Bird's Engineering Mathematics

Why is knowledge of mathematics important in engineering?

A career in any engineering or scientific field will require both basic and advanced mathematics. Without mathematics to determine principles, calculate dimensions and limits, explore variations, prove concepts and so on, there would be no mobile telephones, televisions, stereo systems, video games, microwave ovens, computers or virtually anything electronic. There would be no bridges, tunnels, roads, skyscrapers, automobiles, ships, planes, rockets or most things mechanical. There would be no metals beyond the common ones, such as iron and copper, no plastics, no synthetics. In fact, society would most certainly be less advanced without the use of mathematics throughout the centuries and into the future.

Electrical engineers require mathematics to design, develop, test or supervise the manufacturing and installation of electrical equipment, components or systems for commercial, industrial, military or scientific use.

Mechanical engineers require mathematics to perform engineering duties in planning and designing tools, engines, machines and other mechanically functioning equipment; they oversee installation, operation, maintenance and repair of such equipment as centralised heat, gas, water and steam systems.

Aerospace engineers require mathematics to perform a variety of engineering work in designing, constructing and testing aircraft, missiles and spacecraft; they conduct basic and applied research to evaluate adaptability of materials and equipment to aircraft design and manufacture and recommend improvements in testing equipment and techniques.

Nuclear engineers require mathematics to conduct research on nuclear engineering problems or apply principles and theory of nuclear science to problems concerned with release, control and utilisation of nuclear energy and nuclear waste disposal.

Petroleum engineers require mathematics to devise methods to improve oil and gas well production and determine the need for new or modified tool designs; they oversee drilling and offer technical advice to achieve economical and satisfactory progress.

Industrial engineers require mathematics to design, develop, test, and evaluate integrated systems for managing industrial production processes, including human work factors, quality control, inventory control, logistics and material flow, cost analysis and production coordination.

Environmental engineers require mathematics to design, plan or perform engineering duties in the prevention, control and remediation of environmental health hazards, using various engineering disciplines; their work may include waste treatment, site remediation or pollution control technology.

Civil engineers require mathematics in all levels in civil engineering – structural engineering, hydraulics and geotechnical engineering are all fields that employ mathematical tools such as differential equations, tensor analysis, field theory, numerical methods and operations research.

Knowledge of mathematics is therefore needed by each of the engineering disciplines listed above.

It is intended that this text – *Bird's Engineering Mathematics* – will provide a step by step approach to learning fundamental mathematics needed for your engineering studies.

Now in its ninth edition, *Bird's Engineering Mathematics* has helped thousands of students to succeed in their exams. Mathematical theories are explained in a straightforward manner, supported by practical engineering examples and applications to ensure that readers can relate theory to practice. Some 1,300 engineering situations/problems have been 'flagged-up' to help demonstrate that engineering cannot be fully understood without a good knowledge of mathematics.

The extensive and thorough topic coverage makes this a great text for a range of level 2 and 3 engineering courses – such as for aeronautical, construction, electrical, electronic, mechanical, manufacturing engineering and vehicle technology – including for BTEC First, National and Diploma syllabuses, City & Guilds Technician Certificate and Diploma syllabuses, and even for GCSE and A-level revision.

Its companion website at www.routledge.com/cw/bird provides resources for both students and lecturers, including full solutions for all 2,000 further questions, lists of essential formulae, multiple-choice tests, and illustrations, as well as full solutions to revision tests for course instructors.

John Bird, BSc (Hons), CEng, CMath, CSci, FIMA, FIET, FCollT, is the former Head of Applied Electronics in the Faculty of Technology at Highbury College, Portsmouth, UK. More recently, he has combined freelance lecturing at the University of Portsmouth, with Examiner responsibilities for Advanced Mathematics with City and Guilds and examining for the International Baccalaureate Organisation. He has over 45 years' experience of successfully teaching, lecturing, instructing, training, educating and planning trainee engineers study programmes. He is the author of 146 textbooks on engineering, science and mathematical subjects, with worldwide sales of over one million copies. He is a chartered engineer, a chartered mathematician, a chartered scientist and a Fellow of three professional institutions. He has recently retired from lecturing at the Royal Navy's Defence College of Marine Engineering in the Defence College of Technical Training at H.M.S. Sultan, Gosport, Hampshire, UK, one of the largest engineering training establishments in Europe.

In memory of Elizabeth

Bird's Engineering Mathematics

Ninth Edition

John Bird

LONDON AND NEW YORK

Ninth edition published 2021
by Routledge
2 Park Square, Milton Park, Abingdon, Oxon OX14 4RN

and by Routledge
52 Vanderbilt Avenue, New York, NY 10017

Routledge is an imprint of the Taylor & Francis Group, an informa business

First edition published by Newnes 1999
Eighth edition published by Routledge 2017

British Library Cataloguing-in-Publication Data
A catalogue record for this book is available from the British Library

Library of Congress Cataloging-in-Publication Data
A catalog record has been requested for this book

ISBN: 978-0-367-64379-9 (hbk)
ISBN: 978-0-367-64378-2 (pbk)
ISBN: 978-1-003-12423-8 (ebk)

Typeset in Times
by KnowledgeWorks Global Ltd.

Visit the companion website: www.routledge.com/cw/bird

Contents

Preface

'Bird's Engineering Mathematics 9th Edition' covers a wide range of syllabus requirements. The text is suitable for any course involving engineering mathematics, and in particular for the latest **National Certificate and Diploma courses and City & Guilds syllabuses in Engineering.**

This text will provide a foundation in mathematical principles, which will enable students to solve mathematical, scientific and associated engineering problem. In addition, the material will provide engineering applications and mathematical principles necessary for advancement onto a range of Incorporated Engineer degree profiles. It is widely recognised that a student's ability to use mathematics is a key element in determining subsequent success. First year undergraduates who need some remedial mathematics will also find this book meets their needs.

In *Bird's Engineering Mathematics 9th Edition*, chapters have been re-ordered, examples and problems where ***engineering applications*** occur have been 'flagged up', new multiple-choice questions have been added to each chapter, the text has been added to and simplified, together with other minor modifications.

Throughout the text, theory is introduced in each chapter by an outline of essential definitions, formulae, laws and procedures. The theory is kept to a minimum, for **problem solving** is extensively used to establish and exemplify the theory. It is intended that readers will gain real understanding through seeing problems solved and then through solving similar problems themselves.

For clarity, the text is divided into **eleven topic areas**, these being: number and algebra, trigonometry, areas and volumes, graphs, complex numbers, vectors, differential calculus, integral calculus, differential equations, further number and algebra and statistics.

This new edition covers, in particular, the following syllabuses:

(i) **Mathematics for Technicians**, the core unit for **National Certificate/Diploma** courses in Engineering, to include all or part of the following chapters:
  1. **Algebraic methods**: 2, 5, 11, 13, 14, 27, 29 (1, 4, 6, 8, 9 and 10 for revision)
  2. **Trigonometric methods and areas and volumes**: 17-20, 23-25, 32, 33
  3. **Statistical methods**: 60, 61
  4. **Elementary calculus**: 36, 44, 51

(ii) **Further Mathematics for Technicians,** the optional unit for **National Certificate/Diploma** courses in Engineering, to include all or part of the following chapters:
  1. **Advanced graphical techniques:** 28-30
  2. **Algebraic techniques**: 15, 32, 60, 61
  3. **Trigonometry**: 17-22
  4. **Calculus**: 36-38, 44, 50-52

(iii) **Mathematics contents of City & Guilds Technician Certificate/Diploma courses**

(iv) Any **introductory/access/foundation course** involving Engineering Mathematics **at University, Colleges of Further and Higher education and in schools.**

Each topic considered in the text is presented in a way that assumes in the reader little previous knowledge of that topic.

'Bird's Engineering Mathematics 9th Edition' provides a follow-up to ***'Bird's Basic Engineering Mathematics 8th Edition'*** and a lead into ***'Bird's Higher Engineering Mathematics 9th Edition'***.

This textbook contains over **1050 worked problems**, followed by some **2000 further problems** (all **with answers at the back of the book**). The further problems are contained within some **304 Practice Exercises**; each Exercise follows on directly from the relevant section of work, every two or three pages. In addition, the text contains **575 multiple-choice questions** and **577 line diagrams** to enhance the understanding of the theory. Where at all possible, the problems mirror practical situations found in engineering and science. In fact, some **1300 engineering situations/problems** have been 'flagged-up' to help demonstrate that engineering cannot be fully understood without a good knowledge of mathematics. Look out for the symbol ⚑.

At regular intervals throughout the text are some **19 Revision Tests** to check understanding. For example, Revision Test 1 covers material contained in Chapters 1 to 4, Revision Test 2 covers the material in Chapters 5 to 8 and so on. These Revision Tests do not have answers given since it is envisaged that lecturers could set the tests for students to attempt as part of their course structure. Lecturers may obtain a set of solutions for the Revision Tests in an **Instructor's Manual** available online at www.routledge.com/cw/bird

A list of **Essential Formulae** is included in the text for convenience of reference.

'Learning by Example' is at the heart of ***'Bird's Engineering Mathematics 9th Edition'***.

JOHN BIRD
Formerly Royal Naval Defence College of Marine and Air Engineering, HMS Sultan, University of Portsmouth and Highbury College, Portsmouth

Free Web downloads at www.routledge.com/cw/bird

For students

1. **Full solutions** to the 2000 questions contained in the 304 Practice Exercises
2. **List of Essential Formulae**
3. **Famous Engineers/Scientists** – 25 are mentioned in the text.

For instructors/lecturers

1. **Full solutions** to the 2000 questions contained in the 304 Practice Exercises
2. **Full solutions** and marking scheme to each of the **19 Revision Tests**
3. **Revision Tests – available to run off to be given to students**
4. **List of Essential Formulae**
5. **Illustrations – all 577 available on PowerPoint**
6. **Famous Engineers/Scientists** – 25 are mentioned in the text

Section 1

Number and algebra

Chapter 1

Revision of fractions, decimals and percentages

Why it is important to understand: Revision of fractions, decimals and percentages

Engineers use fractions all the time, examples including stress to strain ratios in mechanical engineering, chemical concentration ratios and reaction rates, and ratios in electrical equations to solve for current and voltage. Fractions are also used everywhere in science, from radioactive decay rates to statistical analysis. Also, engineers and scientists use decimal numbers all the time in calculations. Calculators are able to handle calculations with fractions and decimals; however, there will be times when a quick calculation involving addition, subtraction, multiplication and division of fractions and decimals is needed. Engineers and scientists also use percentages a lot in calculations; for example, percentage change is commonly used in engineering, statistics, physics, finance, chemistry and economics. When you feel able to do calculations with basic arithmetic, fractions, decimals and percentages, with or without the aid of a calculator, then suddenly mathematics doesn't seem quite so difficult.

At the end of this chapter, you should be able to:

- add, subtract, multiply and divide with fractions
- understand practical examples involving ratio and proportion
- add, subtract, multiply and divide with decimals
- understand and use percentages

1.1 Fractions

When 2 is divided by 3, it may be written as $\frac{2}{3}$ or 2/3. $\frac{2}{3}$ is called a **fraction**. The number above the line, i.e. 2, is called the **numerator** and the number below the line, i.e. 3, is called the **denominator**.

When the value of the numerator is less than the value of the denominator, the fraction is called a **proper fraction**; thus $\frac{2}{3}$ is a proper fraction. When the value of the numerator is greater than the denominator, the fraction is called an **improper fraction**. Thus $\frac{7}{3}$ is an improper fraction and can also be expressed as a **mixed number**, that is, an integer and a proper fraction. Thus the improper fraction $\frac{7}{3}$ is equal to the mixed number $2\frac{1}{3}$

When a fraction is simplified by dividing the numerator and denominator by the same number, the process is called **cancelling**. Cancelling by 0 is not permissible.

Problem 1. Simplify: $\frac{1}{3}+\frac{2}{7}$

The lowest common multiple (i.e. LCM) of the two denominators is 3×7, i.e. 21

Expressing each fraction so that their denominators are 21, gives:

$$\frac{1}{3}+\frac{2}{7}=\frac{1}{3}\times\frac{7}{7}+\frac{2}{7}\times\frac{3}{3}=\frac{7}{21}+\frac{6}{21}$$
$$=\frac{7+6}{21}=\mathbf{\frac{13}{21}}$$

Alternatively:

$$\frac{1}{3}+\frac{2}{7}=\frac{\overset{\text{Step (2)}}{(7\times 1)}+\overset{\text{Step (3)}}{(3\times 2)}}{\underset{\text{Step (1)}}{21}}$$

Step 1: the LCM of the two denominators;

Step 2: for the fraction $\frac{1}{3}$, 3 into 21 goes 7 times, $7\times$ the numerator is 7×1;

Step 3: for the fraction $\frac{2}{7}$, 7 into 21 goes 3 times, $3\times$ the numerator is 3×2

Thus $\frac{1}{3}+\frac{2}{7}=\frac{7+6}{21}=\mathbf{\frac{13}{21}}$ as obtained previously.

Problem 2. Find the value of $3\frac{2}{3}-2\frac{1}{6}$

One method is to split the mixed numbers into integers and their fractional parts. Then

$$3\frac{2}{3}-2\frac{1}{6}=\left(3+\frac{2}{3}\right)-\left(2+\frac{1}{6}\right)$$
$$=3+\frac{2}{3}-2-\frac{1}{6}$$
$$=1+\frac{4}{6}-\frac{1}{6}=1\frac{3}{6}=\mathbf{1\frac{1}{2}}$$

Another method is to express the mixed numbers as improper fractions.

Since $3=\frac{9}{3}$, then $3\frac{2}{3}=\frac{9}{3}+\frac{2}{3}=\frac{11}{3}$

Similarly, $2\frac{1}{6}=\frac{12}{6}+\frac{1}{6}=\frac{13}{6}$

Thus $3\frac{2}{3}-2\frac{1}{6}=\frac{11}{3}-\frac{13}{6}=\frac{22}{6}-\frac{13}{6}=\frac{9}{6}=\mathbf{1\frac{1}{2}}$ as obtained previously.

Problem 3. Determine the value of

$$4\frac{5}{8}-3\frac{1}{4}+1\frac{2}{5}$$

$$4\frac{5}{8}-3\frac{1}{4}+1\frac{2}{5}=(4-3+1)+\left(\frac{5}{8}-\frac{1}{4}+\frac{2}{5}\right)$$
$$=2+\frac{5\times 5-10\times 1+8\times 2}{40}$$
$$=2+\frac{25-10+16}{40}$$
$$=2+\frac{31}{40}=\mathbf{2\frac{31}{40}}$$

Problem 4. Find the value of $\frac{3}{7}\times\frac{14}{15}$

Dividing numerator and denominator by 3 gives:

$$\frac{{}^{1}\not{3}}{7}\times\frac{14}{\not{15}_{5}}=\frac{1}{7}\times\frac{14}{5}=\frac{1\times 14}{7\times 5}$$

Dividing numerator and denominator by 7 gives:

$$\frac{1\times \not{14}^{2}}{{}_{1}\not{7}\times 5}=\frac{1\times 2}{1\times 5}=\mathbf{\frac{2}{5}}$$

This process of dividing both the numerator and denominator of a fraction by the same factor(s) is called **cancelling**.

Problem 5. Evaluate: $1\frac{3}{5}\times 2\frac{1}{3}\times 3\frac{3}{7}$

Mixed numbers **must** be expressed as improper fractions before multiplication can be performed. Thus,

$$1\frac{3}{5}\times 2\frac{1}{3}\times 3\frac{3}{7}$$
$$=\left(\frac{5}{5}+\frac{3}{5}\right)\times\left(\frac{6}{3}+\frac{1}{3}\right)\times\left(\frac{21}{7}+\frac{3}{7}\right)$$

$$= \frac{8}{5} \times \frac{1}{3} \times \frac{24}{7} = \frac{8 \times 1 \times 8}{5 \times 1 \times 1}$$

$$= \frac{64}{5} = \mathbf{12\frac{4}{5}}$$

Problem 6. Simplify: $\frac{3}{7} \div \frac{12}{21}$

$$\frac{3}{7} \div \frac{12}{21} = \frac{\frac{3}{7}}{\frac{12}{21}}$$

Multiplying both numerator and denominator by the reciprocal of the denominator gives:

$$\frac{\frac{3}{7}}{\frac{12}{21}} = \frac{\frac{3}{7} \times \frac{21}{12}}{\frac{12}{21} \times \frac{21}{12}} = \frac{\frac{3}{4}}{1} = \mathbf{\frac{3}{4}}$$

This method can be remembered by the rule: invert the second fraction and change the operation from division to multiplication. Thus:

$\frac{3}{7} \div \frac{12}{21} = \frac{3}{7} \times \frac{21}{12} = \mathbf{\frac{3}{4}}$ as obtained previously.

Problem 7. Find the value of $5\frac{3}{5} \div 7\frac{1}{3}$

The mixed numbers must be expressed as improper fractions. Thus,

$$5\frac{3}{5} \div 7\frac{1}{3} = \frac{28}{5} \div \frac{22}{3} = \frac{28}{5} \times \frac{3}{22} = \mathbf{\frac{42}{55}}$$

Problem 8. Simplify:

$$\frac{1}{3} - \left(\frac{2}{5} + \frac{1}{4}\right) \div \left(\frac{3}{8} \times \frac{1}{3}\right)$$

The order of precedence of operations for problems containing fractions is the same as that for integers, i.e. remembered by **BODMAS** (**B**rackets, **O**f, **D**ivision, **M**ultiplication, **A**ddition and **S**ubtraction). Thus,

$$\frac{1}{3} - \left(\frac{2}{5} + \frac{1}{4}\right) \div \left(\frac{3}{8} \times \frac{1}{3}\right)$$

$$= \frac{1}{3} - \frac{4 \times 2 + 5 \times 1}{20} \div \frac{3}{24} \quad \text{(B)}$$

$$= \frac{1}{3} - \frac{13}{20} \times \frac{8}{1} \quad \text{(D)}$$

$$= \frac{1}{3} - \frac{26}{5} \quad \text{(M)}$$

$$= \frac{(5 \times 1) - (3 \times 26)}{15} \quad \text{(S)}$$

$$= \frac{-73}{15} = \mathbf{-4\frac{13}{15}}$$

Problem 9. Determine the value of

$$\frac{7}{6} \text{ of } \left(3\frac{1}{2} - 2\frac{1}{4}\right) + 5\frac{1}{8} \div \frac{3}{16} - \frac{1}{2}$$

$$\frac{7}{6} \text{ of } \left(3\frac{1}{2} - 2\frac{1}{4}\right) + 5\frac{1}{8} \div \frac{3}{16} - \frac{1}{2}$$

$$= \frac{7}{6} \text{ of } 1\frac{1}{4} + \frac{41}{8} \div \frac{3}{16} - \frac{1}{2} \quad \text{(B)}$$

$$= \frac{7}{6} \times \frac{5}{4} + \frac{41}{8} \div \frac{3}{16} - \frac{1}{2} \quad \text{(O)}$$

$$= \frac{7}{6} \times \frac{5}{4} + \frac{41}{8} \times \frac{16}{3} - \frac{1}{2} \quad \text{(D)}$$

$$= \frac{35}{24} + \frac{82}{3} - \frac{1}{2} \quad \text{(M)}$$

$$= \frac{35 + 656}{24} - \frac{1}{2} \quad \text{(A)}$$

$$= \frac{691}{24} - \frac{1}{2} \quad \text{(A)}$$

$$= \frac{691 - 12}{24} \quad \text{(S)}$$

$$= \frac{679}{24} = \mathbf{28\frac{7}{24}}$$

Problem 10. If a storage tank is holding 450 litres when it is three-quarters full, how much will it contain when it is two-thirds full?

If 450 litres is $\frac{3}{4}$ full then $\frac{1}{4}$ full would be $450 \div 3 =$ 150 litres.

Thus, a full tank would have $4 \times 150 = 600$ litres.

$\frac{2}{3}$ of the tank will contain $\frac{2}{3} \times 600 =$ **400 litres**

Now try the following Practice Exercise

Practice Exercise 1 Fractions (Answers on page 701)

Evaluate the following:

1. (a) $\frac{1}{2}+\frac{2}{5}$ (b) $\frac{7}{16}-\frac{1}{4}$
2. (a) $\frac{2}{7}+\frac{3}{11}$ (b) $\frac{2}{9}-\frac{1}{7}+\frac{2}{3}$
3. (a) $10\frac{3}{7}-8\frac{2}{3}$ (b) $3\frac{1}{4}-4\frac{4}{5}+1\frac{5}{6}$
4. (a) $\frac{3}{4}\times\frac{5}{9}$ (b) $\frac{17}{35}\times\frac{15}{119}$
5. (a) $\frac{3}{5}\times\frac{7}{9}\times 1\frac{2}{7}$ (b) $\frac{13}{17}\times 4\frac{7}{11}\times 3\frac{4}{39}$
6. (a) $\frac{3}{8}\div\frac{45}{64}$ (b) $1\frac{1}{3}\div 2\frac{5}{9}$
7. $\frac{1}{2}+\frac{3}{5}\div\frac{8}{15}-\frac{1}{3}$
8. $\frac{7}{15}$ of $\left(15\times\frac{5}{7}\right)+\left(\frac{3}{4}\div\frac{15}{16}\right)$
9. $\frac{1}{4}\times\frac{2}{3}-\frac{1}{3}\div\frac{3}{5}+\frac{2}{7}$
10. $\left(\frac{2}{3}\times 1\frac{1}{4}\right)\div\left(\frac{2}{3}+\frac{1}{4}\right)+1\frac{3}{5}$
11. The movement ratio, M, of a differential pulley is given by the formula: $M=\frac{2R}{R-r}$ where R and r are the radii of the larger and smaller portions of the stepped pulley. Find the movement ratio of such a pulley block having diameters of 140 mm and 120 mm.
12. Three people, P, Q and R contribute to a fund. P provides 3/5 of the total, Q provides 2/3 of the remainder, and R provides £8. Determine (a) the total of the fund, (b) the contributions of P and Q.

1.2 Ratio and proportion

The ratio of one quantity to another is a fraction, and is the number of times one quantity is contained in another quantity **of the same kind**. If one quantity is **directly proportional** to another, then as one quantity doubles, the other quantity also doubles. When a quantity is **inversely proportional** to another, then as one quantity doubles, the other quantity is halved.

Problem 11. A piece of timber 273 cm long is cut into three pieces in the ratio of 3 to 7 to 11. Determine the lengths of the three pieces

The total number of parts is $3+7+11$, that is, 21. Hence 21 parts correspond to 273 cm

$$1 \text{ part corresponds to } \frac{273}{21}=13\,\text{cm}$$

$$3 \text{ parts correspond to } 3\times 13=39\,\text{cm}$$

$$7 \text{ parts correspond to } 7\times 13=91\,\text{cm}$$

$$11 \text{ parts correspond to } 11\times 13=143\,\text{cm}$$

i.e. **the lengths of the three pieces are 39 cm, 91 cm and 143 cm**.

(Check: $39+91+143=273$)

Problem 12. A gear wheel having 80 teeth is in mesh with a 25 tooth gear. What is the gear ratio?

$$\text{Gear ratio}=80:25=\frac{80}{25}=\frac{16}{5}=3.2$$

i.e. gear ratio = **16 : 5** or **3.2 : 1**

Problem 13. An alloy is made up of metals A and B in the ratio 2.5 : 1 by mass. How much of A has to be added to 6 kg of B to make the alloy?

Ratio A : B: :2.5 : 1 (i.e. A is to B as 2.5 is to 1) or $\frac{A}{B}=\frac{2.5}{1}=2.5$

When B = 6 kg, $\frac{A}{6}=2.5$ from which,

$A=6\times 2.5=$ **15 kg**

Problem 14. If 3 people can complete a task in 4 hours, how long will it take 5 people to complete the same task, assuming the rate of work remains constant?

The more the number of people, the more quickly the task is done, hence inverse proportion exists.

3 people complete the task in 4 hours.
1 person takes three times as long, i.e.

$4 \times 3 = 12$ hours,

5 people can do it in one fifth of the time that one person takes, that is $\frac{12}{5}$ hours or **2 hours 24 minutes**.

Now try the following Practice Exercise

Practice Exercise 2 Ratio and proportion (Answers on page 701)

1. Divide 621 cm in the ratio of 3 to 7 to 13.
2. When mixing a quantity of paints, dyes of four different colours are used in the ratio of 7 : 3 : 19 : 5. If the mass of the first dye used is $3\frac{1}{2}$ g, determine the total mass of the dyes used.
3. Determine how much copper and how much zinc is needed to make a 99 kg brass ingot if they have to be in the proportions copper : zinc: :8 : 3 by mass.
4. It takes 21 hours for 12 men to resurface a stretch of road. Find how many men it takes to resurface a similar stretch of road in 50 hours 24 minutes, assuming the work rate remains constant.
5. It takes 3 hours 15 minutes to fly from city A to city B at a constant speed. Find how long the journey takes if
   (a) the speed is $1\frac{1}{2}$ times that of the original speed and
   (b) if the speed is three-quarters of the original speed.
6. A mass of 56 kg is divided into 3 parts in the ratio 3:5:6. Calculate the mass of each part.

1.3 Decimals

The decimal system of numbers is based on the **digits** 0 to 9. A number such as 53.17 is called a **decimal fraction**, a decimal point separating the integer part, i.e. 53, from the fractional part, i.e. 0.17

A number which can be expressed exactly as a decimal fraction is called a **terminating decimal** and those which cannot be expressed exactly as a decimal fraction are called **non-terminating decimals**. Thus, $\frac{3}{2} = 1.5$ is a terminating decimal, but $\frac{4}{3} = 1.33333\ldots$ is a non-terminating decimal. 1.33333... can be written as $1.\dot{3}$, called 'one point-three recurring'.

The answer to a non-terminating decimal may be expressed in two ways, depending on the accuracy required:

(i) correct to a number of **significant figures**, that is, figures which signify something, and

(ii) correct to a number of **decimal places**, that is, the number of figures after the decimal point.

The last digit in the answer is unaltered if the next digit on the right is in the group of numbers 0, 1, 2, 3 or 4, but is increased by 1 if the next digit on the right is in the group of numbers 5, 6, 7, 8 or 9. Thus the non-terminating decimal 7.6183... becomes 7.62, correct to 3 significant figures, since the next digit on the right is 8, which is in the group of numbers 5, 6, 7, 8 or 9. Also 7.6183... becomes 7.618, correct to 3 decimal places, since the next digit on the right is 3, which is in the group of numbers 0, 1, 2, 3 or 4

Problem 15. Evaluate: $42.7 + 3.04 + 8.7 + 0.06$

The numbers are written so that the decimal points are under each other. Each column is added, starting from the right.

$$\begin{array}{r} 42.7 \\ 3.04 \\ 8.7 \\ 0.06 \\ \hline 54.50 \\ \hline \end{array}$$

Thus $\mathbf{42.7 + 3.04 + 8.7 + 0.06 = 54.50}$

Problem 16. Take 81.70 from 87.23

The numbers are written with the decimal points under each other.

$$\begin{array}{r} 87.23 \\ -81.70 \\ \hline 5.53 \\ \hline \end{array}$$

Thus $\mathbf{87.23 - 81.70 = 5.53}$

Problem 17. Find the value of

$$23.4 - 17.83 - 57.6 + 32.68$$

The sum of the positive decimal fractions is

$$23.4+32.68=56.08$$

The sum of the negative decimal fractions is

$$17.83+57.6=75.43$$

Taking the sum of the negative decimal fractions from the sum of the positive decimal fractions gives:

$$56.08-75.43$$

i.e. $-(75.43-56.08)=\mathbf{-19.35}$

Problem 18. Determine the value of 74.3×3.8

When multiplying decimal fractions: (i) the numbers are multiplied as if they are integers, and (ii) the position of the decimal point in the answer is such that there are as many digits to the right of it as the sum of the digits to the right of the decimal points of the two numbers being multiplied together. Thus

(i)
```
   743
    38
 5 944
22 290
28 234
```

(ii) As there are $(1+1)=2$ digits to the right of the decimal points of the two numbers being multiplied together, $(74.\underline{3}\times 3.\underline{8})$, then

$$\mathbf{74.3\times 3.8=282.34}$$

Problem 19. Evaluate $37.81\div 1.7$, correct to (i) 4 significant figures and (ii) 4 decimal places

$$37.81\div 1.7=\frac{37.81}{1.7}$$

The denominator is changed into an integer by multiplying by 10. The numerator is also multiplied by 10 to keep the fraction the same. Thus

$$37.81\div 1.7=\frac{37.81\times 10}{1.7\times 10}=\frac{378.1}{17}$$

The long division is similar to the long division of integers and the first four steps are as shown:

```
        22.24117
    ____________
17 ) 378.100000
     34
     __
      38
      34
      __
       41
       34
       __
        70
        68
        __
         20
```

(i) **$37.81\div 1.7=22.24$, correct to 4 significant figures**, and

(ii) **$37.81\div 1.7=22.2412$, correct to 4 decimal places.**

Problem 20. Convert (a) 0.4375 to a proper fraction and (b) 4.285 to a mixed number

(a) 0.4375 can be written as $\frac{0.4375\times 10000}{10000}$ without changing its value,

i.e. $0.4375=\frac{4375}{10000}$

By cancelling

$$\frac{4375}{10000}=\frac{875}{2000}=\frac{175}{400}=\frac{35}{80}=\frac{7}{16}$$

i.e. $\mathbf{0.4375=\frac{7}{16}}$

(b) Similarly, $\mathbf{4.285}=4\frac{285}{1000}=\mathbf{4\frac{57}{200}}$

Problem 21. Express as decimal fractions: (a) $\frac{9}{16}$ and (b) $5\frac{7}{8}$

(a) To convert a proper fraction to a decimal fraction, the numerator is divided by the denominator. Division by 16 can be done by the long division method, or, more simply, by dividing by 2 and then 8:

$$2\overline{)9.00}\ \text{gives}\ 4.50 \qquad 8\overline{)4.5000}\ \text{gives}\ 0.5625$$

Thus $\mathbf{\frac{9}{16}=0.5625}$

(b) For mixed numbers, it is only necessary to convert the proper fraction part of the mixed number to a decimal fraction. Thus, dealing with the $\frac{7}{8}$ gives:

$$8\overline{)7.000}\ \text{gives}\ 0.875 \qquad \text{i.e.}\ \frac{7}{8}=0.875$$

Thus $\mathbf{5\frac{7}{8}=5.875}$

Now try the following Practice Exercise

Practice Exercise 3 Decimals (Answers on page 701)

In Problems 1 to 6, determine the values of the expressions given:

1. $23.6+14.71-18.9-7.421$
2. $73.84-113.247+8.21-0.068$
3. $3.8\times4.1\times0.7$
4. 374.1×0.006
5. $421.8\div17$, (a) correct to 4 significant figures and (b) correct to 3 decimal places.
6. $\frac{0.0147}{2.3}$, (a) correct to 5 decimal places and (b) correct to 2 significant figures.
7. Convert to proper fractions: (a) 0.65 (b) 0.84 (c) 0.0125 (d) 0.282 and (e) 0.024
8. Convert to mixed numbers: (a) 1.82 (b) 4.275 (c) 14.125 (d) 15.35 and (e) 16.2125

In Problems 9 to 12, express as decimal fractions to the accuracy stated:

9. $\frac{4}{9}$, correct to 5 significant figures.
10. $\frac{17}{27}$, correct to 5 decimal places.
11. $1\frac{9}{16}$, correct to 4 significant figures.
12. $13\frac{31}{37}$, correct to 2 decimal places.
13. Determine the dimension marked x in the length of shaft shown in Fig. 1.1. The dimensions are in millimetres.

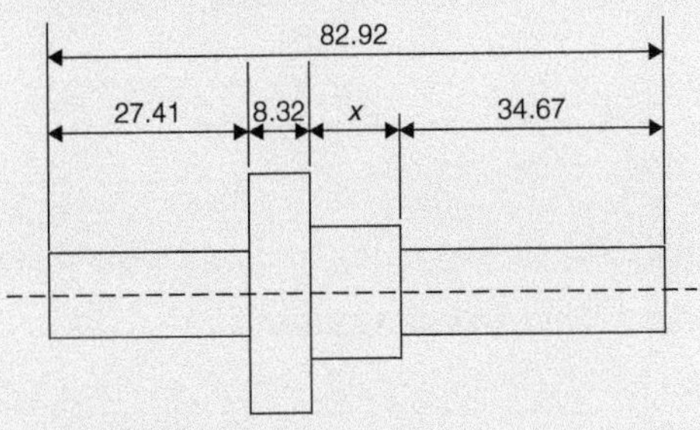

Figure 1.1

14. A tank contains 1800 litres of oil. How many tins containing 0.75 litres can be filled from this tank?

1.4 Percentages

Percentages are used to give a common standard and are fractions having the number 100 as their denominators. For example, 25 per cent means $\frac{25}{100}$ i.e. $\frac{1}{4}$ and is written 25%

Problem 22. Express as percentages: (a) 1.875 and (b) 0.0125

A decimal fraction is converted to a percentage by multiplying by 100. Thus,

(a) 1.875 corresponds to $1.875\times100\%$, i.e. **187.5%**

(b) 0.0125 corresponds to $0.0125\times100\%$, i.e. **1.25%**

Problem 23. Express as percentages:

$$\text{(a)}\ \frac{5}{16} \quad \text{and} \quad \text{(b)}\ 1\frac{2}{5}$$

To convert fractions to percentages, they are (i) converted to decimal fractions and (ii) multiplied by 100

(a) By division, $\frac{5}{16}=0.3125$, hence $\frac{5}{16}$ corresponds to $0.3125\times100\%$, i.e. **31.25%**

(b) Similarly, $1\frac{2}{5}=1.4$ when expressed as a decimal fraction.

Hence $1\frac{2}{5}=1.4\times100\%=\mathbf{140\%}$

Problem 24. It takes 50 minutes to machine a certain part, Using a new type of tool, the time can be reduced by 15%. Calculate the new time taken

$$15\% \text{ of } 50 \text{ minutes} = \frac{15}{100} \times 50 = \frac{750}{100}$$
$$= 7.5 \text{ minutes.}$$

hence the **new time taken** is

$$50 - 7.5 = \mathbf{42.5\ minutes}.$$

Alternatively, if the time is reduced by 15%, then it now takes 85% of the original time, i.e. 85% of $50 = \frac{85}{100} \times 50 = \frac{4250}{100} =$ **42.5 minutes**, as above.

Problem 25. Find 12.5% of £378

12.5% of £378 means $\frac{12.5}{100} \times 378$, since per cent means 'per hundred'.

Hence 12.5% of $£378 = \frac{\cancel{12.5}^1}{\cancel{100}_8} \times 378 = \frac{1}{8} \times 378$
$= \frac{378}{8} =$ **£47.25**

Problem 26. Express 25 minutes as a percentage of 2 hours, correct to the nearest 1%

Working in minute units, 2 hours = 120 minutes.
Hence 25 minutes is $\frac{25}{120}$ths of 2 hours. By cancelling, $\frac{25}{120} = \frac{5}{24}$
Expressing $\frac{5}{24}$ as a decimal fraction gives $0.208\dot{3}$
Multiplying by 100 to convert the decimal fraction to a percentage gives:

$$0.208\dot{3} \times 100 = 20.8\dot{3}\%$$

Thus **25 minutes is 21% of 2 hours,** correct to the nearest 1%

Problem 27. A German silver alloy consists of 60% copper, 25% zinc and 15% nickel. Determine the masses of the copper, zinc and nickel in a 3.74 kilogram block of the alloy

By direct proportion:
100% corresponds to 3.74 kg

1% corresponds to $\frac{3.74}{100} = 0.0374$ kg

60% corresponds to $60 \times 0.0374 = 2.244$ kg

25% corresponds to $25 \times 0.0374 = 0.935$ kg

15% corresponds to $15 \times 0.0374 = 0.561$ kg

Thus, the masses of the copper, zinc and nickel are **2.244 kg, 0.935 kg and 0.561 kg**, respectively.
(Check: $2.244 + 0.935 + 0.561 = 3.74$)

Problem 28. A mixture used for making concrete contains cement, sand and rubble in the proportion 2:5:8. Calculate (a) the mass of sand in 750 kg of mixture (b) the percentage of cement in the mixture.

Since the ratio is 2:3:8, the total number of parts is $2 + 5 + 8 = 15$ parts

(a) Sand is 5 parts out of 15, i.e. $\frac{5}{15}$ of the 750 kg mixture
Hence, **the mass of sand in the mixture** is $\frac{5}{15} \times 750 =$ **250 kg**

(b) Cement is 2 parts out of 15, i.e. $\frac{2}{15}$ of the mixture
As a percentage of the mixture, cement is $\frac{2}{15} \times 100 =$ **13.33%**

Problem 29. A brick being tested for its water absorption properties is found to have a mass of 2.628 kg when dry and 3.127 kg after soaking in water for a day. Calculate its percentage absorption by mass, correct to the nearest percent.

Percentage absorption

$$= \frac{\text{new value-original value}}{\text{original value}} \times 100\%$$
$$= \frac{3.127 - 2.628}{2.628} \times 100\%$$
$$= \frac{0.499}{2.628} \times 100\%$$
$$= 18.99\%$$
$= \mathbf{19\%}$ correct to the nearest percent

Now try the following Practice Exercise

Practice Exercise 4 Percentages (Answers on page 701)

1. Convert to percentages:
(a) 0.057 (b) 0.374 (c) 1.285

2. Express as percentages, correct to 3 significant figures:
(a) $\frac{7}{33}$ (b) $\frac{19}{24}$ (c) $1\frac{11}{16}$

3. Calculate correct to 4 significant figures:
(a) 18% of 2758 tonnes (b) 47% of 18.42 grams (c) 147% of 14.1 seconds

4. When 1600 bolts are manufactured, 36 are unsatisfactory. Determine the percentage unsatisfactory.

5. Express: (a) 140 kg as a percentage of 1 t (b) 47 s as a percentage of 5 min (c) 13.4 cm as a percentage of 2.5 m

6. A block of monel alloy consists of 70% nickel and 30% copper. If it contains 88.2 g of nickel, determine the mass of copper in the block.

7. A drilling machine should be set to 250 rev/min. The nearest speed available on the machine is 268 rev/min. Calculate the percentage over speed.

8. Two kilograms of a compound contains 30% of element A, 45% of element B and 25% of element C. Determine the masses of the three elements present.

9. A concrete mixture contains seven parts by volume of ballast, four parts by volume of sand and two parts by volume of cement. Determine the percentage of each of these three constituents correct to the nearest 1% and the mass of cement in a two tonne dry mix, correct to 1 significant figure.

10. In a sample of iron ore, 18% is iron. How much ore is needed to produce 3600 kg of iron?

11. A screws' dimension is $12.5 \pm 8\%$ mm. Calculate the possible maximum and minimum length of the screw.

12. The output power of an engine is 450 kW. If the efficiency of the engine is 75%, determine the power input.

13. A production run produces 4200 components of which 97% are reliable. Calculate the number of unreliable components

Practice Exercise 5 Multiple-choice questions on fractions, decimals and percentages (Answers on page 702)

Each question has only one correct answer

1. $\frac{3}{4} \div 1\frac{3}{4}$ is equal to:
(a) $\frac{3}{7}$ (b) $1\frac{9}{16}$ (c) $1\frac{5}{16}$ (d) $2\frac{1}{2}$

2. $1\frac{1}{3} + 1\frac{2}{3} \div 2\frac{2}{3} - \frac{1}{3}$ is equal to:
(a) $1\frac{5}{8}$ (b) $\frac{19}{24}$ (c) $2\frac{1}{21}$ (d) $1\frac{2}{7}$

3. The value of
$$\frac{2}{5} \text{ of } \left(4\frac{1}{2} - 3\frac{1}{4}\right) + 5 \div \frac{5}{16} - \frac{1}{4} \text{ is:}$$
(a) $17\frac{7}{20}$ (b) $80\frac{1}{2}$ (c) $16\frac{1}{4}$ (d) 88

4. Which of the fractions $\frac{2}{3}, \frac{3}{5}, \frac{5}{8}, \frac{7}{9}$ is the smallest?
(a) $\frac{2}{3}$ (b) $\frac{3}{5}$ (c) $\frac{5}{8}$ (d) $\frac{7}{9}$

5. $\frac{1}{\pi}\left(1\frac{1}{18} + \frac{1}{6} - \frac{5}{3}\right)$ is equal to:
(a) $\frac{2\pi}{9}$ (b) $\frac{2}{9\pi}$ (c) $-\frac{4}{9\pi}$ (d) $-\frac{16\pi}{9}$

6. Four engineers can complete a task in 5 hours. Assuming the rate of work remains constant, six engineers will complete the task in:
(a) 126 h (b) 4 h 48 min
(c) 3 h 20 min (d) 7 h 30 min

7. How many litres of water needs to be added to 6 litres of a 60% alcohol solution to create a 40% alcohol solution?
(a) 2 litres (b) 3 litres
(c) 6 litres (d) 9 litres

8. A machine can produce 24 identical face masks per hour. The number of machines needed to produce 144 masks in 20 minutes is:
(a) 36 (b) 18 (c) 12 (d) 6

9. A map has a scale of 1:200,000. Two engineering sites are 15 cm apart on the map. The actual distance apart of the two sites is:
(a) 30 km (b) 300 km
(c) 15 km (d) 150 km

10. y varies as the square of x, and $y = 12$ when $x = 2$. When $x = 3$, the value of y is:
(a) 27 (b) 36 (c) 432 (d) 9

11. A simple machine has an effort to load ratio of 5:39. The effort, in newtons, to lift a load of 5.46 kN is:
(a) 700 N (b) 620.45 N
(c) 42588 N (d) 0.7 N

12. In an engineering laboratory, acid and water are mixed in the ratio 2:7. To make 369 ml of the mixture, the amount of acid added is:
(a) 287 ml (b) 184.5 ml
(c) 105.4 ml (d) 82 ml

13. Correct to 2 significant figures, 3.748 is equal to:
(a) 3.75 (b) 3.8 (c) 3.74 (d) 3.7

14. $1\frac{5}{6}$ expressed as a decimal, correct to 4 significant figures is:
(a) 1.83 (b) 0.633 (c) 1.833 (d) 1.8333

15. Expressing $\frac{1}{11}$ as a decimal, correct to 3 significant figures, is equal to:
(a) 0.099 (b) 0.909
(c) 0.0909 (d) 0.009

16. Correct to 2 significant figures, 0.01479 is:
(a) 0.01 (b) 0.014 (c) 0.0148 (d) 0.015

17. The value of 0.02×0.003 is:
(a) 0.06 (b) 0.006
(c) 0.0006 (d) 0.00006

18. $\frac{1524 \div 15.24}{0.5}$ is equal to:
(a) 0.2 (b) 20 (c) 100 (d) 200

19. 11 mm expressed as a percentage of 41 mm is:
(a) 2.68, correct to 3 significant figures
(b) 2.6, correct to 2 significant figures
(c) 26.83, correct to 2 decimal places
(d) 0.2682, correct to 4 decimal places

20. The current in a component in an electrical circuit is calculated as 25 mA using Ohm's law. When measured, the actual current is 25.2 mA. Correct to 2 decimal places, the percentage error in the calculation is:
(a) 0.80% (b) 1.25%
(c) 0.79% (d) 1.26%

21. Given that 12% of P is 48, the value of P is:
(a) 250 (b) 400 (c) 100 (d) 200

22. Given that $12q = 75\%$ of 336, the value of q is:
(a) 21 (b) 48 (c) 28 (d) 252

23. 0.226 as a percentage is:
(a) 2.26% (b) 22.6%
(c) 0.226% (d) 226%

24. A resistor of 47 kΩ has a tolerance of ±5%. The highest possible value is:
(a) 52 kΩ (b) 47.005 kΩ
(c) 42 kΩ (d) 49.35 kΩ

25. The length and width of a rectangle are each increased by 10%. The percentage increase in the area is:
(a) 20 (b) 21 (c) 12.1 (d) 10

COMPANION @ WEBSITE

For fully worked solutions to each of the problems in Practice Exercises 1 to 4 in this chapter, go to the website:
www.routledge.com/cw/bird

Chapter 2

Indices, engineering notation and metric conversions

Why it is important to understand: Indices, engineering notation and metric conversions

Powers and roots are used extensively in mathematics and engineering, so it is important to get a good grasp of what they are and how, and why, they are used. Being able to multiply powers together by adding their indices is particularly useful for disciplines like engineering and electronics, where quantities are often expressed as a value multiplied by some power of ten. In the field of electrical engineering, for example, the relationship between electric current, voltage and resistance in an electrical system is critically important, and yet the typical unit values for these properties can differ by several orders of magnitude. Studying, or working, in an engineering discipline, you very quickly become familiar with powers and roots and laws of indices. In engineering there are many different quantities to get used to, and hence many units to become familiar with. For example, force is measured in Newtons, electric current is measured in amperes and pressure is measured in Pascals. Sometimes the units of these quantities are either very large or very small and hence prefixes are used. For example, 1000 Pascals may be written as 10^3 Pa which is written as 1 kPa in prefix form, the k being accepted as a symbol to represent 1000 or 10^3. Studying, or working, in an engineering discipline, you very quickly become familiar with the standard units of measurement, the prefixes used and engineering notation. In addition, metric conversions are needed all the time in engineering situations. An electronic calculator is essential with engineering notation and with metric conversions.

At the end of this chapter, you should be able to:

- use the laws of indices
- understand standard form
- understand and use engineering notation
- understand and use common prefixes
- understand and use metric conversions for length, area and volume
- convert between metric and imperial units

2.1 Indices

The lowest factors of 2000 are $2\times2\times2\times2\times5\times5\times5$. These factors are written as $2^4\times5^3$, where 2 and 5 are called **bases** and the numbers 4 and 5 are called **indices**.

When an index is an integer it is called a **power**. Thus, 2^4 is called 'two to the power of four', and has a base of 2 and an index of 4. Similarly, 5^3 is called 'five to the power of 3' and has a base of 5 and an index of 3.

Special names may be used when the indices are 2 and 3, these being called '**squared**' and '**cubed**', respectively. Thus 7^2 is called 'seven squared' and 9^3 is called 'nine cubed'. When no index is shown, the power is 1, i.e. 2 means 2^1

Reciprocal

The **reciprocal** of a number is when the index is -1 and its value is given by 1, divided by the base. Thus the reciprocal of 2 is 2^{-1} and its value is $\frac{1}{2}$ or 0.5. Similarly, the reciprocal of 5 is 5^{-1} which means $\frac{1}{5}$ or 0.2

Square root

The **square root** of a number is when the index is $\frac{1}{2}$, and the square root of 2 is written as $2^{1/2}$ or $\sqrt{2}$. The value of a square root is the value of the base which when multiplied by itself gives the number. Since $3 \times 3 = 9$, then $\sqrt{9} = 3$. However, $(-3) \times (-3) = 9$, so $\sqrt{9} = -3$. There are always two answers when finding the square root of a number and this is shown by putting both a + and a − sign in front of the answer to a square root problem. Thus $\sqrt{9} = \pm 3$ and $4^{1/2} = \sqrt{4} = \pm 2$ and so on.

Laws of indices

When simplifying calculations involving indices, certain basic rules or laws can be applied, called the **laws of indices**. These are given below.

(i) When multiplying two or more numbers having the same base, the indices are added. Thus

$$3^2 \times 3^4 = 3^{2+4} = 3^6$$

(ii) When a number is divided by a number having the same base, the indices are subtracted. Thus

$$\frac{3^5}{3^2} = 3^{5-2} = 3^3$$

(iii) When a number which is raised to a power is raised to a further power, the indices are multiplied. Thus

$$(3^5)^2 = 3^{5 \times 2} = 3^{10}$$

(iv) When a number has an index of 0, its value is 1. Thus $3^0 = 1$

(v) A number raised to a negative power is the reciprocal of that number raised to a positive power. Thus $3^{-4} = \frac{1}{3^4}$ Similarly, $\frac{1}{2^{-3}} = 2^3$

(vi) When a number is raised to a fractional power the denominator of the fraction is the root of the number and the numerator is the power.

Thus $8^{2/3} = \sqrt[3]{8^2} = (2)^2 = 4$

and $25^{1/2} = \sqrt[2]{25^1} = \sqrt{25^1} = \pm 5$

(Note that $\sqrt{} \equiv \sqrt[2]{}$)

Problem 1. Evaluate: (a) $5^2 \times 5^3$, (b) $3^2 \times 3^4 \times 3$ and (c) $2 \times 2^2 \times 2^5$

From law (i):

(a) $5^2 \times 5^3 = 5^{(2+3)} = 5^5 = 5 \times 5 \times 5 \times 5 \times 5 = \mathbf{3125}$

(b) $3^2 \times 3^4 \times 3 = 3^{(2+4+1)} = 3^7$

$= 3 \times 3 \times \cdots$ to 7 terms

$= \mathbf{2187}$

(c) $2 \times 2^2 \times 2^5 = 2^{(1+2+5)} = 2^8 = \mathbf{256}$

Problem 2. Find the value of: (a) $\frac{7^5}{7^3}$ and (b) $\frac{5^7}{5^4}$

From law (ii):

(a) $\frac{7^5}{7^3} = 7^{(5-3)} = 7^2 = \mathbf{49}$

(b) $\frac{5^7}{5^4} = 5^{(7-4)} = 5^3 = \mathbf{125}$

Problem 3. Evaluate: (a) $5^2 \times 5^3 \div 5^4$ and (b) $(3 \times 3^5) \div (3^2 \times 3^3)$

From laws (i) and (ii):

(a) $5^2 \times 5^3 \div 5^4 = \frac{5^2 \times 5^3}{5^4} = \frac{5^{(2+3)}}{5^4}$

$= \frac{5^5}{5^4} = 5^{(5-4)} = 5^1 = \mathbf{5}$

(b) $(3 \times 3^5) \div (3^2 \times 3^3) = \frac{3 \times 3^5}{3^2 \times 3^3} = \frac{3^{(1+5)}}{3^{(2+3)}}$

$= \frac{3^6}{3^5} = 3^{(6-5)} = 3^1 = \mathbf{3}$

Problem 4. Simplify: (a) $(2^3)^4$ and (b) $(3^2)^5$, expressing the answers in index form.

From law (iii):

(a) $(2^3)^4 = 2^{3\times4} = \mathbf{2^{12}}$ (b) $(3^2)^5 = 3^{2\times5} = \mathbf{3^{10}}$

Problem 5. Evaluate: $\dfrac{(10^2)^3}{10^4 \times 10^2}$

From the laws of indices:

$$\frac{(10^2)^3}{10^4 \times 10^2} = \frac{10^{(2\times3)}}{10^{(4+2)}} = \frac{10^6}{10^6} = 10^{6-6} = 10^0 = \mathbf{1}$$

Problem 6. Find the value of:

(a) $\dfrac{2^3 \times 2^4}{2^7 \times 2^5}$ and (b) $\dfrac{(3^2)^3}{3 \times 3^9}$

From the laws of indices:

(a) $$\frac{2^3 \times 2^4}{2^7 \times 2^5} = \frac{2^{(3+4)}}{2^{(7+5)}} = \frac{2^7}{2^{12}} = 2^{7-12} = 2^{-5} = \frac{1}{2^5} = \mathbf{\frac{1}{32}}$$

(b) $$\frac{(3^2)^3}{3 \times 3^9} = \frac{3^{2\times3}}{3^{1+9}} = \frac{3^6}{3^{10}} = 3^{6-10} = 3^{-4} = \frac{1}{3^4} = \mathbf{\frac{1}{81}}$$

Now try the following Practice Exercise

Practice Exercise 6 Indices (Answers on page 702)

In Problems 1 to 10, simplify the expressions given, expressing the answers in index form and with positive indices:

1. (a) $3^3 \times 3^4$ (b) $4^2 \times 4^3 \times 4^4$
2. (a) $2^3 \times 2 \times 2^2$ (b) $7^2 \times 7^4 \times 7 \times 7^3$
3. (a) $\dfrac{2^4}{2^3}$ (b) $\dfrac{3^7}{3^2}$
4. (a) $5^6 \div 5^3$ (b) $7^{13}/7^{10}$
5. (a) $(7^2)^3$ (b) $(3^3)^2$
6. (a) $\dfrac{2^2 \times 2^3}{2^4}$ (b) $\dfrac{3^7 \times 3^4}{3^5}$
7. (a) $\dfrac{5^7}{5^2 \times 5^3}$ (b) $\dfrac{13^5}{13 \times 13^2}$
8. (a) $\dfrac{(9 \times 3^2)^3}{(3 \times 27)^2}$ (b) $\dfrac{(16 \times 4)^2}{(2 \times 8)^3}$
9. (a) $\dfrac{5^{-2}}{5^{-4}}$ (b) $\dfrac{3^2 \times 3^{-4}}{3^3}$
10. (a) $\dfrac{7^2 \times 7^{-3}}{7 \times 7^{-4}}$ (b) $\dfrac{2^3 \times 2^{-4} \times 2^5}{2 \times 2^{-2} \times 2^6}$

Here are some further worked problems to help understanding of indices.

Problem 7. Evaluate: $\dfrac{3^3 \times 5^7}{5^3 \times 3^4}$

The laws of indices only apply to terms **having the same base**. Grouping terms having the same base, and then applying the laws of indices to each of the groups independently gives:

$$\frac{3^3 \times 5^7}{5^3 \times 3^4} = \frac{3^3}{3^4} = \frac{5^7}{5^3} = 3^{(3-4)} \times 5^{(7-3)}$$
$$= 3^{-1} \times 5^4 = \frac{5^4}{3^1} = \frac{625}{3} = \mathbf{208\frac{1}{3}}$$

Problem 8. Find the value of:

$$\frac{2^3 \times 3^5 \times (7^2)^2}{7^4 \times 2^4 \times 3^3}$$

$$\frac{2^3 \times 3^5 \times (7^2)^2}{7^4 \times 2^4 \times 3^3} = 2^{3-4} \times 3^{5-3} \times 7^{2\times2-4}$$
$$= 2^{-1} \times 3^2 \times 7^0$$
$$= \frac{1}{2} \times 3^2 \times 1 = \frac{9}{2} = \mathbf{4\frac{1}{2}}$$

Problem 9. Evaluate:

(a) $4^{1/2}$ (b) $16^{3/4}$ (c) $27^{2/3}$ (d) $9^{-1/2}$

(a) $4^{1/2} = \sqrt{4} = \mathbf{\pm 2}$

(b) $16^{3/4} = \sqrt[4]{16^3} = (\pm 2)^3 = \mathbf{\pm 8}$

(Note that it does not matter whether the 4th root of 16 is found first or whether 16 cubed is found first – the same answer will result).

(c) $27^{2/3} = \sqrt[3]{27^2} = (3)^2 = \mathbf{9}$

(d) $9^{-1/2} = \dfrac{1}{9^{1/2}} = \dfrac{1}{\sqrt{9}} = \dfrac{1}{\pm 3} = \pm\dfrac{\mathbf{1}}{\mathbf{3}}$

Problem 10. Evaluate: $\dfrac{4^{1.5} \times 8^{1/3}}{2^2 \times 32^{-2/5}}$

$$4^{1.5} = 4^{3/2} = \sqrt{4^3} = 2^3 = 8$$
$$8^{1/3} = \sqrt[3]{8} = 2,\ 2^2 = 4$$

and $32^{-2/5} = \dfrac{1}{32^{2/5}} = \dfrac{1}{\sqrt[5]{32^2}} = \dfrac{1}{2^2} = \dfrac{1}{4}$

Hence $\dfrac{4^{1.5} \times 8^{1/3}}{2^2 \times 32^{-2/5}} = \dfrac{8 \times 2}{4 \times \frac{1}{4}} = \dfrac{16}{1} = \mathbf{16}$

Alternatively,

$$\frac{4^{1.5} \times 8^{1/3}}{2^2 \times 32^{-2/5}} = \frac{[(2)^2]^{3/2} \times (2^3)^{1/3}}{2^2 \times (2^5)^{-2/5}} = \frac{2^3 \times 2^1}{2^2 \times 2^{-2}}$$
$$= 2^{3+1-2-(-2)} = 2^4 = \mathbf{16}$$

Problem 11. Evaluate: $\dfrac{3^2 \times 5^5 + 3^3 \times 5^3}{3^4 \times 5^4}$

Dividing each term by the HCF (i.e. highest common factor) of the three terms, i.e. $3^2 \times 5^3$, gives:

$$\frac{3^2 \times 5^5 + 3^3 \times 5^3}{3^4 \times 5^4} = \frac{\dfrac{3^2 \times 5^5}{3^2 \times 5^3} + \dfrac{3^3 \times 5^3}{3^2 \times 5^3}}{\dfrac{3^4 \times 5^4}{3^2 \times 5^3}}$$
$$= \frac{3^{(2-2)} \times 5^{(5-3)} + 3^{(3-2)} \times 5^0}{3^{(4-2)} \times 5^{(4-3)}}$$
$$= \frac{3^0 \times 5^2 + 3^1 \times 5^0}{3^2 \times 5^1}$$
$$= \frac{1 \times 25 + 3 \times 1}{9 \times 5} = \frac{\mathbf{28}}{\mathbf{45}}$$

Problem 12. Find the value of:

$$\frac{3^2 \times 5^5}{3^4 \times 5^4 + 3^3 \times 5^3}$$

To simplify the arithmetic, each term is divided by the HCF of all the terms, i.e. $3^2 \times 5^3$. Thus

$$\frac{3^2 \times 5^5}{3^4 \times 5^4 + 3^3 \times 5^3}$$
$$= \frac{\dfrac{3^2 \times 5^5}{3^2 \times 5^3}}{\dfrac{3^4 \times 5^4}{3^2 \times 5^3} + \dfrac{3^3 \times 5^3}{3^2 \times 5^3}}$$
$$= \frac{3^{(2-2)} \times 5^{(5-3)}}{3^{(4-2)} \times 5^{(4-3)} + 3^{(3-2)} \times 5^{(3-3)}}$$
$$= \frac{3^0 \times 5^2}{3^2 \times 5^1 + 3^1 \times 5^0} = \frac{25}{45+3} = \frac{\mathbf{25}}{\mathbf{48}}$$

Problem 13. Simplify: $\dfrac{\left(\frac{4}{3}\right)^3 \times \left(\frac{3}{5}\right)^{-2}}{\left(\frac{2}{5}\right)^{-3}}$ giving the answer with positive indices

A fraction raised to a power means that both the numerator and the denominator of the fraction are raised to that power, i.e. $\left(\dfrac{4}{3}\right)^3 = \dfrac{4^3}{3^3}$

A fraction raised to a negative power has the same value as the inverse of the fraction raised to a positive power.

Thus, $\left(\dfrac{3}{5}\right)^{-2} = \dfrac{1}{\left(\frac{3}{5}\right)^2} = \dfrac{1}{\frac{3^2}{5^2}} = 1 \times \dfrac{5^2}{3^2} = \dfrac{5^2}{3^2}$

Similarly, $\left(\dfrac{2}{5}\right)^{-3} = \left(\dfrac{5}{2}\right)^3 = \dfrac{5^3}{2^3}$

Thus, $\dfrac{\left(\frac{4}{3}\right)^3 \times \left(\frac{3}{5}\right)^{-2}}{\left(\frac{2}{5}\right)^{-3}} = \dfrac{\frac{4^3}{3^3} \times \frac{5^2}{3^2}}{\frac{5^3}{2^3}}$

$$= \frac{4^3}{3^3} \times \frac{5^2}{3^2} \times \frac{2^3}{5^3}$$
$$= \frac{(2^2)^3 \times 2^3}{3^{(3+2)} \times 5^{(3-2)}}$$
$$= \frac{\mathbf{2^9}}{\mathbf{3^5 \times 5}}$$

Now try the following Practice Exercise

Practice Exercise 7 Indices (Answers on page 702)

In Problems 1 and 2, simplify the expressions given, expressing the answers in index form and with positive indices:

1. (a) $\dfrac{3^3 \times 5^2}{5^4 \times 3^4}$ (b) $\dfrac{7^{-2} \times 3^{-2}}{3^5 \times 7^4 \times 7^{-3}}$

2. (a) $\dfrac{4^2 \times 9^3}{8^3 \times 3^4}$ (b) $\dfrac{8^{-2} \times 5^2 \times 3^{-4}}{25^2 \times 2^4 \times 9^{-2}}$

3. Evaluate (a) $\left(\dfrac{1}{3^2}\right)^{-1}$ (b) $81^{0.25}$
   (c) $16^{(-1/4)}$ (d) $\left(\dfrac{4}{9}\right)^{1/2}$

In Problems 4 to 8, evaluate the expressions given.

4. $\dfrac{9^2 \times 7^4}{3^4 \times 7^4 + 3^3 \times 7^2}$

5. $\dfrac{(2^4)^2 - 3^{-2} \times 4^4}{2^3 \times 16^2}$

6. $\dfrac{\left(\dfrac{1}{2}\right)^3 - \left(\dfrac{2}{3}\right)^{-2}}{\left(\dfrac{3}{5}\right)^2}$

7. $\dfrac{\left(\dfrac{4}{3}\right)^4}{\left(\dfrac{2}{9}\right)^2}$

8. $\dfrac{(3^2)^{3/2} \times (8^{1/3})^2}{(3)^2 \times (4^3)^{1/2} \times (9)^{-1/2}}$

2.2 Standard form

A number written with one digit to the left of the decimal point and multiplied by 10 raised to some power is said to be written in **standard form**. Thus: 5837 is written as 5.837×10^3 in standard form, and 0.0415 is written as 4.15×10^{-2} in standard form.

Problem 14. Express in standard form:
(a) 38.71 (b) 3746 (c) 0.0124

For a number to be in standard form, it is expressed with only one digit to the left of the decimal point. Thus:

(a) 38.71 must be divided by 10 to achieve one digit to the left of the decimal point and it must also be multiplied by 10 to maintain the equality, i.e.

$$38.71 = \frac{38.71}{10} \times 10 = \mathbf{3.871 \times 10} \text{ in standard form}$$

(b) $3746 = \dfrac{3746}{1000} \times 1000 = \mathbf{3.746 \times 10^3}$ in standard form

(c) $0.0124 = 0.0124 \times \dfrac{100}{100} = \dfrac{1.24}{100}$

$$= \mathbf{1.24 \times 10^{-2}} \text{ in standard form}$$

Problem 15. Express the following numbers, which are in standard form, as decimal numbers:
(a) 1.725×10^{-2} (b) 5.491×10^4 (c) 9.84×10^0

(a) $1.725 \times 10^{-2} = \dfrac{1.725}{100} = \mathbf{0.01725}$

(b) $5.491 \times 10^4 = 5.491 \times 10000 = \mathbf{54\,910}$

(c) $9.84 \times 10^0 = 9.84 \times 1 = \mathbf{9.84}$ (since $10^0 = 1$)

Problem 16. Express in standard form, correct to 3 significant figures:
(a) $\dfrac{3}{8}$ (b) $19\dfrac{2}{3}$ (c) $741\dfrac{9}{16}$

(a) $\dfrac{3}{8} = 0.375$, and expressing it in standard form gives: $0.375 = \mathbf{3.75 \times 10^{-1}}$

(b) $19\dfrac{2}{3} = 19.\dot{6} = \mathbf{1.97 \times 10}$ in standard form, correct to 3 significant figures

(c) $741\dfrac{9}{16} = 741.5625 = \mathbf{7.42 \times 10^2}$ in standard form, correct to 3 significant figures

Problem 17. Express the following numbers, given in standard form, as fractions or mixed numbers: (a) 2.5×10^{-1} (b) 6.25×10^{-2} (c) 1.354×10^{2}

(a) $2.5\times10^{-1}=\frac{2.5}{10}=\frac{25}{100}=\mathbf{\frac{1}{4}}$

(b) $6.25\times10^{-2}=\frac{6.25}{100}=\frac{625}{10000}=\mathbf{\frac{1}{16}}$

(c) $1.354\times10^{2}=135.4=135\frac{4}{10}=\mathbf{135\frac{2}{5}}$

Now try the following Practice Exercise

Practice Exercise 8 Standard form (Answers on page 702)

In Problems 1 to 4, express in standard form:

1. (a) 73.9 (b) 28.4 (c) 197.72
2. (a) 2748 (b) 33 170 (c) 274 218
3. (a) 0.2401 (b) 0.0174 (c) 0.00923
4. (a) $\frac{1}{2}$ (b) $11\frac{7}{8}$ (c) $130\frac{3}{5}$ (d) $\frac{1}{32}$

In Problems 5 and 6, express the numbers given as integers or decimal fractions:

5. (a) 1.01×10^{3} (b) 9.327×10^{2}
   (c) 5.41×10^{4} (d) 7×10^{0}
6. (a) 3.89×10^{-2} (b) 6.741×10^{-1}
   (c) 8×10^{-3}
7. Write the following statements in standard form:
   (a) The density of aluminium is $2710\,\text{kg m}^{-3}$
   (b) Poisson's ratio for gold is 0.44
   (c) The impedance of free space is 376.73 Ω
   (d) The electron rest energy is 0.511 MeV
   (e) Proton charge-mass ratio is $95\,789\,700\,\text{C kg}^{-1}$
   (f) The normal volume of a perfect gas is $0.02241\,\text{m}^3\,\text{mol}^{-1}$

2.3 Engineering notation and common prefixes

Engineering notation is similar to scientific notation except that the power of ten is always a multiple of 3.

For example, $0.00035=3.5\times10^{-4}$ in scientific notation,

but $0.00035=0.35\times10^{-3}$ or 350×10^{-6} in engineering notation.

Units used in engineering and science may be made larger or smaller by using **prefixes** that denote multiplication or division by a particular amount. The eight most common multiples, with their meaning, are listed in Table 2.1, on page 19, where it is noticed that the prefixes involve powers of ten which are all multiples of 3. For example,

5 MV means $5\times1\,000\,000=5\times10^6$ $=\mathbf{5\,000\,000\ volts}$

3.6 kΩ means $3.6\times1000=3.6\times10^3$ $=\mathbf{3600\ ohms}$

7.5 μC means $7.5\div1\,000\,000=\frac{7.5}{10^6}$ or $7.5\times10^{-6}=\mathbf{0.0000075\ coulombs}$

and **4 mA** means 4×10^{-3} or $=\frac{4}{10^3}$ $=\frac{4}{1000}=\mathbf{0.004\ amperes}$

Similarly,

0.00006 J = 0.06 mJ or **60 μJ**

5 620 000 N = 5620 kN or **5.62 MN**

$\mathbf{47\times10^4\,\Omega}=470000\,\Omega=\mathbf{470\ k\Omega}$ or $\mathbf{0.47\ M\Omega}$

and $\mathbf{12\times10^{-5}\,A}=0.00012\,\text{A}=\mathbf{0.12\ mA}$ or $\mathbf{120\ \mu A}$

A calculator is needed for many engineering calculations, and having a calculator which has an 'ENG' function is most helpful.
For example, to calculate: $3\times10^4\times0.5\times10^{-6}$ volts, input your calculator in the following order: (a) Enter '3' (b) Press $\times10^x$ (c) Enter '4' (d) Press '×' (e) Enter '0.5' (f) Press $\times10^x$ (g) Enter '−6' (h) Press '='

The answer is **0.015 V** or $\left(\mathbf{\frac{7}{200}}\right)$ Now press the 'ENG' button, and the answer changes to $\mathbf{15\times10^{-3}\,V}$

Table 2.1

Prefix	Name	Meaning	
T	tera	multiply by 1 000 000 000 000	(i.e. $\times 10^{12}$)
G	giga	multiply by 1 000 000 000	(i.e. $\times 10^{9}$)
M	mega	multiply by 1 000 000	(i.e. $\times 10^{6}$)
k	kilo	multiply by 1000	(i.e. $\times 10^{3}$)
m	milli	divide by 1000	(i.e. $\times 10^{-3}$)
μ	micro	divide by 1 000 000	(i.e. $\times 10^{-6}$)
n	nano	divide by 1 000 000 000	(i.e. $\times 10^{-9}$)
p	pico	divide by 1 000 000 000 000	(i.e. $\times 10^{-12}$)

The 'ENG' or 'Engineering' button ensures that the value is stated to a power of 10 that is a multiple of 3, enabling you, in this example, to express the answer as **15 mV**

Problem 18. The strength value, V, of a rivet in double shear is given by the formula:

$$V = 2f_s\left(\frac{1}{4}\pi d^2\right)$$

Determine the strength of a 20 mm diameter, d, rivet in double shear, given that $f_s = 100\text{ N/mm}^2$. State the answer in kN, correct to 4 significant figures.

The strength value, $\mathbf{V} = 2f_s\left(\frac{1}{4}\pi d^2\right)$

$= 2(100\text{N/mm}^2)\left(\frac{1}{4}\pi(20)^2\text{ mm}^2\right)$

$= 62831.85$ N $=$ **62.83 kN** correct to 4 significant figures.

Problem 19. The heat h generated in an electrical circuit is given by: $h = I^2Rt$ joules, where I is the current in amperes, R the resistance in ohms and t the time in seconds. Calculate the heat generated when a current of 300 mA is maintained through a 2 kΩ resistor for 30 minutes.

Current, $I = 300$ mA $= 300 \times 10^{-3}$A, resistance $R = 2\text{k}\Omega = 2 \times 10^3\Omega$, and time $t = 30 \times 60$ s

Hence, heat generated,

$h = I^2Rt = (300 \times 10^{-3})^2 \times (2 \times 10^3) \times (30 \times 60)$

i.e. **heat generated, h = 324000 J = 324 kJ**

Now try the following Practice Exercise

Practice Exercise 9 Engineering notation and common prefixes (Answers on page 702)

1. Express the following in engineering notation and in prefix form:

   (a) 100 000 W (b) 0.00054 A
   (c) $15 \times 10^5\,\Omega$ (d) 225×10^{-4} V
   (e) 35 000 000 000 Hz (f) 1.5×10^{-11} F
   (g) 0.000017 A (h) 46200 Ω

2. Rewrite the following as indicated:

   (a) 0.025 mA = ...μA

   (b) 1000 pF = ...nF

   (c) 62×10^4 V = ...kV

   (d) 1 250 000 Ω = ...MΩ

3. Use a calculator to evaluate the following in engineering notation:

   (a) $4.5 \times 10^{-7} \times 3 \times 10^4$

   (b) $\dfrac{(1.6 \times 10^{-5})(25 \times 10^3)}{(100 \times 10^6)}$

4. The mass of a steel bolt is 4.50×10^{-4}kg. Calculate, in grams, the mass of 1600 bolts.

2.4 Metric conversions

Length in metric units

$$\mathbf{1\ m = 100\ cm = 1000\ mm}$$

$$\mathbf{1\ cm = \frac{1}{100}\ m = \frac{1}{10^2}\ m = 10^{-2}\ m}$$

$$\mathbf{1\ mm = \frac{1}{1000}\ m = \frac{1}{10^3}\ m = 10^{-3}\ m}$$

Problem 20. Rewrite 14,700 mm in metres

1 m = 1000 mm hence, $1\text{ mm} = \frac{1}{1000} = \frac{1}{10^3} = 10^{-3}$ m

Hence, $14,700\text{ mm} = 14,700 \times 10^{-3}\text{m} = \mathbf{14.7\ m}$

Problem 21. Rewrite 276 cm in metres

1 m = 100 cm hence, $1\text{ cm} = \frac{1}{100} = \frac{1}{10^2} = 10^{-2}$ m

Hence, $276\text{ cm} = 276 \times 10^{-2}\text{m} = \mathbf{2.76\ m}$

Now try the following Practice Exercise

Practice Exercise 10 Length in metric units (Answers on page 702)

1. State 2.45 m in millimetres
2. State 1.675 m in centimetres
3. State the number of millimetres in 65.8 cm
4. Rewrite 25,400 mm in metres
5. Rewrite 5632 cm in metres
6. State the number of millimetres in 4.356 m
7. How many centimetres are there in 0.875 m?
8. State a length of 465 cm in (a) mm (b) m
9. State a length of 5040 mm in (a) cm (b) m
10. A machine part is measured as 15.0 cm ± 1%. Between what two values would the measurement be? Give the answer in millimetres.

Areas in metric units

Area is a measure of the size or extent of a plane surface.
Area is measured in **square units** such as mm^2, cm^2 and m^2.

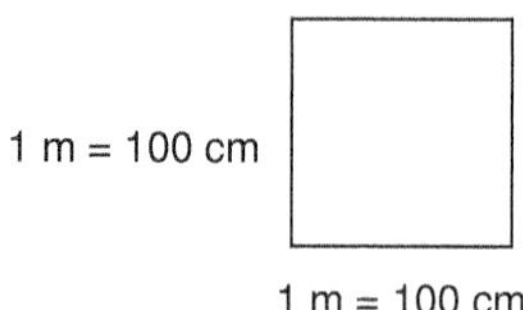

The area of the above square is 1 m^2

$$\mathbf{1\ m^2 = 100\ cm \times 100\ cm = 10000\ cm^2 = 10^4 cm^2}$$

i.e. to change from square metres to square centimetres, multiply by 10^4
Hence, $\mathbf{2.5\ m^2 = 2.5 \times 10^4\ cm^2}$
and $\mathbf{0.75\ m^2 = 0.75 \times 10^4\ cm^2}$

Since $\mathbf{1\ m^2 = 10^4\ cm^2}$ then $\mathbf{1 cm^2 = \frac{1}{10^4} m^2 = 10^{-4} m^2}$
i.e. to change from square centimetres to square metres, multiply by 10^{-4}
Hence, $\mathbf{52\ cm^2 = 52 \times 10^{-4}\ m^2}$
and $\mathbf{643\ cm^2 = 643 \times 10^{-4}\ m^2}$

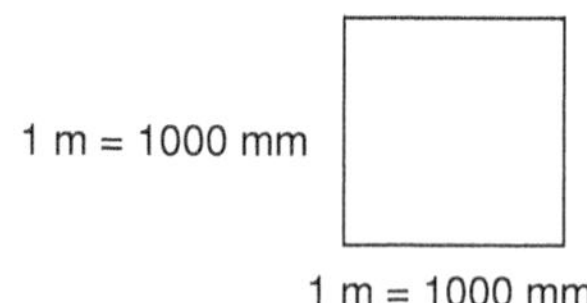

The area of the above square is 1m^2

$$\mathbf{1\ m^2 = 1000\ mm \times 1000\ mm = 1000000\ mm^2 = 10^6 mm^2}$$

i.e. to change from square metres to square millimetres, multiply by 10^6
Hence, $\mathbf{7.5\ m^2 = 7.5 \times 10^6 mm^2}$

and $\mathbf{0.63\,m^2 = 0.63 \times 10^6 mm^2}$
Since $\mathbf{1\,m^2 = 10^6 mm^2}$
then $\mathbf{1\,mm^2 = \frac{1}{10^6} m^2 = 10^{-6}\,m^2}$
i.e. to change from square millimetres to square metres, multiply by 10^{-6}
Hence, $\mathbf{235\,mm^2 = 235 \times 10^{-6} m^2}$
and $\mathbf{47\,mm^2 = 47 \times 10^{-6} m^2}$

1 cm = 10 mm

1 cm = 10 mm

The area of the above square is 1 cm^2

$$\mathbf{1\,cm^2 = 10\,mm \times 10\,mm = 100\,mm^2 = 10^2 mm^2}$$

i.e. to change from square centimetres to square millimetres, multiply by 100 or 10^2
Hence, $\mathbf{3.5\,cm^2 = 3.5 \times 10^2 mm^2 = 350\,mm^2}$
and $\mathbf{0.75\,cm^2 = 0.75 \times 10^2 mm^2 = 75\,mm^2}$
Since $\mathbf{1\,cm^2 = 10^2 mm^2}$
then $\mathbf{1\,mm^2 = \frac{1}{10^2} cm^2 = 10^{-2}\,cm^2}$
i.e. to change from square millimetres to square centimetres, multiply by 10^{-2}
Hence, $\mathbf{250\,mm^2 = 250 \times 10^{-2} cm^2 = 2.5\,cm^2}$
and $\mathbf{85\,mm^2 = 85 \times 10^{-2} cm^2 = 0.85\,cm^2}$

Problem 22. Rewrite 12 m^2 in square centimetres

$1\,m^2 = 10^4 cm^2$ hence, $\mathbf{12\,m^2 = 12 \times 10^4 cm^2}$

Problem 23. Rewrite 50 cm^2 in square metres

$1\,cm^2 = 10^{-4} m^2$ hence, $\mathbf{50\,cm^2 = 50 \times 10^{-4} m^2}$

Problem 24. Rewrite 2.4 m^2 in square millimetres

$1\,m^2 = 10^6 mm^2$ hence, $\mathbf{2.4\,m^2 = 2.4 \times 10^6 mm^2}$

Problem 25. Rewrite 147 mm^2 in square metres

$1\,mm^2 = 10^{-6} m^2$ hence, $\mathbf{147\,mm^2 = 147 \times 10^{-6} m^2}$

Problem 26. Rewrite 34.5 cm^2 in square millimetres

$1\,cm^2 = 10^2 mm^2$
hence, $\mathbf{34.5\,cm^2 = 34.5 \times 10^2 mm^2 = 3450\,mm^2}$

Problem 27. Rewrite 400 mm^2 in square centimetres

$1\,mm^2 = 10^{-2} cm^2$
hence, $\mathbf{400\,mm^2 = 400 \times 10^{-2} cm^2 = 4\,cm^2}$

Problem 28. The top of a small rectangular table is 800 mm long and 500 mm wide. Determine its
area in (a) mm^2 (b) cm^2 (c) m^2

(a) **Area of rectangular table top** $= l \times b = 800 \times 500$
$= \mathbf{400{,}000\,mm^2}$

(b) Since 1 cm = 10 mm then
$1\,cm^2 = 1\,cm \times 1\,cm = 10\,mm \times 10\,mm = 100\,mm^2$

or $1\,mm^2 = \frac{1}{100} = 0.01\,cm^2$

Hence, $\mathbf{400{,}000\,mm^2} = 400{,}000 \times 0.01\,cm^2$
$= \mathbf{4000\,cm^2}$

(c) $1\,cm^2 = 10^{-4} m^2$
hence, $\mathbf{4000\,cm^2} = 4000 \times 10^{-4} m^2 = \mathbf{0.4\,m^2}$

Now try the following Practice Exercise

Practice Exercise 11 Areas in metric units (Answers on page 702)

1. Rewrite 8 m^2 in square centimetres
2. Rewrite 240 cm^2 in square metres
3. Rewrite 3.6 m^2 in square millimetres
4. Rewrite 350 mm^2 in square metres

5. Rewrite $50\,\text{cm}^2$ in square millimetres
6. Rewrite $250\,\text{mm}^2$ in square centimetres
7. A rectangular piece of metal is 720 mm long and 400 mm wide. Determine its area in

   (a) mm^2 (b) cm^2 (c) m^2

Volumes in metric units

The **volume** of any solid is a measure of the space occupied by the solid.
Volume is measured in **cubic units** such as mm^3, cm^3 and m^3.

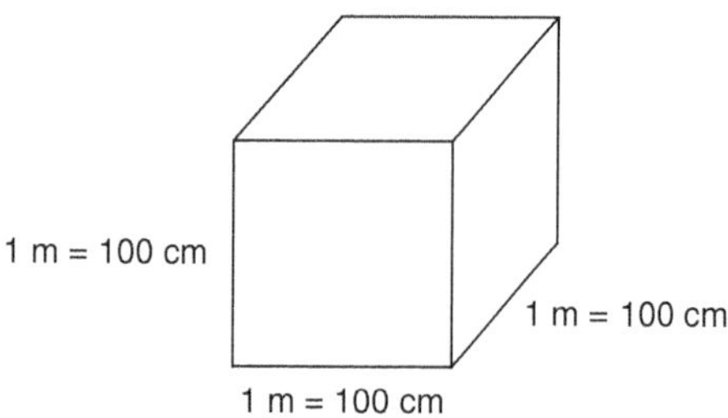

The volume of the cube shown is $1\ \text{m}^3$

$$\mathbf{1\,m^3 = 100\,cm \times 100\,cm \times 100\,cm}$$

$$\mathbf{= 1000000\,cm^2 = 10^6 cm^2}$$

$$\mathbf{1\ litre = 1000\ cm^3}$$

i.e. to change from cubic metres to cubic centimetres, multiply by 10^6
Hence, $\mathbf{3.2\,m^3 = 3.2 \times 10^6 cm^3}$
and $\mathbf{0.43\,m^3 = 0.43 \times 10^6 cm^3}$

Since $\mathbf{1\,m^3 = 10^6 cm^3}$ then
$\mathbf{1\,cm^3 = \frac{1}{10^6} m^3 = 10^{-6}\,m^3}$
i.e. to change from cubic centimetres to cubic metres, multiply by 10^{-6}
Hence, $\mathbf{140\,cm^3 = 140 \times 10^{-6} m^3}$
and $\mathbf{2500\,cm^3 = 2500 \times 10^{-6} m^3}$

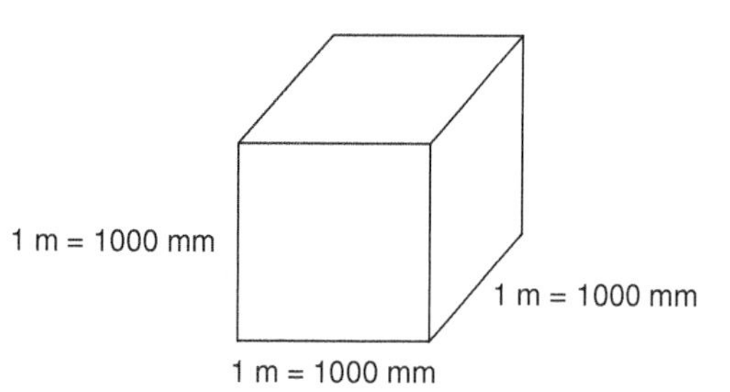

The volume of the cube shown is $1\ \text{m}^3$

$$\mathbf{1\ m^3 = 1000\,mm \times 1000\,mm \times 1000\,mm}$$

$$\mathbf{= 1000000000\,mm^3 = 10^9 mm^3}$$

i.e. to change from cubic metres to cubic millimetres, multiply by 10^9
Hence, $\mathbf{4.5\,m^3 = 4.5 \times 10^9 mm^3}$
and $\mathbf{0.25\,m^3 = 0.25 \times 10^9 mm^3}$

Since $\mathbf{1\,m^3 = 10^9 mm^2}$

then $\mathbf{1\,mm^3 = \frac{1}{10^9} m^3 = 10^{-9}\,m^3}$

i.e. to change from cubic millimetres to cubic metres, multiply by 10^{-9}
Hence, $\mathbf{500\,mm^3 = 500 \times 10^{-9} m^3}$
and $\mathbf{4675\,mm^3 = 4675 \times 10^{-9} m^3}$ or $\mathbf{4.675 \times 10^{-6} m^3}$

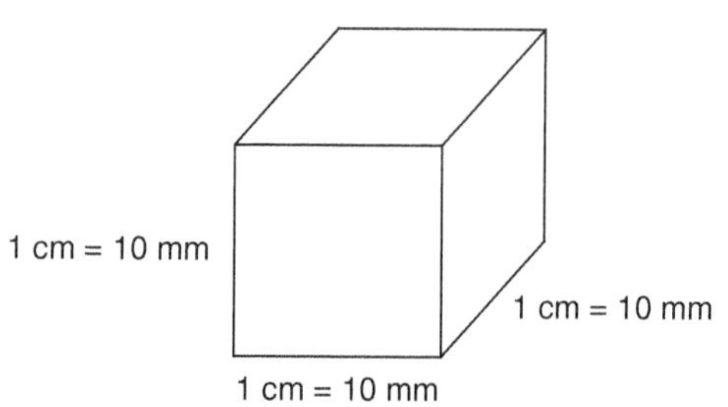

The volume of the cube shown is $1\ \text{cm}^3$

$$\mathbf{1\ cm^3 = 10\,mm \times 10\,mm \times 10\,mm}$$

$$\mathbf{= 1000\,mm^3 = 10^3 mm^3}$$

i.e. to change from cubic centimetres to cubic millimetres, multiply by 1000 or 10^3
Hence, $\mathbf{5\,cm^3 = 5 \times 10^3 mm^3 = 5000\,mm^3}$
and $\mathbf{0.35\,cm^3 = 0.35 \times 10^3 mm^3 = 350\,mm^3}$

Since $\mathbf{1\,cm^3 = 10^3 mm^3}$

then $\mathbf{1\ mm^3 = \frac{1}{10^3} cm^3 = 10^{-3}\,cm^3}$

i.e. to change from cubic millimetres to cubic centimetres, multiply by 10^{-3}
Hence, $\mathbf{650\,mm^3 = 650 \times 10^{-3} cm^3 = 0.65\,cm^3}$
and $\mathbf{75\,mm^3 = 75 \times 10^{-3} cm^3 = 0.075\,cm^3}$

Problem 29. Rewrite $1.5\,\text{m}^3$ in cubic centimetres

$1\,m^3 = 10^6 cm^3$ hence, $\mathbf{1.5\,m^3 = 1.5 \times 10^6 cm^3}$

Problem 30. Rewrite $300\,cm^3$ in cubic metres

$1\,cm^3 = 10^{-6} m^3$ hence, $\mathbf{300\,cm^3 = 300 \times 10^{-6} m^3}$

Problem 31. Rewrite $0.56\,m^3$ in cubic millimetres

$1\,m^3 = 10^9 mm^3$ hence, $\mathbf{0.56\,m^3 = 0.56 \times 10^9 mm^3}$ or $\mathbf{560 \times 10^6 mm^3}$

Problem 32. Rewrite $1250\,mm^3$ in cubic metres

$1\,mm^3 = 10^{-9} m^3$ hence,
$\mathbf{1250\,mm^3 = 1250 \times 10^{-9} m^3}$ or $\mathbf{1.25 \times 10^{-6} m^3}$

Problem 33. Rewrite $8\,cm^3$ in cubic millimetres

$1\,cm^3 = 10^3 mm^3$
hence, $\mathbf{8\,cm^3 = 8 \times 10^3 mm^3 = 8000\,mm^3}$

Problem 34. Rewrite $600\,mm^3$ in cubic centimetres

$1\,mm^3 = 10^{-3} cm^3$
hence, $\mathbf{600\,mm^3 = 600 \times 10^{-3} cm^3 = 0.6\,cm^3}$

Problem 35. A water tank is in the shape of a rectangular prism having length 1.2 m, breadth 50 cm and height 250 mm. Determine the capacity of the tank (a) m^3 (b) cm^3 (c) litres

Capacity means volume. When dealing with liquids, the word capacity is usually used.

(a) Capacity of water tank = $l \times b \times h$
where l = 1.2 m, b = 50 cm and h = 250 mm.
To use this formula, all dimensions **must** be in the same units.
Thus, l = 1.2 m, b = 0.50 m and h = 0.25 m (since 1 m = 100 cm = 1000 mm)
Hence, **capacity of tank** = $1.2 \times 0.50 \times 0.25$
$= \mathbf{0.15\ m^3}$

(b) $1\,m^3 = 10^6 cm^3$
Hence, **capacity** $= 0.15\ m^3 = 0.15 \times 10^6\,cm^3$
$= \mathbf{150{,}000\ cm^3}$

(c) 1 litre = 1000 cm^3
Hence, **150,000** $cm^3 = \dfrac{150{,}000}{1000} = $ **150 litres**

Now try the following Practice Exercise

Practice Exercise 12 Volumes in metric units (Answers on page 702)

1. Rewrite $2.5\,m^3$ in cubic centimetres
2. Rewrite $400\,cm^3$ in cubic metres
3. Rewrite $0.87\,m^3$ in cubic millimetres
4. Change a volume of 2,400,000 cm^3 to cubic metres.
5. Rewrite $1500\,mm^3$ in cubic metres
6. Rewrite $400\,mm^3$ in cubic centimetres
7. Rewrite $6.4\,cm^3$ in cubic millimetres
8. Change a volume of 7500 mm^3 to cubic centimetres.
9. An oil tank is in the shape of a rectangular prism having length 1.5 m, breadth 60 cm and height 200 mm. Determine the capacity of the tank in (a) m^3 (b) cm^3 (c) litres

2.5 Metric - US/imperial conversions

The Imperial System (which uses yards, feet, inches, etc to measure length) was developed over hundreds of years in the UK. Then the French developed the Metric System (metres) in 1670, which soon spread through Europe, even to England itself in 1960. But the USA and a few other countries still prefer feet and inches.
When converting from metric to imperial units, or vice versa, one of the following tables (2.2 to 2.9) should help.

Problem 36. Calculate the number of inches in 350 mm, correct to 2 decimal places

350 mm = 350×0.03937 inches = **13.78 inches** from Table 2.2

Table 2.2 Metric to imperial length

Metric	US or Imperial
1 millimetre, mm	0.03937 inch
1 centimetre, cm = 10 mm	0.3937 inch
1 metre, m = 100 cm	1.0936 yard
1 kilometre, km = 1000 m	0.6214 mile

Problem 37. Calculate the number of inches in 52 cm, correct to 4 significant figures

52 cm = 52 × 0.3937 inches = **20.47 inches** from Table 2.2

Problem 38. Calculate the number of yards in 74 m, correct to 2 decimal places

74 m = 74 × 1.0936 yards = **80.93 yds** from Table 2.2

Problem 39. Calculate the number of miles in 12.5 km, correct to 3 significant figures

12.5 km = 12.5 × 0.6214 miles = **7.77 miles** from Table 2.2

Table 2.3 Imperial to metric length

US or Imperial	Metric
1 inch, in	2.54 cm
1 foot, ft = 12 in	0.3048 m
1 yard, yd = 3 ft	0.9144 m
1 mile = 1760 yd	1.6093 km
1 nautical mile = 2025.4 yd	1.853 km

Problem 40. Calculate the number of centimetres in 35 inches, correct to 1 decimal places

35 inches = 35 × 2.54 cm = **88.9 cm** from Table 2.3

Problem 41. Calculate the number of metres in 66 inches, correct to 2 decimal places

66 inches = $\frac{66}{12}$ feet = $\frac{66}{12} \times 0.3048$ m = **1.68 m** from Table 2.3

Problem 42. Calculate the number of metres in 50 yards, correct to 2 decimal places

50 yards = 50 × 0.9144 m = **45.72 m** from Table 2.3

Problem 43. Calculate the number of kilometres in 7.2 miles, correct to 2 decimal places

7.2 miles = 7.2 × 1.6093 km = **11.59 km** from Table 2.3

Problem 44. Calculate the number of (a) yards (b) kilometres in 5.2 nautical miles

(a) 5.2 nautical miles = 5.2 × 2025.4 yards
= **10532 yards** from Table 2.3

(b) 5.2 nautical miles = 5.2 × 1.853 km
= **9.636 km** from Table 2.3

Table 2.4 Metric to imperial area

Metric	US or Imperial
1 cm^2 = 100 mm^2	0.1550 in^2
1 m^2 = 10,000 cm^2	1.1960 yd^2
1 hectare, ha = 10,000 m^2	2.4711 acres
1 km^2 = 100 ha	0.3861 $mile^2$

Problem 45. Calculate the number of square inches in 47 cm^2, correct to 4 significant figures

47 cm^2 = 47 × 0.1550 in^2 = **7.285** in^2 from Table 2.4

Problem 46. Calculate the number of square yards in 20 m^2, correct to 2 decimal places

20 m^2 = 20 × 1.1960 yd^2 = **23.92 yd^2** from Table 2.4

Problem 47. Calculate the number of acres in 23 hectares of land, correct to 2 decimal places

23 hectares = 23 × 2.4711 acres = **56.84 acres** from Table 2.4

Problem 48. Calculate the number of square miles in a field of 15 km^2 area, correct to 2 decimal places

15 km^2= 15 × 0.3861 mile^2
= **5.79 mile^2** from Table 2.4

Table 2.5 Imperial to metric area

US or Imperial	Metric
1 in^2	6.4516 cm^2
1 ft^2= 144 in^2	0.0929 m^2
1 yd^2= 9 ft^2	0.8361 m^2
1 acre = 4840 yd^2	4046.9 m^2
1 mile^2= 640 acres	2.59 km^2

Problem 49. Calculate the number of square centimetres in 17.5 in^2, correct to the nearest square centimetre

17.5 in^2 = 17.5 × 6.4516 cm^2 = **113 cm^2** from Table 2.5

Problem 50. Calculate the number of square metres in 205 ft^2, correct to 2 decimal places

205 ft^2 = 205 × 0.0929 m^2 = **19.04 m ^{2}** from Table 2.5

Problem 51. Calculate the number of square metres in 11.2 acres, correct to the nearest square metre

11.2 acres = 11.2 × 4046.9 m^2 = **45325 m ^{2}** from Table 2.5

Problem 52. Calculate the number of square kilometres in 12.6 mile^2, correct to 2 decimal places

12.6 mile^2 = 12.6 × 2.59 km^2 = **32.63 km ^{2}** from Table 2.5

Table 2.6 Metric to imperial volume/capacity

Metric	US or Imperial
1 cm^3	0.0610 in^3
1 dm^3 =1000 cm^3	0.0353 ft^3
1 m^3= 1000dm^3	1.3080 yd^3
1 litre = 1 dm^3= 1000 cm^3	2.113 fluid pt = 1.7598 pt

Problem 53. Calculate the number of cubic inches in 123.5 cm^3, correct to 2 decimal places

123.5 cm^3 = 123.5 × 0.0610 cm^3 = **7.53 cm^3** from Table 2.6

Problem 54. Calculate the number of cubic feet in 144 dm^3, correct to 3 decimal places

144 dm^3 = 144 × 0.0353 ft^3 = **5.083 ft^3** from Table 2.6

Problem 55. Calculate the number of cubic yards in 5.75 m^3, correct to 4 significant figures

5.75 m^3 = 5.75 × 1.3080 yd^3 = **7.521 yd^3** from Table 2.6

Problem 56. Calculate the number of US fluid pints in 6.34 litres of oil, correct to 1 decimal place

6.34 litre = 6.34 × 2.113 US fluid pints = **13.4 US fluid pints** from Table 2.6

Problem 57. Calculate the number of cubic centimetres in 3.75 in^3, correct to 2 decimal places

3.75 in^3 = 3.75 × 16.387 cm^3 = **61.45 cm^3** from Table 2.7

Table 2.7 Imperial to metric volume/capacity

US or Imperial	Metric
1 in^3	16.387 cm^3
1 ft^3	0.02832 m^3
1 US fl oz = 1.0408 UK fl oz	0.0296 litres
1 US pint (16 fl oz) = 0.8327 UK pt	0.4732 litre
1 US gal (231 in^3) = 0.8327 UK gal	3.7854 litre

Problem 58. Calculate the number of cubic metres in 210 ft^3, correct to 3 significant figures

210 ft^3 = 210 × 0.02832 m^3 = **5.95 m^3** from Table 2.7

Problem 59. Calculate the number of litres in 4.32 US pints, correct to 3 decimal places

4.32 US pints = 4.32 × 0.4732 litres = **2.044 litres** from Table 2.7

Problem 60. Calculate the number of litres in 8.62 US gallons, correct to 2 decimal places

8.62 US gallons = 8.62 × 3.7854 litre = **32.63 litre** from Table 2.7

Table 2.8 Metric to imperial mass

Metric	US or Imperial
1 g = 1000 mg	0.0353 oz
1 kg = 1000 g	2.2046 lb
1 tonne, t, = 1000 kg	1.1023 short ton
1 tonne, t, = 1000 kg	0.9842 long ton

The British ton is the long ton, which is 2240 pounds, and the US ton is the short ton which is 2000 pounds.

Problem 61. Calculate the number of ounces in a mass of 1346 g, correct to 2 decimal places

1346 g = 1346 × 0.0353 oz = **47.51 oz** from Table 2.8

Problem 62. Calculate the mass, in pounds, in a 210.4 kg mass, correct to 4 significant figures

210.4 kg = 210.4 × 2.2046 lb = **463.8 lb** from Table 2.8

Problem 63. Calculate the number of short tons in 5000 kg, correct to 2 decimal places

5000 kg = 5 t = 5 × 1.1023 short tons = **5.51 short tons** from Table 2.8

Table 2.9 Imperial to metric mass

US or Imperial	Metric
1 oz = 437.5 grain	28.35 g
1 lb = 16 oz	0.4536 kg
1 stone = 14 lb	6.3503 kg
1 hundredweight, cwt = 112 lb	50.802 kg
1 short ton	0.9072 tonne
1 long ton	1.0160 tonne

Problem 64. Calculate the number of grams in 5.63 oz, correct to 4 significant figures

5.63 oz = 5.63 × 28.35 g = **159.6 g** from Table 2.9

Problem 65. Calculate the number of kilograms in 75 oz, correct to 3 decimal places

75 oz = $\frac{75}{16}$ lb = $\frac{75}{16}$ × 0.4536 kg = **2.126 kg** from Table 2.9

Problem 66. Convert 3.25 cwt into (a) pounds (b) kilograms

(a) 3.25 cwt = 3.25 × 112 lb = **364 lb** from Table 2.9

(b) 3.25 cwt = 3.25 × 50.802 kg = **165.1 kg** from Table 2.9

Temperature

To convert from Celsius to Fahrenheit, first multiply by 9/5, then add 32.
To convert from Fahrenheit to Celsius, first subtract 32, then multiply by 5/9

Problem 67. Convert 35° C to degrees Fahrenheit

$F = \frac{9}{5}C + 32$, hence $35°\,C = \frac{9}{5}(35) + 32 = 63 + 32$
$= \mathbf{95°\,F}$

Problem 68. Convert 113° F to degrees Celsius

$C = \frac{5}{9}(F - 32)$, hence $113°\,F = \frac{5}{9}(113 - 32)$
$= \frac{5}{9}(81) = \mathbf{45°\,C}$

Now try the following Practice Exercise

Practice Exercise 13 Metric/imperial conversions (Answers on page 702)

In the following Problems, use the metric/imperial conversions in Tables 2.2 to 2.9

1. Calculate the number of inches in 476 mm, correct to 2 decimal places
2. Calculate the number of inches in 209 cm, correct to 4 significant figures
3. Calculate the number of yards in 34.7 m, correct to 2 decimal places
4. Calculate the number of miles in 29.55 km, correct to 2 decimal places
5. Calculate the number of centimetres in 16.4 inches, correct to 2 decimal places
6. Calculate the number of metres in 78 inches, correct to 2 decimal places
7. Calculate the number of metres in 15.7 yards, correct to 2 decimal places
8. Calculate the number of kilometres in 3.67 miles, correct to 2 decimal places
9. Calculate the number of (a) yards (b) kilometres in 11.23 nautical miles
10. Calculate the number of square inches in 62.5 cm^2, correct to 4 significant figures
11. Calculate the number of square yards in 15.2 m^2, correct to 2 decimal places
12. Calculate the number of acres in 12.5 hectares, correct to 2 decimal places
13. Calculate the number of square miles in 56.7 km^2, correct to 2 decimal places
14. Calculate the number of square centimetres in 6.37 in^2, correct to the nearest square centimetre
15. Calculate the number of square metres in 308.6 ft^2, correct to 2 decimal places
16. Calculate the number of square metres in 2.5 acres, correct to the nearest square metre
17. Calculate the number of square kilometres in 21.3 $mile^2$, correct to 2 decimal places
18. Calculate the number of cubic inches in 200.7 cm^3, correct to 2 decimal places
19. Calculate the number of cubic feet in 214.5 dm^3, correct to 3 decimal places
20. Calculate the number of cubic yards in 13.45 m^3, correct to 4 significant figures
21. Calculate the number of US fluid pints in 15 litres, correct to 1 decimal place
22. Calculate the number of cubic centimetres in 2.15 in^3, correct to 2 decimal places
23. Calculate the number of cubic metres in 175 ft^3, correct to 4 significant figures
24. Calculate the number of litres in 7.75 US pints, correct to 3 decimal places
25. Calculate the number of litres in 12.5 US gallons, correct to 2 decimal places
26. Calculate the number of ounces in 980 g, correct to 2 decimal places
27. Calculate the mass, in pounds, in 55 kg, correct to 4 significant figures

28. Calculate the number of short tons in 4000 kg, correct to 3 decimal places

29. Calculate the number of grams in 7.78 oz, correct to 4 significant figures

30. Calculate the number of kilograms in 57.5 oz, correct to 3 decimal places

31. Convert 2.5 cwt into (a) pounds (b) kilograms

32. Convert 55°C to degrees Fahrenheit

33. Convert 167°F to degrees Celsius

Practice Exercise 14 Multiple-choice questions on indices, engineering notation and metric conversions (Answers on page 703)

Each question has only one correct answer

1. The value of $\frac{2^{-3}}{2^{-4}} - 1$ is equal to:
(a) $-\frac{1}{2}$ (b) 2 (c) $\frac{1}{2}$ (d) 1

2. When $p = 3, q = -\frac{1}{2}$ and $r = -2$, the engineering expression $2p^2q^3r^4$ is equal to:
(a) 1296 (b) −36 (c) 36 (d) 18

3. $16^{-\frac{3}{4}}$ is equal to:
(a) 8 (b) $-\frac{1}{2^3}$ (c) 4 (d) $\frac{1}{8}$

4. $p \times \left(\frac{1}{p^{-2}}\right)^{-3}$ is equivalent to:
(a) $\frac{1}{p^{\frac{3}{2}}}$ (b) p^{-5} (c) p^5 (d) $\frac{1}{p^{-5}}$

5. $(16^{-\frac{1}{4}} - 27^{-\frac{2}{3}})$ is equal to:
(a) −7 (b) $\frac{7}{18}$ (c) $1\frac{8}{9}$ (d) $-8\frac{1}{2}$

6. $\left(\frac{y^2}{y^3} \times y^4\right)^2$ is equivalent to:
(a) y^3 (b) y^{-6} (c) y^6 (d) $\frac{1}{y}$

7. $(3x)^2$ is equivalent to:
(a) $6x^2$ (b) $3x^2$ (c) $9x^2$ (d) $3x^3$

8. $\left(-\frac{2}{5}\right)^3$ is equal to:
(a) $-\frac{8}{125}$ (b) $-\frac{6}{15}$ (c) $\frac{8}{125}$ (d) $\frac{6}{15}$

9. $\left(\frac{1}{4}\right)^{-\frac{1}{2}}$ is equal to:
(a) 2 (b) $\frac{1}{16}$ (c) $\sqrt{2}$ (d) $\frac{1}{2}$

10. 30 to the power of zero is:
(a) 0 (b) −30 (c) 1 (d) 30

11. $125^{-\frac{2}{3}}$ is equal to:
(a) 25 (b) $-\frac{1}{25}$ (c) −25 (d) $\frac{1}{25}$

12. $32^{\frac{1}{5}}$ is equal to:
(a) 2 (b) 18 (c) 16 (d) 160

13. $9^{\frac{3}{2}}$ is equal to:
(a) 18 (b) 27 (c) −18 (d) $\frac{1}{27}$

14. $64^{-\frac{3}{2}}$ is equal to:
(a) $\frac{1}{96}$ (b) $\frac{1}{64}$ (c) 512 (d) $\frac{1}{512}$

15. 0.088 m in centimetres is:
(a) 88 (b) 8.8 (c) 880 (d) 0.88

16. 5 g written in kg is:
(a) 5000 (b) 0.05 (c) 0.005 (d) 0.0005

17. 650 mg written in g is:
(a) 0.65 (b) 65 (c) 6500 (d) 0.06530

18. Converting 500 litres into cubic metres gives:
(a) 0.5 (b) 5 (c) 50 (d) 0.005

19. Converting 1.2 m^2 into cm^2 gives:
(a) 12000 (b) 120
(c) 1200 (d) 0.012

20. 17 cm is the same as:
(a) 1.7 m (b) 0.017 m
(c) 0.17 m (d) 170 m

21. 0.000095 H is the same as:
(a) 9.5×10^{-5} H in engineering notation
(b) 95×10^{-6} in prefix form
(c) 95×10^{-4} H in engineering notation
(d) 95 μH in prefix form

22. Four masses are 1850 mg, 1.65 g, 0.002 kg and 2.5×10^3mg. The lightest of these is:
(a) 1850 mg (b) 0.002 kg
(c) 1.65 g (d) 2.5×10^3 mg

23. 3.485 kg is the same mass as:
(a) 0.003485 g (b) 3.485×10^{-3} g
(c) 348.5 g (d) 3485 g

24. Resistance, R, is given by the formula $R = \frac{\rho \ell}{A}$. When resistivity $\rho = 0.017 \mu\Omega$ m, length $\ell = 5$ km and cross-sectional area $A = 25$ mm^2, the resistance is:
(a) 3.4 mΩ (b) 0.034 Ω
(c) 3.4 kΩ (d) 3.4 Ω

25. Modulus of elasticity E is given by the equation $E = \frac{F\ell}{\pi r^2 x}$, where F is measured in newtons, and ℓ, r and x are measured in metres. If $F = 3.2$ kN, $\ell = 0.56$ m, $r = 5$ mm, and $x = 2$ mm, the value of E is:
(a) 1141 MN/m^2
(b) 11.41 GN/m^2
(c) 114.1 MN/m^2
(d) 1.141×10^8 N/m^2

For fully worked solutions to each of the problems in Practice Exercises 6 to 13 in this chapter, go to the website:

www.routledge.com/cw/bird

COMPANION @ WEBSITE

Chapter 3

Binary, octal and hexadecimal numbers

Why it is important to understand: Binary, octal and hexadecimal numbers

There are infinite ways to represent a number. The four commonly associated with modern computers and digital electronics are decimal, binary, octal, and hexadecimal. All four number systems are equally capable of representing any number. Furthermore, a number can be perfectly converted between the various number systems without any loss of numeric value. At a first look, it seems like using any number system other than decimal is complicated and unnecessary. However, since the job of electrical and software engineers is to work with digital circuits, engineers require number systems that can best transfer information between the human world and the digital circuit world. Thus the way in which a number is represented can make it easier for the engineer to perceive the meaning of the number as it applies to a digital circuit, i.e. the appropriate number system can actually make things less complicated. Binary, octal and hexadecimal numbers are explained in this chapter.

At the end of this chapter, you should be able to:

- recognise a binary number
- convert binary to decimal and vice-versa
- add binary numbers
- recognise an octal number
- convert decimal to binary via octal and vice-versa
- recognise a hexadecimal number
- convert from hexadecimal to decimal and vice-versa
- convert from binary to hexadecimal and vice-versa

3.1 Introduction

Man's earliest number or counting system was probably developed to help determine how many possessions a person had. As daily activities became more complex, numbers became more important in trade, time, distance, and all other phases of human life. Ever since people discovered that it was necessary to count objects, they have been looking for easier ways to do so. The **abacus**, developed by the Chinese, is one of the earliest known calculators; it is still in use in some parts of the world. **Blaise Pascal*** invented the first adding machine in 1642. Twenty years later, an Englishman, **Sir Samuel Morland***, developed a more compact device that could multiply, add and subtract. About 1672, **Gottfried Wilhelm von Leibniz*** perfected a machine that could perform all the basic operations (add, subtract, multiply, divide), as well as extract the square root. Modern electronic digital computers still use von Leibniz's principles.

Computers are now employed wherever repeated calculations or the processing of huge amounts of data is needed. The greatest applications are found in the military, scientific, and commercial fields. They have applications that range from mail sorting and engineering design, to the identification and destruction of enemy targets. The advantages of digital computers include speed, accuracy and man-power savings.

*Who was Leibniz? – **Gottfried Wilhelm Leibniz** (sometimes von Leibniz) (1 July, 1646–14 November, 1716) was a German mathematician and philosopher.

*Who was Pascal? – See page 157.

*Who was Morland? – **Sir Samuel Morland**, 1st Baronet (1625–1695), was an English academic, diplomat, spy, inventor and mathematician of the 17th century, a polymath credited with early developments in relation to computing, hydraulics and steam power.

To find out more go to www.routledge.com/cw/bird

Often computers are able to take over routine jobs and release personnel for more important work that cannot be handled by a computer. People and computers do not normally speak the same language. Methods of translating information into forms that are understandable and usable to both are necessary. Humans generally speak in words and numbers expressed in the decimal number system, while computers only understand coded electronic pulses that represent digital information.

All data in modern computers is stored as series of **bits**, a bit being a **bi**nary digi**t**, and can have one of two values, the numbers 0 and 1. The most basic form of representing computer data is to represent a piece of data as a string of 1's and 0's, one for each bit. This is called a **binary** or base-2 number.

Because binary notation requires so many bits to represent relatively small numbers, two further compact notations are often used, called **octal** and **hexadecimal**. Computer programmers who design sequences of number codes instructing a computer what to do, would have a very difficult task if they were forced to work with nothing but long strings of 1's and 0's, the 'native language' of any digital circuit.

Octal notation represents data as **base-8 numbers** with each digit in an octal number representing three bits. Similarly, hexadecimal notation uses **base-16 numbers**, representing four bits with each digit. Octal numbers use only the digits 0–7, while hexadecimal numbers use all ten base-10 digits (0–9) and the letters A–F (representing the numbers 10–15).

This chapter explains how to convert between the decimal, binary, octal and hexadecimal systems.

3.2 Binary numbers

The system of numbers in everyday use is the **denary** or **decimal** system of numbers, using the digits 0 to 9. It has ten different digits (0, 1, 2, 3, 4, 5, 6, 7, 8 and 9) and is said to have a **radix** or **base** of 10.

The **binary** system of numbers has a radix of 2 and uses only the digits 0 and 1.

(a) Conversion of binary to decimal:

The decimal number 234.5 is equivalent to

$$2\times 10^2+3\times 10^1+4\times 10^0+5\times 10^{-1}$$

i.e. is the sum of term comprising: (a digit) multiplied by (the base raised to some power).

In the binary system of numbers, the base is 2, so 1101.1 is equivalent to:

$$1 \times 2^3 + 1 \times 2^2 + 0 \times 2^1 + 1 \times 2^0 + 1 \times 2^{-1}$$

Thus the decimal number equivalent to the binary number 1101.1 is

$$8 + 4 + 0 + 1 + \frac{1}{2}, \text{ that is } 13.5$$

i.e. $\mathbf{1101.1_2 = 13.5_{10}}$, the suffixes 2 and 10 denoting binary and decimal systems of number respectively.

Problem 1. Convert 11011_2 to a decimal number

From above: $$\begin{aligned} 11011_2 &= 1 \times 2^4 + 1 \times 2^3 + 0 \times 2^2 \\ &\quad + 1 \times 2^1 + 1 \times 2^0 \\ &= 16 + 8 + 0 + 2 + 1 \\ &= \mathbf{27_{10}} \end{aligned}$$

Problem 2. Convert 0.1011_2 to a decimal fraction

$$\begin{aligned} 0.1011_2 &= 1 \times 2^{-1} + 0 \times 2^{-2} + 1 \times 2^{-3} \\ &\quad + 1 \times 2^{-4} \\ &= 1 \times \frac{1}{2} + 0 \times \frac{1}{2^2} + 1 \times \frac{1}{2^3} \\ &\quad + 1 \times \frac{1}{2^4} \\ &= \frac{1}{2} + \frac{1}{8} + \frac{1}{16} \\ &= 0.5 + 0.125 + 0.0625 \\ &= \mathbf{0.6875_{10}} \end{aligned}$$

Problem 3. Convert 101.0101_2 to a decimal number

$$\begin{aligned} 101.0101_2 &= 1 \times 2^2 + 0 \times 2^1 + 1 \times 2^0 \\ &\quad + 0 \times 2^{-1} + 1 \times 2^{-2} \\ &\quad + 0 \times 2^{-3} + 1 \times 2^{-4} \\ &= 4 + 0 + 1 + 0 + 0.25 \\ &\quad + 0 + 0.0625 \\ &= \mathbf{5.3125_{10}} \end{aligned}$$

Now try the following Practice Exercise

Practice Exercise 15 Conversion of binary to decimal numbers (Answers on page 703)

In Problems 1 to 5, convert the binary numbers given to decimal numbers.

1. (a) 110 (b) 1011 (c) 1110 (d) 1001
2. (a) 10101 (b) 11001 (c) 101101 (d) 110011
3. (a) 101010 (b) 111000 (c) 1000001 (d) 10111000
4. (a) 0.1101 (b) 0.11001 (c) 0.00111 (d) 0.01011
5. (a) 11010.11 (b) 10111.011 (c) 110101.0111 (d) 11010101.10111

(b) Conversion of decimal to binary:

An integer decimal number can be converted to a corresponding binary number by repeatedly dividing by 2 and noting the remainder at each stage, as shown below for 39_{10}

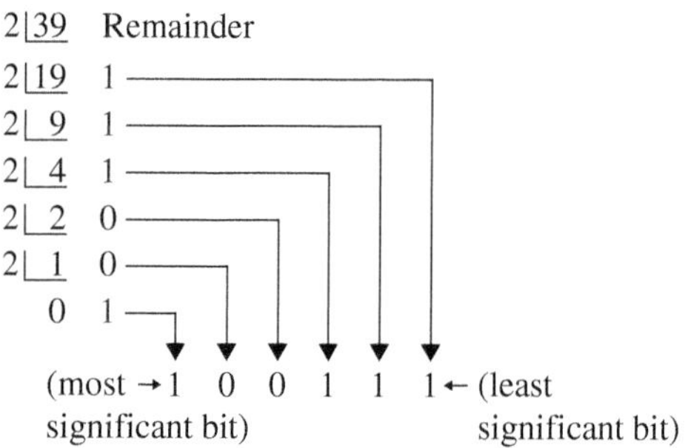

The result is obtained by writing the top digit of the remainder as the least significant bit, (a bit is a **bi**nary digi**t** and the least significant bit is the one on the right). The bottom bit of the remainder is the most significant bit, i.e. the bit on the left.

Thus $39_{10} = 100111_2$

The fractional part of a decimal number can be converted to a binary number by repeatedly multiplying by 2, as shown below for the fraction 0.625

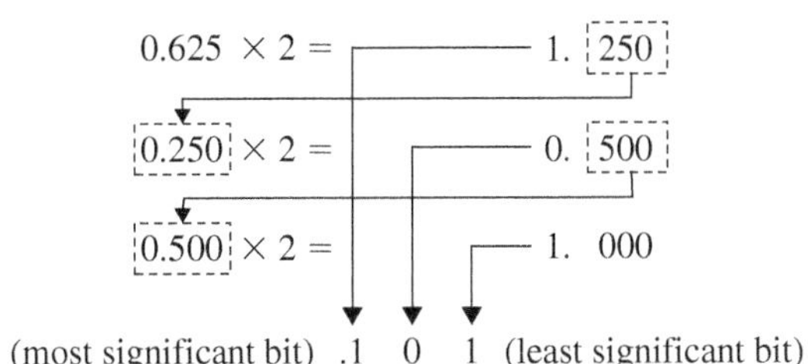

For fractions, the most significant bit of the result is the top bit obtained from the integer part of multiplication by 2. The least significant bit of the result is the bottom bit obtained from the integer part of multiplication by 2

Thus $0.625_{10} = 0.101_2$

Problem 4. Convert 47_{10} to a binary number

From above, repeatedly dividing by 2 and noting the remainder gives:

	Remainder
2 ⌊47	
2 ⌊23	1
2 ⌊11	1
2 ⌊5	1
2 ⌊2	1
2 ⌊1	0
0	1

1 0 1 1 1 1

Thus $47_{10} = 101111_2$

Problem 5. Convert 0.40625_{10} to a binary number

From above, repeatedly multiplying by 2 gives:

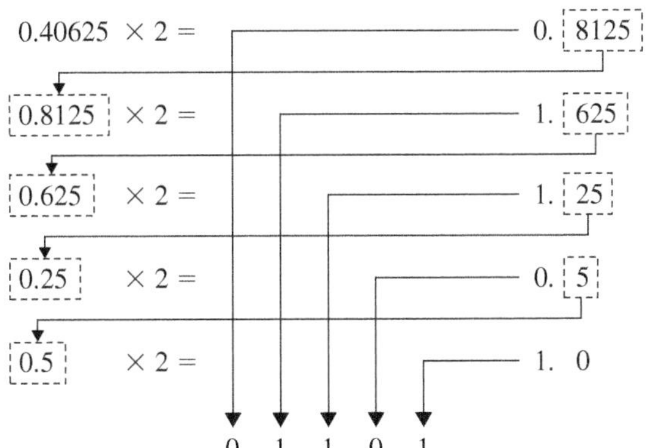

i.e. **$040625_{10} = 0.01101_2$**

Problem 6. Convert 58.3125_{10} to a binary number

The integer part is repeatedly divided by 2, giving:

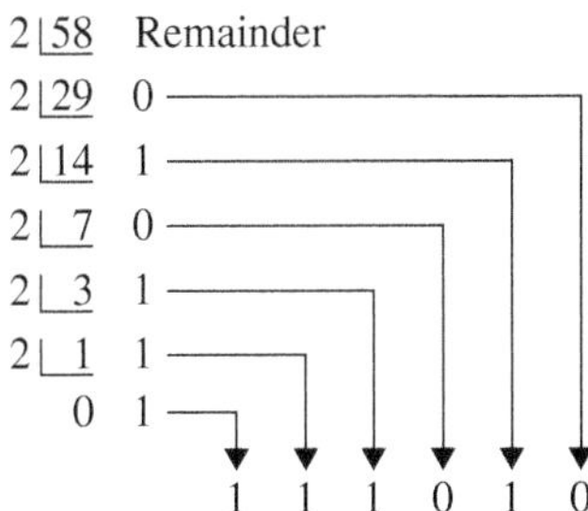

The fractional part is repeatedly multiplied by 2 giving:

0.3125	× 2 =	0.625
0.625	× 2 =	1.25
0.25	× 2 =	0.5
0.5	× 2 =	1.0

.0 1 0 1

Thus $58.3125_{10} = 111010.0101_2$

Now try the following Practice Exercise

Practice Exercise 16 Conversion of decimal to binary numbers (Answers on page 703)

In Problems 1 to 5, convert the decimal numbers given to binary numbers.

1. (a) 5 (b) 15
   (c) 19 (d) 29
2. (a) 31 (b) 42
   (c) 57 (d) 63
3. (a) 47 (b) 60
   (c) 73 (d) 84
4. (a) 0.25 (b) 0.21875
   (c) 0.28125 (d) 0.59375
5. (a) 47.40625 (b) 30.8125
   (c) 53.90625 (d) 61.65625

(c) Binary addition:

Binary addition of two/three bits is achieved according to the following rules:

sum	carry	sum	carry
$0+0=0$	0	$0+0+0=0$	0
$0+1=1$	0	$0+0+1=1$	0
$1+0=1$	0	$0+1+0=1$	0
$1+1=0$	1	$0+1+1=0$	1
		$1+0+0=1$	0
		$1+0+1=0$	1
		$1+1+0=0$	1
		$1+1+1=1$	1

These rules are demonstrated in the following worked problems.

Problem 7. Perform the binary addition: $1001 + 10110$

```
   1001
 +10110
  11111
```

Problem 8. Perform the binary addition: $11111 + 10101$

```
       11111
      +10101
sum   110100
carry  11111
```

Problem 9. Perform the binary addition: $1101001 + 1110101$

```
       1101001
      +1110101
sum   11011110
carry  11   1
```

Problem 10. Perform the binary addition: $1011101 + 1100001 + 110101$

```
       1011101
       1100001
       +110101
sum   11110011
carry  11111 1
```

Now try the following Practice Exercise

Practice Exercise 17 Binary addition (Answers on page 703)

Perform the following binary additions:

1. $10 + 11$
2. $101 + 110$
3. $1101 + 111$
4. $1111 + 11101$
5. $110111 + 10001$
6. $10000101 + 10000101$
7. $11101100 + 111001011$
8. $110011010 + 11100011$
9. $10110 + 1011 + 11011$
10. $111 + 10101 + 11011$
11. $1101 + 1001 + 11101$
12. $100011 + 11101 + 101110$

3.3 Octal numbers

For decimal integers containing several digits, repeatedly dividing by 2 can be a lengthy process. In this case, it is usually easier to convert a decimal number to a binary number via the octal system of numbers. This system has a radix of 8, using the digits 0, 1, 2, 3, 4, 5, 6 and 7. The denary number equivalent to the octal number 4317_8 is

$$4 \times 8^3 + 3 \times 8^2 + 1 \times 8^1 + 7 \times 8^0$$

i.e. $4 \times 512 + 3 \times 64 + 1 \times 8 + 7 \times 1$ or 2255_{10}

An integer decimal number can be converted to a corresponding octal number by repeatedly dividing by 8 and noting the remainder at each stage, as shown below for 493_{10}

```
8 |493   Remainder
8 | 61   5
8 |  7   5
     0   7
         ↓ ↓ ↓
         7 5 5
```

Thus $\mathbf{493_{10} = 755_8}$

The fractional part of a decimal number can be converted to an octal number by repeatedly multiplying by 8, as shown below for the fraction 0.4375_{10}

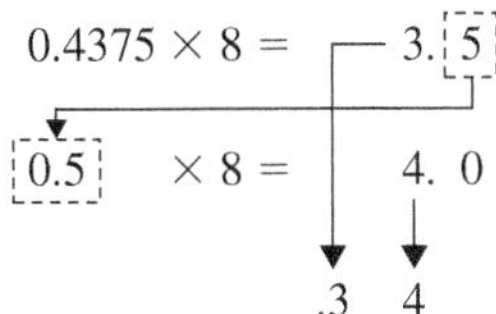

For fractions, the most significant bit is the top integer obtained by multiplication of the decimal fraction by 8, thus

$$0.4375_{10} = 0.34_8$$

The natural binary code for digits 0 to 7 is shown in Table 3.1, and an octal number can be converted to a binary number by writing down the three bits corresponding to the octal digit.

Thus $437_8 = 100\ 011\ 111_2$

and $26.35_8 = 010\ 110.011\ 101_2$

Table 3.1

Octal digit	Natural binary number
0	000
1	001
2	010
3	011
4	100
5	101
6	110
7	111

The '0' on the extreme left does not signify anything, thus $26.35_8 = 10\ 110.011\ 101_2$

Conversion of decimal to binary via octal is demonstrated in the following worked problems.

Problem 11. Convert 3714_{10} to a binary number, via octal

Dividing repeatedly by 8, and noting the remainder gives:

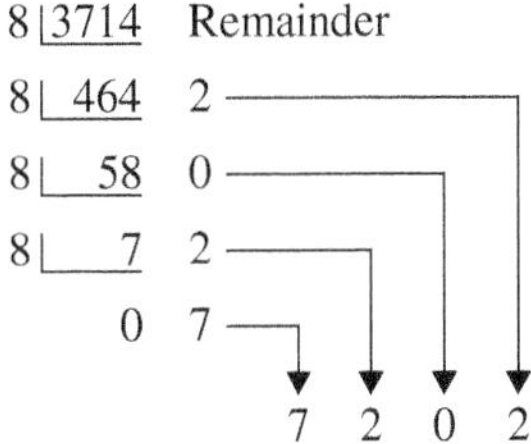

From Table 3.1, $7202_8 = 111\ 010\ 000\ 010_2$

i.e. $\mathbf{3714_{10} = 111\ 010\ 000\ 010_2}$

Problem 12. Convert 0.59375_{10} to a binary number, via octal

Multiplying repeatedly by 8, and noting the integer values, gives:

$0.59375 \times 8 = 4.75$
$0.75 \times 8 = 6.00$
.4 6

Thus $0.59375_{10} = 0.46_8$

From Table 3.1, $0.46_8 = 0.100\ 110_2$

i.e. $\mathbf{0.59375_{10} = 0.100\ 11_2}$

Problem 13. Convert 5613.90625_{10} to a binary number, via octal

The integer part is repeatedly divided by 8, noting the remainder, giving:

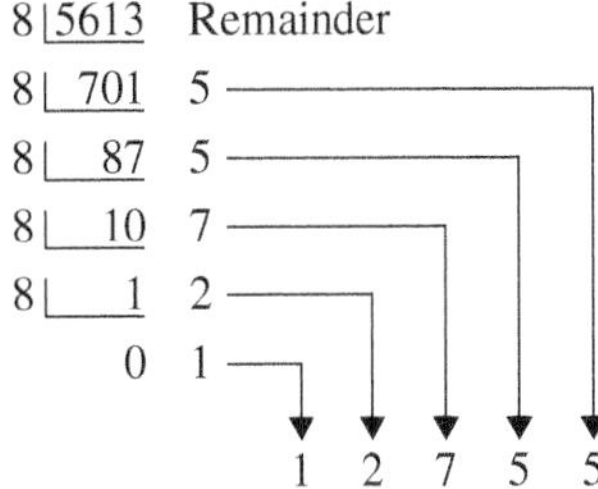

This octal number is converted to a binary number, (see Table 3.1)

$$12755_8 = 001\ 010\ 111\ 101\ 101_2$$

i.e. $5613_{10} = 1\ 010\ 111\ 101\ 101_2$

The fractional part is repeatedly multiplied by 8, and noting the integer part, giving:

$$0.90625 \times 8 = 7.25$$
$$0.25 \quad \times 8 = 2.00$$
$$.7\ 2$$

This octal fraction is converted to a binary number, (see Table 3.1)

$$0.72_8 = 0.111\ 010_2$$

i.e. $0.90625_{10} = 0.111\ 01_2$

Thus, $\mathbf{5613.90625_{10} = 1\ 010\ 111\ 101\ 101.111\ 01_2}$

Problem 14. Convert $11\ 110\ 011.100\ 01_2$ to a decimal number via octal

Grouping the binary number in three's from the binary point gives: $011\ 110\ 011.100\ 010_2$

Using Table 3.1 to convert this binary number to an octal number gives: 363.42_8 and

$$363.42_8 = 3\times 8^2 + 6\times 8^1 + 3\times 8^0 + 4\times 8^{-1} + 2\times 8^{-2}$$
$$= 192 + 48 + 3 + 0.5 + 0.03125$$
$$= \mathbf{243.53125_{10}}$$

Now try the following Practice Exercise

Practice Exercise 18 Conversion between decimal and binary numbers via octal (Answers on page 703)

In Problems 1 to 3, convert the decimal numbers given to binary numbers, via octal.

1. (a) 343 (b) 572 (c) 1265
2. (a) 0.46875 (b) 0.6875 (c) 0.71875
3. (a) 247.09375 (b) 514.4375 (c) 1716.78125
4. Convert the following binary numbers to decimal numbers via octal:

   (a) 111.011 1 (b) 101 001.01

   (c) 1 110 011 011 010.001 1

3.4 Hexadecimal numbers

The hexadecimal system is particularly important in computer programming, since four bits (each consisting of a one or zero) can be succinctly expressed using a single hexadecimal digit. Two hexadecimal digits represent numbers from 0 to 255, a common range used, for example, to specify colours. Thus, in the HTML language of the web, colours are specified using three pairs of hexadecimal digits RRGGBB, where RR is the amount of red, GG the amount of green, and BB the amount of blue.

A **hexadecimal numbering system** has a radix of 16 and uses the following 16 distinct digits:

$$0, 1, 2, 3, 4, 5, 6, 7, 8, 9,\ \text{A, B, C, D, E and F}$$

'A' corresponds to 10 in the denary system, B to 11, C to 12 and so on.

(a) Converting from hexadecimal to decimal:

For example

$$1A_{16} = 1\times 16^1 + A\times 16^0$$
$$= 1\times 16^1 + 10\times 1 = 16 + 10 = 26$$

i.e. $\mathbf{1A_{16} = 26_{10}}$

Similarly,

$$\mathbf{2E_{16}} = 2\times 16^1 + E\times 16^0$$
$$= 2\times 16^1 + 14\times 16^0 = 32 + 14 = \mathbf{46_{10}}$$

and
$$\mathbf{1BF_{16}} = 1\times 16^2 + B\times 16^1 + F\times 16^0$$
$$= 1\times 16^2 + 11\times 16^1 + 15\times 16^0$$
$$= 256 + 176 + 15 = \mathbf{447_{10}}$$

Table 3.2 compares decimal, binary, octal and hexadecimal numbers and shows, for example, that

$$23_{10} = 10111_2 = 27_8 = 17_{16}$$

Problem 15. Convert the following hexadecimal numbers into their decimal equivalents: (a) $7A_{16}$ (b) $3F_{16}$

(a) $7A_{16} = 7\times 16^1 + A\times 16^0 = 7\times 16 + 10\times 1 = 112 + 10 = 122$

Thus $\mathbf{7A_{16} = 122_{10}}$

(b) $3F_{16} = 3 \times 16^1 + F \times 16^0 = 3 \times 16 + 15 \times 1 = 48 + 15 = 63$

Thus, $\mathbf{3F_{16} = 63_{10}}$

Problem 16. Convert the following hexadecimal numbers into their decimal equivalents: (a) $C9_{16}$ (b) BD_{16}

(a) $C9_{16} = C \times 16^1 + 9 \times 16^0 = 12 \times 16 + 9 \times 1 = 192 + 9 = 201$

Thus $\mathbf{C9_{16} = 201_{10}}$

(b) $BD_{16} = B \times 16^1 + D \times 16^0 = 11 \times 16 + 13 \times 1 = 176 + 13 = 189$

Thus, $\mathbf{BD_{16} = 189_{10}}$

Problem 17. Convert $1A4E_{16}$ into a denary number

$1A4E_{16}$

$$= 1 \times 16^3 + A \times 16^2 + 4 \times 16^1 + E \times 16^0$$
$$= 1 \times 16^3 + 10 \times 16^2 + 4 \times 16^1 + 14 \times 16^0$$
$$= 1 \times 4096 + 10 \times 256 + 4 \times 16 + 14 \times 1$$
$$= 4096 + 2560 + 64 + 14 = 6734$$

Thus, $\mathbf{1A4E_{16} = 6734_{10}}$

(b) Converting from decimal to hexadecimal:

This is achieved by repeatedly dividing by 16 and noting the remainder at each stage, as shown below for 26_{10}

	Remainder
16 ⌊ 26	
16 ⌊ 1	$10 \equiv A_{16}$
0	$1 \equiv 1_{16}$

most significant bit → 1 A ← least significant bit

Hence $\mathbf{26_{10} = 1A_{16}}$

Similarly, for 447_{10}

	Remainder
16 ⌊ 447	
16 ⌊ 27	$15 \equiv F_{16}$
16 ⌊ 1	$11 \equiv B_{16}$
0	$1 \equiv 1_{16}$

1 B F

Thus $\mathbf{447_{10} = 1BF_{16}}$

Table 3.2

Decimal	Binary	Octal	Hexadecimal
0	0000	0	0
1	0001	1	1
2	0010	2	2
3	0011	3	3
4	0100	4	4
5	0101	5	5
6	0110	6	6
7	0111	7	7
8	1000	10	8
9	1001	11	9
10	1010	12	A
11	1011	13	B
12	1100	14	C
13	1101	15	D
14	1110	16	E
15	1111	17	F
16	10000	20	10
17	10001	21	11
18	10010	22	12
19	10011	23	13
20	10100	24	14
21	10101	25	15
22	10110	26	16
23	10111	27	17
24	11000	30	18
25	11001	31	19
26	11010	32	1A
27	11011	33	1B
28	11100	34	1C
29	11101	35	1D
30	11110	36	1E
31	11111	37	1F
32	100000	40	20

Problem 18. Convert the following decimal numbers into their hexadecimal equivalents:
(a) 37_{10} (b) 108_{10}

(a)
$$\begin{array}{r|l} 16 & 37 \quad \text{Remainder} \\ 16 & 2 \quad 5 = 5_{16} \\ & 0 \quad 2 = 2_{16} \end{array}$$

2 5

most significant bit → 2, least significant bit → 5

Hence $\mathbf{37_{10} = 25_{16}}$

(b)
$$\begin{array}{r|l} 16 & 108 \quad \text{Remainder} \\ 16 & 6 \quad 12 = C_{16} \\ & 0 \quad 6 = 6_{16} \end{array}$$

6 C

Hence $\mathbf{108_{10} = 6C_{16}}$

Problem 19. Convert the following decimal numbers into their hexadecimal equivalents:
(a) 162_{10} (b) 239_{10}

(a)
$$\begin{array}{r|l} 16 & 162 \quad \text{Remainder} \\ 16 & 10 \quad 2 = 2_{16} \\ & 0 \quad 10 = A_{16} \end{array}$$

A 2

Hence $\mathbf{162_{10} = A2_{16}}$

(b)
$$\begin{array}{r|l} 16 & 239 \quad \text{Remainder} \\ 16 & 14 \quad 15 = F_{16} \\ & 0 \quad 14 = E_{16} \end{array}$$

E F

Hence $\mathbf{239_{10} = EF_{16}}$

Now try the following Practice Exercise

Practice Exercise 19 Hexadecimal numbers (Answers on page 703)

In Problems 1 to 4, convert the given hexadecimal numbers into their decimal equivalents.

1. $E7_{16}$
2. $2C_{16}$
3. 98_{16}
4. $2F1_{16}$

In Problems 5 to 8, convert the given decimal numbers into their hexadecimal equivalents.

5. 54_{10}
6. 200_{10}
7. 91_{10}
8. 238_{10}

(c) Converting from binary to hexadecimal:

The binary bits are arranged in groups of four, starting from right to left, and a hexadecimal symbol is assigned to each group. For example, the binary number 1110011110101001 is initially grouped in

fours as: 1110 0111 1010 1001

and a hexadecimal symbol

assigned to each group as E 7 A 9

from Table 3.2

Hence $\mathbf{1110011110101001_2 = E7A9_{16}}$

Problem 20. Convert the following binary numbers into their hexadecimal equivalents:
(a) 11010110_2 (b) 1100111_2

(a) Grouping bits in fours from the right gives: 1101 0110
and assigning hexadecimal symbols to each group gives: D 6
from Table 3.2

Thus, $\mathbf{11010110_2 = D6_{16}}$

(b) Grouping bits in fours from the right gives: 0110 0111
and assigning hexadecimal symbols to each group gives: 6 7
from Table 3.2

Thus, $\mathbf{1100111_2 = 67_{16}}$

Problem 21. Convert the following binary numbers into their hexadecimal equivalents:
(a) 11001111_2 (b) 110011110_2

(a) Grouping bits in fours from the

right gives: 1100 1111

and assigning hexadecimal

symbols to each group gives: C F
from Table 3.2

Thus, $\mathbf{11001111_2 = CF_{16}}$

(b) Grouping bits in fours from

the right gives: 0001 1001 1110

and assigning hexadecimal

symbols to each group gives: 1 9 E
from Table 3.2

Thus, $\mathbf{110011110_2 = 19E_{16}}$

(d) Converting from hexadecimal to binary:

The above procedure is reversed, thus, for example,

$6CF3_{16} = 0110\ 1100\ 1111\ 0011$

from Table 3.2

i.e. $\mathbf{6CF3_{16} = 110110011110011_2}$

Problem 22. Convert the following hexadecimal numbers into their binary equivalents:
(a) $3F_{16}$ (b) $A6_{16}$

(a) Spacing out hexadecimal digits gives: 3 F

and converting each into

binary gives: 0011 1111
from Table 3.2

Thus, $\mathbf{3F_{16} = 111111_2}$

(b) Spacing out hexadecimal digits

gives: A 6

and converting each into binary

gives: 1010 0110
from Table 3.2

Thus, $\mathbf{A6_{16} = 10100110_2}$

Problem 23. Convert the following hexadecimal numbers into their binary equivalents:
(a) $7B_{16}$ (b) $17D_{16}$

(a) Spacing out hexadecimal

digits gives: 7 B

and converting each into

binary gives: 0111 1011
from Table 3.2

Thus, $\mathbf{7B_{16} = 1111011_2}$

(b) Spacing out hexadecimal

digits gives: 1 7 D

and converting each into

binary gives: 0001 0111 1101
from Table 3.2

Thus, $\mathbf{17D_{16} = 101111101_2}$

Now try the following Practice Exercise

Practice Exercise 20 Hexadecimal numbers (Answers on page 703)

In Problems 1 to 4, convert the given binary numbers into their hexadecimal equivalents.

1. 11010111_2
2. 11101010_2
3. 10001011_2
4. 10100101_2

In Problems 5 to 8, convert the given hexadecimal numbers into their binary equivalents.

5. 37_{16}
6. ED_{16}
7. $9F_{16}$
8. $A21_{16}$

Practice Exercise 21 Multiple-choice questions on binary, octal and hexadecimal numbers (Answers on page 703)

Each question has only one correct answer

1. The base of a hexadecimal number system is:
(a) 6 (b) 8 (c) 16 (d) 10

2. The number of digits in a binary number system is:
(a) 10 (b) 2 (c) 4 (d) 6
3. The number system which uses the alphabet as well as numbers is called:
(a) binary (b) octal
(c) decimal (d) hexadecimal
4. The value of the base of a decimal number system is:
(a) 8 (b) 2 (c) 10 (d) 16
5. The hexadecimal number equivalent of 15 is:
(a) A (b) F (c) D (d) E
6. The symbol D in the hexadecimal system corresponds to:
(a) 8 (b) 16 (c) 13 (d) 14
7. Converting the binary number 1000 to decimal gives:
(a) 2 (b) 4 (c) 6 (d) 8
8. The number of digits required to represent a decimal number 31 in the equivalent binary form is:
(a) 2 (b) 4 (c) 5 (d) 6
9. In decimal, the binary number 1101101 is equal to:
(a) 218 (b) 31 (c) 127 (d) 109
10. In hexadecimal, the decimal number 123 is:
(a) 7B (b) 123 (c) 173 (d) 1111011
11. Converting the binary number 11001 to decimal gives:
(a) 3 (b) 50 (c) 25 (d) 13
12. Converting the hexadecimal number B2 to binary gives:
(a) 10110010 (b) 101101
(c) 11100 (d) 11000010
13. Converting the binary number 11011 to hexadecimal gives:
(a) 1A (b) B1 (c) A1 (d) 1B
14. Converting the decimal number 20 to hexadecimal gives:
(a) 11 (b) 14 (c) 1B (d) 1A
15. Converting the hexadecimal number 2C to decimal gives:
(a) 3A (b) 34 (c) 44 (d) 54
16. The binary addition 1111 + 1111 gives:
(a) 0000 (b) 11111 (c) 11110 (d) 2222
17. Converting the decimal number 21 to binary gives:
(a) 10101 (b) 10001
(c) 10000 (d) 11111
18. The hexadecimal equivalent of the binary number 1110 is:
(a) E (b) 0111 (c) 15 (d) 14
19. The maximum value of a single digit in an octal system is:
(a) 7 (b) 9 (c) 6 (d) 5
20. The binary number 111 in octal is:
(a) 6 (b) 7 (c) 8 (d) 5
21. Converting the octal number 22_8 into its corresponding decimal number is:
(a) 26 (b) 18 (c) 81 (d) 82
22. Which of the following binary numbers is equivalent to the decimal number 24?
(a) 1101111 (b) 11000
(c) 111111 (d) 11001
23. The binary number corresponding to the hexadecimal number B2 is:
(a) 100011 (b) 11011
(c) 10110010 (d) 10010010
24. In binary, the decimal number 45 is:
(a) 11100 (b) 101101
(c) 10100 (d) 110101
25. In decimal, the hexadecimal number 52_{16} is:
(a) 28 (b) 83 (c) 80 (d) 82

For fully worked solutions to each of the problems in Practice Exercises 15 to 20 in this chapter, go to the website:
www.routledge.com/cw/bird

Chapter 4

Calculations and evaluation of formulae

Why it is important to understand: **Calculations and evaluation of formulae**

The availability of electronic pocket calculators, at prices which all can afford, has had a considerable impact on engineering education. Engineers and student engineers now use calculators all the time since calculators are able to handle a very wide range of calculations. You will feel more confident to deal with all aspects of engineering studies if you are able to correctly use a calculator accurately.

At the end of this chapter, you should be able to:

- recognise different types of errors
- determine approximate values of calculations
- use a scientific calculator in a range of calculations
- use conversion tables and charts
- evaluate formulae

4.1 Errors and approximations

(i) In all problems in which the measurement of distance, time, mass or other quantities occurs, an exact answer cannot be given; only an answer which is correct to a stated degree of accuracy can be given. To take account of this an **error due to measurement** is said to exist.

(ii) To take account of measurement errors it is usual to limit answers so that the result given is **not more than one significant figure greater than the least accurate number given in the data**.

(iii) **Rounding-off errors** can exist with decimal fractions. For example, to state that $\pi = 3.142$ is not strictly correct, but '$\pi = 3.142$ correct to 4 significant figures' is a true statement. (Actually, $\pi = 3.14159265\ldots$)

(iv) It is possible, through an incorrect procedure, to obtain the wrong answer to a calculation. This type of error is known as **a blunder**.

(v) An **order of magnitude error** is said to exist if incorrect positioning of the decimal point occurs after a calculation has been completed.

(vi) Blunders and order of magnitude errors can be reduced by determining **approximate values of calculations**. Answers which do not seem feasible must be checked and the calculation must be repeated as necessary.

An engineer will often need to make a quick mental approximation for a calculation. For example, $\dfrac{49.1 \times 18.4 \times 122.1}{61.2 \times 38.1}$ may be approximated to $\dfrac{50 \times 20 \times 120}{60 \times 40}$ and then, by cancelling, $\dfrac{50 \times {}^{1}\cancel{20} \times \cancel{120}^{\cancel{3}1}}{{}_{1}\cancel{60} \times \cancel{40}_{\cancel{2}1}} = 50$. An accurate answer somewhere between 45 and 55 could therefore be expected. Certainly an answer around 500 or 5 would not be expected. Actually, by calculator $\dfrac{49.1 \times 18.4 \times 122.1}{61.2 \times 38.1} = 47.31$, correct to 4 significant figures.

Problem 1. The area A of a triangular metal plate is given by $A = \frac{1}{2}bh$. The base b when measured is found to be 3.26 cm, and the perpendicular height h is 7.5 cm. Determine the area of the triangle.

$$\text{Area of triangle} = \frac{1}{2}bh = \frac{1}{2} \times 3.26 \times 7.5$$
$$= 12.225\,\text{cm}^2 \text{ (by calculator).}$$

The approximate values is $\frac{1}{2} \times 3 \times 8 = 12\,\text{cm}^2$, so there are no obvious blunder or magnitude errors. However, it is not usual in a measurement type problem to state the answer to an accuracy greater than 1 significant figure more than the least accurate number in the data: this is 7.5 cm, so the result should not have more than 3 significant figures.

Thus, **area of triangle = 12.2 cm²**

Problem 2. State which type of error has been made in the following statements:

(a) $72 \times 31.429 = 2262.9$

(b) $16 \times 0.08 \times 7 = 89.6$

(c) $11.714 \times 0.0088 = 0.3247$, correct to 4 decimal places.

(d) $\dfrac{29.74 \times 0.0512}{11.89} = 0.12$, correct to 2 significant figures.

(a) $72 \times 31.429 = 2262.888$ (by calculator), hence a **rounding-off error** has occurred. The answer should have stated:

$72 \times 31.429 = 2262.9$, correct to 5 significant figures or 2262.9, correct to 1 decimal place.

(b) $16 \times 0.08 \times 7 = 16 \times \dfrac{8}{100} \times 7 = \dfrac{32 \times 7}{25}$

$$= \frac{224}{25} = 8\frac{24}{25} = 8.96$$

Hence an **order of magnitude** error has occurred.

(c) 11.714×0.0088 is approximately equal to $12 \times 9 \times 10^{-3}$, i.e. about 108×10^{-3} or 0.108 Thus a **blunder** has been made.

(d) $\dfrac{29.74 \times 0.0512}{11.89} \approx \dfrac{30 \times 5 \times 10^{-2}}{12}$

$$= \frac{150}{12 \times 10^2} = \frac{15}{120} = \frac{1}{8} \text{ or } 0.125$$

hence no order of magnitude error has occurred. However, $\dfrac{29.74 \times 0.0512}{11.89} = 0.128$ correct to 3 significant figures, which equals 0.13 correct to 2 significant figures.

Hence a **rounding-off error** has occurred.

Problem 3. Without using a calculator, determine an approximate value of:

(a) $\dfrac{11.7 \times 19.1}{9.3 \times 5.7}$ (b) $\dfrac{2.19 \times 203.6 \times 17.91}{12.1 \times 8.76}$

(a) $\dfrac{11.7 \times 19.1}{9.3 \times 5.7}$ is approximately equal to $\dfrac{10 \times 20}{10 \times 5}$

i.e. about **4**

(By calculator, $\dfrac{11.7 \times 19.1}{9.3 \times 5.7} = \mathbf{4.22}$, correct to 3 significant figures.)

(b) $\dfrac{2.19 \times 203.6 \times 17.91}{12.1 \times 8.76} \approx \dfrac{2 \times {}^{20}\cancel{200} \times \cancel{20}^{2}}{{}_{1}\cancel{10} \times \cancel{10}_{1}}$

$= 2 \times 20 \times 2$ after cancelling,

i.e. $\dfrac{2.19 \times 203.6 \times 17.91}{12.1 \times 8.76} \approx \mathbf{80}$

(By calculator, $\dfrac{2.19 \times 203.6 \times 17.91}{12.1 \times 8.76} \approx \mathbf{75.3}$, correct to 3 significant figures.)

Now try the following Practice Exercise

Practice Exercise 22 Errors (Answers on page 703)

In Problems 1 to 5 state which type of error, or errors, have been made:

1. $25 \times 0.06 \times 1.4 = 0.21$
2. $137 \times 6.842 = 937.4$
3. $\dfrac{24 \times 0.008}{12.6} = 10.42$

4. For a gas $pV = c$. When pressure $p = 1\,03\,400$ Pa and $V = 0.54\,\text{m}^3$ then $c = 55\,836\,\text{Pa}\,\text{m}^3$.

5. $\dfrac{4.6 \times 0.07}{52.3 \times 0.274} = 0.225$

In Problems 6 to 8, evaluate the expressions approximately, without using a calculator.

6. 4.7×6.3

7. $\dfrac{2.87 \times 4.07}{6.12 \times 0.96}$

8. $\dfrac{72.1 \times 1.96 \times 48.6}{139.3 \times 5.2}$

4.2 Use of calculator

The most modern aid to calculations is the pocket-sized electronic calculator. With one of these, calculations can be quickly and accurately performed, correct to about 9 significant figures.
To help you to become competent at using your calculator check that you agree with the answers to the following problems:

Problem 4. Evaluate the following, correct to 4 significant figures:

(a) $4.7826 + 0.02713$ (b) $17.6941 - 11.8762$
(c) 21.93×0.012981

(a) $4.7826 + 0.02713 = 4.80973 = \mathbf{4.810}$, correct to 4 significant figures

(b) $17.6941 - 11.8762 = 5.8179 = \mathbf{5.818}$, correct to 4 significant figures

(c) $21.93 \times 0.012981 = 0.2846733\ldots = \mathbf{0.2847}$, correct to 4 significant figures

Problem 5. Evaluate the following, correct to 4 decimal places:

(a) $46.32 \times 97.17 \times 0.01258$ (b) $\dfrac{4.621}{23.76}$

(c) $\dfrac{1}{2}(62.49 \times 0.0172)$

(a) $46.32 \times 97.17 \times 0.01258 = 56.6215031\ldots$
$= \mathbf{56.6215}$, correct to 4 decimal places

(b) $\dfrac{4.621}{23.76} = 0.19448653\ldots = \mathbf{0.1945}$, correct to 4 decimal places

(c) $\dfrac{1}{2}(62.49 \times 0.0172) = 0.537414 = \mathbf{0.5374}$, correct to 4 decimal places

Problem 6. Evaluate the following, correct to 3 decimal places:

(a) $\dfrac{1}{52.73}$ (b) $\dfrac{1}{0.0275}$ (c) $\dfrac{1}{4.92} + \dfrac{1}{1.97}$

(a) $\dfrac{1}{52.73} = 0.01896453\ldots = \mathbf{0.019}$, correct to 3 decimal places

(b) $\dfrac{1}{0.0275} = 36.3636363\ldots = \mathbf{36.364}$, correct to 3 decimal places

(c) $\dfrac{1}{4.92} + \dfrac{1}{1.97} = 0.71086624\ldots = \mathbf{0.711}$, correct to 3 decimal places

Problem 7. Evaluate the following, expressing the answers in standard form, correct to 4 significant figures:

(a) $(0.00451)^2$ (b) $631.7 - (6.21 + 2.95)^2$
(c) $46.27^2 - 31.79^2$

(a) $(0.00451)^2 = 2.03401 \times 10^{-5} = \mathbf{2.034 \times 10^{-5}}$, correct to 4 significant figures

(b) $631.7 - (6.21 + 2.95)^2 = 547.7944$
$= 5.477944 \times 10^2 = \mathbf{5.478 \times 10^2}$, correct to 4 significant figures

(c) $46.27^2 - 31.79^2 = 1130.3088 = \mathbf{1.130 \times 10^3}$, correct to 4 significant figures

Problem 8. Evaluate the following, correct to 3 decimal places:

(a) $\dfrac{(2.37)^2}{0.0526}$ (b) $\left(\dfrac{3.60}{1.92}\right)^2 + \left(\dfrac{5.40}{2.45}\right)^2$

(c) $\dfrac{15}{7.6^2 - 4.8^2}$

(a) $\dfrac{(2.37)^2}{0.0526} = 106.785171\ldots = \mathbf{106.785}$, correct to 3 decimal places

(b) $\left(\dfrac{3.60}{1.92}\right)^2 + \left(\dfrac{5.40}{2.45}\right)^2 = 8.37360084\ldots = \mathbf{8.374}$, correct to 3 decimal places

(c) $\dfrac{15}{7.6^2 - 4.8^2} = 0.43202764\ldots = \mathbf{0.432}$, correct to 3 decimal places

Problem 9. Evaluate the following, correct to 4 significant figures:

(a) $\sqrt{5.462}$ (b) $\sqrt{54.62}$ (c) $\sqrt{546.2}$

(a) $\sqrt{5.462} = 2.3370922\ldots = \mathbf{2.337}$, correct to 4 significant figures

(b) $\sqrt{54.62} = 7.39053448\ldots = \mathbf{7.391}$, correct to 4 significant figures

(c) $\sqrt{546.2} = 23.370922\ldots = \mathbf{23.37}$, correct to 4 significant figures

Problem 10. Evaluate the following, correct to 3 decimal places:

(a) $\sqrt{0.007328}$ (b) $\sqrt{52.91} - \sqrt{31.76}$

(c) $\sqrt{1.6291 \times 10^4}$

(a) $\sqrt{0.007328} = 0.08560373 = \mathbf{0.086}$, correct to 3 decimal places

(b) $\sqrt{52.91} - \sqrt{31.76} = 1.63832491\ldots = \mathbf{1.638}$, correct to 3 decimal places

(c) $\sqrt{1.6291 \times 10^4} = \sqrt{16291} = 127.636201\ldots$ $= \mathbf{127.636}$, correct to 3 decimal places

Problem 11. Evaluate the following, correct to 4 significant figures:

(a) 4.72^3 (b) $(0.8316)^4$ (c) $\sqrt{76.21^2 - 29.10^2}$

(a) $4.72^3 = 105.15404\ldots = \mathbf{105.2}$, correct to 4 significant figures

(b) $(0.8316)^4 = 0.47825324\ldots = \mathbf{0.4783}$, correct to 4 significant figures

(c) $\sqrt{76.21^2 - 29.10^2} = 70.4354605\ldots = \mathbf{70.44}$, correct to 4 significant figures

Problem 12. Evaluate the following, correct to 3 significant figures:

(a) $\sqrt{\dfrac{6.09^2}{25.2 \times \sqrt{7}}}$ (b) $\sqrt[3]{47.291}$

(c) $\sqrt{7.213^2 + 6.418^3 + 3.291^4}$

(a) $\sqrt{\dfrac{6.09^2}{25.2 \times \sqrt{7}}} = 0.74583457\ldots = \mathbf{0.746}$, correct to 3 significant figures

(b) $\sqrt[3]{47.291} = 3.61625876\ldots = \mathbf{3.62}$, correct to 3 significant figures

(c) $\sqrt{7.213^2 + 6.418^3 + 3.291^4} = 20.8252991\ldots,$ $= \mathbf{20.8}$ correct to 3 significant figures

Problem 13. Evaluate the following, expressing the answers in standard form, correct to 4 decimal places:

(a) $(5.176 \times 10^{-3})^2$

(b) $\left(\dfrac{1.974 \times 10^1 \times 8.61 \times 10^{-2}}{3.462}\right)^4$

(c) $\sqrt{1.792 \times 10^{-4}}$

(a) $(5.176 \times 10^{-3})^2 = 2.679097\ldots \times 10^{-5}$ $= \mathbf{2.6791 \times 10^{-5}}$, correct to 4 decimal places

(b) $\left(\dfrac{1.974 \times 10^1 \times 8.61 \times 10^{-2}}{3.462}\right)^4 = 0.05808887\ldots$ $= \mathbf{5.8089 \times 10^{-2}}$, correct to 4 decimal places

(c) $\sqrt{1.792 \times 10^{-4}} = 0.0133865\ldots$ $= \mathbf{1.3387 \times 10^{-2}}$, correct to 4 decimal places

Now try the following Practice Exercise

Practice Exercise 23 The use of a calculator (Answers on page 703)

In Problems 1 to 9, use a calculator to evaluate the quantities shown correct to 4 significant figures:

1. (a) 3.249^2 (b) 73.78^2 (c) 311.4^2 (d) 0.0639^2
2. (a) $\sqrt{4.735}$ (b) $\sqrt{35.46}$ (c) $\sqrt{73\,280}$ (d) $\sqrt{0.0256}$

3. (a) $\dfrac{1}{7.768}$ (b) $\dfrac{1}{48.46}$ (c) $\dfrac{1}{0.0816}$
(d) $\dfrac{1}{1.118}$

4. (a) $127.8 \times 0.0431 \times 19.8$
(b) $15.76 \div 4.329$

5. (a) $\dfrac{137.6}{552.9}$ (b) $\dfrac{11.82 \times 1.736}{0.041}$

6. (a) 13.6^3 (b) 3.476^4 (c) 0.124^5

7. (a) $\left(\dfrac{24.68 \times 0.0532}{7.412}\right)^3$
(b) $\left(\dfrac{0.2681 \times 41.2^2}{32.6 \times 11.89}\right)^4$

8. (a) $\dfrac{14.32^3}{21.68^2}$ (b) $\dfrac{4.821^3}{17.33^2 - 15.86 \times 11.6}$

9. (a) $\sqrt{\dfrac{(15.62)^2}{29.21 \times \sqrt{10.52}}}$
(b) $\sqrt{6.921^2 + 4.816^3 - 2.161^4}$

10. Evaluate the following, expressing the answers in engineering notation, correct to 4 significant figures:
(a) $(8.291 \times 10^{-2})^2$ (b) $\sqrt{7.623 \times 10^{-3}}$

4.3 Conversion tables and charts

It is often necessary to make calculations from various conversion tables and charts. Examples include currency exchange rates, imperial to metric unit conversions, train or bus timetables, production schedules and so on.

Problem 14. Currency exchange rates for five countries are shown in Table 4.1

Table 4.1

France	£1 = 1.15 euros
Japan	£1 = 140 yen
Norway	£1 = 11.30 kronor
Switzerland	£1 = 1.20 francs
USA	£1 = 1.25 dollars ($)

Calculate:
(a) how many euros £27.80 will buy in France
(b) the number of Japanese yen which can be bought for £23
(c) the pounds sterling which can be exchanged for 8836.60 Norwegian kronor
(d) the number of American dollars which can be purchased for £90, and
(e) the pounds sterling which can be exchanged for 384.60 Swiss francs.

(a) £1 = 1.15 euros, hence
£27.80 = 27.80 × 1.15 euros = **31.97 euros**

(b) £1 = 140 yen, hence
£23 = 23 × 140 yen = **3220 yen**

(c) £1 = 11.30 kronor, hence
8836.60 kr = £$\dfrac{8836.60}{11.30}$ = **£782**

(d) £1 = 1.25 dollars, hence
£90 = 90 × 1.25 dollars = **$112.50**

(e) £1 = 1.20 Swiss francs, hence
384.60 francs = £$\dfrac{384.60}{1.20}$ = **£320.50**

Problem 15. Some approximate imperial to metric conversions are shown in Table 4.2

Table 4.2

length	1 inch = 2.54 cm 1 mile = 1.61 km
weight	2.2 lb = 1 kg (1 lb = 16 oz)
capacity	1.76 pints = 1 litre (8 pints = 1 gallon)

Use the table to determine:
(a) the number of millimetres in 9.5 inches,
(b) a speed of 50 miles per hour in kilometres per hour,
(c) the number of miles in 300 km,
(d) the number of kilograms in 30 pounds weight,
(e) the number of pounds and ounces in 42 kilograms (correct to the nearest ounce),
(f) the number of litres in 15 gallons, and
(g) the number of gallons in 40 litres.

(a) 9.5 inches $= 9.5 \times 2.54\,\text{cm} = 24.13\,\text{cm}$

$24.13\,\text{cm} = 24.13 \times 10\,\text{mm} = \mathbf{241.3\,mm}$

(b) 50 m.p.h. $= 50 \times 1.61\,\text{km/h} = \mathbf{80.5\,km/h}$

(c) $300\,\text{km} = \frac{300}{1.61}\,\text{miles} = \mathbf{186.3\,miles}$

(d) $30\,\text{lb} = \frac{30}{2.2}\,\text{kg} = \mathbf{13.64\,kg}$

(e) $42\,\text{kg} = 42 \times 2.2\,\text{lb} = 92.4\,\text{lb}$

$0.4\,\text{lb} = 0.4 \times 16\,\text{oz} = 6.4\,\text{oz} = 6\,\text{oz}$, correct to the nearest ounce

Thus $42\,\text{kg} = \mathbf{92\ lb\ 6\,oz}$, correct to the nearest ounce.

(f) 15 gallons $= 15 \times 8$ pints $= 120$ pints

120 pints $= \frac{120}{1.76}$ litres $=$ **68.18 litres**

(g) 40 litres $= 40 \times 1.76$ pints $= 70.4$ pints

70.4 pints $= \frac{70.4}{8}$ gallons $=$ **8.8 gallons**

Now try the following Practice Exercise

Practice Exercise 24 Conversion tables and charts (Answers on page 704)

1. Currency exchange rates listed in a newspaper included the following:

Italy	£1 = 1.17 euro
Japan	£1 = 135 yen
Australia	£1 = 1.70 dollars
Canada	£1 = $1.60
Sweden	£1 = 11.30 kronor

Calculate (a) how many Italian euros £32.50 will buy, (b) the number of Canadian dollars that can be purchased for £74.80, (c) the pounds sterling which can be exchanged for 14 040 yen, (d) the pounds sterling which can be exchanged for 1753.76 Swedish kronor and (e) the Australian dollars which can be bought for £55

2. Below is a list of some metric to imperial conversions.

Length	2.54 cm = 1 inch 1.61 km = 1 mile
Weight	1 kg = 2.2 lb (1 lb = 16 ounces)
Capacity	1 litre = 1.76 pints (8 pints = 1 gallon)

Use the list to determine (a) the number of millimetres in 15 inches, (b) a speed of 35 mph in km/h, (c) the number of kilometres in 235 miles, (d) the number of pounds and ounces in 24 kg (correct to the nearest ounce), (e) the number of kilograms in 15 lb, (f) the number of litres in 12 gallons and (g) the number of gallons in 25 litres.

3. Deduce the following information from the train timetable shown in Table 4.3 on pages 47 and 48:

   (a) At what time should a man catch a train at Mossley Hill to enable him to be in Manchester Piccadilly by 8.15 a.m.?

   (b) A girl leaves Hunts Cross at 8.17 a.m. and travels to Manchester Oxford Road. How long does the journey take? What is the average speed of the journey?

   (c) A man living at Edge Hill has to be at work at Trafford Park by 8.45 a.m. It takes him 10 minutes to walk to his work from Trafford Park station. What time train should he catch from Edge Hill?

4.4 Evaluation of formulae

The statement $v = u + at$ is said to be a **formula** for v in terms of u, a and t.
v, u, a and t are called **symbols** or **variables**.
The single term on the left-hand side of the equation, v, is called the **subject of the formulae**.

Provided values are given for all the symbols in a formula except one, the remaining symbol can be made the subject of the formula and may be evaluated by using a calculator.

Problem 16. In an electrical circuit the voltage V is given by Ohm's law, i.e. $V = IR$. Find, correct to 4 significant figures, the voltage when $I = 5.36$ A and $R = 14.76\,\Omega$.

Table 4.3 Liverpool, Hunt's Cross and Warrington → Manchester

Miles				MX	MO A $$ I	SX ◇ C I	SO ◇ C	SX BHX	BHX	◇ I		BHX	BHX ◇ D I		SX BHX	BHX ◇ E		BHX	◇ C I		SX BHX		◇ I		BHX
0	Liverpool Lime Street	82, 99	d			05 25	05 37		06 03	06 23		06 30	06 54		07 00	07 17		07 30	07 52			08 00	08 23		08 30
$1\frac{1}{2}$	Edge Hill	82, 99	d									06 34			07 04			07 34				08 04			08 34
$3\frac{1}{2}$	Mossley Hill	82	d									06 39			07 09			07 39				08 09			08 39
$4\frac{1}{2}$	West Allerton	82	d									06 41			07 11			07 41				08 11			08 41
$5\frac{1}{2}$	Allerton	82	d									06 43			07 13			07 43				08 13			08 43
–	Liverpool Central	101	d									*06 73*	*06 45*					*07 15*	*07 45*						*08 75*
–	Garston (Merseyside)	101	d									*06 26*	*06 56*					*07 26*	*07 56*						*08 26*
$7\frac{1}{2}$	Hunt's Cross		d			05u38	05u50		06 17			06 47	07u07		07 17			07 47	08u05			08 17			08 47
$8\frac{1}{2}$	Halewood		d						06 20			06 50			07 20			07 50				08 20			08 50
$10\frac{1}{2}$	Hough Green		d						06 24			06 54			07 24			07 54				08 24			08 54
$12\frac{1}{2}$	Widnes		d						06 27			06 57			07 27	07 35		07 57	08 12			08 27			08 57
16	Sankey for Penketh		d	00 02					06 32			07 02			07 32			08 02				08 32			09 02
$18\frac{1}{2}$	Warrington Central		a	00 07		05 50	06 02		06 37	06 46		07 07	07 19		07 37	07 43		08 07	08 20			08 37	08 46		09 07
–			d			05 51	06 03	06 30		06 46	07 00		07 20	07 30		07 43	08 00		08 20		08 30		08 46	09 00	
$20\frac{1}{2}$	Padgate		d					06 33			07 03			07 33			08 03				08 33			09 03	
$21\frac{1}{2}$	Birchwood		d			05 56	06 08	06 36		06 51	07 06		07 25	07 36		07 48	08 06		08 25		08 34		08 51	09 06	
$24\frac{1}{2}$	Glazebrook		d					06 41			07 11			07 41			08 11				08 41			09 11	
$25\frac{1}{2}$	Iriam		d			06 02		06 44			07 14			07 44		07 54	08 14			08 34	08 44			09 14	
28	Flixton		d			06 06		06 48			07 18			07 48			08 15			08 38	08 48			09 18	
$28\frac{1}{2}$	Chassen Road		d			06 08		06 50			07 20			07 50			08 20			08 40	08 50			09 20	
29	Urmston		d		00 03	06 10		06 52			07 22			07 52			08 22			08 42	08 52			09 22	
$30\frac{1}{2}$	Humphrey Park		d		00 13	06 13		06 55			07 25			07 55			08 25			08 45	08 55			09 25	
31	Trafford Park		d			06 15		06 57			07 27			07 57			08 27			08 47	08 57			09 27	
34	Deansgate	81	d		00 23			07 03			07 33			08 03			08 33			08 52	09 03			09 33	
$34\frac{1}{2}$	Manchester Oxford Road	81	a		00 27	06 22	06 22	07 05		07 08	07 35		07 40	08 05		08 08	08 35		08 40	08 54	09 05		09 08	09 35	
–			d		00 27	06 23	06 23			07 09			07 41			08 09	08 37		08 41	08 55			09 09		
35	Manchester Piccadilly	81	a		00 34	06 25	06 25			07 11			07 43			08 11	08 39		08 43	08 57			09 11		
–	Stockport	81, 90	a			06 34	06 34			*07 32*			07 54			*08 32*			08 54	*09 19*			*09 32*		
–	Shetfield	90	a			07 30	07 30						08 42						08 42						

(Continued)

Table 4.3 *(Continued)*

			◇ I			◇ I		BHX	◇ I		◇ I		BHX	◇ I		◇ I		BHX	◇ I		◇ I		BHX	◇ I	
Liverpool Lime Street	82, 99	d	08 54		09 00	09 23		09 30	09 56	10 00	10 23		10 30	10 56	11 00	11 23		11 30	11 56	12 00	12 23		12 30	12 56	13 00
Edge Hill	82, 99	d			09 04			09 34		10 04			10 34		11 04			11 34		12 04			12 34		13 04
Mossley Hill	82	d			09 09			09 39		10 09			10 39		11 09			11 39		12 09			12 39		13 09
West Allerton	82	d			09 11			09 41		10 11			10 41		11 11			11 41		12 11			12 41		13 11
Allerton	82	d			09 13			09 43		10 13			10 43		11 13			11 43		12 13			12 43		13 13
Liverpool Central	101	d	*09 45*					*09 15*	*09 45*				*10 15*	*10 45*				*11 75*	*11 45*				*12 15*	*12 45*	
Garston (Merseyside)	101	d	*09 56*					*09 26*	*09 56*				*10 26*	*10 56*				*11 26*	*11 56*				*12 26*	*12 56*	
Hunt's Cross		d	09u09		09 17			09 47	10u09	10 17			10 47	11u09	11 17			11 47	12u09	12 17			12 47	13u09	13 17
Halewood		d			09 20			09 50		10 20			10 50		11 20			11 50		12 20			12 50		13 20
Hough Green		d			09 24			09 54		10 24			10 54		11 24			11 54		12 24			12 54		13 24
Widnes		d			09 27			09 57		10 27			10 57		11 27			11 57		12 27			12 57		13 27
Sankey for Penketh		d			09 32			10 02		10 32			11 02		11 32			12 02		12 32			13 02		13 32
Warrington Central		a	09 21		09 37	09 46		10 07	10 21	10 37	10 46		11 07	11 21	11 37	11 46		12 07	12 21	12 37	12 46		13 07	13 21	13 37
		d	09 22	09 30		09 46	10 00		10 22		10 46	11 00		11 22		11 46	12 00		12 22		12 46	13 00		13 22	
Padgate		d		09 33			10 03					11 03					12 03					13 03			
Birchwood		d		09 36		09 51	10 06				10 51	11 06				11 51	12 06				12 51	13 04			
Glazebrook		d		09 41			10 11					11 11					12 11					13 11			
Iriam		d		09 44			10 14					11 14					12 14					13 14			
Flexton		d		09 48			10 18					11 18					12 18					13 18			
Chassen Road		d		09 50			10 20					11 20					12 20					13 20			
Urmston		d		09 52			10 22					11 22					12 22					13 22			
Humphrey Park		d		09 55			10 25					11 25					12 25					13 25			
Trafford Park		d		09 57			10 27					11 27					12 27					13 27			
Deansgate	81	d		10 03			10 33					11 33					12 33					13 33			
Manchester Oxford Road	81	a	09 40	10 05		10 08	10 35		10 40		11 08	11 35		11 40		12 08	12 35		12 40		13 08	13 35		13 40	
		d	09 41	10 06		10 09			10 41		11 09			11 41		12 09			12 41		13 09			13 41	
Manchester Piccadilly	81	a	09 43	10 08		10 11			10 43		11 11			11 43		12 11			12 43		13 11			13 43	
Stockport	81, 90	a	09 54	*10 25*		*10 32*			10 54		*11 32*			11 54		*12 32*			12 54		*13 32*			13 54	
Sheffield	90	a	10 42						11 42					12 41					13 42					14 39	

Reproduced with permission of British Rail

$$V = IR = (5.36)(14.76)$$

Hence, **voltage $V = 79.11$ V, correct to 4 significant figures**.

Problem 17. The surface area A of a hollow cone is given by $A = \pi rl$. Determine, correct to 1 decimal place, the surface area when $r = 3.0$ cm and $l = 8.5$ cm.

$$A = \pi rl = \pi\ (3.0)(8.5)\ \text{cm}^2$$

Hence, **surface area $A = 80.1\ \text{cm}^2$**, correct to 1 decimal place.

Problem 18. Velocity v is given by $v = u + at$. If $u = 9.86$ m/s, $a = 4.25\ \text{m/s}^2$ and $t = 6.84$ s, find v, correct to 3 significant figures.

$$v = u + at = 9.86 + (4.25)(6.84)$$
$$= 9.86 + 29.07 = 38.93$$

Hence, **velocity $v = 38.9$ m/s, correct to 3 significant figures**.

Problem 19. The power, P watts, dissipated in an electrical circuit may be expressed by the formula $P = \dfrac{V^2}{R}$. Evaluate the power, correct to 3 significant figures, given that $V = 17.48$ V and $R = 36.12\,\Omega$.

$$P = \frac{V^2}{R} = \frac{(17.48)^2}{36.12} = \frac{305.5504}{36.12}$$

Hence **power, $P = 8.46$ W, correct to 3 significant figures**.

Problem 20. The volume $V\ \text{cm}^3$ of a right circular cone is given by $V = \dfrac{1}{3}\pi r^2 h$. Given that $r = 4.321$ cm and $h = 18.35$ cm, find the volume, correct to 4 significant figures.

$$V = \frac{1}{3}\pi r^2 h = \frac{1}{3}\pi(4.321)^2(18.35)$$
$$= \frac{1}{3}\pi(18.671041)(18.35)$$

Hence **volume, $V = 358.8\ \text{cm}^3$, correct to 4 significant figures**.

Problem 21. Force F newtons is given by the formula $F = \dfrac{Gm_1m_2}{d^2}$, where m_1 and m_2 are masses, d their distance apart and G is a constant. Find the value of the force given that $G = 6.67 \times 10^{-11}$, $m_1 = 7.36$, $m_2 = 15.5$ and $d = 22.6$. Express the answer in standard form, correct to 3 significant figures.

$$F = \frac{Gm_1m_2}{d^2} = \frac{(6.67 \times 10^{-11})(7.36)(15.5)}{(22.6)^2}$$
$$= \frac{(6.67)(7.36)(15.5)}{(10^{11})(510.76)} = \frac{1.490}{10^{11}}$$

Hence **force $F = 1.49 \times 10^{-11}$ newtons, correct to 3 significant figures**.

Problem 22. The time of swing t seconds, of a simple pendulum is given by $t = 2\pi\sqrt{\dfrac{l}{g}}$. Determine the time, correct to 3 decimal places, given that $l = 12.0$ and $g = 9.81$

$$t = 2\pi\sqrt{\frac{l}{g}} = (2)\pi\sqrt{\frac{12.0}{9.81}}$$
$$= (2)\pi\sqrt{1.22324159}$$
$$= (2)\pi(1.106002527)$$

Hence **time $t = 6.950$ seconds, correct to 3 decimal places**.

Problem 23. Resistance, $R\,\Omega$, varies with temperature according to the formula $R = R_0(1 + \alpha t)$. Evaluate R, correct to 3 significant figures, given $R_0 = 14.59$, $\alpha = 0.0043$ and $t = 80$

$$R = R_0(1 + \alpha t) = 14.59[1 + (0.0043)(80)]$$
$$= 14.59(1 + 0.344)$$
$$= 14.59(1.344)$$

Hence, **resistance, $R = 19.6\,\Omega$, correct to 3 significant figures**.

Problem 24. A concrete pillar, which is reinforced with steel rods, supports a compressive axial load, P, of 1 MN. The compressive stress in the steel, σ_S, is given by: $\sigma_S = \frac{-P \times E_S}{(A_S E_S + A_C E_C)}$
Calculate the stress in the steel given that the cross-sectional area of the steel, $A_S = 3 \times 10^{-3}$ m^2, the cross-sectional area of the concrete, $A_C = 0.1$ m^2, and Young's modulus for steel and concrete respectively is $E_S = 2 \times 10^{11}$ N/m^2 and $E_C = 2 \times 10^{10}$ N/m^2. Give the answer in MN correct to 4 significant figures.

Compressive stress in steel, $\sigma_S = \frac{-P \times E_S}{(A_S E_S + A_C E_C)}$

$$= \frac{-1 \times 10^6 \times 2 \times 10^{11}}{(3 \times 10^{-3} \times 2 \times 10^{11} + 0.1 \times 2 \times 10^{10})}$$

$$= \frac{-2 \times 10^{17}}{(6 \times 10^8 + 0.2 \times 10^{10})}$$

$$= \frac{-2 \times 10^{17}}{(2.6 \times 10^9)} = 76923076.92 \text{ N}$$

$$= \mathbf{76.92\ MN}$$

Now try the following Practice Exercise

Practice Exercise 25 Evaluation of formulae (Answers on page 704)

1. A formula used in connection with gases is $R = (PV)/T$. Evaluate R when $P = 1500$, $V = 5$ and $T = 200$
2. The velocity of a body is given by $v = u + at$. The initial velocity u is measured when time t is 15 seconds and found to be 12 m/s. If the acceleration a is 9.81 m/s^2 calculate the final velocity v
3. Find the distance s, given that $s = \frac{1}{2}gt^2$, time $t = 0.032$ seconds and acceleration due to gravity $g = 9.81$ m/s^2
4. The energy stored in a capacitor is given by $E = \frac{1}{2}CV^2$ joules. Determine the energy when capacitance $C = 5 \times 10^{-6}$ farads and voltage $V = 240V$
5. Resistance R_2 is given by $R_2 = R_1(1 + \alpha t)$. Find R_2, correct to 4 significant figures, when $R_1 = 220$, $\alpha = 0.00027$ and $t = 75.6$
6. Density $= \frac{\text{mass}}{\text{volume}}$. Find the density when the mass is 2.462 kg and the volume is 173 cm^3. Give the answer in units of kg/m^3
7. Velocity = frequency × wavelength. Find the velocity when the frequency is 1825 Hz and the wavelength is 0.154 m
8. Evaluate resistance R_T, given $\frac{1}{R_T} = \frac{1}{R_1} + \frac{1}{R_2} + \frac{1}{R_3}$ when $R_1 = 5.5\,\Omega$, $R_2 = 7.42\,\Omega$ and $R_3 = 12.6\,\Omega$
9. Power $= \frac{\text{force} \times \text{distance}}{\text{time}}$. Find the power when a force of 3760 N raises an object a distance of 4.73 m in 35 s
10. The potential difference, V volts, available at battery terminals is given by $V = E - Ir$. Evaluate V when $E = 5.62$, $I = 0.70$ and $R = 4.30$
11. Given force $F = \frac{1}{2}m(v^2 - u^2)$, find F when $m = 18.3$, $v = 12.7$ and $u = 8.24$
12. The current I amperes flowing in a number of cells is given by $I = \frac{nE}{R + nr}$. Evaluate the current when $n = 36$. $E = 2.20$, $R = 2.80$ and $r = 0.50$
13. The time, t seconds, of oscillation for a simple pendulum is given by $t = 2\pi\sqrt{\frac{l}{g}}$. Determine the time when $\pi = 3.142$, $l = 54.32$ and $g = 9.81$
14. Energy, E joules, is given by the formula $E = \frac{1}{2}LI^2$. Evaluate the energy when $L = 5.5$ and $I = 1.2$
15. The current I amperes in an a.c. circuit is given by $I = \frac{V}{\sqrt{R^2 + X^2}}$. Evaluate the current when $V = 250$, $R = 11.0$ and $X = 16.2$
16. Distance s metres is given by the formula $s = ut + \frac{1}{2}\,at^2$. If $u = 9.50$, $t = 4.60$ and $a = -2.50$, evaluate the distance
17. The area, A, of a triangular template is given by $A = \sqrt{s(s-a)(s-b)(s-c)}$ where $s = \frac{a+b+c}{2}$. Evaluate the area given $a = 3.60$ cm, $b = 4.00$ cm and $c = 5.20$ cm

18. Given that $a = 0.290$, $b = 14.86$, $c = 0.042$, $d = 31.8$ and $e = 0.650$, evaluate v, given that $v = \sqrt{\frac{ab}{c} - \frac{d}{e}}$

19. A solid disc of uniform thickness has a diameter of 0.3 m and thickness, $t = 0.08$ m. The density, ρ, of its material of construction is 7860 kg/m^3. The mass moment of inertia about its centroid, I, is given by: $I = \rho\pi R^2 t \times \frac{R^2}{2}$ where R is its radius. Calculate the mass moment of inertia for the disc.

20. If Buoyancy $= V \times \rho \times g$ newtons, where $g = 9.81$ m/s^2 and the density of water, $\rho = 1000$ kg/m^3. A barge of length 30 m and width 8 m floats on an even keel at a depth of 3 m. Calculate the value of its buoyancy.

Practice Exercise 26 Multiple-choice questions on calculations and evaluation of formulae (Answers on page 704)

Each question has only one correct answer

1. Using the 24-hour clock, the time, in hours and minutes, between 0340 and 2222 is:
(a) 26 h 2 min (b) 18 h 42 min
(c) 18 h 18 min (d) 6 h 42 min

2. The symbol for pi is:
(a) Ω (b) ε (c) μ (d) π

3. $106 \times 106 - 94 \times 94$ is equal to:
(a) 2400 (b) 1904 (c) 1906 (d) 2004

4. The reciprocal of 5 is:
(a) 0.5 (b) 5 (c) $\frac{1}{5}$ (d) -5

5. Which of the following is a square number?
(a) 2 (b) 4 (c) 6 (d) 8

6. $(0.5)^2 \times (0.1)^3$ is equal to:
(a) 25 (b) 0.0025 (c) 0.00025 (d) 0.25

7. If $x = \frac{57.06 \times 0.0711}{\sqrt{0.0635}}$ cm, which of the following statements is correct?
(a) $x = 16.09$ cm, correct to 4 significant figures
(b) $x = 16$ cm, correct to 2 significant figures
(c) $x = 1.61 \times 10^1$ cm, correct to 3 decimal places
(d) $x = 16.099$ cm, correct to 3 decimal places

8. $\frac{62.91}{0.12} + \sqrt{\left(\frac{6.36\pi}{2e^{-3} \times \sqrt{73.81}}\right)}$ is equal to:
(a) 529.08 correct to 5 significant figures
(b) 529.082 correct to 3 decimal places
(c) 5.29×10^2
(d) 529.0 correct to 1 decimal place

9. The area A of a triangular piece of land of sides a, b and c may be calculated using

$$A = \sqrt{[s(s-a)(s-b)(s-c)]}$$

where $s = \frac{a+b+c}{2}$

When $a = 15$ m, $b = 11$ m and $c = 8$ m, the area, correct to the nearest square metre, is:
(a) 1836 m^2 (b) 648 m^2
(c) 445 m^2 (d) 43 m^2

10. Evaluating $\frac{1}{10}\sqrt{\left(\frac{2e^{(1-\sqrt{5})}}{3\pi \times \sqrt{12.94}}\right)}$ correct to 4 significant figures gives:
(a) 0.0131 (b) 0.131
(c) 0.1309 (d) 0.01309

For fully worked solutions to each of the problems in Practice Exercises 22 to 25 in this chapter, go to the website:
www.routledge.com/cw/bird

Revision Test 1 Fractions, decimals, percentages, indices, numbering systems and calculations

This Revision Test covers the material contained in Chapters 1 to 4. *The marks for each question are shown in brackets at the end of each question.*

1. Simplify (a) $2\frac{2}{3} \div 3\frac{1}{3}$

   (b) $\dfrac{1}{\left(\frac{4}{7} \times 2\frac{1}{4}\right)} \div \left(\frac{1}{3} + \frac{1}{5}\right) + 2\frac{7}{24}$ (9)

2. A piece of steel, 1.69 m long, is cut into three pieces in the ratio 2 to 5 to 6. Determine, in centimetres, the lengths of the three pieces. (4)

3. Evaluate $\dfrac{576.29}{19.3}$

   (a) correct to 4 significant figures.

   (b) correct to 1 decimal place. (2)

4. Determine, correct to 1 decimal places, 57% of 17.64 g. (2)

5. Express 54.7 mm as a percentage of 1.15 m, correct to 3 significant figures. (3)

6. Evaluate the following:

   (a) $\dfrac{2^3 \times 2 \times 2^2}{2^4}$ (b) $\dfrac{(2^3 \times 16)^2}{(8 \times 2)^3}$

   (c) $\left(\dfrac{1}{4^2}\right)^{-1}$ (d) $(27)^{-\frac{1}{3}}$

   (e) $\dfrac{\left(\frac{3}{2}\right)^{-2} - \frac{2}{9}}{\left(\frac{2}{3}\right)^2}$ (14)

7. Express the following in both standard form and engineering notation:

   (a) 1623 (b) 0.076 (c) $145\frac{2}{5}$ (3)

8. Determine the value of the following, giving the answer in both standard form and engineering notation:

   (a) $5.9 \times 10^2 + 7.31 \times 10^2$

   (b) $2.75 \times 10^{-2} - 2.65 \times 10^{-3}$ (4)

9. Convert the following binary numbers to decimal form:

   (a) 1101 (b) 101101.0101 (5)

10. Convert the following decimal number to binary form:

    (a) 27 (b) 44.1875 (6)

11. Convert the following decimal numbers to binary, via octal:

    (a) 479 (b) 185.2890625 (6)

12. Convert (a) $5F_{16}$ into its decimal equivalent (b) 132_{10} into its hexadecimal equivalent (c) 110101011_2 into its hexadecimal equivalent. (6)

13. Evaluate the following, each correct to 4 significant figures:

    (a) 61.22^2 (b) $\dfrac{1}{0.0419}$ (c) $\sqrt{0.0527}$ (3)

14. Evaluate the following, each correct to 2 decimal places:

    (a) $\left(\dfrac{36.2^2 \times 0.561}{27.8 \times 12.83}\right)^3$

    (b) $\sqrt{\dfrac{14.69^2}{\sqrt{17.42} \times 37.98}}$ (7)

15. If 1.6 km = 1 mile, determine the speed of 45 miles/hour in kilometres per hour. (3)

16. Evaluate B, correct to 3 significant figures, when $W = 7.20$, $v = 10.0$ and $g = 9.81$, given that $B = \dfrac{Wv^2}{2g}$ (3)

17. Rewrite 32 cm^2 in square millimetres. (1)

18. A rectangular tabletop is 1500 mm long and 800 mm wide. Determine its area in
(a) mm^2 (b) cm^2 (c) m^2 (3)

19. Rewrite 0.065 m^3 in cubic millimetres. (1)

20. Rewrite 20000 mm^3 in cubic metres. (1)

21. Rewrite 8.3 cm^3 in cubic millimetres. (1)

22. A petrol tank is in the shape of a rectangular prism having length 1.0 m, breadth 75 cm and height 120 mm. Determine the capacity of the tank in
(a) m^3 (b) cm^3 (c) litres (3)

For lecturers/instructors/teachers, fully worked solutions to each of the problems in Revision Test 1, together with a full marking scheme, are available at the website:
www.routledge.com/cw/bird

COMPANION @ WEBSITE

Chapter 5

Algebra

Why it is important to understand: Algebra

Algebra is one of the most fundamental tools for engineers because it allows them to determine the value of something (length, material constant, temperature, mass and so on) given values that they do know (possibly other length, material properties, mass). Although the types of problems that mechanical, chemical, civil, environmental, electrical engineers deal with vary, all engineers use algebra to solve problems. An example where algebra is frequently used is in simple electrical circuits, where the resistance is proportional to voltage. Using Ohm's law, or $V = I \times R$, an engineer simply multiplies the current in a circuit by the resistance to determine the voltage across the circuit. Engineers and scientists use algebra in many ways, and so frequently that they don't even stop the think about it. Depending on what type of engineer you choose to be, you will use varying degrees of algebra, but in all instances, algebra lays the foundation for the mathematics you will need to become an engineer.

At the end of this chapter, you should be able to:

- understand basic operations in algebra
- understand and use the laws of indices
- use brackets in an algebraic expression
- factorise simple functions
- understand and use the fundamental laws of precedence
- understand direct and inverse proportionality
- apply direct and inverse proportion to practical situations

5.1 Basic operations

Algebra is that part of mathematics in which the relations and properties of numbers are investigated by means of general symbols. For example, the area of a rectangle is found by multiplying the length by the breadth; this is expressed algebraically as $A = l \times b$, where A represents the area, l the length and b the breadth.

The basic laws introduced in arithmetic are generalised in algebra.

Let a, b, c and d represent any four numbers. Then:

(i) $a + (b + c) = (a + b) + c$

(ii) $a(bc) = (ab)c$

(iii) $a + b = b + a$

(iv) $ab = ba$

(v) $a(b + c) = ab + ac$

(vi) $\dfrac{a + b}{c} = \dfrac{a}{c} + \dfrac{b}{c}$

(vii) $(a + b)(c + d) = ac + ad + bc + bd$

Problem 1. Evaluate: $3ab-2bc+abc$ when $a=1$, $b=3$ and $c=5$

Replacing a, b and c with their numerical values gives:

$$3ab-2bc+abc=3\times1\times3-2\times3\times5+1\times3\times5$$
$$=9-30+15=\mathbf{-6}$$

Problem 2. Find the value of $4p^2qr^3$, given the $p=2$, $q=\frac{1}{2}$ and $r=1\frac{1}{2}$

Replacing p, q and r with their numerical values gives:

$$4p^2qr^3=4(2)^2\left(\frac{1}{2}\right)\left(\frac{3}{2}\right)^3$$
$$=4\times2\times2\times\frac{1}{2}\times\frac{3}{2}\times\frac{3}{2}\times\frac{3}{2}=\mathbf{27}$$

Problem 3. Find the sum of: $3x$, $2x$, $-x$ and $-7x$

The sum of the positive term is: $3x+2x=5x$

The sum of the negative terms is: $x+7x=8x$

Taking the sum of the negative terms from the sum of the positive terms gives:

$$5x-8x=\mathbf{-3x}$$

Alternatively

$$3x+2x+(-x)+(-7x)=3x+2x-x-7x$$
$$=\mathbf{-3x}$$

Problem 4. Find the sum of: $4a$, $3b$, c, $-2a$, $-5b$ and $6c$

Each symbol must be dealt with individually.

For the 'a' terms: $+4a-2a=2a$

For the 'b' terms: $+3b-5b=-2b$

For the 'c' terms: $+c+6c=7c$

Thus

$$4a+3b+c+(-2a)+(-5b)+6c$$
$$=4a+3b+c-2a-5b+6c$$
$$=\mathbf{2a-2b+7c}$$

Problem 5. Find the sum of: $5a-2b$, $2a+c$, $4b-5d$ and $b-a+3d-4c$

The algebraic expressions may be tabulated as shown below, forming columns for the a's, b's, c's and d's. Thus:

$$\begin{array}{rrrr} +5a & -2b & & \\ +2a & & +c & \\ & +4b & & -5d \\ -a & +b & -4c & +3d \\ \hline \end{array}$$

Adding gives: $\mathbf{6a+3b-3c-2d}$

Problem 6. Subtract $2x+3y-4z$ from $x-2y+5z$

$$\begin{array}{rrr} x & -2y & +5z \\ 2x & +3y & -4z \\ \hline \end{array}$$

Subtracting gives: $\mathbf{-x-5y+9z}$

(Note that $+5z--4z=+5z+4z=9z$)

An alternative method of subtracting algebraic expressions is to 'change the signs of the bottom line and add'. Hence:

$$\begin{array}{rrr} x & -2y & +5z \\ -2x & -3y & +4z \\ \hline \end{array}$$

Adding gives: $\mathbf{-x-5y+9z}$

Problem 7. Multiply $2a+3b$ by $a+b$

Each term in the first expression is multiplied by a, then each term in the first expression is multiplied by b, and the two results are added. The usual layout is shown below.

$$\begin{array}{rrr} 2a & +3b & \\ a & +b & \\ \hline \end{array}$$

Multiplying by $a\rightarrow$ $2a^2+3ab$

Multiplying by $b\rightarrow$ $+2ab+3b^2$

Adding gives: $\mathbf{2a^2+5ab+3b^2}$

Problem 8. Multiply $3x-2y^2+4xy$ by $2x-5y$

$$\begin{array}{llll} & 3x - 2y^2 + 4xy & & \\ & 2x - 5y & & \\ \text{Multiplying by } 2x\rightarrow & 6x^2 - 4xy^2 + 8x^2y & & \\ \text{Multiplying by } -5y\rightarrow & \quad -20xy^2 & -15xy + 10y^3 & \\ \text{Adding gives:} & \mathbf{6x^2 - 24xy^2 + 8x^2y - 15xy + 10y^3} & & \end{array}$$

Problem 9. Simplify: $2p \div 8pq$

$2p \div 8pq$ means $\frac{2p}{8pq}$. This can be reduced by cancelling as in arithmetic.

Thus: $\frac{2p}{8pq} = \frac{^1\cancel{2} \times \cancel{p}^1}{\cancel{8}_4 \times \cancel{p}_1 \times q} = \mathbf{\frac{1}{4q}}$

Now try the following Practice Exercise

Practice Exercise 27 Basic operations (Answers on page 704)

1. Find the value of $2xy+3yz-xyz$, when $x=2$, $y=-2$ and $z=4$
2. Evaluate $3pq^3r^3$ when $p=\frac{2}{3}$, $q=-2$ and $r=-1$
3. Find the sum of $3a$, $-2a$, $-6a$, $5a$ and $4a$
4. Add together $2a+3b+4c$, $-5a-2b+c$, $4a-5b-6c$
5. Add together $3d+4e$, $-2e+f$, $2d-3f$, $4d-e+2f-3e$
6. From $4x-3y+2z$ subtract $x+2y-3z$
7. Subtract $\frac{3}{2}a-\frac{b}{3}+c$ from $\frac{b}{2}-4a-3c$
8. Multiply $3x+2y$ by $x-y$
9. Multiply $2a-5b+c$ by $3a+b$
10. Simplify (i) $3a \div 9ab$ (ii) $4a^2b \div 2a$

5.2 Laws of indices

The laws of indices are:

(i) $a^m \times a^n = a^{m+n}$ (ii) $\frac{a^m}{a^n} = a^{m-n}$

(iii) $(a^m)^n = a^{mn}$ (iv) $a^{m/n} = \sqrt[n]{a^m}$

(v) $a^{-n} = \frac{1}{a^n}$ (vi) $a^0 = 1$

Problem 10. Simplify: $a^3b^2c \times ab^3c^5$

Grouping like terms gives:

$$a^3 \times a \times b^2 \times b^3 \times c \times c^5$$

Using the first law of indices gives:

$$a^{3+1} \times b^{2+3} \times c^{1+5}$$

i.e. $a^4 \times b^5 \times c^6 = \mathbf{a^4b^5c^6}$

Problem 11. Simplify: $a^{1/2}b^2c^{-2} \times a^{1/6}b^{1/2}c$

Using the first law of indices,

$$a^{1/2}b^2c^{-2} \times a^{(1/6)}b^{(1/2)}c$$
$$= a^{(1/2)+(1/6)} \times b^{2+(1/2)} \times c^{-2+1}$$
$$= \mathbf{a^{2/3}b^{5/2}c^{-1}}$$

Problem 12. Simplify: $\frac{a^3b^2c^4}{abc^{-2}}$ and evaluate when $a=3$, $b=\frac{1}{8}$ and $c=2$

Using the second law of indices,

$$\frac{a^3}{a} = a^{3-1} = a^2, \quad \frac{b^2}{b} = b^{2-1} = b$$

and $\frac{c^4}{c^{-2}} = c^{4-(-2)} = c^6$

Thus $\frac{a^3b^2c^4}{abc^{-2}} = \mathbf{a^2bc^6}$

When $a=3$, $b=\frac{1}{8}$ and $c=2$,

$$a^2bc^6 = (3)^2\left(\frac{1}{8}\right)(2)^6 = (9)\left(\frac{1}{8}\right)(64) = \mathbf{72}$$

Problem 13. Simplify: $\frac{p^{1/2}q^2r^{2/3}}{p^{1/4}q^{1/2}r^{1/6}}$ and evaluate when $p=16$, $q=9$ and $r=4$, taking positive roots only

Using the second law of indices gives:

$$p^{(1/2)-(1/4)}q^{2-(1/2)}r^{(2/3)-(1/6)} = \boldsymbol{p^{1/4}q^{3/2}r^{1/2}}$$

When $p=16$, $q=9$ and $r=4$,

$$\begin{aligned} p^{1/4}q^{3/2}r^{1/2} &= (16)^{1/4}(9)^{3/2}(4)^{1/2} \\ &= (\sqrt[4]{16})(\sqrt{9^3})(\sqrt{4}) \\ &= (2)(3^3)(2) = \mathbf{108} \end{aligned}$$

Problem 14. Simplify: $\dfrac{x^2y^3+xy^2}{xy}$

Algebraic expressions of the form $\dfrac{a+b}{c}$ can be split into $\dfrac{a}{c}+\dfrac{b}{c}$. Thus

$$\begin{aligned} \frac{x^2y^3+xy^2}{xy} &= \frac{x^2y^3}{xy}+\frac{xy^2}{xy} \\ &= x^{2-1}y^{3-1}+x^{1-1}y^{2-1} \\ &= \boldsymbol{xy^2+y} \end{aligned}$$

(since $x^0=1$, from the sixth law of indices).

Problem 15. Simplify: $\dfrac{x^2y}{xy^2-xy}$

The highest common factor (HCF) of each of the three terms comprising the numerator and denominator is xy. Dividing each term by xy gives:

$$\frac{x^2y}{xy^2-xy} = \frac{\dfrac{x^2y}{xy}}{\dfrac{xy^2}{xy}-\dfrac{xy}{xy}} = \boldsymbol{\frac{x}{y-1}}$$

Problem 16. Simplify: $(p^3)^{1/2}(q^2)^4$

Using the third law of indices gives:

$$p^{3\times(1/2)}q^{2\times4} = \boldsymbol{p^{(3/2)}q^8}$$

Problem 17. Simplify: $\dfrac{(mn^2)^3}{(m^{1/2}n^{1/4})^4}$

The brackets indicate that each letter in the bracket must be raised to the power outside. Using the third law of indices gives:

$$\frac{(mn^2)^3}{(m^{1/2}n^{1/4})^4} = \frac{m^{1\times3}n^{2\times3}}{m^{(1/2)\times4}n^{(1/4)\times4}} = \frac{m^3n^6}{m^2n^1}$$

Using the second law of indices gives:

$$\frac{m^3n^6}{m^2n^1} = m^{3-2}n^{6-1} = \boldsymbol{mn^5}$$

Problem 18. Simplify: $(a^3\sqrt{b}\sqrt{c^5})(\sqrt{a}\sqrt[3]{b^2}c^3)$ and evaluate when $a=\dfrac{1}{4}$, $b=64$ and $c=1$

Using the fourth law of indices, the expression can be written as:

$$(a^3b^{1/2}c^{5/2})(a^{1/2}b^{2/3}c^3)$$

Using the first law of indices gives:

$$a^{3+(1/2)}b^{(1/2)+(2/3)}c^{(5/2)+3} = a^{7/2}b^{7/6}c^{11/2}$$

It is usual to express the answer in the same form as the question. Hence

$$a^{7/2}b^{7/6}c^{11/2} = \boldsymbol{\sqrt{a^7}\,\sqrt[6]{b^7}\,\sqrt{c^{11}}}$$

When $a=\dfrac{1}{4}$, $b=64$ and $c=1$,

$$\begin{aligned} \sqrt{a^7}\,\sqrt[6]{b^7}\,\sqrt{c^{11}} &= \sqrt{\left(\frac{1}{4}\right)^7}\left(\sqrt[6]{64^7}\right)\left(\sqrt{1^{11}}\right) \\ &= \left(\frac{1}{2}\right)^7(2)^7(1) = \mathbf{1} \end{aligned}$$

Problem 19. Simplify: $\dfrac{d^2e^2f^{1/2}}{(d^{3/2}ef^{5/2})^2}$ expressing the answer with positive indices only

Using the third law of indices gives:

$$\frac{d^2e^2f^{1/2}}{(d^{3/2}ef^{5/2})^2} = \frac{d^2e^2f^{1/2}}{d^3e^2f^5}$$

Using the second law of indices gives:

$$d^{2-3}e^{2-2}f^{(1/2)-5} = d^{-1}e^0f^{-9/2}$$

$$= d^{-1}f^{(-9/2)} \text{ since } e^0 = 1$$

from the sixth law of indices

$$= \frac{\mathbf{1}}{\mathbf{df^{9/2}}}$$

from the fifth law of indices.

Problem 20. Simplify: $\dfrac{(x^2y^{1/2})(\sqrt{x}\sqrt[3]{y^2})}{(x^5y^3)^{1/2}}$

Using the third and fourth laws of indices gives:

$$\frac{(x^2y^{1/2})(\sqrt{x}\sqrt[3]{y^2})}{(x^5y^3)^{1/2}} = \frac{(x^2y^{1/2})(x^{1/2}y^{2/3})}{x^{5/2}y^{3/2}}$$

Using the first and second laws of indices gives:

$$x^{2+(1/2)-(5/2)}y^{(1/2)+(2/3)-(3/2)} = x^0y^{-1/3}$$

$$= \mathbf{y^{-1/3}}$$

$$\text{or } \frac{\mathbf{1}}{\mathbf{y^{1/3}}} \text{ or } \frac{\mathbf{1}}{\sqrt[3]{\mathbf{y}}}$$

from the fifth and sixth law of indices.

Now try the following Practice Exercise

Practice Exercise 28 The laws of indices (Answers on page 704)

1. Simplify $(x^2y^3z)(x^3yz^2)$ and evaluate when $x=\frac{1}{2}$, $y=2$ and $z=3$
2. Simplify $(a^{3/2}bc^{-3})(a^{1/2}bc^{-1/2}c)$ and evaluate when $a=3$, $b=4$ and $c=2$
3. Simplify: $\dfrac{a^5bc^3}{a^2b^3c^2}$ and evaluate when $a=\frac{3}{2}$, $b=\frac{1}{2}$ and $c=\frac{2}{3}$

In Problems 4 to 10, simplify the given expressions:

4. $\dfrac{x^{1/5}y^{1/2}z^{1/3}}{x^{-1/2}y^{1/3}z^{-1/6}}$
5. $\dfrac{a^2b+a^3b}{a^2b^2}$
6. $\dfrac{p^3q^2}{pq^2-p^2q}$
7. $(a^2)^{1/2}(b^2)^3(c^{1/2})^3$
8. $\dfrac{(abc)^2}{(a^2b^{-1}c^{-3})^3}$
9. $(\sqrt{x}\sqrt{y^3}\sqrt[3]{z^2})(\sqrt{x}\sqrt{y^3}\sqrt{z^3})$
10. $\dfrac{(a^3b^{1/2}c^{-1/2})(ab)^{1/3}}{(\sqrt{a^3}\sqrt{b}c)}$

5.3 Brackets and factorisation

When two or more terms in an algebraic expression contain a common factor, then this factor can be shown outside of a **bracket**. For example

$$ab + ac = a(b + c)$$

which is simply the reverse of law (v) of algebra on page 54, and

$$6px + 2py - 4pz = 2p(3x + y - 2z)$$

This process is called **factorisation**.

Problem 21. Remove the brackets and simplify the expression:

$$(3a + b) + 2(b + c) - 4(c + d)$$

Both b and c in the second bracket have to be multiplied by 2, and c and d in the third bracket by -4 when the brackets are removed. Thus:

$$(3a + b) + 2(b + c) - 4(c + d)$$

$$= 3a + b + 2b + 2c - 4c - 4d$$

Collecting similar terms together gives:

$$\mathbf{3a + 3b - 2c - 4d}$$

Problem 22. Simplify:

$$a^2 - (2a - ab) - a(3b + a)$$

When the brackets are removed, both $2a$ and $-ab$ in the first bracket must be multiplied by -1 and both $3b$ and

a in the second bracket by $-a$. Thus:

$$a^2 - (2a - ab) - a(3b + a)$$
$$= a^2 - 2a + ab - 3ab - a^2$$

Collecting similar terms together gives: $-2a - 2ab$ Since $-2a$ is a common factor, the answer can be expressed as: $\mathbf{-2a(1+b)}$

Problem 23. Simplify: $(a+b)(a-b)$

Each term in the second bracket has to be multiplied by each term in the first bracket. Thus:

$$(a+b)(a-b) = a(a-b) + b(a-b)$$
$$= a^2 - ab + ab - b^2$$
$$= \mathbf{a^2 - b^2}$$

Alternatively

$$\begin{array}{ll} & a + b \\ & a - b \\ \hline \text{Multiplying by } a \rightarrow & a^2 + ab \\ \text{Multiplying by } -b \rightarrow & \quad - ab - b^2 \\ \hline \text{Adding gives:} & a^2 \quad - b^2 \\ \hline \end{array}$$

Problem 24. Simplify: $(3x-3y)^2$

$$(2x - 3y)^2 = (2x - 3y)(2x - 3y)$$
$$= 2x(2x - 3y) - 3y(2x - 3y)$$
$$= 4x^2 - 6xy - 6xy + 9y^2$$
$$= \mathbf{4x^2 - 12xy + 9y^2}$$

Alternatively,

$$\begin{array}{ll} & 2x \ -3y \\ & 2x \ -3y \\ \hline \text{Multiplying by } 2x \rightarrow & 4x^2 - 6xy \\ \text{Multiplying by } -3y \rightarrow & \quad - 6xy \ + 9y^2 \\ \hline \text{Adding gives:} & 4x^2 - 12xy + 9y^2 \\ \hline \end{array}$$

Problem 25. Remove the brackets from the expression: $2[p^2 - 3(q+r) + q^2]$

In this problem there are two brackets and the 'inner' one is removed first.

Hence, $2[p^2 - 3(q+r) + q^2]$

$$= 2[p^2 - 3q - 3r + q^2]$$
$$= \mathbf{2p^2 - 6q - 6r + 2q^2}$$

Problem 26. Remove the brackets and simplify the expression:

$$2a - [3\{2(4a - b) - 5(a + 2b)\} + 4a]$$

Removing the innermost brackets gives:

$$2a - [3\{8a - 2b - 5a - 10b\} + 4a]$$

Collecting together similar terms gives:

$$2a - [3\{3a - 12b\} + 4a]$$

Removing the 'curly' brackets gives:

$$2a - [9a - 36b + 4a]$$

Collecting together similar terms gives:

$$2a - [13a - 36b]$$

Removing the outer brackets gives:

$$2a - 13a - 36b$$

i.e. $\mathbf{-11a + 36b}$ or $\mathbf{36b - 11a}$

(see law (iii), page 54)

Problem 27. Simplify:

$$x(2x - 4y) - 2x(4x + y)$$

Removing brackets gives:

$$2x^2 - 4xy - 8x^2 - 2xy$$

Collecting together similar terms gives:

$$-6x^2 - 6xy$$

Factorising gives:

$$\mathbf{-6x(x + y)}$$

(since $-6x$ is common to both terms).

Problem 28. Factorise: (a) $xy - 3xz$ (b) $4a^2 + 16ab^3$ (c) $3a^2b - 6ab^2 + 15ab$

For each part of this problem, the HCF of the terms will become one of the factors. Thus:

(a) $xy - 3xz = \mathbf{x(y - 3z)}$

(b) $4a^2 + 16ab^3 = \mathbf{4a(a + 4b^3)}$

(c) $3a^2b - 6ab^2 + 15ab = \mathbf{3ab(a - 2b + 5)}$

Problem 29. Factorise: $ax-ay+bx-by$

The first two terms have a common factor of a and the last two terms a common factor of b. Thus:

$$ax-ay+bx-by=a(x-y)+b(x-y)$$

The two newly formed terms have a common factor of $(x-y)$. Thus:

$$a(x-y)+b(x-y)=\mathbf{(x-y)(a+b)}$$

Problem 30. Factorise:
$$2ax-3ay+2bx-3by$$

a is a common factor of the first two terms and b a common factor of the last two terms. Thus:

$$\begin{aligned}&2ax-3ay+2bx-3by\\&\quad=a(2x-3y)+b(2x-3y)\end{aligned}$$

$(2x-3y)$ is now a common factor, thus:

$$\begin{aligned}&a(2x-3y)+b(2x-3y)\\&\quad=\mathbf{(2x-3y)(a+b)}\end{aligned}$$

Alternatively, $2x$ is a common factor of the original first and third terms and $-3y$ is a common factor of the second and fourth terms. Thus:

$$\begin{aligned}&2ax-3ay+2bx-3by\\&\quad=2x(a+b)-3y(a+b)\end{aligned}$$

$(a+b)$ is now a common factor thus:

$$2x(a+b)-3y(a+b)=\mathbf{(a+b)(2x-3y)}$$

as before.

Problem 31. Factorise: x^3+3x^2-x-3

x^2 is a common factor of the first two terms, thus:

$$x^3+3x^2-x-3=x^2(x+3)-x-3$$

-1 is a common factor of the last two terms, thus:

$$x^2(x+3)-x-3=x^2(x+3)-1(x+3)$$

$(x+3)$ is now a common factor, thus:

$$x^2(x+3)-1(x+3)=\mathbf{(x+3)(x^2-1)}$$

Now try the following Practice Exercise

Practice Exercise 29 Brackets and factorisation (Answers on page 704)

In Problems 1 to 9, remove the brackets and simplify where possible:

1. $(x+2y)+(2x-y)$
2. $2(x-y)-3(y-x)$
3. $2(p+3q-r)-4(r-q+2p)+p$
4. $(a+b)(a+2b)$
5. $(p+q)(3p-2q)$
6. (i) $(x-2y)^2$ (ii) $(3a-b)^2$
7. $3a+2[a-(3a-2)]$
8. $2-5[a(a-2b)-(a-b)^2]$
9. $24p-[2\{3(5p-q)-2(p+2q)\}+3q]$

In Problems 10 to 12, factorise:

10. (i) $pb+2pc$ (ii) $2q^2+8qn$
11. (i) $21a^2b^2-28ab$ (ii) $2xy^2+6x^2y+8x^3y$
12. (i) $ay+by+a+b$ (ii) $px+qx+py+qy$ (iii) $2ax+3ay-4bx-6by$

5.4 Fundamental laws and precedence

The **laws of precedence** which apply to arithmetic also apply to algebraic expressions. The order is **B**rackets, **O**f, **D**ivision, **M**ultiplication, **A**ddition and **S**ubtraction (i.e. **BODMAS**).

Problem 32. Simplify: $2a+5a\times 3a-a$

Multiplication is performed before addition and subtraction thus:

$$\begin{aligned}2a+5a\times 3a-a&=2a+15a^2-a\\&=\mathbf{a+15a^2}\text{ or }\mathbf{a(1+15a)}\end{aligned}$$

Problem 33. Simplify: $(a+5a)\times 2a-3a$

The order of precedence is brackets, multiplication, then subtraction. Hence

$$(a+5a)\times 2a-3a = 6a\times 2a-3a$$
$$= \mathbf{12a^2-3a}$$
$$\text{or } \mathbf{3a(4a-1)}$$

Problem 34. Simplify: $a+5a\times(2a-3a)$

The order of precedence is brackets, multiplication, then subtraction. Hence

$$a+5a\times(2a-3a) = a+5a\times -a$$
$$= a+-5a^2$$
$$= \mathbf{a-5a^2} \text{ or } \mathbf{a(1-5a)}$$

Problem 35. Simplify: $a\div 5a+2a-3a$

The order of precedence is division, then addition and subtraction. Hence

$$a\div 5a+2a-3a = \frac{a}{5a}+2a-3a$$
$$= \frac{1}{5}+2a-3a = \mathbf{\frac{1}{5}-a}$$

Problem 36. Simplify:
$$a\div(5a+2a)-3a$$

The order of precedence is brackets, division and subtraction. Hence

$$a\div(5a+2a)-3a = a\div 7a-3a$$
$$= \frac{a}{7a}-3a = \mathbf{\frac{1}{7}-3a}$$

Problem 37. Simplify:
$$3c+2c\times 4c+c\div 5c-8c$$

The order of precedence is division, multiplication, addition and subtraction. Hence:

$$3c+2c\times 4c+c\div 5c-8c$$
$$= 3c+2c\times 4c+\frac{c}{5c}-8c$$
$$= 3c+8c^2+\frac{1}{5}-8c$$
$$= \mathbf{8c^2-5c+\frac{1}{5}} \quad\text{or}\quad \mathbf{c(8c-5)+\frac{1}{5}}$$

Problem 38. Simplify:
$$3c+2c\times 4c+c\div(5c-8c)$$

The order of precedence is brackets, division, multiplication and addition. Hence,

$$3c+2c\times 4c+c\div(5c-8c)$$
$$= 3c+2c\times 4c+c\div -3c$$
$$= 3c+2c\times 4c+\frac{c}{-3c}$$

Now $\dfrac{c}{-3c} = \dfrac{1}{-3}$

Multiplying numerator and denominator by -1 gives:

$$\frac{1\times -1}{-3\times -1} \quad\text{i.e.}\quad -\frac{1}{3}$$

Hence:

$$3c+2c\times 4c+\frac{c}{-3c}$$
$$= 3c+2c\times 4c-\frac{1}{3}$$
$$= \mathbf{3c+8c^2-\frac{1}{3}} \quad\text{or}\quad \mathbf{c(3+8c)-\frac{1}{3}}$$

Problem 39. Simplify:
$$(3c+2c)(4c+c)\div(5c-8c)$$

The order of precedence is brackets, division and multiplication. Hence

$$(3c+2c)(4c+c)\div(5c-8c)$$
$$= 5c\times 5c\div -3c = 5c\times\frac{5c}{-3c}$$
$$= 5c\times -\frac{5}{3} = \mathbf{-\frac{25}{3}c}$$

Problem 40. Simplify:
$$(2a-3)\div 4a+5\times 6-3a$$

The bracket around the $(2a-3)$ shows that both $2a$ and -3 have to be divided by $4a$, and to remove the bracket the expression is written in fraction form.

Hence, $\quad (2a-3)\div 4a+5\times 6-3a$

$$= \frac{2a-3}{4a}+5\times 6-3a$$
$$= \frac{2a-3}{4a}+30-3a$$
$$= \frac{2a}{4a}-\frac{3}{4a}+30-3a$$
$$= \frac{1}{2}-\frac{3}{4a}+30-3a$$
$$= \mathbf{30\frac{1}{2}-\frac{3}{4a}-3a}$$

Problem 41. Simplify:

$$\frac{1}{3} \text{ of } 3p + 4p(3p - p)$$

Applying BODMAS, the expression becomes

$$\frac{1}{3} \text{ of } 3p + 4p \times 2p$$

and changing 'of' to '×' gives:

$$\frac{1}{3} \times 3p + 4p \times 2p$$

i.e. $\boldsymbol{p + 8p^2}$ or $\boldsymbol{p(1 + 8p)}$

Now try the following Practice Exercise

Practice Exercise 30 Fundamental laws of precedence (Answers on page 704)

Simplify the following:

1. $2x \div 4x + 6x$
2. $2x \div (4x + 6x)$
3. $3a - 2a \times 4a + a$
4. $3a - 2a(4a + a)$
5. $2y + 4 \div 6y + 3 \times 4 - 5y$
6. $2y + 4 \div 6y + 3(4 - 5y)$
7. $3 \div y + 2 \div y + 1$
8. $p^2 - 3pq \times 2p \div 6q + pq$
9. $(x + 1)(x - 4) \div (2x + 2)$
10. $\frac{1}{4}$ of $2y + 3y(2y - y)$

5.5 Direct and inverse proportionality

An expression such as $y = 3x$ contains two variables. For every value of x there is a corresponding value of y. The variable x is called the **independent variable** and y is called the **dependent variable**.

When an increase or decrease in an independent variable leads to an increase or decrease of the same proportion in the dependent variable this is termed **direct proportion**. If $y = 3x$ then y is directly proportional to x, which may be written as $y \alpha x$ or $y = kx$, where k is called the **coefficient of proportionality** (in this case, k being equal to 3).

When an increase in an independent variable leads to a decrease of the same proportion in the dependent variable (or vice versa) this is termed **inverse proportion**. If y is inversely proportional to x then $y \alpha \frac{1}{x}$ or $y = k/x$. Alternatively, $k = xy$, that is, for inverse proportionality the product of the variable is constant.

Examples of laws involving direct and inverse proportional in science include:

(i) **Hooke's* law**, which states that within the elastic limit of a material, the strain ε produced is directly proportional to the stress, σ, producing it, i.e. $\varepsilon \alpha \sigma$ or $\varepsilon = k\sigma$

*Who was Hooke? – **Robert Hooke** FRS (28 July 1635 – 3 March 1703) was an English natural philosopher, architect and polymath. To find out more go to www.routledge.com/cw/bird

(ii) **Charles's* law**, which states that for a given mass of gas at constant pressure the volume V is directly proportional to its thermodynamic temperature T, i.e. $V \alpha T$ or $V = kT$

(iii) **Ohm's* law**, which states that the current I flowing through a fixed resistor is directly proportional to the applied voltage V, i.e. $I \alpha V$ or $I = kV$

(iv) **Boyle's* law**, which states that for a gas at constant temperature, the volume V of a fixed mass of a gas is inversely proportional to its absolute pressure p, i.e. $p\alpha(1/V)$ or $p = k/V$, i.e. $pV = k$

Problem 42. If y is directly proportional to x and $y = 2.48$ when $x = 0.4$, determine (a) the coefficient of proportionality and (b) the value of y when $x = 0.65$

(a) $y\alpha x$, i.e. $y = kx$. If $y = 2.48$ when $x = 0.4$, $2.48 = k(0.4)$
Hence the coefficient of proportionality,

$$k = \frac{2.48}{0.4} = \mathbf{6.2}$$

*Who was Charles? – **Jacques Alexandre César Charles** (12 November 1746–7 April 1823) was a French inventor, scientist, mathematician and balloonist. To find out more go to www.routledge.com/cw/bird

(b) $y = kx$, hence, when $x = 0.65$,
$y = (6.2)(0.65) = \mathbf{4.03}$

*Who was Ohm? – **Georg Simon Ohm** (16 March 1789–6 July 1854) was a Bavarian physicist and mathematician. To find out more go to www.routledge.com/cw/bird

*Who was Boyle? – **Robert Boyle** FRS (25 January 1627–31 December 1691) was a natural philosopher, chemist, physicist and inventor. To find out more go to www.routledge.com/cw/bird

Problem 43. Hooke's law states that stress σ is directly proportional to strain ε within the elastic limit of a material. When, for mild steel, the stress is 25×10^6 Pascals, the strain is 0.000125. Determine (a) the coefficient of proportionality and (b) the value of strain when the stress is 18×10^6 Pascals

(a) $\sigma\,\alpha\,\varepsilon$, i.e. $\sigma=k\varepsilon$, from which $k=\sigma/\varepsilon$. Hence the coefficient of proportionality,

$$k=\frac{25\times10^6}{0.000125}=\mathbf{200\times10^9\,pascals}$$

(The coefficient of proportionality k in this case is called **Young's Modulus of Elasticity**.)

(b) Since $\sigma=k\varepsilon$, $\varepsilon=\sigma/k$
Hence when $\sigma=18\times10^6$,

$$\text{strain } \varepsilon=\frac{18\times10^6}{200\times10^9}=\mathbf{0.00009}$$

Problem 44. The electrical resistance R of a piece of wire is inversely proportional to the cross-sectional area A. When $A=5\,\text{mm}^2$, $R=7.02$ ohms. Determine (a) the coefficient of proportionality and (b) the cross-sectional area when the resistance is 4 ohms

(a) $R\,\alpha\,\frac{1}{A}$, i.e. $R=k/A$ or $k=RA$. Hence, when $R=7.2$ and $A=5$, the coefficient of proportionality, $k=(7.2)(5)=\mathbf{36}$

(b) Since $k=RA$ then $A=k/R$
When $R=4$, the cross-sectional area,

$$A=\frac{36}{4}=\mathbf{9\,mm^2}$$

Problem 45. Boyle's law states that at constant temperature, the volume V of a fixed mass of gas is inversely proportional to its absolute pressure p. If a gas occupies a volume of $0.08\,\text{m}^3$ at a pressure of 1.5×10^6 Pascals determine (a) the coefficient of proportionality and (b) the volume if the pressure is changed to 4×10^6 Pascals

(a) $V\,\alpha\,\frac{1}{p}$ i.e. $V=k/p$ or $k=pV$

Hence the coefficient of proportionality,

$$k=(1.5\times10^6)(0.08)=\mathbf{0.12\times10^6}$$

(b) Volume $V=\frac{k}{p}=\frac{0.12\times10^6}{4\times10^6}=\mathbf{0.03\,m^3}$

Now try the following Practice Exercise

Practice Exercise 31 Direct and inverse proportionality (Answers on page 704)

1. If p is directly proportional to q and $p=37.5$ when $q=2.5$, determine (a) the constant of proportionality and (b) the value of p when q is 5.2
2. Charles's law states that for a given mass of gas at constant pressure the volume is directly proportional to its thermodynamic temperature. A gas occupies a volume of 2.25 litres at 300 K. Determine (a) the constant of proportionality, (b) the volume at 420 K, and (c) the temperature when the volume is 2.625 litres
3. Ohm's law states that the current flowing in a fixed resistor is directly proportional to the applied voltage. When 30 volts is applied across a resistor the current flowing through the resistor is 2.4×10^{-3} amperes. Determine (a) the constant of proportionality, (b) the current when the voltage is 52 volts and (c) the voltage when the current is 3.6×10^{-3} amperes
4. If y is inversely proportional to x and $y=15.3$ when $x=0.6$, determine (a) the coefficient of proportionality, (b) the value of y when x is 1.5, and (c) the value of x when y is 27.2
5. Boyle's law states that for a gas at constant temperature, the volume of a fixed mass of gas is inversely proportional to its absolute pressure. If a gas occupies a volume of $1.5\,\text{m}^3$ at a pressure of 200×10^3 Pascals, determine (a) the constant of proportionality, (b) the volume when the pressure is 800×10^3 Pascals and (c) the pressure when the volume is $1.25\,\text{m}^3$

Practice Exercise 32 Multiple-choice questions on algebra (Answers on page 704)

Each question has only one correct answer

1. $\left(\frac{b+2b}{b}\right) \times 3b - 2b$ simplifies to:
(a) 3b (b) 6b (c) $9b^2 - 2b$ (d) 7b

2. $(\sqrt{x})\left(y^{\frac{3}{2}}\right)(x^2y)$ is equivalent to:
(a) $\sqrt{(xy)^5}$ (b) $x^{\sqrt{2}}y^{\frac{5}{2}}$
(c) xy^5 (d) $x\sqrt{y^3}$

3. $q^3 \times \left(\frac{1}{q^{-2}}\right)^{-3}$ is equivalent to:
(a) $\frac{1}{q^{\frac{3}{2}}}$ (b) q^{-3}
(c) q^9 (d) $\frac{1}{q^{-3}}$

4. Given that $3(5t - 2y) = 0$ then:
(a) $15t - 2y = 0$ (b) $15t = 6y$
(c) $5t - 2y = -3$ (d) $8t - 5y = 0$

5. Given that $7x + 8 - 2y = 12y - 4 - x$ then:
(a) $y = \frac{4x+6}{7}$ (b) $x = \frac{7y+2}{4}$
(c) $y = \frac{3x-6}{5}$ (d) $x = \frac{5y-6}{4}$

6. $\frac{2}{5xy} \div \frac{4}{15x^2}$ is equivalent to:
(a) $\frac{2}{3x^2y}$ (b) $\frac{2y}{3x}$
(c) $\frac{8}{75x^3y}$ (d) $\frac{3x}{2y}$

7. $\frac{(3x^2y)^2}{6xy^2}$ simplifies to:
(a) $\frac{x}{y}$ (b) $\frac{3x^2}{2}$ (c) $\frac{3x^3}{2}$ (d) $\frac{x}{2y}$

8. $\frac{(pq^2r)^3}{(p^{-2}qr^2)^2}$ simplifies to:
(a) $\frac{p^3q^4}{r}$ (b) $\frac{p^3}{pr}$ (c) $\frac{p^7q^4}{r}$ (d) $\frac{1}{pqr}$

9. $2x^2 - (x - xy) - x(2y - x)$ simplifies to:
(a) $x(3x - 1 - y)$ (b) $x^2 - 3xy - xy$
(c) $x(xy - y - 1)$ (d) $3x^2 - x + xy$

10. $3d + 2d \times 4d + d \div (6d - 2d)$ simplifies to:
(a) $\frac{25d}{4}$ (b) $\frac{1}{4} + d(3 + 8d)$
(c) $1 + 2d$ (d) $20d^2 + \frac{1}{4}$

11. $5a \div a + 2a - 3a \times a$ simplifies to:
(a) $4a^2$ (b) $5 + 2a - 3a^2$
(c) $5 - a^2$ (d) $\frac{5}{3} - 3a^2$

12. $(2x - 3y)(x + y)$ is equal to:
(a) $2x^2 - 3y^2$ (b) $2x^2 - 5xy - 3y^2$
(c) $2x^2 + 3y^2$ (d) $2x^2 - xy - 3y^2$

13. $(2x - y)^2$ is equal to:
(a) $4x^2 + y^2$ (b) $2x^2 - 2xy + y^2$
(c) $4x^2 - y^2$ (d) $4x^2 - 4xy + y^2$

14. Factorising $2xy^2 + 6x^3y - 8x^3y^2$ gives:
(a) $2x(y^2 + 3x^2 - 4x^2y)$
(b) $2xy(y + 3x^2y - 4x^2y^2)$
(c) $2xy(y + 3x^2 - 4x^2y)$
(d) $96x^7y^5$

15. y varies as the square of x, and $y = 12$ when $x = 2$. When $x = 3$, the value of y is:
(a) 432 (b) 36 (c) 27 (d) 9

For fully worked solutions to each of the problems in Practice Exercises 27 to 31 in this chapter, go to the website:

www.routledge.com/cw/bird

COMPANION @ WEBSITE

Chapter 6

Further algebra

Why it is important to understand: Further algebra

The study of algebra revolves around using and manipulating polynomials. Polynomials are used in engineering, computer programming, software engineering, in management and in business. Mathematicians, statisticians and engineers of all sciences employ the use of polynomials to solve problems; among them are aerospace engineers, chemical engineers, civil engineers, electrical engineers, environmental engineers, industrial engineers, materials engineers, mechanical engineers and nuclear engineers. The factor and remainder theorems are also employed in engineering software and electronic mathematical applications, through which polynomials of higher degrees and longer arithmetic structures are divided without any complexity. The study of polynomial division and the factor and remainder theorems is therefore of some importance in engineering.

At the end of this chapter, you should be able to:

- divide algebraic expressions using polynomial division
- factorise expressions using the factor theorem
- use the remainder theorem to factorise algebraic expressions

6.1 Polynomial division

Before looking at long division in algebra let us revise long division with numbers (we may have forgotten, since calculators do the job for us!)

For example, $\frac{208}{16}$ is achieved as follows:

$$\begin{array}{r|l} & 13 \\ \hline 16 & 208 \\ & 16 \\ \hline & 48 \\ & 48 \\ \hline & \cdot\cdot \end{array}$$

(1) 16 divided into 2 won't go
(2) 16 divided into 20 goes 1
(3) Put 1 above the zero
(4) Multiply 16 by 1 giving 16
(5) Subtract 16 from 20 giving 4
(6) Bring down the 8
(7) 16 divided into 48 goes 3 times
(8) Put the 3 above the 8
(9) $3 \times 16 = 48$
(10) $48 - 48 = 0$

$$\text{Hence} \quad \frac{208}{16} = \mathbf{13} \text{ exactly}$$

Similarly, $\frac{172}{15}$ is laid out as follows:

$$\begin{array}{r|l} & 11 \\ \hline 15 & 172 \\ & 15 \\ \hline & 22 \\ & 15 \\ \hline & 7 \\ \hline \end{array}$$

Hence $\frac{175}{15} = 11$ remainder 7 or $11 + \frac{7}{15} = \mathbf{11\frac{7}{15}}$

Below are some examples of division in algebra, which in some respects, are similar to long division with numbers.

(Note that a **polynomial** is an expression of the form

$$f(x) = a + bx + cx^2 + dx^3 + \cdots$$

and **polynomial division** is sometimes required when resolving into partial fractions — see Chapter 7).

Problem 1. Divide $2x^2 + x - 3$ by $x - 1$

$2x^2 + x - 3$ is called the **dividend** and $x - 1$ the **divisor**. The usual layout is shown below with the dividend and divisor both arranged in descending powers of the symbols.

$$\begin{array}{r|l} & 2x + 3 \\ \hline x - 1 & 2x^2 + x - 3 \\ & 2x^2 - 2x \\ \hline & \quad\;\; 3x - 3 \\ & \quad\;\; 3x - 3 \\ \hline & \quad\;\; \cdot \quad \cdot \end{array}$$

Dividing the first term of the dividend by the first term of the divisor, i.e. $\frac{2x^2}{x}$ gives $2x$, which is put above the first term of the dividend as shown. The divisor is then multiplied by $2x$, i.e. $2x(x - 1) = 2x^2 - 2x$, which is placed under the dividend as shown. Subtracting gives $3x - 3$. The process is then repeated, i.e. the first term of the divisor, x, is divided into $3x$, giving $+3$, which is placed above the dividend as shown. Then $3(x - 1) = 3x - 3$ which is placed under the $3x - 3$. The remainder, on subtraction, is zero, which completes the process.

Thus $\mathbf{(2x^2 + x - 3) \div (x - 1) = (2x + 3)}$

[A check can be made on this answer by multiplying $(2x + 3)$ by $(x - 1)$ which equals $2x^2 + x - 3$]

Problem 2. Divide $3x^3 + x^2 + 3x + 5$ by $x + 1$

$$\begin{array}{r|l} & \text{(1)}\ \ \text{(4)}\ \ \text{(7)} \\ & 3x^2 - 2x + 5 \\ \hline x + 1 & 3x^3 + x^2 + 3x + 5 \\ & 3x^3 + 3x^2 \\ \hline & \quad\;\; -2x^2 + 3x + 5 \\ & \quad\;\; -2x^2 - 2x \\ \hline & \qquad\qquad\;\; 5x + 5 \\ & \qquad\qquad\;\; 5x + 5 \\ \hline & \qquad\qquad\;\; \cdot \quad \cdot \end{array}$$

(1) x into $3x^3$ goes $3x^2$. Put $3x^2$ above $3x^3$
(2) $3x^2(x + 1) = 3x^3 + 3x^2$
(3) Subtract
(4) x into $-2x^2$ goes $-2x$. Put $-2x$ above the dividend
(5) $-2x(x + 1) = -2x^2 - 2x$
(6) Subtract
(7) x into $5x$ goes 5. Put 5 above the dividend
(8) $5(x + 1) = 5x + 5$
(9) Subtract

Thus $\frac{3x^3 + x^2 + 3x + 5}{x + 1} = \mathbf{3x^2 - 2x + 5}$

Problem 3. Simplify: $\frac{x^3 + y^3}{x + y}$

$$\begin{array}{r|l} & \text{(1)}\ \ \text{(4)}\ \ \text{(7)} \\ & x^2 - xy + y^2 \\ \hline x + y & x^3 + 0 + 0 \;\; + y^3 \\ & x^3 + x^2y \\ \hline & \quad -x^2y \qquad + y^3 \\ & \quad -x^2y - xy^2 \\ \hline & \qquad\qquad xy^2 + y^3 \\ & \qquad\qquad xy^2 + y^3 \\ \hline & \qquad\qquad \cdot \quad \cdot \end{array}$$

(1) x into x^3 goes x^2. Put x^2 above x^3 of dividend
(2) $x^2(x + y) = x^3 + x^2y$
(3) Subtract
(4) x into $-x^2y$ goes $-xy$. Put $-xy$ above dividend
(5) $-xy(x + y) = -x^2y - xy^2$
(6) Subtract
(7) x into xy^2 goes y^2. Put y^2 above divided
(8) $y^2(x + y) = xy^2 + y^3$
(9) Subtract

Thus $\frac{x^3 + y^3}{x + y} = \mathbf{x^2 - xy + y^2}$

The zeros shown in the dividend are not normally shown, but are included to clarify the subtraction process and to keep similar terms in their respective columns.

Problem 4. Divide $(x^2 + 3x - 2)$ by $(x - 2)$

$$\begin{array}{r|l} & x+5 \\ \hline x-2 & x^2+3x-2 \\ & x^2-2x \\ \hline & \quad 5x-2 \\ & \quad 5x-10 \\ \hline & \qquad 8 \end{array}$$

Hence $\quad \dfrac{x^2+3x-2}{x-2} = \mathbf{x+5+\dfrac{8}{x-2}}$

Problem 5. Divide $4a^3-6a^2b+5b^3$ by $2a-b$

$$\begin{array}{r|l} & 2a^2-2ab-b^2 \\ \hline 2a-b & 4a^3-6a^2b \qquad +5b^3 \\ & 4a^3-2a^2b \\ \hline & \quad -4a^2b \qquad +5b^3 \\ & \quad -4a^2b+2ab^2 \\ \hline & \qquad -2ab^2+5b^3 \\ & \qquad -2ab^2+b^3 \\ \hline & \qquad\qquad 4b^3 \end{array}$$

Thus

$$\frac{4a^3-6a^2b+5b^3}{2a-b} = \mathbf{2a^2-2ab-b^2+\frac{4b^3}{2a-b}}$$

Now try the following Practice Exercise

Practice Exercise 33 Polynomial division (Answers on page 704)

1. Divide $(2x^2+xy-y^2)$ by $(x+y)$
2. Divide $(3x^2+5x-2)$ by $(x+2)$
3. Determine $(10x^2+11x-6) \div (2x+3)$
4. Find: $\dfrac{14x^2-19x-3}{2x-3}$
5. Divide $(x^3+3x^2y+3xy^2+y^3)$ by $(x+y)$
6. Find $(5x^2-x+4) \div (x-1)$
7. Divide $(3x^3+2x^2-5x+4)$ by $(x+2)$
8. Determine: $\dfrac{5x^4+3x^3-2x+1}{x-3}$

6.2 The factor theorem

There is a simple relationship between the factors of a quadratic expression and the roots of the equation obtained by equating the expression to zero.
For example, consider the quadratic equation $x^2+2x-8=0$
To solve this we may factorise the quadratic expression x^2+2x-8 giving $(x-2)(x+4)$
Hence $(x-2)(x+4)=0$
Then, if the product of two number is zero, one or both of those numbers must equal zero. Therefore,

either $(x-2)=0$, from which, $x=2$
or $\quad (x+4)=0$, from which, $x=-4$

It is clear then that a factor of $(x-2)$ indicates a root of $+2$, while a factor of $(x+4)$ indicates a root of -4. In general, we can therefore say that:

a factor of $(x-a)$ corresponds to a root of $x=a$

In practice, we always deduce the roots of a simple quadratic equation from the factors of the quadratic expression, as in the above example. However, we could reverse this process. If, by trial and error, we could determine that $x=2$ is a root of the equation $x^2+2x-8=0$ we could deduce at once that $(x-2)$ is a factor of the expression x^2+2x-8. We wouldn't normally solve quadratic equations this way – but suppose we have to factorise a cubic expression (i.e. one in which the highest power of the variable is 3). A cubic equation might have three simple linear factors and the difficulty of discovering all these factors by trial and error would be considerable. It is to deal with this kind of case that we use the **factor theorem**. This is just a generalised version of what we established above for the quadratic expression. The factor theorem provides a method of factorising any polynomial, $f(x)$, which has simple factors.
A statement of the **factor theorem** says:

'if $x=a$ is a root of the equation $f(x)=0$, then $(x-a)$ is a factor of $f(x)$'

The following worked problems show the use of the factor theorem.

Problem 6. Factorise: x^3-7x-6 and use it to solve the cubic equation: $x^3-7x-6=0$

Let $f(x)=x^3-7x-6$

If $x=1$, then $f(1)=1^3-7(1)-6=-12$

If $x = 2$, then $f(2) = 2^3 - 7(2) - 6 = -12$

If $x = 3$, then $f(3) = 3^3 - 7(3) - 6 = 0$

If $f(3) = 0$, then $(x - 3)$ is a factor – from the factor theorem.
We have a choice now. We can divide $x^3 - 7x - 6$ by $(x - 3)$ or we could continue our 'trial and error' by substituting further values for x in the given expression – and hope to arrive at $f(x) = 0$.
Let us do both ways. Firstly, dividing out gives:

$$\begin{array}{r l}
 & x^2 + 3x + 2 \\
x - 3 \,\big) & x^3 + 0 \quad - 7x \quad - 6 \\
 & x^3 - 3x^2 \\
 & \quad\;\; 3x^2 - 7x - 6 \\
 & \quad\;\; 3x^2 - 9x \\
 & \qquad\quad\;\; 2x - 6 \\
 & \qquad\quad\;\; 2x - 6 \\
 & \qquad\quad\;\; \cdot \quad \cdot
\end{array}$$

Hence $\dfrac{x^3 - 7x - 6}{x - 3} = x^2 + 3x + 2$

i.e. $x^3 - 7x - 6 = (x - 3)(x^2 + 3x + 2)$

$x^3 + 3x + 2$ factorises 'on sight' as $(x + 1)(x + 2)$

Therefore

$$\mathbf{x^3 - 7x - 6 = (x - 3)(x + 1)(x + 2)}$$

A second method is to continue to substitute values of x into $f(x)$.
Our expression for $f(3)$ was $3^3 - 7(3) - 6$. We can see that if we continue with positive values of x the first term will predominate such that $f(x)$ will not be zero.
Therefore let us try some negative values for x: $f(-1) = (-1)^3 - 7(-1) - 6 = 0$; hence $(x + 1)$ is a factor (as shown above).
Also, $f(-2) = (-2)^3 - 7(-2) - 6 = 0$; hence $(x + 2)$ is a factor (also as shown above).
To solve $x^3 - 7x - 6 = 0$, we substitute the factors, i.e.

$$(x - 3)(x + 1)(x + 2) = 0$$

from which, $\mathbf{x = 3, x = -1}$ **and** $\mathbf{x = -2}$
Note that the values of x, i.e. 3, -1 and -2, are all factors of the constant term, i.e. the 6. This can give us a clue as to what values of x we should consider.

Problem 7. Solve the cubic equation $x^3 - 2x^2 - 5x + 6 = 0$ by using the factor theorem

Let $f(x) = x^3 - 2x^2 - 5x + 6$ and let us substitute simple values of x like 1, 2, 3, -1, -2 and so on.

$$f(1) = 1^3 - 2(1)^2 - 5(1) + 6 = 0,$$
hence $(x - 1)$ is a factor

$$f(2) = 2^3 - 2(2)^2 - 5(2) + 6 \neq 0$$

$$f(3) = 3^3 - 2(3)^2 - 5(3) + 6 = 0,$$
hence $(x - 3)$ is a factor

$$f(-1) = (-1)^3 - 2(-1)^2 - 5(-1) + 6 \neq 0$$

$$f(-2) = (-2)^3 - 2(-2)^2 - 5(-2) + 6 = 0,$$
hence $(x + 2)$ is a factor

Hence, $x^3 - 2x^2 - 5x + 6 = (x - 1)(x - 3)(x + 2)$

Therefore if $x^3 - 2x^2 - 5x + 6 = 0$

then $(x - 1)(x - 3)(x + 2) = 0$

from which, $\mathbf{x = 1, x = 3}$ **and** $\mathbf{x = -2}$

Alternatively, having obtained one factor, i.e. $(x - 1)$ we could divide this into $(x^3 - 2x^2 - 5x + 6)$ as follows:

$$\begin{array}{r l}
 & x^2 - x - 6 \\
x - 1 \,\big) & x^3 - 2x^2 - 5x + 6 \\
 & x^3 - \;\, x^2 \\
 & \quad\; - x^2 \;\, - 5x + 6 \\
 & \quad\; - x^2 \;\, + \;\, x \\
 & \qquad\qquad\; -6x + 6 \\
 & \qquad\qquad\; -6x + 6 \\
 & \qquad\qquad\quad\; \cdot \quad \cdot
\end{array}$$

Hence $x^3 - 2x^2 - 5x + 6$

$$= (x - 1)(x^2 - x - 6)$$

$$= \mathbf{(x - 1)(x - 3)(x + 2)}$$

Summarising, the factor theorem provides us with a method of factorising simple expressions, and an alternative, in certain circumstances, to polynomial division.

Now try the following Practice Exercise

Practice Exercise 34 The factor theorem (Answers on page 705)

Use the factor theorem to factorise the expressions given in problems 1 to 4.

1. $x^2 + 2x - 3$
2. $x^3 + x^2 - 4x - 4$
3. $2x^3 + 5x^2 - 4x - 7$
4. $2x^3 - x^2 - 16x + 15$
5. Use the factor theorem to factorise $x^3 + 4x^2 + x - 6$ and hence solve the cubic equation $x^3 + 4x^2 + x - 6 = 0$
6. Solve the equation $x^3 - 2x^2 - x + 2 = 0$

6.3 The remainder theorem

Dividing a general quadratic expression $(ax^2 + bx + c)$ by $(x - p)$, where p is any whole number, by long division (see Section 6.1) gives:

$$
\begin{array}{r|l}
 & ax + (b + ap) \\
x - p & ax^2 + bx \qquad + c \\
 & ax^2 - apx \\
\hline
 & (b + ap)x + c \\
 & (b + ap)x - (b + ap)p \\
\hline
 & c + (b + ap)p \\
\end{array}
$$

The remainder, $c + (b + ap)p = c + bp + ap^2$ or $ap^2 + bp + c$. This is, in fact, what the **remainder theorem** states, i.e.

'if $(ax^2 + bx + c)$ is divided by $(x - p)$, the remainder will be $ap^2 + bp + c$'

If, in the dividend $(ax^2 + bx + c)$, we substitute p for x we get the remainder $ap^2 + bp + c$

For example, when $(3x^2 - 4x + 5)$ is divided by $(x - 2)$ the remainder is $ap^2 + bp + c$, (where $a = 3$, $b = -4$, $c = 5$ and $p = 2$), i.e. the remainder is:

$$3(2)^2 + (-4)(2) + 5 = 12 - 8 + 5 = \mathbf{9}$$

We can check this by dividing $(3x^2 - 4x + 5)$ by $(x - 2)$ by long division:

$$
\begin{array}{r|l}
 & 3x + 2 \\
x - 2 & 3x^2 - 4x + 5 \\
 & 3x^2 - 6x \\
\hline
 & \qquad 2x + 5 \\
 & \qquad 2x - 4 \\
\hline
 & \qquad\quad 9 \\
\end{array}
$$

Similarly, when $(4x^2 - 7x + 9)$ is divided by $(x + 3)$, the remainder is $ap^2 + bp + c$, (where $a = 4$, $b = -7$, $c = 9$ and $p = -3$) i.e. the remainder is:
$4(-3)^2 + (-7)(-3) + 9 = 36 + 21 + 9 = \mathbf{66}$
Also, when $(x^2 + 3x - 2)$ is divided by $(x - 1)$, the remainder is $1(1)^2 + 3(1) - 2 = \mathbf{2}$
It is not particularly useful, on its own, to know the remainder of an algebraic division. However, if the remainder should be zero then $(x - p)$ is a factor. This is very useful therefore when factorising expressions.
For example, when $(2x^2 + x - 3)$ is divided by $(x - 1)$, the remainder is $2(1)^2 + 1(1) - 3 = 0$, which means that $(x - 1)$ is a factor of $(2x^2 + x - 3)$.
In this case the other factor is $(2x + 3)$, i.e.

$$(2x^2 + x - 3) = (x - 1)(2x - 3).$$

The **remainder theorem** may also be stated for a **cubic equation** as:

'if $(ax^3 + bx^2 + cx + d)$ is divided by $(x - p)$, the remainder will be $ap^3 + bp^2 + cp + d$'

As before, the remainder may be obtained by substituting p for x in the dividend.
For example, when $(3x^3 + 2x^2 - x + 4)$ is divided by $(x - 1)$, the remainder is: $ap^3 + bp^2 + cp + d$ (where $a = 3$, $b = 2$, $c = -1$, $d = 4$ and $p = 1$), i.e. the remainder is:

$$3(1)^3 + 2(1)^2 + (-1)(1) + 4 = 3 + 2 - 1 + 4 = \mathbf{8}.$$

Similarly, when $(x^3 - 7x - 6)$ is divided by $(x - 3)$, the remainder is: $1(3)^3 + 0(3)^2 - 7(3) - 6 = 0$, which mean that $(x - 3)$ is a factor of $(x^3 - 7x - 6)$.
Here are some more examples on the remainder theorem.

Problem 8. Without dividing out, find the remainder when $2x^2 - 3x + 4$ is divided by $(x - 2)$

By the remainder theorem, the remainder is given by: $ap^2 + bp + c$, where $a = 2$, $b = -3$, $c = 4$ and $p = 2$

Hence **the remainder is:**

$$2(2)^2 + (-3)(2) + 4 = 8 - 6 + 4 = \mathbf{6}$$

Problem 9. Use the remainder theorem to determine the remainder when $(3x^3 - 2x^2 + x - 5)$ is divided by $(x + 2)$

By the remainder theorem, the remainder is given by: $ap^3 + bp^2 + cp + d$, where $a = 3$, $b = -2$, $c = 1$, $d = -5$ and $p = -2$

Hence **the remainder is:**

$$3(-2)^3 + (-2)(-2)^2 + (1)(-2) + (-5)$$
$$= -24 - 8 - 2 - 5 = \mathbf{-39}$$

Problem 10. Determine the remainder when $(x^3 - 2x^2 - 5x + 6)$ is divided by (a) $(x - 1)$ and (b) $(x + 2)$. Hence factorise the cubic expression

(a) When $(x^3 - 2x^2 - 5x + 6)$ is divided by $(x - 1)$, the remainder is given by $ap^3 + bp^2 + cp + d$, where $a = 1$, $b = -2$, $c = -5$, $d = 6$ and $p = 1$,

i.e. **the remainder** $= (1)(1)^3 + (-2)(1)^2 + (-5)(1) + 6$
$= 1 - 2 - 5 + 6 = \mathbf{0}$

Hence $(x - 1)$ is a factor of $(x^3 - 2x^2 - 5x + 6)$

(b) When $(x^3 - 2x^2 - 5x + 6)$ is divided by $(x + 2)$, **the remainder is** given by

$$(1)(-2)^3 + (-2)(-2)^2 + (-5)(-2) + 6$$
$$= -8 - 8 + 10 + 6 = \mathbf{0}$$

Hence $(x + 2)$ is also a factor of: $(x^3 - 2x^2 - 5x + 6)$

Therefore $(x - 1)(x + 2)(\quad) = x^3 - 2x^2 - 5x + 6$

To determine the third factor (shown blank) we could

(i) divide $(x^3 - 2x^2 - 5x + 6)$ by $(x - 1)(x + 2)$

or (ii) use the factor theorem where $f(x) = x^3 - 2x^2 - 5x + 6$ and hoping to choose a value of x which makes $f(x) = 0$

or (iii) use the remainder theorem, again hoping to choose a factor $(x - p)$ which makes the remainder zero

(i) Dividing $(x^3 - 2x^2 - 5x + 6)$ by $(x^2 + x - 2)$ gives:

$$\begin{array}{r} x - 3 \\ x^2 + x - 2 \overline{\smash{)}\, x^3 - 2x^2 - 5x + 6} \\ x^3 + x^2 - 2x \\ \hline -3x^2 - 3x + 6 \\ -3x^2 - 3x + 6 \\ \hline \cdot \quad \cdot \quad \cdot \\ \hline \end{array}$$

Thus $(\mathbf{x^3 - 2x^2 - 5x + 6}) = \mathbf{(x - 1)(x + 2)(x - 3)}$

(ii) Using the factor theorem, we let

$$f(x) = x^3 - 2x^2 - 5x + 6$$

Then $f(3) = 3^3 - 2(3)^2 - 5(3) + 6$
$= 27 - 18 - 15 + 6 = 0$

Hence $(x - 3)$ is a factor.

(iii) Using the remainder theorem, when $(x^3 - 2x^2 - 5x + 6)$ is divided by $(x - 3)$, the remainder is given by $ap^3 + bp^2 + cp + d$, where $a = 1$, $b = -2$, $c = -5$, $d = 6$ and $p = 3$.

Hence the remainder is:

$$1(3)^3 + (-2)(3)^2 + (-5)(3) + 6$$
$$= 27 - 18 - 15 + 6 = 0$$

Hence $(x - 3)$ is a factor.

Thus $(\mathbf{x^3 - 2x^2 - 5x + 6}) = \mathbf{(x - 1)(x + 2)(x - 3)}$

Now try the following Practice Exercise

Practice Exercise 35 The remainder theorem (Answers on page 705)

1. Find the remainder when $3x^2 - 4x + 2$ is divided by:
(a) $(x - 2)$ (b) $(x + 1)$

2. Determine the remainder when $x^3 - 6x^2 + x - 5$ is divided by:
(a) $(x + 2)$ (b) $(x - 3)$

3. Use the remainder theorem to find the factors of $x^3 - 6x^2 + 11x - 6$
4. Determine the factors of $x^3 + 7x^2 + 14x + 8$ and hence solve the cubic equation: $x^3 + 7x^2 + 14x + 8 = 0$
5. Determine the value of 'a' if $(x + 2)$ is a factor of $(x^3 - ax^2 + 7x + 10)$
6. Using the remainder theorem, solve the equation: $2x^3 - x^2 - 7x + 6 = 0$

Practice Exercise 36 Multiple-choice questions on further algebra (Answers on page 705)

Each question has only one correct answer

1. The degree of the polynomial equation $6x^5 + 7x^2 - 4x + 2 = 0$ is:
   (a) 1 (b) 2 (c) 3 (d) 5
2. $6x^2 - 5x - 6$ divided by $2x - 3$ gives:
   (a) $2x - 1$ (b) $3x + 2$
   (c) $3x - 2$ (d) $6x + 1$
3. $(x^3 - x^2 - x + 1)$ divided by $(x - 1)$ gives:
   (a) $x^2 - x - 1$ (b) $x^2 + 1$
   (c) $x^2 - 1$ (d) $x^2 + x - 1$
4. The remainder when $(x^2 - 2x + 5)$ is divided by $(x - 1)$ is:
   (a) 4 (b) 2 (c) 6 (d) 8
5. The remainder when $(3x^3 - x^2 + 2x - 5)$ is divided by $(x + 2)$ is:
   (a) -19 (b) 21 (c) -29 (d) -37

For fully worked solutions to each of the problems in Practice Exercises 33 to 35 in this chapter, go to the website:
www.routledge.com/cw/bird

Chapter 7

Partial fractions

Why it is important to understand: Partial fractions

The algebraic technique of resolving a complicated fraction into partial fractions is often needed by electrical and mechanical engineers for not only determining certain integrals in calculus, but for determining inverse Laplace transforms, and for analysing linear differential equations such as resonant circuits and feedback control systems. This chapter explains the techniques in resolving expressions into partial fractions.

At the end of this chapter, you should be able to:

- understand the term 'partial fraction'
- appreciate the conditions needed to resolve a fraction into partial fractions
- resolve into partial fractions a fraction containing linear factors in the denominator
- resolve into partial fractions a fraction containing repeated linear factors in the denominator
- resolve into partial fractions a fraction containing quadratic factors in the denominator

7.1 Introduction to partial fractions

By algebraic addition,

$$\frac{1}{x-2}+\frac{3}{x+1}=\frac{(x+1)+3(x-2)}{(x-2)(x+1)}$$

$$=\frac{4x-5}{x^2-x-2}$$

The reverse process of moving from $\frac{4x-5}{x^2-x-2}$ to $\frac{1}{x-2}+\frac{3}{x+1}$ is called resolving into **partial fractions**.

In order to resolve an algebraic expression into partial fractions:

(i) the denominator must factorise (in the above example, x^2-x-2 factorises as $(x-2)(x+1)$, and

(ii) the numerator must be at least one degree less than the denominator (in the above example $(4x-5)$ is of degree 1 since the highest powered x term is x^1 and (x^2-x-2) is of degree 2)

When the degree of the numerator is equal to or higher than the degree of the denominator, the numerator must be divided by the denominator (see Problems 3 and 4). There are basically three types of partial fraction and the form of partial fraction used is summarised in Table 7.1 where $f(x)$ is assumed to be of less degree than the relevant denominator and A, B and C are constants to be determined.

(In the latter type in Table 7.1, ax^2+bx+c is a quadratic expression which does not factorise without containing surds or imaginary terms.)

Resolving an algebraic expression into partial fractions is used as a preliminary to integrating certain functions (see Chapter 47).

Table 7.1

Type	Denominator containing	Expression	Form of partial fraction
1	Linear factors (see Problems 1 to 4)	$\frac{f(x)}{(x+a)(x-b)(x+c)}$	$\frac{A}{(x+a)}+\frac{B}{(x-b)}+\frac{C}{(x+c)}$
2	Repeated linear factors (see Problems 5 to 7)	$\frac{f(x)}{(x+a)^3}$	$\frac{A}{(x+a)}+\frac{B}{(x+a)^2}+\frac{C}{(x+a)^3}$
3	Quadratic factors (see Problems 8 and 9)	$\frac{f(x)}{(ax^2+bx+c)(x+d)}$	$\frac{Ax+B}{(ax^2+bx+c)}+\frac{C}{(x+d)}$

7.2 Partial fractions with linear factors

Problem 1. Resolve $\frac{11-3x}{x^2+2x-3}$ into partial fractions

The denominator factorises as $(x-1)(x+3)$ and the numerator is of less degree than the denominator. Thus $\frac{11-3x}{x^2+2x-3}$ may be resolved into partial fractions.

Let

$$\frac{11-3x}{x^2+2x-3} \equiv \frac{11-3x}{(x-1)(x+3)} \equiv \frac{A}{(x-1)}+\frac{B}{(x+3)}$$

where A and B are constants to be determined,

i.e. $$\frac{11-3x}{(x-1)(x+3)} \equiv \frac{A(x+3)+B(x-1)}{(x-1)(x+3)}$$

by algebraic addition.

Since the denominators are the same on each side of the identity then the numerators are equal to each other.

Thus, $11-3x \equiv A(x+3)+B(x-1)$

To determine constants A and B, values of x are chosen to make the term in A or B equal to zero.

When $x=1$, then $11-3(1) \equiv A(1+3)+B(0)$

i.e. $8=4A$

i.e. $\mathbf{A=2}$

When $x=-3$, then $11-3(-3) \equiv A(0)+B(-3-1)$

i.e. $20=-4B$

i.e. $\mathbf{B=-5}$

Thus $$\frac{\mathbf{11-3x}}{\mathbf{x^2+2x-3}} \equiv \frac{2}{(x-1)}+\frac{-5}{(x+3)} \equiv \frac{\mathbf{2}}{\mathbf{(x-1)}}-\frac{\mathbf{5}}{\mathbf{(x+3)}}$$

$$\left[\text{Check: } \frac{2}{(x-1)}-\frac{5}{(x+3)} = \frac{2(x+3)-5(x-1)}{(x-1)(x+3)} = \frac{11-3x}{x^2+2x-3}\right]$$

Problem 2. Convert $\frac{2x^2-9x-35}{(x+1)(x-2)(x+3)}$ into the sum of three partial fractions

Let $$\frac{2x^2-9x-35}{(x+1)(x-2)(x+3)} \equiv \frac{A}{(x+1)}+\frac{B}{(x-2)}+\frac{C}{(x+3)}$$

$$\equiv \frac{A(x-2)(x+3)+B(x+1)(x+3)+C(x+1)(x-2)}{(x+1)(x-2)(x+3)}$$

by algebraic addition

Equating the numerators gives:

$$2x^2-9x-35 \equiv A(x-2)(x+3)+B(x+1)(x+3)+C(x+1)(x-2)$$

Let $x=-1$. Then

$$2(-1)^2-9(-1)-35 \equiv A(-3)(2)+B(0)(2)+C(0)(-3)$$

i.e. $-24 = -6\text{A}$

i.e. $\mathbf{A} = \dfrac{-24}{-6} = \mathbf{4}$

Let $x = 2$. Then

$$2(2)^2 - 9(2) - 35 \equiv \text{A}(0)(5) + \text{B}(3)(5) + \text{C}(3)(0)$$

i.e. $-45 = 15\text{B}$

i.e. $\mathbf{B} = \dfrac{-45}{15} = \mathbf{-3}$

Let $x = -3$. Then

$$2(-3)^2 - 9(-3) - 35 \equiv \text{A}(-5)(0) + \text{B}(-2)(0) + \text{C}(-2)(-5)$$

i.e. $10 = 10\text{C}$

i.e. $\mathbf{C = 1}$

Thus
$$\frac{2x^2 - 9x - 35}{(x+1)(x-2)(x+3)} \equiv \frac{\mathbf{4}}{\mathbf{(x+1)}} - \frac{\mathbf{3}}{\mathbf{(x-2)}} + \frac{\mathbf{1}}{\mathbf{(x+3)}}$$

Problem 3. Resolve $\dfrac{x^2+1}{x^2-3x+2}$ into partial fractions

The denominator is of the same degree as the numerator. Thus dividing out gives:

$$\begin{array}{r|l} & 1 \\ x^2 - 3x + 2 & x^2 \qquad + 1 \\ & x^2 - 3x + 2 \\ \hline & 3x - 1 \end{array}$$

For more on polynomial division, see Section 6.1, page 66.

Hence
$$\frac{x^2+1}{x^2-3x+2} \equiv 1 + \frac{3x-1}{x^2-3x+2} \equiv 1 + \frac{3x-1}{(x-1)(x-2)}$$

Let
$$\frac{3x-1}{(x-1)(x-2)} \equiv \frac{\text{A}}{(x-1)} + \frac{\text{B}}{(x-2)} \equiv \frac{\text{A}(x-2) + \text{B}(x-1)}{(x-1)(x-2)}$$

Equating numerators gives:

$$3x - 1 \equiv \text{A}(x-2) + \text{B}(x-1)$$

Let $x = 1$. Then $2 = -\text{A}$

i.e. $\mathbf{A = -2}$

Let $x = 2$. Then $\mathbf{5 = B}$

Hence
$$\frac{3x-1}{(x-1)(x-2)} \equiv \frac{-2}{(x-1)} + \frac{5}{(x-2)}$$

Thus
$$\frac{\mathbf{x^2+1}}{\mathbf{x^2-3x+2}} \equiv \mathbf{1} - \frac{\mathbf{2}}{\mathbf{(x-1)}} + \frac{\mathbf{5}}{\mathbf{(x-2)}}$$

Problem 4. Express $\dfrac{x^3-2x^2-4x-4}{x^2+x-2}$ in partial fractions

The numerator is of higher degree than the denominator. Thus dividing out gives:

$$\begin{array}{r|l} & x - 3 \\ x^2 + x - 2 & x^3 - 2x^2 - 4x - 4 \\ & x^3 + x^2 - 2x \\ \hline & -3x^2 - 2x - 4 \\ & -3x^2 - 3x + 6 \\ \hline & x - 10 \end{array}$$

Thus
$$\frac{x^3-2x^2-4x-4}{x^2+x-2} \equiv x - 3 + \frac{x-10}{x^2+x-2} \equiv x - 3 + \frac{x-10}{(x+2)(x-1)}$$

Let
$$\frac{x-10}{(x+2)(x-1)} \equiv \frac{\text{A}}{(x+2)} + \frac{\text{B}}{(x-1)} \equiv \frac{\text{A}(x-1) + \text{B}(x+2)}{(x+2)(x-1)}$$

Equating the numerators gives:

$$x - 10 \equiv \text{A}(x-1) + \text{B}(x+2)$$

Let $x = -2$. Then $-12 = -3\text{A}$

i.e. $\mathbf{A = 4}$

Let $x = 1$. Then $-9 = 3\text{B}$

i.e. $\mathbf{B = -3}$

Hence $\frac{x-10}{(x+2)(x-1)} \equiv \frac{4}{(x+2)} - \frac{3}{(x-1)}$

Thus $\frac{\mathbf{x^3-2x^2-4x-4}}{\mathbf{x^2+x-2}}$

$$\equiv \mathbf{x-3+\frac{4}{(x+2)}-\frac{3}{(x-1)}}$$

Now try the following Practice Exercise

Practice Exercise 37 Partial fractions with linear factors (Answers on page 705)

Resolve the following into partial fractions:

1. $\frac{12}{x^2-9}$
2. $\frac{4(x-4)}{x^2-2x-3}$
3. $\frac{x^2-3x+6}{x(x-2)(x-1)}$
4. $\frac{3(2x^2-8x-1)}{(x+4)(x+1)(2x-1)}$
5. $\frac{x^2+9x+8}{x^2+x-6}$
6. $\frac{x^2-x-14}{x^2-2x-3}$
7. $\frac{3x^3-2x^2-16x+20}{(x-2)(x+2)}$

7.3 Partial fractions with repeated linear factors

Problem 5. Resolve $\frac{2x+3}{(x-2)^2}$ into partial fractions

The denominator contains a repeated linear factor, $(x-2)^2$

Let $\frac{2x+3}{(x-2)^2} \equiv \frac{A}{(x-2)} + \frac{B}{(x-2)^2}$

$$\equiv \frac{A(x-2)+B}{(x-2)^2}$$

Equating the numerators gives:

$$2x+3 \equiv A(x-2)+B$$

Let $x=2$. Then $7 = A(0)+B$

i.e. $\mathbf{B=7}$

$$2x+3 \equiv A(x-2)+B$$
$$\equiv Ax-2A+B$$

Since an identity is true for all values of the unknown, the coefficients of similar terms may be equated.

Hence, equating the coefficients of x gives: $\mathbf{2=A}$
[Also, as a check, equating the constant terms gives: $3=-2A+B$. When $A=2$ and $B=7$, RHS$=-2(2)+7=3=$LHS]

Hence $\frac{\mathbf{2x+3}}{\mathbf{(x-2)^2}} \equiv \frac{\mathbf{2}}{\mathbf{(x-2)}} + \frac{\mathbf{7}}{\mathbf{(x-2)^2}}$

Problem 6. Express $\frac{5x^2-2x-19}{(x+3)(x-1)^2}$ as the sum of three partial fractions

The denominator is a combination of a linear factor and a repeated linear factor.

Let $\frac{5x^2-2x-19}{(x+3)(x-1)^2}$

$$\equiv \frac{A}{(x+3)} + \frac{B}{(x-1)} + \frac{C}{(x-1)^2}$$

$$\equiv \frac{A(x-1)^2+B(x+3)(x-1)+C(x+3)}{(x+3)(x-1)^2}$$

by algebraic addition

Equating the numerators gives:

$$5x^2-2x-19 \equiv A(x-1)^2+B(x+3)(x-1) + C(x+3) \quad (1)$$

Let $x=-3$. Then

$$5(-3)^2-2(-3)-19 \equiv A(-4)^2+B(0)(-4)+C(0)$$

i.e. $32=16A$

i.e. $\mathbf{A=2}$

Let $x=1$. Then

$$5(1)^2-2(1)-19 \equiv A(0)^2+B(4)(0)+C(4)$$

i.e. $-16 = 4C$

i.e. $\mathbf{C = -4}$

Without expanding the RHS of equation (1) it can be seen that equating the coefficients of x^2 gives: $5 = A + B$, and since $A = 2$, $\mathbf{B = 3}$
[Check: Identity (1) may be expressed as:

$$5x^2 - 2x - 19 \equiv A(x^2 - 2x + 1) + B(x^2 + 2x - 3) + C(x + 3)$$

i.e. $$5x^2 - 2x - 19 \equiv Ax^2 - 2Ax + A + Bx^2 + 2Bx - 3B + Cx + 3C$$

Equating the x term coefficients gives:

$$-2 \equiv -2A + 2B + C$$

When $A = 2$, $B = 3$ and $C = -4$ then
$-2A + 2B + C = -2(2) + 2(3) - 4 = -2 = \text{LHS}$
Equating the constant term gives:

$$-19 \equiv A - 3B + 3C$$

$$\text{RHS} = 2 - 3(3) + 3(-4) = 2 - 9 - 12 = -19 = \text{LHS}]$$

Hence $$\frac{\mathbf{5x^2 - 2x - 19}}{\mathbf{(x+3)(x-1)^2}} \equiv \frac{\mathbf{2}}{\mathbf{(x+3)}} + \frac{\mathbf{3}}{\mathbf{(x-1)}} - \frac{\mathbf{4}}{\mathbf{(x-1)^2}}$$

Problem 7. Resolve $\dfrac{3x^2 + 16x + 15}{(x+3)^3}$ into partial fractions

Let

$$\frac{3x^2 + 16x + 15}{(x+3)^3} \equiv \frac{A}{(x+3)} + \frac{B}{(x+3)^2} + \frac{C}{(x+3)^3} \equiv \frac{A(x+3)^2 + B(x+3) + C}{(x+3)^3}$$

Equating the numerators gives:

$$3x^2 + 16x + 15 \equiv A(x+3)^2 + B(x+3) + C \quad (1)$$

Let $x = -3$. Then

$$3(-3)^2 + 16(-3) + 15 \equiv A(0)^2 + B(0) + C$$

i.e. $\mathbf{-6 = C}$

Identity (1) may be expanded as:

$$3x^2 + 16x + 15 \equiv A(x^2 + 6x + 9) + B(x + 3) + C$$

i.e. $$3x^2 + 16x + 15 \equiv Ax^2 + 6Ax + 9A + Bx + 3B + C$$

Equating the coefficients of x^2 terms gives:

$$\mathbf{3 = A}$$

Equating the coefficients of x terms gives:

$$16 = 6A + B$$

Since $A = 3$, $\mathbf{B = -2}$

[Check: equating the constant terms gives:

$$15 = 9A + 3B + C$$

When $A = 3$, $B = -2$ and $C = -6$,

$$9A + 3B + C = 9(3) + 3(-2) + (-6) = 27 - 6 - 6 = 15 = \text{LHS}]$$

Thus $$\frac{\mathbf{3x^2 + 16x + 15}}{\mathbf{(x+3)^3}} \equiv \frac{\mathbf{3}}{\mathbf{(x+3)}} - \frac{\mathbf{2}}{\mathbf{(x+3)^2}} - \frac{\mathbf{6}}{\mathbf{(x+3)^3}}$$

Now try the following Practice Exercise

Practice Exercise 38 Partial fractions with repeated linear factors (Answers on page 705)

Resolve the following:

1. $\dfrac{4x - 3}{(x+1)^2}$
2. $\dfrac{x^2 + 7x + 3}{x^2(x+3)}$
3. $\dfrac{5x^2 - 30x + 44}{(x-2)^3}$
4. $\dfrac{18 + 21x - x^2}{(x-5)(x+2)^2}$

7.4 Partial fractions with quadratic factors

Problem 8. Express $\dfrac{7x^2 + 5x + 13}{(x^2 + 2)(x + 1)}$ in partial fractions

The denominator is a combination of a quadratic factor, (x^2+2), which does not factorise without introducing imaginary surd terms, and a linear factor, $(x+1)$. Let

$$\frac{7x^2+5x+13}{(x^2+2)(x+1)} \equiv \frac{Ax+B}{(x^2+2)}+\frac{C}{(x+1)}$$

$$\equiv \frac{(Ax+B)(x+1)+C(x^2+2)}{(x^2+2)(x+1)}$$

Equating numerators gives:

$$7x^2+5x+13 \equiv (Ax+B)(x+1)+C(x^2+2) \quad (1)$$

Let $x=-1$. Then

$$7(-1)^2+5(-1)+13 \equiv (Ax+B)(0)+C(1+2)$$

i.e. $15=3C$

i.e. $\mathbf{C=5}$

Identity (1) may be expanded as:

$$7x^2+5x+13 \equiv Ax^2+Ax+Bx+B+Cx^2+2C$$

Equating the coefficients of x^2 terms gives:

$$7=A+C, \text{and since } C=5, \mathbf{A=2}$$

Equating the coefficients of x terms gives:

$$5=A+B, \text{and since } A=2, \mathbf{B=3}$$

[Check: equating the constant terms gives:

$$13=B+2C$$

When $B=3$ and $C=5$, $B+2C=3+10=13=\text{LHS}$]

Hence $$\boldsymbol{\frac{7x^2+5x+13}{(x^2+2)(x+1)} \equiv \frac{2x+3}{(x^2+2)}+\frac{5}{(x+1)}}$$

Problem 9. Resolve $\dfrac{3+6x+4x^2-2x^3}{x^2(x^2+3)}$ into partial fractions

Terms such as x^2 may be treated as $(x+0)^2$, i.e. they are repeated linear factors

Let $$\frac{3+6x+4x^2-2x^3}{x^2(x^2+3)}$$

$$\equiv \frac{A}{x}+\frac{B}{x^2}+\frac{Cx+D}{(x^2+3)}$$

$$\equiv \frac{Ax(x^2+3)+B(x^2+3)+(Cx+D)x^2}{x^2(x^2+3)}$$

Equating the numerators gives:

$$3+6x+4x^2-2x^3 \equiv Ax(x^2+3) + B(x^2+3)+(Cx+D)x^2$$

$$\equiv Ax^3+3Ax+Bx^2+3B + Cx^3+Dx^2$$

Let $x=0$. Then $3=3B$

i.e. $\mathbf{B=1}$

Equating the coefficients of x^3 terms gives:

$$-2=A+C \quad (1)$$

Equating the coefficients of x^2 terms gives:

$$4=B+D$$

Since $B=1$, $\mathbf{D=3}$

Equating the coefficients of x terms gives:

$$6=3A$$

i.e. $\mathbf{A=2}$

From equation (1), since $A=2$, $\mathbf{C=-4}$

Hence $$\boldsymbol{\frac{3+6x+4x^2-2x^3}{x^2(x^2+3)}}$$

$$\equiv \frac{2}{x}+\frac{1}{x^2}+\frac{-4x+3}{x^2+3}$$

$$\boldsymbol{\equiv \frac{2}{x}+\frac{1}{x^2}+\frac{3-4x}{x^2+3}}$$

Now try the following Practice Exercise

Practice Exercise 39 Partial fractions with quadratic factors (Answers on page 705)

Resolve the following:

1. $\dfrac{x^2-x-13}{(x^2+7)(x-2)}$
2. $\dfrac{6x-5}{(x-4)(x^2+3)}$
3. $\dfrac{15+5x+5x^2-4x^3}{x^2(x^2+5)}$
4. $\dfrac{x^3+4x^2+20x-7}{(x-1)^2(x^2+8)}$

5. When solving the differential equation $\frac{d^2\theta}{dt^2} - 6\frac{d\theta}{dt} - 10\theta = 20 - e^{2t}$ by Laplace transforms, for given boundary conditions, the following expression for $\mathcal{L}\{\theta\}$ results:

$$\mathcal{L}\{\theta\} = \frac{4s^3 - \frac{39}{2}s^2 + 42s - 40}{s(s-2)(s^2 - 6s + 10)}$$

Show that the expression can be resolved into partial fractions to give:

$$\mathcal{L}\{\theta\} = \frac{2}{s} - \frac{1}{2(s-2)} + \frac{5s-3}{2(s^2 - 6s + 10)}$$

For fully worked solutions to each of the problems in Practice Exercises 37 to 39 in this chapter, go to the website:

www.routledge.com/cw/bird

COMPANION @ WEBSITE

Chapter 8

Solving simple equations

Why it is important to understand: Solving simple equations

In mathematics, engineering and science, formulae are used to relate physical quantities to each other. They provide rules so that if we know the values of certain quantities, we can calculate the values of others. Equations occur in all branches of engineering. Simple equations always involve one unknown quantity which we try to find when we solve the equation. In reality, we all solve simple equations in our heads all the time without even noticing it. If, for example, you have bought two CDs, each for the same price, and a DVD, and know that you spent £25 in total and that the DVD was £11, then you actually solve the linear equation $2x + 11 = 25$ to find out that the price of each CD was £7. It is probably true to say that there is no branch of engineering, physics, economics, chemistry and computer science which does not require the solution of simple equations. The ability to solve simple equations is another stepping stone on the way to having confidence to handle engineering mathematics.

At the end of this chapter, you should be able to:

- distinguish between an algebraic expression and an algebraic equation
- maintain the equality of a given equation whilst applying arithmetic operations
- solve linear equations in one unknown including those involving brackets and fractions
- form and solve linear equations involved with practical situations

8.1 Expressions, equations and identities

$(3x-5)$ is an example of an **algebraic expression**, whereas $3x-5=1$ is an example of an **equation** (i.e. it contains an 'equals' sign).

An equation is simply a statement that two quantities are equal. For example, 1 m = 1000 mm or $F = \frac{9}{5}C + 32$ or $y = mx + c$.

An **identity** is a relationship that is true for all values of the unknown, whereas an equation is only true for particular values of the unknown. For example, $3x-5=1$ is an equation, since it is only true when $x=2$, whereas $3x \equiv 8x - 5x$ is an identity since it is true for all values of x. (Note '$\equiv$' means 'is identical to').

Simple linear equations (or equations of the first degree) are those in which an unknown quantity is raised only to the power 1.

To **'solve an equation'** means 'to find the value of the unknown'.

Any arithmetic operation may be applied to an equation **as long as the equality of the equation is maintained**.

8.2 Worked problems on simple equations

Problem 1. Solve the equation: $4x = 20$

Dividing each side of the equation by 4 gives: $\frac{4x}{4} = \frac{20}{4}$
(Note that the same operation has been applied to both the left-hand side (LHS) and the right-hand side (RHS) of the equation so the equality has been maintained).
Cancelling gives: $\boldsymbol{x = 5}$, which is the solution to the equation.
Solutions to simple equations should always be checked and this is accomplished by substituting the solution into the original equation. In this case, LHS $= 4(5) = 20 =$ RHS.

Problem 2. Solve: $\frac{2x}{5} = 6$

The LHS is a fraction and this can be removed by multiplying both sides of the equation by 5.

Hence, $5\left(\frac{2x}{5}\right) = 5(6)$

Cancelling gives: $2x = 30$

Dividing both sides of the equation by 2 gives:

$$\frac{2x}{2} = \frac{30}{2} \quad \text{i.e.} \quad \boldsymbol{x = 15}$$

Problem 3. Solve: $a - 5 = 8$

Adding 5 to both sides of the equation gives:

$$a - 5 + 5 = 8 + 5$$

i.e. $\boldsymbol{a = 13}$

The result of the above procedure is to move the '-5' from the LHS of the original equation, across the equals sign, to the RHS, but the sign is changed to $+$

Problem 4. Solve: $x + 3 = 7$

Subtracting 3 from both sides of the equation gives:

$$x + 3 - 3 = 7 - 3$$

i.e. $\boldsymbol{x = 4}$

The result of the above procedure is to move the '$+3$' from the LHS of the original equation, across the equals sign, to the RHS, but the sign is changed to $-$. Thus a term can be moved from one side of an equation to the other as long as a change in sign is made.

Problem 5. Solve: $6x + 1 = 2x + 9$

In such equations the terms containing x are grouped on one side of the equation and the remaining terms grouped on the other side of the equation. As in Problems 3 and 4, changing from one side of an equation to the other must be accompanied by a change of sign.

Thus since $6x + 1 = 2x + 9$

then $6x - 2x = 9 - 1$

$$4x = 8$$

$$\frac{4x}{4} = \frac{8}{4}$$

i.e. $\boldsymbol{x = 2}$

Check: LHS of original equation $= 6(2) + 1 = 13$

RHS of original equation $= 2(2) + 9 = 13$

Hence the solution $x = 2$ is correct.

Problem 6. Solve: $4 - 3p = 2p - 11$

In order to keep the p term positive the terms in p are moved to the RHS and the constant terms to the LHS.

Hence $4 + 11 = 2p + 3p$

$$15 = 5p$$

$$\frac{15}{5} = \frac{5p}{5}$$

Hence $\boldsymbol{3 = p}$ or $\boldsymbol{p = 3}$

Check: LHS $= 4 - 3(3) = 4 - 9 = -5$

RHS $= 2(3) - 11 = 6 - 11 = -5$

Hence the solution $p = 3$ is correct.
If, in this example, the unknown quantities had been grouped initially on the LHS instead of the RHS then:

$$-3p - 2p = -11 - 4$$

i.e. $-5p = -15$

$$\frac{-5p}{-5} = \frac{-15}{-5}$$

and $\boldsymbol{p = 3}$, as before

It is often easier, however, to work with positive values where possible.

Problem 7. Solve: $3(x-2)=9$

Removing the bracket gives: $3x-6=9$

Rearranging gives:

$$3x=9+6$$
$$3x=15$$
$$\frac{3x}{3}=\frac{15}{3}$$

i.e. $\mathbf{x=5}$

Check: LHS $=3(5-2)=3(3)=9=$ RHS
Hence the solution $x=5$ is correct.

Problem 8. Solve:

$$4(2r-3)-2(r-4)=3(r-3)-1$$

Removing brackets gives:

$$8r-12-2r+8=3r-9-1$$

Rearranging gives:

$$8r-2r-3r=-9-1+12-8$$

i.e. $3r=-6$

$$\mathbf{r}=\frac{-6}{3}=\mathbf{-2}$$

Check:

LHS $=4(-4-3)-2(-2-4)=-28+12=-16$

RHS $=3(-2-3)-1=-15-1=-16$

Hence the solution $r=-2$ is correct.

Now try the following Practice Exercise

Practice Exercise 40 Simple equations (Answers on page 705)

Solve the following equations:

1. $2x+5=7$
2. $8-3t=2$
3. $2x-1=5x+11$
4. $7-4p=2p-3$
5. $2a+6-5a=0$
6. $3x-2-5x=2x-4$
7. $20d-3+3d=11d+5-8$
8. $5(f-2)-3(2f+5)+15=0$
9. $2x=4(x-3)$
10. $6(2-3y)-42=-2(y-1)$
11. $2(3g-5)-5=0$
12. $4(3x+1)=7(x+4)-2(x+5)$
13. $10+3(r-7)=16-(r+2)$
14. $8+4(x-1)-5(x-3)=2(5-2x)$

8.3 Further worked problems on simple equations

Problem 9. Solve: $\frac{3}{x}=\frac{4}{5}$

The lowest common multiple (LCM) of the denominators, i.e. the lowest algebraic expression that both x and 5 will divide into, is $5x$.

Multiplying both sides by $5x$ gives:

$$5x\left(\frac{3}{x}\right)=5x\left(\frac{4}{5}\right)$$

Cancelling gives:

$$15=4x \qquad (1)$$
$$\frac{15}{4}=\frac{4x}{4}$$

i.e. $\mathbf{x=\frac{15}{4}}$ or $\mathbf{3\frac{3}{4}}$

Check:

$$\text{LHS}=\frac{3}{3\frac{3}{4}}=\frac{3}{\frac{15}{4}}=3\left(\frac{4}{15}\right)=\frac{12}{15}=\frac{4}{5}=\text{RHS}$$

(Note that when there is only one fraction on each side of an equation 'cross-multiplication' can be applied. In this example, if $\frac{3}{x}=\frac{4}{5}$ then $(3)(5)=4x$, which is a quicker
way of arriving at equation (1) above.)

Problem 10. Solve: $\frac{2y}{5}+\frac{3}{4}+5=\frac{1}{20}-\frac{3y}{2}$

The LCM of the denominators is 20. Multiplying each term by 20 gives:

$$20\left(\frac{2y}{5}\right)+20\left(\frac{3}{4}\right)+20(5)$$
$$=20\left(\frac{1}{20}\right)-20\left(\frac{3y}{2}\right)$$

Cancelling gives:

$$4(2y)+5(3)+100=1-10(3y)$$

i.e. $\quad 8y+15+100=1-30y$

Rearranging gives:

$$8y+30y=1-15-100$$
$$38y=-114$$
$$y=\frac{-114}{38}=\mathbf{-3}$$

Check: LHS $=\dfrac{2(-3)}{5}+\dfrac{3}{4}+5=\dfrac{-6}{5}+\dfrac{3}{4}+5$

$$=\frac{-9}{20}+5=4\frac{11}{20}$$

$$\text{RHS}=\frac{1}{20}-\frac{3(-3)}{2}=\frac{1}{20}+\frac{9}{2}=4\frac{11}{20}$$

Hence the solution $y=-3$ is correct.

Problem 11. Solve: $\dfrac{3}{t-2}=\dfrac{4}{3t+4}$

By 'cross-multiplication': $\quad 3(3t+4)=4(t-2)$

Removing brackets gives: $\quad 9t+12=4t-8$

Rearranging gives: $\quad 9t-4t=-8-12$

i.e. $\quad 5t=-20$

$$t=\frac{-20}{5}=\mathbf{-4}$$

Check: LHS $=\dfrac{3}{-4-2}=\dfrac{3}{-6}=-\dfrac{1}{2}$

$$\text{RHS}=\frac{4}{3(-4)+4}=\frac{4}{-12+4}$$
$$=\frac{4}{-8}=-\frac{1}{2}$$

Hence the solution $t=-4$ is correct.

Problem 12. Solve: $\sqrt{x}=2$

[$\sqrt{x}=2$ is not a 'simple equation' since the power of x is $\frac{1}{2}$ i.e $\sqrt{x}=x^{(1/2)}$; however, it is included here since it occurs often in practice].

Wherever square root signs are involved with the unknown quantity, both sides of the equation must be squared. Hence

$$(\sqrt{x})^2=(2)^2$$

i.e. $\quad \boldsymbol{x=4}$

Problem 13. Solve: $2\sqrt{2}=8$

To avoid possible errors it is usually best to arrange the term containing the square root on its own. Thus

$$\frac{2\sqrt{d}}{2}=\frac{8}{2}$$

i.e. $\quad \sqrt{d}=4$

Squaring both sides gives: $\boldsymbol{d=16}$, which may be checked in the original equation

Problem 14. Solve: $x^2=25$

This problem involves a square term and thus is not a simple equation (it is, in fact, a quadratic equation). However the solution of such an equation is often required and is therefore included here for completeness. Whenever a square of the unknown is involved, the square root of both sides of the equation is taken. Hence

$$\sqrt{x^2}=\sqrt{25}$$

i.e. $\quad x=5$

However, $x=-5$ is also a solution of the equation because $(-5)\times(-5)=+25$. Therefore, whenever the square root of a number is required there are always two answers, one positive, the other negative.
The solution of $x^2=25$ is thus written as $\boldsymbol{x=\pm 5}$

Problem 15. Solve: $\dfrac{15}{4t^2}=\dfrac{2}{3}$

'Cross-multiplying' gives: $\quad 15(3)=2(4t^2)$

i.e. $\quad 45=8t^2$

$$\frac{45}{8}=t^2$$

i.e. $\quad t^2=5.625$

Hence $t=\sqrt{5.625}=\mathbf{\pm 2.372}$, correct to 4 significant figures.

Now try the following Practice Exercise

Practice Exercise 41 Simple equations (Answers on page 705)

Solve the following equations:

1. $2+\frac{3}{4}y=1+\frac{2}{3}y+\frac{5}{6}$
2. $\frac{1}{4}(2x-1)+3=\frac{1}{2}$
3. $\frac{1}{5}(2f-3)+\frac{1}{6}(f-4)+\frac{2}{15}=0$
4. $\frac{1}{3}(3m-6)-\frac{1}{4}(5m+4)+\frac{1}{5}(2m-9)=-3$
5. $\frac{x}{3}-\frac{x}{5}=2$
6. $1-\frac{y}{3}=3+\frac{y}{3}-\frac{y}{6}$
7. $\frac{1}{3n}+\frac{1}{4n}=\frac{7}{24}$
8. $\frac{x+3}{4}=\frac{x-3}{5}+2$
9. $\frac{y}{5}+\frac{7}{20}=\frac{5-y}{4}$
10. $\frac{v-2}{2v-3}=\frac{1}{3}$
11. $\frac{2}{a-3}=\frac{3}{2a+1}$
12. $\frac{x}{4}-\frac{x+6}{5}=\frac{x+3}{2}$
13. $3\sqrt{t}=9$
14. $\frac{3\sqrt{x}}{1-\sqrt{x}}=-6$
15. $10=5\sqrt{\frac{x}{2}-1}$
16. $16=\frac{t^2}{9}$
17. $\sqrt{\frac{y+2}{y-2}}=\frac{1}{2}$
18. $\frac{11}{2}=5+\frac{8}{x^2}$

8.4 Practical problems involving simple equations

Problem 16. A copper wire has a length l of 1.5 km, a resistance R of 5 Ω and a resistivity of 17.2×10^{-6} Ω mm. Find the cross-sectional area, a, of the wire, given that $R=\rho l/a$

Since $R=\rho l/a$

then

$$5\,\Omega=\frac{(17.2\times10^{-6}\,\Omega\,\text{mm})(1500\times10^3\,\text{mm})}{a}$$

From the units given, a is measured in mm^2.

Thus $$5a=17.2\times10^{-6}\times1500\times10^3$$

and $$a=\frac{17.2\times10^{-6}\times1500\times10^3}{5}=\frac{17.2\times1500\times10^3}{10^6\times5}=\frac{17.2\times15}{10\times5}=5.16$$

Hence the cross-sectional area of the wire is 5.16 mm^2

Problem 17. The temperature coefficient of resistance α may be calculated from the formula $R_t=R_0(1+\alpha t)$. Find α given $R_t=0.928$, $R_0=0.8$ and $t=40$

Since $R_t=R_0(1+\alpha t)$ then

$$0.928=0.8[1+\alpha(40)]$$

$$0.928=0.8+(0.8)(\alpha)(40)$$

$$0.928-0.8=32\alpha$$

$$0.128=32\alpha$$

Hence $$\alpha=\frac{0.128}{32}=\mathbf{0.004}$$

Problem 18. The distance s metres travelled in time t seconds is given by the formula: $s=ut+\frac{1}{2}at^2$, where u is the initial velocity in m/s and a is the acceleration in m/s^2. Find the acceleration of the body if it travels 168 m in 6 s, with an initial velocity of 10 m/s

$s = ut + \frac{1}{2}at^2$, and $s = 168$, $u = 10$ and $t = 6$

Hence $$168 = (10)(6) + \frac{1}{2}a(6)^2$$

$$168 = 60 + 18a$$

$$168 - 60 = 18a$$

$$108 = 18a$$

$$a = \frac{108}{18} = 6$$

Hence the acceleration of the body is 6 m/s²

Problem 19. When three resistors in an electrical circuit are connected in parallel the total resistance R_T is given by:

$$\frac{1}{R_T} = \frac{1}{R_1} + \frac{1}{R_2} + \frac{1}{R_3}$$

Find the total resistance when $R_1 = 5\,\Omega$, $R_2 = 10\,\Omega$ and $R_3 = 30\,\Omega$

$$\frac{1}{R_T} = \frac{1}{5} + \frac{1}{10} + \frac{1}{30}$$

$$= \frac{6+3+1}{30} = \frac{10}{30} = \frac{1}{3}$$

Taking the reciprocal of both sides gives: $\boldsymbol{R_T = 3\,\Omega}$

Alternatively, if $\frac{1}{R_T} = \frac{1}{5} + \frac{1}{10} + \frac{1}{30}$ the LCM of the denominators is $30R_T$

Hence

$$30R_T\left(\frac{1}{R_T}\right) = 30R_T\left(\frac{1}{5}\right) + 30R_T\left(\frac{1}{10}\right) + 30R_T\left(\frac{1}{30}\right)$$

Cancelling gives:

$$30 = 6R_T + 3R_T + R_T$$

$$30 = 10R_T$$

$$\boldsymbol{R_T} = \frac{30}{10} = \boldsymbol{3\,\Omega}, \text{ as above.}$$

Now try the following Practice Exercise

Practice Exercise 42 Practical problems involving simple equations (Answers on page 705)

1. A formula used for calculating resistance of a cable is $R = (\rho l)/a$. Given $R = 1.25$, $l = 2500$ and $a = 2 \times 10^{-4}$ find the value of ρ
2. Force F newtons is given by $F = ma$, where m is the mass in kilograms and a is the acceleration in metres per second squared. Find the acceleration when a force of 4 kN is applied to a mass of 500 kg.
3. $PV = mRT$ is the characteristic gas equation. Find the value of m when $P = 100 \times 10^3$, $V = 3.00$, $R = 288$ and $T = 300$
4. When three resistors R_1, R_2 and R_3 are connected in parallel the total resistance R_T is determined from $\frac{1}{R_T} = \frac{1}{R_1} + \frac{1}{R_2} + \frac{1}{R_3}$
   (a) Find the total resistance when $R_1 = 3\,\Omega$, $R_2 = 6\,\Omega$ and $R_3 = 18\,\Omega$.
   (b) Find the value of R_3 given that $R_T = 3\,\Omega$, $R_1 = 5\,\Omega$ and $R_2 = 10\,\Omega$.
5. Ohm's law may be represented by $I = V/R$, where I is the current in amperes, V is the voltage in volts and R is the resistance in ohms. A soldering iron takes a current of 0.30 A from a 240 V supply. Find the resistance of the element.
6. The stress, σ Pascal's, acting on the reinforcing rod in a concrete column is given in the following equation:
   $500 \times 10^{-6}\sigma + 2.67 \times 10^5 = 3.55 \times 10^5$
   Find the value of the stress in MPa.

8.5 Further practical problems involving simple equations

Problem 20. The extension x m of an aluminium tie bar of length l m and cross-sectional area A m² when carrying a load of F newtons is given by the modulus of elasticity $E = Fl/Ax$. Find the extension of the tie bar (in mm) if $E = 70 \times 10^9\,\text{N/m}^2$, $F = 20 \times 10^6\,\text{N}$, $A = 0.1\,\text{m}^2$ and $l = 1.4\,\text{m}$

$E = Fl/Ax$, hence

$$70 \times 10^9 \frac{\text{N}}{\text{m}^2} = \frac{(20 \times 10^6 \text{N})(1.4 \text{m})}{(0.1 \text{m}^2)(x)}$$

(the unit of x is thus metres)

$$70 \times 10^9 \times 0.1 \times x = 20 \times 10^6 \times 1.4$$

$$x = \frac{20 \times 10^6 \times 1.4}{70 \times 10^9 \times 0.1}$$

Cancelling gives: $x = \frac{2 \times 1.4}{7 \times 100} \text{m}$

$$= \frac{2 \times 1.4}{7 \times 100} \times 1000 \text{mm}$$

Hence the extension of the tie bar, $x = 4$ mm

Problem 21. Power in a d.c. circuit is given by $P = \frac{V^2}{R}$ where V is the supply voltage and R is the circuit resistance. Find the supply voltage if the circuit resistance is 1.25 Ω and the power measured is 320 W

Since $P = \frac{V^2}{R}$ then $320 = \frac{V^2}{1.25}$

$$(320)(1.25) = V^2$$

i.e. $V^2 = 400$

Supply voltage, $V = \sqrt{400} = \mathbf{\pm 20\,V}$

Problem 22. A formula relating initial and final states of pressures, P_1 and P_2, volumes V_1 and V_2, and absolute temperatures, T_1 and T_2, of an ideal gas is $\frac{P_1V_1}{T_1} = \frac{P_2V_2}{T_2}$. Find the value of P_2 given $P_1 = 100 \times 10^3$, $V_1 = 1.0$, $V_2 = 0.266$, $T_1 = 423$ and $T_2 = 293$

Since $\frac{P_1V_1}{T_1} = \frac{P_2V_2}{T_2}$

then $\frac{(100 \times 10^3)(1.0)}{423} = \frac{P_2(0.266)}{293}$

'Cross-multiplying' gives:

$$(100 \times 10^3)(1.0)(293) = P_2(0.266)(423)$$

$$P_2 = \frac{(100 \times 10^3)(1.0)(293)}{(0.266)(423)}$$

Hence $\mathbf{P_2 = 260 \times 10^3}$ or $\mathbf{2.6 \times 10^5}$

Problem 23. The stress f in a material of a thick cylinder can be obtained from $\frac{D}{d} = \sqrt{\frac{f+p}{f-p}}$ Calculate the stress, given that $D = 21.5$, $d = 10.75$ and $p = 1800$

Since $\frac{D}{d} = \sqrt{\frac{f+p}{f-p}}$

then $\frac{21.5}{10.75} = \sqrt{\frac{f+1800}{f-1800}}$

i.e. $2 = \sqrt{\frac{f+1800}{f-1800}}$

Squaring both sides gives:

$$4 = \frac{f+1800}{f-1800}$$

$$4(f - 1800) = f + 1800$$

$$4f - 7200 = f + 1800$$

$$4f - f = 1800 + 7200$$

$$3f = 9000$$

$$f = \frac{9000}{3} = 3000$$

Hence **stress, $f = 3000$**

Now try the following Practice Exercise

Practice Exercise 43 Practical problems involving simple equations (Answers on page 705)

1. Given $R_2 = R_1(1 + \alpha t)$, find α given $R_1 = 5.0$, $R_2 = 6.03$ and $t = 51.5$
2. If $v^2 = u^2 + 2as$, find u given $v = 24$, $a = -40$ and $s = 4.05$
3. The relationship between the temperature on a Fahrenheit scale and that on a Celsius scale

is given by $F = \frac{9}{5}C + 32$. Express 113°F in degrees Celsius.

4. If $t = 2\pi\sqrt{w/Sg}$, find the value of S given $w = 1.219$, $g = 9.81$ and $t = 0.3132$

5. Applying the principle of moments to a beam results in the following equation:

$$F \times 3 = (5 - F) \times 7$$

where F is the force in newtons. Determine the value of F.

6. A rectangular laboratory has a length equal to one and a half times its width and a perimeter of 40 m. Find its length and width.

7. The reaction moment M of a cantilever carrying three, point loads is given by:

$$M = 3.5x + 2.0(x - 1.8) + 4.2(x - 2.6)$$

where x is the length of the cantilever in metres. If $M = 55.32$ kN m, calculate the value of x.

Practice Exercise 44 Multiple-choice questions on solving simple equations (Answers on page 705)

(Answers on page 705)

Each question has only one correct answer

1. Solving the equation $-\frac{x}{3} = 3$ for x gives:
(a) 18 (b) 12 (c) −9 (d) 8

2. Solving the equation $-8b = 4.4$ for b gives:
(a) 0.5 (b) −0.55 (c) 5 (d) 0.55

3. Solving the equation $y - \frac{2}{3} = 7$ for y gives:
(a) $7\frac{2}{3}$ (b) $10\frac{1}{2}$ (c) $6\frac{1}{3}$ (d) $-4\frac{2}{3}$

4. Solving the equation $\frac{x}{9} = \frac{7}{72}$ for x gives:
(a) $1\frac{1}{7}$ (b) $9\frac{7}{72}$ (c) $\frac{7}{8}$ (d) $8\frac{65}{72}$

5. The solution of the equation $2x + 1 = x + 3$ is:
(a) 1 (b) 2 (c) 3 (d) 4

6. Solving the equation $3 - 2x = 3x + 8$ for x gives:
(a) −1 (b) 5 (c) 1 (d) −5

7. Given $4 + \frac{x}{2} = \frac{7}{2}$, the value of x is:
(a) 2 (b) 1 (c) −2 (d) −1

8. The value of x in the equation $5x - 27 + 3x = 4 + 9 - 2x$ is:
(a) −4 (b) 5 (c) 4 (d) 6

9. Solving the equation $1 + 3x = 2(x - 1)$ gives:
(a) $x = -1$ (b) $x = -2$
(c) $x = 1$ (d) $x = -3$

10. Solving the equation $17 + 19(x + y) = 19(y - x) - 21$ for x gives:
(a) −1 (b) −2 (c) −3 (d) −4

For fully worked solutions to each of the problems in Practice Exercises 40 to 43 in this chapter, go to the website:
www.routledge.com/cw/bird

COMPANION @ WEBSITE

Revision Test 2 Algebra, partial fractions and simple equations

This Revision Test covers the material contained in Chapters 5 to 8. *The marks for each question are shown in brackets at the end of each question.*

1. Evaluate: $3xy^2z^3 - 2yz$ when $x = \frac{4}{3}$, $y = 2$ and $z = \frac{1}{2}$ (3)

2. Simplify the following:

   (a) $\dfrac{8a^2b\sqrt{c^3}}{(2a)^2\sqrt{b}\sqrt{c}}$

   (b) $3x + 4 \div 2x + 5 \times 2 - 4x$ (6)

3. Remove the brackets in the following expressions and simplify:

   (a) $(2x - y)^2$

   (b) $4ab - [3\{2(4a - b) + b(2 - a)\}]$ (5)

4. Factorise: $3x^2y + 9xy^2 + 6xy^3$ (3)

5. If x is inversely proportional to y and $x = 12$ when $y = 0.4$, determine:

   (a) the value of x when y is 3, and

   (b) the value of y when $x = 2$ (4)

6. Factorise $x^3 + 4x^2 + x - 6$ using the factor theorem. Hence solve the equation: $x^3 + 4x^2 + x - 6 = 0$ (6)

7. Use the remainder theorem to find the remainder when $2x^3 + x^2 - 7x - 6$ is divided by

   (a) $(x - 2)$ (b) $(x + 1)$

   Hence factorise the cubic expression. (7)

8. Simplify $\dfrac{6x^2 + 7x - 5}{2x - 1}$ by dividing out. (5)

9. Resolve the following into partial fractions:

   (a) $\dfrac{x - 11}{x^2 - x - 2}$ (b) $\dfrac{3 - x}{(x^2 + 3)(x + 3)}$

   (c) $\dfrac{x^3 - 6x + 9}{x^2 + x - 2}$ (24)

10. Solve the following equations:

    (a) $3t - 2 = 5t + 4$

    (b) $4(k - 1) - 2(3k + 2) + 14 = 0$

    (c) $\dfrac{a}{2} - \dfrac{2a}{5} = 1$ (d) $\sqrt{\dfrac{s + 1}{s - 1}} = 2$ (13)

11. A rectangular football pitch has its length equal to twice its width and a perimeter of 360 m. Find its length and width. (4)

For lecturers/instructors/teachers, fully worked solutions to each of the problems in Revision Test 2, together with a full marking scheme, are available at the website:

www.routledge.com/cw/bird

Chapter 9

Transposition of formulae

Why it is important to understand: Transposing formulae

As was mentioned in the last chapter, formulae are used frequently in almost all aspects of engineering in order to relate a physical quantity to one or more others. Many well known physical laws are described using formulae - for example, Ohm's law, $V = I \times R$, or Newton's second law of motion, $F = m \times a$. In an everyday situation, imagine you buy 5 identical items for £20. How much did each item cost? If you divide £20 by 5 to get an answer of £4, you are actually applying transposition of a formula. Transposing formulae is a basic skill required in all aspects of engineering. The ability to transpose formulae is yet another stepping stone on the way to having confidence to handle engineering mathematics.

At the end of this chapter, you should be able to:

- define 'subject of the formula'
- transpose equations whose terms are connected by plus and/or minus signs
- transpose equations that involve fractions
- transpose equations that contain a root or power
- transpose equations in which the subject appears in more than one term

9.1 Introduction to transposition of formulae

When a symbol other than the subject is required to be calculated it is usual to rearrange the formula to make a new subject. This rearranging process is called **transposing the formula** or **transposition**.

The rules used for transposition of formulae are the same as those used for the solution of simple equations (see Chapter 8)—basically, **that the equality of an equation must be maintained**.

9.2 Worked problems on transposition of formulae

Problem 1. Transpose $p=q+r+s$ to make r the subject

The aim is to obtain r on its own on the left-hand side (LHS) of the equation. Changing the equation around so that r is on the LHS gives:

$$q+r+s=p \qquad (1)$$

Substracting $(q+s)$ from both sides of the equation gives:

$$q+r+s-(q+s)=p-(q+s)$$

Thus $\quad q+r+s-q-s=p-q-s$

i.e. $\quad \boldsymbol{r=p-q-s} \qquad (2)$

It is shown with simple equations, that a quantity can be moved from one side of an equation to the other with an appropriate change of sign. Thus equation (2) follows immediately from equation (1) above.

Problem 2. If $a+b=w-x+y$, express x as the subject

Rearranging gives:

$$w-x+y=a+b \quad \text{and} \quad -x=a+b-w-y$$

Multiplying both sides by -1 gives:

$$(-1)(-x)=(-1)(a+b-w-y)$$

i.e. $$x=-a-b+w+y$$

The result of multiplying each side of the equation by -1 is to change all the signs in the equation.
It is conventional to express answers with positive quantities first. Hence rather than $x=-a-b+w+y$, $\boldsymbol{x=w+y-a-b}$, since the order of terms connected by + and − signs is immaterial.

Problem 3. Transpose $v=f\lambda$ to make λ the subject

Rearranging gives: $$f\lambda=v$$

Dividing both sides by f gives: $$\frac{f\lambda}{f}=\frac{v}{f}$$

i.e. $$\boldsymbol{\lambda=\frac{v}{f}}$$

Problem 4. When a body falls freely through a height h, the velocity v is given by $v^2=2gh$. Express this formula with h as the subject

Rearranging gives: $$2gh=v^2$$

Dividing both sides by $2g$ gives: $$\frac{2gh}{2g}=\frac{v^2}{2g}$$

i.e. $$\boldsymbol{h=\frac{v^2}{2g}}$$

Problem 5. If $I=\frac{V}{R}$, rearrange to make V the subject

Rearranging gives: $$\frac{V}{R}=I$$

Multiplying both sides by R gives:

$$R\left(\frac{V}{R}\right)=R(I)$$

Hence $$\boldsymbol{V=IR}$$

Problem 6. Transpose: $a=\frac{F}{m}$ for m

Rearranging gives: $\frac{F}{m}=a$

Multiplying both sides by m gives:

$$m\left(\frac{F}{m}\right)=m(a) \quad \text{i.e.} \quad F=ma$$

Rearranging gives: $$ma=F$$

Dividing both sides by a gives: $$\frac{ma}{a}=\frac{F}{a}$$

i.e. $$\boldsymbol{m=\frac{F}{a}}$$

Problem 7. Rearrange the formula: $R=\frac{\rho l}{a}$ to make (i) a the subject, and (ii) l the subject

(i) Rearranging gives: $\frac{\rho l}{a}=R$

Multiplying both sides by a gives:

$$a\left(\frac{\rho l}{a}\right)=a(R) \quad \text{i.e.} \quad \rho l=aR$$

Rearranging gives: $aR=\rho l$
Dividing both sides by R gives:

$$\frac{aR}{R}=\frac{\rho l}{R}$$

i.e. $$\boldsymbol{a=\frac{\rho l}{R}}$$

(ii) Multiplying both sides of $\frac{\rho l}{a}=R$ by a gives:

$$\rho l=aR$$

Dividing both sides by ρ gives: $$\frac{\rho l}{\rho}=\frac{aR}{\rho}$$

i.e. $$\boldsymbol{l=\frac{aR}{\rho}}$$

Now try the following Practice Exercise

Practice Exercise 45 Transposition of formulae (Answers on page 706)

Make the symbol indicated the subject of each of the formulae shown and express each in its simplest form.

1. $a+b=c-d-e$ $\quad(d)$
2. $x+3y=t$ $\quad(y)$
3. $c=2\pi r$ $\quad(r)$

4. $y=mx+c$ (x)

5. $I=PRT$ (T)

6. $I=\frac{E}{R}$ (R)

7. $Q=mc\Delta T$ (c)

8. $S=\frac{a}{1-r}$ (r)

9. $F=\frac{9}{5}C+32$ (C)

10. $pV=mRT$ (R)

9.3 Further worked problems on transposition of formulae

Problem 8. Transpose the formula: $v=u+\frac{ft}{m}$ to make f the subject

Rearranging gives: $u+\frac{ft}{m}=v$ and $\frac{ft}{m}=v-u$

Multiplying each side by m gives:

$$m\left(\frac{ft}{m}\right)=m(v-u) \quad \text{i.e.} \quad ft=m(v-u)$$

Dividing both sides by t gives:

$$\frac{ft}{t}=\frac{m}{t}(v-u) \quad \text{i.e.} \quad \boldsymbol{f=\frac{m}{t}(v-u)}$$

Problem 9. The final length, l_2 of a piece of wire heated through θ°C is given by the formula $l_2=l_1(1+\alpha\theta)$. Make the coefficient of expansion, α, the subject

Rearranging gives: $l_1(1+\alpha\theta)=l_2$

Removing the bracket gives: $l_1+l_1\alpha\theta=l_2$

Rearranging gives: $l_1\alpha\theta=l_2-l_1$

Dividing both sides by $l_1\theta$ gives:

$$\frac{l_1\alpha\theta}{l_1\theta}=\frac{l_2-l_1}{l_1\theta} \quad \text{i.e.} \quad \alpha=\boldsymbol{\frac{l_2-l_1}{l_1\theta}}$$

Problem 10. A formula for the distance moved by a body is given by: $s=\frac{1}{2}(v+u)t$. Rearrange the formula to make u the subject

Rearranging gives: $\frac{1}{2}(v+u)t=s$

Multiplying both sides by 2 gives: $(v+u)t=2s$

Dividing both sides by t gives:

$$\frac{(v+u)t}{t}=\frac{2s}{t}$$

i.e. $$v+u=\frac{2s}{t}$$

Hence $$\boldsymbol{u=\frac{2s}{t}-v} \quad \text{or} \quad \boldsymbol{u=\frac{2s-vt}{t}}$$

Problem 11. A formula for kinetic energy is $k=\frac{1}{2}mv^2$. Transpose the formula to make v the subject

Rearranging gives: $\frac{1}{2}mv^2=k$

Whenever the prospective new subject is a squared term, that term is isolated on the LHS, and then the square root of both sides of the equation is taken.

Multiplying both sides by 2 gives: $mv^2=2k$

Dividing both sides by m gives: $\frac{mv^2}{m}=\frac{2k}{m}$

i.e. $v^2=\frac{2k}{m}$

Taking the square root of both sides gives:

$$\sqrt{v^2}=\sqrt{\frac{2k}{m}}$$

i.e. $$\boldsymbol{v=\sqrt{\frac{2k}{m}}}$$

Problem 12. In a right-angled triangle having sides x, y and hypotenuse z, Pythagoras' theorem states $z^2=x^2+y^2$. Transpose the formula to find x

Rearranging gives: $x^2+y^2=z^2$

and $x^2=z^2-y^2$

Taking the square root of both sides gives:

$$x = \sqrt{z^2 - y^2}$$

Problem 13. Given $t = 2\pi\sqrt{\frac{l}{g}}$ find g in terms of t, l and π

Whenever the prospective new subject is within a square root sign, it is best to isolate that term on the LHS and then to square both sides of the equation.

Rearranging gives: $2\pi\sqrt{\frac{l}{g}} = t$

Dividing both sides by 2π gives: $\sqrt{\frac{l}{g}} = \frac{t}{2\pi}$

Squaring both sides gives: $\frac{l}{g} = \left(\frac{t}{2\pi}\right)^2 = \frac{t^2}{4\pi^2}$

Cross-multiplying, i.e. multiplying each term by $4\pi^2 g$, gives:

$$4\pi^2 l = gt^2$$

or $$gt^2 = 4\pi^2 l$$

Dividing both sides by t^2 gives: $$\frac{gt^2}{t^2} = \frac{4\pi^2 l}{t^2}$$

i.e. $$\boldsymbol{g = \frac{4\pi^2 l}{t^2}}$$

Problem 14. The impedance of an a.c. circuit is given by $Z = \sqrt{R^2 + X^2}$. Make the reactance, X, the subject

Rearranging gives: $\sqrt{R^2 + X^2} = Z$

Squaring both sides gives: $R^2 + X^2 = Z^2$

Rearranging gives: $X^2 = Z^2 - R^2$

Taking the square root of both sides gives:

$$\boldsymbol{X = \sqrt{Z^2 - R^2}}$$

Problem 15. The volume V of a hemisphere is given by $V = \frac{2}{3}\pi r^3$. Find r in terms of V

Rearranging gives: $\frac{2}{3}\pi r^3 = V$

Multiplying both sides by 3 gives: $2\pi r^3 = 3V$

Dividing both sides by 2π gives:

$$\frac{2\pi r^3}{2\pi} = \frac{3V}{2\pi} \quad \text{i.e.} \quad r^3 = \frac{3V}{2\pi}$$

Taking the cube root of both sides gives:

$$\sqrt[3]{r^3} = \sqrt[3]{\frac{3V}{2\pi}} \quad \text{i.e.} \quad \boldsymbol{r = \sqrt[3]{\frac{3V}{2\pi}}}$$

Now try the following Practice Exercise

Practice Exercise 46 Transposition of formulae (Answers on page 706)

Make the symbol indicated the subject of each of the formulae shown and express each in its simplest form.

1. $y = \frac{\lambda(x-d)}{d}$ (x)
2. $A = \frac{3(F-f)}{L}$ (f)
3. $y = \frac{Ml^2}{8EI}$ (E)
4. $R = R_0(1 + \alpha t)$ (t)
5. $\frac{1}{R} = \frac{1}{R_1} + \frac{1}{R_2}$ (R_2)
6. $I = \frac{E-e}{R+r}$ (R)
7. $y = 4ab^2c^2$ (b)
8. $\frac{P_1V_1}{T_1} = \frac{P_2V_2}{T_2}$ (V_2)
9. $\frac{a^2}{x^2} + \frac{b^2}{y^2} = 1$ (x)
10. $t = 2\pi\sqrt{\frac{l}{g}}$ (l)
11. $v^2 = u^2 + 2as$ (u)
12. $A = \frac{\pi R^2 \theta}{360}$ (R)
13. $N = \sqrt{\frac{a+x}{y}}$ (a)

14. $\frac{P_1V_1}{T_1}=\frac{P_2V_2}{T_2}$ (T_2)

15. $Z=\sqrt{R^2+(2\pi fL)^2}$ (L)

16. The lift force, L, on an aircraft is given by: $L=\frac{1}{2}\rho v^2 ac$ where ρ is the density, v is the velocity, a is the area and c is the lift coefficient. Transpose the equation to make the velocity the subject.

17. The angular deflection θ of a beam of electrons due to a magnetic field is given by: $\theta=k\left(\frac{HL}{V^{\frac{1}{2}}}\right)$. Transpose the equation for V.

18. The velocity of a particle moving with simple harmonic motion is: $V=W\sqrt{(a^2-x^2)}$ Rearrange the formula to make x the subject.

19. A radar has a wavelength, λ, of 40 mm. The radar emits and receives electromagnetic waves which have a speed, v, of 300×10^8 m/s. Given that $v=f\lambda$, where f is the frequency in hertz, calculate the frequency.

9.4 Harder worked problems on transposition of formulae

Problem 16. Transpose the formula $p=\frac{a^2x+a^2y}{r}$ to make a the subject

Rearranging gives: $\frac{a^2x+a^2y}{r}=p$

Multiplying both sides by r gives: $a^2x+a^2y=rp$

Factorising the LHS gives: $a^2(x+y)=rp$

Dividing both sides by $(x+y)$ gives:

$$\frac{a^2(x+y)}{(x+y)}=\frac{rp}{(x+y)} \quad \text{i.e.} \quad a^2=\frac{rp}{(x+y)}$$

Taking the square root of both sides gives:

$$\boldsymbol{a=\sqrt{\frac{rp}{x+y}}}$$

Problem 17. Make b the subject of the formula

$$a=\frac{x-y}{\sqrt{bd+be}}$$

Rearranging gives: $\frac{x-y}{\sqrt{bd+be}}=a$

Multiplying both sides by $\sqrt{bd+be}$ gives:

$$x-y=a\sqrt{bd+be}$$

or $$a\sqrt{bd+be}=x-y$$

Dividing both sides by a gives:

$$\sqrt{bd+be}=\frac{x-y}{a}$$

Squaring both sides gives:

$$bd+be=\left(\frac{x-y}{a}\right)^2$$

Factorising the LHS gives:

$$b(d+e)=\left(\frac{x-y}{a}\right)^2$$

Dividing both sides by $(d+e)$ gives:

$$b=\frac{\left(\frac{x-y}{a}\right)^2}{(d+e)}$$

i.e. $$\boldsymbol{b=\frac{(x-y)^2}{a^2(d+e)}}$$

Problem 18. If $a=\frac{b}{1+b}$ make b the subject of the formula

Rearranging gives: $\frac{b}{1+b}=a$

Multiplying both sides by $(1+b)$ gives:

$$b=a(1+b)$$

Removing the bracket gives: $b=a+ab$

Rearranging to obtain terms in b on the LHS gives:

$$b-ab=a$$

Factorising the LHS gives: $b(1-a)=a$

Dividing both sides by $(1-a)$ gives:

$$\boldsymbol{b=\frac{a}{1-a}}$$

Problem 19. Transpose the formula $V=\dfrac{Er}{R+r}$ to make r the subject

Rearranging gives: $\dfrac{Er}{R+r}=V$

Multiplying both sides by $(R+r)$ gives:

$$Er=V(R+r)$$

Removing the bracket gives: $Er=VR+Vr$

Rearranging to obtain terms in r on the LHS gives:

$$Er-Vr=VR$$

Factorising gives: $r(E-V)=VR$

Dividing both sides by $(E-V)$ gives:

$$\mathbf{r=\frac{VR}{E-V}}$$

Problem 20. Given that: $\dfrac{D}{d}=\sqrt{\dfrac{f+p}{f-p}}$ express p in terms of D, d and f

Rearranging gives: $\sqrt{\dfrac{f+p}{f-p}}=\dfrac{D}{d}$

Squaring both sides gives: $\left(\dfrac{f+p}{f-p}\right)=\dfrac{D^2}{d^2}$

Cross-multiplying, i.e. multiplying each term by $d^2(f-p)$, gives:

$$d^2(f+p)=D^2(f-p)$$

Removing brackets gives: $d^2f+d^2p=D^2f-D^2p$

Rearranging, to obtain terms in p on the LHS gives:

$$d^2p+D^2p=D^2f-d^2f$$

Factorising gives: $p(d^2+D^2)=f(D^2-d^2)$

Dividing both sides by (d^2+D^2) gives:

$$\mathbf{p=\frac{f(D^2-d^2)}{(d^2+D^2)}}$$

Problem 21. The mass moment of inertia through the centre of gravity for a compound pendulum, I_G, is given by: $I_G=mk_G^2$. The value of the radius of gyration about G, k_G^2, may be determined from the frequency, f, of a compound pendulum, given by: $f=\dfrac{1}{2\pi}\sqrt{\dfrac{gh}{(k_G^2+h^2)}}$

Given that the distance $h=50$ mm, $g=9.81$ m/s^2 and the frequency of oscillation, $f=1.26$ Hz, calculate the mass moment of inertia I_G when $m=10.5$ kg.

Frequency, $$f=\frac{1}{2\pi}\sqrt{\frac{gh}{(k_G^2+h^2)}}$$

i.e. $$1.26=\frac{1}{2\pi}\sqrt{\frac{9.81\times 0.05}{(k_G^2+0.05^2)}}$$

i.e. $$(1.26)^2=\frac{1}{(2\pi)^2}\times\frac{9.81\times 0.05}{(k_G^2+0.05^2)}$$

from which, $$(k_G^2+0.0025)=\frac{0.4905}{1.5876\times(2\pi)^2}=0.007826$$

$$k_G^2=0.007826-0.0025=0.005326$$

The mass moment of inertia about the centre of gravity,

$$I_G=mk_G^2=10.5\text{ kg}\times 0.005326\text{ m}^2$$

i.e. $$\mathbf{I_G=0.0559\ kg\ m^2}$$

Now try the following Practice Exercise

Practice Exercise 47 Transposition of formulae (Answers on page 706)

Make the symbol indicated the subject of each of the formulae shown in Problems 1 to 7, and express each in its simplest form.

1. $y=\dfrac{a^2m-a^2n}{x}$ (a)
2. $M=\pi(R^4-r^4)$ (R)
3. $x+y=\dfrac{r}{3+r}$ (r)
4. $m=\dfrac{\mu L}{L+rCR}$ (L)

5. $a^2 = \dfrac{b^2 - c^2}{b^2}$ $\quad (b)$

6. $\dfrac{x}{y} = \dfrac{1 + r^2}{1 - r^2}$ $\quad (r)$

7. $\dfrac{p}{q} = \sqrt{\dfrac{a + 2b}{a - 2b}}$ $\quad (b)$

8. A formula for the focal length, f, of a convex lens is $\dfrac{1}{f} = \dfrac{1}{u} + \dfrac{1}{v}$. Transpose the formula to make v the subject and evaluate v when $f = 5$ and $u = 6$

9. The quantity of heat, Q, is given by the formula $Q = mc(t_2 - t_1)$. Make t_2 the subject of the formula and evaluate t_2 when $m = 10$, $t_1 = 15$, $c = 4$ and $Q = 1600$

10. The velocity, v, of water in a pipe appears in the formula $h = \dfrac{0.03Lv^2}{2dg}$. Express v as the subject of the formula and evaluate v when $h = 0.712$, $L = 150$, $d = 0.30$ and $g = 9.81$

11. The sag S at the centre of a wire is given by the formula: $S = \sqrt{\dfrac{3d(l - d)}{8}}$. Make l the subject of the formula and evaluate l when $d = 1.75$ and $S = 0.80$

12. In an electrical alternating current circuit the impedance Z is given by:

$$Z = \sqrt{R^2 + \left(\omega L - \frac{1}{\omega C}\right)^2}$$

Transpose the formula to make C the subject and hence evaluate C when $Z = 130$, $R = 120$, $\omega = 314$ and $L = 0.32$

13. An approximate relationship between the number of teeth, T, on a milling cutter, the diameter of cutter, D, and the depth of cut, d, is given by: $T = \dfrac{12.5D}{D + 4d}$. Determine the value of D when $T = 10$ and $d = 4$ mm.

14. Make λ, the wavelength of X-rays, the subject of the following formula:

$$\frac{\mu}{\rho} = \frac{CZ^4\sqrt{\lambda^5}\,n}{a}$$

15. A simply supported beam of length L has a centrally applied load F and a uniformly distributed load of w per metre length of beam. The reaction at the beam support is given by:

$$R = \frac{1}{2}(F + wL)$$

Rearrange the equation to make w the subject. Hence determine the value of w when $L = 4$ m, $F = 8$ kN and $R = 10$ kN

16. The rate of heat conduction through a slab of material, Q, is given by the formula $Q = \dfrac{kA(t_1 - t_2)}{d}$ where t_1 and t_2 are the temperatures of each side of the material, A is the area of the slab, d is the thickness of the slab, and k is the thermal conductivity of the material. Rearrange the formula to obtain an expression for t_2

17. The slip, s, of a vehicle is given by: $s = \left(1 - \dfrac{r\omega}{v}\right) \times 100\%$ where r is the tyre radius, ω is the angular velocity and v the velocity. Transpose to make r the subject of the formula.

18. The critical load, F newtons, of a steel column may be determined from the formula $L\sqrt{\dfrac{F}{EI}} = n\pi$ where L is the length, EI is the flexural rigidity, and n is a positive integer. Transpose for F and hence determine the value of F when $n = 1$, $E = 0.25 \times 10^{12}$ N/m^2, $I = 6.92 \times 10^{-6}$m^4 and $L = 1.12$ m

19. Bernoulli's equation relates the flow velocity, v, the pressure, p, of the liquid, and the height h of the liquid above some reference level. Given two locations 1 and 2, the equation states:

$$\frac{p_1}{\rho g} + \frac{v_1^2}{2g} + h_1 = \frac{p_2}{\rho g} + \frac{v_2^2}{2g} + h_2$$

where ρ is the density of the liquid. Rearrange the equation to make the velocity of the liquid at location 2, i.e. v_2, the subject.

20. Young's modulus, $E = \dfrac{\text{stress}, \sigma}{\text{strain}, \varepsilon}$ and $\text{strain}, \varepsilon = \dfrac{\text{length increase}, \delta}{\text{length}, \ell}$
For a 50 mm length of a steel bolt under tensile stress, its length increases by 1.25×10^{-4} m. If $E = 2 \times 10^{11}$ N/m^2, calculate the stress.

21. A thermodynamic equation of state, the van der Waals' equation, states that for one mole of gas:

$$\left(P + \frac{a}{V^2}\right)(V - b) = RT$$

Given that constant, $a = 5.083$, volume, $V = 1.300$, constant, $b = 0.0380$, $R = 0.1700$ and $T = 297.0$, determine the value of pressure, P, correct to 3 significant figures.

Practice Exercise 48 Multiple-choice questions on transposition of formulae (Answers on page 707)

Each question has only one correct answer

1. Rearranging the formula $P = \dfrac{Q}{R}$ to make Q the subject gives:
(a) $\dfrac{P}{R}$ (b) $P - R$ (c) PR (d) $\dfrac{R}{P}$

2. Transposing $I = \dfrac{V}{R}$ for resistance R gives:
(a) $I - V$ (b) $\dfrac{V}{I}$ (c) $\dfrac{I}{V}$ (d) VI

3. Rearranging the formula $W = XYZ$ to make Z the subject gives:
(a) $\dfrac{W}{X} - Y$ (b) $W - XY$ (c) WXY (d) $\dfrac{W}{XY}$

4. Transposing $v = f\lambda$ to make wavelength λ the subject gives:
(a) $\dfrac{v}{f}$ (b) $v + f$ (c) $f - v$ (d) $\dfrac{f}{v}$

5. The current I in an a.c. circuit is given by $I = \dfrac{V}{\sqrt{R^2 + X^2}}$ When $R = 4.8$, $X = 10.5$ and $I = 15$, the value of voltage V is:
(a) 173.18 (b) 1.30 (c) 0.98 (d) 229.50

6. The relationship between the temperature in degrees Fahrenheit (F) and the temperature in degrees Celsius (C) is given by:
$$\text{F} = \frac{9}{5}\text{C} + 32$$
135°F is equivalent to:
(a) 43°C (b) 57.2°C
(c) 185.4°C (d) 184°C

7. A formula for the focal length f of a convex lens is $\dfrac{1}{f} = \dfrac{1}{u} + \dfrac{1}{v}$
When $f = 4$ and $u = 6$, v is:
(a) -2 (b) $\dfrac{1}{12}$ (c) 12 (d) $-\dfrac{1}{2}$

8. Electrical resistance $R = \dfrac{\rho\ell}{a}$; transposing this equation for ℓ gives:
(a) $\dfrac{\rho a}{R}$ (b) $\dfrac{R}{a\rho}$ (c) $\dfrac{a}{R\rho}$ (d) $\dfrac{Ra}{\rho}$

9. $PV = mRT$ is the characteristic gas equation. When $P = 100 \times 10^3$, $V = 4.0$, $R = 288$ and $T = 300$, the value of m is:
(a) 4.630 (b) 313600
(c) 0.216 (d) 100592

10. Transposing the formula $R = R_0(1 + \alpha t)$ for t gives:
(a) $\dfrac{R - R_0}{(1 + \alpha)}$ (b) $\dfrac{R - R_0 - 1}{\alpha}$
(c) $\dfrac{R - R_0}{\alpha R_0}$ (d) $\dfrac{R}{R_0 \alpha}$

11. When two resistors R_1 and R_2 are connected in parallel the formula $\dfrac{1}{R_T} = \dfrac{1}{R_1} + \dfrac{1}{R_2}$ is used to determine the total resistance R_T. If $R_1 = 470\,\Omega$ and $R_2 = 2.7$ kΩ, R_T (correct to 3 significant figures) is equal to:
(a) 2.68 Ω (b) 400 Ω
(c) 473 Ω (d) 3170 Ω

12. The height s of a mass projected vertically upwards at time t is given by: $s = ut - \dfrac{1}{2}gt^2$. When $g = 10$, $t = 1.5$ and $s = 3.75$, the value of u is:
(a) 10 (b) -5 (c) $+5$ (d) -10

13. Current I is given by: $I = \dfrac{V}{R + r}$
When $V = 50$ kV, $R = 5$ kΩ and $I = 5$ A, the value of r is:
(a) 9.99 Ω (b) 15 kΩ (c) 9 kΩ (d) 5 kΩ

14. The quantity of heat Q is given by the formula $Q = mc(t_2 - t_1)$. When $m = 5$, $t_1 = 20$, $c = 8$ and $Q = 1200$, the value of t_2 is:
(a) 10 (b) 1.5 (c) 21.5 (d) 50

15. Volume $= \dfrac{\text{mass}}{\text{density}}$. The density (in kg/m^3) when the mass is 2.532 kg and the volume is 162 cm^3 is:
(a) 0.01563 kg/m^3 (b) 410.2 kg/m^3
(c) 15630 kg/m^3 (d) 64.0 kg/m^3

16. Resistance R ohms varies with temperature t according to the formula $R = R_0(1 + \alpha t)$. Given $R = 21\Omega$, $\alpha = 0.004$ and $t = 100$, R_0 has a value of:
(a) 21.4 Ω (b) 15 Ω
(c) 29.4 Ω (d) 0.067 Ω

17. The final length l_2 of a piece of wire heated through θ°C is given by the formula $\ell_2 = \ell_1(1 + \alpha\theta)$ Transposing, the coefficient of expansion, α, is given by:
(a) $\dfrac{\ell_2}{\ell_1} - \dfrac{1}{\theta}$ (b) $\dfrac{\ell_2 - \ell_1}{\ell_1\theta}$
(c) $\ell_2 - \ell_1 - \ell_1\theta$ (d) $\dfrac{\ell_1 - \ell_2}{\ell_1\theta}$

18. Current I in an electrical circuit is given by $I = \dfrac{E - e}{R + r}$. Transposing for R gives:
(a) $\dfrac{E - e - Ir}{I}$ (b) $\dfrac{E - e}{I + r}$
(c) $(E - e)(I + r)$ (d) $\dfrac{E - e}{Ir}$

19. Transposing $t = 2\pi\sqrt{\dfrac{\ell}{g}}$ for g gives:
(a) $\dfrac{(t - 2\pi)^2}{\ell}$ (b) $\left(\dfrac{2\pi}{t}\right)\ell^2$
(c) $\dfrac{\sqrt{\frac{t}{2\pi}}}{\ell}$ (d) $\dfrac{4\pi^2\ell}{t^2}$

20. The volume V_2 of a material when the temperature is increased is given by

$$V_2 = V_1[1 + \gamma(t_2 - t_1)]$$

The value of t_2 when $V_2 = 61.5$ cm^3, $V_1 = 60$ cm^3, $\gamma = 54 \times 10^{-6}$ and $t_1 = 250$ is:
(a) 213 (b) 463 (c) 713 (d) 28028

For fully worked solutions to each of the problems in Practice Exercises 45 to 47 in this chapter, go to the website:
www.routledge.com/cw/bird

COMPANION @ WEBSITE

Chapter 10

Solving simultaneous equations

Why it is important to understand: Solving simultaneous equations

Simultaneous equations arise a great deal in engineering and science, some applications including theory of structures, data analysis, electrical circuit analysis and air traffic control. Systems that consist of a small number of equations can be solved analytically using standard methods from algebra (as explained in this chapter). Systems of large numbers of equations require the use of numerical methods and computers. Matrices are generally used to solve simultaneous equations (as explained in chapter 59). Solving simultaneous equations is an important skill required in all aspects of engineering.

At the end of this chapter, you should be able to:

- solve simultaneous equations in two unknowns by substitution
- solve simultaneous equations in two unknowns by elimination
- solve simultaneous equations involving practical situations

10.1 Introduction to simultaneous equations

Only one equation is necessary when finding the value of a **single unknown quantity** (as with simple equations in Chapter 8). However, when an equation contains **two unknown quantities** it has an infinite number of solutions. When two equations are available connecting the same two unknown values then a unique solution is possible. Similarly, for three unknown quantities it is necessary to have three equations in order to solve for a particular value of each of the unknown quantities and so on.

Equations that have to be solved together to find the unique values of the unknown quantities, which are true for each of the equations, are called **simultaneous equations**.

Two methods of solving simultaneous equations analytically are:

(a) by **substitution**, and (b) by **elimination**.

(A graphical solution of simultaneous equations is shown in Chapter 30 and determinants and matrices are used to solve simultaneous equations in Chapter 59).

10.2 Worked problems on simultaneous equations in two unknowns

Problem 1. Solve the following equations for x and y, (a) by substitution, and (b) by elimination:

$$x + 2y = -1 \qquad (1)$$

$$4x - 3y = 18 \qquad (2)$$

(a) **By substitution**

From equation (1): $x=-1-2y$

Substituting this expression for x into equation (2) gives:

$$4(-1-2y)-3y=18$$

This is now a simple equation in y.

Removing the bracket gives:

$$-4-8y-3y=18$$

$$-11y=18+4=22$$

$$y=\frac{22}{-11}=-2$$

Substituting $y=-2$ into equation (1) gives:

$$x+2(-2)=-1$$

$$x-4=-1$$

$$x=-1+4=3$$

Thus $x=3$ and $y=-2$ is the solution to the simultaneous equations.

(Check: In equation (2), since $x=3$ and $y=-2$, LHS$=4(3)-3(-2)=12+6=18=$RHS.)

(b) **By elimination**

$$x+2y=-1 \qquad (1)$$

$$4x-3y=18 \qquad (2)$$

If equation (1) is multiplied throughout by 4 the coefficient of x will be the same as in equation (2), giving:

$$4x+8y=-4 \qquad (3)$$

Subtracting equation (3) from equation (2) gives:

$$4x-3y=\ 18 \qquad (2)$$

$$4x+8y=-4 \qquad (3)$$

$$0-11y=\ 22$$

Hence $y=\dfrac{22}{-11}=-2$

(Note, in the above subtraction,
$18-(-4)=18+4=22$)

Substituting $y=-2$ into either equation (1) or equation (2) will give $x=3$ as in method (a). The solution $\boldsymbol{x=3}$, $\boldsymbol{y=-2}$ is the only pair of values that satisfies both of the original equations.

Problem 2. Solve, by a substitution method, the simultaneous equations:

$$3x-2y=12 \qquad (1)$$

$$x+3y=-7 \qquad (2)$$

From equation (2), $x=-7-3y$

Substituting for x in equation (1) gives:

$$3(-7-3y)-2y=12$$

i.e. $$-21-9y-2y=12$$

$$-11y=12+21=33$$

Hence $$y=\frac{33}{-11}=-3$$

Substituting $y=-3$ in equation (2) gives:

$$x+3(-3)=-7$$

i.e. $$x-9=-7$$

Hence $$x=-7+9=2$$

Thus $\boldsymbol{x=2, y=-3}$ is the solution of the simultaneous equations.

(Such solutions should always be checked by substituting values into each of the original two equations.)

Problem 3. Use an elimination method to solve the simultaneous equations:

$$3x+4y=\ 5 \qquad (1)$$

$$2x-5y=-12 \qquad (2)$$

If equation (1) is multiplied throughout by 2 and equation (2) by 3, then the coefficient of x will be the same in the newly formed equations. Thus

$2 \times$ equation (1) gives: $\quad 6x+8y=\ 10 \qquad (3)$

$3 \times$ equation (2) gives: $\quad 6x-15y=-36 \qquad (4)$

Equation (3) – equation (4) gives:

$$0+23y=46$$

i.e. $$y=\frac{46}{23}=2$$

(Note $+8y--15y=8y+15y=23y$ and $10-(-36)=10+36=46$. Alternatively, 'change the signs of the bottom line and add').

Substituting $y=2$ in equation (1) gives:

$$3x+4(2)=5$$

from which $3x=5-8=-3$

and $x=-1$

Checking in equation (2), left-hand side $=$ $2(-1)-5(2)=-2-10=-12=$ right-hand side.

Hence $\boldsymbol{x=-1}$ **and** $\boldsymbol{y=2}$ is the solution of the simultaneous equations.
The elimination method is the most common method of solving simultaneous equations.

Problem 4. Solve:

$$7x-2y=26 \quad (1)$$
$$6x+5y=29 \quad (2)$$

When equation (1) is multiplied by 5 and equation (2) by 2 the coefficients of y in each equation are numerically the same, i.e. 10, but are of opposite sign.

$5\times$ equation (1) gives: $35x-10y=130$ (3)

$2\times$ equation (2) gives: $12x+10y=58$ (4)

Adding equation (3) and (4) gives: $47x+0=188$ (5)

Hence $x=\dfrac{188}{47}=4$

[Note that when the signs of common coefficients are **different** the two equations are **added**, and when the signs of common coefficients are the **same** the two equations are **subtracted** (as in Problems 1 and 3)]

Substituting $x=4$ in equation (1) gives:

$$7(4)-2y=26$$
$$28-2y=26$$
$$28-26=2y$$
$$2=2y$$

Hence $y=1$

Checking, by substituting $x=4$ and $y=1$ in equation (2), gives:

$$\text{LHS}=6(4)+5(1)=24+5=29=\text{RHS}$$

Thus the solution is $\boldsymbol{x=4}$, $\boldsymbol{y=1}$, since these values maintain the equality when substituted in both equations.

Now try the following Practice Exercise

Practice Exercise 49 Simultaneous equations (Answers on page 707)

Solve the following simultaneous equations and verify the results.

1. $a+b=7$
   $a-b=3$
2. $2x+5y=7$
   $x+3y=4$
3. $3s+2t=12$
   $4s-t=5$
4. $3x-2y=13$
   $2x+5y=-4$
5. $5x=2y$
   $3x+7y=41$
6. $5c=1-3d$
   $2d+c+4=0$

10.3 Further worked problems on simultaneous equations

Problem 5. Solve:

$$3p=2q \quad (1)$$
$$4p+q+11=0 \quad (2)$$

Rearranging gives:

$$3p-2q=0 \quad (3)$$
$$4p+q=-11 \quad (4)$$

Multiplying equation (4) by 2 gives:

$$8p+2q=-22 \quad (5)$$

Adding equations (3) and (5) gives:

$$11p+0=-22$$

$$p=\frac{-22}{11}=-2$$

Substituting $p=-2$ into equation (1) gives:

$$3(-2)=2q$$

$$-6=2q$$

$$q=\frac{-6}{2}=-3$$

Checking, by substituting $p=-2$ and $q=-3$ into equation (2) gives:

$$\text{LHS}=4(-2)+(-3)+11=-8-3+11$$

$$=0=\text{RHS}$$

Hence the solution is $p=-2, q=-3$

Problem 6. Solve:

$$\frac{x}{8}+\frac{5}{2}=y \quad (1)$$

$$13-\frac{y}{3}=3x \quad (2)$$

Whenever fractions are involved in simultaneous equation it is usual to firstly remove them. Thus, multiplying equation (1) by 8 gives:

$$8\left(\frac{x}{8}\right)+8\left(\frac{5}{2}\right)=8y$$

i.e. $$x+20=8y \quad (3)$$

Multiplying equation (2) by 3 gives:

$$39-y=9x \quad (4)$$

Rearranging equation (3) and (4) gives:

$$x-8y=-20 \quad (5)$$

$$9x+y=39 \quad (6)$$

Multiplying equation (6) by 8 gives:

$$72x+8y=312 \quad (7)$$

Adding equations (5) and (7) gives:

$$73x+0=292$$

$$\boldsymbol{x}=\frac{292}{73}=\mathbf{4}$$

Substituting $x=4$ into equation (5) gives:

$$4-8y=-20$$

$$4+20=8y$$

$$24=8y$$

$$\boldsymbol{y}=\frac{24}{8}=\mathbf{3}$$

Checking: substituting $x=4$, $y=3$ in the original equations, gives:

Equation (1): $$\text{LHS}=\frac{4}{8}+\frac{5}{2}=\frac{1}{2}+2\frac{1}{2}=3$$

$$=y=\text{RHS}$$

Equation (2): $$\text{LHS}=13-\frac{3}{3}=13-1=12$$

$$\text{RHS}=3x=3(4)=12$$

Hence the solution is $x=4, y=3$

Problem 7. Solve:

$$2.5x+0.75-3y=0$$

$$1.6x=1.08-1.2y$$

It is often easier to remove decimal fractions. Thus multiplying equations (1) and (2) by 100 gives:

$$250x+75-300y=0 \quad (1)$$

$$160x=108-120y \quad (2)$$

Rearranging gives:

$$250x-300y=-75 \quad (3)$$

$$160x+120y=108 \quad (4)$$

Multiplying equation (3) by 2 gives:

$$500x-600y=-150 \quad (5)$$

Multiplying equation (4) by 5 gives:

$$800x+600y=540 \quad (6)$$

Adding equations (5) and (6) gives:

$$1300x+0=390$$

$$\boldsymbol{x}=\frac{390}{1300}=\frac{39}{130}=\frac{3}{10}=\mathbf{0.3}$$

Substituting $x=0.3$ into equation (1) gives:

$$250(0.3)+75-300y=0$$
$$75+75=300y$$
$$150=300y$$
$$y=\frac{150}{300}=\mathbf{0.5}$$

Checking $x=0.3$, $y=0.5$ in equation (2) gives:

$$\text{LHS}=160(0.3)=48$$
$$\text{RHS}=108-120(0.5)$$
$$=108-60=48$$

Hence the solution is $x=0.3, y=0.5$

Now try the following Practice Exercise

Practice Exercise 50 Simultaneous equations (Answers on page 707)

Solve the following simultaneous equations and verify the results.

1. $7p+11+2q=0$

   $-1=3q-5p$
2. $\frac{x}{2}+\frac{y}{3}=4$

   $\frac{x}{6}-\frac{y}{9}=0$
3. $\frac{a}{2}-7=-2b$

   $12=5a+\frac{2}{3}b$
4. $\frac{x}{5}+\frac{2y}{3}=\frac{49}{15}$

   $\frac{3x}{7}-\frac{y}{2}+\frac{5}{7}=0$
5. $1.5x-2.2y=-18$

   $2.4x+0.6y=33$
6. $3b-2.5a=0.45$

   $1.6a+0.8b=0.8$

10.4 More difficult worked problems on simultaneous equations

Problem 8. Solve:

$$\frac{2}{x}+\frac{3}{y}=7 \qquad (1)$$

$$\frac{1}{x}-\frac{4}{y}=-2 \qquad (2)$$

In this type of equation the solutions is easier if a substitution is initially made. Let $\frac{1}{x}=a$ and $\frac{1}{y}=b$

Thus equation (1) becomes: $2a+3b=7$ (3)

and equation (2) becomes: $a-4b=-2$ (4)

Multiplying equation (4) by 2 gives:

$$2a-8b=-4 \qquad (5)$$

Subtracting equation (5) from equation (3) gives:

$$0+11b=11$$

i.e. $b=1$

Substituting $b=1$ in equation (3) gives:

$$2a+3=7$$
$$2a=7-3=4$$

i.e. $a=2$

Checking, substituting $a=2$ and $b=1$ in equation (4) gives:

$$\text{LHS}=2-4(1)=2-4=-2=\text{RHS}$$

Hence $a=2$ and $b=1$

However, since $\frac{1}{x}=a$ then $x=\frac{1}{a}=\frac{1}{2}$

and since $\frac{1}{y}=b$ then $y=\frac{1}{b}=\frac{1}{1}=1$

Hence the solutions is $x=\frac{1}{2}, y=1$,

which may be checked in the original equations.

Problem 9. Solve:

$$\frac{1}{2a}+\frac{3}{5b}=4 \quad (1)$$

$$\frac{4}{a}+\frac{1}{2b}=10.5 \quad (2)$$

Let $\frac{1}{a}=x$ and $\frac{1}{b}=y$

then $$\frac{x}{2}+\frac{3}{5}y=4 \quad (3)$$

$$4x+\frac{1}{2}y=10.5 \quad (4)$$

To remove fractions, equation (3) is multiplied by 10 giving:

$$10\left(\frac{x}{2}\right)+10\left(\frac{3}{5}y\right)=10(4)$$

i.e. $$5x+6y=40 \quad (5)$$

Multiplying equation (4) by 2 gives:

$$8x+y=21 \quad (6)$$

Multiplying equation (6) by 6 gives:

$$48x+6y=126 \quad (7)$$

Subtracting equation (5) from equation (7) gives:

$$43x+0=86$$

$$x=\frac{86}{43}=2$$

Substituting $x=2$ into equation (3) gives:

$$\frac{2}{2}+\frac{3}{5}y=4$$

$$\frac{3}{5}y=4-1=3$$

$$y=\frac{5}{3}(3)=5$$

Since $\frac{1}{a}=x$ then $a=\frac{1}{x}=\frac{1}{2}$

and since $\frac{1}{b}=y$ then $b=\frac{1}{y}=\frac{1}{5}$

Hence the solutions is $\boldsymbol{a=\frac{1}{2}, b=\frac{1}{5}}$

which may be checked in the original equations.

Problem 10. Solve:

$$\frac{1}{x+y}=\frac{4}{27} \quad (1)$$

$$\frac{1}{2x-y}=\frac{4}{33} \quad (2)$$

To eliminate fractions, both sides of equation (1) are multiplied by $27(x+y)$ giving:

$$27(x+y)\left(\frac{1}{x+y}\right)=27(x+y)\left(\frac{4}{27}\right)$$

i.e. $$27(1)=4(x+y)$$

$$27=4x+4y \quad (3)$$

Similarly, in equation (2): $33=4(2x-y)$

i.e. $$33=8x-4y \quad (4)$$

Equation (3) + equation (4) gives:

$$60=12x, \quad \text{i.e. } x=\frac{60}{12}=5$$

Substituting $x=5$ in equation (3) gives:

$$27=4(5)+4y$$

from which $4y=27-20=7$

and $y=\frac{7}{4}=1\frac{3}{4}$

Hence $\boldsymbol{x=5,\ y=1\frac{3}{4}}$ is the required solution, which may be checked in the original equations.

Now try the following Practice Exercise

Practice Exercise 51 More difficult simultaneous equations (Answers on page 707)

In Problems 1 to 5, solve the simultaneous equations and verify the results

1. $\frac{3}{x}+\frac{2}{y}=14$

   $\frac{5}{x}-\frac{3}{y}=-2$

2. $\frac{4}{a}-\frac{3}{b}=18$

$\frac{2}{a}+\frac{5}{b}=-4$

3. $\frac{1}{2p}+\frac{3}{5q}=5$

$\frac{5}{p}-\frac{1}{2q}=\frac{35}{2}$

4. $\frac{c+1}{4}-\frac{d+2}{3}+1=0$

$\frac{1-c}{5}+\frac{3-d}{4}+\frac{13}{20}=0$

5. $\frac{3r+2}{5}-\frac{2s-1}{4}=\frac{11}{5}$

$\frac{3+2r}{4}+\frac{5-s}{3}=\frac{15}{4}$

6. If $5x-\frac{3}{y}=1$ and $x+\frac{4}{y}=\frac{5}{2}$ find the value of $\frac{xy+1}{y}$

10.5 Practical problems involving simultaneous equations

There are a number of situations in engineering and science where the solution of simultaneous equations is required. Some are demonstrated in the following worked problems.

Problem 11. The law connecting friction F and load L for an experiment is of the form $F=aL+b$, where a and b are constants. When $F=5.6, L=8.0$ and when $F=4.4, L=2.0$. Find the values of a and b and the value of F when $L=6.5$

Substituting $F=5.6, L=8.0$ into $F=aL+b$ gives:

$$5.6=8.0a+b \quad (1)$$

Substituting $F=4.4, L=2.0$ into $F=aL+b$ gives:

$$4.4=2.0a+b \quad (2)$$

Subtracting equation (2) from equation (1) gives:

$$1.2=6.0a$$

$$a=\frac{1.2}{6.0}=\mathbf{\frac{1}{5}}$$

Substituting $a=\frac{1}{5}$ into equation (1) gives:

$$5.6=8.0\left(\frac{1}{5}\right)+b$$

$$5.6=1.6+b$$

$$5.6-1.6=b$$

i.e. $\mathbf{b=4}$

Checking, substituting $a=\frac{1}{5}$ and $b=4$ in equation (2), gives:

$$\text{RHS}=2.0\left(\frac{1}{5}\right)+4=0.4+4=4.4=\text{LHS}$$

Hence $\mathbf{a=\frac{1}{5}}$ **and** $\mathbf{b=4}$

When $\mathbf{L=6.5}$, $F=aL+b=\frac{1}{5}(6.5)+4$

$=1.3+4$, i.e. $\mathbf{F=5.3}$

Problem 12. The equation of a straight line, of gradient m and intercept on the y-axis c, is $y=mx+c$. If a straight line passes through the point where $x=1$ and $y=-2$, and also through the point where $x=3\frac{1}{2}$ and $y=10\frac{1}{2}$, find the values of the gradient and the y-axis intercept

Substituting $x=1$ and $y=-2$ into $y=mx+c$ gives:

$$-2=m+c \quad (1)$$

Substituting $x=3\frac{1}{2}$ and $y=10\frac{1}{2}$ into $y=mx+c$ gives:

$$10\frac{1}{2}=3\frac{1}{2}m+c \quad (2)$$

Subtracting equation (1) from equation (2) gives:

$$12\frac{1}{2}=2\frac{1}{2}m \quad \text{from which,} \quad m=\frac{12\frac{1}{2}}{2\frac{1}{2}}=\mathbf{5}$$

Substituting $m=5$ into equation (1) gives:

$$-2=5+c$$

$$c=-2-5=\mathbf{-7}$$

Checking, substituting $m=5$ and $c=-7$ in equation (2), gives:

$$\text{RHS}=\left(3\frac{1}{2}\right)(5)+(-7)=17\frac{1}{2}-7$$

$$=10\frac{1}{2}=\text{LHS}$$

Hence the gradient, $m=5$ and the y-axis intercept, $c=-7$

Problem 13. When Kirchhoff's laws* are applied to the electrical circuit shown in Fig. 10.1 the currents I_1 and I_2 are connected by the equations:

$$27 = 1.5I_1 + 8(I_1 - I_2) \qquad (1)$$

$$-26 = 2I_2 - 8(I_1 - I_2) \qquad (2)$$

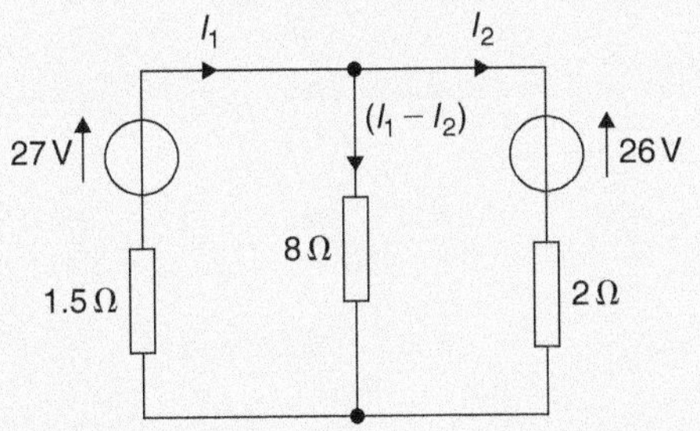

Figure 10.1

Solve the equations to find the values of currents I_1 and I_2

*Who was **Kirchhoff**? – **Gustav Robert Kirchhoff** (12 March 1824–17 October 1887) was a German physicist. Concepts in circuit theory and thermal emission are named 'Kirchhoff's laws' after him, as well as a law of thermochemistry. To find out more go to www.routledge.com/cw/bird

Removing the brackets from equation (1) gives:

$$27 = 1.5I_1 + 8I_1 - 8I_2$$

Rearranging gives:

$$9.5I_1 - 8I_2 = 27 \qquad (3)$$

Removing the brackets from equation (2) gives:

$$-26 = 2I_2 - 8I_1 + 8I_2$$

Rearranging gives:

$$-8I_1 + 10I_2 = -26 \qquad (4)$$

Multiplying equation (3) by 5 gives:

$$47.5I_1 - 40I_2 = 135 \qquad (5)$$

Multiplying equation (4) by 4 gives:

$$-32I_1 + 40I_2 = -104 \qquad (6)$$

Adding equations (5) and (6) gives:

$$15.5I_1 + 0 = 31$$

$$\boldsymbol{I_1} = \frac{31}{15.5} = \mathbf{2}$$

Substituting $I_1 = 2$ into equation (3) gives:

$$9.5(2) - 8I_2 = 27$$

$$19 - 8I_2 = 27$$

$$19 - 27 = 8I_2$$

$$-8 = 8I_2$$

$$\boldsymbol{I_2 = -1}$$

Hence the solution is $\boldsymbol{I_1 = 2}$ **and** $\boldsymbol{I_2 = -1}$

(which may be checked in the original equations).

Problem 14. The distance s metres from a fixed point of a vehicle travelling in a straight line with constant acceleration, a m/s^2, is given by $s = ut + \frac{1}{2}at^2$, where u is the initial velocity in m/s and t the time in seconds. Determine the initial velocity and the acceleration given that $s = 42$ m when $t = 2$ s and $s = 144$ m when $t = 4$ s. Find also the distance travelled after 3 s

Substituting $s = 42, t = 2$ into $s = ut + \frac{1}{2}at^2$ gives:

$$42 = 2u + \frac{1}{2}a(2)^2$$

i.e. $$42 = 2u + 2a \quad (1)$$

Substituting $s = 144, t = 4$ into $s = ut + \frac{1}{2}at^2$ gives:

$$144 = 4u + \frac{1}{2}a(4)^2$$

i.e. $$144 = 4u + 8a \quad (2)$$

Multiplying equation (1) by 2 gives:

$$84 = 4u + 4a \quad (3)$$

Subtracting equation (3) from equation (2) gives:

$$60 = 0 + 4a$$

$$\boldsymbol{a} = \frac{60}{15} = \mathbf{15}$$

Substituting $a = 15$ into equation (1) gives:

$$42 = 2u + 2(15)$$

$$42 - 30 = 2u$$

$$\boldsymbol{u} = \frac{12}{2} = \mathbf{6}$$

Substituting $a = 15, u = 6$ in equation (2) gives:

$$\text{RHS} = 4(6) + 8(15) = 24 + 120 = 144 = \text{LHS}$$

Hence the initial velocity, $u = 6$ m/s and the acceleration, $a = 15$ m/s^2.

Distance travelled after 3 s is given by $s = ut + \frac{1}{2}at^2$ where $t = 3, u = 6$ and $a = 15$

Hence $$s = (6)(3) + \frac{1}{2}(15)(3)^2 = 18 + 67.5$$

i.e. distance travelled after 3 s = 85.5 m.

Problem 15. The resistance $R\,\Omega$ of a length of wire at t°C is given by $R = R_0(1 + \alpha t)$, where R_0 is the resistance at 0°C and α is the temperature coefficient of resistance in /°C. Find the values of α and R_0 if $R = 30\,\Omega$ at 50°C and $R = 35\,\Omega$ at 100°C

Substituting $R = 30, t = 50$ into $R = R_0(1 + \alpha t)$ gives:

$$30 = R_0(1 + 50\alpha) \quad (1)$$

Substituting $R = 35, t = 100$ into $R = R_0(1 + \alpha t)$ gives:

$$35 = R_0(1 + 100\alpha) \quad (2)$$

Although these equations may be solved by the conventional substitution method, an easier way is to eliminate R_0 by division. Thus, dividing equation (1) by equation (2) gives:

$$\frac{30}{35} = \frac{R_0(1 + 50\alpha)}{R_0(1 + 100\alpha)} = \frac{1 + 50\alpha}{1 + 100\alpha}$$

'Cross-multiplying' gives:

$$30(1 + 100\alpha) = 35(1 + 50\alpha)$$

$$30 + 3000\alpha = 35 + 1750\alpha$$

$$3000\alpha - 1750\alpha = 35 - 30$$

$$1250\alpha = 5$$

i.e. $$\alpha = \frac{5}{1250} = \frac{1}{250} \quad \text{or} \quad \mathbf{0.004}$$

Substituting $\alpha = \frac{1}{250}$ into equation (1) gives:

$$30 = R_0\left\{1 + (50)\left(\frac{1}{250}\right)\right\}$$

$$30 = R_0(1.2)$$

$$\boldsymbol{R_0} = \frac{30}{1.2} = \mathbf{25}$$

Checking, substituting $\alpha = \frac{1}{250}$ and $R_0 = 25$ in equation (2) gives:

$$\text{RHS} = 25\left\{1 + (100)\left(\frac{1}{250}\right)\right\}$$

$$= 25(1.4) = 35 = \text{LHS}$$

Thus the solution is $\alpha = 0.004$/°C and $R_0 = 25\,\Omega$.

Problem 16. The molar heat capacity of a solid compound is given by the equation $c = a + bT$, where a and b are constants. When $c = 52, T = 100$ and when $c = 172, T = 400$. Determine the values of a and b

When $c = 52$, $T = 100$, hence

$$52 = a + 100b \quad (1)$$

When $c = 172$, $T = 400$, hence

$$172 = a + 400b \quad (2)$$

Equation (2) – equation (1) gives:

$$120 = 300b$$

from which, $\mathbf{b} = \frac{120}{300} = \mathbf{0.4}$

Substituting $b = 0.4$ in equation (1) gives:

$$52 = a + 100(0.4)$$

$$\mathbf{a} = 52 - 40 = \mathbf{12}$$

Hence $a = 12$ and $b = 0.4$

Now try the following Practice Exercise

Practice Exercise 52 Practical problems involving simultaneous equations (Answers on page 707)

1. In a system of pulleys, the effort P required to raise a load W is given by $P = aW + b$, where a and b are constants
If $W = 40$ when $P = 12$ and $W = 90$ when $P = 22$, find the values of a and b

2. Applying Kirchhoff's laws to an electrical circuit produces the following equations:
$$5 = 0.2I_1 + 2(I_1 - I_2)$$
$$12 = 3I_2 + 0.4I_2 - 2(I_1 - I_2)$$
Determine the values of currents I_1 and I_2

3. Velocity v is given by the formula $v = u + at$. If $v = 20$ when $t = 2$ and $v = 40$ when $t = 7$, find the values of u and a. Hence find the velocity when $t = 3.5$

4. $y = mx + c$ is the equation of a straight line of slope m and y-axis intercept c. If the line passes through the point where $x = 2$ and $y = 2$, and also through the point where $x = 5$ and $y = \frac{1}{2}$, find the slope and y-axis intercept of the straight line

5. The resistance R ohms of copper wire at t°C is given by $R = R_0(1 + \alpha t)$, where R_0 is the resistance at 0°C and α is the temperature coefficient of resistance. If $R = 25.44\,\Omega$ at 30°C and $R = 32.17\,\Omega$ at 100°C, find α and R_0

6. The molar heat capacity of a solid compound is given by the equation $c = a + bT$. When $c = 70, T = 110$ and when $c = 160, T = 290$. Find the values of a and b

7. For a balanced beam, the equilibrium of forces is given by: $R_1 + R_2 = 12.0$ kN
As a result of taking moments:
$0.2R_1 + 7 \times 0.3 + 3 \times 0.6 = 0.8R_2$
Determine the values of the reaction forces R_1 and R_2

8. In an engineering scenario involving reciprocal motion, the following simultaneous equations resulted:
$$\omega\sqrt{(r^2 - 0.07^2)} = 8 \text{ and } \omega\sqrt{(r^2 - 0.25^2)} = 2$$
Calculate the value of radius r and angular velocity ω, each correct to 3 significant figures.

Practice Exercise 53 Multiple-choice questions on solving simultaneous equations (Answers on page 707)

Each question has only one correct answer

1. The solution of the simultaneous equations $3x - 2y = 13$ and $2x + 5y = -4$ is:
(a) $x = -2, y = 3$ (b) $x = 1, y = -5$
(c) $x = 3, y = -2$ (d) $x = -7, y = 2$

2. In a system of pulleys, the effort P required to raise a load W is given by $P = aW + b$, where a and b are constants. If $W = 40$ when $P = 12$ and $W = 90$ when $P = 22$, the values of a and b are:
(a) $a = 5, b = \frac{1}{4}$ (b) $a = 1, b = -28$
(c) $a = \frac{1}{3}, b = -8$ (d) $a = \frac{1}{5}, b = 4$

3. The cost of 8 pencils and 16 pens is £3.52, and the cost of 4 pens and 4 pencils is 96 p. The cost of a pen is:
(a) 28 p (b) 36 p (c) 20 p (d) 32 p

4. A pair of simultaneous equations are:
$x + 2y - 8 = 0$ and $2x + 4y = 16$.
The number of possible solutions is:
(a) 0 (b) 1 (c) 2 (d) infinite

5. Given two simultaneous equations $x-2y=5$ and $2y+5x=7$, then the value of $9x$ is:
(a) 39 (b) 18 (c) −13.5 (d) −3

6. If $x+2y=1$ and $3x-y=-11$ then the value of x is:
(a) −3 (b) 4.2 (c) 2 (d) −1.6

7. If $4x-3y=7$ and $2x+5y=-1$ then the value of y is:
(a) −1 (b) $-\frac{9}{13}$ (c) 2 (d) $\frac{59}{26}$

8. If $5x-3y=17.5$ and $2x+y=4.25$ then the value of $4x+3y$ is:
(a) 7.25 (b) 8.5 (c) 12 (d) 14.75

9. Given that $3x=4y$ and $5x+y-23=0$, then the value of $2x-y$ is:
(a) 2 (b) 5 (c) 1.5 (d) 1

10. Velocity is given by the formula $v=u+at$. If $v=18$ when $t=2$ and $v=38$ when $t=6$, the value of the velocity when $t=7.5$ is:
(a) 45.5 (b) 65 (c) 29.5 (d) 55

For fully worked solutions to each of the problems in Practice Exercises 49 to 52 in this chapter, go to the website:

www.routledge.com/cw/bird

Chapter 11

Solving quadratic equations

Why it is important to understand: Solving quadratic equations

Quadratic equations have many applications in engineering and science; they are used in describing the trajectory of a ball, determining the height of a throw and in the concepts of acceleration, velocity, ballistics and stopping power. In addition, the quadratic equation has been found to be widely evident in a number of natural processes; some of these include the processes by which light is reflected off a lens, water flows down a rocky stream, or even the manner in which fur, spots or stripes develop on wild animals. When traffic policemen arrive at the scene of a road accident, they measure the length of the skid marks and assess the road conditions. They can then use a quadratic equation to calculate the speed of the vehicles and hence reconstruct exactly what happened. The U-shape of a parabola can describe the trajectories of water jets in a fountain and a bouncing ball, or be incorporated into structures like the parabolic reflectors that form the base of satellite dishes and car headlights. Quadratic functions can help plot the course of moving objects and assist in determining minimum and maximum values. Most of the objects we use every day, from cars to clocks, would not exist if someone somewhere hadn't applied quadratic functions to their design. Solving quadratic equations is an important skill required in all aspects of engineering.

At the end of this chapter, you should be able to:

- define a quadratic equation
- solve quadratic equations by factorisation
- solve quadratic equations by 'completing the square'
- solve quadratic equations by formula
- solve quadratic equations involving practical situations
- solve linear and quadratic equations simultaneously

11.1 Introduction to quadratic equations

As stated in Chapter 8, an **equation** is a statement that two quantities are equal and to **'solve an equation'** means 'to find the value of the unknown'. The value of the unknown is called the **root** of the equation.

A **quadratic equation** is one in which the highest power of the unknown quantity is 2. For example, $x^2-3x+1=0$ is a quadratic equation.

There are four methods of **solving quadratic equations**.

These are: (i) **by factorisation** (where possible)
(ii) **by 'completing the square'**
(iii) **by using the 'quadratic formula'**
or (iv) **graphically** (see Chapter 30).

11.2 Solution of quadratic equations by factorisation

Multiplying out $(2x+1)(x-3)$ gives $2x^2-6x+x-3$, i.e. $2x^2-5x-3$. The reverse process of moving from $2x^2-5x-3$ to $(2x+1)(x-3)$ is called **factorising**.
If the quadratic expression can be factorised this provides the simplest method of solving a quadratic equation.

For example, if $\quad 2x^2-5x-3=0$, then,

by factorising: $\quad (2x+1)(x-3)=0$

Hence either $\quad (2x+1)=0 \quad$ i.e. $\quad x=-\dfrac{1}{2}$

or $\quad (x-3)=0 \quad$ i.e. $\quad x=3$

The technique of factorising is often one of 'trial and error'.

Problem 1. Solve the equations:
(a) $x^2+2x-8=0$ (b) $3x^2-11x-4=0$
by factorisation

(a) $x^2+2x-8=0$. The factors of x^2 are x and x. These are placed in brackets thus: $(x \quad)(x \quad)$

The factors of -8 are $+8$ and -1, or -8 and $+1$, or $+4$ and -2, or -4 and $+2$. The only combination to given a middle term of $+2x$ is $+4$ and -2, i.e.

$$x^2+2x-8=(x+4)(x-2)$$

(Note that the product of the two inner terms added to the product of the two outer terms must equal to the middle term, $+2x$ in this case.)

The quadratic equation $x^2+2x-8=0$ thus becomes $(x+4)(x-2)=0$

Since the only way that this can be true is for either the first or the second, or both factors to be zero, then

either $\quad (x+4)=0 \quad$ i.e. $\quad x=-4$

or $\quad (x-2)=0 \quad$ i.e. $\quad x=2$

Hence the roots of $x^2+2x-8=0$ are $x=-4$ and 2

(b) $3x^2-11x-4=0$
The factors of $3x^2$ are $3x$ and x. These are placed in brackets thus: $(3x \quad)(x \quad)$

The factors of -4 are -4 and $+1$, or $+4$ and -1, or -2 and 2

Remembering that the product of the two inner terms added to the product of the two outer terms must equal $-11x$, the only combination to give this is $+1$ and -4, i.e.

$$3x^2-11x-4=(3x+1)(x-4)$$

The quadratic equation $3x^2-11x-4=0$ thus becomes $(3x+1)(x-4)=0$

Hence, either $\quad (3x+1)=0 \quad$ i.e. $\quad \boldsymbol{x=-\dfrac{1}{3}}$

or $\quad (x-4)=0 \quad$ i.e. $\quad \boldsymbol{x=4}$

and both solutions may be checked in the original equation.

Problem 2. Determine the roots of:
(a) $x^2-6x+9=0$, and (b) $4x^2-25=0$, by factorisation

(a) $x^2-6x+9=0$. Hence $(x-3)(x-3)=0$, i.e. $(x-3)^2=0$ (the left-hand side is known as **a perfect square**). Hence $\boldsymbol{x=3}$ is the only root of the equation $x^2-6x+9=0$

(b) $4x^2-25=0$ (the left-hand side is **the difference of two squares**, $(2x)^2$ and $(5)^2$). Thus $(2x+5)(2x-5)=0$

Hence either $\quad (2x+5)=0 \quad$ i.e. $\quad \boldsymbol{x=-\dfrac{5}{2}}$

or $\quad (2x-5)=0 \quad$ i.e. $\quad \boldsymbol{x=\dfrac{5}{2}}$

Problem 3. Solve the following quadratic equations by factorising: (a) $4x^2+8x+3=0$ (b) $15x^2+2x-8=0$.

(a) $4x^2+8x+3=0$. The factors of $4x^2$ are $4x$ and x or $2x$ and $2x$. The factors of 3 are 3 and 1, or -3 and -1. Remembering that the product of the inner terms added to the product of the two outer terms must equal $+8x$, the only combination that is true (by trial and error) is:

$$(4x^2+8x+3)=(2x+3)(2x+1)$$

Hence $\quad (2x+3)(2x+1)=0$ from which, either

$$(2x+3)=0 \quad \text{or} \quad (2x+1)=0$$

Thus, $2x=-3$, from which, $\boldsymbol{x=-\frac{3}{2}}$

or $2x=-1$, from which, $\boldsymbol{x=-\frac{1}{2}}$

which may be checked in the original equation.

(b) $15x^2+2x-8=0$. The factors of $15x^2$ are $15x$ and x or $5x$ and $3x$. The factors of -8 are -4 and $+2$, or 4 and -2, or -8 and $+1$, or 8 and -1. By trial and error the only combination that works is:

$$15x^2+2x-8=(5x+4)(3x-2)$$

Hence $(5x+4)(3x-2)=0$ from which

either $5x+4=0$

or $3x-2=0$

Hence $\boldsymbol{x=-\frac{4}{5}}$ or $\boldsymbol{x=\frac{2}{3}}$

which may be checked in the original equation.

Problem 4. The roots of quadratic equation are $\frac{1}{3}$ and -2. Determine the equation

If the roots of a quadratic equation are α and β then $(x-\alpha)(x-\beta)=0$

Hence if $\alpha=\frac{1}{3}$ and $\beta=-2$, then

$$\left(x-\frac{1}{3}\right)(x-(-2))=0$$

$$\left(x-\frac{1}{3}\right)(x+2)=0$$

$$x^2-\frac{1}{3}x+2x-\frac{2}{3}=0$$

$$x^2+\frac{5}{3}x-\frac{2}{3}=0$$

Hence $\boldsymbol{3x^2+5x-2=0}$

Problem 5. Find the equations of x whose roots are: (a) 5 and -5 (b) 1.2 and -0.4

(a) If 5 and -5 are the roots of a quadratic equation then:

$$(x-5)(x+5)=0$$

i.e. $x^2-5x+5x-25=0$

i.e. $\boldsymbol{x^2-25=0}$

(b) If 1.2 and -0.4 are the roots of a quadratic equation then:

$$(x-1.2)(x+0.4)=0$$

i.e. $x^2-1.2x+0.4x-0.48=0$

i.e. $\boldsymbol{x^2-0.8x-0.48=0}$

Now try the following Practice Exercise

Practice Exercise 54 Solving quadratic equations by factorisation (Answers on page 707)

In Problems 1 to 10, solve the given equations by factorisation.

1. $x^2+4x-32=0$
2. $x^2-16=0$
3. $(x+2)^2=16$
4. $2x^2-x-3=0$
5. $6x^2-5x+1=0$
6. $10x^2+3x-4=0$
7. $x^2-4x+4=0$
8. $21x^2-25x=4$
9. $6x^2-5x-4=0$
10. $8x^2+2x-15=0$

In Problems 11 to 16, determine the quadratic equations in x whose roots are:

11. 3 and 1
12. 2 and -5
13. -1 and -4
14. $2\frac{1}{2}$ and $-\frac{1}{2}$
15. 6 and -6
16. 2.4 and -0.7

11.3 Solution of quadratic equations by 'completing the square'

An expression such as x^2 or $(x+2)^2$ or $(x-3)^2$ is called a perfect square.

If $x^2=3$ then $x=\pm\sqrt{3}$
If $(x+2)^2=5$ then $x+2=\pm\sqrt{5}$ and $x=-2\pm\sqrt{5}$
If $(x-3)^2=8$ then $x-3=\pm\sqrt{8}$ and $x=3\pm\sqrt{8}$

Hence if a quadratic equation can be rearranged so that one side of the equation is a perfect square and the other side of the equation is a number, then the solution of the equation is readily obtained by taking the square roots of each side as in the above examples. The process of rearranging one side of a quadratic equation into a perfect square before solving is called **'completing the square'**.

$$(x+a)^2=x^2+2ax+a^2$$

Thus in order to make the quadratic expression x^2+2ax into a perfect square it is necessary to add (half the coefficient of x)^2 i.e. $\left(\frac{2a}{2}\right)^2$ or a^2

For example, x^2+3x becomes a perfect square by adding $\left(\frac{3}{2}\right)^2$, i.e.

$$x^2+3x+\left(\frac{3}{2}\right)^2=\left(x+\frac{3}{2}\right)^2$$

The method is demonstrated in the following worked problems.

Problem 6. Solve $2x^2+5x=3$ by 'completing the square'

The procedure is as follows:

1. Rearrange the equations so that all terms are on the same side of the equals sign (and the coefficient of the x^2 term is positive).

   Hence $2x^2+5x-3=0$

2. Make the coefficient of the x^2 term unity. In this case this is achieved by dividing throughout by 2. Hence

$$\frac{2x^2}{2}+\frac{5x}{2}-\frac{3}{2}=0$$

i.e. $$x^2+\frac{5}{2}x-\frac{3}{2}=0$$

3. Rearrange the equations so that the x^2 and x terms are on one side of the equals sign and the constant is on the other side, Hence

$$x^2+\frac{5}{2}x=\frac{3}{2}$$

4. Add to both sides of the equation (half the coefficient of x)^2. In this case the coefficient of x is $\frac{5}{2}$

   Half the coefficient squared is therefore $\left(\frac{5}{4}\right)^2$

   Thus, $x^2+\frac{5}{2}x+\left(\frac{5}{4}\right)^2=\frac{3}{2}+\left(\frac{5}{4}\right)^2$

   The LHS is now a perfect square, i.e.

$$\left(x+\frac{5}{4}\right)^2=\frac{3}{2}+\left(\frac{5}{4}\right)^2$$

5. Evaluate the RHS. Thus

$$\left(x+\frac{5}{4}\right)^2=\frac{3}{2}+\frac{25}{16}=\frac{24+25}{16}=\frac{49}{16}$$

6. Taking the square root of both sides of the equation (remembering that the square root of a number gives a $\pm$ answer). Thus

$$\sqrt{\left(x+\frac{5}{4}\right)^2}=\sqrt{\frac{49}{16}}$$

i.e. $$x+\frac{5}{4}=\pm\frac{7}{4}$$

7. Solve the simple equation. Thus

$$x=-\frac{5}{4}\pm\frac{7}{4}$$

i.e. $$x=-\frac{5}{4}+\frac{7}{4}=\frac{2}{4}=\frac{1}{2}$$

and $$x=-\frac{5}{4}-\frac{7}{4}=-\frac{12}{4}=-3$$

Hence $\boldsymbol{x=\frac{1}{2}}$ or $\boldsymbol{-3}$ are the roots of the equation $2x^2+5x=3$

Problem 7. Solve $2x^2+9x+8=0$, correct to 3 significant figures, by 'completing the square'

Making the coefficient of x^2 unity gives:

$$x^2+\frac{9}{2}x+4=0$$

and rearranging gives: $x^2+\frac{9}{2}x=-4$

Adding to both sides (half the coefficient of x)^2 gives:

$$x^2+\frac{9}{2}x+\left(\frac{9}{4}\right)^2=\left(\frac{9}{4}\right)^2-4$$

The LHS is now a perfect square, thus:

$$\left(x+\frac{9}{4}\right)^2=\frac{81}{16}-4=\frac{17}{16}$$

Taking the square root of both sides gives:

$$x+\frac{9}{4}=\sqrt{\frac{17}{16}}=\pm 1.031$$

Hence $x=-\frac{9}{4}\pm 1.031$

i.e. $\boldsymbol{x=-1.22}$ or $\boldsymbol{-3.28}$, correct to 3 significant figures.

Problem 8. By 'completing the square', solve the quadratic equation $4.6y^2+3.5y-1.75=0$, correct to 3 decimal places

Making the coefficient of y^2 unity gives:

$$y^2+\frac{3.5}{4.6}y-\frac{1.75}{4.6}=0$$

and rearranging gives: $y^2+\frac{3.5}{4.6}y=\frac{1.75}{4.6}$

Adding to both sides (half the coefficient of y)^2 gives:

$$y^2+\frac{3.5}{4.6}y+\left(\frac{3.5}{9.2}\right)^2=\frac{1.75}{4.6}+\left(\frac{3.5}{9.2}\right)^2$$

The LHS is now a perfect square, thus:

$$\left(y+\frac{3.5}{9.2}\right)^2=0.5251654$$

Taking the square root of both sides gives:

$$y+\frac{3.5}{9.2}=\sqrt{0.5251654}=\pm 0.7246830$$

Hence, $y=-\frac{3.5}{9.2}\pm 0.7246830$

i.e $\boldsymbol{y=0.344}$ or $\boldsymbol{-1.105}$

Now try the following Practice Exercise

Practice Exercise 55 Solving quadratic equations by completing the square (Answers on page 707)

Solve the following equations by completing the square, each correct to 3 decimal places.

1. $x^2+4x+1=0$
2. $2x^2+5x-4=0$
3. $3x^2-x-5=0$
4. $5x^2-8x+2=0$
5. $4x^2-11x+3=0$

11.4 Solution of quadratic equations by formula

Let the general form of a quadratic equation be given by:

$$ax^2+bx+c=0$$

where a, b and c are constants.

Dividing $ax^2+bx+c=0$ by a gives:

$$x^2+\frac{b}{a}x+\frac{c}{a}=0$$

Rearranging gives:

$$x^2+\frac{b}{a}x=-\frac{c}{a}$$

Adding to each side of the equation the square of half the coefficient of the terms in x to make the LHS a perfect square gives:

$$x^2+\frac{b}{a}x+\left(\frac{b}{2a}\right)^2=\left(\frac{b}{2a}\right)^2-\frac{c}{a}$$

Rearranging gives:

$$\left(x+\frac{b}{a}\right)^2=\frac{b^2}{4a^2}-\frac{c}{a}=\frac{b^2-4ac}{4a^2}$$

Taking the square root of both sides gives:

$$x+\frac{b}{2a}=\sqrt{\frac{b^2-4ac}{4a^2}}=\frac{\pm\sqrt{b^2-4ac}}{2a}$$

Hence $x=-\frac{b}{2a}\pm\frac{\sqrt{b^2-4ac}}{2a}$

i.e. the quadratic formula is: $x=\dfrac{-b\pm\sqrt{b^2-4ac}}{2a}$

(This method of solution is 'completing the square' – as shown in Section 11.3.) Summarising:

if $ax^2+bx+c=0$

then $$x=\frac{-b\pm\sqrt{b^2-4ac}}{2a}$$

This is known as the **quadratic formula**.

Problem 9. Solve (a) $x^2+2x-8=0$ and (b) $3x^2-11x-4=0$ by using the quadratic formula

(a) Comparing $x^2+2x-8=0$ with $ax^2+bx+c=0$ gives $a=1$, $b=2$ and $c=-8$

Substituting these values into the quadratic formula

$$x=\frac{-b\pm\sqrt{b^2-4ac}}{2a} \text{ gives}$$

$$x=\frac{-2\pm\sqrt{2^2-4(1)(-8)}}{2(1)}$$

$$=\frac{-2\pm\sqrt{4+32}}{2}=\frac{-2\pm\sqrt{36}}{2}$$

$$=\frac{-2\pm 6}{2}=\frac{-2+6}{2} \text{ or } \frac{-2-6}{2}$$

Hence $\boldsymbol{x}=\dfrac{4}{2}=\mathbf{2}$ or $\dfrac{-8}{2}=\mathbf{-4}$ (as in Problem 1(a)).

(b) Comparing $3x^2-11x-4=0$ with $ax^2+bx+c=0$ gives $a=3$, $b=-11$ and $c=-4$. Hence,

$$x=\frac{-(-11)\pm\sqrt{(-11)^2-4(3)(-4)}}{2(3)}$$

$$=\frac{-11\pm\sqrt{121+48}}{6}=\frac{11\pm\sqrt{169}}{6}$$

$$=\frac{11\pm 13}{6}=\frac{11+13}{6} \text{ or } \frac{11-13}{6}$$

Hence $\boldsymbol{x}=\dfrac{24}{6}=\mathbf{4}$ or $\dfrac{-2}{6}=-\dfrac{\mathbf{1}}{\mathbf{3}}$ (as in Problem 1(b)).

Problem 10. Solve $4x^2+7x+2=0$ giving the roots correct to 2 decimal places

Comparing $4x^2+7x+2=0$ with $ax^2+bx+c=0$ gives $a=4$, $b=7$ and $c=2$. Hence,

$$x=\frac{-7\pm\sqrt{7^2-4(4)(2)}}{2(4)}$$

$$=\frac{-7\pm\sqrt{17}}{8}=\frac{-7\pm 4.123}{8}$$

$$=\frac{-7\pm 4.123}{8} \text{ or } \frac{-7-4.123}{8}$$

Hence, $\boldsymbol{x=-0.36}$ or $\mathbf{-1.39}$**, correct to 2 decimal places**.

Now try the following Practice Exercise

Practice Exercise 56 Solving quadratic equations by formula (Answers on page 707)

Solve the following equations by using the quadratic formula, correct to 3 decimal places.

1. $2x^2+5x-4=0$
2. $5.76x^2+2.86x-1.35=0$
3. $2x^2-7x+4=0$
4. $4x+5=\dfrac{3}{x}$
5. $(2x+1)=\dfrac{5}{x-3}$

11.5 Practical problems involving quadratic equations

There are many **practical problems** where a quadratic equation has first to be obtained, from given information, before it is solved.

Problem 11. Calculate the diameter of a solid cylinder which has a height of 82.0 cm and a total surface area of 2.0 m^2

Total surface area of a cylinder

= curved surface area + 2 circular ends (from Chapter 25)

$=2\pi rh+2\pi r^2$

(where r = radius and h = height)

Since the total surface area $= 2.0\,\text{m}^2$ and the height $h = 82$ cm or 0.82 m, then

$$2.0 = 2\pi r(0.82) + 2\pi r^2$$

i.e. $2\pi r^2 + 2\pi r(0.82) - 2.0 = 0$

Dividing throughout by 2π gives:

$$r^2 + 0.82r - \frac{1}{\pi} = 0$$

Using the quadratic formula:

$$r = \frac{-0.82 \pm \sqrt{(0.82)^2 - 4(1)\left(-\frac{1}{\pi}\right)}}{2(1)}$$

$$= \frac{-0.82 \pm \sqrt{1.9456}}{2} = \frac{-0.82 \pm 1.3948}{2}$$

$$= 0.2874 \quad \text{or} \quad -1.1074$$

Thus the radius r of the cylinder is 0.2874 m (the negative solution being neglected).
Hence the diameter of the cylinder

$$= 2 \times 0.2874$$

$$= \mathbf{0.5748\,m} \quad \textbf{or} \quad \mathbf{57.5\,cm}$$

correct to 3 significant figures

Problem 12. The height s metres of a mass projected vertically upward at time t seconds is $s = ut - \frac{1}{2}gt^2$. Determine how long the mass will take after being projected to reach a height of 16 m (a) on the ascent and (b) on the descent, when $u = 30$ m/s and $g = 9.81\,\text{m/s}^2$

When height $s = 16$ m, $16 = 30t - \frac{1}{2}(9.81)t^2$

i.e. $4.905t^2 - 30t + 16 = 0$

Using the quadratic formula:

$$t = \frac{-(-30) \pm \sqrt{(-30)^2 - 4(4.905)(16)}}{2(4.905)}$$

$$= \frac{30 \pm \sqrt{586.1}}{9.81} = \frac{30 \pm 24.21}{9.81}$$

$$= 5.53 \quad \text{or} \quad 0.59$$

Hence the mass will reach a height of 16 m after 0.59 s on the ascent and after 5.53 s on the descent.

Problem 13. A shed is 4.0 m long and 2.0 m wide. A concrete path of constant width is laid all the way around the shed. If the area of the path is $9.50\,\text{m}^2$ calculate its width to the nearest centimetre

Figure 11.1 shows a plan view of the shed with its surrounding path of width t metres.

Area of path $= 2(2.0 \times t) + 2t(4.0 + 2t)$

i.e. $9.50 = 4.0t + 8.0t + 4t^2$

or $4t^2 + 12.0t - 9.50 = 0$

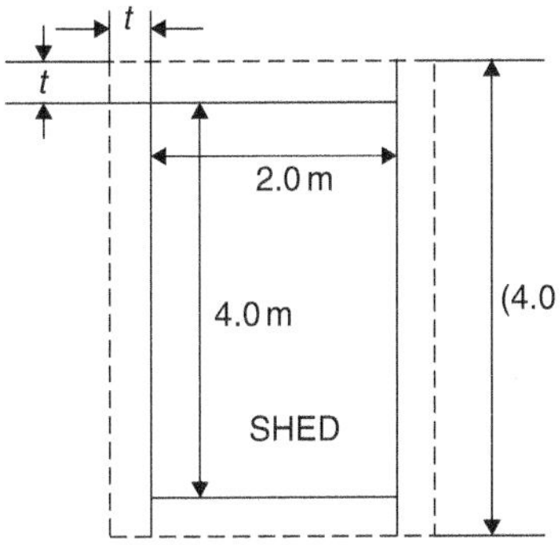

Figure 11.1

Hence $t = \dfrac{-(12.0) \pm \sqrt{(12.0)^2 - 4(4)(-9.50)}}{2(4)}$

$$= \frac{-12.0 \pm \sqrt{296.0}}{8}$$

$$= \frac{-12.0 \pm 17.20465}{8}$$

Hence $t = 0.6506$ m or -3.65058 m

Neglecting the negative result which is meaningless, the width of the path, $\boldsymbol{t} = \mathbf{0.651\,m}$ or **65 cm**, correct to the nearest centimetre.

Problem 14. If the total surface area of a solid cone is $486.2\,\text{cm}^2$ and its slant height is 15.3 cm, determine its base diameter

From Chapter 25, page 258, the total surface area A of a solid cone is given by: $A = \pi rl + \pi r^2$ where l is the slant height and r the base radius.

If $A = 482.2$ and $l = 15.3$, then

$$482.2 = \pi r(15.3) + \pi r^2$$

i.e. $\pi r^2 + 15.3\pi r - 482.2 = 0$

or $r^2 + 15.3r - \dfrac{482.2}{\pi} = 0$

Using the quadratic formula,

$$r=\frac{-15.3\pm\sqrt{(15.3)^2-4\left(\frac{-482.2}{\pi}\right)}}{2}$$

$$=\frac{-15.3\pm\sqrt{848.0461}}{2}$$

$$=\frac{-15.3\pm 29.12123}{2}$$

Hence radius $r=6.9106$ cm (or -22.21 cm, which is meaningless, and is thus ignored).

Thus **the diameter of the base**

$$=2r=2(6.9106)=\mathbf{13.82\,cm}$$

Problem 15. The stress, ρ, set up in a bar of length, ℓ, cross-sectional area, A, by a mass W, falling a distance h is given by the formula:

$$\rho^2-\frac{2W}{A}\rho-\frac{2WEh}{A\ell}=0$$

Given that $E=12500$, $\ell=110$, $h=0.5$, $W=2.25$ and $A=3.20$, calculate the positive value of ρ correct to 3 significant figures.

Since $\rho^2-\frac{2W}{A}\rho-\frac{2WEh}{A\ell}=0$ and $E=12500$, $\ell=110$, $h=0.5$, $W=2.25$ and $A=3.20$

then $\quad \rho^2-\frac{2(2.25)}{3.20}\rho-\frac{2(2.25)(12500)(0.5)}{(3.20)(110)}=0$

i.e. $\quad \rho^2-1.40625\rho-79.90057=0$

Solving using the quadratic formula gives:

$$\rho=\frac{--1.40625\pm\sqrt{(-1.40625)^2-4(1)(-79.90057)}}{2(1)}$$

$$=9.669 \text{ or } -8.263$$

Hence, correct to 3 significant figures, **the positive value of ρ is 9.67**

Now try the following Practice Exercise

Practice Exercise 57 Practical problems involving quadratic equations (Answers on page 707)

1. The angle a rotating shaft turns through in t seconds is given by: $\theta=\omega t+\frac{1}{2}\alpha t^2$. Determine the time taken to complete 4 radians if ω is 3.0 rad/s and α is 0.60 rad/s^2
2. The power P developed in an electrical circuit is given by $P=10I-8I^2$, where I is the current in amperes. Determine the current necessary to produce a power of 2.5 watts in the circuit
3. The sag l metres in a cable stretched between two supports, distance x m apart is given by: $l=\frac{12}{x}+x$. Determine the distance between supports when the sag is 20 m
4. The acid dissociation constant K_a of ethanoic acid is 1.8×10^{-5} mol dm^{-3} for a particular solution. Using the Ostwald dilution law $K_a=\frac{x^2}{v(1-x)}$ determine x, the degree of ionization, given that $v=10$ dm^3
5. A rectangular building is 15 m long by 11 m wide. A concrete path of constant width is laid all the way around the building. If the area of the path is 60.0 m^2, calculate its width correct to the neareast millimetre
6. The total surface area of a closed cylindrical container is 20.0 m^2. Calculate the radius of the cylinder if its height is 2.80 m
7. The bending moment M at a point in a beam is given by $M=\frac{3x(20-x)}{2}$ where x metres is the distance from the point of support. Determine the value of x when the bending moment is 50 Nm
8. A tennis court measures 24 m by 11 m. In the layout of a number of courts an area of ground must be allowed for at the ends and at the sides of each court. If a border of constant width is allowed around each court and the total area of the court and its border is 950 m^2, find the width of the borders
9. Two resistors, when connected in series, have a total resistance of 40 ohms. When connected in parallel their total resistance is 8.4 ohms. If one of the resistors has a resistance R_x ohms:
   (a) show that $R_x^2-40R_x+336=0$ and
   (b) calculated the resistance of each

10. When a ball is thrown vertically upwards its height h varies with time t according to the equation $h = 25t - 4t^2$. Determine the times, correct to 3 significant figures, when the height is 12 m.

11. In an RLC electrical circuit, reactance X is given by: $X = \omega L - \frac{1}{\omega C}$ $X = 220\Omega$, inductance $L = 800\text{mH}$ and capacitance $C = 25\mu\text{F}$. The angular velocity ω is measured in radians per second. Calculate the value of ω.

12. A point of contraflexure from the left-hand end of a 6 m beam is given by the value of x in the following equation:

$$-x^2 + 11.25x - 22.5 = 0$$

Determine the point of contraflexure.

13. The vertical height, h, and the horizontal distance travelled, x, of a projectile fired at an angle of 45° at an initial velocity, v_0, are related by the equation:

$$h = x - \frac{gx^2}{v_0^2}$$

If the projectile has an initial velocity of 120 m/s, calculate the values of x when the projectile is at a height of 200 m, assuming that $g = 9.81$ m/s^2.

11.6 The solution of linear and quadratic equations simultaneously

Sometimes a linear equation and a quadratic equation need to be solved simultaneously. An algebraic method of solution is shown in Problem 16; a graphical solution is shown in Chapter 30, page 326.

Problem 16. Determine the values of x and y which simultaneously satisfy the equations: $y = 5x - 4 - 2x^2$ and $y = 6x - 7$

For a simultaneous solution the values of y must be equal, hence the RHS of each equation is equated.

Thus $5x - 4 - 2x^2 = 6x - 7$

Rearranging gives:

$$5x^2 - 4 - 2x^2 - 6x + 7 = 0$$

i.e. $-x + 3 - 2x^2 = 0$

or $2x^2 + x - 3 = 0$

Factorising gives: $(2x + 3)(x - 1) = 0$

i.e. $x = -\frac{3}{2}$ or $x = 1$

In the equation $y = 6x - 7$

when $x = -\frac{3}{2}$, $y = 6\left(\frac{-3}{2}\right) - 7 = -16$

and when $x = 1$, $y = 6 - 7 = -1$

[Checking the result in $y = 5x - 4 - 2x^2$:

when $x = -\frac{3}{2}$, $y = 5\left(-\frac{3}{2}\right) - 4 - 2\left(-\frac{3}{2}\right)^2$

$$= -\frac{15}{2} - 4 - \frac{9}{2} = -16$$

as above; and when $x = 1$, $y = 5 - 4 - 2 = -1$ as above.]

Hence the simultaneous solutions occur when

$$\boldsymbol{x = -\frac{3}{2}, \quad y = -16}$$

and when $\boldsymbol{x = 1, \quad y = -1}$

Now try the following Practice Exercise

Practice Exercise 58 Solving linear and quadratic equations simultaneously (Answers on page 707)

In Problems 1 to 3 determine the solutions of the simulations equations.

1. $y = x^2 + x + 1$

   $y = 4 - x$

2. $y = 15x^2 + 21x - 11$

   $y = 2x - 1$

3. $2x^2 + y = 4 + 5x$

   $x + y = 4$

Practice Exercise 59 Multiple-choice questions on solving quadratic equations (Answers on page 708)

Each question has only one correct answer

1. A quadratic equation has degree:
(a) 0 (b) 1 (c) 3 (d) 2

2. Which of the following equations is not a quadratic equation?
(a) $x^2+4x-5=0$
(b) $5+2x-x^2=0$
(c) $x^2-16=0$
(d) $x^2+x^3-7=0$

3. The quadratic equation in x whose roots are -2 and $+5$ is:
(a) $x^2-3x-10=0$ (b) $x^2+7x+10=0$
(c) $x^2+3x-10=0$ (d) $x^2-7x-10=0$

4. If the equation $ax^2+bx+c=0$ has equal roots, then c is equal to;
(a) $\frac{b}{2a}$ (b) $-\frac{b^2}{2a}$
(c) $\frac{b^2}{4a}$ (d) $-\frac{b}{2a}$

5. In the equation $8x^2+13x-6=(x+p)(qx-3)$ the values of p and q are:
(a) $p=2, q=8$ (b) $p=3, q=2$
(c) $p=-2, q=4$ (d) $p=1, q=8$

6. The roots of the quadratic equation $8x^2+10x-3=0$ are:
(a) $-\frac{1}{4}$ and $\frac{3}{2}$ (b) 4 and $\frac{2}{3}$
(c) $-\frac{3}{2}$ and $\frac{1}{4}$ (d) $\frac{2}{3}$ and -4

7. For the equation $(4x^2+20x+p)$ to be a perfect square, the value of p will be:
(a) 5 (b) 25 (c) 100 (d) 10

8. The roots of the quadratic equation $2x^2-5x+1=0$, correct to 2 decimal places, are:
(a) -0.22 and -2.28 (b) 2.69 and -0.19
(c) 0.19 and -2.69 (d) 2.28 and 0.22

9. The number of real roots of the equation $x^2-x-6=0$ is:
(a) 4 (b) 3 (c) 2 (d) 1

10. The quadratic equation in x whose roots are 4 and -2 is:
(a) $x^2-2x+8=0$ (b) $x^2-2x-8=0$
(c) $x^2+2x+8=0$ (d) $x^2+2x-8=0$

11. The roots of the quadratic equation $3x^2-7x-6=0$ are:
(a) 2 and -5 (b) 2 and -9
(c) 3 and $-\frac{2}{3}$ (d) 3 and -6

12. The quadratic equation in x whose roots are 1 and 0 is:
(a) $x^2-2x+1=0$ (b) $x^2-x=0$
(c) $x^2+2x+1=0$ (d) $x^2+x=0$

13. For the equation $x^2+kx+9=0$ to be a perfect square, the value of k will be:
(a) 6 (b) -3 (c) -6 (d) 3

14. The roots of the quadratic equation $2x^2+5x+2=0$ are:
(a) 4 and -1 (b) -2 and 2.5
(c) 4 and 1 (d) -2 and -0.5

15. The roots of the quadratic equation $x^2+7x+10=0$ are:
(a) -2 and -5 (b) 2 and 5
(c) 2 and -5 (d) -2 and 5

16. The quadratic equation in x whose roots are -1 and $\frac{1}{2}$ is:
(a) $2x^2+x-1=0$ (b) $2x^2-x-1=0$
(c) $2x^2+x+1=0$ (d) $2x^2-x+1=0$

17. If one of the roots of the quadratic $2x^2+cx-6=0$ is 2, the value of c is:
(a) 2 (b) 1 (c) -2 (d) -1

18. The equation $x^2-x-5=0$ has:
(a) no roots
(b) real and different roots
(c) real and equal roots
(d) imaginary roots

19. The height S metres of a mass thrown vertically upwards at time t seconds is given by $S = 80t - 16t^2$. To reach a height of 50 metres on the descent will take the mass:
(a) 0.73 s (b) 5.56 s
(c) 4.27 s (d) 81.77 s

20. Solving the quadratic equation $x^2 - 2x - 15 = 0$ gives:
(a) $x = 3$ or -5 (b) $x = 0$ or 3
(c) $x = 0$ or -5 (d) $x = -3$ or 5

For fully worked solutions to each of the problems in Practice Exercises 54 to 58 in this chapter, go to the website:
www.routledge.com/cw/bird

COMPANION @ WEBSITE

Chapter 12

Inequalities

Why it is important to understand: Inequalities

In mathematics, an inequality is a relation that holds between two values when they are different. A working knowledge of inequalities can be beneficial to the practising engineer, and inequalities are central to the definitions of all limiting processes, including differentiation and integration. When exact solutions are unavailable, inconvenient, or unnecessary, inequalities can be used to obtain error bounds for numerical approximation. Understanding and using inequalities is important in many branches of engineering.

At the end of this chapter, you should be able to:

- define an inequality
- state simple rules for inequalities
- solve simple inequalities
- solve inequalities involving a modulus
- solve inequalities involving quotients
- solve inequalities involving square functions
- solve quadratic inequalities

12.1 Introduction in inequalities

An **inequality** is any expression involving one of the symbols $<, > \leq$ or $\geq$

$p<q$ means p is less than q
$p>q$ means p is greater than q
$p\leq q$ means p is less than or equal to q
$p\geq q$ means p is greater than or equal to q

Some simple rules

(i) When a quantity is **added or subtracted** to both sides of an inequality, the inequality still remains.

For example, if $p<3$

then $p+2<3+2$ (adding 2 to both sides)

and $p-2<3-2$ (subtracting 2 from both sides)

(ii) When **multiplying or dividing** both sides of an inequality by a **positive** quantity, say 5, the inequality **remains the same**. For example,

$$\text{if } p>4 \quad \text{then } 5p>20 \quad \text{and} \quad \frac{p}{5}>\frac{4}{5}$$

(iii) When **multiplying or dividing** both sides of an inequality by a **negative** quantity, say -3, **the inequality is reversed**. For example,

$$\text{if } p>1 \quad \text{then } -3p<-3 \quad \text{and} \quad \frac{p}{-3}<\frac{1}{-3}$$

(Note $>$ has changed to $<$ in each example.)

To **solve an inequality** means finding all the values of the variable for which the inequality is true. Knowledge of simple equations and quadratic equations are needed in this chapter.

12.2 Simple inequalities

The solution of some simple inequalities, using only the rules given in Section 12.1, is demonstrated in the following worked problems.

Problem 1. Solve the following inequalities:
(a) $3+x>7$ (b) $3t<6$
(c) $z-2\geq 5$ (d) $\frac{p}{3}\leq 2$

(a) Subtracting 3 from both sides of the inequality: $3+x>7$ gives:

$$3+x-3>7-3, \text{ i.e. } \boldsymbol{x>4}$$

Hence, all values of x greater than 4 satisfy the inequality.

(b) Dividing both sides of the inequality: $3t<6$ by 3 gives:

$$\frac{3t}{3}<\frac{6}{3}, \text{ i.e. } \boldsymbol{t<2}$$

Hence, all values of t less than 2 satisfy the inequality.

(c) Adding 2 to both sides of the inequality $z-2\geq 5$ gives:

$$z-2+2\geq 5+2, \text{ i.e. } \boldsymbol{z\geq 7}$$

Hence, all values of z greater than or equal to 7 satisfy the inequality.

(d) Multiplying both sides of the inequality $\frac{p}{3}\leq 2$ by 3 gives:

$$(3)\frac{p}{3}\leq (3)2, \text{ i.e. } \boldsymbol{p\leq 6}$$

Hence, all values of p less than or equal to 6 satisfy the inequality.

Problem 2. Solve the inequality: $4x+1>x+5$

Subtracting 1 from both sides of the inequality: $4x+1>x+5$ gives:

$$4x>x+4$$

Subtracting x from both sides of the inequality: $4x>x+4$ gives:

$$3x>4$$

Dividing both sides of the inequality: $3x>4$ by 3 gives:

$$\boldsymbol{x>\frac{4}{3}}$$

Hence all values of x greater than $\frac{4}{3}$ satisfy the inequality:

$$4x+1>x+5$$

Problem 3. Solve the inequality: $3-4t\leq 8+t$

Subtracting 3 from both sides of the inequality: $3-4t\leq 8+t$ gives:

$$-4t\leq 5+t$$

Subtracting t from both sides of the inequality: $-4t\leq 5+t$ gives:

$$-5t\leq 5$$

Dividing both sides of the inequality $-5t\leq 5$ by -5 gives:

$\boldsymbol{t\geq -1}$ (remembering to reverse the inequality)

Hence, all values of t greater than or equal to -1 satisfy the inequality.

Now try the following Practice Exercise

Practice Exercise 60 Simple inequalities (Answers on page 708)

Solve the following inequalities:

1. (a) $3t>6$ (b) $2x<10$
2. (a) $\frac{x}{2}>1.5$ (b) $x+2\geq 5$
3. (a) $4t-1\leq 3$ (b) $5-x\geq -1$
4. (a) $\frac{7-2k}{4}\leq 1$ (b) $3z+2>z+3$
5. (a) $5-2y\leq 9+y$ (b) $1-6x\leq 5+2x$

12.3 Inequalities involving a modulus

The **modulus** of a number is the size of the number, regardless of sign. Vertical lines enclosing the number denote a modulus.
For example, $|4|=4$ and $|-4|=4$ (the modulus of a number is never negative)
The inequality: $|t|<1$ means that all numbers whose actual size, regardless of sign, is less than 1, i.e. any value between -1 and $+1$
Thus $|t|<1$ **means** $-1<t<1$
Similarly, $|x|>3$ means all numbers whose actual size, regardless of sign, is greater than 3, i.e. any value greater than 3 and any value less than -3

Thus $|x|>3$ **means** $x>3$ **and** $x<-3$
Inequalities involving a modulus are demonstrated in the following worked problems.

Problem 4. Solve the following inequality:

$$|3x+1|<4$$

Since $|3x+1|<4$ then $-4<3x+1<4$

Now $-4<3x+1$ becomes $-5<3x$, i.e. $\mathbf{-\frac{5}{3}<x}$

and $3x+1<4$ becomes $3x<3$, i.e. $\mathbf{x<1}$

Hence, these two results together become $\mathbf{-\frac{5}{3}<x<1}$ and mean that the inequality $|3x+1|<4$ is satisfied for any value of x greater than $-\frac{5}{3}$ but less than 1

Problem 5. Solve the inequality: $|1+2t|\le 5$

Since $|1+2t|\le 5$ then $-5\le 1+2t\le 5$

Now $-5\le 1+2t$ becomes $-6\le 2t$, i.e. $-3\le t$

and $1+2t\le 5$ becomes $2t\le 4$ i.e. $\mathbf{t\le 2}$

Hence, these two results together become: $\mathbf{-3\le t\le 2}$

Problem 6. Solve the inequality: $|3z-4|>2$

$|3z-4|>2$ means $3z-4>2$ and $3z-4<-2$,
i.e. $3z>6$ and $3z<2$,

i.e. the inequality: $|3z-4|>2$ is satisfied when

$$\mathbf{z>2 \text{ and } z<\frac{2}{3}}$$

Now try the following Practice Exercise

Practice Exercise 61 Inequalities involving a modulus (Answers on page 708)

Solve the following inequalities:

1. $|t+1|<4$
2. $|y+3|\le 2$
3. $|2x-1|<4$
4. $|3t-5|>4$
5. $|1-k|\ge 3$

12.4 Inequalities involving quotients

If $\frac{p}{q}>0$ then $\frac{p}{q}$ must be a **positive** value.

For $\frac{p}{q}$ to be positive, **either** p is positive **and** q is positive **or** p is negative **and** q is negative.

i.e. $\frac{+}{+}=+$ and $\frac{-}{-}=+$

If $\frac{p}{q}<0$ then $\frac{p}{q}$ must be a **negative** value.

For $\frac{p}{q}$ to be negative, **either** p is positive **and** q is negative **or** p is negative **and** q is positive.

i.e. $\frac{+}{-}=-$ and $\frac{-}{+}=-$

This reasoning is used when solving inequalities involving quotients as demonstrated in the following worked problems.

Problem 7. Solve the inequality: $\frac{t+1}{3t-6}>0$

Since $\frac{t+1}{3t-6}>0$ then $\frac{t+1}{3t-6}$ must be **positive**.

For $\frac{t+1}{3t-6}$ to be positive,

either (i) $t+1>0$ **and** $3t-6>0$
or (ii) $t+1<0$ **and** $3t-6<0$

(i) If $t+1>0$ then $t>-1$ and if $3t-6>0$ then $3t>6$ and $t>2$
Both of the inequalities $t>-1$ **and** $t>2$ are only true when $t>2$,
i.e. the fraction $\frac{t+1}{3t-6}$ is positive when $\mathbf{t>2}$

(ii) If $t+1<0$ then $t<-1$ and if $3t-6<0$ then $3t<6$ and $t<2$
Both of the inequalities $t<-1$ **and** $t<2$ are only true when $t<-1$,
i.e. the fraction $\frac{t+1}{3t-6}$ is positive when $\mathbf{t<-1}$

Summarising, $\frac{t+1}{3t-6}>0$ when $\mathbf{t>2}$ **or** $\mathbf{t<-1}$

Problem 8. Solve the inequality: $\frac{2x+3}{x+2}\le 1$

Since $\frac{2x+3}{x+2}\le 1$ then $\frac{2x+3}{x+2}-1\le 0$

i.e. $\frac{2x+3}{x+2} - \frac{x+2}{x+2} \le 0$,

i.e. $\frac{2x+3-(x+2)}{x+2} \le 0$ or $\frac{x+1}{x+2} \le 0$

For $\frac{x+1}{x+2}$ to be negative or zero,

either (i) $x+1 \le 0$ **and** $x+2>0$
or (ii) $x+1 \ge 0$ **and** $x+2<0$

(i) If $x+1 \le 0$ then $x \le -1$ and if $x+2>0$ then $x>-2$. (Note that $>$ is used for the denominator, not $\ge$; a zero denominator gives a value for the fraction which is impossible to evaluate.)

Hence, the inequality $\frac{x+1}{x+2} \le 0$ is true when x is greater than -2 and less than or equal to -1, which may be written as $\mathbf{-2 < x \le -1}$

(ii) If $x+1 \ge 0$ then $x \ge -1$ and if $x+2<0$ then $x<-2$ It is not possible to satisfy both $x \ge -1$ and $x<-2$ thus no values of x satisfies (ii).

Summarising, $\frac{2x+3}{x+2} \le 1$ when $\mathbf{-2 < x \le -1}$

Now try the following Practice Exercise

Practice Exercise 62 Inequalities involving quotients (Answers on page 708)

Solve the following inequalitites:

1. $\frac{x+4}{6-2x} \ge 0$
2. $\frac{2t+4}{t-5} > 1$
3. $\frac{3z-4}{z+5} \le 2$
4. $\frac{2-x}{x+3} \ge 4$

12.5 Inequalities involving square functions

The following two general rules apply when inequalities involve square functions:

(i) **if** $x^2 > k$ **then** $x > \sqrt{k}$ **or** $x < -\sqrt{k}$ (1)

(ii) **if** $x^2 < k$ **then** $-\sqrt{k} < x < \sqrt{k}$ (2)

These rules are demonstrated in the following worked problems.

Problem 9. Solve the inequality: $t^2>9$

Since $t^2>9$ then $t^2-9>0$, i.e. $(t+3)(t-3)>0$ by factorising.
For $(t+3)(t-3)$ to be positive,

either (i) $(t+3)>0$ **and** $(t-3)>0$
or (ii) $(t+3)<0$ **and** $(t-3)<0$

(i) If $(t+3)>0$ then $t>-3$ and if $(t-3)>0$ then $t>3$
Both of these are true only when $\mathbf{t > 3}$

(ii) If $(t+3)<0$ then $t<-3$ and if $(t-3)<0$ then $t<3$
Both of these are true only when $\mathbf{t < -3}$

Summarising, $t^2>9$ when $\mathbf{t > 3}$ **or** $\mathbf{t < -3}$

This demonstrates the general rule:

if $x^2 > k$ **then** $x > \sqrt{k}$ **or** $x < -\sqrt{k}$ (1)

Problem 10. Solve the inequality: $x^2>4$

From the general rule stated above in equation (1): if $x^2>4$ then $x>\sqrt{4}$ or $x<-\sqrt{4}$

i.e. the inequality: $x^2>4$ is satisfied when $\mathbf{x>2}$ **or** $\mathbf{x<-2}$

Problem 11. Solve the inequality: $(2z+1)^2>9$

From equation (1), if $(2z+1)^2>9$ then

$$2z+1>\sqrt{9} \quad \text{or} \quad 2z+1<-\sqrt{9}$$
i.e. $$2z+1>3 \quad \text{or} \quad 2z+1<-3$$
i.e. $$2z>2 \quad \text{or} \quad 2z<-4,$$
i.e. $$\mathbf{z>1} \quad \text{or} \quad \mathbf{z<-2}$$

Problem 12. Solve the inequality: $t^2<9$

Since $t^2<9$ then $t^2-9<0$, i.e. $(t+3)(t-3)<0$ by factorising. For $(t+3)(t-3)$ to be negative,

either (i) $(t+3)>0$ **and** $(t-3)<0$
or (ii) $(t+3)<0$ **and** $(t-3)>0$

(i) If $(t+3)>0$ then $t>-3$ and if $(t-3)<0$ then $t<3$

Hence (i) is satisfied when $t > -3$ and $t < 3$ which may be written as: $\mathbf{-3 < t < 3}$

(ii) If $(t+3) < 0$ then $t < -3$ and if $(t-3) > 0$ then $t > 3$
It is not possible to satisfy both $t < -3$ and $t > 3$, thus no values of t satisfies (ii).

Summarising, $t^2 < 9$ when $\mathbf{-3 < t < 3}$ which means that all values of t between -3 and $+3$ will satisfy the inequality.

This demonstrates the general rule:

$$\textbf{if } x^2 < k \textbf{ then } -\sqrt{k} < x < \sqrt{k} \qquad (2)$$

Problem 13. Solve the inequality: $x^2 < 4$

From the general rule stated above in equation (2): if $x^2 < 4$ then $-\sqrt{4} < x < \sqrt{4}$
i.e. the inequality: $x^2 < 4$ is satisfied when:
$$\mathbf{-2 < x < 2}$$

Problem 14. Solve the inequality: $(y-3)^2 \leq 16$

From equation (2), $-\sqrt{16} \leq (y-3) \leq \sqrt{16}$

i.e. $-4 \leq (y-3) \leq 4$

from which, $3-4 \leq y \leq 4+3$,

i.e. $\mathbf{-1 \leq y \leq 7}$

Now try the following Practice Exercise

Practice Exercise 63 Inequalities involving square functions (Answers on page 708)

Solve the following inequalities:

1. $z^2 > 16$
2. $z^2 < 16$
3. $2x^2 \geq 6$
4. $3k^2 - 2 \leq 10$
5. $(t-1)^2 \leq 36$
6. $(t-1)^2 \geq 36$
7. $7 - 3y^2 \leq -5$
8. $(4k+5)^2 > 9$

12.6 Quadratic inequalities

Inequalities involving quadratic expressions are solved using either **factorisation** or **'completing the square'**. For example,

$$x^2 - 2x - 3 \text{ is factorised as } (x+1)(x-3)$$
$$\text{and} \quad 6x^2 + 7x - 5 \text{ is factorised as } (2x-1)(3x+5)$$

If a quadratic expression does not factorise, then the technique of 'completing the square' is used. In general, the procedure for $x^2 + bx + c$ is:

$$x^2 + bx + c \equiv \left(x + \frac{b}{2}\right)^2 + c - \left(\frac{b}{2}\right)^2$$

For example, $x^2 + 4x - 7$ does not factorise; completing the square gives:

$$x^2 + 4x - 7 \equiv (x+2)^2 - 7 - 2^2 \equiv (x+2)^2 - 11$$

Similarly,

$$x^2 + 6x - 5 \equiv (x+3)^2 - 5 - 3^2 \equiv (x-3)^2 - 14$$

Solving quadratic inequalities is demonstrated in the following worked problems.

Problem 15. Solve the inequality: $x^2 + 2x - 3 > 0$

Since $x^2 + 2x - 3 > 0$ then $(x-1)(x+3) > 0$ by factorising. For the product $(x-1)(x+3)$ to be positive,

either (i) $(x-1) > 0$ **and** $(x+3) > 0$
or (ii) $(x-1) < 0$ **and** $(x+3) < 0$

(i) Since $(x-1) > 0$ then $x > 1$ and since $(x+3) > 0$ then $x > -3$
Both of these inequalities are satisfied only when $\mathbf{x > 1}$

(ii) Since $(x-1) < 0$ then $x < 1$ and since $(x+3) < 0$ then $x < -3$
Both of these inequalities are satisfied only when $\mathbf{x < -3}$

Summarising, $x^2 + 2x - 3 > 0$ is satisfied when either
$$\mathbf{x > 1 \text{ or } x < -3}$$

Problem 16. Solve the inequality: $t^2 - 2t - 8 < 0$

Since $t^2 - 2t - 8 < 0$ then $(t-4)(t+2) < 0$ by factorising.

For the product $(t-4)(t+2)$ to be negative,

either (i) $(t-4)>0$ **and** $(t+2)<0$
or (ii) $(t-4)<0$ **and** $(t+2)>0$

(i) Since $(t-4)>0$ then $t>4$ and since $(t+2)<0$ then $t<-2$
It is not possible to satisfy both $t>4$ and $t<-2$, thus no values of t satisfies the inequality (i)

(ii) Since $(t-4)<0$ then $t<4$ and since $(t+2)>0$ then $t>-2$
Hence, (ii) is satisfied when $-2<t<4$

Summarising, $t^2-2t-8<0$ is satisfied when $\mathbf{-2<t<4}$

Problem 17. Solve the inequality:
$$x^2+6x+3<0$$

x^2+6x+3 does not factorise; completing the square gives:

$$x^2+6x+3\equiv(x+3)^2+3-3^2$$
$$\equiv(x+3)^2-6$$

The inequality thus becomes: $(x+3)^2-6<0$ or $(x+3)^2<6$

From equation (2), $-\sqrt{6}<(x+3)<\sqrt{6}$

from which, $(-\sqrt{6}-3)<x<(\sqrt{6}-3)$

Hence, $x^2+6x+3<0$ is satisfied when $\mathbf{-5.45<x<-0.55}$ correct to 2 decimal places.

Problem 18. Solve the inequality:
$$y^2-8y-10\geq 0$$

$y^2-8y-10$ does not factorise; completing the square gives:

$$y^2-8y-10\equiv(y-4)^2-10-4^2$$
$$\equiv(y-4)^2-26$$

The inequality thus becomes: $(y-4)^2-26\geq 0$ or $(y-4)^2\geq 26$

From equation (1), $(y-4)\geq\sqrt{26}$ or $(y-4)\leq-\sqrt{26}$

from which, $\mathbf{y\geq 4+\sqrt{26}}$ **or** $\mathbf{y\leq 4-\sqrt{26}}$

Hence, $y^2-8y-10\geq 0$ is satisfied when $\mathbf{y\geq 9.10}$ **or** $\mathbf{y\leq -1.10}$ correct to 2 decimal places.

Now try the following Practice Exercise

Practice Exercise 64 Quadratic inequalities (Answers on page 708)

Solve the following inequalities:

1. $x^2-x-6>0$
2. $t^2+2t-8\leq 0$
3. $2x^2+3x-2<0$
4. $y^2-y-20\geq 0$
5. $z^2+4z+4\leq 4$
6. $x^2+6x-6\leq 0$
7. $t^2-4t-7\geq 0$
8. $k^2+k-3\geq 0$

Practice Exercise 65 Multiple-choice questions on inequalities (Answers on page 708)

Each question has only one correct answer

1. The solution of the inequality $\dfrac{3t+2}{t+1}\leq 1$ is:
(a) $t\geq -2\frac{1}{2}$ (b) $-1<t\leq -\frac{1}{2}$
(c) $t<-1$ (d) $-\frac{1}{2}<t\leq 1$
2. The solution of the inequality $x^2-x-2<0$ is:
(a) $1<x<-2$ (b) $x>2$
(c) $-1<x<2$ (d) $x<-1$
3. The solution of the inequality $|2x-14|<6$ is:
(a) $4<x<10$ (b) $11<x<17$
(c) $-4<x<10$ (d) $12<x<14$
4. Solving the inequality $7<3+4x\leq 21$ gives:
(a) $4.4<x<5.25$ (b) $1.75<x<7$
(c) $2.33<x<17$ (d) $1<x<4.5$
5. The solution of the inequality $\dfrac{x+1}{3x-9}>0$ is:
(a) $x>3$ (b) $x>3$ or $x<-1$
(c) $x<-1$ (d) $x<-3$ or $x<1$

For fully worked solutions to each of the problems in Practice Exercises 60 to 64 in this chapter, go to the website:
www.routledge.com/cw/bird

COMPANION @ WEBSITE

Chapter 13

Logarithms

Why it is important to understand: Logarithms

All types of engineers use natural and common logarithms. Chemical engineers use them to measure radioactive decay and pH solutions, both of which are measured on a logarithmic scale. The Richter scale which measures earthquake intensity is a logarithmic scale. Biomedical engineers use logarithms to measure cell decay and growth, and also to measure light intensity for bone mineral density measurements. In electrical engineering, a dB (decibel) scale is very useful for expressing attenuations in radio propagation and circuit gains, and logarithms are used for implementing arithmetic operations in digital circuits. Logarithms are especially useful when dealing with the graphical analysis of non-linear relationships and logarithmic scales are used to linearise data to make data analysis simpler. Understanding and using logarithms is clearly important in all branches of engineering.

At the end of this chapter, you should be able to:

- define base, power, exponent and index
- define a logarithm
- distinguish between common and Napierian (i.e. hyperbolic or natural) logarithms
- evaluate logarithms to any base
- state the laws of logarithms
- simplify logarithmic expressions
- solve equations involving logarithms
- solve indicial equations
- sketch graphs of $\log_{10} x$ and $\log_e x$

13.1 Introduction to logarithms

With the use of calculators firmly established, logarithmic tables are no longer used for calculations. However, the theory of logarithms is important, for there are several scientific and engineering laws that involve the rules of logarithms.

From Chapter 5, we know that: $16 = 2^4$

The number 4 is called the **power** or the **exponent** or the **index**. In the expression 2^4, the number 2 is called the **base**.

In another example, we know that: $64 = 8^2$

In this example, 2 is the power, or exponent, or index. The number 8 is the base.

What is a logarithm?

Consider the expression $16 = 2^4$

An alternative, yet equivalent, way of writing this expression is: $\log_2 16 = 4$

This is stated as 'log to the base 2 of 16 equals 4'

We see that the logarithm is the same as the power or index in the original expression. It is the base in the original expression which becomes the base of the logarithm.

The two statements: $16 = 2^4$ and $\log_2 16 = 4$ are equivalent.

If we write either of them, we are automatically implying the other.

In general, if a number y can be written in the form a^x, then the index x is called the 'logarithm of y to the base of a',

i.e. **if $y = a^x$ then $x = \log_a y$**

In another example, if we write down that $64 = 8^2$ then the equivalent statement using logarithms is:

$$\log_8 64 = 2$$

In another example, if we write down that: $\log_3 27 = 3$ then the equivalent statement using powers is:

$$3^3 = 27$$

So the two sets of statements, one involving powers and one involving logarithms, are equivalent.

Common logarithms

From above, if we write down that: $1000 = 10^3$, then $3 = \log_{10} 1000$.

This may be checked using the 'log' button on your calculator.

Logarithms having a base of 10 are called **common logarithms** and $\log_{10}$ is usually abbreviated to lg. The following values may be checked by using a calculator:

$$\lg 27.5 = 1.4393\ldots, \lg 378.1 = 2.5776\ldots$$

and $\lg 0.0204 = -1.6903\ldots$

Napierian logarithms

Logarithms having a base of e (where 'e' is a mathematical constant approximately equal to 2.7183) are called **hyperbolic, Napierian** or **natural logarithms**, and $\log_e$ is usually abbreviated to ln.

The following values may be checked by using a calculator:

$$\ln 3.65 = 1.2947\ldots, \ln 417.3 = 6.0338\ldots$$

and $\ln 0.182 = -1.7037\ldots$

More on Napierian logarithms is explained in Chapter 14.

Here are some worked problems to help you understand logarithms.

Problem 1. Evaluate: $\log_3 9$

Let $x = \log_3 9$ then $3^x = 9$ from the definition of a logarithm,

i.e. $3^x = 3^2$ from which, $x = 2$

Hence, $\mathbf{\log_3 9 = 2}$

Problem 2. Evaluate: $\log_{10} 10$

Let $x = \log_{10} 10$ then $10^x = 10$ from the definition of a logarithm,

i.e. $10^x = 10^1$ from which, $x = 1$

Hence, $\mathbf{\log_{10} 10 = 1}$ (which may be checked by a calculator)

Problem 3. Evaluate: $\log_{16} 8$

Let $x = \log_{16} 8$ then $16^x = 8$ from the definition of a logarithm,

i.e. $(2^4)^x = 2^3$, i.e. $2^{4x} = 2^3$ from the laws of indices,

from which, $4x = 3$ and $x = \frac{3}{4}$

Hence, $\mathbf{\log_{16} 8 = \frac{3}{4}}$

Problem 4. Evaluate: lg 0.001

Let $x = \lg 0.001 = \log_{10} 0.001$ then $10^x = 0.001$

i.e. $10^x = 10^{-3}$ from which $x = -3$

Hence, $\mathbf{\lg 0.001 = -3}$ (which may be checked by a calculator)

Problem 5. Evaluate: $\ln e$

Let $x = \ln e = \log_e e$ then $e^x = e$,

i.e. $e^x = e^1$ from which $x = 1$

Hence, $\mathbf{\ln e = 1}$ (which may be checked by a calculator)

Problem 6. Evaluate: $\log_3 \frac{1}{81}$

Let $x = \log_3 \frac{1}{81}$ then $3^x = \frac{1}{81} = \frac{1}{3^4} = 3^{-4}$ from which, $x = -4$

Hence, $$\log_3 \frac{1}{81} = -4$$

Problem 7. Solve the equation: $\lg x = 3$

If $\lg x = 3$ then $\log_{10} x = 3$

and $x = 10^3$ i.e. $\boldsymbol{x = 1000}$

Problem 8. Solve the equation: $\log_2 x = 5$

If $\log_2 x = 5$ then $\boldsymbol{x = 2^5 = 32}$

Problem 9. Solve the equation: $\log_5 x = -2$

If $\log_5 x = -2$ then $x = 5^{-2} = \frac{1}{5^2} = \boldsymbol{\frac{1}{25}}$

Now try the following Practice Exercise

Practice Exercise 66 Introduction to logarithms (Answers on page 708)

In Problems 1 to 11, evaluate the given expression:

1. $\log_{10} 10000$
2. $\log_2 16$
3. $\log_5 125$
4. $\log_2 \frac{1}{8}$
5. $\log_8 2$
6. $\log_7 343$
7. $\lg 100$
8. $\lg 0.01$
9. $\log_4 8$
10. $\log_{27} 3$
11. $\ln e^2$

In Problems 12 to 18 solve the equations:

12. $\log_{10} x = 4$
13. $\lg x = 5$
14. $\log_3 x = 2$
15. $\log_4 x = -2\frac{1}{2}$
16. $\lg x = -2$
17. $\log_8 x = -\frac{4}{3}$
18. $\ln x = 3$

13.2 Laws of logarithms

There are three laws of logarithms, which apply to any base:

(i) To multiply two numbers:

$$\log (A \times B) = \log A + \log B$$

The following may be checked by using a calculator:

$$\lg 10 = 1$$

Also, $\lg 5 + \lg 2 = 0.69897\ldots + 0.301029\ldots = 1$.
Hence, $\lg (5 \times 2) = \lg 10 = \lg 5 + \lg 2$

(ii) To divide two numbers:

$$\log \left(\frac{A}{B}\right) = \log A - \log B$$

The following may be checked using a calculator:

$$\ln \left(\frac{5}{2}\right) = \ln 2.5 = 0.91629\ldots$$

Also, $\ln 5 - \ln 2 = 1.60943\ldots - 0.69314\ldots = 0.91629\ldots$

Hence, $\ln \left(\frac{5}{2}\right) = \ln 5 - \ln 2$

(iii) To raise a number to a power:

$$\log A^n = n \log A$$

The following may be checked using a calculator:

$$\lg 5^2 = \lg 25 = 1.39794\ldots$$

Also, $2 \lg 5 = 2 \times 0.69897\ldots = 1.39794\ldots$
Hence, $\lg 5^2 = 2 \lg 5$

Here are some worked problems to help you understand the laws of logarithms.

Problem 10. Write $\log 4 + \log 7$ as the logarithm of a single number

$\log 4 + \log 7 = \log (7 \times 4)$

by the first law of logarithms

$= \boldsymbol{\log 28}$

Problem 11. Write $\log 16 - \log 2$ as the logarithm of a single number

$$\log 16 - \log 2 = \left(\frac{16}{2}\right)$$

by the second law of logarithms

$$= \mathbf{\log 8}$$

Problem 12. Write 2 log 3 as the logarithm of a single number

$$2\log 3 = \log 3^2$$

by the third law of logarithms

$$= \mathbf{\log 9}$$

Problem 13. Write $\frac{1}{2}\log 25$ as the logarithm of a single number

$$\frac{1}{2}\log 25 = \log 25^{\frac{1}{2}}$$

by the third law of logarithms

$$= \log\sqrt{25} = \mathbf{\log 5}$$

Problem 14. Simplify: $\log 64 - \log 128 + \log 32$

$64 = 2^6, 128 = 2^7$ and $32 = 2^5$

Hence, $\log 64 - \log 128 + \log 32$

$$= \log 2^6 - \log 2^7 + \log 2^5$$

$$= 6\log 2 - 7\log 2 + 5\log 2$$

by the third law of logarithms

$$= \mathbf{4\log 2}$$

Problem 15. Write $\frac{1}{2}\log 16 + \frac{1}{3}\log 27 - 2\log 5$ as the logarithm of a single number

$$\frac{1}{2}\log 16 + \frac{1}{3}\log 27 - 2\log 5$$

$$= \log 16^{\frac{1}{2}} + \log 27^{\frac{1}{3}} - \log 5^2$$

by the third law of logarithms

$$= \log\sqrt{16} + \log\sqrt[3]{27} - \log 25$$

by the laws of indices

$$= \log 4 + \log 3 - \log 25$$

$$= \log\left(\frac{4 \times 3}{25}\right)$$

by the first and second laws of logarithms

$$= \log\left(\frac{12}{25}\right) = \mathbf{\log 0.48}$$

Problem 16. Evaluate:

$$\frac{\log 25 - \log 125 + \frac{1}{2}\log 625}{3\log 5}$$

$$\frac{\log 25 - \log 125 + \frac{1}{2}\log 625}{3\log 5}$$

$$= \frac{\log 5^2 - \log 5^3 + \frac{1}{2}\log 5^4}{3\log 5}$$

$$= \frac{2\log 5 - 3\log 5 + \frac{4}{2}\log 5}{3\log 5} = \frac{1\log 5}{3\log 5} = \mathbf{\frac{1}{3}}$$

Problem 17. Solve the equation: $\log(x-1) + \log(x+8) = 2\log(x+2)$

LHS $= \log(x-1) + \log(x+8) = \log(x-1)(x+8)$

from the first law of logarithms

$$= \log(x^2 + 7x - 8)$$

RHS $= 2\log(x+2) = \log(x+2)^2$

from the third law of logarithms

$$= \log(x^2 + 4x + 4)$$

Hence, $\log(x^2 + 7x - 8) = \log(x^2 + 4x + 4)$

from which, $x^2 + 7x - 8 = x^2 + 4x + 4,$

i.e. $7x - 8 = 4x + 4,$

i.e. $3x = 12$

and $\mathbf{x = 4}$

Problem 18. Solve the equation: $\frac{1}{2}\log 4 = \log x$

$$\frac{1}{2}\log 4 = \log 4^{\frac{1}{2}}$$

from the third law of logarithms

$$= \log\sqrt{4} \text{ from the laws of indices}$$

Hence, $\frac{1}{2}\log 4 = \log x$

becomes $\log\sqrt{4} = \log x$

i.e. $\log 2 = \log x$

from which, $2 = x$

i.e. the **solution of the equation is:** $\mathbf{x = 2}$

Problem 19. Solve the equation: $\log(x^2-3)-\log x=\log 2$

$\log(x^2-3)-\log x=\log\left(\frac{x^2-3}{x}\right)$ from the second law of logarithms

Hence, $\log\left(\frac{x^2-3}{x}\right)=\log 2$

from which, $\frac{x-3}{x}=2$

Rearranging gives: $x^2-3=2x$

and $x^2-2x-3=0$

Factorising gives: $(x-3)(x+1)=0$

from which, $x=3$ or $x=-1$

$x=-1$ is not a valid solution since the logarithm of a negative number has no real root.
Hence, **the solution of the equation is: $x=3$**

Now try the following Practice Exercise

Practice Exercise 67 Laws of logarithms (Answers on page 708)

In Problems 1 to 11, write as the logarithm of a single number:

1. $\log 2+\log 3$
2. $\log 3+\log 5$
3. $\log 3+\log 4-\log 6$
4. $\log 7+\log 21-\log 49$
5. $2\log 2+\log 3$
6. $2\log 2+3\log 5$
7. $2\log 5-\frac{1}{2}\log 81+\log 36$
8. $\frac{1}{3}\log 8-\frac{1}{2}\log 81+\log 27$
9. $\frac{1}{2}\log 4-2\log 3+\log 45$
10. $\frac{1}{4}\log 16+2\log 3-\log 18$
11. $2\log 2+\log 5-\log 10$

Simplify the expressions given in Problems 12 to 14:

12. $\log 27-\log 9+\log 81$
13. $\log 64+\log 32-\log 128$
14. $\log 8-\log 4+\log 32$

Evaluate the expressions given in Problems 15 and 16:

15. $\dfrac{\frac{1}{2}\log 16-\frac{1}{3}\log 8}{\log 4}$
16. $\dfrac{\log 9-\log 3+\frac{1}{2}\log 81}{2\log 3}$

Solve the equations given in Problems 17 to 22:

17. $\log x^4-\log x^3=\log 5x-\log 2x$
18. $\log 2t^3-\log t=\log 16+\log t$
19. $2\log b^2-3\log b=\log 8b-\log 4b$
20. $\log(x+1)+\log(x-1)=\log 3$
21. $\frac{1}{3}\log 27=\log(0.5a)$
22. $\log(x^2-5)-\log x=\log 4$

13.3 Indicial equations

The laws of logarithms may be used to solve certain equations involving powers — called **indicial equations**. For example, to solve, say, $3^x=27$, logarithms to base of 10 are taken of both sides,

i.e. $\log_{10}3^x=\log_{10}27$

and $x\log_{10}3=\log_{10}27$ by the third law of logarithms.

Rearranging gives $x=\frac{\log_{10}27}{\log_{10}3}$

$=\frac{1.43136\ldots}{0.4771\ldots}=3$

which may be readily checked.

(Note, $(\log 8/\log 2)$ is **not** equal to $\log(8/2)$)

Problem 20. Solve the equation $2^x = 3$, correct to 4 significant figures

Taking logarithms to base 10 of both sides of $2^x = 3$ gives:

$$\log_{10}2^x = \log_{10}3$$

i.e. $$x\log_{10}2 = \log_{10}3$$

Rearranging gives:

$$x = \frac{\log_{10}3}{\log_{10}2} = \frac{0.47712125\ldots}{0.30102999\ldots} = \mathbf{1.585}$$

correct to 4 significant figures.

Problem 21. Solve the equation $2^{x+1} = 3^{2x-5}$ correct to 2 decimal places

Taking logarithms to base 10 of both sides gives:

$$\log_{10}2^{x+1} = \log_{10}3^{2x-5}$$

i.e. $$(x+1)\log_{10}2 = (2x-5)\log_{10}3$$

$$x\log_{10}2 + \log_{10}2 = 2x\log_{10}3 - 5\log_{10}3$$

$$x(0.3010) + (0.3010) = 2x(0.4771) - 5(0.4771)$$

i.e. $$0.3010x + 0.3010 = 0.9542x - 2.3855$$

Hence $$2.3855 + 0.3010 = 0.9542x - 0.3010x$$

$$2.6865 = 0.6532x$$

from which $$\boldsymbol{x} = \frac{2.6865}{0.6532} = \mathbf{4.11}$$

correct to 2 decimal places.

Problem 22. Solve the equation $x^{3.2} = 41.15$, correct to 4 significant figures

Taking logarithms to base 10 of both sides gives:

$$\log_{10}x^{3.2} = \log_{10}41.15$$

$$3.2\log_{10}x = \log_{10}41.15$$

Hence $$\log_{10}x = \frac{\log_{10}41.15}{3.2} = 0.50449$$

Thus $\boldsymbol{x}$ = antilog $0.50449 = 10^{0.50449} = \mathbf{3.195}$ correct to 4 significant figures.

Problem 23. The velocity v of a machine is given by the formula: $v = 15^{\left(\frac{40}{1.25\,b}\right)}$

(a) Transpose the formula to make b the subject.

(b) Evaluate b, correct to 3 significant figures, when $v = 12$

(a) Taking logarithms to the base of 10 of both sides of the equation gives:

$$\log_{10}v = \log_{10}15^{\left(\frac{40}{1.25\,b}\right)}$$

From the third law of logarithms:

$$\log_{10}v = \left(\frac{40}{1.25\,b}\right)\log_{10}15$$

$$= \left(\frac{40}{1.25\,b}\right)(1.1761)$$

correct to 4 decimal places

$$= \frac{47.044}{1.25\,b}$$

from which,

$$1.25\,b\log_{10}v = 47.044$$

and $$\boldsymbol{b} = \frac{\mathbf{47.044}}{\mathbf{1.25\log_{10}}\boldsymbol{v}}$$

(b) When $v = 12$, $$b = \frac{47.044}{1.25\log_{10}12} = 34.874$$

i.e. $\mathbf{b = 34.9}$ correct to 3 significant figures.

Problem 24. A gas follows the polytropic law $PV^{1.25} = C$. Determine the new volume of the gas, given that its original pressure and volume are 101 kPa and 0.35 m^3, respectively, and its final pressure is 1.18 MPa.

If $PV^{1.25} = C$ then $P_1V_1^{1.25} = P_2V_2^{1.25}$

$P_1 = 101$ kPa, $P_2 = 1.18$ MPa and $V_1 = 0.35$ m^3

$$P_1V_1^{1.25} = P_2V_2^{1.25}$$

i.e. $$(101\times10^3)(0.35)^{1.25} = (1.18\times10^6)V_2^{1.25}$$

from which, $$V_2^{1.25} = \frac{(101\times10^3)(0.35)^{1.25}}{(1.18\times10^6)}$$

$$= 0.02304$$

Taking logarithms of both sides of the equation gives:

$$\log_{10} V_2^{1.25} = \log_{10} 0.02304$$

i.e. $$1.25 \log_{10} V_2 = \log_{10} 0.02304$$

from the third law of logarithms

and $$\log_{10} V_2 = \frac{\log_{10} 0.02304}{1.25} = -1.3100$$

from which, **volume,** $\mathbf{V_2} = 10^{-1.3100} = \mathbf{0.049\ m^3}$

Now try the following Practice Exercise

Practice Exercise 68 Indicial equations (Answers on page 708)

Solve the following indicial equations for x, each correct to 4 significant figures:

1. $3^x = 6.4$
2. $2^x = 9$
3. $2^{x-1} = 3^{2x-1}$
4. $x^{1.5} = 14.91$
5. $25.28 = 4.2^x$
6. $4^{2x-1} = 5^{x+2}$
7. $x^{-0.25} = 0.792$
8. $0.027^x = 3.26$
9. The decibel gain n of an amplifier is given by: $n = 10\log_{10}\left(\frac{P_2}{P_1}\right)$ where P_1 is the power input and P_2 is the power output. Find the power gain $\frac{P_2}{P_1}$ when $n = 25$ decibels.
10. A gas follows the polytropic law $PV^{1.26} = C$. Determine the new volume of the gas, given that its original pressure and volume are 101 kPa and 0.42 m^3, respectively, and its final pressure is 1.25 MPa.

13.4 Graphs of logarithmic functions

A graph of $y = \log_{10} x$ is shown in Fig. 13.1 and a graph of $y = \log_e x$ is shown in Fig. 13.2. Both are seen to be of similar shape; in fact, the same general shape occurs for a logarithm to any base.

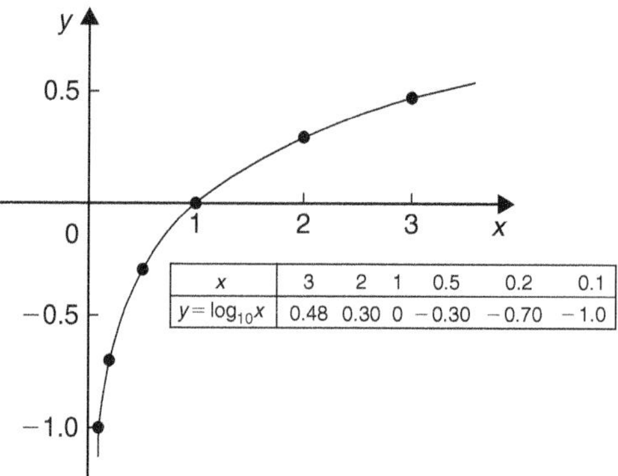

Figure 13.1

In general, with a logarithm to any base a, it is noted that:

(i) $\mathbf{\log_a 1 = 0}$

Let $\log_a = x$, then $a^x = 1$ from the definition of the logarithm.

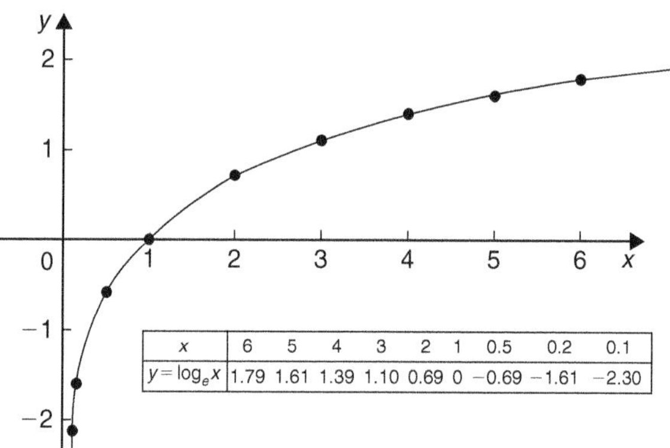

Figure 13.2

If $a^x = 1$ then $x = 0$ from the laws of logarithms.

Hence $\log_a 1 = 0$. In the above graphs it is seen that $\log_{10} 1 = 0$ and $\log_e 1 = 0$

(ii) $\mathbf{\log_a a = 1}$

Let $\log_a a = x$, then $a^x = a$, from the definition of a logarithm.

If $a^x = a$ then $x = 1$

Hence $\log_a a = 1$. (Check with a calculator that $\log_{10} 10 = 1$ and $\log_e e = 1$)

(iii) $\log_a 0 \to -\infty$

Let $\log_a 0 = x$ then $a^x = 0$ from the definition of a logarithm.

If $a^x = 0$, and a is a positive real number, then x must approach minus infinity. (For example, check with a calculator, $2^{-2} = 0.25$, $2^{-20} = 9.54 \times 10^{-7}$, $2^{-200} = 6.22 \times 10^{-61}$ and so on.)

Hence $\log_a 0 \to -\infty$

Practice Exercise 69 Multiple-choice questions on logarithms (Answers on page 708)

Each question has only one correct answer

1. $\log_{16} 8$ is equal to:
   (a) $\frac{1}{2}$ (b) 144 (c) $\frac{3}{4}$ (d) 2
2. Simplifying $(\log 2 + \log 3)$ gives:
   (a) $\log 1$ (b) $\log 5$ (c) $\log 6$ (d) $\log \frac{2}{3}$
3. The value of $\log_{10} \frac{1}{100}$ is:
   (a) −5 (b) 2 (c) 5 (d) −2
4. Solving $(\log 30 - \log 5) = \log x$ for x gives:
   (a) 150 (b) 6 (c) 35 (d) 25
5. Solving $\log_x 81 = 4$ for x gives:
   (a) 3 (b) 9 (c) 20.25 (d) 324
6. $\frac{1}{4}\log 16 - 2\log 5 + \frac{1}{3}\log 27$ is equivalent to:
   (a) $-\log 20$ (b) $\log 0.24$
   (c) $-\log 5$ (d) $\log 3$
7. Solving $\log_{10}(3x+1) = 5$ for x gives:
   (a) 300 (b) 8 (c) 33333 (d) $\frac{4}{3}$
8. Solving $5^{2-x} = \frac{1}{125}$ for x gives:
   (a) $\frac{5}{3}$ (b) 5 (c) −1 (d) $\frac{7}{3}$
9. $\log_{10} 4 + \log_{10} 25$ is equal to:
   (a) 5 (b) 4 (c) 3 (d) 2
10. Simplifying $(2\log 9)$ gives:
    (a) $\log 11$ (b) $\log 18$ (c) $\log 81$ (d) $\log \frac{2}{9}$
11. Solving the equation $\log(x+2) - \log(x-2) = \log 2$ for x gives:
    (a) 2 (b) 3 (c) 4 (d) 6
12. Writing $(3\log 1 + \frac{1}{2}\log 16)$ as a single logarithm gives:
    (a) $\log 24$ (b) $\log 18$ (c) $\log 5$ (d) $\log 4$
13. Solving $(3\log 10) = \log x$ for x gives:
    (a) 1000 (b) 30 (c) 13 (d) 0.3
14. Solving $2^x = 6$ for x, correct to 1 decimal place, gives:
    (a) 2.6 (b) 1.6 (c) 12.0 (d) 5.2
15. The information I contained in a message is given by $I = \log_2\left(\frac{1}{x}\right)$
    If $x = \frac{1}{8}$ then I is equal to:
    (a) 4 (b) 3 (c) 8 (d) 16

For fully worked solutions to each of the problems in Practice Exercises 66 to 68 in this chapter, go to the website:
www.routledge.com/cw/bird

COMPANION @ WEBSITE

Revision Test 3 Simultaneous and quadratic equations, transposition of formulae, inequalities and logarithms

This Revision Test covers the material contained in Chapters 9 to 13. *The marks for each question are shown in brackets at the end of each question.*

1. Solve the following pairs of simultaneous equations:

   (a) $7x-3y=23$
   $2x+4y=-8$

   (b) $3a-8+\frac{b}{8}=0$

   $b+\frac{a}{2}=\frac{21}{4}$ (12)

2. In an engineering process two variables x and y are related by the equation $y=ax+\frac{b}{x}$ where a and b are constants.Evaluate a and b if $y=15$ when $x=1$ and $y=13$ when $x=3$. (4)

3. Transpose the following equations:

   (a) $y=mx+c$ for m

   (b) $x=\frac{2(y-z)}{t}$ for z

   (c) $\frac{1}{R_T}=\frac{1}{R_A}+\frac{1}{R_B}$ for R_A

   (d) $x^2-y^2=3ab$ for y

   (e) $K=\frac{p-q}{1+pq}$ for q (16)

4. The passage of sound waves through walls is governed by the equation:

   $$v=\sqrt{\frac{K+\frac{4}{3}G}{\rho}}$$

   Make the shear modulus G the subject of the formula. (4)

5. Solve the following equations by factorisation:

   (a) $x^2-9=0$ (b) $2x^2-5x-3=0$ (6)

6. Determine the quadratic equation in x whose roots are 1 and -3. (3)

7. Solve the equation $4x^2-9x+3=0$ correct to 3 decimal places. (5)

8. The current i flowing through an electronic device is given by:

   $$i=0.005\,v^2+0.014\,v$$

   where v is the voltage. Calculate the values of v when $i=3\times10^{-3}$. (5)

9. Solve the following inequalities:

   (a) $2-5x\leq 9+2x$ (b) $|3+2t|\leq 6$

   (c) $\frac{x-1}{3x+5}>0$ (d) $(3t+2)^2>16$

   (e) $2x^2-x-3<0$ (14)

10. Evaluate $\log_{16}8$. (3)

11. Solve:

    (a) $\log_3 x=-2$

    (b) $\log 2x^2+\log x=\log 32-\log x$

    (c) $\log(x^2+8)-\log(2x)=\log 3$ (11)

12. Solve the following equations, each correct to 3 significant figures:

    (a) $2^x=5.5$

    (b) $3^{2t-1}=7^{t+2}$ (7)

For lecturers/instructors/teachers, fully worked solutions to each of the problems in Revision Test 3, together with a full marking scheme, are available at the website:

www.routledge.com/cw/bird

Chapter 14

Exponential functions

Why it is important to understand: Exponential functions

Exponential functions are used in engineering, physics, biology and economics. There are many quantities that grow exponentially; some examples are population, compound interest and charge in a capacitor. With exponential growth, the rate of growth increases as time increases. We also have exponential decay; some examples are radioactive decay, atmospheric pressure, Newton's law of cooling and linear expansion. Understanding and using exponential functions is important in many branches of engineering.

At the end of this chapter, you should be able to:

- evaluate exponential functions using a calculator
- state the exponential series for e^x
- plot graphs of exponential functions
- evaluate Napierian logarithms using a calculator
- solve equations involving Napierian logarithms
- appreciate the many examples of laws of growth and decay in engineering and science
- perform calculations involving the laws of growth and decay
- reduce exponential laws to linear form using log-linear graph paper

14.1 Introduction to exponential functions

An exponential function is one which contains e^x, e being a constant called the exponent and having an approximate value of 2.7183. The exponent arises from the natural laws of growth and decay and is used as a base for natural or Napierian logarithms.

The most common method of evaluating an exponential function is by using a scientific notation **calculator**. Use your calculator to check the following values:

$e^1 = 2.7182818$, correct to 8 significant figures,

$e^{-1.618} = 0.1982949$, each correct to 7 significant figures,

$e^{0.12} = 1.1275$, correct to 5 significant figures,

$e^{-1.47} = 0.22993$, correct to 5 decimal places,

$e^{-0.431} = 0.6499$, correct to 4 decimal places,

$e^{9.32} = 11159$, correct to 5 significant figures,

$e^{-2.785} = 0.0617291$, correct to 7 decimal places.

Problem 1. Evaluate the following correct to 4 decimal places, using a calculator:

$$0.0256\left(e^{5.21} - e^{2.49}\right)$$

$0.0256\left(e^{5.21} - e^{2.49}\right) = 0.0256\,(183.094058\ldots - 12.0612761\ldots)$

$= \mathbf{4.3784}$, correct to 4 decimal places.

Problem 2. Evaluate the following correct to 4 decimal places, using a calculator:

$$5\left(\frac{e^{0.25}-e^{-0.25}}{e^{0.25}+e^{-0.25}}\right)$$

$$5\left(\frac{e^{0.25}-e^{-0.25}}{e^{0.25}+e^{-0.25}}\right)$$

$$=5\left(\frac{1.28402541\ldots-0.77880078\ldots}{1.28402541\ldots+0.77880078\ldots}\right)$$

$$=5\left(\frac{0.5052246\ldots}{2.0628262\ldots}\right)$$

$= \mathbf{1.2246}$, correct to 4 decimal places.

Problem 3. The instantaneous voltage v in a capacitive circuit is related to time t by the equation: $v=Ve^{-t/CR}$ where V, C and R are constants. Determine v, correct to 4 significant figures, when $t=50\,\text{ms}$, $C=10\,\mu\text{F}$, $R=47\,\text{k}\Omega$ and $V=300$ volts

$$v=Ve^{-t/CR}=300e^{(-50\times10^{-3})/(10\times10^{-6}\times47\times10^{3})}$$

Using a calculator,

$$v=300e^{-0.1063829\ldots}=300(0.89908025\ldots)$$
$$=\mathbf{269.7\ volts}$$

Now try the following Practice Exercise

Practice Exercise 70 Evaluating exponential functions (Answers on page 708)

1. Evaluate the following, correct to 4 significant figures: (a) $e^{-1.8}$ (b) $e^{-0.78}$ (c) e^{10}
2. Evaluate the following, correct to 5 significant figures:
(a) $e^{1.629}$ (b) $e^{-2.7483}$ (c) $0.62e^{4.178}$

In Problems 3 and 4, evaluate correct to 5 decimal places:

3. (a) $\frac{1}{7}e^{3.4629}$ (b) $8.52e^{-1.2651}$ (c) $\frac{5e^{2.6921}}{3e^{1.1171}}$
4. (a) $\frac{5.6823}{e^{-2.1347}}$ (b) $\frac{e^{2.1127}-e^{-2.1127}}{2}$
(c) $\frac{4(e^{-1.7295}-1)}{e^{3.6817}}$
5. The length of a bar, l, at a temperature θ is given by $l=l_0e^{\alpha\theta}$, where l_0 and α are constants. Evaluate 1, correct to 4 significant figures, where $l_0=2.587, \theta=321.7$ and $\alpha=1.771\times10^{-4}$
6. When a chain of length $2L$ is suspended from two points, $2D$ metres apart, on the same horizontal level: $D=k\left\{\ln\left(\frac{L+\sqrt{L^2+k^2}}{k}\right)\right\}$. Evaluate D when $k=75$ m and $L=180$ m.

14.2 The power series for e^x

The value of e^x can be calculated to any required degree of accuracy since it is defined in terms of the following **power series**:

$$e^x=1+x+\frac{x^2}{2!}+\frac{x^3}{3!}+\frac{x^4}{4!}+\cdots \quad (1)$$

(where $3!=3\times2\times1$ and is called 'factorial 3')
The series is valid for all values of x.
The series is said to **converge**, i.e. if all the terms are added, an actual value for e^x (where x is a real number) is obtained. The more terms that are taken, the closer will be the value of e^x to its actual value. The value of the exponent e, correct to say 4 decimal places, may be determined by substituting $x=1$ in the power series of equation (1). Thus

$$e^1=1+1+\frac{(1)^2}{2!}+\frac{(1)^3}{3!}+\frac{(1)^4}{4!}+\frac{(1)^5}{5!}+\frac{(1)^6}{6!}+\frac{(1)^7}{7!}+\frac{(1)^8}{8!}+\cdots$$
$$=1+1+0.5+0.16667+0.04167+0.00833+0.00139+0.00020+0.00002+\cdots$$
$$=2.71828$$

i.e. $e=2.7183$ correct to 4 decimal places.

The value of $e^{0.05}$, correct to say 8 significant figures, is found by substituting $x=0.05$ in the power series for e^x. Thus

$$e^{0.05} = 1 + 0.05 + \frac{(0.05)^2}{2!} + \frac{(0.05)^3}{3!} + \frac{(0.05)^4}{4!} + \frac{(0.05)^5}{5!} + \cdots$$

$$= 1 + 0.05 + 0.00125 + 0.000020833 + 0.000000260 + 0.000000003$$

and by adding,

$$e^{0.05} = 1.0512711,$$
correct to 8 significant figures

In this example, successive terms in the series grow smaller very rapidly and it is relatively easy to determine the value of $e^{0.05}$ to a high degree of accuracy. However, when x is nearer to unity or larger than unity, a very large number of terms are required for an accurate result.

If in the series of equation (1), x is replaced by $-x$, then

$$e^{-x} = 1 + (-x) + \frac{(-x)^2}{2!} + \frac{(-x)^3}{3!} + \cdots$$

$$e^{-x} = 1 - x + \frac{x^2}{2!} - \frac{x^3}{3!} + \cdots$$

In a similar manner the power series for e^x may be used to evaluate any exponential function of the form ae^{kx}, where a and k are constants. In the series of equation (1), let x be replaced by kx. Then

$$ae^{kx} = a\left\{1 + (kx) + \frac{(kx)^2}{2!} + \frac{(kx)^3}{3!} + \cdots\right\}$$

Thus $5e^{2x} = 5\left\{1 + (2x) + \frac{(2x)^2}{2!} + \frac{(2x)^3}{3!} + \cdots\right\}$

$$= 5\left\{1 + 2x + \frac{4x^2}{2} + \frac{8x^3}{6} + \cdots\right\}$$

i.e. $5e^{2x} = 5\left\{1 + 2x + 2x^2 + \frac{4}{3}x^3 + \cdots\right\}$

Problem 4. Determine the value of $5e^{0.5}$, correct to 5 significant figures by using the power series for e^x

$$e^x = 1 + x + \frac{x^2}{2!} + \frac{x^3}{3!} + \frac{x^4}{4!} + \cdots$$

Hence $e^{0.5} = 1 + 0.5 + \frac{(0.5)^2}{(2)(1)} + \frac{(0.5)^3}{(3)(2)(1)} + \frac{(0.5)^4}{(4)(3)(2)(1)} + \frac{(0.5)^5}{(5)(4)(3)(2)(1)} + \frac{(0.5)^6}{(6)(5)(4)(3)(2)(1)}$

$$= 1 + 0.5 + 0.125 + 0.020833 + 0.0026042 + 0.0002604 + 0.0000217$$

i.e. $e^{0.5} = 1.64872$ correct to 6 significant figures

Hence $\mathbf{5e^{0.5}} = 5(1.64872) = \mathbf{8.2436}$, correct to 5 significant figures.

Problem 5. Expand $e^x(x^2 - 1)$ as far as the term in x^5

The power series for e^x is:

$$e^x = 1 + x + \frac{x^2}{2!} + \frac{x^3}{3!} + \frac{x^4}{4!} + \frac{x^5}{5!} + \cdots$$

Hence:

$$e^x(x^2 - 1) = \left(1 + x + \frac{x^2}{2!} + \frac{x^3}{3!} + \frac{x^4}{4!} + \frac{x^5}{5!} + \cdots\right)(x^2 - 1)$$

$$= \left(x^2 + x^3 + \frac{x^4}{2!} + \frac{x^5}{3!} + \cdots\right) - \left(1 + x + \frac{x^2}{2!} + \frac{x^3}{3!} + \frac{x^4}{4!} + \frac{x^5}{5!} + \cdots\right)$$

Grouping like terms gives:

$$e^x(x^2 - 1) = -1 - x + \left(x^2 - \frac{x^2}{2!}\right) + \left(x^3 - \frac{x^3}{3!}\right) + \left(\frac{x^4}{2!} - \frac{x^4}{4!}\right) + \left(\frac{x^5}{3!} - \frac{x^5}{5!}\right) + \cdots$$

$$= \mathbf{-1 - x + \frac{1}{2}x^2 + \frac{5}{6}x^3 + \frac{11}{24}x^4 + \frac{19}{120}x^5}$$

when expanded as far as the term in x^5

Now try the following Practice Exercise

Practice Exercise 71 Power series for e^x (Answers on page 709)

1. Evaluate $5.6e^{-1}$, correct to 4 decimal places, using the power series for e^x
2. Use the power series for e^x to determine, correct to 4 significant figures, (a) e^2 (b) $e^{-0.3}$ and check your result by using a calculator
3. Expand $(1-2x)e^{2x}$ as far as the term in x^4
4. Expand $(2e^{x^2})(x^{1/2})$ to six terms

14.3 Graphs of exponential functions

Values of e^x and e^{-x} obtained from a calculator, correct to 2 decimal places, over a range $x=-3$ to $x=3$, are shown in the following table.

x	−3.0	−2.5	−2.0	−1.5	−1.0	−0.5	0
e^x	0.05	0.08	0.14	0.22	0.37	0.61	1.00
e^{-x}	20.09	12.18	7.9	4.48	2.72	1.65	1.00

x	0.5	1.0	1.5	2.0	2.5	3.0
e^x	1.65	2.72	4.48	7.39	12.18	20.09
e^{-x}	0.61	0.37	0.22	0.14	0.08	0.05

Figure 14.1 shows graphs of $y=e^x$ and $y=e^{-x}$

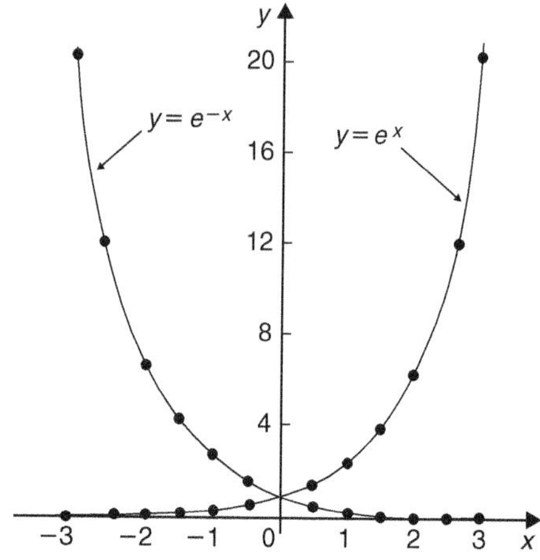

Figure 14.1

Problem 6. Plot a graph of $y=2e^{0.3x}$ over a range of $x=-2$ to $x=3$. Hence determine the value of y when $x=2.2$ and the value of x when $y=1.6$

A table of values is drawn up as shown below.

x	−3	−2	−1	0	1	2	3
$0.3x$	−0.9	−0.6	−0.3	0	0.3	0.6	0.9
$e^{0.3x}$	0.407	0.549	0.741	1.000	1.350	1.822	2.460
$2e^{0.3x}$	0.81	1.10	1.48	2.00	2.70	3.64	4.92

A graph of $y=2e^{0.3x}$ is shown plotted in Fig. 14.2.

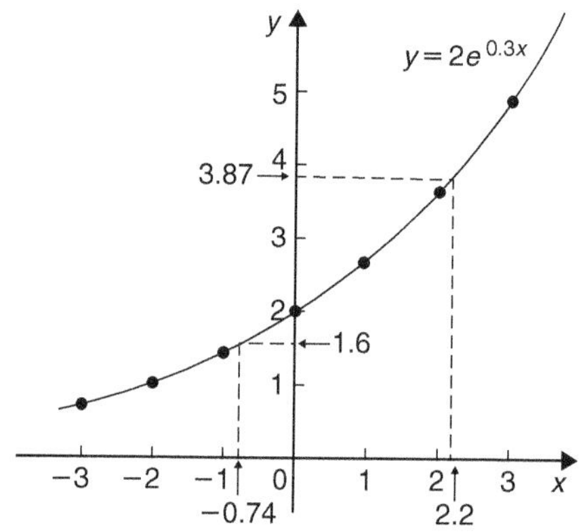

Figure 14.2

From the graph, **when $x=2.2$, $y=3.87$** and **when $y=1.6$, $x=-0.74$**

Problem 7. Plot a graph of $y=\frac{1}{3}e^{-2x}$ over the range $x=-1.5$ to $x=1.5$. Determine from the graph the value of y when $x=-1.2$ and the value of x when $y=1.4$

A table of values is drawn up as shown below.

x	−1.5	−1.0	−0.5	0	0.5	1.0	1.5
$-2x$	3	2	1	0	−1	−2	−3
e^{-2x}	20.086	7.389	2.718	1.00	0.368	0.135	0.050
$\frac{1}{3}e^{-2x}$	6.70	2.46	0.91	0.33	0.12	0.05	0.02

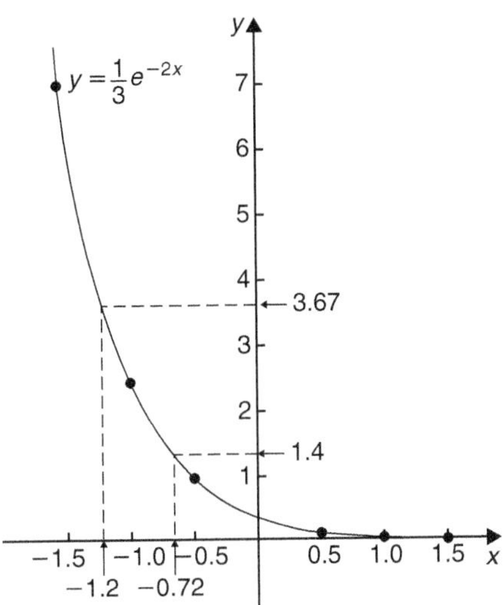

Figure 14.3

A graph of $\frac{1}{3}e^{-2x}$ is shown in Fig. 14.3.
From the graph, **when $x=-1.2$, $y=3.67$** and **when $y=1.4$, $x=-0.72$**

Problem 8. The decay of voltage, v volts, across a capacitor at time t seconds is given by $v=250e^{-t/3}$. Draw a graph showing the natural decay curve over the first 6 seconds. From the graph, find (a) the voltage after 3.4 s, and (b) the time when the voltage is 150 V

A table of values is drawn up as shown below.

t	0	1	2	3
$e^{-t/3}$	1.00	0.7165	0.5134	0.3679
$v=250e^{-t/3}$	250.0	179.1	128.4	91.97

t	4	5	6
$e^{-t/3}$	0.2636	0.1889	0.1353
$v=250e^{-t/3}$	65.90	47.22	33.83

The natural decay curve of $v=250e^{-t/3}$ is shown in Fig. 14.4.
From the graph:

(a) **when time $t=3.4$ s, voltage $v=80$ volts** and

(b) **when voltage $v=150$ volts, time $t=1.5$ seconds**.

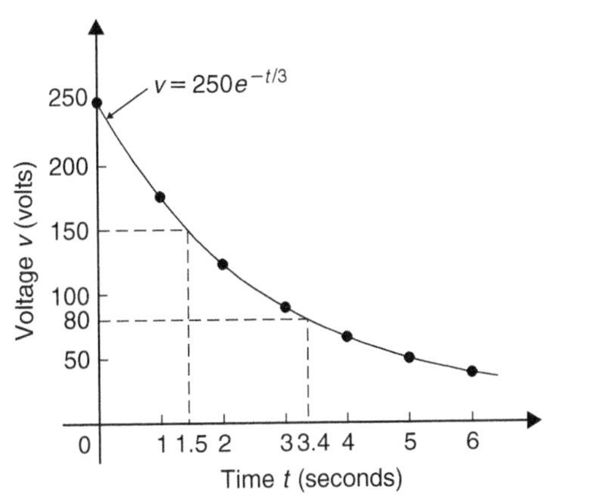

Figure 14.4

Now try the following Practice Exercise

Practice Exercise 72 Exponential graphs (Answers on page 709)

1. Plot a graph of $y=3e^{0.2x}$ over the range $x=-3$ to $x=3$. Hence determine the value of y when $x=1.4$ and the value of x when $y=4.5$
2. Plot a graph of $y=\frac{1}{2}e^{-1.5x}$ over a range $x=-1.5$ to $x=1.5$ and hence determine the value of y when $x=-0.8$ and the value of x when $y=3.5$
3. In a chemical reaction the amount of starting material C cm^3 left after t minutes is given by $C=40e^{-0.006t}$. Plot a graph of C against t and determine (a) the concentration C after 1 hour, and (b) the time taken for the concentration to decrease by half
4. The rate at which a body cools is given by $\theta=250e^{-0.05t}$ where the excess of temperature of a body above its surroundings at time t minutes is $\theta°$ C. Plot a graph showing the natural decay curve for the first hour of cooling. Hence determine (a) the temperature after 25 minutes, and (b) the time when the temperature is 195°C

14.4 Napierian logarithms

Logarithms having a base of 'e' are called **hyperbolic, Napierian** or **natural logarithms** and the Napierian logarithm of x is written as $\log_e x$, or more commonly as $\ln x$. Logarithms were invented by **John Napier***, a Scotsman (1550–1617).

The most common method of evaluating a Napierian logarithm is by a scientific notation **calculator**. Use your calculator to check the following values:

$\ln 4.328 = 1.46510554\ldots$

$= 1.4651$, correct to 4 decimal places

$\ln 1.812 = 0.59443$, correct to 5 significant figures

$\ln 1 = 0$

$\ln 527 = 6.2672$, correct to 5 significant figures

$\ln 0.17 = -1.772$, correct to 4 significant figures

$\ln 0.00042 = -7.77526$, correct to 6 significant figures

$\ln e^3 = 3$

$\ln e^1 = 1$

From the last two examples we can conclude that:

$$\log_e e^x = x$$

This is useful when **solving equations involving exponential functions**. For example, to solve $e^{3x} = 7$, take Napierian logarithms of both sides, which gives:

$$\ln e^{3x} = \ln 7$$

i.e. $3x = \ln 7$

from which $\boldsymbol{x} = \frac{1}{3}\ln 7 = \mathbf{0.6486}$, correct to 4 decimal places.

*Who was Napier? – **John Napier of Merchiston** (1550–4 April 1617) is best known as the discoverer of logarithms. To find out more go to **www.routledge.com/cw/bird**

Problem 9. Evaluate the following, each correct to 5 significant figures:

(a) $\frac{1}{2}\ln 4.7291$ (b) $\frac{\ln 7.8693}{7.8693}$ (c) $\frac{3.17\ln 24.07}{e^{-0.1762}}$

(a) $\frac{1}{2}\ln 4.7291 = \frac{1}{2}(1.5537349\ldots) = \mathbf{0.77687}$, correct to 5 significant figures

(b) $\frac{\ln 7.8693}{7.8693} = \frac{2.06296911\ldots}{7.8693} = \mathbf{0.26215}$, correct to 5 significant figures

(c) $\frac{3.17\ln 24.07}{e^{-0.1762}} = \frac{3.17(3.18096625\ldots)}{0.83845027\ldots}$

$= \mathbf{12.027}$, correct to 5 significant figures.

Problem 10. Evaluate the following:

(a) $\frac{\ln e^{2.5}}{\lg 10^{0.5}}$ (b) $\frac{5e^{2.23}\lg 2.23}{\ln 2.23}$ (correct to 3 decimal places)

(a) $\frac{\ln e^{2.5}}{\lg 10^{0.5}} = \frac{2.5}{0.5} = \mathbf{5}$

(b) $\frac{5e^{2.23}\lg 2.23}{\ln 2.23}$

$= \frac{5(9.29986607\ldots)(0.34830486\ldots)}{0.80200158\ldots}$

$= \mathbf{20.194}$, correct to 3 decimal places.

Problem 11. Solve the equation: $9 = 4e^{-3x}$ to find x, correct to 4 significant figures

Rearranging $9 = 4e^{-3x}$ gives: $\frac{9}{4} = e^{-3x}$

Taking the reciprocal of both sides gives:

$$\frac{4}{9} = \frac{1}{e^{-3x}} = e^{3x}$$

Taking Napierian logarithms of both sides gives:

$$\ln\left(\frac{4}{9}\right) = \ln(e^{3x})$$

Since $\log_e e^{\alpha} = \alpha$, then $\ln\left(\frac{4}{9}\right) = 3x$

Hence, $\boldsymbol{x} = \frac{1}{3}\ln\left(\frac{4}{9}\right) = \frac{1}{3}(-0.81093) = \mathbf{-0.2703}$, correct to 4 significant figures.

Problem 12. Given $32 = 70\left(1 - e^{-\frac{t}{2}}\right)$ determine the value of t, correct to 3 significant figures

Rearranging $32 = 70(1 - e^{-\frac{t}{2}})$ gives:

$$\frac{32}{70} = 1 - e^{-\frac{t}{2}}$$

and $$e^{-\frac{t}{2}} = 1 - \frac{32}{70} = \frac{38}{70}$$

Taking the reciprocal of both sides gives:

$$e^{\frac{t}{2}} = \frac{70}{38}$$

Taking Napierian logarithms of both sides gives:

$$\ln e^{\frac{t}{2}} = \ln\left(\frac{70}{38}\right)$$

i.e. $$\frac{t}{2} = \ln\left(\frac{70}{38}\right)$$

from which, $\boldsymbol{t} = 2\ln\left(\frac{70}{38}\right) = \mathbf{1.22}$, correct to 3 significant figures.

Problem 13. Solve the equation:

$$2.68 = \ln\left(\frac{4.87}{x}\right)$$

to find x

From the definition of a logarithm, since

$2.68 = \ln\left(\frac{4.87}{x}\right)$ then $e^{2.68} = \frac{4.87}{x}$

Rearranging gives: $x = \frac{4.87}{e^{2.68}} = 4.87e^{-2.68}$

i.e. $\boldsymbol{x} = \mathbf{0.3339}$, correct to 4 significant figures.

Problem 14. Solve $\frac{7}{4} = e^{3x}$ correct to 4 significant figures

Take natural logs of both sides gives:

$$\ln\frac{7}{4} = \ln e^{3x}$$

$$\ln\frac{7}{4} = 3x\ln e$$

Since $\ln e = 1$ $$\ln\frac{7}{4} = 3x$$

i.e. $0.55962 = 3x$

i.e. $\boldsymbol{x} = \mathbf{0.1865}$, correct to 4 significant figures.

Problem 15. Solve: $e^{x-1} = 2e^{3x-4}$ correct to 4 significant figures

Taking natural logarithms of both sides gives:

$$\ln\left(e^{x-1}\right) = \ln\left(2e^{3x-4}\right)$$

and by the first law of logarithms,

$$\ln\left(e^{x-1}\right) = \ln 2 + \ln\left(e^{3x-4}\right)$$

i.e. $x - 1 = \ln 2 + 3x - 4$

Rearranging gives: $4 - 1 - \ln 2 = 3x - x$

i.e. $3 - \ln 2 = 2x$

from which, $\boldsymbol{x} = \frac{3 - \ln 2}{2}$

$= \mathbf{1.153}$

Problem 16. Solve, correct to 4 significant figures: $\ln(x-2)^2 = \ln(x-2) - \ln(x+3) + 1.6$

Rearranging gives:

$$\ln(x-2)^2 - \ln(x-2) + \ln(x+3) = 1.6$$

and by the laws of logarithms,

$$\ln\left\{\frac{(x-2)^2(x+3)}{(x-2)}\right\} = 1.6$$

Cancelling gives: $\ln\{(x-2)(x+3)\} = 1.6$

and $(x-2)(x+3) = e^{1.6}$

i.e. $x^2 + x - 6 = e^{1.6}$

or $x^2 + x - 6 - e^{1.6} = 0$

i.e. $x^2 + x - 10.953 = 0$

Using the quadratic formula,

$$x = \frac{-1 \pm \sqrt{1^2 - 4(1)(-10.953)}}{2}$$

$$= \frac{-1 \pm \sqrt{44.812}}{2} = \frac{-1 \pm 6.6942}{2}$$

i.e. $x = 2.847$ or -3.8471

$x = -3.8471$ is not valid since the logarithm of a negative number has no real root.

Hence, **the solution of the equation is: $x = 2.847$**

Now try the following Practice Exercise

Practice Exercise 73 Napierian logarithms (Answers on page 709)

In Problems 1 and 2, evaluate correct to 5 significant figures:

1. (a) $\frac{1}{3}\ln 5.2932$ (b) $\frac{\ln 82.473}{4.829}$

   (c) $\frac{5.62 \ln 321.62}{e^{1.2942}}$

2. (a) $\frac{1.786 \ln e^{1.76}}{\lg 10^{1.41}}$ (b) $\frac{5e^{-0.1629}}{2 \ln 0.00165}$

   (c) $\frac{\ln 4.8629 - \ln 2.4711}{5.173}$

In Problems 3 to 6 solve the given equations, each correct to 4 significant figures.

3. $\ln x = 2.10$
4. $24 + e^{2x} = 45$
5. $5 = e^{x+1} - 7$
6. $1.5 = 4e^{2t}$
7. $7.83 = 2.91e^{-1.7x}$
8. $16 = 24\left(1 - e^{-\frac{t}{2}}\right)$
9. $5.17 = \ln\left(\frac{x}{4.64}\right)$
10. $3.72 \ln\left(\frac{1.59}{x}\right) = 2.43$
11. $5 = 8\left(1 - e^{\frac{-x}{2}}\right)$
12. $\ln(x+3) - \ln x = \ln(x-1)$
13. $\ln(x-1)^2 - \ln 3 = \ln(x-1)$
14. $\ln(x+3) + 2 = 12 - \ln(x-2)$
15. $e^{(x+1)} = 3e^{(2x-5)}$
16. $\ln(x+1)^2 = 1.5 - \ln(x-2) + \ln(x+1)$
17. Transpose: $b = \ln t - a \ln D$ to make t the subject.
18. If $\frac{P}{Q} = 10 \log_{10}\left(\frac{R_1}{R_2}\right)$ find the value of R_1 when $P = 160$, $Q = 8$ and $R_2 = 5$
19. If $U_2 = U_1 e^{\left(\frac{W}{PV}\right)}$ make W the subject of the formula.
20. The work done in an isothermal expansion of a gas from pressure p_1 to p_2 is given by:

    $$w = w_0 \ln\left(\frac{p_1}{p_2}\right)$$

    If the initial pressure $p_1 = 7.0$ kPa, calculate the final pressure p_2 if $w = 3w_0$
21. The velocity v_2 of a rocket is given by: $v_2 = v_1 + C \ln\left(\frac{m_1}{m_2}\right)$ where v_1 is the initial rocket velocity, C is the velocity of the jet exhaust gases, m_1 is the mass of the rocket before the jet engine is fired, and m_2 is the mass of the rocket after the jet engine is switched off. Calculate the velocity of the rocket given $v_1 = 600$ m/s, $C = 3500$ m/s, $m_1 = 8.50 \times 10^4$ kg and $m_2 = 7.60 \times 10^4$ kg.
22. A steel bar is cooled with running water. Its temperature, θ, in degrees Celsius, is given by: $\theta = 17 + 1250e^{-0.17t}$ where t is the time in minutes. Determine the time taken, correct to the nearest minute, for the temperature to fall to 35°C.

14.5 Laws of growth and decay

The laws of exponential growth and decay are of the form $y = Ae^{-kx}$ and $y = A(1 - e^{-kx})$, where A and k are constants. When plotted, the form of each of these equations is as shown in Fig. 14.5. The laws occur frequently in engineering and science and examples of quantities related by a natural law include:

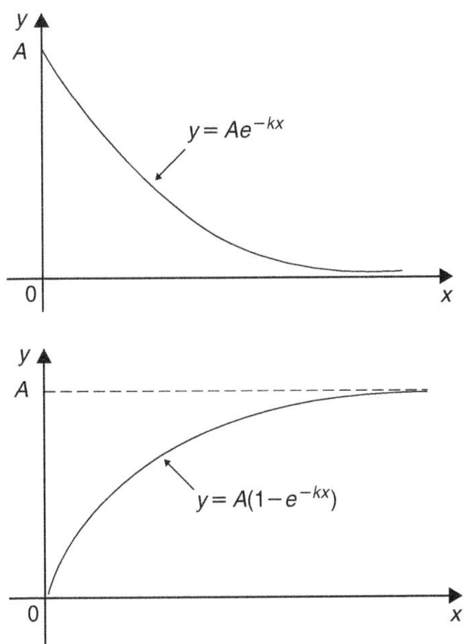

Figure 14.5

(i)	Linear expansion	$l = l_0 e^{\alpha\theta}$
(ii)	Change in electrical resistance with temperature	$R_\theta = R_0 e^{\alpha\theta}$
(iii)	Tension in belts	$T_1 = T_0 e^{\mu\theta}$
(iv)	Newton's law of cooling	$\theta = \theta_0 e^{-kt}$
(v)	Biological growth	$y = y_0 e^{kt}$
(vi)	Discharge of a capacitor	$q = Qe^{-t/CR}$
(vii)	Atmospheric pressure	$p = p_0 e^{-h/c}$
(viii)	Radioactive decay	$N = N_0 e^{-\lambda t}$
(ix)	Decay of current in an inductive circuit	$i = Ie^{-Rt/L}$
(x)	Growth of current in a capacitive circuit	$i = I(1 - e^{-t/CR})$

Problem 17. The resistance R of an electrical conductor at temperature θ°C is given by $R = R_0 e^{\alpha\theta}$, where α is a constant and $R_0 = 5 \times 10^3$ ohms. Determine the value of α, correct to 4 significant figures, when $R = 6 \times 10^3$ ohms and $\theta = 1500$°C. Also, find the temperature, correct to the nearest degree, when the resistance R is 5.4×10^3 ohms

Transposing $R = R_0 e^{\alpha\theta}$ gives $\dfrac{R}{R_0} = e^{\alpha\theta}$

Taking Napierian logarithms of both sides gives:

$$\ln \frac{R}{R_0} = \ln e^{\alpha\theta} = \alpha\theta$$

Hence

$$\alpha = \frac{1}{\theta} \ln \frac{R}{R_0} = \frac{1}{1500} \ln \left(\frac{6 \times 10^3}{5 \times 10^3} \right)$$
$$= \frac{1}{1500}(0.1823215\ldots)$$
$$= 1.215477\ldots \times 10^{-4}$$

Hence $\boldsymbol{\alpha = 1.215 \times 10^{-4}}$, correct to 4 significant figures.

From above, $\ln \dfrac{R}{R_0} = \alpha\theta$ hence $\theta = \dfrac{1}{\alpha} \ln \dfrac{R}{R_0}$

When $R = 5.4 \times 10^3$, $\alpha = 1.215477\ldots \times 10^{-4}$ and $R_0 = 5 \times 10^3$

$$\theta = \frac{1}{1.215477\ldots \times 10^{-4}} \ln \left(\frac{5.4 \times 10^3}{5 \times 10^3} \right)$$
$$= \frac{10^4}{1.215477\ldots}(7.696104\ldots \times 10^{-2})$$
$$= \mathbf{633°C} \quad \text{correct to the nearest degree.}$$

Problem 18. In an experiment involving Newton's law of cooling, the temperature θ(°C) is given by $\theta = \theta_0 e^{-kt}$. Find the value of constant k when $\theta_0 = 56.6$°C, $\theta = 16.5$°C and $t = 83.0$ seconds

Transposing $\theta = \theta_0 e^{-kt}$ gives $\dfrac{\theta}{\theta_0} = e^{-kt}$ from which

$$\frac{\theta_0}{\theta} = \frac{1}{e^{-kt}} = e^{kt}$$

Taking Napierian logarithms of both sides gives:

$$\ln \frac{\theta_0}{\theta} = kt$$

from which,

$$k = \frac{1}{t} \ln \frac{\theta_0}{\theta} = \frac{1}{83.0} \ln \left(\frac{56.6}{16.5} \right)$$
$$= \frac{1}{83.0}(1.2326486\ldots)$$

Hence $\boldsymbol{k = 1.485 \times 10^{-2}}$

Problem 19. The current i amperes flowing in a capacitor at time t seconds is given by $i=8.0(1-e^{-t/CR})$, where the circuit resistance R is 25×10^3 ohms and capacitance C is 16×10^{-6} farads. Determine (a) the current i after 0.5 seconds and (b) the time, to the nearest millisecond, for the current to reach 6.0 A. Sketch the graph of current against time

(a) Current $i=8.0(1-e^{-t/CR})$

$$=8.0[1-e^{0.5/(16\times10^{-6})(25\times10^3)}]$$
$$=8.0(1-e^{-1.25})$$
$$=8.0(1-0.2865047\ldots)$$
$$=8.0(0.7134952\ldots)$$
$$=\mathbf{5.71\ amperes}$$

(b) Transposing $i=8.0(1-e^{-t/CR})$ gives:

$$\frac{i}{8.0}=1-e^{-t/CR}$$

from which, $e^{-t/CR}=1-\frac{i}{8.0}=\frac{8.0-i}{8.0}$

Taking the reciprocal of both sides gives:

$$e^{t/CR}=\frac{8.0}{8.0-i}$$

Taking Napierian logarithms of both sides gives:

$$\frac{t}{CR}=\ln\left(\frac{8.0}{8.0-i}\right)$$

Hence

$$t=CR\ln\left(\frac{8.0}{8.0-i}\right)$$
$$=(16\times10^{-6})(25\times10^3)\ln\left(\frac{8.0}{8.0-6.0}\right)$$

when $i=6.0$ amperes,

i.e. $\boldsymbol{t}=\frac{400}{10^3}\ln\left(\frac{8.0}{2.0}\right)=0.4\ln 4.0$

$$=0.4(1.3862943\ldots)=0.5545\text{ s}$$

$=\mathbf{555\ ms}$, to the nearest millisecond

A graph of current against time is shown in Fig. 14.6.

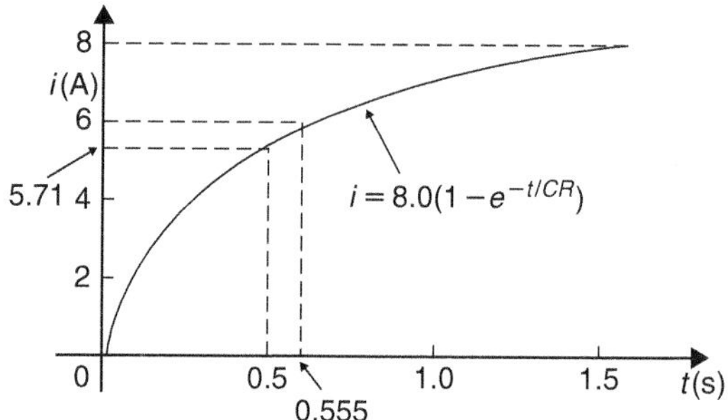

Figure 14.6

Problem 20. The temperature θ_2 of a winding which is being heated electrically at time t is given by: $\theta_2=\theta_1(1-e^{-t/\tau})$ where θ_1 is the temperature (in degrees Celsius) at time $t=0$ and τ is a constant. Calculate:

(a) θ_1, correct to the nearest degree, when θ_2 is 50°C, t is 30 s and τ is 60 s

(b) the time t, correct to 1 decimal place, for θ_2 to be half the value of θ_1

(a) Transposing the formula to make θ_1 the subject gives:

$$\theta_1=\frac{\theta_2}{(1-e^{-t/\tau})}=\frac{50}{1-e^{-30/60}}$$
$$=\frac{50}{1-e^{-0.5}}=\frac{50}{0.393469\ldots}$$

i.e. $\boldsymbol{\theta_1=127°}$**C**, correct to the nearest degree

(b) Transposing to make t the subject of the formula gives:

$$\frac{\theta_2}{\theta_1}=1-e^{-t/\tau}$$

from which, $\quad e^{-t/\tau}=1-\frac{\theta_2}{\theta_1}$

Hence $\quad -\frac{t}{\tau}=\ln\left(1-\frac{\theta_2}{\theta_1}\right)$

i.e. $\quad t=-\tau\ln\left(1-\frac{\theta_2}{\theta_1}\right)$

Since $\quad \theta_2=\frac{1}{2}\theta_1$

$$t=-60\ln\left(1-\frac{1}{2}\right)=-60\ln 0.5$$
$$=41.59\text{ s}$$

Hence the time for the temperature θ_2 to be one half of the value of θ_1 is 41.6 s, correct to 1 decimal place.

Problem 21. The power, P, developed in a flat belt drive system is given by the equation: $P = Tv(1 - e^{-\mu\theta})$ where T is the belt tension, v is the linear velocity, θ is the angle of lap and μ is the coefficient of friction. (a) Transpose the equation for μ (b) Evaluate μ, correct to 3 significant figures, given that $P = 2750$, $T = 1300$, $v = 3$ and $\theta = 2.95$

(a) If $P = Tv(1 - e^{-\mu\theta})$ then $\frac{P}{Tv} = (1 - e^{-\mu\theta})$

and $e^{-\mu\theta} = 1 - \frac{P}{Tv}$

Taking logs of both sides gives:

$$\ln(e^{-\mu\theta}) = \ln\left(1 - \frac{P}{Tv}\right)$$

from which, $-\mu\theta = \ln\left(1 - \frac{P}{Tv}\right)$

and $\boldsymbol{\mu = \frac{1}{-\theta}\ln\left(1 - \frac{P}{Tv}\right)}$

$$\text{or } \boldsymbol{\mu = -\frac{1}{\theta}\ln\left(1 - \frac{P}{Tv}\right)}$$

(b) When $P = 2750$, $T = 1300$, $v = 3$ and $\theta = 2.95$, then

$$\mu = -\frac{1}{\theta}\ln\left(1 - \frac{P}{Tv}\right) = -\frac{1}{2.95}\ln\left(1 - \frac{2750}{1300 \times 3}\right)$$

i.e. $\boldsymbol{\mu = 0.414}$**, correct to 3 significant figures.**

Now try the following Practice Exercise

Practice Exercise 74 The laws of growth and decay (Answers on page 709)

(Answers on page 709)

1. The temperature, T°C, of a cooling object varies with time, t minutes, according to the equation: $T = 150e^{-0.04t}$. Determine the temperature when (a) $t = 0$, (b) $t = 10$ minutes.

2. The pressure p pascals at height h metres above ground level is given by $p = p_0 e^{-h/C}$, where p_0 is the pressure at ground level and C is a constant. Find pressure p when $p_0 = 1.012 \times 10^5$ Pa, height $h = 1420$ m and $C = 71500$

3. The voltage drop, v volts, across an inductor L henrys at time t seconds is given by $v = 200e^{-Rt/L}$, where $R = 150\,\Omega$ and $L = 12.5 \times 10^{-3}$ H. Determine: (a) the voltage when $t = 160 \times 10^{-6}$ s, and (b) the time for the voltage to reach 85 V.

4. The length l metres of a metal bar at temperature t°C is given by $l = l_0 e^{\alpha t}$, when l_0 and α are constants. Determine: (a) the value of l when $l_0 = 1.894$, $\alpha = 2.038 \times 10^{-4}$ and $t = 250$°C, and (b) the value of l_0 when $l = 2.416$, $t = 310$°C and $\alpha = 1.682 \times 10^{-4}$

5. The temperature θ_2°C of an electrical conductor at time t seconds is given by $\theta_2 = \theta_1(1 - e^{-t/T})$, when θ_1 is the initial temperature and T seconds is a constant. Determine (a) θ_2 when $\theta_1 = 159.9$°C, $t = 30$ s and $T = 80$ s, and (b) the time t for θ_2 to fall to half the value of θ_1 if T remains at 80 s

6. A belt is in contact with a pulley for a sector of $\theta = 1.12$ radians and the coefficient of friction between these two surfaces is $\mu = 0.26$. Determine the tension on the taut side of the belt, T newtons, when tension on the slack side is given by $T_0 = 22.7$ newtons, given that these quantities are related by the law $T = T_0 e^{\mu\theta}$. Determine also the value of θ when $T = 28.0$ newtons

7. The instantaneous current i at time t is given by:
$$i = 10e^{-t/CR}$$
when a capacitor is being charged. The capacitance C is 7×10^{-6} farads and the resistance R is 0.3×10^6 ohms. Determine:
(a) the instantaneous current when t is 2.5 seconds, and
(b) the time for the instantaneous current to fall to 5 amperes
Sketch a curve of current against time from $t = 0$ to $t = 6$ seconds

8. The amount of product x (in mol/cm^3) found in a chemical reaction starting with 2.5 mol/cm^3 of reactant is given by $x = 2.5(1 - e^{-4t})$ where t is the time, in minutes, to form product x. Plot a graph at 30 second intervals up to 2.5 minutes and determine x after 1 minute

9. The current i flowing in a capacitor at time t is given by:
$$i = 12.5(1 - e^{-t/CR})$$

where resistance R is 30 kilohms and the capacitance C is 20 microfarads. Determine:

(a) the current flowing after 0.5 seconds, and

(b) the time for the current to reach 10 amperes

10. The amount A after n years of a sum invested P is given by the compound interest law: $A = Pe^{rn/100}$ when the per unit interest rate r is added continuously. Determine, correct to the nearest pound, the amount after 8 years for a sum of £1500 invested if the interest rate is 2% per annum

11. The percentage concentration C of the starting material in a chemical reaction varies with time t according to the equation $C = 100e^{-0.004t}$. Determine the concentration when (a) $t = 0$, (b) $t = 100$ s, (c) $t = 1000$ s.

12. The current i flowing through a diode at room temperature is given by: $i = i_S(e^{40V} - 1)$ amperes. Calculate the current flowing in a silicon diode when the reverse saturation current $i_S = 50$ nA and the forward voltage $V = 0.27$ V

13. A formula for chemical decomposition is given by: $C = A\left(1 - e^{-\frac{t}{10}}\right)$ where t is the time in seconds. Calculate the time, in milliseconds, for a compound to decompose to a value of $C = 0.12$ given $A = 8.5$

14. The mass, m, of pollutant in a water reservoir decreases according to the law $m = m_0 e^{-0.1t}$ where t is the time in days and m_0 is the initial mass. Calculate the percentage decrease in the mass after 60 days, correct to 3 decimal places.

15. A metal bar is cooled with water. Its temperature, in °C, is given by: $\theta = 15 + 1300e^{-0.2t}$ where t is the time in minutes. Calculate how long it will take for the temperature, θ, to decrease to 36°C, correct to the nearest second.

Practice Exercise 75 Multiple-choice questions on exponential functions (Answers on page 709)

Each question has only one correct answer

1. The value of $\frac{\ln 2}{e^2 \lg 2}$, correct to 3 significant figures, is:
(a) 0.0588 (b) 0.312 (c) 17.0 (d) 3.209

2. The value of $\frac{3.67 \ln 21.28}{e^{-0.189}}$, correct to 4 significant figures, is:
(a) 9.289 (b) 13.56
(c) 13.5566 (d) -3.844×10^9

3. A voltage, v, is given by $v = Ve^{-\frac{t}{CR}}$ volts. When $t = 30$ ms, $C = 150$ nF, $R = 50$ MΩ and $V = 250$ V, the value of v, correct to 3 significant figures, is:
(a) 249 V (b) 250 V (c) 4.58 V (d) 0 V

4. The value of $\log_x x^{2k}$ is:
(a) k (b) k^2 (c) $2k$ (d) $2kx$

5. In the equation $5.0 = 3.0 \ln\left(\frac{2.9}{x}\right)$, x has a value correct to 3 significant figures of:
(a) 1.59 (b) 0.392 (c) 0.548 (d) 0.0625

6. Solving the equation $7 + e^{2k-4} = 8$ for k gives:
(a) 2 (b) 3 (c) 4 (d) 5

7. Given that $2e^{x+1} = e^{2x-5}$, the value of x is:
(a) 1 (b) 6.693 (c) 3 (d) 2.231

8. The current i amperes flowing in a capacitor at time t seconds is given by $i = 10(1 - e^{-t/CR})$, where resistance R is 25×10^3 ohms and capacitance C is 16×10^{-6} farads. When current i reaches 7 amperes, the time t is:
(a) -0.48 s (b) 0.14 s
(c) 0.21 s (d) 0.48 s

9. Solving $13 + 2e^{3x} = 31$ for x, correct to 4 significant figures, gives:
(a) 0.4817 (b) 0.09060
(c) 0.3181 (d) 0.7324

10. Chemical decomposition, C, is given by $C = A(1 - e^{-\frac{t}{15}})$ where t is the time in seconds. If $A = 9$, the time, correct to 3 decimal places, for C to reach 0.10 is:
(a) 34.388 s (b) 0.007 s
(c) 0.168 s (d) 15.168 s

For fully worked solutions to each of the problems in Practice Exercises 70 to 74 in this chapter, go to the website:

www.routledge.com/cw/bird

Chapter 15

Number sequences

Why it is important to understand: Number sequences

Number sequences are widely used in engineering applications including computer data structure and sorting algorithms, financial engineering, audio compression and architectural engineering. Thanks to engineers, robots have migrated from factory shop floors – as industrial manipulators, to outer space – as interplanetary explorers, to hospitals – as minimally invasive surgical assistants, to homes – as vacuum cleaners and lawn mowers and to battlefields – as unmanned air, underwater and ground vehicles. Arithmetic progressions (APs) are used in simulation engineering and in the reproductive cycle of bacteria. Some uses of APs in daily life include uniform increase in the speed at regular intervals, completing patterns of objects, calculating simple interest, speed of an aircraft, increase or decrease in the costs of goods, sales and production and so on. Geometric progressions (GPs) are used in compound interest and the range of speeds on a drilling machine. In fact, GPs are used throughout mathematics, and they have many important applications in physics, engineering, biology, economics, computer science, queuing theory and finance. In this chapter, APs, GPs, combinations and permutations are introduced and explained.

At the end of this chapter, you should be able to:

- calculate the n'th term of an AP
- calculate the sum of n terms of an AP
- calculate the n'th term of a GP
- calculate the sum of n terms of a GP
- calculate the sum to infinity of a GP
- understand and perform calculations with combinations and permutations

15.1 Arithmetic progressions

When a sequence has a constant difference between successive terms it is called an **arithmetic progression** (often abbreviated to AP).

Examples include:

(i) 1, 4, 7, 10, 13, ... where the **common difference** is 3

and (ii) $a, a+d, a+2d, a+3d, \ldots$ where the common difference is d.

General expression for the n'th term of an AP

If the 1st term of an AP is 'a' and the common difference is 'd' then

the n'th term is: $a+(n-1)d$

In example (i) above, the 7th term is given by $1+(7-1)3=\mathbf{19}$, which may be readily checked.

Sum of n terms of an AP

The sum S of an AP can be obtained by multiplying the average of all the terms by the number of terms.

The average of all the terms $=\frac{a+l}{2}$, where 'a' is the 1st term and l is the last term, i.e. $l=a+(n-1)d$, for n terms.
Hence the sum of n terms,

$$S_n = n\left(\frac{a+l}{2}\right) = \frac{n}{2}\{a+[a+(n-1)d]\}$$

i.e. $$S_n = \frac{n}{2}[2a+(n-1)d]$$

For example, the sum of the first 7 terms of the series 1, 4, 7, 10, 13, ... is given by

$$S_7 = \frac{7}{2}[2(1)+(7-1)3] \quad \text{since } a=1 \quad \text{and} \quad d=3$$

$$= \frac{7}{2}[2+18] = \frac{7}{2}[20] = \mathbf{70}$$

15.2 Worked problems on arithmetic progressions

Problem 1. Determine: (a) the 9th, and (b) the 16th term of the series 2, 7, 12, 17, ...

2, 7, 12, 17, ... is an arithmetic progression with a common difference, d, of 5

(a) The n'th term of an AP is given by $a+(n-1)d$ Since the first term $a=2$, $d=5$ and $n=9$ then the 9th term is:

$$2+(9-1)5=2+(8)(5)=2+40=\mathbf{42}$$

(b) The 16th term is:

$$2+(16-1)5=2+(15)(5)=2+75=\mathbf{77}$$

Problem 2. The 6th term of an AP is 17 and the 13th term is 38. Determine the 19th term.

The n'th term of an AP is $a+(n-1)d$

The 6th term is: $a+5d=17$ (1)

The 13th term is: $a+12d=38$ (2)

Equation (2) – equation (1) gives: $7d=21$, from which, $d=\frac{21}{7}=3$

Substituting in equation (1) gives: $a+15=17$, from which, $a=2$

Hence the 19th term is:

$$a+(n-1)d = 2+(19-1)3 = 2+(18)(3)$$
$$= 2+54 = \mathbf{56}$$

Problem 3. Determine the number of the term whose value is 22 is the series

$$2\frac{1}{2}, 4, 5\frac{1}{2}, 7, \ldots$$

$2\frac{1}{2}, 4, 5\frac{1}{2}, 7, \ldots$ is an AP

where $a=2\frac{1}{2}$ and $d=1\frac{1}{2}$

Hence if the n'th term is 22 then: $a+(n-1)d=22$

i.e. $$2\frac{1}{2}+(n-1)\left(1\frac{1}{2}\right)=22$$

$$(n-1)\left(1\frac{1}{2}\right) = 22-2\frac{1}{2} = 19\frac{1}{2}$$

$$n-1 = \frac{19\frac{1}{2}}{1\frac{1}{2}} = 13$$

and $$n = 13+1 = 14$$

i.e. **the 14th term of the AP is 22**

Problem 4. Find the sum of the first 12 terms of the series 5, 9, 13, 17, ...

5, 9, 13, 17, ... is an AP where $a=5$ and $d=4$

The sum of n terms of an AP,

$$S_n = \frac{n}{2}[2a+(n-1)d]$$

Hence the sum of the first 12 terms,

$$S_{12} = \frac{12}{2}[2(5)+(12-1)4]$$

$$= 6[10+44] = 6(54) = \mathbf{324}$$

Now try the following Practice Exercise

Practice Exercise 76 Arithmetic progressions (Answers on page 709)

1. Find the 11th term of the series 8, 14, 20, 26, ...
2. Find the 17th term of the series 11, 10.7, 10.4, 10.1, ...
3. The 7th term of a series is 29 and the 11th term is 54. Determine the sixteenth term.
4. Find the 15th term of an arithmetic progression of which the first term is 2.5 and the 10th term is 16.
5. Determine the number of the term which is 29 in the series 7, 9.2, 11.4, 13.6, ...
6. Find the sum of the first 11 terms of the series 4, 7, 10, 13, ...
7. Determine the sum of the series 6.5, 8.0, 9.5, 11.0, ..., 32

15.3 Further worked problems on arithmetic progressions

Problem 5. The sum of 7 terms of an *AP* is 35 and the common difference is 1.2. Determine the 1st term of the series

$n=7$, $d=1.2$ and $S_7=35$

Since the sum of n terms of an *AP* is given by

$$S_n = \frac{n}{2}[2a+(n-1)]d$$

then $$35 = \frac{7}{2}[2a+(7-1)1.2] = \frac{7}{2}[2a+7.2]$$

Hence $$\frac{35\times 2}{7} = 2a+7.2$$

$$10 = 2a+7.2$$

Thus $$2a = 10-7.2 = 2.8$$

from which $$a = \frac{2.8}{2} = 1.4$$

i.e. **the first term, $a=1.4$**

Problem 6. Three numbers are in arithmetic progression. Their sum is 15 and their product is 80. Determine the three numbers

Let the three numbers be $(a-d)$, a and $(a+d)$

Then $(a-d)+a+(a+d)=15$, i.e. $3a=15$, from which, $a=5$

Also, $a(a-d)(a+d)=80$, i.e. $a(a^2-d^2)=80$

Since a = 5, $$5(5^2-d^2) = 80$$
$$125-5d^2 = 80$$
$$125-80 = 5d^2$$
$$45 = 5d^2$$

from which, $d^2=\frac{45}{5}=9$. Hence, $d=\sqrt{9}=\pm 3$

The three numbers are thus $(5-3)$, 5 and $(5+3)$, i.e. **2, 5 and 8**

Problem 7. Find the sum of all the numbers between 0 and 207 which are exactly divisible by 3

The series 3, 6, 9, 12, ... 207 is an *AP* whose first term $a=3$ and common difference $d=3$

The last term is $a+(n-1)d=207$

i.e. $$3+(n-1)3 = 207$$

from which $$(n-1) = \frac{207-3}{3} = 68$$

Hence $$n = 68+1 = 69$$

The sum of all 69 terms is given by

$$S_{69} = \frac{n}{2}[2a+(n-1)d]$$
$$= \frac{69}{2}[2(3)+(69-1)3]$$
$$= \frac{69}{2}[6+204] = \frac{69}{2}(210) = \mathbf{7245}$$

Problem 8. The 1st, 12th and last term of an arithmetic progression are 4, 31.5, and 376.5 respectively. Determine: (a) the number of terms in the series, (b) the sum of all the terms and (c) the 80th term

(a) Let the *AP* be a, $a+d$, $a+2d$, ..., $a+(n-1)d$, where $a=4$

The 12th term is: $a+(12-1)d=31.5$

i.e. $4+11d=31.5$, from which,

$11d=31.5-4=27.5$

Hence $d=\frac{27.5}{11}=2.5$

The last term is $a+(n-1)d$

i.e. $4+(n-1)(2.5)=376.5$

$$(n-1)=\frac{376.5-4}{2.5}$$

$$=\frac{372.5}{2.5}=149$$

Hence the number of terms in the series, $n=149+1=150$

(b) Sum of all the terms,

$$S_{150}=\frac{n}{2}[2a+(n-1)d]$$

$$=\frac{150}{2}[2(4)+(150-1)(2.5)]$$

$$=75[8+(149)(2.5)]$$

$$=85[8+372.5]$$

$$=75(380.5)=\mathbf{28537.5}$$

(c) The 80th term is:

$$a+(n-1)d=4+(80-1)(2.5)$$

$$=4+(79)(2.5)$$

$$=4+197.5=\mathbf{201.5}$$

Problem 9. An oil company bores a hole 120 m deep. Estimate the cost of boring if the cost is £70 for drilling the first metre with an increase in cost of £3 per metre for each succeeding metre

The series is: 70, 73, 76, ... to 120 terms,

i.e. $a=70, d=3$ and $n=120$

Thus, **total cost**,

$$S_n=\frac{n}{2}[2a+(n-1)d]=\frac{120}{2}[2(70)+(120-1)(3)]$$

$$=60[140+357]=60(497)=\mathbf{£29\,820}$$

Now try the following Practice Exercise

Practice Exercise 77 Arithmetic progressions (Answers on page 709)

1. The sum of 15 terms of an arithmetic progression is 202.5 and the common difference is 2. Find the first term of the series
2. Three numbers are in arithmetic progression. Their sum is 9 and their product is 20.25. Determine the three numbers
3. Find the sum of all the numbers between 5 and 250 which are exactly divisible by 4
4. Find the number of terms of the series 5, 8, 11, ... of which the sum is 1025
5. Insert four terms between 5 and 22.5 to form an arithmetic progression
6. The 1st, 10th and last terms of an arithmetic progression are 9, 40.5, and 425.5 respectively. Find (a) the number of terms, (b) the sum of all terms and (c) the 70th term
7. On commencing employment a man is paid a salary of £16 000 per annum and receives annual increments of £480. Determine his salary in the 9th year and calculate the total he will have received in the first 12 years
8. An oil company bores a hole 80 m deep. Estimate the cost of boring if the cost is £30 for drilling the first metre with an increase in cost of £2 per metre for each succeeding metre

15.4 Geometric progressions

When a sequence has a constant ratio between successive terms it is called a **geometric progression** (often abbreviated to *GP*). The constant is called the **common ratio, *r***.

Examples include

(i) 1, 2, 4, 8, ... where the common ratio is 2

and (ii) $a, ar, ar^2, ar^3, \ldots$ where the common ratio is r

General expression for the n'th term of a *GP*

If the first term of a *GP* is 'a' and the common ratio is r, then

the n'th term is: ar^{n-1}

which can be readily checked from the above examples. For example, the 8th term of the *GP* 1, 2, 4, 8, ... is $(1)(2)^7 = \mathbf{128}$, since $a = 1$ and $r = 2$

Sum to *n* terms of a *GP*

Let a *GP* be $a, ar, ar^2, ar^3, \ldots ar^{n-1}$

then the sum of n terms,

$$S_n = a + ar + ar^2 + ar^3 + \cdots + ar^{n-1} \ldots \quad (1)$$

Multiplying throughout by r gives:

$$rS_n = ar + ar^2 + ar^3 + ar^4 + \cdots + ar^{n-1} + ar^n \ldots \quad (2)$$

Subtracting equation (2) from equation (1) gives:

$$S_n - rS_n = a - ar^n$$

i.e. $S_n(1 - r) = a(1 - r^n)$

Thus the sum of n terms,

$$S_n = \frac{a(1 - r^n)}{(1 - r)} \quad \text{which is valid when } r < 1$$

Subtracting equation (1) from equation (2) gives

$$S_n = \frac{a(r^n - 1)}{(r - 1)} \quad \text{which is valid when } r > 1$$

For example, the sum of the first 8 terms of the *GP* 1, 2, 4, 8, 16, ... is given by:

$$S_8 = \frac{1(2^8 - 1)}{(2 - 1)} \quad \text{since } a = 1 \text{ and } r = 2$$

i.e. $S_8 = \frac{1(256 - 1)}{1} = \mathbf{255}$

Sum to infinity of a *GP*

When the common ratio r of a *GP* is less than unity, the sum of n terms,

$$S_n = \frac{a(1 - r^n)}{(1 - r)}, \text{which may be written as}$$

$$S_n = \frac{a}{(1 - r)} - \frac{ar^n}{(1 - r)}$$

Since, $r < 1$, r^n becomes less as n increases,

i.e. $r^n \to 0$ as $n \to \infty$

Hence $\frac{ar^n}{(1 - r)} \to 0$ as $n \to \infty$.

Thus $S_n \to \frac{a}{(1 - r)}$ as $n \to \infty$

The quantity $\frac{a}{(1 - r)}$ is called the **sum to infinity**, S_∞, and is the limiting value of the sum of an infinite number of terms,

i.e. $S_\infty = \frac{a}{(1 - r)}$ which is valid when $-1 < r < 1$

Convergence means that the values of the terms must get progressively smaller and the sum of the terms must reach a limiting value.

For example, the function $y = \frac{1}{x}$ converges to zero as x increases

Similarly, the series $1 + \frac{1}{2} + \frac{1}{4} + \frac{1}{8} + \cdots$ is convergent since the value of the terms is getting smaller and the sum of the terms is approaching a limiting value of 2, i.e. the sum to infinity, $S_\infty = \frac{a}{1 - r} = \frac{1}{1 - \frac{1}{2}} = 2$

15.5 Worked problems on geometric progressions

Problem 10. Determine the 10th term of the series 3, 6, 12, 24, ...

3, 6, 12, 24, ... is a geometric progression with a common ratio r of 2.

The n'th term of a *GP* is ar^{n-1}, where a is the first term. Hence the 10th term is:

$$(3)(2)^{10-1} = (3)(2)^9 = 3(512) = \mathbf{1536}$$

Problem 11. Find the sum of the first 7 terms of the series, $\frac{1}{2}, 1\frac{1}{2}, 4\frac{1}{2}, 13\frac{1}{2}, \ldots$

$$\frac{1}{2}, 1\frac{1}{2}, 4\frac{1}{2}, 13\frac{1}{2}, \ldots$$

is a *GP* with a common ratio $r = 3$

The sum of n terms, $S_n = \frac{a(r^n - 1)}{(r - 1)}$

Hence $S_7 = \frac{\frac{1}{2}(3^7 - 1)}{(3 - 1)} = \frac{\frac{1}{2}(2187 - 1)}{2} = \mathbf{546\frac{1}{2}}$

Problem 12. The first term of a geometric progression is 12 and the 5th term is 55. Determine the 8th term and the 11th term

The 5th term is given by $ar^4 = 55$, where the first term $a = 12$

Hence $r^4 = \frac{55}{a} = \frac{55}{12}$ and

$$r = \sqrt[4]{\frac{55}{12}} = 1.4631719\ldots$$

The 8th term is

$$ar^7 = (12)(1.4631719\ldots)^7 = \mathbf{172.3}$$

The 11th term is

$$ar^{10} = (12)(1.4631719\ldots)^{10} = \mathbf{539.7}$$

Problem 13. Which term of the series:

$$2187, 729, 243, \ldots \text{is} \frac{1}{9}?$$

2187, 729, 243, ... is a *GP* with a common ratio $r = \frac{1}{3}$ and first term $a = 2187$

The n′th term of a *GP* is given by: ar^{n-1}

Hence $\quad \frac{1}{9} = (2187)\left(\frac{1}{3}\right)^{n-1}$ from which

$$\left(\frac{1}{3}\right)^{n-1} = \frac{1}{(9)(2187)} = \frac{1}{3^2 3^7} = \frac{1}{3^9} = \left(\frac{1}{3}\right)^9$$

Thus $(n-1) = 9$, from which, $n = 9 + 1 = 10$

i.e. $\mathbf{\frac{1}{9}}$ **is the 10th term of the** ***GP***.

Problem 14. Find the sum of the first 9 terms of the series: 72.0, 57.6, 46.08, ...

The common ratio,

$$r = \frac{ar}{a} = \frac{57.6}{72.0} = 0.8$$

$$\left(\text{also} \frac{ar^2}{ar} = \frac{46.08}{57.6} = 0.8\right)$$

The sum of 9 terms,

$$S_9 = \frac{a(1-r^n)}{(1-r)} = \frac{72.0(1-0.8^9)}{(1-0.8)}$$

$$= \frac{72.0(1-0.1342)}{0.2} = \mathbf{311.7}$$

Problem 15. Find the sum to infinity of the series $3, 1, \frac{1}{3}, \ldots$

$3, 1, \frac{1}{3}, \ldots$ is a *GP* of common ratio, $r = \frac{1}{3}$

The sum to infinity,

$$S_\infty = \frac{a}{1-r} = \frac{3}{1-\frac{1}{3}} = \frac{3}{\frac{2}{3}} = \frac{9}{2} = \mathbf{4\frac{1}{2}}$$

Now try the following Practice Exercise

Practice Exercise 78 Geometric progressions (Answers on page 709)

1. Find the 10th term of the series 5, 10, 20, 40, ...
2. Determine the sum to the first 7 terms of the series 0.25, 0.75, 2.25, 6.75, ...
3. The first term of a geometric progression is 4 and the 6th term is 128. Determine the 8th and 11th terms.
4. Find the sum of the first 7 terms of the series $2, 5, 12\frac{1}{2}, \ldots$ (correct to 4 significant figures).
5. Determine the sum to infinity of the series 4, 2, 1, ...
6. Find the sum to infinity of the series $2\frac{1}{2}, -1\frac{1}{4}, \frac{5}{8}, \ldots$

15.6 Further worked problems on geometric progressions

Problem 16. In a geometric progression the 6th term is 8 times the 3rd term and the sum of the 7th and 8th terms is 192. Determine: (a) the common ratio, (b) the 1st term, and (c) the sum of the 5th to 11th term, inclusive

(a) Let the *GP* be $a, ar, ar^2, ar^3, \ldots, ar^{n-1}$

The 3rd term $= ar^2$ and the 6th term $= ar^5$

The 6th term is 8 times the 3rd

Hence $ar^5 = 8\,ar^2$ from which, $r^3 = 8$ and $r = \sqrt[3]{8}$

i.e. **the common ratio $r = 2$**

(b) The sum of the 7th and 8th terms is 192. Hence $ar^6 + ar^7 = 192$. Since $r = 2$, then

$$64a + 128a = 192$$
$$192a = 192,$$

from which, ***a*, the first term = 1**

(c) The sum of the 5th to 11th terms (inclusive) is given by:

$$S_{11} - S_4 = \frac{a(r^{11}-1)}{(r-1)} - \frac{a(r^4-1)}{(r-1)}$$
$$= \frac{1(2^{11}-1)}{(2-1)} - \frac{1(2^4-1)}{(2-1)}$$
$$= (2^{11}-1) - (2^4-1)$$
$$= 2^{11} - 2^4 = 2408 - 16 = \mathbf{2032}$$

Problem 17. A hire tool firm finds that their net return from hiring tools is decreasing by 10% per annum. If their net gain on a certain tool this year is £400, find the possible total of all future profits from this tool (assuming the tool lasts for ever)

The net gain forms a series:

$$£400 + £400 \times 0.9 + £400 \times 0.9^2 + \cdots,$$

which is a *GP* with $a = 400$ and $r = 0.9$

The sum to infinity,

$$S_\infty = \frac{a}{(1-r)} = \frac{400}{(1-0.9)}$$

$$= \mathbf{£4000 = total\ future\ profits}$$

Problem 18. If £100 is invested at compound interest of 2% per annum, determine (a) the value after 10 years, (b) the time, correct to the nearest year, it takes to reach more than £150

(a) Let the *GP* be $a, ar, ar^2, \ldots ar^n$

The first term $a = £100$ and

the common ratio $r = 1.02$

Hence the second term is $ar = (100)(1.02) = £102$, which is the value after 1 year, the 3rd term is $ar^2 = (100)(1.02)^2 = £104.04$, which is the value after 2 years and so on.

Thus the value after 10 years

$$= ar^{10} = (100)\,(1.02)^{10} = \mathbf{£\ 121.90}$$

(b) When £150 has been reached, $150 = ar^n$

i.e. $150 = 100(1.02)^n$

and $1.5 = (1.02)^n$

Taking logarithms to base 10 of both sides gives:

$$\lg 1.5 = \lg(1.02)^n = n \lg(1.02),$$

by the laws of logarithms from which,

$$n = \frac{\lg 1.5}{\lg 1.02} = 20.48$$

Hence it will take 21 years to reach more than £150

Problem 19. A drilling machine is to have 6 speeds ranging from 50 rev/min to 750 rev/min. If the speeds form a geometric progression determine their values, each correct to the nearest whole number

Let the *GP* of *n* terms by given by $a, ar, ar^2, \ldots ar^{n-1}$

The 1st term $a = 50$ rev/min.

The 6th term is given by ar^{6-1}, which is 750 rev/min, i.e., $ar^5 = 750$

from which $r^5 = \frac{750}{a} = \frac{750}{50} = 15$

Thus the common ratio, $r = \sqrt[5]{15} = 1.7188$

The 1st term is $a = 50$ rev/min.

the 2nd term is $ar = (50)(1.7188) = 85.94$,

the 3rd term is $ar^2 = (50)(1.7188)^2 = 147.71$,

the 4th term is $ar^3 = (50)(1.7188)^3 = 253.89$,

the 5th term is $ar^4 = (50)(1.7188)^4 = 436.39$,

the 6th term is $ar^5 = (50)(1.7188)^5 = 750.06$

Hence, correct to the nearest whole number, the 6 speeds of the drilling machine are: **50, 86, 148, 254, 436 and 750 rev/min.**

Now try the following Practice Exercise

Practice Exercise 79 Geometric progressions (Answers on page 709)

1. In a geometric progression the 5th term is 9 times the 3rd term and the sum of the 6th and 7th terms is 1944. Determine: (a) the common ratio, (b) the 1st term and (c) the sum of the 4th to 10th terms inclusive.

2. Which term of the series $3, 9, 27, \ldots$ is 59 049?

3. The value of a lathe originally valued at £3000 depreciates 15% per annum. Calculate its value after 4 years. The machine is sold when its value is less than £550. After how many years is the lathe sold?

4. If the population of Great Britain is 65 million and is decreasing at 2.4% per annum, what will be the population in 5 years time?

5. 100 g of a radioactive substance disintegrates at a rate of 3% per annum. How much of the substance is left after 11 years?

6. If £250 is invested at compound interest of 2.2% per annum determine (a) the value after 15 years, (b) the time, correct to the nearest year, it takes to reach £400.

7. A drilling machine is to have 8 speeds ranging from 100 rev/min to 1000 rev/min. If the speeds form a geometric progression determine their values, each correct to the nearest whole number.

15.7 Combinations and permutations

A **combination** is the number of selections of r different items from n distinguishable items when order of selection is ignored. A combination is denoted by ^nC_r or $\binom{n}{r}$

where $$^nC_r = \frac{n!}{r!(n-r)!}$$

where, for example, $4!$ denotes $4\times3\times2\times1$ and is termed 'factorial 4'.

Thus,

$$^5C_3 = \frac{5!}{3!(5-3)!} = \frac{5\times4\times3\times2\times1}{(3\times2\times1)(2\times1)} = \frac{120}{6\times2} = 10$$

For example, the five letters A, B, C, D, E can be arranged in groups of three as follows: ABC, ABD, ABE, ACD, ACE, ADE, BCD, BCE, BDE, CDE, i.e. there are ten groups. The above calculation $^5C_3$ produces the answer of 10 combinations without having to list all of them.

A **permutation** is the number of ways of selecting $r \leq n$ objects from n distinguishable objects when order of selection is important. A permutation is denoted by ^nP_r or $_nP_r$

where $$^nP_r = n(n-1)(n-2)\ldots(n-r+1)$$

or $$^nP_r = \frac{n!}{(n-r)!}$$

Thus, $$^4P_2 = \frac{4!}{(4-2)!} = \frac{4!}{2!} = \frac{4\times3\times2}{2} = 12$$

Problem 20. Evaluate: (a) $^7C_4$ (b) $^{10}C_6$

(a) $$^7C_4 = \frac{7!}{4!(7-4)!} = \frac{7!}{4!3!} = \frac{7\times6\times5\times4\times3\times2}{(4\times3\times2)(3\times2)} = \mathbf{35}$$

(b) $$^{10}C_6 = \frac{10!}{6!(10-6)!} = \frac{10!}{6!4!} = \mathbf{210}$$

Problem 21. Evaluate: (a) $^6P_2$ (b) $^9P_5$

(a) $$^6P_2 = \frac{6!}{(6-2)!} = \frac{6!}{4!} = \frac{6\times5\times4\times3\times2}{4\times3\times2} = \mathbf{30}$$

(b) $$^9P_5 = \frac{9!}{(9-5)!} = \frac{9!}{4!} = \frac{9\times8\times7\times6\times5\times4!}{4!} = \mathbf{15\,120}$$

Now try the following Practice Exercise

Practice Exercise 80 Combinations and permutations (Answers on page 709)

Evaluate the following:

1. (a) ${}^9C_6$ (b) ${}^3C_1$
2. (a) ${}^6C_2$ (b) ${}^8C_5$
3. (a) ${}^4P_2$ (b) ${}^7P_4$
4. (a) ${}^{10}P_3$ (b) ${}^8P_5$

Practice Exercise 81 Multiple-choice questions on number sequences (Answers on page 709)

Each question has only one correct answer

1. The fifth term of an arithmetic progression is 18 and the twelfth term is 46. The eighteenth term is:
   (a) 72 (b) 74 (c) 68 (d) 70
2. The first term of a geometric progression is 9 and the fourth term is 45. The eighth term is:
   (a) 384.7 (b) 150.5 (c) 225 (d) 657.9
3. The value of the permutation ${}^9P_3$ is:
   (a) 84 (b) 504 (c) 3024 (d) 60480
4. The first and last term of an arithmetic progression are 1 and 13. If the sum of its terms is 42, then the number of terms is:
   (a) 5 (b) 6 (c) 7 (d) 8
5. The sum of three consecutive terms of an increasing arithmetic progression is 51 and the product of their terms is 4641. The third term is:
   (a) 9 (b) 13 (c) 17 (d) 21
6. The value of the combination ${}^8C_5$ is:
   (a) 336 (b) 6720 (c) 56 (d) 2016
7. An arithmetic progression is given by 3, 7, 11, 15, ...If the sum of the series is 406, the number of terms is:
   (a) 5 (b) 10 (c) 12 (d) 14
8. The sum of the first 9 terms of the series 1, 3, 9, 27, ... is:
   (a) 3280 (b) 9841 (c) 19682 (d) 6560
9. The sum to infinity of the series 4, 2, 1, ... is:
   (a) 2 (b) 4 (c) 8 (d) 16
10. The 12th term of the series 4, 8, 16, 32, ... is:
   (a) 8192 (b) 4096 (c) 16384 (d) 2048

For fully worked solutions to each of the problems in Practice Exercises 76 to 80 in this chapter, go to the website:
www.routledge.com/cw/bird

COMPANION @ WEBSITE

Chapter 16

The binomial series

Why it is important to understand: The binomial series

There are many applications of the binomial theorem in every part of algebra, and in general with permutations, combinations and probability. It is also used in atomic physics where it is used to count s, p, d and f orbitals. There are applications of the binomial series in financial mathematics to determine the number of stock price paths that leads to a particular stock price at maturity.

At the end of this chapter, you should be able to:

- define a binomial expression
- use Pascal's triangle to expand a binomial expression
- state the general binomial expansion of $(a+x)^n$ and $(1+x)^n$
- use the binomial series to expand expressions of the form $(a+x)^n$ for positive, negative and fractional values of n
- determine the r'th term of a binomial expansion
- use the binomial expansion with practical applications

16.1 Pascal's triangle

A **binomial expression** is one that contains two terms connected by a plus or minus sign. Thus $(p+q)$, $(a+x)^2$, $(2x+y)^3$ are examples of binomial expression. Expanding $(a+x)^n$ for integer values of n from 0 to 6 gives the results shown at the top of page 157.
From the results the following patterns emerge:

(i) 'a' decreases in power moving from left to right.

(ii) 'x' increases in power moving from left to right.

(iii) The coefficients of each term of the expansions are symmetrical about the middle coefficient when n is even and symmetrical about the two middle coefficients when n is odd.

(iv) The coefficients are shown separately in Table 16.1 and this arrangement is known as

$(a+x)^0 =$	1
$(a+x)^1 =$	$a+x$
$(a+x)^2 = (a+x)(a+x) =$	$a^2+2ax+x^2$
$(a+x)^3 = (a+x)^2(a+x) =$	$a^3+3a^2x+3ax^2+x^3$
$(a+x)^4 = (a+x)^3(a+x) =$	$a^4+4a^3x+6a^2x^2+4ax^3+x^4$
$(a+x)^5 = (a+x)^4(a+x) =$	$a^5+5a^4x+10a^3x^2+10a^2x^3+5ax^4+x^5$
$(a+x)^6 = (a+x)^5(a+x) =$	$a^6+6a^5x+15a^4x^2+20a^3x^3+15a^2x^4+6ax^5+x^6$

Pascal's triangle*. A coefficient of a term may be obtained by adding the two adjacent coefficients immediately above in the previous row. This is shown by the triangles in Table 16.1, where, for example, $1+3=4$, $10+5=15$ and so on.

(v) Pascal's triangle method is used for expansions of the form $(a+x)^n$ for integer values of n less than about 8.

Problem 1. Use the Pascal's triangle method to determine the expansion of $(a+x)^7$

From Table 16.1 the row the Pascal's triangle corresponding to $(a+x)^6$ is as shown in (1) below. Adding adjacent coefficients gives the coefficients of $(a+x)^7$ as shown in (2) below.

*Who was Pascal? – **Blaise Pascal** (19 June 1623–19 August 1662) was a French polymath. A child prodigy, he wrote a significant treatise on the subject of projective geometry at the age of sixteen, and later corresponded with Pierre de Fermat on probability theory, strongly influencing the development of modern economics and social science. To find out more go to www.routledge.com/cw/bird

Table 16.1

$(a+x)^0$							1						
$(a+x)^1$						1		1					
$(a+x)^2$					1		2		1				
$(a+x)^3$				1		3		3		1			
$(a+x)^4$			1		4		6		4		1		
$(a+x)^5$		1		5		10		10		5		1	
$(a+x)^6$	1		6		15		20		15		6		1

	1	6	15	20	15	6	1	(1)
1	7	21	35	35	21	7	1	(2)

The first and last terms of the expansion of $(a+x)^7$ and a^7 and x^7 respectively. The powers of 'a' decrease and the powers of 'x' increase moving from left to right. Hence,

$$\boldsymbol{(a+x)^7 = a^7+7a^6x+21a^5x^2+35a^4x^3+35a^3x^4 + 21a^2x^5+7ax^6+x^7}$$

Problem 2. Determine, using Pascal's triangle method, the expansion of $(2p-3q)^5$

Comparing $(2p-3q)^5$ with $(a+x)^5$ shows that $a=2p$ and $x=-3q$

Using Pascal's triangle method:

$$(a+x)^5 = a^5+5a^4x+10a^3x^2+10a^2x^3+\cdots$$

Hence

$$\begin{aligned}(2p-3q)^5 &= (2p)^5+5(2p)^4(-3q)\\ &\quad +10(2p)^3(-3q)^2\\ &\quad +10(2p)^2(-3q)^3\\ &\quad +5(2p)(-3q)^4+(-3q)^5\end{aligned}$$

i.e. $\mathbf{(2p-3q)^5 = 32p^5 - 240p^4q + 720p^3q^2 - 1080p^2q^3 + 810pq^4 - 243q^5}$

Now try the following Practice Exercise

Practice Exercise 82 Pascals triangle (Answers on page 709)

1. Use Pascal's triangle to expand $(x-y)^7$
2. Expand $(2a+3b)^5$ using Pascal's triangle.

16.2 The binomial series

The **binomial series** or **binomial theorem** is a formula for raising a binomial expression to any power without lengthy multiplication. The general binomial expansion of $(a+x)^n$ is given by:

$$(a+x)^n = a^n + na^{n-1}x + \frac{n(n-1)}{2!}a^{n-2}x^2 + \frac{n(n-1)(n-2)}{3!}a^{n-3}x^3 + \cdots + x^n$$

where, for example, 3! denote $3\times2\times1$ and is termed 'factorial 3'.
With the binomial theorem n may be a fraction, a decimal fraction or a positive or negative integer.
In the general expansion of $(a+x)^n$ it is noted that the 4th term is:

$$\frac{n(n-1)(n-2)}{3!}a^{n-3}x^3$$

The number 3 is very evident in this expression.

For any term in a binomial expansion, say the r'th term, $(r-1)$ is very evident. It may therefore be reasoned that **the r'th term of the expansion $(a+x)^n$** is:

$$\frac{n(n-1)(n-2)\ldots \text{to}(r-1)\text{terms}}{(r-1)!}a^{n-(r-1)}x^{r-1}$$

If $a=1$ in the binomial expansion of $(a+x)^n$ then:

$$(1+x)^n = 1 + nx + \frac{n(n-1)}{2!}x^2 + \frac{n(n-1)(n-2)}{3!}x^3 + \cdots$$

which is valid for $-1<x<1$

When x is small compared with 1 then:

$$(1+x)^n \approx 1 + nx$$

16.3 Worked problems on the binomial series

Problem 3. Use the binomial series to determine the expansion of $(2+x)^7$

The binomial expansion is given by:

$$(a+x)^n = a^n + na^{n-1}x + \frac{n(n-1)}{2!}a^{n-2}x^2 + \frac{n(n-1)(n-2)}{3!}a^{n-3}x^3 + \cdots$$

When $a=2$ and $n=7$:

$$(2+x)^7 = 2^7 + 7(2)^6 + \frac{(7)(6)}{(2)(1)}(2)^5x^2 + \frac{(7)(6)(5)}{(3)(2)(1)}(2)^4x^3 + \frac{(7)(6)(5)(4)}{(4)(3)(2)(1)}(2)^3x^4 + \frac{(7)(6)(5)(4)(3)}{(5)(4)(3)(2)(1)}(2)^2x^5 + \frac{(7)(6)(5)(4)(3)(2)}{(6)(5)(4)(3)(2)(1)}(2)x^6 + \frac{(7)(6)(5)(4)(3)(2)(1)}{(7)(6)(5)(4)(3)(2)(1)}x^7$$

i.e. $\mathbf{(2+x)^7 = 128 + 448x + 672x^2 + 560x^3 + 280x^4 + 84x^5 + 14x^6 + x^7}$

Problem 4. Use the binomial series to determine the expansion of $(2a-3b)^5$

The binomial expansion is given by:

$$(a+x)^n = a^n + na^{n-1}x + \frac{n(n-1)}{2!}a^{n-2}x^2 + \frac{n(n-1)(n-2)}{3!}a^{n-3}x^3 + \cdots$$

When $a=2a, x=-3b$ and $n=5$:

$$(2a-3b)^5 = (2a)^5 + 5(2a)^4(-3b) + \frac{(5)(4)}{(2)(1)}(2a)^3(-3b)^2 + \frac{(5)(4)(3)}{(3)(2)(1)}(2a)^2(-3b)^3$$

$$+\frac{(5)(4)(3)(2)}{(4)(3)(2)(1)}(2a)(-3b)^4$$

$$+\frac{(5)(4)(3)(2)(1)}{(5)(4)(3)(2)(1)}(-3b)^5,$$

i.e. $\mathbf{(2a-3b)^5 = 32a^5 - 240a^4b + 720a^3b^2 - 1080a^2b^3 + 810ab^4 - 243b^5}$

Problem 5. Expand $\left(c-\frac{1}{c}\right)^5$ using the binomial series

$$\left(c-\frac{1}{c}\right)^5 = c^5 + 5c^4\left(-\frac{1}{c}\right) + \frac{(5)(4)}{(2)(1)}c^3\left(-\frac{1}{c}\right)^2$$

$$+\frac{(5)(4)(3)}{(3)(2)(1)}c^2\left(-\frac{1}{c}\right)^3$$

$$+\frac{(5)(4)(3)(2)}{(4)(3)(2)(1)}c\left(-\frac{1}{c}\right)^4$$

$$+\frac{(5)(4)(3)(2)(1)}{(5)(4)(3)(2)(1)}\left(-\frac{1}{c}\right)^5$$

i.e. $\mathbf{\left(c-\frac{1}{c}\right)^5 = c^5 - 5c^3 + 10c - \frac{10}{c} + \frac{5}{c^3} - \frac{1}{c^5}}$

Problem 6. Without fully expanding $(3+x)^7$, determine the fifth term

The r'th term of the expansion $(a+x)^n$ is given by:

$$\frac{n(n-1)(n-2)\ldots\text{to }(r-1)\text{ terms}}{(r-1)!}a^{n-(r-1)}x^{r-1}$$

Substituting $n=7$, $a=3$ and $r-1=5-1=4$ gives:

$$\frac{(7)(6)(5)(4)}{(4)(3)(2)(1)}(3)^{7-4}x^4$$

i.e. the fifth term of $(3+x)^7 = 35(3)^3x^4 = \mathbf{945x^4}$

Problem 7. Find the middle term of $\left(2p-\frac{1}{2q}\right)^{10}$

In the expansion of $(a+x)^{10}$ there are $10+1$, i.e. 11 terms. Hence the middle term is the sixth. Using the general expression for the r'th term where $a=2p$, $x=-\frac{1}{2q}$, $n=10$ and $r-1=5$ gives:

$$\frac{(10)(9)(8)(7)(6)}{(5)(4)(3)(2)(1)}(2p)^{10-5}\left(-\frac{1}{2q}\right)^5$$

$$=252(32p^5)\left(-\frac{1}{32q^5}\right)$$

Hence the middle term of $\left(2q-\frac{1}{2q}\right)^{10}$ is $\mathbf{-252\frac{p^5}{q^5}}$

Now try the following Practice Exercise

Practice Exercise 83 The binomial series (Answers on page 710)

1. Use the binomial theorem to expand $(a+2x)^4$
2. Use the binomial theorem to expand $(2-x)^6$
3. Expand $(2x-3y)^4$
4. Determine the expansion of $\left(2x+\frac{2}{x}\right)^5$
5. Expand $(p+2q)^{11}$ as far as the fifth term
6. Determine the sixth term of $\left(3p+\frac{q}{3}\right)^{13}$
7. Determine the middle term of $(2a-5b)^8$

16.4 Further worked problems on the binomial series

Problem 8.

(a) Expand $\frac{1}{(1+2x)^3}$ in ascending powers of x as far as the term in x^3, using the binomial series.
(b) State the limits of x for which the expansion is valid

(a) Using the binomial expansion of $(1+x)^n$, where $n=-3$ and x is replaced by $2x$ gives:

$$\frac{1}{(1+2x)^3} = (1+2x)^{-3}$$

$$= 1 + (-3)(2x) + \frac{(-3)(-4)}{2!}(2x)^2 + \frac{(-3)(-4)(-5)}{3!}(2x)^3 + \cdots$$

$$= \mathbf{1 - 6x + 24x^2 - 80x^3 +}$$

(b) The expansion is valid provided $|2x|<1$,

i.e. $\mathbf{|x|<\frac{1}{2}}$ or $\mathbf{-\frac{1}{2}<x<\frac{1}{2}}$

Problem 9.

(a) Expand $\dfrac{1}{(4-x)^2}$ in ascending powers of x as far as the term in x^3, using the binomial theorem.

(b) What are the limits of x for which the expansion in (a) is true?

(a) $$\frac{1}{(4-x)^2}=\frac{1}{\left[4\left(1-\frac{x}{4}\right)\right]^2}=\frac{1}{4^2\left(1-\frac{x}{4}\right)^2}=\frac{1}{16}\left(1-\frac{x}{4}\right)^{-2}$$

Using the expansion of $(1+x)^n$

$$\frac{1}{(4-x)^2}=\frac{1}{16}\left(1-\frac{x}{4}\right)^{-2}$$
$$=\frac{1}{16}\left[1+(-2)\left(-\frac{x}{4}\right)+\frac{(-2)(-3)}{2!}\left(-\frac{x}{4}\right)^2+\frac{(-2)(-3)(-4)}{3!}\left(-\frac{x}{4}\right)^3+\ldots\right]$$
$$=\mathbf{\frac{1}{16}\left(1+\frac{x}{2}+\frac{3x^2}{16}+\frac{x^3}{16}+\cdots\right)}$$

(b) The expansion in (a) is true provided $\left|\frac{x}{4}\right|<1$,

i.e. $\mathbf{|x|<4}$ or $\mathbf{-4<x<4}$

Problem 10. Use the binomial theorem to expand $\sqrt{4+x}$ in ascending powers of x to four terms. Give the limits of x for which the expansion is valid

$$\sqrt{4+x}=\sqrt{4\left(1+\frac{x}{4}\right)}=\sqrt{4}\sqrt{1+\frac{x}{4}}=2\left(1+\frac{x}{4}\right)^{\frac{1}{2}}$$

Using the expansion of $(1+x)^n$,

$$2\left(1+\frac{x}{4}\right)^{\frac{1}{2}}$$
$$=2\left[1+\left(\frac{1}{2}\right)\left(\frac{x}{4}\right)+\frac{(1/2)(-1/2)}{2!}\left(\frac{x}{4}\right)^2+\frac{(1/2)(-1/2)(-3/2)}{3!}\left(\frac{x}{4}\right)^3+\cdots\right]$$
$$=2\left(1+\frac{x}{8}-\frac{x^2}{128}+\frac{x^3}{1024}-\cdots\right)$$
$$=\mathbf{2+\frac{x}{4}-\frac{x^2}{64}+\frac{x^3}{512}-\cdots}$$

This is valid when $\left|\frac{x}{4}\right|<1$,

i.e. $\mathbf{\left|\frac{x}{4}\right|<4}$ or $\mathbf{-4<x<4}$

Problem 11. Expand $\dfrac{1}{\sqrt{1-2t}}$ in ascending powers of t as far as the term in t^3. State the limits of t for which the expression is valid

$$\frac{1}{\sqrt{1-2t}}$$
$$=(1-2t)^{-\frac{1}{2}}$$
$$=1+\left(-\frac{1}{2}\right)(-2t)+\frac{(-1/2)(-3/2)}{2!}(-2t)^2+\frac{(-1/2)(-3/2)(-5/2)}{3!}(-2t)^3+\cdots$$

using the expansion for $(1+x)^n$

$$=\mathbf{1+t+\frac{3}{2}t^2+\frac{5}{2}t^3+\cdots}$$

The expression is valid when $|2t|<1$,

i.e. $\mathbf{|t|<\frac{1}{2}}$ or $\mathbf{-\frac{1}{2}<t<\frac{1}{2}}$

Problem 12. Simplify $\dfrac{\sqrt[3]{1-3x}\sqrt{1+x}}{\left(1+\frac{x}{2}\right)^3}$ given that powers of x above the first may be neglected

$$\frac{\sqrt[3]{1-3x}\sqrt{1+x}}{\left(1+\frac{x}{2}\right)^3}$$

$$= (1-3x)^{\frac{1}{3}}(1+x)^{\frac{1}{2}}\left(1+\frac{x}{2}\right)^{-3}$$

$$\approx \left[1+\left(\frac{1}{3}\right)(-3x)\right]\left[1+\left(\frac{1}{2}\right)(x)\right]\left[1+(-3)\left(\frac{x}{2}\right)\right]$$

when expanded by the binomial theorem as far as the x term only,

$$= (1-x)\left(1+\frac{x}{2}\right)\left(1-\frac{3x}{2}\right)$$

$$= \left(1-x+\frac{x}{2}-\frac{3x}{2}\right) \text{ when powers of } x \text{ higher than unity are neglected}$$

$$= \mathbf{(1-2x)}$$

Problem 13. Express $\frac{\sqrt{1+2x}}{\sqrt[3]{1-3x}}$ as a power series as far as the term in x^2. State the range of values of x for which the series is convergent

$$\frac{\sqrt{1+2x}}{\sqrt[3]{1-3x}} = (1+2x)^{\frac{1}{2}}(1-3x)^{-\frac{1}{3}}$$

$$(1+2x)^{\frac{1}{2}} = 1+\left(\frac{1}{2}\right)(2x) + \frac{(1/2)(-1/2)}{2!}(2x)^2+\cdots$$

$$= 1+x-\frac{x^2}{2}+\cdots \text{ which is valid for } |2x|<1, \text{ i.e. } |x|<\frac{1}{2}$$

$$(1-3x)^{-\frac{1}{3}} = 1+(-1/3)(-3x) + \frac{(-1/3)(-4/3)}{2!}(-3x)^2+\cdots$$

$$= 1+x+2x^2+\cdots \text{ which is valid for } |3x|<1, \text{ i.e. } |x|<\frac{1}{3}$$

Hence $\frac{\sqrt{1+2x}}{\sqrt[3]{1-3x}}$

$$= (1+2x)^{\frac{1}{2}}(1-3x)^{\frac{1}{3}}$$

$$= \left(1+x-\frac{x^2}{2}+\cdots\right)(1+x+2x^2+\cdots)$$

$$= 1+x+2x^2+x+x^2-\frac{x^2}{2}$$

neglecting terms of higher power than 2

$$= \mathbf{1+2x+\frac{5}{2}x^2}$$

The series is convergent if $\mathbf{-\frac{1}{3}<x<\frac{1}{3}}$

Now try the following Practice Exercise

Practice Exercise 84 The binomial series (Answers on page 710)

In Problems 1 to 5 expand in ascending powers of x as far as the term in x^3, using the binomial theorem. State in each case the limits of x for which the series is valid.

1. $\frac{1}{(1-x)}$
2. $\frac{1}{(1+x)^2}$
3. $\frac{1}{(2+x)^3}$
4. $\sqrt{2+x}$
5. $\frac{1}{\sqrt{1+3x}}$
6. Expand $(2+3x)^{-6}$ to three terms. For what values of x is the expansion valid?
7. When x is very small show that:
   (a) $\frac{1}{(1-x)^2\sqrt{1-x}} \approx 1+\frac{5}{2}x$
   (b) $\frac{(1-2x)}{(1-3x)^4} \approx 1+10x$
   (c) $\frac{\sqrt{1+5x}}{\sqrt[3]{1-2x}} \approx 1+\frac{19}{6}x$

8. If x is very small such that x^2 and higher powers may be neglected, determine the power series for $\dfrac{\sqrt{x+4}\,\sqrt[3]{8-x}}{\sqrt[5]{(1+x)^3}}$

9. Express the following as power series in ascending powers of x as far as the term in x^2. State in each case the range of x for which the series is valid.
(a) $\sqrt{\dfrac{1-x}{1+x}}$ (b) $\dfrac{(1+x)\,\sqrt[3]{(1-3x)^2}}{\sqrt{1+x^2}}$

16.5 Practical problems involving the binomial theorem

Binomial expansions may be used for numerical approximations, for calculations with small variations and in probability theory.

Problem 14. The radius of a cylinder is reduced by 4% and its height is increased by 2%. Determine the approximate percentage change in (a) its volume and (b) its curved surface area, (neglecting the products of small quantities)

Volume of cylinder $=\pi r^2 h$

Let r and h be the original values of radius and height. The new values are $0.96r$ or $(1-0.04)r$ and $1.02\,h$ or $(1+0.02)h$

(a) New volume $=\pi[(1-0.04)r]^2[(1+0.02)h]$
$= \pi r^2 h(1-0.04)^2(1+0.02)$

Now $(1-0.04)^2 = 1-2(0.04)+(0.04)^2$
$=(1-0.08)$, neglecting powers of small terms

Hence new volume
$\approx \pi r^2 h(1-0.08)(1+0.02)$
$\approx \pi r^2 h(1-0.08+0.02)$, neglecting products of small terms
$\approx \pi r^2 h(1-0.06)$ or $0.94\pi r^2 h$, i.e. 94% of the original volume

Hence the volume is reduced by approximately 6%

(b) Curved surface area of cylinder $=2\pi rh$.

New surface area
$=2\pi[(1-0.04)r][(1+0.02)h]$
$=2\pi rh(1-0.04)(1+0.02)$
$\approx 2\pi rh(1-0.04+0.02)$, neglecting products of small terms
$\approx 2\pi rh(1-0.02)$ or $0.98(2\pi rh)$, i.e. 98% of the original surface area

Hence the curved surface area is reduced by approximately 2%

Problem 15. The second moment of area of a rectangle through its centroid is given by $\dfrac{bl^3}{12}$. Determine the approximate change in the second moment of area if b is increased by 3.5% and l is reduced by 2.5%

New values of b and l are $(1+0.035)b$ and $(1-0.025)l$ respectively.
New second moment of area

$$=\frac{1}{12}[(1+0.035)b][(1-0.025)l]^3$$
$$=\frac{bl^3}{12}(1+0.035)(1-0.025)^3$$

$\approx \dfrac{bl^3}{12}(1+0.035)(1-0.075)$, neglecting powers of small terms

$\approx \dfrac{bl^3}{12}(1+0.035-0.075)$, neglecting products of small terms

$\approx \dfrac{bl^3}{12}(1-0.040)$ or $(0.96)\dfrac{bl^3}{12}$, i.e. 96% of the original second moment of area

Hence the second moment of area is reduced by approximately 4%

Problem 16. The resonant frequency of a vibrating shaft is given by: $f=\dfrac{1}{2\pi}\sqrt{\dfrac{k}{I}}$ where k is the stiffness and I is the inertia of the shaft. Use the binomial theorem to determine the approximate percentage error in determining the frequency using the measured values of k and I when the measured value of k is 4% too large and the measured value of I is 2% too small

Let f, k and I be the true values of frequency, stiffness and inertia respectively. Since the measured value of stiffness, k_1, is 4% too large, then

$$k_1 = \frac{104}{100}k = (1+0.04)k$$

The measured value of inertia, I_1, is 2% too small, hence

$$I_1 = \frac{98}{100}I = (1-0.02)I$$

The measured value of frequency,

$$f_1 = \frac{1}{2\pi}\sqrt{\frac{k_1}{I_1}} = \frac{1}{2\pi}k_1^{\frac{1}{2}}I_1^{-\frac{1}{2}}$$

$$= \frac{1}{2\pi}[(1+0.04)k]^{\frac{1}{2}}[(1-0.02)I]^{-\frac{1}{2}}$$

$$= \frac{1}{2\pi}(1+0.04)^{\frac{1}{2}}k^{\frac{1}{2}}(1-0.02)^{-\frac{1}{2}}I^{-\frac{1}{2}}$$

$$= \frac{1}{2\pi}k^{\frac{1}{2}}I^{-\frac{1}{2}}(1+0.04)^{\frac{1}{2}}(1-0.02)^{-\frac{1}{2}}$$

i.e. $f_1 = f(1+0.04)^{\frac{1}{2}}(1-0.02)^{-\frac{1}{2}}$

$$\approx f\left[1+\left(\frac{1}{2}\right)(0.04)\right]\left[1+\left(-\frac{1}{2}\right)(-0.02)\right]$$

$$\approx f(1+0.02)(1+0.01)$$

Neglecting the products of small terms,

$$f_1 \approx (1+0.02+0.01)f \approx 1.03f$$

Thus the percentage error in f based on the measured values of k and I is approximately $[(1.03)(100) - 100]$, i.e. **3% too large**

Problem 17. The kinetic energy, k, in a hammer with mass m as it strikes with velocity v is given by: $k = \frac{1}{2}mv^2$. Determine the approximate change in kinetic energy when the mass of the hammer is increased by 3% and its velocity is reduced by 4%

New value of mass $= (1+0.03)$m and new value of velocity $= (1-0.04)v$

$$\text{New value of } k = \frac{1}{2}mv^2 = \frac{1}{2}(1+0.03)m(1-0.04)^2v^2$$

$$= \frac{1}{2}mv^2[(1+0.03)(1-0.04)^2]$$

$$\approx \frac{1}{2}mv^2[(1+0.03)(1-0.08)]$$

$$\approx \frac{1}{2}mv^2[(1+0.03-0.08)]$$

$$\approx \frac{1}{2}mv^2[(1-0.05)]$$

$$\approx k[(1-0.05)]$$

i.e. **the kinetic energy in the hammer, k, has decreased by approximately 5%**

Now try the following Practice Exercise

Practice Exercise 85 Practical problems involving the binomial theorem (Answers on page 710)

1. Pressure p and volume v are related by $pv^3 = c$, where c is a constant. Determine the approximate percentage change in c when p is increased by 3% and v decreased by 1.2%
2. Kinetic energy is given by $\frac{1}{2}mv^2$. Determine the approximate change in the kinetic energy when mass m is increased by 2.5% and the velocity v is reduced by 3%
3. An error of +1.5% was made when measuring the radius of a sphere. Ignoring the products of small quantities determine the approximate error in calculating (a) the volume, and (b) the surface area
4. The power developed by an engine is given by $I = k$ PLAN, where k is a constant. Determine the approximate percentage change in the power when P and A are each increased by 2.5% and L and N are each decreased by 1.4%
5. The radius of a cone is increased by 2.7% and its height reduced by 0.9%. Determine the approximate percentage change in its volume, neglecting the products of small terms
6. The electric field strength H due to a magnet of length $2l$ and moment M at a point on its

axis distance x from the centre is given by:

$$H=\frac{M}{2l}\left\{\frac{1}{(x-l)^2}-\frac{1}{(x+l)^2}\right\}$$

Show that if l is very small compared with x, then $H\approx\frac{2M}{x^3}$

7. The shear stress τ in a shaft of diameter D under a torque T is given by: $\tau=\frac{kT}{\pi D^3}$. Determine the approximate percentage error in calculating τ if T is measured 3% too small and D 1.5% too large

8. The energy W stored in a flywheel is given by: $W=kr^5N^2$, where k is a constant, r is the radius and N the number of revolutions. Determine the approximate percentage change in W when r is increased by 1.3% and N is decreased by 2%

9. In a series electrical circuit containing inductance L and capacitance C the resonant frequency is given by: $f_r=\frac{1}{2\pi\sqrt{LC}}$. If the values of L and C used in the calculation are 2.6% too large and 0.8% too small respectively, determine the approximate percentage error in the frequency

10. The viscosity η of a liquid is given by: $\eta=\frac{kr^4}{\nu l}$, where k is a constant. If there is an error in r of +2%, in ν of +4% and l of −3%, what is the resultant error in η?

11. A magnetic pole, distance x from the plane of a coil of radius r, and on the axis of the coil, is subject to a force F when a current flows in the coil. The force is given by: $F=\frac{kx}{\sqrt{(r^2+x^2)^5}}$, where k is a constant. Use the binomial theorem to show that when x is small compared to r, then $F\approx\frac{kx}{r^5}-\frac{5kx^3}{2r^7}$

12. The flow of water through a pipe is given by: $G=\sqrt{\frac{(3d)^5H}{L}}$. If d decreases by 2% and H by 1%, use the binomial theorem to estimate the decrease in G

13. The force of a jet of water on a flat plate is given by: $F=2\rho Av^2$ where ρ is the density of water, A is the area of the plate and v is the velocity of the jet. If ρ and A are constants, find the approximate change in F when v is increased by 2.5%

14. The Q-factor at resonance in an R-L-C series circuit is given by: $Q=\frac{1}{R}\sqrt{\frac{L}{C}}$ Determine the approximate error in Q if R is 3% high, L is 4% high and C is 5% low.

Practice Exercise 86 Multiple-choice questions on the binomial series (Answers on page 710)

Each question has only one correct answer

1. $(a+x)^4=a^4+4a^3x+6a^2x^2+4ax^3+x^4$. Using Pascal's triangle, the third term of $(a+x)^5$ is:
(a) $10a^2x^3$ (b) $5a^4x$
(c) $5a^2x^2$ (d) $10a^3x^2$

2. Expanding $(2+x)^4$ using the binomial series gives:
(a) $2+4x+12x^2+8x^3+x^4$
(b) $16+32x+24x^2+8x^3+x^4$
(c) $16+32x+24x^2+48x^3+x^4$
(d) $16+32x+24x^2+16x^3+6x^4$

3. The term without p in the expression $\left(2p-\frac{1}{2p^2}\right)^{12}$ is:
(a) 495 (b) 7920 (c) −495 (d) −7920

4. The second moment of area of a rectangle through its centroid is given by $\frac{bl^3}{12}$. Using the binomial theorem, the approximate percentage change in the second moment of area if b is increased by 3% and l is reduced by 2% is:
(a) +3% (b) +1% (c) −3% (d) −6%

5. When $(2x-1)^5$ is expanded by the binomial series, the fourth term is:
(a) $-40x^2$ (b) $10x^2$ (c) $40x^2$ (d) $-10x^2$

COMPANION @ WEBSITE

For fully worked solutions to each of the problems in Practice Exercises 82 to 85 in this chapter, go to the website:
www.routledge.com/cw/bird

Revision Test 4 Exponential functions, number sequences, and the binomial series

This Revision Test covers the material contained in Chapters 14 to 16. *The marks for each question are shown in brackets at the end of each question.*

1. Evaluate the following, each correct to 4 significant figures:

   (a) $e^{-0.683}$ (b) $\dfrac{5(e^{-2.73}-1)}{e^{1.68}}$ (3)

2. Expand xe^{3x} to six terms. (5)

3. Plot a graph of $y=\frac{1}{2}e^{-1.2x}$ over the range $x=-2$ to $x=+1$ and hence determine, correct to 1 decimal place, (a) the value of y when $x=-0.75$, and (b) the value of x when $y=4.0$. (6)

4. Evaluate the following, each correct to 3 decimal places:

   (a) $\ln 0.0753$ (b) $\dfrac{\ln 3.68-\ln 2.91}{4.63}$ (2)

5. Two quantities x and y are related by the equation $y=ae^{-kx}$, where a and k are constants. Determine, correct to 1 decimal place, the value of y when $a=2.114$, $k=-3.20$ and $x=1.429$. (3)

6. If $\theta_f-\theta_i=\frac{R}{J}\ln\left(\frac{U_2}{U_1}\right)$ find the value of U_2 given that $\theta_f=3.5$, $\theta_i=2.5$, $R=0.315$, $J=0.4$, $U_1=50$. (5)

7. Solve, correct to 4 significant figures:

   (a) $13e^{2x-1}=7e^x$
   (b) $\ln(x+1)^2=\ln(x+1)-\ln(x+2)+2$

   (13)

8. Determine the 20th term of the series 15.6, 15, 14.4, 13.8, … (3)

9. The sum of 13 terms of an arithmetic progression is 286 and the common difference is 3. Determine the first term of the series. (4)

10. Determine the 11th term of the series 1.5, 3, 6, 12, … (2)

11. A machine is to have seven speeds ranging from 25 rev/min to 500 rev/min. If the speeds form a geometric progression, determine their value, each correct to the nearest whole number. (7)

12. Use the binomial series to expand:
    $(2a-3b)^6$ (7)

13. Expand the following in ascending powers of t as far as the term in t^3:

    (a) $\dfrac{1}{1+t}$ (b) $\dfrac{1}{\sqrt{1-3t}}$

    For each case, state the limits for which the expansion is valid. (9)

14. The modulus of rigidity G is given by $G=\dfrac{R^4\theta}{L}$ where R is the radius, θ the angle of twist and L the length. Find the approximate percentage error in G when R is measured 1.5% too large, θ is measure 3% too small and L is measured 1% too small. (6)

For lecturers/instructors/teachers, fully worked solutions to each of the problems in Revision Test 4, together with a full marking scheme, are available at the website:
www.routledge.com/cw/bird

Section 2

Trigonometry

Chapter 17

Introduction to trigonometry

Why it is important to understand: Introduction to trigonometry

There are an enormous number of uses of trigonometry and trigonometric functions. Fields that use trigonometry or trigonometric functions include astronomy (especially for locating apparent positions of celestial objects, in which spherical trigonometry is essential) and hence navigation (on the oceans, in aircraft and in space), music theory, acoustics, optics, analysis of financial markets, electronics, probability theory, statistics, biology, medical imaging (CAT scans and ultrasound), pharmacy, chemistry, number theory (and hence cryptology), seismology, meteorology, oceanography, many physical sciences, land surveying and geodesy (a branch of earth sciences), architecture, phonetics, economics, electrical engineering, mechanical engineering, civil engineering, computer graphics, cartography, crystallography and game development. It is clear that a good knowledge of trigonometry is essential in many fields of engineering.

At the end of this chapter, you should be able to:

- state the theorem of Pythagoras and use it to find the unknown side of a right angled triangle
- define sine, cosine, tangent, secant, cosecant and cotangent of an angle in a right angled triangle
- understand fractional and surd forms of trigonometric ratios
- evaluate trigonometric ratios of angles
- solve right angled triangles
- understand angles of elevation and depression
- appreciate trigonometric approximations for small angles

17.1 Trigonometry

Trigonometry is the branch of mathematics that deals with the measurement of sides and angles of triangles, and their relationship with each other. There are many applications in engineering where knowledge of trigonometry is needed.

17.2 The theorem of Pythagoras

With reference to Fig. 17.1, the side opposite the right angle (i.e. side b) is called the **hypotenuse**. The theorem of Pythagoras* states:

'In any right-angle triangle, the square on the hypotenuse is equal to the sum of the squares on the other two sides.'

Hence $\qquad b^2 = a^2 + c^2$

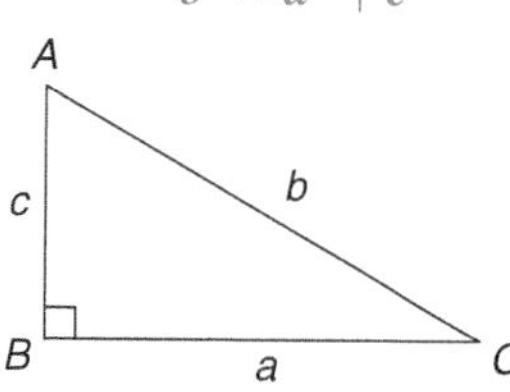

Figure 17.1

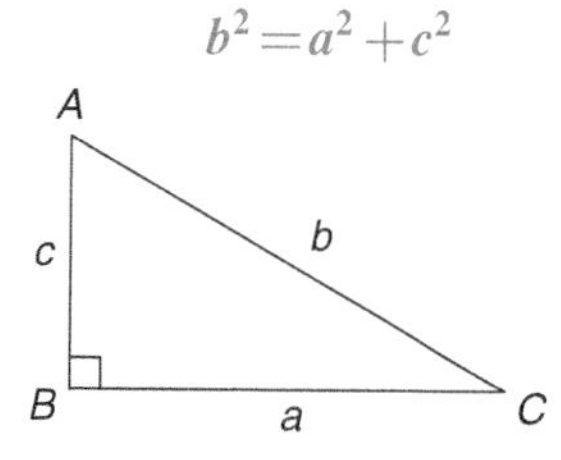

Problem 1. In Fig. 17.2, find the length of EF.

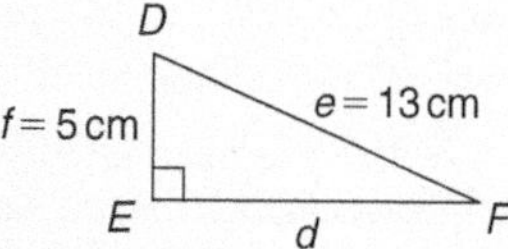

Figure 17.2

*Who was Pythagoras? – **Pythagoras of Samos** (Born about 570 BC and died about 495 BC) was an Ionian Greek philosopher and mathematician. He is best known for the Pythagorean theorem, which states that in a right-angled triangle $a^2 + b^2 = c^2$. To find out more go to www.routledge.com/cw/bird

By Pythagoras' theorem: $\qquad e^2 = d^2 + f^2$

Hence $\qquad 13^2 = d^2 + 5^2$

$$169 = d^2 + 25$$

$$d^2 = 169 - 25 = 144$$

Thus $\qquad d = \sqrt{144} = 12\,\text{cm}$

i.e. $\qquad$ **EF = 12 cm**

Problem 2. Two aircraft leave an airfield at the same time. One travels due north at an average speed of 300 km/h and the other due west at an average speed of 220 km/h. Calculate their distance apart after 4 hours

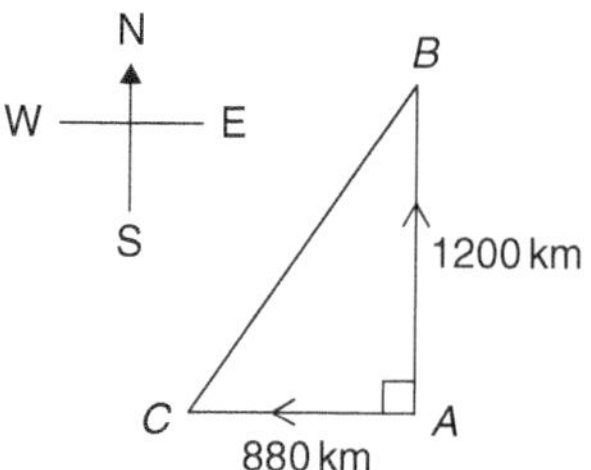

Figure 17.3

After 4 hours, the first aircraft has travelled $4 \times 300 = 1{,}200\,\text{km}$, due north, and the second aircraft has travelled $4 \times 220 = 880\,\text{km}$ due west, as shown in Fig. 17.3. Distance apart after 4 hour $= BC$. From Pythagoras' theorem:

$$BC^2 = 1200^2 + 880^2$$

$$= 1440000 + 774400 \text{ and}$$

$$BC = \sqrt{2214400}$$

Hence distance apart after 4 hours = 1488 km

Now try the following Practice Exercise

Practice Exercise 87 The theorem of Pythagoras (Answers on page 710)

1. In a triangle CDE, $D = 90°$, $CD = 14.83\,\text{mm}$ and $CE = 28.31\,\text{mm}$. Determine the length of DE

2. Triangle PQR is isosceles, Q being a right angle. If the hypotenuse is 38.47 cm find (a) the lengths of sides PQ and QR, and (b) the value of $\angle QPR$

3. A man cycles 24 km due south and then 20 km due east. Another man, starting at the same time and position as the first man, cycles 32 km due east and then 7 km due south. Find the distance between the two men
4. A ladder 3.5 m long is placed against a perpendicular wall with its foot 1.0 m from the wall. How far up the wall (to the nearest centimetre) does the ladder reach? If the foot of the ladder is now moved 30 cm further away from the wall, how far does the top of the ladder fall?
5. Two ships leave a port at the same time. One travels due west at 18.4 km/h and the other due south at 27.6 km/h. Calculate how far apart the two ships are after 4 hours
6. Figure 17.4 shows a bolt rounded off at one end. Determine the dimension h

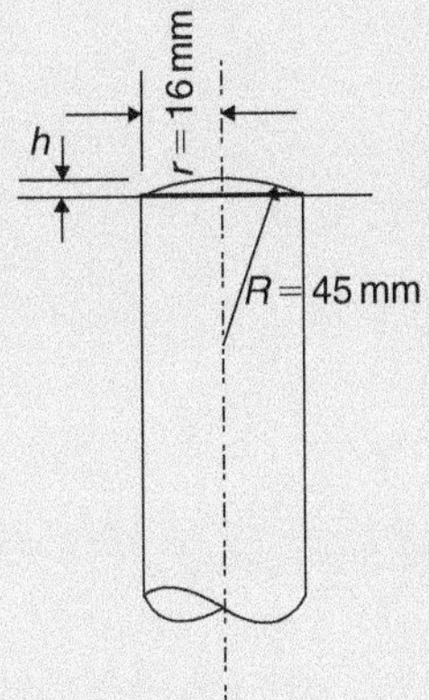

Figure 17.4

7. Figure 17.5 shows a cross-section of a component that is to be made from a round bar. If the diameter of the bar is 74 mm, calculate the dimension x

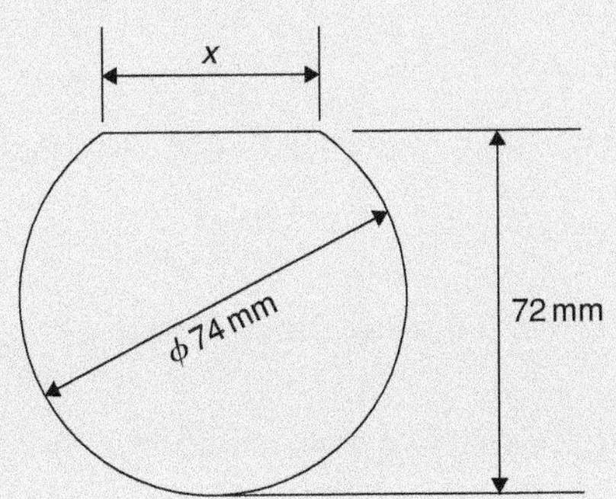

Figure 17.5

17.3 Trigonometric ratios of acute angles

(a) With reference to the right-angled triangle shown in Fig. 17.6:

(i) $\sin\theta = \dfrac{\text{opposite side}}{\text{hypotenuse}}$

i.e. $\mathbf{\sin\theta = \dfrac{b}{c}}$

(ii) $\text{cosine}\,\theta = \dfrac{\text{adjacent side}}{\text{hypotenuse}}$

i.e. $\mathbf{\cos\theta = \dfrac{a}{c}}$

(iii) $\text{tangent}\,\theta = \dfrac{\text{opposite side}}{\text{adjacent side}}$

i.e. $\mathbf{\tan\theta = \dfrac{b}{a}}$

(iv) $\text{secant}\,\theta = \dfrac{\text{hypotenuse}}{\text{adjacent side}}$

i.e. $\mathbf{\sec\theta = \dfrac{c}{a}}$

(v) $\text{cosecant}\,\theta = \dfrac{\text{hypotenuse}}{\text{opposite side}}$

i.e. $\mathbf{\text{cosec}\,\theta = \dfrac{c}{b}}$

(vi) $\text{cotangent}\,\theta = \dfrac{\text{adjacent side}}{\text{opposite side}}$

i.e. $\mathbf{\cot\theta = \dfrac{a}{b}}$

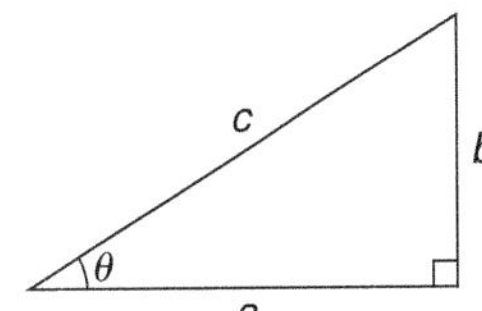

Figure 17.6

(b) From above,

(i) $\dfrac{\sin\theta}{\cos\theta} = \dfrac{\frac{b}{c}}{\frac{a}{c}} = \dfrac{b}{a} = \tan\theta,$

i.e. $\mathbf{\tan\theta = \dfrac{\sin\theta}{\cos\theta}}$

(ii) $\dfrac{\cos\theta}{\sin\theta} = \dfrac{\frac{a}{c}}{\frac{b}{c}} = \dfrac{a}{b} = \cot\theta,$

i.e. $\mathbf{\cot\theta = \dfrac{\cos\theta}{\sin\theta}}$

(iii) $\mathbf{\sec\theta = \dfrac{1}{\cos\theta}}$

(iv) $\mathbf{\text{cosec}\,\theta = \dfrac{1}{\sin\theta}}$

(Note '*s*' and '*c*' go together)

(v) $\mathbf{\cot\theta = \dfrac{1}{\tan\theta}}$

Secants, cosecants and cotangents are called the **reciprocal ratios**.

Problem 3. If $\cos X = \dfrac{9}{41}$ determine the value of the other five trigonometric ratios

Figure 17.7 shows a right-angled triangle *XYZ*.

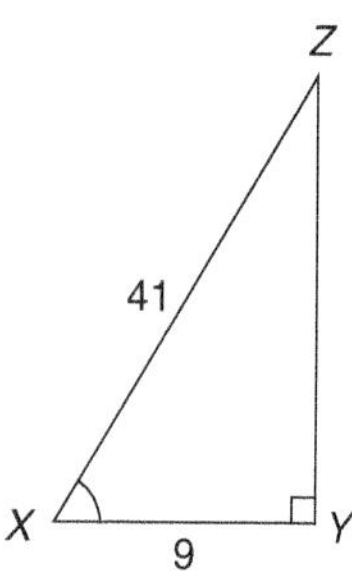

Figure 17.7

Since $\cos X = \dfrac{9}{41}$, then $XY = 9$ units and $XZ = 41$ units.

Using Pythagoras' theorem: $41^2 = 9^2 + YZ^2$ from which $YZ = \sqrt{41^2 - 9^2} = 40$ units.

Thus, $\mathbf{\sin X = \dfrac{40}{41}}$, $\mathbf{\tan X = \dfrac{40}{9} = 4\dfrac{4}{9}}$,

$\mathbf{\text{cosec}\,X = \dfrac{41}{40} = 1\dfrac{1}{40}}$, $\mathbf{\sec X = \dfrac{41}{9} = 4\dfrac{5}{9}}$

and $\mathbf{\cot X = \dfrac{9}{40}}$

Problem 4. If $\sin\theta = 0.625$ and $\cos\theta = 0.500$ determine the values of $\text{cosec}\,\theta, \sec\theta, \tan\theta$ and $\cot\theta$

$$\text{cosec}\,\theta = \frac{1}{\sin\theta} = \frac{1}{0.625} = \mathbf{1.60}$$

$$\sec\theta = \frac{1}{\cos\theta} = \frac{1}{0.500} = \mathbf{2.00}$$

$$\tan\theta = \frac{\sin\theta}{\cos\theta} = \frac{0.625}{0.500} = \mathbf{1.25}$$

$$\cot\theta = \frac{\cos\theta}{\sin\theta} = \frac{0.500}{0.625} = \mathbf{0.80}$$

Problem 5. In Fig. 17.8 point *A* lies at co-ordinate (2, 3) and point *B* at (8, 7). Determine (a) the distance *AB*, (b) the gradient of the straight line *AB*, and (c) the angle *AB* makes with the horizontal

(a) Points *A* and *B* are shown in Fig. 17.8(a).

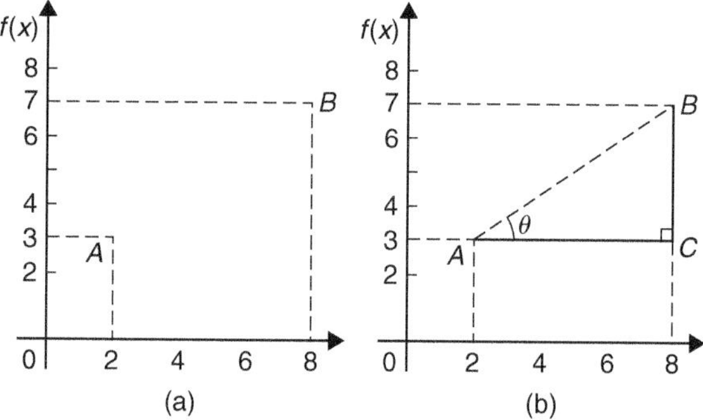

Figure 17.8

In Fig. 17.8(b), the horizontal and vertical lines *AC* and *BC* are constructed. Since *ABC* is a right-angled triangle, and $AC = (8-2) = 6$ and $BC = (7-3) = 4$, then by Pythagoras' theorem:

$$AB^2 = AC^2 + BC^2 = 6^2 + 4^2$$

and $\mathbf{AB} = \sqrt{6^2 + 4^2} = \sqrt{52} = \mathbf{7.211}$ correct to 3 decimal places

(b) The gradient of *AB* is given by $\tan\theta$,

i.e. **gradient** $= \tan\theta = \dfrac{BC}{AC} = \dfrac{4}{6} = \mathbf{\dfrac{2}{3}}$

(c) **The angle *AB* makes with the horizontal** is given by: $\tan^{-1}\dfrac{2}{3} = \mathbf{33.69°}$

Now try the following Practice Exercise

Practice Exercise 88 Trigonometric ratios of acute angles (Answers on page 710)

1. In triangle *ABC* shown in Fig. 17.9, find sin *A*, cos *A*, tan *A*, sin *B*, cos *B* and tan *B*

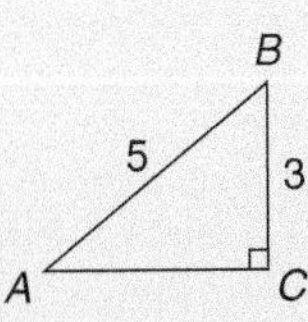

Figure 17.9

2. For the right-angled triangle shown in Fig. 17.10, find:
(a) $\sin\alpha$ (b) $\cos\theta$ (c) $\tan\theta$

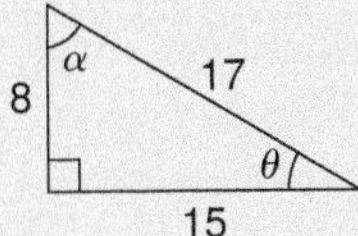

Figure 17.10

3. If $\cos A = \dfrac{12}{13}$ find $\sin A$ and $\tan A$, in fraction form

4. Point P lies at co-ordinate $(-3, 1)$ and point Q at $(5, -4)$. Determine (a) the distance PQ, (b) the gradient of the straight line PQ, and (c) the angle PQ makes with the horizontal

17.4 Fractional and surd forms of trigonometric ratios

In Fig. 17.11, ABC is an equilateral triangle of side 2 units. AD bisects angle A and bisects the side BC. Using Pythagoras' theorem on triangle ABD gives:

$$AD = \sqrt{2^2 - 1^2} = \sqrt{3}$$

Hence, $$\sin 30^\circ = \frac{BD}{AB} = \frac{1}{2}, \ \cos 30^\circ = \frac{AD}{AB} = \frac{\sqrt{3}}{2}$$

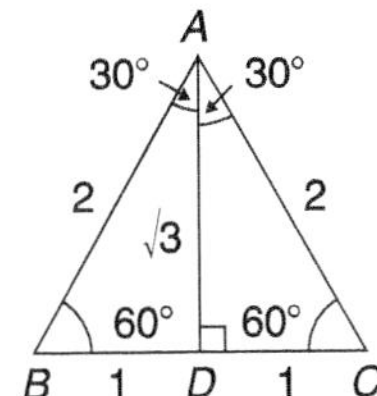

Figure 17.11

and $$\tan 30^\circ = \frac{BD}{AD} = \frac{1}{\sqrt{3}}$$

$$\sin 60^\circ = \frac{AD}{AB} = \frac{\sqrt{3}}{2}, \ \cos 60^\circ = \frac{BD}{AB} = \frac{1}{2}$$

and $$\tan 60^\circ = \frac{AD}{BD} = \sqrt{3}$$

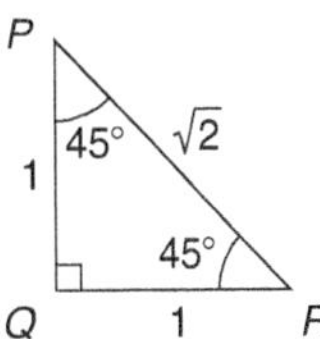

Figure 17.12

In Fig. 17.12, PQR is an isosceles triangle with $PQ = QR = 1$ unit. By Pythagoras' theorem, $PR = \sqrt{1^2 + 1^2} = \sqrt{2}$

Hence,

$$\sin 45^\circ = \frac{1}{\sqrt{2}}, \quad \cos 45^\circ = \frac{1}{\sqrt{2}} \quad \text{and } \tan 45^\circ = 1$$

A quantity that is not exactly expressible as a rational number is called a **surd**. For example, $\sqrt{2}$ and $\sqrt{3}$ are called surds because they cannot be expressed as a fraction and the decimal part may be continued indefinitely. For example,

$$\sqrt{2} = 1.4142135\ldots, \text{ and } \sqrt{3} = 1.7320508\ldots$$

From above,

$$\sin 30^\circ = \cos 60^\circ, \sin 45^\circ = \cos 45^\circ \text{ and}$$
$$\sin 60^\circ = \cos 30^\circ.$$

In general,

$$\sin\theta = \cos(90^\circ - \theta) \quad \text{and} \quad \cos\theta = \sin(90^\circ - \theta)$$

For example, it may be checked by calculator that $\sin 25^\circ = \cos 65^\circ$, $\sin 42^\circ = \cos 48^\circ$ and $\cos 84^\circ\ 10' = \sin 5^\circ\ 50'$ and so on.

Problem 6. Using surd forms, evaluate:
$$\frac{3\tan 60^\circ - 2\cos 30^\circ}{\tan 30^\circ}$$

From above, $\tan 60^\circ = \sqrt{3}$, $\cos 30^\circ = \dfrac{\sqrt{3}}{2}$ and

$\tan 30° = \frac{1}{\sqrt{3}}$, hence

$$\frac{3\tan 60° - 2\cos 30°}{\tan 30°} = \frac{3(\sqrt{3}) - 2\left(\frac{\sqrt{3}}{2}\right)}{\frac{1}{\sqrt{3}}}$$

$$= \frac{3\sqrt{3} - \sqrt{3}}{\frac{1}{\sqrt{3}}} = \frac{2\sqrt{3}}{\frac{1}{\sqrt{3}}}$$

$$= 2\sqrt{3}\left(\frac{\sqrt{3}}{1}\right) = 2(3) = \mathbf{6}$$

Now try the following Practice Exercise

Practice Exercise 89 Fractional and surd forms of trigonometric ratios (Answers on page 710)

Evaluate the following, without using calculators, leaving where necessary in surd form:

1. $3 \sin 30° - 2 \cos 60°$
2. $5 \tan 60° - 3 \sin 60°$
3. $\frac{\tan 60°}{3\tan 30°}$
4. $(\tan 45°)(4 \cos 60° - 2 \sin 60°)$
5. $\frac{\tan 60° - \tan 30°}{1 + \tan 30° \tan 60°}$

17.5 Evaluating trigonometric ratios of any angles

The easiest method of evaluating trigonometric functions of any angle is by using a **calculator**. The following values, correct to 4 decimal places, may be checked:

$$\begin{aligned} \text{sine } 18° &= 0.3090 \\ \text{sine } 172° &= 0.1392 \\ \text{sine } 241.63° &= -0.8799 \\ \text{cosine } 56° &= 0.5592 \\ \text{cosine } 115° &= -0.4226 \\ \text{cosine } 331.78° &= 0.8811 \\ \text{tangent } 29° &= 0.5543 \\ \text{tangent } 178° &= -0.0349 \\ \text{tangent } 296.42° &= -2.0127 \end{aligned}$$

To evaluate, say, sine $42°23'$ using a calculating means finding sine $42\frac{23°}{60}$ since there are 60 minutes in 1 degree.

$$\frac{23}{60} = 0.383\dot{3}, \text{ thus } 42°23' = 42.383\dot{3}°$$

Thus sine $42°23'$ = sine $42.383\dot{3}° = 0.6741$, correct to 4 decimal places.

Similarly, cosine $72°38'$ = cosine $72\frac{38°}{60} = 0.2985$, correct to 4 decimal places.

Most calculators contain only sine, cosine and tangent functions. Thus to evaluate secants, cosecants and cotangents, reciprocals need to be used.

The following values, correct to 4 decimal places, may be checked:

$$\text{secant } 32° = \frac{1}{\cos 32°} = 1.1792$$

$$\text{cosecant } 75° = \frac{1}{\sin 75°} = 1.0353$$

$$\text{cotangent } 41° = \frac{1}{\tan 41°} = 1.1504$$

$$\text{secant } 215.12° = \frac{1}{\cos 215.12°} = -1.2226$$

$$\text{cosecant } 321.62° = \frac{1}{\sin 321.62°} = -1.6106$$

$$\text{cotangent } 263.59° = \frac{1}{\tan 263.59°} = 0.1123$$

If we know the value of a trigonometric ratio and need to find the angle we use the **inverse function** on our calculators.

For example, using shift and sin on our calculator gives $\sin^{-1}$

If, for example, we know the sine of an angle is 0.5 then the value of the angle is given by:

$\sin^{-1} 0.5 = \mathbf{30°}$ (Check that $\sin 30° = 0.5$)

(Note that $\sin^{-1} x$ does not mean $\frac{1}{\sin x}$; also, $\sin^{-1} x$ may also be written as $\arcsin x$)

Similarly, if $\cos\theta = 0.4371$ then

$\theta = \cos^{-1} 0.4371 = \mathbf{64.08°}$

and if $\tan A = 3.5984$ then $A = \tan^{-1} 3.5984 = \mathbf{74.47°}$ each correct to 2 decimal places.

Use your calculator to check the following worked examples.

Problem 7. Determine, correct to 4 decimal places, $\sin 43°39'$

$$\sin 43°39' = \sin 43\frac{39°}{60} = \sin 43.65° = \mathbf{0.6903}$$

This answer can be obtained using the **calculator** as follows:

1. Press sin 2. Enter 43 3. Press °''' 4. Enter 39
5. Press °''' 6. Press) 7. Press =
Answer = **0.6902512...**

Problem 8. Determine, correct to 3 decimal places, $6\cos 62°12'$

$$6\cos 62°12' = 6\cos 62\frac{12°}{60} = 6\cos 62.20° = \mathbf{2.798}$$

This answer can be obtained using the **calculator** as follows:

1. Enter 6 2. Press cos 3. Enter 62 4. Press °'''
5. Enter 12 6. Press °''' 7. Press) 8. Press =
Answer = **2.798319...**

Problem 9. Evaluate correct to 4 decimal places:
(a) sine 168°14′ (b) cosine 271.41°
(c) tangent 98°4′

(a) $\text{sine } 168°14' = \text{sine } 168\frac{14°}{60} = \mathbf{0.2039}$

(b) $\text{cosine } 271.41° = \mathbf{0.0246}$

(c) $\text{tangent } 98°4' = \tan 98\frac{4°}{60} = \mathbf{-7.0558}$

Problem 10. Evaluate, correct to 4 decimal places:
(a) secant 161° (b) secant 302°29′

(a) $\sec 161° = \dfrac{1}{\cos 161°} = \mathbf{-1.0576}$

(b) $\sec 302°29' = \dfrac{1}{\cos 302°29'} = \dfrac{1}{\cos 302\frac{29°}{60}}$
$= \mathbf{1.8620}$

Problem 11. Evaluate, correct to 4 significant figures:
(a) cosecant 279.16° (b) cosecant 49°7′

(a) $\text{cosec } 279.16° = \dfrac{1}{\sin 279.16°} = \mathbf{-1.013}$

(b) $\text{cosec } 49°7' = \dfrac{1}{\sin 49°7'} = \dfrac{1}{\sin 49\frac{7°}{60}} = \mathbf{1.323}$

Problem 12. Evaluate, correct to 4 decimal places:
(a) cotangent 17.49° (b) cotangent 163°52′

(a) $\cot 17.49° = \dfrac{1}{\tan 17.49°} = \mathbf{3.1735}$

(b) $\cot 163°52' = \dfrac{1}{\tan 163°52'} = \dfrac{1}{\tan 163\frac{52°}{60}}$
$= \mathbf{-3.4570}$

Problem 13. Evaluate, correct to 4 significant figures:
(a) sin 1.481 (b) cos (3π/5) (c) tan 2.93

(a) sin 1.481 means the sine of 1.481 radians. Hence a calculator needs to be on the radian function.

Hence $\sin 1.481 = \mathbf{0.9960}$

(b) $\cos(3\pi/5) = \cos 1.884955\ldots = \mathbf{-0.3090}$

(c) $\tan 2.93 = \mathbf{-0.2148}$

Problem 14. Evaluate, correct to 4 decimal places:
(a) secant 5.37 (b) cosecant π/4
(c) cotangent π/24

(a) Again, with no degrees sign, it is assumed that 5.37 means 5.37 radians.

Hence $\sec 5.37 = \dfrac{1}{\cos 5.37} = \mathbf{1.6361}$

(b) $\text{cosec}(\pi/4) = \dfrac{1}{\sin(\pi/4)} = \dfrac{1}{\sin 0.785398\ldots}$
$= \mathbf{1.4142}$

(c) $\cot(5\pi/24) = \dfrac{1}{\tan(5\pi/24)} = \dfrac{1}{\tan 0.654498\ldots}$
$= \mathbf{1.3032}$

Problem 15. Find, in degrees, the acute angle $\sin^{-1} 0.4128$ correct to 2 decimal places.

$\sin^{-1} 0.4128$ means 'the angle whose sine is 0.4128'
Using a calculator:
1. Press shift 2. Press sin 3. Enter 0.4128
4. Press)
5. Press = The answer 24.380848…is displayed
Hence, $\mathbf{\sin^{-1} 0.4128 = 24.38°}$

Problem 16. Find the acute angle $\cos^{-1} 0.2437$ in degrees and minutes

$\cos^{-1} 0.2437$ means 'the angle whose cosine is 0.2437'
Using a calculator:
1. Press shift 2. Press cos 3. Enter 0.2437
4. Press)
5. Press = The answer 75.894979… is displayed
6. Press °’’’ and 75°53′ 41.93″ is displayed
Hence, $\mathbf{\cos^{-1} 0.2437 = 75.89° = 77°54'}$ correct to the nearest minute.

Problem 17. Find the acute angle $\tan^{-1} 7.4523$ in degrees and minutes

$\tan^{-1} 7.4523$ means 'the angle whose tangent is 7.4523'
Using a calculator:
1. Press shift 2. Press tan 3. Enter 7.4523
4. Press) 5. Press = The answer 82.357318… is displayed 6. Press °’’’ and 82°21′ 26.35″ is displayed
Hence, $\mathbf{\tan^{-1} 7.4523 = 82.36° = 82°21'}$ correct to the nearest minute.

Problem 18. Determine the acute angles:
(a) $\sec^{-1} 2.3164$ (b) $\mathrm{cosec}^{-1}\, 1.1784$
(c) $\cot^{-1} 2.1273$

(a) $\sec^{-1} 2.3164 = \cos^{-1}\left(\dfrac{1}{2.3164}\right)$

$= \cos^{-1} 0.4317\ldots$

$= \mathbf{64.42°}$ or $\mathbf{64°25'}$ or **1.124 radians**

(b) $\mathrm{cosec}^{-1}\, 1.1784 = \sin^{-1}\left(\dfrac{1}{1.1784}\right)$

$= \sin^{-1} 0.8486\ldots$

$= \mathbf{58.06°}$ or $\mathbf{58°4'}$ or **1.013 radians**

(c) $\cot^{-1} 2.1273 = \tan^{-1}\left(\dfrac{1}{2.1273}\right)$

$= \tan^{-1} 0.4700\ldots$

$= \mathbf{25.18°}$ or $\mathbf{25°11'}$ or **0.439 radians**.

Problem 19. A locomotive moves around a curve of radius, $r = 500$ m. The angle of banking, θ, is given by: $\theta = \tan^{-1}\left(\dfrac{v^2}{rg}\right)$ where $g = 9.81$ m/s^2 and v is the speed in m/s. Calculate the angle of banking when the speed of the locomotive is 30 km/h.

Angle of banking, $\theta = \tan^{-1}\left(\dfrac{v^2}{rg}\right)$

$v = 30\text{km/h} = 30\dfrac{\text{km}}{\text{h}} \times \dfrac{1\,\text{h}}{3600\,\text{s}} \times \dfrac{1000\,\text{m}}{1\,\text{km}} = \dfrac{30}{3.6}$
$= 8.3333$ m/s

Hence, $\theta = \tan^{-1}\left(\dfrac{8.3333^2\,\text{m}^2/\text{s}^2}{500\,\text{m} \times 9.81\,\text{m/s}^2}\right)$
$= \tan^{-1}(0.014158)$

i.e. **angle of banking,** $\boldsymbol{\theta} = \mathbf{0.811°}$

Problem 20. Evaluate correct to 4 decimal places:
(a) sec (−115°) (b) cosec (−95°17′)

(a) Positive angles are considered by convention to be anticlockwise and negative angles as clockwise. Hence −115° is actually the same as 245° (i.e. 360° − 115°)

Hence $\sec(-115°) = \sec 245° = \dfrac{1}{\cos 245°}$

$= \mathbf{-2.3662}$

(b) $\mathrm{cosec}(-95°47') = \dfrac{1}{\sin\left(-95\dfrac{47°}{60}\right)} = \mathbf{-1.0051}$

Problem 21. In the triangular template EFG in Fig. 17.13, calculate angle G.

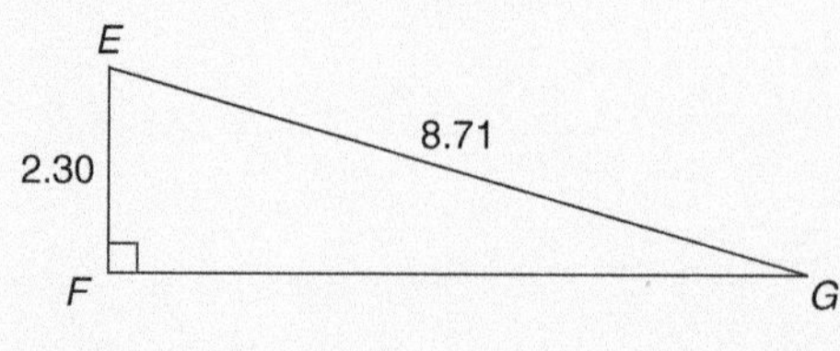

Figure 17.13

With reference to $\angle G$, the two sides of the triangle given are the opposite side EF and the hypotenuse EG; hence, sine is used,

i.e. $\sin G = \frac{2.30}{8.71} = 0.26406429\ldots$

from which, $G = \sin^{-1} 0.26406429\ldots$

i.e. $G = 15.311360\ldots$

Hence, $\angle \mathbf{G} = \mathbf{15.31^\circ}$ or $\mathbf{15^\circ\, 19'}$

Now try the following Practice Exercise

Practice Exercise 90 Evaluating trigonometric ratios (Answers on page 710)

In Problems 1 to 8, evaluate correct to 4 decimal places:

1. (a) sine 27° (b) sine 172.41° (c) sine 302°52′
2. (a) cosine 124° (b) cosine 21.46° (c) cosine 284°10′
3. (a) tangent 145° (b) tangent 310.59° (c) tangent 49°16′
4. (a) secant 73° (b) secant 286.45° (c) secant 155°41′
5. (a) cosecant 213° (b) cosecant 15.62° (c) cosecant 311°50′
6. (a) cotangent 71° (b) cotangent 151.62° (c) cotangent 321°23′
7. (a) sine $\frac{2\pi}{3}$ (b) cos 1.681 (c) tan 3.672
8. (a) sine $\frac{\pi}{8}$ (b) cosec 2.961 (c) cot 2.612

In Problems 9 to 14, determine the acute angle in degrees (correct to 2 decimal places), degrees and minutes, and in radians (correct to 3 decimal places).

9. $\sin^{-1} 0.2341$
10. $\cos^{-1} 0.8271$
11. $\tan^{-1} 0.8106$
12. $\sec^{-1} 1.6214$
13. $\text{cosec}^{-1} 2.4891$
14. $\cot^{-1} 1.9614$
15. In the triangle shown in Fig. 17.14, determine angle θ, correct to 2 decimal places

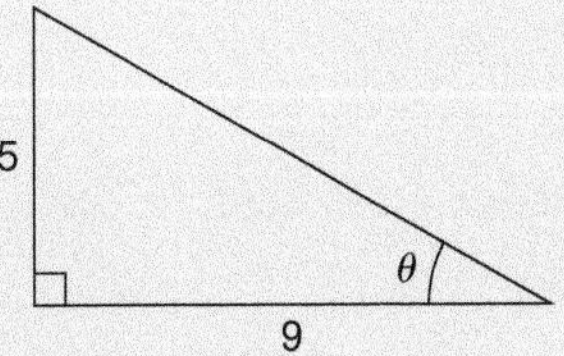

Figure 17.14

16. In the triangle shown in Fig. 17.15, determine angle θ in degrees and minutes

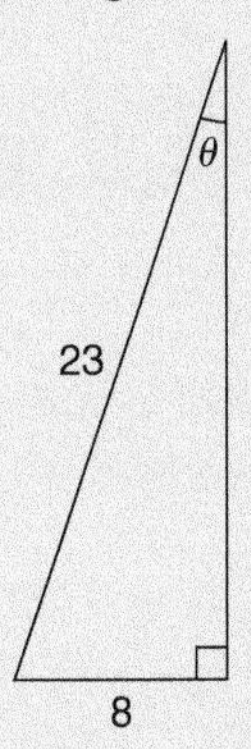

Figure 17.15

In Problems 17 to 20, evaluate correct to 4 significant figures.

17. 4 cos 56°19′ − 3 sin 21°57′
18. $\frac{6.4\,\text{cosec}\,29^\circ 5' - \sec 81^\circ}{2\cot 12^\circ}$
19. Determine the acute angle, in degrees and minutes, correct to the nearest minute, given by: $\sin^{-1}\left(\frac{4.32\sin 42^\circ 16'}{7.86}\right)$
20. If tan x = 1.5276, determine sec x, cosec x, and cot x. (Assume x is an acute angle)

In Problems 21 and 22 evaluate correct to 4 significant figures.

21. $\frac{(\sin 34^\circ 27')(\cos 69^\circ 2')}{(2\tan 53^\circ 39')}$
22. 3 cot 14°15′ sec 23°9′
23. Evaluate correct to 4 decimal places:
(a) sin(−125°) (b) tan(−241°)
(c) cos(−49°15′)
24. Evaluate correct to 5 significant figures:
(a) cosec (−143°) (b) cot(−252°)
(c) sec(−67°22′)

25. A cantilever is subjected to a vertically applied downward load at its free end. The position, β, of the maximum bending stress is given by:

$$\beta = \tan^{-1}\left(-\frac{I_{UU}\tan\theta}{I_{YY}}\right)$$

Evaluate angle β, in degrees, correct to 2 decimal places, when $I_{UU} = 1.381 \times 10^{-5}$ m^4, $\theta = 17.64°$ and $I_{YY} = 1.741 \times 10^{-6}$ m^4.

17.6 Solution of right-angled triangles

To 'solve a right-angled triangle' means 'to find the unknown sides and angles'. This is achieved by using (i) the theorem of Pythagoras, and/or (ii) trigonometric ratios. This is demonstrated in the following problems.

Problem 22. In triangle PQR shown in Fig. 17.16, find the lengths of PQ and PR.

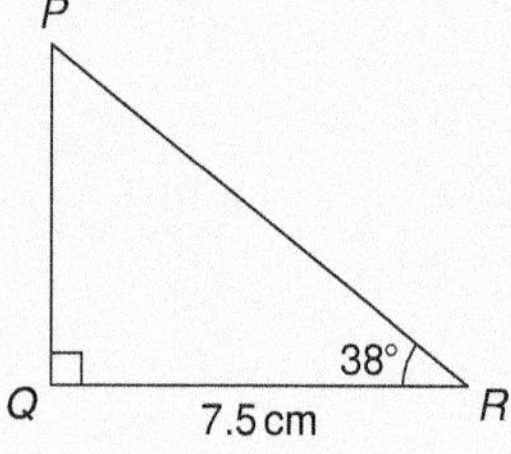

Figure 17.16

$$\tan 38° = \frac{PQ}{QR} = \frac{PQ}{7.5} \text{ hence}$$

$$PQ = 7.5 \tan 38° = 7.5(0.7813)$$

$$= \mathbf{5.860\,cm}$$

$$\cos 38° = \frac{QR}{PR} = \frac{7.5}{PR} \text{ hence}$$

$$PR = \frac{7.5}{\cos 38°} = \frac{7.5}{0.7880} = \mathbf{9.518\,cm}$$

[Check: Using Pythagoras' theorem $(7.5)^2 + (5.860)^2 = 90.59 = (9.518)^2$]

Problem 23. Solve the triangle ABC shown in Fig. 17.17

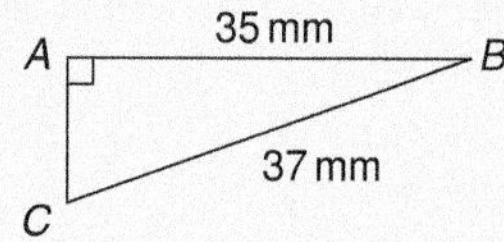

Figure 17.17

To 'solve triangle ABC' means 'to find the length AC and angles B and C'.

$$\sin C = \frac{35}{37} = 0.94595 \text{ hence}$$

$$\angle \mathbf{C} = \sin^{-1} 0.94595 = \mathbf{71.08°} \text{ or } \mathbf{71°5'}$$

$\angle \mathbf{B} = 180° - 90° - 71.08° = \mathbf{18.92°}$ or $\mathbf{18°55'}$ (since angles in a triangle add up to 180°)

$$\sin B = \frac{AC}{37} \text{ hence}$$

$$\mathbf{AC} = 37 \sin 18.92° = 37(0.3242)$$

$$= \mathbf{12.0\,mm},$$

or, using Pythagoras' theorem, $37^2 = 35^2 + AC^2$, from which, $\mathbf{AC} = \sqrt{37^2 - 35^2} = \mathbf{12.0\,mm}$.

Problem 24. Solve triangle XYZ given $\angle X = 90°$, $\angle Y = 23°17'$ and $YZ = 20.0$ mm. Determine also its area

It is always advisable to make a reasonably accurate sketch so as to visualize the expected magnitudes of unknown sides and angles. Such a sketch is shown in Fig. 17.18

$$\angle Z = 180° - 90° - 23°17' = \mathbf{66°43'}$$

$$\sin 23°17' = \frac{XZ}{20.0} \text{ hence } \mathbf{XZ} = 20.0 \sin 23°17'$$

$$= 20.0(0.3953)$$

$$= \mathbf{7.906\,mm}$$

$$\cos 23°17' = \frac{XY}{20.0} \text{ hence } \mathbf{XY} = 20.0 \cos 23°17'$$

$$= 20.0(0.9186)$$

$$= \mathbf{18.37\,mm}$$

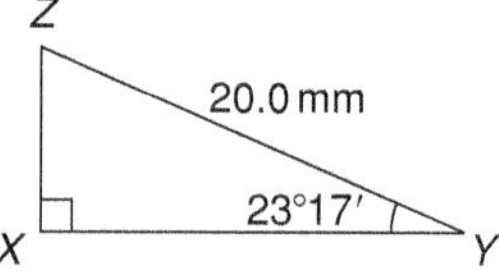

Figure 17.18

[Check: Using Pythagoras' theorem $(18.37)^2 + (7.906)^2 = 400.0 = (20.0)^2$]

Area of triangle XYZ

$$= \frac{1}{2}(\text{base})(\text{perpendicular height})$$

$$= \frac{1}{2}(XY)(XZ) = \frac{1}{2}(18.37)(7.906)$$
$$= \mathbf{72.62\,mm^2}.$$

Now try the following Practice Exercise

Practice Exercise 91 The solution of right-angled triangles (Answers on page 711)

1. Solve triangle ABC in Fig. 17.19(i).

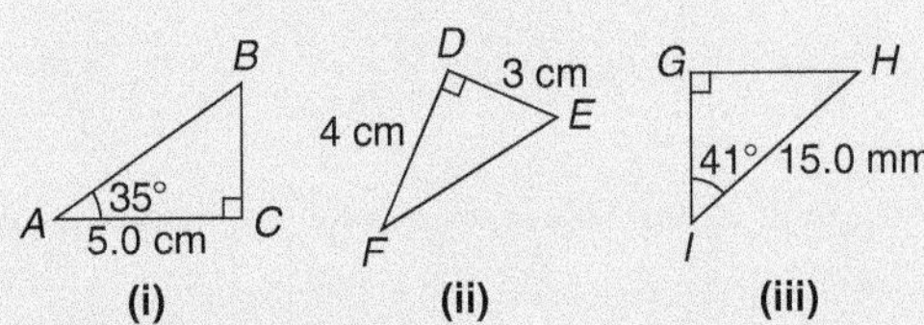

Figure 17.19

2. Solve triangle DEF in Fig. 17.19 (ii).
3. Solve triangle GHI in Fig. 17.19(iii).
4. Solve the triangle JKL in Fig. 17.20 (i) and find its area.

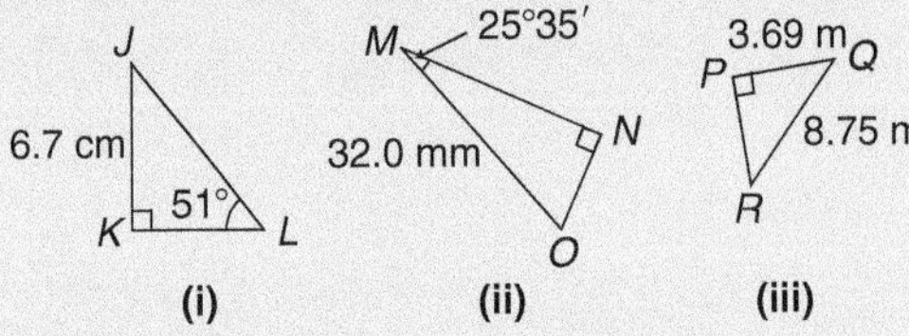

Figure 17.20

5. Solve the triangle MNO in Fig. 17.20(ii) and find its area.
6. Solve the triangle PQR in Fig. 17.20(iii) and find its area.
7. A ladder rests against the top of the perpendicular wall of a building and makes an angle of 73° with the ground. If the foot of the ladder is 2 m from the wall, calculate the height of the building.
8. Determine the length x in Fig. 22.21.

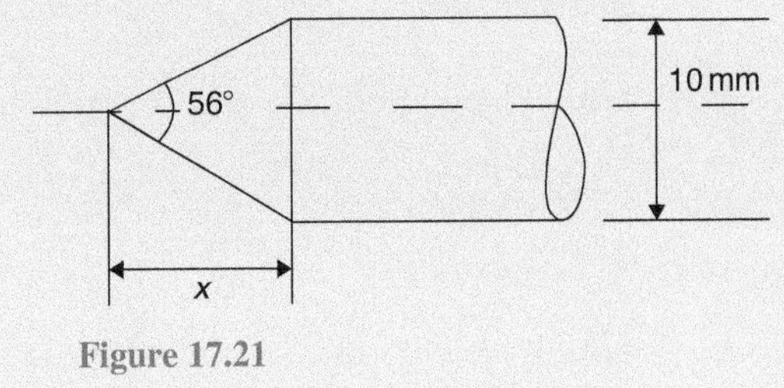

Figure 17.21

17.7 Angle of elevation and depression

(a) If, in Fig. 17.22, BC represents horizontal ground and AB a vertical flagpole, then the **angle of elevation** of the top of the flagpole, A, from the point C is the angle that the imaginary straight line AC must be raised (or elevated) from the horizontal CB, i.e. angle θ. In practice, **surveyors** measure this angle using an instrument called a **theodolite**.

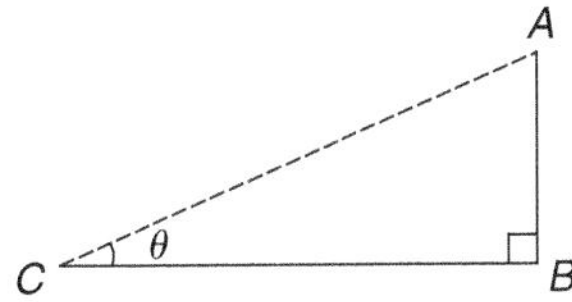

Figure 17.22

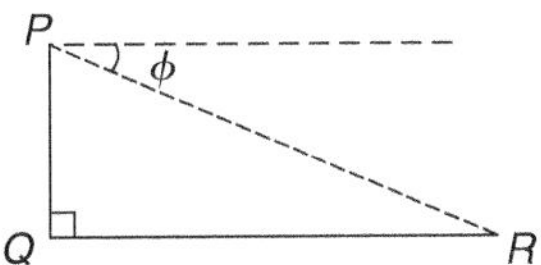

Figure 17.23

(b) If, in Fig. 17.23, PQ represents a vertical cliff and R a ship at sea, then the **angle of depression** of the ship from point P is the angle through which the imaginary straight line PR must be lowered (or depressed) from the horizontal to the ship, i.e. angle ϕ.

(Note, $\angle PRQ$ is also ϕ – alternate angles between parallel lines.)

Problem 25. An electricity pylon stands on horizontal ground. At a point 80 m from the base of the pylon, the angle of elevation of the top of the pylon is 23°. Calculate the height of the pylon to the nearest metre

Figure 17.24 shows the pylon AB and the angle of elevation of A from point C is 23° and

$$\tan 23° = \frac{AB}{BC} = \frac{AB}{80}$$

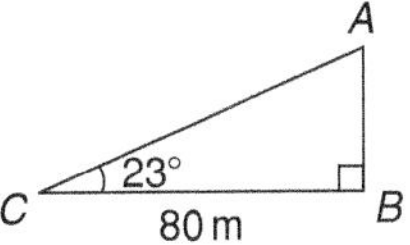

Figure 17.24

Hence, height of pylon $AB = 80 \tan 23°$

$$= 80(0.4245) = 33.96\,\text{m}$$
$$= \textbf{34 m to the nearest metre.}$$

Problem 26. A surveyor measures the angle of elevation of the top of a perpendicular building as 19°. He move 120 m nearer the building and finds the angle of elevation is now 47°. Determine the height of the building

The building PQ and the angles of elevation are shown in Fig. 17.25

In triangle PQS, $\tan 19° = \dfrac{h}{x+120}$

hence $h = \tan 19°(x+120)$

i.e. $h = 0.3443(x+120)$ (1)

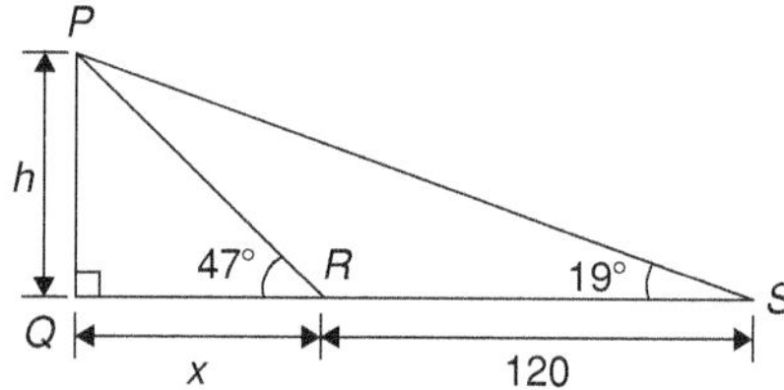

Figure 17.25

In triangle PQR, $\tan 47° = \dfrac{h}{x}$

hence $h = \tan 47°(x)$, i.e. $h = 1.0724x$ (2)

Equating equations (1) and (2) gives:

$$0.3443(x+120) = 1.0724x$$
$$0.3443x + (0.3443)(120) = 1.0724x$$
$$(0.3443)(120) = (1.0724 - 0.3443)x$$
$$41.316 = 0.7281x$$
$$x = \frac{41.316}{0.7281} = 56.74\,\text{m}$$

From equation (2), **height of building,**

$$h = 1.0724x = 1.0724(56.74) = \textbf{60.85 m}.$$

Problem 27. The angle of depression of a ship viewed at a particular instant from the top of a 75 m vertical cliff is 30°. Find the distance of the ship from the base of the cliff at this instant. The ship is sailing away from the cliff at constant speed and 1 minute later its angle of depression from the top of the cliff is 20°. Determine the speed of the ship in km/h

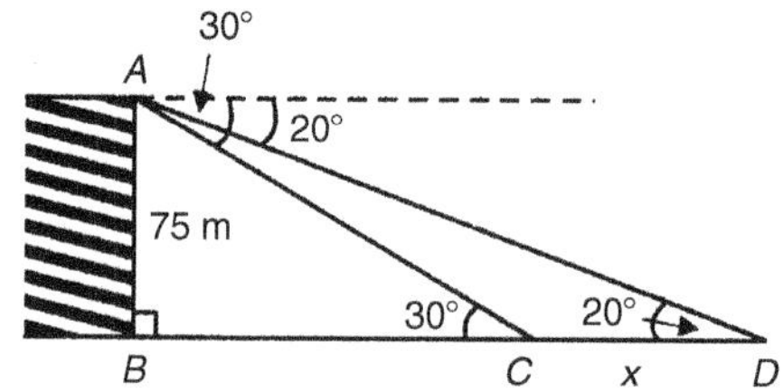

Figure 17.26

Figure 17.26 shows the cliff AB, the initial position of the ship at C and the final position at D. Since the angle of depression is initially 30° then $\angle ACB = 30°$ (alternate angles between parallel lines).

$$\tan 30° = \frac{AB}{BC} = \frac{75}{BC}$$

hence
$$BC = \frac{75}{\tan 30°} = \frac{75}{0.5774}$$
$$= \textbf{129.9 m}$$
$$= \textbf{initial position of ship from base of cliff}$$

In triangle ABD,

$$\tan 20° = \frac{AB}{BD} = \frac{75}{BC+CD} = \frac{75}{129.9+x}$$

Hence

$$129.9 + x = \frac{75}{\tan 20°} = \frac{75}{0.3640} = 206.0\,\text{m}$$

from which,

$$x = 206.0 - 129.9 = 76.1\,\text{m}$$

Thus the ship sails 76.1 m in 1 minute, i.e. 60 s, hence,

$$\textbf{speed of ship} = \frac{\text{distance}}{\text{time}} = \frac{76.1}{60}\,\text{m/s}$$
$$= \frac{76.1 \times 60 \times 60}{60 \times 1{,}000}\,\text{km/h}$$
$$= \textbf{4.57 km/h}$$

Now try the following Practice Exercise

Practice Exercise 92 Angles of elevation and depression (Answers on page 711)

1. If the angle of elevation of the top of a vertical 30 m high aerial is 32°, how far is it to the aerial?

2. From the top of a vertical cliff 80.0 m high the angles of depression of two buoys lying due west of the cliff are 23° and 15°, respectively. How far are the buoys apart?

3. From a point on horizontal ground a surveyor measures the angle of elevation of the top of a flagpole as 18°40′. He moves 50 m nearer to the flagpole and measures the angle of elevation as 26°22′. Determine the height of the flagpole

4. A flagpole stands on the edge of the top of a building. At a point 200 m from the building the angles of elevation of the top and bottom of the pole are 32° and 30° respectively. Calculate the height of the flagpole

5. From a ship at sea, the angle of elevation of the top and bottom of a vertical lighthouse standing on the edge of a vertical cliff are 31° and 26°, respectively. If the lighthouse is 25.0 m high, calculate the height of the cliff

6. From a window 4.2 m above horizontal ground the angle of depression of the foot of a building across the road is 24° and the angle of elevation of the top of the building is 34°. Determine, correct to the nearest centimetre, the width of the road and the height of the building

7. The elevation of a tower from two points, one due east of the tower and the other due west of it are 20° and 24°, respectively, and the two points of observation are 300 m apart. Find the height of the tower to the nearest metre

17.8 Trigonometric approximations for small angles

If angle x is a small angle (i.e. less than about 5°) and is expressed in radians, then the following trigonometric approximations may be shown to be true:

(i) $\sin x \approx x$

(ii) $\tan x \approx x$

(iii) $\cos x \approx 1 - \frac{x^2}{2}$

For example, let $x = 1°$, i.e. $1 \times \frac{\pi}{180} = 0.01745$ radians, correct to 5 decimal places. By calculator, $\sin 1° = 0.01745$ and $\tan 1° = 0.01746$, showing that: $\sin x = x \approx \tan x$ when $x = 0.01745$ radians. Also, $\cos 1° = 0.99985$; when $x = 1°$, i.e. 0.01745 radians,

$$1 - \frac{x^2}{2} = 1 - \frac{0.01745^2}{2} = 0.99985,$$

correct to 5 decimal places, showing that

$$\cos x = 1 - \frac{x^2}{2} \text{ when } x = 0.01745 \text{ radians.}$$

Similarly, let $x = 5°$, i.e. $5 \times \frac{\pi}{180} = 0.08727$ radians, correct to 5 decimal places.

By calculator, $\sin 5° = 0.08716$, thus $\sin x \approx x$,

$\tan 5° = 0.08749$, thus $\tan x \approx x$,

and $\cos 5° = 0.99619$;

since $x = 0.08727$ radians,

$$1 - \frac{x^2}{2} = 1 - \frac{0.08727^2}{2} = 0.99619 \text{ showing that:}$$

$$\cos x = 1 - \frac{x^2}{2} \text{ when } x = 0.0827 \text{ radians.}$$

If $\sin x \approx x$ for small angles, then $\frac{\sin x}{x} \approx 1$, and this relationship can occur in engineering considerations.

Practice Exercise 93 Multiple-choice questions on an introduction to trigonometry (Answers on page 711)

(Answers on page 711)

Each question has only one correct answer

1. In the right-angled triangle ABC shown in Figure 17.27, sine A is given by:
(a) b/a (b) c/b (c) b/c (d) a/b

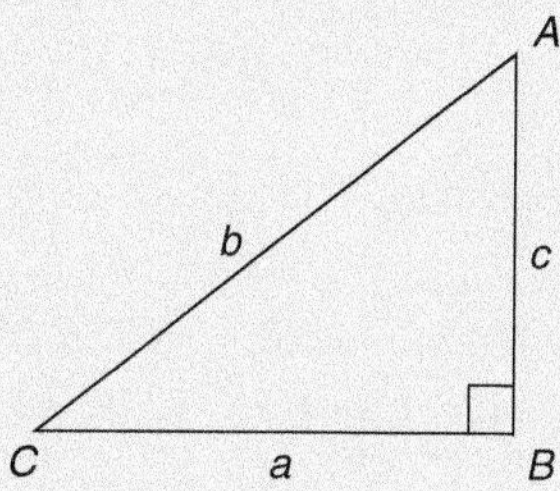

Figure 17.27

2. In the right-angled triangle ABC shown in Figure 17.27, cosine C is given by:
(a) a/b (b) c/b (c) a/c (d) b/a

3. In the right-angled triangle shown in Figure 17.27, tangent A is given by:
(a) b/c (b) a/c (c) a/b (d) c/a

4. Correct to 4 significant figures, the value of sec 161° is:
(a) −1.058 (b) 0.3256
(c) 3.072 (d) −0.9455

5. In the triangular template ABC shown in Figure 17.28, the length AC is:
(a) 6.17 cm (b) 11.17 cm
(c) 9.22 cm (d) 12.40 cm

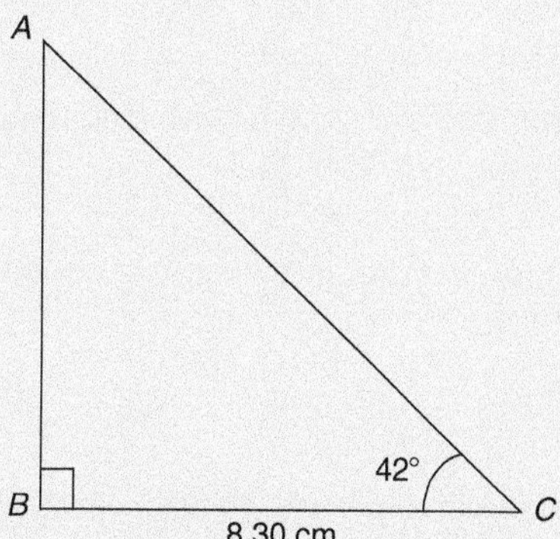

Figure 17.28

6. Correct to 3 decimal places, sin(- 2.6 rad) is:
(a) 0.516 (b) −0.045
(c) −0.516 (d) 0.045

7. For the right-angled triangle PQR shown in Figure 17.29, angle R is equal to:
(a) 41.41° (b) 48.59°
(c) 36.87° (d) 53.13°

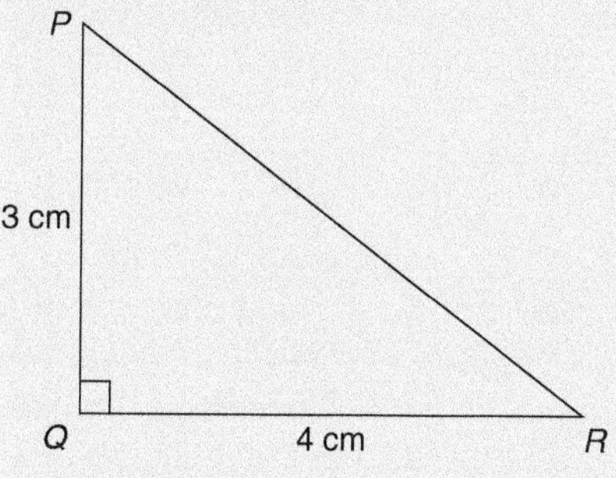

Figure 17.29

8. The area of triangle XYZ in Figure 17.30 is:
(a) 19.35 cm^2 (b) 24.22 cm^2
(c) 38.72 cm^2 (d) 32.16 cm^2

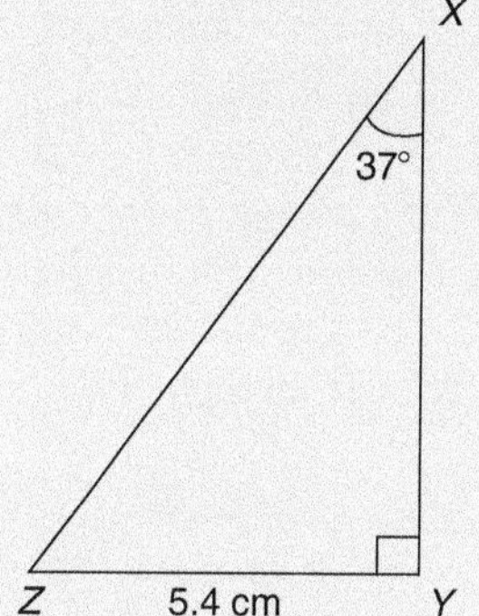

Figure 17.30

9. The length of side XZ in Figure 17.30 is:
(a) 4.313 cm (b) 7.166 cm
(c) 8.973 cm (d) 6.762 cm

10. The value, correct to 3 decimal places, of $\cos\left(\frac{-3\pi}{4}\right)$ is:
(a) 0.999 (b) 0.707
(c) −0.999 (d) −0.707

11. In the right-angled triangle ABC shown in Figure 17.31, secant C is given by:
(a) $\frac{a}{b}$ (b) $\frac{a}{c}$ (c) $\frac{b}{c}$ (d) $\frac{b}{a}$

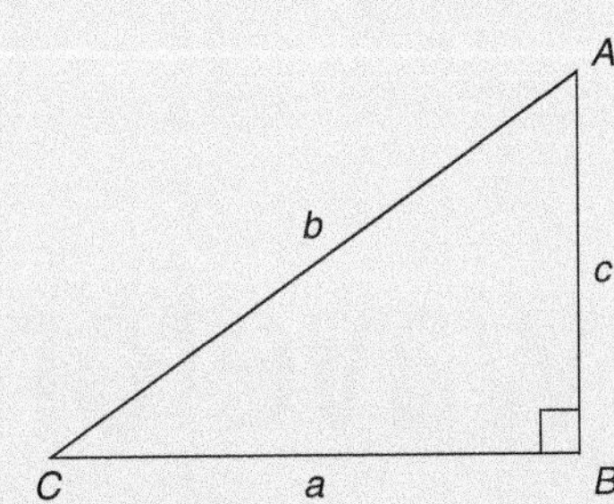

Figure 17.31

12. In the right-angled triangle ABC shown in Figure 17.31, cotangent C is given by:
(a) $\frac{a}{c}$ (b) $\frac{b}{c}$ (c) $\frac{c}{b}$ (d) $\frac{a}{b}$

13. In the right-angled triangle ABC shown in Figure 17.31, cosecant A is given by:
(a) $\frac{c}{a}$ (b) $\frac{b}{a}$ (c) $\frac{a}{b}$ (d) $\frac{b}{c}$

14. A vertical tower stands on level ground. At a point 100 m from the foot of the tower the angle of elevation of the top is 20°. The height of the tower is:
(a) 36.4 m (b) 274.7 m
(c) 34.3 m (d) 94.0 m

15. A 10 m long ladder is placed 2.5 m from the perpendicular wall of a building. How high will the top of the ladder reach on the building?
(a) 7.5 m (b) 9.68 m
(c) 10.31 m (d) 2.74 m

16. A surveyor, standing 120 m from the base of a tower, looks up 60° to see the top of the structure. The height of the tower, correct to the nearest metre is:
(a) 60 m (b) 69 m (c) 208 m (d) 104 m

17. If $\tan A = 1.4276$, sec A is equal to:
(a) 0.8190 (b) 0.5737
(c) 0.7005 (d) 1.743

18. The acute angle $\cot^{-1} 2.562$ is equal to:
(a) 67.03° (b) 21.32°
(c) 22.97° (d) 68.68°

19. The acute angle $\operatorname{cosec}^{-1} 1.429$ is equal to:
(a) 55.02° (b) 45.59°
(c) 44.41° (d) 34.98°

20. The acute angle $\sec^{-1} 2.4178$ is equal to:
(a) 24.43° (b) 22.47°
(c) 0.426 rad (d) 65.57°

For fully worked solutions to each of the problems in Practice Exercises 87 to 92 in this chapter, go to the website:
www.routledge.com/cw/bird

COMPANION @ WEBSITE

Chapter 18

Trigonometric waveforms

Why it is important to understand: Trigonometric waveforms

Trigonometric graphs are commonly used in all areas of science and engineering for modelling many different natural and mechanical phenomena such as waves, engines, acoustics, electronics, populations, UV intensity, growth of plants and animals and so on. Periodic trigonometric graphs mean that the shape repeats itself exactly after a certain amount of time. Anything that has a regular cycle, like the tides, temperatures, rotation of the earth and so on, can be modelled using a sine or cosine curve. The most common periodic signal waveform that is used in electrical and electronic engineering is the sinusoidal waveform. However, an alternating a.c. waveform may not always take the shape of a smooth shape based around the sine and cosine function; a.c. waveforms can also take the shape of square or triangular waves, i.e. complex waves. In engineering, it is therefore important to have some clear understanding of trigonometric waveforms.

At the end of this chapter, you should be able to:

- sketch sine, cosine and tangent waveforms
- determine angles of any magnitude
- understand cycle, amplitude, period, periodic time, frequency, lagging/leading angles with reference to sine and cosine waves
- perform calculations involving sinusoidal form $A\sin(\omega t \pm \alpha)$
- define a complex wave and harmonic analysis

18.1 Graphs of trigonometric functions

By drawing up tables of values from 0° to 360°, graphs of $y = \sin A$, $y = \cos A$ and $y = \tan A$ may be plotted. Values obtained with a calculator (correct to 3 decimal places — which is more than sufficient for plotting graphs), using 30° intervals, are shown below, with the respective graphs shown in Fig. 18.1.

(a) $y = \sin A$

A	0	30°	60°	90°	120°	150°	180°
$\sin A$	0	0.500	0.866	1.000	0.866	0.500	0

A	210°	240°	270°	300°	330°	360°
$\sin A$	−0.500	−0.866	−1.000	−0.866	−0.500	0

(b) $y = \cos A$

A	0	30°	60°	90°	120°	150°	180°
$\cos A$	1.000	0.866	0.500	0	−0.500	−0.866	−1.000

A	210°	240°	270°	300°	330°	360°
$\cos A$	−0.866	−0.500	0	0.500	0.866	1.000

(c) $y = \tan A$

A	0	30°	60°	90°	120°	150°	180°
$\tan A$	0	0.577	1.732	∞	−1.732	−0.577	0

A	210°	240°	270°	300°	330°	360°
$\tan A$	0.577	1.732	∞	−1.732	−0.577	0

From Fig. 18.1 it is seen that:

(i) Sine and cosine graphs oscillate between peak values of ±1

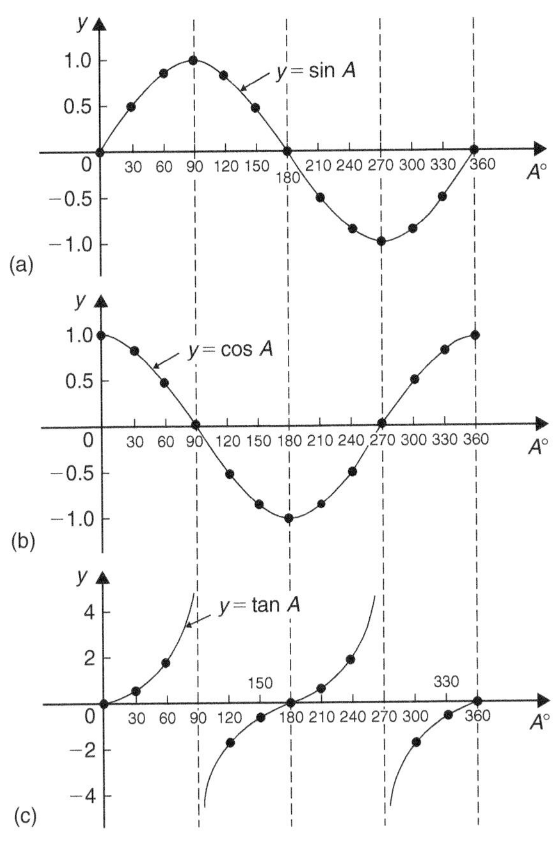

Figure 18.1

(ii) The cosine curve is the same shape as the sine curve but displaced by 90°.

(iii) The sine and cosine curves are continuous and they repeat at intervals of 360°; the tangent curve appears to be discontinuous and repeats at intervals of 180°.

18.2 Angles of any magnitude

Fig. 18.2 shows rectangular axes XX' and YY' intersecting at origin 0. As with graphical work, measurements made to the right and above 0 are positive, while those to the left and downwards are negative. Let $0A$ be free to rotate about 0. By convention, when $0A$ moves anticlockwise angular measurement is considered positive, and vice versa. Let $0A$ be rotated anticlockwise so that θ_1 is any angle in the first quadrant and left perpendicular AB be constructed to form the right-angled triangle $0AB$ in Fig. 18.3. Since all three sides of the triangle are positive, the trigonometric ratios sine, cosine and tangent will all be positive in the first quadrant.

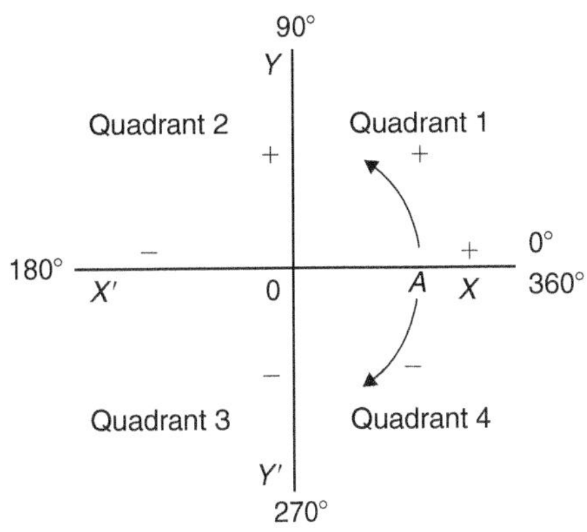

Figure 18.2

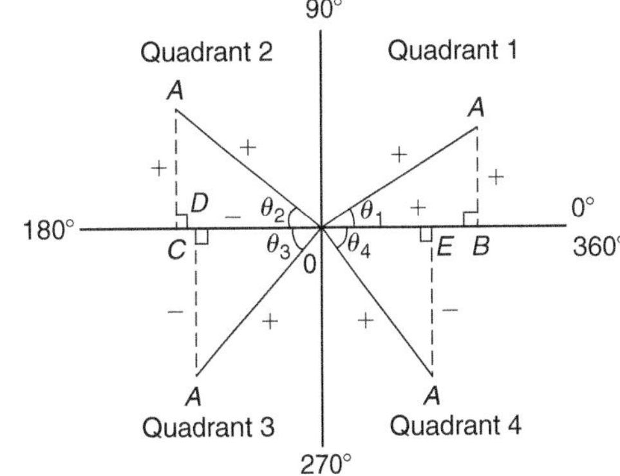

Figure 18.3

(Note: $0A$ is always positive since it is the radius of a circle.)

Let $0A$ be further rotated so that θ_2 is any angle in the second quadrant and let AC be constructed to form the right-angled triangle $0AC$. Then

$$\sin\theta_2 = \frac{+}{+} = + \quad \cos\theta_2 = \frac{-}{+} = -$$

$$\tan\theta_2 = \frac{+}{-} = -$$

Let $0A$ be further rotated so that θ_3 is any angle in the third quadrant and let AD be constructed to form the right-angled triangle $0AD$. Then

$$\sin\theta_3 = \frac{-}{+} = - \quad \cos\theta_3 = \frac{-}{+} = -$$

$$\tan\theta_3 = \frac{-}{-} = +$$

Let $0A$ be further rotated so that θ_4 is any angle in the fourth quadrant and let AE be constructed to form the right-angled triangle $0AE$. Then

$$\sin\theta_4 = \frac{-}{+} = - \qquad \cos\theta_4 = \frac{+}{+} = +$$

$$\tan\theta_4 = \frac{-}{+} = -$$

The above results are summarized in Fig. 18.4. The letters underlined spell the word CAST when starting in the fourth quadrant and moving in an anticlockwise direction.

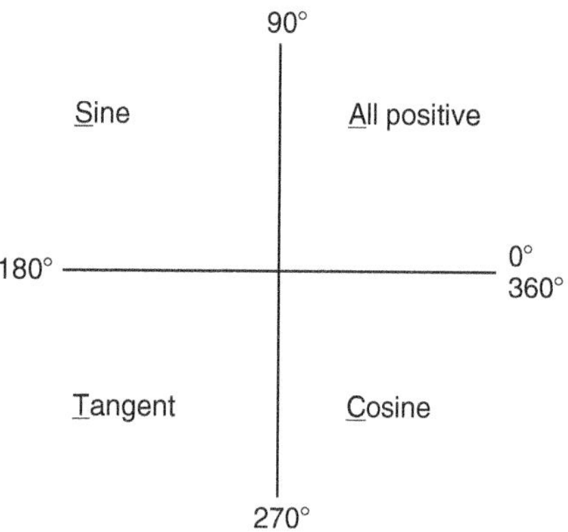

Figure 18.4

In the first quadrant of Fig. 18.1 all of the curves have positive values; in the second only sine is positive; in the third only tangent is positive; in the fourth only cosine is positive — exactly as summarised in Fig. 18.4.

A knowledge of angles of any magnitude is needed when finding, for example, all the angles between 0° and 360° whose sine is, say, 0.3261. If 0.3261 is entered into a calculator and then the inverse sine key pressed (or $\sin^{-1}$ key) the answer 19.03° appears. However, there is a second angle between 0° and 360° which the calculator does not give. Sine is also positive in the second quadrant [either from CAST or from Fig. 18.1(a)]. The other angle is shown in Fig. 18.5 as angle θ where $\theta = 180° - 19.03° = 160.97°$. Thus 19.03° **and** 160.97° are the angles between 0° and 360° whose sine is 0.3261 (check that sin 160.97° = 0.3261 on your calculator).

Be careful! Your calculator only gives you one of these answers. The second answer needs to be deduced from a knowledge of angles of any magnitude, as shown in the following worked problems.

Problem 1. Determine all the angles between 0° and 360° whose sine is −0.4638

The angles whose sine is −0.4638 occurs in the third and fourth quadrants since sine is negative in these quadrants — see Fig. 18.6.

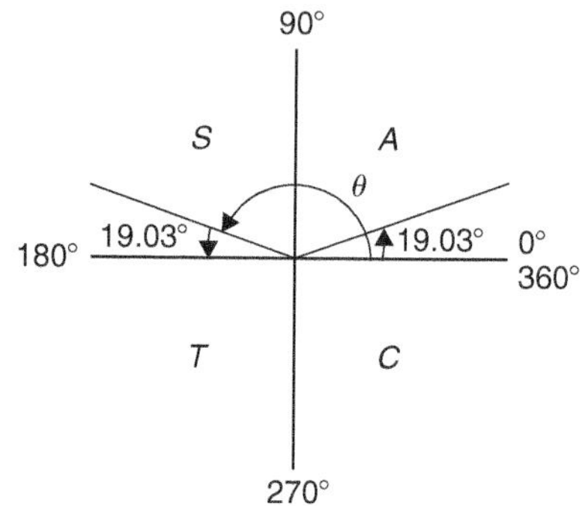

Figure 18.5

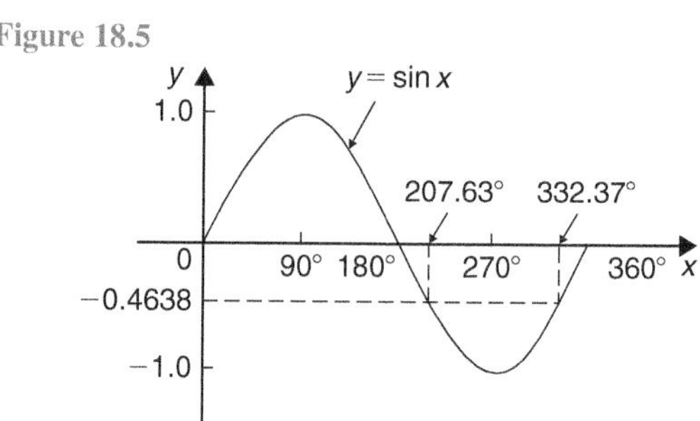

Figure 18.6

From Fig. 18.7, $\theta = \sin^{-1} 0.4638 = 27.63°$. Measured from 0°, the two angles between 0° and 360° whose sine is −0.4638 are 180° + 27.63°, i.e. **207.63°** and 360° − 27.63°, i.e. **332.37°**

(Note that a calculator only gives one answer, i.e. −27.632588°)

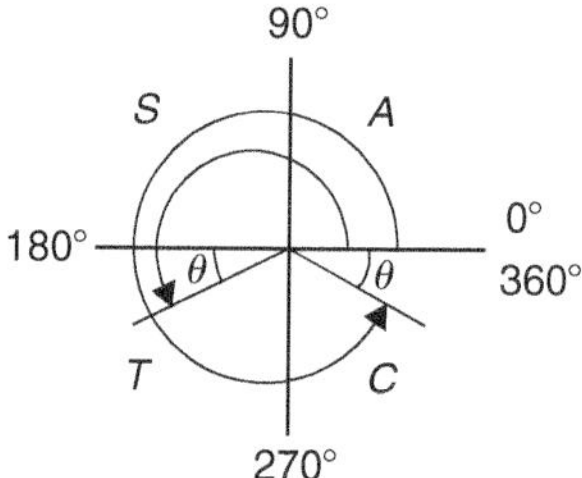

Figure 18.7

Problem 2. Determine all the angles between 0° and 360° whose tangent is 1.7629

A tangent is positive in the first and third quadrants — see Fig. 18.8.

From Fig. 18.9, $\theta = \tan^{-1} 1.7629 = 60.44°$ Measured from 0°, the two angles between 0° and 360° whose tangent is 1.7629 are **60.44°** and 180° + 60.44°, i.e. **240.44°**

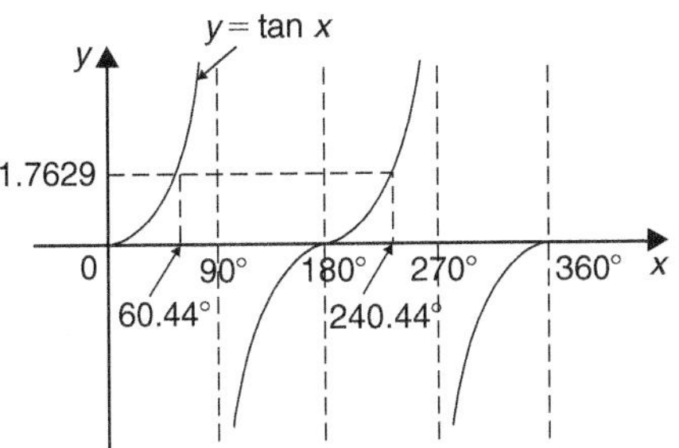

Figure 18.8

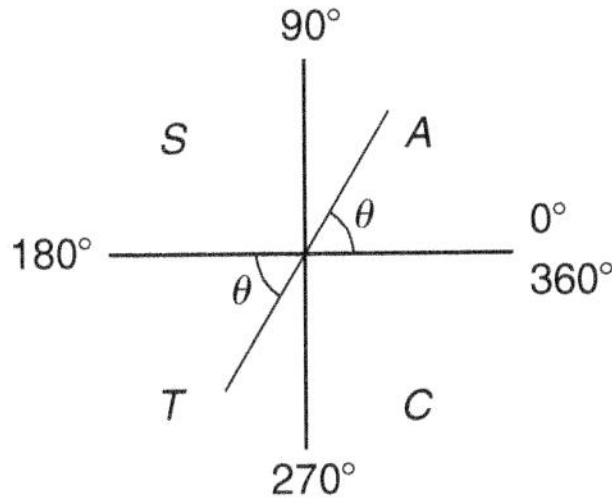

Figure 18.9

Problem 3. Solve the equation $\cos^{-1}(-0.2348) = \alpha$ for angles of α between 0° and 360°

Cosine is positive in the first and fourth quadrants and thus negative in the second and third quadrants — from Fig. 18.5 or from Fig. 18.1(b).
In Fig. 18.10, angle $\theta = \cos^{-1}(0.2348) = 76.42°$

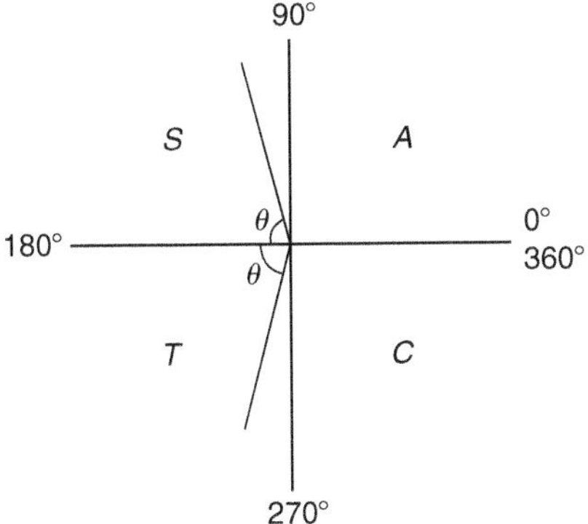

Figure 18.10

Measured from 0°, the two angles whose cosine is -0.2348 are $\alpha = 180° - 76.42°$ i.e. **103.58°** and $\alpha = 180° + 76.42°$, i.e. **256.42°**

Now try the following Practice Exercise

Practice Exercise 94 Evaluating angles of any magnitude (Answers on page 711)

1. Determine all of the angles between 0° and 360° whose sine is:

   (a) 0.6792 (b) −0.1483

2. Solve the following equations for values of x between 0° and 360°:

   (a) $x = \cos^{-1} 0.8739$
   (b) $x = \cos^{-1}(-0.5572)$

3. Find the angles between 0° to 360° whose tangent is:

   (a) 0.9728 (b) −2.3418

In Problems 4 to 6, solve the given equations in the range 0° to 360°, giving the answers in degrees and minutes.

4. $\cos^{-1}(-0.5316) = t$
5. $\sin^{-1}(-0.6250) = \alpha$
6. $\tan^{-1} 0.8314 = \theta$

18.3 The production of a sine and cosine wave

In Fig. 18.11, let OR be a vector 1 unit long and free to rotate anticlockwise about O. In one revolution a circle is produced and is shown with 15° sectors. Each radius arm has a vertical and a horizontal component. For example, at 30°, the vertical component is TS and the horizontal component is OS.
From trigonometric ratios,

$$\sin 30° = \frac{TS}{TO} = \frac{TS}{1}, \quad \text{i.e.} \quad TS = \sin 30°$$

$$\text{and} \quad \cos 30° = \frac{OS}{TO} = \frac{OS}{1}, \quad \text{i.e.} \quad OS = \cos 30°$$

The vertical component TS may be projected across to $T'S'$, which is the corresponding value of 30° on the graph of y against angle $x°$. If all such vertical components as TS are projected on to the graph, then a **sine wave** is produced as shown in Fig. 18.11

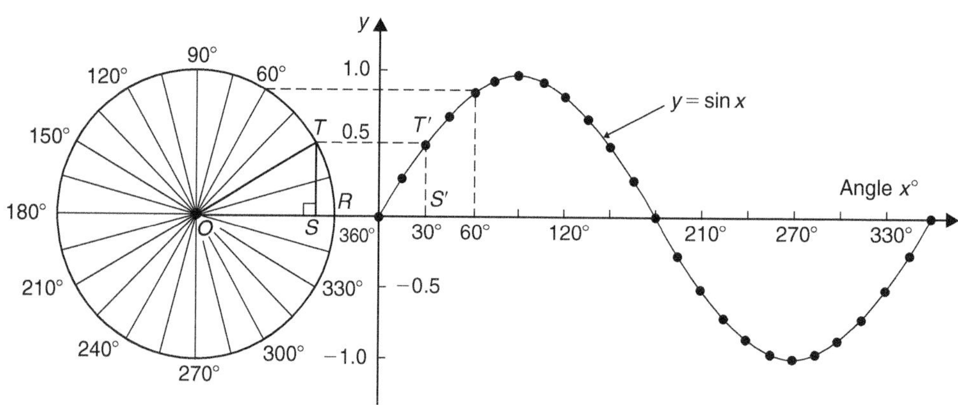

Figure 18.11

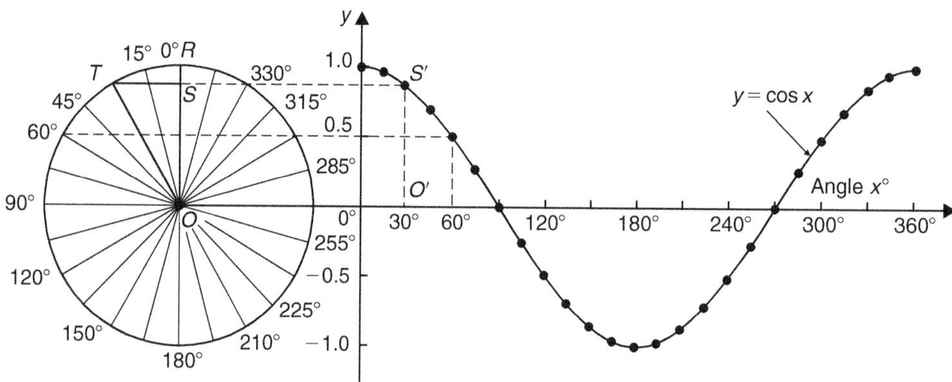

Figure 18.12

If all horizontal components such as OS are projected on to a graph of y against angle $x°$, then a **cosine wave** is produced. It is easier to visualize these projections by redrawing the circle with the radius arm OR initially in a vertical position as shown in Fig. 18.12.

From Figs. 18.11 and 18.12 it is seen that a cosine curve is of the same form as the sine curve but is displaced by 90° (or $\pi/2$ radians).

18.4 Sine and cosine curves

Graphs of sine and cosine waveforms

(i) A graph of $y = \sin A$ is shown by the broken line in Fig. 18.13 and is obtained by drawing up a table of values as in Section 18.1. A similar table may be produced for $y = \sin 2A$.

$A°$	0	30	45	60	90	120
$2A$	0	60	90	120	180	240
$\sin 2A$	0	0.866	1.0	0.866	0	−0.866

$A°$	135	150	180	210	225	240
$2A$	270	300	360	420	450	480
$\sin 2A$	−1.0	−0.866	0	0.866	1.0	0.866

$A°$	270	300	315	330	360
$2A$	540	600	630	660	720
$\sin 2A$	0	−0.866	−1.0	−0.866	0

A graph of $y = \sin 2A$ is shown in Fig. 18.13.

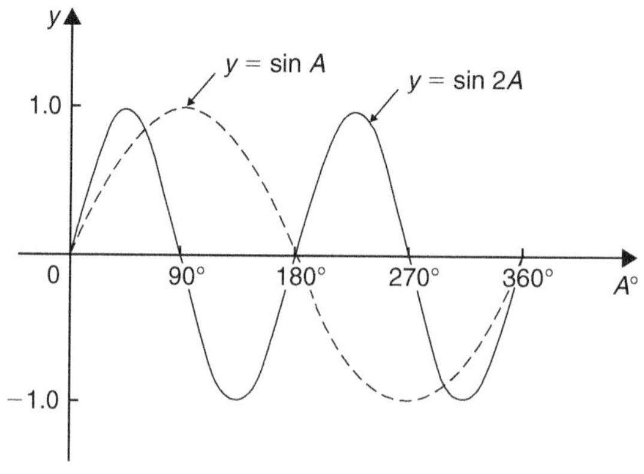

Figure 18.13

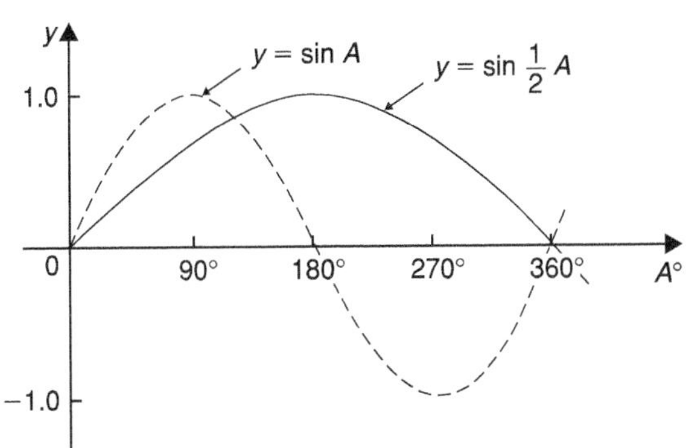

Figure 18.14

(ii) A graph of $y = \sin \frac{1}{2}A$ is shown in Fig. 18.14 using the following table of values.

$A°$	0	30	60	90	120	150	180
$\frac{1}{2}A$	0	15	30	45	60	75	90
$\sin \frac{1}{2}A$	0	0.259	0.500	0.707	0.866	0.966	1.00

$A°$	210	240	270	300	330	360
$\frac{1}{2}A$	105	120	135	150	165	180
$\sin \frac{1}{2}A$	0.966	0.866	0.707	0.500	0.259	0

(iii) A graph of $y = \cos A$ is shown by the broken line in Fig. 18.15 and is obtained by drawing up a table of values. A similar table may be produced for $y = \cos 2A$ with the result as shown.

(iv) A graph of $y = \cos \frac{1}{2}A$ is shown in Fig. 18.16 which may be produced by drawing up a table of values, similar to above.

Periodic functions and period

(i) Each of the graphs shown in Figs. 18.13 to 18.16 will repeat themselves as angle A increases and are thus called **periodic functions**.

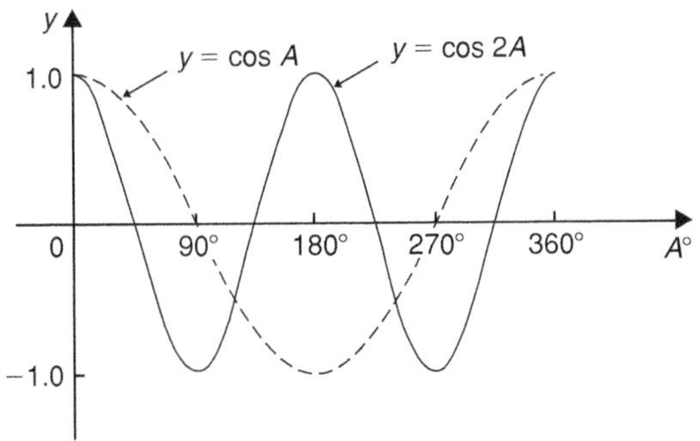

Figure 18.15

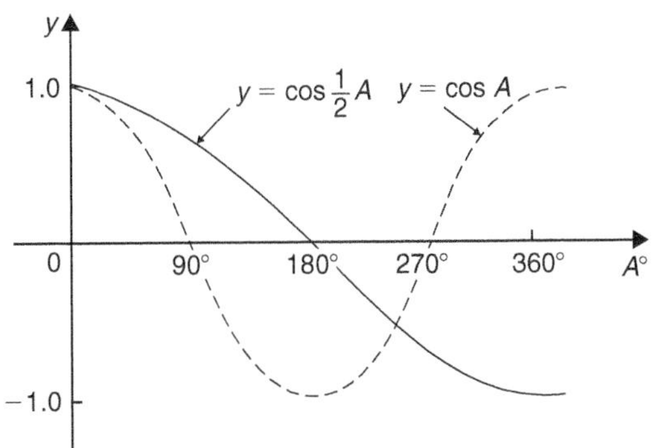

Figure 18.16

(ii) $y = \sin A$ and $y = \cos A$ repeat themselves every 360° (or 2π radians); thus 360° is called the **period** of these waveforms. $y = \sin 2A$ and $y = \cos 2A$ repeat themselves every 180° (or π radians); thus 180° is the period of these waveforms.

(iii) In general, if $y = \sin pA$ or $y = \cos pA$ (where p is a constant) then the period of the waveform is $360°/p$ (or $2\pi/p$ rad). Hence if $y = \sin 3A$ then the period is 360/3, i.e. 120°, and if $y = \cos 4A$ then the period is 360/4, i.e. 90°

Amplitude

Amplitude is the name given to the maximum or peak value of a sine wave. Each of the graphs shown in Figs. 18.13 to 18.16 has an amplitude of +1 (i.e. they oscillate between +1 and −1). However, if $y = 4 \sin A$, each of the values in the table is multiplied by 4 and the maximum value, and thus amplitude, is 4. Similarly, if $y = 5\cos 2A$, the amplitude is 5 and the period is 360°/2, i.e. 180°

Problem 4. Sketch $y = \sin 3A$ between $A = 0°$ and $A = 360°$

Amplitude = 1 and period = 360°/3 = 120°

A sketch of $y = \sin 3A$ is shown in Fig. 18.17.

Problem 5. Sketch $y = 3\sin 2A$ from $A = 0$ to $A = 2\pi$ radians

Amplitude = 3 and period = $2\pi/2 = \pi$ rads (or 180°)

A sketch of $y = 3\sin 2A$ is shown in Fig. 18.18.

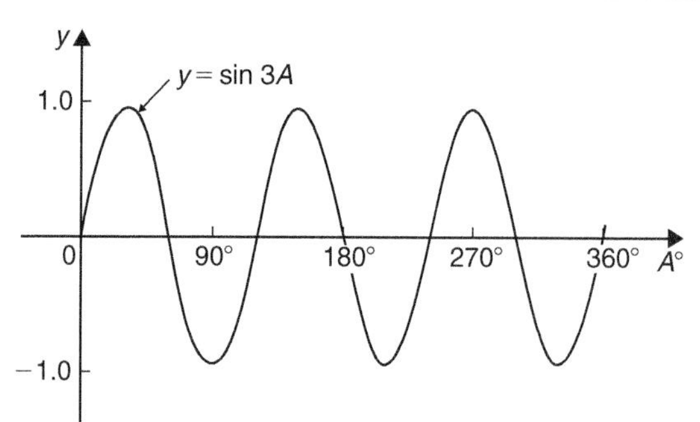

Figure 18.17

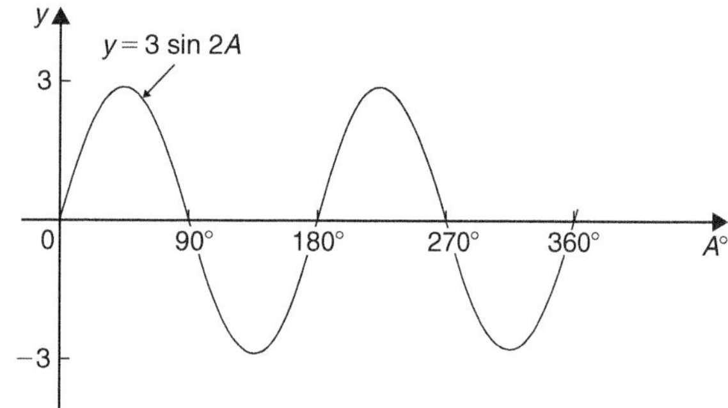

Figure 18.18

Problem 6. Sketch $y = 4 \cos 2x$ form $x = 0°$ to $x = 360°$

Amplitude = 4 and period = 360°/2 = 180° .

A sketch of $y = 4 \cos 2x$ is shown in Fig. 18.19.

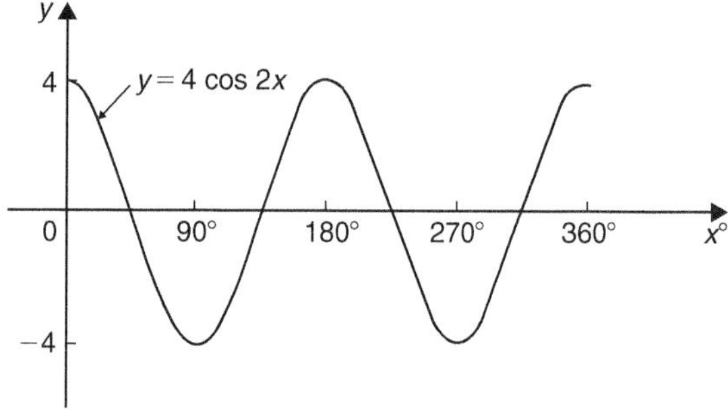

Figure 18.19

Problem 7. Sketch $y = 2 \sin \frac{3}{5}A$ over one cycle

$$\text{Amplitude} = 2; \text{period} = \frac{360°}{\frac{3}{5}} = \frac{360° \times 5}{3} = 600°$$

A sketch of $y = 2 \sin \frac{3}{5}A$ is shown in Fig. 18.20.

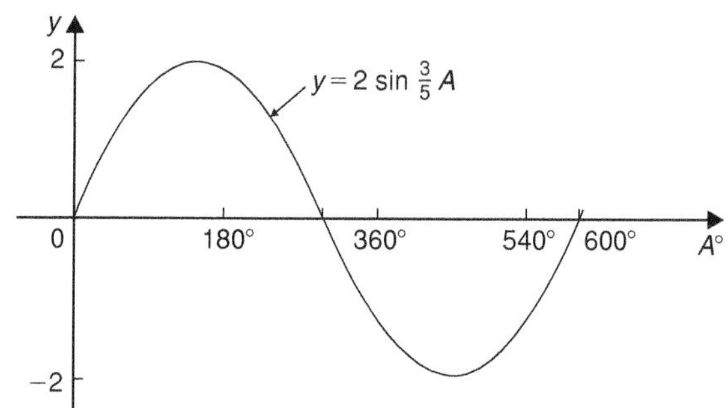

Figure 18.20

Lagging and leading angles

(i) A sine or cosine curve may not always start at 0°. To show this a periodic function is represented by $y = \sin(A \pm \alpha)$ or $y = \cos(A \pm \alpha)$ where α is a phase displacement compared with $y = \sin A$ or $y = \cos A$.

(ii) By drawing up a table of values, a graph of $y = \sin(A - 60°)$ may be plotted as shown in Fig. 18.21. If $y = \sin A$ is assumed to start at 0° then $y = \sin(A - 60°)$ starts 60° later (i.e. has a zero value 60° later). Thus $y = \sin(A - 60°)$ is said to **lag** $y = \sin A$ by 60°

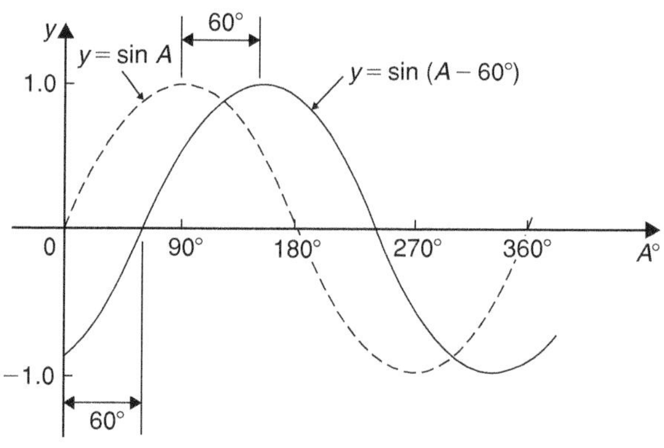

Figure 18.21

(iii) By drawing up a table of values, a graph of $y = \cos(A + 45°)$ may be plotted as shown in Fig. 18.22. If $y = \cos A$ is assumed to start at 0° then $y = \cos(A + 45°)$ starts 45° earlier (i.e. has a maximum value 45° earlier). Thus $y = \cos(A + 45°)$ is said to **lead** $y = \cos A$ by 45°

(iv) Generally, a graph of $y = \sin(A - \alpha)$ lags $y = \sin A$ by angle α, and a graph of $y = \sin(A + \alpha)$ leads $y = \sin A$ by angle α.

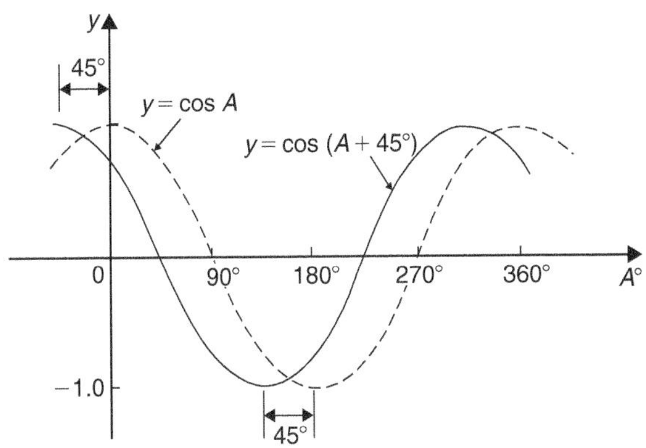

Figure 18.22

(v) A cosine curve is the same shape as a sine curve but starts 90° earlier, i.e. leads by 90°. Hence

$$\cos A = \sin(A + 90°)$$

Problem 8. Sketch $y = 5\sin(A + 30°)$ from $A = 0°$ to $A = 360°$

Amplitude = 5 and period = 360°/1 = 360°.

$5\sin(A + 30°)$ leads $5\sin A$ by 30° (i.e. starts 30° earlier).

A sketch of $y = 5\sin(A + 30°)$ is shown in Fig. 18.23.

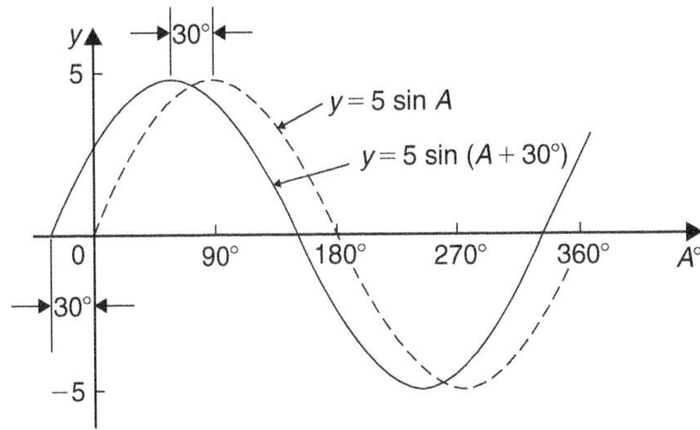

Figure 18.23

Problem 9. Sketch $y = 7\sin(2A - \pi/3)$ in the range $0 \le A \le 360°$

Amplitude = 7 and period = $2\pi/2 = \pi$ radians.

In general, $\boldsymbol{y = \sin(pt - \alpha)}$ **lags** $\boldsymbol{y = \sin pt}$ **by** $\boldsymbol{\alpha/p}$, hence $7\sin(2A - \pi/3)$ lags $7\sin 2A$ by $(\pi/3)/2$, i.e. $\pi/6$ rad or 30°

A sketch of $y = 7\sin(2A - \pi/3)$ is shown in Fig. 18.24.

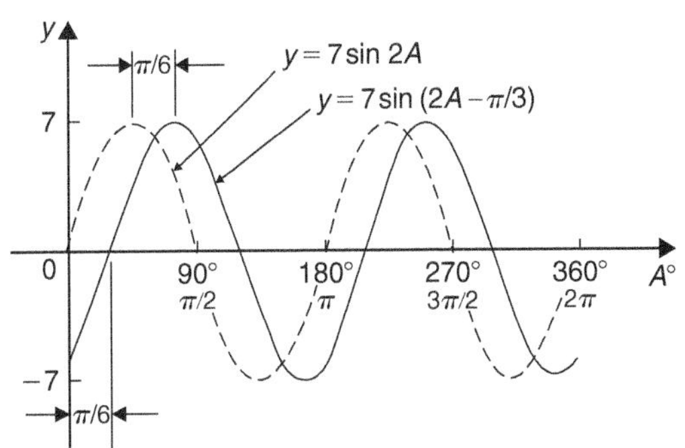

Figure 18.24

Problem 10. Sketch $y = 2\cos(\omega t - 3\pi/10)$ over one cycle

Amplitude = 2 and period = $2\pi/\omega$ rad.

$2\cos(\omega t - 3\pi/10)$ lags $2\cos \omega t$ by $3\pi/10\omega$ seconds.

A sketch of $y = 2\cos(\omega t - 3\pi/10)$ is shown in Fig. 18.25.

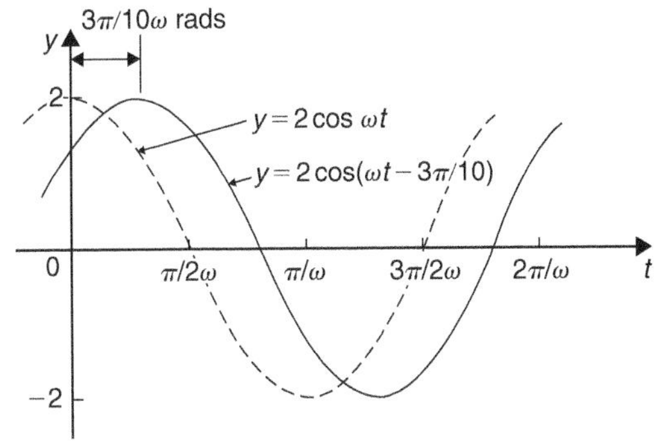

Figure 18.25

Now try the following Practice Exercise

Practice Exercise 95 Sine and cosine curves (Answers on page 711)

In Problems 1 to 7 state the amplitude and period of the waveform and sketch the curve between 0° and 360°.

1. $y = \cos 3A$
2. $y = 2\sin\dfrac{5x}{2}$

3. $y = 3\sin 4t$

4. $y = 5\cos\frac{\theta}{2}$

5. $y = \frac{7}{2}\sin\frac{3x}{8}$

6. $y = 6\sin(t - 45°)$

7. $y = \cos(2\theta + 30°)$

18.5 Sinusoidal form $A\sin(\omega t \pm \alpha)$

In Fig. 18.26, let OR represent a vector that is free to rotate anticlockwise about O at a velocity of ω rad/s. A rotating vector is called a **phasor**. After a time t second OR will have turned through an angle ωt radians (shown as angle TOR in Fig 18.26). If ST is constructed perpendicular to OR, then $\sin\omega t = ST/OT$, i.e. $ST = OT \sin\omega t$.

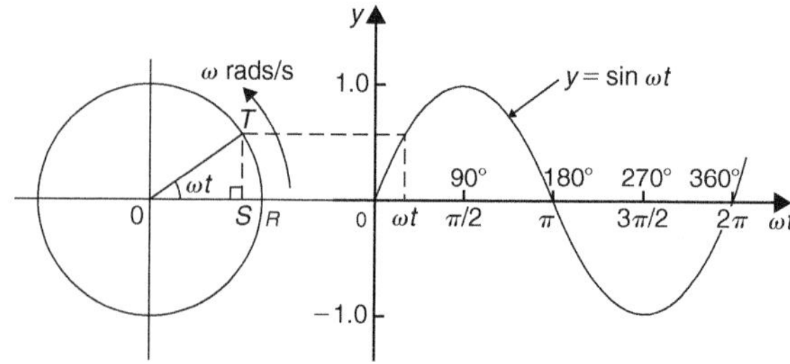

Figure 18.26

If all such vertical components are projected on to a graph of y against ωt, a sine wave results of amplitude OR (as shown in Section 18.3).

If phasor OR makes one revolution (i.e. 2π radians) in T seconds, then the angular velocity, $\omega = 2\pi/T$ rad/s,

from which, $T = 2\pi/\omega$ **seconds**

T is known as the **periodic time**.

The number of complete cycles occurring per second is called the **frequency, f**

$$\text{Frequency} = \frac{\text{number of cycles}}{\text{second}} = \frac{1}{T} = \frac{\omega}{2\pi}\text{Hz}$$

i.e. $$f = \frac{\omega}{2\pi}\text{Hz}$$

Hence **angular velocity,** $\omega = 2\pi f$ **rad/s**

Amplitude is the name given to the maximum or peak value of a sine wave, as explained in Section 18.4. The amplitude of the sine wave shown in Fig. 18.26 has an amplitude of 1.

A sine or cosine wave may not always start at 0°. To show this a periodic function is represented by $y = \sin(\omega t \pm \alpha)$ or $y = \cos(\omega t \pm \alpha)$, where α is a phase displacement compared with $y = \sin A$ or $y = \cos A$. A graph of $y = \sin(\omega t - \alpha)$ **lags** $y = \sin\omega t$ by angle α, and a graph of $y = \sin(\omega t + \alpha)$ **leads** $y = \sin\omega t$ by angle α.

The angle ωt is measured in **radians**

$$\left[\text{i.e. } \left(\omega\frac{\text{rad}}{\text{s}}\right)(ts) = \omega t \text{ radians}\right]$$

hence angle α should also be in radians.

The relationship between degrees and radians is:

$360° = 2\pi$ radians or **$180° = \pi$ radians**

Hence 1 rad $= \frac{180}{\pi} = 57.30°$ and, for example,

$$71° = 71 \times \frac{\pi}{180} = 1.239 \text{ rad}$$

Summarising, given a general sinusoidal function $y = A\sin(\omega t \pm \alpha)$, then:

(i) **A = amplitude**

(ii) **ω = angular velocity $= 2\pi f$ rad/s**

(iii) **$\frac{2\pi}{\omega}$ = periodic time T seconds**

(iv) **$\frac{\omega}{2\pi}$ = frequency, f hertz**

(v) **α = angle of lead or lag (compared with $y = A\sin\omega t$)**

Problem 11. An alternating current is given by $i = 30\sin(100\pi t + 0.27)$ amperes. Find the amplitude, periodic time, frequency and phase angle (in degrees and minutes)

$i = 30\sin(100\pi t + 0.27)A$, hence **amplitude = 30 A**.

Angular velocity $\omega = 100\pi$, hence

$$\textbf{periodic time, } T = \frac{2\pi}{\omega} = \frac{2\pi}{100\pi} = \frac{1}{50} = \mathbf{0.02\,s} \text{ or } \mathbf{20\,ms}$$

Frequency, $f = \frac{1}{T} = \frac{1}{0.02} = \mathbf{50\,Hz}$

Phase angle, $\alpha = 0.27\text{ rad} = \left(0.27 \times \frac{180}{\pi}\right)^\circ$

$= \mathbf{15.47^\circ}$ or $\mathbf{15^\circ 28'}$ **leading**

$i = \mathbf{30\,sin(100\pi t)}$

Problem 12. An oscillating mechanism has a maximum displacement of 2.5 m and a frequency of 60 Hz. At time $t = 0$ the displacement is 90 cm. Express the displacement in the general form $A\sin(\omega t \pm \alpha)$

Amplitude = maximum displacement = 2.5 m

Angular velocity, $\omega = 2\pi f = 2\pi(60) = 120\pi$ rad/s

Hence displacement $= 2.5\sin(120\pi t + \alpha)$ m

When $t = 0$, displacement = 90 cm = 0.90 m

Hence, $0.90 = 2.5\sin(0 + \alpha)$

i.e. $\sin\alpha = \frac{0.90}{2.5} = 0.36$

Hence $\alpha = \sin^{-1}0.36 = 21.10^\circ$

$= 21^\circ 6' = 0.368\text{ rad}$

Thus, **displacement** $= \mathbf{2.5\sin(120\pi t + 0.368)\,m}$

Problem 13. The instantaneous value of voltage in an a.c. circuit at any time t seconds is given by $v = 340\sin(50\pi t - 0.541)$ volts. Determine the:

(a) amplitude, periodic time, frequency and phase angle (in degrees)

(b) value of the voltage when $t = 0$

(c) value of the voltage when $t = 10$ ms

(d) time when the voltage first reaches 200 V and

(e) time when the voltage is a maximum

Sketch one cycle of the waveform

(a) **Amplitude = 340 V**

Angular velocity, $\omega = 50\pi$

Hence **periodic time,** $T = \frac{2\pi}{\omega} = \frac{2\pi}{50\pi} = \frac{1}{25}$

$= \mathbf{0.04}$ **s** or **40 ms**

Frequency $f = \frac{1}{T} = \frac{1}{0.04} = \mathbf{25\,Hz}$

Phase angle $= 0.541\text{ rad} = \left(0.541 \times \frac{180}{\pi}\right)$

$= \mathbf{31^\circ}$ **lagging** $v = 340\sin(50\pi t)$

(b) **When $t = 0$,**

$v = 340\sin(0 - 0.541)$

$= 340\sin(-31^\circ) = \mathbf{-175.1\,V}$

(c) **When $t = 10$ ms,**

then $v = 340\sin\left(50\pi\frac{10}{10^3} - 0.541\right)$

$= 340\sin(1.0298)$

$= 340\sin 59^\circ = \mathbf{291.4\ volts}$

(d) When $v = 200$ volts,

then $200 = 340\sin(50\pi t - 0.541)$

$\frac{200}{340} = \sin(50\pi t - 0.541)$

Hence $(50\pi t - 0.541) = \sin^{-1}\frac{200}{340}$

$= 36.03^\circ$ or 0.6288 rad

$50\pi t = 0.6288 + 0.541$

$= 1.1698$

Hence when $v = 200$ V,

time, $t = \frac{1.1698}{50\pi} = \mathbf{7.447\,ms}$

(e) When the voltage is a maximum, $v = 340$ V

Hence $340 = 340\sin(50\pi t - 0.541)$

$1 = \sin(50\pi t - 0.541)$

$50\pi t - 0.541 = \sin^{-1}1 = 90^\circ$ or 1.5708 rad

$50\pi t = 1.5708 + 0.541 = 2.1118$

Hence time, $t = \frac{2.1118}{50\pi} = \mathbf{13.44\,ms}$

A sketch of $v = 340\sin(50\pi t - 0.541)$ volts is shown in Fig. 18.27.

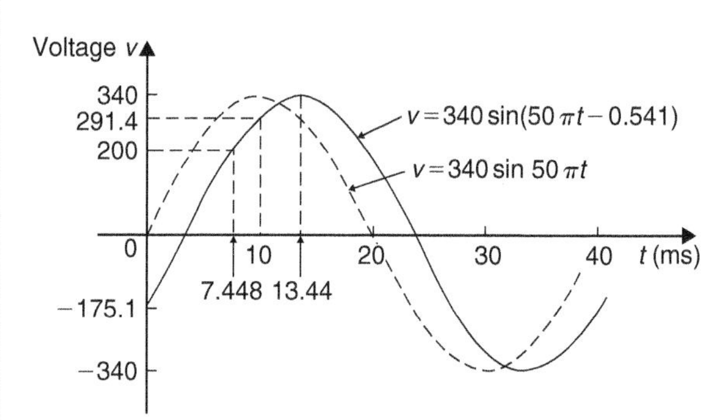

Figure 18.27

Now try the following Practice Exercise

Practice Exercise 96 The sinusoidal form $A\sin(\omega t \pm \alpha)$ (Answers on page 711)

In Problems 1 to 3, find (a) the amplitude, (b) the frequency, (c) the periodic time and (d) the phase angle (stating whether it is leading or lagging $\sin\omega t$) of the alternating quantities given.

1. $i = 40\sin(50\pi t + 0.29)$ mA
2. $y = 75\sin(40t - 0.54)$ cm
3. $v = 300\sin(200\pi t - 0.412)$ V
4. A sinusoidal voltage has a maximum value of 120 V and a frequency of 50 Hz. At time $t = 0$, the voltage is (a) zero and (b) 50 V. Express the instantaneous voltage v in the form $v = A\sin(\omega t \pm \alpha)$
5. An alternating current has a periodic time of 25 ms and a maximum value of 20 A. When time $t = 0$, current $i = -10$ amperes. Express the current i in the form $i = A\sin(\omega t \pm \alpha)$
6. An oscillating mechanism has a maximum displacement of 3.2 m and a frequency of 50 Hz. At time $t = 0$ the displacement is 150 cm. Express the displacement in the general form $A\sin(\omega t \pm \alpha)$
7. The current in an a.c. circuit at any time t seconds is given by:

$$i = 5\sin(100\pi t - 0.432) \text{ amperes}$$

Determine (a) the amplitude, frequency, periodic time and phase angle (in degrees) (b) the value of current at $t = 0$, (c) the value of current at $t = 8$ ms, (d) the time when the current is first a maximum, (e) the time when the current first reaches $3A$.

Sketch one cycle of the waveform showing relevant points

18.6 Waveform harmonics

Let an instantaneous voltage v be represented by $v = V_m \sin 2\pi ft$ volts. This is a waveform which varies sinusoidally with time t, has a frequency f and a maximum value V_m. Alternating voltages are usually assumed to have wave-shapes which are sinusoidal where only one frequency is present. If the waveform is not sinusoidal it is called a **complex wave**, and, whatever its shape, it may be split up mathematically into components called the **fundamental** and a number

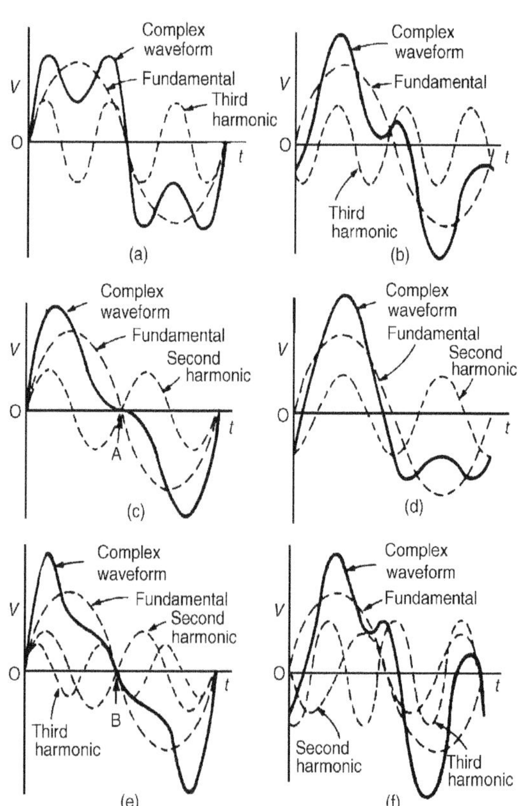

Figure 18.28

of **harmonics**. This process is called **harmonic analysis**. The fundamental (or first harmonic) is sinusoidal and has the supply frequency, f; the other harmonics are also sine waves having frequencies which are integer multiples of f. Thus, if the supply frequency is 50 Hz, then the thid harmonic frequency is 150 Hz, the fifth 250 Hz and so on.

A complex waveform comprising the sum of the fundamental and a third harmonic of about half the amplitude of the fundamental is shown in Fig. 18.28(a), both waveforms being initially in phase with each other. If further odd harmonic waveforms of the appropriate amplitudes are added, a good approximation to a square wave results. In Fig. 18.28(b), the third harmonic is shown having an initial phase displacement from the fundamental. The positive and negative half cycles of each of the complex waveforms shown in Figs. 18.28(a) and (b) are identical in shape, and this is a feature of waveforms containing the fundamental and only odd harmonics.

A complex waveform comprising the sum of the fundamental and a second harmonic of about half the amplitude of the fundamental is shown in Fig. 18.28(c), each waveform being initially in phase with each other. If further even harmonics of appropriate amplitudes are added a good approximation to a triangular wave results. In Fig. 18.28(c), the negative cycle, if reversed, appears as a mirror image of the positive cycle about point A. In Fig. 18.28(d) the second harmonic is shown with an initial phase displacement from the fundamental and the positive and negative half cycles are dissimilar.

A complex waveform comprising the sum of the fundamental, a second harmonic and a third harmonic is shown in Fig. 18.28(e), each waveform being initially 'in-phase'. The negative half cycle, if reversed, appears as a mirror image of the positive cycle about point B. In Fig. 18.28(f), a complex waveform comprising the sum of the fundamental, a second harmonic and a third harmonic are shown with initial phase displacement. The positive and negative half cycles are seen to be dissimilar.

The features mentioned relative to Figs. 18.28(a) to (f) make it possible to recognise the harmonics present in a complex waveform.

Practice Exercise 97 Multiple-choice questions on trigonometric waveforms (Answers on page page 711)

Each question has only one correct answer

1. An alternating current is given by: $i = 15\sin(100\pi t - 0.25)$ amperes. When time $t = 5$ ms, the current i has a value of:
   (a) 0.35 A (b) 0.41 A
   (c) 15 A (d) 14.53 A

2. The displacement x metres of a mass from a fixed point about which it is oscillating is given by $x = 3\cos\omega t - 4\sin\omega t$, where t is the time in seconds. x may be expressed as:
   (a) $5\sin(\omega t + 2.50)$ metres
   (b) $7\sin(\omega t - 36.87°)$ metres
   (c) $5\sin\omega t$ metres
   (d) $-\sin(\omega t - 2.50)$ metres

3. A sinusoidal current is given by: $i = R\sin(\omega t + \alpha)$. Which of the following statements is incorrect?
   (a) R is the average value of the current
   (b) frequency $= \dfrac{\omega}{2\pi}$ Hz
   (c) ω = angular velocity
   (d) periodic time $= \dfrac{2\pi}{\omega}$ s

4. An alternating voltage v is given by $v = 100\sin\left(100\pi t + \dfrac{\pi}{4}\right)$ volts.
   When $v = 50$ volts, the time t is equal to:
   (a) 0.093 s (b) −0.908 ms
   (c) −0.833 ms (d) −0.162

5. An alternating current I at time t seconds is given by $i = 282.9\sin(377t - 0.524)$ mA. For the sinusoidal current, which of the following statements is incorrect?
   (a) The r.m.s. current is 200 mA
   (b) The frequency is 120 Hz
   (c) The current is lagging by 30°
   (d) The periodic time is 16.67 ms

For fully worked solutions to each of the problems in Practice Exercises 94 to 96 in this chapter, go to the website:
www.routledge.com/cw/bird

COMPANION @ WEBSITE

Chapter 19

Cartesian and polar co-ordinates

Why it is important to understand: Cartesian and polar co-ordinates

Applications where polar co-ordinates would be used include terrestrial navigation with sonar-like devices, and those in engineering and science involving energy radiation patterns. Applications where Cartesian co-ordinates would be used include any navigation on a grid and anything involving raster graphics (i.e. bitmap – a dot matrix data structure representing a generally rectangular grid of pixels). The ability to change from Cartesian to polar co-ordinates is vitally important when using complex numbers and their use in a.c. electrical circuit theory and with vector geometry.

At the end of this chapter, you should be able to:

- change from Cartesian to polar co-ordinates
- change from polar to Cartesian co-ordinates
- use a scientific notation calculator to change from Cartesian to polar co-ordinates and vice-versa

19.1 Introduction

There are two ways in which the position of a point in a plane can be represented. These are

(a) by **Cartesian co-ordinates**, (named after Descartes*), i.e. (x, y) and

(b) by **polar co-ordinates**, i.e. (r, θ), where r is a 'radius' from a fixed point and θ is an angle from a fixed point.

19.2 Changing from Cartesian into polar co-ordinates

In Fig. 19.1, if lengths x and y are known, then the length of r can be obtained from Pythagoras' theorem (see Chapter 17) since OPQ is a right-angled triangle.

Hence $\qquad r^2 = (x^2 + y^2)$

from, which $\qquad r = \sqrt{x^2 + y^2}$

*Who was Descartes? – **René Descartes** (31 March 1596–11 February 1650) was a French philosopher, mathematician and writer. He wrote many influential texts including *Meditations on First Philosophy*. Descartes is best known for the philosophical statement *'Cogito ergo sum'* (I think, therefore I am), found in part IV of *Discourse on the Method*. To find out more go to www.routledge.com/cw/bird

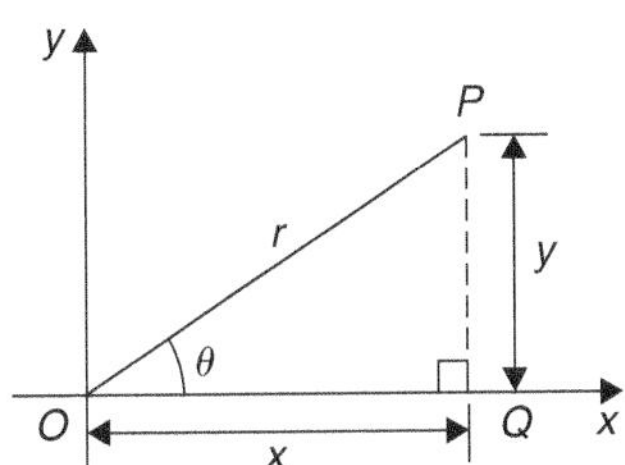

Figure 19.1

For trigonometric ratios (see Chapter 17),

$$\tan\theta = \frac{y}{x}$$

from which $\qquad \boldsymbol{\theta = \tan^{-1}\frac{y}{x}}$

$r = \sqrt{x^2 + y^2}$ and $\theta = \tan^{-1}\frac{y}{x}$ are the two formulae we need to change from Cartesian to polar co-ordinates. The angle θ, which may be expressed in degrees or radians, must **always** be measured from the positive x-axis, i.e. measured from the line OQ in Fig. 19.1. It is suggested that when changing from Cartesian to polar co-ordinates a diagram should always be sketched.

Problem 1. Change the Cartesian co-ordinates (3, 4) into polar co-ordinates

A diagram representing the point (3, 4) is shown in Fig. 19.2.

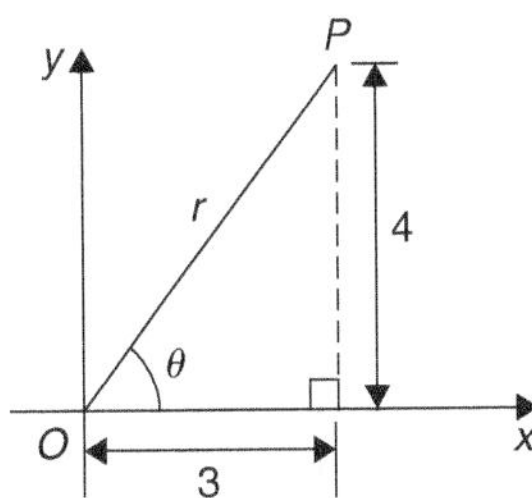

Figure 19.2

From Pythagoras' theorem, $r = \sqrt{3^2 + 4^2} = 5$ (note that -5 has no meaning in this context). By trigonometric ratios, $\theta = \tan^{-1}\frac{4}{3} = 53.13°$ or 0.927 rad
[note that $53.13° = 53.13 \times (\pi/180)$ rad $= 0.927$ rad].

Hence (3, 4) in Cartesian co-ordinates corresponds to (5, 53.13°) or (5, 0.927 rad) in polar co-ordinates.

Problem 2. Express in polar co-ordinates the position (−4, 3)

A diagram representing the point using the Cartesian co-ordinates (−4, 3) is shown in Fig. 19.3.

From Pythagoras' theorem $r=\sqrt{4^2+3^2}=5$

By trigonometric ratios, $\alpha=\tan^{-1}\frac{3}{4}=36.87°$ or 0.644 rad.

Hence $\theta=180°-36.87°=143.13°$

or $\theta=\pi-0.644=2.498$ rad.

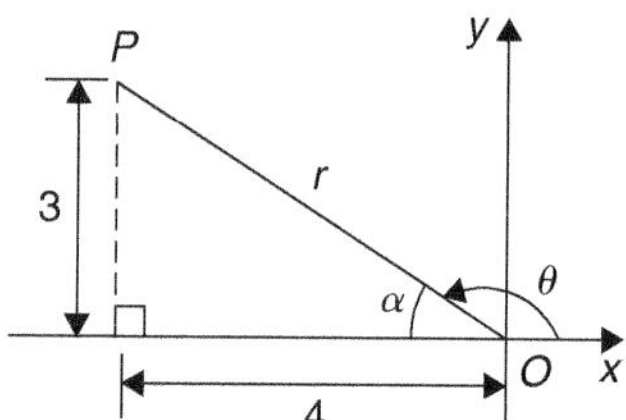

Figure 19.3

Hence the position of point *P* in polar co-ordinate form is (5, 143.13°) or (5, 2.498 rad).

Problem 3. Express (−5, −12) in polar co-ordinates

A sketch showing the position (−5, −12) is shown in Fig. 19.4.

$$r=\sqrt{5^2+12^2}=13$$

and $\alpha=\tan^{-1}\frac{12}{5}=67.38°$ or 1.176 rad.

Hence $\theta=180°+67.38°=247.38°$

or $\theta=\pi+1.176=4.318$ rad.

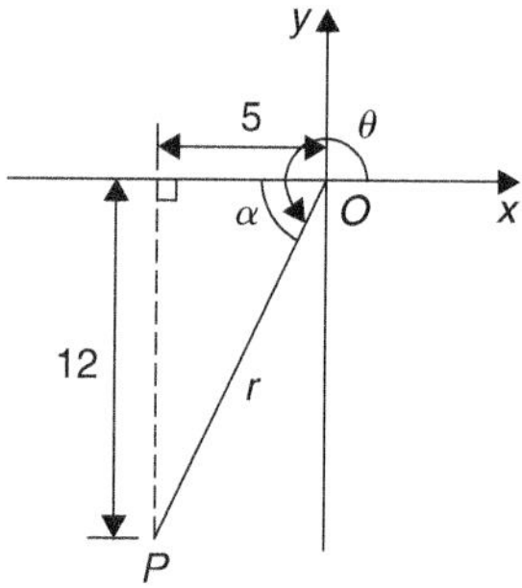

Figure 19.4

Thus (−5, −12) in Cartesian co-ordinates corresponds to (13, 247.38°) or (13, 4.318 rad) in polar co-ordinates.

Problem 4. Express (2, −5) in polar co-ordinates.

A sketch showing the position (2, −5) is shown in Fig. 19.5.

$$r=\sqrt{2^2+5^2}=\sqrt{29}=5.385$$

correct to 3 decimal places

$$\alpha=\tan^{-1}\frac{5}{2}=68.20° \quad \text{or} \quad 1.190 \text{ rad}$$

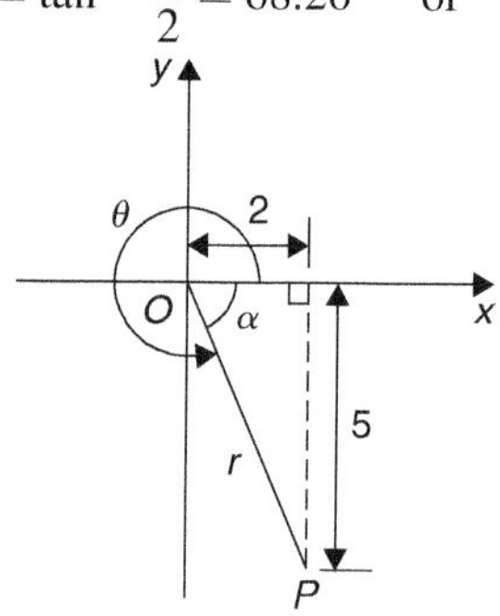

Figure 19.5

Hence $\theta=360°-68.20°=291.80°$

or $\theta=2\pi-1.190=5.093$ rad

Thus (2, −5) in Cartesian co-ordinates corresponds to (5.385, 291.80°) or (5.385, 5.093 rad) in polar co-ordinates.

Now try the following Practice Exercise

Practice Exercise 98 Changing Cartesian into polar co-ordinates (Answers on page 711)

In Problems 1 to 8, express the given Cartesian co-ordinates as polar co-ordinates, correct to 2 decimal places, in both degrees and in radians.

1. (3, 5)
2. (6.18, 2.35)
3. (−2, 4)
4. (−5.4, 3.7)
5. (−7, −3)
6. (−2.4, −3.6)
7. (5, −3)
8. (9.6, −12.4)

19.3 Changing from polar into Cartesian co-ordinates

From the right-angled triangle *OPQ* in Fig. 19.6.

$$\cos\theta = \frac{x}{r} \text{ and } \sin\theta = \frac{y}{r}$$
from trigonometric ratios

Hence $x = r\cos\theta$ and $y = r\sin\theta$

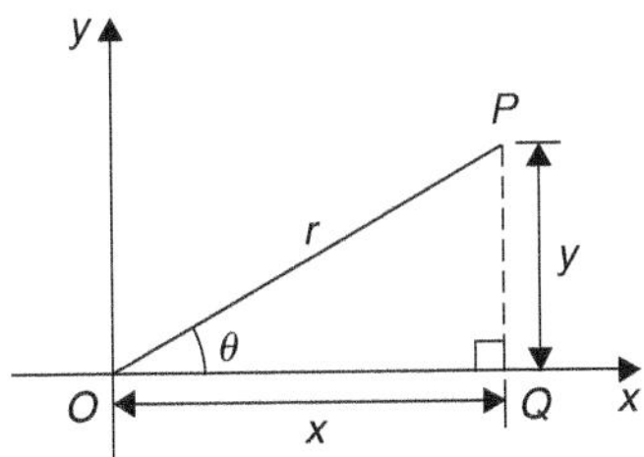

Figure 19.6

If length r and angle θ are known then $x = r\cos\theta$ and $y = r\sin\theta$ are the two formulae we need to change from polar to Cartesian co-ordinates.

Problem 5. Change (4, 32°) into Cartesian co-ordinates

A sketch showing the position (4, 32°) is shown in Fig. 19.7.

Now $x = r\cos\theta = 4\cos 32° = 3.39$
and $y = r\sin\theta = 4\sin 32° = 2.12$

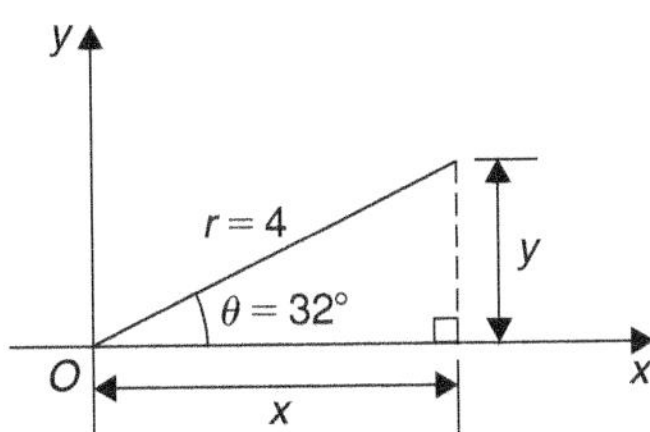

Figure 19.7

Hence (4, 32°) in polar co-ordinates corresponds to (3.39, 2.12) in Cartesian co-ordinates.

Problem 6. Express (6, 137°) in Cartesian co-ordinates

A sketch showing the position (6, 137°) is shown in Fig. 19.8.

$$x = r\cos\theta = 6\cos 137° = -4.388$$

which corresponds to length OA in Fig. 19.8.

$$y = r\sin\theta = 6\sin 137° = 4.092$$

which corresponds to length AB in Fig. 19.8.

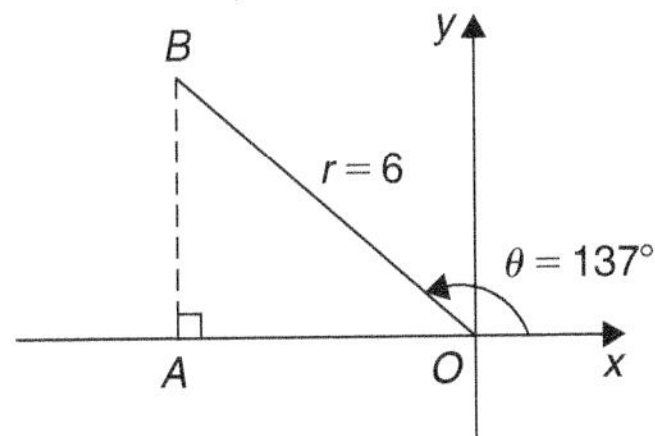

Figure 19.8

Thus (6, 137°) in polar co-ordinates corresponds to (−4.388, 4.092) in Cartesian co-ordinates.

(Note that when changing from polar to Cartesian co-ordinates it is not quite so essential to draw a sketch. Use of $x = r\cos\theta$ and $y = r\sin\theta$ automatically produces the correct signs.)

Problem 7. Express (4.5, 5.16 rad) in Cartesian co-ordinates

A sketch showing the position (4.5, 5.16 rad) is shown in Fig. 19.9.

$$x = r\cos\theta = 4.5\cos 5.16 = 1.948$$

which corresponds to length OA in Fig. 19.9.

$$y = r\sin\theta = 4.5\sin 5.16 = -4.057$$

which corresponds to length AB in Fig. 19.9.

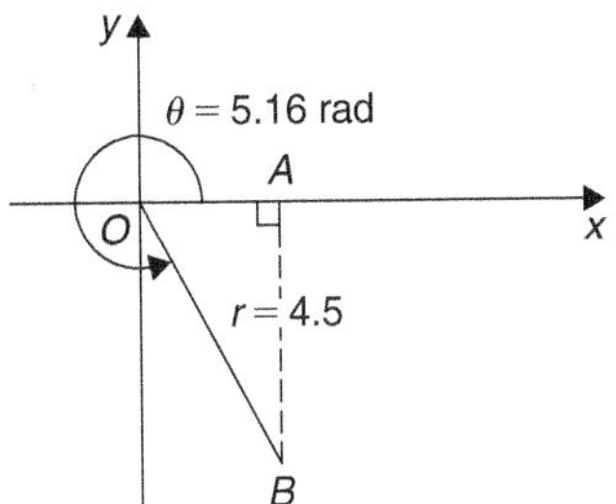

Figure 19.9

Thus (1.948, −4.057) in Cartesian co-ordinates corresponds to (4.5, 5.16 rad) in polar co-ordinates.

Now try the following Practice Exercise

Practice Exercise 99 Changing polar into Cartesian co-ordinates (Answers on page 711)

In Problems 1 to 8, express the given polar co-ordinates as Cartesian co-ordinates, correct to 3 decimal places.

1. (5, 75°)
2. (4.4, 1.12 rad)
3. (7, 140°)
4. (3.6, 2.5 rad)
5. (10.8, 210°)
6. (4, 4 rad)
7. (1.5, 300°)
8. (6, 5.5 rad)
9. Figure 19.10 shows 5 equally spaced holes on an 80 mm pitch circle diameter. Calculate their co-ordinates relative to axes Ox and Oy in (a) polar form, (b) Cartesian form. (c) Calculate also the shortest distance between the centres of two adjacent holes.

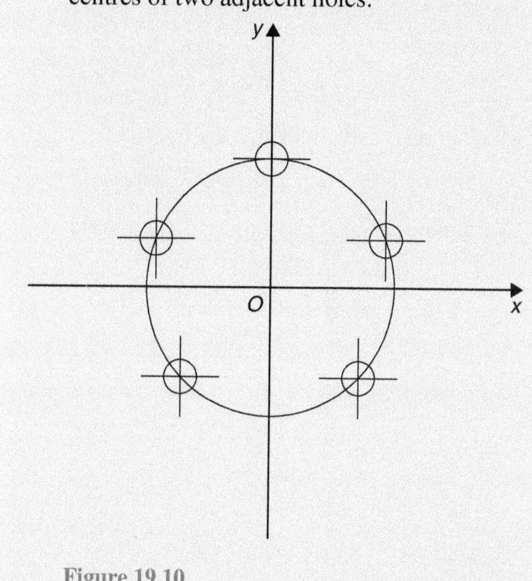

Figure 19.10

19.4 Use of Pol/Rec functions on calculators

Another name for Cartesian co-ordinates is **rectangular** co-ordinates. Many scientific notation calculators possess **Pol** and **Rec** functions. 'Rec' is an abbreviation of 'rectangular' (i.e. Cartesian) and 'Pol' is an abbreviation of 'polar'. Check the operation manual for your particular calculator to determine how to use these two functions. They make changing from Cartesian to polar co-ordinates and vice-versa, so much quicker and easier.

For example, with the Casio fx-991ES PLUS calculator, or similar, to change the Cartesian number (3, 4) into polar form, the following procedure is adopted:

1. Press 'shift'
2. Press 'Pol'
3. Enter 3
4. Enter 'comma' (obtained by 'shift' then))
5. Enter 4
6. Press)
7. Press = The answer is: $r = 5, \theta = 53.13°$

Hence, (3, 4) in Cartesian form is the same as (5, 53.13°) in polar form.

If the angle is required in **radians**, then before repeating the above procedure press 'shift', 'mode' and then 4 to change your calculator to radian mode.

Similarly, to change the polar form number (7, 126°) into Cartesian or rectangular form, adopt the following procedure:

1. Press 'shift'
2. Press 'Rec'
3. Enter 7
4. Enter 'comma'
5. Enter 126 (assuming your calculator is in degrees mode)
6. Press)
7. Press = The answer is: $X = -4.11$**, and scrolling across, Y = 5.66**, correct to 2 decimal places.

Hence, (7, 126°) in polar form is the same as (−4.11, 5.66) in rectangular or Cartesian form.

Now return to Practice Exercises 98 and 99 in this chapter and use your calculator to determine the answers and see how much more quickly they may be obtained.

Practice Exercise 100 Multiple-choice questions on Cartesian and polar co-ordinates (Answers on page 711)

Each question has only one correct answer

1. $(-4, 3)$ in polar co-ordinates is:
   (a) (5, 2.498 rad) (b) (7, 36.87°)
   (c) (5, 36.87°) (d) (5, 323.13°)

2. (7, 141°) in Cartesian co-ordinates is:
   (a) $(5.44, -4.41)$ (b) $(-5.44, -4.41)$
   (c) $(5.44, 4.41)$ (d) $(-5.44, 4.41)$

3. $(-3, -7)$ in polar co-ordinates is:
   (a) (−7.62, −113.20°) (b) (7.62, 246.80°)
   (c) (7.62, 23.20°) (d) (7.62, 203.20°)

4. (6.3, 5.23 rad) in Cartesian co-ordinates is:
   (a) $(3.12, -5.47)$ (b) $(6.27, 0.57)$
   (c) $(-5.47, 3.12)$ (d) $(-3.12, 5.47)$

5. $(-12, -5)$ in polar co-ordinates is:
   (a) (13, 157.38°) (b) (13, 2.75 rad)
   (c) (13, −2.75 rad) (d) (13, 22.62°)

For fully worked solutions to each of the problems in Practice Exercises 98 and 99 in this chapter, go to the website:
www.routledge.com/cw/bird

Revision Test 5 Trigonometry

This Revision Test covers the material contained in Chapters 17 to 19. *The marks for each question are shown in brackets at the end of each question.*

1. Fig. RT5.1 shows a plan view of a kite design. Calculate the lengths of the dimensions shown as a and b. (4)

2. In Fig. RT5.1, evaluate (a) angle θ (b) angle α. (5)

3. Determine the area of the plan view of a kite shown in Fig. RT5.1. (4)

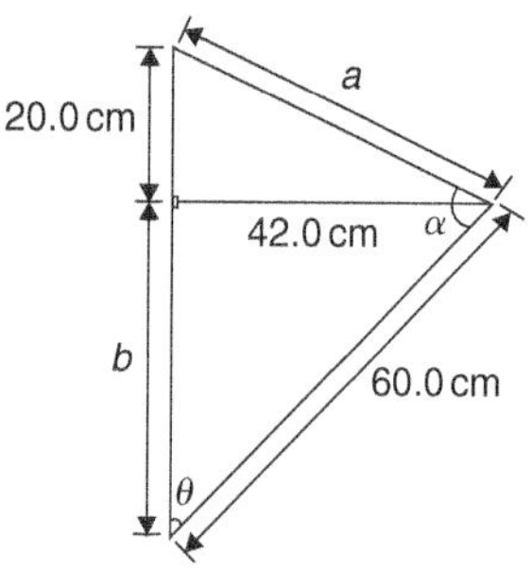

Figure RT5.1

4. If the angle of elevation of the top of a 25 m perpendicular building from point *A* is measured as 27°, determine the distance to the building. Calculate also the angle of elevation at a point *B*, 20 m closer to the building than point *A*. (5)

5. Evaluate, each correct to 4 significant figures: (a) sin 231.78° (b) cos 151°16′ (c) tan $\frac{3\pi}{8}$ (3)

6. Sketch the following curves labelling relevant points: (a) $y = 4\cos(\theta + 45°)$
   (b) $y = 5\sin(2t - 60°)$ (6)

7. Solve the following equations in the range 0° to 360°: (a) $\sin^{-1}(-0.4161) = x$
   (b) $\cot^{-1}(2.4198) = \theta$ (6)

8. The current in an alternating current circuit at any time t seconds is given by: $i = 120\sin(100\pi t + 0.274)$ amperes. Determine:
   (a) the amplitude, periodic time, frequency and phase angle (with reference to $120\sin 100\pi t$)
   (b) the value of current when $t = 0$
   (c) the value of current when $t = 6$ ms
   (d) the time when the current first reaches 80 A

   Sketch one cycle of the oscillation. (17)

9. Change the following Cartesian co-ordinates into polar co-ordinates, correct to 2 decimal places, in both degrees and in radians: (a) (−2.3, 5.4)
   (b) (7.6, −9.2). (6)

10. Change the following polar co-ordinates into Cartesian co-ordinates, correct to 3 decimal places: (a) (6.5, 132°) (b) (3, 3 rad) (4)

For lecturers/instructors/teachers, fully worked solutions to each of the problems in Revision Test 5, together with a full marking scheme, are available at the website:

www.routledge.com/cw/bird

Chapter 20

Triangles and some practical applications

Why it is important to understand: Triangles and some practical applications

As was mentioned earlier, fields that use trigonometry include astronomy, navigation, music theory, acoustics, optics, electronics, probability theory, statistics, biology, medical imaging (CAT scans and ultrasound), pharmacy, chemistry, seismology, meteorology, oceanography, many physical sciences, land surveying, architecture, economics, electrical engineering, mechanical engineering, civil engineering, computer graphics, cartography and crystallography. There are so many examples where triangles are involved in engineering and the ability to solve such triangles is of great importance.

At the end of this chapter, you should be able to:

- state and use the sine rule
- state and use the cosine rule
- use various formulae to determine the area of any triangle
- apply the sine and cosine rules to solving practical trigonometric problems

20.1 Sine and cosine rules

To '**solve a triangle**' means 'to find the values of unknown sides and angles'. If a triangle is **right-angled**, trigonometric ratios and the theorem of Pythagoras* may be used for its solution, as shown in Chapter 17. However, for a **non-right-angled triangle**, trigonometric ratios and Pythagoras' theorem **cannot** be used. Instead, two rules, called the **sine rule** and **cosine rule**, are used.

*Who was Pythagoras? – See page 170. To find out more go to www.routledge.com/cw/bird

Sine rule

With reference to triangle ABC of Fig. 20.1, the **sine rule** states:

$$\frac{a}{\sin A}=\frac{b}{\sin B}=\frac{c}{\sin C}$$

The rule may be used only when:

(i) 1 side and any 2 angles are initially given, or

(ii) 2 sides and an angle (not the included angle) are initially given.

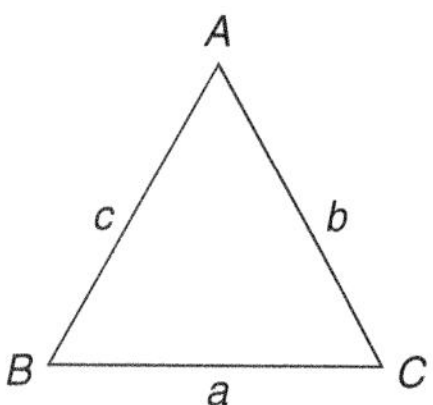

Figure 20.1

Cosine rule

With reference to triangle ABC of Fig. 20.1, the **cosine rule** states:

$$a^2 = b^2 + c^2 - 2bc\cos A$$
$$\text{or} \quad b^2 = a^2 + c^2 - 2ac\cos B$$
$$\text{or} \quad c^2 = a^2 + b^2 - 2ab\cos C$$

The rule may be used only when:

(i) 2 sides and the included angle are initially given, or

(ii) 3 sides are initially given.

20.2 Area of any triangle

The **area of any triangle** such as ABC of Fig. 20.1 is given by:

(i) $\frac{1}{2}\times$ **base** $\times$ **perpendicular height, or**

(ii) $\frac{1}{2}ab\sin C$ **or** $\frac{1}{2}ac\sin B$ **or** $\frac{1}{2}bc\sin A$**, or**

(iii) $\sqrt{s(s-a)(s-b)(s-c)}$

where $s = \dfrac{a+b+c}{2}$

This latter formula is called **Heron's formula** (or sometimes **Hero's formula**).

20.3 Worked problems on the solution of triangles and their areas

Problem 1. In a triangle XYZ, $\angle X = 51°$, $\angle Y = 67°$ and $YZ = 15.2$ cm. Solve the triangle and find its area

The triangle XYZ is shown in Fig. 20.2. Since the angles in a triangle add up to 180°, then $z = 180° - 51° - 67° = \mathbf{62°}$.

Applying the sine rule:

$$\frac{15.2}{\sin 51°} = \frac{y}{\sin 67°} = \frac{z}{\sin 62°}$$

Using $\dfrac{15.2}{\sin 51°} = \dfrac{y}{\sin 67°}$ and transposing gives:

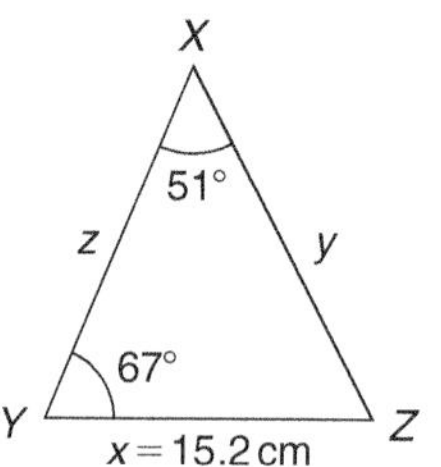

Figure 20.2

$$y = \frac{15.2\sin 67°}{\sin 51°} = \mathbf{18.00\ cm} = \mathbf{XZ}$$

Using $\dfrac{15.2}{\sin 51°} = \dfrac{z}{\sin 62°}$ and transposing gives:

$$z = \frac{15.2\sin 62°}{\sin 51°} = \mathbf{17.27\ cm} = \mathbf{XY}$$

Area of triangle $\mathbf{XYZ} = \frac{1}{2}xy\sin Z$

$$= \tfrac{1}{2}(15.2)(18.00)\sin 62° = \mathbf{120.8\ cm^2}$$

$$(\text{or area } = \tfrac{1}{2}xz\sin Y = \tfrac{1}{2}(15.2)(17.27)\sin 67° = \mathbf{120.8\ cm^2})$$

It is always worth checking with triangle problems that the longest side is opposite the largest angle and vice-versa. In this problem, Y is the largest angle and XZ is the longest of the three sides.

Problem 2. Solve the triangle ABC given $B = 78°51'$, $AC = 22.31$ mm and $AB = 17.92$ mm. Find also its area

Triangle ABC is shown in Fig. 20.3.

Applying the sine rule:

$$\frac{22.31}{\sin 78°51'} = \frac{17.92}{\sin C}$$

from which, $\sin C = \dfrac{17.92\sin 78°51'}{22.31} = 0.7881$

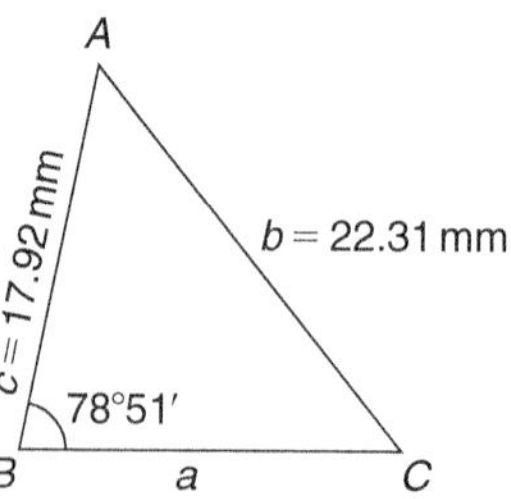

Figure 20.3

Hence $C = \sin^{-1} 0.7881 = 52°0'$ or $128°0'$ (see Chapters 17 and 18).

Since $B = 78°51'$, C cannot be $128°0'$, since $128°0' + 78°51'$ is greater than $180°$.

Thus only $C = 52°0'$ is valid.

Angle $A = 180° - 78°51' - 52°0' = 49°9'$

Applying the sine rule:

$$\frac{a}{\sin 49°9'} = \frac{22.31}{\sin 78°51'}$$

from which, $$a = \frac{22.31 \sin 49°9'}{\sin 78°51'} = 17.20\,\text{mm}$$

Hence $A = \mathbf{49°9'}$, $C = \mathbf{52°0'}$ and $BC = \mathbf{17.20\,mm}$.

Area of triangle $ABC = \frac{1}{2}ac \sin B$

$$= \frac{1}{2}(17.20)(17.92)\sin 78°51' = \mathbf{151.2\,mm^2}$$

Problem 3. Solve the triangle PQR and find its area given that $QR = 36.5$ mm, $PR = 26.6$ mm and $\angle Q = 36°$

Triangle PQR is shown in Fig. 20.4.

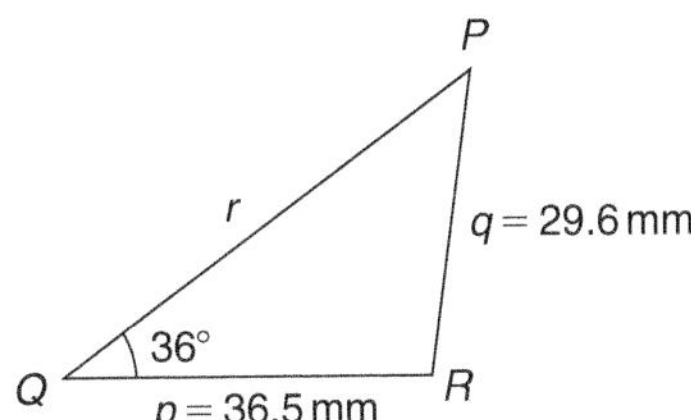

Figure 20.4

Applying the sine rule:

$$\frac{29.6}{\sin 36°} = \frac{36.5}{\sin P}$$

from which, $$\sin P = \frac{36.5 \sin 36°}{29.6} = 0.7248$$

Hence $P = \sin^{-1} 0.7248 = 46°27'$ or $133°33'$

When $P = 46°27'$ and $Q = 36°$

then $R = 180° - 46°27' - 36° = 97°33'$

When $P = 133°33'$ and $Q = 36°$

then $R = 180° - 133°33' - 36° = 10°27'$

Thus, in this problem, there are **two** separate sets of results and both are feasible solutions. Such a situation is called the **ambiguous case**.

Case 1. $P = 46°27'$, $Q = 36°$, $R = 97°33'$, $p = 36.5$ mm and $q = 29.6$ mm

From the sine rule:

$$\frac{r}{\sin 97°33'} = \frac{29.6}{\sin 36°}$$

from which, $$r = \frac{29.6 \sin 97°33'}{\sin 36°} = \mathbf{49.92\,mm}$$

$$\textbf{Area} = \frac{1}{2}pq \sin R = \frac{1}{2}(36.5)(29.6)\sin 97°33' = \mathbf{535.5\,mm^2}$$

Case 2. $P = 133°33'$, $Q = 36°$, $R = 10°27'$, $p = 36.5$ mm and $q = 29.6$ mm

From the sine rule:

$$\frac{r}{\sin 10°27'} = \frac{29.6}{\sin 36°}$$

from which, $$r = \frac{29.6 \sin 10°2'}{\sin 36°} = \mathbf{9.134\,mm}$$

$$\text{Area} = \frac{1}{2}pq \sin R = \frac{1}{2}(36.5)(29.6)\sin 10°27' = \mathbf{97.98\,mm^2}$$

Triangle PQR for case 2 is shown in Fig. 20.5.

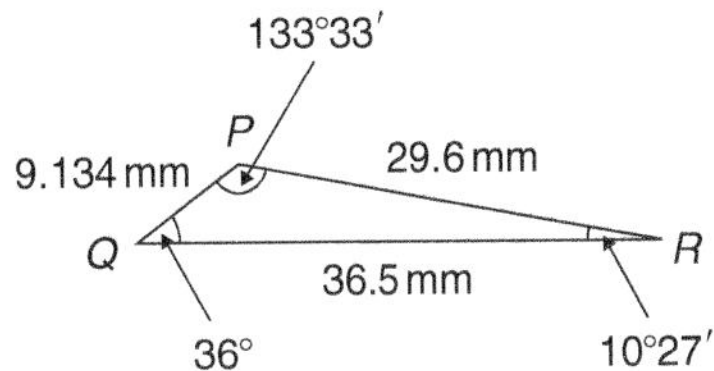

Figure 20.5

Now try the following Practice Exercise

Practice Exercise 101 The solution of triangles and their areas (Answers on page 712)

In Problems 1 and 2, use the sine rule to solve the triangles ABC and find their areas.

1. $A = 29°$, $B = 68°$, $b = 27$ mm
2. $B = 71°26'$, $C = 56°32'$, $b = 8.60$ cm

In Problems 3 and 4, use the sine rule to solve the triangles DEF and find their areas.

3. $d = 17\,\text{cm}, f = 22\,\text{cm}, F = 26°$

4. $d = 32.6\,\text{mm}, e = 25.4\,\text{mm}, D = 104°22'$

In Problems 5 and 6, use the sine rule to solve the triangles *JKL* and find their areas.

5. $j = 3.85\,\text{cm}, k = 3.23\,\text{cm}, K = 36°$

6. $k = 46\,\text{mm}, l = 36\,\text{mm}, L = 35°$

20.4 Further worked problems on the solution of triangles and their areas

Problem 4. Solve triangle *DEF* and find its area given that $EF = 35.0\,\text{mm}$, $DE = 25.0\,\text{mm}$ and $\angle E = 64°$

Triangle *DEF* is shown in Fig. 20.6.

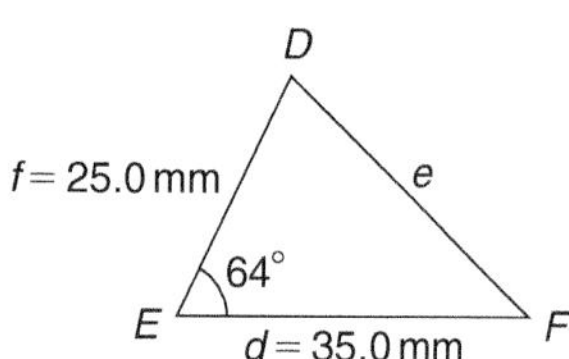

Figure 20.6

Applying the cosine rule:

$$e^2 = d^2 + f^2 - 2df\cos E$$

i.e. $$e^2 = (35.0)^2 + (25.0)^2 - [2(35.0)(25.0)\cos 64°]$$
$$= 1225 + 625 - 767.1 = 1083$$

from which, $\boldsymbol{e} = \sqrt{1083} = \mathbf{32.91\,mm}$

Applying the sine rule:

$$\frac{32.91}{\sin 64°} = \frac{25.0}{\sin F}$$

from which, $$\sin F = \frac{25.0\sin 64°}{32.91} = 0.6828$$

Thus $$\angle F = \sin^{-1} 0.6828$$
$$= 43°4' \quad \text{or} \quad 136°56'$$

$F = 136°56'$ is not possible in this case since $136°56' + 64°$ is greater than 180°. Thus only $\boldsymbol{F = 43°4'}$ is valid.

$$\angle \boldsymbol{D} = 180° - 64° - 43°4' = \mathbf{72°56'}$$

Area of triangle $DEF = \frac{1}{2}df\sin E$

$$= \tfrac{1}{2}(35.0)(25.0)\sin 64° = \mathbf{393.2\,mm^2}$$

Problem 5. A triangle *ABC* has sides $a = 9.0\,\text{cm}$, $b = 7.5\,\text{cm}$ and $c = 6.5\,\text{cm}$. Determine its three angles and its area

Triangle *ABC* is shown in Fig. 20.7. It is usual first to calculate the largest angle to determine whether the triangle is acute or obtuse. In this case the largest angle is *A* (i.e. opposite the longest side).

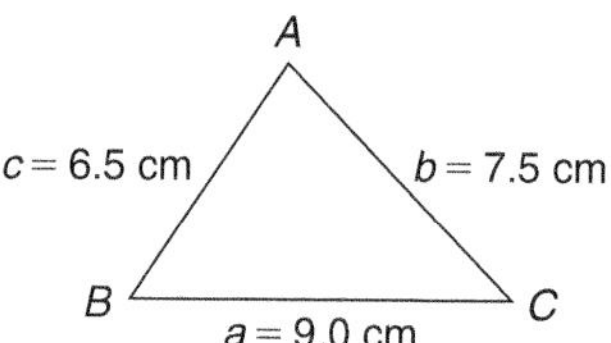

Figure 20.7

Applying the cosine rule:

$$a^2 = b^2 + c^2 - 2bc\cos A$$

from which,

$$2bc\cos A = b^2 + c^2 - a^2$$

and $$\cos A = \frac{b^2 + c^2 - a^2}{2bc}$$

$$= \frac{7.5^2 + 6.5^2 - 9.0^2}{2(7.5)(6.5)} = 0.1795$$

Hence $A = \cos^{-1} 0.1795 = \mathbf{79.66°}$ (or 280.33°, which is obviously impossible). The triangle is thus acute angled since cos *A* is positive. (If cos *A* had been negative, angle *A* would be obtuse, i.e. lie between 90° and 180°.)

Applying the sine rule:

$$\frac{9.0}{\sin 79.66°} = \frac{7.5}{\sin B}$$

from which, $$\sin B = \frac{7.5\sin 79.66°}{9.0} = 0.8198$$

Hence $$\boldsymbol{B} = \sin^{-1} 0.8198 = \mathbf{55.06°}$$

and $$\boldsymbol{C} = 180° - 79.66° - 55.06° = \mathbf{45.28°}$$

Area $=\sqrt{s(s-a)(s-b)(s-c)}$, where

$$s=\frac{a+b+c}{2}=\frac{9.0+7.5+6.5}{2}=11.5\,\text{cm}$$

Hence

$$\textbf{area}=\sqrt{11.5(11.5-9.0)(11.5-7.5)(11.5-6.5)}$$

$$=\sqrt{11.5(2.5)(4.0)(5.0)}=\mathbf{23.98\,cm^2}$$

Alternatively,

$$\text{area}=\tfrac{1}{2}ab\sin C=\tfrac{1}{2}(9.0)(7.5)\sin 45.28^\circ$$

$$=\mathbf{23.98\,cm^2}$$

Problem 6. Solve triangle XYZ, shown in Fig. 20.8, and find its area given that $Y=128^\circ$, $XY=7.2$ cm and $YZ=4.5$ cm

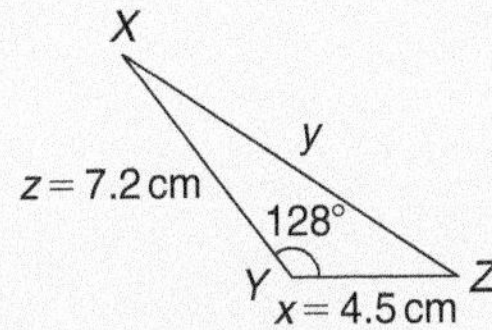

Figure 20.8

Applying the cosine rule:

$$y^2=x^2+z^2-2xz\cos Y$$

$$=4.5^2+7.2^2-[2(4.5)(7.2)\cos 128^\circ]$$

$$=20.25+51.84-[-39.89]$$

$$=20.25+51.84+39.89=112.0$$

$$y=\sqrt{112.0}=\mathbf{10.58\,cm}$$

Applying the sine rule:

$$\frac{10.58}{\sin 128^\circ}=\frac{7.2}{\sin Z}$$

from which, $\sin Z=\dfrac{7.2\sin 128^\circ}{10.58}=0.5363$

Hence $\mathbf{Z}=\sin^{-1}0.5363=\mathbf{32.43^\circ}$ (or 147.57° which, here, is impossible).

$$\boldsymbol{X}=180^\circ-128^\circ-32.43^\circ=\mathbf{19.57^\circ}$$

$$\textbf{Area}=\tfrac{1}{2}xz\sin Y=\tfrac{1}{2}(4.5)(7.2)\sin 128^\circ$$

$$=\mathbf{12.77\,cm^2}$$

Now try the following Practice Exercise

Practice Exercise 102 The solution of triangles and their areas (Answers on page 712)

In Problems 1 and 2, use the cosine and sine rules to solve the triangles PQR and find their areas.

1. $q=12$ cm, $r=16$ cm, $P=54^\circ$
2. $q=3.25$ m, $r=4.42$ m, $P=105^\circ$

In Problems 3 and 4, use the cosine and sine rules to solve the triangles XYZ and find their areas.

3. $x=10.0$ cm, $y=8.0$ cm, $z=7.0$ cm
4. $x=21$ mm, $y=34$ mm, $z=42$ mm

20.5 Practical situations involving trigonometry

There are a number of **practical situations** where the use of trigonometry is needed to find unknown sides and angles of triangles. This is demonstrated in the following worked problems.

Problem 7. A room 8.0 m wide has a span roof which slopes at 33° on one side and 40° on the other. Find the length of the roof slopes, correct to the nearest centimetre

A section of the roof is shown in Fig. 20.9.

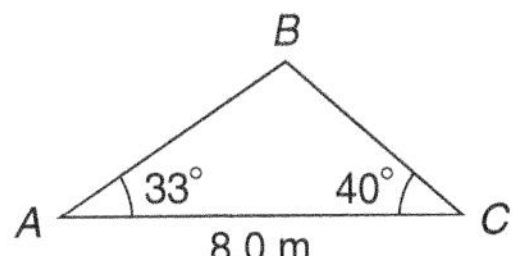

Figure 20.9

Angle at ridge, $B=180^\circ-33^\circ-40^\circ=107^\circ$

From the sine rule:

$$\frac{8.0}{\sin 107^\circ}=\frac{a}{\sin 33^\circ}$$

from which, $a=\dfrac{8.0\sin 33^\circ}{\sin 107^\circ}=4.556\,\text{m}$

Also from the sine rule:

$$\frac{8.0}{\sin 107^\circ}=\frac{c}{\sin 40^\circ}$$

from which, $c=\dfrac{8.0\sin 40^\circ}{\sin 107^\circ}=5.377\,\text{m}$

Hence the roof slopes are 4.56 m and 5.38 m, correct to the nearest centimetre.

Problem 8. A man leaves a point walking at 6.5 km/h in a direction E 20° N (i.e. a bearing of 70°). A cyclist leaves the same point at the same time in a direction E 40° S (i.e. a bearing of 130°) travelling at a constant speed. Find the average speed of the cyclist if the walker and cyclist are 80 km apart after 5 hours

After 5 hours the walker has travelled $5 \times 6.5 = 32.5$ km (shown as AB in Fig. 20.10). If AC is the distance the cyclist travels in 5 hours then $BC = 80$ km.

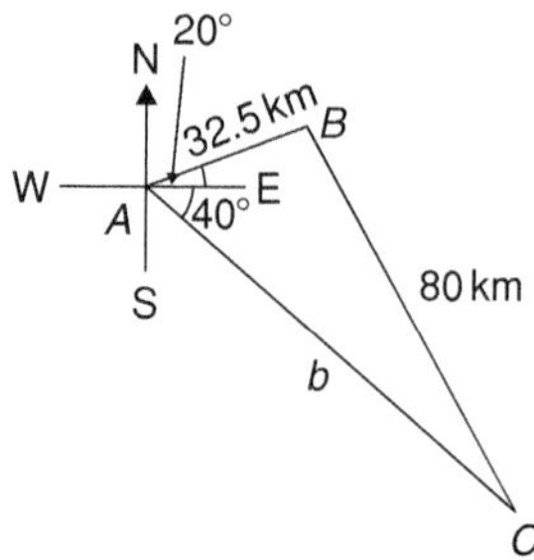

Figure 20.10

Applying the sine rule:

$$\frac{80}{\sin 60°} = \frac{32.5}{\sin C}$$

from which, $$\sin C = \frac{32.5 \sin 60°}{80} = 0.3518$$

Hence $C = \sin^{-1} 0.3518 = 20.60°$ (or 159.40°, which is impossible in this case).

$$B = 180° - 60° - 20.60° = 99.40°$$

Applying the sine rule again:

$$\frac{80}{\sin 60°} = \frac{b}{\sin 99.40°}$$

from which, $$b = \frac{80 \sin 99.40°}{\sin 60°} = 91.14 \text{km}$$

Since the cyclist travels 91.14 km in 5 hours then

$$\textbf{average speed} = \frac{\text{distance}}{\text{time}} = \frac{91.14}{5} = \mathbf{18.23\ km/h}$$

Problem 9. Two voltage phasors are shown in Fig. 20.11. If $V_1 = 40$ V and $V_2 = 100$ V determine the value of their resultant (i.e. length OA) and the angle the resultant makes with V_1

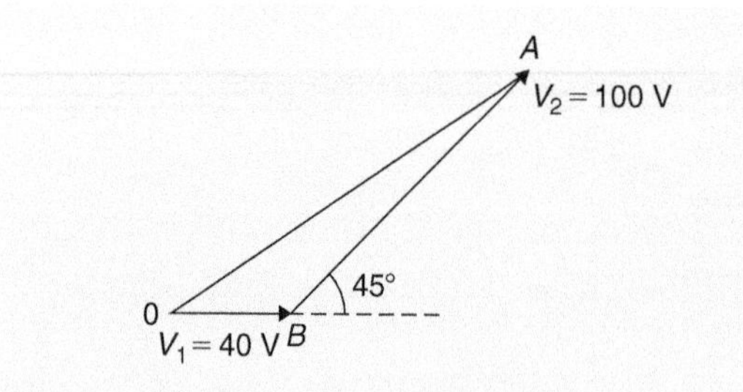

Figure 20.11

Angle $OBA = 180° - 45° = 135°$

Applying the cosine rule:

$$\begin{aligned} OA^2 &= V_1^2 + V_2^2 - 2V_1V_2 \cos OBA \\ &= 40^2 + 100^2 - \{2(40)(100)\cos 135°\} \\ &= 1600 + 10000 - \{-5657\} \\ &= 1600 + 10000 + 5657 = 17257 \end{aligned}$$

The resultant

$$OA = \sqrt{17257} = 131.4 \text{V}$$

Applying the sine rule:

$$\frac{131.4}{\sin 135°} = \frac{100}{\sin AOB}$$

from which, $$\sin AOB = \frac{100 \sin 135°}{131.4} = 0.5381$$

Hence angle $AOB = \sin^{-1} 0.5381 = 32.55°$ (or 147.45°, which is impossible in this case).

Hence the resultant voltage is 131.4 volts at 32.55° to V_1

Problem 10. In Fig.20.12, PR represents the inclined jib of a crane and is 10.0 m long. PQ is 4.0 m long. Determine the inclination of the jib to the vertical and the length of tie QR

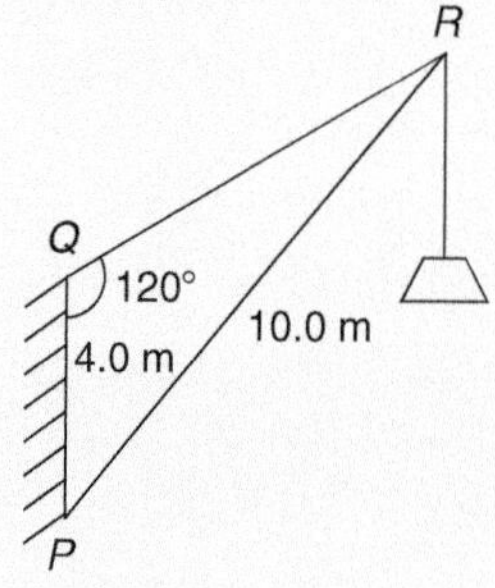

Figure 20.12

Applying the sine rule:

$$\frac{PR}{\sin 120°} = \frac{PQ}{\sin R}$$

from which, $\sin R = \dfrac{PQ \sin 120°}{PR}$

$$= \frac{(4.0)\sin 120°}{10.0} = 0.3464$$

Hence $\angle R = \sin^{-1} 0.3464 = 20.27°$ (or 159.73°, which is impossible in this case).

$\angle P = 180° - 120° - 20.27° = $ **39.73°, which is the inclination of the jib to the vertical**.

Applying the sine rule:

$$\frac{10.0}{\sin 120°} = \frac{QR}{\sin 39.73°}$$

from which, **length of tie, QR** $= \dfrac{10.0 \sin 39.73°}{\sin 120°}$

$$= \mathbf{7.38\,m}$$

Now try the following Practice Exercise

Practice Exercise 103 Practical situations involving trigonometry (Answers on page 712)

1. A ship P sails at a steady speed of 45 km/h in a direction of W 32° N (i.e. a bearing of 302°) from a port. At the same time another ship Q leaves the port at a steady speed of 35 km/h in a direction N 15° E (i.e. a bearing of 015°). Determine their distance apart after 4 hours

2. Two sides of a triangular plot of land are 52.0 m and 34.0 m, respectively. If the area of the plot is 620 m^2 find (a) the length of fencing required to enclose the plot and (b) the angles of the triangular plot

3. A jib crane is shown in Fig. 20.13. If the tie rod PR is 8.0 long and PQ is 4.5 m long determine (a) the length of jib RQ and (b) the angle between the jib and the tie rod

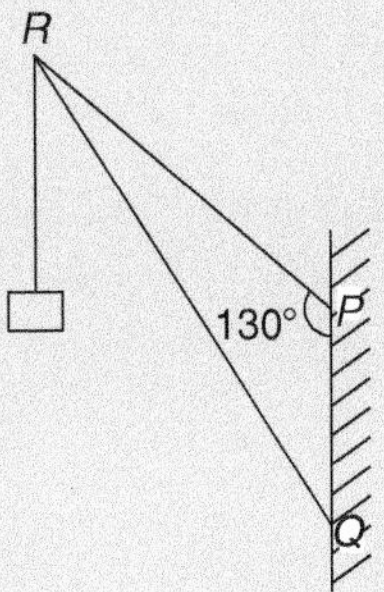

Figure 20.13

4. A building site is in the form of a quadrilateral as shown in Fig. 20.14, and its area is 1510 m^2. Determine the length of the perimeter of the site

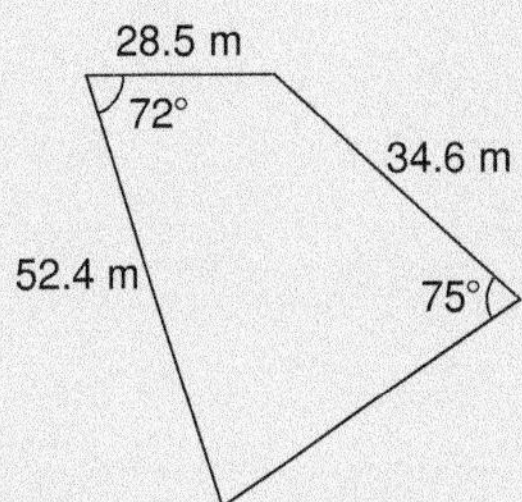

Figure 20.14

5. Determine the length of members BF and EB in the roof truss shown in Fig. 20.15

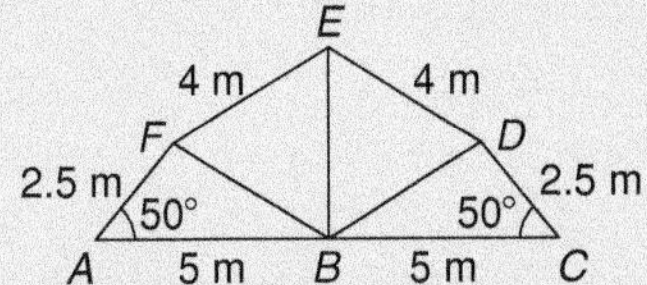

Figure 20.15

6. A laboratory 9.0 m wide has a span roof that slopes at 36° on one side and 44° on the other. Determine the lengths of the roof slopes

7. PQ and QR are the phasors representing the alternating currents in two branches of a circuit. Phasor PQ is 20.0 A and is horizontal. Phasor QR (which is joined to the end of PQ to form triangle PQR) is 14.0 A and is at an angle of 35° to the horizontal. Determine the resultant phasor PR and the angle it makes with phasor PQ

20.6 Further practical situations involving trigonometry

Problem 11. A vertical aerial stands on horizontal ground. A surveyor positioned due east of the aerial measures the elevation of the top as 48°. He moves due south 30.0 m and measures the elevation as 44°. Determine the height of the aerial

In Fig. 20.16, DC represents the aerial, A is the initial position of the surveyor and B his final position.

From triangle ACD, $\tan 48° = \dfrac{DC}{AC}$, from which

$$AC = \frac{DC}{\tan 48°}$$

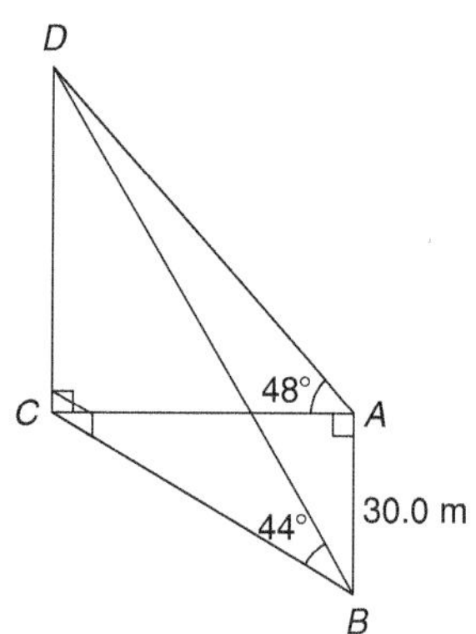

Figure 20.16

Similarly, from triangle BCD, $BC = \dfrac{DC}{\tan 44°}$

For triangle ABC, using Pythagoras' theorem:

$$BC^2 = AB^2 + AC^2$$

$$\left(\frac{DC}{\tan 44°}\right)^2 = (30.0)^2 + \left(\frac{DC}{\tan 48°}\right)^2$$

$$DC^2\left(\frac{1}{\tan^2 44°} - \frac{1}{\tan^2 48°}\right) = 30.0^2$$

$$DC^2(1.072323 - 0.810727) = 30.0^2$$

$$DC^2 = \frac{30.0^2}{0.261596} = 3440.4$$

Hence, height of aerial, $\boldsymbol{DC} = \sqrt{3340.4}$
$= \mathbf{58.65\,m}$.

Problem 12. A crank mechanism of a petrol engine is shown in Fig. 20.17. Arm OA is 10.0 cm long and rotates clockwise about 0. The connecting rod AB is 30.0 cm long and end B is constrained to move horizontally

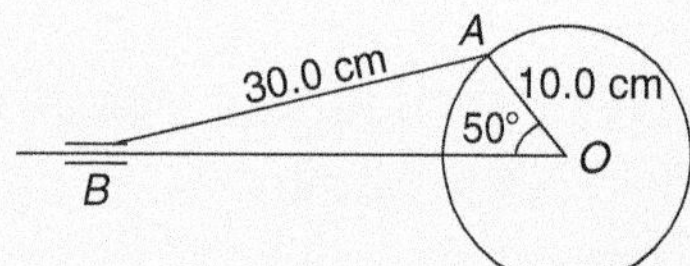

Figure 20.17

(a) For the position shown in Fig. 20.17 determine the angle between the connecting rod AB and the horizontal and the length of OB.

(b) How far does B move when angle AOB changes from 50° to 120°?

(a) Applying the sine rule:

$$\frac{AB}{\sin 50°} = \frac{AO}{\sin B}$$

from which, $\sin B = \dfrac{AO \sin 50°}{AB}$

$$= \frac{10.0 \sin 50°}{30.0} = 0.2553$$

Hence $B = \sin^{-1} 0.2553 = 14.78°$ (or 165.22°, which is impossible in this case).

Hence the connecting rod *AB* makes an angle of 14.78° with the horizontal.

Angle $OAB = 180° - 50° - 14.78° = 115.22°$

Applying the sine rule:

$$\frac{30.0}{\sin 50°} = \frac{OB}{\sin 115.22°}$$

from which, $\boldsymbol{OB} = \dfrac{30.0 \sin 115.22°}{\sin 50°}$

$= \mathbf{35.43\,cm}$

(b) Figure 20.18 shows the initial and final positions of the crank mechanism. In triangle $OA'B'$, applying the sine rule:

$$\frac{30.0}{\sin 120°} = \frac{10.0}{\sin A'B'O}$$

from which, $\sin A'B'O = \dfrac{10.0 \sin 120°}{30.0}$

$= 0.2887$

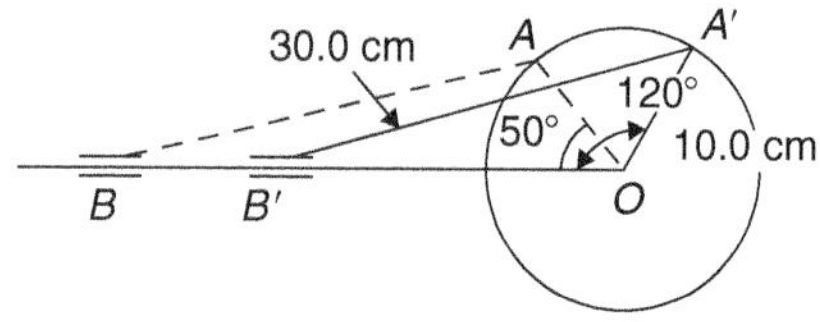

Figure 20.18

Hence $A'B'O = \sin^{-1} 0.2887 = 16.78°$
(or 163.22° which is impossible in this case).

Angle $OA'B' = 180° - 120° - 16.78° = 43.22°$

Applying the sine rule:

$$\frac{30.0}{\sin 120°} = \frac{OB'}{\sin 43.22°}$$

from which, $$OB' = \frac{30.0 \sin 43.22°}{\sin 120°} = 23.72\,\text{cm}$$

Since $OB = 35.43$ cm and $OB' = 23.72$ cm then

$$BB' = 35.43 - 23.72 = 11.71\,\text{cm}$$

Hence *B* moves 11.71 cm when angle *AOB* changes from 50° to 120°

Problem 13. The area of a field is in the form of a quadrilateral *ABCD* as shown in Fig. 20.19. Determine its area

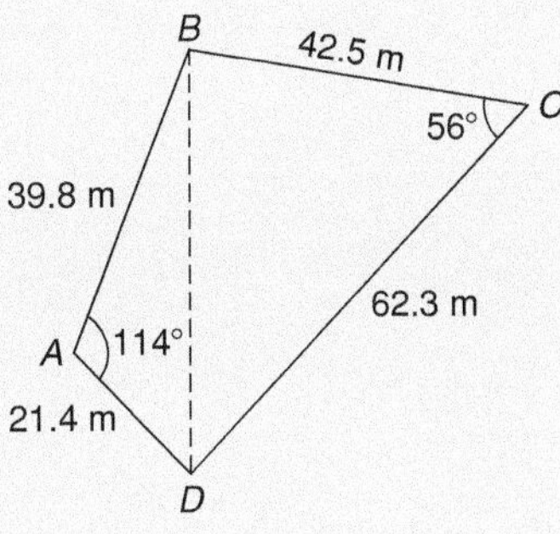

Figure 20.19

A diagonal drawn from *B* to *D* divides the quadrilateral into two triangles.

Area of quadrilateral *ABCD*

$= \text{area of triangle } ABD + \text{area of triangle } BCD$

$= \frac{1}{2}(39.8)(21.4)\sin 114° + \frac{1}{2}(42.5)(62.3)\sin 56°$

$= 389.04 + 1097.5 = \mathbf{1487\,m^2}$.

Now try the following Practice Exercise

Practice Exercise 104 Practical situations involving trigonometry (Answers on page 712)

1. Three forces acting on a fixed point are represented by the sides of a triangle of dimensions 7.2 cm, 9.6 cm and 11.0 cm. Determine the angles between the lines of action of the three forces

2. A vertical aerial *AB*, 9.60 m high, stands on ground which is inclined 12° to the horizontal. A stay connects the top of the aerial *A* to a point *C* on the ground 10.0 m downhill from *B*, the foot of the aerial. Determine (a) the length of the stay and (b) the angle the stay makes with the ground

3. A reciprocating engine mechanism is shown in Fig. 20.20. The crank *AB* is 12.0 cm long and the connecting rod *BC* is 32.0 cm long. For the position shown determine the length of *AC* and the angle between the crank and the connecting rod

4. From Fig. 20.20, determine how far *C* moves, correct to the nearest millimetre when angle *CAB* changes from 40° to 160°, *B* moving in an anticlockwise direction

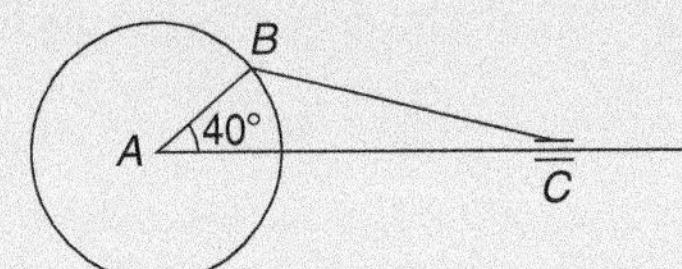

Figure 20.20

5. A surveyor, standing W 25° S of a tower measures the angle of elevation of the top of the tower as 46°30′. From a position E 23° S from the tower the elevation of the top is 37°15′. Determine the height of the tower if the distance between the two observations is 75 m

6. Calculate, correct to 3 significant figures, the co-ordinates *x* and *y* to locate the hole centre at *P* shown in Fig. 20.21

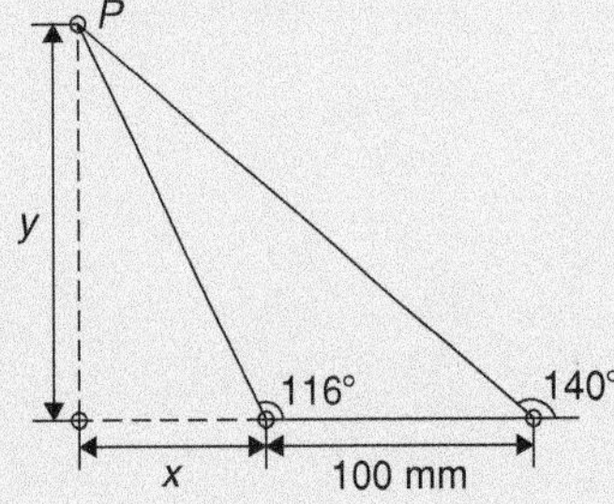

Figure 20.21

7. An idler gear, 30 mm in diameter, has to be fitted between a 70 mm diameter driving gear and a 90 mm diameter driven gear as shown in Fig. 20.22. Determine the value of angle θ between the centre lines

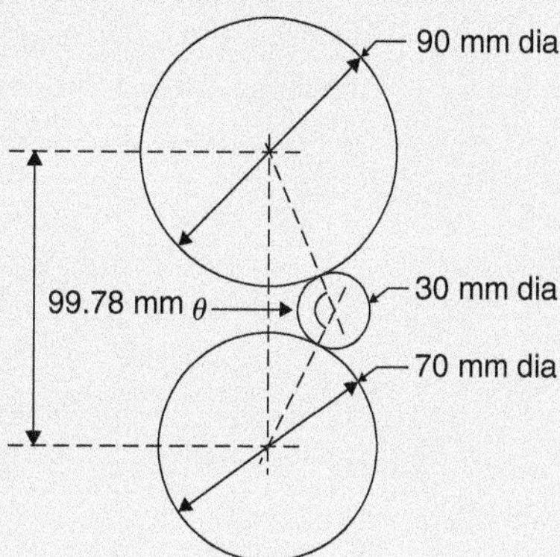

Figure 20.22

8. 16 holes are equally spaced on a pitch circle of 70 mm diameter. Determine the length of the chord joining the centres of two adjacent holes

Practice Exercise 105 Multiple-choice questions on triangles and some practical applications (Answers on page 712)

Each question has only one correct answer

1. In a non-right-angled triangular plate XYZ, $XY = 12$ cm, $YZ = 18$ cm and $\angle XZY = 30°$. The value of sin X is equal to:
(a) $\frac{1}{3}$ (b) $\frac{1}{2}$ (c) $\frac{2}{3}$ (d) $\frac{3}{4}$
2. In the triangular template DEF shown in Figure 20.23, angle F is equal to:
(a) 43.5° (b) 28.6°
(c) 116.4° (d) 101.5°

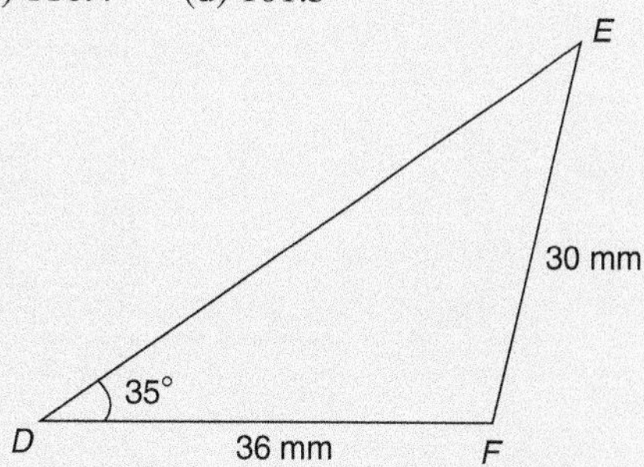

Figure 20.23

3. The length of side DE in Figure 20.23 is:
(a) 51.25 mm (b) 46.85 mm
(c) 50.55 mm (d) 73.70 mm
4. The area of the triangular template DEF shown in Figure 20.23 is:
(a) 529.2 mm^2 (b) 258.5 mm^2
(c) 483.7 mm^2 (d) 371.7 mm^2
5. A triangle has sides a = 9.0 cm, b = 8.0 cm and c = 6.0 cm. Angle A is equal to:
(a) 82.42° (b) 56.49°
(c) 78.58° (d) 79.87°
6. The area of a triangle PQR is given by:
(a) $\frac{1}{2}pr\cos Q$
(b) $\sqrt{[(s-p)(s-q)(s-r)]}$ where $s = \frac{p+q+r}{2}$
(c) $\frac{1}{2}rq\sin P$
(d) $\frac{1}{2}pq\sin Q$
7. In triangle ABC in Figure 20.24, length AC is:
(a) 18.15 cm (b) 14.90 cm
(c) 13.16 cm (d) 14.04 cm

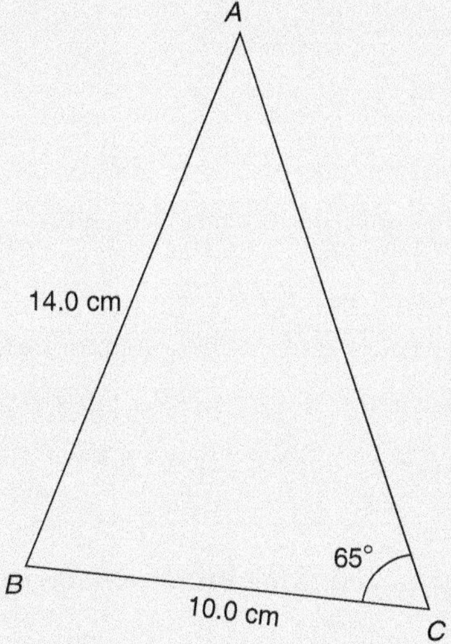

Figure 20.24

8. The area of triangle ABC in Figure 20.24 is:
(a) 70.0 cm^2 (b) 18.51 cm^2
(c) 67.51 cm^2 (d) 71.85 cm^2

9. In triangle ABC in Figure 20.25, the length AC is:
(a) 18.79 cm (b) 70.89 cm
(c) 22.89 cm (d) 16.10 cm

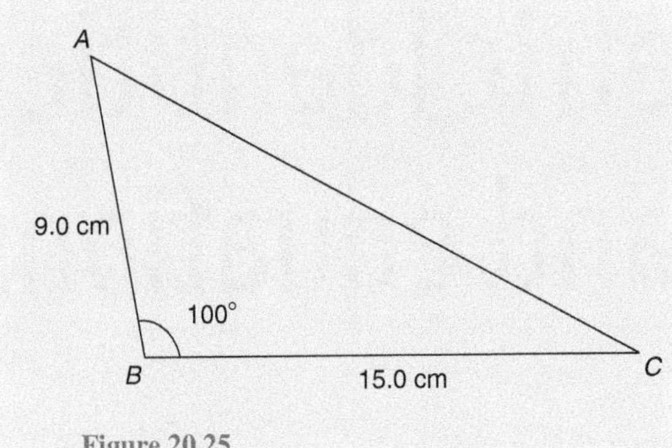

Figure 20.25

10. The area of triangle ABC in Figure 20.25 is:
(a) 53.07 cm^2
(b) 66.47 cm^2
(c) 11.72 cm^2
(d) 138.78 cm^2

For fully worked solutions to each of the problems in Practice Exercises 101 to 104 in this chapter, go to the website:
www.routledge.com/cw/bird

Chapter 21

Trigonometric identities and equations

Why it is important to understand: Trigonometric identities and equations

In engineering, trigonometric identities occur often, examples being in the more advanced areas of calculus to generate derivatives and integrals, with tensors/vectors and with differential equations and partial differential equations. One of the skills required for more advanced work in mathematics, especially in calculus, is the ability to use identities to write expressions in alternative forms. In software engineering, working, say, on the next big blockbuster film, trigonometric identities are needed for computer graphics; an RF engineer working on the next-generation mobile phone will also need trigonometric identities. In addition, identities are needed in electrical engineering when dealing with a.c. power, and wave addition/subtraction and the solutions of trigonometric equations often require knowledge of trigonometric identities.

At the end of this chapter, you should be able to:

- state simple trigonometric identities
- prove simple identities
- solve equations of the form $b\sin A + c = 0$
- solve equations of the form $a\sin^2 A + c = 0$
- solve equations of the form $a\sin^2 A + b\sin A + c = 0$
- solve equations requiring trigonometric identities

21.1 Trigonometric identities

A trigonometric identity is a relationship that is true for all values of the unknown variable.

$$\tan\theta = \frac{\sin\theta}{\cos\theta} \quad \cot\theta = \frac{\cos\theta}{\sin\theta} \quad \sec\theta = \frac{1}{\cos\theta}$$

$$\operatorname{cosec}\theta = \frac{1}{\sin\theta} \quad \text{and} \quad \cot\theta = \frac{1}{\tan\theta}$$

are examples of trigonometric identities from Chapter 17.

Applying Pythagoras' theorem to the right-angled triangle shown in Fig. 21.1 gives:

$$a^2 + b^2 = c^2 \qquad (1)$$

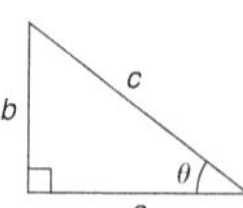

Figure 21.1

Dividing each term of equation (1) by c^2 gives:

$$\frac{a^2}{c^2}+\frac{b^2}{c^2}=\frac{c^2}{c^2}$$

i.e. $$\left(\frac{a}{c}\right)^2+\left(\frac{b}{c}\right)^2=1$$

$$(\cos\theta)^2+(\sin\theta)^2=1$$

Hence $$\mathbf{\cos^2\theta+\sin^2\theta=1} \quad (2)$$

Dividing each term of equation (1) by a^2 gives:

$$\frac{a^2}{a^2}+\frac{b^2}{a^2}=\frac{c^2}{a^2}$$

i.e. $$1+\left(\frac{b}{a}\right)^2=\left(\frac{c}{a}\right)^2$$

Hence $$\mathbf{1+\tan^2\theta=\sec^2\theta} \quad (3)$$

Dividing each term of equation (1) by b^2 gives:

$$\frac{a^2}{b^2}+\frac{b^2}{b^2}=\frac{c^2}{b^2}$$

i.e. $$\left(\frac{a}{b}\right)^2+1=\left(\frac{c}{b}\right)^2$$

Hence $$\mathbf{\cot^2\theta+1=\operatorname{cosec}^2\theta} \quad (4)$$

Equations (2), (3) and (4) are three further examples of trigonometric identities.

21.2 Worked problems on trigonometric identities

Problem 1. Prove the identity $\sin^2\theta\cot\theta\sec\theta=\sin\theta$

With trigonometric identities it is necessary to start with the left-hand side (LHS) and attempt to make it equal to the right-hand side (RHS) or vice-versa. It is often useful to change all of the trigonometric ratios into sines and cosines where possible. Thus

$$\text{LHS}=\sin^2\theta\cot\theta\sec\theta$$
$$=\sin^2\theta\left(\frac{\cos\theta}{\sin\theta}\right)\left(\frac{1}{\cos\theta}\right)$$
$$=\sin\theta\,(\text{by cancelling})=\text{RHS}$$

Problem 2. Prove that:

$$\frac{\tan x+\sec x}{\sec x\left(1+\dfrac{\tan x}{\sec x}\right)}=1$$

$$\text{LHS}=\frac{\tan x+\sec x}{\sec x\left(1+\dfrac{\tan x}{\sec x}\right)}$$
$$=\frac{\dfrac{\sin x}{\cos x}+\dfrac{1}{\cos x}}{\left(\dfrac{1}{\cos x}\right)\left(1+\dfrac{\frac{\sin x}{\cos x}}{\frac{1}{\cos x}}\right)}$$
$$=\frac{\dfrac{\sin x+1}{\cos x}}{\left(\dfrac{1}{\cos x}\right)\left[1+\left(\dfrac{\sin x}{\cos x}\right)\left(\dfrac{\cos x}{1}\right)\right]}$$
$$=\frac{\dfrac{\sin x+1}{\cos x}}{\left(\dfrac{1}{\cos x}\right)[1+\sin x]}$$
$$=\left(\frac{\sin x+1}{\cos x}\right)\left(\frac{\cos x}{1+\sin x}\right)$$
$$=1\text{ (by cancelling)}=\text{RHS}$$

Problem 3. Prove that: $\dfrac{1+\cot\theta}{1+\tan\theta}=\cot\theta$

$$\text{LHS}=\frac{1+\cot\theta}{1+\tan\theta}=\frac{1+\dfrac{\cos\theta}{\sin\theta}}{1+\dfrac{\sin\theta}{\cos\theta}}=\frac{\dfrac{\sin\theta+\cos\theta}{\sin\theta}}{\dfrac{\cos\theta+\sin\theta}{\cos\theta}}$$
$$=\left(\frac{\sin\theta+\cos\theta}{\sin\theta}\right)\left(\frac{\cos\theta}{\cos\theta+\sin\theta}\right)$$
$$=\frac{\cos\theta}{\sin\theta}=\cot\theta=\text{RHS}$$

Problem 4. Show that: $\cos^2\theta-\sin^2\theta=1-2\sin^2\theta$

From equation (2), $\cos^2\theta+\sin^2\theta=1$, from which, $\cos^2\theta=1-\sin^2\theta$

Hence, $$\text{LHS}=\cos^2\theta-\sin^2\theta$$
$$=(1-\sin^2\theta)-\sin^2\theta$$
$$=1-\sin^2\theta-\sin^2\theta$$
$$=1-2\sin^2\theta=\text{RHS}$$

Problem 5. Prove that:

$$\sqrt{\frac{1-\sin x}{1+\sin x}} = \sec x - \tan x$$

$$\text{LHS} = \sqrt{\frac{1-\sin x}{1+\sin x}}$$

$$= \sqrt{\frac{(1-\sin x)(1-\sin x)}{(1+\sin x)(1-\sin x)}}$$

$$= \sqrt{\frac{(1-\sin x)^2}{(1-\sin^2 x)}}$$

Since $\cos^2 x + \sin^2 x = 1$ then $1 - \sin^2 x = \cos^2 x$

$$\text{LHS} = \sqrt{\frac{(1-\sin x)^2}{(1-\sin^2 x)}} = \sqrt{\frac{(1-\sin x)^2}{\cos^2 x}}$$

$$= \frac{1-\sin x}{\cos x} = \frac{1}{\cos x} - \frac{\sin x}{\cos x}$$

$$= \sec x - \tan x = \text{RHS}$$

Now try the following Practice Exercise

Practice Exercise 106 Trigonometric identities (Answers on page 712)

Prove the following trigonometric identities:

1. $\sin x \cot x = \cos x$
2. $\dfrac{1}{\sqrt{1-\cos^2\theta}} = \operatorname{cosec}\theta$
3. $2\cos^2 A - 1 = \cos^2 A - \sin^2 A$
4. $\dfrac{\cos x - \cos^3 x}{\sin x} = \sin x \cos x$
5. $(1+\cot\theta)^2 + (1-\cot\theta)^2 = 2\operatorname{cosec}^2\theta$
6. $\dfrac{\sin^2 x(\sec x + \operatorname{cosec} x)}{\cos x \tan x} = 1 + \tan x$

21.3 Trigonometric equations

Equations which contain trigonometric ratios are called **trigonometric equations**. There are usually an infinite number of solutions to such equations; however, solutions are often restricted to those between 0° and 360°. A knowledge of angles of any magnitude is essential in the solution of trigonometric equations and calculators cannot be relied upon to give all the solutions (as shown in Chapter 18). Figure 21.2 shows a summary for angles of any magnitude.

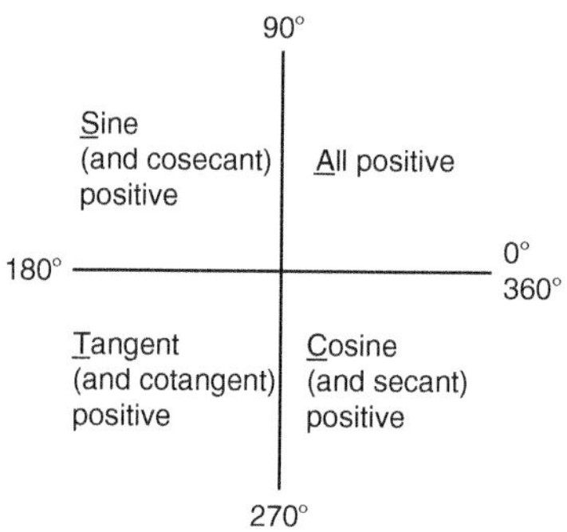

Figure 21.2

Equations of the type $a\sin^2 A + b\sin A + c = 0$

(i) **When $a=0$, $b\sin A + c = 0$, hence $\sin A = -\dfrac{c}{b}$ and $A = \sin^{-1}\left(-\dfrac{c}{b}\right)$**

There are two values of A between 0° and 360° that satisfy such an equation, provided $-1 \le \dfrac{c}{b} \le 1$ (see Problems 6 to 9).

(ii) **When $b=0$, $a\sin^2 A + c = 0$, hence $\sin^2 A = -\dfrac{c}{a}$,**

$\sin A = \sqrt{-\dfrac{c}{a}}$ and $A = \sin^{-1}\sqrt{-\dfrac{c}{a}}$

If either a or c is a negative number, then the value within the square root sign is positive. Since when a square root is taken there is a positive and negative answer there are four values of A between 0° and 360° which satisfy such an equation, provided $-1 \le \dfrac{c}{b} \le 1$ (see Problems 10 and 11).

(iii) **When a, b and c are all non-zero:**

$a\sin^2 A + b\sin A + c = 0$ is a quadratic equation in which the unknown is $\sin A$. The solution of a quadratic equation is obtained either by factorising (if possible) or by using the quadratic formula:

$$\sin A = \frac{-b \pm \sqrt{b^2 - 4ac}}{2a}$$

(see Problems 12 and 13).

(iv) Often the trigonometric identities $\cos^2 A + \sin^2 A = 1$, $1 + \tan^2 A = \sec^2 A$ and $\cot^2 A + 1 = \operatorname{cosec}^2 A$ need to be used to reduce equations to one of the above forms (see Problems 14 to 16).

21.4 Worked problems (i) on trigonometric equations

Problem 6. Solve the trigonometric equation: $5\sin\theta + 3 = 0$ for values of θ from 0° to 360°

$5\sin\theta + 3 = 0$, from which
$\sin\theta = -3/5 = -0.6000$

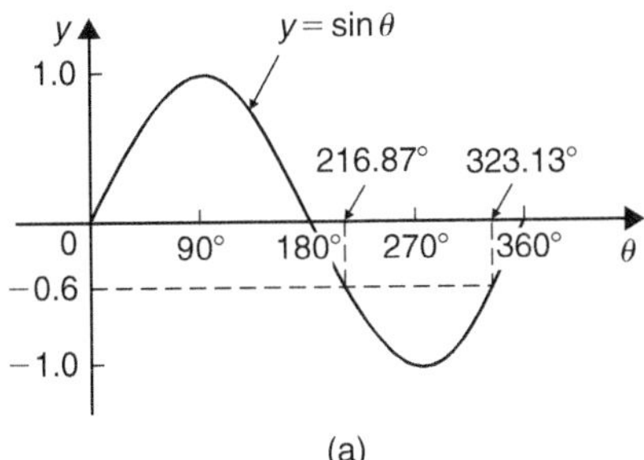

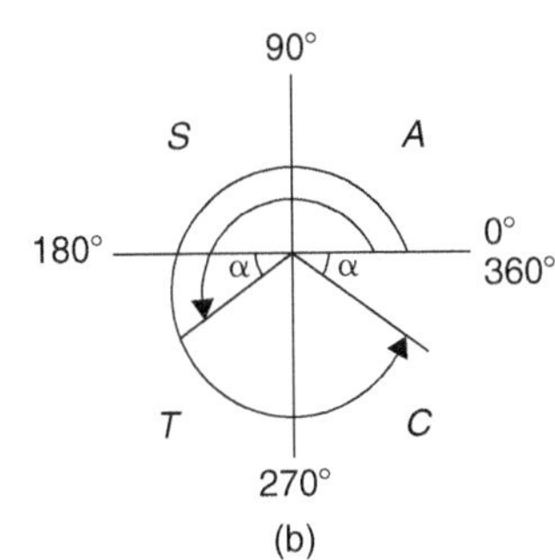

Figure 21.3

Hence $\theta = \sin^{-1}(-0.6000)$. Sine is negative in the third and fourth quadrants (see Fig. 21.3). The acute angle $\sin^{-1}(0.6000) = 36.87°$ (shown as α in Fig. 21.3(b)). Hence $\theta = 180° + 36.87°$, i.e. **216.87°** or $\theta = 360° - 36.87°$, i.e. **323.13°**

Problem 7. Solve: $1.5\tan x - 1.8 = 0$ for $0° \le x \le 360°$

$1.5\tan x - 1.8 = 0$, from which

$$\tan x = \frac{1.8}{1.5} = 1.2000$$

Hence $x = \tan^{-1} 1.2000$

Tangent is positive in the first and third quadrants (see Fig. 21.4).

The acute angle $\tan^{-1} 1.2000 = 50.19°$

Hence, $\mathbf{x = 50.19°}$ or $180° + 50.19° = \mathbf{230.19°}$

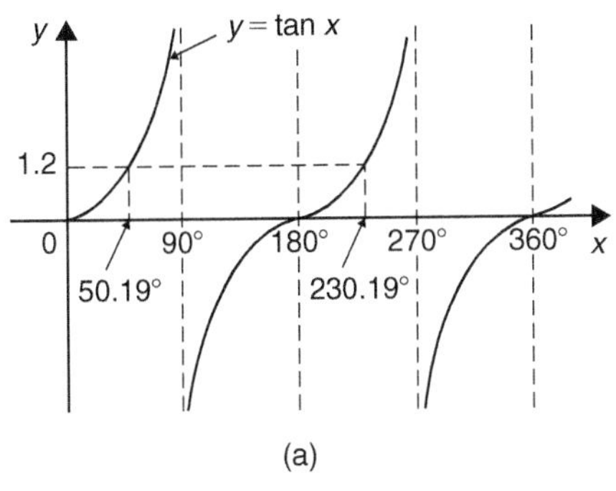

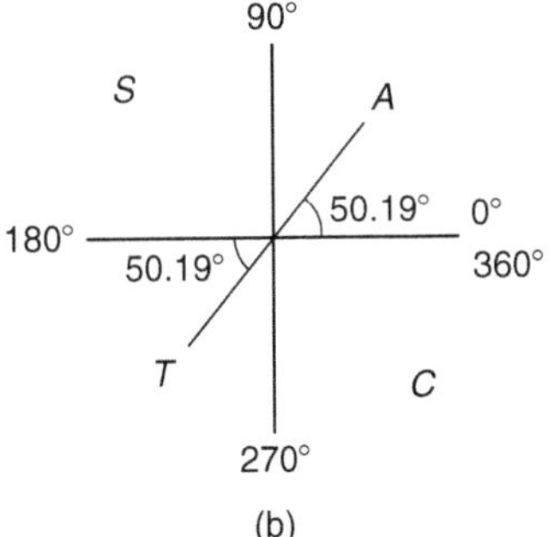

Figure 21.4

Problem 8. Solve for θ in the range $0° \le \theta \le 360°$ for $2\sin\theta = \cos\theta$

Dividing both sides by $\cos\theta$ gives: $\dfrac{2\sin\theta}{\cos\theta} = 1$

From Section 21.1, $\tan\theta = \dfrac{\sin\theta}{\cos\theta}$

hence $2\tan\theta = 1$

Dividing by 2 gives: $\tan\theta = \dfrac{1}{2}$

from which, $\theta = \tan^{-1}\dfrac{1}{2}$

Since tangent is positive in the first and third quadrants, $\theta = $ **26.57° and 206.57°**

Problem 9. Solve: $4\sec t = 5$ for values of t between 0° and 360°

$4\sec t = 5$, from which $\sec t = \frac{5}{4} = 1.2500$

Hence $t = \sec^{-1} 1.2500$

Secant (=1/cosine) is positive in the first and fourth quadrants (see Fig. 21.5). The acute angle $\sec^{-1} 1.2500 = 36.87°$. Hence

$$t = \mathbf{36.87°} \text{ or } 360° - 36.87° = \mathbf{323.13°}$$

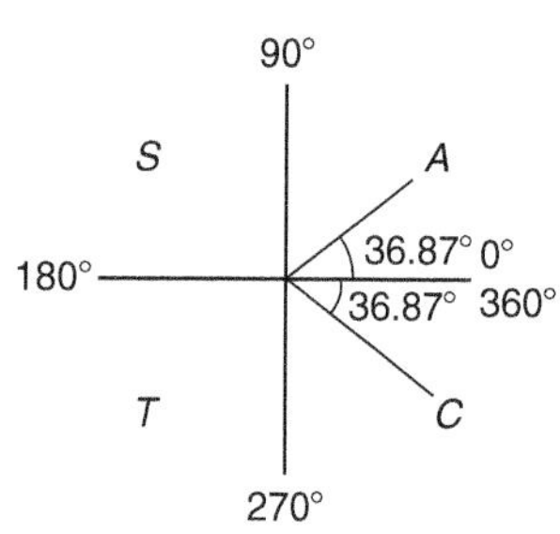

Figure 21.5

Now try the following Practice Exercise

Practice Exercise 107 Trigonometric equations (Answers on page 712)

Solve the following equations for angles between 0° and 360°

1. $4-7\sin\theta=0$
2. $3\,\text{cosec}\,A+5.5=0$
3. $4(2.32-5.4\cot t)=0$

In Problems 4 to 6, solve for θ in the range $0° \le \theta \le 360°$

4. $\sec\theta=2$
5. $\cot\theta=0.6$
6. $\text{cosec}\,\theta=1.5$

In Problems 7 to 9, solve for x in the range $-180° \le x \le 180°$

7. $\sec x=-1.5$
8. $\cot x=1.2$
9. $\text{cosec}\,x=-2$

In Problems 10 and 11, solve for θ in the range $0° \le \theta \le 360°$

10. $3\sin\theta=2\cos\theta$
11. $5\cos\theta=-\sin\theta$

21.5 Worked problems (ii) on trigonometric equations

Problem 10. Solve: $2-4\cos^2 A=0$ for values of A in the range $0° < A < 360°$

$2-4\cos^2 A=0$, from which $\cos^2 A=\dfrac{2}{4}=0.5000$

Hence $\cos A=\sqrt{0.5000}=\pm 0.7071$ and $A=\cos^{-1}(\pm 0.7071)$

Cosine is positive in quadrant one and four and negative in quadrants two and three. Thus in this case there are four solutions, one in each quadrant (see Fig. 21.6).The acute angle $\cos^{-1} 0.7071=45°$.

Hence $\quad$ $\mathbf{A=45°, 135°, 225° \text{ or } 315°}$

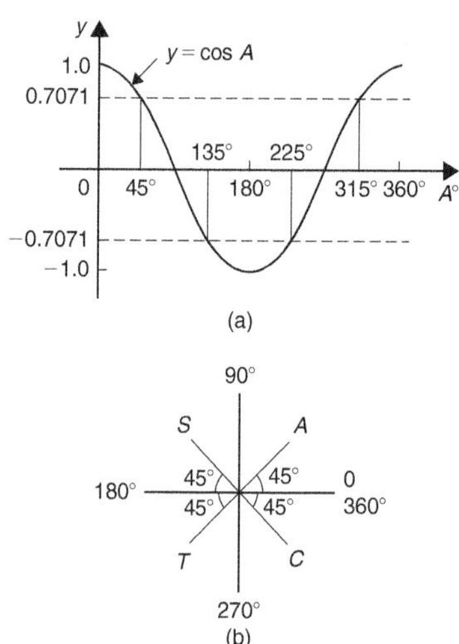

Figure 21.6

Problem 11. Solve: $\frac{1}{2}\cot^2 y=1.3$ for $0° < y < 360°$

$\frac{1}{2}\cot^2 y=1.3$, from which, $\cot^2 y=2(1.3)=2.6$

Hence $\cot y=\sqrt{2.6}=\pm 1.6125$, and $y=\cot^{-1}(\pm 1.6125)$. There are four solutions, one in each quadrant. The acute angle $\cot^{-1} 1.6125=31.81°$.

Hence $\mathbf{y=31.81°, 148.19°, 211.81° \text{ or } 328.19°}$

Now try the following Practice Exercise

Practice Exercise 108 Trigonometric equations (Answers on page 712)

Solve the following equations for angles between 0° and 360°

1. $5\sin^2 y=3$
2. $\cos^2\theta=0.25$

3. $\tan^2 x = 3$

4. $5 + 3\,\text{cosec}^2 D = 8$

5. $2\cot^2\theta = 5$

21.6 Worked problems (iii) on trigonometric equations

Problem 12. Solve the equation: $8\sin^2\theta + 2\sin\theta - 1 = 0$, for all values of θ between 0° and 360°

Factorising $8\sin^2\theta + 2\sin\theta - 1 = 0$ gives $(4\sin\theta - 1)(2\sin\theta + 1) = 0$

Hence $4\sin\theta - 1 = 0$, from which,

$\sin\theta = \frac{1}{4} = 0.2500$, or $2\sin\theta + 1 = 0$, from which, $\sin\theta = -\frac{1}{2} = -0.5000$
(Instead of factorising, the quadratic formula can, of course, be used). $\theta = \sin^{-1} 0.2500 = 14.48°$ or $165.52°$, since sine is positive in the first and second quadrants, or $\theta = \sin^{-1}(-0.5000) = 210°$ or $330°$, since sine is negative in the third and fourth quadrants. Hence

$$\theta = \mathbf{14.48°, 165.52°, 210° \text{ or } 330°}$$

Problem 13. Solve: $6\cos^2\theta + 5\cos\theta - 6 = 0$ for values of θ from 0° to 360°

Factorising $6\cos^2\theta + 5\cos\theta - 6 = 0$ gives $(3\cos\theta - 2)(2\cos\theta + 3) = 0$.

Hence $3\cos\theta - 2 = 0$, from which,

$\cos\theta = \frac{2}{3} = 0.6667$, or $2\cos\theta + 3 = 0$, from which, $\cos = -\frac{3}{2} = -1.5000$

The minimum value of a cosine is -1, hence the latter expression has no solution and is thus neglected. Hence

$$\theta = \cos^{-1} 0.6667 = \mathbf{48.18° \text{ or } 311.82°}$$

since cosine is positive in the first and fourth quadrants.

Now try the following Practice Exercise

Practice Exercise 109 Trigonometric equations (Answers on page 712)

Solve the following equations for angles between 0° and 360°

1. $15\sin^2 A + \sin A - 2 = 0$

2. $8\tan^2\theta + 2\tan\theta = 15$

3. $2\,\text{cosec}^2 t - 5\,\text{cosec}\, t = 12$

4. $2\cos^2\theta + 9\cos\theta - 5 = 0$

21.7 Worked problems (iv) on trigonometric equations

Problem 14. Solve: $5\cos^2 t + 3\sin t - 3 = 0$ for values of t from 0° to 360°

Since $\cos^2 t + \sin^2 t = 1$, $\cos^2 t = 1 - \sin^2 t$.
Substituting for $\cos^2 t$ in $5\cos^2 t + 3\sin t - 3 = 0$ gives

$$5(1 - \sin^2 t) + 3\sin t - 3 = 0$$
$$5 - 5\sin^2 t + 3\sin t - 3 = 0$$
$$-5\sin^2 t + 3\sin t + 2 = 0$$
$$5\sin^2 t - 3\sin t - 2 = 0$$

Factorising gives $(5\sin t + 2)(\sin t - 1) = 0$. Hence $5\sin t + 2 = 0$, from which, $\sin t = -\frac{2}{5} = -0.4000$, or $\sin t - 1 = 0$, from which, $\sin t = 1$.
$t = \sin^{-1}(-0.4000) = 203.58°$ or $336.42°$, since sine is negative in the third and fourth quadrants, or $t = \sin^{-1} 1 = 90°$. Hence

$$t = \mathbf{90°, 203.58° \text{ or } 336.42°}$$

as shown in Fig. 21.7.

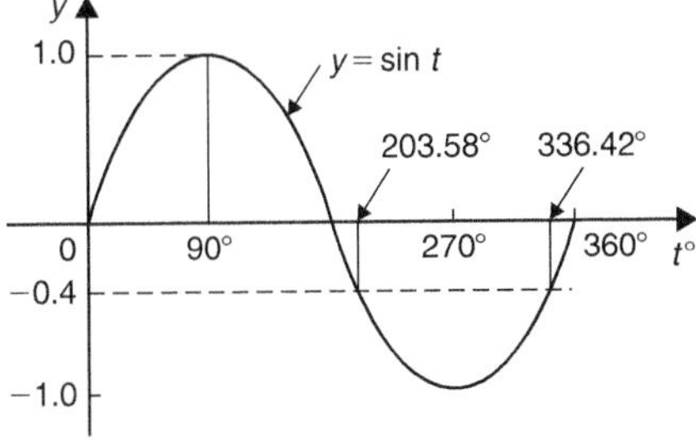

Figure 21.7

Problem 15. Solve: $18\sec^2 A - 3\tan A = 21$ for values of A between 0° and 360°

$1+\tan^2 A = \sec^2 A$. Substituting for $\sec^2 A$ in $18\sec^2 A - 3\tan A = 21$ gives

$$18(1+\tan^2 A) - 3\tan A = 21$$

i.e. $$18 + 18\tan^2 A - 3\tan A - 21 = 0$$

$$18\tan^2 A - 3\tan A - 3 = 0$$

Factorising gives $(6\tan A - 3)(3\tan A + 1) = 0$

Hence $6\tan A - 3 = 0$, from which, $\tan A = \frac{3}{6} = 0.5000$ or $3\tan A + 1 = 0$, from which, $\tan A = -\frac{1}{3} = -0.3333$. Thus $A = \tan^{-1}(0.5000) = 26.57°$ or $206.57°$, since tangent is positive in the first and third quadrants, or $A = \tan^{-1}(-0.3333) = 161.57°$ or $341.57°$, since tangent is negative in the second and fourth quadrants. Hence

$$\mathbf{A = 26.57°, 161.57°, 206.57° \text{ or } 341.57°}$$

Problem 16. Solve: $3\,\text{cosec}^2\theta - 5 = 4\cot\theta$ in the range $0 < \theta < 360°$

$\cot^2\theta + 1 = \text{cosec}^2\theta$, Substituting for $\text{cosec}^2\theta$ in $3\,\text{cosec}^2\theta - 5 = 4\cot\theta$ gives:

$$3(\cot^2\theta + 1) - 5 = 4\cot\theta$$

$$3\cot^2\theta + 3 - 5 = 4\cot\theta$$

$$3\cot^2\theta - 4\cot\theta - 2 = 0$$

Since the left-hand side does not factorise the quadratic formula is used.

Thus, $$\cot\theta = \frac{-(-4) \pm \sqrt{(-4)^2 - 4(3)(-2)}}{2(3)}$$

$$= \frac{4 \pm \sqrt{16+24}}{6} = \frac{4 \pm \sqrt{40}}{6}$$

$$= \frac{10.3246}{6} \text{ or } -\frac{2.3246}{6}$$

Hence $\cot\theta = 1.7208$ or -0.3874
$\theta = \cot^{-1} 1.7208 = 30.17°$ or $210.17°$, since cotangent is positive in the first and third quadrants, or $\theta = \cot^{-1}(-0.3874) = 111.18°$ or $291.18°$, since cotangent is negative in the second and fourth quadrants.

Hence, $\theta = \mathbf{30.17°, 111.18°, 210.17° \text{ or } 291.18°}$

Now try the following Practice Exercise

Practice Exercise 110 Trigonometric equations (Answers on page 712)

Solve the following equations for angles between 0° and 360°

1. $2\cos^2\theta + \sin\theta = 1$
2. $4\cos^2 t + 5\sin t = 3$
3. $2\cos\theta - 4\sin^2\theta = 0$
4. $3\cos\theta + 2\sin^2\theta = 3$
5. $12\sin^2\theta - 6 = \cos\theta$
6. $16\sec x - 2 = 14\tan^2 x$
7. $4\cot^2 A - 6\,\text{cosec}\,A + 6 = 0$
8. $5\sec t + 2\tan^2 t = 3$
9. $2.9\cos^2 a - 7\sin a + 1 = 0$
10. $3\,\text{cosec}^2\beta = 8 - 7\cot\beta$
11. $\cot\theta = \sin\theta$
12. $\tan\theta + 3\cot\theta = 5\sec\theta$

Practice Exercise 111 Multiple-choice questions on trigonometric identities and equations (Answers on page 713)

Each question has only one correct answer

1. Which of the following trigonometric identities is true?
   (a) $\text{cosec}\,\theta = \frac{1}{\cos\theta}$ (b) $\cot\theta = \frac{1}{\sin\theta}$
   (c) $\frac{\sin\theta}{\cos\theta} = \tan\theta$ (d) $\sec\theta = \frac{1}{\sin\theta}$
2. The solutions of the equation $2\tan x - 7 = 0$ for $0° \leq x \leq 360°$ are:
   (a) 74.05° and 254.05°
   (b) 105.95° and 254.05°
   (c) 74.05° and 285.95°
   (d) 254.05° and 285.95°

3. The angles between 0° and 360° whose tangent is – 1.7624 are:
(a) 60.43° and 240.43°
(b) 119.57° and 240.43°
(c) 119.57° and 299.57°
(d) 150.43° and 299.57°

4. In the range $0° \leq \theta \leq 360°$ the solutions of the trigonometric equation $9\tan^2\theta - 12\tan\theta + 4 = 0$ are:
(a) 33.69° and 213.69°
(b) 33.69°, 146.31°, 213.69° and 326.31°
(c) 146.31° and 213.69°
(d) 146.69° and 326.31°

5. The solution of the equation $3 - 5\cos^2 A = 0$ for values of A in the range $0° \leq A \leq 360°$ are:
(a) 39.23° and 320.77°
(b) 39.23°, 140.77°, 219.23° and 320.77°
(c) 140.77° and 219.23°
(d) 53.13°, 126.87°, 233.13° and 306.87°

6. The values of θ that are true for the equation $5\sin\theta + 2 = 0$ in the range $\theta = 0°$ to $\theta = 360°$ are:
(a) 23.58° and 336.42°
(b) 23.58° and 203.58°
(c) 156.42° and 336.42°
(d) 203.58° and 336.42°

7. The values of x that are true for the equation $3\sec x = 5$ in the range $x = 0°$ to $x = 360°$ are:
(a) 36.87° and 143.23°
(b) 53.13° and 306.87°
(c) 53.13°, 126.87°, 233.13 and 306.87°
(d) 36.87°, 143.13°, 216.87° and 323.13°

8. Solving the equation $4\cos^2 x + 3\sin x - 3 = 0$ for values of x from 0° to 360° gives:
(a) 57.95° and 302.05°
(b) 14.48°, 165.52° and 270°
(c) 14.48°, 90° or 165.52°
(d) 90°, 194.48° and 345.52°

For fully worked solutions to each of the problems in Practice Exercises 106 to 110 in this chapter, go to the website:
www.routledge.com/cw/bird

COMPANION @ WEBSITE

Chapter 22

Compound angles

Why it is important to understand: Compound angles

It is often necessary to rewrite expressions involving sines, cosines and tangents in alternative forms. To do this, formulae known as trigonometric identities are used as explained previously. Compound angle (or sum and difference) formulae and double angles are further commonly used identities. Compound angles are required for example in the analysis of acoustics (where a beat is an interference between two sounds of slightly different frequencies), and with phase detectors (which is a frequency mixer, analogue multiplier or logic circuit that generates a voltage signal which represents the difference in phase between two signal inputs). Many rational functions of sine and cosine are difficult to integrate without compound angle formulae.

At the end of this chapter, you should be able to:

- state compound angle formulae for $\sin(A \pm B)$, $\cos(A \pm B)$ and $\tan(A \pm B)$
- convert $a\sin\omega t + b\cos\omega t$ into $R\sin(\omega t + \alpha)$
- derive double angle formulae
- change products of sines and cosines into sums or differences
- change sums or differences of sines and cosines into products

22.1 Compound angle formulae

An electric current i may be expressed as $i = 5\sin(\omega t - 0.33)$ amperes. Similarly, the displacement x of a body from a fixed point can be expressed as $x = 10\sin(2t + 0.67)$ metres. The angles $(\omega t - 0.33)$ and $(2t + 0.67)$ are called **compound angles** because they are the sum or difference of two angles.
The **compound angle formulae** for sines and cosines of the sum and difference of two angles A and B are:

$$\sin(A+B) = \sin A\cos B + \cos A\sin B$$

$$\sin(A-B) = \sin A\cos B - \cos A\sin B$$

$$\cos(A+B) = \cos A\cos B - \sin A\sin B$$

$$\cos(A-B) = \cos A\cos B + \sin A\sin B$$

(Note, $\sin(A+B)$ is **not** equal to $(\sin A + \sin B)$ and so on.)
The formulae stated above may be used to derive two further compound angle formulae:

$$\tan(A+B) = \frac{\tan A + \tan B}{1 - \tan A\tan B}$$

$$\tan(A-B) = \frac{\tan A - \tan B}{1 + \tan A\tan B}$$

The compound-angle formulae are true for all values of A and B, and by substituting values of A and B into the formulae they may be shown to be true.

Problem 1. Expand and simplify the following expressions:
(a) $\sin(\pi+\alpha)$ (b) $-\cos(90°+\beta)$
(c) $\sin(A-B)-\sin(A+B)$

(a) $\sin(\pi+\alpha)=\sin\pi\cos\alpha+\cos\pi\sin\alpha$ (from the formula for $\sin(A+B)$)
$=(0)(\cos\alpha)+(-1)\sin\alpha$
$=-\sin\alpha$

(b) $-\cos(90°+\beta)$
$=-[\cos 90°\cos\beta-\sin 90°\sin\beta]$
$=-[(0)(\cos\beta)-(1)\sin\beta]=\sin\beta$

(c) $\sin(A-B)-\sin(A+B)$
$=[\sin A\cos B-\cos A\sin B]$
$-[\sin A\cos B+\cos A\sin B]$
$=\mathbf{-2\cos A\sin B}$

Problem 2. Prove that:

$$\cos(y-\pi)+\sin\left(y+\frac{\pi}{2}\right)=0$$

$$\cos(y-\pi)=\cos y\cos\pi+\sin y\sin\pi$$
$$=(\cos y)(-1)+(\sin y)(0)$$
$$=-\cos y$$

$$\sin\left(y+\frac{\pi}{2}\right)=\sin y\cos\frac{\pi}{2}+\cos y\sin\frac{\pi}{2}$$
$$=(\sin y)(0)+(\cos y)(1)=\cos y$$

Hence $\cos(y-\pi)+\sin\left(y+\frac{\pi}{2}\right)$
$$=(-\cos y)+(\cos y)=\mathbf{0}$$

Problem 3. Show that

$$\tan\left(x+\frac{\pi}{4}\right)\tan\left(x-\frac{\pi}{4}\right)=-1$$

$$\tan\left(x+\frac{\pi}{4}\right)=\frac{\tan x+\tan\frac{\pi}{4}}{1-\tan x\tan\frac{\pi}{4}}$$

(from the formula for $\tan(A+B)$)

$$=\frac{\tan x+1}{1-(\tan x)(1)}=\left(\frac{1+\tan x}{1-\tan x}\right)$$

since $\tan\frac{\pi}{4}=1$

$$\tan\left(x-\frac{\pi}{4}\right)=\frac{\tan x-\tan\frac{\pi}{4}}{1+\tan x\tan\frac{\pi}{4}}=\left(\frac{\tan x-1}{1+\tan x}\right)$$

Hence, $\tan\left(x+\frac{\pi}{4}\right)\tan\left(x-\frac{\pi}{4}\right)$

$$=\left(\frac{1+\tan x}{1-\tan x}\right)\left(\frac{\tan x-1}{1+\tan x}\right)$$

$$=\frac{\tan x-1}{1-\tan x}=\frac{-(1-\tan x)}{1-\tan x}=\mathbf{-1}$$

Problem 4. If $\sin P=0.8142$ and $\cos Q=0.4432$ evaluate, correct to 3 decimal places: (a) $\sin(P-Q)$, (b) $\cos(P+Q)$ and (c) $\tan(P+Q)$, using the compound angle formulae

Since $\sin P=0.8142$ then
$P=\sin^{-1}0.8142=54.51°$

Thus $\cos P=\cos 54.51°=0.5806$ and
$\tan P=\tan 54.51°=1.4025$

Since $\cos Q=0.4432$, $Q=\cos^{-1}0.4432=63.69°$

Thus $\sin Q=\sin 63.69°=0.8964$ and
$\tan Q=\tan 63.69°=2.0225$

(a) $\sin(P-Q)$
$=\sin P\cos Q-\cos P\sin Q$
$=(0.8142)(0.4432)-(0.5806)(0.8964)$
$=0.3609-0.5204=\mathbf{-0.160}$

(b) $\cos(P+Q)$
$=\cos P\cos Q-\sin P\sin Q$
$=(0.5806)(0.4432)-(0.8142)(0.8964)$
$=0.2573-0.7298=\mathbf{-0.473}$

(c) $\tan(P+Q)$

$$=\frac{\tan P+\tan Q}{1-\tan P\tan Q}=\frac{(1.4025)+(2.0225)}{1-(1.4025)(2.0225)}$$

$$=\frac{3.4250}{-1.8366}=\mathbf{-1.865}$$

Problem 5. Solve the equation:
$4\sin(x-20°)=5\cos x$ for values of x between 0° and 90°

$$4\sin(x-20°)=4[\sin x\cos 20°-\cos x\sin 20°]$$
from the formula for $\sin(A-B)$
$$=4[\sin x(0.9397)-\cos x(0.3420)]$$
$$=3.7588\sin x-1.3680\cos x$$

Since $4\sin(x-20°)=5\cos x$ then
$3.7588\sin x-1.3680\cos x=5\cos x$
Rearranging gives:

$$3.7588\sin x=5\cos x+1.3680\cos x$$
$$=6.3680\cos x$$

and $$\frac{\sin x}{\cos x}=\frac{6.3680}{3.7588}=1.6942$$

i.e. $\tan x=1.6942$, and
$x=\tan^{-1}1.6942=59.449°$ or **59°27′**

$$[\text{Check: LHS}=4\sin(59.449°-20°)$$
$$=4\sin 39.449°=2.542$$
$$\text{RHS}=5\cos x=5\cos 59.449°=2.542]$$

Now try the following Practice Exercise

Practice Exercise 112 Compound angle formulae (Answers on page 713)

1. Reduce the following to the sine of one angle:
   (a) $\sin 37°\cos 21°+\cos 37°\sin 21°$
   (b) $\sin 7t\cos 3t-\cos 7t\sin 3t$
2. Reduce the following to the cosine of one angle:
   (a) $\cos 71°\cos 33°-\sin 71°\sin 33°$
   (b) $\cos\frac{\pi}{3}\cos\frac{\pi}{4}+\sin\frac{\pi}{3}\sin\frac{\pi}{4}$
3. Show that:
   (a) $\sin\left(x+\frac{\pi}{3}\right)+\sin\left(x+\frac{2\pi}{3}\right)=\sqrt{3}\cos x$
   (b) $-\sin\left(\frac{3\pi}{2}-\phi\right)=\cos\phi$
4. Prove that:
   (a) $\sin\left(\theta+\frac{\pi}{4}\right)-\sin\left(\theta-\frac{3\pi}{4}\right)=\sqrt{2}(\sin\theta+\cos\theta)$
   (b) $\frac{\cos(270°+\theta)}{\cos(360°-\theta)}=\tan\theta$
5. Given $\cos A=0.42$ and $\sin B=0.73$ evaluate (a) $\sin(A-B)$, (b) $\cos(A-B)$, (c) $\tan(A+B)$, correct to 4 decimal places.

In Problems 6 and 7, solve the equations for values of θ between 0° and 360°.

6. $3\sin(\theta+30°)=7\cos\theta$
7. $4\sin(\theta-40°)=2\sin\theta$

22.2 Conversion of $a\sin\omega t+b\cos\omega t$ into $R\sin(\omega t+\alpha)$

(i) $R\sin(\omega t+\alpha)$ represents a sine wave of maximum value R, periodic time $2\pi/\omega$, frequency $\omega/2\pi$ and leading $R\sin\omega t$ by angle α. (See Chapter 18.)

(ii) $R\sin(\omega t+\alpha)$ may be expanded using the compound-angle formula for $\sin(A+B)$, where $A=\omega t$ and $B=\alpha$. Hence

$$R\sin(\omega t+\alpha)$$
$$=R[\sin\omega t\cos\alpha+\cos\omega t\sin\alpha]$$
$$=R\sin\omega t\cos\alpha+R\cos\omega t\sin\alpha$$
$$=(R\cos\alpha)\sin\omega t+(R\sin\alpha)\cos\omega t$$

(iii) If $a=R\cos\alpha$ and $b=R\sin\alpha$, where a and b are constants, then

$R\sin(\omega t+\alpha)=a\sin\omega t+b\cos\omega t$, i.e. a sine and cosine function of the same frequency when added produce a sine wave of the same frequency (which is further demonstrated in Chapter 35).

(iv) Since $a=R\cos\alpha$, then $\cos\alpha=\frac{a}{R}$ and since $b=R\sin\alpha$, then $\sin\alpha=\frac{b}{R}$

If the values of a and b are known then the values of R and α may be calculated. The relationship between constants a, b, R and α are shown in Fig. 22.1.
From Fig. 22.1, by Pythagoras' theorem:

$$R=\sqrt{a^2+b^2}$$

Figure 22.1

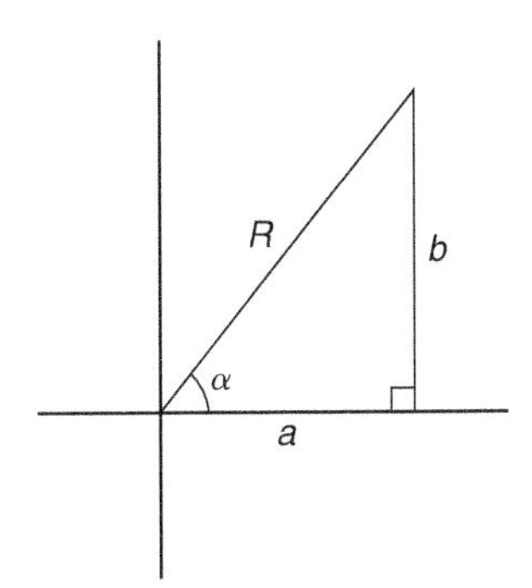

and from trigonometric ratios:

$$\alpha = \tan^{-1}\frac{b}{a}$$

Problem 6. Find an expression for $3\sin\omega t + 4\cos\omega t$ in the form $R\sin(\omega t + \alpha)$ and sketch graphs of $3\sin\omega t$, $4\cos\omega t$ and $R\sin(\omega t+\alpha)$ on the same axes

Let $3\sin\omega t + 4\cos\omega t = R\sin(\omega t+\alpha)$

then $\quad 3\sin\omega t + 4\cos\omega t$

$$= R[\sin\omega t\cos\alpha + \cos\omega t\sin\alpha]$$

$$= (R\cos\alpha)\sin\omega t + (R\sin\alpha)\cos\omega t$$

Equating coefficients of $\sin\omega t$ gives:

$$3 = R\cos\alpha, \text{ from which, } \cos\alpha = \frac{3}{R}$$

Equating coefficients of $\cos\omega t$ gives:

$$4 = R\sin\alpha, \text{ from which, } \sin\alpha = \frac{4}{R}$$

There is only one quadrant where both $\sin\alpha$ **and** $\cos\alpha$ are positive, and this is the first, as shown in Fig. 22.2. From Fig. 22.2, by Pythagoras' theorem:

$$R = \sqrt{3^2 + 4^2} = 5$$

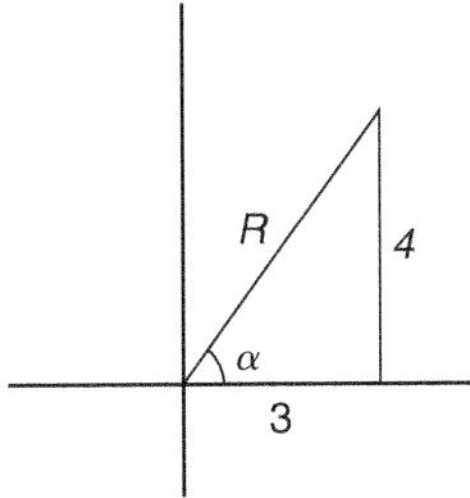

Figure 22.2

From trigonometric ratios:

$$a = \tan^{-1}\frac{4}{3} = 53.13° \quad \text{or} \quad 0.927 \text{ radians}$$

Hence, $\mathbf{3\sin\omega t + 4\cos\omega t = 5\sin(\omega t + 0.927)}$

A sketch of $3\sin\omega t$, $4\cos\omega t$ and $5\sin(\omega t + 0.927)$ is shown in Fig. 22.3.

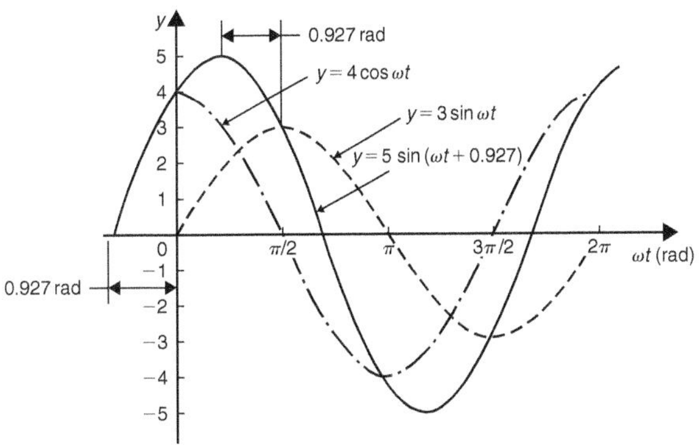

Figure 22.3

Two periodic functions of the same frequency may be combined by

(a) plotting the functions graphically and combining ordinates at intervals, or

(b) by resolution of phasors by drawing or calculation.

Problem 6, together with Problems 7 and 8 following, demonstrate a third method of combining waveforms.

Problem 7. Express: $4.6\sin\omega t - 7.3\cos\omega t$ in the form $R\sin(\omega t+\alpha)$

Let $\quad 4.6\sin\omega t - 7.3\cos\omega t = R\sin(\omega t+\alpha)$

then $\quad 4.6\sin\omega t - 7.3\cos\omega t$

$$= R[\sin\omega t\cos\alpha + \cos\omega t\sin\alpha]$$

$$= (R\cos\alpha)\sin\omega t + (R\sin\alpha)\cos\omega t$$

Equating coefficients of $\sin\omega t$ gives:

$$4.6 = R\cos\alpha, \text{ from which, } \cos\alpha = \frac{4.6}{R}$$

Equating coefficients of $\cos\omega t$ gives:

$$-7.3 = R\sin\alpha, \text{ from which } \sin\alpha = \frac{-7.3}{R}$$

There is only one quadrant where cosine is positive **and** sine is negative, i.e. the fourth quadrant, as shown in Fig. 22.4. By Pythagoras' theorem:

$$R = \sqrt{4.6^2 + (-7.3)^2} = 8.628$$

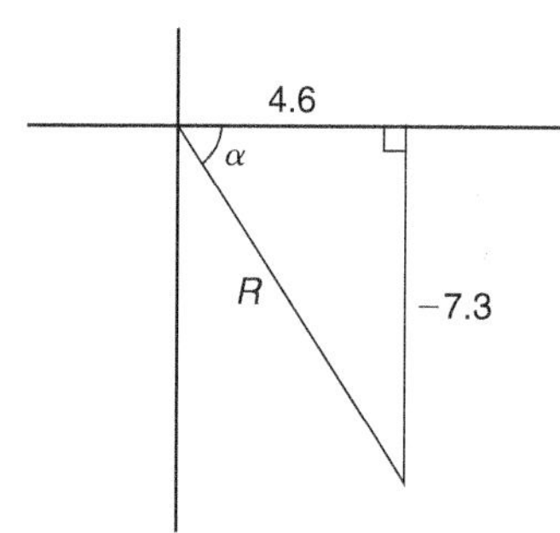

Figure 22.4

By trigonometric ratios:

$$\alpha = \tan^{-1}\left(\frac{-7.3}{4.6}\right)$$
$$= -57.78^\circ \text{ or } -1.008 \text{ radians.}$$

Hence,

$$\mathbf{4.6 \sin \omega t - 7.3 \cos \omega t = 8.628 \sin(\omega t - 1.008)}$$

Problem 8. Express: $-2.7 \sin \omega t - 4.1 \cos \omega t$ in the form $R \sin(\omega t + \alpha)$

Let $-2.7 \sin \omega t - 4.1 \cos \omega t = R \sin(\omega t + \alpha)$

$$= R[\sin \omega t \cos \alpha + \cos \omega t \sin \alpha]$$
$$= (R \cos \alpha) \sin \omega t + (R \sin \alpha) \cos \omega t$$

Equating coefficients gives:

$$-2.7 = R\cos\alpha, \text{from which, } \cos\alpha = \frac{-2.7}{R}$$

and $-4.1 = R\sin\alpha, \text{from which, } \sin\alpha = \dfrac{-4.1}{R}$

There is only one quadrant in which both cosine **and** sine are negative, i.e. the third quadrant, as shown in Fig. 22.5.
From Fig. 22.5,

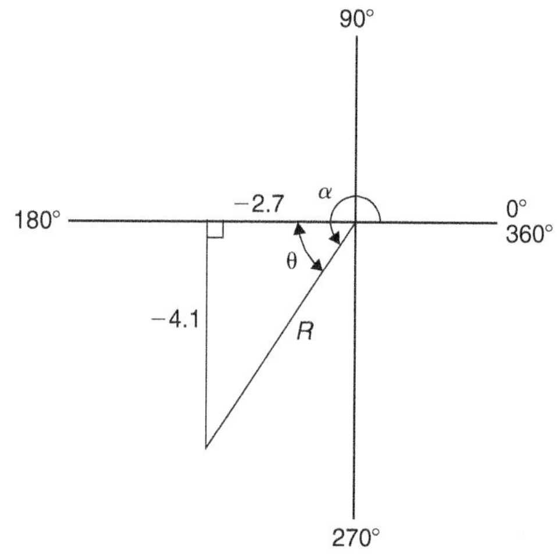

Figure 22.5

$$R = \sqrt{(-2.7)^2 + (-4.1)^2} = 4.909$$

and $\theta = \tan^{-1}\dfrac{4.1}{2.7} = 56.63^\circ$

Hence $\alpha = 180^\circ + 56.63^\circ = 236.63^\circ$ or 4.130 radians. Thus,

$$\mathbf{-2.7 \sin \omega t - 4.1 \cos \omega t = 4.909 \sin(\omega t - 4.130)}$$

An angle of 236.63° is the same as -123.37° or -2.153 radians.
Hence $-2.7 \sin \omega t - 4.1 \cos \omega t$ may be expressed also as **4.909** $\sin(\omega t$ **− 2.153**$)$, which is preferred since it is the **principal value** (i.e. $-\pi \le \alpha \le \pi$).

Problem 9. Express: $3 \sin \theta + 5 \cos \theta$ in the form $R \sin(\theta + \alpha)$, and hence solve the equation $3 \sin \theta + 5 \cos \theta = 4$, for values of θ between 0° and 360°

Let $3 \sin \theta + 5 \cos \theta = R \sin(\theta + \alpha)$

$$= R[\sin \theta \cos \alpha + \cos \theta \sin \alpha]$$
$$= (R \cos \alpha) \sin \theta + (R \sin \alpha) \cos \theta$$

Equating coefficients gives:

$$3 = R\cos\alpha, \text{from which, } \cos\alpha = \frac{3}{R}$$

and $5 = R\sin\alpha, \text{from which, } \sin\alpha = \dfrac{5}{R}$

Since both $\sin \alpha$ and $\cos \alpha$ are positive, R lies in the first quadrant, as shown in Fig. 22.6.

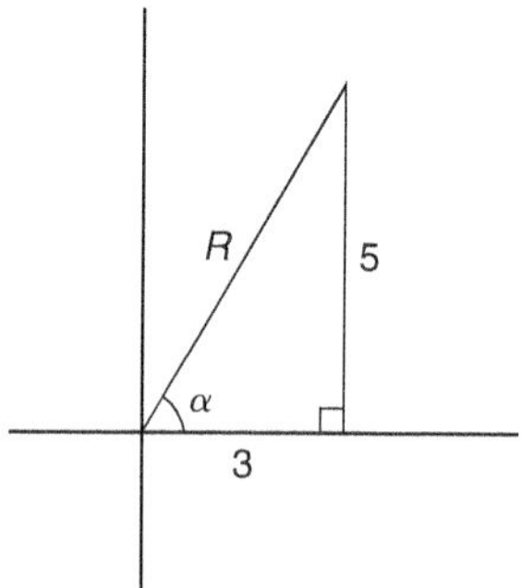

Figure 22.6

From Fig. 22.6, $R=\sqrt{3^2+5^2}=5.831$ and $\alpha=\tan^{-1}\dfrac{5}{3}=59.04°$

Hence $\quad 3\sin\theta+5\cos\theta=5.831\sin(\theta+59.04°)$

However $\quad 3\sin\theta+5\cos\theta=4$

Thus $\quad 5.831\sin(\theta+59.04°)=4,$

from which $\quad (\theta+59.04°)=\sin^{-1}\left(\dfrac{4}{5.831}\right)$

i.e. $\quad \theta+59.04°=43.32° \quad \text{or} \quad 136.68°$

Hence $\theta=43.32°-59.04°=-15.72° \quad \text{or}$

$\theta=136.68°-59.04°=77.64°$

Since $-15.72°$ is the same as $-15.72°+360°$, i.e. 344.28°, then the solutions are $\boldsymbol{\theta=77.64°}$ **or 344.28°**, which may be checked by substituting into the original equation.

Problem 10. Solve the equation: $3.5\cos A-5.8\sin A=6.5$ for $0°\leq A\leq 360°$

Let $\quad 3.5\cos A-5.8\sin A=R\sin(A+\alpha)$

$=R[\sin A\cos\alpha+\cos A\sin\alpha]$

$=(R\cos\alpha)\sin A+(R\sin\alpha)\cos A$

Equating coefficients gives:

$3.5=R\sin\alpha$, from which, $\sin\alpha=\dfrac{3.5}{R}$

and $-5.8=R\cos\alpha$, from which, $\cos\alpha=\dfrac{-5.8}{R}$

There is only one quadrant in which both sine is positive **and** cosine is negative, i.e. the second, as shown in Fig. 22.7.

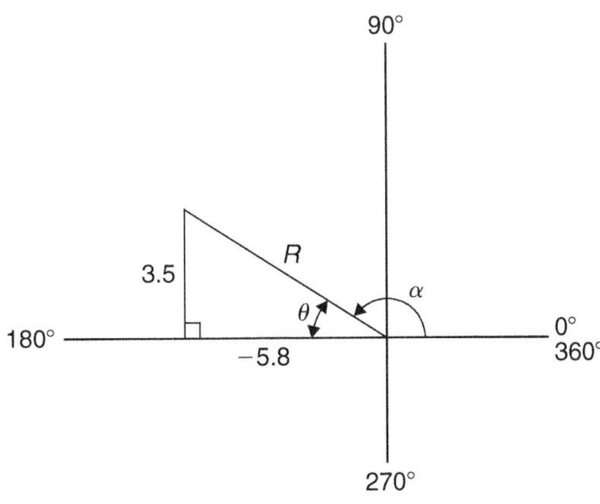

Figure 22.7

From Fig. 22.7, $R=\sqrt{3.5^2+(-5.8)^2}=6.774$ and $\theta=\tan^{-1}\dfrac{3.5}{5.8}=31.12°$

Hence $\alpha=180°-31.12°=148.88°$

Thus $\quad 3.5\cos A-5.8\sin A$

$=6.774\sin(A+148.88°)=6.5$

Hence, $\quad \sin(A+148.88°)=\dfrac{6.5}{6.774}$

from which, $\quad (A+148.88°)=\sin^{-1}\dfrac{6.5}{6.774}$

$=73.65° \text{ or } 106.35°$

Thus, $\quad A=73.65°-148.88°=-75.23°$

$\equiv(-75.23°+360°)=284.77°$

or $\quad A=106.35°-148.88°=-42.53°$

$\equiv(-42.53°+360°)=317.47°$

The solutions are thus $\boldsymbol{A=284.77°}$ **or 317.47°**, which may be checked in the original equation.

Now try the following Practice Exercise

Practice Exercise 113 The conversion of $a\sin\omega t+b\cos\omega t$ into $R\sin(\omega t+\alpha)$ (Answers on page 713)

In Problems 1 to 4, change the functions into the form $R\sin(\omega t\pm\alpha)$.

1. $5\sin\omega t+8\cos\omega t$
2. $4\sin\omega t-3\cos\omega t$
3. $-7\sin\omega t+4\cos\omega t$
4. $-3\sin\omega t-6\cos\omega t$
5. Solve the following equations for values of θ between 0° and 360°:
(a) $2\sin\theta+4\cos\theta=3$
(b) $12\sin\theta-9\cos\theta=7$
6. Solve the following equations for $0°<A<360°$:
(a) $3\cos A+2\sin A=2.8$
(b) $12\cos A-4\sin A=11$
7. Solve the following equations for values of θ between 0° and 360°:
(a) $3\sin\theta+4\cos\theta=3$
(b) $2\cos\theta+\sin\theta=2$

8. Solve the following equations for values of θ between 0° and 360°:
(a) $6\cos\theta + \sin\theta = \sqrt{3}$
(b) $2\sin 3\theta + 8\cos 3\theta = 1$

9. The third harmonic of a wave motion is given by $4.3\cos 3\theta - 6.9\sin 3\theta$.
Express this in the form $R\sin(3\theta \pm \alpha)$

10. The displacement x metres of a mass from a fixed point about which it is oscillating is given by $x = 2.4\sin\omega t + 3.2\cos\omega t$, where t is the time in seconds. Express x in the form $R\sin(\omega t + \alpha)$

11. Two voltages, $v_1 = 5\cos\omega t$ and $v_2 = -8\sin\omega t$ are inputs to an analogue circuit. Determine an expression for the output voltage if this is given by $(v_1 + v_2)$

12. The motion of a piston moving in a cylinder can be described by:
$x = (5\cos 2t + 5\sin 2t)$ cm.
Express x in the form $R\sin(\omega t + \alpha)$.

22.3 Double angles

(i) If, in the compound-angle formula for $\sin(A+B)$, we let $B = A$ then

$$\mathbf{\sin 2A = 2\sin A\cos A}$$

Also, for example, $\sin 4A = 2\sin 2A\cos 2A$ and $\sin 8A = 2\sin 4A\cos 4A$ and so on.

(ii) If, in the compound-angle formula for $\cos(A+B)$, we let $B = A$ then

$$\mathbf{\cos 2A = \cos^2 A - \sin^2 A}$$

Since $\cos^2 A + \sin^2 A = 1$, then $\cos^2 A = 1 - \sin^2 A$, and $\sin^2 A = 1 - \cos^2 A$, and two further formula for $\cos 2A$ can be produced.

Thus $\cos 2A = \cos^2 A - \sin^2 A = (1 - \sin^2 A) - \sin^2 A$

i.e. $\mathbf{\cos 2A = 1 - 2\sin^2 A}$

and $\cos 2A = \cos^2 A - \sin^2 A = \cos^2 A - (1 - \cos^2 A)$

i.e. $\mathbf{\cos 2A = 2\cos^2 A - 1}$

Also, for example, $\cos 4A = \cos^2 2A - \sin^2 2A$ or $1 - 2\sin^2 2A$ or $2\cos^2 2A - 1$ and $\cos 6A = \cos^2 3A - \sin^2 3A$ or $1 - 2\sin^2 3A$ or $2\cos^2 3A - 1$ and so on.

(iii) If, in the compound-angle formula for $\tan(A+B)$, we let $B = A$ then

$$\mathbf{\tan 2A = \frac{2\tan A}{1 - \tan^2 A}}$$

Also, for example, $\tan 4A = \dfrac{2\tan 2A}{1 - \tan^2 2A}$

and $\tan 5A = \dfrac{2\tan\frac{5}{2}A}{1 - \tan^2\frac{5}{2}A}$ and so on.

Problem 11. $I_3 \sin 3\theta$ is the third harmonic of a waveform. Express the third harmonic in terms of the first harmonic $\sin\theta$, when $I_3 = 1$

When $I_3 = 1$,

$$I_3\sin 3\theta = \sin 3\theta = \sin(2\theta + \theta)$$

$$= \sin 2\theta\cos\theta + \cos 2\theta\sin\theta$$

from the $\sin(A+B)$ formula

$$= (2\sin\theta\cos\theta)\cos\theta + (1 - 2\sin^2\theta)\sin\theta,$$

from the double angle expansions

$$= 2\sin\theta\cos^2\theta + \sin\theta - 2\sin^3\theta$$

$$= 2\sin\theta(1 - \sin^2\theta) + \sin\theta - 2\sin^3\theta,$$

$(\text{since } \cos^2\theta = 1 - \sin^2\theta)$

$$= 2\sin\theta - 2\sin^3\theta + \sin\theta - 2\sin^3\theta$$

i.e. $\mathbf{\sin 3\theta = 3\sin\theta - 4\sin^3\theta}$

Problem 12. Prove that: $\dfrac{1 - \cos 2\theta}{\sin 2\theta} = \tan\theta$

$$\text{LHS} = \frac{1 - \cos 2\theta}{\sin 2\theta} = \frac{1 - (1 - 2\sin^2\theta)}{2\sin\theta\cos\theta}$$

$$= \frac{2\sin^2\theta}{2\sin\theta\cos\theta} = \frac{\sin\theta}{\cos\theta} = \tan\theta = \text{RHS}$$

Problem 13. Prove that:

$$\cot 2x + \operatorname{cosec} 2x = \cot x$$

$$\text{LHS} = \cot 2x + \text{cosec}\, 2x$$
$$= \frac{\cos 2x}{\sin 2x} + \frac{1}{\sin 2x} = \frac{\cos 2x + 1}{\sin 2x}$$
$$= \frac{(2\cos^2 x - 1) + 1}{\sin 2x} = \frac{2\cos^2 x}{\sin 2x}$$
$$= \frac{2\cos^2 x}{2\sin x \cos x} = \frac{\cos x}{\sin x} = \cot x = \text{RHS}.$$

Problem 14. Solve the equation $\cos 2\theta + 3\sin\theta = 2$ for θ in the range $0° \le \theta \le 360°$

Replacing the double angle term with the relationship $\cos 2\theta = 1 - 2\sin^2\theta$ gives:

$$1 - 2\sin^2\theta + 3\sin\theta = 2$$

Rearranging gives:

$$-2\sin^2\theta + 3\sin\theta - 1 = 0 \text{ or } 2\sin^2\theta - 3\sin\theta + 1 = 0$$

which is a quadratic in $\sin\theta$.
Using the quadratic formula or by factorising gives:

$$(2\sin\theta - 1)(\sin\theta - 1) = 0$$

from which, $2\sin\theta - 1 = 0$ or $\sin\theta - 1 = 0$

and $\sin\theta = \frac{1}{2}$ or $\sin\theta = 1$

from which, $\boldsymbol{\theta = 30°}$ **or** $\boldsymbol{150°}$ **or** $\boldsymbol{90°}$

Now try the following Practice Exercise

Practice Exercise 114 Double angles (Answers on page 713)

1. The power p in an electrical circuit is given by $p = \frac{v^2}{R}$. Determine the power in terms of V, R and $\cos 2t$ when $v = V\cos t$
2. Prove the following identities:
   (a) $1 - \frac{\cos 2\phi}{\cos^2\phi} = \tan^2\phi$
   (b) $\frac{1 + \cos 2t}{\sin^2 t} = 2\cot^2 t$
   (c) $\frac{(\tan 2x)(1 + \tan x)}{\tan x} = \frac{2}{1 - \tan x}$
   (d) $2\,\text{cosec}\, 2\theta \cos 2\theta = \cot\theta - \tan\theta$
3. If the third harmonic of a waveform is given by $V_3 \cos 3\theta$, express the third harmonic in terms of the first harmonic $\cos\theta$, when $V_3 = 1$

In Problems 4 to 8, solve for θ in the range $-180° \le \theta \le 180°$:

4. $\cos 2\theta = \sin\theta$
5. $3\sin 2\theta + 2\cos\theta = 0$
6. $\sin 2\theta + \cos\theta = 0$
7. $\cos 2\theta + 2\sin\theta = -3$
8. $\tan\theta + \cot\theta = 2$

22.4 Changing products of sines and cosines into sums or differences

(i) $\sin(A + B) + \sin(A - B) = 2\sin A\cos B$ (from the formulae in Section 22.1), i.e.

$$\boldsymbol{\sin A \cos B = \frac{1}{2}[\sin(A + B) + \sin(A - B)]} \quad (1)$$

(ii) $\sin(A + B) - \sin(A - B) = 2\cos A \sin B$ i.e.

$$\boldsymbol{\cos A \sin B = \frac{1}{2}[\sin(A + B) - \sin(A - B)]} \quad (2)$$

(iii) $\cos(A + B) + \cos(A - B) = 2\cos A\cos B$ i.e.

$$\boldsymbol{\cos A \cos B = \frac{1}{2}[\cos(A + B) + \cos(A - B)]} \quad (3)$$

(iv) $\cos(A + B) - \cos(A - B) = -2\sin A\sin B$ i.e.

$$\boldsymbol{\sin A \sin B = -\frac{1}{2}[\cos(A + B) - \cos(A - B)]} \quad (4)$$

Problem 15. Express: $\sin 4x \cos 3x$ as a sum or difference of sines and cosines

From equation (1),

$$\sin 4x\cos 3x = \frac{1}{2}[\sin(4x + 3x) + \sin(4x - 3x)]$$
$$= \boldsymbol{\frac{1}{2}(\sin 7x + \sin x).}$$

Problem 16. Express: $2\cos 5\theta \sin 2\theta$ as a sum or difference of sines or cosines

From equation (2),

$$2\cos 5\theta \sin 2\theta$$
$$= 2\left\{\frac{1}{2}[\sin(5\theta+2\theta)-\sin(5\theta-2\theta)]\right\}$$
$$= \mathbf{\sin 7\theta - \sin 3\theta}$$

Problem 17. Express: $3\cos 4t\cos t$ as a sum or difference of sines or cosines

From equation (3),

$$3\cos 4t\cos t = 3\left\{\frac{1}{2}[\cos(4t+t)+\cos(4t-t)]\right\}$$
$$= \mathbf{\frac{3}{2}(\cos 5t + \cos 3t)}$$

Thus, if the integral $\int 3\cos 4t\cos t\,dt$ was required, then

$$\int 3\cos 4t\cos t\,dt = \int \frac{3}{2}(\cos 5t + \cos 3t)dt$$
$$= \mathbf{\frac{3}{2}\left[\frac{\sin 5t}{5} + \frac{\sin 3t}{3}\right] + c}$$

Problem 18. In an alternating current circuit, voltage $v=5\sin\omega t$ and current $i=10\sin(\omega t-\pi/6)$. Find an expression for the instantaneous power p at time t given that $p=vi$, expressing the answer as a sum or difference of sines and cosines

$$p = vi = (5\sin\omega t)[10\sin(\omega t-\pi/6)]$$
$$= 50\sin\omega t\sin(\omega t-\pi/6).$$

From equation (4),

$$50\sin\omega t\sin(\omega t-\pi/6)$$
$$= (50)\left[-\frac{1}{2}\{\cos(\omega t+\omega t-\pi/6) - \cos[\omega t-(\omega t-\pi/6)]\}\right]$$
$$= -25[\cos(2\omega t-\pi/6)-\cos\pi/6]$$

i.e. instantaneous power,

$$\mathbf{p = 25[\cos\pi/6 - \cos(2\omega t-\pi/6)]}$$

Now try the following Practice Exercise

Practice Exercise 115 Changing products of sines and cosines into sums or differences (Answers on page 713)

In Problems 1 to 5, express as sums or differences:

1. $\sin 7t\cos 2t$
2. $\cos 8x\sin 2x$
3. $2\sin 7t\sin 3t$
4. $4\cos 3\theta\cos\theta$
5. $3\sin\frac{\pi}{3}\cos\frac{\pi}{6}$
6. Determine $\int 2\sin 3t\cos t\,dt$
7. Evaluate $\int_0^{\pi/2} 4\cos 5x\cos 2x\,dx$
8. Solve the equation: $2\sin 2\phi\sin\phi=\cos\phi$ in the range $\phi=0$ to $\phi=180°$

22.5 Changing sums or differences of sines and cosines into products

In the compound-angle formula let $(A+B)=X$ and $(A-B)=Y$

Solving the simultaneous equations gives

$$A=\frac{X+Y}{2} \text{ and } B=\frac{X-Y}{2}$$

Thus $\sin(A+B)+\sin(A-B)=2\sin A\cos B$ becomes

$$\sin X + \sin Y = 2\sin\left(\frac{X+Y}{2}\right)\cos\left(\frac{X-Y}{2}\right) \quad (5)$$

Similarly,

$$\sin X - \sin Y = 2\cos\left(\frac{X+Y}{2}\right)\sin\left(\frac{X-Y}{2}\right) \quad (6)$$

$$\cos X + \cos Y = 2\cos\left(\frac{X+Y}{2}\right)\cos\left(\frac{X-Y}{2}\right) \quad (7)$$

$$\cos X - \cos Y = -2\sin\left(\frac{X+Y}{2}\right)\sin\left(\frac{X-Y}{2}\right) \quad (8)$$

Problem 19. Express: $\sin 5\theta + \sin 3\theta$ as a product

From equation (5),

$$\sin 5\theta + \sin 3\theta = 2\sin\left(\frac{5\theta+3\theta}{2}\right)\cos\left(\frac{5\theta-3\theta}{2}\right)$$
$$= \mathbf{2\sin 4\theta\cos\theta}$$

Problem 20. Express: $\sin 7x - \sin x$ as a product

From equation (6),

$$\sin 7x - \sin x = 2\cos\left(\frac{7x+x}{2}\right)\sin\left(\frac{7x-x}{2}\right)$$

$$= \mathbf{2\cos 4x \sin 3x}$$

Problem 21. Express: $\cos 2t - \cos 5t$ as a product

From equation (8),

$$\cos 2t - \cos 5t = -2\sin\left(\frac{2t+5t}{2}\right)\sin\left(\frac{2t-5t}{2}\right)$$

$$= -2\sin\frac{7}{2}t\sin\left(-\frac{3}{2}t\right)$$

$$= \mathbf{2\sin\frac{7}{2}t\sin\frac{3}{2}t}$$

$$\left[\text{since } \sin\left(-\frac{3}{2}t\right) = -\sin\frac{3}{2}t\right]$$

Problem 22. Show that $\dfrac{\cos 6x + \cos 2x}{\sin 6x + \sin 2x} = \cot 4x$

From equation (7), $\cos 6x + \cos 2x = 2\cos 4x\cos 2x$

From equation (5), $\sin 6x + \sin 2x = 2\sin 4x\cos 2x$

Hence $$\frac{\cos 6x + \cos 2x}{\sin 6x + \sin 2x} = \frac{2\cos 4x\cos 2x}{2\sin 4x\cos 2x}$$

$$= \frac{\cos 4x}{\sin 4x} = \mathbf{\cot 4x}$$

Problem 23. Solve the equation $\cos 4\theta + \cos 2\theta = 0$ for θ in the range $0° \leq \theta \leq 360°$

From equation (7),

$$\cos 4\theta + \cos 2\theta = 2\cos\left(\frac{4\theta+2\theta}{2}\right)\cos\left(\frac{4\theta-2\theta}{2}\right)$$

Hence, $$2\cos 3\theta\cos\theta = 0$$

Dividing by 2 gives: $$\cos 3\theta\cos\theta = 0$$

Hence, either $$\cos 3\theta = 0 \text{ or } \cos\theta = 0$$

Thus, $$3\theta = \cos^{-1} 0 \text{ or } \theta = \cos^{-1} 0$$

from which, $$3\theta = 90° \text{ or } 270° \text{ or } 450° \text{ or } 630° \text{ or } 810° \text{ or } 990°$$

and $$\boldsymbol{\theta = 30°, 90°, 150°, 210°, 270° \text{ or } 330°}$$

Now try the following Practice Exercise

Practice Exercise 116 Changing sums or differences of sines and cosines into products (Answers on page 713)

In Problems 1 to 5, express as products:

1. $\sin 3x + \sin x$
2. $\frac{1}{2}(\sin 9\theta - \sin 7\theta)$
3. $\cos 5t + \cos 3t$
4. $\frac{1}{8}(\cos 5t - \cos t)$
5. $\frac{1}{2}\left(\cos\frac{\pi}{3} + \cos\frac{\pi}{4}\right)$
6. Show that: (a) $\dfrac{\sin 4x - \sin 2x}{\cos 4x + \cos 2x} = \tan x$

   (b) $\frac{1}{2}[\sin(5x-\alpha) - \sin(x+\alpha)] = \cos 3x \sin(2x-\alpha)$

In Problems 7 and 8, solve for θ in the range $0° \leq \theta \leq 180°$.

7. $\cos 6\theta + \cos 2\theta = 0$
8. $\sin 3\theta - \sin\theta = 0$

In Problems 9 and 10, solve in the range $0° \leq \theta \leq 360°$.

9. $\cos 2x = 2\sin x$
10. $\sin 4t + \sin 2t = 0$
11. Two waves, $s_1 = 10\sin\omega_1 t$ and $s_2 = 10\sin\omega_2 t$ meet. If $\omega_1 = 390$ rad/s and $\omega_2 = 410$ rad/s, find the equation for their resultant wave, i.e. $s_1 + s_2$.

Practice Exercise 117 Multiple-choice questions on compound angles (Answers on page 713)

Each question has only one correct answer

1. The compound angle formula for $\cos(X - Y)$ is:
   (a) $\sin X \cos Y + \cos X \sin Y$
   (b) $\cos X \cos Y - \sin X \sin Y$
   (c) $\cos X \cos Y + \sin X \sin Y$
   (d) $\sin X \cos Y - \cos X \sin Y$
2. The trigonometric expression $\cos^2\theta$ - $\sin^2\theta$ is equivalent to:
   (a) $2\sin^2\theta - 1$ (b) $1 + 2\sin^2\theta$
   (c) $2\sin^2\theta + 1$ (d) $1 - 2\sin^2\theta$
3. $\cos(x + \pi) + \sin(x - \pi/2)$ is equivalent to:
   (a) 0 (b) $-2\cos x$ (c) $2\sin x$ (d) $2\cos x$
4. Expressing $(4\sin x - 3\cos x)$ in the form $R\sin(x + \alpha)$ gives:
   (a) $25\sin(x - 36.87°)$ (b) $5\sin(x + 36.87°)$
   (c) $5\sin(x - 36.87°)$ (d) $5\sin(x + 143.13°)$
5. sin 5x cos 3x is equivalent to:
   (a) $\frac{1}{2}[\sin 8x - \sin 2x]$ (b) $\frac{1}{2}[\cos 8x + \sin 2x]$
   (c) $\frac{1}{2}[\cos 8x - \cos 2x]$ (d) $\frac{1}{2}[\sin 8x + \sin 2x]$
6. $\cos 3t - \cos 7t$ is equivalent to:
   (a) $2\sin 5t \sin 2t$ (b) $2\cos 5t \sin 2t$
   (c) $2\sin 5t \cos 2t$ (d) $2\cos 5t \cos 2t$

For fully worked solutions to each of the problems in Practice Exercises 112 to 116 in this chapter, go to the website:
www.routledge.com/cw/bird

Revision Test 6 Further trigonometry

This Revision Test covers the material contained in Chapter 20 to 22. *The marks for each question are shown in brackets at the end of each question.*

1. A triangular plot of land ABC is shown in Fig. RT 6.1. Solve the triangle and determine its area. (10)

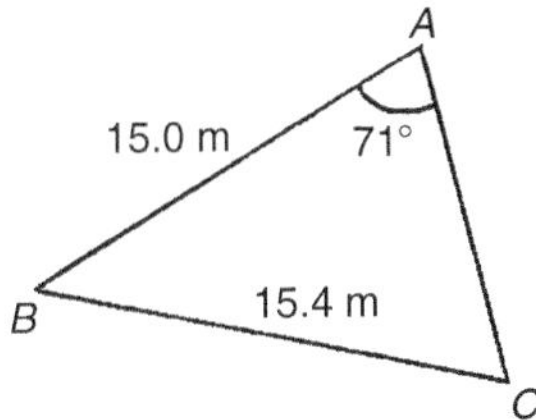

Figure RT 6.1

2. Figure RT 6.2 shows a roof truss PQR with rafter $PQ = 3$ m. Calculate the length of (a) the roof rise PP', (b) rafter PR, and (c) the roof span QR.

   Find also (d) the cross-sectional area of the roof truss. (11)

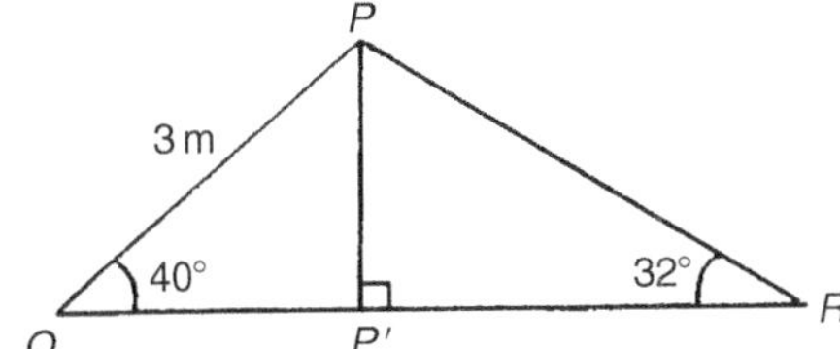

Figure RT 6.2

3. Prove the following identities:

   (a) $\sqrt{\dfrac{1 - \cos^2\theta}{\cos^2\theta}} = \tan\theta$

   (b) $\cos\left(\dfrac{3\pi}{2} + \phi\right) = \sin\phi$ (6)

4. Solve the following trigonometric equations in the range $0° \leq x \leq 360°$:

   (a) $4\cos x + 1 = 0$ (b) $3.25 \operatorname{cosec} x = 5.25$
   (c) $5\sin^2 x + 3\sin x = 4$ (13)

5. Solve the equation $5\sin(\theta - \pi/6) = 8\cos\theta$ for values $0 \leq \theta \leq 2\pi$. (8)

6. Express $5.3\cos t - 7.2\sin t$ in the form $R\sin(t + \alpha)$. Hence solve the equation $5.3\cos t - 7.2\sin t = 4.5$ in the range $0 \leq t \leq 2\pi$. (12)

For lecturers/instructors/teachers, fully worked solutions to each of the problems in Revision Test 6, together with a full marking scheme, are available at the website:
www.routledge.com/cw/bird

COMPANION @ WEBSITE

Section 3

Areas and volumes

Chapter 23

Areas of common shapes

Why it is important to understand: Areas of common shapes

To paint, wallpaper or panel a wall, you must know the total area of the wall so you can buy the appropriate amount of finish. When designing a new building, or seeking planning permission, it is often necessary to specify the total floor area of the building. In construction, calculating the area of a gable end of a building is important when determining the amount of bricks and mortar to order. When using a bolt, the most important thing is that it is long enough for your particular application and it may also be necessary to calculate the shear area of the bolt connection. Ridge vents allow a home to properly vent, while disallowing rain or other forms of precipitation to leak into the attic or crawlspace underneath the roof. Equal amounts of cool air and warm air flowing through the vents is paramount for proper heat exchange. Calculating how much surface area is available on the roof aids in determining how long the ridge vent should run. A race track is an oval shape, and it is sometimes necessary to find the perimeter of the inside of a race track. Arches are everywhere, from sculptures and monuments to pieces of architecture and strings on musical instruments; finding the height of an arch or its cross-sectional area is often required. Determining the cross-sectional areas of beam structures is vitally important in design engineering. There are thus a large number of situations in engineering where determining area is important.

At the end of this chapter, you should be able to:

- state the SI unit of area
- identify common polygons – triangle, quadrilateral, pentagon, hexagon, heptagon and octagon
- identify common quadrilaterals – rectangle, square, parallelogram, rhombus and trapezium
- calculate areas of quadrilaterals and circles
- appreciate that areas of similar shapes are proportional to the squares of the corresponding linear dimensions

23.1 Introduction

Area is a measure of the size or extent of a plane surface.
Area is measured in **square units** such as mm^2, cm^2 and m^2.
This chapter deals with finding areas of common shapes.
In engineering it is often important to be able to calculate simple areas of various shapes.
In everyday life its important to be able to measure area to, say, lay a carpet, or to order sufficient paint for a decorating job or to order sufficient bricks for a new wall.
On completing this chapter you will be able to recognise common shapes and be able to find their areas.

23.2 Properties of quadrilaterals

Polygon

A **polygon** is a closed plane figure bounded by straight lines. A polygon, which has:

(i) 3 sides is called a **triangle**
(ii) 4 sides is called a **quadrilateral**
(iii) 5 sides is called a **pentagon**
(iv) 6 sides is called a **hexagon**
(v) 7 sides is called a **heptagon**
(vi) 8 sides is called an **octagon**

There are five types of **quadrilateral**, these being:

(i) rectangle
(ii) square
(iii) parallelogram
(iv) rhombus
(v) trapezium

(The properties of these are given below.)
If the opposite corners of any quadrilateral are joined by a straight line, two triangles are produced. Since the sum of the angles of a triangle is 180°, the sum of the angles of a quadrilateral is 360°.
In a **rectangle**, shown in Fig. 23.1:

(i) all four angles are right angles,
(ii) opposite sides are parallel and equal in length, and
(iii) diagonals AC and BD are equal in length and bisect one another.

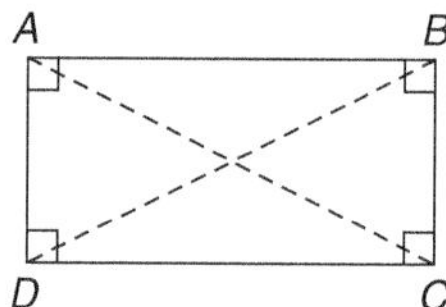

Figure 23.1

In a **square**, shown in Fig. 23.2:

(i) all four angles are right angles,
(ii) opposite sides are parallel,
(iii) all four sides are equal in length, and
(iv) diagonals PR and QS are equal in length and bisect one another at right angles.

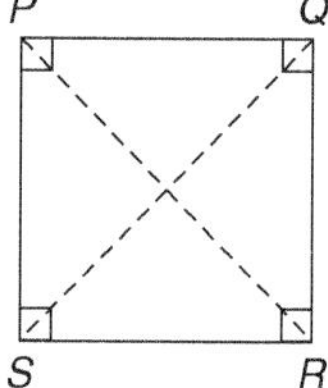

Figure 23.2

In a **parallelogram**, shown in Fig. 23.3:

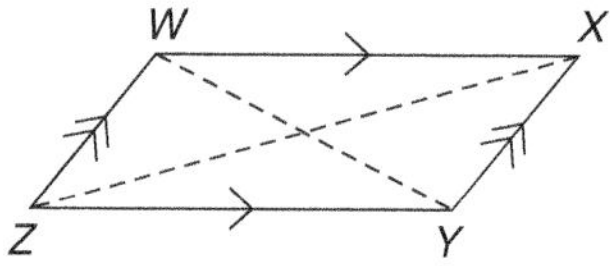

Figure 23.3

(i) opposite angles are equal,
(ii) opposite sides are parallel and equal in length, and
(iii) diagonals WY and XZ bisect one another.

In a **rhombus**, shown in Fig. 23.4:

(i) opposite angles are equal,
(ii) opposite angles are bisected by a diagonal,
(iii) opposite sides are parallel,
(iv) all four sides are equal in length, and
(v) diagonals AC and BD bisect one another at right angles.

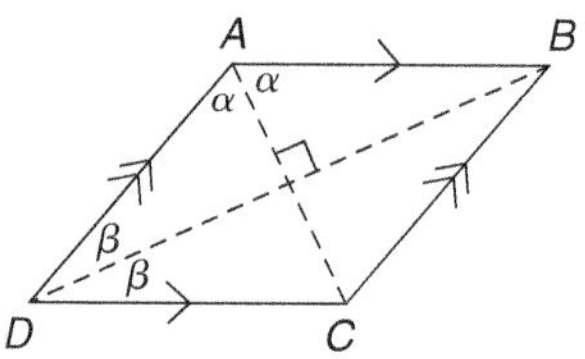

Figure 23.4

In a **trapezium**, shown in Fig. 23.5:
(i) only one pair of sides is parallel

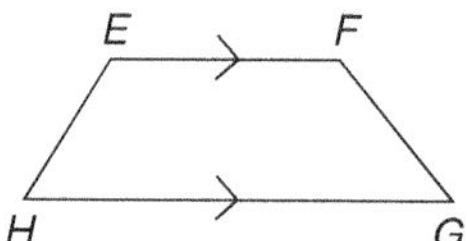

Figure 23.5

23.3 Areas of common shapes

Table 23.1 summarises the areas of common plane figures.

Table 23.1

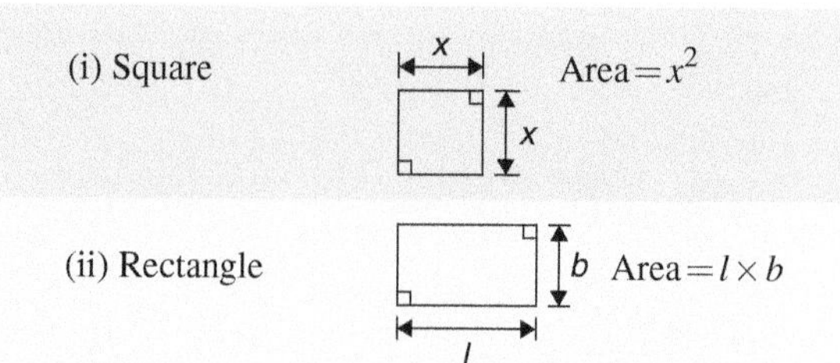

(i) Square	x / x	Area $= x^2$
(ii) Rectangle	b / l	Area $= l \times b$

Table 23.1 *(Continued)*

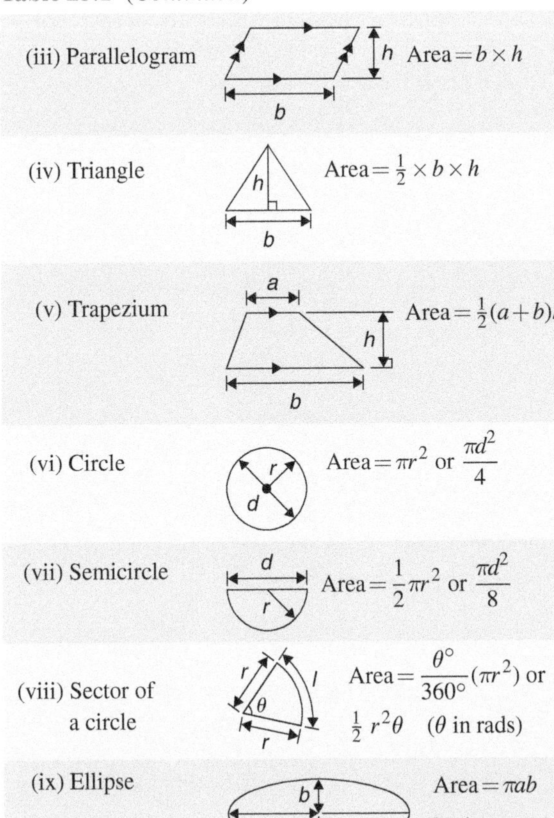

(iii) Parallelogram	Area $= b \times h$
(iv) Triangle	Area $= \frac{1}{2} \times b \times h$
(v) Trapezium	Area $= \frac{1}{2}(a+b)h$
(vi) Circle	Area $= \pi r^2$ or $\frac{\pi d^2}{4}$
(vii) Semicircle	Area $= \frac{1}{2}\pi r^2$ or $\frac{\pi d^2}{8}$
(viii) Sector of a circle	Area $= \frac{\theta°}{360°}(\pi r^2)$ or $\frac{1}{2} r^2\theta$ (θ in rads)
(ix) Ellipse	Area $= \pi ab$ Perimeter $\approx \pi(a+b)$

23.4 Worked problems on areas of common shapes

Problem 1. State the types of quadrilateral shown in Fig. 23.6 and determine the angles marked a to l

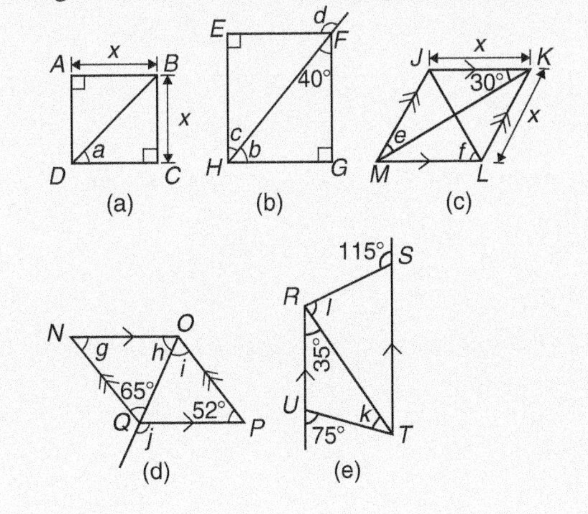

Figure 23.6

(a) ***ABCD* is a square**
The diagonals of a square bisect each of the right angles, hence

$$a = \frac{90°}{2} = \mathbf{45°}$$

(b) ***EFGH* is a rectangle**
In triangle FGH, $40° + 90° + b = 180°$ (angles in a triangle add up to 180°) from which, $\mathbf{b = 50°}$. Also $\mathbf{c = 40°}$ (alternate angles between parallel lines EF and HG).
(Alternatively, b and c are complementary, i.e. add up to 90°)
$d = 90° + c$ (external angle of a triangle equals the sum of the interior opposite angles), hence

$$d = 90° + 40° = \mathbf{130°}$$

(c) ***JKLM* is a rhombus**
The diagonals of a rhombus bisect the interior angles and opposite internal angles are equal. Thus $\angle JKM = \angle MKL = \angle JMK = \angle LMK = 30°$, hence, $\mathbf{e = 30°}$
In triangle KLM, $30° + \angle KLM + 30° = 180°$ (angles in a triangle add up to 180°), hence $\angle KLM = 120°$.
The diagonal JL bisects $\angle KLM$, hence

$$f = \frac{120°}{2} = \mathbf{60°}$$

(d) ***NOPQ* is a parallelogram**
$\mathbf{g = 52°}$ (since opposite interior angles of a parallelogram are equal).
In triangle NOQ, $g + h + 65° = 180°$ (angles in a triangle add up to 180°), from which,

$$h = 180° - 65° - 52° = \mathbf{63°}$$

$\mathbf{i = 65°}$ (alternate angles between parallel lines NQ and OP).
$j = 52° + i = 52° + 65° = \mathbf{117°}$ (external angle of a triangle equals the sum of the interior opposite angles).

(e) ***RSTU* is a trapezium**
$35° + k = 75°$ (external angle of a triangle equals the sum of the interior opposite angles), hence $\mathbf{k = 40°}$
$\angle STR = 35°$ (alternate angles between parallel lines RU and ST).
$l + 35° = 115°$ (external angle of a triangle equals the sum of the interior opposite angles), hence

$$l = 115° - 35° = \mathbf{80°}$$

Problem 2. A rectangular metal tray is 820 mm long and 400 mm wide. Find its area in (a) mm^2, (b) cm^2 and (c) m^2

(a) Area = length × width = 820 × 400 = **328 000 mm²**

(b) $1\,cm^2 = 100\,mm^2$. Hence

$$328\,000\,mm^2 = \frac{328\,000}{100}\,cm^2 = \mathbf{3280\,cm^2}$$

(c) $1\,m^2 = 10\,000\,cm^2$. Hence

$$3280\,cm^2 = \frac{3280}{10000}\,m^2 = \mathbf{0.3280\,m^2}$$

Problem 3. Find (a) the cross-sectional area of the girder shown in Fig. 23.7(a) and (b) the area of the path shown in Fig. 23.7(b)

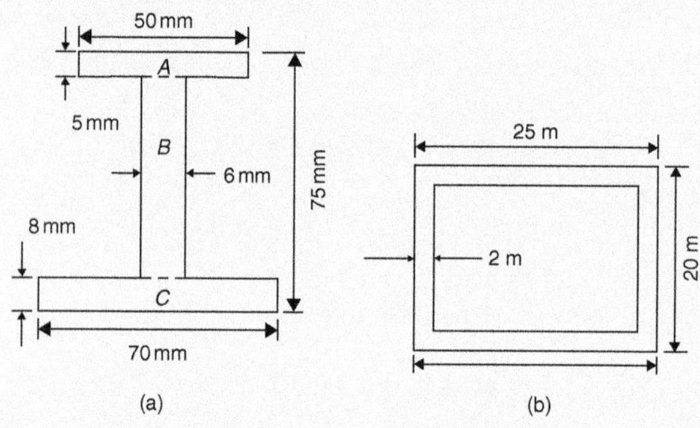

Figure 23.7

(a) The girder may be divided into three separate rectangles as shown.

$$\text{Area of rectangle } A = 50 \times 5 = 250\,mm^2$$
$$\text{Area of rectangle } B = (75 - 8 - 5) \times 6 = 62 \times 6 = 372\,mm^2$$
$$\text{Area of rectangle } C = 70 \times 8 = 560\,mm^2$$

Total area of girder = 250 + 372 + 560 = **1182 mm²** or **11.82 cm²**

(b) Area of path = area of large rectangle − area of small rectangle

$$= (25 \times 20) - (21 \times 16) = 500 - 336 = \mathbf{164\,m^2}$$

Problem 4. Find the area of the aluminium sheet in the form of a parallelogram shown in Fig. 23.8 (dimensions are in mm)

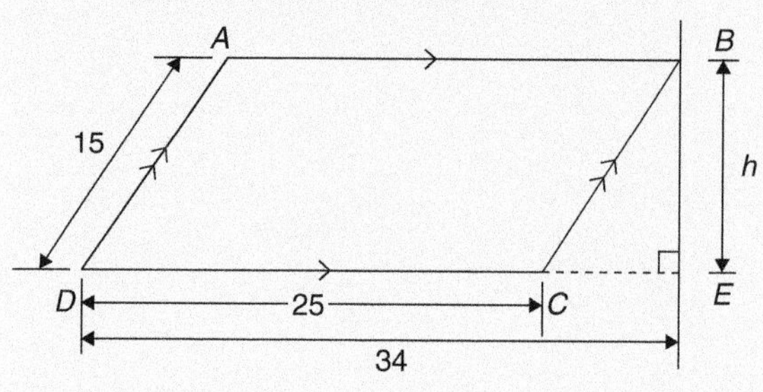

Figure 23.8

Area of parallelogram = base × perpendicular height. The perpendicular height h is found using Pythagoras' theorem.

$$BC^2 = CE^2 + h^2$$

i.e. $$15^2 = (34 - 25)^2 + h^2$$

$$h^2 = 15^2 - 9^2 = 225 - 81 = 144$$

Hence, $h = \sqrt{144} = 12$ mm (−12 can be neglected). Hence, area of $ABCD = 25 \times 12 =$ **300 mm²**

Problem 5. Figure 23.9 shows the gable end of a building. Determine the area of brickwork in the gable end

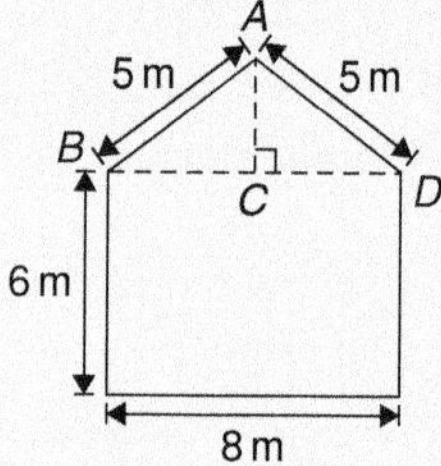

Figure 23.9

The shape is that of a rectangle and a triangle.

Area of rectangle = 6 × 8 = 48 m²

Area of triangle = $\frac{1}{2}$ × base × height.

$CD = 4$ m, $AD = 5$ m, hence $AC = 3$ m (since it is a 3, 4, 5 triangle).

Hence, area of triangle $ABD = \frac{1}{2} \times 8 \times 3 = 12\,m^2$

Total area of brickwork = 48 + 12 = **60 m²**

Problem 6. Determine the area of the metal shape shown in Fig. 23.10

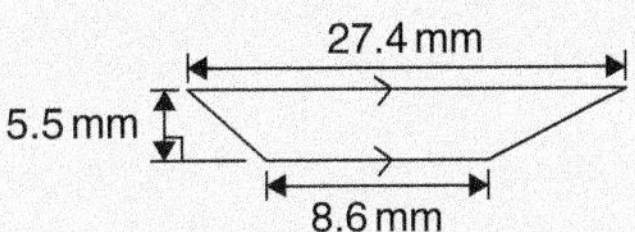

Figure 23.10

The shape shown is a trapezium.
Area of trapezium

$= \frac{1}{2}$ (sum of parallel sides)(perpendicular distance between them)

$= \frac{1}{2}(27.4 + 8.6)(5.5)$

$= \frac{1}{2} \times 36 \times 5.5 = \mathbf{99\ mm^2}$

Now try the following Practice Exercise

Practice Exercise 118 Areas of plane figures (Answers on page 713)

1. A rectangular plate is 85 mm long and 42 mm wide. Find its area in square centimetres.

2. A rectangular field has an area of 1.2 hectares and a length of 150 m. Find (a) its width and (b) the length of a diagonal (1 hectare = 10 000 m^2).

3. Determine the area of each of the angle iron sections shown in Fig. 23.11.

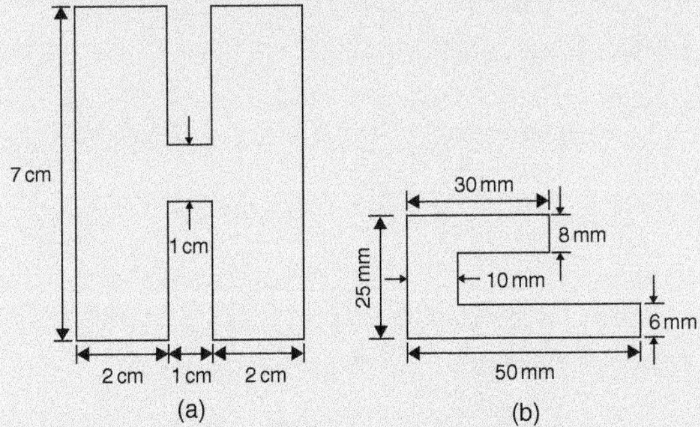

Figure 23.11

4. The outside measurements of a picture frame are 100 cm by 50 cm. If the frame is 4 cm wide, find the area of the wood used to make the frame.

5. A rectangular garden measures 40 m by 15 m. A 1 m flower border is made round the two shorter sides and one long side. A circular swimming pool of diameter 8 m is constructed in the middle of the garden. Find, correct to the nearest square metre, the area remaining.

6. The area of a metal trapezium shape is 13.5 cm^2 and the perpendicular distance between its parallel sides is 3 cm. If the length of one of the parallel sides is 5.6 cm, find the length of the other parallel side.

7. Find the angles p, q, r, s and t in Fig. 23.12(a) to (c).

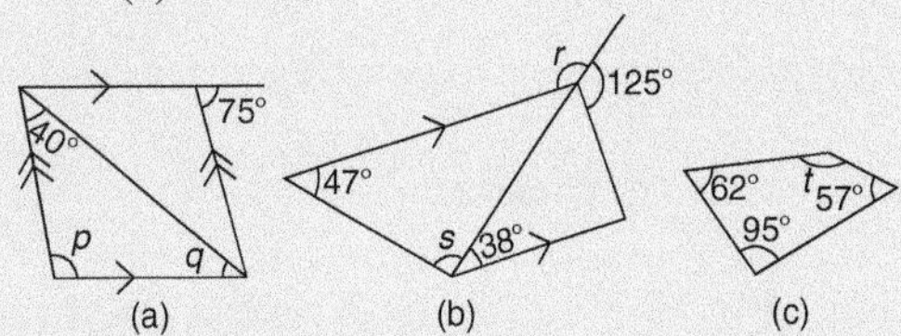

Figure 23.12

8. Name the types of quadrilateral shown in Fig. 23.13(a) to (d), and determine (a) the area, and (b) the perimeter of each.

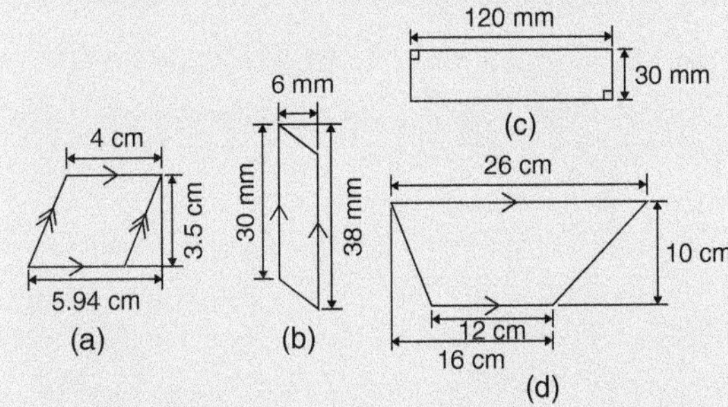

Figure 23.13

9. Calculate the area of the steel plate shown in Fig. 23.14

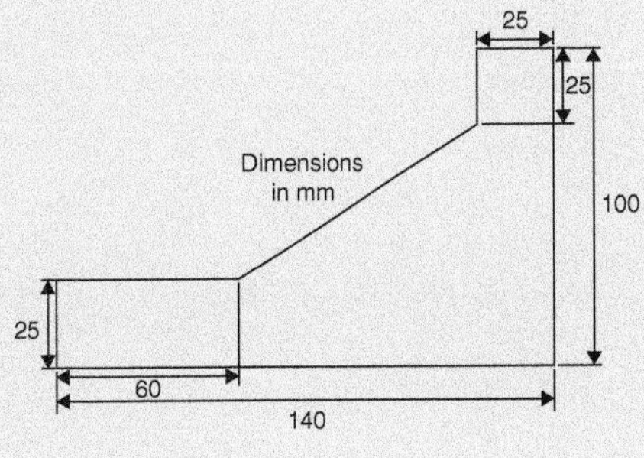

Figure 23.14

23.5 Further worked problems on areas of plane figures

Problem 7. Find the areas of the circles having (a) a radius of 5 cm, (b) a diameter of 15 mm, (c) a circumference of 70 mm

Area of a circle $= \pi r^2$ or $\dfrac{\pi d^2}{4}$

(a) Area $= \pi r^2 = \pi(5)^2 = 25\pi = \mathbf{78.54\ cm^2}$

(b) Area $= \dfrac{\pi d^2}{4} = \dfrac{\pi(15)^2}{4} = \dfrac{225\pi}{4} = \mathbf{176.7\ mm^2}$

(c) Circumference, $c = 2\pi r$, hence

$$r = \frac{c}{2\pi} = \frac{70}{2\pi} = \frac{35}{\pi}\ \text{mm}$$

$$\text{Area of circle} = \pi r^2 = \pi\left(\frac{35}{\pi}\right)^2 = \frac{35^2}{\pi}$$

$$= \mathbf{389.9\ mm^2} \text{ or } \mathbf{3.899\ cm^2}$$

Problem 8. Calculate the areas of the following sectors of circles having:

(a) radius 6 cm with angle subtended at centre 50°
(b) diameter 80 mm with angle subtended at centre 107°42′
(c) radius 8 cm with angle subtended at centre 1.15 radians

$$\text{Area of sector of a circle} = \frac{\theta^2}{360}(\pi r^2)$$

$$\text{or } \frac{1}{2}r^2\theta \quad (\theta \text{ in radians}).$$

(a) Area of sector

$$= \frac{50}{360}(\pi 6^2) = \frac{50 \times \pi \times 36}{360} = 5\pi$$

$$= \mathbf{15.71\ cm^2}$$

(b) If diameter = 80 mm, then radius, $r = 40$ mm, and area of sector

$$= \frac{107°42'}{360}(\pi 40^2) = \frac{107\frac{42}{60}}{360}(\pi 40^2)$$

$$= \frac{107.7}{360}(\pi 40^2) = \mathbf{1504\ mm^2} \text{ or } \mathbf{15.04\ cm^2}$$

(c) Area of sector $= \frac{1}{2}r^2\theta = \frac{1}{2} \times 8^2 \times 1.15$

$$= \mathbf{36.8\ cm^2}$$

Problem 9. A hollow shaft has an outside diameter of 5.45 cm and an inside diameter of 2.25 cm. Calculate the cross-sectional area of the shaft

The cross-sectional area of the shaft is shown by the shaded part in Fig. 23.15 (often called an **annulus**).

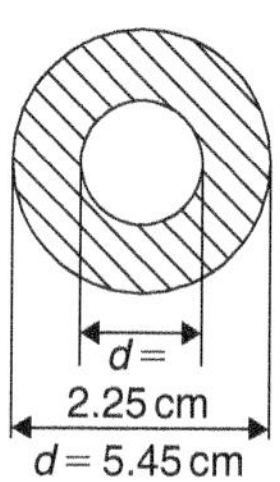

Figure 23.15

Area of shaded part = area of large circle − area of small circle

$$= \frac{\pi D^2}{4} - \frac{\pi d^2}{4} = \frac{\pi}{4}(D^2 - d^2)$$

$$= \frac{\pi}{4}(5.45^2 - 2.25^2) = \mathbf{19.35\ cm^2}$$

Problem 10. The major axis of an ellipse is 15.0 cm and the minor axis is 9.0 cm. Find its area and approximate perimeter

If the major axis = 15.0 cm, then the semi-major axis = 7.5 cm.
If the minor axis = 9.0 cm, then the semi-minor axis = 4.5 cm.
Hence, from Table 23.1(ix),

$$\textbf{area} = \pi ab = \pi(7.5)(4.5) = \mathbf{106.0\ cm^2}$$

and **perimeter** $\approx \pi(a + b) = \pi(7.5 + 4.5)$

$$= 12.0\pi = \mathbf{37.7\ cm}$$

Now try the following Practice Exercise

Practice Exercise 119 Areas of plane figures (Answers on page 713)

1. Determine the area of a circular template having (a) a radius of 4 cm (b) a diameter of 30 mm (c) a circumference of 200 mm

2. A washer in the form of an annulus has an outside diameter of 60 mm and an inside diameter of 20 mm. Determine its area

3. If the area of a circular coin is 320 mm^2, find (a) its diameter, and (b) its circumference

4. Calculate the areas of the following sectors of circles:
   (a) radius 9 cm, angle subtended at centre 75°
   (b) diameter 35 mm, angle subtended at centre 48°37′
   (c) diameter 5 cm, angle subtended at centre 2.19 radians

5. Determine the area of the shaded template shown in Fig. 23.16

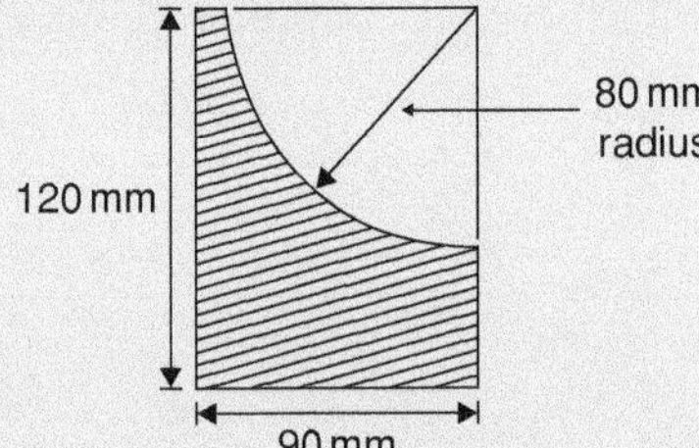

Figure 23.16

6. An archway consists of a rectangular opening topped by a semi-circular arch as shown in Fig. 23.17. Determine the area of the opening if the width is 1 m and the greatest height is 2 m

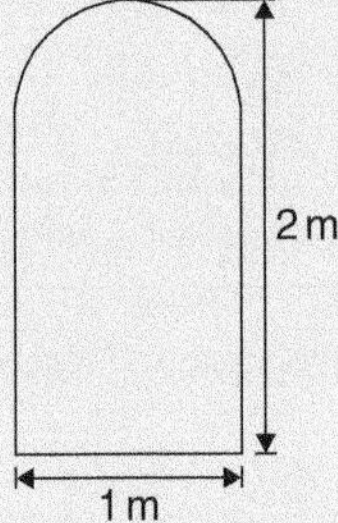

Figure 23.17

7. The major axis of a metal elliptical shape is 200 mm and the minor axis 100 mm. Determine the area and approximate perimeter of the ellipse

8. If fencing costs £15 per metre, find the cost (to the nearest pound) of enclosing an elliptical plot of land which has major and minor diameter lengths of 120 m and 80 m

9. A cycling track is in the form of an ellipse, the axes being 250 m and 150 m respectively for the inner boundary, and 270 m and 170 m for the outer boundary. Calculate the area of the track

10. A base plate is in the form of a quadrant of a circle of radius 0.5 m. Calculate the area and perimeter of the plate

11. A rectangular gasket 350 mm by 200 mm has four holes cut in it, each of diameter 60 mm. Calculate the area of the gasket in square centimetres

23.6 Worked problems on areas of composite figures

Problem 11. Calculate the area of a regular metal octagonal shape, if each side is 5 cm and the width across the flats is 12 cm

An octagon is an 8-sided polygon. If radii are drawn from the centre of the polygon to the vertices then 8 equal triangles are produced (see Fig. 23.18).

$$\text{Area of one triangle} = \frac{1}{2} \times \text{base} \times \text{height}$$

$$= \frac{1}{2} \times 5 \times \frac{12}{2} = 15\ \text{cm}^2$$

$$\text{Area of octagon} = 8 \times 15 = \mathbf{120\ cm^2}$$

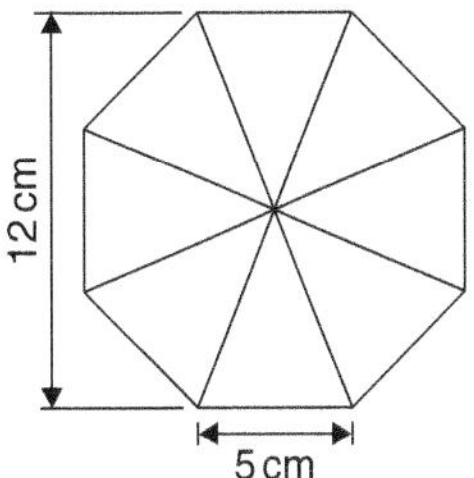

Figure 23.18

Problem 12. Determine the area of a regular hexagonal sheet of metal that has sides 8 cm long

A hexagon is a 6-sided polygon which may be divided into 6 equal triangles as shown in Fig. 23.19. The angle subtended at the centre of each triangle is 360°/6 = 60°.

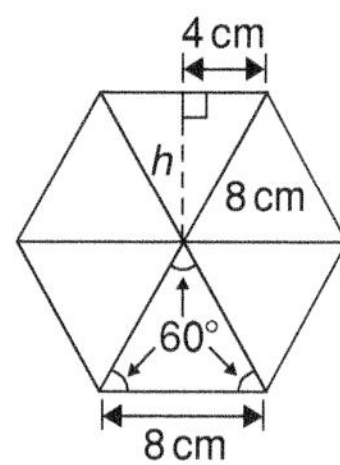

Figure 23.19

The other two angles in the triangle add up to 120° and are equal to each other. Hence each of the triangles is equilateral with each angle 60° and each side 8 cm.

$$\text{Area of one triangle} = \frac{1}{2} \times \text{base} \times \text{height}$$

$$= \frac{1}{2} \times 8 \times h$$

h is calculated using Pythagoras' theorem:

$$8^2 = h^2 + 4^2$$

from which, $h = \sqrt{8^2 - 4^2} = 6.928\,\text{cm}$

Hence area of one triangle

$$= \frac{1}{2} \times 8 \times 6.928 = 27.71\,\text{cm}^2$$

Area of hexagon $= 6 \times 27.71 =$ **166.3 cm²**

Problem 13. Figure 23.20 shows a plan of a floor of a building that is to be carpeted. Calculate the area of the floor in square metres. Calculate the cost, correct to the nearest pound, of carpeting the floor with carpet costing £16.80 per m^2, assuming 30% extra carpet is required due to wastage in fitting

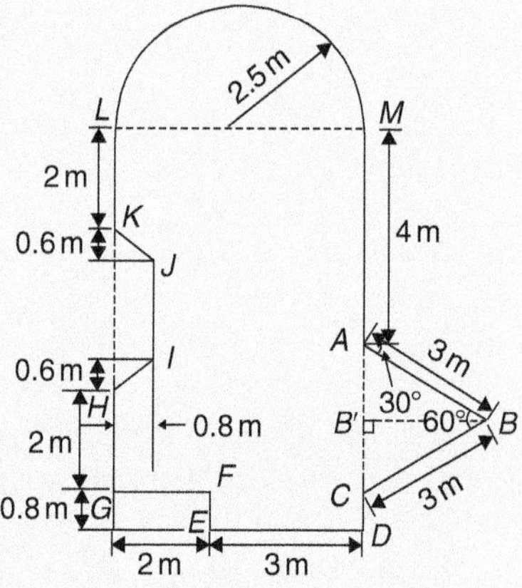

Figure 23.20

Area of floor plan

= area of triangle ABC + area of semicircle

+ area of rectangle $CGLM$

+ area of rectangle $CDEF$

− area of trapezium $HIJK$

Triangle ABC is equilateral since $AB = BC = 3$ m and hence angle $B'CB = 60°$

$\sin B'CB = BB'/3$, i.e. $BB' = 3\sin 60° = 2.598$ m

$$\text{Area of triangle } ABC = \tfrac{1}{2}(AC)(BB')$$

$$= \tfrac{1}{2}(3)(2.598) = 3.897\,\text{m}^2$$

Area of semicircle $= \frac{1}{2}\pi r^2 = \frac{1}{2}\pi(2.5)^2 = 9.817\,\text{m}^2$

Area of $CGLM = 5 \times 7 = 35\,\text{m}^2$

Area of $CDEF = 0.8 \times 3 = 2.4\,\text{m}^2$

Area of $HIJK = \frac{1}{2}(KH + IJ)(0.8)$

Since $MC = 7$ m then $LG = 7$ m, hence $JI = 7 - 5.2 = 1.8$ m

Hence area of $HIJK = \frac{1}{2}(3 + 1.8)(0.8) = 1.92\,\text{m}^2$

Total floor area $= 3.897 + 9.817 + 35 + 2.4 - 1.92$

$= 49.194\,\text{m}^2$

To allow for 30% wastage, amount of carpet required $= 1.3 \times 49.194 = 63.95\,\text{m}^2$

Cost of carpet at £16.80 per m² $= 63.95 \times 16.80 =$ **£1074, correct to the nearest pound**.

Now try the following Practice Exercise

Practice Exercise 120 Areas of composite figures (Answers on page 713)

1. Calculate the area of a regular metal octagonal template if each side is 20 mm and the width across the flats is 48.3 mm
2. Determine the area of a regular wooden hexagonal feature which has sides 25 mm
3. A plot of land is in the shape shown in Fig. 23.21. Determine: (a) its area in hectares (1 ha $= 10^4\,\text{m}^2$), and (b) the length of fencing

required, to the nearest metre, to completely enclose the plot of land

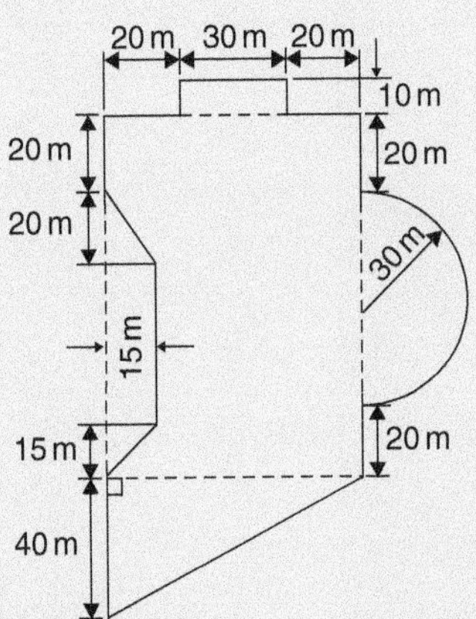

Figure 23.21

4. If paving slabs are produced in 250 mm × 250 mm squares, determine the number of slabs required to cover an area of 2 m^2

5. Calculate the number of turns (to the nearest whole number) on a solenoid which is made by winding 25 m of fine copper wire around a cylindrical former of diameter 26 mm.

6. Determine the area of the largest regular hexagon that can be cut from a circular sheet of aluminium of radius 400 mm. Give the answer in square metres, correct to 3 significant figures.

23.7 Areas of similar shapes

The areas of similar shapes are proportional to the squares of corresponding linear dimensions.

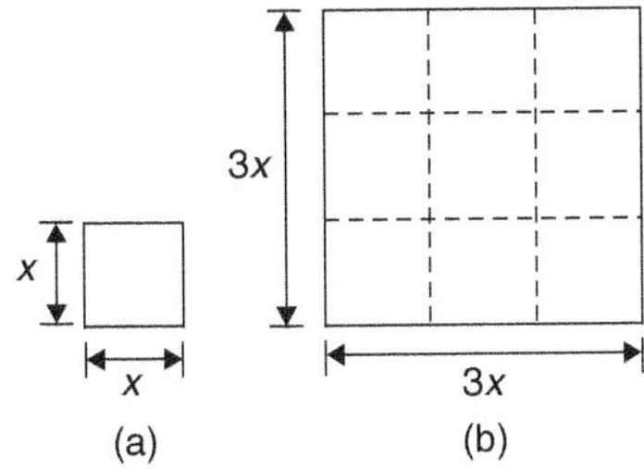

Figure 23.22

For example, Fig. 23.22 shows two squares, one of which has sides three times as long as the other.

$$\text{Area of Fig. 23.22(a)} = (x)(x) = x^2$$

$$\text{Area of Fig. 23.22(b)} = (3x)(3x) = 9x^2$$

Hence Fig. 23.22(b) has an area $(3)^2$, i.e. 9 times the area of Fig. 23.22(a).

Problem 14. A rectangular garage is shown on a building plan having dimensions 10 mm by 20 mm. If the plan is drawn to a scale of 1 to 250, determine the true area of the garage in square metres

$$\text{Area of garage on the plan} = 10\text{ mm} \times 20\text{ mm} = 200\text{ mm}^2$$

Since the areas of similar shapes are proportional to the squares of corresponding dimensions then:

$$\text{true area of garage} = 200 \times (250)^2 = 12.5 \times 10^6\text{ mm}^2 = \frac{12.5 \times 10^6}{10^6}\text{ m}^2 = \mathbf{12.5\ m^2}$$

Now try the following Practice Exercise

Practice Exercise 121 Areas of similar shapes (Answers on page 714)

1. The area of a park on a map is 500 mm^2. If the scale of the map is 1 to 40 000 determine the true area of the park in hectares (1 hectare = 10^4 m^2)

2. A model of a boiler is made having an overall height of 75 mm corresponding to an overall height of the actual boiler of 6 m. If the area of metal required for the model is 12 500 mm^2 determine, in square metres, the area of metal required for the actual boiler

3. The scale of an Ordnance Survey map is 1:2500. A circular sports field has a diameter of 8 cm on the map. Calculate its area in hectares, giving your answer correct to 3 significant figures. (1 hectare = 10^4 m^2)

Practice Exercise 122 Multiple-choice questions on area of common shapes (Answers on page 714)

(Answers on page 714)

Each question has only one correct answer

1. If the circumference of a circle is 100 mm, its area is:

   (a) 314.2 cm^2 (b) 7.96 cm^2
   (c) 31.83 mm^2 (d) 78.54 cm^2

2. The outside measurements of a rectangular picture frame are 80 cm by 30 cm. If the frame is 3 cm wide, the area of the metal used to make the frame is:

   (a) 624 cm^2 (b) 2079 cm^2
   (c) 660 mm^2 (d) 588 cm^2

3. A square piece of metal plate has sides x cm. A square of side $(x-3)$ is machined out of the centre. In terms of x, the remaining area of metal in square centimetres is:

   (a) $2x^2-9$ (b) $6x-9$
   (c) $2x^2-6x-9$ (d) $6x+9$

4. The interior angle of a regular hexagon is:

   (a) 60° (b) 360°
   (c) 150° (d) 120°

5. The length of each side of an equilateral triangle is 2 cm. The area of the triangle is:

   (a) 2 cm^2 (b) $\sqrt{3}$ cm^2
   (c) 4 cm^2 (d) $\sqrt{5}$ cm^2

6. The area of the path shown shaded in Figure 23.23 is:

   (a) 300 m^2 (b) 234 m^2
   (c) 124 m^2 (d) 66 m^2

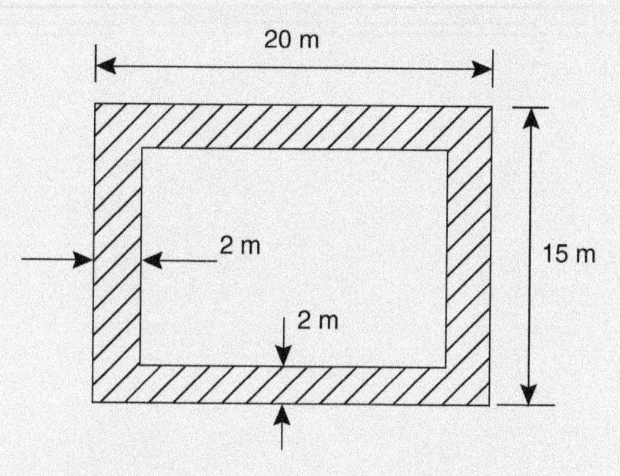

Figure 23.23

7. The area of a semi-circular faceplate with a diameter of 10 cm is:

   (a) 12.5π cm^2 (b) 25π cm^2
   (c) 50π cm^2 (d) 100π cm^2

8. A steel shape is a square of area 120 cm^2. The length of the diagonal of the square is:

   (a) 4.68 cm (b) 10.95 cm
   (c) 15.49 cm (d) 169.71 cm

9. A hollow shaft has an outside diameter of 8 cm and an inside diameter of 4 cm. The cross-sectional area of the shaft is:

   (a) 20π cm^2 (b) 48π cm^2
   (c) 12π cm^2 (d) 24π cm^2

10. A rectangular building is shown on a building plan having dimensions 20 mm by 10 mm. If the plan is drawn to a scale of 1 to 300, the true area of the building in m^2 is:

    (a) 60000 m^2 (b) 1800 m^2
    (c) 0.06 m^2 (d) 18 m^2

COMPANION @ WEBSITE

For fully worked solutions to each of the problems in Practice Exercises 118 to 121 in this chapter, go to the website:

www.routledge.com/cw/bird

Chapter 24

The circle and its properties

Why it is important to understand: The circle and its properties

A circle is one of the fundamental shapes of geometry; it consists of all the points that are equidistant from a central point. Knowledge of calculations involving circles is needed with crank mechanisms, with determinations of latitude and longitude, with pendulums and even in the design of paper clips. The floodlit area at a football ground, the area an automatic garden sprayer sprays and the angle of lap of a belt drive all rely on calculations involving the arc of a circle. The ability to handle calculations involving circles and their properties is clearly essential in several branches of engineering design.

At the end of this chapter, you should be able to:

- define a circle
- state some properties of a circle – including radius, circumference, diameter, semicircle, quadrant, tangent, sector, chord, segment and arc
- appreciate the angle in a semicircle is a right angle
- define a radian, and change radians to degrees, and vice versa
- determine arc length, area of a circle and area of a sector of a circle
- state the equation of a circle
- sketch a circle given its equation

24.1 Introduction

A **circle** is a plane figure enclosed by a curved line, every point on which is equidistant from a point within, called the **centre**.

24.2 Properties of circles

(i) The distance from the centre to the curve is called the **radius**, r, of the circle (see OP in Fig. 24.1).

(ii) The boundary of a circle is called the **circumference**, c.

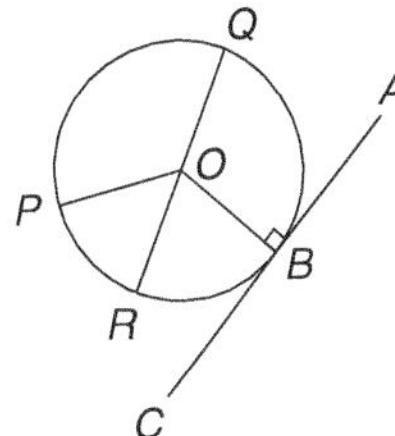

Figure 24.1

(iii) Any straight line passing through the centre and touching the circumference at each end is called the **diameter**, d (see QR in Fig. 24.1). Thus $\boldsymbol{d=2r}$.

(iv) The ratio $\dfrac{\text{circumference}}{\text{diameter}}=$ a constant for any circle.

This constant is denoted by the Greek letter π (pronounced 'pie'), where $\pi = 3.14159$, correct to 5 decimal places.

Hence $c/d = \pi$ or $\boldsymbol{c = \pi d}$ or $\boldsymbol{c = 2\pi r}$.

(v) A **semicircle** is one half of the whole circle.

(vi) A **quadrant** is one quarter of a whole circle.

(vii) A **tangent** to a circle is a straight line that meets the circle in one point only and does not cut the circle when produced. *AC* in Fig. 24.1 is a tangent to the circle since it touches the curve at point *B* only. If radius *OB* is drawn, then angle *ABO* is a right angle.

(viii) A **sector** of a circle is the part of a circle between radii (for example, the portion *OXY* of Fig. 24.2 is a sector). If a sector is less than a semicircle it is called a **minor sector**, if greater than a semicircle it is called a **major sector**.

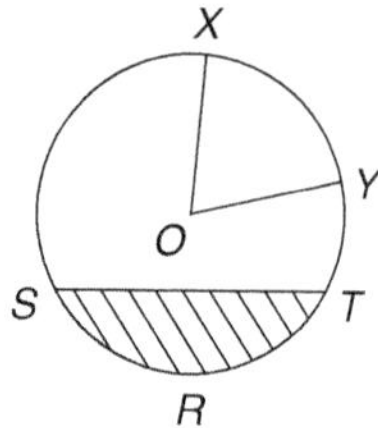

Figure 24.2

(ix) A **chord** of a circle is any straight line that divides the circle into two parts and is terminated at each end by the circumference. *ST*, in Fig. 24.2 is a chord.

(x) A **segment** is the name given to the parts into which a circle is divided by a chord. If the segment is less than a semicircle it is called a **minor segment** (see shaded area in Fig. 24.2). If the segment is greater than a semicircle it is called a **major segment** (see the unshaded area in Fig. 24.2).

(xi) An **arc** is a portion of the circumference of a circle. The distance *SRT* in Fig. 24.2 is called a **minor arc** and the distance *SXYT* is called a **major arc**.

(xii) The angle at the centre of a circle, subtended by an arc, is double the angle at the circumference subtended by the same arc. With reference to Fig. 24.3,

Angle *AOC* = 2 × angle *ABC*

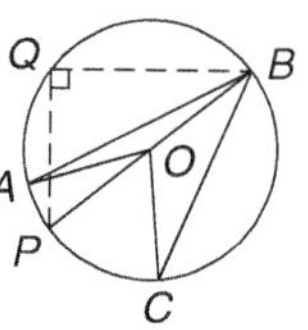

Figure 24.3

(xiii) The angle in a semicircle is a right angle (see angle *BQP* in Fig. 24.3).

Problem 1. Find the circumference of a circular metal plate of radius 12.0 cm

Circumference,

$$c = 2 \times \pi \times \text{radius} = 2\pi r = 2\pi(12.0)$$

$$= \mathbf{75.40\ cm}$$

Problem 2. If the diameter of a circular disc is 75 mm, find its circumference

Circumference,

$$c = \pi \times \text{diameter} = \pi d = \pi(75) = \mathbf{235.6\ mm}$$

Problem 3. Determine the radius of a circular plate if its perimeter is 112 cm

Perimeter = circumference, $c = 2\pi r$

Hence **radius** $\boldsymbol{r} = \dfrac{c}{2\pi} = \dfrac{112}{2\pi} = \mathbf{17.83\ cm}$

Problem 4. In the crank mechanism shown in Fig. 24.4, *AB* is a tangent to the circle at *B*. If the circle radius is 40 mm and *AB* = 150 mm, calculate the length *AO*

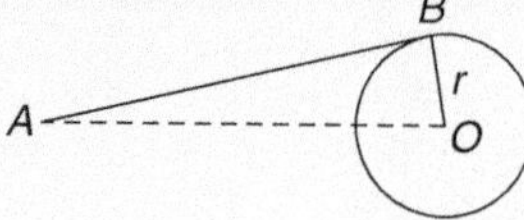

Figure 24.4

A tangent to a circle is at right angles to a radius drawn from the point of contact, i.e. *ABO* = 90°. Hence, using Pythagoras' theorem (see page 170):

$$AO^2 = AB^2 + OB^2$$

from which, $\boldsymbol{AO} = \sqrt{AB^2 + OB^2}$

$$= \sqrt{150^2 + 40^2} = \mathbf{155.2\ mm}$$

Now try the following Practice Exercise

Practice Exercise 123 Properties of circles (Answers on page 714)

1. Calculate the length of the circumference of a circular template of radius 7.2 cm.
2. If the diameter of a circular ceramic plate is 82.6 mm, calculate the circumference of the circle.
3. Determine the radius of a circular medal whose circumference is 16.52 cm.
4. Find the diameter of a circular plate whose perimeter is 149.8 cm.
5. A crank mechanism is shown in Fig. 24.5, where XY is a tangent to the circle at point X. If the circle radius OX is 10 cm and length OY is 40 cm, determine the length of the connecting rod XY.

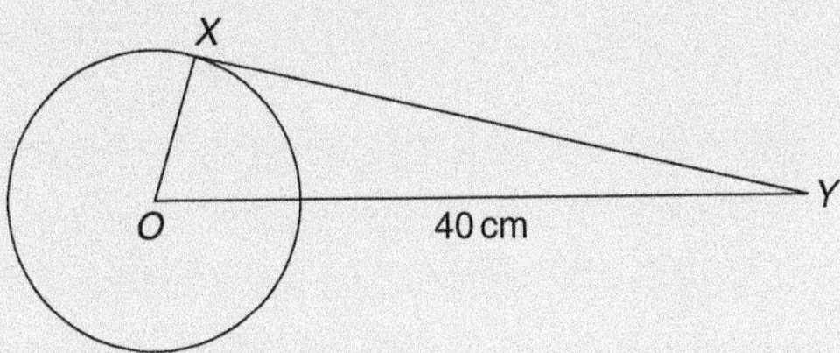

Figure 24.5

6. If the circumference of the earth is 40 000 km at the equator, calculate its diameter.
7. Calculate the length of wire in the paper clip shown in Fig. 24.6. The dimensions are in millimetres.

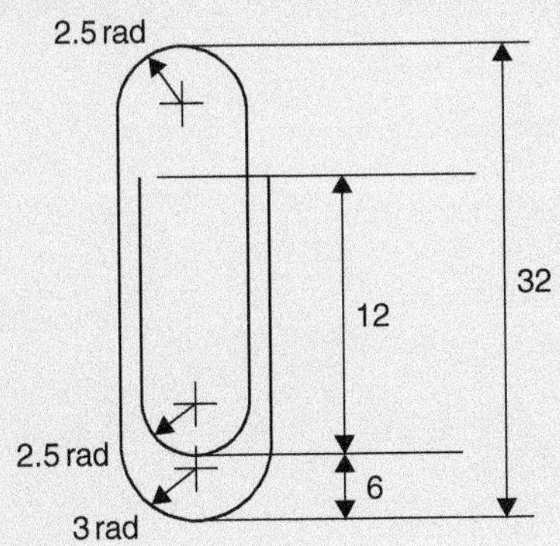

Figure 24.6

24.3 Radians and degrees

One **radian** is defined as the angle subtended at the centre of a circle by an arc equal in length to the radius.

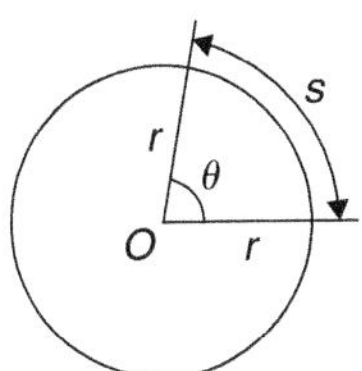

Figure 24.7

With reference to Fig. 24.7, for arc length s,

$$\theta \text{ radians} = \frac{s}{r}$$

When $s =$ whole circumference $(= 2\pi r)$ then

$$\theta = \frac{s}{r} = \frac{2\pi r}{r} = 2\pi$$

i.e. $\mathbf{2\pi}$ **radians** $= \mathbf{360^\circ}$ or $\boldsymbol{\pi}$ **radians** $= \mathbf{180^\circ}$

Thus, **1 rad** $= \dfrac{\mathbf{180^\circ}}{\pi} = \mathbf{57.30^\circ}$, correct to 2 decimal places.

Since π rad $= 180^\circ$, then $\dfrac{\pi}{2} = 90^\circ$, $\dfrac{\pi}{3} = 60^\circ$, $\dfrac{\pi}{4} = 45^\circ$ and so on.

Problem 5. Convert to radians: (a) 125° (b) $69^\circ\,47'$

(a) Since $180^\circ = \pi$ rad then $1^\circ = \pi/180$ rad, therefore

$$125^\circ = 125\left(\frac{\pi}{180}\right)^c = \mathbf{2.182\ radians}$$

(Note that ^{c} means 'circular measure' and indicates radian measure.)

(b) $69^\circ 47' = 69\dfrac{47^\circ}{60} = 69.783^\circ$

$$69.783^\circ = 69.783\left(\frac{\pi}{180}\right)^c = \mathbf{1.218\ radians}$$

Problem 6. Convert to degrees and minutes: (a) 0.749 radians (b) $3\pi/4$ radians

(a) Since π rad = 180° then 1 rad = 180°/π, therefore

$$0.749 \text{ rad} = 0.749\left(\frac{180}{\pi}\right)^\circ = 42.915^\circ$$

$0.915^\circ = (0.915 \times 60)' = 55'$, correct to the nearest minute, hence

$$\mathbf{0.749 \text{ radians} = 42^\circ 55'}$$

(b) Since $1 \text{ rad} = \left(\frac{180}{\pi}\right)^\circ$ then

$$\frac{3\pi}{4} \text{ rad} = \frac{3\pi}{4}\left(\frac{180}{\pi}\right)^\circ$$
$$= \frac{3}{4}(180)^\circ = \mathbf{135^\circ}$$

Problem 7. Express in radians, in terms of π: (a) 150° (b) 270° (c) 37.5°

Since 180° = π rad then 1° = 180/π, hence

(a) $150^\circ = 150\left(\frac{\pi}{180}\right) \text{ rad} = \mathbf{\frac{5\pi}{6} \text{ rad}}$

(b) $270^\circ = 270\left(\frac{\pi}{180}\right) \text{ rad} = \mathbf{\frac{3\pi}{2} \text{ rad}}$

(c) $37.5^\circ = 37.5\left(\frac{\pi}{180}\right) \text{ rad} = \frac{75\pi}{360} \text{ rad} = \mathbf{\frac{5\pi}{24} \text{ rad}}$

Now try the following Practice Exercise

Practice Exercise 124 Radians and degrees (Answers on page 714)

1. Convert to radians in terms of π: (a) 30° (b) 75° (c) 225°
2. Convert to radians: (a) 48° (b) 84°51′ (c) 232°15′
3. Convert to degrees: (a) $\frac{5\pi}{6}$ rad (b) $\frac{4\pi}{9}$ rad (c) $\frac{7\pi}{12}$ rad
4. Convert to degrees and minutes: (a) 0.0125 rad (b) 2.69 rad (c) 7.241 rad
5. A car engine speed is 1000 rev/min. Convert this speed into rad/s.

24.4 Arc length and area of circles and sectors

Arc length

From the definition of the radian in the previous section and Fig. 24.7,

$$\textbf{arc length, } s = r\theta \qquad (1)$$

where θ is in radians

Area of circle

From Chapter 23, for any circle, area = $\pi \times (\text{radius})^2$,

i.e. $\quad \textbf{area} = \pi r^2$

Since, $r = \frac{d}{2}$, then $\quad \textbf{area} = \pi r^2$ or $\frac{\pi d^2}{4}$

Area of sector

$$\textbf{Area of a sector} = \frac{\theta}{360}(\pi r^2) \quad \text{when } \theta \text{ is in degrees}$$

$$= \frac{\theta}{2\pi}(\pi r^2) = \frac{1}{2}r^2\theta \qquad (2)$$

when θ is in radians

24.5 Worked problems on arc length and area of circles and sectors

Problem 8. A hockey pitch has a semicircle of radius 14.63 m around each goal net. Find the area enclosed by the semicircle, correct to the nearest square metre

Area of a semicircle $= \frac{1}{2}\pi r^2$

When $r = 14.63$ m, area $= \frac{1}{2}\pi(14.63)^2$

i.e. **area of semicircle = 336 m²**

Problem 9. Find the area of a circular metal plate, correct to the nearest square millimetre, having a diameter of 35.0 mm

Area of a circle $= \pi r^2 = \frac{\pi d^2}{4}$

When $d = 35.0$ mm, $\quad$ area $= \frac{\pi(35.0)^2}{4}$

i.e. **area of circular plate = 962 mm²**

Problem 10. Find the area of a circular copper disc having a circumference of 60.0 mm

Circumference, $c = 2\pi r$

from which, radius, $r = \frac{c}{2\pi} = \frac{60.0}{2\pi} = \frac{30.0}{\pi}$

Area of a circle $= \pi r^2$

i.e. **area** $= \pi\left(\frac{30.0}{\pi}\right)^2 = \mathbf{286.5\,mm^2}$

Problem 11. Find the length of arc of a circle of radius 5.5 cm when the angle subtended at the centre is 1.20 radians

From equation (1), length of arc, $s = r\theta$, where θ is in radians, hence

$$s = (5.5)(1.20) = \mathbf{6.60\,cm}$$

Problem 12. A circular-section solid bar of linear taper is subjected to an axial pull, W, of 0.1 MN, as shown in Figure 24.8 It may be shown that the bar extension, u, is given by: $u = \frac{Wl}{A_1 E}\left(\frac{d_1}{d_2}\right)$

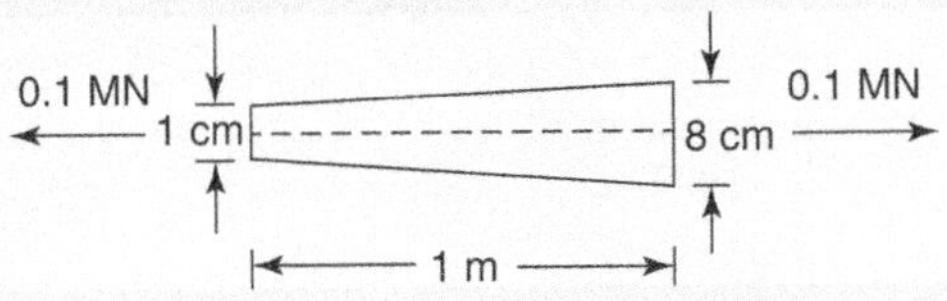

Figure 24.8

If $E = 2 \times 10^{11}$ N/m^2, $d_1 = 1$ cm, $d_2 = 8$ cm and $l = 1$ m, by how much will the bar extend?

Bar extension, $u = \frac{W1}{A_1 E}\left(\frac{d_1}{d_2}\right)$

Area of circle of diameter 1 cm, i.e. 0.01 m,

$$A_1 = \pi r^2 = \pi\left(\frac{d_1}{2}\right)^2 = \pi\left(\frac{0.01}{2}\right)^2 = 7.854 \times 10^{-5}\ \mathrm{m}^2$$

Hence, bar extension,

$$u = \frac{W1}{A_1 E}\left(\frac{d_1}{d_2}\right) = \frac{0.1 \times 10^6 \times 1}{7.854 \times 10^{-5} \times 2 \times 10^{11}}\left(\frac{0.01}{0.08}\right)$$
$$= 7.958 \times 10^{-4}\ \mathrm{m} = 0.796 \times 10^3\ \mathrm{m}$$

i.e. **bar extension, u = 0.796 mm**

Problem 13. Determine the diameter and circumference of a circle if an arc of length 4.75 cm subtends an angle of 0.91 radians

Since $s = r\theta$ then $r = \frac{s}{\theta} = \frac{4.75}{0.91} = 5.22$ cm.

Diameter $= 2 \times$ radius $= 2 \times 5.22 =$ **10.44 cm.**

Circumference, $c = \pi d = \pi(10.44) =$ **32.80 cm**.

Problem 14. If an angle of 125° is subtended by an arc of a circle of radius 8.4 cm, find the length of (a) the minor arc, and (b) the major arc, correct to 3 significant figures

Since $180° = \pi$ rad then $1° = \left(\frac{\pi}{180}\right)$ rad

and $125° = 125\left(\frac{\pi}{180}\right)$ rad

Length of minor arc,

$$s = r\theta = (8.4)(125)\left(\frac{\pi}{180}\right) = \mathbf{18.3\,cm}$$

correct to 3 significant figures.

Length of major arc = (circumference – minor arc) $= 2\pi(8.4) - 18.3 =$ **34.5 cm**, correct to 3 significant figures.
(Alternatively, major arc $= r\theta = 8.4(360 - 125)(\pi/180)$ $=$ **34.5 cm**.)

Problem 15. Determine the angle, in degrees and minutes, subtended at the centre of a circle of diameter 42 mm by an arc of length 36 mm. Calculate also the area of the minor sector formed

Since length of arc, $s = r\theta$ then $\theta = s/r$

Radius, $r = \frac{\text{diameter}}{2} = \frac{42}{2} = 21$ mm

hence $\theta = \frac{s}{r} = \frac{36}{21} = 1.7143$ radians

1.7143 rad $= 1.7143 \times (180/\pi)° = 98.22° =$ **98°13′** = angle subtended at centre of circle.
From equation (2),

$$\textbf{area of sector} = \frac{1}{2}r^2\theta = \frac{1}{2}(21)^2(1.7143)$$
$$= \mathbf{378\,mm^2}$$

Problem 16. A football stadiums floodlights can spread its illumination over an angle of 45° to a distance of 55 m. Determine the maximum area that is floodlit

$$\textbf{Floodlit area} = \text{area of sector} = \frac{1}{2}r^2\theta$$

$$= \frac{1}{2}(55)^2\left(45 \times \frac{\pi}{180}\right) \text{ from equation (2)}$$

$$= \mathbf{1188\ m^2}$$

Problem 17. An automatic garden spray produces a spray to a distance of 1.8 m and revolves through an angle α which may be varied. If the desired spray catchment area is to be 2.5 m^2, to what should angle α be set, correct to the nearest degree

Area of sector $= \frac{1}{2}r^2\theta$, hence $2.5 = \frac{1}{2}(1.8)^2\alpha$ from

$$\text{which, } \alpha = \frac{2.5 \times 2}{1.8^2} = 1.5432 \text{ radians}$$

$$1.5432 \text{ rad} = \left(1.5432 \times \frac{180}{\pi}\right)^\circ = 88.42^\circ$$

Hence **angle α = 88°**, correct to the nearest degree.

Problem 18. The angle of a tapered groove is checked using a 20 mm diameter roller as shown in Fig. 24.9. If the roller lies 2.12 mm below the top of the groove, determine the value of angle θ

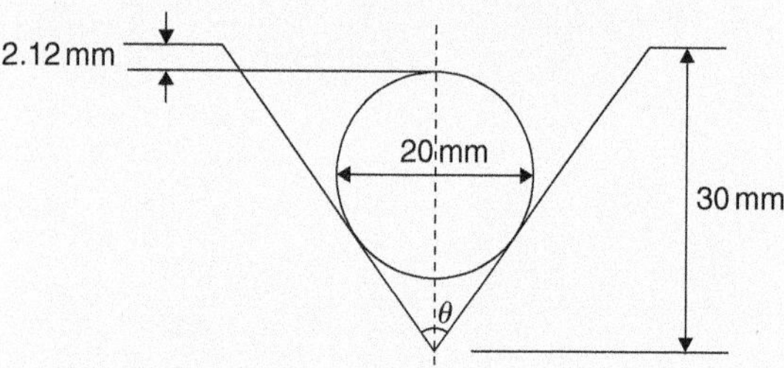

Figure 24.9

In Fig. 24.10, triangle ABC is right-angled at C (see Section 24.2(vii), page 248).
Length BC = 10 mm (i.e. the radius of the circle), and $AB = 30 - 10 - 2.12 = 17.88$ mm from Fig. 24.9

$$\text{Hence} \quad \sin\frac{\theta}{2} = \frac{10}{17.88} \text{ and}$$

$$\frac{\theta}{2} = \sin^{-1}\left(\frac{10}{17.88}\right) = 34^\circ$$

and **angle θ = 68°**

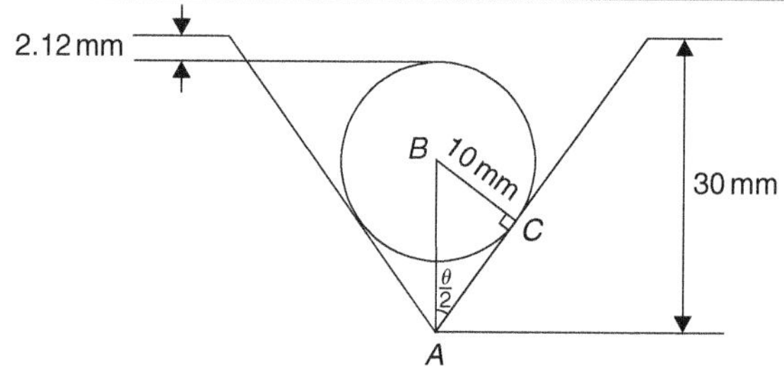

Figure 24.10

Now try the following Practice Exercise

Practice Exercise 125 Arc length and area of circles and sectors (Answers on page 714)

1. Calculate the area of a circular DVD of radius 6.0 cm, correct to the nearest square centimetre.
2. The diameter of a circular medal is 55.0 mm. Determine its area, correct to the nearest square millimetre.
3. The perimeter of a circular badge is 150 mm. Find its area, correct to the nearest square millimetre.
4. Find the area of the sector, correct to the nearest square millimetre, of a circle having a radius of 35 mm, with angle subtended at centre of 75°.
5. A washer in the form of an annulus has an outside diameter of 49.0 mm and an inside diameter of 15.0 mm. Find its area correct to 4 significant figures.
6. Find the area, correct to the nearest square metre, of a 2 m wide path surrounding a circular plot of land 200 m in diameter.
7. A rectangular park measures 50 m by 40 m. A 3 m flower bed is made round the two longer sides and one short side. A circular fish pond of diameter 8.0 m is constructed in the centre of the park. It is planned to grass the remaining area. Find, correct to the nearest square metre, the area of grass.
8. Find the length of an arc of a circle of radius 8.32 cm when the angle subtended at the centre is 2.14 radians. Calculate also the area of the minor sector formed.

9. If the angle subtended at the centre of a circle of diameter 82 mm is 1.46 rad, find the lengths of the (a) minor arc (b) major arc.

10. A pendulum of length 1.5 m swings through an angle of 10° in a single swing. Find, in centimetres, the length of the arc traced by the pendulum bob.

11. Determine the length of the radius and circumference of a circle if an arc length of 32.6 cm subtends an angle of 3.76 radians.

12. Determine the angle of lap, in degrees and minutes, if 180 mm of a belt drive are in contact with a pulley of diameter 250 mm.

13. A helicopter landing pad is circular with a diameter of 5 m. Calculate its area.

14. A square plate of side 50 mm has 5 holes cut out of it. If the 5 holes are all circles of diameter 10 mm calculate the remaining area, correct to the nearest whole number.

15. A circular plastic plate has a radius of 60 mm. A sector subtends an angle of 1.6 radians at the centre. Calculate the area of the sector.

16. The end plate of a steel boiler is a flat circular plate of diameter 2.5 m. Holes are drilled in it as follows: two 500 mm diameter tube holes and 24 water tube holes each of diameter 40 mm. Calculate the remaining area of the end plate in square metres, correct to 3 significant figures.

17. A wheel of diameter 600 mm rolls without slipping along a horizontal surface. If the wheel rotates 4.5 revolutions calculate how far the wheel moves horizontally. Give the answer in metres correct to 3 significant figures.

18. A tight open belt passes directly over two pulleys, of diameters 1.2 m and 1.8 m respectively having their centres 2.4 m apart. Calculate the length of the belt, correct to 3 significant figures.

19. A rubber gasket for a cylinder is a circular annulus of area 4084 mm^2. If the outer diameter is 140 mm calculate the inner radius of the gasket, correct to the nearest whole number.

20. Determine the number of complete revolutions a motorcycle wheel will make in travelling 2 km, if the wheel's diameter is 85.1 cm.

21. The floodlights at a sports ground spread its illumination over an angle of 40° to a distance of 48 m. Determine (a) the angle in radians, and (b) the maximum area that is floodlit.

22. Find the area swept out in 50 minutes by the minute hand of a large floral clock, if the hand is 2 m long.

23. Determine (a) the shaded area in Fig. 24.11, (b) the percentage of the whole sector that the area of the shaded portion represents.

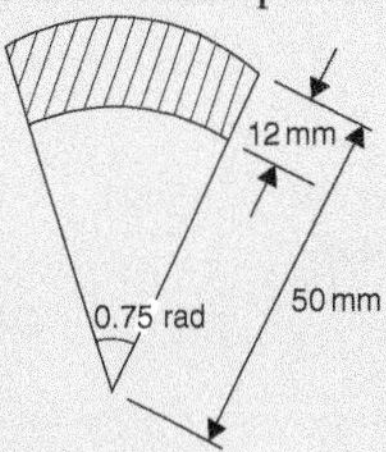

Figure 24.11

24. Determine the length of steel strip required to make the clip shown in Fig. 24.12.

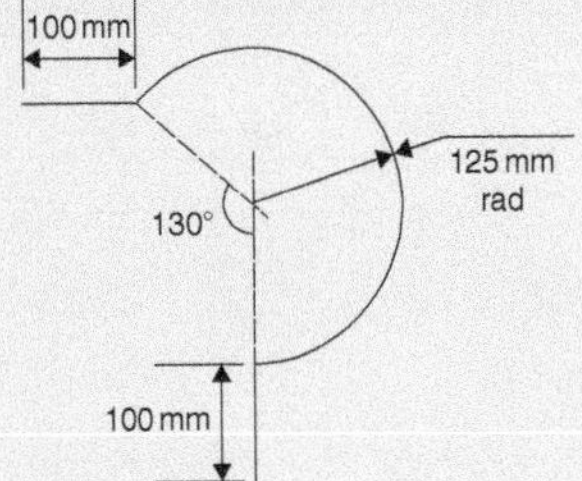

Figure 24.12

25. A 50° tapered hole is checked with a 40 mm diameter ball as shown in Fig. 24.13. Determine the length shown as x.

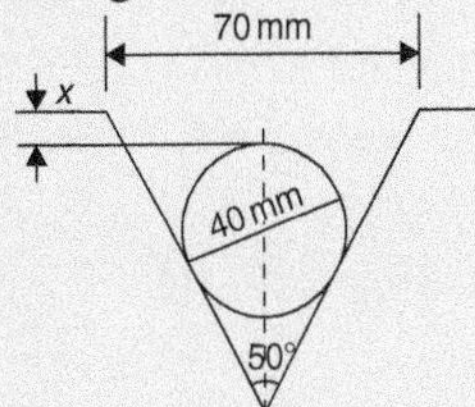

Figure 24.13

24.6 The equation of a circle

The simplest equation of a circle, centre at the origin, radius r, is given by:

$$x^2+y^2=r^2$$

For example, Fig. 24.14 shows a circle $x^2+y^2=9$

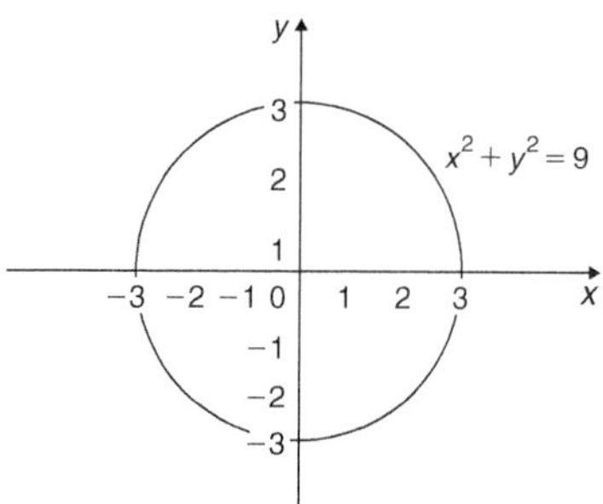

Figure 24.14

More generally, the equation of a circle, centre (a, b), radius r, is given by:

$$(x-a)^2+(y-b)^2=r^2 \quad (3)$$

Figure 24.15 shows a circle $(x-2)^2+(y-3)^2=4$

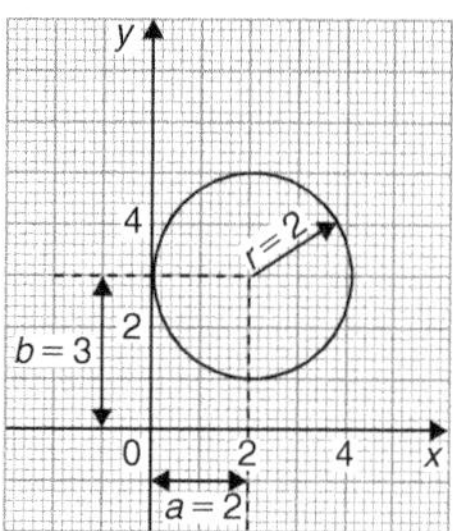

Figure 24.15

The general equation of a circle is:

$$x^2+y^2+2ex+2fy+c=0 \quad (4)$$

Multiplying out the bracketed terms in equation (3) gives:

$$x^2-2ax+a^2+y^2-2by+b^2=r^2$$

Comparing this with equation (4) gives:

$$2e=-2a, \quad \text{i.e.} \quad a=-\frac{2e}{2}$$

$$\text{and} \quad 2f=-2b, \quad \text{i.e.} \quad b=-\frac{2f}{2}$$

$$\text{and} \quad c=a^2+b^2-r^2, \quad \text{i.e.} \quad r=\sqrt{a^2+b^2-c}$$

Thus, for example, the equation

$$x^2+y^2-4x-6y+9=0$$

represents a circle with centre

$$a=-\left(\tfrac{-4}{2}\right), b=-\left(\tfrac{-6}{2}\right)$$

i.e. at (2, 3) and radius $r=\sqrt{2^2+3^2-9}=2$

Hence $x^2+y^2-4x-6y+9=0$ is the circle shown in Fig. 24.15, which may be checked by multiplying out the brackets in the equation

$$(x-2)^2+(y-3)^2=4$$

Problem 19. Determine: (a) the radius, and (b) the co-ordinates of the centre of the circle given by the equation: $x^2+y^2+8x-2y+8=0$

$x^2+y^2+8x-2y+8=0$ is of the form shown in equation (4),

where $\quad a=-\left(\tfrac{8}{2}\right)=-4, \quad b=-\left(\tfrac{-2}{2}\right)=1$

and $\quad r=\sqrt{(-4)^2+1^2-8}=\sqrt{9}=3$

Hence $x^2+y^2+8x-2y+8=0$ represents a circle **centre (−4, 1)** and **radius 3**, as shown in Fig. 24.16.

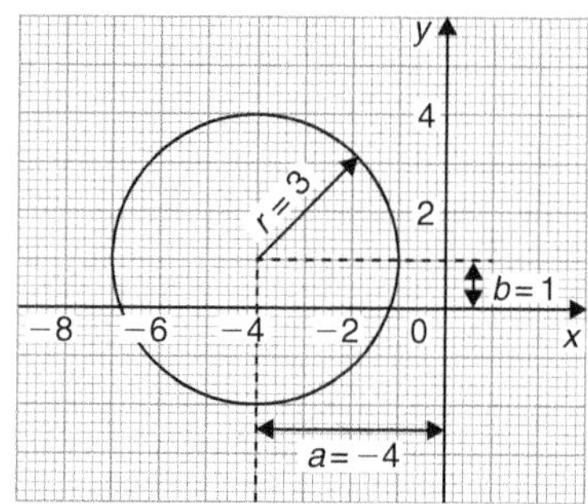

Figure 24.16

Alternatively, $x^2+y^2+8x-2y+8=0$ may be rearranged as:

$$(x+4)^2+(y-1)^2-9=0$$

i.e. $\quad (x+4)^2+(y-1)^2=3^2$

which represents a circle, **centre (−4, 1)** and **radius 3**, as stated above.

Problem 20. Sketch the circle given by the equation: $x^2+y^2-4x+6y-3=0$

The equation of a circle, centre (a,b), radius r is given by:

$$(x-a)^2+(y-b)^2=r^2$$

The general equation of a circle is

$$x^2+y^2+2ex+2fy+c=0$$

From above $a=-\frac{2e}{2}, b=-\frac{2f}{2}$

and $r=\sqrt{a^2+b^2-c}$

Hence if $x^2+y^2-4x+6y-3=0$

then $a=-\left(\frac{-4}{2}\right)=2, \quad b=-\left(\frac{6}{2}\right)=-3$

and $r=\sqrt{2^2+(-3)^2-(-3)}=\sqrt{16}=4$

Thus **the circle has centre (2, −3)** and **radius 4**, as shown in Fig. 24.17.

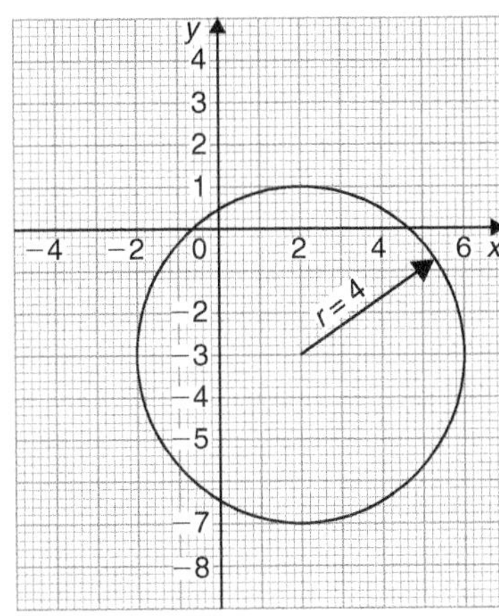

Figure 24.17

Alternatively, $x^2+y^2-4x+6y-3=0$ may be rearranged as:

$$(x-2)^2+(y+3)^2-3-13=0$$

i.e. $$(x-2)^2+(y+3)^2=4^2$$

which represents a circle, **centre (2, −3)** and **radius 4**, as stated above.

Now try the following Practice Exercise

Practice Exercise 126 The equation of a circle (Answers on page 714)

1. Determine: (a) the radius, and (b) the co-ordinates of the centre of the circle given by the equation $x^2+y^2-6x+8y+21=0$
2. Sketch the circle given by the equation $x^2+y^2-6x+4y-3=0$
3. Sketch the curve $x^2+(y-1)^2-25=0$
4. Sketch the curve $x=6\sqrt{1-\left(\frac{y}{6}\right)^2}$

Practice Exercise 127 Multiple-choice questions on the circle and its properties (Answers on page 714)

Each question has only one correct answer

1. An arc of a circle of length 5.0 cm subtends an angle of 2 radians. The circumference of the circle is:

   (a) 2.5 cm (b) 10.0 cm
   (c) 5.0 cm (d) 15.7 cm

2. A pendulum of length 1.2 m swings through an angle of 12° in a single swing. The length of arc traced by the pendulum bob is:

   (a) 14.40 cm (b) 25.13 cm
   (c) 10.00 cm (d) 45.24 cm

3. A wheel on a car has a diameter of 800 mm. If the car travels 5 miles, the number of complete revolutions the wheel makes (given 1 km = $\frac{5}{8}$ mile) is:

   (a) 1989 (b) 1591
   (c) 3183 (d) 10000

4. The equation of a circle is
   $$x^2+2x+y^2-10y+22=0$$
   The co-ordinates of its centre are:

   (a) (1, 5) (b) (−2, 4)
   (c) (−1, 5) (d) (2, 4)

5. A semicircle has a radius of 2 cm. Its perimeter is:

   (a) $(8+2\pi)$ cm (b) 2π cm

   (c) $(4+2\pi)$ cm (d) $(4\pi+2)$ cm

6. The equation of a circle is
   $$x^2+y^2-2x+4y-4=0$$
   Which of the following statements is correct?

   (a) The circle has centre $(1, -2)$ and radius 4

   (b) The circle has centre $(-1, 2)$ and radius 2

   (c) The circle has centre $(-1, -2)$ and radius 4

   (d) The circle has centre $(1, -2)$ and radius 3

7. The area of the sector of a circle of radius 9 cm is $27\pi \text{cm}^2$. The angle of the sector, in degrees, is:

   (a) 120° (b) 60°

   (c) 11° (d) 240°

8. The length of the arc of a circle of radius 5 cm when the angle subtended at the centre is 32° is:

   (a) 3.49 cm (b) 2.79 cm

   (c) 0.44 cm (d) 6.98 cm

9. A sector of a circle subtends an angle of 1.2 radians and has an arc length of 6 cm. The circumference of the circle is:

   (a) 10π cm (b) 5π cm

   (c) 15π cm (d) 25π cm

10. A circle has its centre at co-ordinates $(3, -2)$ and has a radius of 4. The equation of the circle is:

    (a) $(x+3)^2+(y-2)^2=42$

    (b) $(x-3)^2+(y+2)^2=16$

    (c) $x^2+y^2=52$

    (d) $(x-3)^2+(y-2)^2=16$

For fully worked solutions to each of the problems in Practice Exercises 123 to 126 in this chapter, go to the website:

www.routledge.com/cw/bird

Chapter 25

Volumes and surface areas of common solids

Why it is important to understand: Volumes and surface areas of common solids

There are many practical applications where volumes and surface areas of common solids are required. Examples include determining capacities of oil, water, petrol and fish tanks, ventilation shafts and cooling towers, determining volumes of blocks of metal, ball-bearings, boilers and buoys and calculating the cubic metres of concrete needed for a path. Finding the surface areas of loudspeaker diaphragms and lampshades provide further practical examples. Understanding these calculations is essential for the many practical applications in engineering, construction, architecture and science.

At the end of this chapter, you should be able to:

- state the SI unit of volume
- calculate the volumes and surface areas of cuboids, cylinders, prisms, pyramids, cones and spheres
- calculate volumes and surface areas of frusta of pyramids and cones
- calculate the frustum and zone of a sphere
- calculate volumes of regular solids using the prismoidal rule
- appreciate that volumes of similar bodies are proportional to the cubes of the corresponding linear dimensions

25.1 Introduction

The **volume** of any solid is a measure of the space occupied by the solid.

Volume is measured in **cubic units** such as mm^3, cm^3 and m^3.

This chapter deals with finding volumes of common solids; in engineering it is often important to be able to calculate volume or capacity to estimate, say, the amount of liquid, such as water, oil or petrol, in differing shaped containers.

A **prism** is a solid with a constant cross-section and with two ends parallel. The shape of the end is used to describe the prism. For example, there are rectangular prisms (called cuboids), triangular prisms and circular prisms (called cylinders).

On completing this chapter you will be able to calculate the volumes and surface areas of rectangular and other prisms, cylinders, pyramids, cones and spheres, together with frusta of pyramids and cones. Also, volumes of similar shapes are considered.

25.2 Volumes and surface areas of regular solids

A summary of volumes and surface areas of regular solids is shown in Table 25.1.

Table 25.1

(i) Rectangular prism (or cuboid)

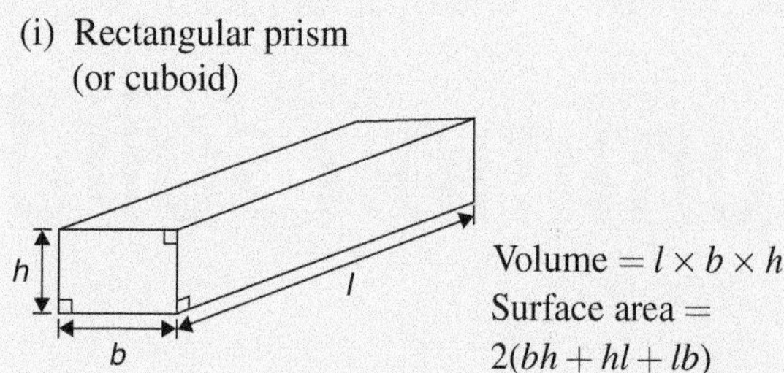

Volume $= l \times b \times h$
Surface area $= 2(bh + hl + lb)$

(ii) Cylinder

Volume $= \pi r^2 h$
Total surface area $= 2\pi rh + 2\pi r^2$

(iii) Pyramid

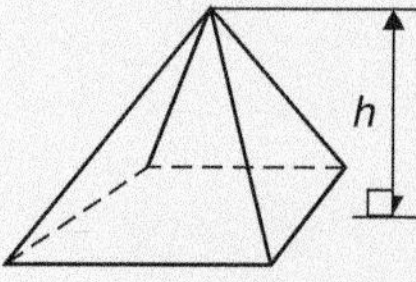

Volume $= \dfrac{1}{3} \times A \times h$
where A = area of base
and h = perpendicular height

Total surface area = (sum of areas of triangles forming sides) + (area of base)

(iv) Cone

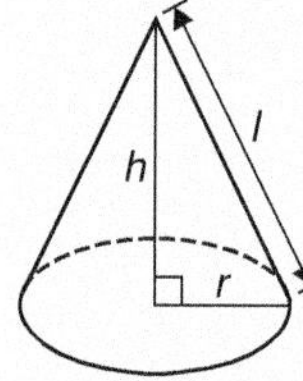

Volume $= \dfrac{1}{3}\pi r^2 h$
Curved surface area $= \pi rl$
Total surface area $= \pi rl + \pi r^2$

(v) Sphere

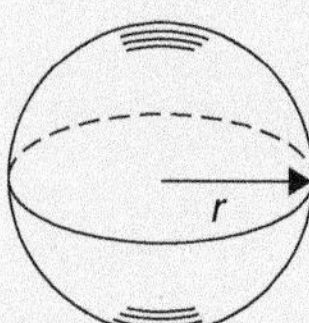

Volume $= \dfrac{4}{3}\pi r^3$
Surface area $= 4\pi r^2$

25.3 Worked problems on volumes and surface areas of regular solids

Problem 1. A water tank is the shape of a rectangular prism having length 2 m, breadth 75 cm and height 50 cm. Determine the capacity of the tank in (a) m^3 (b) cm^3 (c) litres

Volume of rectangular prism $= l \times b \times h$ (see Table 25.1)

(a) Volume of tank $= 2 \times 0.75 \times 0.5 =$ **0.75 m³**

(b) $1\,m^3 = 10^6\,cm^3$, hence
$0.75\,m^3 = 0.75 \times 10^6\,cm^3 =$ **750 000 cm³**

(c) 1 litre $= 1000\,cm^3$, hence
$750\,000\,cm^3 = \dfrac{750\,000}{1000}$ litres $=$ **750 litres**

Problem 2. Find the volume and total surface area of a cylindrical container of length 15 cm and diameter 8 cm

Volume of cylinder $= \pi r^2 h$ (see Table 25.1)

Since diameter $= 8$ cm, then radius $r = 4$ cm

Hence volume $= \pi \times 4^2 \times 15 =$ **754 cm³**

Total surface area (i.e. including the two ends)

$$= 2\pi rh + 2\pi r^2 = (2 \times \pi \times 4 \times 15) + (2 \times \pi \times 4^2) = \mathbf{477.5\ cm^2}$$

Problem 3. Determine the volume (in cm^3) of the wooden shape shown in Fig. 25.1.

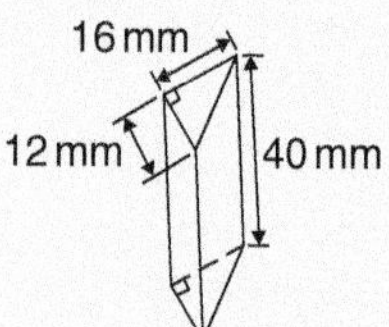

Figure 25.1

The solid shown in Fig. 25.1 is a triangular prism. The volume V of any prism is given by: $V = Ah$, where A is the cross-sectional area and h is the perpendicular height.

$$\text{Hence volume} = \frac{1}{2} \times 16 \times 12 \times 40$$

$$= 3840\,\text{mm}^3 = \mathbf{3.840\,cm^3}$$

(since $1\,\text{cm}^3 = 1000\,\text{mm}^3$)

Problem 4. Calculate the volume and total surface area of the solid metal prism shown in Fig. 25.2.

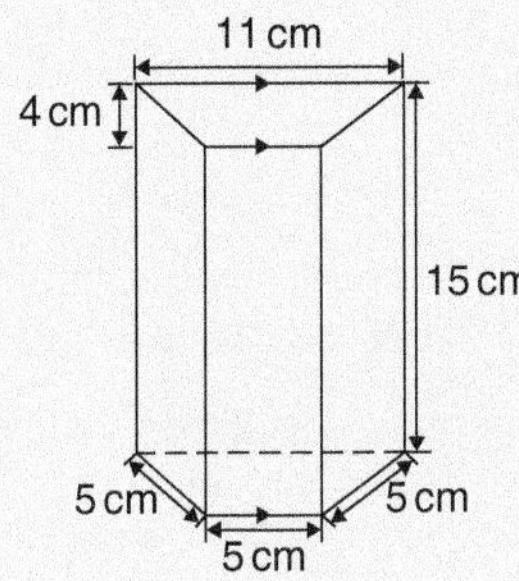

Figure 25.2

The solid shown in Fig. 25.2 is a trapezoidal prism.

$$\text{Volume} = \text{cross-sectional area} \times \text{height}$$

$$= \frac{1}{2}(11+5)4 \times 15 = 32 \times 15 = \mathbf{480\,cm^3}$$

$$\text{Surface area} = \text{sum of two trapeziums} + 4\,\text{rectangles}$$

$$= (2 \times 32) + (5 \times 15) + (11 \times 15) + 2(5 \times 15)$$

$$= 64 + 75 + 165 + 150 = \mathbf{454\,cm^2}$$

Problem 5. Determine the volume and the total surface area of the square wooden pyramid shown in Fig. 25.3 if its perpendicular height is 12 cm.

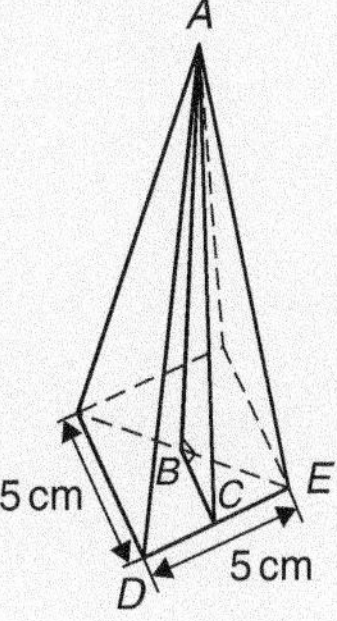

Figure 25.3

Volume of pyramid

$$= \frac{1}{3}(\text{area of base}) \times (\text{perpendicular height})$$

$$= \frac{1}{3}(5 \times 5) \times 12 = \mathbf{100\,cm^3}$$

The total surface area consists of a square base and 4 equal triangles.

Area of triangle ADE

$$= \frac{1}{2} \times \text{base} \times \text{perpendicular height}$$

$$= \frac{1}{2} \times 5 \times AC$$

The length AC may be calculated using Pythagoras' theorem on triangle ABC, where $AB = 12\,\text{cm}$, $BC = \frac{1}{2} \times 5 = 2.5\,\text{cm}$

$$\text{Hence,} \quad AC = \sqrt{AB^2 + BC^2} = \sqrt{12^2 + 2.5^2} = 12.26\,\text{cm}$$

$$\text{Hence area of triangle } ADE = \frac{1}{2} \times 5 \times 12.26 = 30.65\,\text{cm}^2$$

$$\text{Total surface area of pyramid} = (5 \times 5) + 4(30.65) = \mathbf{147.6\,cm^2}$$

Problem 6. Determine the volume and total surface area of a metal cone of radius 5 cm and perpendicular height 12 cm

The cone is shown in Fig. 25.4.

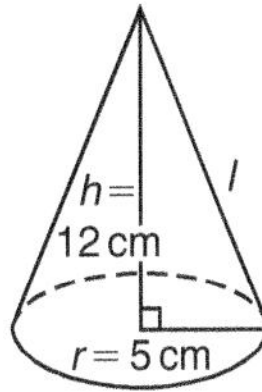

Figure 25.4

$$\text{Volume of cone} = \frac{1}{3}\pi r^2 h = \frac{1}{3} \times \pi \times 5^2 \times 12 = \mathbf{314.2\,cm^3}$$

$$\text{Total surface area} = \text{curved surface area} + \text{area of base} = \pi r l + \pi r^2$$

From Fig. 25.4, slant height l may be calculated using Pythagoras' theorem

$$l = \sqrt{12^2 + 5^2} = 13\,\text{cm}$$

Hence total surface area $= (\pi \times 5 \times 13) + (\pi \times 5^2)$

$$= \mathbf{282.7\,cm^2}$$

Problem 7. Find the volume and surface area of a spherical bearing of diameter 8 cm

Since diameter = 8 cm, then radius, $r = 4$ cm.

$$\text{Volume of sphere} = \frac{4}{3}\pi r^3 = \frac{4}{3} \times \pi \times 4^3 = \mathbf{268.1\,cm^3}$$

$$\text{Surface area of sphere} = 4\pi r^2 = 4 \times \pi \times 4^2 = \mathbf{201.1\,cm^2}$$

Now try the following Practice Exercise

Practice Exercise 128 Volumes and surface areas of regular solids (Answers on page 714)

1. A rectangular block of metal has dimensions of 40 mm by 25 mm by 15 mm. Determine its volume. Find also its mass if the metal has a density of 9 g/cm^3
2. Determine the maximum capacity, in litres, of a fish tank measuring 50 cm by 40 cm by 2.5 m (1 litre = 1000 cm^3)
3. Determine how many cubic metres of concrete are required for a 120 m long path, 150 mm wide and 80 mm deep
4. Calculate the volume of a metal tube whose outside diameter is 8 cm and whose inside diameter is 6 cm, if the length of the tube is 4 m
5. The volume of a metal cylinder is 400 cm^3. If its radius is 5.20 cm, find its height. Determine also its curved surface area
6. If a solid metal cone has a diameter of 80 mm and a perpendicular height of 120 mm calculate its volume in cm^3 and its curved surface area
7. A cylinder is cast from a rectangular piece of alloy 5 cm by 7 cm by 12 cm. If the length of the cylinder is to be 60 cm, find its diameter
8. Find the volume and the total surface area of a regular hexagonal bar of metal of length 3 m if each side of the hexagon is 6 cm
9. A square pyramid feature has a perpendicular height of 4 cm. If a side of the base is 2.4 cm long find the volume and total surface area of the pyramid
10. A spherical bearing has a diameter of 6 cm. Determine its volume and surface area
11. Find the total surface area of a hemispherical paperweight of diameter 50 mm
12. How long will it take a tap dripping at a rate of 800 mm^3/s to fill a 3-litre can?
13. A lorry has a closed cylindrical chemical storage tank of length 3.6 m and diameter 1.3 m. Calculate for the tank its (a) surface area, in m^2, and (b) storage capacity, in m^3 and in litres. (Note that 1 litre = 1000 cm^3).
14. A feed to a process line is in the form of a cone of radius 1.8 m and perpendicular height 2.5 m. The feed supplies flour to a bread mixing line. Calculate the maximum volume of flour that the feed can hold.
15. How many cylindrical jars of radius 80 mm and height 250 mm can be filled completely from a rectangular vat 2.4 m long, 60 cm wide and 950 mm high?

25.4 Further worked problems on volumes and surface areas of regular solids

Problem 8. A wooden section is shown in Fig. 25.5. Find (a) its volume (in m^3), and (b) its total surface area.

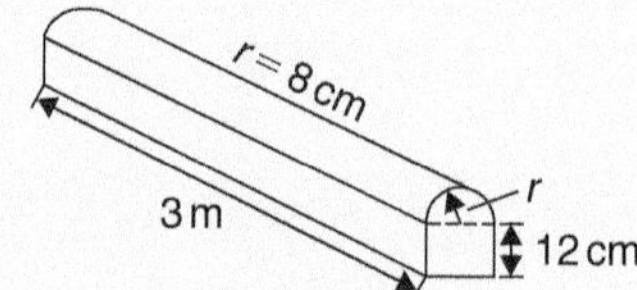

Figure 25.5

The section of wood is a prism whose end comprises a rectangle and a semicircle. Since the radius of the semicircle is 8 cm, the diameter is 16 cm.
Hence the rectangle has dimensions 12 cm by 16 cm.
Area of end $= (12 \times 16) + \frac{1}{2}\pi 8^2 = 292.5\,\text{cm}^2$
Volume of wooden section

$$= \text{area of end} \times \text{perpendicular height}$$
$$= 292.5 \times 300 = 87750\,\text{cm}^3 = \frac{87750\,\text{m}^3}{10^6}$$
$$= \mathbf{0.08775\,m^3}$$

The total surface area comprises the two ends (each of area 292.5 cm^2), three rectangles and a curved surface (which is half a cylinder), hence

$$\text{total surface area} = (2 \times 292.5) + 2(12 \times 300) + (16 \times 300) + \tfrac{1}{2}(2\pi \times 8 \times 300)$$
$$= 585 + 7200 + 4800 + 2400\pi$$
$$= \mathbf{20\,125\,cm^2} \quad \text{or} \quad \mathbf{2.0125\,m^2}$$

Problem 9. A glass pyramid has a rectangular base 3.60 cm by 5.40 cm. Determine the volume and total surface area of the pyramid if each of its sloping edges is 15.0 cm

The pyramid is shown in Fig. 25.6. To calculate the volume of the pyramid the perpendicular height EF is required. Diagonal BD is calculated using Pythagoras' theorem,

i.e. $$BD = \sqrt{3.60^2 + 5.40^2} = 6.490\,\text{cm}$$

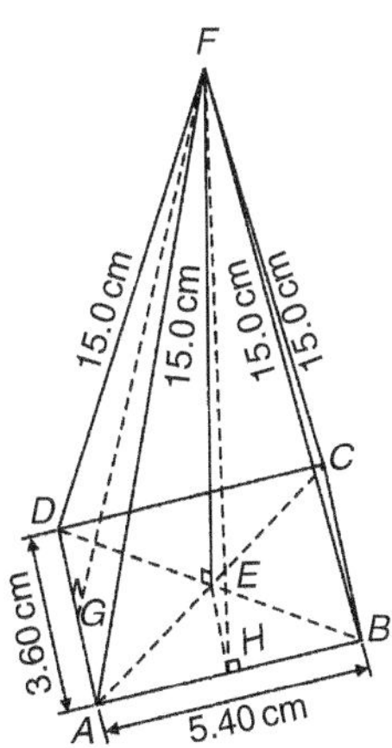

Figure 25.6

$$\text{Hence } EB = \frac{1}{2}BD = \frac{6.490}{2} = 3.245\,\text{cm}$$

Using Pythagoras' theorem on triangle BEF gives

$$BF^2 = EB^2 + EF^2$$

$$\text{from which, } EF = \sqrt{BF^2 - EB^2}$$
$$= \sqrt{15.0^2 - 3.245^2} = 14.64\,\text{cm}$$

Volume of pyramid

$$= \tfrac{1}{3}(\text{area of base})(\text{perpendicular height})$$
$$= \tfrac{1}{3}(3.60 \times 5.40)(14.64) = \mathbf{94.87\,cm^3}$$

Area of triangle ADF (which equals triangle BCF) $= \frac{1}{2}(AD)(FG)$, where G is the midpoint of AD. Using Pythagoras' theorem on triangle FGA gives:

$$FG = \sqrt{15.0^2 - 1.80^2} = 14.89\,\text{cm}$$

$$\text{Hence area of triangle } ADF = \tfrac{1}{2}(3.60)(14.89)$$
$$= 26.80\,\text{cm}^2$$

Similarly, if H is the mid-point of AB, then

$$FH = \sqrt{15.0^2 - 2.70^2} = 14.75\,\text{cm},$$

hence area of triangle ABF (which equals triangle CDF)

$$= \tfrac{1}{2}(5.40)(14.75) = 39.83\,\text{cm}^2$$

Total surface area of pyramid

$$= 2(26.80) + 2(39.83) + (3.60)(5.40)$$
$$= 53.60 + 79.66 + 19.44$$
$$= \mathbf{152.7\,cm^2}$$

Problem 10. Calculate the volume and total surface area of a solid metal hemisphere of diameter 5.0 cm

$$\text{Volume of hemisphere} = \tfrac{1}{2}(\text{volume of sphere})$$
$$= \frac{2}{3}\pi r^3 = \frac{2}{3}\pi\left(\frac{5.0}{2}\right)^3$$
$$= \mathbf{32.7\,cm^3}$$

Total surface area

$$= \text{curved surface area} + \text{area of circle}$$
$$= \tfrac{1}{2}(\text{surface area of sphere}) + \pi r^2$$
$$= \tfrac{1}{2}(4\pi r^2) + \pi r^2$$
$$= 2\pi r^2 + \pi r^2 = 3\pi r^2 = 3\pi \left(\tfrac{5.0}{2}\right)^2$$
$$= \mathbf{58.9\ cm^2}$$

Problem 11. A rectangular piece of metal having dimensions 4 cm by 3 cm by 12 cm is melted down and recast into a pyramid having a rectangular base measuring 2.5 cm by 5 cm. Calculate the perpendicular height of the pyramid

$$\text{Volume of rectangular prism of metal} = 4 \times 3 \times 12$$
$$= 144\,\text{cm}^3$$

Volume of pyramid

$$= \tfrac{1}{3}(\text{area of base})(\text{perpendicular height})$$

Assuming no waste of metal,

$$144 = \tfrac{1}{3}(2.5 \times 5)(\text{height})$$

i.e. $\quad \text{perpendicular height} = \dfrac{144 \times 3}{2.5 \times 5} = \mathbf{34.56\ cm}$

Problem 12. A rivet consists of a cylindrical head, of diameter 1 cm and depth 2 mm, and a shaft of diameter 2 mm and length 1.5 cm. Determine the volume of metal in 2000 such rivets

Radius of cylindrical head $= \tfrac{1}{2}$ cm $= 0.5$ cm and

height of cylindrical head $= 2$ mm $= 0.2$ cm

Hence, volume of cylindrical head

$$= \pi r^2 h = \pi(0.5)^2(0.2) = 0.1571\,\text{cm}^3$$

Volume of cylindrical shaft

$$= \pi r^2 h = \pi \left(\frac{0.2}{2}\right)^2 (1.5) = 0.0471\,\text{cm}^3$$

$$\text{Total volume of 1 rivet} = 0.1571 + 0.0471$$
$$= 0.2042\,\text{cm}^3$$

Volume of metal in 2000 such rivets

$$= 2000 \times 0.2042 = \mathbf{408.4\ cm^3}$$

Problem 13. A solid metal cylinder of radius 6 cm and height 15 cm is melted down and recast into a shape comprising a hemisphere surmounted by a cone. Assuming that 8% of the metal is wasted in the process, determine the height of the conical portion, if its diameter is to be 12 cm

$$\text{Volume of cylinder} = \pi r^2 h = \pi \times 6^2 \times 15$$
$$= 540\pi\,\text{cm}^3$$

If 8% of metal is lost then 92% of 540π gives the volume of the new shape (shown in Fig. 25.7).

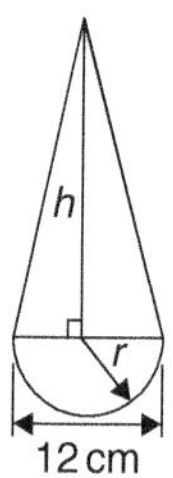

Figure 25.7

Hence the volume of (hemisphere + cone)

$$= 0.92 \times 540\pi\,\text{cm}^3,$$

i.e. $\quad \tfrac{1}{2}\left(\tfrac{4}{3}\pi r^3\right) + \tfrac{1}{3}\pi r^2 h = 0.92 \times 540\pi$

Dividing throughout by π gives:

$$\tfrac{2}{3}r^3 + \tfrac{1}{3}r^2 h = 0.92 \times 540$$

Since the diameter of the new shape is to be 12 cm, then radius $r = 6$ cm,

hence $\quad \tfrac{2}{3}(6)^3 + \tfrac{1}{3}(6)^2 h = 0.92 \times 540$

$$144 + 12h = 496.8$$

i.e. height of conical portion,

$$\boldsymbol{h} = \frac{496.8 - 144}{12} = \mathbf{29.4\ cm}$$

Problem 14. A block of copper having a mass of 50 kg is drawn out to make 500 m of wire of uniform cross-section. Given that the density of copper is 8.91 g/cm^3, calculate (a) the volume of copper, (b) the cross-sectional area of the wire, and (c) the diameter of the cross-section of the wire

(a) A density of 8.91 g/cm^3 means that 8.91 g of copper has a volume of 1 cm^3, or 1 g of copper has a volume of (1/8.91) cm^3

Hence 50 kg, i.e. 50 000 g, has a volume

$$\frac{50000}{8.91}\text{cm}^3 = \mathbf{5612\,cm^3}$$

(b) Volume of wire

$$= \text{area of circular cross-section} \times \text{length of wire.}$$

Hence $5612\,\text{cm}^3 = \text{area} \times (500 \times 100\,\text{cm})$,

from which, $\text{area} = \dfrac{5612}{500 \times 100}\,\text{cm}^2$

$= \mathbf{0.1122\,cm^2}$

(c) Area of circle $= \pi r^2$ or $\dfrac{\pi d^2}{4}$, hence

$0.1122 = \dfrac{\pi d^2}{4}$ from which

$$d = \sqrt{\frac{4 \times 0.1122}{\pi}} = 0.3780\,\text{cm}$$

i.e. diameter of cross-section is 3.780 mm

Problem 15. A boiler consists of a cylindrical section of length 8 m and diameter 6 m, on one end of which is surmounted a hemispherical section of diameter 6 m, and on the other end a conical section of height 4 m and base diameter 6 m. Calculate the volume of the boiler and the total surface area

The boiler is shown in Fig. 25.8

Volume of hemisphere, P

$$= \tfrac{2}{3}\pi r^3 = \tfrac{2}{3} \times \pi \times 3^3 = 18\pi\,\text{m}^3$$

Volume of cylinder, Q

$$= \pi r^2 h = \pi \times 3^2 \times 8 = 72\pi\,\text{m}^3$$

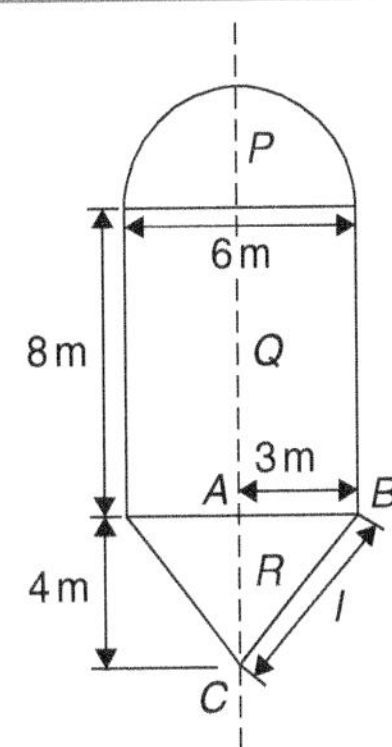

Figure 25.8

Volume of cone, R

$$= \tfrac{1}{3}\pi r^2 h = \tfrac{1}{3} \times \pi \times 3^2 \times 4 = 12\pi\,\text{m}^3$$

Total volume of boiler $= 18\pi + 72\pi + 12\pi$

$= 102\pi = \mathbf{320.4\,m^3}$

Surface area of hemisphere, P

$$= \tfrac{1}{2}(4\pi r^2) = 2 \times \pi \times 3^2 = 18\pi\,\text{m}^2$$

Curved surface area of cylinder, Q

$$= 2\pi r h = 2 \times \pi \times 3 \times 8 = 48\pi\,\text{m}^2$$

The slant height of the cone, l, is obtained by Pythagoras' theorem on triangle ABC, i.e.

$$l = \sqrt{4^2 + 3^2} = 5$$

Curved surface area of cone, R

$$= \pi r l = \pi \times 3 \times 5 = 15\pi\,\text{m}^2$$

Total surface area of boiler $= 18\pi + 48\pi + 15\pi$

$= 81\pi = \mathbf{254.5\,m^2}$

Problem 16. A 3 km tunnel is to be dug under a hill. Initially the tunnel is cut by a boring machine which cuts a circular profile having a radius of 5 m. The floor of the tunnel is then laid to a maximum depth of 2 m. Calculate (a) the surface area of the tunnel wall, and (b) the volume of earth material to be removed.

A view of the entrance to the tunnel is shown in Fig. 25.9

(a) The radius of the circle is 5 m (shown as AB and BC) and the depth of the floor is 2 m (shown as ED).
Hence the distance shown as BE must be 5 − 2 = 3 m since BD is the radius equal to 5 m.

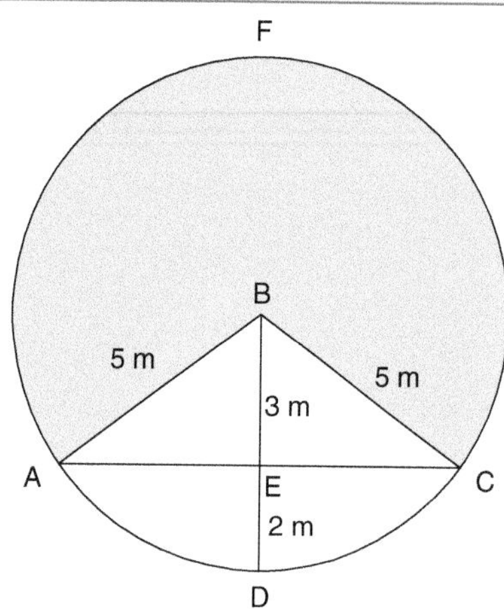

Figure 25.9

$\angle CBE = \cos^{-1}\frac{3}{5} = 53.13°$ and therefore $\angle ABC = 2 \times 53.13° = 106.26$

The major arc from A to F to C

$$= 360° - 106.26° = 253.74°$$

The length of the major arc $AFC = r \times \theta$ where θ is in radians

Converting 253.74° to radians gives

$$253.74 \times \frac{\pi}{180} \text{ rad}$$

Length of major arc AFC

$$= r \times \theta = 5 \times \left(253.74 \times \frac{\pi}{180}\right) = 22.143 \text{ m}$$

Hence, **the surface area of the tunnel wall** = $22.143 \times 3000 =$ **66429 m^2**

(b) In triangle $BCE, BC^2 = BE^2 + EC^2$ by Pythagoras from which, $EC = \sqrt{BC^2 - BE^2} = \sqrt{5^2 - 3^2} = 4$ m and hence, $AC = 2 \times 4 = 8$ m

Area of triangle ABC

$$= \frac{1}{2} \times \text{base} \times \text{height} = \frac{1}{2} \times 8 \times 3 = 12 \text{ m}^2$$

Area of major sector AFC (shown shaded)

$$= \frac{253.74}{360} \times \pi(5)^2 = 55.3575 \text{ m}^2$$

The total area of the tunnel opening = area of major sector AFC+ area of triangle ABC

$$= 55.358 + 12 = 67.358 \text{ m}^2$$

Therefore, **the volume of material to be removed** $= 67.358 \times 3000 =$ **202074 m^3**

Now try the following Practice Exercise

Practice Exercise 129 Volumes and surface areas of regular solids (Answers on page 714)

1. Determine the mass of a hemispherical copper container whose external and internal radii are 12 cm and 10 cm, assuming that 1 cm^3 of copper weighs 8.9 g
2. If the volume of a spherical ball bearing is 566 cm^3, find its radius
3. A metal plumb bob comprises a hemisphere surmounted by a cone. If the diameter of the hemisphere and cone are each 4 cm and the total length is 5 cm, find its total volume
4. A marquee is in the form of a cylinder surmounted by a cone. The total height is 6 m and the cylindrical portion has a height of 3.5 m, with a diameter of 15 m. Calculate the surface area of material needed to make the marquee assuming 12% of the material is wasted in the process
5. Determine (a) the volume and (b) the total surface area of the following metal solids:
   (i) a cone of radius 8.0 cm and perpendicular height 10 cm
   (ii) a sphere of diameter 7.0 cm
   (iii) a hemisphere of radius 3.0 cm
   (iv) a 2.5 cm by 2.5 cm square pyramid of perpendicular height 5.0 cm
   (v) a 4.0 cm by 6.0 cm rectangular pyramid of perpendicular height 12.0 cm
   (vi) a 4.2 cm by 4.2 cm square pyramid whose sloping edges are each 15.0 cm
   (vii) a pyramid having an octagonal base of side 5.0 cm and perpendicular height 20 cm
6. The volume of a spherical bearing is 325 cm^3. Determine its diameter
7. A metal sphere weighing 24 kg is melted down and recast into a solid cone of base radius 8.0 cm. If the density of the metal is 8000 kg/m^3 determine (a) the diameter of the metal sphere and (b) the perpendicular height of the cone, assuming that 15% of the metal is lost in the process

8. Find the volume of a regular hexagonal pyramid feature if the perpendicular height is 16.0 cm and the side of base is 3.0 cm

9. A buoy consists of a hemisphere surmounted by a cone. The diameter of the cone and hemisphere is 2.5 m and the slant height of the cone is 4.0 m. Determine the volume and surface area of the buoy

10. A petrol container is in the form of a central cylindrical portion 5.0 m long with a hemispherical section surmounted on each end. If the diameters of the hemisphere and cylinder are both 1.2 m determine the capacity of the tank in litres (1 litre = 1000 cm^3)

11. Figure 25.10 shows a metal rod section. Determine its volume and total surface area

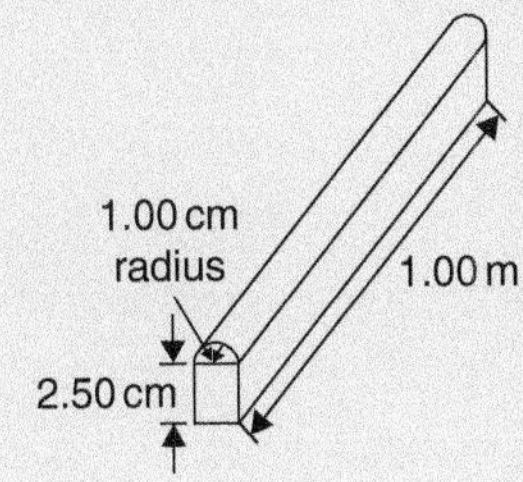

Figure 25.10

12. Find the volume (in cm^3) of the die-casting shown in Fig. 25.11. The dimensions are in millimetres

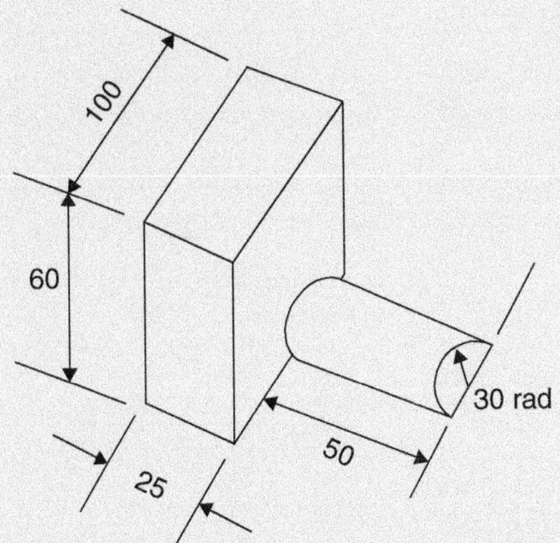

Figure 25.11

13. The cross-section of part of a circular ventilation shaft is shown in Fig. 25.12, ends AB and CD being open. Calculate (a) the volume of the air, correct to the nearest litre, contained in the part of the system shown, neglecting the sheet metal thickness, (given 1 litre = 1000 cm^3), (b) the cross-sectional area of the sheet metal used to make the system, in square metres, and (c) the cost of the sheet metal if the material costs £11.50 per square metre, assuming that 25% extra metal is required due to wastage

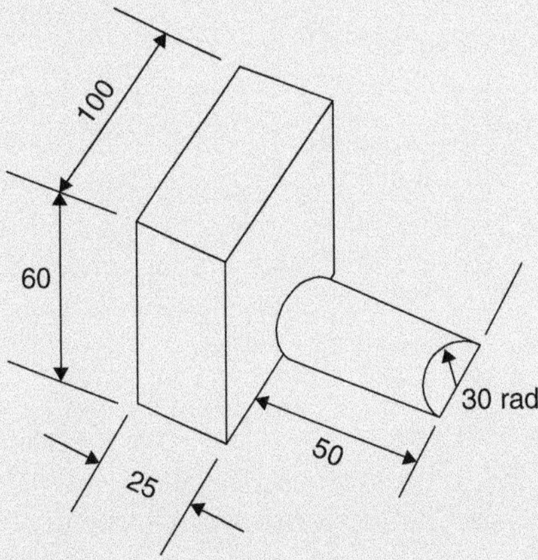

Figure 25.12

14. A spherical chemical storage tank has an internal diameter of 5.6 m. Calculate the storage capacity of the tank, correct to the nearest cubic metre. If 1 litre = 1000 cm^3, determine the tank capacity in litres.

15. In level ground, a drainage trench is excavated 0.5 m wide. The bottom is flat, and the sides are vertical, but the depth varies uniformly over a length of 20 m from a depth of 0.5 m at one end to 1.0 m at the other. Calculate the material dug out of the trench.

16. Each edge of a hexagonal steel plate is 0.32 m in length. The plate is 15 mm thick and has a density of 7800 kg/m^3. Calculate the mass of the steel plate.

17. A steel washer has external and internal diameters of 20 mm and 12 mm respectively and is 1 mm thick. The washers are stamped from sheet steel, a sheet measuring 1 m by 0.5 m by 1 mm thick. Assume the density of the steel as 7900 kg/m^3 The pattern of the stampings is shown in Figure 25.13.

Figure 25.13

Calculate

(a) the circular cross-sectional area of each washer
(b) the maximum number of washers which can be produced from one sheet of metal
(c) the mass of material wasted per sheet of steel
(d) the percentage of steel wasted per sheet
(e) the mass, in grams, of one washer
(f) the percentage profit per 1000 washers produced, assuming one sheet of steel costs £12, production and overhead costs are £300 per 100,000 washers, and washers are sold at £2 per 100.
Express the answers correct to 4 significant figures.

25.5 Volumes and surface areas of frusta of pyramids and cones

The **frustum** of a pyramid or cone is the portion remaining when a part containing the vertex is cut off by a plane parallel to the base.

The **volume of a frustum of a pyramid or cone** is given by the volume of the whole pyramid or cone minus the volume of the small pyramid or cone cut off.

The **surface area of the sides of a frustum of a pyramid or cone** is given by the surface area of the whole pyramid or cone minus the surface area of the small pyramid or cone cut off. This gives the lateral surface area of the frustum. If the total surface area of the frustum is required then the surface area of the two parallel ends are added to the lateral surface area.

There is an alternative method for finding the volume and surface area of a **frustum of a cone**. With reference to Fig. 25.14:

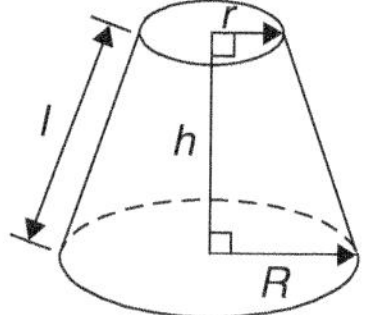

Figure 25.14

$$\textbf{Volume} = \frac{1}{3}\pi h(R^2 + Rr + r^2)$$

$$\textbf{Curved surface area} = \pi l(R + r)$$

$$\textbf{Total surface area} = \pi l(R + r) + \pi r^2 + \pi R^2$$

Problem 17. Determine the volume of a concrete frustum of a cone if the diameter of the ends are 6.0 cm and 4.0 cm and its perpendicular height is 3.6 cm

Method 1

A section through the vertex of a complete cone is shown in Fig. 25.15

Using similar triangles

$$\frac{AP}{DP} = \frac{DR}{BR}$$

Hence $$\frac{AP}{2.0} = \frac{3.6}{1.0}$$

from which $$AP = \frac{(2.0)(3.6)}{1.0} = 7.2\,\text{cm}$$

The height of the large cone $= 3.6 + 7.2 = 10.8$ cm.

Volume of frustum of cone

= volume of large cone

− volume of small cone cut off

$$= \frac{1}{3}\pi(3.0)^2(10.8) - \frac{1}{3}\pi(2.0)^2(7.2)$$

$$= 101.79 - 30.16 = \mathbf{71.6\,cm^3}$$

Method 2

From above, volume of the frustum of a cone

$$= \frac{1}{3}\pi h(R^2 + Rr + r^2)$$

where $R = 3.0$ cm,

$r = 2.0$ cm and $h = 3.6$ cm

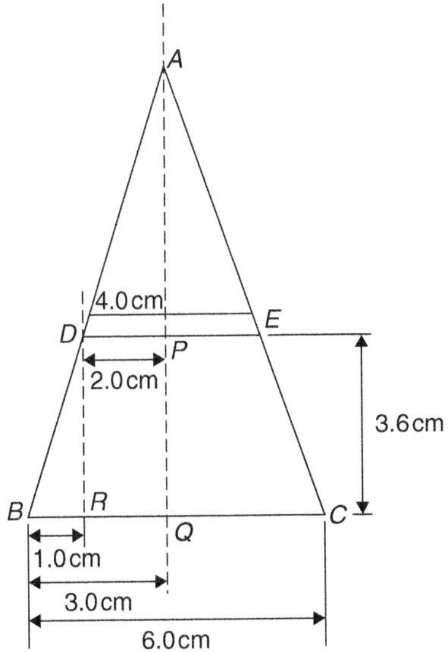

Figure 25.15

Hence volume of frustum

$$= \frac{1}{3}\pi(3.6)[(3.0)^2 + (3.0)(2.0) + (2.0)^2]$$

$$= \frac{1}{3}\pi(3.6)(19.0) = \mathbf{71.6\,cm^3}$$

Problem 18. Find the total surface area of the concrete frustum of the cone in Problem 17

Method 1

Curved surface area of frustum = curved surface area of large cone − curved surface area of small cone cut off.
From Fig. 25.15, using Pythagoras' theorem:

$$AB^2 = AQ^2 + BQ^2 \quad \text{from which}$$

$$AB = \sqrt{10.8^2 + 3.0^2} = 11.21\,\text{cm}$$

and $AD^2 = AP^2 + DP^2$ from which

$$AD = \sqrt{7.2^2 + 2.0^2} = 7.47\,\text{cm}$$

Curved surface area of large cone

$$= \pi rl = \pi(BQ)(AB) = \pi(3.0)(11.21)$$

$$= 105.65\,\text{cm}^2$$

and curved surface area of small cone

$$= \pi(DP)(AD) = \pi(2.0)(7.47) = 46.94\,\text{cm}^2$$

Hence, curved surface area of frustum

$$= 105.65 - 46.94$$

$$= 58.71\,\text{cm}^2$$

Total surface area of the concrete frustum

$$= \text{curved surface area} + \text{area of two circular ends}$$

$$= 58.71 + \pi(2.0)^2 + \pi(3.0)^2$$

$$= 58.71 + 12.57 + 28.27 = \mathbf{99.6\,cm^2}$$

Method 2

From page 266, total surface area of frustum

$$= \pi l(R + r) + \pi r^2 + \pi R^2$$

where $l = BD = 11.21 - 7.47 = 3.74\,\text{cm}$, $R = 3.0\,\text{cm}$ and $r = 2.0\,\text{cm}$.

Hence total surface area of frustum

$$= \pi(3.74)(3.0 + 2.0) + \pi(2.0)^2 + \pi(3.0)^2$$

$$= \mathbf{99.6\,cm^2}$$

Problem 19. A storage hopper is in the shape of a frustum of a pyramid. Determine its volume if the ends of the frustum are squares of sides 8.0 m and 4.6 m, respectively, and the perpendicular height between its ends is 3.6 m

The frustum is shown shaded in Fig. 25.16(a) as part of a complete pyramid. A section perpendicular to the base through the vertex is shown in Fig. 25.16(b)

By similar triangles: $\dfrac{CG}{BG} = \dfrac{BH}{AH}$

Height $CG = BG\left(\dfrac{BH}{AH}\right) = \dfrac{(2.3)(3.6)}{1.7} = 4.87\,\text{m}$

Height of complete pyramid $= 3.6 + 4.87 = 8.47\,\text{m}$

Volume of large pyramid $= \frac{1}{3}(8.0)^2(8.47)$

$$= 180.69\,\text{m}^3$$

Volume of small pyramid cut off

$$= \frac{1}{3}(4.6)^2(4.87) = 34.35\,\text{m}^3$$

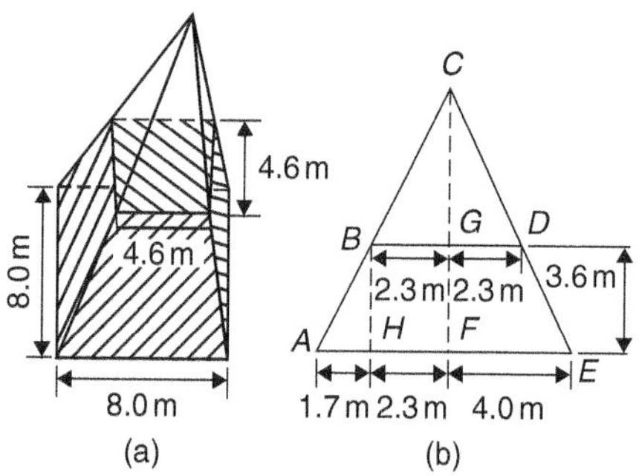

Figure 25.16

Hence volume of storage hopper

$$= 180.69 - 34.35 = \mathbf{146.3\,m^3}$$

Problem 20. Determine the lateral surface area of the storage hopper in Problem 19

The lateral surface area of the storage hopper consists of four equal trapeziums.
From Fig. 25.17, area of trapezium $PRSU$

$$= \frac{1}{2}(PR + SU)(QT)$$

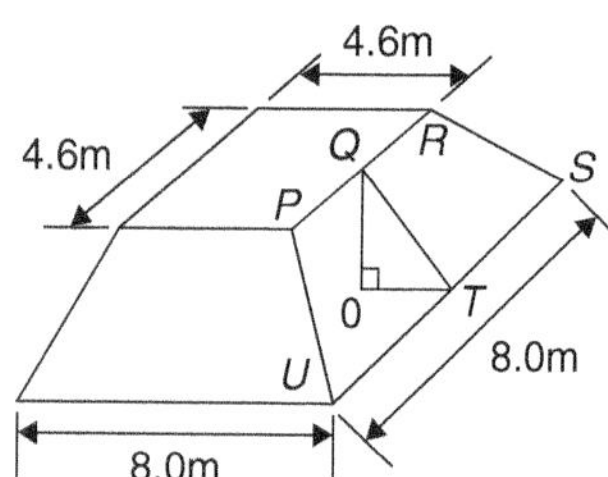

Figure 25.17

$OT = 1.7$ m (same as AH in Fig. 25.16(b)) and $OQ = 3.6$ m.
By Pythagoras' theorem,

$$QT = \sqrt{OQ^2 + OT^2} = \sqrt{3.6^2 + 1.7^2} = 3.98\,\text{m}$$

$$\text{Area of trapezium } PRSU = \frac{1}{2}(4.6 + 8.0)(3.98)$$
$$= 25.07\,\text{m}^2$$

$$\text{Lateral surface area of hopper} = 4(25.07)$$
$$= \mathbf{100.3\,m^2}$$

Problem 21. A lampshade is in the shape of a frustum of a cone. The vertical height of the shade is 25.0 cm and the diameters of the ends are 20.0 cm and 10.0 cm, respectively. Determine the area of the material needed to form the lampshade, correct to 3 significant figures

The curved surface area of a frustum of a cone = $\pi l(R + r)$ from page 266.
Since the diameters of the ends of the frustum are 20.0 cm and 10.0 cm, then from Fig. 25.18,

$$r = 5.0\,\text{cm}, R = 10.0\,\text{cm}$$

and $\quad l = \sqrt{25.0^2 + 5.0^2} = 25.50\,\text{cm},$

from Pythagoras' theorem.

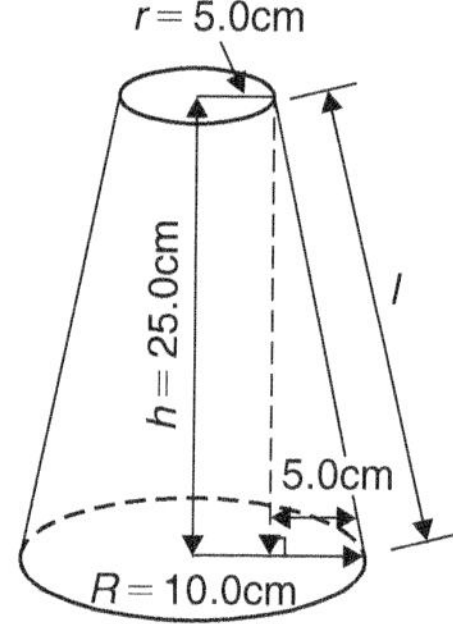

Figure 25.18

Hence curved surface area

$$= \pi(25.50)(10.0 + 5.0) = 1201.7\,\text{cm}^2$$

i.e. the area of material needed to form the lampshade is **1200 cm²**, correct to 3 significant figures.

Problem 22. A cooling tower is in the form of a cylinder surmounted by a frustum of a cone as shown in Fig. 25.19. Determine the volume of air space in the tower if 40% of the space is used for pipes and other structures

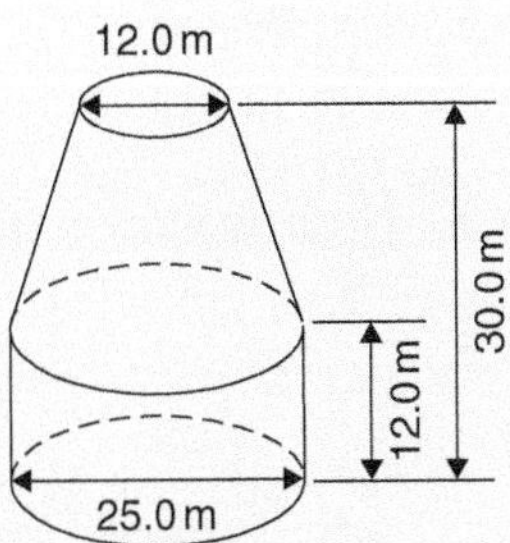

Figure 25.19

Volume of cylindrical portion

$$= \pi r^2 h = \pi \left(\frac{25.0}{2}\right)^2 (12.0) = 5890\,\text{m}^3$$

Volume of frustum of cone

$$= \frac{1}{3}\pi h(R^2 + Rr + r^2)$$

where $h = 30.0 - 12.0 = 18.0$ m,

$R = 25.0/2 = 12.5$ m and $r = 12.0/2 = 6.0$ m

Hence volume of frustum of cone

$$= \frac{1}{3}\pi(18.0)[(12.5)^2 + (12.5)(6.0) + (6.0)^2]$$

$$= 5038\,\text{m}^3$$

Total volume of cooling tower $= 5890 + 5038$

$$= 10928\,\text{m}^3$$

If 40% of space is occupied then volume of air space $= 0.6 \times 10928 =$ **6557 m³**

Now try the following Practice Exercise

Practice Exercise 130 Volumes and surface areas of frusta of pyramids and cones (Answers on page 714)

1. The radii of the faces of a solid metal frustum of a cone are 2.0 cm and 4.0 cm and the thickness of the frustum is 5.0 cm. Determine its volume and total surface area

2. A frustum of a wooden pyramid has square ends, the squares having sides 9.0 cm and 5.0 cm, respectively. Calculate the volume and total surface area of the frustum if the perpendicular distance between its ends is 8.0 cm

3. A cooling tower is in the form of a frustum of a cone. The base has a diameter of 32.0 m, the top has a diameter of 14.0 m and the vertical height is 24.0 m. Calculate the volume of the tower and the curved surface area

4. A loudspeaker diaphragm is in the form of a frustum of a cone. If the end diameters are 28.0 cm and 6.00 cm and the vertical distance between the ends is 30.0 cm, find the area of material needed to cover the curved surface of the speaker

5. A rectangular prism of metal having dimensions 4.3 cm by 7.2 cm by 12.4 cm is melted down and recast into a frustum of a square pyramid, 10% of the metal being lost in the process. If the ends of the frustum are squares of side 3 cm and 8 cm respectively, find the thickness of the frustum

6. Determine the volume and total surface area of a bucket consisting of an inverted frustum of a cone, of slant height 36.0 cm and end diameters 55.0 cm and 35.0 cm

7. A cylindrical tank of diameter 2.0 m and perpendicular height 3.0 m is to be replaced by a tank of the same capacity but in the form of a frustum of a cone. If the diameters of the ends of the frustum are 1.0 m and 2.0 m, respectively, determine the vertical height required

8. A concrete pillar is made up of two parts. The top section is a cylinder of radius 125 mm and height 750 mm. The base is a frustum of a cone with base radius 300 mm, top radius 125 mm and height 240 mm. Calculate the volume of the pillar in cubic centimetres and in cubic metres, each correct to 4 significant figures

25.6 The frustum and zone of a sphere

Volume of sphere $= \frac{4}{3}\pi r^3$ and the surface area of sphere $= 4\pi r^2$

A **frustum of a sphere** is the portion contained between two parallel planes. In Fig. 25.20, *PQRS* is a frustum of the sphere. A **zone of a sphere** is the curved surface of a frustum. With reference to Fig. 25.20:

Surface area of a zone of a sphere $= 2\pi rh$

Volume of frustum of sphere

$$= \frac{\pi h}{6}\left(h^2 + 3r_1^2 + 3r_2^2\right)$$

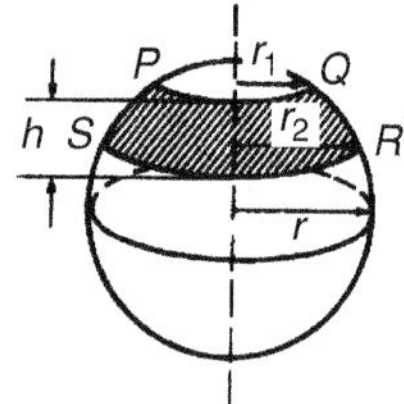

Figure 25.20

Problem 23. Determine the volume of a frustum of a sphere of diameter 49.74 cm if the diameter of the ends of the frustum are 24.0 cm and 40.0 cm, and the height of the frustum is 7.00 cm

From above, volume of frustum of a sphere

$$= \frac{\pi h}{6}(h^2 + 3r_1^2 + 3r_2^2)$$

where $h = 7.00$ cm, $r_1 = 24.0/2 = 12.0$ cm and $r_2 = 40.0/2 = 20.0$ cm.
Hence volume of frustum

$$= \frac{\pi(7.00)}{6}[(7.00)^2 + 3(12.0)^2 + 3(20.0)^2]$$

$$= \mathbf{6161\,cm^3}$$

Problem 24. Determine for the frustum of Problem 23 the curved surface area of the frustum

The curved surface area of the frustum = surface area of zone $= 2\pi rh$ (from above), where r = radius of sphere $= 49.74/2 = 24.87$ cm and $h = 7.00$ cm. Hence, surface area of zone $= 2\pi(24.87)(7.00) = \mathbf{1094\,cm^2}$

Problem 25. The diameters of the ends of the frustum of a sphere are 14.0 cm and 26.0 cm respectively, and the thickness of the frustum is 5.0 cm. Determine, correct to 3 significant figures (a) the volume of the frustum of the sphere, (b) the radius of the sphere and (c) the area of the zone formed

The frustum is shown shaded in the cross-section of Fig. 25.21

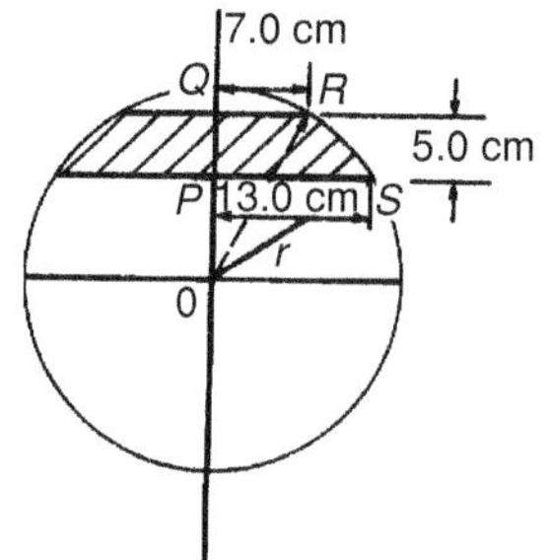

Figure 25.21

(a) Volume of frustum of sphere

$$= \frac{\pi h}{6}(h^2 + 3r_1^2 + 3r_2^2)$$

from above, where $h = 5.0$ cm, $r_1 = 14.0/2 = 7.0$ cm and $r_2 = 26.0/2 = 13.0$ cm.
Hence volume of frustum of sphere

$$= \frac{\pi(5.0)}{6}[(5.0)^2 + 3(7.0)^2 + 3(13.0)^2]$$

$$= \frac{\pi(5.0)}{6}[25.0 + 147.0 + 507.0]$$

$= \mathbf{1780\,cm^3}$ correct to 3 significant figures.

(b) The radius, r, of the sphere may be calculated using Fig. 25.21. Using Pythagoras' theorem:

$$OS^2 = PS^2 + OP^2$$

i.e. $$r^2 = (13.0)^2 + OP^2 \qquad (1)$$

$$OR^2 = QR^2 + OQ^2$$

i.e. $$r^2 = (7.0)^2 + OQ^2$$

However $OQ = QP + OP = 5.0 + OP$, therefore

$$r^2 = (7.0)^2 + (5.0 + OP)^2 \qquad (2)$$

Equating equations (1) and (2) gives:

$$(13.0)^2 + OP^2 = (7.0)^2 + (5.0 + OP)^2$$

$$169.0 + OP^2 = 49.0 + 25.0 + 10.0(OP) + OP^2$$

$$169.0 = 74.0 + 10.0(OP)$$

Hence

$$OP = \frac{169.0 - 74.0}{10.0} = 9.50\,\text{cm}$$

Substituting $OP = 9.50\,\text{cm}$ into equation (1) gives:

$$r^2 = (13.0)^2 + (9.50)^2$$

from which $r = \sqrt{13.0^2 + 9.50^2}$

i.e. **radius of sphere, $r = 16.1\,\text{cm}$**

(c) Area of zone of sphere

$$= 2\pi rh = 2\pi(16.1)(5.0)$$

$= \mathbf{506\,cm^2}$, correct to 3 significant figures.

Problem 26. A frustum of a sphere of diameter 12.0 cm is formed by two parallel planes, one through the diameter and the other distance h from the diameter. The curved surface area of the frustum is required to be $\frac{1}{4}$ of the total surface area of the sphere. Determine (a) the volume and surface area of the sphere, (b) the thickness h of the frustum, (c) the volume of the frustum and (d) the volume of the frustum expressed as a percentage of the sphere

(a) Volume of sphere,

$$V = \frac{4}{3}\pi r^3 = \frac{4}{3}\pi\left(\frac{12.0}{2}\right)^3$$

$$= \mathbf{904.8\,cm^3}$$

Surface area of sphere

$$= 4\pi r^2 = 4\pi\left(\frac{12.0}{2}\right)^2$$

$$= \mathbf{452.4\,cm^2}$$

(b) Curved surface area of frustum

$$= \tfrac{1}{4} \times \text{surface area of sphere}$$

$$= \tfrac{1}{4} \times 452.4 = 113.1\,\text{cm}^2$$

From above,

$$113.1 = 2\pi rh = 2\pi\left(\frac{12.0}{2}\right)h$$

Hence thickness of frustum

$$h = \frac{113.1}{2\pi(6.0)} = \mathbf{3.0\,cm}$$

(c) Volume of frustum,

$$V = \frac{\pi h}{6}(h^2 + 3r_1^2 + 3r_2^2)$$

where $h = 3.0\,\text{cm}$, $r_2 = 6.0\,\text{cm}$ and

$$r_1 = \sqrt{OQ^2 - OP^2}$$

from Fig. 25.22,

i.e. $r_1 = \sqrt{6.0^2 - 3.0^2} = 5.196\,\text{cm}$

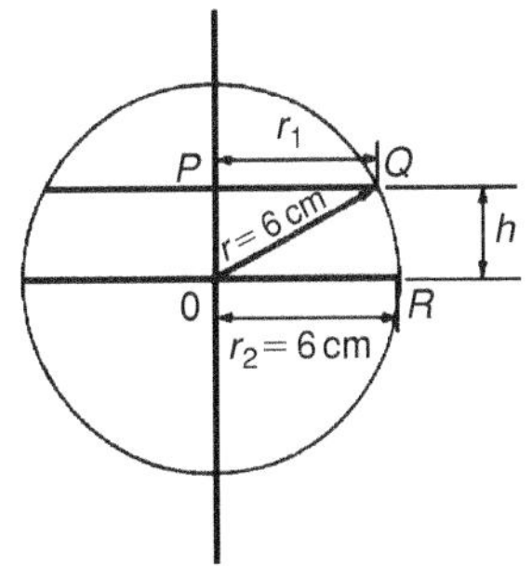

Figure 25.22

Hence volume of frustum

$$= \frac{\pi(3.0)}{6}[(3.0)^2 + 3(5.196)^2 + 3(6.0)^2]$$

$$= \frac{\pi}{2}[9.0 + 81 + 108.0] = \mathbf{311.0\,cm^3}$$

(d) $$\frac{\text{Volume of frustum}}{\text{Volume of sphere}} = \frac{311.0}{904.8} \times 100\%$$

$$= \mathbf{34.37\%}$$

Problem 27. A spherical storage tank is filled with liquid to a depth of 20 cm. If the internal diameter of the vessel is 30 cm, determine the number of litres of liquid in the container (1 litre $= 1000\,\text{cm}^3$)

The liquid is represented by the shaded area in the section shown in Fig. 25.23. The volume of liquid comprises a hemisphere and a frustum of thickness 5 cm. Hence volume of liquid

$$= \frac{2}{3}\pi r^3 + \frac{\pi h}{6}[h^2 + 3r_1^2 + 3r_2^2]$$

where $r_2 = 30/2 = 15\,\text{cm}$ and

$$r_1 = \sqrt{15^2 - 5^2} = 14.14\,\text{cm}$$

Volume of liquid

$$= \frac{2}{3}\pi(15)^3 + \frac{\pi(5)}{6}[5^2 + 3(14.14)^2 + 3(15)^2]$$

$$= 7069 + 3403 = 10470\,\text{cm}^3$$

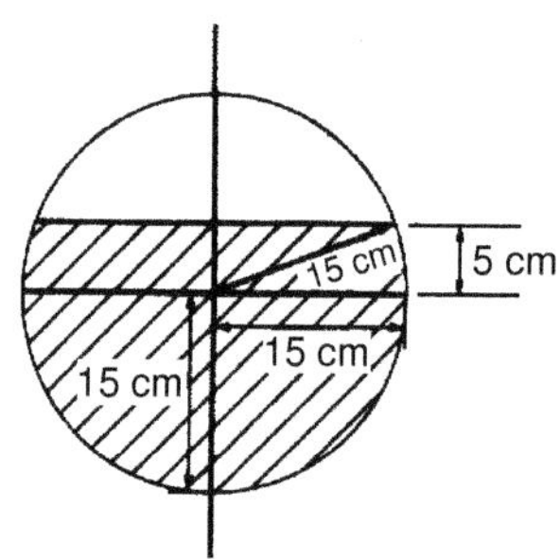

Figure 25.23

Since 1 litre $= 1000\,\text{cm}^3$, the number of litres of liquid

$$= \frac{10470}{1000} = \mathbf{10.47\ litres}$$

Now try the following Practice Exercise

Practice Exercise 131 Frustums and zones of spheres (Answers on page 715)

1. Determine the volume and surface area of a frustum of a sphere of diameter 47.85 cm, if the radii of the ends of the frustum are 14.0 cm and 22.0 cm and the height of the frustum is 10.0 cm
2. Determine the volume (in cm^3) and the surface area (in cm^2) of a frustum of a sphere if the diameter of the ends are 80.0 mm and 120.0 mm and the thickness is 30.0 mm
3. A sphere has a radius of 6.50 cm. Determine its volume and surface area. A frustum of the sphere is formed by two parallel planes, one through the diameter and the other at a distance h from the diameter. If the curved surface area of the frustum is to be $\frac{1}{5}$ of the surface area of the sphere, find the height h and the volume of the frustum
4. A sphere has a diameter of 32.0 mm. Calculate the volume (in cm^3) of the frustum of the sphere contained between two parallel planes distances 12.0 mm and 10.00 mm from the centre and on opposite sides of it
5. A spherical storage tank is filled with liquid to a depth of 30.0 cm. If the inner diameter of the vessel is 45.0 cm determine the number of litres of liquid in the container (1 litre $= 1000\,\text{cm}^3$)

25.7 Prismoidal rule

The prismoidal rule applies to a solid of length x divided by only three equidistant plane areas, A_1, A_2 and A_3 as shown in Fig. 25.24 and is merely an extension of Simpson's rule (see Chapter 26) – but for volumes.

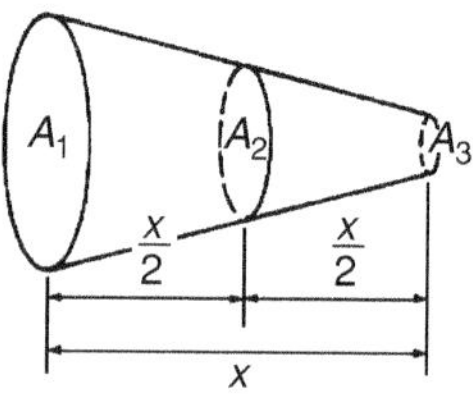

Figure 25.24

With reference to Fig. 25.24

$$\textbf{Volume, } V = \frac{x}{6}[A_1 + 4A_2 + A_3]$$

The prismoidal rule gives precise values of volume for regular solids such as pyramids, cones, spheres and prismoids.

Problem 28. A container is in the shape of a frustum of a cone. Its diameter at the bottom is 18 cm and at the top 30 cm. If the depth is 24 cm determine the capacity of the container, correct to the nearest litre, by the prismoidal rule.
(1 litre $= 1000\,\text{cm}^3$)

The container is shown in Fig. 25.25. At the mid-point, i.e. at a distance of 12 cm from one end, the radius r_2 is $(9 + 15)/2 = 12$ cm, since the sloping side changes uniformly.

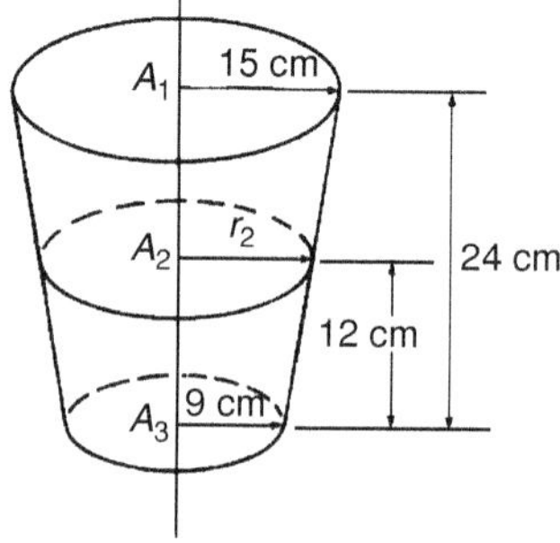

Figure 25.25

Volume of container by the prismoidal rule

$$= \frac{x}{6}[A_1 + 4A_2 + A_3]$$

from above, where $x = 24\,\text{cm}$, $A_1 = \pi(15)^2\,\text{cm}^2$, $A_2 = \pi(12)^2\,\text{cm}^2$ and $A_3 = \pi(9)^2\,\text{cm}^2$
Hence volume of container

$$= \frac{24}{6}[\pi(15)^2 + 4\pi(12)^2 + \pi(9)^2]$$
$$= 4[706.86 + 1809.56 + 254.47]$$
$$= 11\,080\,\text{cm}^3 = \frac{11\,080}{1000}\text{ litres}$$
$$= \textbf{11 litres, correct to the nearest litre}$$

(Check: Volume of frustum of cone

$$= \tfrac{1}{3}\pi h[R^2 + Rr + r^2] \quad \text{from Section 25.5}$$
$$= \tfrac{1}{3}\pi(24)[(15)^2 + (15)(9) + (9)^2]$$
$$= 11\,080\,\text{cm}^3 \text{(as shown above)}$$

Problem 29. A frustum of a sphere of radius 13 cm is formed by two parallel planes on opposite sides of the centre, each at distance of 5 cm from the centre. Determine the volume of the frustum (a) by using the prismoidal rule, and (b) by using the formula for the volume of a frustum of a sphere

The frustum of the sphere is shown by the section in Fig. 25.26.

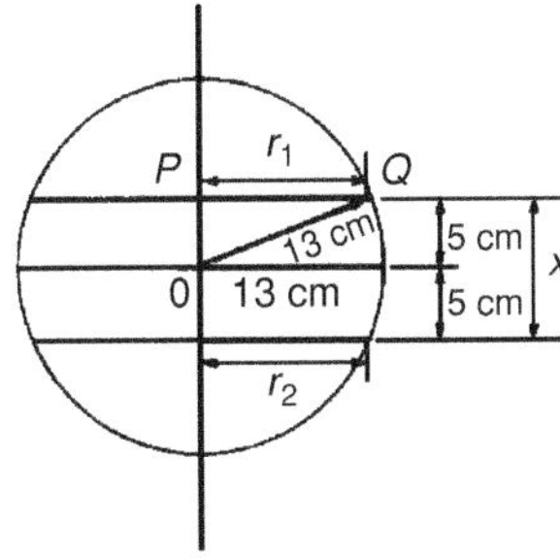

Figure 25.26

Radius $r_1 = r_2 = PQ = \sqrt{13^2 - 5^2} = 12\,\text{cm}$, by Pythagoras' theorem.

(a) Using the prismoidal rule, volume of frustum,

$$V = \frac{x}{6}[A_1 + 4A_2 + A_3]$$
$$= \frac{10}{6}[\pi(12)^2 + 4\pi(13)^2 + \pi(12)^2]$$
$$= \frac{10\pi}{6}[144 + 676 + 144] = \mathbf{5047\,cm^3}$$

(b) Using the formula for the volume of a frustum of a sphere:

$$\text{Volume} \quad V = \frac{\pi h}{6}(h^2 + 3r_1^2 + 3r_2^2)$$
$$= \frac{\pi(10)}{6}[10^2 + 3(12)^2 + 3(12)^2]$$
$$= \frac{10\pi}{6}(100 + 432 + 432)$$
$$= \mathbf{5047\,cm^3}$$

Problem 30. A hole is to be excavated in the form of a prismoid. The bottom is to be a rectangle 16 m long by 12 m wide; the top is also a rectangle, 26 m long by 20 m wide. Find the volume of earth to be removed, correct to 3 significant figures, if the depth of the hole is 6.0 m

The prismoid is shown in Fig. 25.27. Let A_1 represent the area of the top of the hole, i.e. $A_1 = 20 \times 26 = 520\,\text{m}^2$. Let A_3 represent the area of the bottom of the hole, i.e. $A_3 = 16 \times 12 = 192\,\text{m}^2$. Let A_2 represent the rectangular area through the middle of the hole parallel to areas A_1 and A_3. The length of this rectangle is $(26 + 16)/2 = 21\,\text{m}$ and the width is $(20 + 12)/2 = 16\,\text{m}$, assuming the sloping edges are uniform. Thus area $A_2 = 21 \times 16 = 336\,\text{m}^2$.

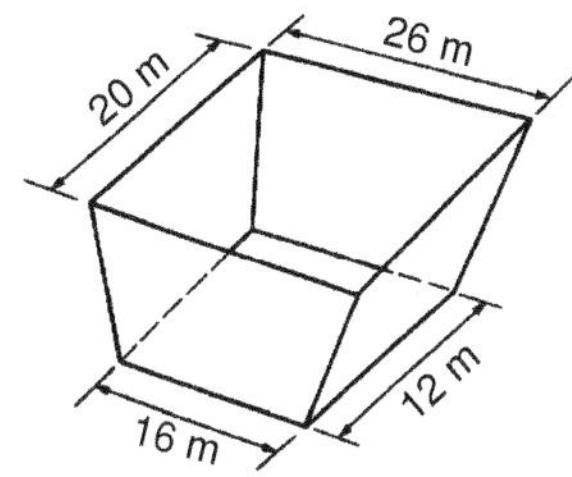

Figure 25.27

Using the prismoidal rule,

$$\text{volume of hole} = \frac{x}{6}[A_1 + 4A_2 + A_3]$$
$$= \frac{6}{6}[520 + 4(336) + 192]$$
$$= 2056\,\text{m}^3 = \mathbf{2060\,m^3},$$

correct to 3 significant figures.

Problem 31. The roof of a building is in the form of a frustum of a pyramid with a square base of side 5.0 m. The flat top is a square of side 1.0 m and all the sloping sides are pitched at the same angle. The vertical height of the flat top above the level of the eaves is 4.0 m. Calculate, using the prismoidal rule, the volume enclosed by the roof

Let area of top of frustum be $A_1 = (1.0)^2 = 1.0\,\text{m}^2$
Let area of bottom of frustum be $A_3 = (5.0)^2 = 25.0\,\text{m}^2$
Let area of section through the middle of the frustum parallel to A_1 and A_3 be A_2. The length of the side of the square forming A_2 is the average of the sides forming A_1 and A_3, i.e. $(1.0 + 5.0)/2 = 3.0\,\text{m}$.
Hence $A_2 = (3.0)^2 = 9.0\,\text{m}^2$.
Using the prismoidal rule,

$$\text{volume of frustum} = \frac{x}{6}[A_1 + 4A_2 + A_3]$$

$$= \frac{4.0}{6}[1.0 + 4(9.0) + 25.0]$$

Hence, **volume enclosed by roof = 41.3 m^3**

Now try the following Practice Exercise

Practice Exercise 132 The prismoidal rule (Answers on page 715)

1. Use the prismoidal rule to find the volume of a frustum of a spherical ball contained between two parallel planes on opposite sides of the centre each of radius 7.0 cm and each 4.0 cm from the centre
2. Determine the volume of a metal cone of perpendicular height 16.0 cm and base diameter 10.0 cm by using the prismoidal rule
3. A bucket is in the form of a frustum of a cone. The diameter of the base is 28.0 cm and the diameter of the top is 42.0 cm. If the height is 32.0 cm, determine the capacity of the bucket (in litres) using the prismoidal rule (1 litre = 1000 cm^3)
4. Determine the capacity of a water reservoir, in litres, the top being a 30.0 m by 12.0 m rectangle, the bottom being a 20.0 m by 8.0 m rectangle and the depth being 5.0 m (1 litre = 1000 cm^3)

25.8 Volumes of similar shapes

The volumes of similar bodies are proportional to the cubes of corresponding linear dimensions.
For example, Fig. 25.28 shows two cubes, one of which has sides three times as long as those of the other.

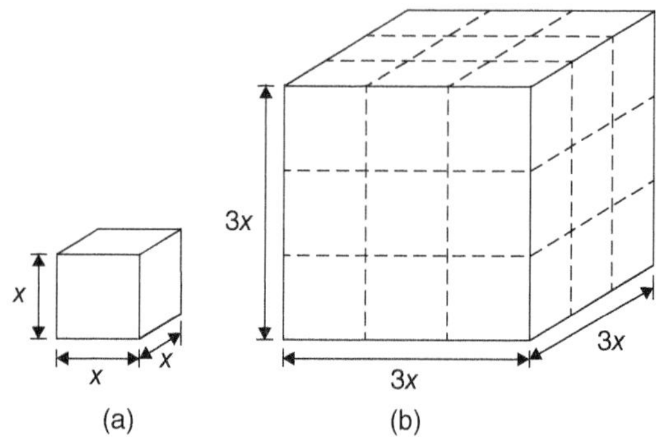

Figure 25.28

Volume of Fig. 25.28(a) $= (x)(x)(x) = x^3$
Volume of Fig. 25.28(b) $= (3x)(3x)(3x) = 27x^3$
Hence Fig. 25.28(b) has a volume $(3)^3$, i.e. 27 times the volume of Fig. 25.28(a).

Problem 32. A car has a mass of 1000 kg. A model of the car is made to a scale of 1 to 50. Determine the mass of the model if the car and its model are made of the same material

$$\frac{\text{Volume of model}}{\text{Volume of car}} = \left(\frac{1}{50}\right)^3$$

since the volume of similar bodies are proportional to the cube of corresponding dimensions.
Mass = density × volume, and since both car and model are made of the same material then:

$$\frac{\text{Mass of model}}{\text{Mass of car}} = \left(\frac{1}{50}\right)^3$$

$$\text{Hence mass of model} = (\text{mass of car})\left(\frac{1}{50}\right)^3$$

$$= \frac{1000}{50^3}$$

$$= \mathbf{0.008\,kg} \text{ or } \mathbf{8\,g}$$

Now try the following Practice Exercise

Practice Exercise 133 Volumes of similar shapes (Answers on page 715)

1. The diameter of two spherical bearings are in the ratio 2:5. What is the ratio of their volumes?
2. An engineering component has a mass of 400 g. If each of its dimensions are reduced by 30% determine its new mass

Practice Exercise 134 Multiple-choice questions on volumes and surface areas of common solids (Answers on page 715)

Each question has only one correct answer

1. A water tank is in the shape of a rectangular prism having length 1.5 m, breadth 60 cm and height 300 mm. If 1 litre = 1000 cm^3, the capacity of the tank is:

   (a) 27 litre (b) 2.7 litre

   (c) 2700 litre (d) 270 litre
2. A hollow shaft has an outside diameter of 6.0 cm and an inside diameter of 4.0 cm. The cross- sectional area of the shaft is:

   (a) 6283 mm^2 (b) 1257 mm^2

   (c) 1571 mm^2 (d) 628 mm^2
3. The surface area of a sphere of diameter 40 mm is:

   (a) 50.27 cm^2 (b) 33.51 cm^2

   (c) 268.08 cm^2 (d) 201.06 cm^2
4. The total surface area of a cylinder of length 20 cm and diameter 6 cm is:

   (a) 56.55 cm^2 (b) 433.54 cm^2

   (c) 980.18 cm^2 (d) 226.19 cm^2
5. The length of each edge of a cube is d. The surface area of the cube is given by:

   (a) $7d^2$ (b) $5d^2$

   (c) $6d^2$ (d) $4d^2$
6. The total surface area of a solid hemisphere of diameter 6.0 cm is:

   (a) 84.82 cm^2 (b) 339.3 cm^2

   (c) 226.2 cm^2 (d) 56.55 cm^2
7. A cylindrical copper pipe, 1.8 m long, has an outside diameter of 300 mm and an inside diameter of 180 mm. The volume of copper in the pipe, in cubic metres is:

   (a) 0.3257 m^2 (b) 0.0814 m^2

   (c) 8.143 m^2 (d) 814.3 m^2
8. A cylinder and a cone have the same base radius and the same height. The ratio of the volume of the cylinder to that of the cone is:

   (a) 3:2 (b) 2:1

   (c) 3:1 (d) 2:3
9. A cube has a volume of 27 cm^3. The length of one edge is:

   (a) 2.25 cm (b) 6.25 cm

   (c) 9 cm (d) 3 cm
10. A vehicle has a mass of 2000 kg. A model of the vehicle is made to a scale of 1 to 100. If the vehicle and model are made of the same material, the mass of the model is:

    (a) 2 g (b) 20 kg

    (c) 200 g (d) 20 g

For fully worked solutions to each of the problems in Practice Exercises 128 to 133 in this chapter, go to the website:
www.routledge.com/cw/bird

Chapter 26

Irregular areas and volumes and mean values of waveforms

Why it is important to understand: Irregular areas and volumes and mean values of waveforms

Surveyors, farmers and landscapers often need to determine the area of irregularly shaped pieces of land to work with the land properly. There are many applications in business, economics and the sciences, including all aspects of engineering, where finding the areas of irregular shapes, the volumes of solids and the lengths of irregular shaped curves are important. Typical earthworks include roads, railway beds, causeways, dams and canals. Other common earthworks are land grading to reconfigure the topography of a site or to stabilise slopes. Engineers need to concern themselves with issues of geotechnical engineering (such as soil density and strength) and with quantity estimation to ensure that soil volumes in the cuts match those of the fills, while minimizing the distance of movement. Simpson's rule is a staple of scientific data analysis and engineering; it is widely used, for example, by naval architects to numerically determine hull offsets and cross-sectional areas to determine volumes and centroids of ships or lifeboats. There are therefore plenty of examples where irregular areas and volumes need to be determined by engineers.

At the end of this chapter, you should be able to:

- use the trapezoidal rule to determine irregular areas
- use the mid-ordinate rule to determine irregular areas
- use Simpson's rule to determine irregular areas
- estimate the volume of irregular solids
- determine the mean values of waveforms

26.1 Area of irregular figures

Area of irregular plane surfaces may be approximately determined by using (a) a planimeter, (b) the trapezoidal rule, (c) the mid-ordinate rule and (d) Simpson's rule. Such methods may be used, for example, by engineers estimating areas of indicator diagrams of steam engines, surveyors estimating areas of plots of land or naval architects estimating areas of water planes or transverse sections of ships.

(a) A **planimeter** is an instrument for directly measuring small areas bounded by an irregular curve.

(b) **Trapezoidal rule**
To determine the areas *PQRS* in Fig. 26.1:

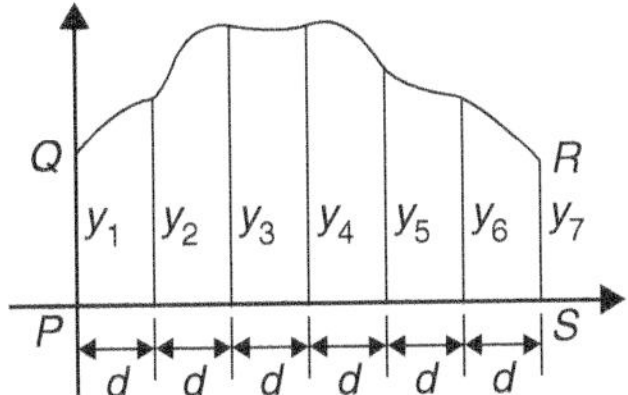

Figure 26.1

(i) Divide base *PS* into any number of equal intervals, each of width d (the greater the number of intervals, the greater the accuracy).
(ii) Accurately measure ordinates y_1, y_2, y_3, etc.
(iii) Area *PQRS*

$$\approx d\left[\frac{y_1+y_7}{2}+y_2+y_3+y_4+y_5+y_6\right]$$

In general, the trapezoidal rule states:

$$\text{Area} \approx \begin{pmatrix}\text{width of}\\ \text{interval}\end{pmatrix}\left[\frac{1}{2}\begin{pmatrix}\text{first + last}\\ \text{ordinate}\end{pmatrix}+\begin{pmatrix}\text{sum of}\\ \text{remaining}\\ \text{ordinates}\end{pmatrix}\right]$$

(c) **Mid-ordinate rule**
To determine the area *ABCD* of Fig. 26.2:

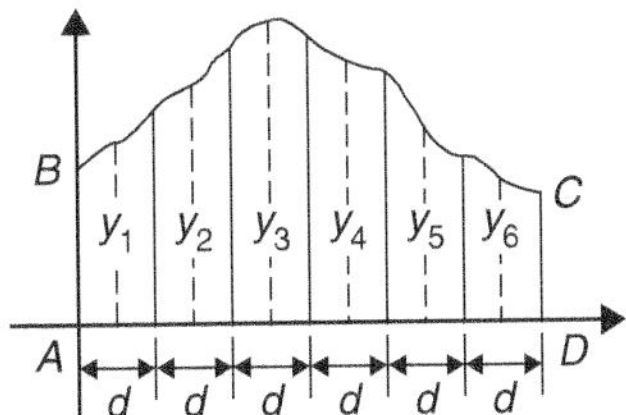

Figure 26.2

(i) Divide base *AD* into any number of equal intervals, each of width d (the greater the number of intervals, the greater the accuracy).
(ii) Erect ordinates in the middle of each interval (shown by broken lines in Fig. 26.2).
(iii) Accurately measure ordinates y_1, y_2, y_3, etc.
(iv) Area *ABCD*

$$\approx d(y_1+y_2+y_3+y_4+y_5+y_6).$$

In general, the mid-ordinate rule states:

$$\text{Area} \approx \begin{pmatrix}\text{width of}\\ \text{interval}\end{pmatrix}\begin{pmatrix}\text{sum of}\\ \text{mid-ordinates}\end{pmatrix}$$

(d) **Simpson's rule***

To determine the area *PQRS* of Fig. 26.1:

(i) Divide base *PS* into an **even** number of intervals, each of width d (the greater the number of intervals, the greater the accuracy).
(ii) Accurately measure ordinates y_1, y_2, y_3, etc.
(iii) Area *PQRS*

$$\approx \frac{d}{3}[(y_1+y_7)+4(y_2+y_4+y_6) + 2(y_3+y_5)]$$

*Who was Simpson? – **Thomas Simpson** FRS (20 August 1710–14 May 1761) was the British mathematician who invented Simpson's rule to approximate definite integrals. To find out more go to www.routledge.com/cw/bird

In general, Simpson's rule states:

$$\textbf{Area} \approx \frac{1}{3}\left(\begin{array}{c}\textbf{width of}\\ \textbf{interval}\end{array}\right)\left[\left(\begin{array}{c}\textbf{first+last}\\ \textbf{ordinate}\end{array}\right) + 4\left(\begin{array}{c}\textbf{sum of even}\\ \textbf{ordinates}\end{array}\right) + 2\left(\begin{array}{c}\textbf{sum of remaining}\\ \textbf{odd ordinates}\end{array}\right)\right]$$

Problem 1. A car starts from rest and its speed is measured every second for 6 s:

Time t(s)	0	1	2	3	4	5	6
Speed v (m/s)	0	2.5	5.5	8.75	12.5	17.5	24.0

Determine the distance travelled in 6 seconds (i.e. the area under the v/t graph), by (a) the trapezoidal rule, (b) the mid-ordinate rule, and (c) Simpson's rule

A graph of speed/time is shown in Fig. 26.3.

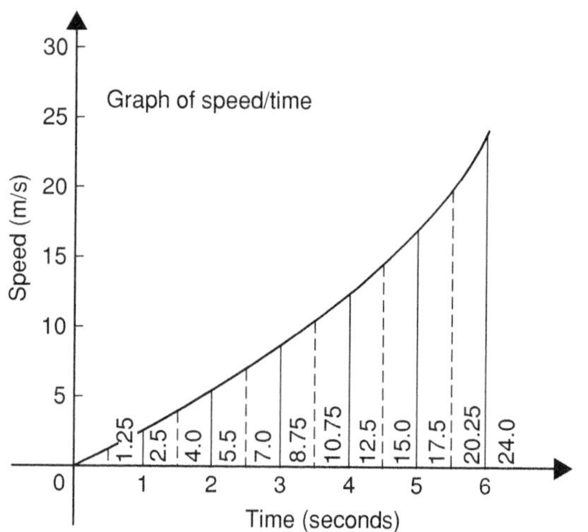

Figure 26.3

(a) **Trapezodial rule** (see para. (b) above).

The time base is divided into 6 strips each of width 1 s, and the length of the ordinates measured. Thus

$$\textbf{area} \approx (1)\left[\left(\frac{0+24.0}{2}\right) + 2.5 + 5.5 + 8.75 + 12.5 + 17.5\right] = \textbf{58.75 m}$$

(b) **Mid-ordinate rule** (see para. (c) above).

The time base is divided into 6 strips each of width 1 second. Mid-ordinates are erected as shown in Fig. 26.3 by the broken lines. The length of each mid-ordinate is measured. Thus

$$\textbf{area} \approx (1)[1.25 + 4.0 + 7.0 + 10.75 + 15.0 + 20.25] = \textbf{58.25 m}$$

(c) **Simpson's rule** (see para. (d) above).

The time base is divided into 6 strips each of width 1 s, and the length of the ordinates measured. Thus

$$\textbf{area} \approx \frac{1}{3}(1)[(0+24.0) + 4(2.5+8.75 + 17.5) + 2(5.5+12.5)] = \textbf{58.33 m}$$

Problem 2. A river is 15 m wide. Soundings of the depth are made at equal intervals of 3 m across the river and are as shown below.

Depth (m)	0	2.2	3.3	4.5	4.2	2.4	0

Calculate the cross-sectional area of the flow of water at this point using Simpson's rule

From para. (d) above,

$$\text{Area} \approx \frac{1}{3}(3)[(0+0) + 4(2.2+4.5+2.4) + 2(3.3+4.2)]$$

$$= (1)[0 + 36.4 + 15] = \textbf{51.4 m}^{\mathbf{2}}$$

Now try the following Practice Exercise

Practice Exercise 135 Areas of irregular figures (Answers on page 715)

1. Plot a graph of $y=3x-x^2$ by completing a table of values of y from $x=0$ to $x=3$. Determine the area enclosed by the curve, the x-axis and ordinate $x=0$ and $x=3$ by (a) the trapezoidal rule, (b) the mid-ordinate rule and (c) by Simpson's rule
Use 6 intervals in each case.
2. Plot the graph of $y=2x^2+3$ between $x=0$ and $x=4$. Estimate the area enclosed by the curve, the ordinates $x=0$ and $x=4$, and the x-axis by an approximate method.

3. The velocity of a car at one second intervals is given in the following table:

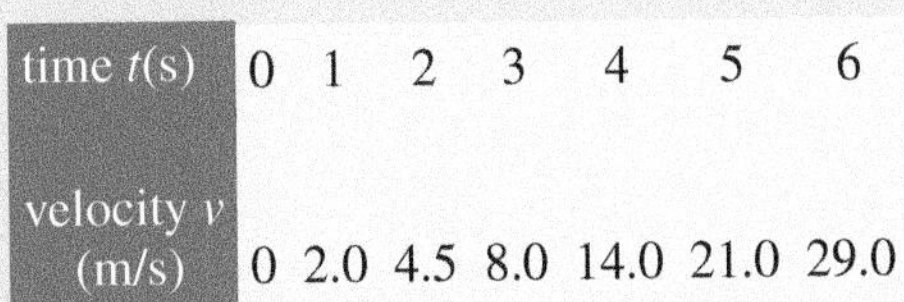

time t(s)	0	1	2	3	4	5	6
velocity v (m/s)	0	2.0	4.5	8.0	14.0	21.0	29.0

Determine the distance travelled in 6 seconds (i.e. the area under the v/t graph) using Simpson's rule

4. The shape of a piece of land is shown in Fig. 26.4. To estimate the area of the land, a surveyor takes measurements at intervals of 50 m, perpendicular to the straight portion with the results shown (the dimensions being in metres). Estimate the area of the land in hectares (1 ha $= 10^4\,\text{m}^2$)

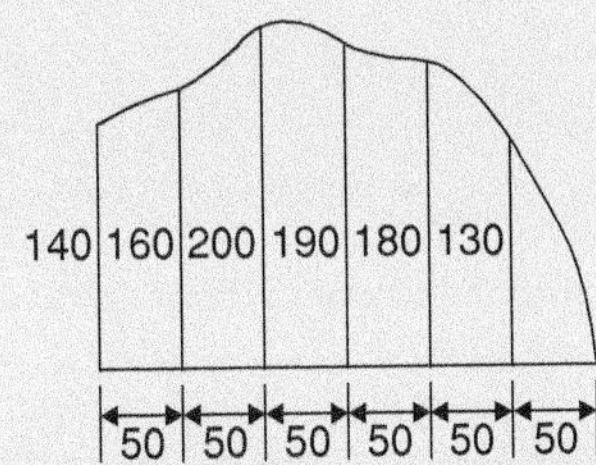

Figure 26.4

5. The deck of a ship is 35 m long. At equal intervals of 5 m the width is given by the following table:

Width (m)	0	2.8	5.2	6.5	5.8	4.1	3.0	2.3

Estimate the area of the deck

26.2 Volumes of irregular solids

If the cross-sectional areas $A_1, A_2, A_3, \ldots$ of an irregular solid bounded by two parallel planes are known at equal intervals of width d (as shown in Fig. 26.5), then by Simpson's rule:

$$\text{Volume, } V \approx \frac{d}{3}\left[\begin{array}{r}(A_1+A_7)+4(A_2+A_4+A_6)\\ +2(A_3+A_5)\end{array}\right]$$

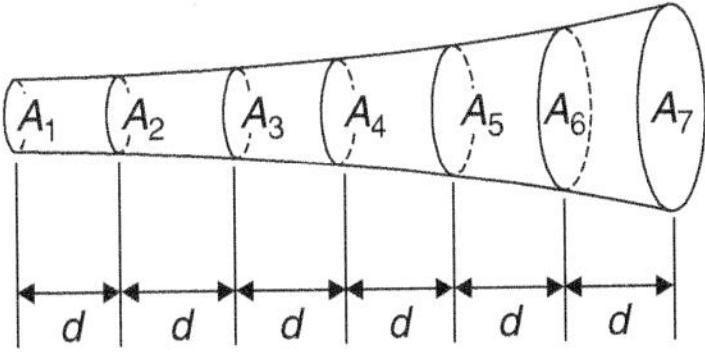

Figure 26.5

Problem 3. A tree trunk is 12 m in length and has a varying cross-section. The cross-sectional areas at intervals of 2 m measured from one end are:

0.52, 0.55, 0.59, 0.63, 0.72, 0.84, 0.97 m²

Estimate the volume of the tree trunk

A sketch of the tree trunk is similar to that shown in Fig. 26.5, where $d = 2\,\text{m}$, $A_1 = 0.52\,\text{m}^2$, $A_2 = 0.55\,\text{m}^2$ and so on.

Using Simpson's rule for volumes gives:

$$\text{Volume} \approx \frac{2}{3}[(0.52+0.97)+4(0.55+0.63+0.84)+2(0.59+0.72)]$$

$$= \frac{2}{3}[1.49+8.08+2.62] = \mathbf{8.13\,m^3}$$

Problem 4. The areas of seven horizontal cross-sections of a water reservoir at intervals of 10 m are:

210, 250, 320, 350, 290, 230, 170 m²

Calculate the capacity of the reservoir in litres

Using Simpson's rule for volumes gives:

$$\text{Volume} \approx \frac{10}{3}[(210+170)+4(250+350+230)+2(320+290)]$$

$$= \frac{10}{3}[380+3320+1220]$$

$$= \mathbf{16400\,m^3}$$

$$16400\,\text{m} = 16400 \times 10^6\,\text{cm}^3$$

Since 1 litre $= 1000\,\text{cm}^3$, capacity of reservoir

$$= \frac{16400 \times 10^6}{1000}\ \text{litres}$$

$$= 16400000 = \mathbf{1.64 \times 10^7\ litres}$$

Now try the following Practice Exercise

Practice Exercise 136 Volumes of irregular solids (Answers on page 715)

1. The areas of equidistantly spaced sections of the underwater form of a small boat are as follows:

   1.76, 2.78, 3.10, 3.12, 2.61, 1.24, 0.85 m^2

   Determine the underwater volume if the sections are 3 m apart.

2. To estimate the amount of earth to be removed when constructing a cutting the cross-sectional area at intervals of 8 m were estimated as follows:

   0, 2.8, 3.7, 4.5, 4.1, 2.6, 0 m^2

   Estimate the volume of earth to be excavated.

3. The circumference of a 12 m long log of timber of varying circular cross-section is measured at intervals of 2 m along its length and the results are:

Distance from one end (m)	Circumference (m)
0	2.80
2	3.25
4	3.94
6	4.32
8	5.16
10	5.82
12	6.36

   Estimate the volume of the timber in cubic metres.

26.3 The mean or average value of a waveform

The mean or average value, y, of the waveform shown in Fig. 26.6 is given by:

$$y = \frac{\textbf{area under curve}}{\textbf{length of base, } b}$$

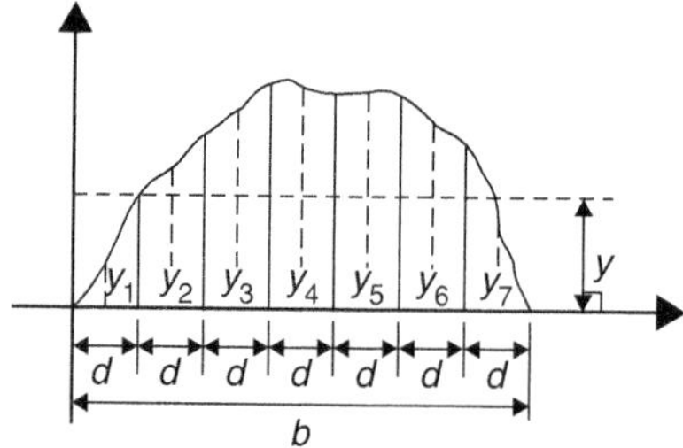

Figure 26.6

If the mid-ordinate rule is used to find the area under the curve, then:

$$y \approx \frac{\textbf{sum of mid-ordinates}}{\textbf{number of mid-ordinates}}$$

$$\left(= \frac{y_1 + y_2 + y_3 + y_4 + y_5 + y_6 + y_7}{7} \text{ for Fig. 26.6}\right)$$

For a **sine wave**, the mean or average value:

(i) over one complete cycle is zero (see Fig. 26.7(a)),

(ii) over half a cycle is **0.637 × maximum value, or $2/\pi$ × maximum value,**

(iii) of a full-wave rectified waveform (see Fig. 26.7(b)) is **0.637 × maximum value,**

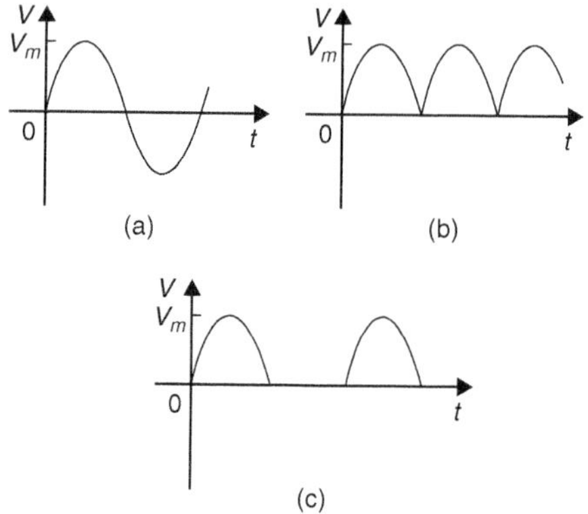

Figure 26.7

(iv) of a half-wave rectified waveform (see Fig. 26.7(c)) is **0.318 × maximum value, or $\frac{1}{\pi}$ × maximum value,**

Problem 5. Determine the average values over half a cycle of the periodic waveforms shown in Fig. 26.8

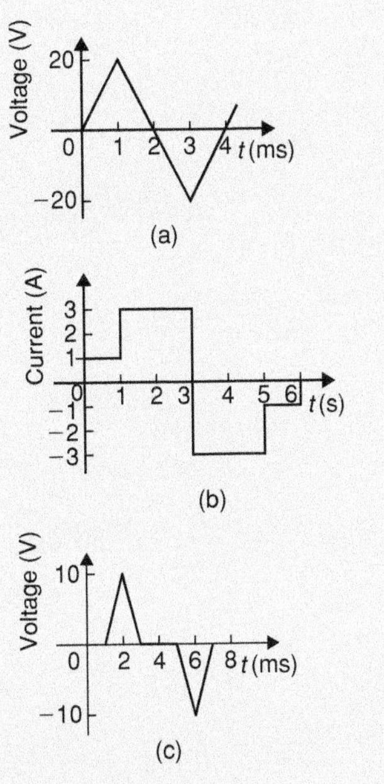

Figure 26.8

(a) Area under triangular waveform (a) for a half cycle is given by:

$$\text{Area} = \frac{1}{2}(\text{base})(\text{perpendicular height})$$
$$= \frac{1}{2}(2 \times 10^{-3})(20)$$
$$= 20 \times 10^{-3}\,\text{Vs}$$

Average value of waveform

$$= \frac{\text{area under curve}}{\text{length of base}}$$
$$= \frac{20 \times 10^{-3}\,\text{Vs}}{2 \times 10^{-3}\,\text{s}} = \mathbf{10\,V}$$

(b) Area under waveform (b) for a half cycle

$$= (1 \times 1) + (3 \times 2) = 7\,\text{As}$$

Average value of waveform

$$= \frac{\text{area under curve}}{\text{length of base}}$$
$$= \frac{7\,\text{As}}{3\,\text{s}} = \mathbf{2.33\,A}$$

(c) A half cycle of the voltage waveform (c) is completed in 4 ms.

Area under curve

$$= \frac{1}{2}\{(3-1)10^{-3}\}(10)$$
$$= 10 \times 10^{-3}\,\text{Vs}$$

Average value of waveform

$$= \frac{\text{area under curve}}{\text{length of base}}$$
$$= \frac{10 \times 10^{-3}\,\text{Vs}}{4 \times 10^{-3}\,\text{s}} = \mathbf{2.5\,V}$$

Problem 6. Determine the mean value of current over one complete cycle of the periodic waveforms shown in Fig. 26.9

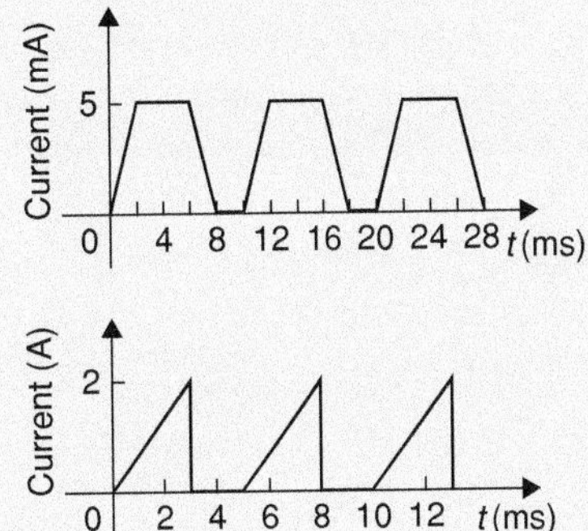

Figure 26.9

(a) One cycle of the trapezoidal waveform (a) is completed in 10 ms (i.e. the periodic time is 10 ms).
Area under curve = area of trapezium

$$= \frac{1}{2}(\text{sum of parallel sides})(\text{perpendicular distance between parallel sides})$$
$$= \frac{1}{2}\{(4+8) \times 10^{-3}\}(5 \times 10^{-3})$$
$$= 30 \times 10^{-6}\,\text{As}$$

Mean value over one cycle

$$= \frac{\text{area under curve}}{\text{length of base}}$$

$$= \frac{30 \times 10^{-6}\,\text{As}}{10 \times 10^{-3}\,\text{s}} = \mathbf{3\,mA}$$

(b) One cycle of the sawtooth waveform (b) is completed in 5 ms.

Area under curve

$$= \frac{1}{2}(3 \times 10^{-3})(2) = 3 \times 10^{-3}\,\text{As}$$

Mean value over one cycle

$$= \frac{\text{area under curve}}{\text{length of base}}$$

$$= \frac{3 \times 10^{-3}\,\text{As}}{5 \times 10^{-3}\,\text{s}} = \mathbf{0.6\,A}$$

Problem 7. The power used in a manufacturing process during a 6 hour period is recorded at intervals of 1 hour as shown below

Time (h)	0	1	2	3	4	5	6
Power (kW)	0	14	29	51	45	23	0

Plot a graph of power against time and, by using the mid-ordinate rule, determine (a) the area under the curve and (b) the average value of the power

The graph of power/time is shown in Fig. 26.10.

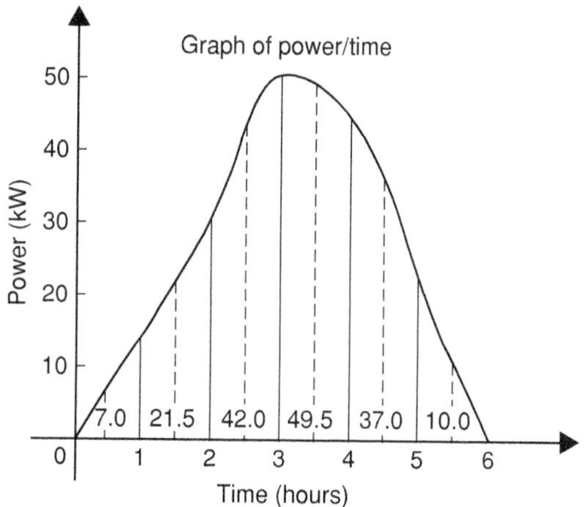

Figure 26.10

(a) The time base is divided into 6 equal intervals, each of width 1 hour. Mid-ordinates are erected (shown by broken lines in Fig. 26.10) and measured. The values are shown in Fig. 26.10.

Area under curve

$$\approx \text{(width of interval)(sum of mid-ordinates)}$$

$$= (1)[7.0 + 21.5 + 42.0 + 49.5 + 37.0 + 10.0]$$

$$= \mathbf{167\,kWh}$$

(i.e. a measure of electrical energy)

(b) Average value of waveform

$$= \frac{\text{area under curve}}{\text{length of base}}$$

$$= \frac{167\,\text{kWh}}{6\,\text{h}} = \mathbf{27.83\,kW}$$

Alternatively, average value

$$= \frac{\text{Sum of mid-ordinates}}{\text{number of mid-ordinate}}$$

Problem 8. Figure 26.11 shows a sinusoidal output voltage of a full-wave rectifier. Determine, using the mid-ordinate rule with 6 intervals, the mean output voltage

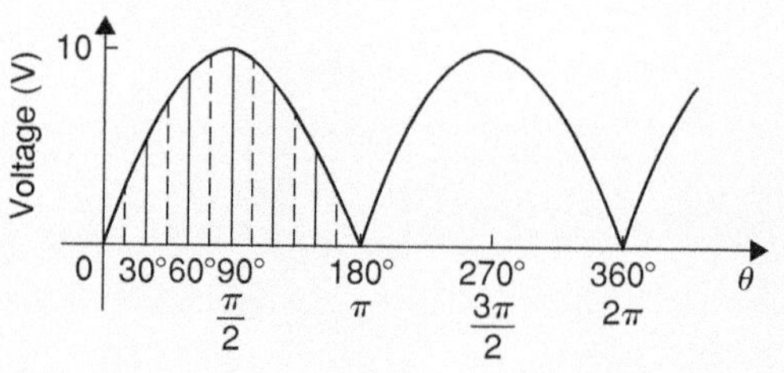

Figure 26.11

One cycle of the output voltage is completed in π radians or 180°. The base is divided into 6 intervals, each of width 30°. The mid-ordinate of each interval will lie at 15°, 45°, 75°, etc.

At 15° the height of the mid-ordinate is $10\sin 15° = 2.588\,\text{V}$

At 45° the height of the mid-ordinate is $10\sin 45° = 7.071\,\text{V}$ and so on.

The results are tabulated below:

Mid-ordinate	Height of mid-ordinate
15°	10 sin 15° = 2.588 V
45°	10 sin 45° = 7.071 V
75°	10 sin 75° = 9.659 V
105°	10 sin 105° = 9.659 V
135°	10 sin 135° = 7.071 V
165°	10 sin 165° = 2.588 V
Sum of mid-ordinates = 38.636 V	

Mean or average value of output voltage

$$\approx \frac{\text{sum of mid-ordinates}}{\text{number of mid-ordinate}}$$

$$= \frac{38.636}{6} = \mathbf{6.439\,V}$$

(With a larger number of intervals a more accurate answer may be obtained.)

For a sine wave the actual mean value is $0.637 \times$ maximum value, which in this problem gives 6.37 V

Problem 9. An indicator diagram for a steam engine is shown in Fig. 26.12. The base line has been divided into 6 equally spaced intervals and the lengths of the 7 ordinates measured with the results shown in centimetres. Determine (a) the area of the indicator diagram using Simpson's rule, and (b) the mean pressure in the cylinder given that 1 cm represents 100 kPa

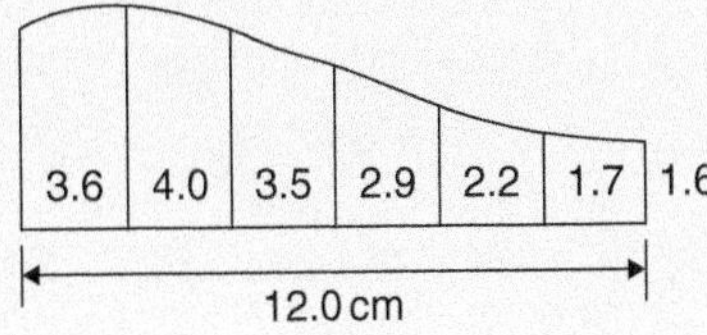

Figure 26.12

(a) The width of each interval is $\frac{12.0}{6}$ cm. Using Simpson's rule,

$$\text{area} \approx \frac{1}{3}(2.0)[(3.6+1.6)+4(4.0+2.9+1.7)+2(3.5+2.2)]$$

$$= \frac{2}{3}[5.2+34.4+11.4]$$

$$= \mathbf{34\,cm^2}$$

(b) Mean height of ordinates

$$= \frac{\text{area of diagram}}{\text{length of base}} = \frac{34}{12} = 2.83\,\text{cm}$$

Since 1 cm represents 100 kPa, the mean pressure in the cylinder

$$= 2.83\,\text{cm} \times 100\,\text{kPa/cm} = \mathbf{283\,kPa}$$

Now try the following Practice Exercise

Practice Exercise 137 Mean or average values of waveforms (Answers on page 715)

1. Determine the mean value of the periodic waveforms shown in Fig. 26.13 over a half cycle

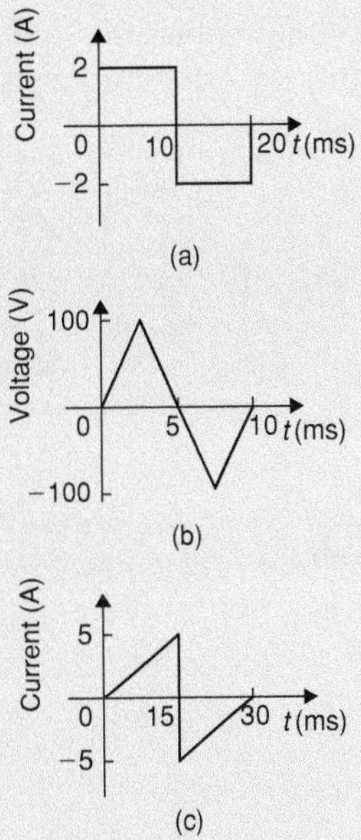

Figure 26.13

2. Find the average value of the periodic waveform shown in Fig. 26.14 over one complete cycle

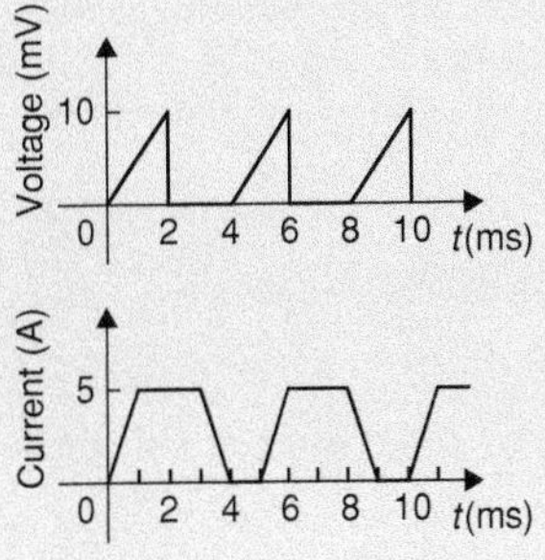

Figure 26.14

3. An alternating current has the following values at equal intervals of 5 ms

Time (ms)	0	5	10	15	20	25	30
Current (A)	0	0.9	2.6	4.9	5.8	3.5	0

Plot a graph of current against time and estimate the area under the curve over the 30 ms period using the mid-ordinate rule and determine its mean value

4. Determine, using an approximate method, the average value of a sine wave of maximum value 50 V for (a) a half cycle and (b) a complete cycle

5. An indicator diagram of a steam engine is 12 cm long. Seven evenly spaced ordinates, including the end ordinates, are measured as follows:

5.90, 5.52, 4.22, 3.63, 3.32, 3.24, 3.16 cm

Determine the area of the diagram and the mean pressure in the cylinder if 1 cm represents 90 kPa

Practice Exercise 138 Multiple-choice questions on irregular areas and volumes and mean values of waveforms (Answers on page 715)

Each question has only one correct answer

1. The speed of a car at 1 second intervals is given in the following table:

Time t (s)	0	1	2	3	4	5	6
Speed v(m/s)	0	2.5	5.0	9.0	15.0	22.0	30.0

The distance travelled in 6 s (i.e. the area under the v/t graph) using the trapezoidal rule is:
(a) 83.5 m (b) 68 m
(c) 68.5 m (d) 204 m

2. The mean value of a sine wave over half a cycle is:

(a) 0.318 × maximum value
(b) 0.707 × maximum value
(c) the peak value
(d) 0.637 × maximum value

3. An indicator diagram for a steam engine is as shown in Figure 26.15. The base has been divided into 6 equally spaced intervals and the lengths of the 7 ordinates measured, with the results shown in centimetres. Using Simpson's rule, the area of the indicator diagram is:
(a) 17.9 cm^2 (b) 32 cm^2
(c) 16 cm^2 (d) 96 cm^2

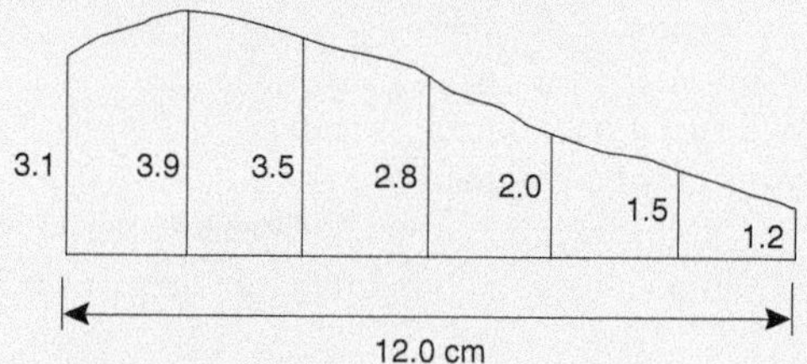

Figure 26.15

4. The mean value over half a cycle of a sine wave is 7.5 volts. The peak value of the sine wave is:
(a) 15 V (b) 23.58 V
(c) 11.77 V (d) 10.61 V

5. The areas of seven horizontal cross-sections of a water reservoir at intervals of 9 m are 70, 160, 210, 280, 250, 220 and 170 m^2. Given that 1 litre = 1000 m^2, the capacity of the reservoir is:
(a) 11.4 × 10^6 litres (b) 34.2 × 10^6 litres
(c) 11.4 × 10^3 litres (d) 32.4 × 10^3 litres

For fully worked solutions to each of the problems in Practice Exercises 135 to 137 in this chapter, go to the website:
www.routledge.com/cw/bird

Revision Test 7 Areas and volumes

This Revision Test covers the material contained in Chapters 23 to 26. *The marks for each question are shown in brackets at the end of each question.*

1. A swimming pool is 55 m long and 10 m wide. The perpendicular depth at the deep end is 5 m and at the shallow end is 1.5 m, the slope from one end to the other being uniform. The inside of the pool needs two coats of a protective paint before it is filled with water. Determine how many litres of paint will be needed if 1 litre covers $10\,m^2$. (7)

2. A steel template is of the shape shown in Fig. RT7.1, the circular area being removed. Determine the area of the template, in square centimetres, correct to 1 decimal place. (7)

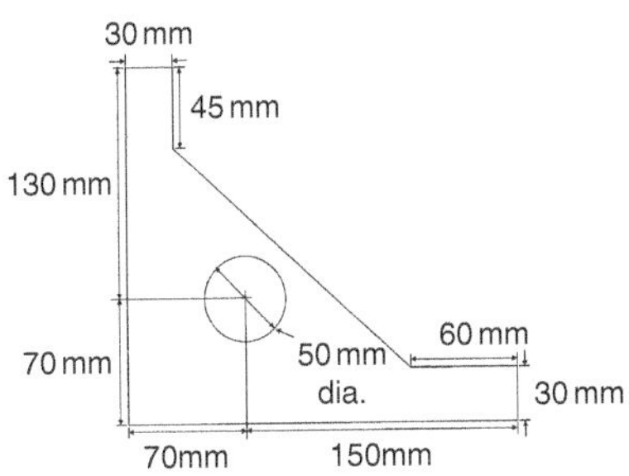

Figure RT7.1

3. The area of a plot of land on a map is $400\,mm^2$. If the scale of the map is 1 to 50 000, determine the true area of the land in hectares ($1\text{ hectare} = 10^4\,m^2$). (3)

4. Determine the shaded area in Fig. RT7.2, correct to the nearest square centimetre. (3)

5. Determine the diameter of a circle whose circumference is 178.4 cm. (2)

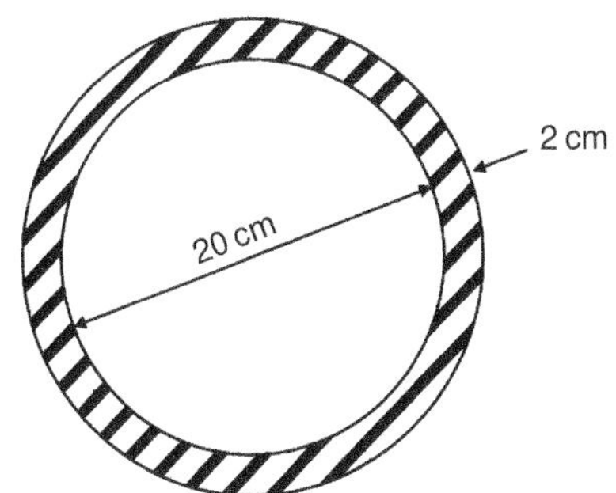

Figure RT7.2

6. Convert:

   (a) $125°47'$ to radians

   (b) 1.724 radians to degrees and minutes (2)

7. Calculate the length of metal strip needed to make the clip shown in Fig. RT7.3. (6)

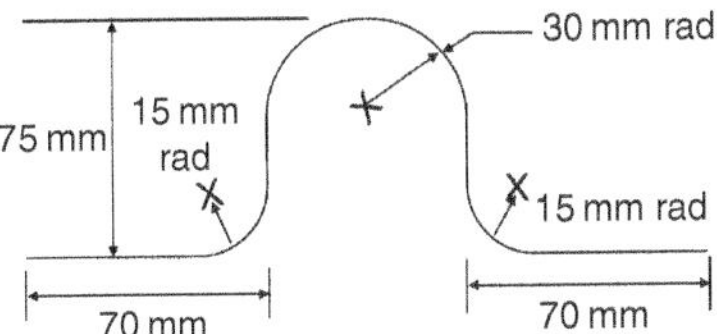

Figure RT7.3

8. A lorry has wheels of radius 50 cm. Calculate the number of complete revolutions a wheel makes (correct to the nearest revolution) when travelling 3 miles (assume 1 mile = 1.6 km). (5)

9. The equation of a circle is:

$$x^2 + y^2 + 12x - 4y + 4 = 0.$$

   Determine (a) the diameter of the circle, and (b) the co-ordinates of the centre of the circle. (5)

10. Determine the volume (in cubic metres) and the total surface area (in square metres) of a solid metal cone of base radius 0.5 m and perpendicular height 1.20 m. Give answers correct to 2 decimal places. (5)

11. Calculate the total surface area of a 10 cm by 15 cm rectangular pyramid of height 20 cm. (5)

12. A water container is of the form of a central cylindrical part 3.0 m long and diameter 1.0 m, with a hemispherical section surmounted at each end as shown in Fig. RT7.4. Determine the maximum capacity of the container, correct to the nearest litre. ($1\text{ litre} = 1000\,cm^3$.) (5)

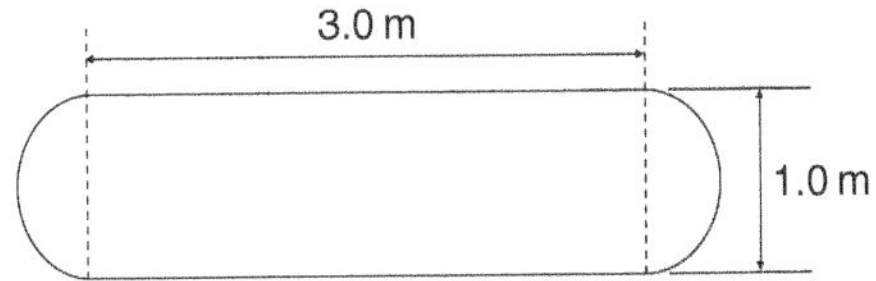

Figure RT7.4

13. Find the total surface area of a bucket consisting of an inverted frustum of a cone, of slant height 35.0 cm and end diameters 60.0 cm and 40.0 cm. (4)

14. A boat has a mass of 20 000 kg. A model of the boat is made to a scale of 1 to 80. If the model is made of the same material as the boat, determine the mass of the model (in grams). (3)

15. Plot a graph of $y=3x^2+5$ from $x=1$ to $x=4$. Estimate, correct to 2 decimal places, using 6 intervals, the area enclosed by the curve, the ordinates $x=1$ and $x=4$, and the x-axis by (a) the trapezoidal rule, (b) the mid-ordinate rule, and (c) Simpson's rule. (12)

16. A vehicle starts from rest and its velocity is measured every second for 6 seconds, with the following results:

Time t (s)	0	1	2	3	4	5	6
Velocity v (m/s)	0	1.2	2.4	3.7	5.2	6.0	9.2

Using Simpson's rule, calculate (a) the distance travelled in 6 s (i.e. the area under the v/t graph) and (b) the average speed over this period. (6)

For lecturers/instructors/teachers, fully worked solutions to each of the problems in Revision Test 7, together with a full marking scheme, are available at the website:

www.routledge.com/cw/bird

Section 4

Graphs

Chapter 27

Straight line graphs

Why it is important to understand: Straight line graphs

Graphs have a wide range of applications in engineering and in physical sciences because of their inherent simplicity. A graph can be used to represent almost any physical situation involving discrete objects and the relationship among them. If two quantities are directly proportional and one is plotted against the other, a straight line is produced. Examples include an applied force on the end of a spring plotted against spring extension, the speed of a flywheel plotted against time and strain in a wire plotted against stress (Hooke's law). In engineering, the straight line graph is the most basic graph to draw and evaluate.

At the end of this chapter, you should be able to:

- understand rectangular axes, scales and co-ordinates
- plot given co-ordinates and draw the best straight line graph
- determine the gradient of a straight line graph
- estimate the vertical-axis intercept
- state the equation of a straight line graph
- plot straight line graphs involving practical engineering examples

27.1 Introduction to graphs

A **graph** is a pictorial representation of information showing how one quantity varies with another related quantity.

The most common method of showing a relationship between two sets of data is to use **Cartesian** (named after Descartes*) **or rectangular axes** as shown in Fig. 27.1.

The points on a graph are called **co-ordinates**. Point A in Fig. 27.1 has the co-ordinates $(3, 2)$, i.e. 3 units in the x direction and 2 units in the y direction. Similarly, point B has co-ordinates $(-4, 3)$ and C has co-ordinates $(-3,-2)$. The origin has co-ordinates (0, 0).

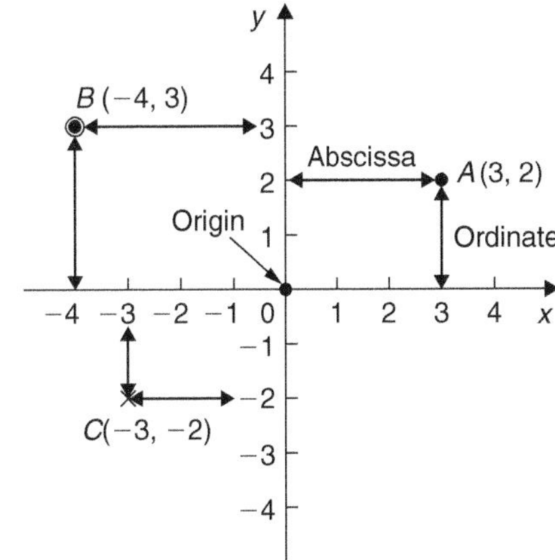

Figure 27.1

The horizontal distance of a point from the vertical axis is called the **abscissa** and the vertical

*Who was Descartes? – See page 197. To find out more go to www.routledge.com/cw/bird

distance from the horizontal axis is called the **ordinate**.

27.2 The straight line graph

Let a relationship between two variables x and y be $y=3x+2$

When $x=0$, $y=3(0)+2=2$.
When $x=1$, $y=3(1)+2=5$.
When $x=2$, $y=3(2)+2=8$ and so on.

Thus co-ordinates (0, 2), (1, 5) and (2, 8) have been produced from the equation by selecting arbitrary values of x, and are shown plotted in Fig. 27.2. When the points are joined together, a **straight line graph** results.

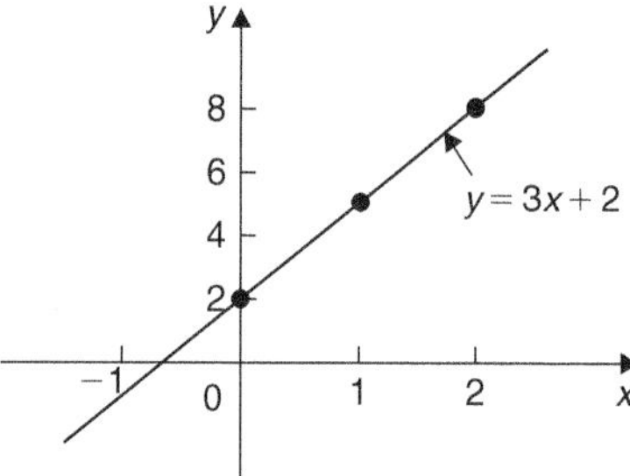

Figure 27.2

The **gradient** or **slope** of a straight line is the ratio of the change in the value of y to the change in the value of x between any two points on the line. If, as x increases, (→), y also increases (↑), then the gradient is positive.
In Fig. 27.3(a),

$$\text{the gradient of } AC = \frac{\text{change in } y}{\text{change in } x} = \frac{CB}{BA}$$
$$= \frac{7-3}{3-1} = \frac{4}{2} = 2$$

If as x increases (→), y decreases (↓), then the gradient is negative.
In Fig. 27.3(b),

$$\text{the gradient of } DF = \frac{\text{change in } y}{\text{change in } x} = \frac{FE}{ED}$$
$$= \frac{11-2}{-3-0} = \frac{9}{-3} = -3$$

Figure 27.3(c) shows a straight line graph $y=3$. Since the straight line is horizontal the gradient is zero.
The value of y when $x=0$ is called the **y-axis intercept**. In Fig. 27.3(a) the y-axis intercept is 1 and in Fig. 27.3(b) is 2.

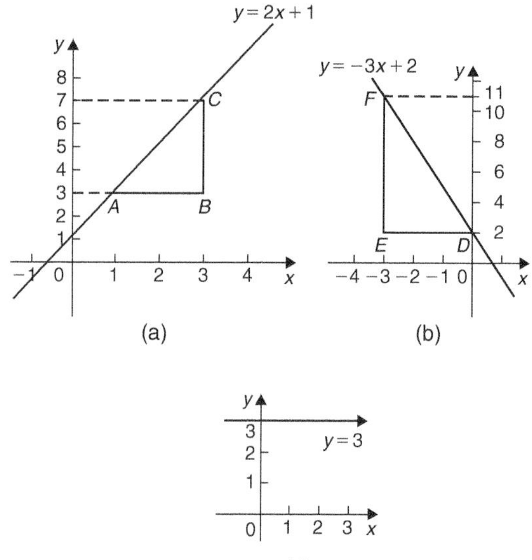

Figure 27.3

If the equation of a graph is of the form $y=mx+c$, where m and c are constants, **the graph will always be a straight line**, m representing the gradient and c the y-axis intercept.

Thus $y=5x+2$ represents a straight line of gradient 5 and y-axis intercept 2. Similarly, $y=-3x-4$ represents a straight line of gradient -3 and y-axis intercept -4.

Summary of general rules to be applied when drawing graphs

(i) Give the graph a title clearly explaining what is being illustrated.

(ii) Choose scales such that the graph occupies as much space as possible on the graph paper being used.

(iii) Choose scales so that interpolation is made as easy as possible. Usually scales such as 1 cm = 1 unit, or 1 cm = 2 units, or 1 cm = 10 units are used. Awkward scales such as 1 cm = 3 units or 1 cm = 7 units should not be used.

(iv) The scales need not start at zero, particularly when starting at zero produces an accumulation of points within a small area of the graph paper.

(v) The co-ordinates, or points, should be clearly marked. This may be done either by a cross, or a dot and circle, or just by a dot (see Fig. 27.1).

(vi) A statement should be made next to each axis explaining the numbers represented with their appropriate units.

(vii) Sufficient numbers should be written next to each axis without cramping.

Problem 1. Plot the graph $y=4x+3$ in the range $x=-3$ to $x=+4$. From the graph, find (a) the value of y when $x=2.2$, and (b) the value of x when $y=-3$

Whenever an equation is given and a graph is required, a table giving corresponding values of the variable is necessary. The table is achieved as follows:

When $x=-3,\ y=4x+3=4(-3)+3$
$$=-12+3=-9$$

When $x=-2,\ y=4(-2)+3$
$$=-8+3=-5 \text{ and so on.}$$

Such a table is shown below:

x	−3	−2	−1	0	1	2	3	4
y	−9	−5	−1	3	7	11	15	19

The co-ordinates (−3, −9), (−2, −5), (−1, −1) and so on, are plotted and joined together to produce the straight line shown in Fig. 27.4. (Note that the scales used on the x and y axes do not have to be the same.)

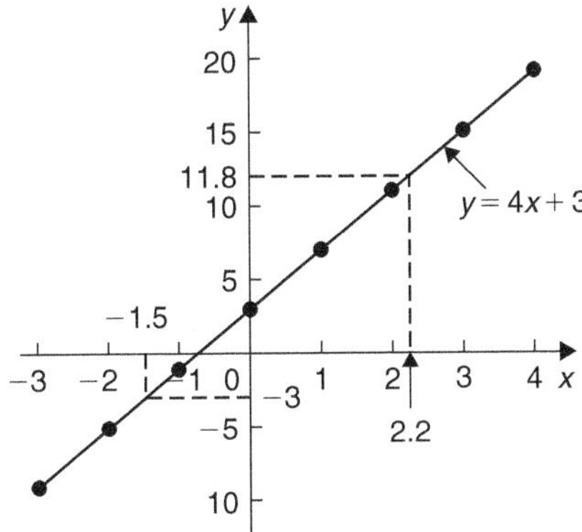

Figure 27.4

From the graph:

(a) when $x=2.2$, $\mathbf{y=11.8}$, and

(b) when $y=-3$, $\mathbf{x=-1.5}$

Problem 2. Plot the following graphs on the same axes between the range $x=-4$ to $x=+4$, and determine the gradient of each.

(a) $y=x$ (b) $y=x+2$

(c) $y=x+5$ (d) $y=x-3$

A table of co-ordinates is produced for each graph.

(a) $y=x$

x	−4	−3	−2	−1	0	1	2	3	4
y	−4	−3	−2	−1	0	1	2	3	4

(b) $y=x+2$

x	−4	−3	−2	−1	0	1	2	3	4
y	−2	−1	0	1	2	3	4	5	6

(c) $y=x+5$

x	−4	−3	−2	−1	0	1	2	3	4
y	1	2	3	4	5	6	7	8	9

(d) $y=x-3$

x	−4	−3	−2	−1	0	1	2	3	4
y	−7	−6	−5	−4	−3	−2	−1	0	1

The co-ordinates are plotted and joined for each graph. The results are shown in Fig. 27.5. Each of the straight

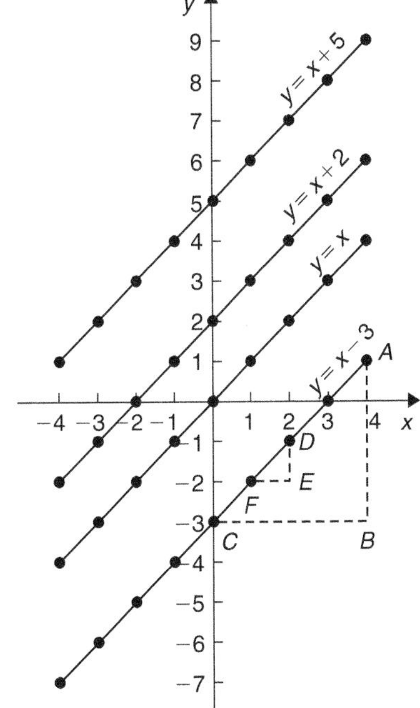

Figure 27.5

lines produced are parallel to each other, i.e. the slope or gradient is the same for each.

To find the gradient of any straight line, say, $y=x-3$ a horizontal and vertical component needs to be constructed. In Fig. 27.5, AB is constructed vertically at $x=4$ and BC constructed horizontally at $y=-3$.

$$\text{The gradient of } AC = \frac{AB}{BC} = \frac{1-(-3)}{4-0} = \frac{4}{4} = 1$$

i.e. the gradient of the straight line $y=x-3$ is 1. The actual positioning of AB and BC is unimportant for the gradient is also given by, for example,

$$\frac{DE}{EF} = \frac{-1-(-2)}{2-1} = \frac{1}{1} = 1$$

The slope or gradient of each of the straight lines in Fig. 27.5 is thus 1 since they are all parallel to each other.

Problem 3. Plot the following graphs on the same axes between the values $x=-3$ to $x=+3$ and determine the gradient and y-axis intercept of each.
(a) $y=3x$ (b) $y=3x+7$
(c) $y=-4x+4$ (d) $y=-4x-5$

A table of co-ordinates is drawn up for each equation.

(a) $y=3x$

x	−3	−2	−1	0	1	2	3
y	−9	−6	−3	0	3	6	9

(b) $y=3x+7$

x	−3	−2	−1	0	1	2	3
y	−2	1	4	7	10	13	16

(c) $y=-4x+4$

x	−3	−2	−1	0	1	2	3
y	16	12	8	4	0	−4	−8

(d) $y=-4x-5$

x	−3	−2	−1	0	1	2	3
y	7	3	−1	−5	−9	−13	−17

Each of the graphs is plotted as shown in Fig. 27.6, and each is a straight line. $y=3x$ and $y=3x+7$ are parallel to each other and thus have the same gradient. The gradient of AC is given by:

$$\frac{CB}{BA} = \frac{16-7}{3-0} = \frac{9}{3} = 3$$

Hence the gradient of both $y=3x$ and $y=3x+7$ is 3.

$y=-4x+4$ and $y=-4x-5$ are parallel to each other and thus have the same gradient. The gradient of DF is given by:

$$\frac{FE}{ED} = \frac{-5-(-17)}{0-3} = \frac{12}{-3} = -4$$

Hence the gradient of both $y=-4x+4$ and $y=-4x-5$ is -4

The y-axis intercept means the value of y where the straight line cuts the y-axis. From Fig. 27.6,

$y=3x$ cuts the y-axis at $y=0$

$y=3x+7$ cuts the y-axis at $y=+7$

$y=-4x+4$ cuts the y-axis at $y=+4$

and $y=-4x-5$ cuts the y-axis at $y=-5$

Some general conclusions can be drawn from the graphs shown in Figs. 27.4, 27.5 and 27.6.

When an equation is of the form $y=mx+c$, where m and c are constants, then

(i) a graph of y against x produces a straight line,

(ii) m represents the slope or gradient of the line, and

(iii) c represents the y-axis intercept.

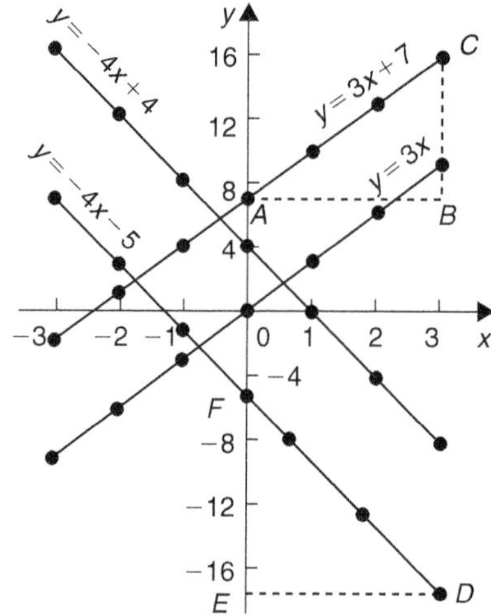

Figure 27.6

Thus, given an equation such as $y=3x+7$, it may be deduced 'on sight' that its gradient is $+3$ and its y-axis intercept is $+7$, as shown in Fig. 27.6. Similarly, if $y=-4x-5$, then the gradient is -4 and the y-axis intercept is -5, as shown in Fig. 27.6.
When plotting a graph of the form $y=mx+c$, only two co-ordinates need be determined. When the co-ordinates are plotted a straight line is drawn between the two points. Normally, three co-ordinates are determined, the third one acting as a check.

Problem 4. The following equations represent straight lines. Determine, without plotting graphs, the gradient and y-axis intercept for each.

(a) $y=3$ (b) $y=2x$
(c) $y=5x-1$ (d) $2x+3y=3$

(a) $y=3$ (which is of the form $y=0x+3$) represents a horizontal straight line intercepting the y-axis at **3**. Since the line is horizontal its **gradient is zero**.

(b) $y=2x$ is of the form $y=mx+c$, where c is zero. Hence **gradient = 2** and **y-axis intercept = 0** (i.e. the origin).

(c) $y=5x-1$ is of the form $y=mx+c$. Hence **gradient = 5** and **y-axis intercept = −1**.

(d) $2x+3y=3$ is not in the form $y=mx+c$ as it stands. Transposing to make y the subject gives $3y=3-2x$, i.e.

$$y=\frac{3-2x}{3}=\frac{3}{3}-\frac{2x}{3}$$

i.e. $$y=-\frac{2x}{3}+1$$

which is of the form $y=mx+c$

Hence **gradient** $=-\frac{2}{3}$ and **y-axis intercept = +1**

Problem 5. Without plotting graphs, determine the gradient and y-axis intercept values of the following equations:

(a) $y=7x-3$ (b) $3y=-6x+2$
(c) $y-2=4x+9$ (d) $\frac{y}{3}=\frac{x}{2}-\frac{1}{5}$
(e) $2x+9y+1=0$

(a) $y=7x-3$ is of the form $y=mx+c$, hence **gradient, $m=7$** and **y-axis intercept, $c=-3$**

(b) Rearranging $3y=-6x+2$ gives

$$y=-\frac{6x}{3}+\frac{2}{3}$$

i.e. $$y=-2x+\frac{2}{3}$$

which is of the form $y=mx+c$. Hence **gradient $m=-2$** and **y-axis intercept, $c=\frac{2}{3}$**

(c) Rearranging $y-2=4x+9$ gives $y=4x+11$, hence **gradient = 4** and **y-axis intercept = 11**

(d) Rearranging $\frac{y}{3}=\frac{x}{2}-\frac{1}{5}$ gives

$$y=3\left(\frac{x}{2}-\frac{1}{5}\right)=\frac{3}{2}x-\frac{3}{5}$$

Hence **gradient** $=\frac{3}{2}$ and **y-axis intercept** $=-\frac{3}{5}$

(e) Rearranging $2x+9y+1=0$ gives

$$9y=-2x-1$$

i.e. $$y=-\frac{2}{9}x-\frac{1}{9}$$

Hence **gradient** $=-\frac{2}{9}$

and **y-axis intercept** $=-\frac{1}{9}$

Problem 6. Determine the gradient of the straight line graph passing through the co-ordinates:
(a) $(-2, 5)$ and $(3, 4)$ (b) $(-2, -3)$ and $(-1, 3)$

A straight line graph passing through co-ordinates (x_1, y_1) and (x_2, y_2) has a gradient given by:

$$m=\frac{y_2-y_1}{x_2-x_1} \text{ (see Fig. 27.7)}$$

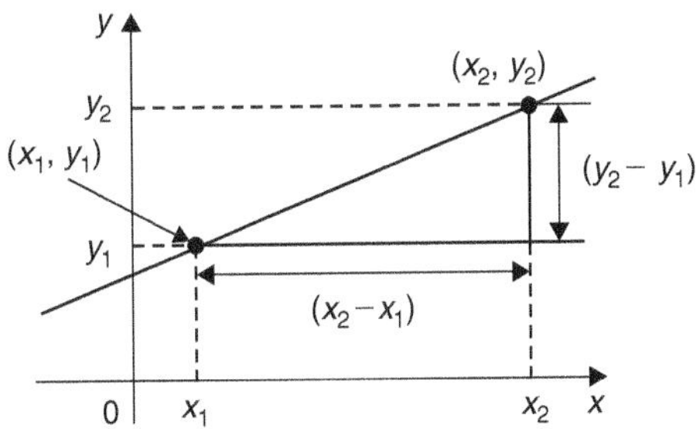

Figure 27.7

(a) A straight line passes through (−2, 5) and (3, 4), hence $x_1=-2$, $y_1=5$, $x_2=3$ and $y_2=4$, hence gradient

$$m=\frac{y_2-y_1}{x_2-x_1}=\frac{4-5}{3-(-2)}=-\frac{1}{5}$$

(b) A straight line passes through (−2,−3) and (−1, 3), hence $x_1=-2$, $y_1=-3$, $x_2=-1$ and $y_2=3$, hence gradient,

$$m=\frac{y_2-y_1}{x_2-x_1}=\frac{3-(-3)}{-1-(-2)}$$
$$=\frac{3+3}{-1+2}=\frac{6}{1}=\mathbf{6}$$

Problem 7. Plot the graph $3x+y+1=0$ and $2y-5=x$ on the same axes and find their point of intersection

Rearranging $3x+y+1=0$ gives: $y=-3x-1$

Rearranging $2y-5=x$ gives: $2y=x+5$ and $y=\frac{1}{2}x+2\frac{1}{2}$

Since both equations are of the form $y=mx+c$ both are straight lines. Knowing an equation is a straight line means that only two co-ordinates need to be plotted and a straight line drawn through them. A third co-ordinate is usually determined to act as a check.

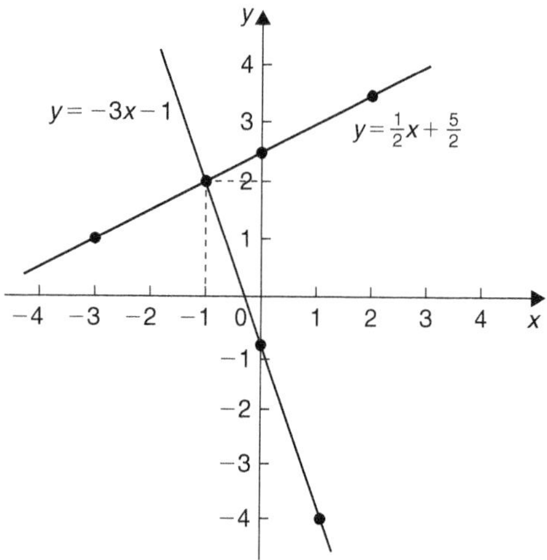

Figure 27.8

A table of values is produced for each equation as shown below.

x	1	0	−1
$-3x-1$	−4	−1	2

x	2	0	−3
$\frac{1}{2}x+2\frac{1}{2}$	$3\frac{1}{2}$	$2\frac{1}{2}$	1

The graphs are plotted as shown in Fig. 27.8.

The two straight lines are seen to intersect at (−1, 2)

Now try the following Practice Exercise

Practice Exercise 139 Straight line graphs (Answers on page 715)

1. Corresponding values obtained experimentally for two quantities are:

x	−2.0	−0.5	0	1.0	2.5	3.0	5.0
y	−13.0	−5.5	−3.0	2.0	9.5	12.0	22.0

Use a horizontal scale for x of 1 cm $=\frac{1}{2}$ unit and a vertical scale for y of 1 cm $=2$ units and draw a graph of x against y. Label the graph and each of its axes. By interpolation, find from the graph the value of y when x is 3.5

2. The equation of a line is $4y=2x+5$. A table of corresponding values is produced and is shown below. Complete the table and plot a graph of y against x. Find the gradient of the graph.

x	−4	−3	−2	−1	0	1	2	3	4
y		−0.25			1.25				3.25

3. Determine the gradient and intercept on the y-axis for each of the following equations:

(a) $y=4x-2$ (b) $y=-x$

(c) $y=-3x-4$ (d) $y=4$

4. Find the gradient and intercept on the y-axis for each of the following equations:

(a) $2y-1=4x$ (b) $6x-2y=5$

(c) $3(2y-1)=\frac{x}{4}$

5. Determine the gradient and y-axis intercept for each of the following equations and sketch the graphs:

   (a) $y=6x-3$ (b) $y=3x$ (c) $y=7$

   (d) $2x+3y+5=0$

6. Determine the gradient of the straight line graphs passing through the co-ordinates:

   (a) (2, 7) and (−3, 4)
   (b) (−4, −1) and (−5, 3)
   (c) $\left(\frac{1}{4}, -\frac{3}{4}\right)$ and $\left(-\frac{1}{2}, \frac{5}{8}\right)$

7. State which of the following equations will produce graphs which are parallel to one another:

   (a) $y-4=2x$ (b) $4x=-(y+1)$

   (c) $x=\frac{1}{2}(y+5)$ (d) $1+\frac{1}{2}y=\frac{3}{2}x$

   (e) $2x=\frac{1}{2}(7-y)$

8. Draw a graph of $y-3x+5=0$ over a range of $x=-3$ to $x=4$. Hence determine (a) the value of y when $x=1.3$ and (b) the value of x when $y=-9.2$

9. Draw on the same axes the graphs of $y=3x-5$ and $3y+2x=7$. Find the co-ordinates of the point of intersection. Check the result obtained by solving the two simultaneous equations algebraically.

10. Plot the graphs $y=2x+3$ and $2y=15-2x$ on the same axes and determine their point of intersection.

27.3 Practical problems involving straight line graphs

When a set of co-ordinate values are given or are obtained experimentally and it is believed that they follow a law of the form $y=mx+c$, then if a straight line can be drawn reasonably close to most of the co-ordinate values when plotted, this verifies that a law of the form $y=mx+c$ exists. From the graph, constants m (i.e. gradient) and c (i.e. y-axis intercept) can be determined. This technique is called **determination of law** (see also Chapter 28).

Problem 8. The temperature in degrees Celsius* and the corresponding values in degrees Fahrenheit are shown in the table below. Construct rectangular axes, choose a suitable scale and plot a graph of degrees Celsius (on the horizontal axis) against degrees Fahrenheit (on the vertical scale).

°C	10	20	40	60	80	100
°F	50	68	104	140	176	212

From the graph find (a) the temperature in degrees Fahrenheit at 55°C, (b) the temperature in degrees Celsius at 167°F, (c) the Fahrenheit temperature at 0°C, and (d) the Celsius temperature at 230°F

The co-ordinates (10, 50), (20, 68), (40, 104) and so on are plotted as shown in Fig. 27.9. When the

*Who was Celsius? – **Anders Celsius** (27 November 1701–25 April 1744) was the Swedish astronomer that proposed the Celsius temperature scale in 1742 which takes his name. To find out more go to www.routledge.com/cw/bird

co-ordinates are joined, a straight line is produced. Since a straight line results there is a linear relationship between degrees Celsius and degrees Fahrenheit.

(a) To find the Fahrenheit temperature at 55°C a vertical line *AB* is constructed from the horizontal axis to meet the straight line at *B*. The point where the horizontal line *BD* meets the vertical axis indicates the equivalent Fahrenheit temperature.

Hence 55°C is equivalent to 131°F

This process of finding an equivalent value in between the given information in the above table is called **interpolation**.

(b) To find the Celsius temperature at 167°F, a horizontal line *EF* is constructed as shown in Fig. 27.9. The point where the vertical line *FG* cuts the horizontal axis indicates the equivalent Celsius temperature.

Hence 167°F is equivalent to 75°C

(c) If the graph is assumed to be linear even outside of the given data, then the graph may be extended at both ends (shown by broken line in Fig. 27.9). From Fig. 27.9, **0°C corresponds to 32°F**

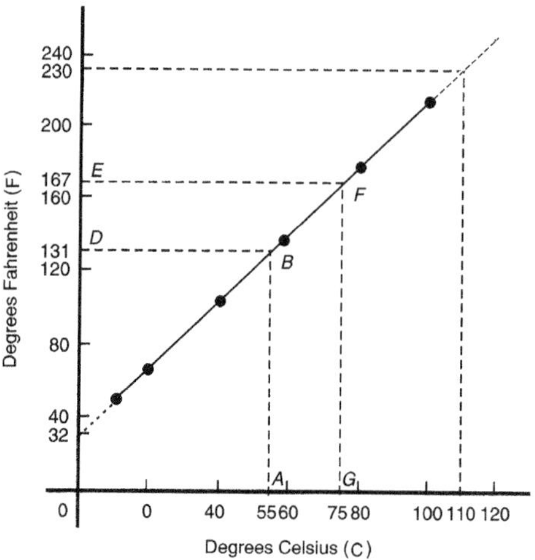

Figure 27.9

(d) **230°F is seen to correspond to 110°C**

The process of finding equivalent values outside of the given range is called **extrapolation**.

Problem 9. In an experiment on Charles's* law, the value of the volume of gas, *V*m^3, was measured for various temperatures *T*°C. Results are shown below.

*V*m^3	25.0	25.8	26.6	27.4	28.2	29.0
T°C	60	65	70	75	80	85

Plot a graph of volume (vertical) against temperature (horizontal) and from it find (a) the temperature when the volume is 28.6 m^3, and (b) the volume when the temperature is 67°C

If a graph is plotted with both the scales starting at zero then the result is as shown in Fig. 27.10. All of the points lie in the top right-hand corner of the graph, making interpolation difficult. A more accurate graph is obtained if the temperature axis starts at 55°C and the volume axis starts at 24.5 m^3. The axes corresponding to these values is shown by the broken lines in Fig. 27.10 and are called **false axes**, since the origin is not now at zero. A magnified version of this relevant part of the graph is shown in Fig. 27.11. From the graph:

(a) when the volume is 28.6 m^3, the equivalent temperature is **82.5°C**, and

(b) when the temperature is 67°C, the equivalent volume is **26.1 m^3**.

*Who was Charles? – **Jacques Alexandre César Charles** (12 November 1746–7 April 1823) was a French inventor, scientist, mathematician and balloonist. Charles' law describes how gases expand when heated. To find out more go to www.routledge.com/cw/bird

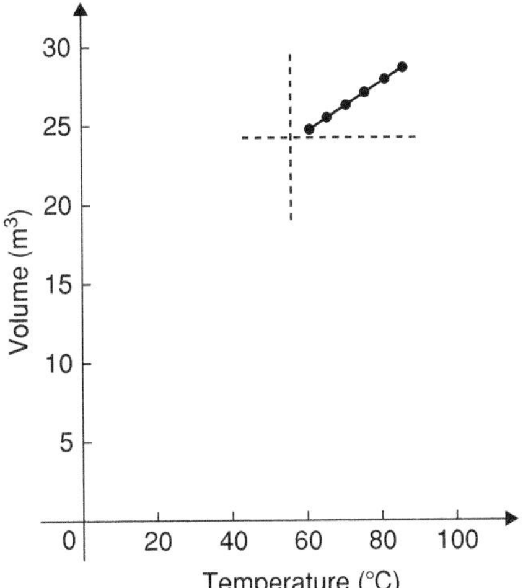

Figure 27.10

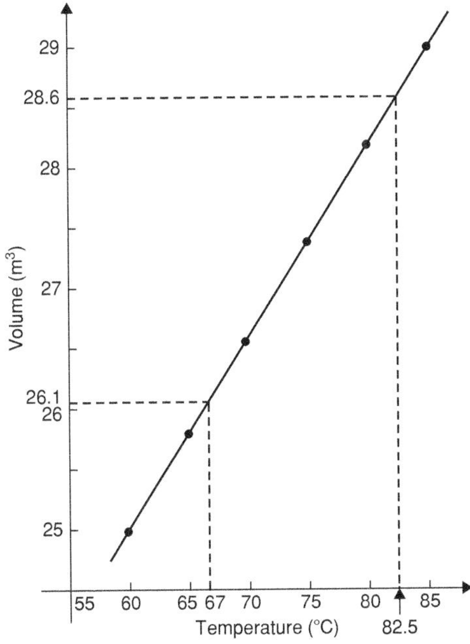

Figure 27.11

Problem 10. In an experiment demonstrating Hooke's* law, the strain in an aluminium wire was measured for various stresses. The results were:

Stress N/mm^2	4.9	8.7	15.0
Strain	0.00007	0.00013	0.00021
Stress N/mm^2	18.4	24.2	27.3
Strain	0.00027	0.00034	0.00039

Plot a graph of stress (vertically) against strain (horizontally). Find:

(a) Young's Modulus of Elasticity for aluminium which is given by the gradient of the graph,

(b) the value of the strain at a stress of 20 N/mm^2, and

(c) the value of the stress when the strain is 0.00020

The co-ordinates (0.00007, 4.9), (0.00013, 8.7) and so on, are plotted as shown in Fig. 27.12. The graph produced is the best straight line which can be drawn corresponding to these points. (With experimental results it is unlikely that all the points will lie exactly on a straight line.) The graph, and each of its axes, are labelled. Since the straight line passes through the origin, then stress is directly proportional to strain for the given range of values.

*Who was Hooke? – **Robert Hooke** FRS (28 July 1635–3 March 1703) was an English natural philosopher, architect and polymath who, amongst other things, discovered the law of elasticity. To find out more go to www.routledge.com/cw/bird

(a) The gradient of the straight line AC is given by

$$\frac{AB}{BC} = \frac{28-7}{0.00040-0.00010} = \frac{21}{0.00030}$$
$$= \frac{21}{3\times 10^{-4}} = \frac{7}{10^{-4}}$$
$$= 7\times 10^{4} = 70\,000\ \text{N/mm}^2$$

Thus Young's* Modulus of Elasticity for aluminium is 70 000 N/mm^2.

Since $1\ \text{m}^2 = 10^6\ \text{mm}^2$, $70\,000\ \text{N/mm}^2$ is equivalent to $70\,000 \times 10^6\ \text{N/m}^2$, i.e. **$70 \times 10^9$ N/m^2 (or Pascals)**.

From Fig. 27.12:

(b) the value of the strain at a stress of 20 N/mm^2 is **0.000285**, and

(c) the value of the stress when the strain is 0.00020 is **14 N/mm^2**.

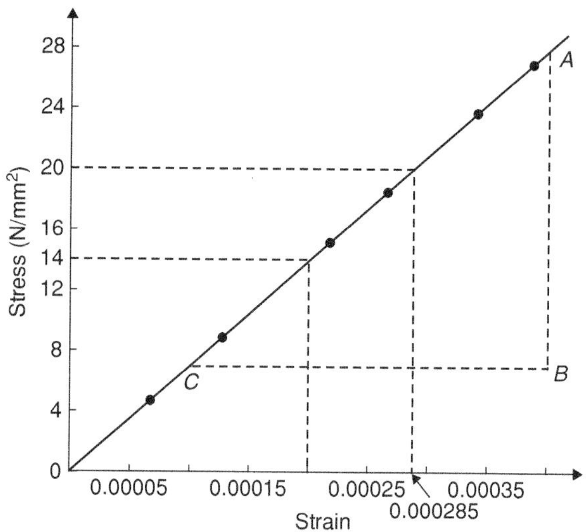

Figure 27.12

*Who was Young? – **Thomas Young** (13 June 1773–10 May 1829) was an English polymath. He is famous for having partly deciphered Egyptian hieroglyphics (specifically the Rosetta Stone). Young made notable scientific contributions to the fields of vision, light, solid mechanics, energy, physiology, language, musical harmony and Egyptology. To find out more go to www.routledge.com/cw/bird

Problem 11. The following values of resistance R ohms and corresponding voltage V volts are obtained from a test on a filament lamp.

R ohms	30	48.5	73	107	128
V volts	16	29	52	76	94

Choose suitable scales and plot a graph with R representing the vertical axis and V the horizontal axis. Determine (a) the gradient of the graph, (b) the R-axis intercept value, (c) the equation of the graph, (d) the value of resistance when the voltage is 60 V, and (e) the value of the voltage when the resistance is 40 ohms. (f) If the graph were to continue in the same manner, what value of resistance would be obtained at 110 V?

The co-ordinates (16, 30), (29, 48.5) and so on, are shown plotted in Fig. 27.13 where the best straight line is drawn through the points.

(a) The slope or gradient of the straight line AC is given by:

$$\frac{AB}{BC} = \frac{135-10}{100-0} = \frac{125}{100} = \mathbf{1.25}$$

(Note that the vertical line AB and the horizontal line BC may be constructed anywhere along the length of the straight line. However, calculations are made easier if the horizontal line BC is carefully chosen, in this case, 100)

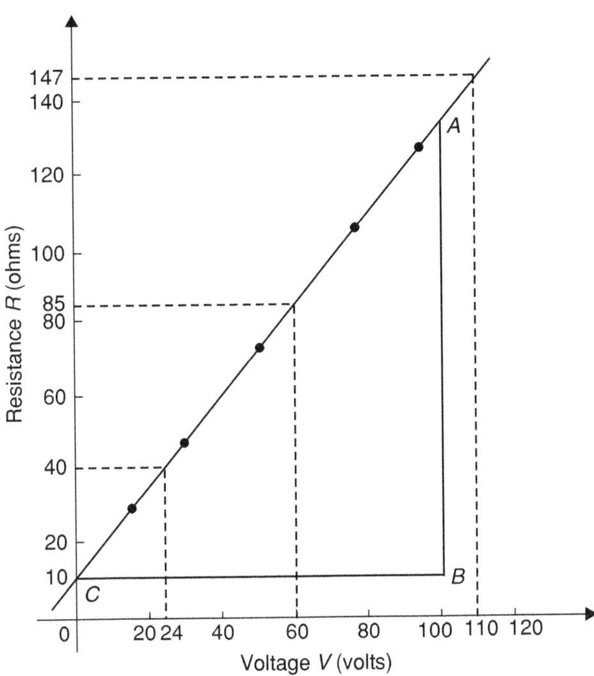

Figure 27.13

(b) The R-axis intercept is at $\boldsymbol{R=10}$ **ohms** (by extrapolation).

(c) The equation of a straight line is $y=mx+c$, when y is plotted on the vertical axis and x on the horizontal axis. m represents the gradient and c the y-axis intercept. In this case, R corresponds to y, V corresponds to x, $m=1.25$ and $c=10$. Hence the equation of the graph is $\boldsymbol{R=(1.25V+10)\,\Omega}$

From Fig. 27.13,

(d) when the voltage is 60 V, the resistance is **85 Ω**

(e) when the resistance is 40 ohms, the voltage is **24 V**, and

(f) by extrapolation, when the voltage is 110 V, the resistance is **147 Ω**.

Problem 12. Experimental tests to determine the breaking stress σ of rolled copper at various temperatures t gave the following results

Stress σ N/cm^2	8.46	8.04	7.78
Temperature t °C	70	200	280

Stress σ N/cm^2	7.37	7.08	6.63
Temperature t °C	410	500	640

Show that the values obey the law $\sigma=at+b$, where a and b are constants and determine approximate values for a and b. Use the law to determine the stress at 250°C and the temperature when the stress is 7.54 N/cm^2

The co-ordinates (70, 8.46), (200, 8.04) and so on, are plotted as shown in Fig. 27.14. Since the graph is a straight line then the values obey the law $\sigma=at+b$, and the gradient of the straight line is:

$$a=\frac{AB}{BC}=\frac{8.36-6.76}{100-600}=\frac{1.60}{-500}=\mathbf{-0.0032}$$

Vertical axis intercept, $\boldsymbol{b=8.68}$

Hence the law of the graph is: $\sigma=\mathbf{-0.0032}\boldsymbol{t}\mathbf{+8.68}$

When the temperature is 250°C, stress σ is given by:

$$\sigma=-0.0032(250)+8.68$$
$$=\mathbf{7.88\ N/cm^2}$$

Rearranging $\sigma=-0.0032t+8.68$ gives:

$$0.0032t=8.68-\sigma,$$

i.e. $$t=\frac{8.68-\sigma}{0.0032}$$

Hence when the stress $\sigma=7.54$ N/cm^2, temperature

$$t=\frac{8.68-7.54}{0.0032}=\mathbf{356.3°C}$$

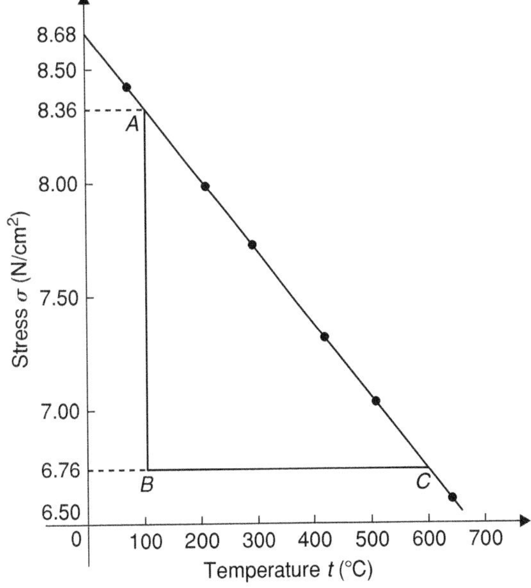

Figure 27.14

Now try the following Practice Exercise

Practice Exercise 140 Practical problems involving straight line graphs (Answers on page 715)

1. The resistance R ohms of a copper winding is measured at various temperatures t °C and the results are as follows:

R ohms	112	120	126	131	134
t °C	20	36	48	58	64

Plot a graph of R (vertically) against t (horizontally) and find from it (a) the temperature when the resistance is 122 Ω and (b) the resistance when the temperature is 52°C

2. The speed of a motor varies with armature voltage as shown by the following experimental results:

n (rev/min)	285	517	615	750	917	1050
V volts	60	95	110	130	155	175

Plot a graph of speed (horizontally) against voltage (vertically) and draw the best straight line through the points. Find from the graph: (a) the speed at a voltage of 145 V, and (b) the voltage at a speed of 400 rev/min

3. The following table gives the force F newtons which, when applied to a lifting machine, overcomes a corresponding load of L newtons

Force F newtons	25	47	64	120	149	187
Load L newtons	50	140	210	430	550	700

Choose suitable scales and plot a graph of F (vertically) against L (horizontally). Draw the best straight line through the points. Determine from the graph:
(a) the gradient, (b) the F-axis intercept, (c) the equation of the graph, (d) the force applied when the load is 310 N, and (e) the load that a force of 160 N will overcome. (f) If the graph were to continue in the same manner, what value of force will be needed to overcome a 800 N load?

4. The following table gives the results of tests carried out to determine the breaking stress σ of rolled copper at various temperature, t:

Stress σ (N/cm^2)	8.51	8.07	7.80
Temperature t (°C)	75	220	310

Stress σ (N/cm^2)	7.47	7.23	6.78
Temperature t (°C)	420	500	650

Plot a graph of stress (vertically) against temperature (horizontally). Draw the best straight line through the plotted co-ordinates. Determine the slope of the graph and the vertical axis intercept

5. The velocity v of a body after varying time intervals t was measured as follows:

t (seconds)	2	5	8	11	15	18
v (m/s)	16.9	19.0	21.1	23.2	26.0	28.1

Plot v vertically and t horizontally and draw a graph of velocity against time. Determine from the graph (a) the velocity after 10 s, (b) the time at 20 m/s and (c) the equation of the graph

6. The mass m of a steel joint varies with length L as follows:

mass, m (kg)	80	100	120	140	160
length, L (m)	3.00	3.74	4.48	5.23	5.97

Plot a graph of mass (vertically) against length (horizontally). Determine the equation of the graph

7. The crushing strength of mortar varies with the percentage of water used in its preparation, as shown below

Crushing strength, F (tonnes)	1.64	1.36	1.07	0.78	0.50	0.22
% of water used, w%	6	9	12	15	18	21

Plot a graph of F (vertically) against w (horizontally).

(a) Interpolate and determine the crushing strength when 10% of water is used.

(b) Assuming the graph continues in the same manner extrapolate and determine the percentage of water used when the crushing strength is 0.15 tonnes.

(c) What is the equation of the graph?

8. In an experiment demonstrating Hooke's law, the strain in a copper wire was measured for various stresses. The results were:

Stress (Pascals)	10.6×10^6	18.2×10^6	24.0×10^6
Strain	0.00011	0.00019	0.00025

Stress (Pascals)	30.7×10^6	39.4×10^6
Strain	0.00032	0.00041

Plot a graph of stress (vertically) against strain (horizontally). Determine (a) Young's Modulus of Elasticity for copper, which is given by the gradient of the graph, (b) the value of strain at a stress of 21×10^6 Pa, (c) the value of stress when the strain is 0.00030

9. An experiment with a set of pulley blocks gave the following results:

Effort, E (newtons)	9.0	11.0	13.6	17.4	20.8	23.6
Load, L (newtons)	15	25	38	57	74	88

Plot a graph of effort (vertically) against load (horizontally) and determine: (a) the gradient, (b) the vertical axis intercept, (c) the law of the graph, (d) the effort when the load is 30 N and (e) the load when the effort is 19 N

10. The variation of pressure p in a vessel with temperature T is believed to follow a law of the form $p = aT + b$, where a and b are constants. Verify this law for the results given below and determine the approximate values of a and b. Hence determine the pressures at temperatures of 285 K and 310 K and the temperature at a pressure of 250 kPa

Pressure, p kPa	244	247	252	258	262	267
Temperature, T K	273	277	282	289	294	300

Practice Exercise 141 Multiple-choice questions on straight line graphs (Answers on page 715)

Each question has only one correct answer

1. A graph of the line $x - y = 0$ passes through:

   (a) (2, 3) (b) (3, 40)
   (c) (5, 6) (d) (0, 0)

2. A graph of the line $x + y = 5$ intersects the x-axis at the point:

   (a) (5, 0) (b) (0, 5)
   (c) (−5, 0) (d) (0, −5)

3. The equation for the line with gradient $\frac{1}{2}$ and y-axis intercept 3 is:

(a) $x+2y=6$ (b) $2x-y=3$

(c) $2y-x=6$ (d) $\frac{1}{2}x-y=3$

4. A graph of resistance against voltage for an electrical circuit is shown in Figure 27.15. The equation relating resistance R and voltage V is:

(a) $R=1.45$ V $+ 40$ (b) $R=0.8$ V $+ 20$

(c) $R=1.45$ V $+ 20$ (d) $R=1.25$ V $+ 20$

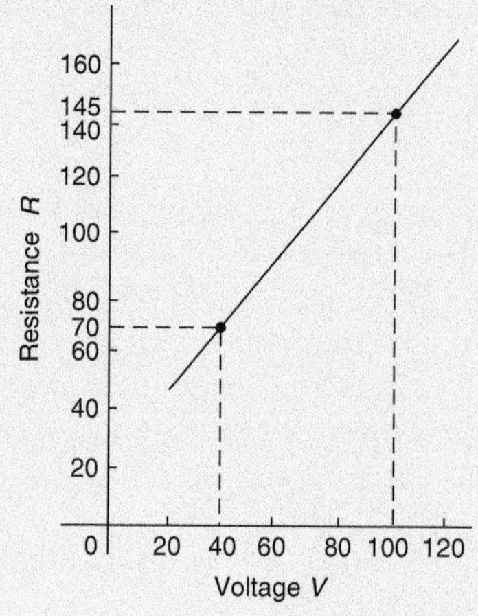

Figure 27.15

5. Two points on rectangular axes are P at $(-1,6)$ and Q at $(5,2)$. The gradient of the straight line between points P and Q is:

(a) 1 (b) −2/3 (c) −1 (d) 2/3

6. A graph of y against x, two engineering quantities, produces a straight line. A table of values is shown below:

x	2	−1	p
y	9	3	5

The value of p is:

(a) $-\frac{1}{2}$ (b) −2 (c) 3 (d) 0

7. Here are four equations in x and y. When x is plotted against y, in each case a straight line results.

(i) $y+3=3x$ (ii) $y+3x=3$

(iii) $\frac{y}{2}-\frac{3}{2}=x$ (iv) $\frac{y}{3}=x+\frac{2}{3}$

Which of these equations are parallel to each other ?

(a) (i) and (ii) (b) (i) and (iv)

(c) (ii) and (iii) (d) (ii) and (iv)

8. Which of the straight lines shown in Figure 27.16 has the equation $y+4=2x$?

(a) (iv) (b) (ii) (c) (iii) (d) (i)

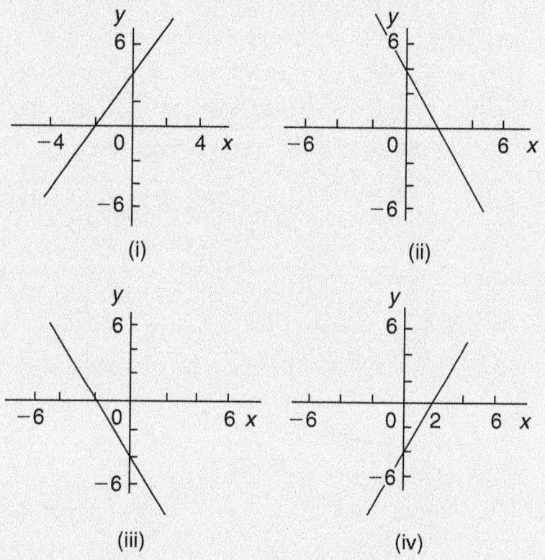

Figure 27.16

9. A graph relating effort E (plotted vertically) against load L (plotted horizontally) for a set of pulleys is given by $L+30=6E$. The gradient of the graph is:

(a) $\frac{1}{6}$ (b) 5 (c) 6 (d) $\frac{1}{5}$

10. The y-intercept of the equation $5x-3y=30$ is at:

(a) 10 (b) 2 (c) −10 (d) 6

11. The graph of the linear equation $2x+3y=6$ intersects the y-axis at the point:

(a) (2, 0) (b) (0, 3) (c) (3, 0) (d) (0, 2)

12. The mirror image about the y-axis of a point (3, 5) is:

    (a) (3, 5) (b) (−3, 5)
    (c) (3, −5) (d) (−3, −5)

13. The perpendicular distance of a point A at (5, 8) from the y-axis is:

    (a) 5 (b) 8 (c) 3 (d) 13

14. The graph of the equation $3x + 5y = 7$ is a:

    (a) vertical line
    (b) straight line
    (c) horizontal line
    (d) line passing through the origin

15. In an experiment demonstrating Hooke's law, the strain in a copper wire was measured for various stresses. The results included

Stress (megapascals)	18.24	24.00	39.36
Strain	0.00019	0.00025	0.00041

When stress is plotted vertically against strain horizontally a straight-line graph results. Young's modulus of elasticity for copper, which is given by the gradient of the graph, is:

(a) 96000 Pa (b) 1.04×10^{-11} Pa
(c) 96 Pa (d) 96×10^{9} Pa

For fully worked solutions to each of the problems in Practice Exercises 139 and 140 in this chapter, go to the website:
www.routledge.com/cw/bird

Chapter 28

Reduction of non-linear laws to linear form

Why it is important to understand: Reduction of non-linear laws to linear form

Graphs are important tools for analysing and displaying data between two experimental quantities. Many times situations occur in which the relationship between the variables is not linear. By manipulation, a straight line graph may be plotted to produce a law relating the two variables. Sometimes this involves using the laws of logarithms. The relationship between the resistance of wire and its diameter is not a linear one. Similarly, the periodic time of oscillations of a pendulum does not have a linear relationship with its length, and the head of pressure and the flow velocity are not linearly related. There are thus plenty of examples in engineering where determination of law is needed.

At the end of this chapter, you should be able to:

- understand what is meant by determination of law
- prepare co-ordinates for a non-linear relationship between two variables
- plot prepared co-ordinates and draw a straight line graph
- determine the gradient and vertical-axis intercept of a straight line graph
- state the equation of a straight line graph
- plot straight line graphs involving practical engineering examples
- determine straight line laws involving logarithms: $y = ax^n$, $y = ab^x$ and $y = ae^{bx}$
- plot straight line graphs involving logarithms

28.1 Determination of law

Frequently, the relationship between two variables, say x and y, is not a linear one, i.e. when x is plotted against y a curve results. In such cases the non-linear equation may be modified to the linear form, $y = mx + c$, so that the constants, and thus the law relating the variables can be determined. This technique is called '**determination of law**'.

Some examples of the reduction of equations to linear form include:

(i) $y = ax^2 + b$ compares with $Y = mX + c$, where $m = a$, $c = b$ and $X = x^2$.
Hence y is plotted vertically against x^2 horizontally to produce a straight line graph of gradient 'a' and y-axis intercept 'b'.

(ii) $y = \frac{a}{x} + b$

y is plotted vertically against $\frac{1}{x}$ horizontally to produce a straight line graph of gradient 'a' and y-axis intercept 'b'.

(iii) $y = ax^2 + bx$

Dividing both sides by x gives $\frac{y}{x} = ax + b$.

Comparing with $Y = mX + c$ shows that $\frac{y}{x}$ is plotted vertically against x horizontally to produce a straight line graph of gradient 'a' and $\frac{y}{x}$ axis intercept 'b'.

Problem 1. Experimental values of x and y, shown below, are believed to be related by the law $y = ax^2 + b$. By plotting a suitable graph verify this law and determine approximate values of a and b

x	1	2	3	4	5
y	9.8	15.2	24.2	36.5	53.0

If y is plotted against x a curve results and it is not possible to determine the values of constants a and b from the curve. Comparing $y = ax^2 + b$ with $Y = mX + c$ shows that y is to be plotted vertically against x^2 horizontally. A table of values is drawn up as shown below.

x	1	2	3	4	5
x^2	1	4	9	16	25
y	9.8	15.2	24.2	36.5	53.0

A graph of y against x^2 is shown in Fig. 28.1, with the best straight line drawn through the points. Since a straight line graph results, the law is verified.

From the graph, gradient

$$a = \frac{AB}{BC} = \frac{53 - 17}{25 - 5} = \frac{36}{20} = \mathbf{1.8}$$

and the y-axis intercept,

$$\mathbf{b = 8.0}$$

Hence the law of the graph is:

$$\mathbf{y = 1.8x^2 + 8.0}$$

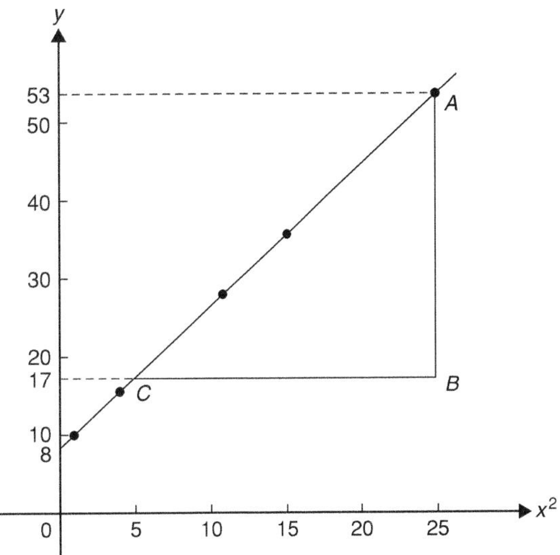

Figure 28.1

Problem 2. Values of load L newtons and distance d metres obtained experimentally are shown in the following table

Load, L N	32.3	29.6	27.0	23.2	18.3	12.8	10.0	6.4
distance, d m	0.75	0.37	0.24	0.17	0.12	0.09	0.08	0.07

Verify that load and distance are related by a law of the form $L = \frac{a}{d} + b$ and determine approximate values of a and b. Hence calculate the load when the distance is 0.20 m and the distance when the load is 20 N.

Comparing $L = \frac{a}{d} + b$ i.e. $L = a\left(\frac{1}{d}\right) + b$ with $Y = mX + c$ shows that L is to be plotted vertically against $\frac{1}{d}$ horizontally. Another table of values is drawn up as shown below.

L	32.3	29.6	27.0	23.2	18.3	12.8	10.0	6.4
d	0.75	0.37	0.24	0.17	0.12	0.09	0.08	0.07
$\frac{1}{d}$	1.33	2.70	4.17	5.88	8.33	11.11	12.50	14.29

A graph of L against $\frac{1}{d}$ is shown in Fig. 28.2. A straight line can be drawn through the points, which verifies that load and distance are related by a law of the form $L = \frac{a}{d} + b$

Gradient of straight line,

$$\mathbf{a} = \frac{AB}{BC} = \frac{31 - 11}{2 - 12} = \frac{20}{-10} = \mathbf{-2}$$

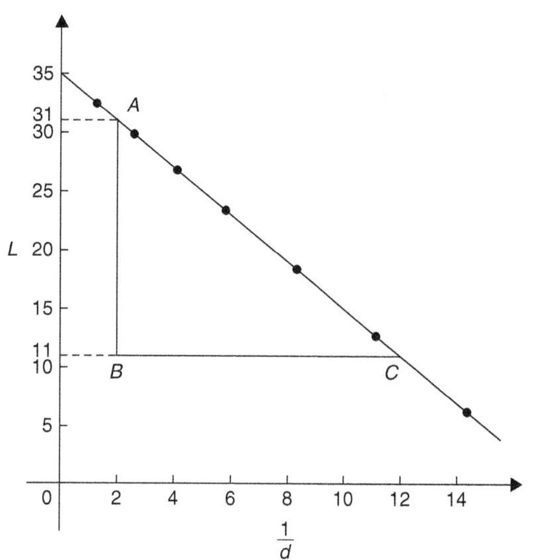

Figure 28.2

L-axis intercept,

$$\mathbf{b = 35}$$

Hence the law of the graph is

$$\mathbf{L = -\frac{2}{d} + 35}$$

When the distance $d = 0.20$ m, load

$$L = \frac{-2}{0.20} + 35 = \mathbf{25.0\,N}$$

Rearranging $L = -\frac{2}{d} + 35$ gives:

$$\frac{2}{d} = 35 - L \quad \text{and} \quad d = \frac{2}{35 - L}$$

Hence when the load $L = 20$ N, distance

$$d = \frac{2}{35 - 20} = \frac{2}{15} = \mathbf{0.13\,m}$$

Problem 3. The solubility s of potassium chlorate is shown by the following table:

t °C	10	20	30	40	50	60	80	100
s	4.9	7.6	11.1	15.4	20.4	26.4	40.6	58.0

The relationship between s and t is thought to be of the form $s = 3 + at + bt^2$. Plot a graph to test the supposition and use the graph to find approximate values of a and b. Hence calculate the solubility of potassium chlorate at 70°C

Rearranging $s = 3 + at + bt^2$ gives $s - 3 = at + bt^2$ and $\frac{s-3}{t} = a + bt$ or $\frac{s-3}{t} = bt + a$ which is of the form $Y = mX + c$, showing that $\frac{s-3}{t}$ is to be plotted vertically and t horizontally. Another table of values is drawn up as shown below.

t	10	20	30	40	50	60	80	100
s	4.9	7.6	11.1	15.4	20.4	26.4	40.6	58.0
$\frac{s-3}{t}$	0.19	0.23	0.27	0.31	0.35	0.39	0.47	0.55

A graph of $\frac{s-3}{t}$ against t is shown plotted in Fig. 28.3.

A straight line fits the points, which shows that s and t are related by

$$s = 3 + at + bt^2$$

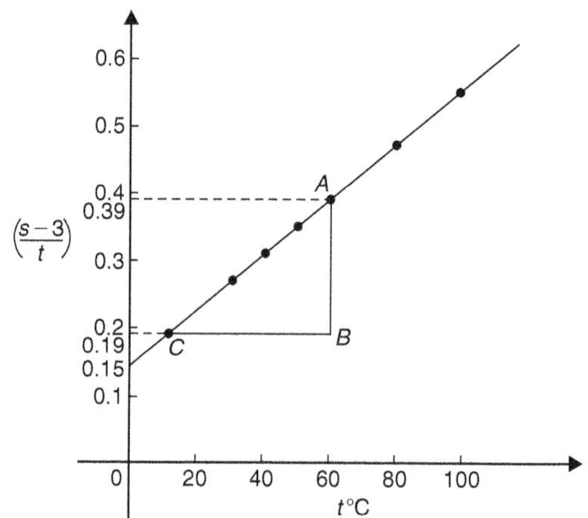

Figure 28.3

Gradient of straight line,

$$b = \frac{AB}{BC} = \frac{0.39 - 0.19}{60 - 10} = \frac{0.20}{50} = \mathbf{0.004}$$

Vertical axis intercept,

$$\mathbf{a = 0.15}$$

Hence the law of the graph is:

$$\mathbf{s = 3 + 0.15t + 0.004t^2}$$

The solubility of potassium chlorate at 70°C is given by

$$s = 3 + 0.15(70) + 0.004(70)^2$$
$$= 3 + 10.5 + 19.6 = \mathbf{33.1}$$

Now try the following Practice Exercise

Practice Exercise 142 Reducing non-linear laws to linear form (Answers on page 715)

In Problems 1 to 5, x and y are two related variables and all other letters denote constants. For the stated laws to be verified it is necessary to plot graphs of the variables in a modified form. State for each (a) what should be plotted on the vertical axis, (b) what should be plotted on the horizontal axis, (c) the gradient and (d) the vertical axis intercept.

1. $y = d + cx^2$
2. $y - a = b\sqrt{x}$
3. $y - e = \frac{f}{x}$
4. $y - cx = bx^2$
5. $y = \frac{a}{x} + bx$
6. In an experiment the resistance of wire is measured for wires of different diameters with the following results:

R ohms	1.64	1.14	0.89	0.76	0.63
d mm	1.10	1.42	1.75	2.04	2.56

It is thought that R is related to d by the law $R = (a/d^2) + b$, where a and b are constants. Verify this and find the approximate values for a and b. Determine the cross-sectional area needed for a resistance reading of 0.50 ohms.

7. Corresponding experimental values of two quantities x and y are given below

x	1.5	3.0	4.5	6.0	7.5	9.0
y	11.5	25.0	47.5	79.0	119.5	169.0

By plotting a suitable graph verify that y and x are connected by a law of the form $y = kx^2 + c$, where k and c are constants. Determine the law of the graph and hence find the value of x when y is 60.0

8. Experimental results of the safe load L kN, applied to girders of varying spans, d m, are show below

Span, *d* m	2.0	2.8	3.6	4.2	4.8
Load, *L* kN	475	339	264	226	198

It is believed that the relationship between load and span is $L = c/d$, where c is a constant. Determine (a) the value of constant c and (b) the safe load for a span of 3.0 m

9. The following laboratory results give corresponding values of two quantities x and y which are believed to be related by a law of the form $y = ax^2 + bx$ where a and b are constants

y	33.86	55.54	72.80	84.10	111.4	168.1
x	3.4	5.2	6.5	7.3	9.1	12.4

Verify the law and determine approximate values of a and b.

Hence determine (i) the value of y when x is 8.0 and (ii) the value of x when y is 146.5

28.2 Determination of law involving logarithms

Examples of reduction of equations to linear form involving logarithms include:

(i) $\boldsymbol{y = ax^n}$

Taking logarithms to a base of 10 of both sides gives:

$$\lg y = \lg(ax^n) = \lg a + \lg x^n$$

i.e. $$\lg y = n \lg x + \lg a$$

by the laws of logarithms which compares with

$$Y = mX + c$$

and shows that $\lg y$ is plotted vertically against $\lg x$ horizontally to produce a straight line graph of gradient n and $\lg y$-axis intercept $\lg a$

(ii) $\boldsymbol{y = ab^x}$

Taking logarithms to a base of 10 of the both sides gives:

$$\lg y = \lg(ab^x)$$

i.e. $$\lg y = \lg a + \lg b^x$$

i.e. $$\lg y = x \lg b + \lg a$$

by the laws of logarithms

or $$\lg y = (\lg b)x + \lg a$$

which compares with

$$Y = mX + c$$

and shows that $\lg y$ is plotted vertically against x horizontally to produce a straight line graph of gradient $\lg b$ and $\lg y$-axis intercept $\lg a$

(iii) $\boldsymbol{y = ae^{bx}}$

Taking logarithms to a base of e of both sides gives:

$$\ln y = \ln(ae^{bx})$$

i.e. $\ln y = \ln a + \ln e^{bx}$

i.e. $\ln y = \ln a + bx \ln e$

i.e. $\ln y = bx + \ln a$

(since $\ln e = 1$), which compares with

$$Y = mX + c$$

and shows that $\ln y$ is plotted vertically against x horizontally to produce a straight line graph of gradient b and $\ln y$-axis intercept $\ln a$.

Problem 4. The current flowing in, and the power dissipated by, a resistor are measured experimentally for various values and the results are as shown below

Current, I amperes	2.2	3.6	4.1	5.6	6.8
Power, P watts	116	311	403	753	1110

Show that the law relating current and power is of the form $P = RI^n$, where R and n are constants, and determine the law

Taking logarithms to a base of 10 of both sides of $P = RI^n$ gives:

$$\lg P = \lg(RI^n) = \lg R + \lg I^n = \lg R + n \lg I$$

by the laws of logarithms

i.e. $\lg P = n \lg I + \lg R$

which is of the form

$$Y = mX + c$$

showing that $\lg P$ is to be plotted vertically against $\lg I$ horizontally.

A table of values for $\lg I$ and $\lg P$ is drawn up as shown below

I	2.2	3.6	4.1	5.6	6.8
$\lg I$	0.342	0.556	0.613	0.748	0.833
P	116	311	403	753	1110
$\lg P$	2.064	2.493	2.605	2.877	3.045

A graph of $\lg P$ against $\lg I$ is shown in Fig. 28.4 and since a straight line results the law $P = RI^n$ is verified.

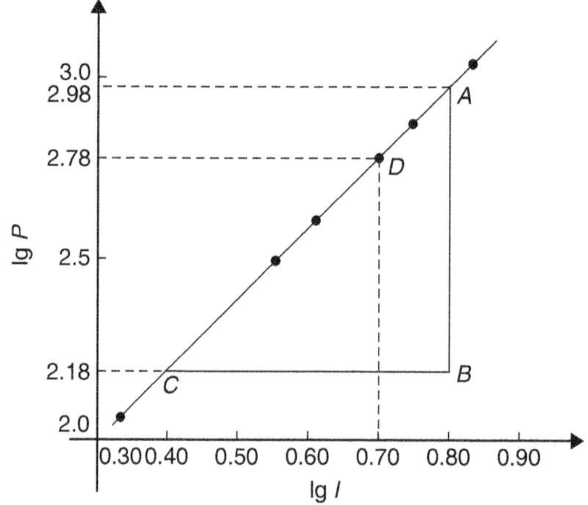

Figure 28.4

Gradient of straight line,

$$n = \frac{AB}{BC} = \frac{2.98 - 2.18}{0.8 - 0.4} = \frac{0.80}{0.4} = \mathbf{2}$$

It is not possible to determine the vertical axis intercept on sight since the horizontal axis scale does not start at zero. Selecting any point from the graph, say point D, where $\lg I = 0.70$ and $\lg P = 2.78$, and substituting values into

$$\lg P = n \lg I + \lg R$$

gives: $2.78 = (2)(0.70) + \lg R$

from which $\lg R = 2.78 - 1.40 = 1.38$

Hence $R = \text{antilog } 1.38 \; (= 10^{1.38}) = \mathbf{24.0}$

Hence the law of the graph is $P = 24.0I^2$

Problem 5. The periodic time, T, of oscillation of a pendulum is believed to be related to it length, l, by a law of the form $T = kl^n$, where k and n are constants. Values of T were measured for various lengths of the pendulum and the results are as shown below

Periodic time, T s	1.0	1.3	1.5	1.8	2.0	2.3
Length, l m	0.25	0.42	0.56	0.81	1.0	1.32

Show that the law is true and determine the approximate values of k and n. Hence find the periodic time when the length of the pendulum is 0.75 m

From para (i), if $T = kl^n$ then

$$\lg T = n \lg l + \lg k$$

and comparing with

$$Y = mX + c$$

shows that $\lg T$ is plotted vertically against $\lg l$ horizontally. A table of values for $\lg T$ and $\lg l$ is drawn up as shown below

T	1.0	1.3	1.5	1.8	2.0	2.3
$\lg T$	0	0.114	0.176	0.255	0.301	0.362
l	0.25	0.42	0.56	0.81	1.0	1.32
$\lg l$	−0.602	−0.377	−0.252	−0.092	0	0.121

A graph of $\lg T$ against $\lg l$ is shown in Fig. 28.5 and the law $T = kl^n$ is true since a straight line results.

From the graph, gradient of straight line,

$$n = \frac{AB}{BC} = \frac{0.25 - 0.05}{-0.10 - (-0.50)} = \frac{0.20}{0.40} = \frac{1}{2}$$

Vertical axis intercept, $\lg k = 0.30$. Hence

$k = \text{antilog } 0.30 \,(= 10^{0.30}) = \mathbf{2.0}$

Hence the law of the graph is:

$$\mathbf{T = 2.0 l^{1/2}} \quad \text{or} \quad \mathbf{T = 2.0\sqrt{l}}$$

When length $l = 0.75$ m then

$$T = 2.0\sqrt{0.75} = \mathbf{1.73\,s}$$

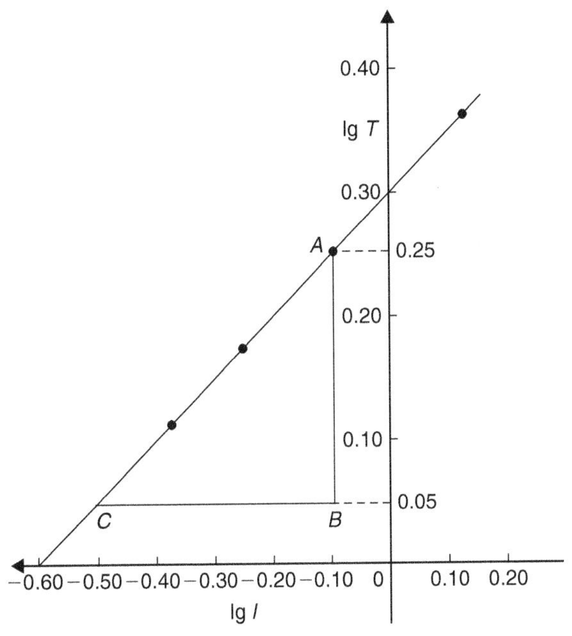

Figure 28.5

Problem 6. In a test lab, quantities x and y are believed to be related by a law of the form $y = ab^x$, where a and b are constants. Values of x and corresponding values of y are:

x	0	0.6	1.2	1.8	2.4	3.0
y	5.0	9.67	18.7	36.1	69.8	135.0

Verify the law and determine the approximate values of a and b. Hence determine (a) the value of y when x is 2.1 and (b) the value of x when y is 100

From para (ii), if $y = ab^x$ then

$$\lg y = (\lg b)x + \lg a$$

and comparing with

$$Y = mX + c$$

shows that $\lg y$ is plotted vertically and x horizontally. Another table is drawn up as shown below

x	0	0.6	1.2	1.8	2.4	3.0
y	5.0	9.67	18.7	36.1	69.8	135.0
$\lg y$	0.70	0.99	1.27	1.56	1.84	2.13

A graph of $\lg y$ against x is shown in Fig. 28.6 and since a straight line results, the law $y = ab^x$ is verified.

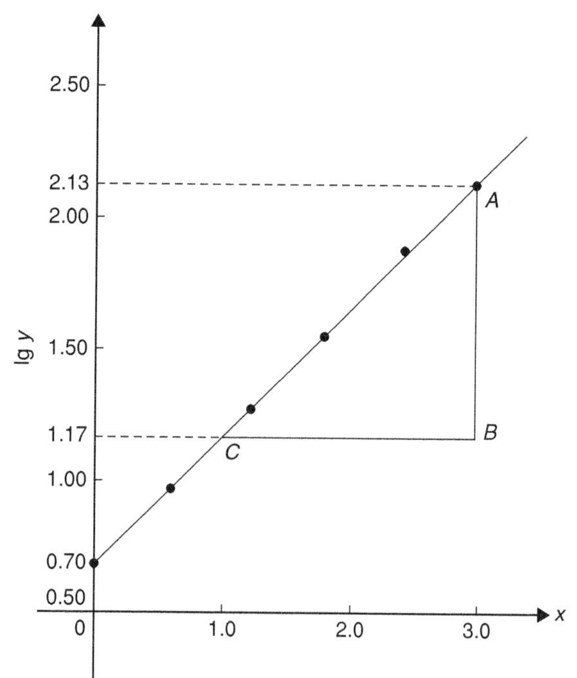

Figure 28.6

Gradient of straight line,

$$\lg b = \frac{AB}{BC} = \frac{2.13 - 1.17}{3.0 - 1.0} = \frac{0.96}{2.0} = 0.48$$

Hence $b =$ antilog 0.48 $(= 10^{0.48}) = \mathbf{3.0}$, correct to 2 significant figures.
Vertical axis intercept,

$$\lg a = 0.70, \text{ from which}$$

$$a = \text{antilog } 0.70 (= 10^{0.70})$$

$$= \mathbf{5.0}, \text{ correct to 2 significant figures.}$$

Hence the law of the graph is $y = 5.0(3.0)^x$

(a) When $x = 2.1$, $y = 5.0(3.0)^{2.1} = \mathbf{50.2}$

(b) When $y = 100$, $100 = 5.0(3.0)x$

from which $100/5.0 = (3.0)^x$

i.e. $20 = (3.0)^x$

Taking logarithms of both sides gives

$$\lg 20 = \lg(3.0)^x = x \lg 3.0$$

Hence $x = \frac{\lg 20}{\lg 3.0} = \frac{1.3010}{0.4771} = \mathbf{2.73}$

Problem 7. The current i mA flowing in a capacitor which is being discharged varies with time t ms as shown below:

i mA	203	61.14	22.49	6.13	2.49	0.615
t ms	100	160	210	275	320	390

Show that these results are related by a law of the form $i = Ie^{t/T}$, where I and T are constants. Determine the approximate values of I and T

Taking Napierian logarithms of both sides of $i = Ie^{t/T}$ gives

$$\ln i = \ln(Ie^{t/T}) = \ln I + \ln e^{t/T} = \ln I + \frac{t}{T} \ln e$$

i.e. $\ln i = \ln I + \frac{t}{T}$ (since $\ln e = 1$)

or $\ln i = \left(\frac{1}{T}\right) t + \ln I$

which compares with $y = mx + c$, showing that $\ln i$ is plotted vertically against t horizontally. (For methods of evaluating Napierian logarithms see Chapter 14.) Another table of values is drawn up as shown below

t	100	160	210	275	320	390
i	203	61.14	22.49	6.13	2.49	0.615
$\ln i$	5.31	4.11	3.11	1.81	0.91	−0.49

A graph of $\ln i$ againt t is shown in Fig. 28.7 and since a straight line results the law $i = Ie^{t/T}$ is verified.
Gradient of straight line,

$$\frac{1}{T} = \frac{AB}{BC} = \frac{5.30 - 1.30}{100 - 300} = \frac{4.0}{-200} = -0.02$$

Hence $T = \frac{1}{-0.02} = \mathbf{-50}$

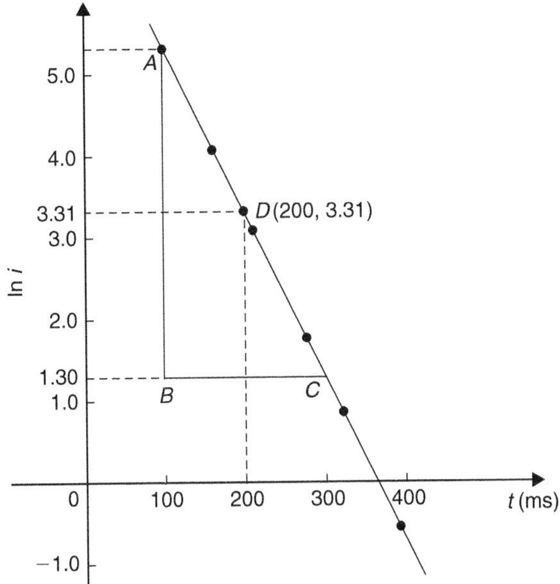

Figure 28.7

Selecting any point on the graph, say point D, where $t = 200$ and $\ln i = 3.31$, and substituting into

$$\ln i = \left(\frac{1}{T}\right)t + \ln I$$

gives: $$3.31 = -\frac{1}{50}(200) + \ln I$$

from which, $$\ln I = 3.31 + 4.0 = 7.31$$

and $$I = \text{antilog } 7.31 (= e^{7.31}) = 1495$$

or **1500** correct to 3 significant figures.

Hence the law of the graph is, $i = 1500\, e^{-t/50}$

Now try the following Practice Exercise

Practice Exercise 143 Reducing non-linear laws to linear form (Answers on page 716)

In Problem 1 to 3, x and y are two related variables and all other letters denote constants. For the stated laws to be verified it is necessary to plot graphs of the variables in a modified form. State for each (a) what should be plotted on the vertical axis, (b) what should be plotted on the horizontal axis, (c) the gradient and (d) the vertical axis intercept.

1. $y = ba^x$
2. $y = kx^l$
3. $\frac{y}{m} = e^{nx}$
4. The luminosity I of a lamp varies with the applied voltage V and the relationship between I and V is thought to be $I = kV^n$. Experimental results obtained are:

I candelas	1.92	4.32	9.72
V volts	40	60	90

I candelas	15.87	23.52	30.72
V volts	115	140	160

Verify that the law is true and determine the law of the graph. Determine also the luminosity when 75 V is applied cross the lamp

5. The head of pressure h and the flow velocity v are measured and are believed to be connected by the law $v = ah^b$, where a and b are constants. The results are as shown below:

h	10.6	13.4	17.2	24.6	29.3
v	9.77	11.0	12.44	14.88	16.24

Verify that the law is true and determine values of a and b

6. Experimental values of x and y are measured in a laboratory as follows:

x	0.4	0.9	1.2	2.3	3.8
y	8.35	13.47	17.94	51.32	215.20

The law relating x and y is believed to be of the form $y = ab^x$, where a and b are constants. Determine the approximate values of a and b. Hence find the value of y when x is 2.0 and the value of x when y is 100

7. The activity of a mixture of radioactive isotope is believed to vary according to the law $R = R_0 t^{-c}$, where R_0 and c are constants. Experimental results are shown below

R	9.72	2.65	1.15	0.47	0.32	0.23
t	2	5	9	17	22	28

Verify that the law is true and determine approximate values of R_0 and c

8. Determine the scientific law of the form $y = ae^{kx}$ which relates the following values

y	0.0306	0.285	0.841	5.21	173.2	1181
x	−4.0	5.3	9.8	17.4	32.0	40.0

9. The tension T in a belt passing round a pulley wheel and in contact with the pulley over an angle of θ radius is given by $T = T_0 e^{\mu\theta}$, where T_0 and μ are constants. Experimental results obtained are:

T newtons	47.9	52.8	60.3	70.1	80.9
θ radians	1.12	1.48	1.97	2.53	3.06

Determine approximate values of T_0 and μ. Hence find the tension when θ is 2.25 radians and the value of θ when the tension is 50.0 newtons

Practice Exercise 144 Multiple-choice questions on reduction of non-linear laws to linear form (Answers on page 716)

Each question has only one correct answer

Questions 1 to 4 relate to the following information:
x and y are two related engineering variables and p and q are constants. For the law $y - p = \frac{q}{x}$ to be verified it is necessary to plot a graph of the variables.

1. On the vertical axis is plotted:

   (a) y (b) p (c) q (d) x

2. On the horizontal axis is plotted:

   (a) x (b) $\frac{q}{x}$ (c) $\frac{1}{x}$ (d) p

3. The gradient of the graph is:

   (a) y (b) p (c) x (d) q

4. The vertical axis intercept is:

   (a) y (b) p (c) q (d) x

5. The relationship between two related engineering variables x and y is $y - cx = bx^2$ where b and c are constants. To produce a straight-line graph, it is necessary to plot:

   (a) x vertically against y horizontally

   (b) y vertically against x^2 horizontally

   (c) $\frac{y}{x}$ vertically against x horizontally

   (d) y vertically against x horizontally

6. The relationship between two related scientific variables x and y is $y = px^q$, where p and q are constants. To produce a straight-line graph, it is necessary to plot:

   (a) x vertically against y horizontally

   (b) lg y vertically against lg x horizontally

   (c) lg y vertically against ln x^q horizontally

   (d) y vertically against x horizontally

7. The relationship between two related engineering variables x and y is $y = cd^x$ where c and d are constants. To produce a straight-line graph, it is necessary to plot:

   (a) x vertically against y horizontally

   (b) ly y vertically against ln x horizontally

   (c) y vertically against x horizontally

   (d) lg y vertically against x horizontally

8. The relationship between two related scientific variables x and y is $\frac{y}{g} = e^{hx}$ where g and h are constants. To produce a straight-line graph, it is necessary to plot:

   (a) ln y vertically against x horizontally

   (b) ln y vertically against ln x horizontally

   (c) x vertically against y horizontally

   (d) y vertically against x horizontally

COMPANION @ WEBSITE **For fully worked solutions to each of the problems in Practice Exercises 142 and 143 in this chapter, go to the website:**
www.routledge.com/cw/bird

Chapter 29

Graphs with logarithmic scales

Why it is important to understand: Graphs with logarithmic scales

As mentioned in previous chapters, graphs are important tools for analysing and displaying data between two experimental quantities and that many times situations occur in which the relationship between the variables is not linear. By manipulation, a straight line graph may be plotted to produce a law relating the two variables. Knowledge of logarithms may be used to simplify plotting the relation between one variable and another. In particular, we consider those situations in which one of the variables requires scaling because the range of its data values is very large in comparison to the range of the other variable. Log-log and log-linear graph paper is available to make the plotting process easier.

At the end of this chapter, you should be able to:

- understand logarithmic scales
- understand log-log and log-linear graph paper
- plot a graph of the form $y = ax^n$ using log-log graph paper and determine constants 'a' and 'n'
- plot a graph of the form $y = ab^x$ using log-linear graph paper and determine constants 'a' and 'b'
- plot a graph of the form $y = ae^{kx}$ using log-linear graph paper and determine constants 'a' and 'k'

29.1 Logarithmic scales

Graph paper is available where the scale markings along the horizontal and vertical axes are proportional to the logarithms of the numbers. Such graph paper is called **log–log graph paper**.

A **logarithmic scale** is shown in Fig. 29.1 where distance between, say 1 and 2, is proportional to lg 2–lg 1, i.e. 0.3010 of the total distance from 1 to 10. Similarly, the distance between 7 and 8 is proportional to lg 8–lg 7, i.e. 0.05799 of the total distance from 1 to 10. Thus the distance between markings progressively decreases as the numbers increase from 1 to 10.

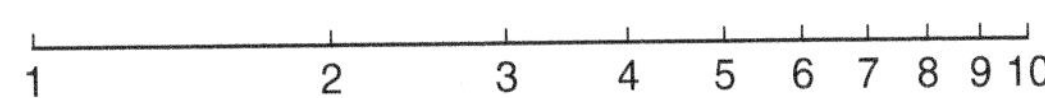

Figure 29.1

With log–log graph paper the scale markings are from 1 to 9, and this pattern can be repeated several times. The number of times the pattern of markings is repeated on an axis signifies the number of **cycles**. When the vertical axis has, say, 3 sets of values from 1 to 9, and the

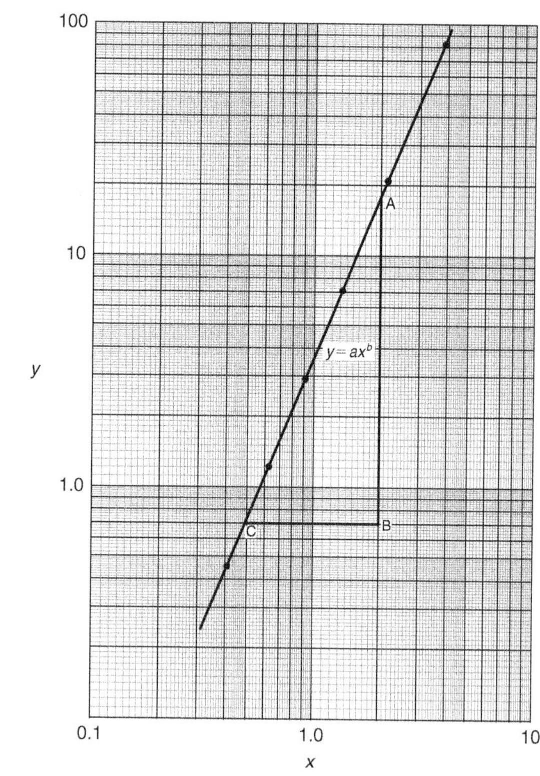

Figure 29.2

horizontal axis has, say, 2 sets of values from 1 to 9, then this log–log graph paper is called 'log 3 cycle × 2 cycle' (see Fig. 29.2). Many different arrangements, are available ranging from 'log 1 cycle × 1 cycle' through to 'log 5 cycle × 5 cycle'.
To depict a set of values, say, from 0.4 to 161, on an axis of log–log graph paper, 4 cycles are required, from 0.1 to 1, 1 to 10, 10 to 100 and 100 to 1000.

29.2 Graphs of the form $y = ax^n$

Taking logarithms to a base of 10 of both sides of $y = ax^n$ gives:

$$\lg y = \lg(ax^n)$$
$$= \lg a + \lg x^n$$

i.e. $\lg y = n \lg x + \lg a$

which compares with $Y = mX + c$

Thus, by plotting lg y vertically against lg x horizontally, a straight line results, i.e. the equation $y = ax^n$ is reduced to linear form. With log–log graph paper available x and y may be plotted directly, without having first to determine their logarithms, as shown in Chapter 28.

Problem 1. Experimental values of two related quantities x and y are shown below:

x	0.41	0.63	0.92	1.36	2.17	3.95
y	0.45	1.21	2.89	7.10	20.79	82.46

The law relating x and y is believed to be $y = ax^b$, where a and b are constants. Verify that this law is true and determine the approximate values of a and b

If $y = ax^b$ then $\lg y = b \lg x + \lg a$, from above, which is of the form $Y = mX + c$, showing that to produce a straight line graph lg y is plotted vertically against lg x horizontally. x and y may be plotted directly on to log–log graph paper as shown in Fig. 29.2. The values of y range from 0.45 to 82.46 and 3 cycles are needed (i.e. 0.1 to 1, 1 to 10 and 10 to 100). The values of x range from 0.41 to 3.95 and 2 cycles are needed (i.e. 0.1 to 1 and 1 to 10). Hence 'log 3 cycle × 2 cycle' is used as shown in Fig. 29.2 where the axes are marked and the points plotted. Since the points lie on a straight line the law $y = ax^b$ is verified.

To evaluate constants a and b:

Method 1. Any two points on the straight line, say points A and C, are selected, and AB and BC are measure (say in centimetres).

Then, gradient, $b = \frac{AB}{BC} = \frac{11.5 \text{ units}}{5 \text{ units}} = \mathbf{2.3}$

Since $\lg y = b \lg x + \lg a$, when $x = 1$, $\lg x = 0$ and $\lg y = \lg a$

The straight line crosses the ordinate $x = 1.0$ at $y = 3.5$

Hence $\lg a = \lg 3.5$, i.e. $\mathbf{a = 3.5}$

Method 2. Any two points on the straight line, say points A and C, are selected. A has co-ordinates (2, 17.25) and C has co-ordinates (0.5, 0.7)

Since $\quad y = ax^b$ then $17.25 = a(2)^b \quad (1)$

and $\quad 0.7 = a(0.5)^b \quad (2)$

i.e. two simultaneous equations are produced and may be solved for a and b.

Dividing equation (1) by equation (2) to eliminate a gives:

$$\frac{17.25}{0.7} = \frac{(2)^b}{(0.5)^b} = \left(\frac{2}{0.5}\right)^b$$

i.e. $\quad 24.643 = (4)^b$

Taking logarithms of both sides gives

$$\lg 24.643 = b \lg 4$$

and
$$b = \frac{\lg 24.643}{\lg 4}$$
$$= 2.3, \text{ correct to 2 significant figures.}$$

Substituting $b = 2.3$ in equation (1) gives:

$$17.25 = a(2)^{2.3}$$

and
$$a = \frac{17.25}{(2)^{2.3}} = \frac{17.25}{4.925}$$
$$= 3.5, \text{ correct to 2 significant figures.}$$

Hence the law of the graph is: $y = 3.5x^{2.3}$

Problem 2. The power dissipated by a resistor was measured for varying values of current flowing in the resistor and the results are as shown:

Current, I amperes	1.4	4.7	6.8	9.1	11.2	13.1
Power, P watts	49	552	1156	2070	3136	4290

Prove that the law relating current and power is of the form $P = RI^n$, where R and n are constants, and determine the law. Hence calculate the power when the current is 12 amperes and the current when the power is 1000 watts

Since $P = RI^n$ then $\lg P = n \lg I + \lg R$, which is of the form $Y = mX + c$, showing that to produce a straight line graph $\lg P$ is plotted vertically against $\lg I$ horizontally. Power values range from 49 to 4290, hence 3 cycles of log–log graph paper are needed (10 to 100, 100 to 1000 and 1000 to 10 000). Current values range from 1.4 to 11.2, hence 2 cycles of log–log graph paper are needed (1 to 10 and 10 to 100). Thus 'log 3 cycles × 2 cycles' is used as shown in Fig. 29.3 (or, if not available, graph paper having a larger number of cycles per axis can be used). The co-ordinates are plotted and a straight line results which proves that the law relating current and power is of the form $P = RI^n$. Gradient of straight line,

$$n = \frac{AB}{BC} = \frac{14 \text{ units}}{7 \text{ units}} = 2$$

At point C, $I = 2$ and $P = 100$. Substituting these values into $P = RI^n$ gives: $100 = R(2)^2$. Hence $R = 100/(2)^2 = 25$ which may have been found from the intercept on the $I = 1.0$ axis in Fig. 29.3.

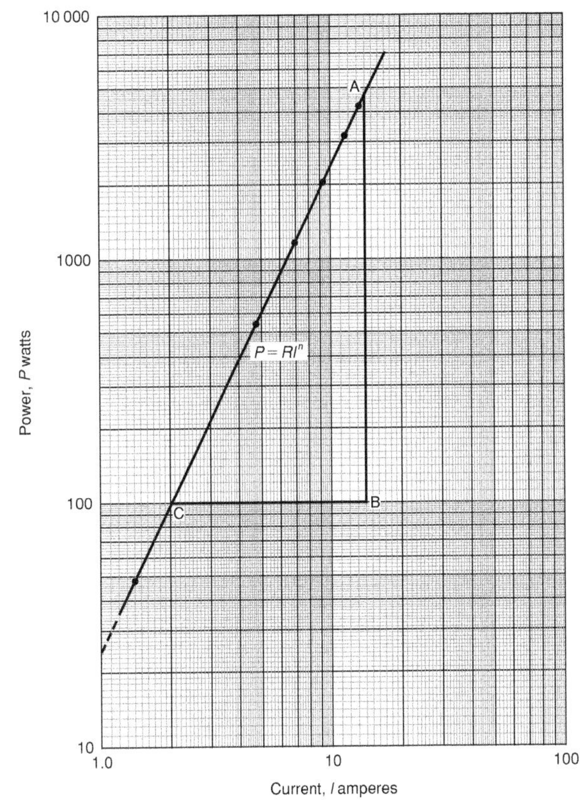

Figure 29.3

Hence the law of the graph is $P = 25I^2$

When current $I = 12$, power $P = 25(12)^2 =$ **3600 watts** (which may be read from the graph).

When power $P = 1000$, $1000 = 25I^2$

Hence
$$I^2 = \frac{1000}{25} = 40$$

from which,
$$I = \sqrt{40} = \mathbf{6.32\ A}$$

Problem 3. The pressure p and volume v of a gas are believed to be related by a law of the form $p = cv^n$, where c and n are constants. Experimental values of p and corresponding values of v obtained in a laboratory are:

p pascals	2.28×10^5	8.04×10^5	20.3×10^6
v m^3	3.2×10^{-2}	1.3×10^{-2}	6.7×10^{-3}

p pascals	5.05×10^6	1.82×10^7
v m^3	3.5×10^{-3}	1.4×10^{-3}

Verify that the law is true and determine approximate values of c and n

Since $p=cv^n$, then $\lg p=n \lg v+\lg c$, which is of the form $Y=mX+c$, showing that to produce a straight line graph $\lg p$ is plotted vertically against $\lg v$ horizontally. The co-ordinates are plotted on 'log 3 cycle × 2 cycle' graph paper as shown in Fig. 29.4. With the data expressed in standard form, the axes are marked in standard form also. Since a straight line results the law $p=cv^n$ is verified.

The straight line has a negative gradient and the value of the gradient is given by:

$$\frac{AB}{BC}=\frac{14 \text{ units}}{10 \text{ units}}=1.4,$$

hence $n=\mathbf{-1.4}$

Selecting any point on the straight line, say point C, having co-ordinates $(2.63\times10^{-2}, 3\times10^5)$, and substituting these values in $p=cv^n$ gives:

$$3\times10^5=c(2.63\times10^{-2})^{-1.4}$$

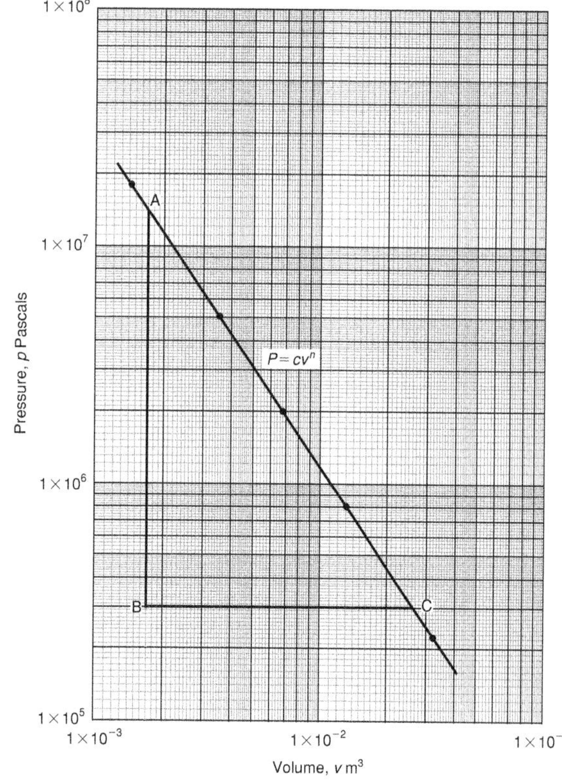

Figure 29.4

Hence
$$c=\frac{3\times10^5}{(2.63\times10^{-2})^{-1.4}}=\frac{3\times10^5}{(0.0263)^{-1.4}}$$
$$=\frac{3\times10^5}{1.63\times10^2}$$
$$=\mathbf{1840}, \text{ correct to 3 significant figures.}$$

Hence the law of the graph is:

$$\mathbf{p=1840v^{-1.4}} \text{ or } \mathbf{pv^{1.4}=1840}$$

Now try the following Practice Exercise

Practice Exercise 145 Graphs of the form $y=ax^n$ (Answers on page 716)

1. Engineering quantities x and y are believed to be related by a law of the form $y=ax^n$, where a and n are constants. Experimental values of x and corresponding values of y are:

x	0.8	2.3	5.4	11.5	21.6	42.9
y	8	54	250	974	3028	10 410

Show that the law is true and determine the values of a and n. Hence determine the value of y when x is 7.5 and the value of x when y is 5000.

2. Show from the following results of voltage V and admittance Y of an electrical circuit that the law connecting the quantities is of the form $V=kY^n$, and determine the values of k and n.

Voltage, V volts	2.88	2.05	1.60	1.22	0.96
Admittance, Y siemens	0.52	0.73	0.94	1.23	1.57

3. Scientific quantities x and y are believed to be related by a law of the form $y=mx^n$. The values of x and corresponding values of y are:

x	0.5	1.0	1.5	2.0	2.5	3.0
y	0.53	3.0	8.27	16.97	29.65	46.77

Verify the law and find the values of m and n.

29.3 Graphs of the form $y = ab^x$

Taking logarithms to a base of 10 of both sides of $y=ab^x$ gives:

$$\lg y = \lg(ab^x) = \lg a + \lg b^x = \lg a + x \lg b$$

i.e. $$\lg y = (\lg b)x + \lg a$$

which compares with $Y = mX + c$

Thus, by plotting lg y vertically against x horizontally a straight line results, i.e. the graph $y = ab^x$ is reduced to linear form. In this case, graph paper having a linear horizontal scale and a logarithmic vertical scale may be used. This type of graph paper is called **log–linear graph paper**, and is specified by the number of cycles of the logarithmic scale. For example, graph paper having 3 cycles on the logarithmic scale is called 'log 3 cycle × linear' graph paper.

Problem 4. Experimental engineering values of quantities x and y are believed to be related by a law of the form $y=ab^x$, where a and b are constants. The values of x and corresponding values of y are:

x	0.7	1.4	2.1	2.9	3.7	4.3
y	18.4	45.1	111	308	858	1850

Verify the law and determine the approximate values of a and b. Hence evaluate (i) the value of y when x is 2.5, and (ii) the value of x when y is 1200

Since $y=ab^x$ then lg y=(lg b)x+lg a (from above), which is of the form $Y=mX+c$, showing that to produce a straight line graph lg y is plotted vertically against x horizontally. Using log-linear graph paper, values of x are marked on the horizontal scale to cover the range 0.7 to 4.3. Values of y range from 18.4 to 1850 and 3 cycles are needed (i.e. 10 to 100, 100 to 1000 and 1000 to 10 000). Thus 'log 3 cycles × linear' graph paper is used as shown in Fig. 29.5. A straight line is drawn through the co-ordinates, hence the law $y=ab^x$ is verified.

Gradient of straight line, lg $b=AB/BC$. Direct measurement (say in centimetres) is not made with log-linear graph paper since the vertical scale is logarithmic and the horizontal scale is linear. Hence

$$\frac{AB}{BC} = \frac{\lg 1000 - \lg 100}{3.82 - 2.02} = \frac{3-2}{1.80}$$

$$= \frac{1}{1.80} = 0.5556$$

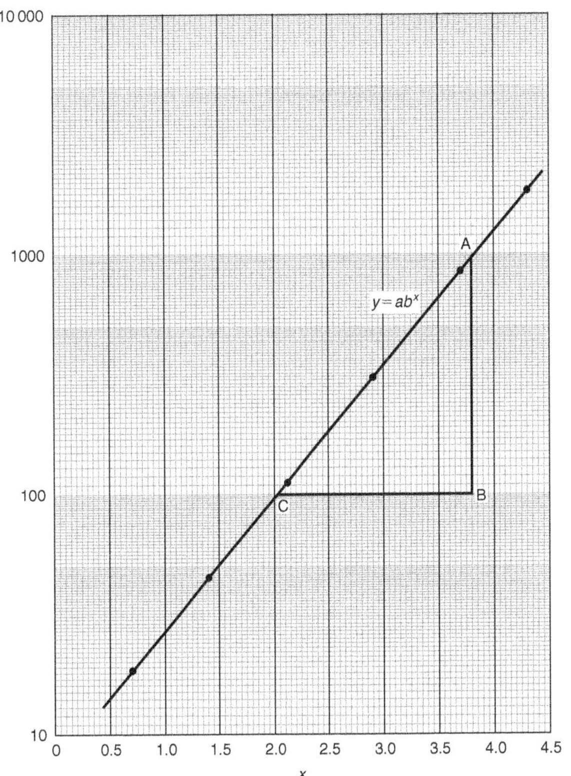

Figure 29.5

Hence b=antilog 0.5556(=$10^{0.5556}$)=**3.6**, correct to 2 significant figures.

Point A has co-ordinates (3.82, 1000).

Substituting these values into $y=ab^x$ gives:

$$1000 = a(3.6)^{3.82}$$

i.e. $$a = \frac{1000}{(3.6)^{3.82}}$$

$$= \mathbf{7.5}, \text{ correct to 2 significant figures.}$$

Hence the law of the graph is: $y = 7.5(3.6)^x$

(i) When $x=2.5$, $y=7.5(3.6)^{2.5}=\mathbf{184}$

(ii) When $y=1200$, $1200=7.5(3.6)^x$, hence

$$(3.6)^x = \frac{1200}{7.5} = 160$$

Taking logarithms gives: $x \lg 3.6 = \lg 160$

i.e. $$x = \frac{\lg 160}{\lg 3.6} = \frac{2.2041}{0.5563}$$

$$= \mathbf{3.96}$$

Now try the following Practice Exercise

Practice Exercise 146 Graphs of the form $y = ab^x$ (Answers on page 716)

1. Experimental scientific values of p and corresponding values of q are shown below:

p	−13.2	−27.9	−62.2	−383.2	−1581	−2931
q	0.30	0.75	1.23	2.32	3.17	3.54

Show that the law relating p and q is $p = ab^q$, where a and b are constants. Determine (i) values of a and b, and state the law, (ii) the value of p when q is 2.0 and (iii) the value of q when p is −2000.

29.4 Graphs of the form $y = ae^{kx}$

Taking logarithms to a base of e of both sides of $y = ae^{kx}$ gives:

$$\ln y = \ln(ae^{kx}) = \ln a + \ln e^{kx} = \ln a + kx \ln e$$

i.e. $\ln y = kx + \ln a$ (since $\ln e = 1$) which compares with $Y = mX + c$

Thus, by plotting $\ln y$ vertically against x horizontally, a straight line results, i.e. the equation $y = ae^{kx}$ is reduced to linear form. In this case, graph paper having a linear horizontal scale and a logarithmic vertical scale may be used.

Problem 5. The data given below is believed to be related by an exponential law of the form $y = ae^{kx}$, where a and b are constants. Verify that the law is true and determine approximate values of a and b. Also determine the value of y when x is 3.8 and the value of x when y is 85

x	−1.2	0.38	1.2	2.5	3.4	4.2	5.3
y	9.3	22.2	34.8	71.2	117	181	332

Since $y = ae^{kx}$ then $\ln y = kx + \ln a$ (from above), which is of the form $Y = mX + c$, showing that to produce a straight line graph $\ln y$ is plotted vertically against x horizontally. The value of y range from 9.3 to 332 hence 'log 3 cycle × linear' graph paper is used. The plotted co-ordinates are shown in Fig. 29.6 and since a straight line passes through the points the law $y = ae^{kx}$ is verified.

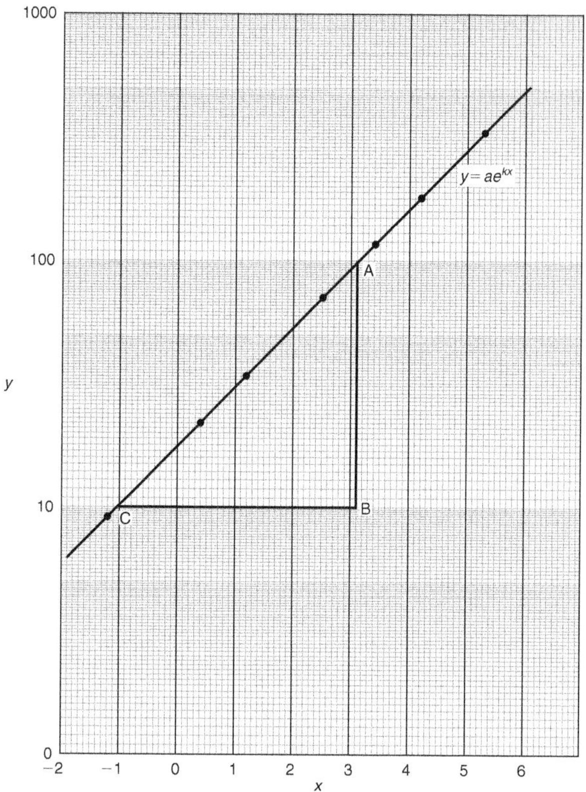

Figure 29.6

Gradient of straight line,

$$k = \frac{AB}{BC} = \frac{\ln 100 - \ln 10}{3.12 - (-1.08)} = \frac{2.3026}{4.20}$$
$$= 0.55, \text{ correct to 2 significant figures.}$$

Since $\ln y = kx + \ln a$, when $x = 0$, $\ln y = \ln a$, i.e. $y = a$
The vertical axis intercept value at $x = 0$ is 18, hence $a = 18$.

The law of the graph is thus: $y = 18e^{0.55x}$

When x is 3.8,

$y = 18e^{0.55(3.8)} = 18e^{2.09} = 18(8.0849) = \mathbf{146}$

When y is 85, $85 = 18e^{0.55x}$

Hence, $\quad e^{0.55x} = \frac{85}{18} = 4.7222$

and $\quad 0.55x = \ln 4.7222 = 1.5523$

Hence $\quad \mathbf{x} = \frac{1.5523}{0.55} = \mathbf{2.82}$

Problem 6. The voltage, v volts, across an inductor is believed to be related to time, t ms, by the law $v = Ve^{t/T}$, where V and T are constants. Experimental results obtained are:

v volts	883	347	90	55.5	18.6	5.2
t ms	10.4	21.6	37.8	43.6	56.7	72.0

Show that the law relating voltage and time is as stated and determine the approximate values of V and T. Find also the value of voltage after 25 ms and the time when the voltage is 30.0 V

Since $v = Ve^{t/T}$ then $\ln v = \frac{1}{T}t + \ln V$
which is of the form $Y = mX + c$
Using 'log 3 cycle × linear' graph paper, the points are plotted as shown in Fig. 29.7.
Since the points are joined by a straight line the law $v = Ve^{t/T}$ is verified.
Gradient of straight line,

$$\frac{1}{T} = \frac{AB}{BC} = \frac{\ln 100 - \ln 10}{36.5 - 64.2} = \frac{2.3026}{-27.7}$$

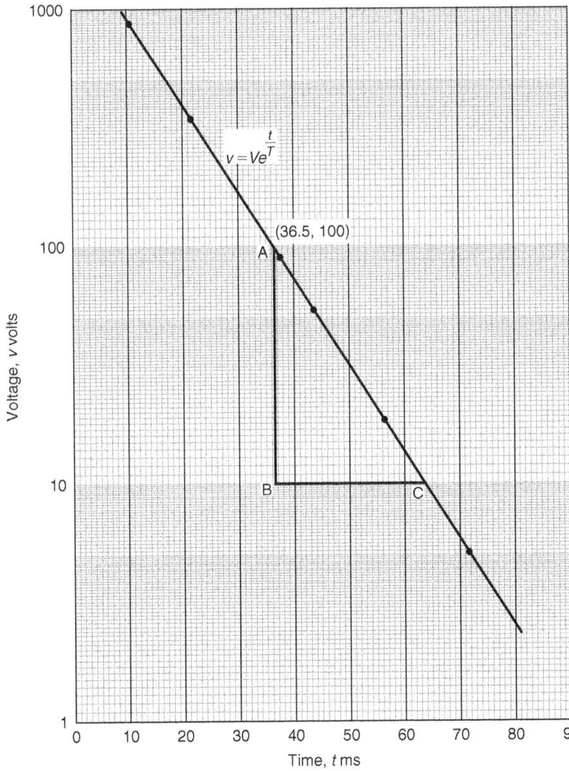

Figure 29.7

Hence $T = \frac{-27.7}{2.3026} = \mathbf{-12.0}$, correct to 3 significant figures.
Since the straight line does not cross the vertical axis at $t = 0$ in Fig. 29.7, the value of V is determined by selecting any point, say A, having co-ordinates (36.5, 100) and substituting these values into $v = Ve^{t/T}$. Thus

$$100 = Ve^{36.5/-12.0}$$

i.e. $$V = \frac{100}{e^{-36.5/12.0}} = \mathbf{2090\ volts},$$

correct to 3 significant figures.

Hence the law of the graph is: $v = 2090e^{-t/12.0}$

When time $t = 25$ ms, voltage $v = 2090e^{-25/12.0}$
$= \mathbf{260\ V}$

When the voltage is 30.0 volts, $30.0 = 2090e^{-t/12.0}$ hence

$$e^{-t/12.0} = \frac{30.0}{2090} \text{ and } e^{t/12.0} = \frac{2090}{30.0} = 69.67$$

Taking Napierian logarithms gives:

$$\frac{t}{12.0} = \ln 69.67 = 4.2438$$

from which, time $t = (12.0)(4.2438) = \mathbf{50.9\ ms}$.

Now try the following Practice Exercise

Practice Exercise 147 Reducing exponential laws to linear form (Answers on page 716)

1. Atmospheric pressure p is measured at varying altitudes h and the results are as shown below:

Altitude, h m	500	1500	3000	5000	8000
pressure, p kPa	73.39	68.42	61.60	53.56	43.41

Show that the quantities are related by the law $p = ae^{kh}$, where a and k are constants. Determine, the values of a and k and state the law. Find also the atmospheric pressure at 10 000 m.

2. At particular times, t minutes, measurements are made of the temperature, θ°C, of a

cooling liquid and the following results are obtained:

Temperature θ°C	92.2	55.9	33.9	20.6	12.5
Time t minutes	10	20	30	40	50

Prove that the quantities follow a law of the form $\theta = \theta_0 e^{kt}$, where θ_0 and k are constants, and determine the approximate value of θ_0 and k.

Practice Exercise 148 Multiple-choice questions on graphs with logarithmic scales (Answers on page 716)

Each question has only one correct answer

Questions 1 to 3 relate to the following information.
A straight-line graph is plotted for the equation $y = ax^n$, where y and x are the variables and a and n are constants.

1. On the vertical axis is plotted:

   (a) y (b) x (c) lg y (d) a

2. On the horizontal axis is plotted:

   (a) lg x (b) x (c) x^n (d) a

3. The gradient of the graph is given by:

   (a) y (b) a (c) x (d) n

4. To depict a set of values from 0.05 to 275, the minimum number of cycles required on logarithmic graph paper is:

   (a) 5 (b) 3 (c) 4 (d) 2

Questions 5 to 7 relate to the following information.
A straight-line graph is plotted for the equation $y = ae^{bx}$, where y and x are the variables and a and b are constants.

5. On the vertical axis is plotted:

   (a) y (b) x (c) ln y (d) a

6. The gradient of the graph is given by:

   (a) y (b) a (c) x (d) b

7. The vertical axis intercept is given by:

   (a) b (b) ln a (c) x (d) ln y

Questions 8 to 10 relate to the following information.
A straight-line graph is plotted for the equation $y = ab^x$, where y and x are the variables and a and b are constants.

8. On the horizontal axis is plotted:

   (a) lg x (b) x (c) b^x (d) a

9. The gradient of the graph is given by:

   (a) y (b) a (c) b (d) lg b

10. The vertical axis intercept is given by:

    (a) b (b) lg a (c) x (d) lg y

COMPANION WEBSITE

For fully worked solutions to each of the problems in Practice Exercises 145 to 147 in this chapter, go to the website:
www.routledge.com/cw/bird

Chapter 30

Graphical solution of equations

Why it is important to understand: Graphical solution of equations

It has been established in previous chapters that the solution of linear, quadratic, simultaneous and cubic equations occur often in engineering and science and may be solved using algebraic means. Being able to solve equations graphically provides another method to aid understanding and interpretation of equations. Architects, surveyors and a variety of engineers in fields such as biomedical, chemical, electrical, mechanical and nuclear, all use equations which need solving by one means or another.

At the end of this chapter, you should be able to:

- solve two simultaneous equations graphically
- solve a quadratic equation graphically
- solve a linear and simultaneous equation simultaneously by graphical means
- solve a cubic equation graphically

30.1 Graphical solution of simultaneous equations

Linear simultaneous equations in two unknowns may be solved graphically by:

(i) plotting the two straight lines on the same axes, and

(ii) noting their point of intersection.

The co-ordinates of the point of intersection give the required solution.

Problem 1. Solve graphically the simultaneous equations:

$$2x - y = 4$$

$$x + y = 5$$

Rearranging each equation into $y = mx + c$ form gives:

$$y = 2x - 4 \quad (1)$$

$$y = -x + 5 \quad (2)$$

Only three co-ordinates need be calculated for each graph since both are straight lines.

x	0	1	2
$y=2x-4$	−4	−2	0

x	0	1	2
$y=-x+5$	5	4	3

Each of the graphs is plotted as shown in Fig. 30.1. The point of intersection is at (3, 2) and since this is the only point which lies simultaneously on both lines then $\mathbf{x=3, y=2}$ is the solution of the simultaneous equations.

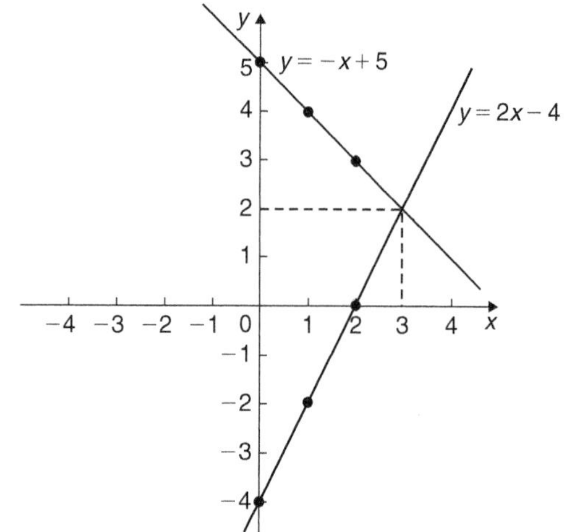

Figure 30.1

Problem 2. Solve graphically the equations:

$$1.20x + y = 1.80$$
$$x - 5.0y = 8.50$$

Rearranging each equation into $y=mx+c$ form gives:

$$y = -1.20x + 1.80 \quad (1)$$

$$y = \frac{x}{5.0} - \frac{8.5}{5.0}$$

i.e. $$y = 0.20x - 1.70 \quad (2)$$

Three co-ordinates are calculated for each equation as shown below:

x	0	1	2
$y=-1.20x+1.80$	1.80	0.60	−0.60

x	0	1	2
$y=0.20x-1.70$	−1.70	−1.50	−1.30

The two lines are plotted as shown in Fig. 30.2. The point of intersection is (2.50, −1.20). Hence the solution of the simultaneous equation is $\mathbf{x=2.50, y=-1.20}$

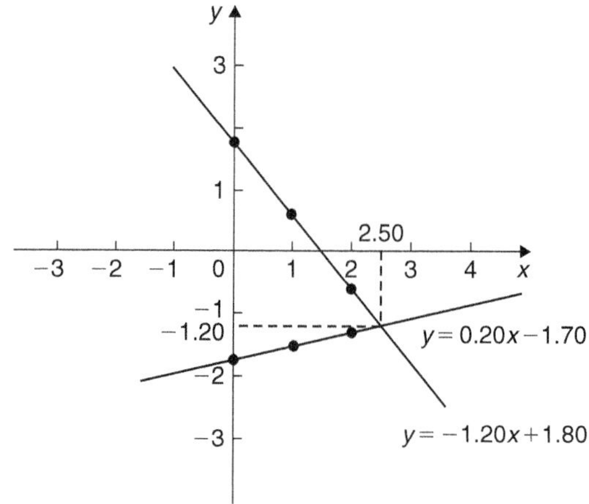

Figure 30.2

(It is sometimes useful initially to sketch the two straight lines to determine the region where the point of intersection is. Then, for greater accuracy, a graph having a smaller range of values can be drawn to 'magnify' the point of intersection.)

Now try the following Practice Exercise

Practice Exercise 149 Graphical solution of simultaneous equations (Answers on page 716)

In Problems 1 to 5, solve the simultaneous equations graphically.

1. $x+y=2$

   $3y-2x=1$

2. $y=5-x$

   $x-y=2$

3. $3x+4y=5$

   $2x-5y+12=0$

4. $1.4x-7.06=3.2y$

   $2.1x-6.7y=12.87$

5. $3x-2y=0$

   $4x+y+11=0$

6. The friction force F Newtons and load L Newton's are connected by a law of the form $F=aL+b$, where a and b are constants. When $F=4$ Newtons, $L=6$ Newtons and when $F=2.4$ Newtons, $L=2$ Newtons. Determine graphically the values of a and b

30.2 Graphical solution of quadratic equations

A general **quadratic equation** is of the form $y=ax^2+bx+c$, where a, b and c are constants and a is not equal to zero.

A graph of a quadratic equation always produces a shape called a **parabola**. The gradient of the curve between 0 and A and between B and C in Fig. 30.3 is positive, whilst the gradient between A and B is negative. Points such as A and B are called **turning points**. At A the gradient is zero and, as x increases, the gradient of the curve changes from positive just before A to negative just after. Such a point is called a **maximum value**. At B the gradient is also zero, and, as x increases, the gradient of the curve changes from negative just before B to positive just after. Such a point is called a **minimum value**.

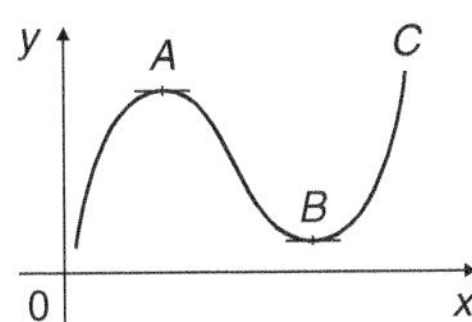

Figure 30.3

Quadratic graphs

(i) $y=ax^2$

Graphs of $y=x^2$, $y=3x^2$ and $y=\frac{1}{2}x^2$ are shown in Fig. 30.4.

All have minimum values at the origin (0, 0).

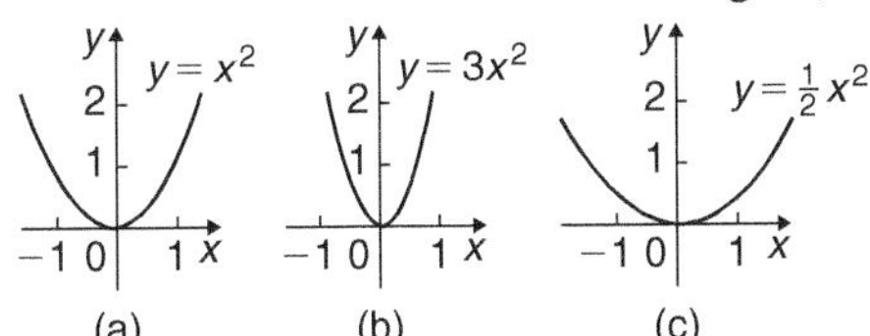

Figure 30.4

Graphs of $y=-x^2$, $y=-3x^2$ and $y=-\frac{1}{2}x^2$ are shown in Fig. 30.5.

All have maximum values at the origin (0, 0).

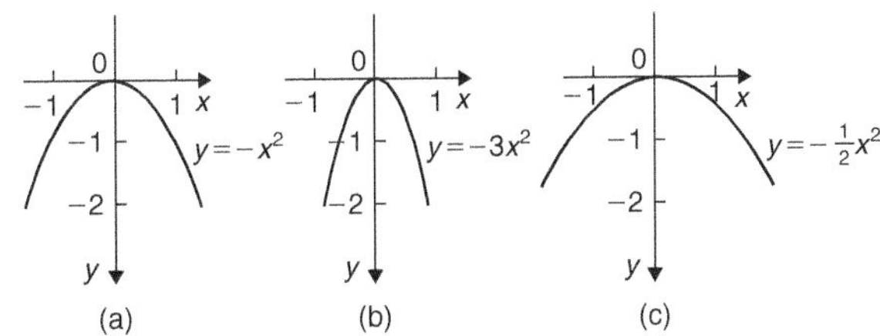

Figure 30.5

When $y=ax^2$,

(a) curves are symmetrical about the y-axis,
(b) the magnitude of 'a' affects the gradient of the curve, and
(c) the sign of 'a' determines whether it has a maximum or minimum value.

(ii) $y=ax^2+c$

Graphs of $y=x^2+3$, $y=x^2-2$, $y=-x^2+2$ and $y=-2x^2-1$ are shown in Fig. 30.6.

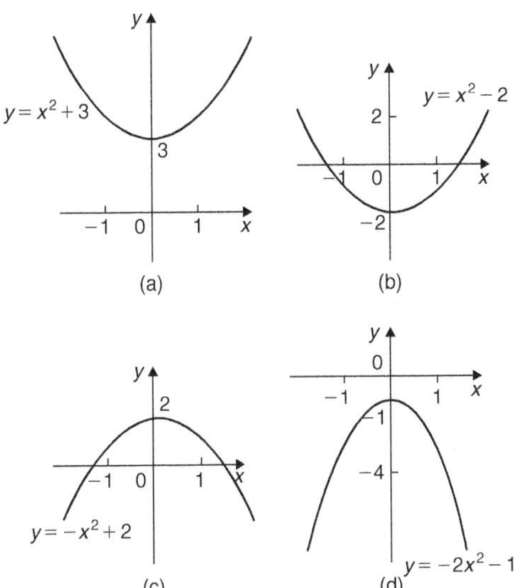

Figure 30.6

When $y=ax^2+c$,

(a) curves are symmetrical about the y-axis,
(b) the magnitude of 'a' affects the gradient of the curve, and
(c) the constant 'c' is the y-axis intercept.

(iii) $y = ax^2 + bx + c$

Whenever 'b' has a value other than zero the curve is displaced to the right or left of the y-axis. When b/a is positive, the curve is displaced $b/2a$ to the left of the y-axis, as shown in Fig. 30.7(a). When b/a is negative the curve is displaced $b/2a$ to the right of the y-axis, as shown in Fig. 30.7(b).

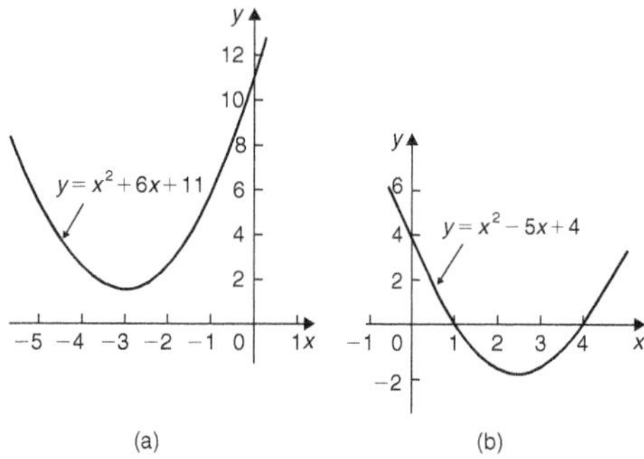

Figure 30.7

Quadratic equations of the form $ax^2 + bx + c = 0$ may be solved graphically by:

(i) plotting the graph $y = ax^2 + bx + c$, and

(ii) noting the points of intersection on the x-axis (i.e. where $y = 0$)

The x values of the points of intersection give the required solutions since at these points both $y = 0$ and $ax^2 + bx + c = 0$. The number of solutions, or roots of a quadratic equation, depends on how many times the curve cuts the x-axis and there can be no real roots (as in Fig. 30.7(a)) or one root (as in Figs 30.4 and 30.5) or two roots (as in Fig. 30.7(b)).

Problem 3. Solve the quadratic equation $4x^2 + 4x - 15 = 0$ graphically given that the solutions lie in the range $x = -3$ to $x = 2$. Determine also the co-ordinates and nature of the turning point of the curve

Let $y = 4x^2 + 4x - 15$. A table of values is drawn up as shown below:

x	-3	-2	-1	0	1	2
$4x^2$	36	16	4	0	4	16
$4x$	-12	-8	-4	0	4	8
-15	-15	-15	-15	-15	-15	-15
$y = 4x^2 + 4x - 15$	9	-7	-15	-15	-7	9

A graph of $y = 4x^2 + 4x - 15$ is shown in Fig. 30.8. The only points where $y = 4x^2 + 4x - 15$ and $y = 0$ are the points marked A and B. This occurs at $\boldsymbol{x = -2.5}$ **and** $\boldsymbol{x = 1.5}$ and these are the solutions of the quadratic equation $4x^2 + 4x - 15 = 0$. (By substituting $x = -2.5$ and $x = 1.5$ into the original equation the solutions may be checked.) The curve has a turning point at $(-0.5, -16)$ and the nature of the point is a **minimum**.

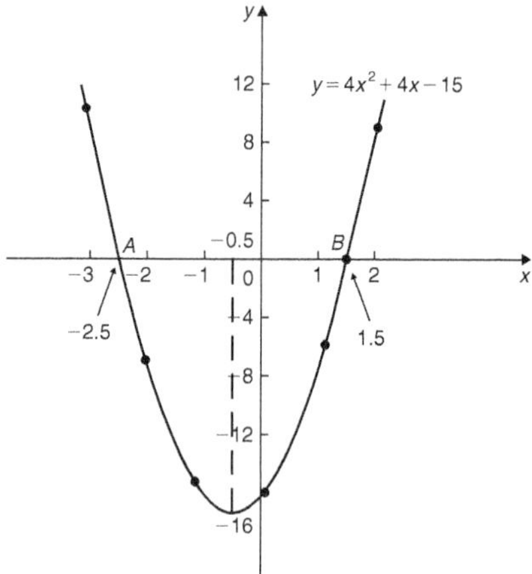

Figure 30.8

An alternative graphical method of solving $4x^2 + 4x - 15 = 0$ is to rearrange the equation as $4x^2 = -4x + 15$ and then plot two separate graphs — in this case $y = 4x^2$ and $y = -4x + 15$. Their points of intersection give the roots of equation $4x^2 = -4x + 15$, i.e. $4x^2 + 4x - 15 = 0$. This is shown in Fig. 30.9, where the roots are $x = -2.5$ and $x = 1.5$ as before.

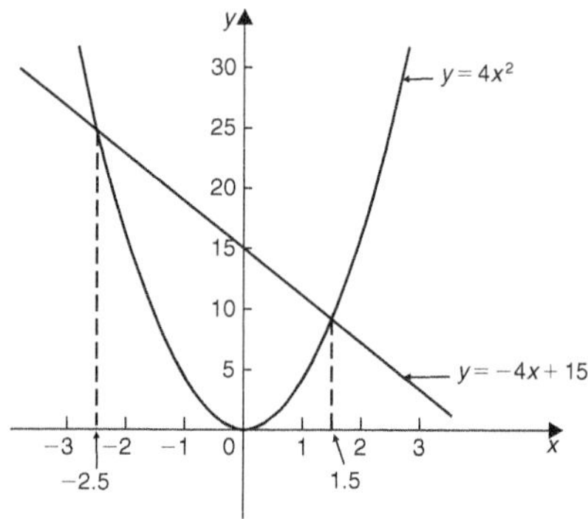

Figure 30.9

Problem 4. Solve graphically the quadratic equation $-5x^2 + 9x + 7.2 = 0$ given that the solutions lie between $x = -1$ and $x = 3$. Determine also the co-ordinates of the turning point and state its nature

Let $y=-5x^2+9x+7.2$. A table of values is drawn up as shown to the right. A graph of $y=-5x^2+9x+7.2$ is shown plotted in Fig. 30.10. The graph crosses the x-axis (i.e. where $y=0$) at $\mathbf{x=-0.6}$ **and** $\mathbf{x=2.4}$ and these are the solutions of the quadratic equation $-5x^2+9x+7.2=0$. The turning point is a **maximum** having co-ordinates **(0.9, 11.25)**.

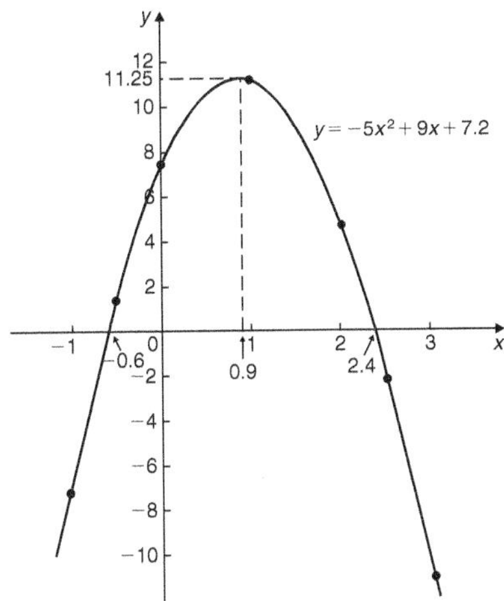

Figure 30.10

x	−1	−0.5	0	1
$-5x^2$	−5	−1.25	0	−5
$+9x$	−9	−4.5	0	9
$+7.2$	7.2	7.2	7.2	7.2
$y=-5x^2+9x+7.2$	−6.8	1.45	7.2	11.2

x	2	2.5	3
$-5x^2$	−20	−31.25	−45
$+9x$	18	22.5	27
$+7.2$	7.2	7.2	7.2
$y=-5x^2+9x+7.2$	5.2	−1.55	−10.8

Problem 5. Plot a graph of: $y=2x^2$ and hence solve the equations: (a) $2x^2-8=0$ and (b) $2x^2-x-3=0$

A graph of $y=2x^2$ is shown in Fig. 30.11.

(a) Rearranging $2x^2-8=0$ gives $2x^2=8$ and the solution of this equation is obtained from the points of intersection of $y=2x^2$ and $y=8$, i.e. at co-ordinates (−2, 8) and (2, 8), shown as A and B, respectively, in Fig. 30.11. Hence the solutions of $2x^2-8=0$ and $\mathbf{x=-2}$ **and** $\mathbf{x=+2}$

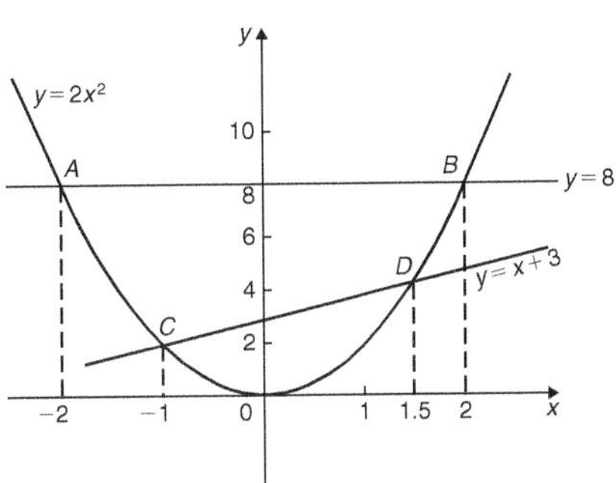

Figure 30.11

(b) Rearranging $2x^2-x-3=0$ gives $2x^2=x+3$ and the solution of this equation is obtained from the points of intersection of $y=2x^2$ and $y=x+3$, i.e. at C and D in Fig. 30.11. Hence the solutions of $2x^2-x-3=0$ are $\mathbf{x=-1}$ **and** $\mathbf{x=1.5}$

Problem 6. Plot the graph of $y=-2x^2+3x+6$ for values of x from $x=-2$ to $x=4$. Use the graph to find the roots of the following equations:

(a) $-2x^2+3x+6=0$
(b) $-2x^2+3x+2=0$
(c) $-2x^2+3x+9=0$
(d) $-2x^2+x+5=0$

A table of values is drawn up as shown below.

x	−2	−1	0	1	2	3	4
$-2x^2$	−8	−2	0	−2	−8	−18	−32
$+3x$	−6	−3	0	3	6	9	12
$+6$	6	6	6	6	6	6	6
y	−8	1	6	7	4	−3	−14

A graph of $y=-2x^2+3x+6$ is shown in Fig. 30.12.

(a) The parabola $y=-2x^2+3x+6$ and the straight line $y=0$ intersect at A and B, where $\mathbf{x=-1.13}$ **and** $\mathbf{x=2.63}$ and these are the roots of the equation $-2x^2+3x+6=0$

(b) Comparing

$$y=-2x^2+3x+6 \quad (1)$$

$$\text{with} \quad 0=-2x^2+3x+2 \quad (2)$$

shows that if 4 is added to both sides of equation (2), the right-hand side of both equations will be the same.

Hence $4=-2x^2+3x+6$. The solution of this equation is found from the points of intersection of

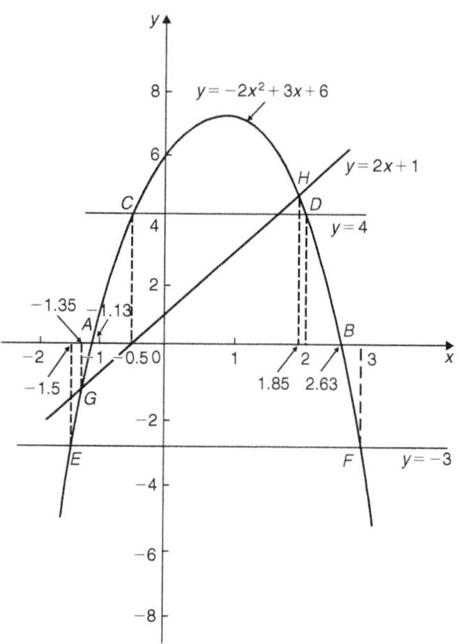

Figure 30.12

the line $y=4$ and the parabola $y=-2x^2+3x+6$, i.e. points C and D in Fig. 30.12. Hence the roots of $-2x^2+3x+2=0$ are $\boldsymbol{x=-0.5}$ **and** $\boldsymbol{x=2}$

(c) $-2x^2+3x+9=0$ may be rearranged as $-2x^2+3x+6=-3$, and the solution of this equation is obtained from the points of intersection of the line $y=-3$ and the parabola $y=-2x^2+3x+6$, i.e. at points E and F in Fig. 30.12. Hence the roots of $-2x^2+3x+9=0$ are $\boldsymbol{x=-1.5}$ **and** $\boldsymbol{x=3}$

(d) Comparing

$$y=-2x^2+3x+6 \qquad (3)$$

with $$0=-2x^2+x+5 \qquad (4)$$

shows that if $2x+1$ is added to both sides of equation (4) the right-hand side of both equations will be the same. Hence equation (4) may be written as $2x+1=-2x^2+3x+6$. The solution of this equation is found from the points of intersection of the line $y=2x+1$ and the parabola $y=-2x^2+3x+6$, i.e. points G and H in Fig. 30.12. Hence the roots of $-2x^2+x+5=0$ are $\boldsymbol{x=-1.35}$ **and** $\boldsymbol{x=1.85}$

Now try the following Practice Exercise

Practice Exercise 150 Solving quadratic equations graphically (Answers on page 716)

1. Sketch the following graphs and state the nature and co-ordinates of their turning points:

   (a) $y=4x^2$ (b) $y=2x^2-1$
   (c) $y=-x^2+3$ (d) $y=-\frac{1}{2}x^2-1$

Solve graphically the quadratic equations in Problems 2 to 5 by plotting the curves between the given limits. Give answers correct to 1 decimal place.

2. $4x^2-x-1=0$; $x=-1$ to $x=1$
3. $x^2-3x=27$; $x=-5$ to $x=8$
4. $2x^2-6x-9=0$; $x=-2$ to $x=5$
5. $2x(5x-2)=39.6$; $x=-2$ to $x=3$
6. Solve the quadratic equation $2x^2+7x+6=0$ graphically, given that the solutions lie in the range $x=-3$ to $x=1$. Determine also the nature and co-ordinates of its turning point
7. Solve graphically the quadratic equation $10x^2-9x-11.2=0$, given that the roots lie between $x=-1$ and $x=2$
8. Plot a graph of $y=3x^2$ and hence solve the equations (a) $3x^2-8=0$ and (b) $3x^2-2x-1=0$
9. Plot the graphs $y=2x^2$ and $y=3-4x$ on the same axes and find the co-ordinates of the points of intersection. Hence determine the roots of the equation $2x^2+4x-3=0$
10. Plot a graph of $y=10x^2-13x-30$ for values of x between $x=-2$ and $x=3$. Solve the equation $10x^2-13x-30=0$ and from the graph determine: (a) the value of y when x is 1.3, (b) the values of x when y is 10 and (c) the roots of the equation $10x^2-15x-18=0$

30.3 Graphical solution of linear and quadratic equations simultaneously

The solution of **linear and quadratic equations simultaneously** may be achieved graphically by: (i) plotting the straight line and parabola on the same axes, and (ii) noting the points of intersection. The co-ordinates of the points of intersection give the required solutions.

Problem 7. Determine graphically the values of x and y which simultaneously satisfy the equations: $y=2x^2-3x-4$ and $y=2-4x$

$y=2x^2-3x-4$ is a parabola and a table of values is drawn up as shown below:

x	−2	−1	0	1	2	3
$2x^2$	8	2	0	2	8	18
$-3x$	6	3	0	−3	−6	−9
−4	−4	−4	−4	−4	−4	−4
y	10	1	−4	−5	−2	5

$y=2-4x$ is a straight line and only three co-ordinates need be calculated:

x	0	1	2
y	2	−2	−6

The two graphs are plotted in Fig. 30.13 and the points of intersection, shown as A and B, are at co-ordinates (−2, 10) and (1.5, −4). Hence the simultaneous solutions occur when $\boldsymbol{x=-2, y=10}$ and when $\boldsymbol{x=1.5, y=-4}$

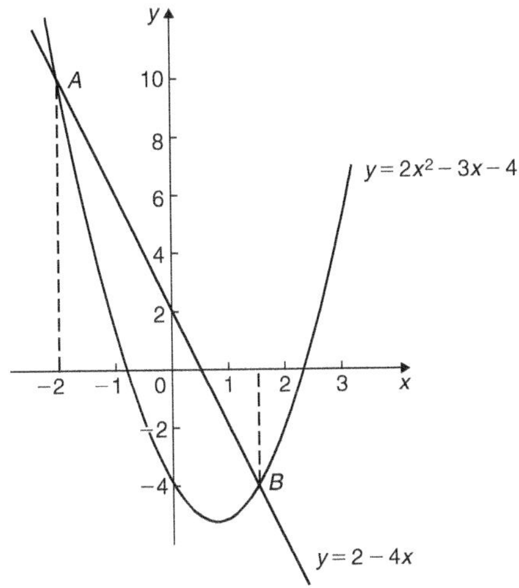

Figure 30.13

(These solutions may be checked by substituting into each of the original equations.)

Now try the following Practice Exercise

Practice Exercise 151 Solving linear and quadratic equations simultaneously (Answers on page 716)

1. Determine graphically the values of x and y which simultaneously satisfy the equations $y=2(x^2-2x-4)$ and $y+4=3x$
2. Plot the graph of $y=4x^2-8x-21$ for values of x from −2 to +4. Use the graph to find the roots of the following equations:
(a) $4x^2-8x-21=0$ (b) $4x^2-8x-16=0$
(c) $4x^2-6x-18=0$

30.4 Graphical solution of cubic equations

A **cubic equation** of the form $ax^3+bx^2+cx+d=0$ may be solved graphically by: (i) plotting the graph $y=ax^3+bx^2+cx+d$, and (ii) noting the points of intersection on the x-axis (i.e. where $y=0$). The x-values of the points of intersection give the required solution since at these points both $y=0$ and $ax^3+bx^2+cx+d=0$

The number of solutions, or roots of a cubic equation depends on how many times the curve cuts the x-axis and there can be one, two or three possible roots, as shown in Fig. 30.14.

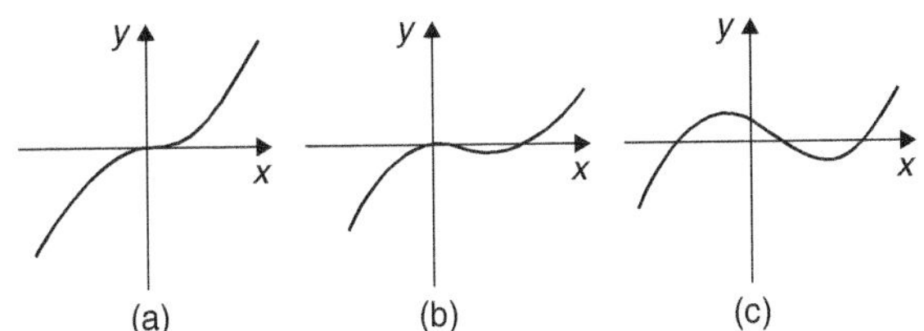

Figure 30.14

Problem 8. Solve graphically the cubic equation $4x^3-8x^2-15x+9=0$ given that the roots lie between $x=-2$ and $x=3$. Determine also the co-ordinates of the turning points and distinguish between them

Let $y=4x^3-8x^2-15x+9$. A table of values is drawn up as shown below:

x	−2	−1	0	1	2	3
$4x^3$	−32	−4	0	4	32	108
$-8x^2$	−32	−8	0	−8	−32	−72
$-15x$	30	15	0	−15	−30	−45
+9	9	9	9	9	9	9
y	−25	12	9	−10	−21	0

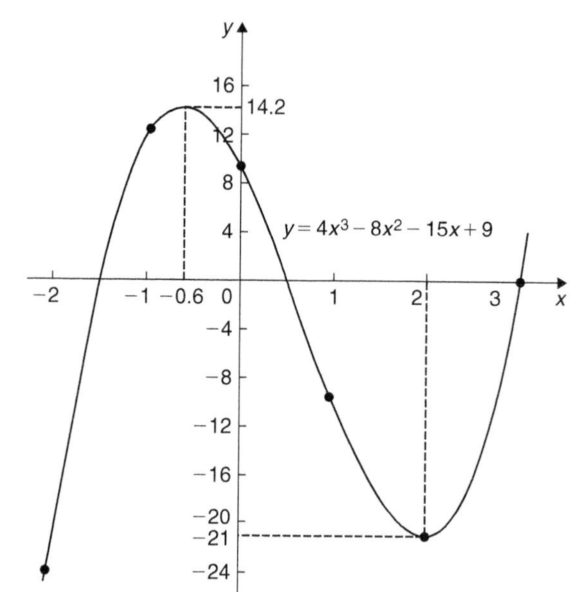

Figure 30.15

A graph of $y = 4x^3 - 8x^2 - 15x + 9$ is shown in Fig. 30.15.

The graph crosses the x-axis (where $y = 0$) at $\boldsymbol{x = -1.5}$, $\boldsymbol{x = 0.5}$ **and** $\boldsymbol{x = 3}$ and these are the solutions to the cubic equation $4x^3 - 8x^2 - 15x + 9 = 0$
The turning points occur at (**−0.6, 14.2**), which is a **maximum**, and (**2, −21**), which is a **minimum**.

Problem 9. Plot the graph of $y = 2x^3 - 7x^2 + 4x + 4$ for values of x between $x = -1$ and $x = 3$. Hence determine the roots of the equation:

$$2x^3 - 7x^2 + 4x + 4 = 0$$

A table of values is drawn up as shown below.

x	−1	0	1	2	3
$2x^3$	−2	0	2	16	54
$-7x^2$	−7	0	−7	−28	−63
$+4x$	−4	0	4	8	12
$+4$	4	4	4	4	4
y	−9	4	3	0	7

A graph of $y = 2x^3 - 7x^2 + 4x + 4$ is shown in Fig. 30.16. The graph crosses the x-axis at $x = -0.5$ and touches the x-axis at $x = 2$. Hence the solutions of the equation $2x^3 - 7x^2 + 4x + 4 = 0$ are $\boldsymbol{x = -0.5}$ **and** $\boldsymbol{x = 2}$

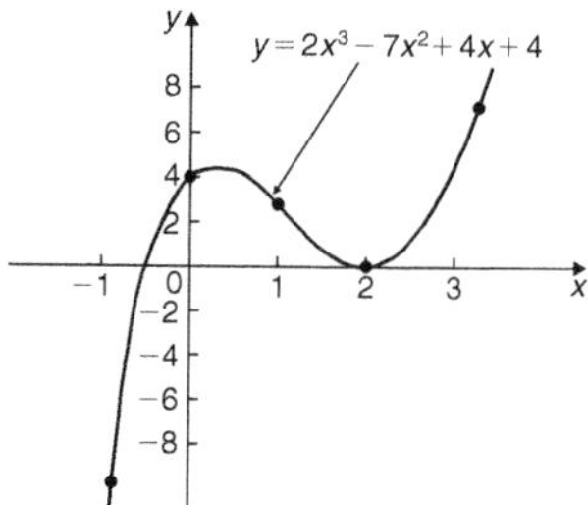

Figure 30.16

Now try the following Practice Exercise

Practice Exercise 152 Solving cubic equations (Answers on page 716)

1. Plot the graph $y = 4x^3 + 4x^2 - 11x - 6$ between $x = -3$ and $x = 2$ and use the graph to solve the cubic equation
$$4x^3 + 4x^2 - 11x - 6 = 0$$
2. By plotting a graph of $y = x^3 - 2x^2 - 5x + 6$ between $x = -3$ and $x = 4$ solve the equation $x^3 - 2x^2 - 5x + 6 = 0$. Determine also the co-ordinates of the turning points and distinguish between them

In Problems 3 to 6, solve graphically the cubic equations given, each correct to 2 significant figures.

3. $x^3 - 1 = 0$
4. $x^3 - x^2 - 5x + 2 = 0$
5. $x^3 - 2x^2 = 2x - 2$
6. $2x^3 - x^2 - 9.08x + 8.28 = 0$
7. Show that the cubic equation $8x^3 + 36x^2 + 54x + 27 = 0$ has only one real root and determine its value

Practice Exercise 153 Multiple-choice questions on graphical solution of equations (Answers on page 716)

Each question has only one correct answer

1. The equation of the graph shown in Figure 30.17 is:

   (a) $x(x+1) = \frac{15}{4}$ (b) $4x^2 - 4x - 15 = 0$

   (c) $x^2 - 4x - 5 = 0$ (d) $4x^2 + 4x - 15 = 0$

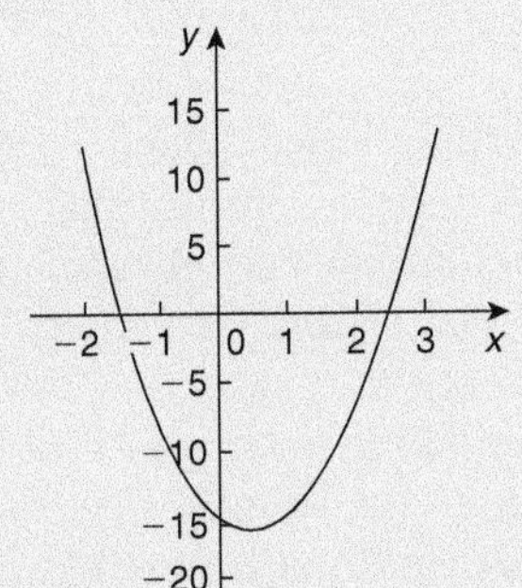

Figure 30.17

2. The numbers of real roots to the equation $ax^2 + bx + c = 0$ may be as many as:

   (a) 1 (b) 2 (c) 3 (d) 4

3. The numbers of real roots to the equation $ax^3 + bx^2 + cx + d = 0$ may be as many as:

   (a) 1 (b) 2 (c) 3 (d) 4

4. A parabola cuts the x-axis at $x = -3$ and at $x = 2$. The equation of the graph is:

   (a) $y = x^2 - x + 6$ (b) $y = x^2 + x + 6$

   (c) $y = x^2 - x - 6$ (d) $y = x^2 + x - 6$

5. Graphs of $y = 2x^2 - 3x + 1$ and $y = 2x - 1$ are drawn on the same axes. The two graphs intersect where:

   (a) $x = 0.5$ and $x = 2$ (b) $x = 1$ and $x = 2$

   (c) $x = 0.5$ and $x = 1$ (d) $x = 1$ and $x = 3$

For fully worked solutions to each of the problems in Practice Exercises 149 to 152 in this chapter, go to the website:

www.routledge.com/cw/bird

COMPANION @ WEBSITE

Chapter 31

Functions and their curves

Why it is important to understand: Functions and their curves

Graphs and diagrams provide a simple and powerful approach to a variety of problems that are typical to computer science in general, and software engineering in particular; graphical transformations have many applications in software engineering problems. Periodic functions are used throughout engineering and science to describe oscillations, waves and other phenomena that exhibit periodicity. Engineers use many basic mathematical functions to represent, say, the input/output of systems – linear, quadratic, exponential, sinusoidal and so on – and knowledge of these are needed to determine how these are used to generate some of the more unusual input/output signals such as the square wave, saw-tooth wave and fully-rectified sine wave. Understanding of continuous and discontinuous functions, odd and even functions and inverse functions are helpful in this – it's all part of the 'language of engineering'.

At the end of this chapter, you should be able to:

- recognise standard curves and their equations – straight line, quadratic, cubic, trigonometric, circle, ellipse, hyperbola, rectangular hyperbola, logarithmic function, exponential function and polar curves
- perform simple graphical transformations
- define a periodic function
- define continuous and discontinuous functions
- define odd and even functions
- define inverse functions

31.1 Standard curves

When a mathematical equation is known, co-ordinates may be calculated for a limited range of values, and the equation may be represented pictorially as a graph, within this range of calculated values. Sometimes it is useful to show all the characteristic features of an equation, and in this case a sketch depicting the equation can be drawn, in which all the important features are shown, but the accurate plotting of points is less important. This technique is called 'curve sketching' and can involve the use of differential calculus, with, for example, calculations involving turning points.

If, say, y depends on, say, x, then y is said to be a function of x and the relationship is expressed as $y=f(x)$; x is called the independent variable and y is the dependent variable.

In engineering and science, corresponding values are obtained as a result of tests or experiments.

Here is a brief resumé of standard curves, some of which have been met earlier in this text.

(i) Straight line (see Chapter 27, page 289)
The general equation of a straight line is $y=mx+c$, where m is the gradient and c is the y-axis intercept.
Two examples are shown in Fig. 31.1.

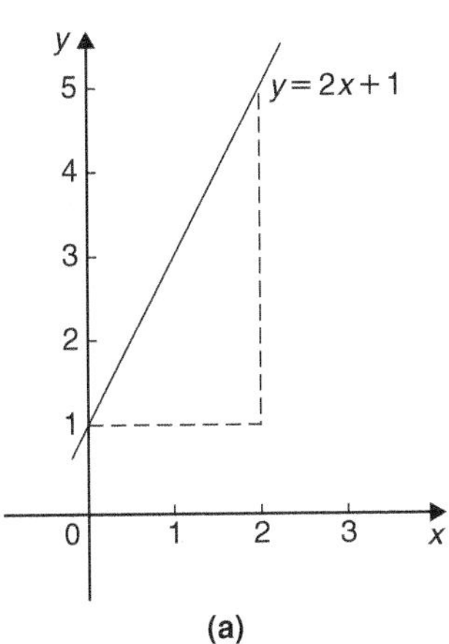

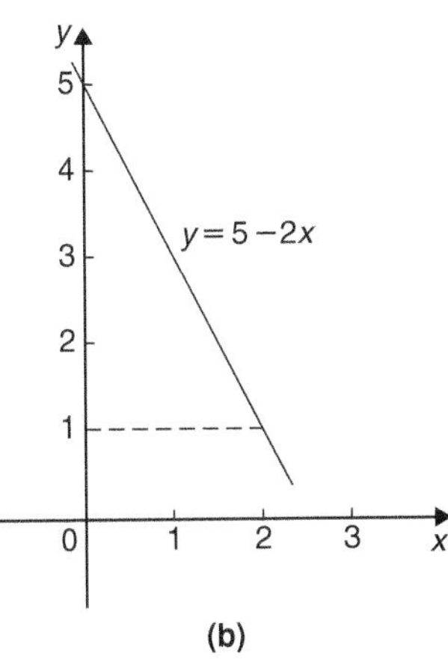

Figure 31.1

(ii) Quadratic graphs (see Chapter 30, page 323)
The general equation of a quadratic graph is $y = ax^2 + bx + c$, and its shape is that of a parabola.
The simplest example of a quadratic graph, $y = x^2$, is shown in Fig. 31.2.

(iii) Cubic equations (see Chapter 30, page 327)
The general equation of a cubic graph is $y = ax^3 + bx^2 + cx + d$.

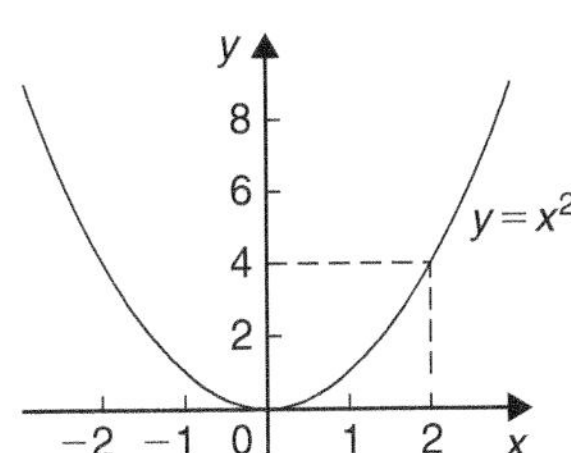

Figure 31.2

The simplest example of a cubic graph, $y = x^3$, is shown in Fig. 31.3.

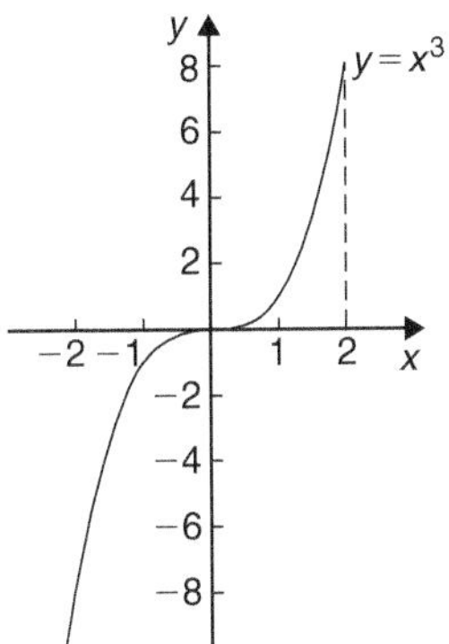

Figure 31.3

(iv) Trigonometric functions (see Chapter 18, page 184)
Graphs of $y = \sin\theta$, $y = \cos\theta$ and $y = \tan\theta$ are shown in Fig. 31.4

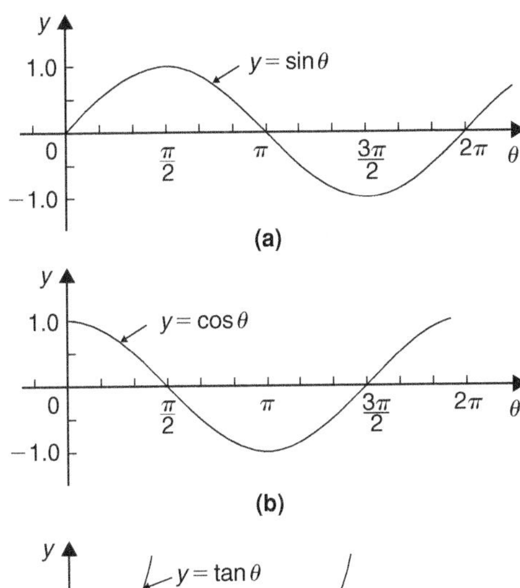

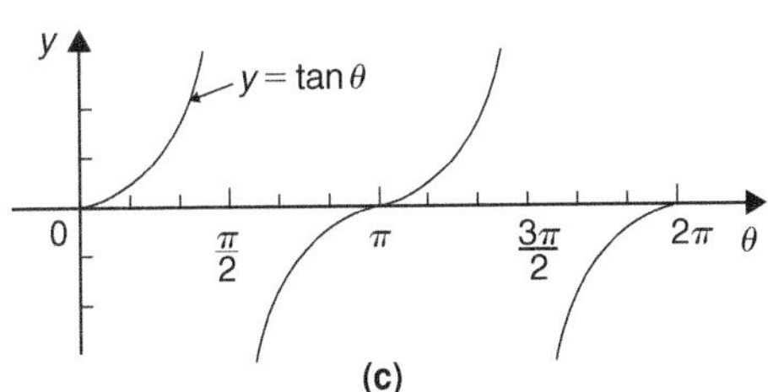

Figure 31.4

(v) Circle (see Chapter 24, page 254)
The simplest equation of a circle is $x^2 + y^2 = r^2$, with centre at the origin and radius r, as shown in Fig. 31.5.

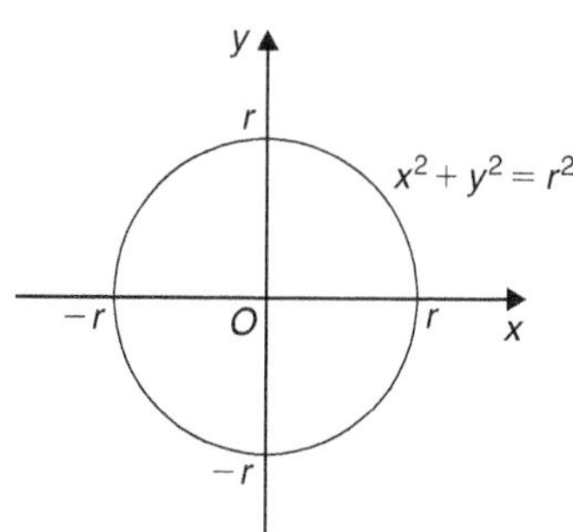

Figure 31.5

More generally, the equation of a circle, centre (a, b), radius r, is given by:

$$(x-a)^2+(y-b)^2=r^2 \qquad (1)$$

Figure 31.6 shows a circle

$$(x-2)^2+(y-3)^2=4$$

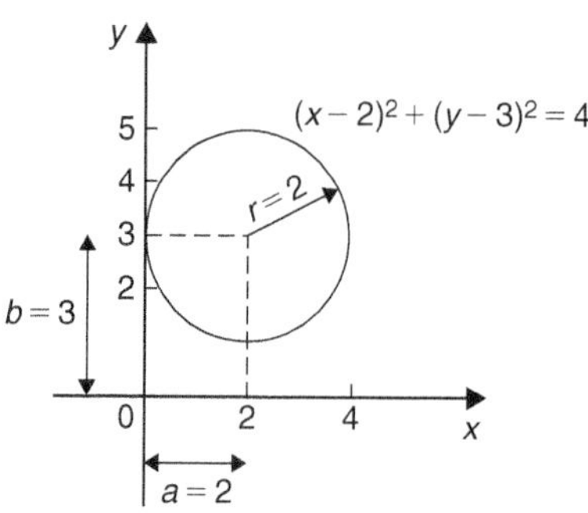

Figure 31.6

(vi) Ellipse

The equation of an ellipse is:

$$\frac{x^2}{a^2}+\frac{y^2}{b^2}=1$$

and the general shape is as shown in Fig. 31.7.

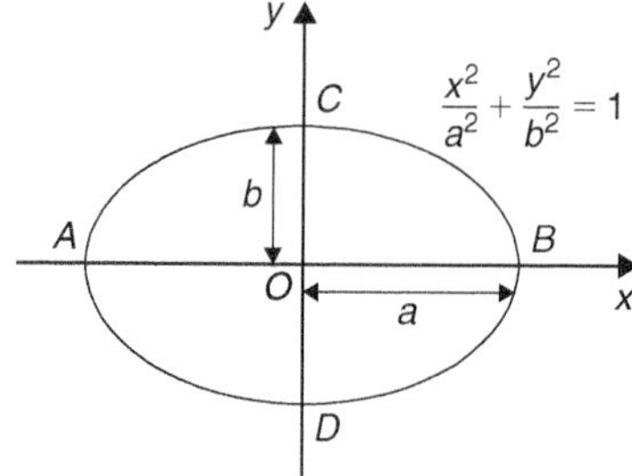

Figure 31.7

The length AB is called the **major axis** and CD the **minor axis**. In the above equation, 'a' is the semi-major axis and 'b' is the semi-minor axis. (Note that if $b=a$, the equation becomes

$$\frac{x^2}{a^2}+\frac{y^2}{a^2}=1,$$

i.e. $x^2+y^2=a^2$, which is a circle of radius a)

(vii) Hyperbola

The equation of a hyperbola is $\frac{x^2}{a^2}-\frac{y^2}{b^2}=1$ and the general shape is shown in Fig. 31.8. The curve is seen to be symmetrical about both the x- and y-axes. The distance AB in Fig. 31.8 is given by $2a$.

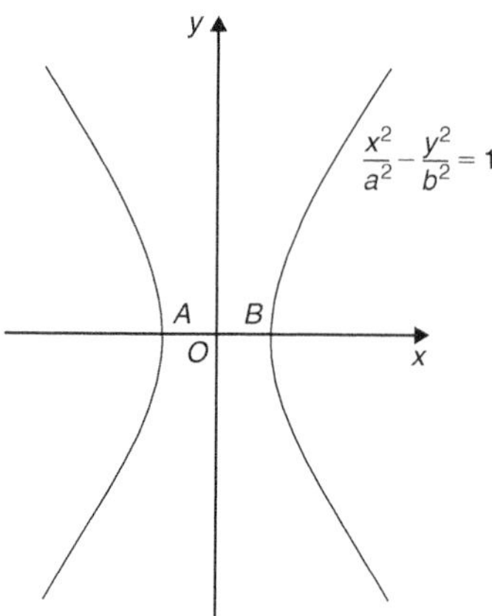

Figure 31.8

(viii) Rectangular hyperbola

The equation of a rectangular hyperbola is $xy=c$ or $y=\frac{c}{x}$ (where c is constant) and the general shape is shown in Fig. 31.9.

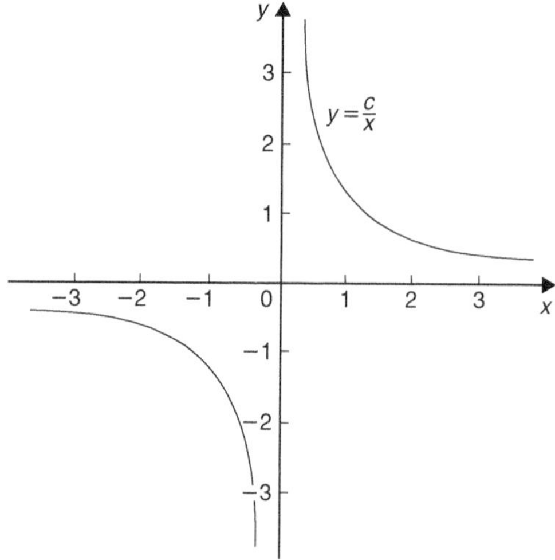

Figure 31.9

(ix) **Logarithmic function** (see Chapter 13, page 132)
$y = \ln x$ and $y = \lg x$ are both of the general shape shown in Fig. 31.10.

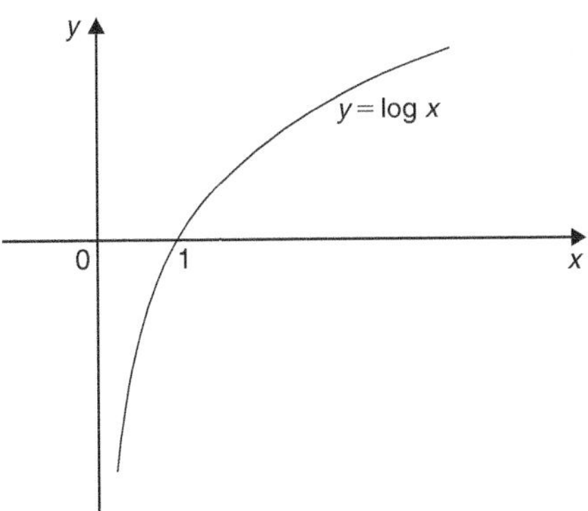

Figure 31.10

(x) **Exponential functions** (see Chapter 14, page 138)
$y = e^x$ is of the general shape shown in Fig. 31.11.

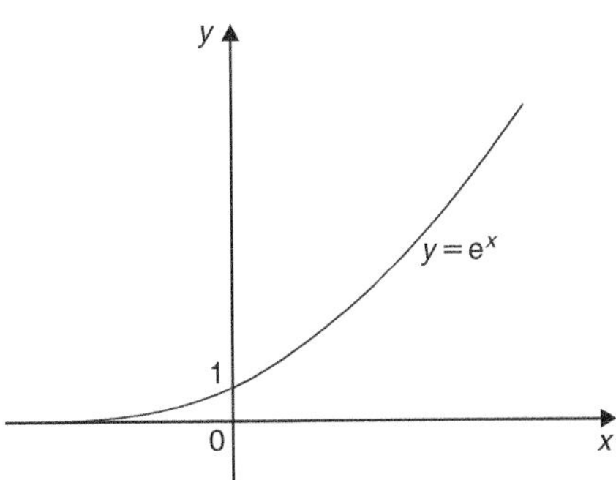

Figure 31.11

(xi) **Polar curves**
The equation of a polar curve is of the form $r = f(\theta)$. An example of a polar curve, $r = a\sin\theta$, is shown in Fig. 31.12.

31.2 Simple transformations

From the graph of $y = f(x)$ it is possible to deduce the graphs of other functions which are transformations of $y = f(x)$. For example, knowing the graph of $y = f(x)$, can help us draw the graphs of $y = af(x)$, $y = f(x) + a$, $y = f(x + a)$, $y = f(ax)$, $y = -f(x)$ and $y = f(-x)$

(i) $y = af(x)$
For each point (x_1, y_1) on the graph of $y = f(x)$ there exists a point (x_1, ay_1) on the graph of

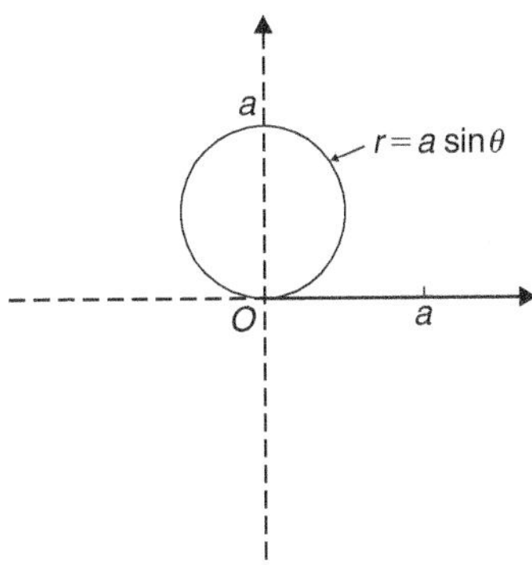

Figure 31.12

$y = af(x)$. Thus the graph of $y = af(x)$ can be obtained by stretching $y = f(x)$ parallel to the y-axis by a scale factor 'a'. Graphs of $y = x + 1$ and $y = 3(x + 1)$ are shown in Fig. 31.13(a) and graphs of $y = \sin\theta$ and $y = 2\sin\theta$ are shown in Fig. 31.13(b).

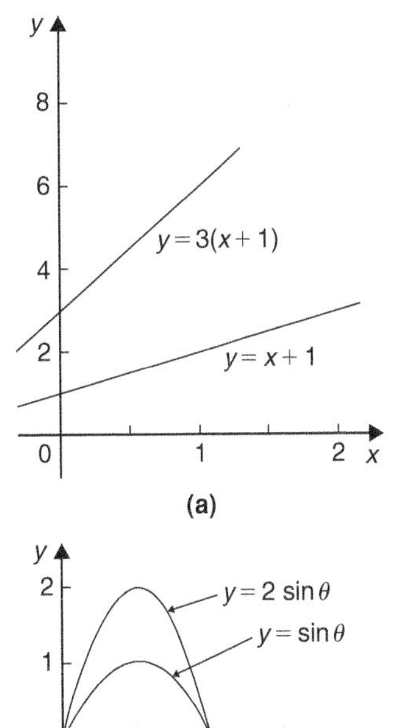

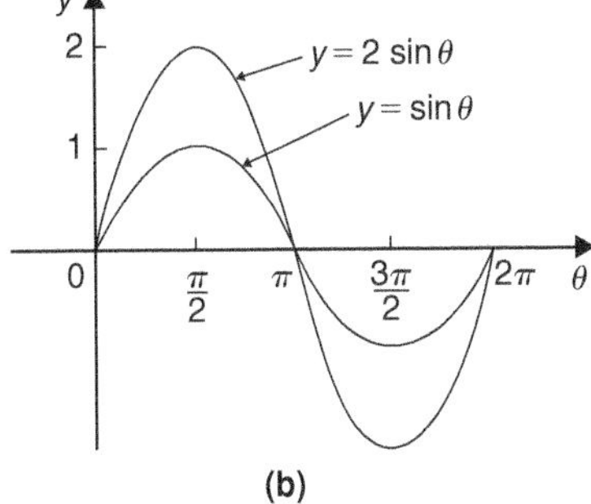

Figure 31.13

(ii) $y = f(x) + a$
The graph of $y = f(x)$ is translated by 'a' units parallel to the y-axis to obtain $y = f(x) + a$. For example, if $f(x) = x$, $y = f(x) + 3$ becomes $y = x + 3$, as shown in Fig. 31.14(a). Similarly, if $f(\theta) = \cos\theta$, then $y = f(\theta) + 2$ becomes

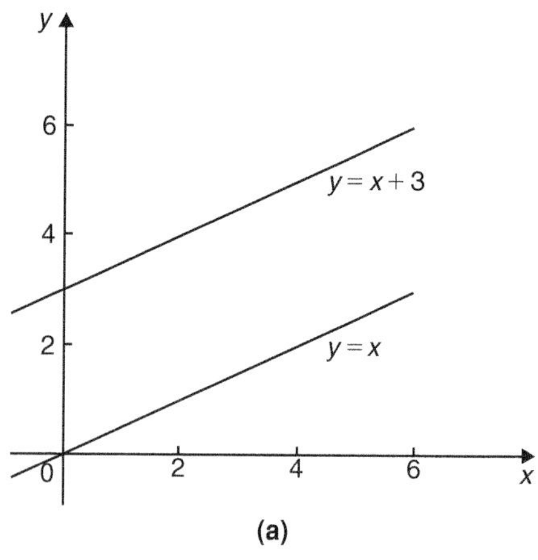

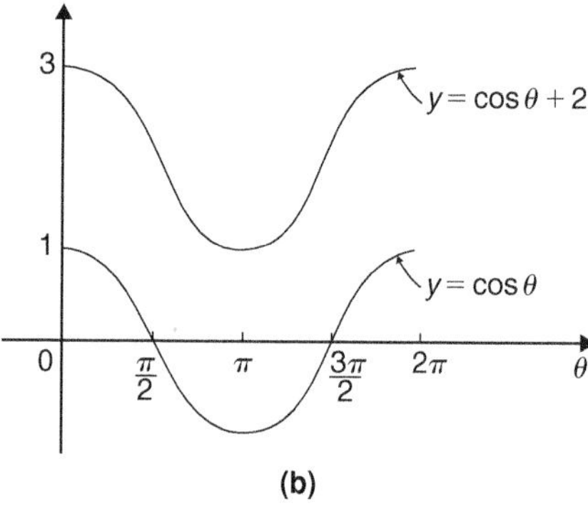

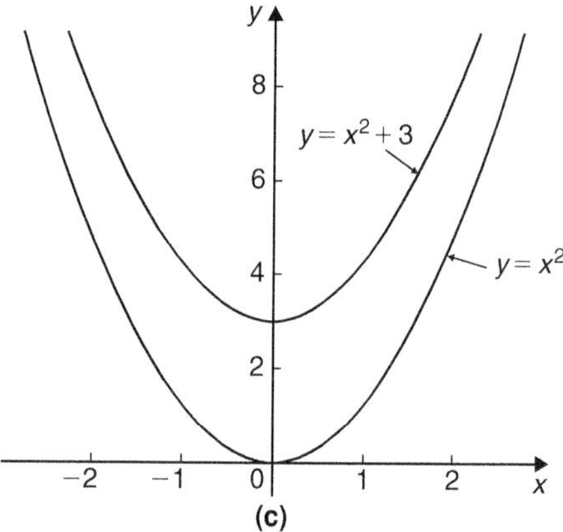

Figure 31.14

$y = \cos\theta + 2$, as shown in Fig. 31.14(b). Also, if $f(x) = x^2$, then $y = f(x) + 3$ becomes $y = x^2 + 3$, as shown in Fig. 31.14(c).

(iii) $y = f(x + a)$

The graph of $y = f(x)$ is translated by 'a' units parallel to the x-axis to obtain $y = f(x + a)$. If 'a' > 0 it moves $y = f(x)$ in the negative direction on the x-axis (i.e. to the left), and if 'a' < 0 it moves $y = f(x)$ in the positive direction on the x-axis (i.e. to the right). For example, if $f(x) = \sin x$, $y = f\left(x - \frac{\pi}{3}\right)$ becomes $y = \sin\left(x - \frac{\pi}{3}\right)$ as shown in Fig. 31.15(a) and $y = \sin\left(x + \frac{\pi}{4}\right)$ is shown in Fig. 31.15(b).

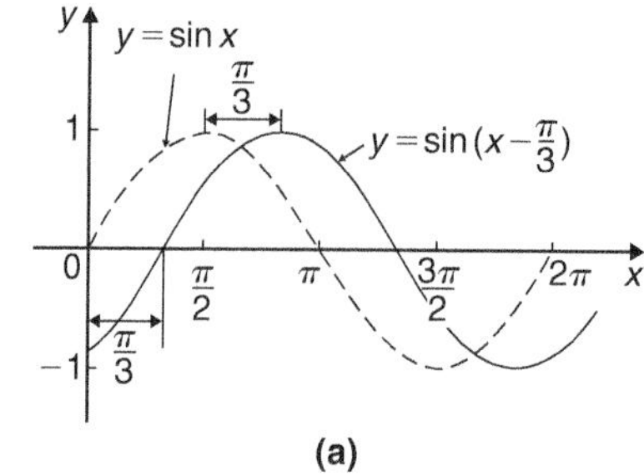

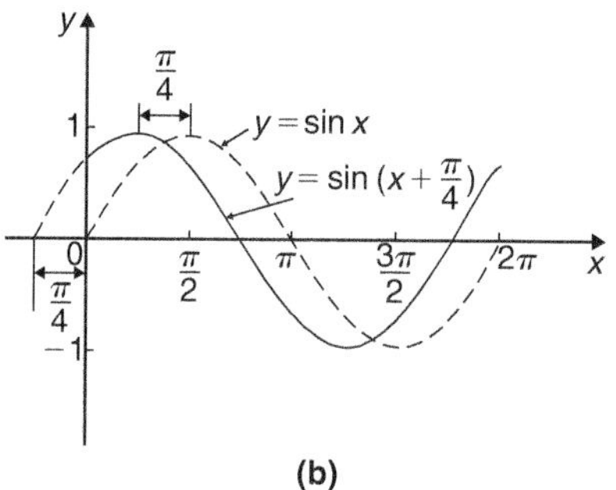

Figure 31.15

Similarly graphs of $y = x^2$, $y = (x - 1)^2$ and $y = (x + 2)^2$ are shown in Fig. 31.16.

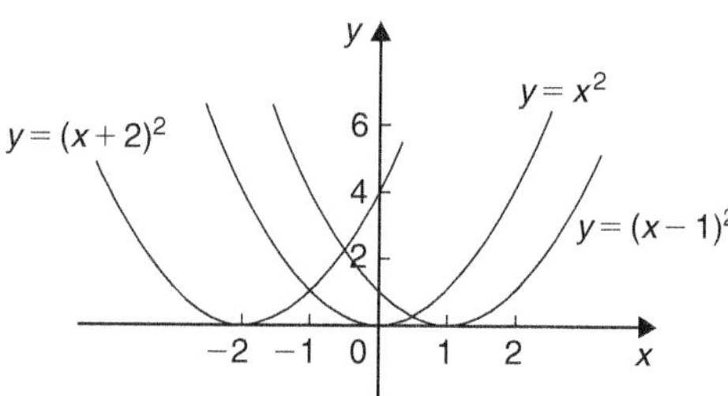

Figure 31.16

(iv) $y = f(ax)$

For each point (x_1, y_1) on the graph of $y = f(x)$, there exists a point $\left(\frac{x_1}{a}, y_1\right)$ on the graph of $y = f(ax)$. Thus the graph of $y = f(ax)$ can be obtained by stretching $y = f(x)$ parallel to the x-axis by a scale factor $\frac{1}{a}$

For example, if $f(x)=(x-1)^2$, and $a=\frac{1}{2}$, then $f(ax)=\left(\frac{x}{2}-1\right)^2$

Both of these curves are shown in Fig. 31.17(a).

Similarly, $y=\cos x$ and $y=\cos 2x$ are shown in Fig. 31.17(b).

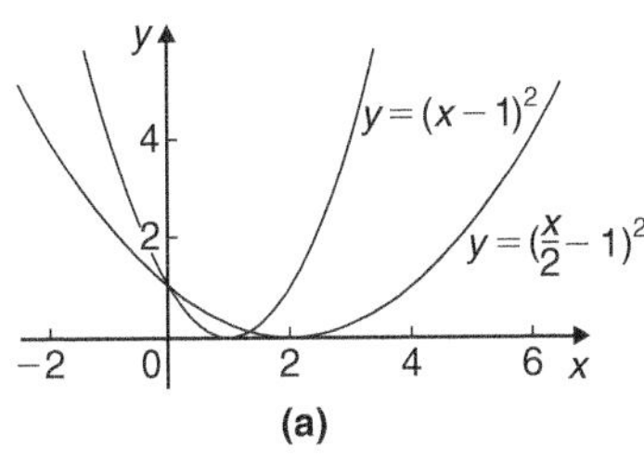

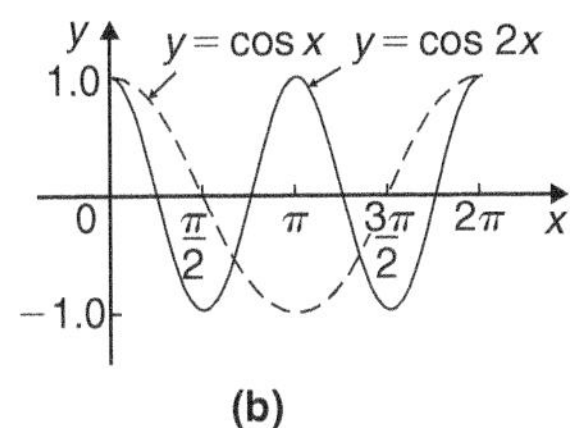

Figure 31.17

(v) $y=-f(x)$

The graph of $y=-f(x)$ is obtained by reflecting $y=f(x)$ in the x-axis. For example, graphs of $y=e^x$ and $y=-e^x$ are shown in Fig. 31.18(a), and graphs of $y=x^2+2$ and $y=-(x^2+2)$ are shown in Fig. 31.18(b).

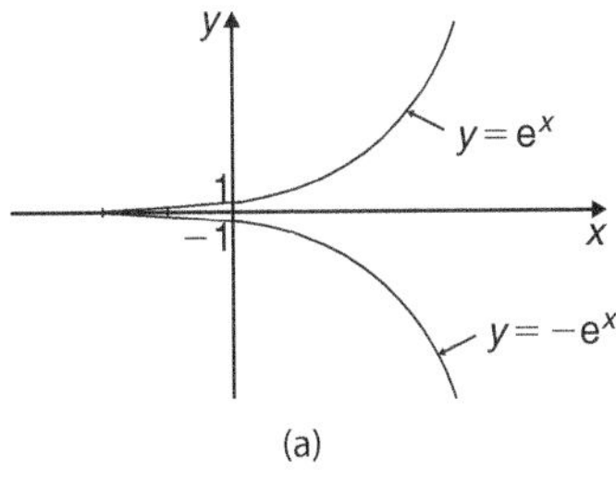

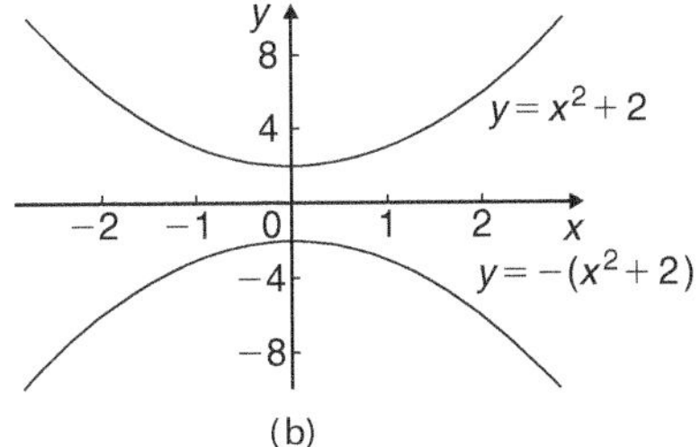

Figure 31.18

(vi) $y=f(-x)$

The graph of $y=f(-x)$ is obtained by reflecting $y=f(x)$ in the y-axis. For example, graphs of $y=x^3$ and $y=(-x)^3=-x^3$ are shown in Fig. 31.19(a) and graphs of $y=\ln x$ and $y=-\ln x$ are shown in Fig. 31.19(b).

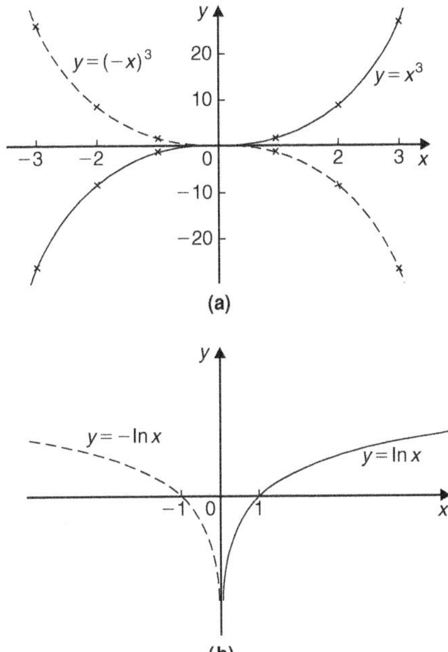

Figure 31.19

Problem 1. Sketch the following graphs, showing relevant points:
(a) $y=(x-4)^2$ (b) $y=x^3-8$

(a) In Fig. 31.20 a graph of $y=x^2$ is shown by the broken line. The graph of $y=(x-4)^2$ is of the form $y=f(x+a)$. Since $a=-4$, then $y=(x-4)^2$ is translated 4 units to the right of $y=x^2$, parallel to the x-axis.
(See section (iii) above.)

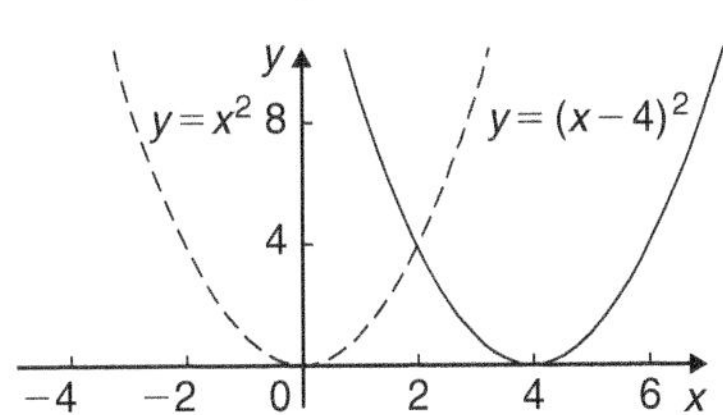

Figure 31.20

(b) In Fig. 31.21 a graph of $y=x^3$ is shown by the broken line. The graph of $y=x^3-8$ is of the form $y=f(x)+a$. Since $a=-8$, then $y=x^3-8$ is translated 8 units down from $y=x^3$, parallel to the y-axis.
(See section (ii) above.)

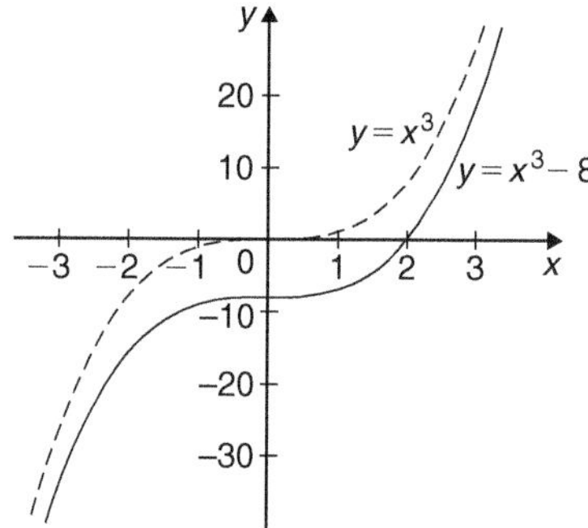

Figure 31.21

Problem 2. Sketch the following graphs, showing relevant points:
(a) $y=5-(x+2)^3$ (b) $y=1+3\sin 2x$

(a) Fig. 31.22(a) shows a graph of $y=x^3$. Fig. 31.22(b) shows a graph of $y=(x+2)^3$ (see $f(x+a)$, section (iii) above). Fig. 31.22(c) shows a graph of $y=-(x+2)^3$ (see $-f(x)$, section (v) above). Fig. 31.22(d) shows the graph of $y=5-(x+2)^3$ (see $f(x)+a$, section (ii) above).

(b) Fig. 31.23(a) shows a graph of $y=\sin x$. Fig. 31.23(b) shows a graph of $y=\sin 2x$ (see $f(ax)$, section (iv) above).

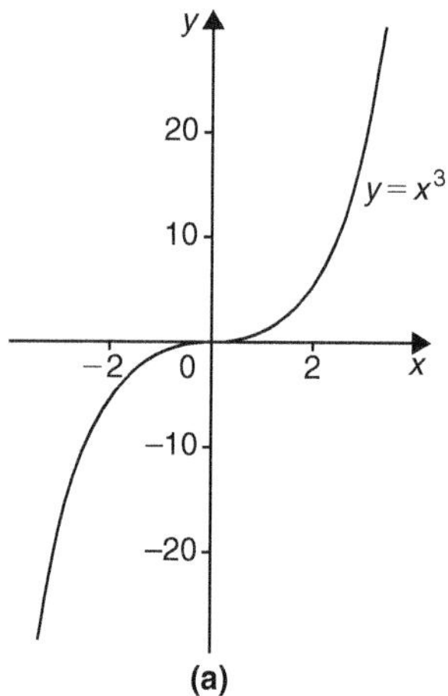

Figure 31.22

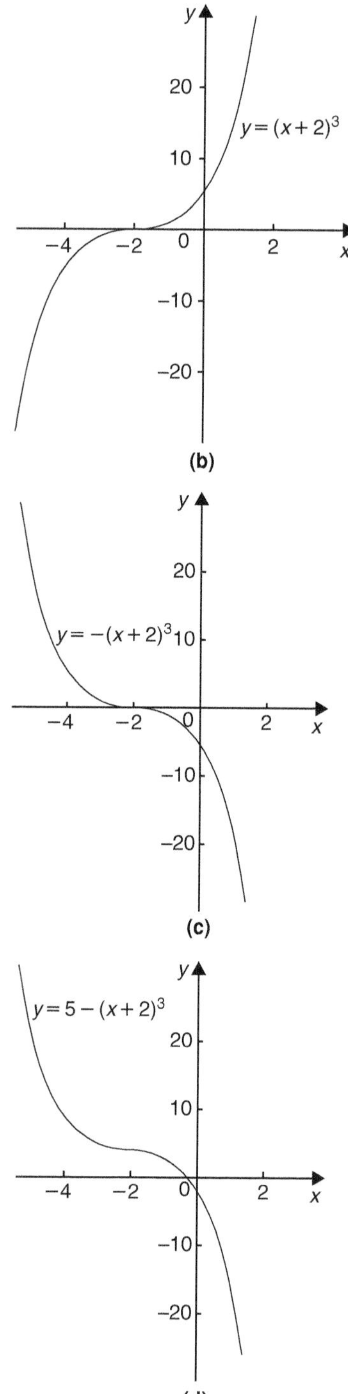

Figure 31.22 (*Continued*)

Fig. 31.23(c) shows a graph of $y=3\sin 2x$ (see $a\,f(x)$, section (i) above). Fig. 31.23(d) shows a graph of $y=1+3\sin 2x$ (see $f(x)+a$, section (ii) above).

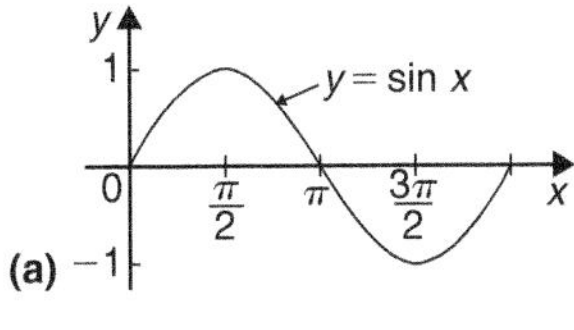

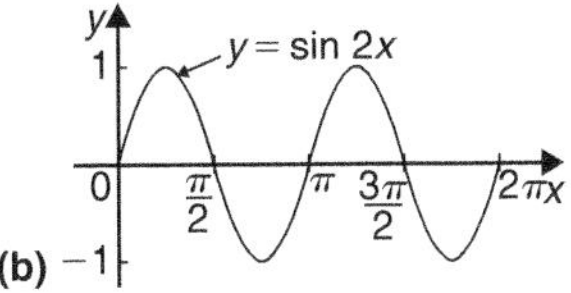

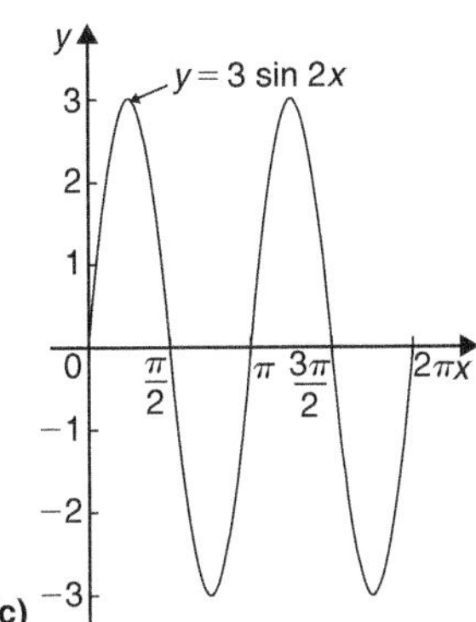

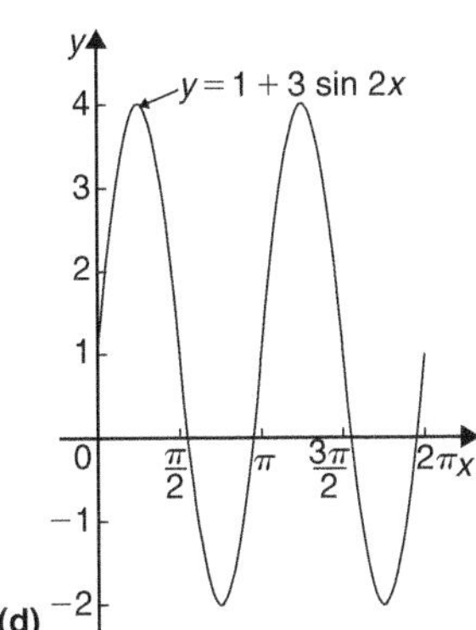

Figure 31.23

Now try the following Practice Exercise

Practice Exercise 154 Simple transformations with curve sketching (Answers on page 717)

Sketch the following graphs, showing relevant points:

1. $y=3x-5$
2. $y=-3x+4$
3. $y=x^2+3$
4. $y=(x-3)^2$
5. $y=(x-4)^2+2$
6. $y=x-x^2$
7. $y=x^3+2$
8. $y=1+2\cos 3x$
9. $y=3-2\sin\left(x+\frac{\pi}{4}\right)$
10. $y=2\ln x$

31.3 Periodic functions

A function $f(x)$ is said to be **periodic** if $f(x+T)=f(x)$ for all values of x, where T is some positive number. T is the interval between two successive repetitions and is called the period of the function $f(x)$. For example, $y=\sin x$ is periodic in x with period 2π since $\sin x=\sin(x+2\pi)=\sin(x+4\pi)$ and so on. Similarly, $y=\cos x$ is a periodic function with period 2π since $\cos x=\cos(x+2\pi)=\cos(x+4\pi)$ and so on. In general, if $y=\sin\omega t$ or $y=\cos\omega t$ then the period of the waveform is $2\pi/\omega$. The function shown in Fig. 31.24 is also periodic of period 2π and is defined by:

$$f(x)=\begin{cases}-1, \text{ when } -\pi\le x\le 0\\ 1, \text{ when } 0\le x\le\pi\end{cases}$$

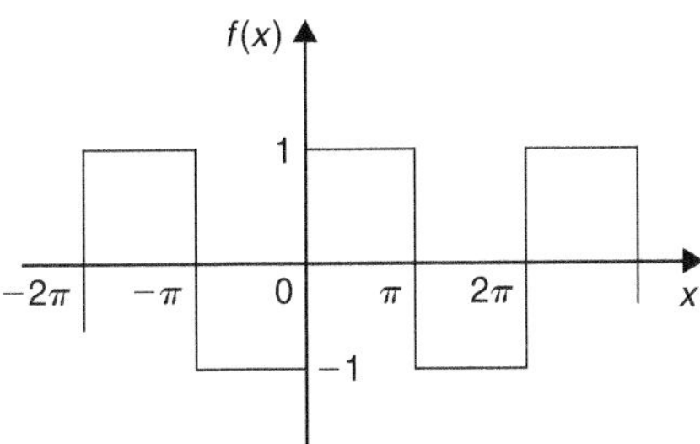

Figure 31.24

31.4 Continuous and discontinuous functions

If a graph of a function has no sudden jumps or breaks it is called a **continuous function**, examples being the graphs of sine and cosine functions. However, other graphs make finite jumps at a point or points in the interval. The square wave shown in Fig. 31.24 has **finite discontinuities** as $x=\pi$, 2π, 3π and so on, and is therefore a **discontinuous function**. $y=\tan x$ is another example of a discontinuous function.

31.5 Even and odd functions

Even functions

A function $y = f(x)$ is said to be even if $f(-x) = f(x)$ for all values of x. Graphs of even functions are always symmetrical about the y-axis (i.e. is a mirror image). Two examples of even functions are $y = x^2$ and $y = \cos x$ as shown in Fig. 31.25.

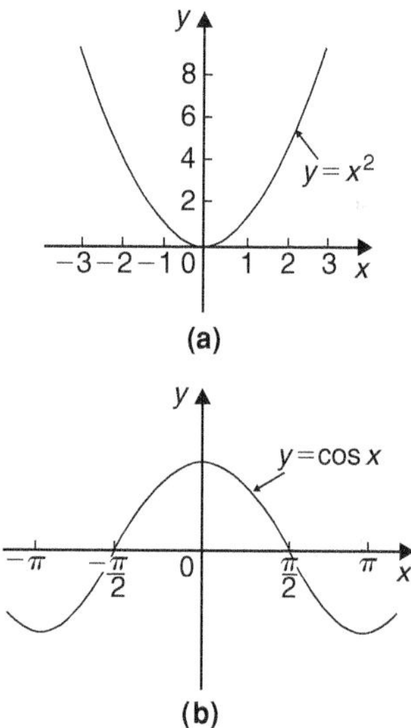

Figure 31.25

Odd functions

A function $y = f(x)$ is said to be odd if $f(-x) = -f(x)$ for all values of x. Graphs of odd functions are always symmetrical about the origin. Two examples of odd functions are $y = x^3$ and $y = \sin x$ as shown in Fig. 31.26. Many functions are neither even nor odd, two such examples being shown in Fig. 31.27.

Problem 3. Sketch the following functions and state whether they are even or odd functions:

(a) $y = \tan x$

(b) $$f(x) = \begin{cases} 2, \text{ when } 0 \le x \le \dfrac{\pi}{2} \\ -2, \text{ when } \dfrac{\pi}{2} \le x \le \dfrac{3\pi}{2} \\ 2, \text{ when } \dfrac{3\pi}{2} \le x \le 2\pi \end{cases}$$

and is periodic of period 2π

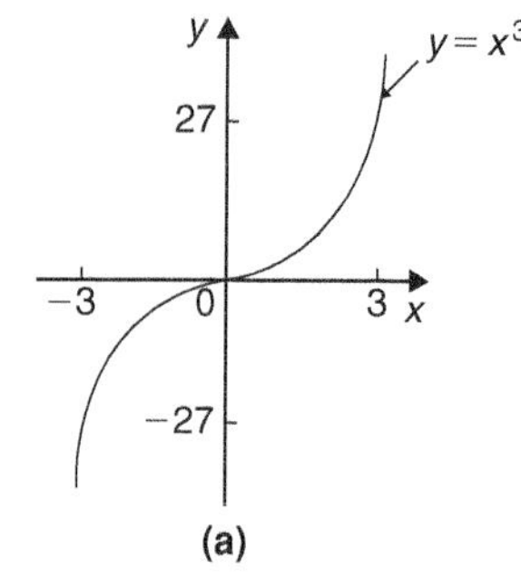

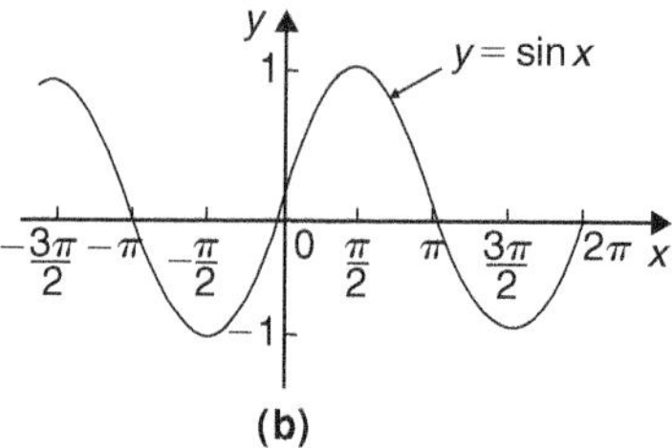

Figure 31.26

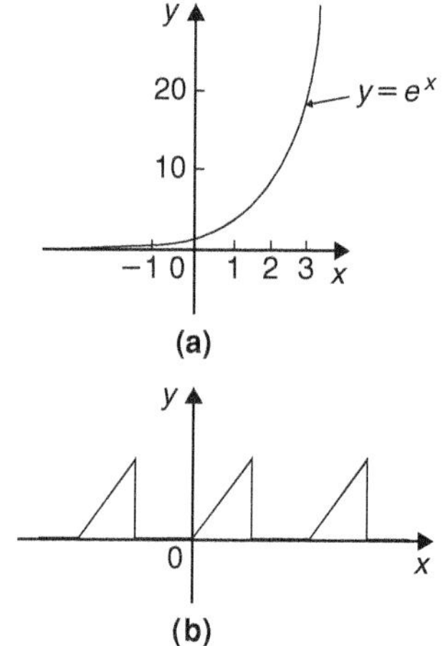

Figure 31.27

(a) A graph of $y = \tan x$ is shown in Fig. 31.28(a) and is symmetrical about the origin and is thus an **odd function** (i.e. $\tan(-x) = -\tan x$).

(b) A graph of $f(x)$ is shown in Fig. 31.28(b) and is symmetrical about the $f(x)$ axis hence the function is an **even** one, ($f(-x) = f(x)$).

Problem 4. Sketch the following graphs and state whether the functions are even, odd or neither even nor odd:

(a) $y = \ln x$

(b) $f(x) = x$ in the range $-\pi$ to π and is periodic of period 2π

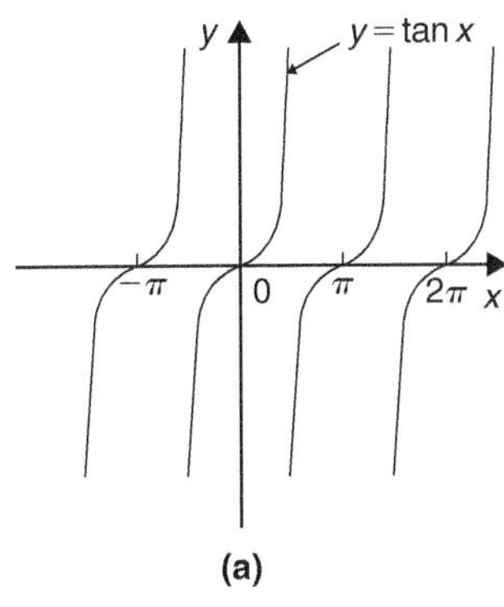

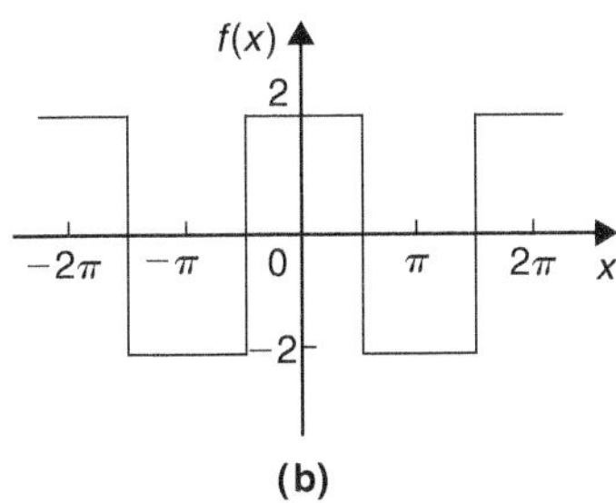

Figure 31.28

(a) A graph of $y = \ln x$ is shown in Fig. 31.29(a) and the curve is neither symmetrical about the y-axis nor symmetrical about the origin and is thus **neither even nor odd**.

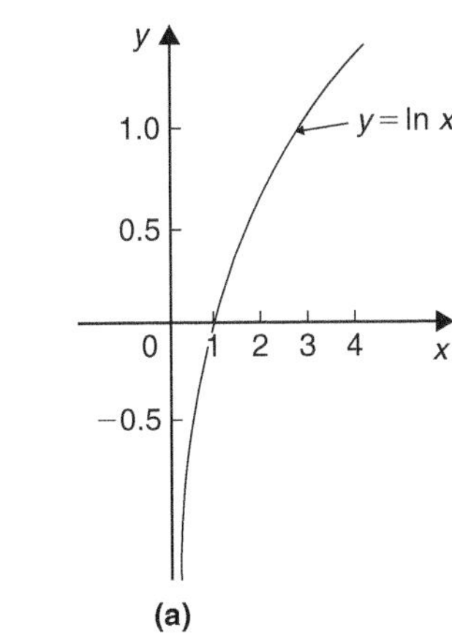

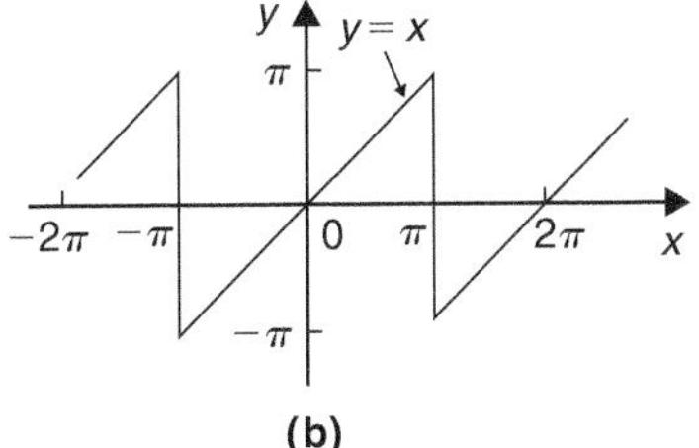

Figure 31.29

(b) A graph of $y = x$ in the range $-\pi$ to π is shown in Fig. 31.29(b) and is symmetrical about the origin and is thus an **odd function**.

Now try the following Practice Exercise

Practice Exercise 155 Even and odd functions (Answers on page 718)

In Problems 1 and 2 determine whether the given functions are even, odd or neither even nor odd.

1. (a) x^4 (b) $\tan 3x$ (c) $2e^{3t}$ (d) $\sin^2 x$
2. (a) $5t^3$ (b) $e^x + e^{-x}$ (c) $\dfrac{\cos\theta}{\theta}$ (d) e^x
3. State whether the following functions, which are periodic of period 2π, are even or odd:

   (a) $f(\theta) = \begin{cases} \theta, \text{ when } -\pi \le \theta \le 0 \\ -\theta, \text{ when } 0 \le \theta \le \pi \end{cases}$

   (b) $f(x) = \begin{cases} x, \text{ when } -\dfrac{\pi}{2} \le x \le \dfrac{\pi}{2} \\ 0, \text{ when } \dfrac{\pi}{2} \le x \le \dfrac{3\pi}{2} \end{cases}$

31.6 Inverse functions

If y is a function of x, the graph of y against x can be used to find x when any value of y is given. Thus the graph also expresses that x is a function of y. Two such functions are called **inverse functions**.

In general, given a function $y = f(x)$, its inverse may be obtained by inter-changing the roles of x and y and then transposing for y. The inverse function is denoted by $y = f^{-1}(x)$

For example, if $y = 2x + 1$, the inverse is obtained by

(i) transposing for x, i.e. $x = \dfrac{y-1}{2} = \dfrac{y}{2} - \dfrac{1}{2}$

and (ii) interchanging x and y, giving the inverse as $y = \dfrac{x}{2} - \dfrac{1}{2}$

Thus if $f(x) = 2x + 1$, then $f^{-1}(x) = \dfrac{x}{2} - \dfrac{1}{2}$

A graph of $f(x) = 2x + 1$ and its inverse $f^{-1}(x) = \dfrac{x}{2} - \dfrac{1}{2}$ is shown in Fig. 31.30 and $f^{-1}(x)$ is seen to be a reflection of $f(x)$ in the line $y = x$.

Similarly, if $y = x^2$, the inverse is obtained by

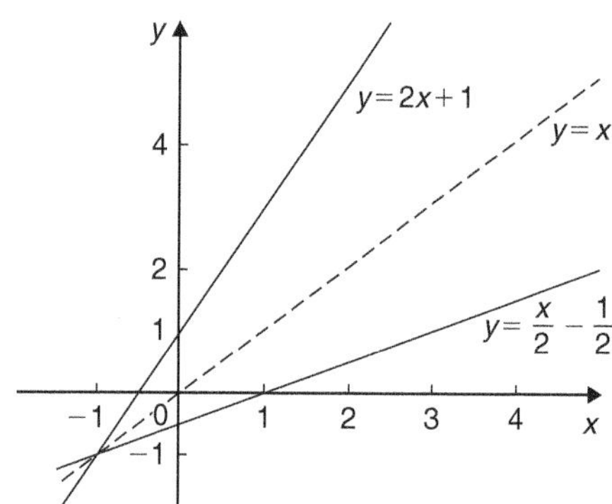

Figure 31.30

(i) transposing for x, i.e. $x = \pm\sqrt{y}$

and (ii) interchanging x and y, giving the inverse $y = \pm\sqrt{x}$

Hence the inverse has two values for every value of x. Thus $f(x) = x^2$ does not have a single inverse. In such a case the domain of the original function may be restricted to $y = x^2$ for $x > 0$. Thus the inverse is then $y = +\sqrt{x}$. A graph of $f(x) = x^2$ and its inverse $f^{-1}(x) = \sqrt{x}$ for $x > 0$ is shown in Fig. 31.31 and, again, $f^{-1}(x)$ is seen to be a reflection of $f(x)$ in the line $y = x$

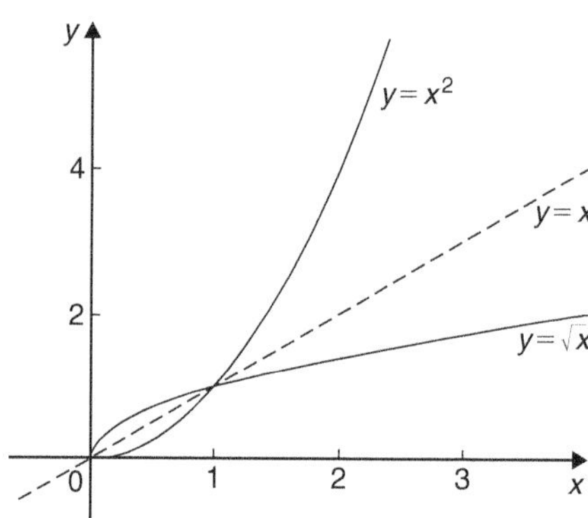

Figure 31.31

It is noted from the latter example, that not all functions have a single inverse. An inverse, however, can be determined if the range is restricted.

Problem 5. Determine the inverse for each of the following functions:
(a) $f(x) = x - 1$ (b) $f(x) = x^2 - 4$ $(x > 0)$
(c) $f(x) = x^2 + 1$

(a) If $y = f(x)$, then $y = x - 1$
Transposing for x gives $x = y + 1$
Interchanging x and y gives $y = x + 1$
Hence if $f(x) = x - 1$, then $\boldsymbol{f^{-1}(x) = x + 1}$

(b) If $y = f(x)$, then $y = x^2 - 4$ $(x > 0)$
Transposing for x gives $x = \sqrt{y + 4}$
Interchanging x and y gives $y = \sqrt{x + 4}$
Hence if $f(x) = x^2 - 4$ $(x > 0)$ then
$\boldsymbol{f^{-1}(x) = \sqrt{x + 4}}$ **if** $\boldsymbol{x > -4}$

(c) If $y = f(x)$, then $y = x^2 + 1$
Transposing for x gives $x = \sqrt{y - 1}$
Interchanging x and y gives $y = \sqrt{x - 1}$, which has two values. **Hence there is no single inverse of** $\boldsymbol{f(x) = x^2 + 1}$, since the domain of $f(x)$ is not restricted.

Inverse trigonometric functions

If $y = \sin x$, then x is the angle whose sine is y. Inverse trigonometrical functions are denoted either by prefixing the function with 'arc' or by using $^{-1}$. Hence transposing $y = \sin x$ for x gives $x = \arcsin y$ or $\sin^{-1} y$. Interchanging x and y gives the inverse $y = \arcsin x$ or $\sin^{-1} x$.
Similarly, $y = \cos^{-1} x$, $y = \tan^{-1} x$, $y = \sec^{-1} x$, $y = \text{cosec}^{-1} x$ and $y = \cot^{-1} x$ are all inverse trigonometric functions. The angle is always expressed in radians.
Inverse trigonometric functions are periodic so it is necessary to specify the smallest or principal value of the angle. For $y = \sin^{-1} x$, $\tan^{-1} x$, $\text{cosec}^{-1} x$ and $\cot^{-1} x$, the principal value is in the range $-\frac{\pi}{2} < y < \frac{\pi}{2}$. For $y = \cos^{-1} x$ and $\sec^{-1} x$ the principal value is in the range $0 < y < \pi$
Graphs of the six inverse trigonometric functions are shown in Fig. 31.32.

Problem 6. Determine the principal values of
(a) arcsin 0.5 (b) arctan(−1)
(c) $\arccos\left(-\frac{\sqrt{3}}{2}\right)$ (d) $\text{arccosec}\,(\sqrt{2})$

Using a calculator,

(a) $\arcsin 0.5 \equiv \sin^{-1} 0.5 = 30° = \boldsymbol{\frac{\pi}{6}}$ **rad** or **0.5236 rad**

(b) $\arctan(-1) \equiv \tan^{-1}(-1) = -45°$
$= \boldsymbol{-\frac{\pi}{4}}$ **rad** or **−0.7854 rad**

(c) $\arccos\left(-\frac{\sqrt{3}}{2}\right) \equiv \cos^{-1}\left(-\frac{\sqrt{3}}{2}\right) = 150°$
$= \boldsymbol{\frac{5\pi}{6}}$ **rad** or **2.6180 rad**

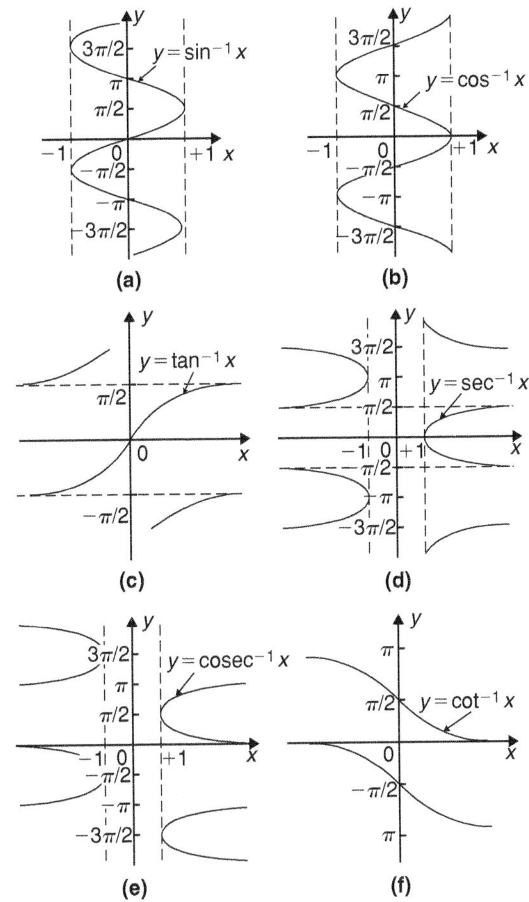

Figure 31.32

(d) $\text{arccosec}(\sqrt{2}) = \arcsin\left(\frac{1}{\sqrt{2}}\right) \equiv \sin^{-1}\left(\frac{1}{\sqrt{2}}\right)$

$= 45° = \frac{\pi}{4}$ **rad** or **0.7854 rad**

Problem 7. Evaluate (in radians), correct to 3 decimal places: $\sin^{-1} 0.30 + \cos^{-1} 0.65$

$$\sin^{-1} 0.30 = 17.4576° = 0.3047 \text{ rad}$$
$$\cos^{-1} 0.65 = 49.4584° = 0.8632 \text{ rad}$$

Hence $\sin^{-1} 0.30 + \cos^{-1} 0.65 = 0.3047 + 0.8632$
$= \mathbf{1.168}$ **rad**, correct to 3 decimal places

Now try the following Practice Exercise

Practice Exercise 156 Inverse functions (Answers on page 718)

Determine the inverse of the functions given in Problems 1 to 4.

1. $f(x) = x + 1$
2. $f(x) = 5x - 1$
3. $f(x) = x^3 + 1$
4. $f(x) = \frac{1}{x} + 2$

Determine the principal value of the inverse functions in Problems 5 to 11.

5. $\sin^{-1}(-1)$
6. $\cos^{-1} 0.5$
7. $\tan^{-1} 1$
8. $\cot^{-1} 2$
9. $\text{cosec}^{-1} 2.5$
10. $\sec^{-1} 1.5$
11. $\sin^{-1}\left(\frac{1}{\sqrt{2}}\right)$
12. Evaluate x, correct to 3 decimal places: $x = \sin^{-1}\frac{1}{3} + \cos^{-1}\frac{4}{5} - \tan^{-1}\frac{8}{9}$
13. Evaluate y, correct to 4 significant figures: $y = 3\sec^{-1}\sqrt{2} - 4\,\text{cosec}^{-1}\sqrt{2} + 5\cot^{-1} 2$

Practice Exercise 157 Multiple-choice questions on functions and their curves (Answers on page 718)

Each question has only one correct answer

1. The graph of $y = \tan 3\theta$ is:
   (a) a continuous, periodic, even function
   (b) a discontinuous, non-periodic, odd function
   (c) a discontinuous, periodic, odd function
   (d) a continuous, non-periodic, even function

2. $\frac{x^2}{a^2}+\frac{b^2}{y^2}=1$ is the equation of:
(a) a hyperbola (b) an ellipse
(c) a circle (d) a rectangular hyperbola

3. $\frac{x^2}{a^2}-\frac{b^2}{y^2}=1$ is the equation of:
(a) a hyperbola (b) a circle
(c) an ellipse (d) a rectangular hyperbola

4. $(x-a)^2+(y-b)^2=r^2$ is the equation of:
(a) a hyperbola (b) a circle
(c) an ellipse (d) a rectangular hyperbola

5. $xy=c$ is the equation of:
(a) a hyperbola (b) a circle
(c) an ellipse (d) a rectangular hyperbola

For fully worked solutions to each of the problems in Practice Exercises 154 to 156 in this chapter, go to the website:
www.routledge.com/cw/bird

Revision Test 8 Graphs

This Revision Test covers the material contained in Chapters 27 to 31. *The marks for each question are shown in brackets at the end of each question.*

1. Determine the gradient and intercept on the y-axis for the following equations:

   (a) $y=-5x+2$ (b) $3x+2y+1=0$ (5)

2. The equation of a line is $2y=4x+7$. A table of corresponding values is produced and is as shown below. Complete the table and plot a graph of y against x. Determine the gradient of the graph.

x	−3	−2	−1	0	1	2	3
y	−2.5					7.5	

(6)

3. Plot the graphs $y=3x+2$ and $\frac{y}{2}+x=6$ on the same axes and determine the co-ordinates of their point of intersection. (7)

4. The velocity v of a body over varying time intervals t was measured as follows:

t seconds	2	5	7
v m/s	15.5	17.3	18.5

t seconds	10	14	17
v m/s	20.3	22.7	24.5

Plot a graph with velocity vertical and time horizontal. Determine from the graph (a) the gradient, (b) the vertical axis intercept, (c) the equation of the graph, (d) the velocity after 12.5 s, and (e) the time when the velocity is 18 m/s. (9)

5. The following experimental values of x and y are believed to be related by the law $y=ax^2+b$, where a and b are constants. By plotting a suitable graph verify this law and find the approximate values of a and b.

x	2.5	4.2	6.0	8.4	9.8	11.4
y	15.4	32.5	60.2	111.8	150.1	200.9

(9)

6. Determine the law of the form $y=ae^{kx}$ which relates the following values:

y	0.0306	0.285	0.841
x	−4.0	5.3	9.8

y	5.21	173.2	1181
x	17.4	32.0	40.0

(9)

7. State the minimum number of cycles on logarithmic graph paper needed to plot a set of values ranging from 0.073 to 490. (2)

8. Plot a graph of $y=2x^2$ from $x=-3$ to $x=+3$ and hence solve the equations:

   (a) $2x^2-8=0$ (b) $2x^2-4x-6=0$ (9)

9. Plot the graph of $y=x^3+4x^2+x-6$ for values of x between $x=-4$ and $x=2$. Hence determine the roots of the equation $x^3+4x^2+x-6=0$. (7)

10. Sketch the following graphs, showing the relevant points:

    (a) $y=(x-2)^2$ (b) $y=3-\cos 2x$

    (c) $f(x)=\begin{cases} -1 & -\pi\le x\le -\frac{\pi}{2} \\ x & -\frac{\pi}{2}\le x\le \frac{\pi}{2} \\ 1 & \frac{\pi}{2}\le x\le \pi \end{cases}$

    (10)

11. Determine the inverse of $f(x)=3x+1$. (3)

12. Evaluate, correct to 3 decimal places:

    $2\tan^{-1}1.64+\sec^{-1}2.43-3\,\text{cosec}^{-1}3.85$ (4)

For lecturers/instructors/teachers, fully worked solutions to each of the problems in Revision Test 8, together with a full marking scheme, are available at the website:
www.routledge.com/cw/bird

COMPANION @ WEBSITE

Section 5

Complex numbers

Chapter 32

Complex numbers

Why it is important to understand: Complex numbers

Complex numbers are used in many scientific fields, including engineering, electromagnetism, quantum physics and applied mathematics, such as chaos theory. Any physical motion which is periodic, such as an oscillating beam, string, wire, pendulum, electronic signal or electromagnetic wave can be represented by a complex number function. This can make calculations with the various components simpler than with real numbers and sines and cosines. In control theory, systems are often transformed from the time domain to the frequency domain using the Laplace transform. In fluid dynamics, complex functions are used to describe potential flow in two dimensions. In electrical engineering, the Fourier transform is used to analyse varying voltages and currents. Complex numbers are used in signal analysis and other fields for a convenient description for periodically varying signals. This use is also extended into digital signal processing and digital image processing, which utilize digital versions of Fourier analysis (and wavelet analysis) to transmit, compress, restore and otherwise process digital audio signals, still images and video signals. Knowledge of complex numbers is clearly absolutely essential for further studies in so many engineering disciplines.

At the end of this chapter, you should be able to:

- define a complex number
- solve quadratic equations with imaginary roots
- use an Argand diagram to represent a complex number pictorially
- add, subtract, multiply and divide Cartesian complex numbers
- solve complex equations
- convert a Cartesian complex number into polar form, and vice-versa
- multiply and divide polar form complex numbers
- apply complex numbers to practical applications

32.1 Cartesian complex numbers

There are several applications of complex numbers in science and engineering, in particular in electrical alternating current theory and in mechanical vector analysis.

There are two main forms of complex number – **Cartesian form** (named after **Descartes***) **and polar form** – and both are explained in this chapter.

If we can add, subtract, multiply and divide complex numbers in both forms and represent the numbers on an Argand diagram then a.c. theory and vector analysis become considerably easier.

---

*Who was Descartes? – See page 197. To find out more go to www.routledge.com/cw/bird

(i) If the quadratic equation $x^2+2x+5=0$ is solved using the quadratic formula then:

$$x=\frac{-2\pm\sqrt{(2)^2-(4)(1)(5)}}{2(1)}$$
$$=\frac{-2\pm\sqrt{-16}}{2}=\frac{-2\pm\sqrt{(16)(-1)}}{2}$$
$$=\frac{-2\pm\sqrt{16}\sqrt{-1}}{2}=\frac{-2\pm 4\sqrt{-1}}{2}$$
$$=-1\pm 2\sqrt{-1}$$

It is not possible to evaluate $\sqrt{-1}$ in real terms. However, if an operator j is defined as $j=\sqrt{-1}$ then the solution may be expressed as $x=-1\pm j2$.

(ii) $-1+j2$ and $-1-j2$ are known as **complex numbers**. Both solutions are of the form $a+jb$, 'a' being termed the **real part** and jb the **imaginary part**. A complex number of the form $a+jb$ is called a **Cartesian complex number**.

(iii) In pure mathematics the symbol i is used to indicate $\sqrt{-1}$ (i being the first letter of the word imaginary). However i is the symbol of electric current in engineering, and to avoid possible confusion the next letter in the alphabet, j, is used to represent $\sqrt{-1}$

Problem 1. Solve the quadratic equation:
$$x^2+4=0$$

Since $x^2+4=0$ then $x^2=-4$ and $x=\sqrt{-4}$

i.e., $x=\sqrt{(-1)(4)}=\sqrt{-1}\sqrt{4}=j(\pm 2)$

$=\pm \mathbf{j2}$, (since $j=\sqrt{-1}$)

(Note that $\pm j2$ may also be written as $\pm\mathbf{2j}$)

Problem 2. Solve the quadratic equation:
$$2x^2+3x+5=0$$

Using the quadratic formula,

$$x=\frac{-3\pm\sqrt{(3)^2-4(2)(5)}}{2(2)}$$
$$=\frac{-3\pm\sqrt{-31}}{4}=\frac{-3\pm\sqrt{-1}\sqrt{31}}{4}$$
$$=\frac{-3\pm j\sqrt{31}}{4}$$

Hence $\mathbf{x=-\frac{3}{4}+j\frac{\sqrt{31}}{4}}$ or $\mathbf{-0.750\pm j1.392}$, correct to 3 decimal places.

(Note, a graph of $y=2x^2+3x+5$ does not cross the x-axis and hence $2x^2+3x+5=0$ has no real roots.)

Problem 3. Evaluate

(a) j^3 (b) j^4 (c) j^{23} (d) $\frac{-4}{j^9}$

(a) $j^3=j^2\times j=(-1)\times j=\mathbf{-j}$, since $j^2=-1$

(b) $j^4=j^2\times j^2=(-1)\times(-1)=\mathbf{1}$

(c) $j^{23}=j\times j^{22}=j\times(j^2)^{11}=j\times(-1)^{11}$
$=j\times(-1)=\mathbf{-j}$

(d) $j^9=j\times j^8=j\times(j^2)^4=j\times(-1)^4$
$=j\times 1=\mathbf{j}$

Hence $$\frac{-4}{j^9}=\frac{-4}{j}=\frac{-4}{j}\times\frac{-j}{-j}=\frac{4j}{-j^2}$$
$$=\frac{4j}{-(-1)}=\mathbf{4j}\text{ or }\mathbf{j4}$$

Now try the following Practice Exercise

Practice Exercise 158 Introduction to Cartesian complex numbers (Answers on page 718)

In Problems 1 to 9, solve the quadratic equations.

1. $x^2+25=0$
2. $x^2-2x+2=0$
3. $x^2-4x+5=0$
4. $x^2-6x+10=0$
5. $2x^2-2x+1=0$
6. $x^2-4x+8=0$
7. $25x^2-10x+2=0$
8. $2x^2+3x+4=0$
9. $4t^2-5t+7=0$
10. Evaluate (a) j^8 (b) $-\frac{1}{j^7}$ (c) $\frac{4}{2j^{13}}$

32.2 The Argand diagram

A complex number may be represented pictorially on rectangular or Cartesian axes. The horizontal (or x) axis is used to represent the real axis and the vertical (or y) axis is used to represent the imaginary axis. Such a diagram is called an **Argand diagram***. In Fig. 32.1, the point A represents the complex number $(3+j2)$ and is obtained by plotting the co-ordinates $(3, j2)$ as in graphical work. Fig. 32.1 also shows the Argand points B, C and D representing the complex numbers $(-2+j4)$, $(-3-j5)$ and $(1-j3)$ respectively.

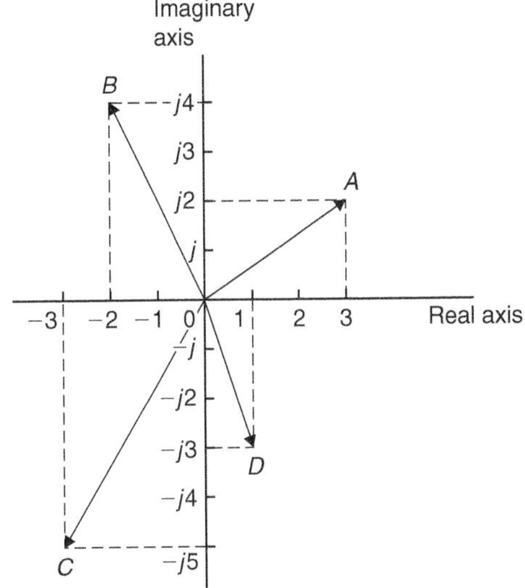

Figure 32.1

32.3 Addition and subtraction of complex numbers

Two complex numbers are added/subtracted by adding/subtracting separately the two real parts and the two imaginary parts.

For example, if $Z_1 = a+jb$ and $Z_2 = c+jd$,

then $Z_1 + Z_2 = (a+jb)+(c+jd)$
$= (a+c)+j(b+d)$

and $Z_1 - Z_2 = (a+jb)-(c+jd)$
$= (a-c)+j(b-d)$

Thus, for example,

$$(2+j3)+(3-j4) = 2+j3+3-j4 = \mathbf{5-j1}$$

and $$(2+j3)-(3-j4) = 2+j3-3+j4 = \mathbf{-1+j7}$$

The addition and subtraction of complex numbers may be achieved graphically as shown in the Argand diagram of Fig. 32.2. $(2+j3)$ is represented by vector $\boldsymbol{OP}$ and $(3-j4)$ by vector $\boldsymbol{OQ}$. In Fig. 32.2(a), by vector addition, (i.e. the diagonal of the parallelogram), $\boldsymbol{OP}+\boldsymbol{OQ}=\boldsymbol{OR}$. R is the point $(5, -j1)$

Hence $(2+j3)+(3-j4) = \mathbf{5-j1}$

In Fig. 32.2(b), vector $\boldsymbol{OQ}$ is reversed (shown as OQ') since it is being subtracted. (Note $\boldsymbol{OQ}=3-j4$ and $\boldsymbol{OQ'}=-(3-j4)=-3+j4$)

$\boldsymbol{OP}-\boldsymbol{OQ}=\boldsymbol{OP}+\boldsymbol{OQ'}=\boldsymbol{OS}$ is found to be the Argand point $(-1, j7)$

Hence $(2+j3)-(3-j4) = \mathbf{-1+j7}$

*Who was Argand? – **Jean-Robert Argand** (18 July 1786–13 August 1822) was a highly influential mathematician. He privately published a landmark essay on the representation of imaginary quantities which became known as the **Argand diagram**. To find out more go to www.routledge.com/cw/bird

Problem 4. Given $Z_1 = 2+j4$ and $Z_2 = 3-j$ determine (a) Z_1+Z_2, (b) Z_1-Z_2, (c) Z_2-Z_1 and show the results on an Argand diagram

(a) $Z_1+Z_2 = (2+j4)+(3-j)$
$= (2+3)+j(4-1) = \mathbf{5+j3}$

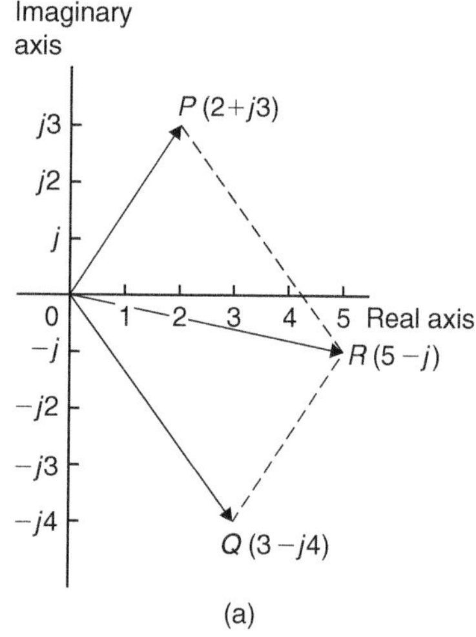

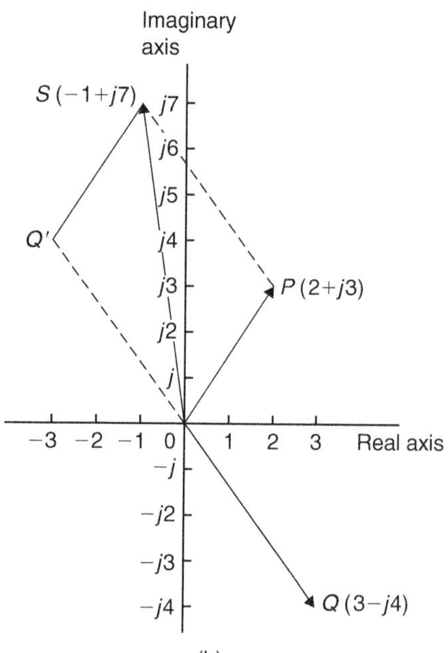

Figure 32.2

(b) $Z_1 - Z_2 = (2 + j4) - (3 - j)$
$= (2-3) + j(4-(-1)) = \mathbf{-1 + j5}$

(c) $Z_2 - Z_1 = (3 - j) - (2 + j4)$
$= (3-2) + j(-1-4) = \mathbf{1 - j5}$

Each result is shown in the Argand diagram of Fig. 32.3.

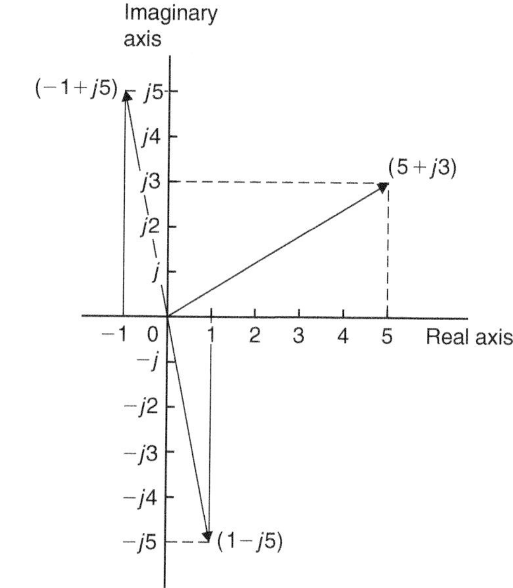

Figure 32.3

32.4 Multiplication and division of complex numbers

(i) **Multiplication of complex numbers** is achieved by assuming all quantities involved are real and then using $j^2 = -1$ to simplify.

Hence $(a + jb)(c + jd)$

$$= ac + a(jd) + (jb)c + (jb)(jd)$$
$$= ac + jad + jbc + j^2bd$$
$$= (ac - bd) + j(ad + bc), \quad \text{since } j^2 = -1$$

Thus $(3 + j2)(4 - j5)$

$$= 12 - j15 + j8 - j^2 10$$
$$= (12 - (-10)) + j(-15 + 8)$$
$$= \mathbf{22 - j7}$$

(ii) The **complex conjugate** of a complex number is obtained by changing the sign of the imaginary part. Hence the complex conjugate of $a + jb$ is $a - jb$. The product of a complex number and its complex conjugate is always a real number.

For example,

$$(3 + j4)(3 - j4) = 9 - j12 + j12 - j^2 16$$
$$= 9 + 16 = 25$$

[$(a + jb)(a - jb)$ may be evaluated 'on sight' as $a^2 + b^2$]

(iii) **Division of complex numbers** is achieved by multiplying both numerator and denominator by the complex conjugate of the denominator.

For example,

$$\frac{2 - j5}{3 + j4} = \frac{2 - j5}{3 + j4} \times \frac{(3 - j4)}{(3 - j4)}$$

$$= \frac{6 - j8 - j15 + j^2 20}{3^2 + 4^2}$$

$$= \frac{-14-j23}{25} = \frac{\mathbf{-14}}{\mathbf{25}} - j\frac{\mathbf{23}}{\mathbf{25}}$$

or $\mathbf{-0.56 - j0.92}$

Problem 5. If $Z_1 = 1-j3$, $Z_2 = -2+j5$ and $Z_3 = -3-j4$, determine in $a+jb$ form:

(a) Z_1Z_2 (b) $\frac{Z_1}{Z_3}$

(c) $\frac{Z_1Z_2}{Z_1+Z_2}$ (d) $Z_1Z_2Z_3$

(a) $Z_1Z_2 = (1-j3)(-2+j5)$

$= -2+j5+j6-j^2 15$

$= (-2+15)+j(5+6)$, since $j^2=-1$,

$= \mathbf{13+j11}$

(b) $\frac{Z_1}{Z_3} = \frac{1-j3}{-3-j4} = \frac{1-j3}{-3-j4} \times \frac{-3+j4}{-3+j4}$

$= \frac{-3+j4+j9-j^2 12}{3^2+4^2}$

$= \frac{9+j13}{25} = \frac{\mathbf{9}}{\mathbf{25}} + j\frac{\mathbf{13}}{\mathbf{25}}$

or $\mathbf{0.36 + j0.52}$

(c) $\frac{Z_1Z_2}{Z_1+Z_2} = \frac{(1-j3)(-2+j5)}{(1-j3)+(-2+j5)}$

$= \frac{13+j11}{-1+j2}$, from part (a),

$= \frac{13+j11}{-1+j2} \times \frac{-1-j2}{-1-j2}$

$= \frac{-13-j26-j11-j^2 22}{1^2+2^2}$

$= \frac{9-j37}{5} = \frac{\mathbf{9}}{\mathbf{5}} - j\frac{\mathbf{37}}{\mathbf{5}}$ or $\mathbf{1.8 - j7.4}$

(d) $Z_1Z_2Z_3 = (13+j11)(-3-j4)$, since

$Z_1Z_2 = 13+j11$, from part (a)

$= -39-j52-j33-j^2 44$

$= (-39+44)-j(52+33) = \mathbf{5-j85}$

Problem 6. Evaluate:

(a) $\frac{2}{(1+j)^4}$ (b) $j\left(\frac{1+j3}{1-j2}\right)^2$

(a) $(1+j)^2 = (1+j)(1+j) = 1+j+j+j^2$

$= 1+j+j-1 = j2$

$(1+j)^4 = [(1+j)^2]^2 = (j2)^2 = j^2 4 = -4$

Hence $\frac{2}{(1+j)^4} = \frac{2}{-4} = -\frac{\mathbf{1}}{\mathbf{2}}$

(b) $\frac{1+j3}{1-j2} = \frac{1+j3}{1-j2} \times \frac{1+j2}{1+j2}$

$= \frac{1+j2+j3+j^2 6}{1^2+2^2} = \frac{-5+j5}{5}$

$= -1+j1 = -1+j$

$\left(\frac{1+j3}{1-j2}\right)^2 = (-1+j)^2 = (-1+j)(-1+j)$

$= 1-j-j+j^2 = -j2$

Hence $j\left(\frac{1+j3}{1-j2}\right)^2 = j(-j2) = -j^2 2 = \mathbf{2}$,

since $j^2=-1$

Now try the following Practice Exercise

Practice Exercise 159 Operations involving Cartesian complex numbers (Answers on page 718)

1. Evaluate (a) $(3+j2)+(5-j)$ and (b) $(-2+j6)-(3-j2)$ and show the results on an Argand diagram
2. Write down the complex conjugates of (a) $3+j4$, (b) $2-j$
3. If $z=2+j$ and $w=3-j$ evaluate:
   (a) $z+w$
   (b) $w-z$
   (c) $3z-2w$
   (d) $5z+2w$
   (e) $j(2w-3z)$
   (f) $2jw-jz$

In Problems 4 to 8 evaluate in $a+jb$ form given $Z_1=1+j2$, $Z_2=4-j3$, $Z_3=-2+j3$ and $Z_4=-5-j$

4. (a) $Z_1+Z_2-Z_3$ (b) $Z_2-Z_1+Z_4$

5. (a) Z_1Z_2 (b) Z_3Z_4

6. (a) $Z_1Z_3+Z_4$ (b) $Z_1Z_2Z_3$

7. (a) $\dfrac{Z_1}{Z_2}$ (b) $\dfrac{Z_1+Z_3}{Z_2-Z_4}$

8. (a) $\dfrac{Z_1Z_3}{Z_1+Z_3}$ (b) $Z_2+\dfrac{Z_1}{Z_4}+Z_3$

9. Evaluate (a) $\dfrac{1-j}{1+j}$ (b) $\dfrac{1}{1+j}$

10. Show that: $\dfrac{-25}{2}\left(\dfrac{1+j2}{3+j4}-\dfrac{2-j5}{-j}\right)$
$=57+j24$

32.5 Complex equations

If two complex numbers are equal, then their real parts are equal and their imaginary parts are equal. Hence **if $a+jb=c+jd$, then $a=c$ and $b=d$**

Problem 7. Solve the complex equations:
(a) $2(x+jy)=6-j3$
(b) $(1+j2)(-2-j3)=a+jb$

(a) $2(x+jy)=6-j3$ hence $2x+j2y=6-j3$

Equating the real parts gives:

$$2x=6, \text{ i.e. } \mathbf{x=3}$$

Equating the imaginary parts gives:

$$2y=-3, \text{ i.e. } \mathbf{y=-\frac{3}{2}}$$

(b) $(1+j2)(-2-j3)=a+jb$

$-2-j3-j4-j^2 6=a+jb$

Hence $4-j7=a+jb$

Equating real and imaginary terms gives:

$$\mathbf{a=4} \text{ and } \mathbf{b=-7}$$

Problem 8. Solve the equations:
(a) $(2-j3)=\sqrt{a+jb}$
(b) $(x-j2y)+(y-j3x)=2+j3$

(a) $(2-j3)=\sqrt{a+jb}$

Hence $(2-j3)^2=a+jb$

i.e. $(2-j3)(2-j3)=a+jb$

Hence $4-j6-j6+j^2 9=a+jb$

and $-5-j12=a+jb$

Thus $\mathbf{a=-5}$ and $\mathbf{b=-12}$

(b) $(x-j2y)+(y-j3x)=2+j3$

Hence $(x+y)+j(-2y-3x)=2+j3$

Equating real and imaginary parts gives:

$$x+y=2 \quad (1)$$

and $$-3x-2y=3 \quad (2)$$

i.e. two stimulaneous equations to solve.

Multiplying equation (1) by 2 gives:

$$2x+2y=4 \quad (3)$$

Adding equations (2) and (3) gives:

$$-x=7, \text{ i.e. } \mathbf{x=-7}$$

From equation (1), $\mathbf{y=9}$, which may be checked in equation (2).

Now try the following Practice Exercise

Practice Exercise 160 Complex equations (Answers on page 718)

In Problems 1 to 4 solve the complex equations.

1. $(2+j)(3-j2)=a+jb$
2. $\dfrac{2+j}{1-j}=j(x+jy)$
3. $(2-j3)=\sqrt{a+jb}$
4. $(x-j2y)-(y-jx)=2+j$
5. If $Z=R+j\omega L+1/j\omega C$, express Z in $(a+jb)$ form when $R=10$, $L=5$, $C=0.04$ and $\omega=4$

32.6 The polar form of a complex number

(i) Let a complex number Z be $x+jy$ as shown in the Argand diagram of Fig. 32.4. Let distance OZ be r and the angle OZ makes with the positive real axis be θ.

From trigonometry, $x=r\cos\theta$ and $y=r\sin\theta$

Hence $Z=x+jy=r\cos\theta+jr\sin\theta = r(\cos\theta+j\sin\theta)$

$Z=r(\cos\theta+j\sin\theta)$ is usually abbreviated to $Z=r\angle\theta$ which is known as the **polar form** of a complex number.

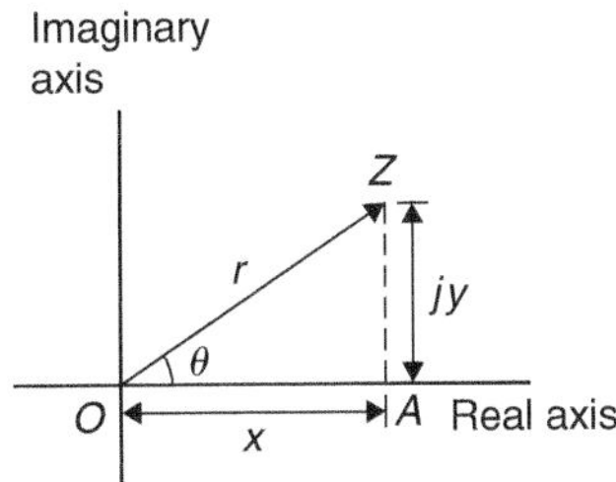

Figure 32.4

(ii) r is called the **modulus** (or magnitude) of Z and is written as mod Z or $|Z|$

r is determined using Pythagoras' theorem on triangle OAZ in Fig. 32.4,

i.e. $$r=\sqrt{x^2+y^2}$$

(iii) θ is called the **argument** (or amplitude) of Z and is written as arg Z.

By trigonometry on triangle OAZ,

$$\arg Z=\theta=\tan^{-1}\frac{y}{x}$$

(iv) Whenever changing from Cartesian form to polar form, or vice-versa, a sketch is invaluable for determining the quadrant in which the complex number occurs

Problem 9. Determine the modulus and argument of the complex number $Z=2+j3$, and express Z in polar form

$Z=2+j3$ lies in the first quadrant as shown in Fig. 32.5.

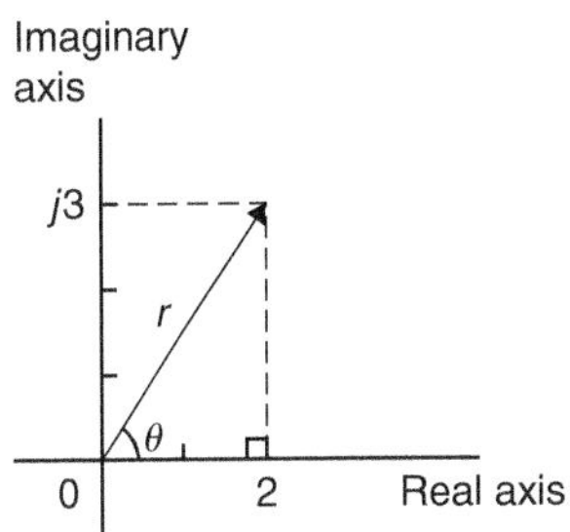

Figure 32.5

Modulus, $|Z|=r=\sqrt{2^2+3^2}=\sqrt{\mathbf{13}}$ or **3.606**, correct to 3 decimal places.

Argument, $\arg Z=\theta=\tan^{-1}\frac{3}{2}$
$= \mathbf{56.31^\circ}$ or $\mathbf{56^\circ 19'}$

In polar form, $2+j3$ is written as $\mathbf{3.606\angle 56.31^\circ}$ or $\mathbf{3.606\angle 56^\circ 19'}$

Problem 10. Express the following complex numbers in polar form:

(a) $3+j4$ (b) $-3+j4$

(c) $-3-j4$ (d) $3-j4$

(a) $3+j4$ is shown in Fig. 32.6 and lies in the first quadrant.

Modulus, $r=\sqrt{3^2+4^2}=5$ and

argument $\theta=\tan^{-1}\frac{4}{3}=53.13^\circ$ or $53^\circ 8'$

Hence $\mathbf{3+j4=5\angle 53.13^\circ}$

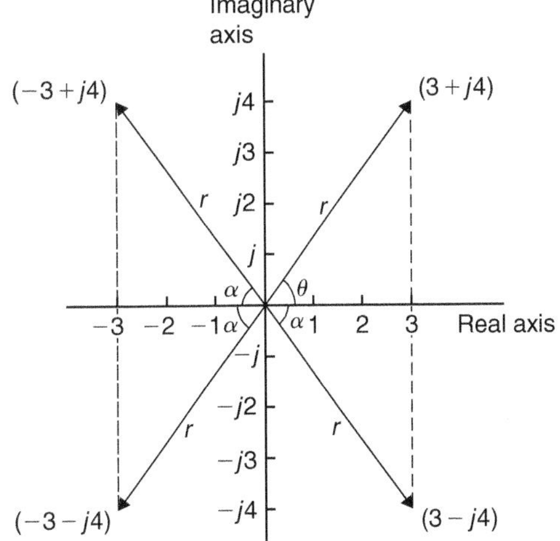

Figure 32.6

(b) $-3+j4$ is shown in Fig. 32.6 and lies in the second quadrant.

Modulus, $r=5$ and angle $\alpha=53.13°$, from part (a).

Argument $=180°-53.13°=126.87°$ (i.e. the argument must be measured from the positive real axis).

Hence $\mathbf{-3+j4 = 5\angle 126.87°}$

(c) $-3-j4$ is shown in Fig. 32.6 and lies in the third quadrant.

Modulus, $r=5$ and $\alpha=53.13°$, as above.

Hence the argument $=180°+53.13°=233.13°$, which is the same as $-126.87°$

Hence $\mathbf{(-3-j4)=5\angle 233.13°}$ or $\mathbf{5\angle -126.87°}$

(By convention the **principal value** is normally used, i.e. the numerically least value, such that $-\pi<\theta<\pi$)

(d) $3-j4$ is shown in Fig. 32.6 and lies in the fourth quadrant.

Modulus, $r=5$ and angle $\alpha=53.13°$, as above.

Hence $\mathbf{(3-j4)=5\angle -53.13°}$

Problem 11. Convert (a) $4\angle 30°$ (b) $7\angle -145°$ into $a+jb$ form, correct to 4 significant figures

(a) $4\angle 30°$ is shown in Fig. 32.7(a) and lies in the first quadrant.

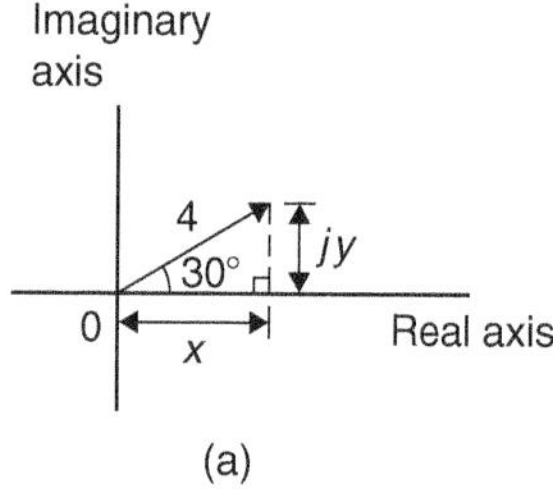

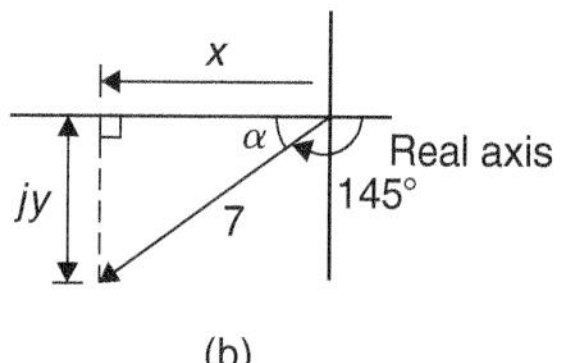

Figure 32.7

Using trigonometric ratios,

$x=4\cos 30°=3.464$ and $y=4\sin 30°=2.000$

Hence $\mathbf{4\angle 30° = 3.464+j2.000}$

(b) $7\angle -145°$ is shown in Fig. 32.7(b) and lies in the third quadrant.

Angle $\alpha=180°-145°=35°$

Hence $x=7\cos 35°=5.734$

and $y=7\sin 35°=4.015$

Hence $\mathbf{7\angle -145° = -5.734-j4.015}$

Alternatively

$$7\angle -145° = 7\cos(-145°)+j7\sin(-145°) = \mathbf{-5.734-j4.015}$$

Calculator

Using the **'Pol'** and **'Rec'** functions on a calculator enables changing from Cartesian to polar and vice versa to be achieved more quickly.

Since complex numbers are used with vectors and with electrical engineering a.c. theory, it is essential that the calculator can be used quickly and accurately.

32.7 Multiplication and division in polar form

If $Z_1=r_1\angle\theta_1$ **and** $Z_2=r_2\angle\theta_2$ **then:**

(i) $Z_1Z_2=r_1r_2\angle(\theta_1+\theta_2)$ **and**

(ii) $\dfrac{Z_1}{Z_2}=\dfrac{r_1}{r_2}\angle(\theta_1-\theta_2)$

Problem 12. Determine, in polar form:
(a) $8\angle 25°\times 4\angle 60°$
(b) $3\angle 16°\times 5\angle -44°\times 2\angle 80°$

(a) $8\angle 25°\times 4\angle 60°=(8\times 4)\angle(25°+60°)=\mathbf{32\angle 85°}$

(b) $3\angle 16°\times 5\angle -44°\times 2\angle 80°$
$=(3\times 5\times 2)\angle[16°+(-44°)+80°]=\mathbf{30\angle 52°}$

Problem 13. Evaluate in polar form:

(a) $\dfrac{16\angle 75°}{2\angle 15°}$ (b) $\dfrac{10\angle\frac{\pi}{4}\times 12\angle\frac{\pi}{2}}{6\angle -\frac{\pi}{3}}$

(a) $\dfrac{16\angle 75^\circ}{2\angle 15^\circ} = \dfrac{16}{2}\angle(75^\circ - 15^\circ) = \mathbf{8\angle 60^\circ}$

(b) $\dfrac{10\angle\frac{\pi}{4} \times 12\angle\frac{\pi}{2}}{6\angle-\frac{\pi}{3}} = \dfrac{10\times 2}{6}\angle\left(\frac{\pi}{4}+\frac{\pi}{2}-\left(-\frac{\pi}{3}\right)\right)$

$$= \mathbf{20\angle\frac{13\pi}{12}} \text{ or } \mathbf{20\angle-\frac{11\pi}{12}} \text{ or}$$

$$\mathbf{20\angle 195^\circ} \text{ or } \mathbf{20\angle -165^\circ}$$

Problem 14. Evaluate, in polar form:

$$2\angle 30^\circ + 5\angle -45^\circ - 4\angle 120^\circ$$

Addition and subtraction in polar form is not possible directly. Each complex number has to be converted into Cartesian form first.

$$\begin{aligned}
2\angle 30^\circ &= 2(\cos 30^\circ + j\sin 30^\circ)\\
&= 2\cos 30^\circ + j2\sin 30^\circ = 1.732 + j1.000\\
5\angle -45^\circ &= 5(\cos(-45^\circ) + j\sin(-45^\circ))\\
&= 5\cos(-45^\circ) + j5\sin(-45^\circ)\\
&= 3.536 - j3.536\\
4\angle 120^\circ &= 4(\cos 120^\circ + j\sin 120^\circ)\\
&= 4\cos 120^\circ + j4\sin 120^\circ\\
&= -2.000 + j3.464
\end{aligned}$$

Hence $2\angle 30^\circ + 5\angle -45^\circ - 4\angle 120^\circ$

$$= (1.732 + j1.000) + (3.536 - j3.536) - (-2.000 + j3.464)$$

$= 7.268 - j6.000$, which lies in the fourth quadrant

$$= \sqrt{7.268^2 + 6.000^2}\angle\tan^{-1}\left(\frac{-6.000}{7.268}\right)$$

$$= \mathbf{9.425\angle -39.54^\circ} \text{ or } \mathbf{9.425\angle -39^\circ 32'}$$

Now try the following Practice Exercise

Practice Exercise 161 Polar form (Answers on page 718)

1. Determine the modulus and argument of (a) $2+j4$ (b) $-5-j2$ (c) $j(2-j)$

In Problems 2 and 3 express the given Cartesian complex numbers in polar form, leaving answers in surd form.

2. (a) $2+j3$ (b) -4 (c) $-6+j$
3. (a) $-j3$ (b) $(-2+j)^3$ (c) $j^3(1-j)$

In Problems 4 and 5 convert the given polar complex numbers into $(a+jb)$ form giving answers correct to 4 significant figures.

4. (a) $5\angle 30^\circ$ (b) $3\angle 60^\circ$ (c) $7\angle 45^\circ$
5. (a) $6\angle 125^\circ$ (b) $4\angle\pi$ (c) $3.5\angle -120^\circ$

In Problems 6 to 8, evaluate in polar form.

6. (a) $3\angle 20^\circ \times 15\angle 45^\circ$
   (b) $2.4\angle 65^\circ \times 4.4\angle -21^\circ$
7. (a) $6.4\angle 27^\circ \div 2\angle -15^\circ$
   (b) $5\angle 30^\circ \times 4\angle 80^\circ \div 10\angle -40^\circ$
8. (a) $4\angle\frac{\pi}{6} + 3\angle\frac{\pi}{8}$
   (b) $2\angle 120^\circ + 5.2\angle 58^\circ - 1.6\angle -40^\circ$

32.8 Applications of complex numbers

There are several applications of complex numbers in science and engineering, in particular in electrical alternating current theory and in mechanical vector analysis.

The effect of multiplying a phasor by j is to rotate it in a positive direction (i.e. anticlockwise) on an Argand diagram through 90° without altering its length. Similarly, multiplying a phasor by $-j$ rotates the phasor through −90°. These facts are used in a.c. theory since certain quantities in the phasor diagrams lie at 90° to each other. For example, in the R–L series circuit shown in Fig. 32.8(a), V_L leads I by 90° (i.e. I lags V_L by 90°) and may be written as jV_L, the vertical axis being regarded as the imaginary axis of an Argand diagram. Thus $V_R + jV_L = V$ and since $V_R = IR$, $V = IX_L$ (where X_L is the inductive reactance, $2\pi fL$ ohms) and $V = IZ$ (where Z is the impedance) then $R + jX_L = Z$
Similarly, for the R–C circuit shown in Fig. 32.8(b), V_C lags I by 90° (i.e. I leads V_C by 90°) and $V_R - jV_C = V$, from which $R - jX_C = Z$ (where X_C is the capacitive reactance $\dfrac{1}{2\pi fC}$ ohms)

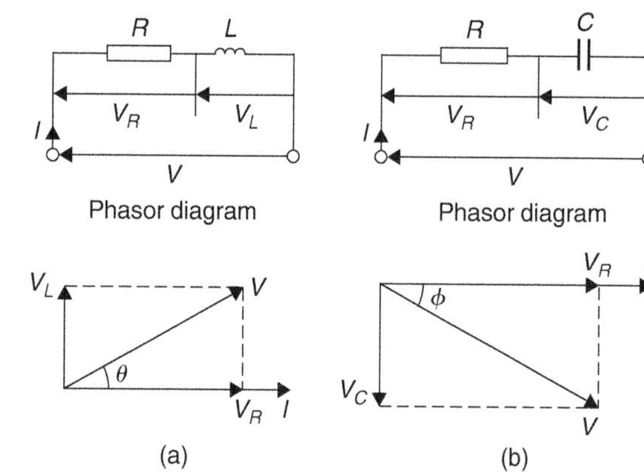

Figure 32.8

Problem 15. Determine the resistance and series inductance (or capacitance) for each of the following impedances, assuming a frequency of 50 Hz:

(a) $(4.0+j7.0)\,\Omega$ (b) $-j20\Omega$
(c) $15\angle-60°\,\Omega$

(a) Impedance, $Z=(4.0+j7.0)\,\Omega$ hence,

resistance = 4.0 Ω and reactance = 7.0 Ω.

Since the imaginary part is positive, the reactance is inductive,

i.e. $X_L = 7.0\,\Omega$.

Since $X_L = 2\pi f L$ then **inductance**,

$$L = \frac{X_L}{2\pi f} = \frac{7.0}{2\pi(50)} = \mathbf{0.0223\ H} \text{ or } \mathbf{22.3\ mH}$$

(b) Impedance, $Z=-j20$, i.e. $Z=(0-j20)\,\Omega$, hence **resistance = 0** and reactance = 20 Ω. Since the imaginary part is negative, the reactance is capacitive, i.e. $X_C=20\,\Omega$ and since $X_C=\frac{1}{2\pi f C}$ then:

$$\textbf{capacitance}, C = \frac{1}{2\pi f X_C} = \frac{1}{2\pi(50)(20)}\text{F}$$

$$= \frac{10^6}{2\pi(50)(20)}\mu\text{F} = \mathbf{159.2\ \mu F}$$

(c) Impedance, Z

$$= 15\angle-60° = 15[\cos(-60°)+j\sin(-60°)]$$

$$= 7.50 - j12.99\,\Omega.$$

Hence **resistance = 7.50 Ω** and capacitive reactance, $X_C = 12.99\,\Omega$

Since $X_C = \frac{1}{2\pi f C}$ then **capacitance**,

$$C = \frac{1}{2\pi f X_C} = \frac{10^6}{2\pi(50)(12.99)}\mu\text{F}$$

$$= \mathbf{245\ \mu F}$$

Problem 16. An alternating voltage of 240 V, 50 Hz is connected across an impedance of $(60-j100)\,\Omega$. Determine (a) the resistance (b) the capacitance (c) the magnitude of the impedance and its phase angle and (d) the current flowing

(a) Impedance $Z=(60-j100)\,\Omega$.

Hence **resistance = 60Ω**

(b) Capacitive reactance $X_C=100\,\Omega$ and since $X_C=\frac{1}{2\pi f C}$ then

$$\text{capacitance,}\quad C = \frac{1}{2\pi f X_C} = \frac{1}{2\pi(50)(100)}$$

$$= \frac{10^6}{2\pi(50)(100)}\mu\text{F}$$

$$= \mathbf{31.83\,\mu F}$$

(c) Magnitude of impedance,

$$|Z| = \sqrt{60^2+(-100)^2} = \mathbf{116.6\Omega}$$

$$\text{Phase angle, arg } Z = \tan^{-1}\left(\frac{-100}{60}\right) = \mathbf{-59.04°}$$

(d) Current flowing, $I = \frac{V}{Z} = \frac{240°\angle 0°}{116.6\angle-59.04°}$

$$= \mathbf{2.058\angle 59.04°\,A}$$

The circuit and phasor diagrams are as shown in Fig. 32.8(b).

Problem 17. For the parallel circuit shown in Fig. 32.9, determine the value of current I and its phase relative to the 240 V supply, using complex numbers

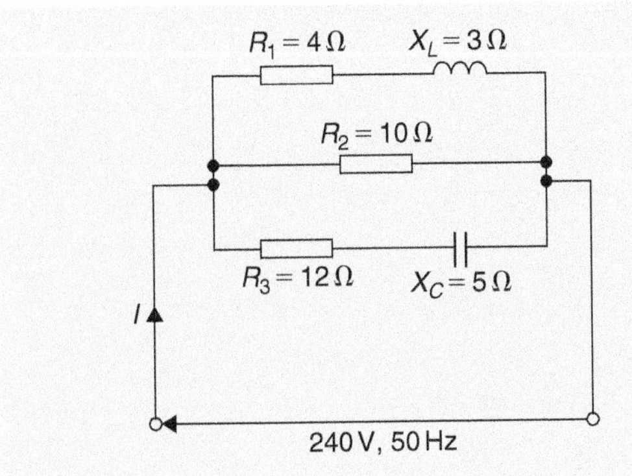

Figure 32.9

Current $I = \frac{V}{Z}$. Impedance Z for the three-branch parallel circuit is given by:

$$\frac{1}{Z} = \frac{1}{Z_1} + \frac{1}{Z_2} + \frac{1}{Z_3}$$

where $Z_1 = 4 + j3$, $Z_2 = 10$ and $Z_3 = 12 - j5$

$$\text{Admittance,} \quad Y_1 = \frac{1}{Z_1} = \frac{1}{4 + j3}$$

$$= \frac{1}{4 + j3} \times \frac{4 - j3}{4 - j3} = \frac{4 - j3}{4^2 + 3^2}$$

$$= 0.160 - j0.120 \text{ siemens}$$

$$\text{Admittance, } Y_2 = \frac{1}{Z_2} = \frac{1}{10} = 0.10 \text{ siemens}$$

$$\text{Admittance,} \quad Y_3 = \frac{1}{Z_3} = \frac{1}{12 - j5}$$

$$= \frac{1}{12 - j5} \times \frac{12 + j5}{12 + j5} = \frac{12 + j5}{12^2 + 5^2}$$

$$= 0.0710 + j0.0296 \text{ siemens}$$

$$\text{Total admittance,} \quad Y = Y_1 + Y_2 + Y_3$$

$$= (0.160 - j0.120) + (0.10) + (0.0710 + j0.0296)$$

$$= 0.331 - j0.0904$$

$$= 0.343\angle{-15.28^\circ} \text{ siemens}$$

$$\text{Current } I = \frac{V}{Z} = VY$$

$$= (240\angle 0^\circ)(0.343\angle{-15.28^\circ})$$

$$= \mathbf{82.32\angle{-15.28^\circ}\ A}$$

Problem 18. Determine the magnitude and direction of the resultant of the three coplanar forces given below, when they act at a point:

Force A, 10 N acting at 45° from the positive horizontal axis,
Force B, 8 N acting at 120° from the positive horizontal axis,
Force C, 15 N acting at 210° from the positive horizontal axis.

The space diagram is shown in Fig. 32.10. The forces may be written as complex numbers.

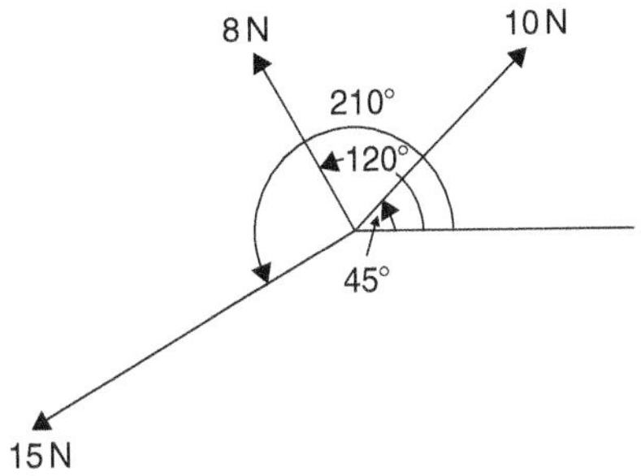

Figure 32.10

Thus force A, $f_A = 10\angle 45^\circ$, force B, $f_B = 8\angle 120^\circ$ and force C, $f_C = 15\angle 210^\circ$.

The resultant force

$$= f_A + f_B + f_C$$

$$= 10\angle 45^\circ + 8\angle 120^\circ + 15\angle 210^\circ$$

$$= 10(\cos 45^\circ + j\sin 45^\circ) + 8(\cos 120^\circ + j\sin 120^\circ) + 15(\cos 210^\circ + j\sin 210^\circ)$$

$$= (7.071 + j7.071) + (-4.00 + j6.928) + (-12.99 - j7.50)$$

$$= -9.919 + j6.499$$

Magnitude of resultant force

$$= \sqrt{(-9.919)^2 + 6.499^2} = \mathbf{11.86\,N}$$

Direction of resultant force

$$= \tan^{-1}\left(\frac{6.499}{-9.919}\right) = \mathbf{146.77^\circ}$$

(since $-9.919 + j6.499$ lies in the second quadrant).

Now try the following Practice Exercise

Practice Exercise 162 Applications of complex numbers (Answers on page 719)

1. Determine the resistance R and series inductance L (or capacitance C) for each of the following impedances assuming the frequency to be 50 Hz.

   (a) $(3+j8)\,\Omega$ (b) $(2-j3)\,\Omega$
   (c) $j14\,\Omega$ (d) $8\angle-60°\,\Omega$

2. Two impedances, $Z_1=(3+j6)\,\Omega$ and $Z_2=(4-j3)\,\Omega$ are connected in series to a supply voltage of 120 V. Determine the magnitude of the current and its phase angle relative to the voltage

3. If the two impedances in Problem 2 are connected in parallel determine the current flowing and its phase relative to the 120 V supply voltage

4. A series circuit consists of a 12 Ω resistor, a coil of inductance 0.10 H and a capacitance of 160 μF. Calculate the current flowing and its phase relative to the supply voltage of 240 V, 50 Hz. Determine also the power factor of the circuit, given that power factor = $\cos\phi$, where ϕ is the phase angle.

5. For the circuit shown in Fig. 32.11, determine the current I flowing and its phase relative to the applied voltage

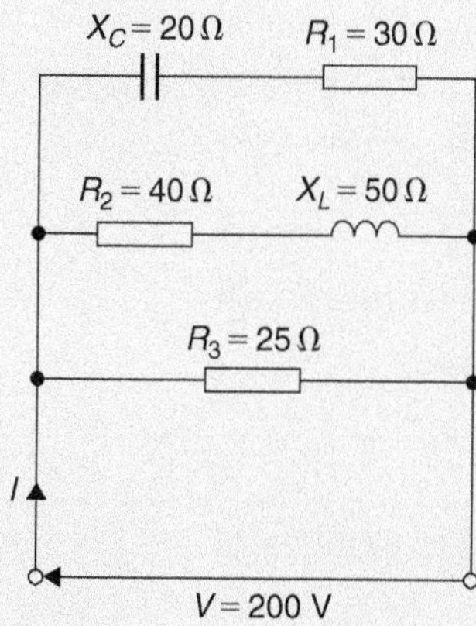

Figure 32.11

6. Determine, using complex numbers, the magnitude and direction of the resultant of the coplanar forces given below, which are acting at a point. Force A, 5 N acting horizontally, Force B, 9 N acting at an angle of 135° to force A, Force C, 12 N acting at an angle of 240° to force A

7. A delta-connected impedance Z_A is given by:

$$Z_A=\frac{Z_1Z_2+Z_2Z_3+Z_3Z_1}{Z_2}$$

   Determine Z_A in both Cartesian and polar form given $Z_1=(10+j0)\,\Omega$, $Z_2=(0-j10)\,\Omega$ and $Z_3=(10+j10)\,\Omega$

8. In the hydrogen atom, the angular momentum, p, of the de Broglie wave is given by: $p\psi=-\left(\frac{jh}{2\pi}\right)(\pm jm\psi)$

   Determine an expression for p

9. An aircraft P flying at a constant height has a velocity of $(400+j300)$ km/h. Another aircraft Q at the same height has a velocity of $(200-j600)$ km/h. Determine (a) the velocity of P relative to Q, and (b) the velocity of Q relative to P. Express the answers in polar form, correct to the nearest km/h

10. Three vectors are represented by P, $2\angle30°$, Q, $3\angle90°$ and R, $4\angle-60°$. Determine in polar form the vectors represented by (a) $P+Q+R$, (b) $P-Q-R$

11. In a Schering bridge circuit, $Z_x=(R_X-jX_{C_X})$, $Z_2=-jX_{C_2}$,

   $Z_3=\frac{(R_3)(-jX_{C_3})}{(R_3-jX_{C_3})}$ and $Z_4=R_4$ where $X_C=\frac{1}{2\pi fC}$

   At balance: $(Z_X)(Z_3)=(Z_2)(Z_4)$.

   Show that at balance $R_X=\frac{C_3R_4}{C_2}$ and $C_X=\frac{C_2R_3}{R_4}$

12. An amplifier has a transfer function, T, given by: $T = \dfrac{500}{1 + j\omega\left(5 \times 10^{-4}\right)}$ where ω is the angular frequency. The gain of the amplifier is given by the modulus of T and the phase is given by the argument of T. If $\omega = 2000$ rad/s, determine the gain and the phase (in degrees).

13. The sending end current of a transmission line is given by: $I_S = \frac{V_S}{Z_0}\tanh PL$. Calculate the value of the sending current, in polar form, given $V_S = 200$V, $Z_0 = 560 + j420\,\Omega$, $P = 0.20$ and $L = 10$

Practice Exercise 163 Multiple-choice questions on complex numbers (Answers on page 719)

Each question has only one correct answer

1. $\dfrac{5}{j^6}$ is equivalent to:
(a) $j5$ (b) -5 (c) $-j5$ (d) 5

2. Which of the following statements is true?
(a) $j^2 = 1$ (b) $j^3 = \sqrt{-1}$
(c) $j^4 = -1$ (d) $j^5 = j$

3. If $x^2 + 1 = 0$ then the value of x is:
(a) -1 (b) j (c) $\pm j$ (d) $-j$

4. In $a + jb$ form, $(6 + j2) - (3 - j2)$ is equal to:
(a) 3 (b) $9 + j4$ (c) 9 (d) $3 + j4$

5. The product $(j2)(j3)$ is:
(a) -6 (b) $j6$ (c) $-j6$ (d) 6

6. $(j2)^3$ is equivalent to:
(a) $j6$ (b) $j8$ (c) $-j6$ (d) $-j8$

7. $(-4 - j3)$ in polar form is:
(a) $5\angle -143.13°$ (b) $5\angle 126.87°$
(c) $5\angle 143.13°$ (d) $5\angle -126.87°$

8. $(1 + j)^4$ is equivalent to:
(a) 4 (b) $-j4$ (c) -4 (d) $j4$

9. The product of $(2 - j3)$ and $(5 + j4)$ is:
(a) $-2 + j7$ (b) $22 - j7$
(c) $-2 + j7$ (d) $22 + j7$

10. j^{13} is equivalent to:
(a) j (b) -1 (c) $-j$ (d) 1

11. $(3 - j5)^2$ is equal to:
(a) 34 (b) $-16 - j30$ (c) 16 (d) $16 + j30$

12. The product of $(4 + j)$ and $(4 - j)$ is:
(a) 16 (b) $\sqrt{17}$ (c) 17 (d) 15

13. In Cartesian form, $\dfrac{3 - j4}{4 - j}$ is equivalent to:
(a) $\dfrac{11 - j7}{15}$ (b) $\dfrac{16 - j13}{17}$
(c) $\dfrac{8 - j7}{17}$ (d) $\dfrac{8 - j7}{15}$

14. In Cartesian form, $\dfrac{20}{3 + j}$ is equivalent to:
(a) $\dfrac{15}{2} - j\dfrac{5}{2}$ (b) $6 + j2$
(c) $\dfrac{15}{2} + j\dfrac{5}{2}$ (d) $6 - j2$

15. $\dfrac{(1 - j)^2}{(-j)^3}$ is equal to:
(a) -2 (b) $-j2$ (c) $2 - j2$ (d) $j2$

16. In Cartesian form, $\dfrac{1}{2 + j}$ is equivalent to:
(a) $\dfrac{2}{5} - j\dfrac{1}{5}$ (b) $\dfrac{2}{3} - j\dfrac{1}{3}$
(c) $\dfrac{15}{2} + j\dfrac{5}{2}$ (d) $\dfrac{2}{3} + j\dfrac{1}{3}$

17. The expression $(-1 + j)^3$ is equivalent to:
(a) $-1 - j$ (b) $2 + j2$ (c) $-3j$ (d) $-2 - j2$

18. The total impedance, Z_T ohms, in an electrical circuit is given by $Z_T = \dfrac{Z_1 \times Z_2}{Z_1 + Z_2}$
When $Z_1 = (1 + j3)\Omega$ and $Z_2 = (1 - j3)\Omega$, the total impedance is:
(a) 2Ω (b) 1.6Ω (c) 5Ω (d) 0.625Ω

19. The expression $(2 - j3)^2$ is equivalent to:
(a) $-5 + j0$ (b) $13 - j12$
(c) $-5 - j12$ (d) $13 + j0$

20. $2\angle\dfrac{\pi}{3} + 3\angle\dfrac{\pi}{6}$ in polar form is:
(a) $5\angle\dfrac{\pi}{2}$ (b) $4.84\angle 0.84$
(c) $6\angle 0.55$ (d) $4.84\angle 0.73$

For fully worked solutions to each of the problems in Practice Exercises 158 to 162 in this chapter, go to the website:
www.routledge.com/cw/bird

COMPANION @ WEBSITE

Chapter 33

De Moivre's theorem

Why it is important to understand: De Moivre's theorem

There are many, many examples of the use of complex numbers in engineering and science. De Moivre's theorem has several uses, including finding powers and roots of complex numbers, solving polynomial equations, calculating trigonometric identities and for evaluating the sums of trigonometric series. The theorem is also used to calculate exponential and logarithmic functions of complex numbers. De Moivre's theorem has applications in electrical engineering and physics.

At the end of this chapter, you should be able to:

- state de Moivre's theorem
- calculate powers of complex numbers
- calculate roots of complex numbers

33.1 Introduction

From multiplication of complex numbers in polar form,

$$(r\angle\theta) \times (r\angle\theta) = r^2\angle 2\theta$$

Similarly, $(r\angle\theta)\times(r\angle\theta)\times(r\angle\theta) = r^3\angle 3\theta$ and so on.

In general, **de Moivre's theorem*** states:

$$[r\angle\theta]^n = r^n\angle n\theta$$

The theorem is true for all positive, negative and fractional values of n. The theorem is used to determine powers and roots of complex numbers.

33.2 Powers of complex numbers

For example, $[3\angle 20°]^4 = 3^4\angle(4\times 20°) = 81\angle 80°$ by de Moivre's theorem.

Problem 1. Determine, in polar form:
(a) $[2\angle 35°]^5$ (b) $(-2+j3)^6$

*Who was de Moivre? – **Abraham de Moivre** (26 May 1667–27 November 1754) was a French mathematician famous for **de Moivre's formula**, which links complex numbers and trigonometry, and for his work on the normal distribution and probability theory. To find out more go to www.routledge.com/cw/bird

(a) $[2\angle 35^\circ]^5 = 2^5\angle(5\times 35^\circ)$, from De Moivre's theorem

$= \mathbf{32\angle 175^\circ}$

(b) $(-2+j3) = \sqrt{(-2)^2+(3)^2}\angle\tan^{-1}\left(\dfrac{3}{-2}\right)$

$= \sqrt{13}\angle 123.69^\circ$, since $-2+j3$ lies in the second quadrant

$(-2+j3)^6 = [\sqrt{3}\angle 123.69^\circ]^6$

$= \sqrt{13^6}\angle(6\times 123.69^\circ)$, by De Moivre's theorem

$= 2197\angle 742.14^\circ$

$= 2197\angle 382.14^\circ$

(since $742.14 \equiv 742.14^\circ - 360^\circ = 382.14^\circ$)

$= \mathbf{2197\angle 22.14^\circ}$

(since $382.14^\circ \equiv 382.14^\circ - 360^\circ = 22.14^\circ$)

Problem 2. Determine the value of $(-7+j5)^4$, expressing the result in polar and rectangular forms

$$(-7+j5) = \sqrt{(-7)^2+5^2}\angle\tan^{-1}\left(\frac{5}{-7}\right)$$

$$= \sqrt{74}\angle 144.46^\circ$$

(Note, by considering the Argand diagram, $-7+j5$ must represent an angle in the second quadrant and **not** in the fourth quadrant.)

Applying de Moivre's theorem:

$(-7+j5)^4 = [\sqrt{74}\angle 144.46^\circ]^4$

$= \sqrt{74^4}\angle 4\times 144.46^\circ$

$= 5476\angle 577.84^\circ$

$= \mathbf{5476\angle 217.84^\circ}$ or

$\mathbf{5476\angle 217^\circ 15'}$ in polar form.

Since $r\angle\theta = r\cos\theta + jr\sin\theta$,

$5476\angle 217.84^\circ = 5476\cos 217.84^\circ + j5476\sin 217.84^\circ$

$= -4325 - j3359$

i.e. $\mathbf{(-7+j5)^4 = -4325 - j3359}$ in rectangular form.

Now try the following Practice Exercise

Practice Exercise 164 Powers of complex numbers (Answers on page 719)

1. Determine in polar form (a) $[1.5\angle 15^\circ]^5$ (b) $(1+j2)^6$
2. Determine in polar and Cartesian forms (a) $[3\angle 41^\circ]^4$ (b) $(-2-j)^5$
3. Convert $(3-j)$ into polar form and hence evaluate $(3-j)^7$, giving the answer in polar form

In Problems 4 to 7, express in both polar and rectangular forms:

4. $(6+j5)^3$
5. $(3-j8)^5$
6. $(-2+j7)^4$
7. $(-16-j9)^6$

33.3 Roots of complex numbers

The **square root** of a complex number is determined by letting $n=\frac{1}{2}$ in De Moivre's theorem,

i.e. $\sqrt{r\angle\theta} = [r\angle\theta]^{1/2} = r^{1/2}\angle\frac{1}{2}\theta = \sqrt{r}\angle\frac{\theta}{2}$

There are two square roots of a real number, equal in size but opposite in sign.

Problem 3. Determine the two square roots of the complex number $(5+j12)$ in polar and Cartesian forms and show the roots on an Argand diagram

$$(5+112) = \sqrt{5^2+12^2}\angle\tan^{-1}\left(\frac{12}{5}\right) = 13\angle 67.38^\circ$$

When determining square roots two solutions result. To obtain the second solution one way is to express $13\angle 67.38^\circ$ also as $13\angle(67.38^\circ+360^\circ)$, i.e.

$13\angle 427.38°$. When the angle is divided by 2 an angle less than 360° is obtained.

Hence

$$\sqrt{5^2+12^2} = \sqrt{13}\angle 67.38° \text{ and } \sqrt{13}\angle 427.38°$$
$$= [13\angle 67.38°]^{1/2} \text{ and } [13\angle 427.38°]^{1/2}$$
$$= 13^{1/2}\angle\left(\tfrac{1}{2}\times 67.38°\right) \text{ and } 13^{1/2}\angle\left(\tfrac{1}{2}\times 427.38°\right)$$
$$= \sqrt{13}\angle 33.69° \text{ and } \sqrt{13}\angle 213.69°$$
$$= 3.61\angle 33.69° \text{ and } 3.61\angle 213.69°$$

Thus, in polar form, the two roots are: $3.61\angle 33.69°$ and $3.61\angle -146.31°$

$$\sqrt{13}\angle 33.69° = \sqrt{13}(\cos 33.69° + j\sin 33.69°) = 3.0 + j2.0$$
$$\sqrt{13}\angle 213.69° = \sqrt{13}(\cos 213.69° + j\sin 213.69°) = -3.0 - j2.0$$

Thus, in Cartesian form the two roots are: $\pm(3.0 + j2.0)$

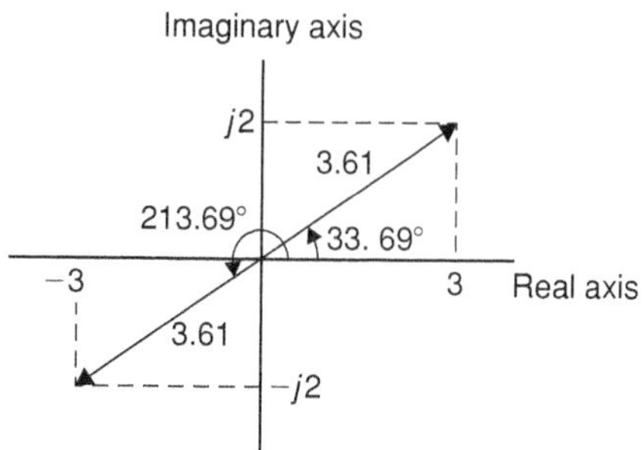

Figure 33.1

From the Argand diagram shown in Fig. 33.1 the two roots are seen to be 180° apart, which is always true when finding square roots of complex numbers.

In general, **when finding the nth root of complex number, there are n solutions.** For example, there are three solutions to a cube root, five solutions to a fifth root and so on. In the solutions to the roots of a complex number, the modulus, r, is always the same, but the arguments, θ, are different. It is shown in Problem 3 that arguments are symmetrically spaced on an Argand diagram and are $\frac{360°}{n}$ apart, where n is the number of the roots required. Thus if one of the solutions to the cube roots of a complex number is, say, $5\angle 20°$, the other two roots are symmetrically spaced $\frac{360°}{3}$, i.e. 120° from this root, and the three roots are $5\angle 20°$, $5\angle 140°$ and $5\angle 260°$.

Problem 4. Find the roots of $(5+j3)]^{1/2}$ in rectangular form, correct to 4 significant figures

$$(5+j3) = \sqrt{34}\angle 30.96°$$

Applying de Moivre's theorem:

$$(5+j3)^{1/2} = \sqrt{34}^{1/2}\angle\tfrac{1}{2}\times 30.96°$$
$$= \mathbf{2.415\angle 15.48°} \quad \text{or} \quad \mathbf{2.415\angle 15°29'}$$

The second root may be obtained as shown above, i.e. having the same modulus but displaced $\frac{360°}{2}$ from the first root.

Thus, $(5+j3)^{1/2} = 2.415\angle(15.48° + 180°)$
$$= \mathbf{2.415\angle 195.48°}$$

In rectangular form:

$$2.415\angle 15.48° = 2.415\cos 15.48° + j2.415\sin 15.48°$$
$$= \mathbf{2.327 + j0.6446}$$

and

$$2.415\angle 195.48° = 2.415\cos 195.48° + j2.415\sin 195.48°$$
$$= \mathbf{-2.327 - j0.6446}$$

Hence $(5+j3)]^{1/2} = \mathbf{2.415\angle 15.48°}$ **and** $\mathbf{2.415\angle 195.48°}$ or $\pm\mathbf{(2.327 + j0.6446)}$.

Problem 5. Express the roots of $(-14+j3)^{-2/5}$ in polar form

$$(-14+j3)=\sqrt{205}\angle 167.905^\circ$$

$$(-14+j3)^{-2/5}=\sqrt{205^{-2/5}}\angle\left[\left(-\tfrac{2}{5}\right)\times 167.905^\circ\right]$$
$$=0.3449\angle -67.162^\circ \text{ or}$$
$$0.3449\angle -67^\circ 10'$$

There are five roots to this complex number,

$$\left(x^{-2/5}=\frac{1}{x^{2/5}}=\frac{1}{\sqrt[5]{x^2}}\right)$$

The roots are symmetrically displaced from one another $\frac{360^\circ}{5}$, i.e. 72° apart round an Argand diagram.

Thus the required roots are **0.3449∠−67°10′, 0.3449∠4°50′, 0.3449∠76°50′, 0.3449∠148°50′** and **0.3449∠220°50′**

Now try the following Practice Exercise

Practice Exercise 165 Roots of complex numbers (Answers on page 719)

In Problems 1 to 3 determine the two square roots of the given complex numbers in Cartesian form and show the results on an Argand diagram.

1. (a) $1+j$ (b) j
2. (a) $3-j4$ (b) $-1-j2$
3. (a) $7\angle 60^\circ$ (b) $12\angle\frac{3\pi}{2}$

In Problems 4 to 7, determine the moduli and arguments of the complex roots.

4. $(3+j4)^{1/3}$
5. $(-2+j)^{1/4}$
6. $(-6-j5)^{1/2}$
7. $(4-j3)^{-2/3}$
8. For a transmission line, the characteristic impedance Z_0 and the propagation coefficient γ are given by:

$$Z_0=\sqrt{\frac{R+j\omega L}{G+j\omega C}} \quad \text{and}$$

$$\gamma=\sqrt{(R+j\omega L)(G+j\omega C)}$$

Given $R=25\,\Omega$, $L=5\times 10^{-3}$ H, $G=80\times 10^{-6}$ S, $C=0.04\times 10^{-6}$ F and $\omega=2000\pi$ rad/s, determine, in polar form, Z_0 and γ

Practice Exercise 166 Multiple-choice questions on De Moivre's theorem (Answers on page 719)

Each question has only one correct answer

1. The two square roots of $(-3+j4)$ are:
(a) $\pm(1+j2)$ (b) $\pm(0.71+j2.12)$
(c) $\pm(1-j2)$ (d) $\pm(0.71-j2.12)$
2. $[2\angle 30^\circ]^4$ in Cartesian form is:
(a) $(0.50+j0.06)$ (b) $(-8+j13.86)$
(c) $(-4+j6.93)$ (d) $(13.86+j8)$
3. The first square root of the complex number $(6-j8)$, correct to 2 decimal places, is:
(a) $3.16\angle -7.29^\circ$ (b) $5\angle 26.57^\circ$
(c) $3.16\angle -26.57^\circ$ (d) $5\angle -26.57^\circ$
4. The first of the five roots of $(-4+j3)^{-\frac{2}{5}}$ is:
(a) $1.90\angle -57.25^\circ$ (b) $0.525\angle -9.75^\circ$
(c) $1.90\angle -21.25^\circ$ (d) $0.525\angle -57.25^\circ$
5. The first of the three roots of $(5-j12)^{\frac{2}{3}}$ is:
(a) $6.61\angle 44.92^\circ$ (b) $5.53\angle -44.92^\circ$
(c) $6.61\angle -44.92^\circ$ (d) $5.53\angle 44.92^\circ$

For fully worked solutions to each of the problems in Practice Exercises 164 and 165 in this chapter, go to the website:
www.routledge.com/cw/bird

COMPANION @ WEBSITE

Section 6

Vectors

Chapter 34

Vectors

Why it is important to understand: Vectors

Vectors are an important part of the language of science, mathematics and engineering. They are used to discuss multivariable calculus, electrical circuits with oscillating currents, stress and strain in structures and materials, flows of atmospheres and fluids and they have many other applications. Resolving a vector into components is a precursor to computing things with or about a vector quantity. Because position, velocity, acceleration, force, momentum and angular momentum are all vector quantities, resolving vectors into components is a most important skill required in any engineering studies.

At the end of this chapter, you should be able to:

- distinguish between scalars and vectors
- recognise how vectors are represented
- add vectors using the nose-to-tail method
- add vectors using the parallelogram method
- resolve vectors into their horizontal and vertical components
- add vectors by calculation – horizontal and vertical components, complex numbers
- perform vector subtraction
- understand relative velocity
- understand **i**, **j**, **k** notation

34.1 Introduction

This chapter initially explains the difference between scalar and vector quantities and shows how a vector is drawn and represented.

Any object that is acted upon by an external force will respond to that force by moving in the line of the force. However, if two or more forces act simultaneously, the result is more difficult to predict; the ability to add two or more vectors then becomes important.

This chapter thus shows how vectors are added and subtracted, both by drawing and by calculation, and finding the resultant of two or more vectors has many uses in engineering. (Resultant means the single vector which would have the same effect as the individual vectors.) Relative velocities and vector **i**, **j**, **k** notation are also briefly explained.

34.2 Scalars and vectors

The time taken to fill a water tank may be measured as, say, 50 s. Similarly, the temperature in a room may be measured as, say, 16°C, or the mass of a bearing may be measured as, say, 3 kg.

Quantities such as time, temperature and mass are entirely defined by a numerical value and are called **scalars** or **scalar quantities**.

Not all quantities are like this. Some are defined by more than just size; some also have direction.
For example, the velocity of a car is 90 km/h due west, or a force of 20 N acts vertically downwards, or an acceleration of $10\,m/s^2$ acts at 50° to the horizontal.
Quantities such as velocity, force and acceleration, which **have both a magnitude and a direction**, are called **vectors**.

Now try the following Practice Exercise

Practice Exercise 167 Scalar and vector quantities (Answers on page 719)

1. State the difference between scalar and vector quantities.

In problems 2 to 9, state whether the quantities given are scalar or vector:

2. A temperature of 70°C
3. $5\,m^3$ volume
4. A downward force of 20 N
5. 500 J of work
6. $30\,cm^2$ area
7. A south-westerly wind of 10 knots
8. 50 m distance
9. An acceleration of $15\,m/s^2$ at 60° to the horizontal

34.3 Drawing a vector

A vector quantity can be represented graphically by a line, drawn so that:

(a) the **length** of the line denotes the magnitude of the quantity, and

(b) the **direction** of the line denotes the direction in which the vector quantity acts.

An arrow is used to denote the sense, or direction, of the vector.
The arrow end of a vector is called the 'nose' and the other end the 'tail'. For example, a force of 9 N acting at 45° to the horizontal is shown in Fig. 34.1. Note that an angle of **+45°** is drawn from the horizontal and moves **anticlockwise**.

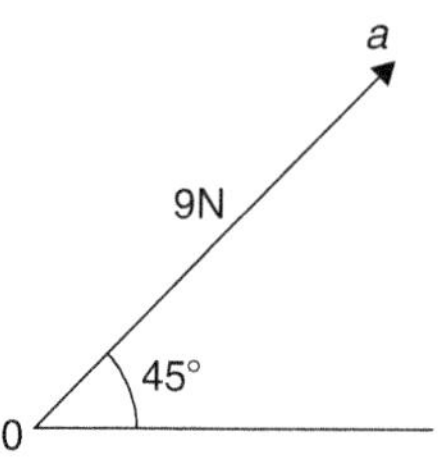

Figure 34.1

A velocity of 20 m/s at −60° is shown in Fig. 34.2. Note that an angle of **−60°** is drawn from the horizontal and moves **clockwise**.

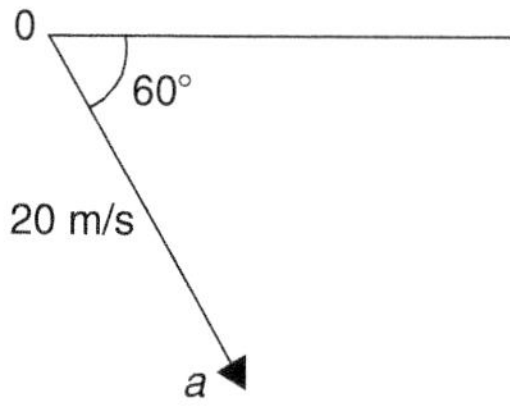

Figure 34.2

Representing a vector

There are a number of ways of representing vector quantities. These include:

1. Using **bold print**
2. $\overrightarrow{AB}$ where an arrow above two capital letters denotes the sense of direction, where A is the starting point and B the end point of the vector
3. $\overline{AB}$ or $\bar{a}$, i.e. a line over the top of letters
4. $\underline{a}$, i.e. an underlined letter

The force of 9 N at 45° shown in Fig. 34.1 may be represented as:

$$\boldsymbol{Oa} \text{ or } \overrightarrow{0a} \text{ or } \overline{0a}$$

The magnitude of the force is Oa.
Similarly, the velocity of 20 m/s at −60° shown in Fig. 34.2 may be represented as:

$$\boldsymbol{Ob} \text{ or } \overrightarrow{0b} \text{ or } \overline{0b}$$

The magnitude of the velocity is Ob.
In this chapter a vector quantity is denoted by **bold print**.

34.4 Addition of vectors by drawing

Adding two or more vectors by drawing assumes that a ruler, pencil and protractor are available. Results obtained by drawing are naturally not as accurate as those obtained by calculation.

(a) Nose-to-tail method

Two force vectors, F_1 and F_2, are shown in Fig. 34.3. When an object is subjected to more than one force, the resultant of the forces is found by the addition of vectors.

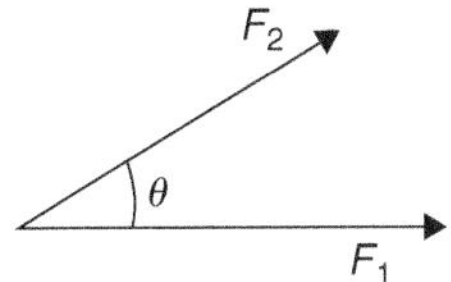

Figure 34.3

To add forces F_1 and F_2:

(i) Force F_1 is drawn to scale horizontally, shown as ***Oa*** in Fig. 34.4.

(ii) From the nose of F_1, force F_2 is drawn at angle θ to the horizontal, shown as ***ab***

(iii) The resultant force is given by length ***Ob***, which may be measured.

This procedure is called the '**nose-to-tail**' or '**triangle**' method.

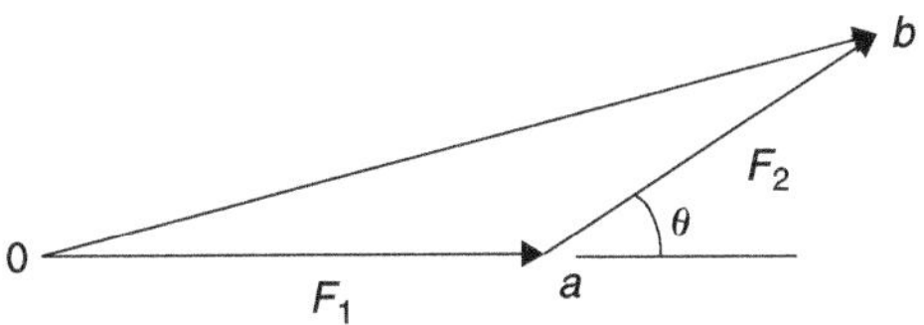

Figure 34.4

(b) Parallelogram method

To add the two force vectors, F_1 and F_2, of Fig. 34.3:

(i) A line *cb* is constructed which is parallel to and equal in length to ***Oa*** (see Fig. 34.5).

(ii) A line *ab* is constructed which is parallel to and equal in length to ***Oc***

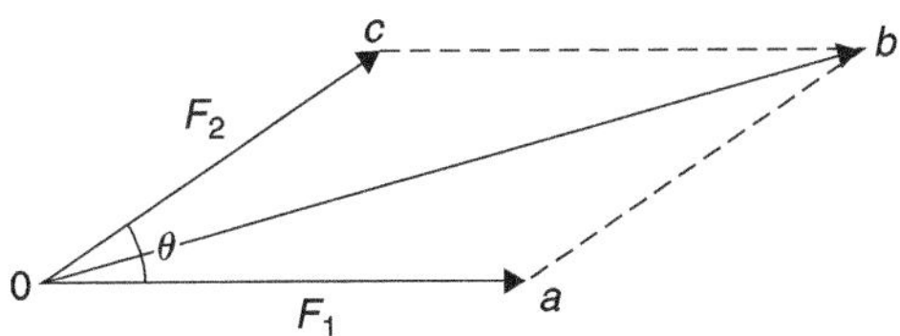

Figure 34.5

(iii) The resultant force is given by the diagonal of the parallelogram, i.e. length ***Ob***

This procedure is called the '**parallelogram**' method.

Problem 1. A force of 5 N is inclined at an angle of 45° to a second force of 8 N, both forces acting at a point. Find the magnitude of the resultant of these two forces and the direction of the resultant with respect to the 8 N force by: (a) the 'nose-to-tail' method, and (b) the 'parallelogram' method.

The two forces are shown in Fig. 34.6. (Although the 8 N force is shown horizontal, it could have been drawn in any direction.)

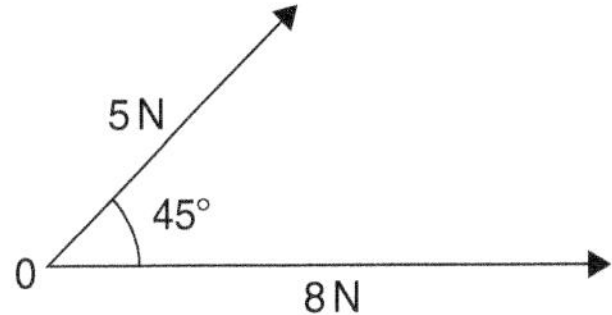

Figure 34.6

(a) 'Nose-to tail' method

(i) The 8 N force is drawn horizontally 8 units long, shown as ***Oa*** in Fig. 34.7.

(ii) From the nose of the 8 N force, the 5 N force is drawn 5 units long at an angle of 45° to the horizontal, shown as ***ab***

(iii) The resultant force is given by length *Ob* and is measured as **12 N** and angle θ is measured as **17°**

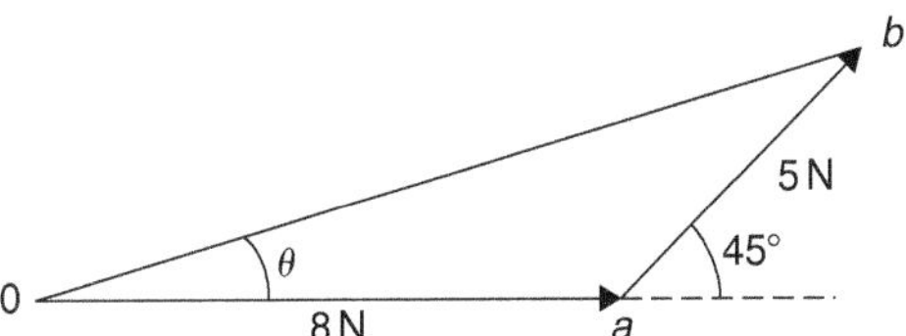

Figure 34.7

(b) 'Parallelogram' method

(i) In Fig. 34.8, a line is constructed which is parallel to and equal in length to the 8 N force.

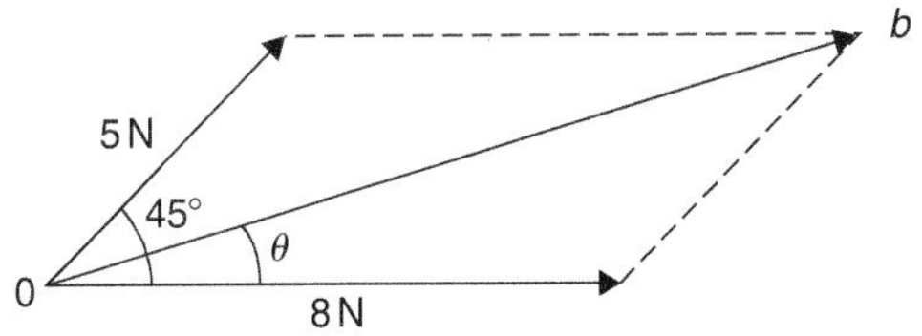

Figure 34.8

(ii) A line is constructed which is parallel to and equal in length to the 5 N force.

(iii) The resultant force is given by the diagonal of the parallelogram, i.e. length Ob, and is measured as **12 N** and angle θ is measured as **17°**

Thus, **the resultant of the two force vectors in Fig. 34.6 is 12 N at 17° to the 8 N force.**

Problem 2. Forces of 15 N and 10 N are at an angle of 90° to each other as shown in Fig. 34.9. Find, by drawing, the magnitude of the resultant of these two forces and the direction of the resultant with respect to the 15 N force.

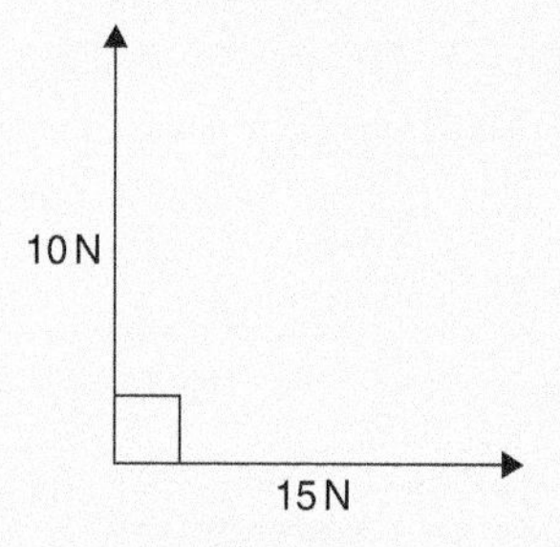

Figure 34.9

Using the 'nose-to-tail' method:

(i) The 15 N force is drawn horizontally 15 units long as shown in Fig. 34.10.

(ii) From the nose of the 15 N force, the 10 N force is drawn 10 units long at an angle of 90° to the horizontal as shown.

(iii) The resultant force is shown as R and is measured as **18 N** and angle θ is measured as **34°**

Thus, **the resultant of the two force vectors is 18 N at 34° to the 15 N force.**

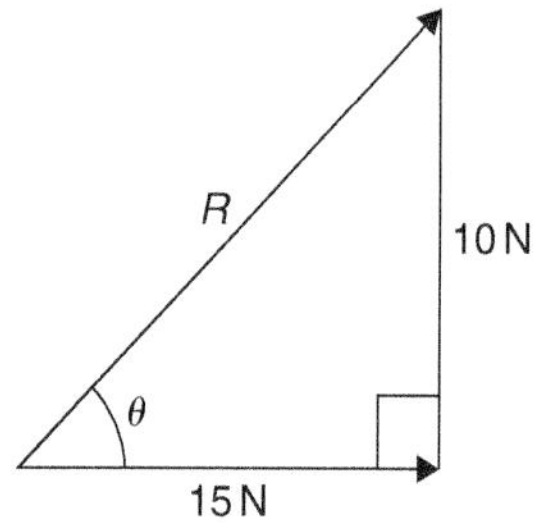

Figure 34.10

Problem 3. Velocities of 10 m/s, 20 m/s and 15 m/s act as shown in Fig. 34.11. Determine, by drawing, the magnitude of the resultant velocity and its direction relative to the horizontal.

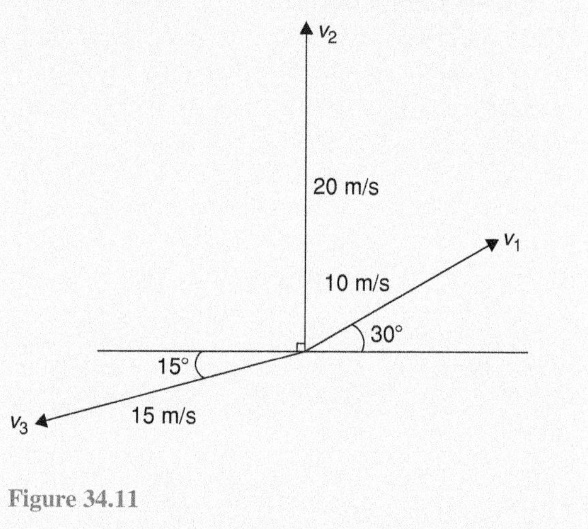

Figure 34.11

When more than 2 vectors are being added the 'nose-to-tail' method is used.

The order in which the vectors are added does not matter. In this case the order taken is v_1, then v_2, then v_3. However, if a different order is taken the same result will occur.

(i) v_1 is drawn 10 units long at an angle of 30° to the horizontal, shown as ***Oa*** in Fig. 34.12.

(ii) From the nose of v_1, v_2 is drawn 20 units long at an angle of 90° to the horizontal, shown as ***ab***

(iii) From the nose of v_2, v_3 is drawn 15 units long at an angle of 195° to the horizontal, shown as ***br***

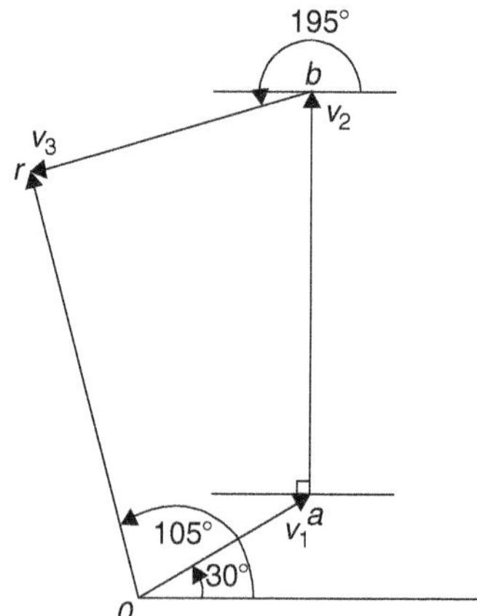

Figure 34.12

(iv) The resultant velocity is given by length ***Or*** and is measured as **22 m/s** and the angle measured to the horizontal is **105°**

Thus, **the resultant of the three velocities is 22 m/s at 105° to the horizontal**.

Worked Problems 1 to 3 have demonstrated how vectors are added to determine their resultant and their direction. However, drawing to scale is time-consuming and not highly accurate. The following sections demonstrate how to determine resultant vectors by calculation using horizontal and vertical components and, where possible, by Pythagoras' theorem.

34.5 Resolving vectors into horizontal and vertical components

A force vector F is shown in Fig. 34.13 at angle θ to the horizontal. Such a vector can be resolved into two components such that the vector addition of the components is equal to the original vector.

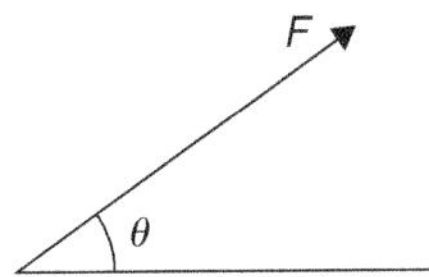

Figure 34.13

The two components usually taken are a **horizontal component** and a **vertical component**.
If a right-angled triangle is constructed as shown in Fig. 34.14, then Oa is called the horizontal component of F and ab is called the vertical component of F.

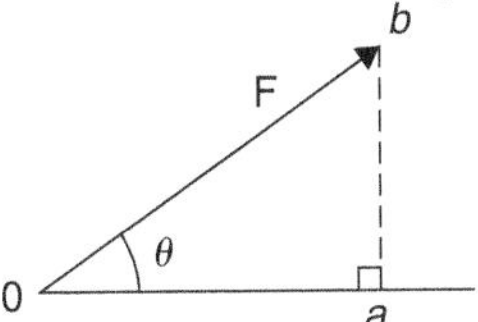

Figure 34.14

From trigonometry (see Chapter 17, page 171),

$$\cos\theta = \frac{\text{Oa}}{\text{Ob}} \text{ from which, } Oa = Ob\cos\theta = F\cos\theta,$$

i.e. **the horizontal component of $F = F\cos\theta$**

and $$\sin\theta = \frac{\text{ab}}{\text{Ob}} \text{ from which, } ab = Ob\sin\theta = F\sin\theta,$$

i.e. **the vertical component of $F = F\sin\theta$**

Problem 4. Resolve the force vector of 50 N at an angle of 35° to the horizontal into its horizontal and vertical components.

The **horizontal component** of the 50 N force,
$Oa = 50\cos 35° = \mathbf{40.96\,N}$
The **vertical component** of the 50 N force,
$ab = 50\sin 35° = \mathbf{28.68\,N}$
The horizontal and vertical components are shown in Fig. 34.15.

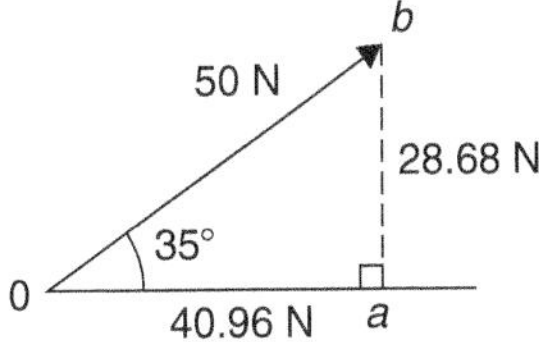

Figure 34.15

(Checking: By Pythagoras, $0b = \sqrt{40.96^2 + 28.68^2}$
$= 50\text{N}$

and $$\theta = \tan^{-1}\left(\frac{28.68}{40.96}\right) = 35°$$

Thus, the vector addition of components 40.96 N and 28.68 N is 50 N at 35°)

Problem 5. Resolve the velocity vector of 20 m/s at an angle of −30° to the horizontal into horizontal and vertical components.

The **horizontal component** of the 20 m/s velocity,
$Oa = 20\cos(-30°) = \mathbf{17.32\,m/s}$

The **vertical component** of the 20 m/s velocity,
$ab = 20\sin(-30°) = \mathbf{-10\,m/s}$

The horizontal and vertical components are shown in Fig. 34.16.

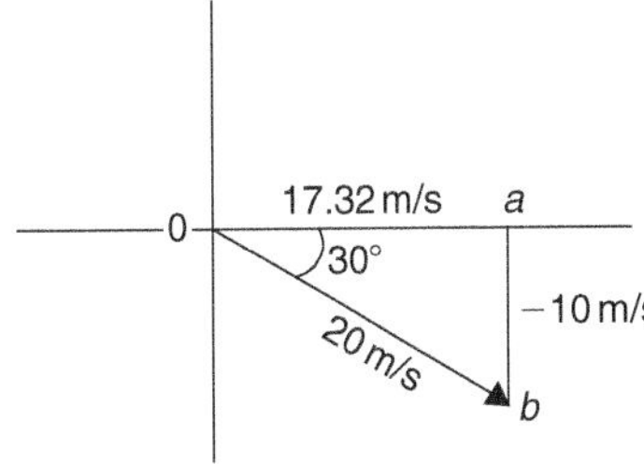

Figure 34.16

Problem 6. Resolve the displacement vector of 40 m at an angle of 120° into horizontal and vertical components.

The **horizontal component** of the 40 m displacement, $Oa = 40\cos 120° = \mathbf{-20.0\,m}$

The **vertical component** of the 40 m displacement, $ab = 40\sin 120° = \mathbf{34.64\,m}$

The horizontal and vertical components are shown in Fig. 34.17.

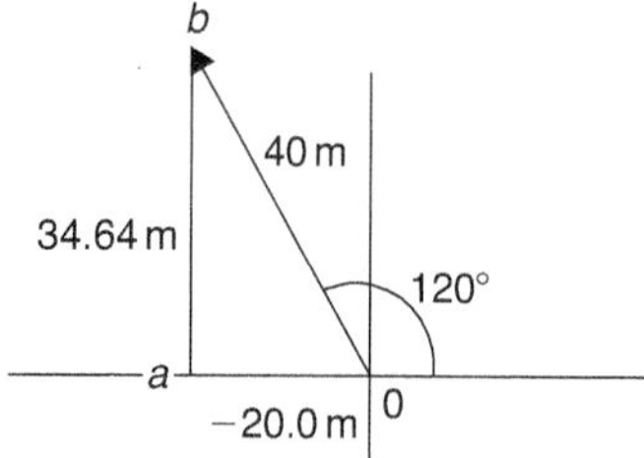

Figure 34.17

34.6 Addition of vectors by calculation

Two force vectors, F_1 and F_2, are shown in Fig. 34.18, F_1 being at an angle of θ_1 and F_2 being at an angle of θ_2

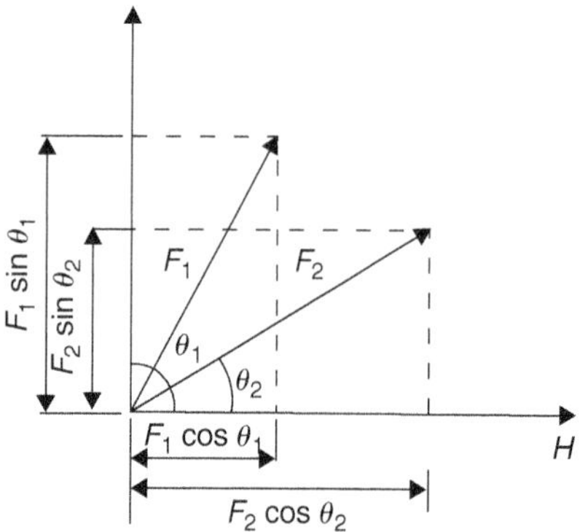

Figure 34.18

A method of adding two vectors together is to use horizontal and vertical components.

The horizontal component of force F_1 is $F_1\cos\theta_1$ and the horizontal component of force F_2 is $F_2\cos\theta_2$

The total horizontal component of the two forces, $\boldsymbol{H = F_1\cos\theta_1 + F_2\cos\theta_2}$

The vertical component of force F_1 is $F_1\sin\theta_1$ and the vertical component of force F_2 is $F_2\sin\theta_2$

The total vertical component of the two forces, $\mathbf{V = F_1\sin\theta_1 + F_2\sin\theta_2}$

Since we have H and V, the resultant of F_1 and F_2 is obtained by using the theorem of Pythagoras. From Fig. 34.19, $0b^2 = H^2 + V^2$,

i.e. **resultant** $= \sqrt{\mathbf{H^2 + V^2}}$ at an angle given by $\theta = \tan^{-1}\left(\frac{\mathbf{V}}{\mathbf{H}}\right)$

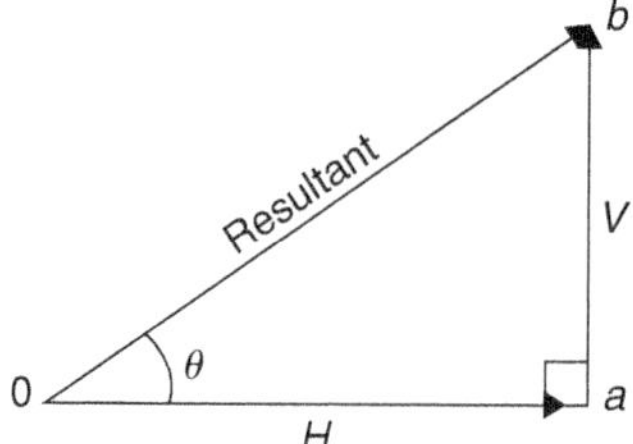

Figure 34.19

Problem 7. A force of 5 N is inclined at an angle of 45° to a second force of 8 N, both forces acting at a point. Calculate the magnitude of the resultant of these two forces and the direction of the resultant with respect to the 8 N force.

The two forces are shown in Fig. 34.20.

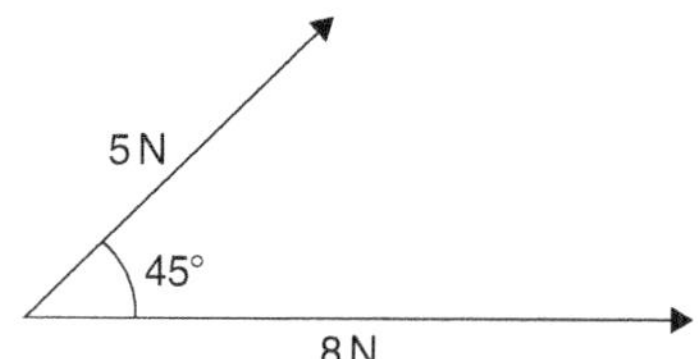

Figure 34.20

The horizontal component of the 8 N force is $8\cos 0°$ and the horizontal component of the 5 N force is $5\cos 45°$

The total horizontal component of the two forces,

$$H = 8\cos 0° + 5\cos 45° = 8 + 3.5355 = \mathbf{11.5355}$$

The vertical component of the 8 N force is $8\sin 0°$ and the vertical component of the 5 N force is $5\sin 45°$

The total vertical component of the two forces,

$$V = 8\sin 0° + 5\sin 45° = 0 + 3.5355 = \mathbf{3.5355}$$

From Fig. 34.21, magnitude of resultant vector

$$= \sqrt{H^2 + V^2}$$

$$= \sqrt{11.5355^2 + 3.5355^2} = \mathbf{12.07\,N}$$

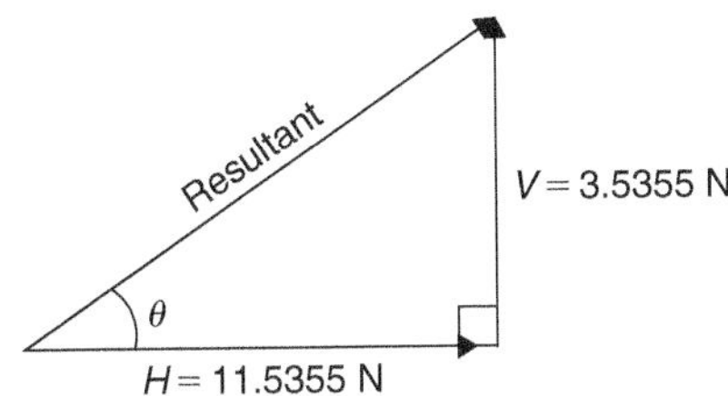

Figure 34.21

The direction of the resultant vector,

$$\theta = \tan^{-1}\left(\frac{V}{H}\right) = \tan^{-1}\left(\frac{3.5355}{11.5355}\right)$$

$$= \tan^{-1} 0.30648866\ldots = \mathbf{17.04^\circ}$$

Thus, **the resultant of the two forces is a single vector of 12.07 N at 17.04° to the 8 N vector**.
Perhaps an easier and quicker method of calculating the magnitude and direction of the resultant is to use **complex numbers** (see Chapter 32).

In this example, the **resultant**

$$= 8\angle 0^\circ + 5\angle 45^\circ$$

$$= (8\cos 0^\circ + j8\sin 0^\circ) + (5\cos 45^\circ + j5\sin 45^\circ)$$

$$= (8 + j0) + (3.536 + j3.536)$$

$$= (11.536 + j3.536)\,\text{N} \quad \text{or} \quad \mathbf{12.07\angle 17.04^\circ N}$$

as obtained above using horizontal and vertical components.

Problem 8. Forces of 15 N and 10 N are at an angle of 90° to each other as shown in Fig. 34.22. Calculate the magnitude of the resultant of these two forces and its direction with respect to the 15 N force.

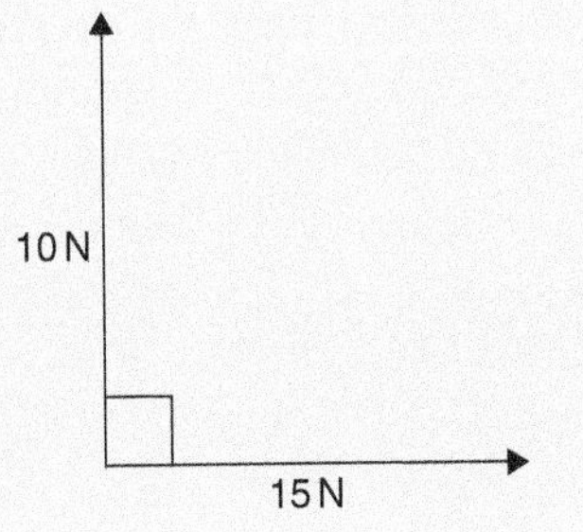

Figure 34.22

The horizontal component of the 15 N force is $15\cos 0^\circ$ and the horizontal component of the 10 N force is $10\cos 90^\circ$
The total horizontal component of the two forces,

$$\boldsymbol{H} = 15\cos 0^\circ + 10\cos 90^\circ = 15 + 0 = \mathbf{15}$$

The vertical component of the 15 N force is $15\sin 0^\circ$ and the vertical component of the 10 N force is $10\sin 90^\circ$
The total vertical component of the two forces,

$$\boldsymbol{V} = 15\sin 0^\circ + 10\sin 90^\circ = 0 + 10 = \mathbf{10}$$

Magnitude of resultant vector

$$= \sqrt{H^2 + V^2} = \sqrt{15^2 + 10^2} = \mathbf{18.03\,N}$$

The direction of the resultant vector,

$$\theta = \tan^{-1}\left(\frac{V}{H}\right) = \tan^{-1}\left(\frac{10}{15}\right) = \mathbf{33.69^\circ}$$

Thus, **the resultant of the two forces is a single vector of 18.03 N at 33.69° to the 15 N vector**.
There is an alternative method of calculating the resultant vector in this case.
If we used the triangle method, then the diagram would be as shown in Fig. 34.23.

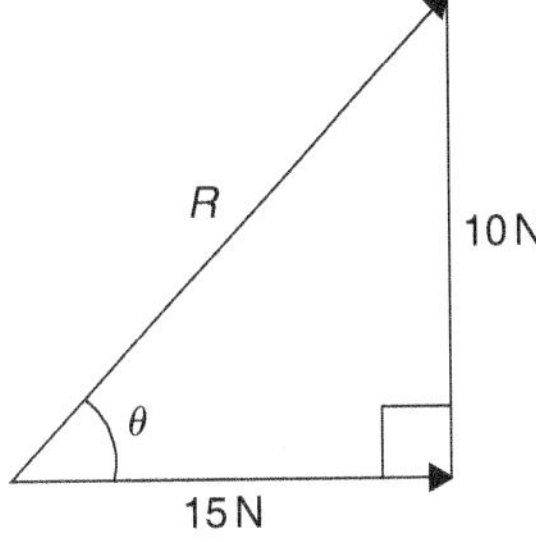

Figure 34.23

Since a right-angled triangle results then we could use **Pythagoras' theorem** without needing to go through the procedure for horizontal and vertical components. In fact, the horizontal and vertical components are 15 N and 10 N respectively.
This is, of course, a special case. Pythagoras can only be used when there is an angle of 90° between vectors. This is demonstrated in the next worked problem.

Problem 9. Calculate the magnitude and direction of the resultant of the two acceleration vectors shown in Fig. 34.24.

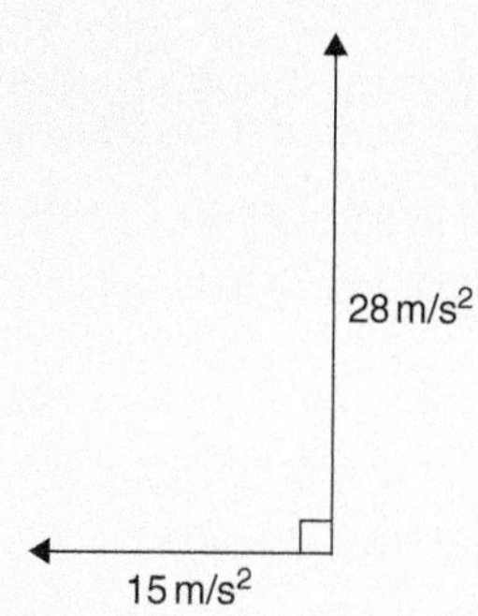

Figure 34.24

The 15 m/s^2 acceleration is drawn horizontally, shown as ***Oa*** in Fig. 34.25.
From the nose of the 15 m/s^2 acceleration, the 28 m/s^2 acceleration is drawn at an angle of 90° to the horizontal, shown as ***ab***

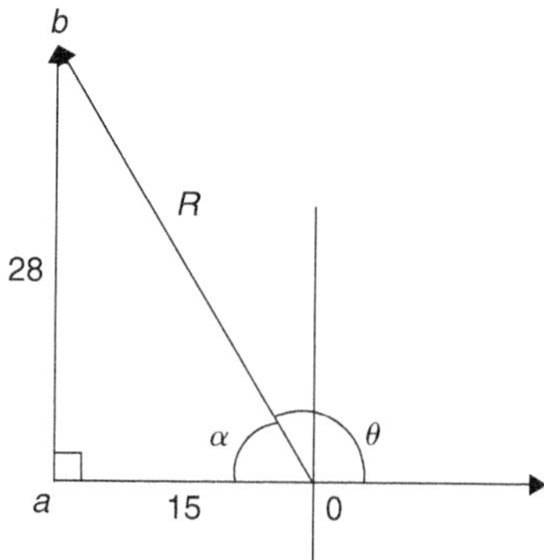

Figure 34.25

The resultant acceleration, R, is given by length ***Ob***
Since a right-angled triangle results, the theorem of Pythagoras may be used.

$$0b = \sqrt{15^2 + 28^2} = \mathbf{31.76\ m/s^2}$$

and $$\alpha = \tan^{-1}\left(\frac{28}{15}\right) = \mathbf{61.82°}$$

Measuring from the horizontal,

$$\theta = 180° - 61.82° = \mathbf{118.18°}$$

Thus, **the resultant of the two accelerations is a single vector of 31.76 m/s^2 at 118.18° to the horizontal.**

Problem 10. Velocities of 10 m/s, 20 m/s and 15 m/s act as shown in Fig. 34.26. Calculate the magnitude of the resultant velocity and its direction relative to the horizontal.

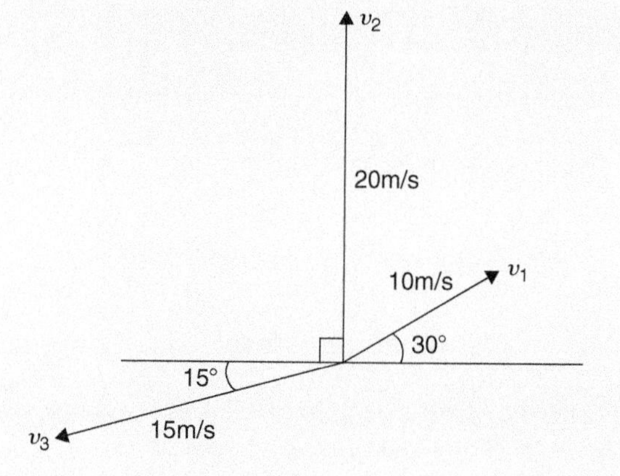

Figure 34.26

The horizontal component of the 10 m/s velocity is $10\cos 30° = 8.660\,\text{m/s}$,

the horizontal component of the 20 m/s velocity is $20\cos 90° = 0\,\text{m/s}$,

and the horizontal component of the 15 m/s velocity is $15\cos 195° = -14.489\,\text{m/s}$.

The total horizontal component of the three velocities,

$$H = 8.660 + 0 - 14.489 = \mathbf{-5.829\ m/s}$$

The vertical component of the 10 m/s velocity is $10\sin 30° = 5\,\text{m/s}$,

the vertical component of the 20 m/s velocity is $20\sin 90° = 20\,\text{m/s}$,

and the vertical component of the 15 m/s velocity is $15\sin 195° = -3.882\,\text{m/s}$.

The total vertical component of the three velocities,

$$V = 5 + 20 - 3.882 = \mathbf{21.118\ m/s}$$

From Fig. 34.27, magnitude of resultant vector,
$R = \sqrt{H^2 + V^2} = \sqrt{5.829^2 + 21.118^2} = \mathbf{21.91\ m/s}$

The direction of the resultant vector,
$$\alpha = \tan^{-1}\left(\frac{V}{H}\right) = \tan^{-1}\left(\frac{21.118}{5.829}\right) = \mathbf{74.57°}$$

Measuring from the horizontal,

$$\theta = 180° - 74.57° = \mathbf{105.43°}$$

Thus, **the resultant of the three velocities is a single vector of 21.91 m/s at 105.43° to the horizontal.**

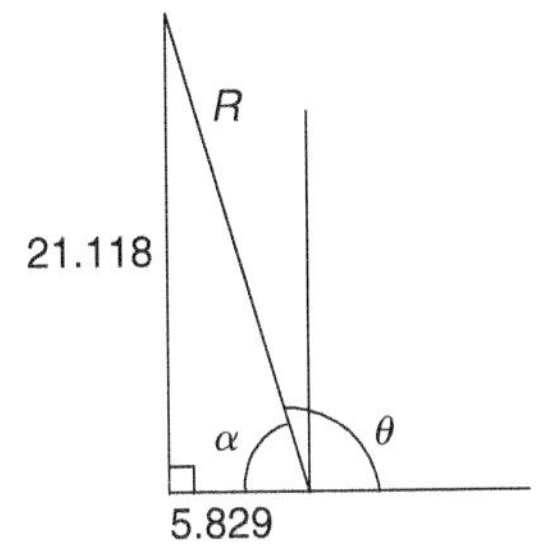

Figure 34.27

Using **complex numbers**, from Fig. 34.26,

$$\textbf{resultant} = 10\angle 30^\circ + 20\angle 90^\circ + 15\angle 195^\circ$$
$$= (10\cos 30^\circ + j10\sin 30^\circ) + (20\cos 90^\circ + j20\sin 90^\circ) + (15\cos 195^\circ + j15\sin 195^\circ)$$
$$= (8.660 + j5.000) + (0 + j20.000) + (-14.489 - j3.882)$$
$$= (-5.829 + j21.118)\,\text{N} \quad \text{or} \quad \mathbf{21.91\angle 105.43^\circ\ N}$$

as obtained above using horizontal and vertical components.

The method used to add vectors by calculation will not be specified – the choice is yours, but probably the quickest and easiest method is by using complex numbers.

Problem 11. A belt-driven pulley is attached to a shaft as shown in Figure 34.28. Calculate the magnitude and direction of the resultant force acting on the shaft.

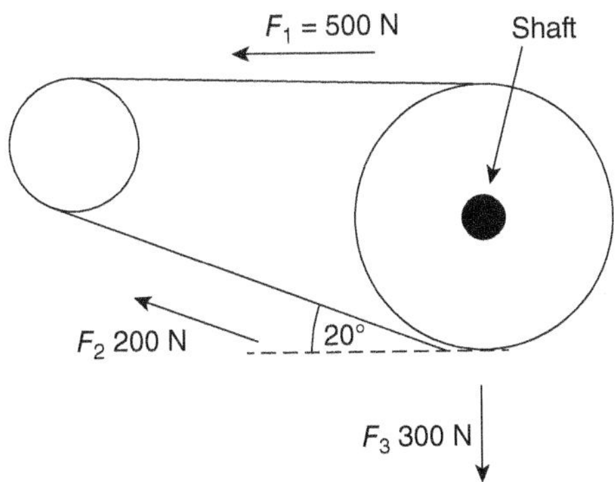

Figure 34.28

The three force vectors are shown in the sketch of Figure 34.29.

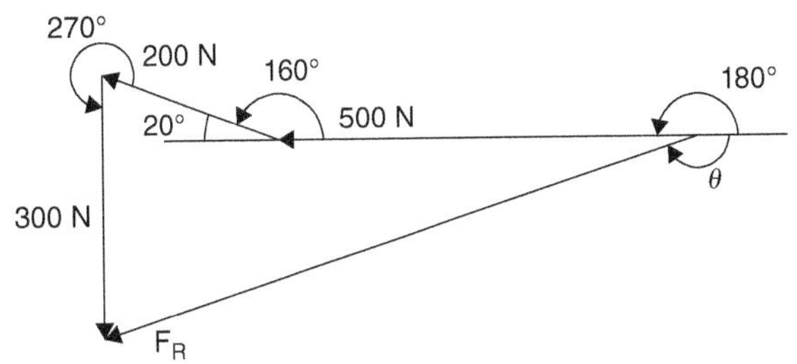

Figure 34.29

Using complex numbers, **the resultant force,**

$$F_R = 500\angle 180^\circ + 200\angle 160^\circ + 300\angle 270^\circ$$
$$= (-500 + j0) + (-187.94 + j68.40) + (0 - j300)$$
$$= -687.94 - -j231.60$$
$$= \mathbf{725.9\angle - 161.39^\circ N}$$

Thus, the resultant force F_R is 725.9 N acting at – 161.39° to the horizontal

---

Now try the following Practice Exercise

Practice Exercise 168 Addition of vectors by calculation (Answers on page 719)

1. A force of 7 N is inclined at an angle of 50° to a second force of 12 N, both forces acting at a point. Calculate the magnitude of the resultant of the two forces, and the direction of the resultant with respect to the 12 N force
2. Velocities of 5 m/s and 12 m/s act at a point at 90° to each other. Calculate the resultant velocity and its direction relative to the 12 m/s velocity
3. Calculate the magnitude and direction of the resultant of the two force vectors shown in Fig. 34.30

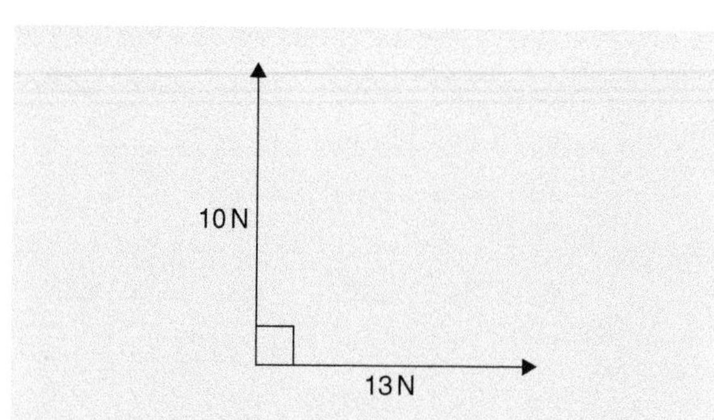

Figure 34.30

4. Calculate the magnitude and direction of the resultant of the two force vectors shown in Fig. 34.31

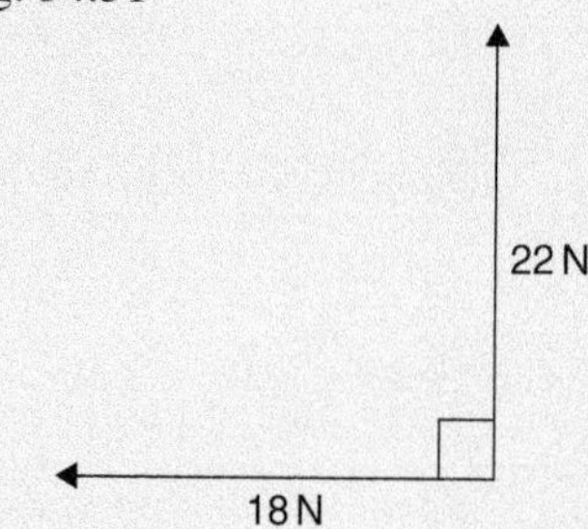

Figure 34.31

5. A displacement vector s_1 is 30 m at 0°. A second displacement vector s_2 is 12 m at 90°. Calculate magnitude and direction of the resultant vector $s_1 + s_2$

6. Three forces of 5 N, 8 N and 13 N act as shown in Fig. 34.32. Calculate the magnitude and direction of the resultant force

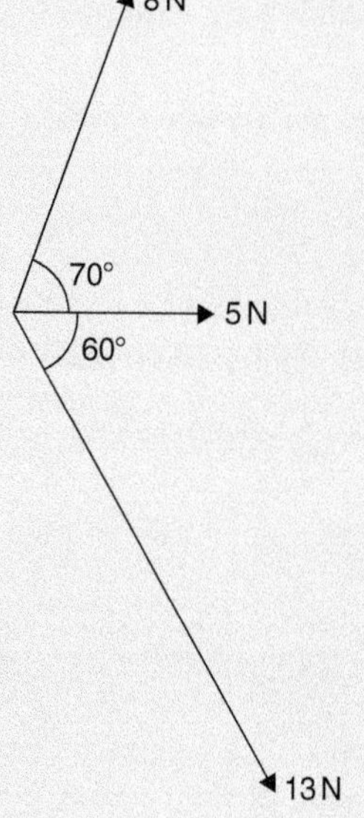

Figure 34.32

7. If velocity $v_1 = 25$ m/s at 60° and $v_2 = 15$ m/s at −30°, calculate the magnitude and direction of $v_1 + v_2$

8. Calculate the magnitude and direction of the resultant vector of the force system shown in Fig. 34.33

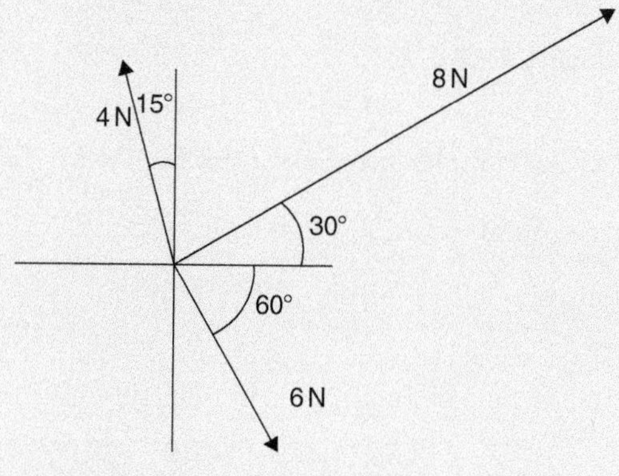

Figure 34.33

9. Calculate the magnitude and direction of the resultant vector of the system shown in Fig. 34.34

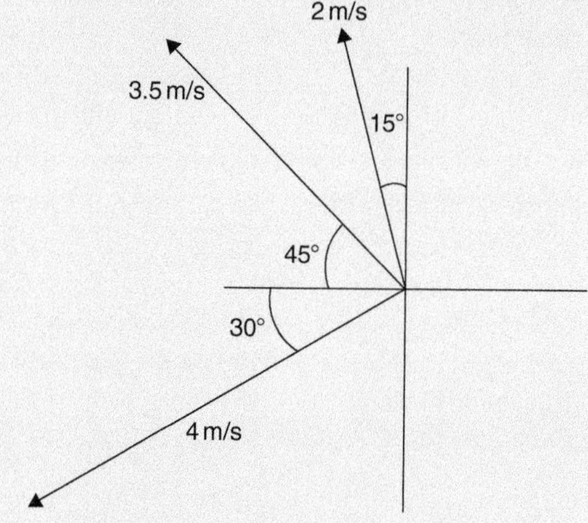

Figure 34.34

10. An object is acted upon by two forces of magnitude 10 N and 8 N at an angle of 60° to each other. Determine the resultant force on the object

11. A ship heads in a direction of E 20° S at a speed of 20 knots while the current is 4 knots in a direction of N 30° E. Determine the speed and actual direction of the ship

34.7 Vector subtraction

In Fig. 34.35, a force vector $\boldsymbol{F}$ is represented by $\boldsymbol{oa}$. The vector $(-\boldsymbol{oa})$ can be obtained by drawing a vector from o in the opposite sense to $\boldsymbol{oa}$ but having the same magnitude, shown as $\boldsymbol{ob}$ in Fig. 34.35, i.e. $\boldsymbol{ob} = (-\boldsymbol{oa})$.

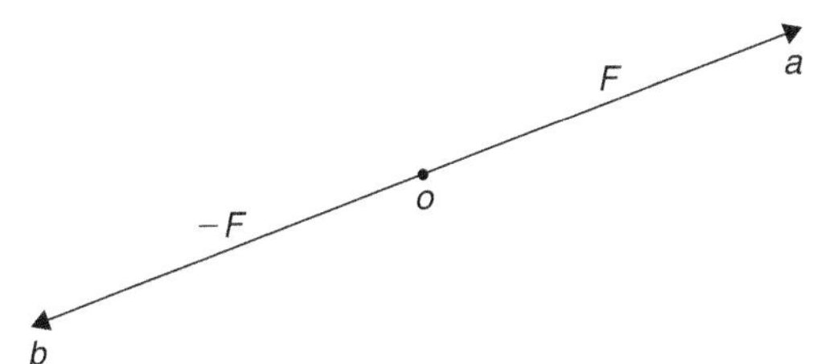

Figure 34.35

For two vectors acting at a point, as shown in Fig. 34.36(a), the resultant of vector addition is: $\boldsymbol{os} = \boldsymbol{oa} + \boldsymbol{ob}$

Figure 34.36(b) shows vectors $\boldsymbol{ob} + (-\boldsymbol{oa})$, that is, $\boldsymbol{ob} - \boldsymbol{oa}$ and the vector equation is $\boldsymbol{ob} - \boldsymbol{oa} = \boldsymbol{od}$. Comparing $\boldsymbol{od}$ in Fig. 34.36(b) with the broken line ab in Fig. 34.36(a) shows that the second diagonal of the 'parallelogram' method of vector addition gives the magnitude and direction of vector subtraction of $\boldsymbol{oa}$ from $\boldsymbol{ob}$

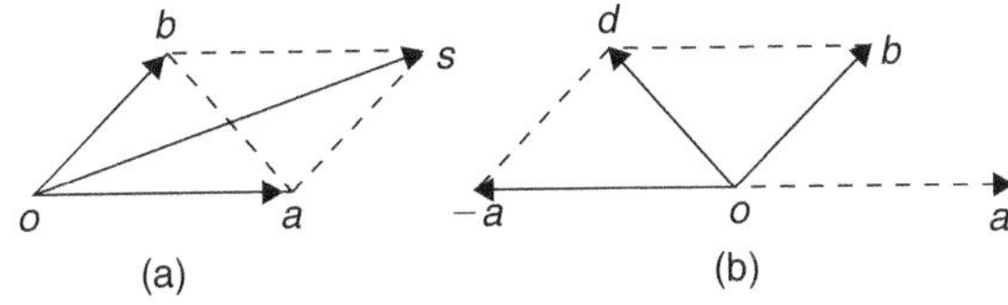

Figure 34.36

Problem 12. Accelerations of $a_1 = 1.5\,\text{m/s}^2$ at 90° and $a_2 = 2.6\,\text{m/s}^2$ at 145° act at a point. Find $\boldsymbol{a_1} + \boldsymbol{a_2}$ and $\boldsymbol{a_1} - \boldsymbol{a_2}$ by (i) drawing a scale vector diagram, and (ii) by calculation.

(i) The scale vector diagram is shown in Fig. 34.37. By measurement,

$$\boldsymbol{a_1} + \boldsymbol{a_2} = \mathbf{3.7\,m/s^2\ at\ 126^\circ}$$
$$\boldsymbol{a_1} - \boldsymbol{a_2} = \mathbf{2.1\,m/s^2\ at\ 0^\circ}$$

(ii) Resolving horizontally and vertically gives:

Horizontal component of $\boldsymbol{a_1} + \boldsymbol{a_2}$,
$H = 1.5\cos 90^\circ + 2.6\cos 145^\circ = \mathbf{-2.13}$

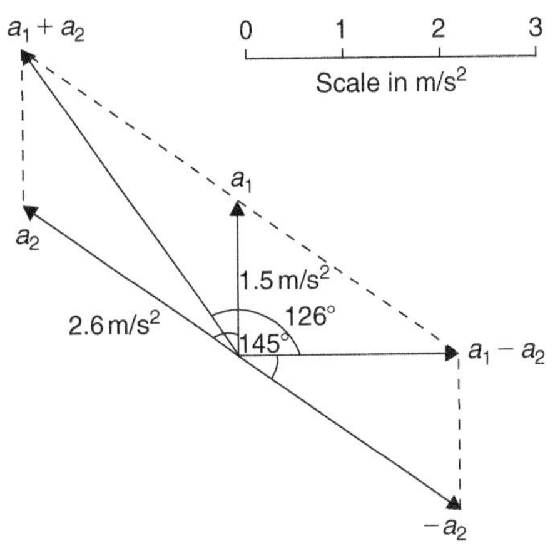

Figure 34.37

Vertical component of $\boldsymbol{a_1} + \boldsymbol{a_2}$,
$V = 1.5\sin 90^\circ + 2.6\sin 145^\circ = \mathbf{2.99}$

From Fig. 34.38, magnitude of $\boldsymbol{a_1} + \boldsymbol{a_2}$,
$R = \sqrt{(-2.13)^2 + 2.99^2} = \mathbf{3.67\,m/s^2}$

In Fig. 34.38, $\alpha = \tan^{-1}\left(\frac{2.99}{2.13}\right) = 54.53^\circ$ and

$$\theta = 180^\circ - 54.53^\circ = \mathbf{125.47^\circ}$$

Thus, $\quad \boldsymbol{a_1} + \boldsymbol{a_2} = \mathbf{3.67\,m/s^2\ at\ 125.47^\circ}$

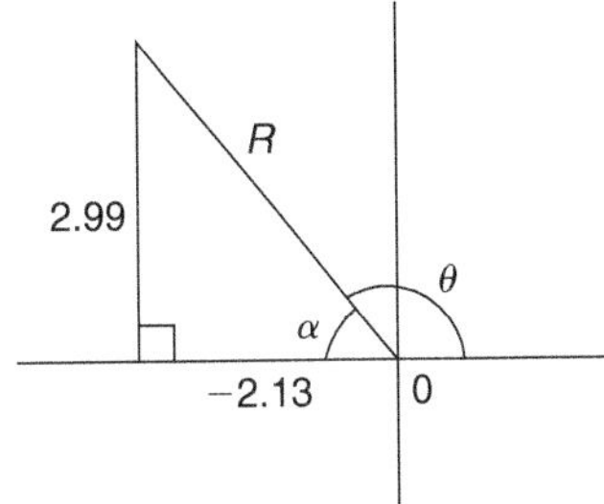

Figure 34.38

Horizontal component of $\boldsymbol{a_1} - \boldsymbol{a_2}$
$= 1.5\cos 90^\circ - 2.6\cos 145^\circ = \mathbf{2.13}$

Vertical component of $\boldsymbol{a_1} - \boldsymbol{a_2}$
$= 1.5\sin 90^\circ - 2.6\sin 145^\circ = 0$

Magnitude of $\boldsymbol{a_1} - \boldsymbol{a_2} = \sqrt{2.13^2 + 0^2} = \mathbf{2.13\,m/s^2}$

Direction of $\boldsymbol{a_1} - \boldsymbol{a_2} = \tan^{-1}\left(\frac{0}{2.13}\right) = \mathbf{0^\circ}$

Thus, $\boldsymbol{a_1} - \boldsymbol{a_2} = \mathbf{2.13\,m/s^2\ at\ 0^\circ}$

Problem 13. Calculate the resultant of (i) $v_1 - v_2 + v_3$ and (ii) $v_2 - v_1 - v_3$ when $v_1 = 22$ units at 140°, $v_2 = 40$ units at 190° and $v_3 = 15$ units at 290°.

(i) The vectors are shown in Fig. 34.39.

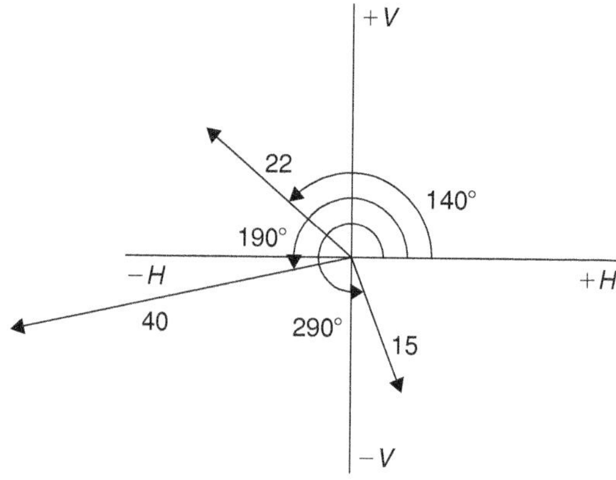

Figure 34.39

The horizontal component of

$$v_1 - v_2 + v_3 = (22\cos 140°) - (40\cos 190°) + (15\cos 290°)$$
$$= (-16.85) - (-39.39) + (5.13)$$
$$= \mathbf{27.67\ units}$$

The vertical component of

$$v_1 - v_2 + v_3 = (22\sin 140°) - (40\sin 190°) + (15\sin 290°)$$
$$= (14.14) - (-6.95) + (-14.10)$$
$$= \mathbf{6.99\ units}$$

The magnitude of the resultant,

$$R = \sqrt{27.67^2 + 6.99^2} = \mathbf{28.54\ units}$$

The direction of the resultant $R = \tan^{-1}\left(\frac{6.99}{27.67}\right)$

$$= 14.18°$$

Thus, $v_1 - v_2 + v_3 = \mathbf{28.54\ units\ at\ 14.18°}$

Using **complex numbers**,

$$v_1 - v_2 + v_3 = 22\angle 140° - 40\angle 190° + 15\angle 290°$$
$$= (-16.853 + j14.141) - (-39.392 - j6.946) + (5.130 - j14.095)$$
$$= 27.669 + j6.992 = \mathbf{28.54\angle 14.18°}$$

(ii) The horizontal component of

$$v_2 - v_1 - v_3 = (40\cos 190°) - (22\cos 140°) - (15\cos 290°)$$
$$= (-39.39) - (-16.85) - (5.13)$$
$$= \mathbf{-27.67\ units}$$

The vertical component of

$$v_2 - v_1 - v_3 = (40\sin 190°) - (22\sin 140°) - (15\sin 290°)$$
$$= (-6.95) - (14.14) - (-14.10)$$
$$= \mathbf{-6.99\ units}$$

From Fig. 34.40:
the magnitude of the resultant,
$R = \sqrt{(-27.67)^2 + (-6.99)^2} = \mathbf{28.54\ units}$

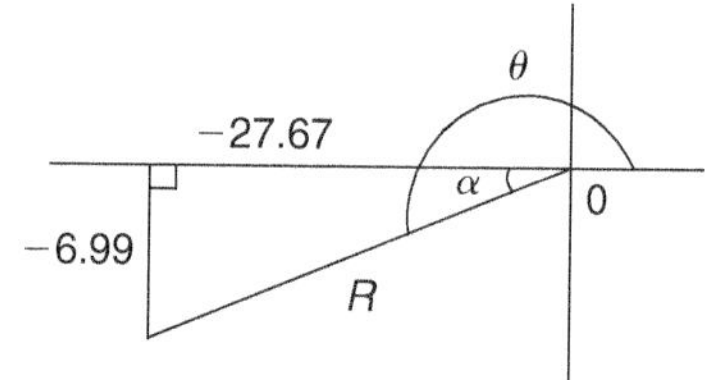

Figure 34.40

and $\alpha = \tan^{-1}\left(\frac{6.99}{27.67}\right) = 14.18°$, from which,

$$\theta = 180° + 14.18° = 194.18°$$

Thus, $v_2 - v_1 - v_3 = \mathbf{28.54\ units\ at\ 194.18°}$

Using **complex numbers**,

$$v_2 - v_1 - v_3 = 40\angle 190° - 22\angle 140° - 15\angle 290°$$
$$= (-39.392 - j6.946) - (-16.853 + j14.141) - (5.130 - j14.095)$$

$$= -27.669 - j6.992$$

$$= \mathbf{28.54\angle -165.82^\circ} \text{ or } \mathbf{28.54\angle 194.18^\circ}$$

This result is as expected, since $\boldsymbol{v_2} - \boldsymbol{v_1} - \boldsymbol{v_3} = -(\boldsymbol{v_1} - \boldsymbol{v_2} + \boldsymbol{v_3})$ and the vector 28.54 units at 194.18° is minus times (i.e. is 180° out of phase with) the vector 28.54 units at 14.18°

Now try the following Practice Exercise

Practice Exercise 169 Vector subtraction (Answers on page 720)

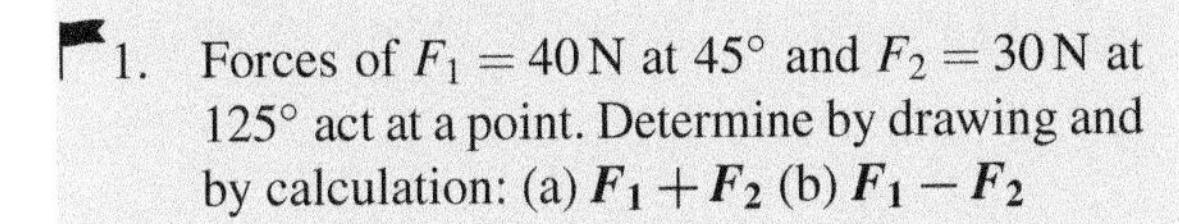

1. Forces of $F_1 = 40$ N at 45° and $F_2 = 30$ N at 125° act at a point. Determine by drawing and by calculation: (a) $\boldsymbol{F_1} + \boldsymbol{F_2}$ (b) $\boldsymbol{F_1} - \boldsymbol{F_2}$

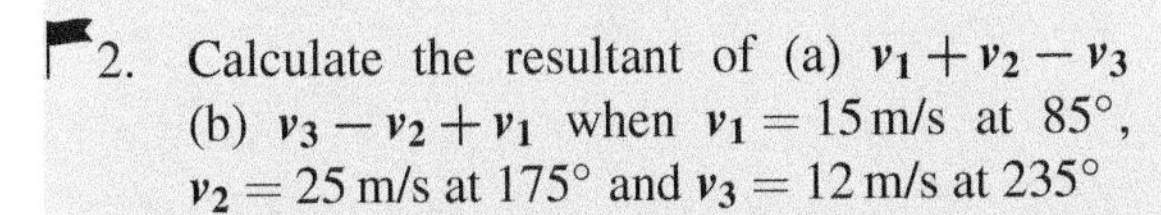

2. Calculate the resultant of (a) $\boldsymbol{v_1} + \boldsymbol{v_2} - \boldsymbol{v_3}$ (b) $\boldsymbol{v_3} - \boldsymbol{v_2} + \boldsymbol{v_1}$ when $\boldsymbol{v_1} = 15$ m/s at 85°, $\boldsymbol{v_2} = 25$ m/s at 175° and $\boldsymbol{v_3} = 12$ m/s at 235°

34.8 Relative velocity

For relative velocity problems, some fixed datum point needs to be selected. This is often a fixed point on the earth's surface. In any vector equation, only the start and finish points affect the resultant vector of a system. Two different systems are shown in Fig. 34.41, but in each of the systems, the resultant vector is ***ad***

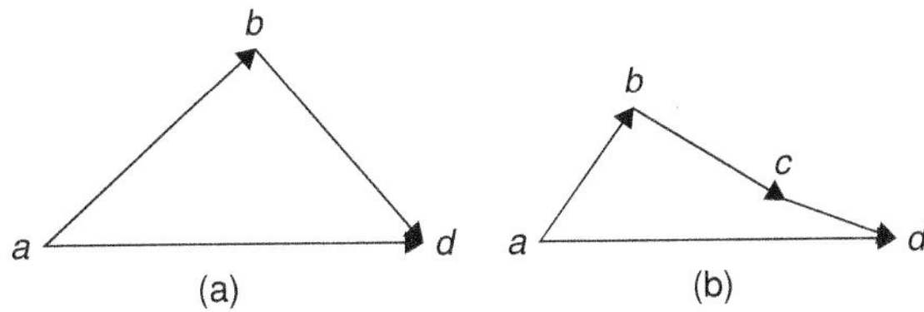

Figure 34.41

The vector equation of the system shown in Fig. 34.41(a) is:

$$\boldsymbol{ad} = \boldsymbol{ab} + \boldsymbol{bd}$$

and that for the system shown in Fig. 34.41(b) is:

$$\boldsymbol{ad} = \boldsymbol{ab} + \boldsymbol{bc} + \boldsymbol{cd}$$

Thus in vector equations of this form, only the first and last letters, 'a' and 'd', respectively, fix the magnitude and direction of the resultant vector. This principle is used in relative velocity problems.

Problem 14. Two cars, P and Q, are travelling towards the junction of two roads which are at right angles to one another. Car P has a velocity of 45 km/h due east and car Q a velocity of 55 km/h due south. Calculate (i) the velocity of car P relative to car Q, and (ii) the velocity of car Q relative to car P.

(i) The directions of the cars are shown in Fig. 34.42(a), called a **space diagram**. The velocity diagram is shown in Fig. 34.42(b), in which ***pe*** is taken as the velocity of car P relative to point e on the earth's surface. The velocity of P relative to Q is vector ***pq*** and the vector equation is ***pq*** = ***pe*** + ***eq***. Hence the vector directions are as shown, ***eq*** being in the opposite direction to ***qe***

From the geometry of the vector triangle, the magnitude of $\boldsymbol{pq} = \sqrt{45^2 + 55^2} = 71.06$ km/h and the direction of $\boldsymbol{pq} = \tan^{-1}\left(\frac{55}{45}\right) = 50.71°$

i.e. **the velocity of car P relative to car Q is 71.06 km/h at 50.71°**

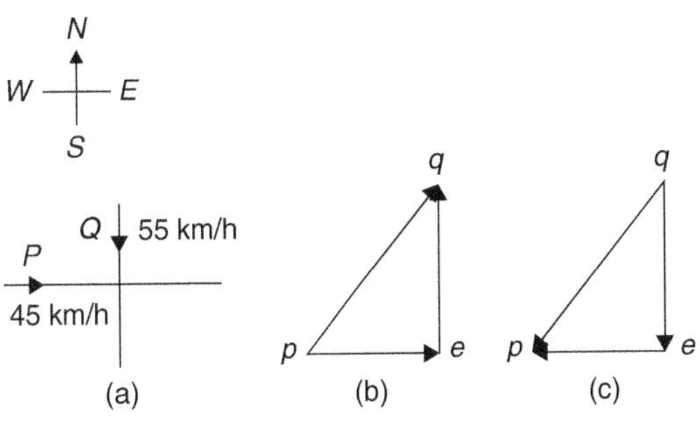

Figure 34.42

(ii) The velocity of car Q relative to car P is given by the vector equation ***qp*** = ***qe*** + ***ep*** and the vector diagram is as shown in Fig. 34.42(c), having ***ep*** opposite in direction to ***pe***

From the geometry of this vector triangle, the magnitude of $\boldsymbol{qp} = \sqrt{45^2 + 55^2} = 71.06$ km/h and the direction of $\boldsymbol{qp} = \tan^{-1}\left(\frac{55}{45}\right) = 50.71°$ but must lie in the third quadrant, i.e. the required angle is: $180° + 50.71° = 230.71°$, i.e. **the velocity of car Q relative to car P is 71.06 km/h at 230.71°**

Now try the following Practice Exercise

Practice Exercise 170 Relative velocity (Answers on page 720)

1. A car is moving along a straight horizontal road at 79.2 km/h and rain is falling vertically downwards at 26.4 km/h. Find the velocity of the rain relative to the driver of the car
2. Calculate the time needed to swim across a river 142 m wide when the swimmer can swim at 2 km/h in still water and the river is flowing at 1 km/h. At what angle to the bank should the swimmer swim?
3. A ship is heading in a direction N 60° E at a speed which in still water would be 20 km/h. It is carried off course by a current of 8 km/h in a direction of E 50° S. Calculate the ship's actual speed and direction

34.9 *i*, *j* and *k* notation

A method of completely specifying the direction of a vector in space relative to some reference point is to use three unit vectors, $\boldsymbol{i}, \boldsymbol{j}$ and $\boldsymbol{k}$, mutually at right angles to each other, as shown in Fig. 34.43.

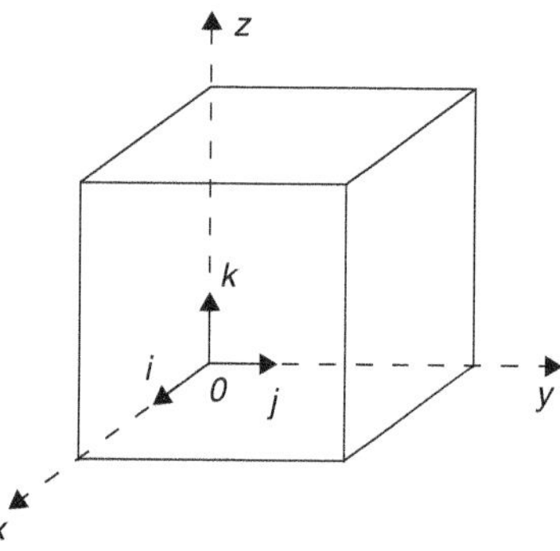

Figure 34.43

Calculations involving vectors given in $\boldsymbol{i}, \boldsymbol{j}$ and $\boldsymbol{k}$ notation are carried out in exactly the same way as standard algebraic calculations, as shown in the worked example below.

Problem 15. Determine:

$$(3i + 2j + 2k) - (4i - 3j + 2k)$$

$$\begin{aligned}(3i + 2j + 2k) - (4i - 3j + 2k) &= 3i + 2j + 2k \\ &\quad - 4i + 3j - 2k \\ &= -i + 5j\end{aligned}$$

Problem 16. Given $\boldsymbol{p} = 3i + 2k$, $\boldsymbol{q} = 4i - 2j + 3k$ and $\boldsymbol{r} = -3i + 5j - 4k$ determine: (a) $-\boldsymbol{r}$ (b) $3\boldsymbol{p}$ (c) $2\boldsymbol{p} + 3\boldsymbol{q}$ (d) $-\boldsymbol{p} + 2\boldsymbol{r}$ (e) $0.2\boldsymbol{p} + 0.6\boldsymbol{q} - 3.2\boldsymbol{r}$

(a) $-\boldsymbol{r} = -(-3i + 5j - 4k) = \mathbf{+3i - 5j + 4k}$

(b) $3\boldsymbol{p} = 3(3i + 2k) = \mathbf{9i + 6k}$

(c) $$\begin{aligned}2\boldsymbol{p} + 3\boldsymbol{q} &= 2(3i + 2k) + 3(4i - 2j + 3k) \\ &= 6i + 4k + 12i - 6j + 9k \\ &= \mathbf{18i - 6j + 13k}\end{aligned}$$

(d) $$\begin{aligned}-\boldsymbol{p} + 2\boldsymbol{r} &= -(3i + 2k) + 2(-3i + 5j - 4k) \\ &= -3i - 2k + (-6i + 10j - 8k) \\ &= -3i - 2k - 6i + 10j - 8k \\ &= \mathbf{-9i + 10j - 10k}\end{aligned}$$

(e) $$\begin{aligned}&0.2\boldsymbol{p} + 0.6\boldsymbol{q} - 3.2\boldsymbol{r} \\ &= 0.2(3i + 2k) + 0.6(4i - 2j + 3k) \\ &\qquad -3.2(-3i + 5j - 4k) \\ &= 0.6i + 0.4k + 2.4i - 1.2j + 1.8k \\ &\qquad +9.6i - 16j + 12.8k \\ &= \mathbf{12.6i - 17.2j + 15k}\end{aligned}$$

Now try the following Practice Exercise

Practice Exercise 171 The *i*, *j*, *k* notation (Answers on page 720)

Given that $\boldsymbol{p} = 2i + 0.5j - 3k$, $\boldsymbol{q} = -i + j + 4k$ and $\boldsymbol{r} = 6j - 5k$, evaluate and simplify the following vectors in $\boldsymbol{i}, \boldsymbol{j}, \boldsymbol{k}$ form:

1. $-\boldsymbol{q}$
2. $2\boldsymbol{p}$

3. $\boldsymbol{q}+\boldsymbol{r}$
4. $-\boldsymbol{q}+2\boldsymbol{p}$
5. $3\boldsymbol{q}+4\boldsymbol{r}$
6. $\boldsymbol{q}-2\boldsymbol{p}$
7. $\boldsymbol{p}+\boldsymbol{q}+\boldsymbol{r}$
8. $\boldsymbol{p}+2\boldsymbol{q}+3\boldsymbol{r}$
9. $2\boldsymbol{p}+0.4\boldsymbol{q}+0.5\boldsymbol{r}$
10. $7\boldsymbol{r}-2\boldsymbol{q}$

Practice Exercise 172 Multiple-choice questions on vectors (Answers on page 720)

Each question has only one correct answer

1. Which of the following quantities is considered a vector?
(a) time (b) temperature
(c) force (d) mass

2. A force vector of 40 N is at an angle of 72° to the horizontal. Its vertical component is:
(a) 40 N (b) 12.36 N
(c) 38.04 N (d) 123.11 N

3. A displacement vector of 30 m is at an angle of −120° to the horizontal. Its horizontal component is:
(a) −15 m (b) −25.98 m
(c) 15 m (d) 25.98 m

4. Two voltage phasors are shown in Figure 34.44. If $V_1 = 40$ volts and $V_2 = 100$ volts, the resultant (i.e. length OA) is:
(a) 131.4 volts at 32.55° to V_1
(b) 105.0 volts at 32.55° to V_1
(c) 131.4 volts at 68.30° to V_1
(d) 105.0 volts at 42.31° to V_1

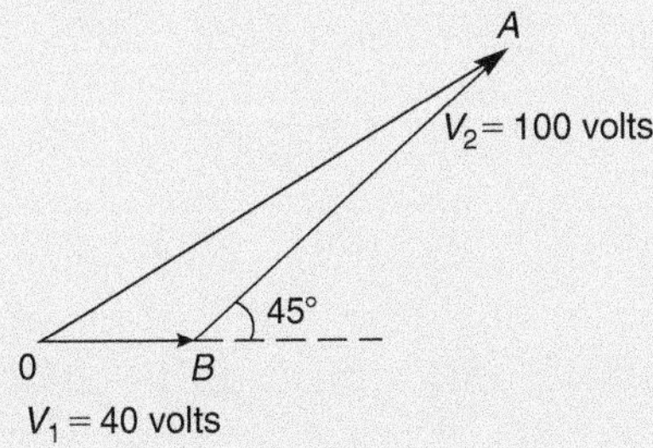

Figure 34.44

5. A force of 4 N is inclined at an angle of 45° to a second force of 7 N, both forces acting at a point, as shown in Figure 34.45. The magnitude of the resultant of these two forces and the direction of the resultant with respect to the 7 N force is:
(a) 3 N at 45°
(b) 5 N at 146°
(c) 11 N at 135°
(d) 10.2 N at 16°

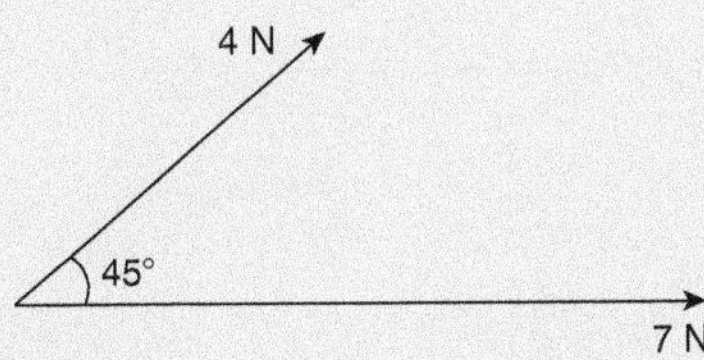

Figure 34.45

Questions 6 and 7 relate to the following information.
Two voltage phasors $\boldsymbol{V}_1$ and $\boldsymbol{V}_2$ are shown in Figure 34.46.

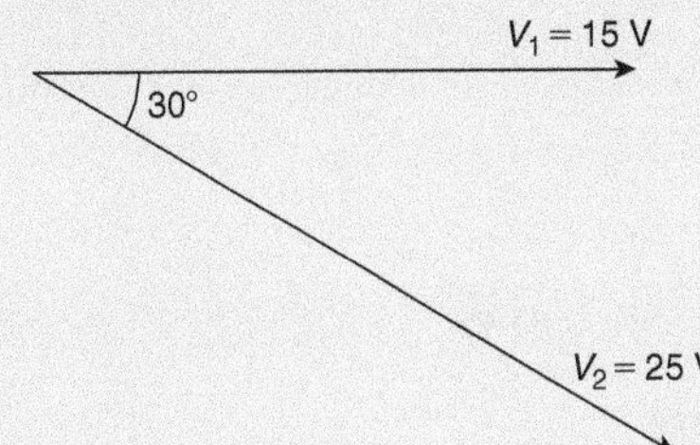

Figure 34.46

6. The resultant $\boldsymbol{V}_1 + \boldsymbol{V}_2$ is given by:
(a) 14.16 V at 62° to V_1
(b) 38.72 V at −19° to V_1
(c) 38.72 V at 161° to V_1
(d) 14.16 V at 118° to V_1

7. The resultant $\boldsymbol{V}_1 - \boldsymbol{V}_2$ is given by:
(a) 38.72 V at −19° to V_1
(b) 14.16 V at 62° to V_1
(c) 38.72 V at 161° to V_1
(d) 14.16 V at 118° to V_1

8. The magnitude of the resultant of velocities of 3 m/s at 20° and 7 m/s at 120° when acting simultaneously at a point is:
(a) 7.12 m/s (b) 10 m/s
(c) 21 m/s (d) 4 m/s

9. Three forces of 2 N, 3 N and 4 N act as shown in Figure 34.47. The magnitude of the resultant force is:
(a) 8.08 N (b) 7.17 N
(c) 9 N (d) 1 N

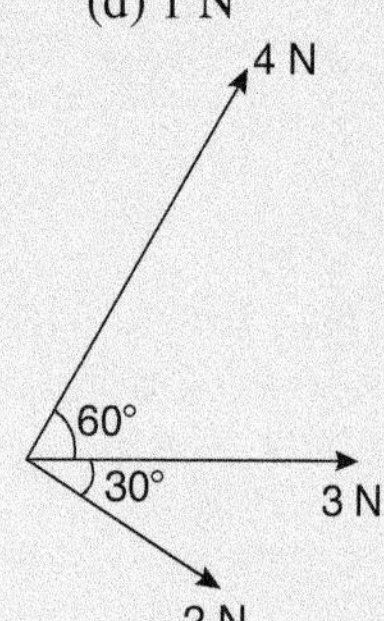

Figure 34.47

10. The following three forces are acting upon an object in space:

$$\boldsymbol{p} = 4\boldsymbol{i} - 3\boldsymbol{j} + 2\boldsymbol{k} \quad \boldsymbol{q} = -5\boldsymbol{k}$$
$$\boldsymbol{r} = -2\boldsymbol{i} + 7\boldsymbol{j} + 6\boldsymbol{k}$$

The resultant of the forces $\boldsymbol{p} + \boldsymbol{q} - \boldsymbol{r}$ is:
(a) $-3\boldsymbol{i} + 4\boldsymbol{j} + 4\boldsymbol{k}$
(b) $2\boldsymbol{i} + 4\boldsymbol{j} + 3\boldsymbol{k}$
(c) $6\boldsymbol{i} - 10\boldsymbol{j} - 9\boldsymbol{k}$
(d) $2\boldsymbol{i} + 4\boldsymbol{j} - 3\boldsymbol{k}$

For fully worked solutions to each of the problems in Practice Exercises 167 to 171 in this chapter, go to the website:
www.routledge.com/cw/bird

Chapter 35

Methods of adding alternating waveforms

Why it is important to understand: Methods of adding alternating waveforms

In electrical engineering, a phasor is a rotating vector representing a quantity such as an alternating current or voltage that varies sinusoidally. Sometimes it is necessary when studying sinusoidal quantities to add together two alternating waveforms, for example in an a.c. series circuit that are not in-phase with each other. Electrical engineers, electronics engineers, electronic engineering technicians and aircraft engineers all use phasor diagrams to visualise complex constants and variables. So, given oscillations to add and subtract, the required rotating vectors are constructed, called a phasor diagram, and graphically the resulting sum and/or difference oscillation are added or calculated. Phasors may be used to analyse the behaviour of electrical and mechanical systems that have reached a kind of equilibrium called sinusoidal steady state. Hence, discovering different methods of combining sinusoidal waveforms is of some importance in certain areas of engineering.

At the end of this chapter, you should be able to:

- determine the resultant of two phasors by graph plotting
- determine the resultant of two or more phasors by drawing
- determine the resultant of two phasors by the sine and cosine rules
- determine the resultant of two or more phasors by horizontal and vertical components
- determine the resultant of two or more phasors by complex numbers

35.1 Combination of two periodic functions

There are a number of instances in engineering and science where waveforms have to be combined and where it is required to determine the single phasor (called the resultant) that could replace two or more separate phasors. Uses are found in electrical alternating current theory, in mechanical vibrations, in the addition of forces and with sound waves.

There are a number of methods of determining the resultant waveform. These include:

(a) by drawing the waveforms and adding graphically

(b) by drawing the phasors and measuring the resultant

(c) by using the cosine and sine rules

(d) by using horizontal and vertical components

(e) by using complex numbers.

35.2 Plotting periodic functions

This may be achieved by sketching the separate functions on the same axes and then adding (or subtracting) ordinates at regular intervals. This is demonstrated in the following worked problems.

Problem 1. Plot the graph of $y_1 = 3\sin A$ from $A = 0°$ to $A = 360°$. On the same axes plot $y_2 = 2\cos A$. By adding ordinates, plot $y_R = 3\sin A + 2\cos A$ and obtain a sinusoidal expression for this resultant waveform

$y_1 = 3\sin A$ and $y_2 = 2\cos A$ are shown plotted in Fig. 35.1. Ordinates may be added at, say, 15° intervals.

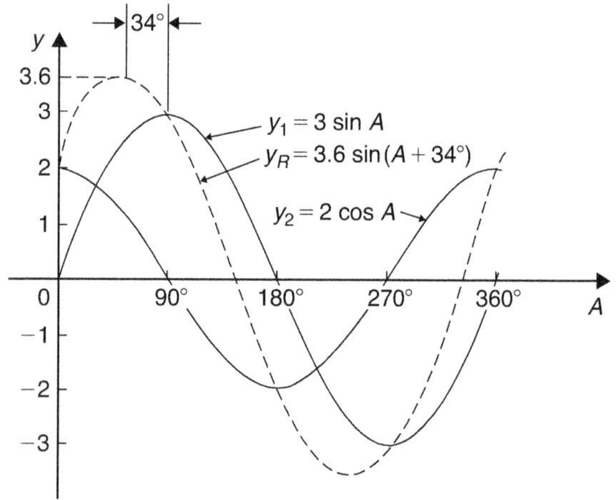

Figure 35.1

For example,

at 0°, $\quad y_1 + y_2 = 0 + 2 = 2$

at 15°, $\quad y_1 + y_2 = 0.78 + 1.93 = 2.71$

at 120°, $\quad y_1 + y_2 = 2.60 + -1 = 1.60$

at 210°, $\quad y_1 + y_2 = -1.50 - 1.73 = -3.23$ and so on

The resultant waveform, shown by the broken line, has the same period, i.e. 360°, and thus the same frequency as the single phasors. The maximum value, or amplitude, of the resultant is 3.6. The resultant waveform **leads** $y_1 = 3\sin A$ by 34° or $34 \times \frac{\pi}{180}$ rad $= 0.593$ rad.

The sinusoidal expression for the resultant waveform is:

$$\mathbf{y_R = 3.6\sin(A + 34°)} \quad \text{or} \quad \mathbf{y_R = 3.6\sin(A + 0.593)}$$

Problem 2. Plot the graphs of $y_1 = 4\sin\omega t$ and $y_2 = 3\sin(\omega t - \pi/3)$ on the same axes, over one cycle. By adding ordinates at intervals plot $y_R = y_1 + y_2$ and obtain a sinusoidal expression for the resultant waveform

$y_1 = 4\sin\omega t$ and $y_2 = 3\sin(\omega t - \pi/3)$ are shown plotted in Fig. 35.2. Ordinates are added at 15° intervals and the resultant is shown by the broken line. The amplitude of the resultant is 6.1 and it **lags** y_1 by 25° or 0.436 rad.

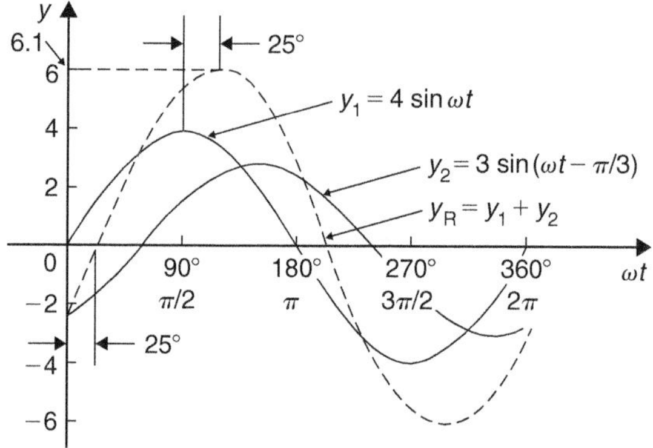

Figure 35.2

Hence, the sinusoidal expression for the resultant waveform is:

$$\mathbf{y_R = 6.1\sin(\omega t - 0.436)}$$

Problem 3. Determine a sinusoidal expression for $y_1 - y_2$ when $y_1 = 4\sin\omega t$ and $y_2 = 3\sin(\omega t - \pi/3)$

y_1 and y_2 are shown plotted in Fig. 35.3. At 15° intervals y_2 is subtracted from y_1. For example:

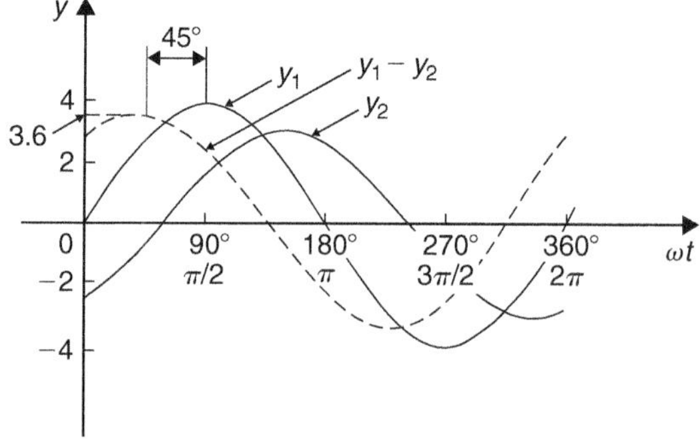

Figure 35.3

at 0°, $y_1 - y_2 = 0 - (-2.6) = +2.6$

at 30°, $y_1 - y_2 = 2 - (-1.5) = +3.5$

at 150°, $y_1 - y_2 = 2 - 3 = -1$ and so on.

The amplitude, or peak value of the resultant (shown by the broken line), is 3.6 and it leads y_1 by 45° or 0.79 rad. Hence, $\boldsymbol{y_1 - y_2 = 3.6\sin(\omega t + 0.79)}$

Problem 4. Two alternating currents are given by: $i_1 = 20\sin\omega t$ amperes and $i_2 = 10\sin\left(\omega t + \frac{\pi}{3}\right)$ amperes. By drawing the waveforms on the same axes and adding, determine the sinusoidal expression for the resultant $i_1 + i_2$

i_1 and i_2 are shown plotted in Fig. 35.4. The resultant waveform for $i_1 + i_2$ is shown by the broken line. It has the same period, and hence frequency, as i_1 and i_2. The amplitude or peak value is 26.5 A.

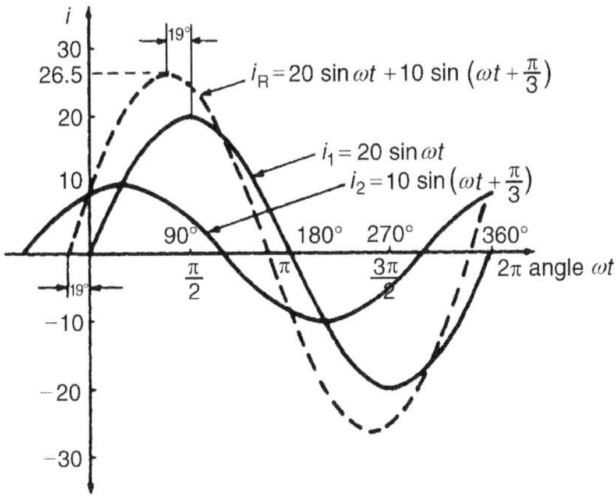

Figure 35.4

The resultant waveform leads the waveform of $i_1 = 20\sin\omega t$ by 19° or 0.33 rad.
Hence, the sinusoidal expression for the resultant $i_1 + i_2$ is given by:

$$\boldsymbol{i_R = i_1 + i_2 = 26.5\sin(\omega t + 0.33)\text{A}}$$

Now try the following Practice Exercise

Practice Exercise 173 Plotting periodic functions (Answers on page 720)

1. Plot the graph of $y = 2\sin A$ from $A = 0°$ to $A = 360°$. On the same axes plot $y = 4\cos A$. By adding ordinates at intervals plot $y = 2\sin A + 4\cos A$ and obtain a sinusoidal expression for the waveform
2. Two alternating voltages are given by $v_1 = 10\sin\omega t$ volts and $v_2 = 14\sin(\omega t + \pi/3)$ volts. By plotting v_1 and v_2 on the same axes over one cycle obtain a sinusoidal expression for (a) $v_1 + v_2$ (b) $v_1 - v_2$
3. Express $12\sin\omega t + 5\cos\omega t$ in the form $A\sin(\omega t \pm \alpha)$ by drawing and measurement

35.3 Determining resultant phasors by drawing

The resultant of two periodic functions may be found from their relative positions when the time is zero. For example, if $y_1 = 4\sin\omega t$ and $y_2 = 3\sin(\omega t - \pi/3)$ then each may be represented as phasors as shown in Fig. 35.5, y_1 being 4 units long and drawn horizontally and y_2 being 3 units long, lagging y_1 by $\pi/3$ radians or 60°. To determine the resultant of $y_1 + y_2$, y_1 is drawn horizontally as shown in Fig. 35.6 and y_2 is joined to the end of y_1 at 60° to the horizontal. The resultant is given by y_R. This is the same as the diagonal of a parallelogram that is shown completed in Fig. 35.7.

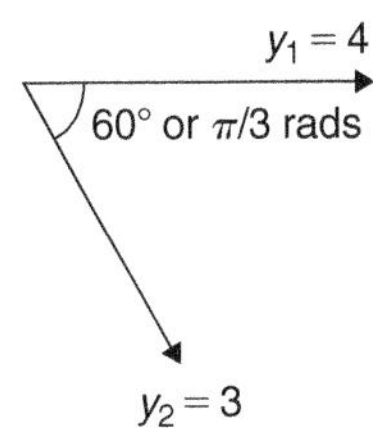

Figure 35.5

Resultant y_R, in Figs 35.6 and 35.7, may be determined by drawing the phasors and their directions to scale and measuring using a ruler and protractor.

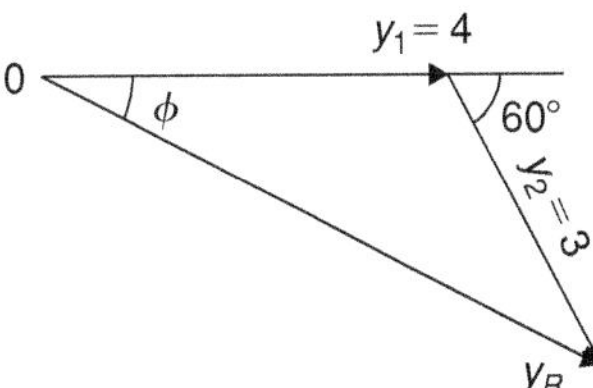

Figure 35.6

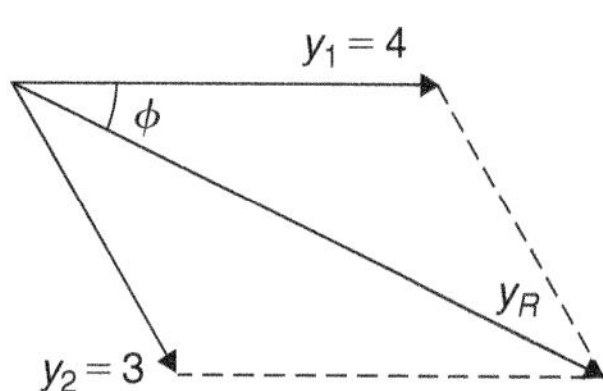

Figure 35.7

In this example, y_R is measured as 6 units long and angle ϕ is measured as 25°.

$$25° = 25 \times \frac{\pi}{180} \text{ radians} = 0.44 \text{ rad}$$

Hence, summarising, by drawing: $\mathbf{y_R = y_1 + y_2 =}$ $4\sin\omega t + 3\sin(\omega t - \pi/3) = \mathbf{6\sin(\omega t - 0.44)}$

If the resultant phasor $y_R = y_1 - y_2$ is required, then y_2 is still 3 units long but is drawn in the opposite direction, as shown in Fig. 35.8.

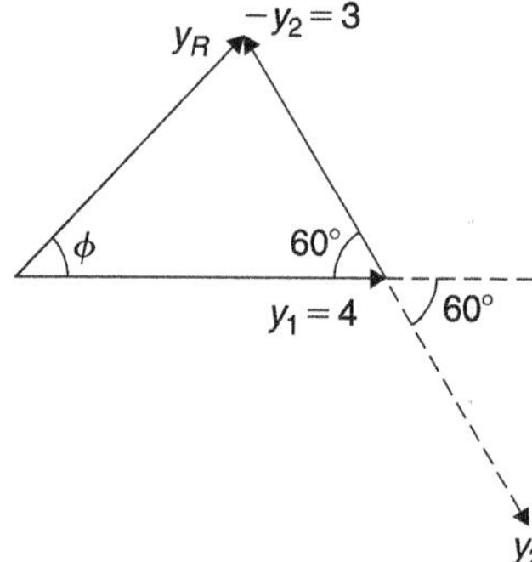

Figure 35.8

It may be shown that $y_R = y_1 - y_2 = \mathbf{3.6\ sin(\omega t + 0.80)}$

Problem 5. Two alternating currents are given by: $i_1 = 20\sin\omega t$ amperes and $i_2 = 10\sin\left(\omega t + \frac{\pi}{3}\right)$ amperes. Determine $i_1 + i_2$ by drawing phasors

The relative positions of i_1 and i_2 at time $t = 0$ are shown as phasors in Fig. 35.9, where $\frac{\pi}{3}$ rad $= 60°$.

The phasor diagram in Fig. 35.10 is drawn to scale with a ruler and protractor.

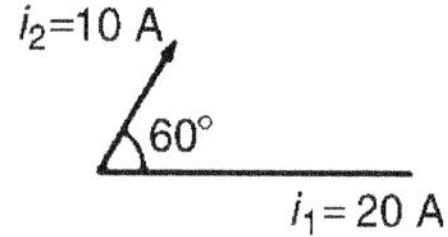

Figure 35.9

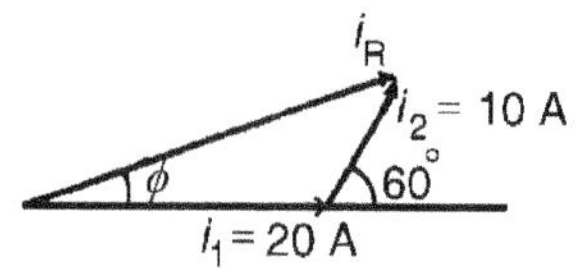

Figure 35.10

The resultant i_R is shown and is measured as 26 A and angle ϕ as 19° or 0.33 rad leading i_1.

Hence, by drawing and measuring:

$$\mathbf{i_R = i_1 + i_2 = 26\sin(\omega t + 0.33)A}$$

Problem 6. For the currents in Problem 5, determine $i_1 - i_2$ by drawing phasors

At time $t = 0$, current i_1 is drawn 20 units long horizontally as shown by Oa in Fig. 35.11. Current i_2 is shown, drawn 10 units long and leading by 60°. The current $-i_2$ is drawn in the opposite direction, shown as ab in Fig. 35.11. The resultant i_R is given by Ob lagging by angle ϕ.

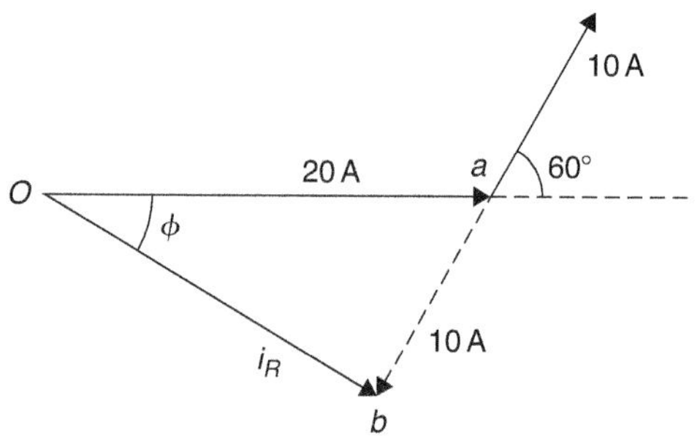

Figure 35.11

By measurement, $i_R = 17$ A and $\phi = 30°$ or 0.52 rad

Hence, by drawing phasors:

$$\mathbf{i_R = i_1 - i_2 = 17\sin(\omega t - 0.52)A}$$

Now try the following Practice Exercise

Practice Exercise 174 Determining resultant phasors by drawing (Answers on page 720)

1. Determine a sinusoidal expression for $2\sin\theta + 4\cos\theta$ by drawing phasors

2. If $v_1 = 10\sin\omega t$ volts and $v_2 = 14\sin(\omega t + \pi/3)$ volts, determine by drawing phasors sinusoidal expressions for (a) $v_1 + v_2$ (b) $v_1 - v_2$

3. Express $12\sin\omega t + 5\cos\omega t$ in the form $R\sin(\omega t \pm \alpha)$ by drawing phasors

35.4 Determining resultant phasors by the sine and cosine rules

As stated earlier, the resultant of two periodic functions may be found from their relative positions when the time is zero. For example, if $y_1 = 5\sin\omega t$ and $y_2 = 4\sin(\omega t - \pi/6)$ then each may be represented by phasors as shown in Fig. 35.12, y_1 being 5 units long and drawn horizontally and y_2 being 4 units long, lagging y_1 by $\pi/6$ radians or 30°. To determine the resultant of $y_1 + y_2$, y_1 is drawn horizontally as shown in Fig. 35.13 and y_2 is joined to the end of y_1 at $\pi/6$ radians, i.e. 30° to the horizontal. The resultant is given by y_R

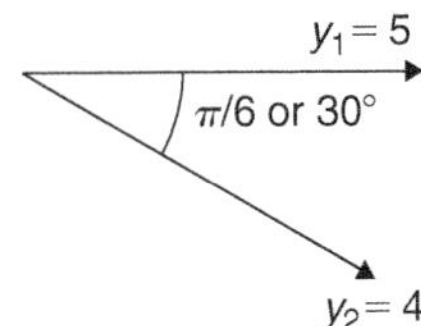

Figure 35.12

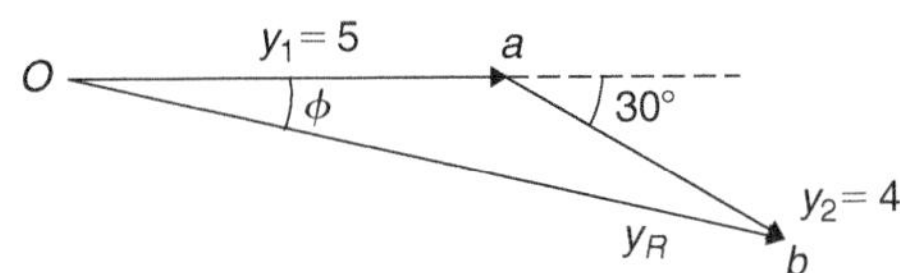

Figure 35.13

Using the cosine rule (from Chapter 20) on triangle *Oab* of Fig. 35.13 gives:

$$y_R{}^2 = 5^2 + 4^2 - [2(5)(4)\cos 150°]$$
$$= 25 + 16 - (-34.641) = 75.641$$

from which, $\quad y_R = \sqrt{75.641} = 8.697$

Using the sine rule, $\quad \dfrac{8.697}{\sin 150°} = \dfrac{4}{\sin\phi}$

from which, $\quad \sin\phi = \dfrac{4\sin 150°}{8.697} = 0.22996$

and $\quad \phi = \sin^{-1} 0.22996$

$$= 13.29° \text{ or } 0.232\,\text{rad}$$

Hence, $\mathbf{y_R = y_1 + y_2} = 5\sin\omega t + 4\sin(\omega t - \pi/6)$

$$= \mathbf{8.697\sin(\omega t - 0.232)}$$

Problem 7. Given $y_1 = 2\sin\omega t$ and $y_2 = 3\sin(\omega t + \pi/4)$, obtain an expression, by calculation, for the resultant, $y_R = y_1 + y_2$

When time $t = 0$, the position of phasors y_1 and y_2 are as shown in Fig. 35.14(a). To obtain the resultant, y_1 is drawn horizontally, 2 units long, y_2 is drawn 3 units long at an angle of $\pi/4$ rads or 45° and joined to the end of y_1 as shown in Fig. 35.14(b).

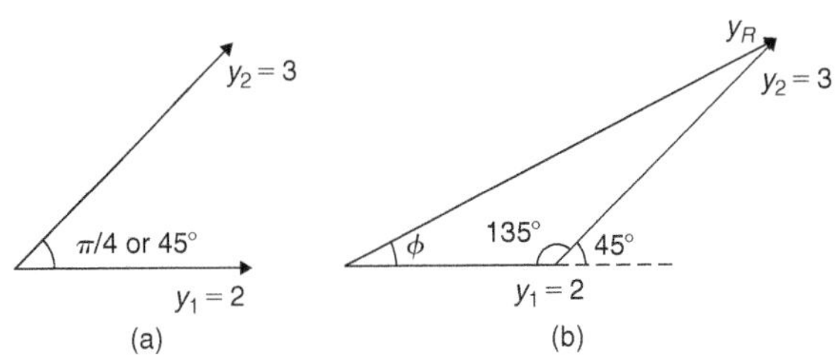

Figure 35.14

From Fig. 35.14(b), and using the cosine rule:

$$y_R^2 = 2^2 + 3^2 - [2(2)(3)\cos 135°]$$
$$= 4 + 9 - [-8.485] = 21.49$$

Hence, $\quad y_R = \sqrt{21.49} = 4.6357$

Using the sine rule: $\quad \dfrac{3}{\sin\phi} = \dfrac{4.6357}{\sin 135°}$

from which, $\quad \sin\phi = \dfrac{3\sin 135°}{4.6357} = 0.45761$

Hence, $\quad \phi = \sin^{-1} 0.45761 = 27.23°$ or 0.475 rad.

Thus, by calculation, $\quad \mathbf{y_R = 4.635\sin(\omega t + 0.475)}$

Problem 8. Determine

$$20\sin\omega t + 10\sin\left(\omega t + \frac{\pi}{3}\right)A$$

using the cosine and sine rules

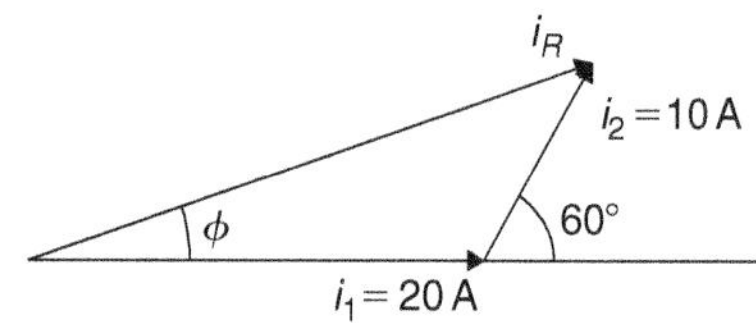

Figure 35.15

From the phasor diagram of Fig. 35.15, and using the cosine rule:

$$i_R{}^2 = 20^2 + 10^2 - [2(20)(10)\cos 120°] = 700$$

Hence, $\quad i_R = \sqrt{700} = \mathbf{26.46\ A}$

Using the sine rule gives : $\dfrac{10}{\sin\phi} = \dfrac{26.46}{\sin 120°}$

from which, $\quad \sin\phi = \dfrac{10\sin 120°}{26.46}$

$$= 0.327296$$

and $\quad \phi = \sin^{-1} 0.327296 = 19.10°$

$$= 19.10 \times \frac{\pi}{180} = \mathbf{0.333\ rad}$$

Hence, by cosine and sine rules,

$$\boldsymbol{i_R = i_1 + i_2 = 26.46\sin(\omega t + 0.333)\text{A}}$$

Now try the following Practice Exercise

Practice Exercise 175 Resultant phasors by the sine and cosine rules (Answers on page 720)

1. Determine, using the cosine and sine rules, a sinusoidal expression for:
$$y = 2\sin A + 4\cos A$$

2. Given $v_1 = 10\sin\omega t$ volts and $v_2 = 14\sin(\omega t + \pi/3)$ volts use the cosine and sine rules to determine sinusoidal expressions for (a) $v_1 + v_2$ (b) $v_1 - v_2$

In Problems 3 to 5, express the given expressions in the form $A\sin(\omega t \pm \alpha)$ by using the cosine and sine rules.

3. $12\sin\omega t + 5\cos\omega t$

4. $7\sin\omega t + 5\sin\left(\omega t + \dfrac{\pi}{4}\right)$

5. $6\sin\omega t + 3\sin\left(\omega t - \dfrac{\pi}{6}\right)$

35.5 Determining resultant phasors by horizontal and vertical components

If a right-angled triangle is constructed as shown in Fig. 35.16, then Oa is called the horizontal component of F and ab is called the vertical component of F.

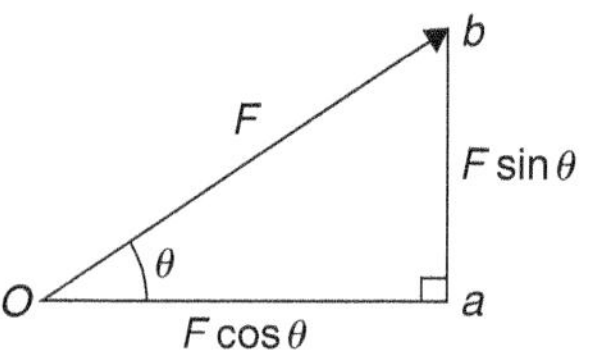

Figure 35.16

From trigonometry (see Chapter 17, page 171)

$\cos\theta = \dfrac{Oa}{Ob}$ from which, $Oa = Ob\cos\theta = F\cos\theta$,

i.e. **the horizontal component of F, $H = F\cos\theta$**

and $\quad \sin\theta = \dfrac{ab}{Ob}$ from which,

$$ab = Ob\sin\theta = F\sin\theta$$

i.e. **the vertical component of F, $V = F\sin\theta$**

Determining resultant phasors by horizontal and vertical components is demonstrated in the following worked problems.

Problem 9. Two alternating voltages are given by $v_1 = 15\sin\omega t$ volts and $v_2 = 25\sin(\omega t - \pi/6)$ volts. Determine a sinusoidal expression for the resultant $v_R = v_1 + v_2$ by finding horizontal and vertical components

The relative positions of v_1 and v_2 at time $t = 0$ are shown in Fig. 35.17(a) and the phasor diagram is shown in Fig. 35.17(b).

The horizontal component of v_R,
$H = 15\cos 0° + 25\cos(-30°) = Oa + ab = \mathbf{36.65\ V}$
The vertical component of v_R,
$V = 15\sin 0° + 25\sin(-30°) = Oc = \mathbf{-12.50\ V}$

Hence, $\quad v_R = Oc = \sqrt{36.65^2 + (-12.50)^2}$

by Pythagoras' theorem

$$= \mathbf{38.72\ volts}$$

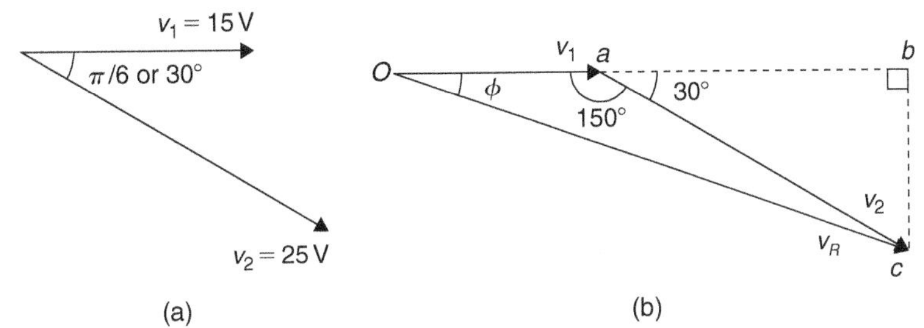

Figure 35.17

$$\tan\phi = \frac{V}{H} = \frac{-12.50}{36.65} = -1.3441$$

from which, $\phi = \tan^{-1}(-0.3411) = -18.83^\circ$ or $-0.329\,\text{rad}$

Hence, $\boldsymbol{v_R = v_1 + v_2 = 38.72\sin(\omega t - 0.329)\,V}$

Problem 10. For the voltages in Problem 9, determine the resultant $v_R = v_1 - v_2$ using horizontal and vertical components

The horizontal component of v_R,
$H = 15\cos 0^\circ - 25\cos(-30^\circ) = \boldsymbol{-6.65\,V}$

The vertical component of v_R,
$V = 15\sin 0^\circ - 25\sin(-30^\circ) = \boldsymbol{12.50\,V}$

Hence, $v_R = \sqrt{(-6.65)^2 + (-12.50)^2}$ by Pythagoras' theorem

$= \mathbf{14.16\ volts}$

$$\tan\phi = \frac{V}{H} = \frac{12.50}{-6.65} = -1.8797$$

from which, $\phi = \tan^{-1}(-1.8797) = 118.01^\circ$ or 2.06 rad (i.e. in the second quadrant)

Hence, $\boldsymbol{v_R = v_1 - v_2 = 14.16\sin(\omega t + 2.06)\,V}$

The phasor diagram is shown in Fig. 35.18.

Problem 11. Determine
$$20\sin\omega t + 10\sin\left(\omega t + \frac{\pi}{3}\right)A$$
using horizontal and vertical components

From the phasors shown in Fig. 35.19:
Total horizontal component,
$$H = 20\cos 0^\circ + 10\cos 60^\circ = \mathbf{25.0}$$

Total vertical component,

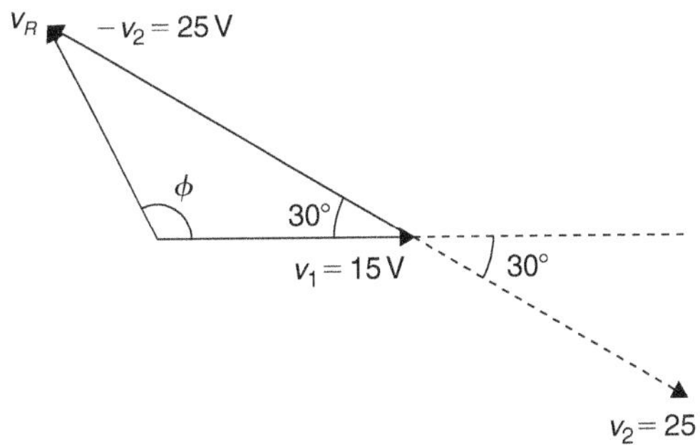

Figure 35.18

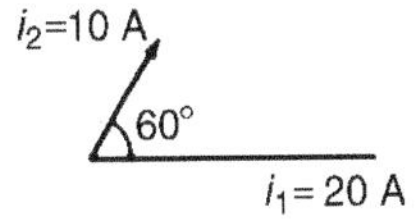

Figure 35.19

$$V = 20\sin 0^\circ + 10\sin 60^\circ = \mathbf{8.66}$$

By Pythagoras, the resultant, $i_R = \sqrt{[25.0^2 + 8.66^2]}$
$= \mathbf{26.46\ A}$

Phase angle, $\phi = \tan^{-1}\left(\frac{8.66}{25.0}\right) = \mathbf{19.11^\circ}$ or $\mathbf{0.333\ rad}$

Hence, by using horizontal and vertical components,

$$20\sin\omega t + 10\sin\left(\omega t + \frac{\pi}{3}\right) = \boldsymbol{26.46\sin(\omega t + 0.333)A}$$

Now try the following Practice Exercise

Practice Exercise 176 Resultant phasors by horizontal and vertical components (Answers on page 720)

In Problems 1 to 4, express the combination of periodic functions in the form $A\sin(\omega t \pm \alpha)$ by horizontal and vertical components:

1. $7\sin\omega t + 5\sin\left(\omega t + \frac{\pi}{4}\right) A$

2. $6\sin\omega t + 3\sin\left(\omega t - \frac{\pi}{6}\right) V$

3. $i = 25\sin\omega t - 15\sin\left(\omega t + \frac{\pi}{3}\right) A$

4. $x = 9\sin\left(\omega t + \frac{\pi}{3}\right) - 7\sin\left(\omega t - \frac{3\pi}{8}\right) m$

5. The voltage drops across two components when connected in series across an a.c. supply are: $v_1 = 200\sin 314.2t$ and $v_2 = 120\sin(314.2t - \pi/5)$ volts respectively. Determine the
   (a) voltage of the supply (given by $v_1 + v_2$) in the form $A\sin(\omega t \pm \alpha)$
   (b) frequency of the supply

6. If the supply to a circuit is $v = 20\sin 628.3t$ volts and the voltage drop across one of the components is $v_1 = 15\sin(628.3t - 0.52)$ volts, calculate the
   (a) voltage drop across the remainder of the circuit, given by $v - v_1$, in the form $A\sin(\omega t \pm \alpha)$
   (b) supply frequency
   (c) periodic time of the supply

7. The voltages across three components in a series circuit when connected across an a.c. supply are: $v_1 = 25\sin\left(300\pi t + \frac{\pi}{6}\right)$ volts, $v_2 = 40\sin\left(300\pi t - \frac{\pi}{4}\right)$ volts and $v_3 = 50\sin\left(300\pi t + \frac{\pi}{3}\right)$ volts. Calculate the
   (a) supply voltage, in sinusoidal form, in the form $A\sin(\omega t \pm \alpha)$
   (b) frequency of the supply
   (c) periodic time

8. In an electrical circuit, two components are connected in series. The voltage across the first component is given by 80 $\sin(\omega t + \pi/3)$ volts, and the voltage across the second component is given by 150 $\sin(\omega t - \pi/4)$ volts. Determine the total supply voltage to the two components. Give the answer in sinusoidal form

35.6 Determining resultant phasors by complex numbers

As stated earlier, the resultant of two periodic functions may be found from their relative positions when the time is zero. For example, if $y_1 = 5\sin\omega t$ and $y_2 = 4\sin(\omega t - \pi/6)$ then each may be represented by phasors as shown in Fig. 35.20, y_1 being 5 units long and drawn horizontally and y_2 being 4 units long, lagging y_1 by $\pi/6$ radians or 30°. To determine the resultant of $y_1 + y_2$, y_1 is drawn horizontally as shown in Fig. 35.21 and y_2 is joined to the end of y_1 at $\pi/6$ radians, i.e. 30° to the horizontal. The resultant is given by y_R

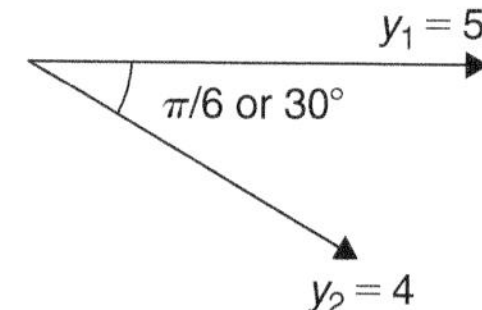

Figure 35.20

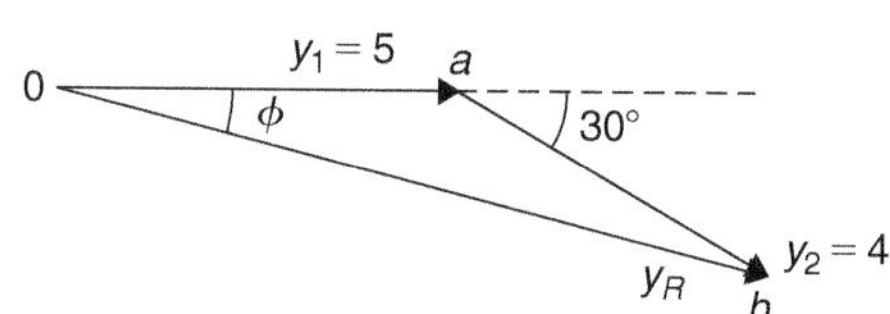

Figure 35.21

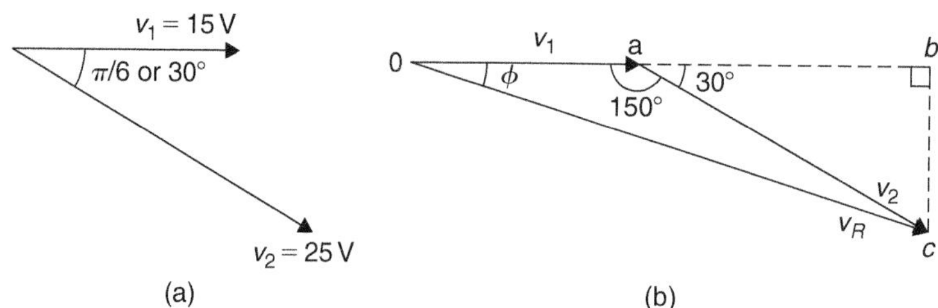

Figure 35.22

From complex numbers, Chapter 32, in polar form,

$$y_R = 5\angle 0 + 4\angle -\frac{\pi}{6}$$
$$= 5\angle 0^\circ + 4\angle -30^\circ$$
$$= (5 + j0) + (3.464 - j2.0)$$
$$= 8.464 - j2.0 = 8.70\angle -13.29^\circ$$
$$= 8.70\angle -0.23\,\text{rad}$$

Hence, by using complex numbers, the resultant in sinusoidal form is:

$$y_1 + y_2 = 5\sin\omega t + 4\sin(\omega t - \pi/6)$$
$$= \mathbf{8.70\sin(\omega t - 0.23)}$$

Problem 12. Two alternating voltages are given by $v_1 = 15\sin\omega t$ volts and $v_2 = 25\sin(\omega t - \pi/6)$ volts. Determine a sinusoidal expression for the resultant $v_R = v_1 + v_2$ by using complex numbers.

The relative positions of v_1 and v_2 at time $t = 0$ are shown in Fig. 35.22(a) and the phasor diagram is shown in Fig. 35.22(b).

In polar form, $v_R = v_1 + v_2 = 15\angle 0 + 25\angle -\frac{\pi}{6}$

$$= 15\angle 0^\circ + 25\angle -30^\circ$$
$$= (15 + j0) + (21.65 - j12.5)$$
$$= 36.65 - j12.5 = 38.72\angle -18.83^\circ$$
$$= 38.72\angle -0.329\,\text{rad}$$

Hence, by using complex numbers, the resultant in sinusoidal form is:

$$v_R = v_1 + v_2 = 15\sin\omega t + 25\sin(\omega t - \pi/6)$$
$$= \mathbf{38.72\sin(\omega t - 0.329)\ volts}$$

Problem 13. For the voltages in Problem 12, determine the resultant $v_R = v_1 - v_2$ using complex numbers.

In polar form, $v_R = v_1 - v_2 = 15\angle 0 - 25\angle -\frac{\pi}{6}$

$$= 15\angle 0^\circ - 25\angle -30^\circ$$
$$= (15 + j0) - (21.65 - j12.5)$$
$$= -6.65 + j12.5 = 14.16\angle 118.01^\circ$$
$$= 14.16\angle 2.06\,\text{rad}$$

Hence, by using complex numbers, the resultant in sinusoidal form is:

$$v_1 - v_2 = 15\sin\omega t - 25\sin(\omega t - \pi/6)$$
$$= \mathbf{14.16\sin(\omega t - 2.06)V}$$

Problem 14. Determine $20\sin\omega t + 10\sin\left(\omega t + \frac{\pi}{6}\right) A$ using complex numbers.

From the phasors shown in Fig. 35.23, the resultant may be expressed in polar form as:

$$i_R = 20\angle 0^\circ + 10\angle 60^\circ$$

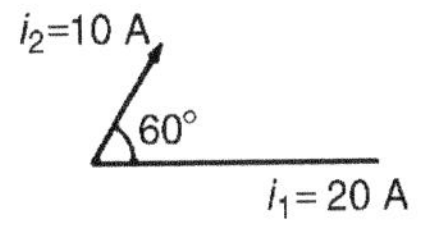

Figure 35.23

i.e. $i_R = (20 + j0) + (5 + j8.66)$

$$= (25 + j8.66)$$
$$= \mathbf{26.46\angle 19.11^\circ A} \text{ or } \mathbf{26.46\angle 0.333\ rad\ A}$$

Hence, by using complex numbers, the resultant in sinusoidal form is:

$$\mathbf{i_R = i_1 + i_2 = 26.46\sin(\omega t + 0.333)A}$$

Problem 15. If the supply to a circuit is $v = 30\sin 100\pi t$ volts and the voltage drop across one of the components is $v_1 = 20\sin(100\pi t - 0.59)$ volts, calculate the

(a) voltage drop across the remainder of the circuit, given by $v - v_1$, in the form $A\sin(\omega t \pm \alpha)$

(b) supply frequency

(c) periodic time of the supply

(d) r.m.s. value of the supply voltage

(a) Supply voltage, $v = v_1 + v_2$ where v_2 is the voltage across the remainder of the circuit

Hence, $v_2 = v - v_1$

$$= 30\sin 100\pi t - 20\sin(100\pi t - 0.59)$$
$$= 30\angle 0 - 20\angle -0.59\,\text{rad}$$
$$= (30 + j0) - (16.619 - j11.127)$$
$$= 13.381 + j11.127$$
$$= 17.40\angle 0.694\,\text{rad}$$

Hence, by using complex numbers, the resultant in sinusoidal form is:

$$v - v_1 = 30\sin 100\pi t - 20\sin(100\pi t - 0.59)$$
$$= \mathbf{17.40\sin(100\pi t + 0.694)\ volts}$$

(b) **Supply frequency,** $f = \frac{\omega}{2\pi} = \frac{100\pi}{2\pi} = \mathbf{50\ Hz}$

(c) **Periodic time,** $T = \frac{1}{f} = \frac{1}{50} = \mathbf{0.02\,s}$ or $\mathbf{20\,ms}$

(d) **R.m.s. value of supply voltage** $= \mathbf{0.707 \times 30}$
$= \mathbf{21.21\ volts}$

Now try the following Practice Exercise

Practice Exercise 177 Resultant phasors by complex numbers (Answers on page 720)

In Problems 1 to 5, express the combination of periodic functions in the form $A\sin(\omega t \pm \alpha)$ by using complex numbers:

1. $8\sin\omega t + 5\sin\left(\omega t + \frac{\pi}{4}\right)$
2. $6\sin\omega t + 9\sin\left(\omega t - \frac{\pi}{6}\right)$
3. $v = 12\sin\omega t - 5\sin\left(\omega t - \frac{\pi}{4}\right)$
4. $x = 10\sin\left(\omega t + \frac{\pi}{3}\right) - 8\sin\left(\omega t - \frac{3\pi}{8}\right)$
5. The voltage drops across two components when connected in series across an a.c. supply are: $v_1 = 240\sin 314.2t$ and $v_2 = 150\sin(314.2t - \pi/5)$ volts respectively. Determine the
   (a) voltage of the supply (given by $v_1 + v_2$) in the form $A\sin(\omega t \pm \alpha)$
   (b) frequency of the supply
6. If the supply to a circuit is $v = 25\sin 200\pi t$ volts and the voltage drop across one of the components is $v_1 = 18\sin(200\pi t - 0.43)$ volts, calculate the
   (a) voltage drop across the remainder of the circuit, given by $v - v_1$, in the form $A\sin(\omega t \pm \alpha)$
   (b) supply frequency
   (c) periodic time of the supply
7. The voltages drops across three components in a series circuit when connected across an a.c. supply are:
   $$v_1 = 20\sin\left(300\pi t - \frac{\pi}{6}\right)\ \text{volts},$$
   $$v_2 = 30\sin\left(300\pi t + \frac{\pi}{4}\right)\ \text{volts, and}$$
   $$v_3 = 60\sin\left(300\pi t - \frac{\pi}{3}\right)\ \text{volts.}$$
   Calculate the
   (a) supply voltage, in sinusoidal form, in the form $A\sin(\omega t \pm \alpha)$
   (b) frequency of the supply
   (c) periodic time
   (d) r.m.s. value of the supply voltage
8. Measurements made at a substation at peak demand of the current in the red, yellow and blue phases of a transmission system are: $I_{red} = 1248\angle -15°$A, $I_{yellow} = 1120\angle -135°$A and $I_{blue} = 1310\angle 95°$A. Determine the current in the neutral cable if the sum of the currents flows through it.
9. When a mass is attached to a spring and oscillates in a vertical plane, the displacement, s, at time t is given by:
   $$s = (75\sin\omega t + 40\cos\omega t)\ \text{mm}$$
   Express this in the form $s = A\sin(\omega t + \phi)$
10. The single waveform resulting from the addition of the alternating voltages $9\sin 120t$ and $12\cos 120t$ volts is shown on an oscilloscope. Calculate the amplitude of the waveform and its phase angle measured with respect to $9\sin 120t$ and expressed in both radians and degrees.

Practice Exercise 178 Multiple-choice questions on methods of adding alternating waveforms (Answers on page 720)

Each question has only one correct answer

Questions 1 and 2 relate to the following information.

Two alternating voltages are given by: $v_1 = 2\sin\omega t$ and $v_2 = 3\sin\left(\omega t + \frac{\pi}{4}\right)$ volts.

1. Which of the phasor diagrams shown in Figure 35.24 represents $v_R = v_1 + v_2$?
(a) (i) (b) (ii) (c) (iii) (d) (iv)

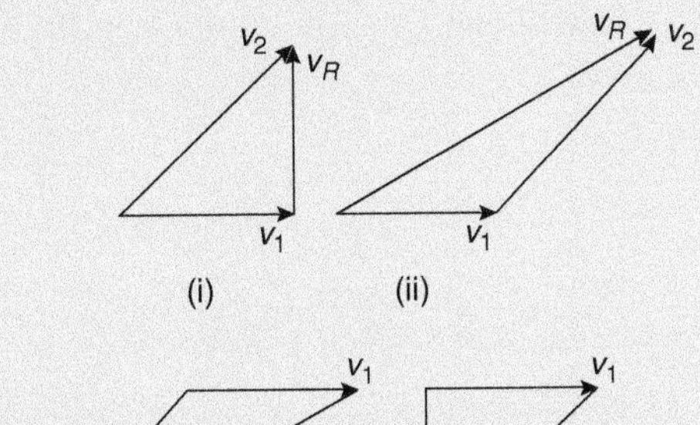

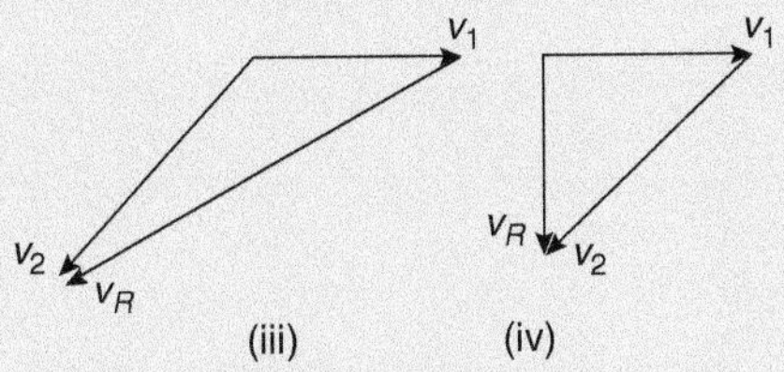

Figure 35.24

2. Which of the phasor diagrams shown represents $v_R = v_1 - v_2$?
(a) (i) (b) (ii) (c) (iii) (d) (iv)

3. Two alternating currents are given by $i_1 = 4\sin\omega t$ amperes and $i_2 = 3\sin\left(\omega t + \frac{\pi}{6}\right)$ amperes. The sinusoidal expression for the resultant $i_1 + i_2$ is:
(a) $5\sin(\omega t + 30)A$
(b) $5\sin(\omega t + \frac{\pi}{6})A$
(c) $6.77\sin(\omega t + 0.22)A$
(d) $6.77\sin(\omega t + 12.81)A$

4. Two alternating voltages are given by $v_1 = 5\sin\omega t$ volts and $v_2 = 12\sin\left(\omega t - \frac{\pi}{3}\right)$ volts. The sinusoidal expression for the resultant $v_1 + v_2$ is:
(a) $13\sin(\omega t - 60)V$
(b) $15.13\sin(\omega t - 0.76)V$
(c) $13\sin(\omega t - 1.05)V$
(d) $15.13\sin(\omega t - 43.37)V$

5. Two alternating currents are given by $i_1 = 3\sin\omega t$ amperes and $i_2 = 4\sin\left(\omega t + \frac{\pi}{4}\right)$ amperes. The sinusoidal expression for the resultant $i_1 - i_2$ is:
(a) $2.83\sin(\omega t - 1.51)A$
(b) $6.48\sin(\omega t + 0.45)A$
(c) $1.00\sin(\omega t - 176.96)$
(d) $6.48\sin(\omega t - 25.89)A$

For fully worked solutions to each of the problems in Practice Exercises 173 to 177 in this chapter, go to the website:
www.routledge.com/cw/bird

COMPANION @ WEBSITE

Revision Test 9 Complex numbers and vectors

This Revision Test covers the material contained in Chapters 32 to 35. *The marks for each question are shown in brackets at the end of each question.*

1. Solve the quadratic equation $x^2 - 2x + 5 = 0$ and show the roots on an Argand diagram. (8)

2. If $Z_1 = 2 + j5$, $Z_2 = 1 - j3$ and $Z_3 = 4 - j$ determine, in both Cartesian and polar forms, the value of: $\frac{Z_1 Z_2}{Z_1 + Z_2} + Z_3$, correct to 2 decimal places. (8)

3. Three vectors are represented by $\boldsymbol{A}$, $4.2\angle 45°$, $\boldsymbol{B}$, $5.5\angle -32°$ and $\boldsymbol{C}$, $2.8\angle 75°$. Determine in polar form the resultant $\boldsymbol{D}$, where $\boldsymbol{D} = \boldsymbol{B} + \boldsymbol{C}\ \boldsymbol{A}$. (8)

4. Two impedances, $Z_1 = (2 + j7)$ ohms and $Z_2 = (3 + j4)$ ohms are connected in series to a supply voltage V of $150\angle 0°$ V. Determine the magnitude of the current I and its phase angle relative to the voltage. (6)

5. Determine in both polar and rectangular forms:
   (a) $[2.37\angle 35°]^4$ (b) $[3.2 - j4.8]^5$ (c) $\sqrt{-1 - j3}$ (15)

6. State whether the following are scalar or a vector quantities:
   (a) A temperature of 50°C
   (b) A downward force of 80 N
   (c) 300 J of work
   (d) A south-westerly wind of 15 knots
   (e) 70 m distance
   (f) An acceleration of $25\,\text{m/s}^2$ at 30° to the horizontal (6)

7. Calculate the resultant and direction of the force vectors shown in Fig. RT9.1, correct to 2 decimal places. (7)

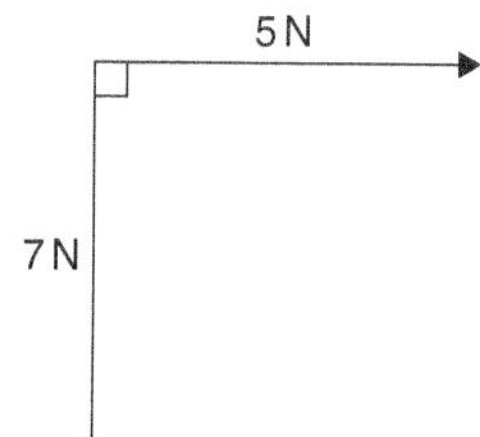

Figure RT9.1

8. Four coplanar forces act at a point A as shown in Fig. RT9.2. Determine the value and direction of the resultant force by (a) drawing (b) by calculation. (11)

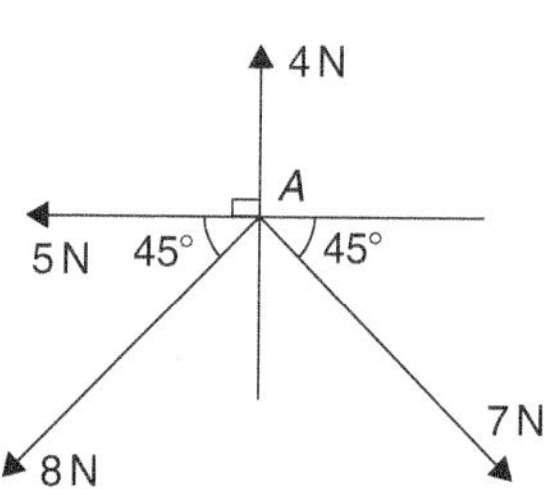

Figure RT9.2

9. The instantaneous values of two alternating voltages are given by:

$$v_1 = 150\sin\left(\omega t + \frac{\pi}{3}\right) \text{ volts and}$$

$$v_2 = 90\sin\left(\omega t - \frac{\pi}{6}\right) \text{ volts}$$

Plot the two voltages on the same axes to scales of 1 cm = 50 volts and 1 cm = $\frac{\pi}{6}$ rad. Obtain a sinusoidal expression for the resultant $v_1 + v_2$ in the form $R\sin(\omega t + \alpha)$: (a) by adding ordinates at intervals and (b) by calculation. (13)

10. If velocity $v_1 = 26$ m/s at 52° and $v_2 = 17$ m/s at −28° calculate the magnitude and direction of $v_1 + v_2$, correct to 2 decimal places, using complex numbers. (10)

11. Given $\boldsymbol{a} = -3\boldsymbol{i} + 3\boldsymbol{j} + 5\boldsymbol{k}$, $\boldsymbol{b} = 2\boldsymbol{i} - 5\boldsymbol{j} + 7\boldsymbol{k}$ and $\boldsymbol{c} = 3\boldsymbol{i} + 6\boldsymbol{j} - 4\boldsymbol{k}$, determine the following:
    (i) $-4\boldsymbol{b}$
    (ii) $\boldsymbol{a} + \boldsymbol{b} - \boldsymbol{c}$
    (iii) $5\boldsymbol{b} - 3\boldsymbol{c}$ (8)

COMPANION WEBSITE

For lecturers/instructors/teachers, fully worked solutions to each of the problems in Revision Test 9, together with a full marking scheme, are available at the website:

www.routledge.com/cw/bird

Section 7

Differential calculus

Chapter 36

Introduction to differentiation

Why it is important to understand: **Introduction to differentiation**

There are many practical situations engineers have to analyse which involve quantities that are varying. Typical examples include the stress in a loaded beam, the temperature of an industrial chemical, the rate at which the speed of a vehicle is increasing or decreasing, the current in an electrical circuit or the torque on a turbine blade. Differential calculus, or differentiation, is a mathematical technique for analysing the way in which functions change. There are many methods and rules of differentiation which are individually covered in the following chapters. A good knowledge of algebra, in particular, laws of indices, is essential. This chapter explains how to differentiate the five most common functions, providing an important base for future chapters.

At the end of this chapter, you should be able to:

- state that calculus comprises two parts - differential and integral calculus
- understand functional notation
- describe the gradient of a curve and limiting value
- differentiate simple functions from first principles
- differentiate $y = ax^n$ by the general rule
- differentiate sine and cosine functions
- differentiate exponential and logarithmic functions

36.1 Introduction to calculus

Calculus is a branch of mathematics involving or leading to calculations dealing with continuously varying functions.
Calculus is a subject that falls into two parts:

(i) **differential calculus** (or **differentiation**) and

(ii) **integral calculus** (or **integration**).

Differentiation is used in calculations involving velocity and acceleration, rates of change and maximum and minimum values of curves.

36.2 Functional notation

In an equation such as $y = 3x^2 + 2x - 5$, y is said to be a function of x and may be written as $y = f(x)$.

An equation written in the form $f(x) = 3x^2 + 2x - 5$ is termed **functional notation**. The value of $f(x)$ when $x = 0$ is denoted by $f(0)$, and the value of $f(x)$ when $x = 2$ is denoted by $f(2)$ and so on. Thus when $f(x) = 3x^2 + 2x - 5$, then

$$f(0) = 3(0)^2 + 2(0) - 5 = -5$$

$$\text{and} \quad f(2) = 3(2)^2 + 2(2) - 5 = 11 \text{ and so on.}$$

Problem 1. If $f(x) = 4x^2 - 3x + 2$ find: $f(0), f(3), f(-1)$ and $f(3) - f(-1)$

$$f(x) = 4x^2 - 3x + 2$$

$$f(0) = 4(0)^2 - 3(0) + 2 = \mathbf{2}$$

$$f(3) = 4(3)^2 - 3(3) + 2 = 36 - 9 + 2 = \mathbf{29}$$

$$f(-1) = 4(-1)^2 - 3(-1) + 2 = 4 + 3 + 2 = \mathbf{9}$$

$$f(3) - f(-1) = 29 - 9 = \mathbf{20}$$

Problem 2. Given that $f(x) = 5x^2 + x - 7$ determine:
(i) $f(2) \div f(1)$ (iii) $f(3+a) - f(3)$
(ii) $f(3+a)$ (iv) $\dfrac{f(3+a) - f(3)}{a}$

$$f(x) = 5x^2 + x - 7$$

(i) $f(2) = 5(2)^2 + 2 - 7 = 15$

$f(1) = 5(1)^2 + 1 - 7 = -1$

$$f(2) \div f(1) = \frac{15}{-1} = \mathbf{-15}$$

(ii) $f(3+a) = 5(3+a)^2 + (3+a) - 7$

$= 5(9 + 6a + a^2) + (3+a) - 7$

$= 45 + 30a + 5a^2 + 3 + a - 7$

$= \mathbf{41 + 31a + 5a^2}$

(iii) $f(3) = 5(3)^2 + 3 - 7 = 41$

$f(3+a) - f(3) = (41 + 31a + 5a^2) - (41)$

$= \mathbf{31a + 5a^2}$

(iv) $\dfrac{f(3+a) - f(3)}{a} = \dfrac{31a + 5a^2}{a} = \mathbf{31 + 5a}$

Now try the following Practice Exercise

Practice Exercise 179 Functional notation (Answers on page 720)

1. If $f(x) = 6x^2 - 2x + 1$ find $f(0)$, $f(1)$, $f(2)$, $f(-1)$ and $f(-3)$
2. If $f(x) = 2x^2 + 5x - 7$ find $f(1), f(2), f(-1)$, $f(2) - f(-1)$
3. Given $f(x) = 3x^3 + 2x^2 - 3x + 2$ prove that $f(1) = \frac{1}{7}f(2)$
4. If $f(x) = -x^2 + 3x + 6$ find $f(2)$, $f(2+a)$, $f(2+a) - f(2)$ and $\dfrac{f(2+a) - f(2)}{a}$

36.3 The gradient of a curve

(a) If a tangent is drawn at a point P on a curve, then the gradient of this tangent is said to be the **gradient of the curve** at P. In Fig. 36.1, the gradient of the curve at P is equal to the gradient of the tangent PQ.

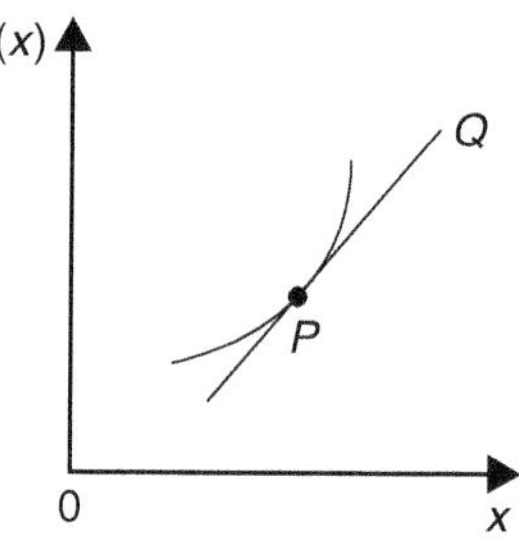

Figure 36.1

(b) For the curve shown in Fig. 36.2, let the points A and B have co-ordinates (x_1, y_1) and (x_2, y_2), respectively. In functional notation, $y_1 = f(x_1)$ and $y_2 = f(x_2)$ as shown.

The gradient of the chord AB

$$= \frac{BC}{AC} = \frac{BD - CD}{ED} = \frac{f(x_2) - f(x_1)}{(x_2 - x_1)}$$

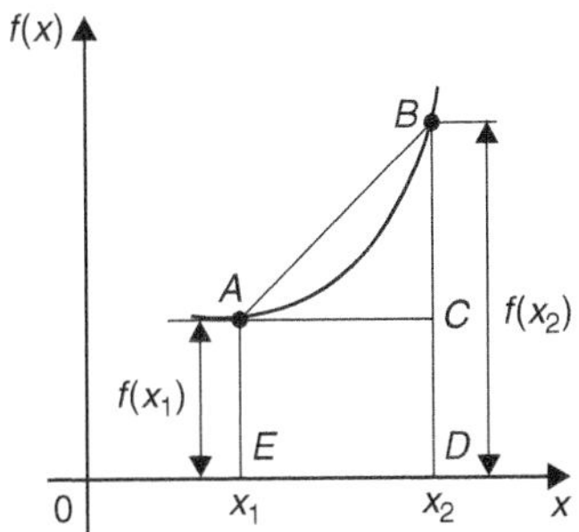

Figure 36.2

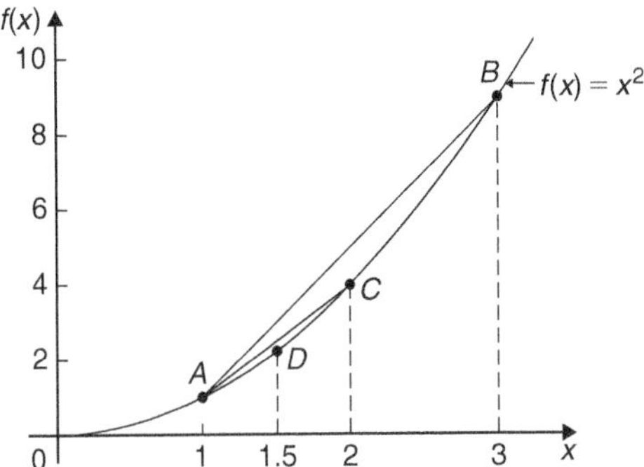

Figure 36.3

(c) For the curve $f(x) = x^2$ shown in Fig. 36.3:

(i) the gradient of chord AB

$$= \frac{f(3) - f(1)}{3 - 1} = \frac{9 - 1}{2} = 4$$

(ii) the gradient of chord AC

$$= \frac{f(2) - f(1)}{2 - 1} = \frac{4 - 1}{1} = 3$$

(iii) the gradient of chord AD

$$= \frac{f(1.5) - f(1)}{1.5 - 1} = \frac{2.25 - 1}{0.5} = 2.5$$

(iv) if E is the point on the curve $(1.1, f(1.1))$ then the gradient of chord AE

$$= \frac{f(1.1) - f(1)}{1.1 - 1}$$

$$= \frac{1.21 - 1}{0.1} = 2.1$$

(v) if F is the point on the curve $(1.01, f(1.01))$ then the gradient of chord AF

$$= \frac{f(1.01) - f(1)}{1.01 - 1}$$

$$= \frac{1.0201 - 1}{0.01} = 2.01$$

Thus as point B moves closer and closer to point A the gradient of the chord approaches nearer and nearer to the value 2. This is called the **limiting value** of the gradient of the chord AB and when B coincides with A the chord becomes the tangent to the curve.

Now try the following Practice Exercise

Practice Exercise 180 The gradient of a curve (Answers on page 720)

1. Plot the curve $f(x) = 4x^2 - 1$ for values of x from $x = -1$ to $x = +4$. Label the co-ordinates $(3, f(3))$ and $(1, f(1))$ as J and K, respectively. Join points J and K to form the chord JK. Determine the gradient of chord JK. By moving J nearer and nearer to K determine the gradient of the tangent of the curve at K.

36.4 Differentiation from first principles

(i) In Fig. 36.4, A and B are two points very close together on a curve, δx (delta x) and δy (delta y) representing small increments in the x and y directions, respectively.

Gradient of chord $AB = \dfrac{\delta y}{\delta x}$

However, $\delta y = f(x + \delta x) - f(x)$

Hence $\dfrac{\delta y}{\delta x} = \dfrac{f(x + \delta x) - f(x)}{\delta x}$

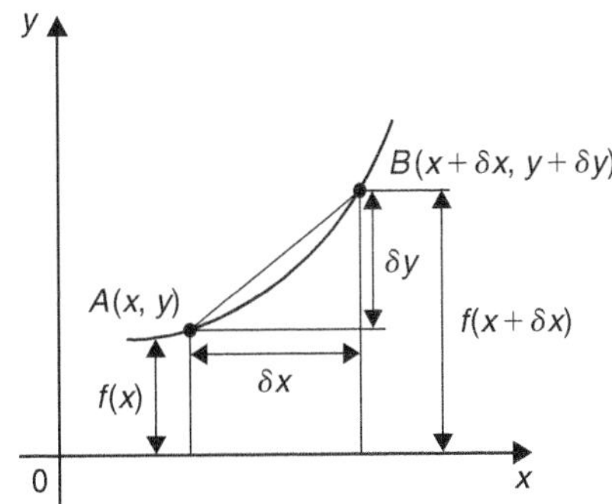

Figure 36.4

As δx approaches zero, $\dfrac{\delta y}{\delta x}$ approaches a limiting value and the gradient of the chord approaches the gradient of the tangent at A.

(ii) When determining the gradient of a tangent to a curve there are two notations used. The gradient of the curve at A in Fig. 36.4 can either be written as:

$$\lim_{\delta x \to 0} \frac{\delta y}{\delta x} \text{ or } \lim_{\delta x \to 0}\left\{\frac{f(x+\delta x)-f(x)}{\delta x}\right\}$$

In **Leibniz* notation**, $\frac{dy}{dx} = \lim_{\delta x \to 0} \frac{\delta y}{\delta x}$

In **functional notation**,

$$f'(x) = \lim_{\delta x \to 0}\left\{\frac{f(x+\delta x)-f(x)}{\delta x}\right\}$$

(iii) $\frac{dy}{dx}$ is the same as $f'(x)$ and is called the **differential coefficient** or the **derivative**. The process of finding the differential coefficient is called **differentiation**.

*Who was **Leibniz**? – **Gottfried Wilhelm Leibniz** (sometimes von Leibniz) (1 July, 1646 – 14 November, 1716) was a German mathematician and philosopher. Leibniz developed infinitesimal calculus and invented the Leibniz wheel. To find out more go to www.routledge.com/cw/bird

Summarising, the differential coefficient,

$$\frac{dy}{dx} = f'(x) = \lim_{\delta x \to 0} \frac{\delta y}{\delta x} = \lim_{\delta x \to 0}\left\{\frac{f(x+\delta x)-f(x)}{\delta x}\right\}$$

Problem 3. Differentiate from first principles $f(x) = x^2$ and determine the value of the gradient of the curve at $x = 2$

To 'differentiate from first principles' means 'to find $f'(x)$' by using the expression

$$f'(x) = \lim_{\delta x \to 0}\left\{\frac{f(x+\delta x)-f(x)}{\delta x}\right\}$$

$$f(x) = x^2$$

Substituting $(x+\delta x)$ for x gives $f(x+\delta x) = (x+\delta x)^2 = x^2 + 2x\delta x + \delta x^2$, hence

$$f'(x) = \lim_{\delta x \to 0}\left\{\frac{(x^2+2x\delta x+\delta x^2)-(x^2)}{\delta x}\right\}$$

$$= \lim_{\delta x \to 0}\left\{\frac{2x\delta x+\delta x^2}{\delta x}\right\} = \lim_{\delta x \to 0}\{2x+\delta x\}$$

As $\delta x \to 0$, $[2x+\delta x] \to [2x+0]$. Thus $f'(x) = \mathbf{2x}$, i.e. the differential coefficient of x^2 is $2x$. At $x = 2$, the gradient of the curve, $f'(x) = 2(2) = \mathbf{4}$

Problem 4. Find the differential coefficient of $y = 5x$

By definition, $\frac{dy}{dx} = f'(x)$

$$= \lim_{\delta x \to 0}\left\{\frac{f(x+\delta x)-f(x)}{\delta x}\right\}$$

The function being differentiated is $y = f(x) = 5x$. Substituting $(x+\delta x)$ for x gives:

$f(x+\delta x) = 5(x+\delta x) = 5x + 5\delta x$. Hence

$$\frac{dy}{dx} = f'(x) = \lim_{\delta x \to 0}\left\{\frac{(5x+5\delta x)-(5x)}{\delta x}\right\}$$

$$= \lim_{\delta x \to 0}\left\{\frac{5\delta x}{\delta x}\right\} = \lim_{\delta x \to 0}\{5\}$$

Since the term δx does not appear in [5] the limiting value as $\delta x \to 0$ of [5] is 5. Thus $\frac{dy}{dx} = \mathbf{5}$, i.e. the differential coefficient of $5x$ is 5. The equation $y = 5x$

represents a straight line of gradient 5 (see Chapter 27). The 'differential coefficient' (i.e. $\frac{dy}{dx}$ or $f'(x)$) means 'the gradient of the curve', and since the slope of the line $y = 5x$ is 5 this result can be obtained by inspection. Hence, in general, if $y = kx$ (where k is a constant), then the gradient of the line is k and $\frac{dy}{dx}$ or $f'(x) = k$.

Problem 5. Find the derivative of $y = 8$

$y = f(x) = 8$. Since there are no x-values in the original equation, substituting $(x+\delta x)$ for x still gives $f(x+\delta x) = 8$. Hence

$$\frac{dy}{dx} = f'(x) = \underset{\delta x \to 0}{\text{limit}}\left\{\frac{f(x+\delta x) - f(x)}{\delta x}\right\}$$

$$= \underset{\delta x \to 0}{\text{limit}}\left\{\frac{8-8}{\delta x}\right\} = 0$$

Thus, when $y = 8$, $\boldsymbol{\frac{dy}{dx} = 0}$

The equation $y = 8$ represents a straight horizontal line and the gradient of a horizontal line is zero, hence the result could have been determined by inspection. 'Finding the derivative' means 'finding the gradient', hence, in general, for any horizontal line if $y = k$ (where k is a constant) then $\frac{dy}{dx} = 0$

Problem 6. Differentiate from first principles $f(x) = 2x^3$

Substituting $(x+\delta x)$ for x gives

$$f(x+\delta x) = 2(x+\delta x)^3$$
$$= 2(x+\delta x)(x^2 + 2x\delta x + \delta x^2)$$
$$= 2(x^3 + 3x^2\delta x + 3x\delta x^2 + \delta x^3)$$
$$= 2x^3 + 6x^2\delta x + 6x\delta x^2 + 2\delta x^3$$

$$\frac{dy}{dx} = f'(x) = \underset{\delta x \to 0}{\text{limit}}\left\{\frac{f(x+\delta x) - f(x)}{\delta x}\right\}$$

$$= \underset{\delta x \to 0}{\text{limit}}\left\{\frac{(2x^3 + 6x^2\delta x + 6x\delta x^2 + 2\delta x^3) - (2x^3)}{\delta x}\right\}$$

$$= \underset{\delta x \to 0}{\text{limit}}\left\{\frac{6x^2\delta x + 6x\delta x^2 + 2\delta x^3}{\delta x}\right\}$$

$$= \underset{\delta x \to 0}{\text{limit}}\{6x^2 + 6x\delta x + 2\delta x^2\}$$

Hence $\boldsymbol{f'(x) = 6x^2}$, i.e. the differential coefficient of $2x^3$ is $6x^2$

Problem 7. Find the differential coefficient of $y = 4x^2 + 5x - 3$ and determine the gradient of the curve at $x = -3$

$$y = f(x) = 4x^2 + 5x - 3$$
$$f(x+\delta x) = 4(x+\delta x)^2 + 5(x+\delta x) - 3$$
$$= 4(x^2 + 2x\delta x + \delta x^2) + 5x + 5\delta x - 3$$
$$= 4x^2 + 8x\delta x + 4\delta x^2 + 5x + 5\delta x - 3$$

$$\frac{dy}{dx} = f'(x) = \underset{\delta x \to 0}{\text{limit}}\left\{\frac{f(x+\delta x) - f(x)}{\delta x}\right\}$$

$$= \underset{\delta x \to 0}{\text{limit}}\left\{\frac{(4x^2 + 8x\delta x + 4\delta x^2 + 5x + 5\delta x - 3) - (4x^2 + 5x - 3)}{\delta x}\right\}$$

$$= \underset{\delta x \to 0}{\text{limit}}\left\{\frac{8x\delta x + 4\delta x^2 + 5\delta x}{\delta x}\right\}$$

$$= \underset{\delta x \to 0}{\text{limit}}\{8x + 4\delta x + 5\}$$

i.e. $\boldsymbol{\frac{dy}{dx} = f'(x) = 8x + 5}$

At $x = -3$, the gradient of the curve

$$= \frac{dy}{dx} = f'(x) = 8(-3) + 5 = \mathbf{-19}$$

Differentiation from first principles can be a lengthy process and it would not be convenient to go through this procedure every time we want to differentiate a function. In reality we do not have to, because a set of general rules have evolved from the above procedure, which we consider in the following section.

Now try the following Practice Exercise

Practice Exercise 181 Differentiation from first principles (Answers on page 720)

In Problems 1 to 12, differentiate from first principles.

1. $y = x$
2. $y = 7x$
3. $y = 4x^2$

4. $y = 5x^3$
5. $y = -2x^2 + 3x - 12$
6. $y = 23$
7. $f(x) = 9x$
8. $f(x) = \dfrac{2x}{3}$
9. $f(x) = 9x^2$
10. $f(x) = -7x^3$
11. $f(x) = x^2 + 15x - 4$
12. $f(x) = 4$
13. Determine $\dfrac{d}{dx}(4x^3)$ from first principles
14. Find $\dfrac{d}{dx}(3x^2 + 5)$ from first principles

36.5 Differentiation of $y = ax^n$ by the general rule

From differentiation by first principles, a general rule for differentiating ax^n emerges where a and n are any constants. This rule is:

$$\text{if } y = ax^n \textbf{ then } \frac{dy}{dx} = anx^{n-1}$$

$$\text{or, if } f(x) = ax^n \textbf{ then } f'(x) = anx^{n-1}$$

(Each of the results obtained in worked problems 3 to 7 may be deduced by using this general rule.)
When differentiating, results can be expressed in a number of ways.

For example:

(i) if $y = 3x^2$ then $\dfrac{dy}{dx} = 6x$,

(ii) if $f(x) = 3x^2$ then $f'(x) = 6x$,

(iii) the differential coefficient of $3x^2$ is $6x$,

(iv) the derivative of $3x^2$ is $6x$, and

(v) $\dfrac{d}{dx}(3x^2) = 6x$

Problem 8. Using the general rule, differentiate the following with respect to x:
(a) $y = 5x^7$ (b) $y = 3\sqrt{x}$ (c) $y = \dfrac{4}{x^2}$

(a) Comparing $y = 5x^7$ with $y = ax^n$ shows that $a = 5$ and $n = 7$. Using the general rule,

$$\frac{dy}{dx} = anx^{n-1} = (5)(7)x^{7-1} = \mathbf{35x^6}$$

(b) $y = 3\sqrt{x} = 3x^{\frac{1}{2}}$. Hence $a = 3$ and $n = \dfrac{1}{2}$

$$\frac{dy}{dx} = anx^{n-1} = (3)\frac{1}{2}x^{\frac{1}{2}-1}$$

$$= \frac{3}{2}x^{-\frac{1}{2}} = \frac{3}{2x^{\frac{1}{2}}} = \mathbf{\frac{3}{2\sqrt{x}}}$$

(c) $y = \dfrac{4}{x^2} = 4x^{-2}$. Hence $a = 4$ and $n = -2$

$$\frac{dy}{dx} = anx^{n-1} = (4)(-2)x^{-2-1}$$

$$= -8x^{-3} = \mathbf{-\frac{8}{x^3}}$$

Problem 9. Find the differential coefficient of $y = \dfrac{2}{5}x^3 - \dfrac{4}{x^3} + 4\sqrt{x^5} + 7$

$$y = \frac{2}{5}x^3 - \frac{4}{x^3} + 4\sqrt{x^5} + 7$$

i.e. $$y = \frac{2}{5}x^3 - 4x^{-3} + 4x^{5/2} + 7$$

$$\frac{dy}{dx} = \left(\frac{2}{5}\right)(3)x^{3-1} - (4)(-3)x^{-3-1}$$

$$+ (4)\left(\frac{5}{2}\right)x^{(5/2)-1} + 0$$

$$= \frac{6}{5}x^2 + 12x^{-4} + 10x^{3/2}$$

i.e. $$\mathbf{\frac{dy}{dx} = \frac{6}{5}x^2 + \frac{12}{x^4} + 10\sqrt{x^3}}$$

Problem 10. If $f(t) = 5t + \dfrac{1}{\sqrt{t^3}}$ find $f'(t)$

$$f(t) = 5t + \frac{1}{\sqrt{t^3}} = 5t + \frac{1}{t^{\frac{3}{2}}} = 5t^1 + t^{-\frac{3}{2}}$$

Hence $$f'(t) = (5)(1)t^{1-1} + \left(-\frac{3}{2}\right)t^{-\frac{3}{2}-1}$$

$$= 5t^0 - \frac{3}{2}t^{-\frac{5}{2}}$$

i.e. $$f'(t) = 5 - \frac{3}{2t^{\frac{5}{2}}} = \mathbf{5 - \frac{3}{2\sqrt{t^5}}}$$

Problem 11. Differentiate $y = \dfrac{(x+2)^2}{x}$ with respect to x

$$y = \frac{(x+2)^2}{x} = \frac{x^2+4x+4}{x}$$

$$= \frac{x^2}{x} + \frac{4x}{x} + \frac{4}{x}$$

i.e. $$y = x + 4 + 4x^{-1}$$

Hence $$\frac{dy}{dx} = 1 + 0 + (4)(-1)x^{-1-1}$$

$$= 1 - 4x^{-2} = \mathbf{1 - \frac{4}{x^2}}$$

Now try the following Practice Exercise

Practice Exercise 182 Differentiation of $y = ax^n$ by the general rule (Answers on page 721)

In Problems 1 to 8, determine the differential coefficients with respect to the variable.

1. $y = 7x^4$
2. $y = \sqrt{x}$
3. $y = \sqrt{t^3}$
4. $y = 6 + \dfrac{1}{x^3}$
5. $y = 3x - \dfrac{1}{\sqrt{x}} + \dfrac{1}{x}$
6. $y = \dfrac{5}{x^2} - \dfrac{1}{\sqrt{x^7}} + 2$
7. $y = 3(t-2)^2$
8. $y = (x+1)^3$
9. Using the general rule for ax^n check the results of Problems 1 to 12 of Exercise 181, page 401
10. Differentiate $f(x) = 6x^2 - 3x + 5$ and find the gradient of the curve at (a) $x = -1$, and (b) $x = 2$
11. Find the differential coefficient of $y = 2x^3 + 3x^2 - 4x - 1$ and determine the gradient of the curve at $x = 2$
12. Determine the derivative of $y = -2x^3 + 4x + 7$ and determine the gradient of the curve at $x = -1.5$

36.6 Differentiation of sine and cosine functions

Figure 36.5(a) shows a graph of $y = \sin\theta$. The gradient is continually changing as the curve moves from O to A to B to C to D. The gradient, given by $\dfrac{dy}{d\theta}$, may be plotted in a corresponding position below $y = \sin\theta$, as shown in Fig. 36.5(b).

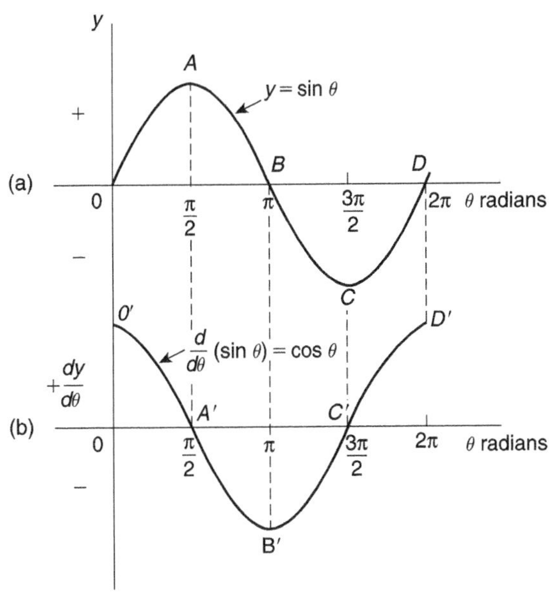

Figure 36.5

(i) At 0, the gradient is positive and is at its steepest. Hence 0′ is the maximum positive value.

(ii) Between 0 and A the gradient is positive but is decreasing in value until at A the gradient is zero, shown as A'.

(iii) Between A and B the gradient is negative but is increasing in value until at B the gradient is at its steepest. Hence B' is a maximum negative value.

(iv) If the gradient of $y = \sin\theta$ is further investigated between B and C and C and D then the resulting graph of $\dfrac{dy}{d\theta}$ is seen to be a cosine wave.

Hence the rate of change of $\sin\theta$ is $\cos\theta$, i.e.

if $\quad y=\sin\theta$ **then** $\dfrac{dy}{d\theta}=\cos\theta$

It may also be shown that:

if $\quad y=\sin a\theta,\ \dfrac{dy}{d\theta}=a\cos a\theta$

(where a is a constant)

and if $\quad y=\sin(a\theta+\alpha),\ \dfrac{dy}{d\theta}=a\cos(a\theta+\alpha)$

(where a and α are constants).

If a similar exercise is followed for $y=\cos\theta$ then the graphs of Fig. 36.6 result, showing $\dfrac{dy}{d\theta}$ to be a graph of $\sin\theta$, but displaced by π radians. If each point on the curve $y=\sin\theta$ (as shown in Fig. 36.5(a)) were to be made negative, (i.e. $+\dfrac{\pi}{2}$ is made $-\dfrac{\pi}{2}$, $-\dfrac{3\pi}{2}$ is made $+\dfrac{3\pi}{2}$ and so on) then the graph shown in Fig. 36.6(b) would result. This latter graph therefore represents the curve of $-\sin\theta$.

Thus, if $y=\cos\theta,\ \dfrac{dy}{d\theta}=-\sin\theta$

It may also be shown that:

if $\quad y=\cos a\theta,\ \dfrac{dy}{d\theta}=-a\sin a\theta$

(where a is a constant)

and if $\quad y=\cos(a\theta+\alpha),\ \dfrac{dy}{d\theta}=-a\sin(a\theta+\alpha)$

(where a and α are constants).

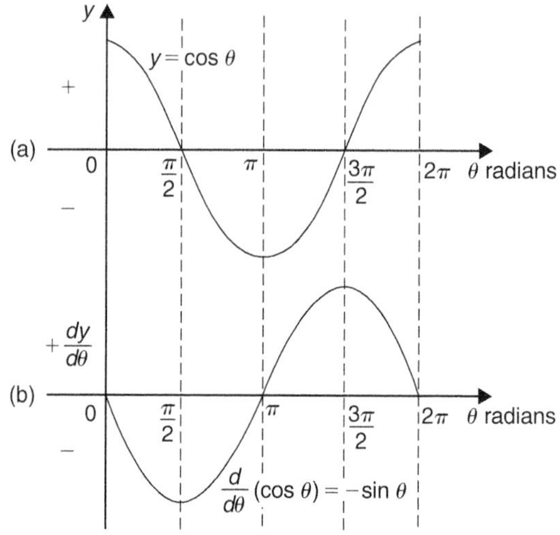

Figure 36.6

Problem 12. Differentiate the following with respect to the variable: (a) $y=2\sin 5\theta$ (b) $f(t)=3\cos 2t$

(a) $y=2\sin 5\theta$

$\dfrac{dy}{d\theta}=(2)(5)\cos 5\theta=\mathbf{10\cos 5\theta}$

(b) $f(t)=3\cos 2t$

$f'(t)=(3)(-2)\sin 2t=\mathbf{-6\sin 2t}$

Problem 13. Find the differential coefficient of $y=7\sin 2x-3\cos 4x$

$$y=7\sin 2x-3\cos 4x$$

$$\frac{dy}{dx}=(7)(2)\cos 2x-(3)(-4)\sin 4x$$

$$=\mathbf{14\cos 2x+12\sin 4x}$$

Problem 14. Differentiate the following with respect to the variable:

(a) $f(\theta)=5\sin(100\pi\theta-0.40)$
(b) $f(t)=2\cos(5t+0.20)$

(a) If $f(\theta)=5\sin(100\pi\theta-0.40)$

$f'(\theta)=5[100\pi\cos(100\pi\theta-0.40)]$

$=\mathbf{500\pi\cos(100\pi\theta-0.40)}$

(b) If $f(t)=2\cos(5t+0.20)$

$f'(t)=2[-5\sin(5t+0.20)]$

$=\mathbf{-10\sin(5t+0.20)}$

Problem 15. An alternating voltage is given by: $v=100\sin 200t$ volts, where t is the time in seconds. Calculate the rate of change of voltage when (a) $t=0.005$ s and (b) $t=0.01$ s

$v=100\sin 200t$ volts. The rate of change of v is given by $\dfrac{dv}{dt}$

$$\frac{dv}{dt}=(100)(200)\cos 200t=20000\cos 200t$$

(a) When $t=0.005$ s,

$\dfrac{dv}{dt}=20\,000\cos(200)(0.005)=20\,000\cos 1$

cos 1 means 'the cosine of 1 radian' (make sure your calculator is on radians — not degrees).

Hence $\frac{dv}{dt}$ = **10 806 volts per second**

(b) When $t = 0.01$ s,

$\frac{dv}{dt} = 20\,000\cos(200)(0.01) = 20\,000\cos 2.$

Hence $\frac{dv}{dt}$ = **−8323 volts per second**

Now try the following Practice Exercise

Practice Exercise 183 Differentiation of sine and cosine functions (Answers on page 721)

1. Differentiate with respect to x: (a) $y = 4\sin 3x$ (b) $y = 2\cos 6x$
2. Given $f(\theta) = 2\sin 3\theta - 5\cos 2\theta$, find $f'(\theta)$
3. An alternating current is given by $i = 5\sin 100t$ amperes, where t is the time in seconds. Determine the rate of change of current when $t = 0.01$ seconds
4. $v = 50\sin 40t$ volts represents an alternating voltage where t is the time in seconds. At a time of 20×10^{-3} seconds, find the rate of change of voltage
5. If $f(t) = 3\sin(4t + 0.12) - 2\cos(3t - 0.72)$ determine $f'(t)$

36.7 Differentiation of e^{ax} and ln ax

A graph of $y = e^x$ is shown in Fig. 36.7(a). The gradient of the curve at any point is given by $\frac{dy}{dx}$ and is continually changing. By drawing tangents to the curve at many points on the curve and measuring the gradient of the tangents, values of $\frac{dy}{dx}$ for corresponding values of x may be obtained. These values are shown graphically in Fig. 36.7(b). The graph of $\frac{dy}{dx}$ against x is identical to the original graph of $y = e^x$. It follows that:

$$\textbf{if } y = e^x, \textbf{ then } \frac{dy}{dx} = e^x$$

It may also be shown that

$$\textbf{if } y = e^{ax}, \textbf{ then } \frac{dy}{dx} = ae^{ax}$$

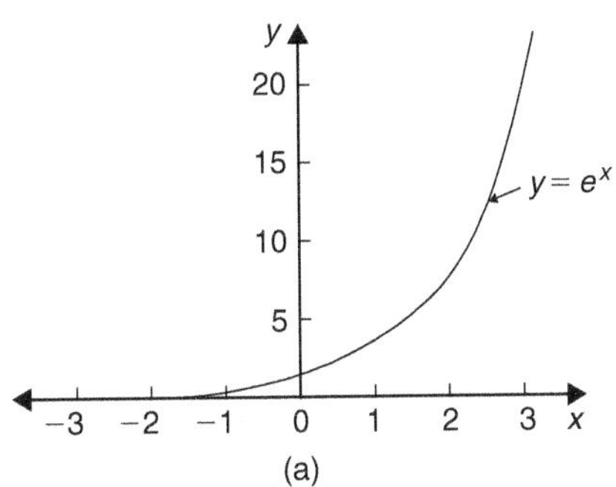

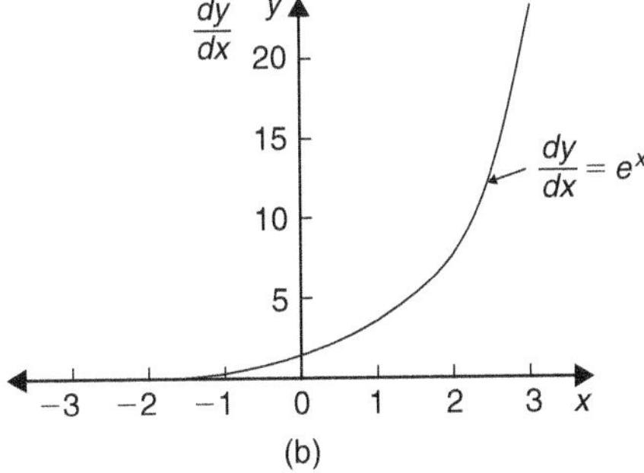

Figure 36.7

Therefore if $y = 2e^{6x}$, then $\frac{dy}{dx} = (2)(6e^{6x}) = \mathbf{12e^{6x}}$

A graph of $y = \ln x$ is shown in Fig. 36.8(a). The gradient of the curve at any point is given by $\frac{dy}{dx}$ and is continually changing. By drawing tangents to the curve at many points on the curve and measuring the gradient of the tangents, values of $\frac{dy}{dx}$ for corresponding values of x may be obtained. These values are shown graphically in Fig. 36.8(b). The graph of $\frac{dy}{dx}$ against x is the graph of $\frac{dy}{dx} = \frac{1}{x}$

It follows that: **if $y = \ln x$, then $\frac{dy}{dx} = \frac{1}{x}$**

It may also be shown that

$$\textbf{if } y = \ln ax, \textbf{ then } \frac{dy}{dx} = \frac{1}{x}$$

(Note that in the latter expression 'a' does not appear in the $\frac{dy}{dx}$ term).

Thus if $y = \ln 4x$, then $\frac{dy}{dx} = \mathbf{\frac{1}{x}}$

Problem 16. Differentiate the following with respect to the variable: (a) $y = 3e^{2x}$ (b) $f(t) = \frac{4}{3e^{5t}}$

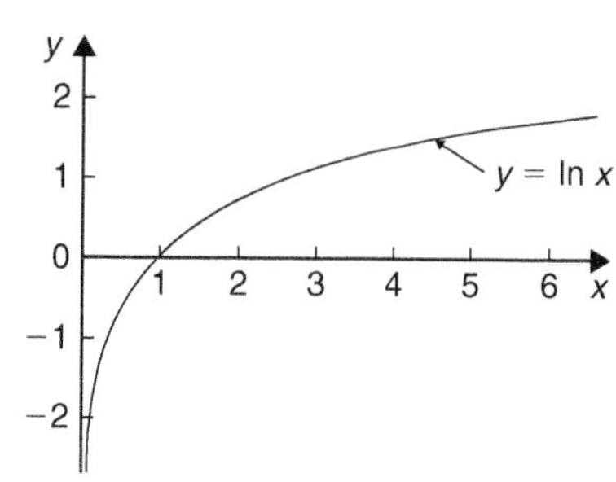

(a)

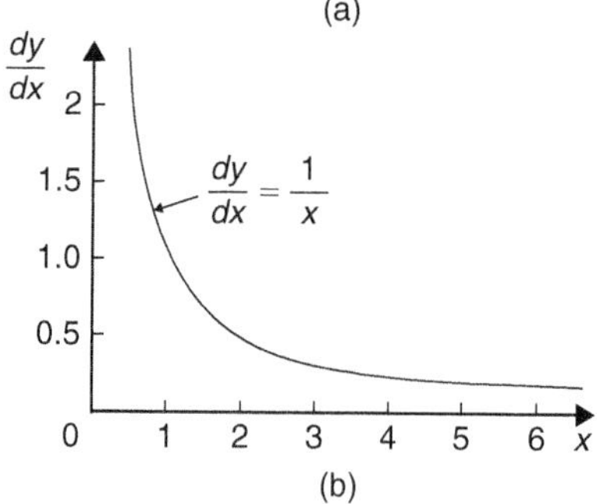

(b)

Figure 36.8

(a) If $y = 3e^{2x}$ then $\frac{dy}{dx} = (3)(2e^{2x}) = \mathbf{6e^{2x}}$

(b) If $f(t) = \frac{4}{3e^{5t}} = \frac{4}{3}e^{-5t}$, then

$$f'(t) = \frac{4}{3}(-5e^{-5t}) = -\frac{20}{3}e^{-5t} = -\frac{\mathbf{20}}{\mathbf{3e^{5t}}}$$

Problem 17. Differentiate $y = 5 \ln 3x$

If $y = 5 \ln 3x$, then $\frac{dy}{dx} = (5)\left(\frac{1}{x}\right) = \frac{\mathbf{5}}{\mathbf{x}}$

Now try the following Practice Exercise

Practice Exercise 184 Differentiation of e^{ax} and ln ax (Answers on page 721)

1. Differentiate with respect to x:
   (a) $y = 5e^{3x}$ (b) $y = \frac{2}{7e^{2x}}$

2. Given $f(\theta) = 5 \ln 2\theta - 4 \ln 3\theta$, determine $f'(\theta)$

3. If $f(t) = 4 \ln t + 2$, evaluate $f'(t)$ when $t = 0.25$

4. Evaluate $\frac{dy}{dx}$ when $x = 1$, given $y = 3e^{4x} - \frac{5}{2e^{3x}} + 8 \ln 5x$. Give the answer correct to 3 significant figures

Practice Exercise 185 Multiple-choice questions on the introduction to differentiation (Answers on page 721)

Each question has only one correct answer

1. Differentiating $y = 4x^5$ gives:
   (a) $\frac{dy}{dx} = \frac{2}{3}x^6$ (b) $\frac{dy}{dx} = 20x^4$
   (c) $\frac{dy}{dx} = 4x^6$ (d) $\frac{dy}{dx} = 5x^4$

2. Differentiating $i = 3\sin 2t - 2\cos 3t$ with respect to t gives:
   (a) $3\cos 2t + 2\sin 3t$ (b) $6(\sin 2t - \cos 3t)$
   (c) $\frac{3}{2}\cos 2t + \frac{2}{3}\sin 3t$ (d) $6(\cos 2t + \sin 3t)$

3. Given $y = 3e^x + 2\ln 3x$, $\frac{dy}{dx}$ is equal to:
   (a) $6e^x + \frac{2}{3x}$ (b) $3e^x + \frac{2}{x}$
   (c) $6e^x + \frac{2}{x}$ (d) $3e^x + \frac{2}{3}$

4. Given $f(t) = 3t^4 - 2$, $f'(t)$ is equal to:
   (a) $12t^3 - 2$ (b) $\frac{3}{4}t^5 - 2t + c$
   (c) $12t^3$ (d) $3t^5 - 2$

5. An alternating current is given by $i = 4\sin 150t$ amperes, where t is the time in seconds. The rate of change of current at $t = 0.025$s is:
   (a) −492.3 A/s (b) 3.99 A/s
   (c) −3.28 A/s (d) 598.7 A/s

6. If $y = 5x + \frac{1}{2x^3}$ then $\frac{dy}{dx}$ is equal to:
   (a) $5 + \frac{1}{6x^2}$ (b) $5 - \frac{3}{2x^4}$
   (c) $5 - \frac{6}{x^4}$ (d) $5x^2 - \frac{1}{6x^2}$

7. If y $=\sqrt{t^5}$ then $\dfrac{dy}{dt}$ is equal to:

(a) $-\dfrac{5}{2\sqrt{t^7}}$ (b) $\dfrac{2}{5\sqrt[5]{t^3}}$ (c) $\dfrac{5}{2}\sqrt{t^3}$ (d) $\dfrac{5}{2}t^2$

8. Given that $y=\cos 2x-3e^{-3x}+\ln 4x$ then $\dfrac{dy}{dx}$ is equal to:

(a) $2\sin 2x+3e^{-3x}+\dfrac{1}{4x}$

(b) $-2\sin 2x-3e^{-3x}+\dfrac{1}{x}$

(c) $2\sin 2x-3e^{-3x}+\dfrac{1}{4x}$

(d) $-2\sin 2x+3e^{-3x}+\dfrac{1}{x}$

9. Given that $y=2(x-1)^2$, then $\dfrac{dy}{dx}$ is equal to:

(a) $4(x-1)$
(b) $4x-1$
(c) $4x+4$
(d) $4x$

10. An alternating voltage is given by $v=10\sin 300t$ volts, where t is the time in seconds. The rate of change of voltage when $t=0.01$ s is:
(a) −2996 V/s (b) 157 V/s
(c) −2970 V/s (d) 0.523 V/s

For fully worked solutions to each of the problems in Practice Exercises 179 to 184 in this chapter, go to the website:
www.routledge.com/cw/bird

COMPANION @ WEBSITE

Chapter 37

Methods of differentiation

Why it is important to understand: Methods of differentiation

Calculus is one of the most powerful mathematical tools used by engineers; this chapter continues with explaining the basic techniques involved. As was mentioned in the last chapter, engineers have to analyse varying quantities; further such examples include the voltage on a transmission line, the rate of growth of a bacteriological culture and the rate at which the charge on a capacitor is changing. This chapter explains how to differentiate common functions, products, quotients and function of a function – all important methods providing a basis for further study in later chapters.

At the end of this chapter, you should be able to:

- differentiate common functions
- differentiate a product using the product rule
- differentiate a quotient using the quotient rule
- differentiate a function of a function
- differentiate successively

37.1 Differentiation of common functions

The **standard derivatives** summarised below were derived in Chapter 36 and are true for all real values of x.

y or $f(x)$	$\frac{dy}{dx}$ or $f'(x)$
ax^n	anx^{n-1}
$\sin ax$	$a\cos ax$
$\cos ax$	$-a\sin ax$
e^{ax}	ae^{ax}
$\ln ax$	$\frac{1}{x}$

The **differential coefficient of a sum or difference** is the sum or difference of the differential coefficients of the separate terms.

Thus, if $f(x)=p(x)+q(x)-r(x)$, (where f, p, q and r are functions), then $f'(x)=p'(x)+q'(x)-r'(x)$

Differentiation of common functions is demonstrated in the following worked problems.

Problem 1. Find the differential coefficients of:

(a) $y=12x^3$ (b) $y=\frac{12}{x^3}$

If $y=ax^n$ then $\frac{dy}{dx}=anx^{n-1}$

(a) Since $y=12x^3$, $a=12$ and $n=3$ thus $\frac{dy}{dx}=(12)(3)x^{3-1}=\mathbf{36x^2}$

(b) $y = \frac{12}{x^3}$ is rewritten in the standard ax^n form as $y = 12x^{-3}$ and in the general rule $a = 12$ and $n = -3$

Thus $\frac{dy}{dx} = (12)(-3)x^{-3-1}$

$= -36x^{-4} = -\frac{\mathbf{36}}{\mathbf{x^4}}$

Problem 2. Differentiate: (a) $y = 6$ (b) $y = 6x$

(a) $y = 6$ may be written as $y = 6x^0$, i.e. in the general rule $a = 6$ and $n = 0$.

Hence $\frac{dy}{dx} = (6)(0)x^{0-1} = \mathbf{0}$

In general, **the differential coefficient of a constant is always zero**.

(b) Since $y = 6x$, in the general rule $a = 6$ and $n = 1$

Hence $\frac{dy}{dx} = (6)(1)x^{1-1} = 6x^0 = \mathbf{6}$

In general, the differential coefficient of kx, where k is a constant, is always k.

Problem 3. Find the derivatives of:
(a) $y = 3\sqrt{x}$ (b) $y = \frac{5}{\sqrt[3]{x^4}}$

(a) $y = 3\sqrt{x}$ is rewritten in the standard differential form as $y = 3x^{1/2}$

In the general rule, $a = 3$ and $n = \frac{1}{2}$

Thus $\frac{dy}{dx} = (3)\left(\frac{1}{2}\right)x^{\frac{1}{2}-1} = \frac{3}{2}x^{-\frac{1}{2}}$

$= \frac{3}{2x^{1/2}} = \frac{\mathbf{3}}{\mathbf{2\sqrt{x}}}$

(b) $y = \frac{5}{\sqrt[3]{x^4}} = \frac{5}{x^{4/3}} = 5x^{-4/3}$

In the general rule, $a = 5$ and $n = -\frac{4}{3}$

Thus $\frac{dy}{dx} = (5)\left(-\frac{4}{3}\right)x^{(-4/3)-1}$

$= \frac{-20}{3}x^{-7/3} = \frac{-20}{3x^{7/3}} = \frac{\mathbf{-20}}{\mathbf{3\sqrt[3]{x^7}}}$

Problem 4. Differentiate:
$y = 5x^4 + 4x - \frac{1}{2x^2} + \frac{1}{\sqrt{x}} - 3$ with respect to x

$y = 5x^4 + 4x - \frac{1}{2x^2} + \frac{1}{\sqrt{x}} - 3$ is rewritten as

$y = 5x^4 + 4x - \frac{1}{2}x^{-2} + x^{-1/2} - 3$

When differentiating a sum, each term is differentiated in turn.

Thus $\frac{dy}{dx} = (5)(4)x^{4-1} + (4)(1)x^{1-1} - \frac{1}{2}(-2)x^{-2-1}$

$+ (1)\left(-\frac{1}{2}\right)x^{(-1/2)-1} - 0$

$= 20x^3 + 4 + x^{-3} - \frac{1}{2}x^{-3/2}$

i.e. $\frac{\mathbf{dy}}{\mathbf{dx}} = \mathbf{20x^3 + 4 + \frac{1}{x^3} - \frac{1}{2\sqrt{x^3}}}$

Problem 5. Find the differential coefficients of:
(a) $y = 3 \sin 4x$ (b) $f(t) = 2 \cos 3t$ with respect to the variable

(a) When $y = 3 \sin 4x$ then $\frac{dy}{dx} = (3)(4 \cos 4x)$

$= \mathbf{12 \cos 4x}$

(b) When $f(t) = 2 \cos 3t$ then

$f'(t) = (2)(-3 \sin 3t) = \mathbf{-6 \sin 3t}$

Problem 6. Determine the derivatives of:
(a) $y = 3e^{5x}$ (b) $f(\theta) = \frac{2}{e^{3\theta}}$ (c) $y = 6 \ln 2x$

(a) When $y = 3e^{5x}$ then $\frac{dy}{dx} = (3)(5)e^{5x} = \mathbf{15e^{5x}}$

(b) $f(\theta) = \frac{2}{e^{3\theta}} = 2e^{-3\theta}$, thus

$f'(\theta) = (2)(-3)e^{-3\theta} = -6e^{-3\theta} = \frac{\mathbf{-6}}{\mathbf{e^{3\theta}}}$

(c) When $y = 6 \ln 2x$ then $\frac{dy}{dx} = 6\left(\frac{1}{x}\right) = \frac{\mathbf{6}}{\mathbf{x}}$

Problem 7. Find the gradient of the curve $y = 3x^4 - 2x^2 + 5x - 2$ at the points $(0, -2)$ and $(1, 4)$

The gradient of a curve at a given point is given by the corresponding value of the derivative. Thus, since $y = 3x^4 - 2x^2 + 5x - 2$ then

the gradient $= \frac{dy}{dx} = 12x^3 - 4x + 5$

At the point $(0, -2)$, $x = 0$.

Thus the gradient $= 12(0)^3 - 4(0) + 5 = \mathbf{5}$

At the point $(1, 4)$, $x = 1$

Thus the gradient $= 12(1)^3 - 4(1) + 5 = \mathbf{13}$

Problem 8. Determine the co-ordinates of the point on the graph $y = 3x^2 - 7x + 2$ where the gradient is -1

The gradient of the curve is given by the derivative.

When $y = 3x^2 - 7x + 2$ then $\frac{dy}{dx} = 6x - 7$

Since the gradient is -1 then $6x - 7 = -1$, from which, $x = 1$

When $x = 1$, $y = 3(1)^2 - 7(1) + 2 = -2$

Hence the gradient is -1 at the point $(1, -2)$

Now try the following Practice Exercise

Practice Exercise 186 Differentiating common functions (Answers on page 721)

In Problems 1 to 6 find the differential coefficients of the given functions with respect to the variable.

1. (a) $5x^5$ (b) $2.4x^{3.5}$ (c) $\frac{1}{x}$
2. (a) $\frac{-4}{x^2}$ (b) 6 (c) $2x$
3. (a) $2\sqrt{x}$ (b) $3\sqrt[3]{x^5}$ (c) $\frac{4}{\sqrt{x}}$
4. (a) $\frac{-3}{\sqrt[3]{x}}$ (b) $(x-1)^2$ (c) $2\sin 3x$
5. (a) $-4\cos 2x$ (b) $2e^{6x}$ (c) $\frac{3}{e^{5x}}$
6. (a) $4\ln 9x$ (b) $\frac{e^x - e^{-x}}{2}$ (c) $\frac{1-\sqrt{x}}{x}$
7. Find the gradient of the curve $y = 2t^4 + 3t^3 - t + 4$ at the points $(0, 4)$ and $(1, 8)$
8. Find the co-ordinates of the point on graph $y = 5x^2 - 3x + 1$ where the gradient is 2
9. (a) Differentiate
   $$y = \frac{2}{\theta^2} + 2\ln 2\theta - 2(\cos 5\theta + 3\sin 2\theta) - \frac{2}{e^{3\theta}}$$
   (b) Evaluate $\frac{dy}{d\theta}$ when $\theta = \frac{\pi}{2}$, correct to 4 significant figures
10. Evaluate $\frac{ds}{dt}$, correct to 3 significant figures, when $t = \frac{\pi}{6}$ given $s = 3\sin t - 3 + \sqrt{t}$
11. A mass, m, is held by a spring with a stiffness constant k. The potential energy, p, of the system is given by: $p = \frac{1}{2}kx^2 - mgx$ where x is the displacement and g is acceleration due to gravity. The system is in equilibrium if $\frac{dp}{dx} = 0$. Determine the expression for x for system equilibrium
12. The current i flowing in an inductor of inductance 100 mH is given by: $i = 5\sin 100t$ amperes, where t is the time t in seconds. The voltage v across the inductor is given by: $v = L\frac{di}{dt}$ volts. Determine the voltage when $t = 10$ ms

37.2 Differentiation of a product

When $y = uv$ and u and v are both functions of x,

then $$\frac{dy}{dx} = u\frac{dv}{dx} + v\frac{du}{dx}$$

This is known as the **product rule**.

Problem 9. Find the differential coefficient of: $y = 3x^2 \sin 2x$

$3x^2 \sin 2x$ is a product of two terms $3x^2$ and $\sin 2x$. Let $u = 3x^2$ and $v = \sin 2x$

Using the product rule:

$$\frac{dy}{dx} = u \quad \frac{dv}{dx} \quad + \quad v \quad \frac{du}{dx}$$

gives: $$\frac{dy}{dx} = (3x^2)(2\cos 2x) + (\sin 2x)(6x)$$

i.e. $$\frac{dy}{dx} = 6x^2\cos 2x + 6x\sin 2x$$

$$= \mathbf{6x(x\cos 2x + \sin 2x)}$$

Note that the differential coefficient of a product is **not** obtained by merely differentiating each term and multiplying the two answers together. The product rule formula **must** be used when differentiating products.

Problem 10. Find the rate of change of y with respect to x given: $y = 3\sqrt{x}\ln 2x$

The rate of change of y with respect to x is given by $\frac{dy}{dx}$

$y = 3\sqrt{x}\ln 2x = 3x^{1/2}\ln 2x$, which is a product.

Let $u = 3x^{1/2}$ and $v = \ln 2x$

$$\text{Then } \frac{dy}{dx} = u\frac{dv}{dx} + v\frac{du}{dx}$$

$$= (3x^{1/2})\left(\frac{1}{x}\right) + (\ln 2x)\left[3\left(\frac{1}{2}\right)x^{(1/2)-1}\right]$$

$$= 3x^{(1/2)-1} + (\ln 2x)\left(\frac{3}{2}\right)x^{-1/2}$$

$$= 3x^{-1/2}\left(1 + \frac{1}{2}\ln 2x\right)$$

i.e. $$\boldsymbol{\frac{dy}{dx} = \frac{3}{\sqrt{x}}\left(1 + \frac{1}{2}\ln 2x\right)}$$

Problem 11. Differentiate: $y = x^3 \cos 3x \ln x$

Let $u = x^3 \cos 3x$ (i.e. a product) and $v = \ln x$

$$\text{Then } \frac{dy}{dx} = u\frac{dv}{dx} + v\frac{du}{dx}$$

$$\text{where } \frac{du}{dx} = (x^3)(-3\sin 3x) + (\cos 3x)(3x^2)$$

$$\text{and } \frac{dv}{dx} = \frac{1}{x}$$

$$\text{Hence } \frac{dy}{dx} = (x^3\cos 3x)\left(\frac{1}{x}\right) + (\ln x)[-3x^3\sin 3x + 3x^2\cos 3x]$$

$$= x^2\cos 3x + 3x^2\ln x(\cos 3x - x\sin 3x)$$

i.e. $$\boldsymbol{\frac{dy}{dx} = x^2\{\cos 3x + 3\ln x(\cos 3x - x\sin 3x)\}}$$

Problem 12. Determine the rate of change of voltage, given $v = 5t \sin 2t$ volts, when $t = 0.2$ s

Rate of change of voltage

$$= \frac{dv}{dt} = (5t)(2\cos 2t) + (\sin 2t)(5)$$

$$= 10t\cos 2t + 5\sin 2t$$

When $t = 0.2$,

$$\frac{dv}{dt} = 10(0.2)\cos 2(0.2) + 5\sin 2(0.2)$$

$$= 2\cos 0.4 + 5\sin 0.4$$

(where cos 0.4 means the cosine of 0.4 radians = 0.92106)

$$\text{Hence } \frac{dv}{dt} = 2(0.92106) + 5(0.38942)$$

$$= 1.8421 + 1.9471 = 3.7892$$

i.e. **the rate of change of voltage when $t = 0.2$ s is 3.79 volts/s, correct to 3 significant figures.**

Now try the following Practice Exercise

Practice Exercise 187 Differentiating products (Answers on page 721)

In Problems 1 to 8 differentiate the given products with respect to the variable.

1. $x\sin x$
2. x^2e^{2x}
3. $x^2\ln x$
4. $2x^3\cos 3x$
5. $\sqrt{x^3}\ln 3x$
6. $e^{3t}\sin 4t$
7. $e^{4\theta}\ln 3\theta$
8. $e^t\ln t\cos t$
9. Evaluate $\frac{di}{dt}$, correct to 4 significant figure, when $t = 0.1$, and $i = 15t\sin 3t$
10. Evaluate $\frac{dz}{dt}$, correct to 4 significant figures, when $t = 0.5$, given that $z = 2e^{3t}\sin 2t$

37.3 Differentiation of a quotient

When $y = \frac{u}{v}$ and u and v are both functions of x

$$\text{then } \frac{dy}{dx} = \frac{v\frac{du}{dx} - u\frac{dv}{dx}}{v^2}$$

This is known as the **quotient rule**.

Problem 13. Find the differential coefficient of:

$$y = \frac{4\sin 5x}{5x^4}$$

$\dfrac{4\sin 5x}{5x^4}$ is a quotient. Let $u = 4\sin 5x$ and $v = 5x^4$

(Note that v is **always** the denominator and u the numerator)

$$\frac{dy}{dx} = \frac{v\frac{du}{dx} - u\frac{dv}{dx}}{v^2}$$

where $\dfrac{du}{dx} = (4)(5)\cos 5x = 20\cos 5x$

and $\dfrac{dv}{dx} = (5)(4)x^3 = 20x^3$

Hence
$$\frac{dy}{dx} = \frac{(5x^4)(20\cos 5x) - (4\sin 5x)(20x^3)}{(5x^4)^2}$$
$$= \frac{100x^4\cos 5x - 80x^3\sin 5x}{25x^8}$$
$$= \frac{20x^3[5x\cos 5x - 4\sin 5x]}{25x^8}$$

i.e. $$\boldsymbol{\frac{dy}{dx} = \frac{4}{5x^5}(5x\cos 5x - 4\sin 5x)}$$

Note that the differential coefficient is **not** obtained by merely differentiating each term in turn and then dividing the numerator by the denominator. The quotient formula **must** be used when differentiating quotients.

Problem 14. Determine the differential coefficient of: $y = \tan ax$

$y = \tan ax = \dfrac{\sin ax}{\cos ax}$. Differentiation of $\tan ax$ is thus treated as a quotient with $u = \sin ax$ and $v = \cos ax$

$$\frac{dy}{dx} = \frac{v\frac{du}{dx} - u\frac{dv}{dx}}{v^2}$$
$$= \frac{(\cos ax)(a\cos ax) - (\sin ax)(-a\sin ax)}{(\cos ax)^2}$$
$$= \frac{a\cos^2 ax + a\sin^2 ax}{(\cos ax)^2}$$
$$= \frac{a(\cos^2 ax + \sin^2 ax)}{\cos^2 ax}$$
$$= \frac{a}{\cos^2 ax} \text{ since } \cos^2 ax + \sin^2 ax = 1$$

(see Chapter 21)

Hence $\boldsymbol{\dfrac{dy}{dx} = a\sec^2 ax}$ since $\sec^2 ax = \dfrac{1}{\cos^2 ax}$

(see Chapter 17)

Problem 15. Find the derivative of: $y = \sec ax$

$y = \sec ax = \dfrac{1}{\cos ax}$ (i.e. a quotient), Let $u = 1$ and $v = \cos ax$

$$\frac{dy}{dx} = \frac{v\frac{du}{dx} - u\frac{dv}{dx}}{v^2}$$
$$= \frac{(\cos ax)(0) - (1)(-a\sin ax)}{(\cos ax)^2}$$
$$= \frac{a\sin ax}{\cos^2 ax} = a\left(\frac{1}{\cos ax}\right)\left(\frac{\sin ax}{\cos ax}\right)$$

i.e. $$\boldsymbol{\frac{dy}{dx} = a\sec ax\tan ax}$$

Problem 16. Differentiate: $y = \dfrac{te^{2t}}{2\cos t}$

The function $\dfrac{te^{2t}}{2\cos t}$ is a quotient, whose numerator is a product.

Let $u = te^{2t}$ and $v = 2\cos t$ then

$\dfrac{du}{dt} = (t)(2e^{2t}) + (e^{2t})(1)$ and $\dfrac{dv}{dt} = -2\sin t$

Hence
$$\frac{dy}{dt} = \frac{v\frac{du}{dt} - u\frac{dv}{dt}}{v^2}$$
$$= \frac{(2\cos t)[2te^{2t} + e^{2t}] - (te^{2t})(-2\sin t)}{(2\cos t)^2}$$
$$= \frac{4te^{2t}\cos t + 2e^{2t}\cos t + 2te^{2t}\sin t}{4\cos^2 t}$$
$$= \frac{2e^{2t}[2t\cos t + \cos t + t\sin t]}{4\cos^2 t}$$

i.e. $$\boldsymbol{\frac{dy}{dt} = \frac{e^{2t}}{2\cos^2 t}(2t\cos t + \cos t + t\sin t)}$$

Problem 17. Determine the gradient of the curve $y = \dfrac{5x}{2x^2 + 4}$ at the point $\left(\sqrt{3}, \dfrac{\sqrt{3}}{2}\right)$

Let $y = 5x$ and $v = 2x^2 + 4$

$$\frac{dy}{dx} = \frac{v\frac{du}{dx} - u\frac{dv}{dx}}{v^2} = \frac{(2x^2+4)(5) - (5x)(4x)}{(2x^2+4)^2}$$

$$= \frac{10x^2 + 20 - 20x^2}{(2x^2+4)^2} = \frac{20 - 10x^2}{(2x^2+4)^2}$$

At the point $\left(\sqrt{3}, \frac{\sqrt{3}}{2}\right)$, $x = \sqrt{3}$,

hence the gradient $= \frac{dy}{dx} = \frac{20 - 10(\sqrt{3})^2}{[2(\sqrt{3})^2 + 4]^2}$

$$= \frac{20 - 30}{100} = -\frac{\mathbf{1}}{\mathbf{10}}$$

Now try the following Practice Exercise

Practice Exercise 188 Differentiating quotients (Answers on page 721)

In Problems 1 to 7, differentiate the quotients with respect to the variable.

1. $\frac{\sin x}{x}$
2. $\frac{2\cos 3x}{x^3}$
3. $\frac{2x}{x^2+1}$
4. $\frac{\sqrt{x}}{\cos x}$
5. $\frac{3\sqrt{\theta^3}}{2\sin 2\theta}$
6. $\frac{\ln 2t}{\sqrt{t}}$
7. $\frac{2xe^{4x}}{\sin x}$
8. Find the gradient of the curve $y = \frac{2x}{x^2 - 5}$ at the point (2, –4)
9. Evaluate $\frac{dy}{dx}$ at $x = 2.5$, correct to 3 significant figures, given $y = \frac{2x^2+3}{\ln 2x}$

37.4 Function of a function

It is often easier to make a substitution before differentiating.

If y is a function of x then $\frac{dy}{dx} = \frac{dy}{du} \times \frac{du}{dx}$

This is known as the **'function of a function'** rule (or sometimes the **chain rule**).

For example, if $y = (3x - 1)^9$ then, by making the substitution $u = (3x - 1)$, $y = u^9$, which is of the 'standard' from.

Hence $\frac{dy}{du} = 9u^8$ and $\frac{du}{dx} = 3$

Then $\frac{dy}{dx} = \frac{dy}{du} \times \frac{du}{dx} = (9u^8)(3) = 27u^8$

Rewriting u as $(3x - 1)$ gives: $\mathbf{\frac{dy}{dx} = 27(3x - 1)^8}$

Since y is a function of u, and u is a function of x, then y is a function of a function of x.

Problem 18. Differentiate: $y = 3\cos(5x^2 + 2)$

Let $u = 5x^2 + 2$ then $y = 3\cos u$

Hence $\frac{du}{dx} = 10x$ and $\frac{dy}{du} = -3\sin u$

Using the function of a function rule,

$$\frac{dy}{dt} = \frac{dy}{du} \times \frac{du}{dt} = (-3\sin u)(10x) = -30x\sin u$$

Rewriting u as $5x^2 + 2$ gives:

$$\mathbf{\frac{dy}{dx} = -30x\sin(5x^2 + 2)}$$

Problem 19. Find the derivative of: $y = (4t^3 - 3t)^6$

Let $u = 4t^3 - 3t$, then $y = u^6$

Hence $\frac{du}{dt} = 12t^2 - 3$ and $\frac{dy}{dt} = 6u^5$

Using the function of a function rule,

$$\frac{dy}{dt} = \frac{dy}{du} \times \frac{du}{dt} = (6u^5)(12t^2 - 3)$$

Rewriting u as $(4t^3 - 3t)$ gives:

$$\frac{dy}{dt} = 6(4t^3 - 3t)^5(12t^2 - 3)$$

$$= \mathbf{18(4t^2 - 1)(4t^3 - 3t)^5}$$

Problem 20. Determine the differential coefficient of: $y=\sqrt{3x^2+4x-1}$

$y=\sqrt{3x^2+4x-1}=(3x^2+4x-1)^{1/2}$

Let $u=3x^2+4x-1$ then $y=u^{1/2}$

Hence $\frac{du}{dx}=6x+4$ and $\frac{dy}{du}=\frac{1}{2}u^{-1/2}=\frac{1}{2\sqrt{u}}$

Using the function of a function rule,

$$\frac{dy}{dx}=\frac{dy}{du}\times\frac{du}{dx}=\left(\frac{1}{2\sqrt{u}}\right)(6x+4)=\frac{3x+2}{\sqrt{u}}$$

i.e. $$\boldsymbol{\frac{dy}{dx}=\frac{3x+2}{\sqrt{3x^2+4x-1}}}$$

Problem 21. Differentiate: $y=3\tan^4 3x$

Let $u=\tan 3x$ then $y=3u^4$

Hence $\frac{du}{dx}=3\sec^2 3x$, (from Problem 14),

and $\frac{dy}{du}=12u^3$

Then $$\frac{dy}{dx}=\frac{dy}{du}\times\frac{du}{dx}=(12u^3)(3\sec^2 3x)$$

$$=12(\tan 3x)^3(3\sec^2 3x)$$

i.e. $$\boldsymbol{\frac{dy}{dx}=36\tan^3 3x\sec^2 3x}$$

Problem 22. Find the differential coefficient of: $y=\frac{2}{(2t^3-5)^4}$

$y=\frac{2}{(2t^3-5)^4}=2(2t^3-5)^{-4}$. Let $u=(2t^3-5)$, then $y=2u^{-4}$

Hence $\frac{du}{dt}=6t^2$ and $\frac{dy}{du}=-8u^{-5}=\frac{-8}{u^5}$

Then $$\frac{dy}{dt}=\frac{dy}{du}\times\frac{du}{dt}=\left(\frac{-8}{u^5}\right)(6t^2)=\boldsymbol{\frac{-48t^2}{(2t^3-5)^5}}$$

Now try the following Practice Exercise

Practice Exercise 189 Differentiating a function of a function (Answers on page 722)

In Problems 1 to 9, find the differential coefficients with respect to the variable.

1. $(2x-1)^6$
2. $(2x^3-5x)^5$
3. $2\sin(3\theta-2)$
4. $2\cos^5\alpha$
5. $\frac{1}{(x^3-2x+1)^5}$
6. $5e^{2t+1}$
7. $2\cot(5t^2+3)$
8. $6\tan(3y+1)$
9. $2e^{\tan\theta}$
10. Differentiate: $\theta\sin\left(\theta-\frac{\pi}{3}\right)$ with respect to θ and evaluate, correct to 3 significant figures, when $\theta=\frac{\pi}{2}$
11. The extension, x metres, of an un-damped vibrating spring after t seconds is given by:

$$x=0.54\cos(0.3t-0.15)+3.2$$

Calculate the speed of the spring, given by $\frac{dx}{dt}$, when (a) $t=0$, (b) $t=2$ s

37.5 Successive differentiation

When a function $y=f(x)$ is differentiated with respect to x the differential coefficient is written as $\frac{dy}{dx}$ or $f'(x)$. If the expression is differentiated again, the second differential coefficient is obtained and is written as $\frac{d^2y}{dx^2}$ (pronounced dee two y by dee x squared) or $f''(x)$ (pronounced f double–dash x). By successive differentiation further higher derivatives such as $\frac{d^3y}{dx^3}$ and $\frac{d^4y}{dx^4}$ may be obtained.

Thus if $y=3x^4$,

$$\frac{dy}{dx}=12x^3,\ \frac{d^2y}{dx^2}=36x^2,$$

$$\frac{d^3y}{dx^3}=72x,\ \frac{d^4y}{dx^4}=72 \text{ and } \frac{d^5y}{dx^5}=0$$

Problem 23. If $f(x)=2x^5-4x^3+3x-5$, find $f''(x)$

$$f(x)=2x^5-4x^3+3x-5$$
$$f'(x)=10x^4-12x^2+3$$
$$f''(x)=40x^3-24x=\mathbf{4x(10x^2-6)}$$

Problem 24. If $y=\cos x-\sin x$, evaluate x, in the range $0\le x\le\frac{\pi}{2}$, when $\frac{d^2y}{dx^2}$ is zero

Since $y=\cos x-\sin x$, $\frac{dy}{dx}=-\sin x-\cos x$ and

$\frac{d^2y}{dx^2}=-\cos x+\sin x$

When $\frac{d^2y}{dx^2}$ is zero, $-\cos x+\sin x=0$,

i.e. $\sin x=\cos x$ or $\frac{\sin x}{\cos x}=1$

Hence $\tan x=1$ and $\boldsymbol{x}=\tan^{-1}1=\mathbf{45°}$ **or** $\boldsymbol{\frac{\pi}{4}}$ **rads** in the range $0\le x\le\frac{\pi}{2}$

Problem 25. Given $y=2xe^{-3x}$ show that
$$\frac{d^2y}{dx^2}+6\frac{dy}{dx}+9y=0$$

$y=2xe^{-3x}$ (i.e. a product)

Hence
$$\frac{dy}{dx}=(2x)(-3e^{-3x})+(e^{-3x})(2)$$
$$=-6xe^{-3x}+2e^{-3x}$$
$$\frac{d^2y}{dx^2}=[(-6x)(-3e^{-3x})+(e^{-3x})(-6)]+(-6e^{-3x})$$
$$=18xe^{-3x}-6e^{-3x}-6e^{-3x}$$

i.e.
$$\frac{d^2y}{dx^2}=18xe^{-3x}-12e^{-3x}$$

Substituting values into $\frac{d^2y}{dx^2}+6\frac{dy}{dx}+9y$ gives:

$$(18xe^{-3x}-12e^{-3x})+6(-6xe^{-3x}+2e^{-3x})+9(2xe^{-3x})$$
$$=18xe^{-3x}-12e^{-3x}-36xe^{-3x}+12e^{-3x}+18xe^{-3x}=0$$

Thus when $y=2xe^{-3x}$, $\frac{d^2y}{dx^2}+6\frac{dy}{dx}+9y=0$

Problem 26. Evaluate $\frac{d^2y}{d\theta^2}$ when $\theta=0$ given: $y=4\sec 2\theta$

Since $y=4\sec 2\theta$, then

$$\frac{dy}{d\theta}=(4)(2)\sec 2\theta\tan 2\theta \text{ (from Problem 15)}$$
$$=8\sec 2\theta\tan 2\theta \text{ (i.e. a product)}$$
$$\frac{d^2y}{d\theta^2}=(8\sec 2\theta)(2\sec^2 2\theta)+(\tan 2\theta)[(8)(2)\sec 2\theta\tan 2\theta]$$
$$=16\sec^3 2\theta+16\sec 2\theta\tan^2 2\theta$$

When $\theta=0$,

$$\frac{d^2y}{d\theta^2}=16\sec^3 0+16\sec 0\tan^2 0$$
$$=16(1)+16(1)(0)=\mathbf{16}$$

Now try the following Practice Exercise

Practice Exercise 190 Successive differentiation (Answers on page 722)

1. If $y=3x^4+2x^3-3x+2$ find (a) $\frac{d^2y}{dx^2}$ (b) $\frac{d^3y}{dx^3}$
2. (a) Given $f(t)=\frac{2}{5}t^2-\frac{1}{t^3}+\frac{3}{t}-\sqrt{t}+1$ determine $f''(t)$
   (b) Evaluate $f''(t)$ when $t=1$

In Problems 3 and 4, find the second differential coefficient with respect to the variable.

3. The charge q on the plates of a capacitor is given by $q=CVe^{-\frac{t}{CR}}$, where t is the time, C is the capacitance and R the resistance. Determine (a) the rate of change of charge, which is given by $\frac{dq}{dt}$, (b) the rate of change of current, which is given by $\frac{d^2q}{dt^2}$
4. (a) $3\sin 2t+\cos t$ (b) $2\ln 4\theta$
5. (a) $2\cos^2 x$ (b) $(2x-3)^4$

6. Evaluate $f''(\theta)$ when $\theta = 0$ given $f(\theta) = 2\sec 3\theta$

7. Show that the differential equation $\frac{d^2y}{dx^2} - 4\frac{dy}{dx} + 4y = 0$ is satisfied when $y = xe^{2x}$

8. Show that, if P and Q are constants and $y = P\cos(\ln t) + Q\sin(\ln t)$, then

$$t^2\frac{d^2y}{dt^2} + t\frac{dy}{dt} + y = 0$$

9. The displacement, s, of a mass in a vibrating system is given by: $s = (1+t)e^{-\omega t}$ where ω is the natural frequency of vibration. Show that: $\frac{d^2s}{dt^2} + 2\omega\frac{ds}{dt} + \omega^2 s = 0$

10. An equation for the deflection of a cantilever is:
$y = \frac{Px^2}{2EI}\left(L - \frac{x}{3}\right) + \frac{Wx^2}{12EI}\left(3L^2 + 2xL + \frac{x^2}{2}\right)$
where P, W, L, E and I are all constants. If the bending moment $M = -EI\frac{d^2y}{dx^2}$ determine an equation for M in terms of P, W, L and x.

Practice Exercise 191 Multiple-choice questions on methods of differentiation (Answers on page 722)

Each question has only one correct answer

1. The gradient of the curve $y = -2x^3 + 3x + 5$ at $x = 2$ is:
(a) -21 (b) 27 (c) -16 (d) -5

2. Given $y = 4x^3 - 5\cos x - e^{-x} + \ln x - 25$ then $\frac{dy}{dx}$ is equal to:
(a) $12x^2 + 5\sin x - e^{-x} + \frac{1}{x}$
(b) $12x^2 - 5\sin x - e^{-x} + \frac{1}{x}$
(c) $x^4 - 5\sin x - e^{-x} + x(1 + \ln x) + 25x + c$
(d) $12x^2 + 5\sin x + e^{-x} + \frac{1}{x}$

3. If $y = 5\sqrt{x^3} - 2$, then $\frac{dy}{dx}$ is equal to:
(a) $\frac{15}{2}\sqrt{x}$ (b) $2\sqrt{x^5} - 2x + c$
(c) $\frac{5}{2}\sqrt{x} - 2$ (d) $5\sqrt{x} - 2x$

4. The gradient of the curve $y = 4x^2 - 7x + 3$ at the point (1, 0) is
(a) 3 (b) 1 (c) 0 (d) -7

5. A displacement s is given by $s = 30\sin 200t$ mm, where t is the time in seconds. The velocity when $t = 5$ ms is:
(a) 6.00 m/s (b) 16.21 mm/s
(c) 3.24 m/s (d) 30.00 mm/s

6. If $f(t) = e^{2t}\ln 2t, f'(t)$ is equal to:
(a) $\frac{2e^{2t}}{t}$ (b) $e^{2t}\left(\frac{1}{t} + 2\ln 2t\right)$
(c) $\frac{e^{2t}}{2t}$ (d) $\frac{e^{2t}}{2t} + 2e^{2t}\ln 2t$

7. If $y = 3x^2 - \ln 5x$ then $\frac{d^2y}{dx^2}$ is equal to:
(a) $6 + \frac{1}{5x^2}$ (b) $6x - \frac{1}{x}$
(c) $6 - \frac{1}{5x}$ (d) $6 + \frac{1}{x^2}$

8. If $s = 2t\ln t$ then $\frac{ds}{dt}$ is equal to:
(a) $2 + 2\ln t$ (b) $2t + \frac{2}{t}$
(c) $2 - 2\ln t$ (d) $2 + 2e^t$

9. Given $y = (2x^2 - 3)^2$ then:
(a) $\frac{dy}{dx} = 16x^3$ (b) $\frac{dy}{dx} = 16x^3 - 24x$
(c) $\frac{dy}{dx} = 4x^2 - 6$ (d) $\frac{dy}{dx} = 16x - 18$

10. If a function is given by $y = 4x^3 - 4\sqrt{x}$, then $\frac{d^2y}{dx^2}$ is given by:
(a) $24x + x^{\frac{1}{2}}$ (b) $12x^2 - 2x^{-\frac{1}{2}}$
(c) $24x + x^{-\frac{3}{2}}$ (d) $12x^2 - x^{\frac{1}{2}}$

For fully worked solutions to each of the problems in Practice Exercises 186 to 190 in this chapter, go to the website:
www.routledge.com/cw/bird

Chapter 38

Some applications of differentiation

Why it is important to understand: Some applications of differentiation

In the previous two chapters some basic differentiation techniques were explored, sufficient to allow us to look at some applications of differential calculus. Some practical rates of change problems are initially explained, followed by some practical velocity and acceleration problems. Determining maximum and minimum points and points of inflexion on curves, together with some practical maximum and minimum problems follow. Tangents and normals to curves and errors and approximations complete this initial look at some applications of differentiation. In general, with these applications, the differentiation tends to be straightforward.

At the end of this chapter, you should be able to:

- determine rates of change using differentiation
- solve velocity and acceleration problems
- understand turning points
- determine the turning points on a curve and determine their nature
- solve practical problems involving maximum and minimum values
- determine points of inflexion on a curve
- determine tangents and normals to a curve
- determine small changes in functions

38.1 Rates of change

If a quantity y depends on and varies with a quantity x then the rate of change of y with respect to x is $\frac{dy}{dx}$

Thus, for example, the rate of change of pressure p with height h is $\frac{dp}{dh}$

A rate of change with respect to time is usually just called 'the rate of change', the 'with respect to time' being assumed. Thus, for example, a rate of change of current, i, is $\frac{di}{dt}$ and a rate of change of temperature, θ, is $\frac{d\theta}{dt}$ and so on.

Problem 1. The length l metres of a certain metal rod at temperature θ°C is given by: $l = 1 + 0.00005\theta + 0.0000004\theta^2$. Determine the rate of change of length, in mm/°C, when the temperature is (a) 100°C and (b) 400°C

The rate of change of length means $\frac{dl}{d\theta}$

Since length $l = 1 + 0.00005\theta + 0.0000004\theta^2$,

then $\frac{dl}{d\theta} = 0.00005 + 0.0000008\theta$

(a) When $\theta = 100°C$,

$$\frac{dl}{d\theta} = 0.00005 + (0.0000008)(100)$$
$$= 0.00013\,\text{m/°C} = \mathbf{0.13\,mm/°C}$$

(b) When $\theta = 400°C$,

$$\frac{dl}{d\theta} = 0.00005 + (0.0000008)(400)$$
$$= 0.00037\,\text{m/°C} = \mathbf{0.37\,mm/°C}$$

Problem 2. The luminous intensity I candelas of a lamp at varying voltage V is given by: $I = 4\times 10^{-4}\,V^2$. Determine the voltage at which the light is increasing at a rate of 0.6 candelas per volt

The rate of change of light with respect to voltage is given by $\frac{dI}{dV}$

Since $I = 4\times 10^{-4}V^2$, $\frac{dI}{dV} = (4\times 10^{-4})(2)V$

$$= 8\times 10^{-4}\,V$$

When the light is increasing at 0.6 candelas per volt then $+0.6 = 8\times 10^{-4}$ V, from which, voltage

$$V = \frac{0.6}{8\times 10^{-4}} = 0.075\times 10^{+4} = \mathbf{750\ volts}$$

Problem 3. Newtons law of cooling is given by: $\theta = \theta_0 e^{-kt}$, where the excess of temperature at zero time is θ_0°C and at time t seconds is θ°C. Determine the rate of change of temperature after 40 s, given that $\theta_0 = 16°C$ and $k = -0.03$

The rate of change of temperture is $\frac{d\theta}{dt}$

Since $\theta = \theta_0 e^{-kt}$ then $\frac{d\theta}{dt} = (\theta_0)(-k)e^{-kt}$

$$= -k\theta_0 e^{-kt}$$

When $\theta_0 = 16$, $k = -0.03$ and $t = 40$ then

$$\frac{d\theta}{dt} = -(-0.03)(16)e^{-(-0.03)(40)}$$
$$= 0.48e^{1.2} = \mathbf{1.594°C/s}$$

Problem 4. The displacement s cm of the end of a stiff spring at time t seconds is given by: $s = ae^{-kt}\sin 2\pi ft$. Determine the velocity of the end of the spring after 1 s, if $a = 2$, $k = 0.9$ and $f = 5$

Velocity $v = \frac{ds}{dt}$ where $s = ae^{-kt}\sin 2\pi ft$ (i.e. a product)

Using the product rule,

$$\frac{ds}{dt} = (ae^{-kt})(2\pi f\cos 2\pi ft) + (\sin 2\pi ft)(-ake^{-kt})$$

When $a = 2$, $k = 0.9$, $f = 5$ and $t = 1$,

$$\textbf{velocity, } v = (2e^{-0.9})(2\pi 5\cos 2\pi 5) + (\sin 2\pi 5)(-2)(0.9)e^{-0.9}$$
$$= 25.5455\cos 10\pi - 0.7318\sin 10\pi$$
$$= 25.5455(1) - 0.7318(0)$$
$$= \mathbf{25.55\ cm/s}$$

(Note that $\cos 10\pi$ means 'the cosine of 10π radians', *not* degrees, and $\cos 10\pi \equiv \cos 2\pi = 1$)

Now try the following Practice Exercise

Practice Exercise 192 Rates of change (Answers on page 722)

1. An alternating current, i amperes, is given by $i = 10\sin 2\pi ft$, where f is the frequency in hertz and t the time in seconds. Determine the rate of change of current when $t = 20$ ms, given that $f = 150$ Hz.
2. The luminous intensity, I candelas, of a lamp is given by $I = 6\times 10^{-4}\,V^2$, where V is the voltage. Find (a) the rate of change of luminous intensity with voltage when $V = 200$ volts, and (b) the voltage at which the light is increasing at a rate of 0.3 candelas per volt.
3. The voltage across the plates of a capacitor at any time t seconds is given by $v = Ve^{-t/CR}$, where V, C and R are constants. Given $V = 300$ volts, $C = 0.12\times 10^{-6}$ farads and $R = 4\times 10^6$ ohms find (a) the initial rate of change of voltage, and (b) the rate of change of voltage after 0.5 s.

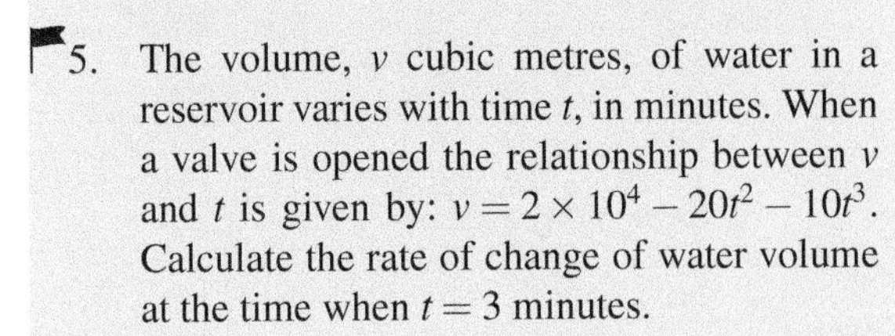

4. The pressure p of the atmosphere at height h above ground level is given by $p = p_0 e^{-h/c}$, where p_0 is the pressure at ground level and c is a constant. Determine the rate of change of pressure with height when $p_0 = 1.013 \times 10^5$ Pascals and $c = 6.05 \times 10^4$ at 1450 metres.

5. The volume, v cubic metres, of water in a reservoir varies with time t, in minutes. When a valve is opened the relationship between v and t is given by: $v = 2 \times 10^4 - 20t^2 - 10t^3$. Calculate the rate of change of water volume at the time when $t = 3$ minutes.

38.2 Velocity and acceleration

When a car moves a distance x metres in a time t seconds along a straight road, if the **velocity** v is constant then $v = \frac{x}{t}$ m/s, i.e. the gradient of the distance/time graph shown in Fig. 38.1 is constant.

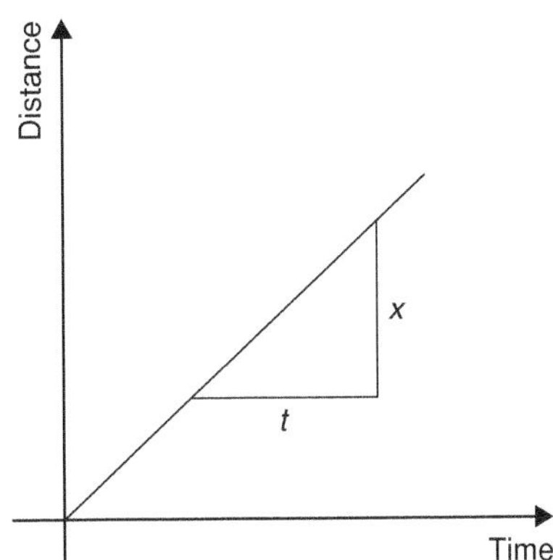

Figure 38.1

If, however, the velocity of the car is not constant then the distance/time graph will not be a straight line. It may be as shown in Fig. 38.2.

The average velocity over a small time δt and distance δx is given by the gradient of the chord AB, i.e. the average velocity over time δt is $\frac{\delta x}{\delta t}$. As $\delta t \to 0$, the chord AB becomes a tangent, such that at point A, the velocity is given by: $v = \frac{dx}{dt}$

Hence the velocity of the car at any instant is given by the gradient of the distance/time graph. If an expression

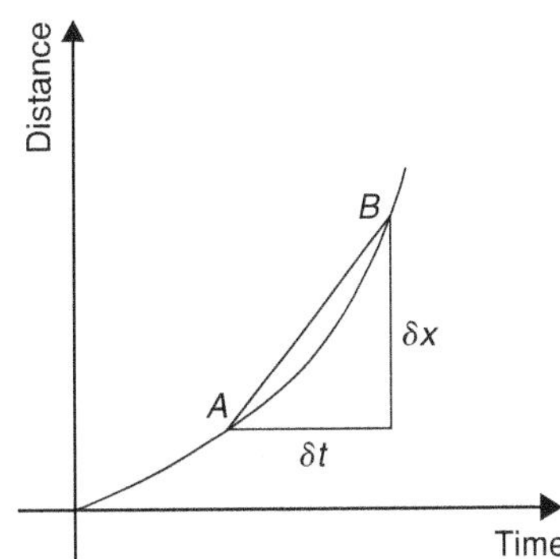

Figure 38.2

for the distance x is known in terms of time t then the velocity is obtained by differentiating the expression.

The **acceleration** $\boldsymbol{a}$ of the car is defined as the rate of change of velocity. A velocity/time graph is shown in Fig. 38.3. If δv is the change in v and δt the corresponding change in time, then $a = \frac{\delta v}{\delta t}$. As $\delta t \to 0$, the chord CD becomes a tangent, such that at point C, the acceleration is given by: $a = \frac{dv}{dt}$

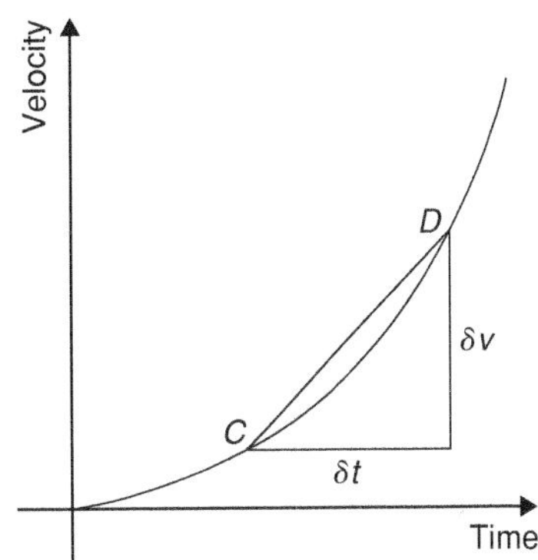

Figure 38.3

Hence the acceleration of the car at any instant is given by the gradient of the velocity/time graph. If an expression for velocity is known in terms of time t then the acceleration is obtained by differentiating the expression.

Acceleration $\quad a = \frac{dv}{dt}$

However, $\quad v = \frac{dx}{dt}$

Hence $\quad a = \frac{d}{dt}\left(\frac{dx}{dt}\right) = \frac{d^2x}{dt^2}$

The acceleration is given by the second differential coefficient of distance x with respect to time t

Summarising, if a body moves a distance x metres in a time t seconds then:

(i) distance $x = f(t)$

(ii) velocity $v = f'(t)$ or $\dfrac{dx}{dt}$, which is the gradient of the distance/time graph

(iii) acceleration $a = \dfrac{dv}{dt} = f''$ or $\dfrac{d^2x}{dt^2}$, which is the gradient of the velocity/time graph.

Problem 5. The distance x metres moved by a car in a time t seconds is given by: $x = 3t^3 - 2t^2 + 4t - 1$. Determine the velocity and acceleration when (a) $t = 0$, and (b) $t = 1.5$ s

Distance $\quad x = 3t^3 - 2t^2 + 4t - 1$ m.

Velocity $\quad v = \dfrac{dx}{dt} = 9t^2 - 4t + 4$ m/s

Acceleration $\quad a = \dfrac{d^2x}{dt^2} = 18t - 4$ m/s^2

(a) When time $t = 0$,

velocity $v = 9(0)^2 - 4(0) + 4 = \mathbf{4\ m/s}$

and acceleration $a = 18(0) - 4 = \mathbf{-4\ m/s^2}$ (i.e. a deceleration)

(b) When time $t = 1.5$ s,

velocity $v = 9(1.5)^2 - 4(1.5) + 4 = \mathbf{18.25\ m/s}$

and acceleration $a = 18(1.5) - 4 = \mathbf{23\ m/s^2}$

Problem 6. Supplies are dropped from a helicopter and the distance fallen in a time t seconds is given by: $x = \frac{1}{2}gt^2$, where $g = 9.8$ m/s^2. Determine the velocity and acceleration of the supplies after it has fallen for 2 seconds

Distance $\quad x = \dfrac{1}{2}gt^2 = \dfrac{1}{2}(9.8)t^2 = 4.9t^2$ m

Velocity $\quad v = \dfrac{dv}{dt} = 9.8t$ m/s

and acceleration $\quad a = \dfrac{d^2x}{dt^2} = 9.8$ m/s^2

When time $t = 2$ s,
velocity $v = (9.8)(2) = \mathbf{19.6\ m/s}$
and **acceleration** $a = \mathbf{9.8\ m/s^2}$ (which is acceleration due to gravity).

Problem 7. The distance x metres travelled by a vehicle in time t seconds after the brakes are applied is given by: $x = 20t - \frac{5}{3}t^2$. Determine (a) the speed of the vehicle (in km/h) at the instant the brakes are applied, and (b) the distance the car travels before it stops

(a) Distance, $x = 20t - \dfrac{5}{3}t^2$

Hence velocity $v = \dfrac{dx}{dt} = 20 - \dfrac{10}{3}t$

At the instant the brakes are applied, time $= 0$

Hence

$$\textbf{velocity } v = 20\,\text{m/s} = \frac{20 \times 60 \times 60}{1000}\,\text{km/h} = \mathbf{72\ km/h}$$

(Note: changing from m/s to km/h merely involves multiplying by 3.6)

(b) When the car finally stops, the velocity is zero, i.e. $v = 20 - \dfrac{10}{3}t = 0$, from which, $20 = \dfrac{10}{3}t$, giving $t = 6$ s. Hence the distance travelled before the car stops is given by:

$$x = 20t - \frac{5}{3}t^2 = 20(6) - \frac{5}{3}(6)^2 = 120 - 60 = \mathbf{60\ m}$$

Problem 8. The angular displacement θ radians of a flywheel varies with time t seconds and follows the equation: $\theta = 9t^2 - 2t^3$. Determine (a) the angular velocity and acceleration of the flywheel when time, $t = 1$ s, and (b) the time when the angular acceleration is zero

(a) Angular displacement $\theta = 9t^2 - 2t^3$ rad.

Angular velocity $\omega = \dfrac{d\theta}{dt} = 18t - 6t^2$ rad/s.

When time $t = 1$ s,

$\omega = 18(1) - 6(1)^2 = \mathbf{12\ rad/s}$.

Angular acceleration $\alpha = \dfrac{d^2\theta}{dt^2} = 18 - 12t$ rad/s^2.

When time $t = 1$ s, $\alpha = 18 - 12(1) = \mathbf{6\ rad/s^2}$

(b) When the angular acceleration is zero, $18 - 12t = 0$, from which, $18 = 12t$, giving time, $t = \mathbf{1.5\ s}$

Problem 9. The displacement x cm of the slide valve of an engine is given by: $x = 2.2\cos 5\pi t + 3.6\sin 5\pi t$. Evaluate the velocity (in m/s) when time $t = 30$ ms

Displacement $x = 2.2\cos 5\pi t + 3.6\sin 5\pi t$

$$\text{Velocity } v = \frac{dx}{dt} = (2.2)(-5\pi)\sin 5\pi t + (3.6)(5\pi)\cos 5\pi t$$

$$= -11\pi\sin 5\pi t + 18\pi\cos 5\pi t \text{ cm/s}$$

When time $t = 30$ ms,

$$\text{velocity} = -11\pi\sin(5\pi \times 30 \times 10^{-3}) + 18\pi\cos(5\pi \times 30 \times 10^{-3})$$

$$= -11\pi\sin 0.4712 + 18\pi\cos 0.4712$$

$$= -11\pi\sin 27^\circ + 18\pi\cos 27^\circ$$

$$= -15.69 + 50.39$$

$= 34.7$ cm/s $=$ **0.347 m/s**

Now try the following Practice Exercise

Practice Exercise 193 Velocity and acceleration (Answers on page 722)

1. A missile fired from ground level rises x metres vertically upwards in t seconds and $x = 100t - \frac{25}{2}t^2$. Find (a) the initial velocity of the missile, (b) the time when the height of the missile is a maximum, (c) the maximum height reached, (d) the velocity with which the missile strikes the ground.
2. The distance s metres travelled by a car in t seconds after the brakes are applied is given by $s = 25t - 2.5t^2$. Find (a) the speed of the car (in km/h) when the brakes are applied, (b) the distance the car travels before it stops.
3. The equation $\theta = 10\pi + 24t - 3t^2$ gives the angle θ, in radians, through which a wheel turns in t seconds. Determine (a) the time the wheel takes to come to rest, (b) the angle turned through in the last second of movement.
4. At any time t seconds the distance x metres of a particle moving in a straight line from a fixed point is given by: $x = 4t + \ln(1 - t)$. Determine (a) the initial velocity and acceleration, (b) the velocity and acceleration after 1.5 s, and (c) the time when the velocity is zero.
5. The angular displacement θ of a rotating disc is given by: $\theta = 6\sin\frac{t}{4}$, where t is the time in seconds. Determine (a) the angular velocity of the disc when t is 1.5 s, (b) the angular acceleration when t is 5.5 s, and (c) the first time when the angular velocity is zero.
6. $x = \frac{20t^3}{3} - \frac{23t^2}{2} + 6t + 5$ represents the distance, x metres, moved by a body in t seconds. Determine (a) the velocity and acceleration at the start, (b) the velocity and acceleration when $t = 3$ s, (c) the values of t when the body is at rest, (d) the value of t when the acceleration is 37 m/s^2, and (e) the distance travelled in the third second.
7. A particle has a displacement s given by: $s = 30t + 27t^2 - 3t^3$ metres, where time t is in seconds. Determine the time at which the acceleration will be zero.
8. The angular displacement θ radians of the spoke of a wheel is given by the expression: $\theta = \frac{1}{2}t^4 - \frac{1}{3}t^3$ where t is the time in seconds. Calculate (a) the angular velocity after 2 seconds, (b) the angular acceleration after 3 seconds, (c) the time when the angular acceleration is zero.
9. A mass of 4000 kg moves along a straight line so that the distance s metres travelled in a time t seconds is given by: $s = 4t^2 + 3t + 2$
   Kinetic energy is given by the formula $\frac{1}{2}mv^2$ joules, where m is the mass in kg and v is the velocity in m/s. Calculate the kinetic energy after 0.75 s, giving the answer in kJ.
10. The height, h metres, of a liquid in a chemical storage tank is given by: $h = 100\left(1 - e^{-\frac{t}{40}}\right)$ where t seconds is the time from which the tank starts to be filled. Calculate the speed in m/s at which the liquid level rises in the tank after 2 seconds.
11. The position of a missile, neglecting gravitational force, is described by: $s = (40t)\ln(1 + t)$ where s is the displacement of the missile, in metres, and t is the time, in seconds. 10 seconds after the missile has been fired, calculate (a) the displacement, (b) the velocity, and (c) the acceleration.

38.3 Turning points

In Fig. 38.4, the gradient (or rate of change) of the curve changes from positive between O and P to negative between P and Q, and then positive again between Q and R. At point P, the gradient is zero and, as x increases, the gradient of the curve changes from positive just before P to negative just after. Such a point is called a **maximum point** and appears as the 'crest of a wave'. At point Q, the gradient is also zero and, as x increases, the gradient of the curve changes from negative just before Q to positive just after. Such a point is called a **minimum point**, and appears as the 'bottom of a valley'. Points such as P and Q are given the general name of **turning points**.

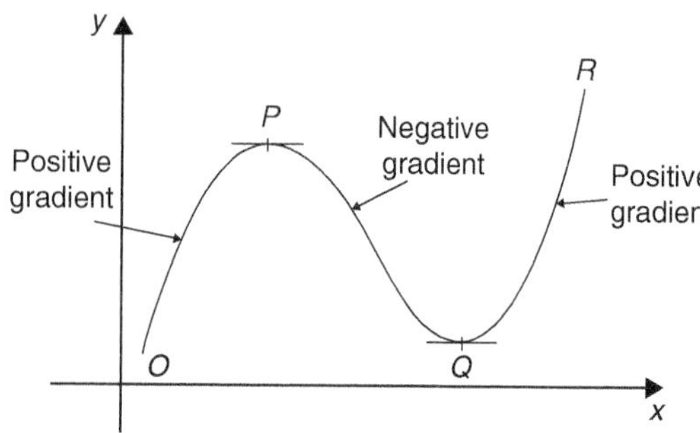

Figure 38.4

It is possible to have a turning point, the gradient on either side of which is the same. Such a point is given the special name of a **point of inflexion**, and examples are shown in Fig. 38.5.

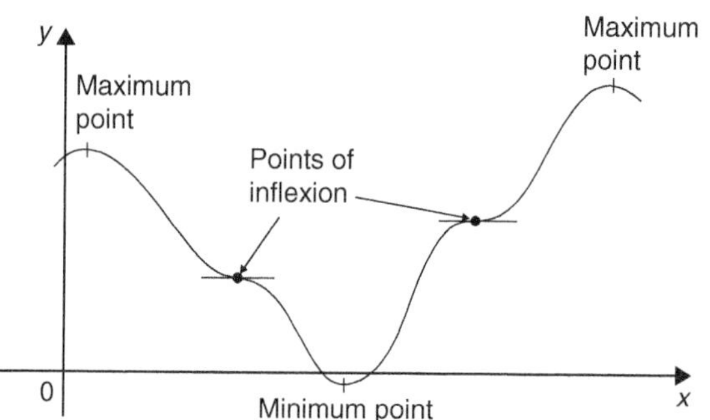

Figure 38.5

Maximum and minimum points and points of inflexion are given the general term of **stationary points**.

Procedure for finding and distinguishing between stationary points.

(i) Given $y=f(x)$, determine $\frac{dy}{dx}$ (i.e. $f'(x)$)

(ii) Let $\frac{dy}{dx}=0$ and solve for the values of x

(iii) Substitute the values of x into the original equation, $y=f(x)$, to find the corresponding y-ordinate values. This establishes the co-ordinates of the stationary points.

To determine the nature of the stationary points: Either

(iv) Find $\frac{d^2y}{dx^2}$ and substitute into it the values of x found in (ii).

If the result is: (a) positive — the point is a minimum one,
(b) negative — the point is a maximum one,
(c) zero — the point is a point of inflexion (see section 38.5)

or

(v) Determine the sign of the gradient of the curve just before and just after the stationary points. If the sign change for the gradient of the curve is:
(a) positive to negative — the point is a maximum one
(b) negative to positive — the point is a minimum one
(c) positive to positive or negative to negative — the point is a point of inflexion (see section 38.5).

Problem 10. Locate the turning point on the curve $y=3x^2-6x$ and determine its nature by examining the sign of the gradient on either side

Following the above procedure:

(i) Since $y=3x^2-6x$, $\frac{dy}{dx}=6x-6$

(ii) At a turning point, $\frac{dy}{dx}=0$, hence $6x-6=0$, from which, $x=1$

(iii) When $x=1$, $y=3(1)^2-6(1)=-3$

Hence the co-ordinates of the turning point is (1, −3)

(v) If x is slightly less than 1, say, 0.9, then $\frac{dy}{dx}=6(0.9)-6=-0.6$, i.e. negative

If x is slightly greater than 1, say, 1.1, then $\frac{dy}{dx}=6(1.1)-6=0.6$, i.e. positive

Since the gradient of the curve is negative just before the turning point and positive just after (i.e. $-\smile+$), **(1, −3) is a minimum point**

Problem 11. Find the maximum and minimum values of the curve $y=x^3-3x+5$ by (a) examining the gradient on either side of the turning points, and (b) determining the sign of the second derivative

Since $y=x^3-3x+5$ then $\frac{dy}{dx}=3x^2-3$

For a maximum or minimum value $\frac{dy}{dx}=0$

Hence $3x^2-3=0$

from which, $3x^2=3$

and $x=\pm 1$

When $x=1$, $y=(1)^3-3(1)+5=3$

When $x=-1$, $y=(-1)^3-3(-1)+5=7$

Hence (1, 3) and (−1, 7) are the co-ordinates of the turning points.

(a) Considering the point (1, 3):

If x is slightly less than 1, say 0.9, then $\frac{dy}{dx}=3(0.9)^2-3$, which is negative.

If x is slightly more than 1, say 1.1, then $\frac{dy}{dx}=3(1.1)^2-3$, which is positive.

Since the gradient changes from negative to positive, **the point (1, 3) is a minimum point**.

Considering the point (−1, 7):

If x is slightly less than −1, say −1.1, then $\frac{dy}{dx}=3(-1.1)^2-3$, which is positive.

If x is slightly more than −1, say −0.9, then $\frac{dy}{dx}=3(-0.9)^2-3$, which is negative.

Since the gradient changes from positive to negative, **the point (−1, 7) is a maximum point**.

(b) Since $\frac{dy}{dx}=3x^2-3$, then $\frac{d^2y}{dx^2}=6x$

When $x=1$, $\frac{d^2y}{dx^2}$ is positive, hence (1, 3) is a **minimum value**.

When $x=-1$, $\frac{d^2y}{dx^2}$ is negative, hence (−1, 7) is a **maximum value**.

Thus the maximum value is 7 and the minimum value is 3.

It can be seen that the second differential method of determining the nature of the turning points is, in this case, quicker when investigating the gradient.

Problem 12. Locate the turning point on the following curve and determine whether it is a maximum or minimum point: $y=4\theta+e^{-\theta}$

Since $y=4\theta+e^{-\theta}$ then $\frac{dy}{d\theta}=4-e^{-\theta}=0$ for a maximum or minimum value.

Hence $4=e^{-\theta}$ and $\frac{1}{4}=e^{\theta}$

giving $\theta=\ln\frac{1}{4}=-1.3863$

When $\theta=-1.3863$,

$$y=4(-1.3863)+e^{-(-1.3863)}=5.5452+4.0000$$
$$=-1.5452$$

Thus (−1.3863, −1.5452) are the co-ordinates of the turning point.

$$\frac{d^2y}{d\theta^2}=e^{-\theta}$$

When $\theta=-1.3863$, $\frac{d^2y}{d\theta^2}=e^{+1.3863}=4.0$, which is positive, hence
(−1.3863, −1.5452) is a minimum point.

Problem 13. Determine the co-ordinates of the maximum and minimum values of the graph $y=\frac{x^3}{3}-\frac{x^2}{2}-6x+\frac{5}{3}$ and distinguish between them. Sketch the graph

Following the given procedure:

(i) Since $y = \frac{x^3}{3} - \frac{x^2}{2} - 6x + \frac{5}{3}$ then

$$\frac{dy}{dx} = x^2 - x - 6$$

(ii) At a turning point, $\frac{dy}{dx} = 0$

Hence $\quad x^2 - x - 6 = 0$

i.e. $\quad (x+2)(x-3) = 0$

from which $\quad x = -2$ or $x = 3$

(iii) When $x = -2$

$$y = \frac{(-2)^3}{3} - \frac{(-2)^2}{2} - 6(-2) + \frac{5}{3} = 9$$

When $x = 3, y = \frac{(3)^3}{3} - \frac{(3)^2}{2} - 6(3) + \frac{5}{3}$

$$= -11\frac{5}{6}$$

Thus the co-ordinates of the turning points are (−2, 9) and $\left(3, -11\frac{5}{6}\right)$

(iv) Since $\frac{dy}{dx} = x^2 - x - 6$ then $\frac{d^2y}{dx^2} = 2x - 1$

When $x = -2$, $\frac{d^2y}{dx^2} = 2(-2) - 1 = -5$, which is negative.

Hence (−2, 9) is a maximum point.

When $x = 3$, $\frac{d^2y}{dx^2} = 2(3) - 1 = 5$, which is positive.

Hence $\left(3, -11\frac{5}{6}\right)$ is a minimum point.

Knowing (−2, 9) is a maximum point (i.e. crest of a wave), and $\left(3, -11\frac{5}{6}\right)$ is a minimum point (i.e. bottom of a valley) and that when $x = 0$, $y = \frac{5}{3}$, a sketch may be drawn as shown in Fig. 38.6.

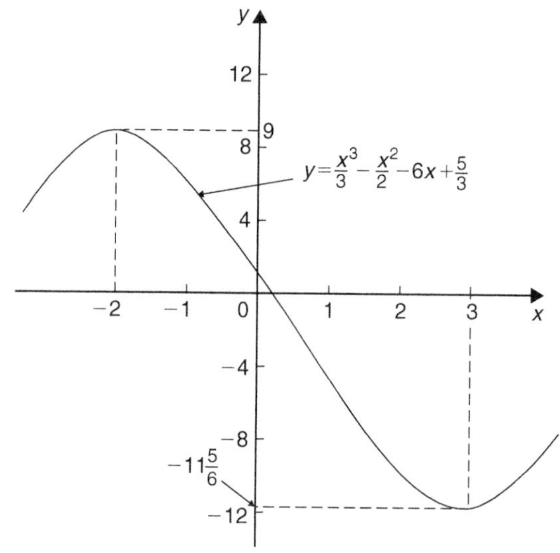

Figure 38.6

Problem 14. Determine the turning points on the curve $y = 4\sin x - 3\cos x$ in the range $x = 0$ to $x = 2\pi$ radians, and distinguish between them. Sketch the curve over one cycle

Since $y = 4\sin x - 3\cos x$ then

$\frac{dy}{dx} = 4\cos x + 3\sin x = 0$, for a turning point,

from which, $4\cos x = -3\sin x$ and $\frac{-4}{3} = \frac{\sin x}{\cos x}$

$$= \tan x$$

Hence $\quad x = \tan^{-1}\left(\frac{-4}{3}\right) = 126.87°$ or $306.87°$, since tangent is negative in the second and fourth quadrants.

When $\quad x = 126.87°$,

$$y = 4\sin 126.87° - 3\cos 126.87° = 5$$

When $\quad x = 306.87°$

$$y = 4\sin 306.87° - 3\cos 306.87° = -5$$

$$126.87° = \left(125.87° \times \frac{\pi}{180}\right) \text{ radians}$$

$$= 2.214 \text{ rad}$$

$$306.87° = \left(306.87° \times \frac{\pi}{180}\right) \text{ radians}$$

$$= 5.356 \text{ rad}$$

Hence (2.214, 5) and (5.356, −5) are the co-ordinates of the turning points.

$$\frac{d^2y}{dx^2} = -4\sin x + 3\cos x$$

When $x = 2.214$ rad,

$\frac{d^2y}{dx^2} = -4\sin 2.214 + 3\cos 2.214$, which is negative.

Hence (2.214, 5) is a maximum point.

When $x = 5.356$ rad,

$\frac{d^2y}{dx^2} = -4\sin 5.356 + 3\cos 5.356$, which is positive.

Hence (5.356, −5) is a minimum point.

A sketch of $y = 4\sin x - 3\cos x$ is shown in Fig. 38.7.

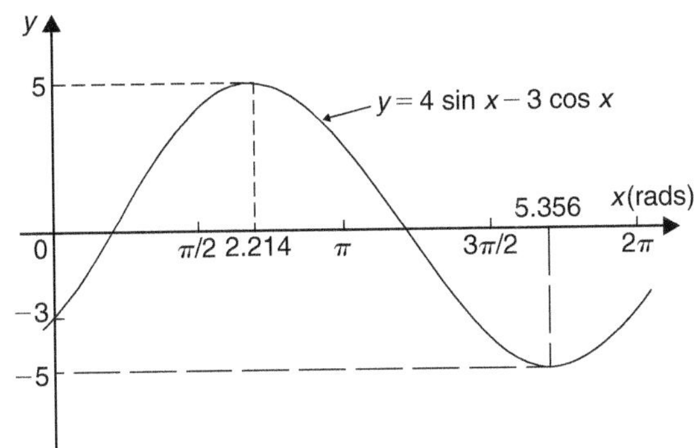

Figure 38.7

Now try the following Practice Exercise

Practice Exercise 194 Turning points (Answers on page 722)

In Problems 1 to 11, find the turning points and distinguish between them.

1. $y = x^2 - 6x$
2. $y = 8 + 2x - x^2$
3. $y = x^2 - 4x + 3$
4. $y = 3 + 3x^2 - x^3$
5. $y = 3x^2 - 4x + 2$
6. $x = \theta(6 - \theta)$
7. $y = 4x^3 + 3x^2 - 60x - 12$
8. $y = 5x - 2\ln x$
9. $y = 2x - e^x$
10. $y = t^3 - \frac{t^2}{2} - 2t + 4$
11. $x = 8t + \frac{1}{2t^2}$
12. Determine the maximum and minimum values on the graph $y = 12\cos\theta - 5\sin\theta$ in the range $\theta = 0$ to $\theta = 360°$. Sketch the graph over one cycle showing relevant points.
13. Show that the curve $y = \frac{2}{3}(t-1)^3 + 2t(t-2)$ has a maximum value of $\frac{2}{3}$ and a minimum value of −2.

38.4 Practical problems involving maximum and minimum values

There are many **practical problems** involving maximum and minimum values which occur in science and engineering. Usually, an equation has to be determined from given data, and rearranged where necessary, so that it contains only one variable. Some examples are demonstrated in Problems 15 to 20.

Problem 15. A rectangular area is formed having a perimeter of 40 cm. Determine the length and breadth of the rectangle if it is to enclose the maximum possible area

Let the dimensions of the rectangle be x and y. Then the perimeter of the rectangle is $(2x + 2y)$. Hence

$$2x + 2y = 40, \quad \text{or} \quad x + y = 20 \qquad (1)$$

Since the rectangle is to enclose the maximum possible area, a formula for area A must be obtained in terms of one variable only.

Area $A = xy$. From equation (1), $x = 20 - y$

Hence, area $A = (20 - y)y = 20y - y^2$

$\frac{dA}{dy} = 20 - 2y = 0$ for a turning point, from which, $y = 10$ cm.

$\dfrac{d^2A}{dy^2}=-2$, which is negative, giving a maximum point.

When $y=10$ cm, $x=10$ cm, from equation (1).

Hence the length and breadth of the rectangle are each 10 cm, i.e. a square gives the maximum possible area. When the perimeter of a rectangle is 40 cm, the maximum possible area is $10\times 10=$ **100 cm^2**.

Problem 16. A rectangular sheet of metal having dimensions 20 cm by 12 cm has squares removed from each of the four corners and the sides bent upwards to form an open box. Determine the maximum possible volume of the box

The squares to be removed from each corner are shown in Fig. 38.8, having sides x cm. When the sides are bent upwards the dimensions of the box will be: length $(20-2x)$ cm, breadth $(12-2x)$ cm and height, x cm.

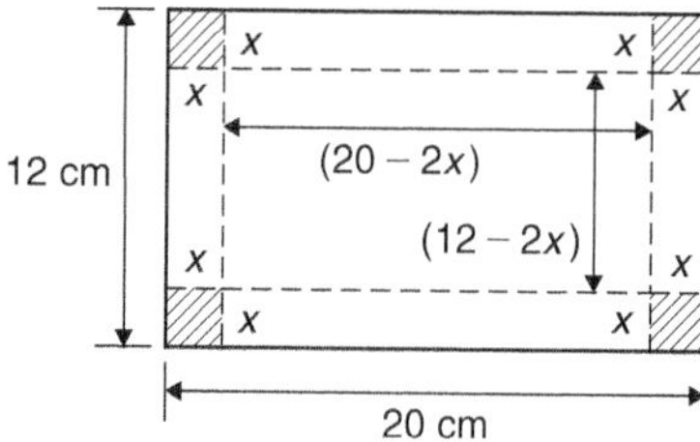

Figure 38.8

Volume of box, $V=(20-2x)(12-2x)(x)$

$$=240x-64x^2+4x^3$$

$\dfrac{dV}{dx}=240-128x+12x^2=0$ for a turning point.

Hence $4(60-32x+3x^2)=0$, i.e. $3x^2-32x+60=0$

Using the quadratic formula,

$$x=\frac{32\pm\sqrt{(-32)^2-4(3)(60)}}{2(3)}$$

$$=8.239\text{ cm or }2.427\text{ cm.}$$

Since the breadth is $(12-2x)$ cm then $x=8.239$ cm is not possible and is neglected.

Hence $x=2.427$ cm.

$$\frac{d^2V}{dx^2}=-128+24x$$

When $x=2.427$, $\dfrac{d^2V}{dx^2}$ is negative, giving a maximum value.

The dimensions of the box are:

length $=20-2(2.427)=15.146$ cm,

breadth $=12-2(2.427)=7.146$ cm, and

height $=2.427$ cm.

Maximum volume $=(15.146)(7.146)(2.427)$

$=$ **262.7 cm^3**

Problem 17. Determine the height and radius of a cylinder of volume 200 cm^3 which has the least surface area

Let the cylinder have radius r and perpendicular height h.

From Chapter 25, Volume of cylinder, $V=\pi r^2h=200$ (1)

Surface area of cylinder, $A=2\pi rh+2\pi r^2$

Least surface area means minimum surface area and a formula for the surface area in terms of one variable only is required.

From equation (1), $h=\dfrac{200}{\pi r^2}$ (2)

Hence surface area,

$$A=2\pi r\left(\frac{200}{\pi r^2}\right)+2\pi r^2$$

$$=\frac{400}{r}+2\pi r^2=400r^{-1}+2\pi r^2$$

$\dfrac{dA}{dr}=\dfrac{-400}{r^2}+4\pi r=0$, for a turning point.

Hence $\quad 4\pi r=\dfrac{400}{r^2}$

and $\quad r^3=\dfrac{400}{4\pi}$

from which, $\quad r=\sqrt[3]{\dfrac{100}{\pi}}=3.169$ cm.

$$\frac{d^2A}{dr^2}=\frac{800}{r^3}+4\pi$$

When $r=3.169$ cm, $\dfrac{d^2A}{dr^2}$ is positive, giving a minimum value.

From equation (2), when $r=3.169$ cm,

$$h=\frac{200}{\pi(3.169)^2}=6.339\text{ cm.}$$

Hence for the least surface area, a cylinder of volume 200 cm^3 has a radius of 3.169 cm and height of 6.339 cm.

Problem 18. Determine the area of the largest piece of rectangular ground that can be enclosed by 100 m of fencing, if part of an existing straight wall is used as one side

Let the dimensions of the rectangle be x and y as shown in Fig. 38.9, where PQ represents the straight wall.

From Fig. 38.9, $\quad x + 2y = 100 \quad (1)$

Area of rectangle, $\quad A = xy \quad (2)$

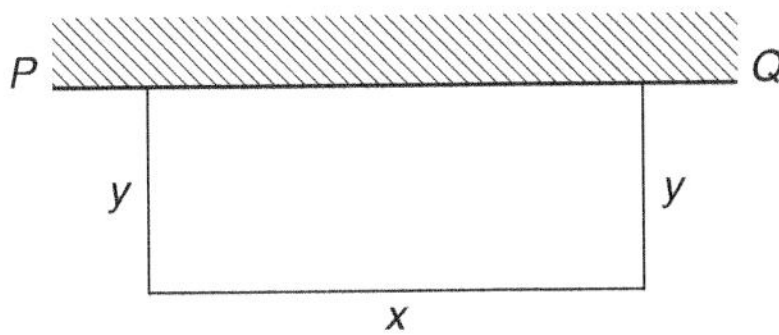

Figure 38.9

Since the maximum area is required, a formula for area A is needed in terms of one variable only.
From equation (1), $x = 100 - 2y$
Hence, area $A = xy = (100 - 2y)y = 100y - 2y^2$
$\frac{dA}{dy} = 100 - 4y = 0$, for a turning point, from which, $y = 25$ m.
$\frac{d^2A}{dy^2} = -4$, which is negative, giving a maximum value.
When $y = 25$ m, $x = 50$ m from equation (1).
Hence the **maximum possible area**

$$= xy = (50)(25) = \mathbf{1250\ m^2}$$

Problem 19. An open rectangular box with square ends is fitted with an overlapping lid which covers the top and the front face. Determine the maximum volume of the box if 6 m^2 of metal are used in its construction

A rectangular box having square ends of side x and length y is shown in Fig. 38.10.
Surface area of box, A, consists of two ends and five faces (since the lid also covers the front face).

Hence $\quad A = 2x^2 + 5xy = 6 \quad (1)$

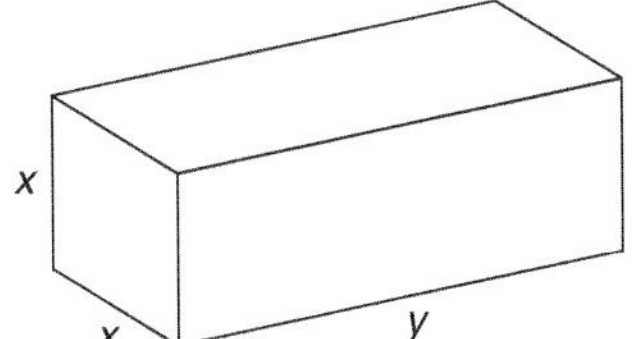

Figure 38.10

Since it is the maximum volume required, a formula for the volume in terms of one variable only is needed.
Volume of box, $V = x^2y$
From equation (1),

$$y = \frac{6 - 2x^2}{5x} = \frac{6}{5x} - \frac{2x}{5} \quad (2)$$

$$\text{Hence volume } V = x^2y = x^2\left(\frac{6}{5x} - \frac{2x}{5}\right) = \frac{6x}{5} - \frac{2x^3}{5}$$

$\frac{dV}{dx} = \frac{6}{5} - \frac{6x^2}{5} = 0$ for a maximum or minimum value.

Hence $6 = 6x^2$, giving $x = 1$ m ($x = -1$ is not possible, and is thus neglected).

$$\frac{d^2V}{dx^2} = \frac{-12x}{5}$$

When $x = 1$, $\frac{d^2V}{dx^2}$ is negative, giving a maximum value.

From equation (2), when $x = 1$, $y = \frac{6}{5(1)} - \frac{2(1)}{5} = \frac{4}{5}$

Hence the maximum volume of the box is given by $V = x^2y = (1)^2\left(\frac{4}{5}\right) = \mathbf{\frac{4}{5}\ m^3}$

Problem 20. Find the diameter and height of a cylinder of maximum volume which can be cut from a sphere of radius 12 cm

A cylinder of radius r and height h is shown enclosed in a sphere of radius $R = 12$ cm in Fig. 38.11.

Volume of cylinder, $\quad V = \pi r^2 h \quad (1)$

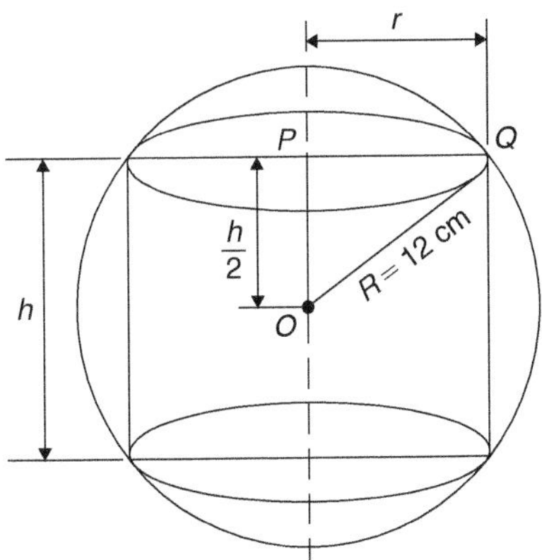

Figure 38.11

Using the right-angled triangle OPQ shown in Fig. 38.11,

$$r^2 + \left(\frac{h}{2}\right)^2 = R^2 \text{ by Pythagoras' theorem,}$$

i.e. $$r^2 + \frac{h^2}{4} = 144 \qquad (2)$$

Since the maximum volume is required, a formula for the volume V is needed in terms of one variable only.

From equation (2), $r^2 = 144 - \dfrac{h^2}{4}$

Substituting into equation (1) gives:

$$V = \pi\left(144 - \frac{h^2}{4}\right)h = 144\pi h - \frac{\pi h^3}{4}$$

$\dfrac{dV}{dh} = 144\pi - \dfrac{3\pi h^2}{4} = 0$, for a maximum or minimum value.

Hence $144\pi = \dfrac{3\pi h^2}{4}$, from which,

$$h = \sqrt{\frac{(144)(4)}{3}} = 13.86\,\text{cm}.$$

$$\frac{d^2V}{dh^2} = \frac{-6\pi h}{4}$$

When $h = 13.86$, $\dfrac{d^2V}{dh^2}$ is negative, giving a maximum value.

From equation (2),

$r^2 = 144 - \dfrac{h^2}{4} = 144 - \dfrac{13.86^2}{4}$, from which, radius $r = 9.80\,\text{cm}$

Diameter of cylinder $= 2r = 2(9.80) = 19.60\,\text{cm}$.

Hence the cylinder having the maximum volume that can be cut from a sphere of radius 12 cm is one in which the diameter is 19.60 cm and the height is 13.86 cm.

Now try the following Practice Exercise

Practice Exercise 195 Practical maximum and minimum problems (Answers on page 722)

1. The speed, v, of a car (in m/s) is related to time t s by the equation $v = 3 + 12t - 3t^2$. Determine the maximum speed of the car in km/h.
2. Determine the maximum area of a rectangular piece of land that can be enclosed by 1200 m of fencing.
3. A shell is fired vertically upwards and its vertical height, x metres, is given by: $x = 24t - 3t^2$, where t is the time in seconds. Determine the maximum height reached.
4. A lidless box with square ends is to be made from a thin sheet of metal. Determine the least area of the metal for which the volume of the box is $3.5\,\text{m}^3$.
5. A closed cylindrical container has a surface area of $400\,\text{cm}^2$. Determine the dimensions for maximum volume.
6. Calculate the height of a cylinder of maximum volume that can be cut from a cone of height 20 cm and base radius 80 cm.
7. The power developed in a resistor R by a battery of emf E and internal resistance r is given by $P = \dfrac{E^2R}{(R+r)^2}$. Differentiate P with respect to R and show that the power is a maximum when $R = r$.
8. Find the height and radius of a closed cylinder of volume $125\,\text{cm}^3$ which has the least surface area.
9. Resistance to motion, F, of a moving vehicle, is given by: $F = \dfrac{5}{x} + 100x$. Determine the minimum value of resistance.
10. An electrical voltage E is given by: $E = (15\sin 50\pi t + 40\cos 50\pi t)$ volts, where t is the time in seconds. Determine the maximum value of voltage.
11. The fuel economy E of a car, in miles per gallon, is given by:

    $$E = 21 + 2.10 \times 10^{-2}v^2 - 3.80 \times 10^{-6}v^4$$

    where v is the speed of the car in miles per hour.
    Determine, correct to 3 significant figures, the most economical fuel consumption, and the speed at which it is achieved.
12. The horizontal range of a projectile, x, launched with velocity u at an angle θ to the horizontal is given by: $x = \dfrac{2u^2\sin\theta\cos\theta}{g}$

To achieve maximum horizontal range, determine the angle the projectile should be launched at.

13. The signalling range, x, of a submarine cable is given by the formula: $x = r^2 \ln\left(\frac{1}{r}\right)$ where r is the ratio of the radii of the conductor and cable. Determine the value of r for maximum range.

14. The equation $y = 2e^{-t} \sin 2t$ represents the motion of a particle performing damped vibrations, y being the displacement from its mean position at time t. Calculate, correct to 3 significant figures, the maximum displacement.

38.5 Points of inflexion

As mentioned earlier in the chapter, it is possible to have a turning point, the gradient on either side of which is the same. This is called a **point of inflexion**.

Procedure to determine points of inflexion:

(i) Given $y = f(x)$, determine $\frac{dy}{dx}$ and $\frac{d^2y}{dx^2}$

(ii) Solve the equation $\frac{d^2y}{dx^2} = 0$

(iii) Test whether there is a change of sign occurring in $\frac{d^2y}{dx^2}$. This is achieved by substituting into the expression for $\frac{d^2y}{dx^2}$ firstly a value of x just less than the solution and then a value just greater than the solution.

(iv) A point of inflexion has been found if $\frac{d^2y}{dx^2} = 0$ **and** there is a change of sign.

This procedure is demonstrated in the following worked problems.

Problem 21. Determine the point(s) of inflexion (if any) on the graph of the function

$$y = x^3 - 6x^2 + 9x + 5$$

Find also any other turning points. Hence, sketch the graph.

Using the above procedure:

(i) Given $y = x^3 - 6x^2 + 9x + 5$,

$\frac{dy}{dx} = 3x^2 - 12x + 9$ and $\frac{d^2y}{dx^2} = 6x - 12$

(ii) Solving the equation $\frac{d^2y}{dx^2} = 0$ gives: $6x - 12 = 0$ from which, $6x = 12$ and $\boldsymbol{x = 2}$

Hence, if there is a point of inflexion, it occurs at $x = 2$

(iii) Taking a value just less than 2, say, 1.9:

$\frac{d^2y}{dx^2} = 6x - 12 = 6(1.9) - 12$, which is negative.

Taking a value just greater than 2, say, 2.1:

$\frac{d^2y}{dx^2} = 6x - 12 = 6(2.1) - 12$, which is positive.

(iv) Since a change of sign has occurred a point of inflexion exists at $x = 2$

When $x = 2$, $y = 2^3 - 6(2)^2 + 9(2) + 5 = 7$

i.e. **a point of inflexion occurs at the co-ordinates (2, 7)**

From above, $\frac{dy}{dx} = 3x^2 - 12x + 9 = 0$ for a turning point

i.e. $\quad x^2 - 4x + 3 = 0$

Using the quadratic formula or factorising (or by calculator), the solution is: $x = 1$ or $x = 3$

Since $y = x^3 - 6x^2 + 9x + 5$, then

when $x = 1$, $y = 1^3 - 6(1)^2 + 9(1) + 5 = 9$

and when $x = 3$, $y = 3^3 - 6(3)^2 + 9(3) + 5 = 5$

Hence, there are turning points at (1, 9) and at (3, 5)

Since $\frac{d^2y}{dx^2} = 6x - 12$, when $x = 1$, $\frac{d^2y}{dx^2} = 6(1) - 12$ which is negative – hence a maximum point and when $x = 3$, $\frac{d^2y}{dx^2} = 6(3) - 12$ which is positive – hence a minimum point

Thus, **(1, 9) is a maximum point and (3, 5) is a minimum point**

A sketch of the graph $y = x^3 - 6x^2 + 9x + 5$ is shown in Fig. 38.12

Problem 22. Determine the point(s) of inflexion (if any) on the graph of the function

$$y = x^4 - 24x^2 + 5x + 60$$

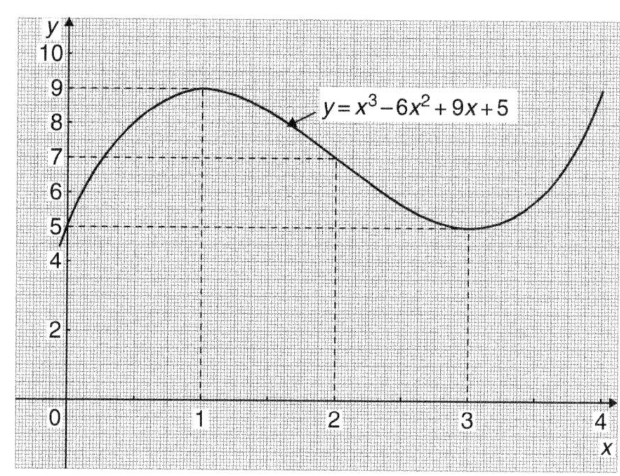

Figure 38.12

Using the above procedure:

(i) Given $y = x^4 - 24x^2 + 5x + 60$,
$\frac{dy}{dx} = 4x^3 - 48x + 5$ and $\frac{d^2y}{dx^2} = 12x^2 - 48$

(ii) Solving the equation $\frac{d^2y}{dx^2} = 0$ gives:
$12x^2 - 48 = 0$ from which, $12x^2 = 48$ and $x^2 = 4$ from which, $x = \sqrt{4} = \pm 2$
Hence, if there are points of inflexion, they occur at $x = 2$ and at $x = -2$

(iii) Taking a value just less than 2, say, 1.9:
$\frac{d^2y}{dx^2} = 12x^2 - 48 = 12(1.9)^2 - 48$, which is negative.
Taking a value just greater than 2, say, 2.1:
$\frac{d^2y}{dx^2} = 12x^2 - 48 = 12(2.1)^2 - 48$, which is positive.
Taking a value just less than −2, say, −2.1:
$\frac{d^2y}{dx^2} = 12x^2 - 48 = 12(-2.1)^2 - 48$, which is positive.
Taking a value just greater than −2, say, −1.9:
$\frac{d^2y}{dx^2} = 12x^2 - 48 = 12(-1.9)^2 - 48$, which is negative.

(iv) Since changes of signs have occurred, points of inflexion exist at $x = 2$ and $x = -2$
When $x = 2$,
$y = 2^4 - 24(2)^2 + 5(2) + 60 = -10$
When $x = -2$,
$y = (-2)^4 - 24(-2)^2 + 5(-2) + 60 = -30$

i.e. **points of inflexion occur at the co-ordinates (2, −10) and at (−2, −30)**

Now try the following Practice Exercise

Practice Exercise 196 Further problems on points of inflexion (Answers on page 723)

1. Find the points of inflexion (if any) on the graph of the function:
$$y = \frac{1}{3}x^3 - \frac{1}{2}x^2 - 2x + \frac{1}{12}$$
2. Find the points of inflexion (if any) on the graph of the function:
$$y = 4x^3 + 3x^2 - 18x - \frac{5}{8}$$
3. Find the point(s) of inflexion on the graph of the function $y = x + \sin x$ for $0 < x < 2\pi$.
4. Find the point(s) of inflexion on the graph of the function $y = 3x^3 - 27x^2 + 15x + 17$.
5. Find the point(s) of inflexion on the graph of the function $y = 2xe^{-x}$.
6. The displacement, s, of a particle is given by: $s = 3t^3 - 9t^2 + 10$. Determine the maximum, minimum and point of inflexion of s.

38.6 Tangents and normals

Tangents

The equation of the tangent to a curve $y = f(x)$ at the point (x_1, y_1) is given by:
$$y - y_1 = m(x - x_1)$$
where $m = \frac{dy}{dx}$ = gradient of the curve at (x_1, y_1)

Problem 23. Find the equation of the tangent to the curve $y = x^2 - x - 2$ at the point (1, −2)

Gradient, $m = \frac{dy}{dx} = 2x - 1$

At the point (1, −2), $x = 1$ and $m = 2(1) - 1 = 1$
Hence the equation of the tangent is:

$$y - y_1 = m(x - x_1)$$
i.e. $y - -2 = 1(x - 1)$
i.e. $y + 2 = x - 1$
or $\mathbf{y = x - 3}$

The graph of $y=x^2-x-2$ is shown in Fig. 38.13. The line AB is the tangent to the curve at the point C, i.e. $(1, -2)$, and the equation of this line is $y=x-3$

Normals

The normal at any point on a curve is the line that passes through the point and is at right angles to the tangent. Hence, in Fig. 38.13, the line CD is the normal.

It may be shown that if two lines are at right angles then the product of their gradients is -1. Thus if m is the gradient of the tangent, then the gradient of the normal is $-\frac{1}{m}$

Hence the equation of the normal at the point (x_1, y_1) is given by:

$$y-y_1=-\frac{1}{m}(x-x_1)$$

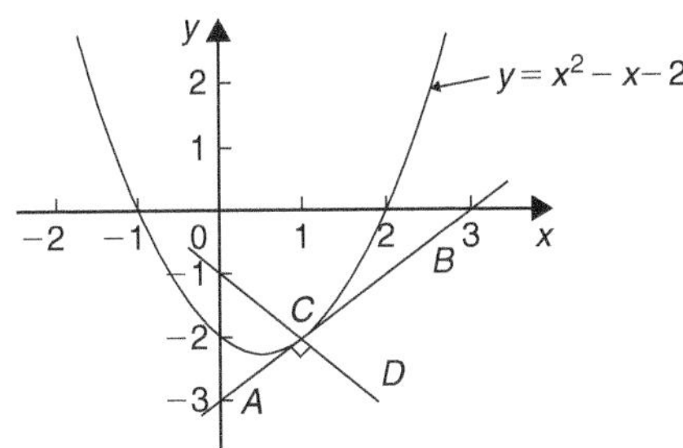

Figure 38.13

Problem 24. Find the equation of the normal to the curve $y=x^2-x-2$ at the point $(1, -2)$

$m=1$ from Problem 23, hence the equation of the normal is $y-y_1=-\frac{1}{m}(x-x_1)$

i.e. $y--2=-\frac{1}{1}(x-1)$

i.e. $y+2=-x+1$ or $\mathbf{y=-x-1}$

Thus the line CD in Fig. 38.13 has the equation $y=-x-1$

Problem 25. Determine the equations of the tangent and normal to the curve $y=\frac{x^3}{5}$ at the point $(-1, -\frac{1}{5})$

Gradient m of curve $y=\frac{x^3}{5}$ is given by

$$m=\frac{dy}{dx}=\frac{3x^2}{5}$$

At the point $(-1, -\frac{1}{5})$, $x=-1$ and

$$m=\frac{3(-1)^2}{5}=\frac{3}{5}$$

Equation of the tangent is:

$$y-y_1=m(x-x_1)$$

i.e. $y--\frac{1}{5}=\frac{3}{5}(x--1)$

i.e. $y+\frac{1}{5}=\frac{3}{5}(x+1)$

vor $5y+1=3x+3$

or $\mathbf{5y-3x=2}$

Equation of the normal is:

$$y-y_1=-\frac{1}{m}(x-x_1)$$

i.e. $y--\frac{1}{5}=\frac{-1}{\left(\frac{3}{5}\right)}(x--1)$

i.e. $y+\frac{1}{5}=-\frac{5}{3}(x+1)$

i.e. $y+\frac{1}{5}=-\frac{5}{3}x-\frac{5}{3}$

Multiplying each term by 15 gives:

$$15y+3=-25x-25$$

Hence **equation of the normal** is:

$$\mathbf{15y+25x+28=0}$$

Now try the following Practice Exercise

Practice Exercise 197 Tangents and normals (Answers on page 723)

For the following curves, at the points given, find (a) the equation of the tangent, and (b) the equation of the normal.

1. $y=2x^2$ at the point (1, 2)
2. $y=3x^2-2x$ at the point (2, 8)
3. $y=\frac{x^3}{2}$ at the point $\left(-1,-\frac{1}{2}\right)$
4. $y=1+x-x^2$ at the point (−2, −5)
5. $\theta=\frac{1}{t}$ at the point $\left(3,\frac{1}{3}\right)$

38.7 Small changes

If y is a function of x, i.e. $y=f(x)$, and the approximate change in y corresponding to a small change δx in x is required, then:

$$\frac{\delta y}{\delta x}\approx\frac{dy}{dx}$$

and $\delta y\approx\frac{dy}{dx}\cdot\delta x$ or $\delta y\approx f'(x)\cdot\delta x$

Problem 26. Given $y=4x^2-x$, determine the approximate change in y if x changes from 1 to 1.02

Since $y=4x^2-x$, then $\frac{dy}{dx}=8x-1$

Approximate change in y,

$\delta y\approx\frac{dy}{dx}\cdot\delta x\approx(8x-1)\delta x$

When $x=1$ and $\delta x=0.02$, $\boldsymbol{\delta y}\approx[8(1)-1](0.02)$

$$\approx\mathbf{0.14}$$

[Obviously, in this case, the exact value of δy may be obtained by evaluating y when $x=1.02$, i.e. $y=4(1.02)^2-1.02=3.1416$ and then subtracting from it the value of y when $x=1$, i.e. $y=4(1)^2-1=3$, giving $\boldsymbol{\delta y}=3.1416-3=\mathbf{0.1416}$. Using $\delta y=\frac{dy}{dx}\cdot\delta x$ above gave 0.14, which shows that the formula gives the approximate change in y for a small change in x]

Problem 27. The time of swing T of a pendulum is given by $T=k\sqrt{l}$, where k is a constant. Determine the percentage change in the time of swing if the length of the pendulum l changes from 32.1 cm to 32.0 cm

If $T=k\sqrt{l}=kl^{1/2}$

then $\frac{dT}{dl}=k\left(\frac{1}{2}l^{-1/2}\right)=\frac{k}{2\sqrt{l}}$

Approximate change in T,

$$\delta T\approx\frac{dT}{dl}\delta l\approx\left(\frac{k}{2\sqrt{l}}\right)\delta l\approx\left(\frac{k}{2\sqrt{l}}\right)(-0.1)$$

(negative since l decreases)

Percentage change

$$=\left(\frac{\text{approximate change in } T}{\text{original value of } T}\right)100\%$$

$$=\frac{\left(\frac{k}{2\sqrt{l}}\right)(-0.1)}{k\sqrt{l}}\times 100\%$$

$$=\left(\frac{-0.1}{2l}\right)100\%=\left(\frac{-0.1}{2(32.1)}\right)100\%$$

$$=\mathbf{-0.156\%}$$

Hence the percentage change in the time of swing is a decrease of 0.156%

Problem 28. A circular template has a radius of 10 cm (±0.02). Determine the possible error in calculating the area of the template. Find also the percentage error

Area of circular template, $A=\pi r^2$, hence

$\frac{dA}{dr}=2\pi r$

Approximate change in area,

$\delta A\approx\frac{dA}{dr}\cdot\delta r\approx(2\pi r)\delta r$

When $r=10$ cm and $\delta r=0.02$,

$\delta A=(2\pi 10)(0.02)\approx 0.4\pi\text{ cm}^2$

i.e. **the possible error in calculating the template area is approximately 1.257 cm^2.**

Percentage error $\approx\left(\frac{0.4\pi}{\pi(10)^2}\right)100\%=\mathbf{0.40\%}$

Now try the following Practice Exercise

Practice Exercise 198 Small changes (Answers on page 723)

1. Determine the change in y if x changes from 2.50 to 2.51 when (a) $y=2x-x^2$ (b) $y=\frac{5}{x}$

2. The pressure p and volume v of a mass of gas are related by the equation $pv=50$. If the pressure increases from 25.0 to 25.4, determine the approximate change in the volume of the gas. Find also the percentage change in the volume of the gas.

3. Determine the approximate increase in (a) the volume, and (b) the surface area of a cube of side x cm if x increases from 20.0 cm to 20.05 cm.

4. The radius of a sphere decreases from 6.0 cm to 5.96 cm. Determine the approximate change in (a) the surface area, and (b) the volume.

5. The rate of flow of a liquid through a tube is given by Poiseuilles's equation as: $Q=\frac{p\pi r^4}{8\eta L}$ where Q is the rate of flow, p is the pressure difference between the ends of the tube, r is the radius of the tube, L is the length of the tube and η is the coefficient of viscosity of the liquid. η is obtained by measuring Q, p, r and L. If Q can be measured accurate to $\pm 0.5\%$, p accurate to $\pm 3\%$, r accurate to $\pm 2\%$ and L accurate to $\pm 1\%$, calculate the maximum possible percentage error in the value of η.

Practice Exercise 199 Multiple-choice questions on applications of differentiation (Answers on page 723)

Each question has only one correct answer

1. The length ℓ metres of a certain metal rod at temperature t°C is given by:

$$\ell=1+4\times 10^{-5}t+4\times 10^{-7}t^2$$

The rate of change of length, in mm/°C, when the temperature is 400°C, is:
(a) 3.6×10^{-4} (b) 1.00036
(c) 0.36 (d) 3.2×10^{-4}

2. An alternating current, i amperes, is given by $i=100\sin 2\pi ft$ amperes, where f is the frequency in hertz and t is the time in seconds. The rate of change of current when $t=12$ ms and $f=50$ Hz is:
(a) 31348 A/s (b) −58.78 A/s
(c) 627.0 A/s (d) −25416 A/s

3. For the curve shown in Figure 38.14, which of the following statements is incorrect?
(a) P is a turning point
(b) Q is a minimum point
(c) R is a maximum value
(d) Q is a stationary value

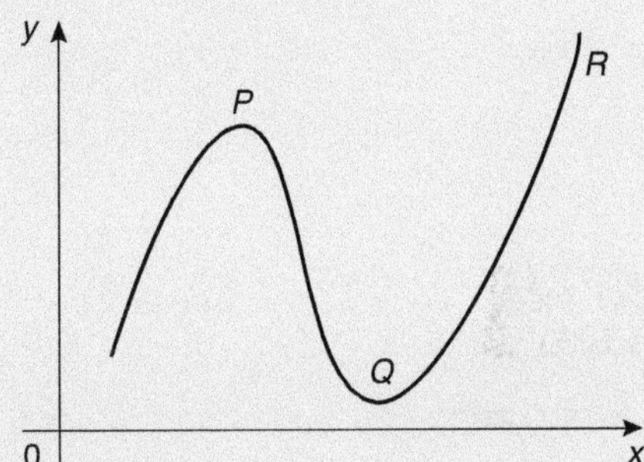

Figure 38.14

4. The equation of a curve is $y=2x^3-6x+1$. The maximum value of the curve is:
(a) −3 (b) 5 (c) 1 (d) −6

5. The resistance to motion F of a moving vehicle is given by $F=\frac{5}{x}+100x$. The minimum value of resistance is:
(a) −44.72 (b) 0.2236
(c) 44.72 (d) −0.2236

6. The vertical displacement, s, of a prototype model in a tank is given by $s=40\sin(0.1t)$ mm, where t is the time in seconds. The vertical velocity of the model, in mm/s, is given by:
(a) $-\cos 0.1t$ (b) $400\cos 0.1t$
(c) $-400\cos 0.1t$ (d) $4\cos 0.1t$

7. The current i in a circuit at time t seconds is given by $i=0.20(1-e^{-20t})$ A. When time $t=0.1$ s, the rate of change of current is:
(a) −1.022 A/s (b) 0.541 A/s
(c) 0.173 A/s (d) 0.373 A/s

8. The velocity of a car (in m/s) is related to time t seconds by the equation

$$v = 4.5 + 18t - 4.5t^2$$

The maximum speed of the car, in km/h, is:
(a) 81 (b) 6.25 (c) 22.5 (d) 77

9. The equation of a curve is $y = \frac{1}{3}x^3 + \frac{1}{2}x^2 - 6x$
The maximum value of the curve is:
(a) 13.5 (b) −3 (c) −7.33 (d) 2

10. The equation of the tangent to the curve $y = x^2 - 3x + 4$ at the point $(2, -1)$ is:
(a) $y = x - 1$
(b) $y = x - 3$
(c) $y = -6x - 12$
(d) $y = x + 3$

For fully worked solutions to each of the problems in Practice Exercises 192 to 198 in this chapter, go to the website:
www.routledge.com/cw/bird

Chapter 39

Solving equations by Newton's method

Why it is important to understand: Solving equations by Newton's method

There are many, many different types of equations used in every branch of engineering and science. There are straightforward methods for solving simple, quadratic and simultaneous equations; however, there are many other types of equations than these three. Great progress has been made in the engineering and scientific disciplines regarding the use of iterative methods for linear systems. In engineering it is important that we can solve any equation; iterative methods, such as the Newton–Raphson method, help us do that.

At the end of this chapter, you should be able to:

- define iterative methods
- state the Newton-Raphson formula
- use Newton's method to solve equations

39.1 Introduction to iterative methods

Many equations can only be solved graphically or by methods of successive approximation to the roots, called **iterative methods**. Three methods of successive approximations are (i) by using the Newton-Raphson formula, given in Section 39.2, (ii) the bisection methods, and (iii) an algebraic methods. The latter two methods are discussed in *Bird's Higher Engineering Mathematics, ninth edition*.

Each successive approximation method relies on a reasonably good first estimate of the value of a root being made. One way of determining this is to sketch a graph of the function, say $y=f(x)$, and determine the approximate values of roots from the points where the graph cuts the x-axis. Another way is by using a functional notation method. This method uses the property that the value of the graph of $f(x)=0$ changes sign for values of x just before and just after the value of a root. For example, one root of the equation $x^2-x-6=0$ is $x=3$

Using functional notation:

$$f(x)=x^2-x-6$$

$$f(2)=2^2-2-6=-4$$

$$f(4)=4^2-4-=+6$$

It can be seen from these results that the value of $f(x)$ changes from -4 at $f(2)$ to $+6$ at $f(4)$, indicating that a root lies between 2 and 4. This is shown more clearly in Fig. 39.1.

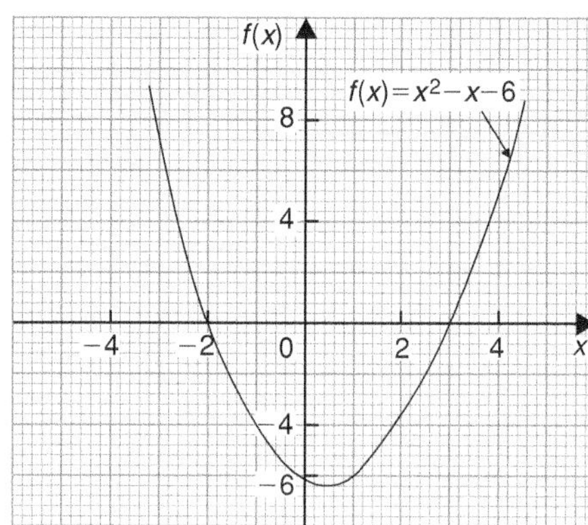

Figure 39.1

39.2 The Newton–Raphson method

The Newton–Raphson formula*, often just referred to as **Newton's method,** may be stated as follows:

*Who were Newton and Raphson? – **Sir Isaac Newton** PRS MP (25 December 1642–20 March 1727) was an English polymath. Newton showed that the motions of objects are governed by the same set of natural laws, by demonstrating the consistency between Kepler's laws of planetary motion and his theory of gravitation. To find out more go to www.routledge.com/cw/bird

Joseph Raphson was an English mathematician known best for the Newton–Raphson method for approximating the roots of an equation. To find out more go to www.routledge.com/cw/bird

if r_1 is the approximate value of a real root of the equation $f(x) = 0$, then a closer approximation to the root r_2 is given by:

$$r_2 = r_1 - \frac{f(r_1)}{f'(r_1)}$$

The advantages of Newton's method over other methods of successive approximations is that it can be used for any type of mathematical equation (i.e. ones containing trigonometric, exponential, logarithmic, hyperbolic and algebraic functions), and it is usually easier to apply than other methods. The method is demonstrated in the following worked problems.

39.3 Worked problems on the Newton–Raphson method

Problem 1. Use Newton's method to determine the positive root of the quadratic equation $5x^2 + 11x - 17 = 0$, correct to 3 significant figures. Check the value of the root by using the quadratic formula

The functional notation method is used to determine the first approximation to the root:

$$f(x) = 5x^2 + 11x - 17$$
$$f(0) = 5(0)^2 + 11(0) - 17 = -17$$
$$f(1) = 5(1)^2 + 11(1) - 17 = -1$$
$$f(2) = 5(2)^2 + 11(2) - 17 = 25$$

This shows that the value of the root is close to $x = 1$

Let the first approximation to the root, r_1, be 1. Newton's formula states that a closer approximation,

$$r_2 = r_1 - \frac{f(r_1)}{f'(r_1)}$$

$$f(x) = 5x^2 + 11x - 17, \quad \text{thus,}$$
$$f(r) = 5(r_1)^2 + 11(r_1) - 17$$
$$= 5(1)^2 + 11(1) - 17 = -1$$

$f'(x)$ is the differential coefficient of $f(x)$, i.e. $f'(x) = 10x + 11$ (see Chapter 36). Thus $f'(r_1) = 10(r_1) + 11 = 10(1) + 11 = 21$

By Newton's formula, a better approximation to the root is:

$$r_2 = 1 - \frac{-1}{21} = 1 - (-0.048) = 1.05,$$

correct to 3 significant figures

A still better approximation to the root, r_3, is given by:

$$r_3 = r_2 - \frac{f(r_2)}{f'(r_2)}$$
$$= 1.05 - \frac{[5(1.05)^2 + 11(1.05) - 17]}{[10(1.05) + 11]}$$
$$= 1.05 - \frac{0.0625}{21.5}$$
$$= 1.05 - 0.003 = 1.047,$$

i.e. 1.05, correct to 3 significant figures.
Since the value of r_2 and r_3 are the same when expressed to the required degree of accuracy, the required root is **1.05**, correct to 3 significant figures.
Checking, using the quadratic equation formula,

$$x = \frac{-11 \pm \sqrt{121 - 4(5)(-17)}}{(2)(5)}$$
$$= \frac{-11 \pm 21.47}{10}$$

The positive root is 1.047, i.e. **1.05**, correct to 3 significant figures.

Problem 2. Taking the first approximation as 2, determine the root of the equation $x^2 - 3\sin x + 2\ln(x+1) = 3.5$, correct to 3 significant figures, by using Newton's method

Newton's formula state that $r_2 = r_1 - \frac{f(r_1)}{f'(r_1)}$, where r_1 is a first approximation to the root and r_2 is a better approximation to the root.

Since $\quad fx = x^2 - 3\sin x + 2\ln(x+1) - 3.5$

$$f(r_1) = f(2) = 2^2 - 3\sin 2 + 2\ln 3 - 3.5,$$

where sin 2 means the sin of 2 radians

$$= 4 - 2.7279 + 2.1972 - 3.5 = -0.0307$$

$$f'x = 2x - 3\cos x + \frac{2}{x+1}$$

$$f'(r_1) = f'(2) = 2(2) - 3\cos 2 + \frac{2}{3}$$
$$= 4 + 1.2484 + 0.6667 = 5.9151$$

Hence,

$$r_2 = r_1 - \frac{f(r_1)}{f'(r_1)}$$
$$= 2 - \frac{-0.0307}{5.9151} = 2.005 \text{ or } 2.01,$$

correct to 3 significant figures.

A still better approximation to the root, r_3, is given by:

$$r_3 = r_2 - \frac{f(r_2)}{f'(r_2)}$$
$$= 2.005 - \frac{[(2.005)^2 - 3\sin 2.005 + 2\ln 3.005 - 3.5]}{\left[2(2.005) - 3\cos 2.005 + \frac{2}{2.005+1}\right]}$$
$$= 2.005 - \frac{(-0.00104)}{5.9376} = 2.005 + 0.000175$$

i.e. $r_3 = 2.01$, correct to 3 significant figures.
Since the values of r_2 and r_3 are the same when expressed to the required degree of accuracy, then the required root is **2.01**, correct to 3 significant figures.

Problem 3. Use Newton's method to find the positive root of:

$$(x+4)^3 - e^{1.92x} + 5\cos\frac{x}{3} = 9,$$

correct to 3 significant figures

The function notational method is used to determine the approximate value of the root:

$$f(x) = (x+4)^3 - e^{1.92x} + 5\cos\frac{x}{3} - 9$$
$$f(0) = (0+4)^3 - e^0 + 5\cos 0 - 9 = 59$$
$$f(1) = 5^3 - e^{1.92} + 5\cos\frac{1}{3} - 9 \approx 114$$
$$f(2) = 6^3 - e^{3.84} + 5\cos\frac{2}{3} - 9 \approx 164$$
$$f(3) = 7^3 - e^{5.76} + 5\cos 1 - 9 \approx 19$$
$$f(4) = 8^3 - e^{7.68} + 5\cos\frac{4}{3} - 9 \approx -1660$$

From these results, let a first approximation to the root be $r_1 = 3$.
Newton's formula states that a better approximation to the root,

$$r_2 = r_1 - \frac{f(r_1)}{f'(r_1)}$$
$$f(r_1) = f(3) = 7^3 - e^{5.76} + 5\cos 1 - 9$$
$$= 19.35$$
$$f'(x) = 3(x+4)^2 - 1.92e^{1.92x} - \frac{5}{3}\sin\frac{x}{3}$$
$$f'(r_1) = f'(3) = 3(7)^2 - 1.92e^{5.76} - \frac{5}{3}\sin 1$$
$$= -463.7$$

Thus, $\quad r_3 = 3 - \dfrac{19.35}{-463.7} = 3 + 0.042 = 3.042 = 3.04,$

correct to 3 significant figures.

$$\text{Similarly,} \quad r_3 = 3.042 - \frac{f(3.042)}{f'(3.042)}$$
$$= 3.042 - \frac{-1.146}{(-513.1)}$$
$$= 3.042 - 0.0022 = 3.0398 = 3.04,$$

correct to 3 significant figures.

Since r_2 and r_3 are the same when expressed to the required degree of accuracy, then the required root is **3.04**, correct to 3 significant figures.

Now try the following Practice Exercise

Practice Exercise 200 Newton's method (Answers on page 723)

In Problems 1 to 7, use **Newton's method** to solve the equations given to the accuracy stated.

1. $x^2 - 2x - 13 = 0$, correct to 3 decimal places
2. $3x^3 - 10x = 14$, correct to 4 significant figures
3. $x^4 - 3x^3 + 7x - 12 = 0$, correct to 3 decimal places
4. $3x^4 - 4x^3 + 7x = 12$, correct to 3 decimal places
5. $3 \ln x + 4x = 5$, correct to 3 decimal places
6. $x^3 = 5\cos 2x$, correct to 3 significant figures
7. $300e^{-2\theta} + \dfrac{\theta}{2} = 6$, correct to 3 significant figures
8. A Fourier analysis of the instantaneous value of a waveform can be represented by: $y = \left(t + \dfrac{\pi}{4}\right) + \sin t + \dfrac{1}{8}\sin 3t$

   Use Newton's method to determine the value of t near to 0.04, correct to 4 decimal places, when the amplitude, y, is 0.880
9. A damped oscillation of a system is given by the equation: $y = -7.4e^{0.5t} \sin 3t$. Determine the value of t near to 4.2, correct to 3 significant figure, when the magnitude y of the oscillation is zero
10. The critical speeds of oscillation, λ, of a loaded beam are given by the equation:

    $$\lambda^3 - 3.250\lambda^2 + \lambda - 0.063 = 0$$

    Determine the value of λ which is approximately equal to 3.0 by Newton's method, correct to 4 decimal places.

Practice Exercise 201 Multiple-choice questions on solving equations using Newton's method (Answers on page 723)

Each question has only one correct answer

1. The motion of a particle in an electrostatic field is described by the equation:

   $$y = x^3 + 3x^2 + 5x - 28$$

   When $x = 2$, y is approximately zero. Using one iteration of the Newton-Raphson method, a better approximation (correct to 2 decimal places) is:
   (a) 1.89 (b) 2.07 (c) 2.11 (d) 1.93
2. A first approximation indicates that a root of the equation $x^3 - 9x + 1 = 0$ lies close to 3. Using the Newton-Raphson method a second approximation, correct to 4 decimal places is:
   (a) 2.9444 (b) 3.0580
   (c) 2.9420 (c) 3.0556
3. Using Newton's method to find the real root of the equation $3x^3 - 7x = 15$, correct to 3 decimal places, and starting at $x = 2$, gives:
   (a) 1.987 (b) 2.157 (c) 2.132 (d) 2.172
4. Using Newton's method to find the real root of the equation $2\ln x + 3x - 7 = 0$, correct to 3 decimal places, and starting at $x = 2$, gives:
   (a) 2.096 (b) 1.893 (c) 1.904 (d) 2.034
5. Using Newton's method to find the real root of the equation $200e^{-2t} + t = 6$, correct to 3 decimal places, and starting at $x = 2$, gives:
   (a) 1.950 (b) 2.050 (c) 1.947 (d) 2.143

For fully worked solutions to each of the problems in Practice Exercise 200 in this chapter, go to the website:
www.routledge.com/cw/bird

Chapter 40

Maclaurin's series

Why it is important to understand: Maclaurin's series

One of the simplest kinds of function to deal with, in either algebra or calculus, is a polynomial (i.e. an expression of the form $a+bx+cx^2+dx^3+\ldots$). Polynomials are easy to substitute numerical values into, and they are easy to differentiate. One useful application of Maclaurin's series is to approximate, to a polynomial, functions which are not already in polynomial form. In the simple theory of flexure of beams, the slope, bending moment, shearing force, load and other quantities are functions of a derivative of y with respect to x. The elastic curve of a transversely loaded beam can be represented by the Maclaurin series. Substitution of the values of the derivatives gives a direct solution of beam problems. Another application of Maclaurin series is in relating inter-atomic potential functions. At this stage, not all of the above applications would have been met or understood; however, sufficient to say that Maclaurin's series has a number of applications in engineering and science.

At the end of this chapter, you should be able to:

- determine simple derivatives
- derive Maclaurin's theorem
- appreciate the conditions of Maclaurin's series
- use Maclaurin's series to determine the power series for simple trigonometric, logarithmic and exponential functions

40.1 Introduction

Some mathematical functions may be represented as power series, containing terms in ascending powers of the variable. For example,

$$e^x = 1 + x + \frac{x^2}{2!} + \frac{x^3}{3!} + \cdots$$

$$\sin x = x - \frac{x^3}{3!} + \frac{x^5}{5!} - \frac{x^7}{7!} + \cdots$$

and $$\cos x = 1 - \frac{x^2}{2!} + \frac{x^4}{4!} - \frac{x^6}{6!} + \cdots$$

Using a series, called **Maclaurin's* series**, mixed functions containing, say, algebraic, trigonometric and exponential functions, may be expressed solely as algebraic functions, and differentiation and integration can often be more readily performed.

*Who was Maclaurin? – **Colin Maclaurin** (February 1698–14 June 1746) was a Scottish mathematician who made important contributions to geometry and algebra. **The Maclaurin series** are named after him. To find out more go to www.routledge.com/cw/bird

To expand a function using Maclaurin's theorem, some knowledge of differentiation is needed. Here is a revision of derivatives of the main functions needed in this chapter.

y or $f(x)$	$\frac{dy}{dx}$ or $f'(x)$
ax^n	anx^{n-1}
$\sin ax$	$a\cos ax$
$\cos ax$	$-a\sin ax$
e^{ax}	ae^{ax}
$\ln ax$	$\frac{1}{x}$

Given a general function $f(x)$, then $f'(x)$ is the first derivative, $f''(x)$ is the second derivative and so on. Also, $f(0)$ means the value of the function when $x=0, f'(0)$ means the value of the first derivative when $x=0$ and so on.

40.2 Derivation of Maclaurin's theorem

Let the power series for $f(x)$ be

$$f(x) = a_0 + a_1x + a_2x^2 + a_3x^3 + a_4x^4 + a_5x^5 + \cdots \quad (1)$$

where $a_0, a_1, a_2, \ldots$ are constants.

When $x=0$, $\boldsymbol{f(0) = a_0}$

Differentiating equation (1) with respect to x gives:

$$f'(x) = a_1 + 2a_2x + 3a_3x^2 + 4a_4x^3 + 5a_5x^4 + \cdots \quad (2)$$

When $x=0$, $\boldsymbol{f'(0) = a_1}$

Differentiating equation (2) with respect to x gives:

$$f''(x) = 2a_2 + (3)(2)a_3x + (4)(3)a_4x^2 + (5)(4)a_5x^3 + \cdots \quad (3)$$

When $x=0$, $f''(0) = 2a_2 = 2!a_2$, i.e. $\boldsymbol{a_2 = \frac{f''(0)}{2!}}$

Differentiating equation (3) with respect to x gives:

$$f'''(x) = (3)(2)a_3 + (4)(3)(2)a_4x + (5)(4)(3)a_5x^2 + \cdots \quad (4)$$

When $x=0$, $f'''(0) = (3)(2)a_3 = 3!a_3$, i.e. $\boldsymbol{a_3 = \frac{f'''(0)}{3!}}$

Continuing the same procedure gives $a_4 = \dfrac{f^{iv}(0)}{4!}$, $a_5 = \dfrac{f^{v}(0)}{5!}$ and so on.

Substituting for $a_0, a_1, a_2, \ldots$ in equation (1) gives:

$$f(x) = f(0) + f'(0)x + \frac{f''(0)}{2!}x^2 + \frac{f'''(0)}{3!}x^3 + \cdots$$

i.e. $$f(x) = f(0) + xf'(0) + \frac{x^2}{2!}f''(0) + \frac{x^3}{3!}f'''(0) + \cdots \quad (5)$$

Equation (5) is a mathematical statement called **Maclaurin's theorem** or **Maclaurin's series**.

40.3 Conditions of Maclaurin's series

Maclaurin's series may be used to represent any function, say $f(x)$, as a power series provided that at $x = 0$ the following three conditions are met:

(a) $f(0) \neq \infty$

For example, for the function $f(x) = \cos x$, $f(0) = \cos 0 = 1$, thus $\cos x$ meets the condition. However, if $f(x) = \ln x$, $f(0) = \ln 0 = -\infty$, thus $\ln x$ does not meet this condition.

(b) $f'(0), f''(0), f'''(0), \ldots \neq \infty$

For example, for the function $f(x) = \cos x$, $f'(0) = -\sin 0 = 0$, $f''(0) = -\cos 0 = -1$ and so on; thus $\cos x$ meets this condition. However, if $f(x) = \ln x$, $f'(0) = \frac{1}{0} = \infty$, thus $\ln x$ does not meet this condition

(c) **The resultant Maclaurin's series must be convergent**.

In general, this means that the values of the terms, or groups of terms, must get progressively smaller and the sum of the terms must reach a limiting value.

For example, the series $1 + \frac{1}{2} + \frac{1}{4} + \frac{1}{8} + \cdots$ is convergent since the value of the terms is getting smaller and the sum of the terms is approaching a limiting value of 2.

40.4 Worked problems on Maclaurin's series

Problem 1. Determine the first four non-zero terms of the power series for $\cos x$

The values of $f(0), f'(0), f''(0), \ldots$ in Maclaurin's series are obtained as follows:

$f(x) = \cos x \qquad f(0) = \cos 0 = 1$

$f'(x) = -\sin x \qquad f'(0) = -\sin 0 = 0$

$f''(x) = -\cos x \qquad f''(0) = -\cos 0 = -1$

$f'''(x) = \sin x \qquad f'''(0) = \sin 0 = 0$

$f^{iv}(x) = \cos x \qquad f^{iv}(0) = \cos 0 = 1$

$f^{v}(x) = -\sin x \qquad f^{v}(0) = -\sin 0 = 0$

$f^{vi}(x) = -\cos x \qquad f^{vi}(0) = -\cos 0 = -1$

Substituting these values into equation (5) gives:

$$f(x) = \cos x = 1 + x(0) + \frac{x^2}{2!}(-1) + \frac{x^3}{3!}(0) + \frac{x^4}{4!}(1) + \frac{x^5}{5!}(0) + \frac{x^6}{6!}(-1) + \cdots$$

i.e. $$\mathbf{\cos x = 1 - \frac{x^2}{2!} + \frac{x^4}{4!} - \frac{x^6}{6!} + \cdots}$$

Problem 2. Determine the power series for $\cos 2\theta$

Replacing x with 2θ in the series obtained in Problem 1 gives:

$$\cos 2\theta = 1 - \frac{(2\theta)^2}{2!} + \frac{(2\theta)^4}{4!} - \frac{(2\theta)^6}{6!} + \cdots$$

$$= 1 - \frac{4\theta^2}{2} + \frac{16\theta^4}{24} - \frac{64\theta^6}{720} + \cdots$$

i.e. $$\mathbf{\cos 2\theta = 1 - 2\theta^2 + \frac{2}{3}\theta^4 - \frac{4}{45}\theta^6 + \cdots}$$

Problem 3. Using Maclaurin's series, find the first 4 (non zero) terms for the function $f(x) = \sin x$

$$f(x) = \sin x \qquad f(0) = \sin 0 = 0$$

$$f''(x) = -\sin x \qquad f''(0) = -\sin 0 = 0$$

$$f'''(x) = -\cos x \qquad f'''(0) = -\cos 0 = -1$$

$$f^{iv}(x) = \sin x \qquad f^{iv}(0) = \sin 0 = 0$$

$$f^{v}(x) = \cos x \qquad f^{v}(0) = \cos 0 = 1$$

$$f^{vi}(x) = -\sin x \qquad f^{vi}(0) = -\sin 0 = 0$$

$$f^{vii}(x) = -\cos x \qquad f^{vii}(0) = -\cos 0 = -1$$

Substituting the above values into Maclaurin's series of equation (5) gives:

$$\sin x = 0 + x(1) + \frac{x^2}{2!}(0) + \frac{x^3}{3!}(-1) + \frac{x^4}{4!}(0) + \frac{x^5}{5!}(1) + \frac{x^6}{6!}(0) + \frac{x^7}{7!}(-1) + \cdots$$

i.e. $\mathbf{\sin x = x - \frac{x^3}{3!} + \frac{x^5}{5!} - \frac{x^7}{7!} + \cdots}$

Problem 4. Using Maclaurin's series, find the first five terms for the expansion of the function $f(x) = e^{3x}$

$$f(x) = e^{3x} \qquad f(0) = e^0 = 1$$

$$f'(x) = 3e^{3x} \qquad f'(0) = 3e^0 = 3$$

$$f''(x) = 9e^{3x} \qquad f''(0) = 9e^0 = 9$$

$$f'''(x) = 27e^{3x} \qquad f'''(0) = 27e^0 = 27$$

$$f^{iv}(x) = 81e^{3x} \qquad f^{iv}(0) = 81e^0 = 81$$

Substituting the above values into Maclaurin's series of equation (5) gives:

$$e^{3x} = 1 + x(3) + \frac{x^2}{2!}(9) + \frac{x^3}{3!}(27) + \frac{x^4}{4!}(81) + \cdots$$

$$e^{3x} = 1 + 3x + \frac{9x^2}{2!} + \frac{27x^3}{3!} + \frac{81x^4}{4!} + \cdots$$

i.e. $\mathbf{e^{3x} = 1 + 3x + \frac{9x^2}{2} + \frac{9x^3}{2} + \frac{27x^4}{8} + \cdots}$

Problem 5. Determine the power series for $\tan x$ as far as the term in x^3

$$f(x) = \tan x$$

$$f(0) = \tan 0 = 0$$

$$f'(x) = \sec^2 x$$

$$f'(0) = \sec^2 0 = \frac{1}{\cos^2 0} = 1$$

$$f''(x) = (2\sec x)(\sec x \tan x)$$

$$= 2\sec^2 x \tan x$$

$$f''(0) = 2\sec^2 0 \tan 0 = 0$$

$$f'''(x) = (2\sec^2 x)(\sec^2 x) + (\tan x)(4\sec x \sec x \tan x), \text{ by the product rule,}$$

$$= 2\sec^4 x + 4\sec^2 x \tan^2 x$$

$$f'''(0) = 2\sec^4 0 + 4\sec^2 0 \tan^2 0 = 2$$

Substituting these values into equation (5) gives:

$$f(x) = \tan x = 0 + (x)(1) + \frac{x^2}{2!}(0) + \frac{x^3}{3!}(2)$$

i.e. $\mathbf{\tan x = x + \frac{1}{3}x^3}$

Problem 6. Expand $\ln(1+x)$ to five terms.

$$f(x) = \ln(1+x) \qquad f(0) = \ln(1+0) = 0$$

$$f'(x) = \frac{1}{(1+x)} \qquad f'(0) = \frac{1}{1+0} = 1$$

$$f''(x) = \frac{-1}{(1+x)^2} \qquad f''(0) = \frac{-1}{(1+0)^2} = -1$$

$$f'''(x) = \frac{2}{(1+x)^3} \qquad f'''(0) = \frac{2}{(1+0)^3} = 2$$

$$f^{iv}(x) = \frac{-6}{(1+x)^4} \qquad f^{iv}(0) = \frac{-6}{(1+0)^4} = -6$$

$$f^{v}(x) = \frac{24}{(1+x)^5} \qquad f^{v}(0) = \frac{24}{(1+0)^5} = 24$$

Substituting these values into equation (5) gives:

$$f(x) = \ln(1+x) = 0 + x(1) + \frac{x^2}{2!}(-1) + \frac{x^3}{3!}(2) + \frac{x^4}{4!}(-6) + \frac{x^5}{5!}(24)$$

i.e. $\mathbf{\ln(1+x) = x - \frac{x^2}{2} + \frac{x^3}{3} - \frac{x^4}{4} + \frac{x^5}{5} - \cdots}$

Problem 7. Expand $\ln(1-x)$ to five terms.

Replacing x by $-x$ in the series for $\ln(1+x)$ in Problem 6 gives:

$$\ln(1-x) = (-x) - \frac{(-x)^2}{2} + \frac{(-x)^3}{3} - \frac{(-x)^4}{4} + \frac{(-x)^5}{5} - \cdots$$

i.e. $\mathbf{\ln(1-x) = -x - \frac{x^2}{2} - \frac{x^3}{3} - \frac{x^4}{4} - \frac{x^5}{5} - \cdots}$

Problem 8. Determine the power series for $\ln\left(\frac{1+x}{1-x}\right)$

$\ln\left(\frac{1+x}{1-x}\right) = \ln(1+x) - \ln(1-x)$ by the laws of logarithms, and from Problems 6 and 7,

$$\ln\left(\frac{1+x}{1-x}\right) = \left(x - \frac{x^2}{2} + \frac{x^3}{3} - \frac{x^4}{4} + \frac{x^5}{5} - \cdots\right) - \left(-x - \frac{x^2}{2} - \frac{x^3}{3} - \frac{x^4}{4} - \frac{x^5}{5} - \cdots\right)$$

$$= 2x + \frac{2}{3}x^3 + \frac{2}{5}x^5 + \cdots$$

i.e. $\mathbf{\ln\left(\frac{1+x}{1-x}\right) = 2\left(x + \frac{x^3}{3} + \frac{x^5}{5} + \cdots\right)}$

Problem 9. Use Maclaurin's series to find the expansion of $(2+x)^4$

$f(x) = (2+x)^4 \qquad f(0) = 2^4 = 16$

$f'(x) = 4(2+x)^3 \qquad f'(0) = 4(2)^3 = 32$

$f''(x) = 12(2+x)^2 \qquad f''(0) = 12(2)^2 = 48$

$f'''(x) = 24(2+x)^1 \qquad f'''(0) = 24(2) = 48$

$f^{iv}(x) = 24 \qquad f^{iv}(0) = 24$

Substituting in equation (5) gives:

$$\mathbf{(2+x)^4}$$

$$= f(0) + xf'(0) + \frac{x^2}{2!}f''(0) + \frac{x^3}{3!}f'''(0) + \frac{x^4}{4!}f^{iv}(0)$$

$$= 16 + (x)(32) + \frac{x^2}{2!}(48) + \frac{x^3}{3!}(48) + \frac{x^4}{4!}(24)$$

$$\mathbf{= 16 + 32x + 24x^2 + 8x^3 + x^4}$$

(This expression could also have been obtained by applying the binomial theorem.)

Problem 10. Expand $e^{\frac{x}{2}}$ as far as the term in x^4

$f(x) = e^{\frac{x}{2}} \qquad f(0) = e^0 = 1$

$f'(x) = \frac{1}{2}e^{\frac{x}{2}} \qquad f'(0) = \frac{1}{2}e^0 = \frac{1}{2}$

$f''(x) = \frac{1}{4}e^{\frac{x}{2}} \qquad f''(0) = \frac{1}{4}e^0 = \frac{1}{4}$

$f'''(x) = \frac{1}{8}e^{\frac{x}{2}} \qquad f'''(0) = \frac{1}{8}e^0 = \frac{1}{8}$

$f^{iv}(x) = \frac{1}{16}e^{\frac{x}{2}} \qquad f^{iv}(0) = \frac{1}{16}e^0 = \frac{1}{16}$

Substituting in equation (5) gives:

$$e^{\frac{x}{2}} = f(0) + xf'(0) + \frac{x^2}{2!}f''(0) + \frac{x^3}{3!}f'''(0) + \frac{x^4}{4!}f^{iv}(0) + \cdots$$

$$= 1 + (x)\left(\frac{1}{2}\right) + \frac{x^2}{2!}\left(\frac{1}{4}\right) + \frac{x^3}{3!}\left(\frac{1}{8}\right) + \frac{x^4}{4!}\left(\frac{1}{16}\right) + \cdots$$

i.e. $\mathbf{e^{\frac{x}{2}} = 1 + \frac{1}{2}x + \frac{1}{8}x^2 + \frac{1}{48}x^3 + \frac{1}{384}x^4 + \cdots}$

Problem 11. Produce a power series for $\cos^2 2x$ as far as the term in x^6

From double angle formulae, $\cos 2A = 2\cos^2 A - 1$ (see Chapter 22).

from which, $\cos^2 A = \frac{1}{2}(1+\cos 2A)$

and $\cos^2 2x = \frac{1}{2}(1+\cos 4x)$

From Problem 1,

$$\cos x = 1 - \frac{x^2}{2!} + \frac{x^4}{4!} - \frac{x^6}{6!} + \cdots$$

hence
$$\cos 4x = 1 - \frac{(4x)^2}{2!} + \frac{(4x)^4}{4!} - \frac{(4x)^6}{6!} + \cdots$$
$$= 1 - 8x^2 + \frac{32}{3}x^4 - \frac{256}{45}x^6 + \cdots$$

Thus $\cos^2 2x = \frac{1}{2}(1+\cos 4x)$

$$= \frac{1}{2}\left(1 + 1 - 8x^2 + \frac{32}{3}x^4 - \frac{256}{45}x^6 + \cdots\right)$$

i.e. $\mathbf{\cos^2 2x = 1 - 4x^2 + \frac{16}{3}x^4 - \frac{128}{45}x^6 + \cdots}$

Now try the following Practice Exercise

Practice Exercise 202 Maclaurin's series (Answers on page 723)

1. Determine the first four terms of the power series for $\sin 2x$ using Maclaurin's series.
2. Use Maclaurin's series to produce a power series for e^{2x} as far as the term in x^4.
3. Use Maclaurin's theorem to determine the first three terms of the power series for $\ln(1+e^x)$.
4. Determine the power series for $\cos 4t$ as far as the term in t^6.
5. Expand $e^{\frac{3}{2}x}$ in a power series as far as the term in x^3.
6. Develop, as far as the term in x^4, the power series for $\sec 2x$.
7. Expand $e^{2\theta}\cos 3\theta$ as far as the term in θ^2 using Maclaurin's series.
8. Determine the first three terms of the series for $\sin^2 x$ by applying Maclaurin's theorem.
9. Use Maclaurin's series to determine the expansion of $(3+2t)^4$.

Practice Exercise 203 Multiple-choice questions on Maclaurin's series (Answers on page 723)

Each question has only one correct answer

1. Using Maclaurin's series, the first 4 terms of the power series for $\cos\theta$ are:

   (a) $1 + \frac{\theta^2}{2!} - \frac{\theta^4}{4!} + \frac{\theta^6}{6!}$
   (b) $1 - \frac{\theta^2}{2!} + \frac{\theta^4}{4!} - \frac{\theta^6}{6!}$
   (c) $1 - \frac{\theta^2}{2} + \frac{\theta^4}{4} - \frac{\theta^6}{6}$
   (d) $\theta - \frac{\theta^3}{3!} + \frac{\theta^5}{5!} - \frac{\theta^7}{7!}$

2. Using Maclaurin's series, the first 4 terms of the power series for $\sin t$ are:

   (a) $t - \frac{t^2}{3} + \frac{t^4}{5} - \frac{t^6}{7}$
   (b) $t + \frac{t^2}{2!} - \frac{t^4}{4!} + \frac{t^6}{6!}$
   (c) $1 - \frac{t^2}{2!} + \frac{t^4}{4!} - \frac{t^6}{6!}$
   (d) $t - \frac{t^3}{3!} + \frac{t^5}{5!} - \frac{t^7}{7!}$

3. Using Maclaurin's series, the first 4 terms of the power series for e^θ are:

   (a) $1 - \theta + \frac{\theta^2}{2!} - \frac{\theta^3}{3!}$
   (b) $\theta + \frac{\theta^2}{2!} - \frac{\theta^3}{3!} + \frac{\theta^4}{4!}$
   (c) $1 + \theta + \frac{\theta^2}{2!} + \frac{\theta^3}{3!}$
   (d) $1 + \theta + \frac{\theta^2}{2} + \frac{\theta^3}{3}$

4. Using Maclaurin's series, the fourth term of the power series for e^{2t} is:

(a) $\frac{4}{3}t^3$ (b) $\frac{2}{3}t^4$

(c) $2t^2$ (d) $\frac{2}{3}t^3$

5. Using Maclaurin's series, the third term of the power series for $\cos 3x$ is:

(a) $\frac{81}{80}x^6$ (b) $\frac{27}{8}x^4$

(c) $\frac{81}{40}x^5$ (d) $\frac{1}{8}x^4$

For fully worked solutions to each of the problems in Practice Exercises 202 in this chapter, go to the website:

www.routledge.com/cw/bird

COMPANION @ WEBSITE

Revision Test 10 Differentiation

This Revision Test covers the material contained in Chapters 36 to 40. *The marks for each question are shown in brackets at the end of each question.*

1. Differentiate the following with respect to the variable:
   (a) $y=5+2\sqrt{x^3}-\dfrac{1}{x^2}$ (b) $s=4e^{2\theta}\sin 3\theta$
   (c) $y=\dfrac{3\ln 5t}{\cos 2t}$ (d) $x=\dfrac{2}{\sqrt{t^2-3t+5}}$
   (15)

2. If $f(x)=2.5x^2-6x+2$ find the co-ordinates at the point at which the gradient is -1. (5)

3. The displacement s cm of the end of a stiff spring at time t seconds is given by: $s=ae^{-kt}\sin 2\pi f t$. Determine the velocity and acceleration of the end of the spring after 2 seconds if $a=3$, $k=0.75$ and $f=20$. (10)

4. Find the co-ordinates of the turning points on the curve $y=3x^3+6x^2+3x-1$ and distinguish between them.
   Find also the point(s) of inflexion. (14)

5. The heat capacity C of a gas varies with absolute temperature θ as shown:

   $$C=26.50+7.20\times 10^{-3}\theta-1.20\times 10^{-6}\theta^2$$

   Determine the maximum value of C and the temperature at which it occurs. (7)

6. Determine for the curve $y=2x^2-3x$ at the point (2, 2): (a) the equation of the tangent (b) the equation of the normal. (7)

7. A rectangular block of metal with a square cross-section has a total surface area of $250\,\text{cm}^2$. Find the maximum volume of the block of metal. (7)

8. The solution to a differential equation associated with the path taken by a projectile for which the resistance to motion is proportional to the velocity is given by:

   $$y=2.5(e^x-e^{-x})+x-25$$

   Use Newton's method to determine the value of x, correct to 2 decimal places, for which the value of y is zero. (10)

9. Use Maclaurin's series to determine the first five terms of the power series for $3e^{2x}$. (7)

10. Show, using Maclaurin's series, that the first four terms of the power series for $\cos 3x$ is given by: $\cos 3x=1-\dfrac{9}{2}x^2+\dfrac{27}{8}x^4-\dfrac{81}{80}x^6+\cdots$ (9)

11. Use Maclaurin's series to determine a power series for $e^{2x}\cos 3x$ as far as the term in x^2. (9)

For lecturers/instructors/teachers, fully worked solutions to each of the problems in Revision Test 10, together with a full marking scheme, are available at the website:
www.routledge.com/cw/bird

Chapter 41

Differentiation of parametric equations

Why it is important to understand: **Differentiation of parametric equations**

Rather than using a single equation to define two variables with respect to one another, parametric equations exist as a set that relates the two variables to one another with respect to a third variable. Some curves are easier to describe using a pair of parametric equations. The co-ordinates x and y of the curve are given using a third variable t, such as $x = f(t)$ and $y = g(t)$, where t is referred to as the parameter. Hence, for a given value of t, a point (x, y) is determined. For example, let t be the time and x and y are the positions of a particle; the parametric equations then describe the path of the particle at different times. Parametric equations are useful in defining three-dimensional curves and surfaces, such as determining the velocity or acceleration of a particle following a three-dimensional path. CAD systems use parametric versions of equation. Sometimes in engineering, differentiation of parametric equations is necessary, for example, when determining the radius of curvature of part of the surface when finding the surface tension of a liquid. Knowledge of standard differentials and the function of a function rule from previous chapters are needed to be able to differentiate parametric equations.

At the end of this chapter, you should be able to:

- recognise parametric equations – ellipse, parabola, hyperbola, rectangular hyperbola, cardioids, asteroid and cycloid
- differentiate parametric equations

41.1 Introduction to parametric equations

Certain mathematical functions can be expressed more simply by expressing, say, x and y separately in terms of a third variable. For example, $y=r\sin\theta$, $x=r\cos\theta$. Then, any value given to θ will produce a pair of values for x and y, which may be plotted to provide a curve of $y=f(x)$.

The third variable, θ, is called a **parameter** and the two expressions for y and x are called **parametric equations**.

The above example of $y=r\sin\theta$ and $x=r\cos\theta$ are the parametric equations for a circle. The equation of any point on a circle, centre at the origin and of radius r is given by: $x^2+y^2=r^2$, as shown in Chapter 24.

To show that $y=r\sin\theta$ and $x=r\cos\theta$ are suitable parametric equations for such a circle:

Left hand side of equation

$$= x^2 + y^2$$
$$= (r\cos\theta)^2 + (r\sin\theta)^2$$
$$= r^2\cos^2\theta + r^2\sin^2\theta$$
$$= r^2(\cos^2\theta + \sin^2\theta)$$
$$= r^2 = \text{right hand side}$$

(since $\cos^2\theta + \sin^2\theta = 1$, as shown in Chapter 21).

41.2 Some common parametric equations

The following are some of the most common parametric equations, and Fig. 41.1 shows typical shapes of these curves.

(a) Ellipse $x = a\cos\theta,\ y = b\sin\theta$
(b) Parabola $x = at^2,\ y = 2at$
(c) Hyperbola $x = a\sec\theta,\ y = b\tan\theta$
(d) Rectangular hyperbola $x = ct,\ y = \frac{c}{t}$
(e) Cardioid $x = a(2\cos\theta - \cos 2\theta),\ y = a(2\sin\theta - \sin 2\theta)$
(f) Astroid $x = a\cos^3\theta,\ y = a\sin^3\theta$
(g) Cycloid $x = a(\theta - \sin\theta),\ y = a(1 - \cos\theta)$

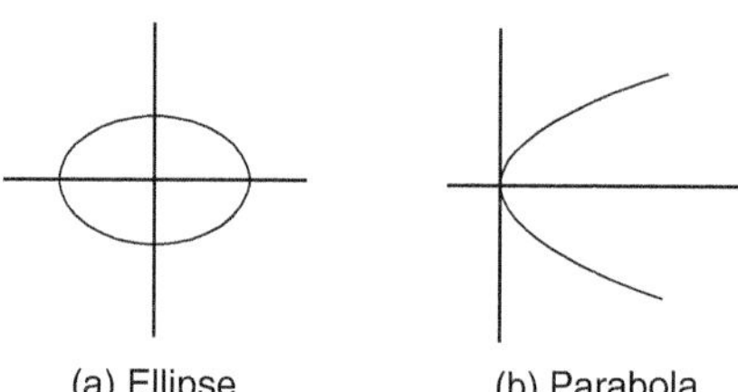

(a) Ellipse (b) Parabola

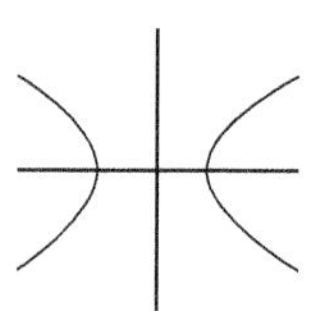

(c) Hyperbola

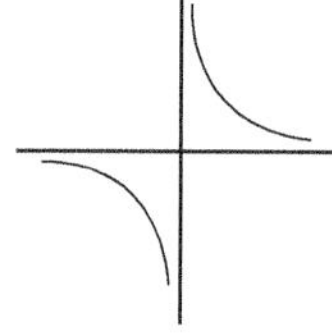

(d) Rectangular hyperbola

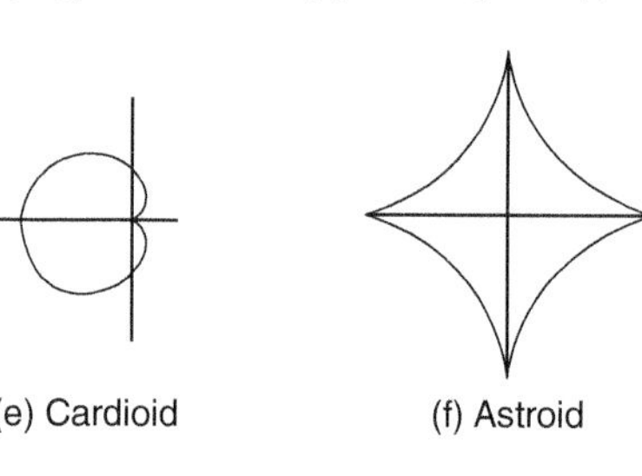

(e) Cardioid (f) Astroid

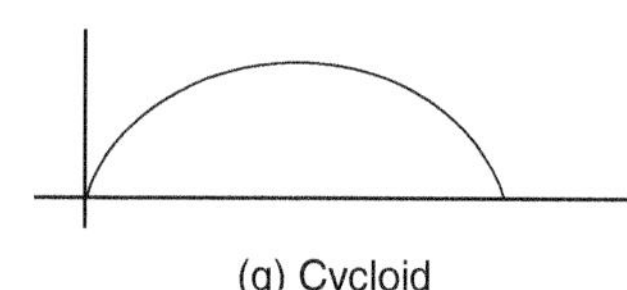

(g) Cycloid

Figure 41.1

41.3 Differentiation in parameters

When x and y are given in terms of a parameter say θ, then by the function of a function rule of differentiation (from Chapter 37):

$$\frac{dy}{dx} = \frac{dy}{d\theta} \times \frac{d\theta}{dx}$$

It may be shown that this can be written as:

$$\frac{dy}{dx} = \frac{\frac{dy}{d\theta}}{\frac{dx}{d\theta}} \qquad (1)$$

For the second differential,

$$\frac{d^2y}{dx^2} = \frac{d}{dx}\left(\frac{dy}{dx}\right) = \frac{d}{d\theta}\left(\frac{dy}{dx}\right) \cdot \frac{d\theta}{dx}$$

or

$$\frac{d^2y}{dx^2} = \frac{\frac{d}{d\theta}\left(\frac{dy}{dx}\right)}{\frac{dx}{d\theta}} \qquad (2)$$

Problem 1. Given $x = 5\theta - 1$ and $y = 2\theta(\theta - 1)$, determine $\frac{dy}{dx}$ in terms of θ

$x = 5\theta - 1$, hence $\frac{dx}{d\theta} = 5$

$y = 2\theta(\theta - 1) = 2\theta^2 - 2\theta$,

hence $\frac{dy}{d\theta} = 4\theta - 2 = 2(2\theta - 1)$

From equation (1),

$$\frac{dy}{dx} = \frac{\frac{dy}{d\theta}}{\frac{dx}{d\theta}} = \mathbf{\frac{2(2\theta - 1)}{5}} \text{ or } \mathbf{\frac{2}{5}(2\theta - 1)}$$

Problem 2. The parametric equations of a function are given by $y = 3\cos 2t$, $x = 2\sin t$. Determine expressions for (a) $\frac{dy}{dx}$ (b) $\frac{d^2y}{dx^2}$

(a) $y=3\cos 2t$, hence $\frac{dy}{dt}=-6\sin 2t$

$x=2\sin t$, hence $\frac{dx}{dt}=2\cos t$

From equation (1),

$$\frac{dy}{dx}=\frac{\frac{dy}{dt}}{\frac{dx}{dt}}=\frac{-6\sin 2t}{2\cos t}=\frac{-6(2\sin t\cos t)}{2\cos t}$$

from double angles, Chapter 22

i.e. $\boldsymbol{\frac{dy}{dx}=-6\sin t}$

(b) From equation (2),

$$\frac{d^2y}{dx^2}=\frac{\frac{d}{dt}\left(\frac{dy}{dx}\right)}{\frac{dx}{dt}}=\frac{\frac{d}{dt}(-6\sin t)}{2\cos t}=\frac{-6\cos t}{2\cos t}$$

i.e. $\boldsymbol{\frac{d^2y}{dx^2}=-3}$

Problem 3. The equation of a tangent drawn to a curve at point (x_1, y_1) is given by:

$$y-y_1=\frac{dy_1}{dx_1}(x-x_1)$$

Determine the equation of the tangent drawn to the parabola $x=2t^2$, $y=4t$ at the point t

At point t, $x_1=2t^2$, hence $\frac{dx_1}{dt}=4t$

and $y_1=4t$, hence $\frac{dy_1}{dt}=4$

From equation (1),

$$\frac{dy}{dx}=\frac{\frac{dy}{dt}}{\frac{dx}{dt}}=\frac{4}{4t}=\frac{1}{t}$$

Hence, the equation of the tangent is:

$$\boldsymbol{y-4t=\frac{1}{t}(x-2t^2)}$$

Problem 4. The parametric equations of a cycloid are $x=4(\theta-\sin\theta)$, $y=4(1-\cos\theta)$. Determine (a) $\frac{dy}{dx}$ (b) $\frac{d^2y}{dx^2}$

(a) $x=4(\theta-\sin\theta)$

hence $\frac{dx}{d\theta}=4-4\cos\theta=4(1-\cos\theta)$

$y=4(1-\cos\theta)$, hence $\frac{dy}{d\theta}=4\sin\theta$

From equation (1),

$$\frac{dy}{dx}=\frac{\frac{dy}{d\theta}}{\frac{dx}{d\theta}}=\frac{4\sin\theta}{4(1-\cos\theta)}=\boldsymbol{\frac{\sin\theta}{(1-\cos\theta)}}$$

(b) From equation (2),

$$\frac{d^2y}{dx^2}=\frac{\frac{d}{d\theta}\left(\frac{dy}{dx}\right)}{\frac{dx}{d\theta}}=\frac{\frac{d}{d\theta}\left(\frac{\sin\theta}{1-\cos\theta}\right)}{4(1-\cos\theta)}$$

$$=\frac{\frac{(1-\cos\theta)(\cos\theta)-(\sin\theta)(\sin\theta)}{(1-\cos\theta)^2}}{4(1-\cos\theta)}$$

$$=\frac{\cos\theta-\cos^2\theta-\sin^2\theta}{4(1-\cos\theta)^3}$$

$$=\frac{\cos\theta-(\cos^2\theta+\sin^2\theta)}{4(1-\cos\theta)^3}$$

$$=\frac{\cos\theta-1}{4(1-\cos\theta)^3}$$

$$=\frac{-(1-\cos\theta)}{4(1-\cos\theta)^3}=\boldsymbol{\frac{-1}{4(1-\cos\theta)^2}}$$

Now try the following Practice Exercise

Practice Exercise 204 Differentiation of parametric equations (Answers on page 723)

1. Given $x=3t-1$ and $y=t(t-1)$, determine $\frac{dy}{dx}$ in terms of t
2. A parabola has parametric equations: $x=t^2$, $y=2t$. Evaluate $\frac{dy}{dx}$ when $t=0.5$
3. The parametric equations for an ellipse are $x=4\cos\theta$, $y=\sin\theta$. Determine (a) $\frac{dy}{dx}$ (b) $\frac{d^2y}{dx^2}$

4. Evaluate $\frac{dy}{dx}$ at $\theta=\frac{\pi}{6}$ radians for the hyperbola whose parametric equations are $x=3\sec\theta$, $y=6\tan\theta$

5. The parametric equations for a rectangular hyperbola are $x=2t$, $y=\frac{2}{t}$. Evaluate $\frac{dy}{dx}$ when $t=0.40$

   The equation of a tangent drawn to a curve at point (x_1, y_1) is given by:

$$y-y_1=\frac{dy_1}{dx_1}(x-x_1)$$

   Use this in Problems 6 and 7.

6. Determine the equation of the tangent drawn to the ellipse $x=3\cos\theta$, $y=2\sin\theta$ at $\theta=\frac{\pi}{6}$

7. Determine the equation of the tangent drawn to the rectangular hyperbola $x=5t$, $y=\frac{5}{t}$ at $t=2$

41.4 Further worked problems on differentiation of parametric equations

Problem 5. The equation of the normal drawn to a curve at point (x_1, y_1) is given by:

$$y-y_1=-\frac{1}{\frac{dy_1}{dx_1}}(x-x_1)$$

Determine the equation of the normal drawn to the astroid $x=2\cos^3\theta$, $y=2\sin^3\theta$ at the point $\theta=\frac{\pi}{4}$

$x=2\cos^3\theta$, hence $\frac{dx}{d\theta}=-6\cos^2\theta\sin\theta$

by the function of a function rule

$y=2\sin^3\theta$, hence $\frac{dy}{d\theta}=6\sin^2\theta\cos\theta$

From equation (1),

$$\frac{dy}{dx}=\frac{\frac{dy}{d\theta}}{\frac{dx}{d\theta}}=\frac{6\sin^2\theta\cos\theta}{-6\cos^2\theta\sin\theta}=-\frac{\sin\theta}{\cos\theta}=-\tan\theta$$

When $\theta=\frac{\pi}{4}$, $\frac{dy}{dx}=-\tan\frac{\pi}{4}=-1$

$x_1=2\cos^3\frac{\pi}{4}=0.7071$ and $y_1=2\sin^3\frac{\pi}{4}=0.7071$

Hence, **the equation of the normal is:**

$$y-0.7071=-\frac{1}{-1}(x-0.7071)$$

i.e. $y-0.7071=x-0.7071$

i.e. $\boldsymbol{y=x}$

Problem 6. The parametric equations for a hyperbola are $x=2\sec\theta$, $y=4\tan\theta$. Evaluate (a) $\frac{dy}{dx}$ (b) $\frac{d^2y}{dx^2}$, correct to 4 significant figures, when $\theta=1$ radian

(a) $x=2\sec\theta$, hence $\frac{dx}{d\theta}=2\sec\theta\tan\theta$

$y=4\tan\theta$, hence $\frac{dy}{d\theta}=4\sec^2\theta$

From equation (1),

$$\frac{dy}{dx}=\frac{\frac{dy}{d\theta}}{\frac{dx}{d\theta}}=\frac{4\sec^2\theta}{2\sec\theta\tan\theta}=\frac{2\sec\theta}{\tan\theta}$$

$$=\frac{2\left(\frac{1}{\cos\theta}\right)}{\left(\frac{\sin\theta}{\cos\theta}\right)}=\frac{2}{\sin\theta}\quad\text{or}\quad 2\operatorname{cosec}\theta$$

When $\theta=1$ rad, $\frac{dy}{dx}=\frac{2}{\sin 1}=\mathbf{2.377}$, correct to 4 significant figures.

(b) From equation (2),

$$\frac{d^2y}{dx^2}=\frac{\frac{d}{d\theta}\left(\frac{dy}{dx}\right)}{\frac{dx}{d\theta}}=\frac{\frac{d}{d\theta}(2\operatorname{cosec}\theta)}{2\sec\theta\tan\theta}$$

$$=\frac{-2\operatorname{cosec}\theta\cot\theta}{2\sec\theta\tan\theta}$$

$$=\frac{-\left(\frac{1}{\sin\theta}\right)\left(\frac{\cos\theta}{\sin\theta}\right)}{\left(\frac{1}{\cos\theta}\right)\left(\frac{\sin\theta}{\cos\theta}\right)}$$

$$= -\left(\frac{\cos\theta}{\sin^2\theta}\right)\left(\frac{\cos^2\theta}{\sin\theta}\right)$$

$$= -\frac{\cos^3\theta}{\sin^3\theta} = -\cot^3\theta$$

When $\theta = 1$ rad, $\frac{d^2y}{dx^2} = -\cot^3 1 = -\frac{1}{(\tan 1)^3}$

$= \mathbf{-0.2647}$, correct to 4 significant figures.

Problem 7. When determining the surface tension of a liquid, the radius of curvature, ρ, of part of the surface is given by:

$$\rho = \frac{\sqrt{\left[1 + \left(\frac{dy}{dx}\right)^2\right]^3}}{\frac{d^2y}{dx^2}}$$

Find the radius of curvature of the part of the surface having the parametric equations $x = 3t^2$, $y = 6t$ at the point $t = 2$

$x = 3t^2$, hence $\frac{dx}{dt} = 6t$

$y = 6t$, hence $\frac{dy}{dt} = 6$

From equation (1), $\frac{dy}{dx} = \frac{\frac{dy}{dt}}{\frac{dx}{dt}} = \frac{6}{6t} = \frac{1}{t}$

From equation (2),

$$\frac{d^2y}{dx^2} = \frac{\frac{d}{dt}\left(\frac{dy}{dx}\right)}{\frac{dx}{dt}} = \frac{\frac{d}{dt}\left(\frac{1}{t}\right)}{6t} = \frac{-\frac{1}{t^2}}{6t} = -\frac{1}{6t^3}$$

Hence, radius of curvature, $\rho = \frac{\sqrt{\left[1 + \left(\frac{dy}{dx}\right)^2\right]^3}}{\frac{d^2y}{dx^2}}$

$$= \frac{\sqrt{\left[1 + \left(\frac{1}{t}\right)^2\right]^3}}{-\frac{1}{6t^3}}$$

When $t = 2$, $\rho = \frac{\sqrt{\left[1 + \left(\frac{1}{2}\right)^2\right]^3}}{-\frac{1}{6(2)^3}} = \frac{\sqrt{(1.25)^3}}{-\frac{1}{48}}$

$= -48\sqrt{(1.25)^3} = \mathbf{-67.08}$

Now try the following Practice Exercise

Practice Exercise 205 Differentiation of parametric equations (Answers on page 723)

1. A cycloid has parametric equations $x = 2(\theta - \sin\,\theta)$, $y = 2(1 - \cos\,\theta)$. Evaluate, at $\theta = 0.62$ rad, correct to 4 significant figures, (a)s $\frac{dy}{dx}$ (b) $\frac{d^2y}{dx^2}$

   The equation of the normal drawn to a curve at point (x_1, y_1) is given by:

   $$y - y_1 = -\frac{1}{\frac{dy_1}{dx_1}}(x - x_1)$$

   Use this in Problems 2 and 3.

2. Determine the equation of the normal drawn to the parabola $x = \frac{1}{4}t^2$, $y = \frac{1}{2}t$ at $t = 2$
3. Find the equation of the normal drawn to the cycloid $x = 2(\theta - \sin\theta)$, $y = 2(1 - \cos\theta)$ at $\theta = \frac{\pi}{2}$ rad.
4. Determine the value of $\frac{d^2y}{dx^2}$, correct to 4 significant figures, at $\theta = \frac{\pi}{6}$ rad for the cardioid $x = 5(2\theta - \cos 2\theta)$, $y = 5(2\sin\theta - \sin 2\theta)$.

5. The radius of curvature, ρ, of part of a surface when determining the surface tension of a liquid is given by:

$$\rho = \frac{\left[1+\left(\frac{dy}{dx}\right)^2\right]^{3/2}}{\frac{d^2y}{dx^2}}$$

Find the radius of curvature (correct to 4 significant figures) of the part of the surface having parametric equations:

(a) $x=3t,\ y=\frac{3}{t}$ at the point $t=\frac{1}{2}$

(b) $x=4\cos^3 t,\ y=4\sin^3 t$ at $t=\frac{\pi}{6}$ rad

Practice Exercise 206 Multiple-choice questions on differentiation of parametric equations (Answers on page 723)

Each question has only one correct answer

1. Given $x=3t-1$ and $y=3t(t-1)$ then which of the following is correct?

(a) $\frac{dy}{dx}=2t-1$ (b) $\frac{d^2y}{dx^2}=2$

(c) $\frac{d^2y}{dx^2}=\frac{1}{2}$ (d) $\frac{dy}{dx}=\frac{1}{2t-1}$

2. Given $x=2\sin t$ and $y=4\cos 2t$ then:

(a) $\frac{dy}{dx}=-8\sin t$ (b) $\frac{d^2y}{dx^2}=4$

(c) $\frac{d^2y}{dx^2}=-\tan t$ (d) $\frac{dy}{dx}=-\frac{4\sin 2t}{\cos t}$

3. The parametric equations for a hyperbola are $x=\sec\theta, y=3\tan\theta$. The value of $\frac{dy}{dx}$ when $\theta=1.5$ rad, correct to 2 decimal places is:

(a) 0.33 (b) 114.61 (c) 3.01 (d) 42.52

4. The parametric equations for a hyperbola are $x=3\sec\theta, y=2\tan\theta$. The value of $\frac{d^2y}{dx^2}$ when $\theta=1.2$ rad, correct to 3 decimal places is:

(a) 0.259 (b) −0.013
(c) −0.086 (d) −0.778

5. If the parametric equations of a cycloid are $x=5(\theta-\sin\theta), x=5(1-\cos\theta)$ then $\frac{dy}{dx}$ is equal to:

(a) $\frac{\sin\theta}{1-\cos\theta}$ (b) $\frac{1-\cos\theta}{\sin\theta}$

(c) $\frac{\sin\theta}{1+\cos\theta}$ (d) $\frac{1+\cos\theta}{\sin\theta}$

For fully worked solutions to each of the problems in Practice Exercises 204 and 205 in this chapter, go to the website:

www.routledge.com/cw/bird

Chapter 42

Differentiation of implicit functions

Why it is important to understand: **Differentiation of implicit functions**

Differentiation of implicit functions is another special technique, but it occurs often enough to be important. It is needed for more complicated problems involving different rates of change. Up to this chapter we have been finding derivatives of functions of the form $y = f(x)$; unfortunately not all functions fall into this form. However, implicit differentiation is nothing more than a special case of the function of a function (or chain rule) for derivatives. Engineering applications where implicit differentiation is needed are found in optics, electronics, control and even some thermodynamics.

At the end of this chapter, you should be able to:

- recognise implicit functions
- differentiate simple implicit functions
- differentiate implicit functions containing products and quotients

42.1 Implicit functions

When an equation can be written in the form $y=f(x)$ it is said to be an **explicit function** of x. Examples of explicit functions include

$$y = 2x^3 - 3x + 4, \quad y = 2x\ln x$$

and
$$y = \frac{3e^x}{\cos x}$$

In these examples y may be differentiated with respect to x by using standard derivatives, the product rule and the quotient rule of differentiation respectively.

Sometimes with equations involving, say, y and x, it is impossible to make y the subject of the formula. The equation is then called an **implicit function** and examples of such functions include $y^3 + 2x^2 = y^2 - x$ and $\sin y = x^2 + 2xy$

42.2 Differentiating implicit functions

It is possible to **differentiate an implicit function** by using the **function of a function rule**, which may be stated as

$$\frac{du}{dx} = \frac{du}{dy} \times \frac{dy}{dx}$$

Thus, to differentiate y^3 with respect to x, the substitution $u = y^3$ is made, from which, $\frac{du}{dy} = 3y^2$

Hence, $\frac{d}{dx}(y^3)=(3y^2)\times\frac{dy}{dx}$, by the function of a function rule.

A simple rule for differentiating an implicit function is summarised as:

$$\frac{d}{dx}[f(y)]=\frac{d}{dy}[f(y)]\times\frac{dy}{dx} \quad (1)$$

Problem 1. Differentiate the following functions with respect to x:

(a) $2y^4$ (b) $\sin 3t$

(a) Let $u=2y^4$, then, by the function of a function rule:

$$\frac{du}{dx}=\frac{du}{dy}\times\frac{dy}{dx}=\frac{d}{dy}(2y^4)\times\frac{dy}{dx}$$

$$=\mathbf{8y^3\frac{dy}{dx}}$$

(b) Let $u=\sin 3t$, then, by the function of a function rule:

$$\frac{du}{dx}=\frac{du}{dt}\times\frac{dt}{dx}=\frac{d}{dt}(\sin 3t)\times\frac{dt}{dx}$$

$$=\mathbf{3\cos 3t\frac{dt}{dx}}$$

Problem 2. Differentiate the following functions with respect to x:

(a) $4\ln 5y$ (b) $\frac{1}{5}e^{3\theta-2}$

(a) Let $u=4\ln 5y$, then, by the function of a function rule:

$$\frac{du}{dx}=\frac{du}{dy}\times\frac{dy}{dx}=\frac{d}{dy}(4\ln 5y)\times\frac{dy}{dx}$$

$$=\mathbf{\frac{4}{y}\frac{dy}{dx}}$$

(b) Let $u=\frac{1}{5}e^{3\theta-2}$, then, by the function of a function rule:

$$\frac{du}{dx}=\frac{du}{d\theta}\times\frac{d\theta}{dx}=\frac{d}{d\theta}\left(\frac{1}{5}e^{3\theta-2}\right)\times\frac{d\theta}{dx}$$

$$=\mathbf{\frac{3}{5}e^{3\theta-2}\frac{d\theta}{dx}}$$

Now try the following Practice Exercise

Practice Exercise 207 Differentiating implicit functions (Answers on page 724)

In Problems 1 and 2 differentiate the given functions with respect to x.

1. (a) $3y^5$ (b) $2\cos 4\theta$ (c) $\sqrt{k}$
2. (a) $\frac{5}{2}\ln 3t$ (b) $\frac{3}{4}e^{2y+1}$ (c) $2\tan 3y$
3. Differentiate the following with respect to y: (a) $3\sin 2\theta$ (b) $4\sqrt{x^3}$ (c) $\frac{2}{e^t}$
4. Differentiate the following with respect to u: (a) $\frac{2}{(3x+1)}$ (b) $3\sec 2\theta$ (c) $\frac{2}{\sqrt{y}}$

42.3 Differentiating implicit functions containing products and quotients

The product and quotient rules of differentiation must be applied when differentiating functions containing products and quotients of two variables.

For example, $\frac{d}{dx}(x^2y)=(x^2)\frac{d}{dx}(y)+(y)\frac{d}{dx}(x^2)$, by the product rule

$$=(x^2)\left(1\frac{dy}{dx}\right)+y(2x)$$

by using equation (1)

$$=\mathbf{x^2\frac{dy}{dx}+2xy}$$

Problem 3. Determine $\frac{d}{dx}(2x^3y^2)$

In the product rule of differentiation let $u=2x^3$ and $v=y^2$

$$\text{Thus } \frac{d}{dx}(2x^3y^2)=(2x^3)\frac{d}{dx}(y^2)+(y^2)\frac{d}{dx}(2x^3)$$

$$=(2x^3)\left(2y\frac{dy}{dx}\right)+(y^2)(6x^2)$$

$$=4x^3y\frac{dy}{dx}+6x^2y^2$$

$$=\mathbf{2x^2y\left(2x\frac{dy}{dx}+3y\right)}$$

Problem 4. Find $\frac{d}{dx}\left(\frac{3y}{2x}\right)$

In the quotient rule of differentiation let $u=3y$ and $v=2x$

$$\text{Thus } \frac{d}{dx}\left(\frac{3y}{2x}\right)=\frac{(2x)\frac{d}{dx}(3y)-(3y)\frac{d}{dx}(2x)}{(2x)^2}$$

$$=\frac{(2x)\left(3\frac{dy}{dx}\right)-(3y)(2)}{4x^2}$$

$$=\frac{6x\frac{dy}{dx}-6y}{4x^2}=\mathbf{\frac{3}{2x^2}\left(x\frac{dy}{dx}-y\right)}$$

Problem 5. Differentiate $z=x^2+3x\cos 3y$ with respect to y

$$\frac{dz}{dy}=\frac{d}{dy}(x^2)+\frac{d}{dy}(3x\cos 3y)$$

$$=2x\frac{dx}{dy}+\left[(3x)(-3\sin 3y)+(\cos 3y)\left(3\frac{dx}{dy}\right)\right]$$

$$=\mathbf{2x\frac{dx}{dy}-9x\sin 3y+3\cos 3y\frac{dx}{dy}}$$

Now try the following Practice Exercise

Practice Exercise 208 Differentiating implicit functions involving products and quotients (Answers on page 724)

1. Determine $\frac{d}{dx}(3x^2y^3)$
2. Find $\frac{d}{dx}\left(\frac{2y}{5x}\right)$
3. Determine $\frac{d}{du}\left(\frac{3u}{4v}\right)$
4. Given $z=3\sqrt{y}\cos 3x$ find $\frac{dz}{dx}$
5. Determine $\frac{dz}{dy}$ given $z=2x^3\ln y$

42.4 Further implicit differentiation

An implicit function such as $3x^2+y^2-5x+y=2$, may be differentiated term by term with respect to x. This gives:

$$\frac{d}{dx}(3x^2)+\frac{d}{dx}(y^2)-\frac{d}{dx}(5x)+\frac{d}{dx}(y)=\frac{d}{dx}(2)$$

$$\text{i.e.}\qquad 6x+2y\frac{dy}{dx}-5+1\frac{dy}{dx}=0,$$

using equation (1) and standard derivatives.

An expression for the derivative $\frac{dy}{dx}$ in terms of x and y may be obtained by rearranging this latter equation. Thus:

$$(2y+1)\frac{dy}{dx}=5-6x$$

$$\text{from which, }\quad \mathbf{\frac{dy}{dx}=\frac{5-6x}{2y+1}}$$

Problem 6. Given $2y^2-5x^4-2-7y^3=0$, determine $\frac{dy}{dx}$

Each term in turn is differentiated with respect to x:

$$\text{Hence}\quad \frac{d}{dx}(2y^2)-\frac{d}{dx}(5x^4)-\frac{d}{dx}(2)-\frac{d}{dx}(7y^3)=\frac{d}{dx}(0)$$

$$\text{i.e.}\qquad 4y\frac{dy}{dx}-20x^3-0-21y^2\frac{dy}{dx}=0$$

Rearranging gives:

$$(4y-21y^2)\frac{dy}{dx}=20x^3$$

$$\text{i.e.}\qquad \mathbf{\frac{dy}{dx}=\frac{20x^3}{(4y-21y^2)}}$$

Problem 7. Determine the values of $\frac{dy}{dx}$ when $x=4$ given that $x^2+y^2=25$

Differentiating each term in turn with respect to x gives:

$$\frac{d}{dx}(x^2)+\frac{d}{dx}(y^2)=\frac{d}{dx}(25)$$

i.e. $$2x+2y\frac{dy}{dx}=0$$

Hence $$\frac{dy}{dx}=-\frac{2x}{2y}=-\frac{x}{y}$$

Since $x^2+y^2=25$, when $x=4$, $y=\sqrt{(25-4^2)}=\pm 3$

Thus when $x=4$ and $y=\pm 3$, $\boldsymbol{\frac{dy}{dx}}=-\frac{4}{\pm 3}=\boldsymbol{\pm\frac{4}{3}}$

$x^2+y^2=25$ is the equation of a circle, centre at the origin and radius 5, as shown in Fig. 42.1. At $x=4$, the two gradients are shown.

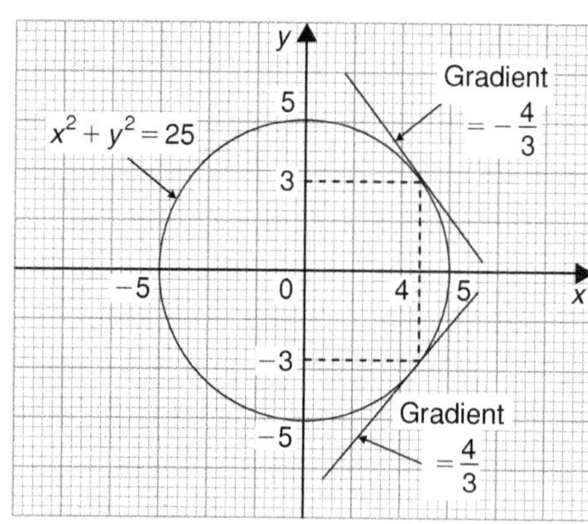

Figure 42.1

Above, $x^2+y^2=25$ was differentiated implicitly; actually, the equation could be transposed to $y=\sqrt{(25-x^2)}$ and differentiated using the function of a function rule. This gives

$$\frac{dy}{dx}=\frac{1}{2}(25-x^2)^{-\frac{1}{2}}(-2x)=-\frac{x}{\sqrt{(25-x^2)}}$$

and when $x=4$, $\frac{dy}{dx}=-\frac{4}{\sqrt{(25-4^2)}}=\pm\frac{4}{3}$ as obtained above.

Problem 8.

(a) Find $\frac{dy}{dx}$ in terms of x and y given

$$4x^2+2xy^3-5y^2=0$$

(b) Evalate $\frac{dy}{dx}$ when $x=1$ and $y=2$

(a) Differentiating each term in turn with respect to x gives:

$$\frac{d}{dx}(4x^2)+\frac{d}{dx}(2xy^3)-\frac{d}{dx}(5y^2)=\frac{d}{dx}(0)$$

i.e. $$8x+\left[(2x)\left(3y^2\frac{dy}{dx}\right)+(y^3)(2)\right]-10y\frac{dy}{dx}=0$$

i.e. $$8x+6xy^2\frac{dy}{dx}+2y^3-10y\frac{dy}{dx}=0$$

Rearranging gives:

$$8x+2y^3=(10y-6xy^2)\frac{dy}{dx}$$

and $$\boldsymbol{\frac{dy}{dx}}=\frac{8x+2y^3}{10y-6xy^2}=\boldsymbol{\frac{4x+y^3}{y(5-3xy)}}$$

(b) When $x=1$ and $y=2$,

$$\boldsymbol{\frac{dy}{dx}}=\frac{4(1)+(2)^3}{2[5-(3)(1)(2)]}=\frac{12}{-2}=\boldsymbol{-6}$$

Problem 9. Find the gradients of the tangents drawn to the circle $x^2+y^2-2x-2y=3$ at $x=2$

The gradient of the tangent is given by $\frac{dy}{dx}$

Differentiating each term in turn with respect to x gives:

$$\frac{d}{dx}(x^2)+\frac{d}{dx}(y^2)-\frac{d}{dx}(2x)-\frac{d}{dx}(2y)=\frac{d}{dx}(3)$$

i.e. $$2x+2y\frac{dy}{dx}-2-2\frac{dy}{dx}=0$$

Hence $$(2y-2)\frac{dy}{dx}=2-2x$$

from which $$\frac{dy}{dx}=\frac{2-2x}{2y-2}=\frac{1-x}{y-1}$$

The value of y when $x=2$ is determined from the original equation

Hence $(2)^2+y^2-2(2)-2y=3$

i.e. $4+y^2-4-2y=3$

or $y^2-2y-3=0$

Factorising gives: $(y+1)(y-3)=0$, from which $y=-1$ or $y=3$

When $x=2$ and $y=-1$,

$$\frac{dy}{dx}=\frac{1-x}{y-1}=\frac{1-2}{-1-1}=\frac{-1}{-2}=\frac{1}{2}$$

When $x = 2$ and $y = 3$,

$$\frac{dy}{dx} = \frac{1-2}{3-1} = -\frac{1}{2}$$

Hence the gradients of the tangents are $\pm\frac{1}{2}$

The circle having the given equation has its centre at (1, 1) and radius $\sqrt{5}$ (see Chapter 24) and is shown in Fig. 42.2 with the two gradients of the tangents.

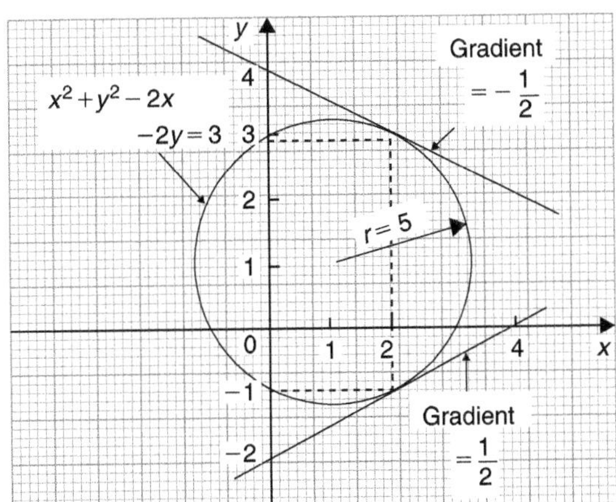

Figure 42.2

Problem 10. Pressure p and volume v of a gas are related by the law $pv^{\gamma} = k$, where γ and k are constants. Show that the rate of change of pressure $\frac{dp}{dt} = -\gamma \frac{p}{v}\frac{dv}{dt}$

Since $pv^{\gamma} = k$, then $p = \frac{k}{v^{\gamma}} = kv^{-\gamma}$

$$\frac{dp}{dt} = \frac{dp}{dv} \times \frac{dv}{dt}$$

by the function of a function rule

$$\frac{dp}{dv} = \frac{d}{dv}(kv^{-\gamma})$$

$$= -\gamma kv^{-\gamma-1} = \frac{-\gamma k}{v^{\gamma+1}}$$

$$\frac{dp}{dt} = \frac{-\gamma k}{v^{\gamma+1}} \times \frac{dv}{dt}$$

Since $\quad k = pv^{\gamma}$

$$\frac{dp}{dt} = \frac{-\gamma(pv^{\gamma})}{v^{r+1}}\frac{dv}{dt} = \frac{-\gamma pv^{\gamma}}{v^{\gamma}v^{1}}\frac{dv}{dt}$$

i.e. $$\boldsymbol{\frac{dp}{dt} = -\gamma \frac{p}{v}\frac{dv}{dt}}$$

Now try the following Practice Exercise

Practice Exercise 209 Implicit differentiation (Answers on page 724)

In Problems 1 and 2 determine $\frac{dy}{dx}$

1. $x^2 + y^2 + 4x - 3y + 1 = 0$
2. $2y^3 - y + 3x - 2 = 0$
3. Given $x^2 + y^2 = 9$ evaluate $\frac{dy}{dx}$ when $x = \sqrt{5}$ and $y = 2$

In Problems 4 to 7, determine $\frac{dy}{dx}$

4. $x^2 + 2x \sin 4y = 0$
5. $3y^2 + 2xy - 4x^2 = 0$
6. $2x^2y + 3x^3 = \sin y$
7. $3y + 2x \ln y = y^4 + x$
8. If $3x^2 + 2x^2y^3 - \frac{5}{4}y^2 = 0$ evaluate $\frac{dy}{dx}$ when $x = \frac{1}{2}$ and $y = 1$
9. Determine the gradients of the tangents drawn to the circle $x^2 + y^2 = 16$ at the point where $x = 2$. Give the answer correct to 4 significant figures
10. Find the gradients of the tangents drawn to the ellipse $\frac{x^2}{4} + \frac{y^2}{9} = 2$ at the point where $x = 2$
11. Determine the gradient of the curve $3xy + y^2 = -2$ at the point (1, −2)

Practice Exercise 210 Multiple-choice questions on differentiation of implicit functions (Answers on page 724)

Each question has only one correct answer

1. The function $\frac{d}{dx}(3x^2y^5)$ is equal to:

   (a) $6xy^5$ (b) $3xy^4\left(2y\frac{dx}{dy}+5x\right)$

   (c) $15x^2y^4+6xy^5$ (d) $3xy^4\left(5x\frac{dy}{dx}+2y\right)$

2. Given $3y^2+1-4x^5-8y^3=0$ then $\frac{dy}{dx}$ is equal to:

   (a) $\frac{10x^4}{3y(1-4y)}$ (b) $\frac{3y-12y^2}{24x^4}$

   (c) $20x^4-6y-24y^2$ (d) $\frac{20x^4-1}{6y-24y^2}$

3. $\frac{d}{dx}\left(\frac{5y}{3x}\right)$ is equal to:

   (a) $\frac{5(x-y)}{3x^2}$ (b) $\frac{5\left(x\frac{dx}{dy}-y\right)}{3x^2}$

   (c) $\frac{5\left(x\frac{dy}{dx}-y\right)}{3x^2}$ (d) $\frac{5y}{3}$

4. Given $5y^2+3x^4-6y^3+4=0$ then the value of $\frac{dy}{dx}$ when $x=2$ and $y=1$ is equal to:

   (a) -52 (b) 12 (c) -48 (d) 0.194

5. The gradient of the curve $y^2+5xy=-3$ at the point $(2, -1)$ is:

   (a) 7 (b) -2.5 (c) 0.625 (d) -0.125

For fully worked solutions to each of the problems in Practice Exercises 207 to 209 in this chapter, go to the website:

www.routledge.com/cw/bird

Chapter 43

Logarithmic differentiation

Why it is important to understand: **Logarithmic differentiation**

Logarithmic differentiation is a means of differentiating algebraically complicated functions or functions for which the ordinary rules of differentiation do not apply. The technique is performed in cases where it is easier to differentiate the logarithm of a function rather than the function itself. Logarithmic differentiation relies on the function of a function rule (i.e. chain rule) as well as properties of logarithms (in particular, the natural logarithm or logarithm to the base e) to transform products into sums and divisions into subtractions, and can also be applied to functions raised to the power of variables of functions. Logarithmic differentiation occurs often enough in engineering calculations to make it an important technique.

At the end of this chapter, you should be able to:

- state the laws of logarithms
- differentiate simple logarithmic functions
- differentiate an implicit function involving logarithms
- differentiate more difficult logarithmic functions involving products and quotients
- differentiate functions of the form $y = [f(x)]^x$

43.1 Introduction to logarithmic differentiation

With certain functions containing more complicated products and quotients, differentiation is often made easier if the logarithm of the function is taken before differentiating. This technique, called **'logarithmic differentiation'** is achieved with a knowledge of (i) the laws of logarithms, (ii) the differential coefficients of logarithmic functions and (iii) the differentiation of implicit functions.

43.2 Laws of logarithms

Three laws of logarithms may be expressed as:

(i) $\log(A \times B) = \log A + \log B$

(ii) $\log\left(\frac{A}{B}\right) = \log A - \log B$

(iii) $\log A^n = n \log A$

In calculus, Napierian logarithms (i.e. logarithms to a base of 'e') are invariably used. Thus for two functions $f(x)$ and $g(x)$ the laws of logarithms may be expressed as:

(i) $\ln[f(x) \cdot g(x)] = \ln f(x) + \ln g(x)$

(ii) $\ln\left(\frac{f(x)}{g(x)}\right) = \ln f(x) - \ln g(x)$

(iii) $\ln[f(x)]^n = n \ln f(x)$

Taking Napierian logarithms of both sides of the equation $y=\frac{f(x)\cdot g(x)}{h(x)}$ gives :

$$\ln y = \ln\left(\frac{f(x)\cdot g(x)}{h(x)}\right)$$

which may be simplified using the above laws of logarithms, giving;

$$\ln y = \ln f(x) + \ln g(x) - \ln h(x)$$

This latter form of the equation is often easier to differentiate.

43.3 Differentiation of logarithmic functions

The differential coefficient of the logarithmic function $\ln x$ is given by:

$$\frac{d}{dx}(\ln x)=\frac{1}{x}$$

More generally, it may be shown that:

$$\frac{d}{dx}[\ln f(x)]=\frac{f'(x)}{f(x)} \qquad (1)$$

For example, if $y=\ln(3x^2+2x-1)$ then,

$$\frac{dy}{dx}=\frac{6x+2}{3x^2+2x-1}$$

Similarly, if $y=\ln(\sin 3x)$ then

$$\frac{dy}{dx}=\frac{3\cos 3x}{\sin 3x}=3\cot 3x \quad \text{from Chapter 17}$$

Now try the following Practice Exercise

Practice Exercise 211 Differentiating logarithmic functions (Answers on page 724)

Differentiate the following using the laws for logarithms.

1. $\ln(4x-10)$
2. $\ln(\cos 3x)$
3. $\ln(3x^3+x)$
4. $\ln(5x^2+10x-7)$
5. $\ln 8x$
6. $\ln(x^2-1)$
7. $3\ln 4x$
8. $2\ln(\sin x)$
9. $\ln(4x^3-6x^2+3x)$

43.4 Differentiation of further logarithmic functions

As explained in Chapter 42, by using the function of a function rule:

$$\frac{d}{dx}(\ln y)=\left(\frac{1}{y}\right)\frac{dy}{dx} \qquad (2)$$

Differentiation of an expression such as $y=\frac{(1+x)^2\sqrt{(x-1)}}{x\sqrt{(x+2)}}$ may be achieved by using the product and quotient rules of differentiation; however the working would be rather complicated. With logarithmic differentiation the following procedure is adopted:

(i) Take Napierian logarithms of both sides of the equation.

$$\text{Thus } \ln y = \ln\left\{\frac{(1+x)^2\sqrt{(x-1)}}{x\sqrt{(x+2)}}\right\} = \ln\left\{\frac{(1+x)^2(x-1)^{\frac{1}{2}}}{x(x+2)^{\frac{1}{2}}}\right\}$$

(ii) Apply the laws of logarithms.

Thus $\ln y = \ln(1+x)^2+\ln(x-1)^{\frac{1}{2}} - \ln x - \ln(x+2)^{\frac{1}{2}}$, by laws (i) and (ii) of Section 43.2

i.e. $\ln y = 2\ln(1+x)+\frac{1}{2}\ln(x-1) - \ln x - \frac{1}{2}\ln(x+2)$, by law (iii) of Section 43.2.

(iii) Differentiate each term in turn with respect to x using equations (1) and (2).

$$\text{Thus } \frac{1}{y}\frac{dy}{dx}=\frac{2}{(1+x)}+\frac{\frac{1}{2}}{(x-1)}-\frac{1}{x}-\frac{\frac{1}{2}}{(x+2)}$$

(iv) Rearrange the equation to make $\frac{dy}{dx}$ the subject.

$$\text{Thus} \quad \frac{dy}{dx}=y\left\{\frac{2}{(1+x)}+\frac{1}{2(x-1)}-\frac{1}{x}-\frac{1}{2(x+2)}\right\}$$

(v) Substitute for y in terms of x.

$$\text{Thus} \quad \frac{\mathbf{dy}}{\mathbf{dx}}=\frac{\mathbf{(1+x)^2(\sqrt{(x-1)}}}{\mathbf{x\sqrt{(x+2)}}}\left\{\frac{\mathbf{2}}{\mathbf{(1+x)}}+\frac{\mathbf{1}}{\mathbf{2(x-1)}}-\frac{\mathbf{1}}{\mathbf{x}}-\frac{\mathbf{1}}{\mathbf{2(x+2)}}\right\}$$

Problem 1. Use logarithmic differentiation to differentiate $y=\frac{(x+1)(x-2)^3}{(x-3)}$

Following the above procedure:

(i) Since $y=\frac{(x+1)(x-2)^3}{(x-3)}$

then $\ln y=\ln\left\{\frac{(x+1)(x-2)^3}{(x-3)}\right\}$

(ii) $\ln y=\ln(x+1)+\ln(x-2)^3-\ln(x-3)$, by laws (i) and (ii) of Section 43.2,

i.e. $\ln y=\ln(x+1)+3\ln(x-2)-\ln(x-3)$, by law (iii) of Section 43.2.

(iii) Differentiating with respect to x gives:

$$\frac{1}{y}\frac{dy}{dx}=\frac{1}{(x+1)}+\frac{3}{(x-2)}-\frac{1}{(x-3)}$$

by using equations (1) and (2).

(iv) Rearranging gives:

$$\frac{dy}{dx}=y\left\{\frac{1}{(x+1)}+\frac{3}{(x-2)}-\frac{1}{(x-3)}\right\}$$

(v) Substituting for y gives:

$$\frac{\mathbf{dy}}{\mathbf{dx}}=\frac{\mathbf{(x+1)(x-2)^3}}{\mathbf{(x-3)}}\left\{\frac{\mathbf{1}}{\mathbf{(x+1)}}+\frac{\mathbf{3}}{\mathbf{(x-2)}}-\frac{\mathbf{1}}{\mathbf{(x-3)}}\right\}$$

Problem 2. Differentiate $y=\frac{\sqrt{(x-2)^3}}{(x+1)^2(2x-1)}$ with respect to x and evalualte $\frac{dy}{dx}$ when $x=3$

Using logarithmic differentiation and following the above procedure:

(i) Since $y=\frac{\sqrt{(x-2)^3}}{(x+1)^2(2x-1)}$

then $\ln y=\ln\left\{\frac{\sqrt{(x-2)^3}}{(x+1)^2(2x-1)}\right\}$

$$=\ln\left\{\frac{(x-2)^{\frac{3}{2}}}{(x+1)^2(2x-1)}\right\}$$

(ii) $\ln y=\ln(x-2)^{\frac{3}{2}}-\ln(x+1)^2-\ln(2x-1)$

i.e. $\ln y=\frac{3}{2}\ln(x-2)-2\ln(x+1)-\ln(2x-1)$

(iii) $\frac{1}{y}\frac{dy}{dx}=\frac{\frac{3}{2}}{(x-2)}-\frac{2}{(x+1)}-\frac{2}{(2x-1)}$

(iv) $\frac{dy}{dx}=y\left\{\frac{3}{2(x-2)}-\frac{2}{(x+1)}-\frac{2}{(2x-1)}\right\}$

(v) $$\frac{dy}{dx}=\frac{\sqrt{\mathbf{(x-2)^3}}}{\mathbf{(x+1)^2(2x-1)}}\left\{\frac{\mathbf{3}}{\mathbf{2(x-2)}}-\frac{\mathbf{2}}{\mathbf{(x+1)}}-\frac{\mathbf{2}}{\mathbf{(2x-1)}}\right\}$$

When $x=3$, $\frac{dy}{dx}=\frac{\sqrt{(1)^3}}{(4)^2(5)}\left(\frac{3}{2}-\frac{2}{4}-\frac{2}{5}\right)$

$$=\pm\frac{1}{80}\left(\frac{3}{5}\right)=\pm\frac{\mathbf{3}}{\mathbf{400}}\text{ or }\pm\mathbf{0.0075}$$

Problem 3. Given $y=\frac{3e^{2\theta}\sec 2\theta}{\sqrt{(\theta-2)}}$ determine $\frac{dy}{d\theta}$

Using logarithmic differentiation and following the procedure:

(i) Since $y=\frac{3e^{2\theta}\sec 2\theta}{\sqrt{(\theta-2)}}$

then $\ln y=\ln\left\{\frac{3e^{2\theta}\sec 2\theta}{\sqrt{(\theta-2)}}\right\}$

$$=\ln\left\{\frac{3e^{2\theta}\sec 2\theta}{(\theta-2)^{\frac{1}{2}}}\right\}$$

(ii) $\ln y = \ln 3e^{2\theta} + \ln \sec 2\theta - \ln(\theta - 2)^{\frac{1}{2}}$

i.e. $\ln y = \ln 3 + \ln e^{2\theta} + \ln \sec 2\theta - \frac{1}{2}\ln(\theta - 2)$

i.e. $\ln y = \ln 3 + 2\theta + \ln \sec 2\theta - \frac{1}{2}\ln(\theta - 2)$

(iii) Differentiating with respect to θ gives:

$$\frac{1}{y}\frac{dy}{d\theta} = 0 + 2 + \frac{2\sec 2\theta \tan 2\theta}{\sec 2\theta} - \frac{\frac{1}{2}}{(\theta - 2)}$$

from equations (1) and (2).

(iv) Rearranging gives:

$$\frac{dy}{d\theta} = y\left\{2 + 2\tan 2\theta - \frac{1}{2(\theta - 2)}\right\}$$

(v) Substituting for y gives:

$$\boldsymbol{\frac{dy}{d\theta} = \frac{3e^{2\theta}\sec 2\theta}{\sqrt{(\theta - 2)}}\left\{2 + 2\tan 2\theta - \frac{1}{2(\theta - 2)}\right\}}$$

Problem 4. Differentiate $y = \dfrac{x^3 \ln 2x}{e^x \sin x}$ with respect to x

Using logarithmic differentiation and following the procedure gives:

(i) $\ln y = \ln\left\{\dfrac{x^3 \ln 2x}{e^x \sin x}\right\}$

(ii) $\ln y = \ln x^3 + \ln(\ln 2x) - \ln(e^x) - \ln(\sin x)$
i.e. $\ln y = 3\ln x + \ln(\ln 2x) - x - \ln(\sin x)$

(iii) $\dfrac{1}{y}\dfrac{dy}{dx} = \dfrac{3}{x} + \dfrac{\frac{1}{x}}{\ln 2x} - 1 - \dfrac{\cos x}{\sin x}$

(iv) $\dfrac{dy}{dx} = y\left\{\dfrac{3}{x} + \dfrac{1}{x\ln 2x} - 1 - \cot x\right\}$

(v) $\boldsymbol{\dfrac{dy}{dx} = \dfrac{x^3 \ln 2x}{e^x \sin x}\left\{\dfrac{3}{x} + \dfrac{1}{x\ln 2x} - 1 - \cot x\right\}}$

Now try the following Practice Exercise

Practice Exercise 212 Differentiating logarithmic functions (Answers on page 724)

In Problems 1 to 6, use logarithmic differentiation to differentiate the given functions with respect to the variable.

1. $y = \dfrac{(x-2)(x+1)}{(x-1)(x+3)}$

2. $y = \dfrac{(x+1)(2x+1)^3}{(x-3)^2(x+2)^4}$

3. $y = \dfrac{(2x-1)\sqrt{(x+2)}}{(x-3)\sqrt{(x+1)^3}}$

4. $y = \dfrac{e^{2x}\cos 3x}{\sqrt{(x-4)}}$

5. $y = 3\theta \sin\theta \cos\theta$

6. $y = \dfrac{2x^4 \tan x}{e^{2x}\ln 2x}$

7. Evaluate $\dfrac{dy}{dx}$ when $x = 1$ given
$y = \dfrac{(x+1)^2\sqrt{(2x-1)}}{\sqrt{(x+3)^3}}$

8. Evaluate $\dfrac{dy}{d\theta}$, correct to 3 significant figures, when $\theta = \dfrac{\pi}{4}$ given $y = \dfrac{2e^{\theta}\sin\theta}{\sqrt{\theta^5}}$

43.5 Differentiation of $[f(x)]^x$

Whenever an expression to be differentiated contains a term raised to a power which is itself a function of the variable, then logarithmic differentiation must be used. For example, the differentiation of expressions such as x^x, $(x+2)^x$, $\sqrt[x]{(x-1)}$ and x^{3x+2} can only be achieved using logarithmic differentiation.

Problem 5. Determine $\dfrac{dy}{dx}$ given $y = x^x$

Taking Napierian logarithms of both sides of $y = x^x$ gives:
$\ln y = \ln x^x = x\ln x$, by law (iii) of Section 43.2.
Differentiating both sides with respect to x gives:
$\dfrac{1}{y}\dfrac{dy}{dx} = (x)\left(\dfrac{1}{x}\right) + (\ln x)(1)$, using the product rule

i.e. $\frac{1}{y}\frac{dy}{dx} = 1 + \ln x$

from which, $\frac{dy}{dx} = y(1 + \ln x)$

i.e. $\boldsymbol{\frac{dy}{dx} = x^x(1 + \ln x)}$

Problem 6. Evaluate $\frac{dy}{dx}$ when $x = -1$ given $y = (x+2)^x$

Taking Napierian logarithms of both sides of $y = (x+2)^x$ gives:

$\ln y = \ln(x+2)^x = x\ln(x+2)$, by law (iii) of Section 43.2.

Differentiating both sides with respect to x gives:

$$\frac{1}{y}\frac{dy}{dx} = (x)\left(\frac{1}{x+2}\right) + [\ln(x+2)](1),$$

by the product rule.

$$\text{Hence} \quad \frac{dy}{dx} = y\left(\frac{x}{x+2} + \ln(x+2)\right)$$

$$= \boldsymbol{(x+2)^x\left\{\frac{x}{x+2} + \ln(x+2)\right\}}$$

$$\text{When } x = -1, \frac{dy}{dx} = (1)^{-1}\left(\frac{-1}{1} + \ln 1\right)$$

$$= (+1)(-1) = \boldsymbol{-1}$$

Problem 7. Determine (a) the differential coefficient of $y = \sqrt[x]{(x-1)}$ and (b) evaluate $\frac{dy}{dx}$ when $x = 2$

(a) $y = \sqrt[x]{(x-1)} = (x-1)^{\frac{1}{x}}$, since by the laws of indices $\sqrt[n]{a^m} = a^{\frac{m}{n}}$

Taking Napierian logarithms of both sides gives:

$$\ln y = \ln(x-1)^{\frac{1}{x}} = \frac{1}{x}\ln(x-1),$$

by law (iii) of Section 43.2.

Differentiating each side with respect to x gives:

$$\frac{1}{y}\frac{dy}{dx} = \left(\frac{1}{x}\right)\left(\frac{1}{x-1}\right) + [\ln(x-1)]\left(\frac{-1}{x^2}\right)$$

by the product rule.

$$\text{Hence} \quad \frac{dy}{dx} = y\left\{\frac{1}{x(x-1)} - \frac{\ln(x-1)}{x^2}\right\}$$

$$\text{i.e.} \quad \boldsymbol{\frac{dy}{dx} = \sqrt[x]{(x-1)}\left\{\frac{1}{x(x-1)} - \frac{\ln(x-1)}{x^2}\right\}}$$

(b) When $x = 2$, $\frac{dy}{dx} = \sqrt[2]{(1)}\left\{\frac{1}{2(1)} - \frac{\ln(1)}{4}\right\}$

$$= \pm 1\left\{\frac{1}{2} - 0\right\} = \boldsymbol{\pm\frac{1}{2}}$$

Problem 8. Differentiate x^{3x+2} with respect to x

Let $y = x^{3x+2}$
Taking Napierian logarithms of both sides gives:

$\ln y = \ln x^{3x+2}$

i.e. $\ln y = (3x+2)\ln x$, by law (iii) of Section 43.2 Differentiating each term with respect to x gives:

$$\frac{1}{y}\frac{dy}{dx} = (3x+2)\left(\frac{1}{x}\right) + (\ln x)(3),$$

by the product rule.

$$\text{Hence} \quad \frac{dy}{dx} = y\left\{\frac{3x+2}{x} + 3\ln x\right\}$$

$$= x^{3x+2}\left\{\frac{3x+2}{x} + 3\ln x\right\}$$

$$= \boldsymbol{x^{3x+2}\left\{3 + \frac{2}{x} + 3\ln x\right\}}$$

Now try the following Practice Exercise

Practice Exercise 213 Differentiating $[f(x)]^x$ type functions (Answers on page 725)

In Problems 1 to 4, differentiate with respect to x

1. $y = x^{2x}$
2. $y = (2x-1)^x$

3. $y=\sqrt[x]{(x+3)}$

4. $y=3x^{4x+1}$

5. Show that when $y=2x^x$ and $x=1$, $\frac{dy}{dx}=2$

6. Evaluate $\frac{d}{dx}\{\sqrt[x]{(x-2)}\}$ when $x=3$

7. Show that if $y=\theta^\theta$ and $\theta=2$, $\frac{dy}{d\theta}=6.77$, correct to 3 significant figures.

Practice Exercise 214 Multiple-choice questions on logarithmic differentiation (Answers on page 725)

(Answers on page 725)

Each question has only one correct answer

1. Given that $y=\ln(4x^3-3x^2)$ then $\frac{dy}{dx}$ is equal to:
(a) $\frac{6x}{4x^3+3x^2}$ (b) $\frac{6(2x-1)}{x(4x-3)}$
(c) $\frac{12x^2}{4x^3+3x^2}$ (d) $\frac{1}{4x^3+3x^2}$

2. The value of $\frac{d}{d\theta}(2\ln\cos\theta)$ when $\theta=\frac{\pi}{6}$ is:
(a) −3.464 (b) −7.464 (c) −9.774 (d) 4

3. If $y=3x^{2x}$ then $\frac{dy}{dx}$ is equal to:
(a) $6x^{2x}(1+\ln x)$ (b) $\ln 3+2x\ln x$
(c) $2x(3x)^x$ (d) $3x^{2x}(2x\ln 3x)$

4. The value of $\frac{d}{dx}\ln(x^3-2x^2+1)$ when $x=2$ is:
(a) 1 (b) 12 (c) 5 (d) 4

5. Given that $y=\sqrt[x]{(x-2)}$, the value of $\frac{dy}{dx}$ when $x=3$ is:
(a) $-\frac{1}{3}$ (b) $-\frac{1}{9}$ (c) $\frac{1}{3}$ (d) $\frac{1}{9}$

For fully worked solutions to each of the problems in Practice Exercises 211 to 213 in this chapter, go to the website:
www.routledge.com/cw/bird

Revision Test 11 Further differentiation

This Revision Test covers the material contained in Chapters 41 to 43. *The marks for each question are shown in brackets at the end of each question.*

1. A cycloid has parametric equations given by: $x=5(\theta-\sin\theta)$ and $y=5(1-\cos\theta)$. Evaluate (a) $\frac{dy}{dx}$ (b) $\frac{d^2y}{dx^2}$ when $\theta=1.5$ radians. Give answers correct to 3 decimal places. (8)

2. Determine the equation of (a) the tangent, and (b) the normal, drawn to an ellipse $x=4\cos\theta$, $y=\sin\theta$ at $\theta=\frac{\pi}{3}$ (8)

3. Determine expressions for $\frac{dz}{dy}$ for each of the following functions:
   (a) $z=5y^2\cos x$ (b) $z=x^2+4xy-y^2$ (5)

4. If $x^2+y^2+6x+8y+1=0$, find $\frac{dy}{dx}$ in terms of x and y (4)

5. Determine the gradient of the tangents drawn to the hyperbola $x^2-y^2=8$ at $x=3$ (4)

6. Use logarithmic differentiation to differentiate $y=\frac{(x+1)^2\sqrt{(x-2)}}{(2x-1)\sqrt[3]{(x-3)^4}}$ with respect to x (6)

7. Differentiate $y=\frac{3e^{\theta}\sin 2\theta}{\sqrt{\theta^5}}$ and hence evaluate $\frac{dy}{d\theta}$, correct to 2 decimal places, when $\theta=\frac{\pi}{3}$ (9)

8. Evaluate $\frac{d}{dt}[\sqrt{(2t+1)}]$ when $t=2$, correct to 4 significant figures (6)

For lecturers/instructors/teachers, fully worked solutions to each of the problems in Revision Test 11, together with a full marking scheme, are available at the website:
www.routledge.com/cw/bird

COMPANION @ WEBSITE

Section 8

Integral calculus

Chapter 44

Standard integration

Why it is important to understand: Standard integration

Engineering is all about problem solving and many problems in engineering can be solved using calculus. Physicists, chemists, engineers, and many other scientific and technical specialists use calculus in their everyday work; it is a technique of fundamental importance. Both integration and differentiation have numerous applications in engineering and science and some typical examples include determining areas, mean and r.m.s. values, volumes of solids of revolution, centroids, second moments of area, differential equations and Fourier series. Besides the standard integrals covered in this chapter, there are a number of other methods of integration covered in future chapters. For any further studies in engineering, differential and integral calculus are unavoidable.

At the end of this chapter, you should be able to:

- understand that integration is the reverse process of differentiation
- determine integrals of the form ax^n where n is fractional, zero, or a positive or negative integer
- integrate standard functions - $\cos ax, \sin ax, \sec^2 ax, \text{cosec}^2 ax, \text{cosec}\, ax \cot ax, \sec ax \tan ax, e^{ax}, \frac{1}{x}$
- evaluate definite integrals

44.1 The process of integration

The process of integration reverses the process of differentiation. In differentiation, if $f(x)=2x^2$ then $f'(x)=4x$. Thus the integral of $4x$ is $2x^2$, i.e. integration is the process of moving from $f'(x)$ to $f(x)$. By similar reasoning, the integral of $2t$ is t^2.

Integration is a process of summation or adding parts together and an elongated S, shown as $\int$, is used to replace the words 'the integral of'. Hence, from above, $\int 4x=2x^2$ and $\int 2t$ is t^2.

In differentiation, the differential coefficient $\frac{dy}{dx}$ indicates that a function of x is being differentiated with respect to x, the dx indicating that it is 'with respect to x'. In integration the variable of integration is shown by adding d(the variable) after the function to be integrated.

Thus $\int 4x\,dx$ means 'the integral of $4x$ with respect to x',

and $\int 2t\,dt$ means 'the integral of $2t$ with respect to t'

As stated above, the differential coefficient of $2x^2$ is $4x$, hence $\int 4x\,dx=2x^2$. However, the differential coefficient of $2x^2+7$ is also $4x$. Hence $\int 4x\,dx$ is also equal to $2x^2+7$. To allow for the possible presence of a constant, whenever the process of integration is performed, a constant 'c' is added to the result.

Thus $\int 4x\,dx = 2x^2 + c$ and $\int 2t\,dt = t^2 + c$

'c' is called the **arbitrary constant of integration.**

44.2 The general solution of integrals of the form ax^n

The general solution of integrals of the form $\int ax^n dx$, where a and n are constants is given by:

$$\int ax^n dx = \frac{ax^{n+1}}{n+1} + c$$

This rule is true when n is fractional, zero, or a positive or negative integer, with the exception of $n = -1$.

Using this rule gives:

(i) $\int 3x^4 dx = \frac{3x^{4+1}}{4+1} + c = \mathbf{\frac{3}{5}x^5 + c}$

(ii) $\int \frac{2}{x^2} dx = \int 2x^{-2} dx = \frac{2x^{-2+1}}{-2+1} + c$

$= \frac{2x^{-1}}{-1} + c = \mathbf{\frac{-2}{x} + c}$

and

(iii) $\int \sqrt{x}\,dx = \int x^{1/2} dx = \frac{x^{\frac{1}{2}+1}}{\frac{1}{2}+1} + c$

$= \frac{x^{\frac{3}{2}}}{\frac{3}{2}} + c = \mathbf{\frac{2}{3}\sqrt{x^3} + c}$

Each of these three results may be checked by differentiation.

(a) The integral of a constant k is $kx + c$. For example,

$$\int 8\,dx = 8x + c$$

(b) When a sum of several terms is integrated the result is the sum of the integrals of the separate terms. For example,

$$\int (3x + 2x^2 - 5)dx$$

$$= \int 3x\,dx + \int 2x^2 dx - \int 5\,dx$$

$$= \mathbf{\frac{3x^2}{2} + \frac{2x^3}{3} - 5x + c}$$

44.3 Standard integrals

Since integration is the reverse process of differentiation the **standard integrals** listed in Table 44.1 may be deduced and readily checked by differentiation.

Table 44.1 Standard integrals

(i)	$\int ax^n dx = \frac{ax^{n+1}}{n+1} + c$ (except when $n = -1$)
(ii)	$\int \cos ax\,dx = \frac{1}{a}\sin ax + c$
(iii)	$\int \sin ax\,dx = -\frac{1}{a}\cos ax + c$
(iv)	$\int \sec^2 ax\,dx = \frac{1}{a}\tan ax + c$
(v)	$\int \text{cosec}^2 ax\,dx = -\frac{1}{a}\cot ax + c$
(vi)	$\int \text{cosec}\,ax \cot ax\,dx = -\frac{1}{a}\text{cosec}\,ax + c$
(vii)	$\int \sec ax \tan ax\,dx = \frac{1}{a}\sec ax + c$
(viii)	$\int e^{ax} dx = \frac{1}{a}e^{ax} + c$
(ix)	$\int \frac{1}{x} dx = \ln x + c$

Problem 1. Determine:

(a) $\int 5x^2 dx$ (b) $\int 2t^3 dt$

The standard integral, $\int ax^n dx = \frac{ax^{n+1}}{n+1} + c$

(a) When $a = 5$ and $n = 2$ then

$$\int 5x^2 dx = \frac{5x^{2+1}}{2+1} + c = \mathbf{\frac{5x^3}{3} + c}$$

(b) When $a = 2$ and $n = 3$ then

$$\int 2t^3 dt = \frac{2t^{3+1}}{3+1} + c = \frac{2t^4}{4} + c = \mathbf{\frac{1}{2}t^4 + c}$$

Each of these results may be checked by differentiating them.

Problem 2. Determine: $\int \left(4+\frac{3}{7}x-6x^2\right) dx$

$\int \left(4+\frac{3}{7}x-6x^2\right) dx$ may be written as

$\int 4\,dx+\int \frac{3}{7}x\,dx-\int 6x^2\,dx$

i.e. each term is integrated separately. (This splitting up of terms only applies, however, for addition and subtraction.)

Hence

$$\int \left(4+\frac{3}{7}x-6x^2\right)dx$$
$$=4x+\left(\frac{3}{7}\right)\frac{x^{1+1}}{1+1}-(6)\frac{x^{2+1}}{2+1}+c$$
$$=4x+\left(\frac{3}{7}\right)\frac{x^2}{2}-(6)\frac{x^3}{3}+c$$
$$=\mathbf{4x+\frac{3}{14}x^2-2x^3+c}$$

Note that when an integral contains more than one term there is no need to have an arbitrary constant for each; just a single constant at the end is sufficient.

Problem 3. Determine:

(a) $\int \frac{2x^3-3x}{4x}dx$ (b) $\int (1-t)^2 dt$

(a) Rearranging into standard integral form gives:

$$\int \frac{2x^3-3x}{4x}dx=\int \left(\frac{2x^3}{4x}-\frac{3x}{4x}\right)dx$$
$$=\int \left(\frac{x^2}{2}-\frac{3}{4}\right)dx=\left(\frac{1}{2}\right)\frac{x^{2+1}}{2+1}-\frac{3}{4}x+c$$
$$=\left(\frac{1}{2}\right)\frac{x^3}{3}-\frac{3}{4}x+c=\mathbf{\frac{1}{6}x^3-\frac{3}{4}x+c}$$

(b) Rearranging $\int (1-t)^2 dt$ gives:

$$\int (1-2t+t^2)dt=t-\frac{2t^{1+1}}{1+1}+\frac{t^{2+1}}{2+1}+c$$
$$=t-\frac{2t^2}{2}+\frac{t^3}{3}+c$$
$$=\mathbf{t-t^2+\frac{1}{3}t^3+c}$$

This problem shows that functions often have to be rearranged into the standard form of $\int ax^n dx$ before it is possible to integrate them.

Problem 4. Determine: $\int \frac{3}{x^2}dx$

$\int \frac{3}{x^2}dx=\int 3x^{-2}$. Using the standard integral, $\int ax^n dx$ when $a=3$ and $n=-2$ gives:

$$\int 3x^{-2}dx=\frac{3x^{-2+1}}{-2+1}+c=\frac{3x^{-1}}{-1}+c$$
$$=-3x^{-1}+c=\mathbf{\frac{-3}{x}+c}$$

Problem 5. Determine: $\int 3\sqrt{x}dx$

For fractional powers it is necessary to appreciate $\sqrt[n]{a^m}=a^{\frac{m}{n}}$

$$\int 3\sqrt{x}dx=\int 3x^{1/2}dx=\frac{3x^{\frac{1}{2}+1}}{\frac{1}{2}+1}+c$$
$$=\frac{3x^{\frac{3}{2}}}{\frac{3}{2}}+c=2x^{\frac{3}{2}}+c=\mathbf{2\sqrt{x^3}+c}$$

Problem 6. Determine: $\int \frac{-5}{9\sqrt[4]{t^3}}dt$

$$\int \frac{-5}{9\sqrt[4]{t^3}}dt=\int \frac{-5}{9t^{\frac{3}{4}}}dt=\int \left(-\frac{5}{9}\right)t^{-\frac{3}{4}}dt$$
$$=\left(-\frac{5}{9}\right)\frac{t^{-\frac{3}{4}+1}}{-\frac{3}{4}+1}+c$$
$$=\left(-\frac{5}{9}\right)\frac{t^{\frac{1}{4}}}{\frac{1}{4}}+c=\left(-\frac{5}{9}\right)\left(\frac{4}{1}\right)t^{1/4}+c$$
$$=\mathbf{-\frac{20}{9}\sqrt[4]{t}+c}$$

Problem 7. Determine: $\int \frac{(1+\theta)^2}{\sqrt{\theta}}d\theta$

$$\int \frac{(1+\theta)^2}{\sqrt{\theta}}d\theta = \int \frac{(1+2\theta+\theta^2)}{\sqrt{\theta}}d\theta$$
$$= \int \left(\frac{1}{\theta^{\frac{1}{2}}} + \frac{2\theta}{\theta^{\frac{1}{2}}} + \frac{\theta^2}{\theta^{\frac{1}{2}}}\right)d\theta$$
$$= \int \left(\theta^{-\frac{1}{2}} + 2\theta^{1-\left(\frac{1}{2}\right)} + \theta^{2-\left(\frac{1}{2}\right)}\right)d\theta$$
$$= \int \left(\theta^{-\frac{1}{2}} + 2\theta^{\frac{1}{2}} + \theta^{\frac{3}{2}}\right)d\theta$$
$$= \frac{\theta^{\left(-\frac{1}{2}\right)+1}}{-\frac{1}{2}+1} + \frac{2\theta^{\left(\frac{1}{2}\right)+1}}{\frac{1}{2}+1} + \frac{\theta^{\left(\frac{3}{2}\right)+1}}{\frac{3}{2}+1} + c$$
$$= \frac{\theta^{\frac{1}{2}}}{\frac{1}{2}} + \frac{2\theta^{\frac{3}{2}}}{\frac{3}{2}} + \frac{\theta^{\frac{5}{2}}}{\frac{5}{2}} + c$$
$$= 2\theta^{\frac{1}{2}} + \frac{4}{3}\theta^{\frac{3}{2}} + \frac{2}{5}\theta^{\frac{5}{2}} + c$$
$$= \mathbf{2\sqrt{\theta} + \frac{4}{3}\sqrt{\theta^3} + \frac{2}{5}\sqrt{\theta^5} + c}$$

Problem 8. Determine:
(a) $\int 4\cos 3x\,dx$ (b) $\int 5\sin 2\theta\,d\theta$

(a) From Table 44.1 (ii),

$$\int 4\cos 3x\,dx = (4)\left(\frac{1}{3}\right)\sin 3x + c = \mathbf{\frac{4}{3}\sin 3x + c}$$

(b) From Table 44.1(iii),

$$\int 5\sin 2\theta\,d\theta = (5)\left(-\frac{1}{2}\right)\cos 2\theta + c = \mathbf{-\frac{5}{2}\cos 2\theta + c}$$

Problem 9. Determine: (a) $\int 7\sec^2 4t\,dt$
(b) $3\int \text{cosec}^2 2\theta\,d\theta$

(a) From Table 44.1(iv),

$$\int 7\sec^2 4t\,dt = (7)\left(\frac{1}{4}\right)\tan 4t + c = \mathbf{\frac{7}{4}\tan 4t + c}$$

(b) From Table 44.1(v),

$$3\int \text{cosec}^2 2\theta\,d\theta = (3)\left(-\frac{1}{2}\right)\cot 2\theta + c = \mathbf{-\frac{3}{2}\cot 2\theta + c}$$

Problem 10. Determine: (a) $\int 5e^{3x}dx$
(b) $\int \frac{2}{3e^{4t}}dt$

(a) From Table 44.1(viii),

$$\int 5e^{3x}dx = (5)\left(\frac{1}{3}\right)e^{3x} + c = \mathbf{\frac{5}{3}e^{3x} + c}$$

(b) $$\int \frac{2}{3e^{4t}}dt = \int \frac{2}{3}e^{-4t}dt = \left(\frac{2}{3}\right)\left(-\frac{1}{4}\right)e^{-4t} + c = -\frac{1}{6}e^{-4t} + c = \mathbf{-\frac{1}{6e^{4t}} + c}$$

Problem 11. Determine:
(a) $\int \frac{3}{5x}dx$ (b) $\int \left(\frac{2m^2+1}{m}\right)dm$

(a) $$\int \frac{3}{5x}dx = \int \left(\frac{3}{5}\right)\left(\frac{1}{x}\right)dx = \mathbf{\frac{3}{5}\ln x + c}$$
(from Table 44.1(ix))

(b) $$\int \left(\frac{2m^2+1}{m}\right)dm = \int \left(\frac{2m^2}{m} + \frac{1}{m}\right)dm$$
$$= \int \left(2m + \frac{1}{m}\right)dm$$
$$= \frac{2m^2}{2} + \ln m + c$$
$$= \mathbf{m^2 + \ln m + c}$$

Now try the following Practice Exercise

Practice Exercise 215 Standard integrals (Answers on page 725)

Determine the following integrals:

1. (a) $\int 4\,dx$ (b) $\int 7x\,dx$
2. (a) $\int \frac{2}{5}x^2 dx$ (b) $\int \frac{5}{6}x^3 dx$

3. (a) $\int\left(\frac{3x^2-5x}{x}\right)dx$ (b) $\int(2+\theta)^2 d\theta$

4. (a) $\int\frac{4}{3x^2}dx$ (b) $\int\frac{3}{4x^4}dx$

5. (a) $2\int\sqrt{x^3}dx$ (b) $\int\frac{1}{4}\sqrt[4]{x^5}dx$

6. (a) $\int\frac{-5}{\sqrt{t^3}}dt$ (b) $\int\frac{3}{7\sqrt[5]{x^4}}dx$

7. (a) $\int 3\cos 2x\,dx$ (b) $\int 7\sin 3\theta\,d\theta$

8. (a) $\int\frac{3}{4}\sec^2 3x\,dx$ (b) $\int 2\,\text{cosec}^2 4\theta\,d\theta$

9. (a) $5\int\cot 2t\,\text{cosec}\,2t\,dt$

   (b) $\int\frac{4}{3}\sec 4t\tan 4t\,dt$

10. (a) $\int\frac{3}{4}e^{2x}dx$ (b) $\frac{2}{3}\int\frac{dx}{e^{5x}}$

11. (a) $\int\frac{2}{3x}dx$ (b) $\int\left(\frac{u^2-1}{u}\right)du$

12. (a) $\int\frac{(2+3x)^2}{\sqrt{x}}dx$ (b) $\int\left(\frac{1}{t}+2t\right)^2 dt$

44.4 Definite integrals

Integrals containing an arbitrary constant c in their results are called **indefinite integrals** since their precise value cannot be determined without further information. **Definite integrals** are those in which limits are applied. If an expression is written as $[x]_a^b$, 'b' is called the upper limit and 'a' the lower limit.
The operation of applying the limits is defined as: $[x]_a^b=(b)-(a)$
The increase in the value of the integral x^2 as x increases from 1 to 3 is written as $\int_1^3 x^2\,dx$

Applying the limits gives:

$$\int_1^3 x^2 dx=\left[\frac{x^3}{3}+c\right]_1^3=\left(\frac{3^3}{3}+c\right)-\left(\frac{1^3}{3}+c\right)$$
$$=(9+c)-\left(\frac{1}{3}+c\right)=\mathbf{8\frac{2}{3}}$$

Note that the 'c' term always cancels out when limits are applied and it need not be shown with definite integrals.

Problem 12. Evaluate: (a) $\int_1^2 3x\,dx$
(b) $\int_{-2}^3(4-x^2)dx$

(a) $$\int_1^2 3x\,dx=\left[\frac{3x^2}{2}\right]_1^2=\left\{\frac{3}{2}(2)^2\right\}-\left\{\frac{3}{2}(1)^2\right\}$$
$$=6-1\frac{1}{2}=\mathbf{4\frac{1}{2}}$$

(b) $$\int_{-2}^3(4-x^2)dx=\left[4x-\frac{x^3}{3}\right]_{-2}^3$$
$$=\left\{4(3)-\frac{(3)^3}{3}\right\}-\left\{4(-2)-\frac{(-2)^3}{3}\right\}$$
$$=\{12-9\}-\left\{-8-\frac{-8}{3}\right\}$$
$$=\{3\}-\left\{-5\frac{1}{3}\right\}=\mathbf{8\frac{1}{3}}$$

Problem 13. Evaluate: $\int_1^4\left(\frac{\theta+2}{\sqrt{\theta}}\right)d\theta$, taking positive square roots only

$$\int_1^4\left(\frac{\theta+2}{\sqrt{\theta}}\right)d\theta=\int_1^4\left(\frac{\theta}{\theta^{\frac{1}{2}}}+\frac{2}{\theta^{\frac{1}{2}}}\right)d\theta$$
$$=\int_1^4\left(\theta^{\frac{1}{2}}+2\theta^{-\frac{1}{2}}\right)d\theta$$
$$=\left[\frac{\theta^{\left(\frac{1}{2}\right)+1}}{\frac{1}{2}+1}+\frac{2\theta^{\left(-\frac{1}{2}\right)+1}}{-\frac{1}{2}+1}\right]_1^4$$

$$= \left[\frac{\theta^{\frac{3}{2}}}{\frac{3}{2}} + \frac{2\theta^{\frac{1}{2}}}{\frac{1}{2}}\right]_1^4 = \left[\frac{2}{3}\sqrt{\theta^3} + 4\sqrt{\theta}\right]_1^4$$

$$= \left\{\frac{2}{3}\sqrt{(4)^3} + 4\sqrt{4}\right\} - \left\{\frac{2}{3}\sqrt{(1)^3} + 4\sqrt{1}\right\}$$

$$= \left\{\frac{16}{3} + 8\right\} - \left\{\frac{2}{3} + 4\right\}$$

$$= 5\frac{1}{3} + 8 - \frac{2}{3} - 4 = \mathbf{8\frac{2}{3}}$$

Problem 14. Evaluate: $\int_0^{\pi/2} 3\sin 2x\,dx$

$$\int_0^{\frac{\pi}{2}} 3\sin 2x\,dx$$

$$= \left[(3)\left(-\frac{1}{2}\right)\cos 2x\right]_0^{\frac{\pi}{2}} = \left[-\frac{3}{2}\cos 2x\right]_0^{\frac{\pi}{2}}$$

$$= \left\{-\frac{3}{2}\cos 2\left(\frac{\pi}{2}\right)\right\} - \left\{-\frac{3}{2}\cos 2(0)\right\}$$

$$= \left\{-\frac{3}{2}\cos\pi\right\} - \left\{-\frac{3}{2}\cos 0\right\}$$

$$= \left\{-\frac{3}{2}(-1)\right\} - \left\{-\frac{3}{2}(1)\right\} = \frac{3}{2} + \frac{3}{2} = \mathbf{3}$$

Problem 15. Evaluate: $\int_1^2 4\cos 3t\,dt$

$$\int_1^2 4\cos 3t\,dt = \left[(4)\left(\frac{1}{3}\right)\sin 3t\right]_1^2 = \left[\frac{4}{3}\sin 3t\right]_1^2$$

$$= \left\{\frac{4}{3}\sin 6\right\} - \left\{\frac{4}{3}\sin 3\right\}$$

Note that limits of trigonometric functions are always expressed in radians—thus, for example, sin 6 means the sine of 6 radians $= -0.279415\ldots$

Hence $$\int_1^2 4\cos 3t\,dt = \left\{\frac{4}{3}(-0.279415\ldots)\right\} - \left\{\frac{4}{3}(-0.141120\ldots)\right\}$$

$$= (-0.37255) - (0.18816) = \mathbf{-0.5607}$$

Problem 16. Evaluate:

(a) $\int_1^2 4e^{2x}\,dx$ (b) $\int_1^4 \frac{3}{4u}\,du$,

each correct to 4 significant figures

(a) $$\int_1^2 4e^{2x}\,dx = \left[\frac{4}{2}e^{2x}\right]_1^2$$

$$= 2[e^{2x}]_1^2 = 2[e^4 - e^2]$$

$$= 2[54.5982 - 7.3891] = \mathbf{94.42}$$

(b) $$\int_1^4 \frac{3}{4u}\,du = \left[\frac{3}{4}\ln u\right]_1^4 = \frac{3}{4}[\ln 4 - \ln 1]$$

$$= \frac{3}{4}[1.3863 - 0] = \mathbf{1.040}$$

Now try the following Practice Exercise

Practice Exercise 216 Definite integrals (Answers on page 725)

In Problems 1 to 8, evaluate the definite integrals (where necessary, correct to 4 significant figures).

1. (a) $\int_1^4 5x^2\,dx$ (b) $\int_{-1}^1 -\frac{3}{4}t^2\,dt$
2. (a) $\int_{-1}^2 (3 - x^2)\,dx$ (b) $\int_1^3 (x^2 - 4x + 3)\,dx$
3. (a) $\int_0^{\pi} \frac{3}{2}\cos\theta\,d\theta$ (b) $\int_0^{\frac{\pi}{2}} 4\cos\theta\,d\theta$
4. (a) $\int_{\frac{\pi}{6}}^{\frac{\pi}{3}} 2\sin 2\theta\,d\theta$ (b) $\int_0^2 3\sin t\,dt$
5. (a) $\int_0^1 5\cos 3x\,dx$ (b) $\int_0^{\frac{\pi}{6}} 3\sec^2 2x\,dx$
6. (a) $\int_1^2 \mathrm{cosec}^2 4t\,dt$

   (b) $\int_{\frac{\pi}{4}}^{\frac{\pi}{2}} (3\sin 2x - 2\cos 3x)\,dx$
7. (a) $\int_0^1 3e^{3t}\,dt$ (b) $\int_{-1}^2 \frac{2}{3e^{2x}}\,dx$
8. (a) $\int_2^3 \frac{2}{3x}\,dx$ (b) $\int_1^3 \frac{2x^2 + 1}{x}\,dx$

9. The entropy change ΔS, for an ideal gas is given by:

$$\Delta S = \int_{T_1}^{T_2} C_v \frac{dT}{T} - R\int_{V_1}^{V_2} \frac{dV}{V}$$

where T is the thermodynamic temperature, V is the volume and $R = 8.314$. Determine the entropy change when a gas expands from 1 litre to 3 litres for a temperature rise from 100 K to 400 K given that:

$$C_v = 45 + 6 \times 10^{-3}T + 8 \times 10^{-6}T^2$$

10. The p.d. between boundaries a and b of an electric field is given by:

$$V = \int_a^b \frac{Q}{2\pi r \varepsilon_0 \varepsilon_r} dr$$

If $a = 10$, $b = 20$, $Q = 2 \times 10^{-6}$ coulombs, $\varepsilon_0 = 8.85 \times 10^{-12}$ and $\varepsilon_r = 2.77$, show that $V = 9$ kV.

11. The average value of a complex voltage waveform is given by:

$$V_{AV} = \frac{1}{\pi}\int_0^{\pi} (10\sin\omega t + 3\sin 3\omega t + 2\sin 5\omega t)d(\omega t)$$

Evaluate V_{AV} correct to 2 decimal places

12. The volume of liquid in a tank is given by: $v = \int_{t_1}^{t_2} q dt$. Determine the volume of a chemical, given $q = (5 - 0.05t + 0.003t^2)\text{m}^3/\text{s}$, $t_1 = 0$ and $t_2 = 16$s

Practice Exercise 217 Multiple-choice questions on standard integration (Answers on page 725)

Each question has only one correct answer

1. $\int (5 - 3t^2)dt$ is equal to:
(a) $5 - t^3 + c$ (b) $-3t^3 + c$
(c) $-6t + c$ (d) $5t - t^3 + c$

2. Evaluating $\int_{-1}^{2} (3x^2 + 4x^3)\, dx$ gives:
(a) 54 (b) 22 (c) 24 (d) 26

3. $\int \left(\frac{5x - 1}{x}\right) dx$ is equal to:
(a) $5x - \ln x + c$ (b) $\frac{5x^2 - x}{\frac{x^2}{2}}$
(c) $\frac{5x^2}{2} + \frac{1}{x^2} + c$ (d) $5x + \frac{1}{x^2} + c$

4. $\int \frac{2}{9}t^3\, dt$ is equal to:
(a) $\frac{t^4}{18} + c$ (b) $\frac{2}{3}t^2 + c$
(c) $\frac{2}{9}t^4 + c$ (d) $\frac{2}{9}t^3 + c$

5. $\int (5\sin 3t - 3\cos 5t)dt$ is equal to:
(a) $-5\cos 3t + 3\sin 5t + c$
(b) $15(\cos 3t + \sin 3t) + c$
(c) $-\frac{5}{3}\cos 3t - \frac{3}{5}\sin 5t + c$
(d) $\frac{3}{5}\cos 3t - \frac{5}{3}\sin 5t + c$

6. $\int (\sqrt{x} - 3)dx$ is equal to:
(a) $\frac{3}{2}\sqrt{x^3} - 3x + c$ (b) $\frac{2}{3}\sqrt{x^3} + c$
(c) $\frac{1}{2\sqrt{x}} + c$ (d) $\frac{2}{3}\sqrt{x^3} - 3x + c$

7. Evaluating $\int_0^{\pi/3} 3\sin 3x\, dx$ gives:
(a) 1.503 (b) 2 (c) −18 (d) 6

8. $\int \left(1 + \frac{4}{e^{2x}}\right) dx$ is equal to:
(a) $\frac{8}{e^{2x}} + c$ (b) $x - \frac{2}{e^{2x}} + c$
(c) $x + \frac{4}{e^{2x}} + c$ (d) $x - \frac{8}{e^{2x}} + c$

9. Evaluating $\int_1^2 2e^{3t}\, dt$, correct to 4 significant figures, gives:
(a) 2300 (b) 255.6 (c) 766.7 (d) 282.3

10. $\int_{\frac{\pi}{2}}^{\frac{3\pi}{2}} (2\sin\theta - 3\cos\theta)\, d\theta$ is equal to:
(a) 0 (b) 1 (c) 2 (d) 6

For fully worked solutions to each of the problems in Practice Exercises 215 and 216 in this chapter, go to the website:
www.routledge.com/cw/bird

COMPANION WEBSITE

Chapter 45

Integration using algebraic substitutions

Why it is important to understand: Integration using algebraic substitutions

As intimated in the previous chapter, most complex engineering problems cannot be solved without calculus. Calculus has widespread applications in science, economics and engineering and can solve many problems for which algebra alone is insufficient. For example, calculus is needed to calculate the force exerted on a particle a specific distance from an electrically charged wire, and is needed for computations involving arc length, centre of mass, work and pressure. Sometimes the integral is not a standard one; in these cases it may be possible to replace the variable of integration by a function of a new variable. A change in variable can reduce an integral to a standard form, and this is demonstrated in this chapter.

At the end of this chapter, you should be able to:

- appreciate when an algebraic substitution is required to determine an integral
- integrate functions which require an algebraic substitution
- determine definite integrals where an algebraic substitution is required
- appreciate that it is possible to change the limits when determining a definite integral

45.1 Introduction

Functions that require integrating are not always in the 'standard form' shown in Chapter 44. However, it is often possible to change a function into a form which can be integrated by using either:

(i) an algebraic substitution (see Section 45.2),

(ii) trigonometric substitutions (see Chapter 46),

(iii) partial fractions (see Chapter 47),

(iv) the $t=\tan\frac{\theta}{2}$ substitution (see Chapter 48), or

(v) integration by parts (see Chapter 49).

45.2 Algebraic substitutions

With **algebraic substitutions**, the substitution usually made is to let u be equal to $f(x)$ such that $f(u)\,du$ is a standard integral. It is found that integrals of the forms:

$$k\int [f(x)]^n f'(x)dx \text{ and } k\int \frac{f'(x)^n}{[f(x)]}dx$$

(where k and n are constants) can both be integrated by substituting u for $f(x)$.

45.3 Worked problems on integration using algebraic substitutions

Problem 1. Determine: $\int \cos(3x+7)\,dx$

$\int \cos(3x+7)\,dx$ is not a standard integral of the form shown in Table 44.1, page 470, thus an algebraic substitution is made.

Let $u=3x+7$ then $\dfrac{du}{dx}=3$ and rearranging gives $dx=\dfrac{du}{3}$

Hence $\displaystyle\int \cos(3x+7)\,dx = \int (\cos u)\frac{du}{3} = \int \frac{1}{3}\cos u\,du,$

which is a standard integral

$$= \frac{1}{3}\sin u + c$$

Rewriting u as $(3x+7)$ gives:

$$\int \cos(3x+7)\,dx = \mathbf{\frac{1}{3}\sin(3x+7)+c},$$

which may be checked by differentiating it.

Problem 2. Find: $\int (2x-5)^7\,dx$

$(2x-5)$ may be multiplied by itself 7 times and then each term of the result integrated. However, this would be a lengthy process, and thus an algebraic substitution is made.

Let $u=(2x-5)$ then $\dfrac{du}{dx}=2$ and $dx=\dfrac{du}{2}$

Hence

$$\int (2x-5)^7\,dx = \int u^7\frac{du}{2} = \frac{1}{2}\int u^7\,du = \frac{1}{2}\left(\frac{u^8}{8}\right)+c = \frac{1}{16}u^8+c$$

Rewriting u as $(2x-5)$ gives:

$$\int \mathbf{(2x-5)^7\,dx = \frac{1}{16}(2x-5)^8+c}$$

Problem 3. Find: $\displaystyle\int \frac{4}{(5x-3)}\,dx$

Let $u=(5x-3)$ then $\dfrac{du}{dx}=5$ and $dx=\dfrac{du}{5}$

Hence $\displaystyle\int \frac{4}{(5x-3)}\,dx = \int \frac{4}{u}\frac{du}{5} = \frac{4}{5}\int \frac{1}{u}\,du$

$$= \frac{4}{5}\ln u + c$$

$$= \mathbf{\frac{4}{5}\ln(5x-3)+c}$$

Problem 4. Evaluate: $\int_0^1 2e^{6x-1}\,dx$, correct to 4 significant figures

Let $u=6x-1$ then $\dfrac{du}{dx}=6$ and $dx=\dfrac{du}{6}$

Hence $\displaystyle\int 2e^{6x-1}\,dx = \int 2e^u\frac{du}{6} = \frac{1}{3}\int e^u\,du$

$$= \frac{1}{3}e^u + c = \frac{1}{3}e^{6x-1}+c$$

Thus $\displaystyle\int_0^1 2e^{6x-1}\,dx = \frac{1}{3}\left[e^{6x-1}\right]_0^1$

$$= \frac{1}{3}[e^5 - e^{-1}] = \mathbf{49.35},$$

correct to 4 significant figures.

Problem 5. Determine: $\int 3x(4x^2+3)^5\,dx$

Let $u=(4x^2+3)$ then $\dfrac{du}{dx}=8x$ and $dx=\dfrac{du}{8x}$

Hence

$$\int 3x(4x^2+3)^5\,dx = \int 3x(u)^5\frac{du}{8x}$$

$$= \frac{3}{8}\int u^5\,du, \text{ by cancelling}$$

The original variable 'x' has been completely removed and the integral is now only in terms of u and is a standard integral.

Hence $\displaystyle\frac{3}{8}\int u^5\,du = \frac{3}{8}\left(\frac{u^6}{6}\right)+c = \frac{1}{16}u^6+c$

$$= \mathbf{\frac{1}{16}(4x^2+3)^6+c}$$

Problem 6. Evaluate: $\int_0^{\pi/6} 24\sin^5\theta\cos\theta\, d\theta$

Let $u = \sin\theta$ then $\frac{du}{d\theta} = \cos\theta$ and $d\theta = \frac{du}{\cos\theta}$

Hence $\int 24\sin^5\theta\cos\theta\, d\theta$

$$= \int 24u^5\cos\theta\frac{du}{\cos\theta}$$

$$= 24\int u^5 du, \text{by cancelling}$$

$$= 24\frac{u^6}{6} + c = 4u^6 + c = 4(\sin\theta)^6 + c$$

$$= 4\sin^6\theta + c$$

Thus $\int_0^{\pi/6} 24\sin^5\theta\cos\theta\, d\theta$

$$= \left[4\sin^6\theta\right]_0^{\pi/6} = 4\left[\left(\sin\frac{\pi}{6}\right)^6 - (\sin 0)^6\right]$$

$$= 4\left[\left(\frac{1}{2}\right)^6 - 0\right] = \mathbf{\frac{1}{16}} \text{ or } \mathbf{0.0625}$$

Now try the following Practice Exercise

Practice Exercise 218 Integration using algebraic substitutions (Answers on page 475)

In Problems 1 to 6, integrate with respect to the variable.

1. $2\sin(4x+9)$
2. $3\cos(2\theta-5)$
3. $4\sec^2(3t+1)$
4. $\frac{1}{2}(5x-3)^6$
5. $\frac{-3}{(2x-1)}$
6. $3e^{3\theta+5}$

In Problems 7 to 10, evaluate the definite integrals correct to 4 significant figures.

7. $\int_0^1 (3x+1)^5 dx$
8. $\int_0^2 x\sqrt{2x^2+1}\, dx$
9. $\int_0^{\pi/3} 2\sin\left(3t+\frac{\pi}{4}\right) dt$
10. $\int_0^1 3\cos(4x-3)\, dx$
11. The mean time to failure, M years, for a set of components is given by: $M = \int_0^4 (1-0.25t)^{1.5} dt$. Determine the mean time to failure.

45.4 Further worked problems on integration using algebraic substitutions

Problem 7. Find: $\int \frac{x}{2+3x^2}dx$

Let $u = 2+3x^2$ then $\frac{du}{dx} = 6x$ and $dx = \frac{du}{6x}$

Hence $\int \frac{x}{2+3x^2}dx$

$$= \int \frac{x}{u}\frac{du}{6x} = \frac{1}{6}\int \frac{1}{u}du, \text{ by cancelling,}$$

$$= \frac{1}{6}\ln u + x$$

$$= \mathbf{\frac{1}{6}\ln(2+3x^2) + c}$$

Problem 8. Determine: $\int \frac{2x}{\sqrt{4x^2-1}}dx$

Let $u = 4x^2 - 1$ then $\frac{du}{dx} = 8x$ and $dx = \frac{du}{8x}$

Hence $\int \frac{2x}{\sqrt{4x^2-1}}dx$

$$= \int \frac{2x}{\sqrt{u}}\frac{du}{8x} = \frac{1}{4}\int \frac{1}{\sqrt{u}}du, \text{ by cancelling}$$

$$= \frac{1}{4}\int u^{-1/2} du$$

$$= \frac{1}{4}\left[\frac{u^{(-1/2)+1}}{-\frac{1}{2}+1}\right] + c = \frac{1}{4}\left[\frac{u^{1/2}}{\frac{1}{2}}\right] + c$$

$$= \frac{1}{2}\sqrt{u} + c = \mathbf{\frac{1}{2}\sqrt{4x^2-1}+c}$$

Problem 9. Show that:

$$\int \tan\theta\, d\theta = \ln(\sec\theta) + c$$

$$\int \tan\theta\, d\theta = \int \frac{\sin\theta}{\cos\theta} d\theta$$

Let u = cos θ

then $\frac{du}{d\theta} = -\sin\theta$ and $d\theta = \frac{-du}{\sin\theta}$

Hence

$$\int \frac{\sin\theta}{\cos\theta} d\theta = \int \frac{\sin\theta}{u}\left(\frac{-du}{\sin\theta}\right)$$

$$= -\int \frac{1}{u} du = -\ln u + c$$

$$= -\ln(\cos\theta) + c$$

$$= \ln(\cos\theta)^{-1} + c,$$

by the laws of logarithms

Hence $\int \tan\theta\, d\theta = \ln(\sec\theta) + c$,

since $(\cos\theta)^{-1} = \frac{1}{\cos\theta} = \sec\theta$

Problem 10. The distance, d, a body has fallen in a resistive medium when its velocity, v, is 0.3 m/s is given by the integral: $d = \int_0^2 \left(\frac{v}{9.81 - 0.3v^2}\right) dv$

Evaluate distance d, correct to 4 significant figures.

Distance, $d = \int_0^2 \left(\frac{v}{9.81 - 0.3v^2}\right) dv$

Let $u = 9.81 - 0.3v^2$ then $\frac{du}{dx} = -0.6v$ and $dx = \frac{du}{-0.6v}$

Hence, $\int \left(\frac{v}{9.81 - 0.3v^2}\right) dv$

$$= \int \left(\frac{v}{u}\right)\frac{du}{-0.6v} = \frac{1}{-0.6}\int \frac{1}{u} du = -\frac{1}{0.6}\ln u + c$$

Since $u = 9.81 - 0.3v^2$ then

$$\int \left(\frac{v}{9.81 - 0.3v^2}\right) dv = -\frac{1}{0.6}\ln\left(9.81 - 0.3v^2\right) + c$$

Therefore, distance,

$$d = \int_0^2 \left(\frac{v}{9.81 - 0.3v^2}\right) dv = -\frac{1}{0.6}[\ln(9.81 - 0.3v^2)]_0^2$$

$$= -\frac{1}{0.6}[\ln(9.81 - 0.3(2)^2) - \ln(9.81 - 0.3(0)^2)]$$

$$= -\frac{1}{0.6}[\ln(8.61) - \ln(9.81)] = 0.2175 \text{ m}$$

i.e. **the distance a body has fallen in the resistive medium is 0.2175 m**

45.5 Change of limits

When evaluating definite integrals involving substitutions it is sometimes more convenient to **change the limits** of the integral as shown in Problems 11 and 12.

Problem 11. Evaluate: $\int_1^3 5x\sqrt{2x^2+7}\,dx$, taking positive values of square roots only

Let $u = 2x^2 + 7$, then $\frac{du}{dx} = 4x$ and $dx = \frac{du}{4x}$

It is possible in this case to change the limits of integration. Thus when $x = 3$, $u = 2(3)^2 + 7 = 25$ and when $x = 1$, $u = 2(1)^2 + 7 = 9$

Hence $\int_{x=1}^{x=3} 5x\sqrt{2x^2+7}\,dx$

$$= \int_{u=9}^{u=25} 5x\sqrt{u}\,\frac{du}{4x} = \frac{5}{4}\int_9^{25}\sqrt{u}\,du$$

$$= \frac{5}{4}\int_9^{25} u^{1/2} du$$

Thus the limits have been changed, and it is unnecessary to change the integral back in terms of x.

Thus $\int_{x=1}^{x=3} 5x\sqrt{2x^2+7}\,dx$

$$= \frac{5}{4}\left[\frac{u^{3/2}}{3/2}\right]_9^{25} = \frac{5}{6}\left[\sqrt{u^3}\right]_9^{25}$$

$$= \frac{5}{6}[\sqrt{25^3} - \sqrt{9^3}] = \frac{5}{6}(125 - 27) = \mathbf{81\frac{2}{3}}$$

Problem 12. Evaluate: $\int_0^2 \frac{3x}{\sqrt{2x^2+1}}\,dx$, taking positive values of square roots only

Let $u=2x^2+1$ then $\frac{du}{dx}=4x$ and $dx=\frac{du}{4x}$

$$\text{Hence} \quad \int_0^2 \frac{3x}{\sqrt{2x^2+1}}\,dx = \int_{x=0}^{x=2} \frac{3x}{\sqrt{u}}\frac{du}{4x}$$

$$= \frac{3}{4}\int_{x=0}^{x=2} u^{-1/2}du$$

Since $u=2x^2+1$, when $x=2$, $u=9$ and when $x=0$, $u=1$

$$\text{Thus } \frac{3}{4}\int_{x=0}^{x=2} u^{-1/2}\,du = \frac{3}{4}\int_{u=1}^{u=9} u^{-1/2}\,du,$$

i.e. the limits have been changed

$$= \frac{3}{4}\left[\frac{u^{1/2}}{\frac{1}{2}}\right]_1^9 = \frac{3}{2}[\sqrt{9}-\sqrt{1}] = \mathbf{3},$$

taking positive values of square roots only.

Now try the following Practice Exercise

Practice Exercise 219 Integration using algebraic substitutions (Answers on page 725)

In Problems 1 to 7, integrate with respect to the variable.

1. $2x(2x^2-3)^5$
2. $5\cos^5 t \sin t$
3. $3\sec^2 3x \tan 3x$
4. $2t\sqrt{3t^2-1}$
5. $\frac{\ln\theta}{\theta}$
6. $3\tan 2t$
7. $\frac{2e^t}{\sqrt{e^t+4}}$

In Problems 8 to 10, evaluate the definite integrals correct to 4 significant figures.

8. $\int_0^1 3xe^{(2x^2-1)}dx$
9. $\int_0^{\pi/2} 3\sin^4\theta\cos\theta\,d\theta$
10. $\int_0^1 \frac{3x}{(4x^2-1)^5}dx$
11. The electrostatic potential on all parts of a conducting circular disc of radius r is given by the equation:
$$V = 2\pi\sigma\int_0^9 \frac{R}{\sqrt{R^2+r^2}}dR$$
Solve the equation by determining the integral.
12. In the study of a rigid rotor the following integration occurs:
$$Z_r = \int_0^\infty (2J+1)\,e^{\frac{-J(J+1)h^2}{8\pi^2 IkT}}\,dJ$$
Determine Z_r for constant temperature T assuming h, I and k are constants.
13. In electrostatics,
$$E = \int_0^\pi \left\{\frac{a^2\sigma\sin\theta}{2\varepsilon\sqrt{(a^2-x^2-2ax\cos\theta)}}d\theta\right\}$$
where a, σ and ε are constants, x is greater than a, and x is independent of θ. Show that $E=\frac{a^2\sigma}{\varepsilon x}$
14. The time taken, t hours, for a vehicle to reach a velocity of 130 km/h with an initial speed of 60 km/h is given by: $t=\int_{60}^{130}\frac{dv}{650-3v}$ where v is the velocity in km/h. Determine t, correct to the nearest second.

Practice Exercise 220 Multiple-choice questions on integration using algebraic substitutions (Answers on page 726)

Each question has only one correct answer

1. The value of $\int_0^{\pi/6} 2\sin\left(3t+\frac{\pi}{2}\right) dt$ is:

   (a) 6 (b) $-\frac{2}{3}$ (c) -6 (d) $\frac{2}{3}$

2. The indefinite integral $\int x\left(x^2-3\right)^3 dx$ is equal to:

   (a) $\frac{1}{4}\left(x^2-3\right)^4+c$ (b) $2\left(x^2-3\right)^4+c$
   (c) $\frac{3}{2}\left(x^2-3\right)^2+c$ (d) $\frac{1}{8}\left(x^2-3\right)^4+c$

3. Evaluating $\int_1^3 \frac{5}{4x-3} dx$, correct to 3 decimal places, gives:

   (a) 2.747 (b) 1.758
   (c) −0.025 (d) 11.513

4. The value of $\int_0^{\pi/3} 16\cos^4\theta\sin\theta\, d\theta$ is:

   (a) −0.1 (b) 3.1 (c) 0.1 (d) −3.1

5. The value of $\int_1^2 \frac{12x}{\sqrt{3x^2-1}} dx$, correct to 3 decimal places, is:

   (a) −1.249 (b) 0.852
   (c) 7.610 (d) 10.228

For fully worked solutions to each of the problems in Practice Exercises 218 and 219 in this chapter, go to the website:

www.routledge.com/cw/bird

COMPANION @ WEBSITE

Chapter 46

Integration using trigonometric substitutions

Why it is important to understand: Integration using trigonometric substitutions

Calculus is the most powerful branch of mathematics. It is capable of computing many quantities accurately which cannot be calculated using any other branch of mathematics. Many integrals are not 'standard' ones that we can determine from a list of results. Some need substitutions to rearrange them into a standard form. There are a number of trigonometric substitutions that may be used for certain integrals to change them into a form that can be integrated. These are explained in this chapter which provides another piece of the integral calculus jigsaw.

At the end of this chapter, you should be able to:

- integrate functions of the form $\sin^2 x$, $\cos^2 x$, $\tan^2 x$ and $\cot^2 x$
- integrate functions having powers of sines and cosines
- integrate functions that are products of sines and cosines
- integrate using the $\sin\theta$ substitution
- integrate using the $\tan\theta$ substitution

46.1 Introduction

Table 46.1 gives a summary of the integrals that require the use of **trigonometric substitutions**, and their application is demonstrated in Problems 1 to 20.

46.2 Worked problems on integration of $\sin^2 x$, $\cos^2 x$, $\tan^2 x$ and $\cot^2 x$

Problem 1. Evaluate: $\int_0^{\frac{\pi}{4}} 2\cos^2 4t\,dt$

Since $\cos 2t = 2\cos^2 t - 1$ (from Chapter 22),

then $\cos^2 t = \frac{1}{2}(1+\cos 2t)$ and

$$\cos^2 4t = \frac{1}{2}(1+\cos 8t)$$

Hence $\int_0^{\frac{\pi}{4}} 2\cos^2 4t\,dt$

$$= 2\int_0^{\frac{\pi}{4}} \frac{1}{2}(1+\cos 8t)\,dt = \left[t + \frac{\sin 8t}{8}\right]_0^{\frac{\pi}{4}}$$

$$= \left[\frac{\pi}{4} + \frac{\sin 8\left(\frac{\pi}{4}\right)}{8}\right] - \left[0 + \frac{\sin 0}{8}\right]$$

$$= \frac{\pi}{4} \text{ or } \mathbf{0.7854}$$

Table 46.1 Integrals using trigonometric substitutions

$f(x)$	$\int f(x)dx$	Method	See problem
1. $\cos^2 x$	$\frac{1}{2}\left(x+\frac{\sin 2x}{2}\right)+c$	Use $\cos 2x=2\cos^2 x-1$	1
2. $\sin^2 x$	$\frac{1}{2}\left(x-\frac{\sin 2x}{2}\right)+c$	Use $\cos 2x=1-2\sin^2 x$	2
3. $\tan^2 x$	$\tan x-x+c$	Use $1+\tan^2 x=\sec^2 x$	3
4. $\cot^2 x$	$-\cot x-x+c$	Use $\cot^2 x+1=\text{cosec}^2 x$	4
5. $\cos^m x\sin^n x$	(a) If either m or n is odd (but not both), use $\cos^2 x+\sin^2 x=1$		5, 6
	(b) If both m and n are even, use either $\cos 2x=2\cos^2 x-1$ or $\cos 2x=1-2\sin^2 x$		7, 8
6. $\sin A\cos B$		Use $\frac{1}{2}[\sin(A+B)+\sin(A-B)]$	9
7. $\cos A\sin B$		Use $\frac{1}{2}[\sin(A+B)-\sin(A-B)]$	10
8. $\cos A\cos B$		Use $\frac{1}{2}[\cos(A+B)+\cos(A-B)]$	11
9. $\sin A\sin B$		Use $-\frac{1}{2}[\cos(A+B)-\cos(A-B)]$	12
10. $\frac{1}{\sqrt{a^2-x^2}}$	$\sin^{-1}\frac{x}{a}+c$	Use $x=a\sin\theta$ substitution	13, 14
11. $\sqrt{a^2-x^2}$	$\frac{a^2}{2}\sin^{-1}\frac{x}{a}+\frac{x}{2}\sqrt{a^2-x^2}+c$	Use $x=a\sin\theta$ substitution	15, 16
12. $\frac{1}{a^2+x^2}$	$\frac{1}{a}\tan^{-1}\frac{x}{a}+c$	Use $x=a\tan\theta$ substitution	17–19

Problem 2. Determine: $\int \sin^2 3x\,dx$

Since $\cos 2x = 1-2\sin^2 x$ (from Chapter 22),

then $\sin^2 x = \frac{1}{2}(1-\cos 2x)$ and

$$\sin^2 3x = \frac{1}{2}(1-\cos 6x)$$

Hence
$$\int \sin^2 3x\,dx = \int \frac{1}{2}(1-\cos 6x)\,dx = \mathbf{\frac{1}{2}\left(x-\frac{\sin 6x}{6}\right)+c}$$

Problem 3. Find: $3\int \tan^2 4x\,dx$

Since $1+\tan^2 x=\sec^2 x$, then $\tan^2 x=\sec^2 x-1$ and $\tan^2 4x=\sec^2 4x-1$

Hence $3\int \tan^2 4x\,dx = 3\int (\sec^2 4x - 1)\,dx$

$$= \mathbf{3\left(\frac{\tan 4x}{4} - x\right) + c}$$

Problem 4. Evaluate: $\int_{\frac{\pi}{6}}^{\frac{\pi}{3}} \frac{1}{2}\cot^2 2\theta\, d\theta$

Since $\cot^2\theta + 1 = \text{cosec}^2\theta$, then

$\cot^2\theta = \text{cosec}^2\theta - 1$ and $\cot^2 2\theta = \text{cosec}^2 2\theta - 1$

Hence $\int_{\frac{\pi}{6}}^{\frac{\pi}{3}} \frac{1}{2}\cot^2 2\theta\, d\theta$

$$= \frac{1}{2}\int_{\frac{\pi}{6}}^{\frac{\pi}{3}} (\text{cosec}^2 2\theta - 1)\, d\theta = \frac{1}{2}\left[\frac{-\cot 2\theta}{2} - \theta\right]_{\frac{\pi}{6}}^{\frac{\pi}{3}}$$

$$= \frac{1}{2}\left[\left(\frac{-\cot 2\left(\frac{\pi}{3}\right)}{2} - \frac{\pi}{3}\right) - \left(\frac{-\cot 2\left(\frac{\pi}{6}\right)}{2} - \frac{\pi}{6}\right)\right]$$

$$= \frac{1}{2}[(--0.2887 - 1.0472) - (-0.2887 - 0.5236)]$$

$$= \mathbf{0.0269}$$

Problem 5. In a thin-walled closed tube of uniform thickness, t, the second moment of area about its neutral axis is given by:

$$I = 2 \times \int_0^{0.206} t\,ds_1 (s_1 \sin\alpha)^2$$
$$+ 0.1 \times t \times 0.05^2 \times 2 + \int_0^{\pi} (tR\,d\phi)(R\cos\phi)^2$$

Given that angle $\alpha = 14.04°$ and radius $R = 0.05$ m, determine I in terms of t

$$I = 2 \times \int_0^{0.206} t\,ds_1 (s_1 \sin\alpha)^2$$
$$+ 0.1 \times t \times 0.05^2 \times 2 + \int_0^{\pi} (tR\,d\phi)(R\cos\phi)^2$$
$$= 2t \times \int_0^{0.206} (s_1 \sin 14.04°)^2 ds_1 + 5 \times 10^{-4} t$$
$$+ t(0.05)^3 \int_0^{\pi} \cos^2\phi\, d\phi$$
$$= 0.1177t \times \int_0^{0.206} s_1^2 ds_1 + 5 \times 10^{-4} t$$
$$+ t(0.05)^3 \int_0^{\pi} \left(\frac{1 + \cos 2\phi}{2}\right) d\phi$$

since from 1 of Table 46.1, $\cos 2\phi = 2\cos^2\phi - 1$ from which, $\cos^2\phi = \frac{1 + \cos 2\phi}{2}$

$$= 0.1177t \times \left[\frac{s_1^3}{3}\right]_0^{0.206} + 5 \times 10^{-4} t$$
$$+ \frac{t(0.05)^3}{2}\left[\phi + \frac{\sin 2\phi}{2}\right]_0^{\pi}$$
$$= 0.1177t \times \left[\frac{(0.206)^3}{3} - \frac{(0)^3}{3}\right] + 5 \times 10^{-4} t$$
$$+ \frac{t(0.05)^3}{2}\left[\left(\pi + \frac{\sin 2\pi}{2}\right) - (0)\right]$$
$$= 3.430 \times 10^{-4} t + 5 \times 10^{-4} t + 1.963 \times 10^{-4} t$$

Hence, **the second moment of area,**

$$\mathbf{I = 1.039 \times 10^{-3} t}$$

Now try the following Practice Exercise

Practice Exercise 221 Integration of $\sin^2 x$, $\cos^2 x$, $\tan^2 x$ and $\cot^2 x$ (Answers on page 726)

In Problems 1 to 4, integrate with respect to the variable.

1. $\sin^2 2x$
2. $3\cos^2 t$
3. $5\tan^2 3\theta$
4. $2\cot^2 2t$

In Problems 5 to 8, evaluate the definite integrals, correct to 4 significant figures.

5. $\int_0^{\pi/3} 3\sin^2 3x\,dx$
6. $\int_0^{\pi/4} \cos^2 4x\,dx$
7. $\int_0^{0.5} 2\tan^2 2t\,dt$
8. $\int_{\pi/6}^{\pi/3} \cot^2\theta\,d\theta$

46.3 Worked problems on integration of powers of sines and cosines

Problem 6. Determine: $\int \sin^5 \theta \, d\theta$

Since $\cos^2\theta + \sin^2\theta = 1$ then $\sin^2\theta = (1 - \cos^2\theta)$.

Hence $\int \sin^5 \theta \, d\theta$

$$= \int \sin\theta(\sin^2\theta)^2 \, d\theta = \int \sin\theta(1-\cos^2\theta)^2 d\theta$$
$$= \int \sin\theta(1 - 2\cos^2\theta + \cos^4\theta)d\theta$$
$$= \int (\sin\theta - 2\sin\theta\cos^2\theta + \sin\theta\cos^4\theta)d\theta$$
$$= \mathbf{-\cos\theta + \frac{2\cos^3\theta}{3} - \frac{\cos^5\theta}{5} + c}$$

[Whenever a power of a cosine is multiplied by a sine of power 1, or vice-versa, the integral may be determined by inspection as shown.

In general, $\displaystyle\int \cos^n\theta \sin\theta \, d\theta = \frac{-\cos^{n+1}\theta}{(n+1)} + c$

and $\displaystyle\int \sin^n\theta \cos\theta \, d\theta = \frac{\sin^{n+1}\theta}{(n+1)} + c$

Alternatively, an algebraic substitution may be used as shown in Problem 6, Chapter 45, page 478]

Problem 7. Evaluate: $\int_0^{\frac{\pi}{2}} \sin^2 x \cos^3 x \, dx$

$$\int_0^{\frac{\pi}{2}} \sin^2 x\cos^3 x \, dx = \int_0^{\frac{\pi}{2}} \sin^2 x \cos^2 x \cos x \, dx$$
$$= \int_0^{\frac{\pi}{2}} (\sin^2 x)(1-\sin^2 x)(\cos x)\, dx$$
$$= \int_0^{\frac{\pi}{2}} (\sin^2 x\cos x - \sin^4 x\cos x)\, dx$$
$$= \left[\frac{\sin^3 x}{3} - \frac{\sin^5 x}{5}\right]_0^{\frac{\pi}{2}}$$
$$= \left[\frac{\left(\sin\frac{\pi}{2}\right)^3}{3} - \frac{\left(\sin\frac{\pi}{2}\right)^5}{5}\right] - [0-0]$$
$$= \frac{1}{3} - \frac{1}{5} = \mathbf{\frac{2}{15}} \text{ or } \mathbf{0.1333}$$

Problem 8. Evaluate: $\int_0^{\frac{\pi}{4}} 4\cos^4\theta \, d\theta$, correct to 4 significant figures

$$\int_0^{\frac{\pi}{4}} 4\cos^4\theta \, d\theta = 4\int_0^{\frac{\pi}{4}} (\cos^2\theta)^2 d\theta$$
$$= 4\int_0^{\frac{\pi}{4}} \left[\frac{1}{2}(1+\cos 2\theta)\right]^2 d\theta$$
$$= \int_0^{\frac{\pi}{4}} (1 + 2\cos 2\theta + \cos^2 2\theta)\, d\theta$$
$$= \int_0^{\frac{\pi}{4}} \left[1 + 2\cos 2\theta + \frac{1}{2}(1+\cos 4\theta)\right] d\theta$$
$$= \int_0^{\frac{\pi}{4}} \left[\frac{3}{2} + 2\cos 2\theta + \frac{1}{2}\cos 4\theta\right] d\theta$$
$$= \left[\frac{3\theta}{2} + \sin 2\theta + \frac{\sin 4\theta}{8}\right]_0^{\frac{\pi}{4}}$$
$$= \left[\frac{3}{2}\left(\frac{\pi}{4}\right) + \sin\frac{2\pi}{4} + \frac{\sin 4(\pi/4)}{8}\right] - [0]$$
$$= \frac{3\pi}{8} + 1$$
$$= \mathbf{2.178}, \text{correct to 4 significant figures.}$$

Problem 9. Find: $\int \sin^2 t \cos^4 t \, dt$

$$\int \sin^2 t\cos^4 t \, dt = \int \sin^2 t(\cos^2 t)^2 \, dt$$
$$= \int \left(\frac{1-\cos 2t}{2}\right)\left(\frac{1+\cos 2t}{2}\right)^2 dt$$
$$= \frac{1}{8}\int (1-\cos 2t)(1 + 2\cos 2t + \cos^2 2t)\, dt$$
$$= \frac{1}{8}\int (1 + 2\cos 2t + \cos^2 2t - \cos 2t - 2\cos^2 2t - \cos^3 2t)\, dt$$

$$= \frac{1}{8}\int (1 + \cos 2t - \cos^2 2t - \cos^3 2t)\,dt$$

$$= \frac{1}{8}\int \left[1 + \cos 2t - \left(\frac{1+\cos 4t}{2}\right) - \cos 2t(1 - \sin^2 2t)\right] dt$$

$$= \frac{1}{8}\int \left(\frac{1}{2} - \frac{\cos 4t}{2} + \cos 2t \sin^2 2t\right) dt$$

$$= \mathbf{\frac{1}{8}\left(\frac{t}{2} - \frac{\sin 4t}{8} + \frac{\sin^3 2t}{6}\right) + c}$$

Now try the following Practice Exercise

Practice Exercise 222 Integration of powers of sines and cosines (Answers on page 726)

Integrate the following with respect to the variable:

1. $\sin^3\theta$
2. $2\cos^3 2x$
3. $2\sin^3 t\cos^2 t$
4. $\sin^3 x\cos^4 x$
5. $2\sin^4 2\theta$
6. $\sin^2 t\cos^2 t$

46.4 Worked problems on integration of products of sines and cosines

Problem 10. Determine: $\int \sin 3t \cos 2t\,dt$

$$\int \sin 3t\cos 2t\,dt$$

$$= \int \frac{1}{2}[\sin(3t+2t) + \sin(3t-2t)]\,dt,$$

from 6 of Table 46.1, which follows from Section 22.4, page 229,

$$= \frac{1}{2}\int (\sin 5t + \sin t)\,dt$$

$$= \mathbf{\frac{1}{2}\left(\frac{-\cos 5t}{5} - \cos t\right) + c}$$

Problem 11. Find: $\int \frac{1}{3}\cos 5x \sin 2x\,dx$

$$\int \frac{1}{3}\cos 5x\sin 2x\,dx$$

$$= \frac{1}{3}\int \frac{1}{2}[\sin(5x+2x) - \sin(5x-2x)]\,dx,$$

from 7 of Table 46.1

$$= \frac{1}{6}\int (\sin 7x - \sin 3x)\,dx$$

$$= \mathbf{\frac{1}{6}\left(\frac{-\cos 7x}{7} + \frac{\cos 3x}{3}\right) + c}$$

Problem 12. Evaluate: $\int_0^1 2\cos 6\theta\cos\theta\,d\theta$, correct to 4 decimal places

$$\int_0^1 2\cos 6\theta\cos\theta\,d\theta$$

$$= 2\int_0^1 \frac{1}{2}[\cos(6\theta+\theta) + \cos(6\theta-\theta)]\,d\theta,$$

from 8 of Table 46.1

$$= \int_0^1 (\cos 7\theta + \cos 5\theta)\,d\theta = \left[\frac{\sin 7\theta}{7} + \frac{\sin 5\theta}{5}\right]_0^1$$

$$= \left(\frac{\sin 7}{7} + \frac{\sin 5}{5}\right) - \left(\frac{\sin 0}{7} + \frac{\sin 0}{5}\right)$$

'sin 7' means 'the sine of 7 radians' ($\equiv 401.07°$) and $\sin 5 \equiv 286.48°$.

Hence $\int_0^1 2\cos 6\theta\cos\theta\,d\theta$

$$= (0.09386 + -0.19178) - (0)$$

$$= \mathbf{-0.0979}, \text{correct to 4 decimal places}$$

Problem 13. Find: $3\int \sin 5x \sin 3x\,dx$

$$3\int \sin 5x\sin 3x\,dx$$

$$= 3\int -\frac{1}{2}[\cos(5x+3x) - \cos(5x-3x)]\,dx,$$

from 9 of Table 46.1

$$= -\frac{3}{2}\int (\cos 8x - \cos 2x)\,dx$$

$$= \mathbf{-\frac{3}{2}\left(\frac{\sin 8x}{8} - \frac{\sin 2x}{2}\right) + c} \quad \text{or}$$

$$\mathbf{\frac{3}{16}(4\sin 2x - \sin 8x) + c}$$

Now try the following Practice Exercise

Practice Exercise 223 Integration of products of sines and cosines (Answers on page 726)

In Problems 1 to 4, integrate with respect to the variable.

1. $\sin 5t \cos 2t$
2. $2\sin 3x \sin x$
3. $3\cos 6x \cos x$
4. $\frac{1}{2}\cos 4\theta \sin 2\theta$

In Problems 6 to 9, evaluate the definite integrals.

5. $\int_0^{\pi/2} \cos 4x \cos 3x\, dx$
6. $\int_0^1 2\sin 7t \cos 3t\, dt$
7. $-4\int_0^{\pi/3} \sin 5\theta \sin 2\theta\, d\theta$
8. $\int_1^2 3\cos 8t \sin 3t\, dt$

46.5 Worked problems on integration using the sin θ substitution

Problem 14. Determine: $\int \frac{1}{\sqrt{a^2 - x^2}}\, dx$

Let $x = a\sin\theta$, then $\frac{dx}{d\theta} = a\cos\theta$ and $dx = a\cos\theta\, d\theta$.

Hence $\int \frac{1}{\sqrt{a^2-x^2}}\,dx$

$$= \int \frac{1}{\sqrt{a^2 - a^2\sin^2\theta}}\, a\cos\theta\, d\theta$$

$$= \int \frac{a\cos\theta\, d\theta}{\sqrt{a^2(1-\sin^2\theta)}}$$

$$= \int \frac{a\cos\theta\, d\theta}{\sqrt{a^2\cos^2\theta}}, \text{ since } \sin^2\theta + \cos^2\theta = 1$$

$$= \int \frac{a\cos\theta\, d\theta}{a\cos\theta} = \int d\theta = \theta + c$$

Since $x = a\sin\theta$, then $\sin\theta = \frac{x}{a}$ and $\theta = \sin^{-1}\frac{x}{a}$

Hence $\int \frac{1}{\sqrt{a^2-x^2}}\,dx = \mathbf{\sin^{-1}\frac{x}{a} + c}$

Problem 15. Evaluate: $\int_0^3 \frac{1}{\sqrt{9-x^2}}\,dx$

From Problem 14, $\int_0^3 \frac{1}{\sqrt{9-x^2}}\,dx$

$$= \left[\sin^{-1}\frac{x}{3}\right]_0^3 \quad \text{since } a = 3$$

$$= (\sin^{-1}1 - \sin^{-1}0) = \frac{\pi}{2} \text{ or } \mathbf{1.5708}$$

Problem 16. Find: $\int \sqrt{a^2 - x^2}\,dx$

Let $x = a\sin\theta$ then $\frac{dx}{d\theta} = a\cos\theta$ and $dx = a\cos\theta\, d\theta$

Hence $\int \sqrt{a^2 - x^2}\,dx$

$$= \int \sqrt{a^2 - a^2\sin^2\theta}\,(a\cos\theta\, d\theta)$$

$$= \int \sqrt{a^2(1-\sin^2\theta)}\,(a\cos\theta\, d\theta)$$

$$= \int \sqrt{a^2\cos^2\theta}\,(a\cos\theta\, d\theta)$$

$$= \int (a\cos\theta)(a\cos\theta\, d\theta)$$

$$= a^2\int \cos^2\theta\, d\theta = a^2\int\left(\frac{1+\cos 2\theta}{2}\right)d\theta$$

(since $\cos 2\theta = 2\cos^2\theta - 1$)

$$= \frac{a^2}{2}\left(\theta + \frac{\sin 2\theta}{2}\right) + c$$

$$= \frac{a^2}{2}\left(\theta + \frac{2\sin\theta\cos\theta}{2}\right) + c$$

since from Chapter 22, $\sin 2\theta = 2\sin\theta\cos\theta$

$$= \frac{a^2}{2}[\theta + \sin\theta\cos\theta] + c$$

Since $x = a\sin\theta$, then $\sin\theta = \frac{x}{a}$ and $\theta = \sin^{-1}\frac{x}{a}$

Also, $\cos^2\theta+\sin^2\theta=1$, from which,

$$\cos\theta=\sqrt{1-\sin^2\theta}=\sqrt{1-\left(\frac{x}{a}\right)^2}$$

$$=\sqrt{\frac{a^2-x^2}{a^2}}=\frac{\sqrt{a^2-x^2}}{a}$$

Thus $\int\sqrt{a^2-x^2}\,dx=\frac{a^2}{2}[\theta+\sin\theta\cos\theta]+c$

$$=\frac{a^2}{2}\left[\sin^{-1}\frac{x}{a}+\left(\frac{x}{a}\right)\frac{\sqrt{a^2-x^2}}{a}\right]+c$$

$$=\mathbf{\frac{a^2}{2}\sin^{-1}\frac{x}{a}+\frac{x}{2}\sqrt{a^2-x^2}+c}$$

Problem 17. Evaluate: $\int_0^4\sqrt{16-x^2}\,dx$

From Problem 16, $\int_0^4\sqrt{16-x^2}\,dx$

$$=\left[\frac{16}{2}\sin^{-1}\frac{x}{4}+\frac{x}{2}\sqrt{16-x^2}\right]_0^4$$

$$=\left[8\sin^{-1}1+2\sqrt{0}\right]-\left[8\sin^{-1}0+0\right]$$

$$=8\sin^{-1}1=8\left(\frac{\pi}{2}\right)$$

$$=\mathbf{4\pi}\text{ or }\mathbf{12.57}$$

Now try the following Practice Exercise

Practice Exercise 224 Integration using the sine θ substitution (Answers on page 726)

1. Determine: $\int\frac{5}{\sqrt{4-t^2}}\,dt$
2. Determine: $\int\frac{3}{\sqrt{9-x^2}}\,dx$
3. Determine: $\int\sqrt{4-x^2}\,dx$
4. Determine: $\int\sqrt{16-9t^2}\,dt$
5. Evaluate: $\int_0^4\frac{1}{\sqrt{16-x^2}}\,dx$
6. Evaluate: $\int_0^1\sqrt{9-4x^2}\,dx$

46.6 Worked problems on integration using the tan θ substitution

Problem 18. Determine: $\int\frac{1}{(a^2+x^2)}\,dx$

Let $x=a\tan\theta$ then $\frac{dx}{d\theta}=a\sec^2\theta$ and $dx=a\sec^2\theta\,d\theta$

Hence $\int\frac{1}{(a^2+x^2)}\,dx$

$$=\int\frac{1}{(a^2+a^2\tan^2\theta)}(a\sec^2\theta\,d\theta)$$

$$=\int\frac{a\sec^2\theta\,d\theta}{a^2(1+\tan^2\theta)}$$

$$=\int\frac{a\sec^2\theta\,d\theta}{a^2\sec^2\theta}\quad\text{since }1+\tan^2\theta=\sec^2\theta$$

$$=\int\frac{1}{a}\,d\theta=\frac{1}{a}(\theta)+c$$

Since $x=a\tan\theta$, $\theta=\tan^{-1}\frac{x}{a}$

Hence $\mathbf{\int\frac{1}{(a^2+x^2)}\,dx=\frac{1}{a}\tan^{-1}\frac{x}{a}+c}$

Problem 19. Evaluate: $\int_0^2\frac{1}{(4+x^2)}\,dx$

From Problem 18, $\int_0^2\frac{1}{(4+x^2)}\,dx$

$$=\frac{1}{2}\left[\tan^{-1}\frac{x}{2}\right]_0^2\quad\text{since }a=2$$

$$=\frac{1}{2}(\tan^{-1}1-\tan^{-1}0)=\frac{1}{2}\left(\frac{\pi}{4}-0\right)$$

$$=\frac{\pi}{8}\text{ or }\mathbf{0.3927}$$

Problem 20. Evaluate: $\int_0^1\frac{5}{(3+2x^2)}\,dx$, correct to 4 decimal places

$$\int_0^1 \frac{5}{(3+2x^2)}dx = \int_0^1 \frac{5}{2[(3/2)+x^2]}dx$$

$$= \frac{5}{2}\int_0^1 \frac{1}{[\sqrt{3/2}]^2+x^2}dx$$

$$= \frac{5}{2}\left[\frac{1}{\sqrt{3/2}}\tan^{-1}\frac{x}{\sqrt{3/2}}\right]_0^1$$

$$= \frac{5}{2}\sqrt{\frac{2}{3}}\left[\tan^{-1}\sqrt{\frac{2}{3}} - \tan^{-1}0\right]$$

$$= (2.0412)[0.6847 - 0]$$

$$= \mathbf{1.3976}, \text{correct to 4 decimal places.}$$

Now try the following Practice Exercise

Practice Exercise 225 Integration using the $\tan\theta$ substitution (Answers on page 726)

1. Determine: $\int \frac{3}{4+t^2}dt$
2. Determine: $\int \frac{5}{16+9\theta^2}d\theta$
3. Evaluate: $\int_0^1 \frac{3}{1+t^2}dt$
4. Evaluate: $\int_0^3 \frac{5}{4+x^2}dx$

Practice Exercise 226 Multiple-choice questions on integration using trigonometric substitutions (Answers on page 726)

Each question has only one correct answer

1. The value of $\int_0^{\pi/3} 4\cos^2 2\theta\, d\theta$, correct to 3 decimal places, is:
   (a) 1.661 (b) 3.464
   (c) −1.167 (d) −2.417
2. $\int_0^{\pi/2} 2\sin^3 t\, dt$ is equal to:
   (a) 1.33 (b) −0.25 (c) −1.33 (d) 0.25
3. The integral $\int \sqrt{(4-x^2)}\, dx$ is equal to:
   (a) $2\cos^{-1}\frac{x}{2} + \frac{2\sqrt{(4-x^2)}}{x} + c$
   (b) $2\sin^{-1}\frac{x}{2} + \frac{x}{2}\sqrt{(4-x^2)} + c$
   (c) $2\sin^{-1}\frac{x}{2} + \frac{x}{2} + c$
   (d) $4\sin^{-1}\frac{x}{2} + x\sqrt{(4-x^2)} + c$
4. Evaluating $\int_0^{\frac{\pi}{2}} 4\cos 6x \sin 3x\, dx$, correct to 3 decimal places, gives:
   (a) −0.888 (b) 4.000
   (c) −0.444 (d) 0.888
5. The value of $\int_0^3 \frac{5}{9+x^2}dx$, correct to 3 decimal places, is:
   (a) 2.596 (b) 75.000 (c) 0.536 (d) 1.309

For fully worked solutions to each of the problems in Practice Exercises 221 to 225 in this chapter, go to the website:
www.routledge.com/cw/bird

COMPANION @ WEBSITE

Revision Test 12 Integration

This Revision Test covers the material contained in Chapters 44 to 46. *The marks for each question are shown in brackets at the end of each question.*

1. Determine:

   (a) $\int 3\sqrt{t^5}\,dt$

   (b) $\int \frac{2}{\sqrt[3]{x^2}}\,dx$

   (c) $\int (2+\theta)^2 d\theta$ (9)

2. Evaluate the following integrals, each correct to 4 significant figures:

   (a) $\int_0^{\pi/3} 3\sin 2t\,dt$

   (b) $\int_1^2 \left(\frac{2}{x^2}+\frac{1}{x}+\frac{3}{4}\right)dx$ (10)

3. Determine the following integrals:

   (a) $\int 5(6t+5)^7 dt$

   (b) $\int \frac{3\ln x}{x}\,dx$

   (c) $\int \frac{2}{\sqrt{(2\theta-1)}}\,d\theta$ (9)

4. Evaluate the following definite integrals:

   (a) $\int_0^{\pi/2} 2\sin\left(2t+\frac{\pi}{3}\right)dt$

   (b) $\int_0^1 3xe^{4x^2-3}\,dx$ (10)

5. Determine the following integrals:

   (a) $\int \cos^3 x\sin^2 x\,dx$

   (b) $\int \frac{2}{\sqrt{9-4x^2}}\,dx$ (8)

6. Evaluate the following definite integrals, correct to 4 significant figures:

   (a) $\int_0^{\pi/2} 3\sin^2 t\,dt$

   (b) $\int_0^{\pi/3} 3\cos 5\theta\sin 3\theta\,d\theta$

   (c) $\int_0^2 \frac{5}{4+x^2}\,dx$ (14)

COMPANION @ WEBSITE

For lecturers/instructors/teachers, fully worked solutions to each of the problems in Revision Test 12, together with a full marking scheme, are available at the website:

www.routledge.com/cw/bird

Chapter 47

Integration using partial fractions

Why it is important to understand: **Integration using partial fractions**

Sometimes expressions which at first sight look impossible to integrate using standard techniques may in fact be integrated by first expressing them as simpler partial fractions and then using earlier learned techniques. As explained in Chapter 7, the algebraic technique of resolving a complicated fraction into partial fractions is often needed by electrical and mechanical engineers for not only determining certain integrals in calculus, but for determining inverse Laplace transforms and for analysing linear differential equations like resonant circuits and feedback control systems.

At the end of this chapter, you should be able to:

- integrate functions using partial fractions with linear factors
- integrate functions using partial fractions with repeated linear factors
- integrate functions using partial fractions with quadratic factors

47.1 Introduction

The process of expressing a fraction in terms of simpler fractions—called **partial fractions**—is discussed in Chapter 7, with the forms of partial fractions used being summarised in Table 7.1, page 74.

Certain functions have to be resolved into partial fractions before they an be integrated, as demonstrated in the following worked problems.

47.2 Integration using partial fractions with linear factors

Problem 1. Determine: $\displaystyle\int \frac{11-3x}{x^2+2x-3}\,dx$

As shown in Problem 1, page 74:

$$\frac{11-3x}{x^2+2x-3} \equiv \frac{2}{(x-1)} - \frac{5}{(x+3)}$$

Hence $\displaystyle\int \frac{11-3x}{x^2+2x-3}dx$

$$= \int \left\{ \frac{2}{(x-1)} - \frac{5}{(x+3)} \right\} dx$$

$$= \mathbf{2\ln(x-1) - 5\ln(x+3) + c}$$

(by algebraic substitutions—see Chapter 45)

or $\ln\left\{\frac{(x-1)^2}{(x+3)^5}\right\}+c$ by the laws of logarithms

Problem 2. Find: $\displaystyle\int \frac{2x^2-9x-35}{(x+1)(x-2)(x+3)}\,dx$

It was shown in Problem 2, page 74:

$$\frac{2x^2-9x-35}{(x+1)(x-2)(x+3)} \equiv \frac{4}{(x+1)} - \frac{3}{(x-2)} + \frac{1}{(x+3)}$$

Hence $\displaystyle\int \frac{2x^2-9x-35}{(x+1)(x-2)(x+3)}\,dx$

$$\equiv \int\left\{\frac{4}{(x+1)} - \frac{3}{(x-2)} + \frac{1}{(x+3)}\right\}dx$$

$$= \mathbf{4\ln(x+1) - 3\ln(x-2) + \ln(x+3) + c}$$

or $\ln\left\{\frac{(x+1)^4(x+3)}{(x-2)^3}\right\}+c$

Problem 3. Determine: $\displaystyle\int \frac{x^2+1}{x^2-3x+2}\,dx$

By dividing out (since the numerator and denominator are of the same degree) and resolving into partial fractions it was shown in Problem 3, page 75:

$$\frac{x^2+1}{x^2-3x+2} \equiv 1 - \frac{2}{(x-1)} + \frac{5}{(x-2)}$$

Hence $\displaystyle\int \frac{x^2+1}{x^2-3x+2}\,dx$

$$\equiv \int\left\{1 - \frac{2}{(x-1)} + \frac{5}{(x-2)}\right\}dx$$

$$= \mathbf{x - 2\ln(x-1) + 5\ln(x-2) + c}$$

or $x + \ln\left\{\frac{(x-2)^5}{(x-1)^2}\right\}+c$

Problem 4. Evaluate:

$\displaystyle\int_2^3 \frac{x^3-2x^2-4x-4}{x^2+x-2}\,dx$, correct to 4 significant figures

By dividing out and resolving into partial fractions, it was shown in Problem 4, page 75:

$$\frac{x^3-2x^2-4x-4}{x^2+x-2} \equiv x-3+\frac{4}{(x+2)} - \frac{3}{(x-1)}$$

Hence $\displaystyle\int_2^3 \frac{x^3-2x^2-4x-4}{x^2+x-2}\,dx$

$$\equiv \int_2^3\left\{x-3+\frac{4}{(x+2)} - \frac{3}{(x-1)}\right\}dx$$

$$= \left[\frac{x^2}{2} - 3x + 4\ln(x+2) - 3\ln(x-1)\right]_2^3$$

$$= \left(\frac{9}{2} - 9 + 4\ln 5 - 3\ln 2\right)$$

$$-(2-6+4\ln 4 - 3\ln 1)$$

$= \mathbf{-1.687}$, correct to 4 significant figures

Now try the following Practice Exercise

Practice Exercise 227 Integration using partial fractions with linear factors (Answers on page 726)

In Problems 1 to 5, integrate with respect to x

1. $\displaystyle\int \frac{12}{(x^2-9)}\,dx$

2. $\displaystyle\int \frac{4(x-4)}{(x^2-2x-3)}\,dx$

3. $\displaystyle\int \frac{3(2x^2-8x-1)}{(x+4)(x+1)(2x-1)}\,dx$

4. $\displaystyle\int \frac{x^2+9x+8}{x^2+x-6}\,dx$

5. $\displaystyle\int \frac{3x^3-2x^2-16x+20}{(x-2)(x+2)}\,dx$

In Problems 6 and 7, evaluate the definite integrals correct to 4 significant figures.

6. $\displaystyle\int_3^4 \frac{x^2-3x+6}{x(x-2)(x-1)}\,dx$

7. $\displaystyle\int_4^6 \frac{x^2-x-14}{x^2-2x-3}\,dx$

8. The velocity, v, of an object in a medium at time t seconds is given by: $t=\int_{20}^{80}\frac{dv}{v(2v-1)}$ Evaluate t, in milliseconds, correct to 2 decimal places.

47.3 Integration using partial fractions with repeated linear factors

Problem 5. Determine: $\int\frac{2x+3}{(x-2)^2}dx$

It was shown in Problem 5, page 76:

$$\frac{2x+3}{(x-2)^2}\equiv\frac{2}{(x-2)}+\frac{7}{(x-2)^2}$$

Thus $\int\frac{2x+3}{(x-2)^2}dx$

$$\equiv\int\left\{\frac{2}{(x-2)}+\frac{7}{(x-2)^2}\right\}dx$$

$$=\mathbf{2\ln(x-2)-\frac{7}{(x-2)}+c}$$

$\left[\int\frac{7}{(x-2)^2}dx\right.$ is determined using the algebraic substitution $u=(x-2)$, see Chapter 45$\left.\right]$

Problem 6. Find: $\int\frac{5x^2-2x-19}{(x+3)(x-1)^2}dx$

It was shown in Problem 6, page 76:

$$\frac{5x^2-2x-19}{(x+3)(x-1)^2}\equiv\frac{2}{(x+3)}+\frac{3}{(x-1)}-\frac{4}{(x-1)^2}$$

Hence $\int\frac{5x^2-2x-19}{(x+3)(x-1)^2}dx$

$$\equiv\int\left\{\frac{2}{(x+3)}+\frac{3}{(x-1)}-\frac{4}{(x-1)^2}\right\}dx$$

$$=\mathbf{2\ln(x+3)+3\ln(x-1)+\frac{4}{(x-1)}+c}$$

$$\text{or}\quad\mathbf{\ln(x+3)^2(x-1)^3+\frac{4}{(x-1)}+c}$$

Problem 7. Evaluate:

$\int_{-2}^{1}\frac{3x^2+16x+15}{(x+3)^3}dx$, correct to 4 significant figures

It was shown in Problem 7, page 77:

$$\frac{3x^2+16x+15}{(x+3)^3}\equiv\frac{3}{(x+3)}-\frac{2}{(x+3)^2}-\frac{6}{(x+3)^3}$$

Hence $\int\frac{3x^2+16x+15}{(x+3)^3}dx$

$$\equiv\int_{-2}^{1}\left\{\frac{3}{(x+3)}-\frac{2}{(x+3)^2}-\frac{6}{(x+3)^3}\right\}dx$$

$$=\left[3\ln(x+3)+\frac{2}{(x+3)}+\frac{3}{(x+3)^2}\right]_{-2}^{1}$$

$$=\left(3\ln4+\frac{2}{4}+\frac{3}{16}\right)-\left(3\ln1+\frac{2}{1}+\frac{3}{1}\right)$$

$=\mathbf{-0.1536}$, correct to 4 significant figures.

Now try the following Practice Exercise

Practice Exercise 228 Integration using partial fractions with repeated linear factors (Answers on page 727)

In Problems 1 and 2, integrate with respect to x.

1. $\int\frac{4x-3}{(x+1)^2}dx$
2. $\int\frac{5x^2-30x+44}{(x-2)^3}dx$

In Problems 3 and 4, evaluate the definite integrals correct to 4 significant figures.

3. $\int_{1}^{2}\frac{x^2+7x+3}{x^2(x+3)}dx$
4. $\int_{6}^{7}\frac{18+21x-x^2}{(x-5)(x+2)^2}dx$

47.4 Integration using partial fractions with quadratic factors

Problem 8. Find: $\int \frac{3+6x+4x^2-2x^3}{x^2(x^2+3)}dx$

It was shown in Problem 9, page 78:

$$\frac{3+6x+4x^2-2x^2}{x^2(x^2+3)} \equiv \frac{2}{x}+\frac{1}{x^2}+\frac{3-4x}{(x^2+3)}$$

Thus $\int \frac{3+6x+4x^2-2x^3}{x^2(x^2+3)}dx$

$$\equiv \int\left(\frac{2}{x}+\frac{1}{x^2}+\frac{3-4x}{(x^2+3)}\right)dx$$

$$= \int\left\{\frac{2}{x}+\frac{1}{x^2}+\frac{3}{(x^2+3)}-\frac{4x}{(x^2+3)}\right\}dx$$

$$\int \frac{3}{(x^2+3)}dx = 3\int \frac{1}{x^2+(\sqrt{3})^2}dx$$

$$= \frac{3}{\sqrt{3}}\tan^{-1}\frac{x}{\sqrt{3}}$$

from 12, Table 46.1, page 483.

$\int \frac{4x}{x^2+3}dx$ is determined using the algebraic substitutions $u=(x^2+3)$.

Hence $\int\left\{\frac{2}{x}+\frac{1}{x^2}+\frac{3}{(x^2+3)}-\frac{4x}{(x^2+3)}\right\}dx$

$$= 2\ln x - \frac{1}{x}+\frac{3}{\sqrt{3}}\tan^{-1}\frac{x}{\sqrt{3}} - 2\ln(x^2+3)+c$$

$$= \mathbf{\ln\left(\frac{x}{x^2+3}\right)^2 - \frac{1}{x} + \sqrt{3}\tan^{-1}\frac{x}{\sqrt{3}} + c}$$

Problem 9. Determine: $\int \frac{1}{(x^2-a^2)}dx$

Let $\frac{1}{(x^2-a^2)} \equiv \frac{A}{(x-a)}+\frac{B}{(x+a)}$

$$\equiv \frac{A(x+a)+B(x-a)}{(x+a)(x-a)}$$

Equating the numerators gives:

$$1 \equiv A(x+a)+B(x-a)$$

Let $x=a$, then $A=\frac{1}{2a}$

and let $x=-a$,

then $B=-\frac{1}{2a}$

Hence $\int \frac{1}{(x^2-a^2)}dx \equiv \int \frac{1}{2a}\left[\frac{1}{(x-a)}-\frac{1}{(x+a)}\right]dx$

$$= \frac{1}{2a}[\ln(x-a)-\ln(x+a)]+c$$

$$= \mathbf{\frac{1}{2a}\ln\left(\frac{x-a}{x+a}\right)+c}$$

Problem 10. Evaluate: $\int_3^4 \frac{3}{(x^2-4)}dx$, correct to 3 significant figures

From Problem 9,

$$\int_3^4 \frac{3}{(x^2-4)}dx = 3\left[\frac{1}{2(2)}\ln\left(\frac{x-2}{x+2}\right)\right]_3^4$$

$$= \frac{3}{4}\left[\ln\frac{2}{6}-\ln\frac{1}{5}\right]$$

$$= \frac{3}{4}\ln\frac{5}{3} = \mathbf{0.383}, \text{ correct to 3 significant figures.}$$

Problem 11. Determine: $\int \frac{1}{(a^2-x^2)}dx$

Using partial fractions, let

$$\frac{1}{(a^2-x^2)} \equiv \frac{1}{(a-x)(a+x)} \equiv \frac{A}{(a-x)}+\frac{B}{(a+x)}$$

$$\equiv \frac{A(a+x)+B(a-x)}{(a-x)(a+x)}$$

Then $1 \equiv A(a+x)+B(a-x)$

Let $x=a$ then $A=\frac{1}{2a}$. Let $x=-a$ then $B=\frac{1}{2a}$

Hence $\int \frac{1}{(a^2-x^2)}dx$

$$= \int \frac{1}{2a}\left[\frac{1}{(a-x)}+\frac{1}{(a+x)}\right]dx$$

$$= \frac{1}{2a}[-\ln(a-x)+\ln(a+x)]+c$$

$$= \frac{\mathbf{1}}{\mathbf{2a}}\,\mathbf{ln}\left(\frac{\boldsymbol{a+x}}{\boldsymbol{a-x}}\right)+\boldsymbol{c}$$

Problem 12. Evaluate: $\int_0^2 \frac{5}{(9-x^2)}\,dx$, correct to 4 decimal places

From Problem 11,

$$\int_0^2 \frac{5}{(9-x^2)}\,dx = 5\left[\frac{1}{2(3)}\ln\left(\frac{3+x}{3-x}\right)\right]_0^2$$

$$= \frac{5}{6}\left[\ln\frac{5}{1}-\ln 1\right] = \mathbf{1.3412}$$

correct to 4 decimal places

Now try the following Practice Exercise

Practice Exercise 229 Integration using partial fractions with quadratic factors (Answers on page 727)

1. Determine: $\int \frac{x^2-x-13}{(x^2+7)(x-2)}\,dx$

In Problems 2 to 4, evaluate the definite integrals, correct to 4 significant figures.

2. $\int_5^6 \frac{6x-5}{(x-4)(x^2+3)}\,dx$
3. $\int_1^2 \frac{4}{(16-x^2)}\,dx$
4. $\int_4^5 \frac{2}{(x^2-9)}\,dx$

Practice Exercise 230 Multiple-choice questions on integration using partial fractions (Answers on page 727)

Each question has only one correct answer

1. $\int_3^4 \frac{3x+4}{(x-2)(x+3)}\,dx$ is equal to:
   (a) 1.540 (b) 1.001
   (c) 1.232 (d) 0.847
2. $\int_5^6 \frac{8}{x^2-16}\,dx$ is equal to:
   (a) −0.588 (b) 6.388
   (c) 0.588 (d) −0.748
3. $\int_2^3 \frac{3}{x^2+x-2}\,dx$ is equal to:
   (a) 3 ln 2.5 (b) $\frac{1}{3}$lg 1.6
   (c) ln 40 (d) ln 1.6
4. $\int_4^5 \frac{2(x-2)}{(x-3)^2}\,dx$ is equal to:
   (a) 0.386 (b) 2.386
   (c) −2.500 (d) 1.000
5. $\int_4^5 \frac{1}{x^2-9}\,dx$ is equal to:
   (a) −0.0933 (b) 0.827
   (c) 0.0933 (d) −0.827

For fully worked solutions to each of the problems in Practice Exercises 227 to 229 in this chapter, go to the website:
www.routledge.com/cw/bird

COMPANION WEBSITE

Chapter 48

The $t = \tan\frac{\theta}{2}$ substitution

Why it is important to understand: **The $t = \tan\theta/2$ substitution**

Sometimes, with an integral containing sin θ and/or cos θ, it is possible, after making a substitution $t = \tan\theta/2$, to obtain an integral which can be determined using partial fractions. This is explained in this chapter where we continue to build the picture of integral calculus, each step building from the previous. A simple substitution can make things so much easier.

At the end of this chapter, you should be able to:

- develop formulae for sin θ, cos θ and $d\theta$ in terms of t, where $t = \tan\theta/2$
- integrate functions using $t = \tan\theta/2$ substitution

48.1 Introduction

Integrals of the form $\int \frac{1}{a\cos\theta + b\sin\theta + c}d\theta$, where a, b and c are constants, may be determined by using the substitution $t = \tan\frac{\theta}{2}$. The reason is explained below.

If angle A in the right-angled triangle ABC shown in Fig. 48.1 is made equal to $\frac{\theta}{2}$ then, since

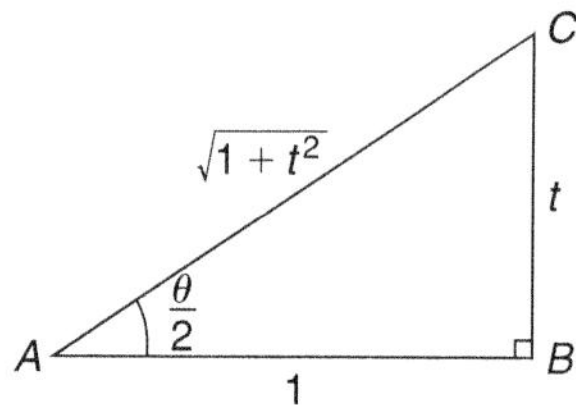

Figure 48.1

tangent $= \frac{\text{opposite}}{\text{adjacent}}$, if $BC = t$ and $AB = 1$, then $\tan\frac{\theta}{2} = t$

By Pythagoras' theorem, $AC = \sqrt{1+t^2}$

Therefore $\sin\frac{\theta}{2} = \frac{t}{\sqrt{1+t^2}}$ and $\cos\frac{\theta}{2} = \frac{1}{\sqrt{1+t^2}}$

Since sin $2x = 2\sin x\cos x$ (from double angle formulae, Chapter 22), then

$$\sin\theta = 2\sin\frac{\theta}{2}\cos\frac{\theta}{2}$$

$$= 2\left(\frac{t}{\sqrt{1+t^2}}\right)\left(\frac{1}{\sqrt{1+t^2}}\right)$$

i.e. $$\sin\theta = \frac{2t}{(1+t^2)} \qquad (1)$$

Since $\cos 2x = \cos^2 x - \sin^2 x$

then $\cos\theta = \cos^2\frac{\theta}{2} - \sin^2\frac{\theta}{2}$

$$= \left(\frac{1}{\sqrt{1+t^2}}\right)^2 - \left(\frac{t}{\sqrt{1+t^2}}\right)^2$$

i.e. $$\cos\theta = \frac{1-t^2}{1+t^2} \qquad (2)$$

Also, since $t = \tan\frac{\theta}{2}$

$\frac{dt}{d\theta} = \frac{1}{2}\sec^2\frac{\theta}{2} = \frac{1}{2}\left(1+\tan^2\frac{\theta}{2}\right)$ from trigonometric identities,

i.e. $$\frac{dt}{d\theta} = \frac{1}{2}(1+t^2)$$

from which, $$d\theta = \frac{2\,dt}{1+t^2} \qquad (3)$$

Equations (1), (2) and (3) are used to determine integrals of the form $\int \frac{1}{a\cos\theta + b\sin\theta + c}d\theta$ where a, b or c may be zero.

48.2 Worked problems on the $t = \tan\frac{\theta}{2}$ substitution

Problem 1. Determine: $\int \frac{d\theta}{\sin\theta}$

If $t = \tan\frac{\theta}{2}$ then $\sin\theta = \frac{2t}{1+t^2}$ and $d\theta = \frac{2\,dt}{1+t^2}$ from equations (1) and (3).

Thus
$$\int \frac{d\theta}{\sin\theta} = \int \frac{1}{\sin\theta}d\theta$$
$$= \int \frac{1}{\frac{2t}{1+t^2}}\left(\frac{2\,dt}{1+t^2}\right)$$
$$= \int \frac{1}{t}dt = \ln t + c$$

Hence $$\int \frac{d\theta}{\sin\theta} = \mathbf{\ln\left(\tan\frac{\theta}{2}\right) + c}$$

Problem 2. Determine: $\int \frac{dx}{\cos x}$

If $t = \tan\frac{x}{2}$ then $\cos x = \frac{1-t^2}{1+t^2}$ and $dx = \frac{2\,dt}{1+t^2}$ from equations (2) and (3).

Thus
$$\int \frac{dx}{\cos x} = \int \frac{1}{\frac{1-t^2}{1+t^2}}\left(\frac{2\,dt}{1+t^2}\right)$$
$$= \int \frac{2}{1-t^2}dt$$

$\frac{2}{1-t^2}$ may be resolved into partial fractions (see Chapter 7).

Let
$$\frac{2}{1-t^2} = \frac{2}{(1-t)(1+t)}$$
$$= \frac{A}{(1-t)} + \frac{B}{(1+t)}$$
$$= \frac{A(1+t)+B(1-t)}{(1-t)(1+t)}$$

Hence $2 = A(1+t) + B(1-t)$

When $t = 1$, $2 = 2A$, from which, $A = 1$

When $t = -1$, $2 = 2B$, from which, $B = 1$

Hence
$$\int \frac{2dt}{1-t^2} = \int \frac{1}{(1-t)} + \frac{1}{(1+t)}dt$$
$$= -\ln(1-t) + \ln(1+t) + c$$
$$= \ln\left\{\frac{(1+t)}{(1-t)}\right\} + c$$

Thus
$$\int \frac{dx}{\cos x} = \mathbf{\ln\left\{\frac{1+\tan\frac{x}{2}}{1-\tan\frac{x}{2}}\right\} + c}$$

Note that since $\tan\frac{\pi}{4} = 1$, the above result may be written as:

$$\int \frac{dx}{\cos x} = \ln\left\{\frac{\tan\frac{\pi}{4} + \tan\frac{x}{2}}{1 - \tan\frac{\pi}{4}\tan\frac{x}{2}}\right\} + c$$
$$= \mathbf{\ln\left\{\tan\left(\frac{\pi}{4} + \frac{x}{2}\right)\right\} + c}$$

from compound angles, Chapter 22

Problem 3. Determine: $\int \frac{dx}{1+\cos x}$

If $t = \tan\frac{x}{2}$ then $\cos x = \frac{1-t^2}{1+t^2}$ and $dx = \frac{2dt}{1+t^2}$ from equations (2) and (3).

Thus $\int \frac{dx}{1+\cos x} = \int \frac{1}{1+\cos x}dx$

$$= \int \frac{1}{1+\frac{1-t^2}{1+t^2}}\left(\frac{2\,dt}{1+t^2}\right)$$

$$= \int \frac{1}{\frac{(1+t^2)+(1-t^2)}{1+t^2}}\left(\frac{2\,dt}{1+t^2}\right)$$

$$= \int dt$$

Hence $\int \frac{dx}{1+\cos x} = t + c = \mathbf{\tan\frac{x}{2} + c}$

Problem 4. Determine: $\int \frac{d\theta}{5+4\cos\theta}$

If $t = \tan\frac{\theta}{2}$ then $\cos\theta = \frac{1-t^2}{1+t^2}$ and $d\theta = \frac{2\,dt}{1+t^2}$ from equations (2) and (3).

Thus $\int \frac{d\theta}{5+4\cos\theta} = \int \frac{\left(\frac{2\,dt}{1+t^2}\right)}{5+4\left(\frac{1-t^2}{1+t^2}\right)}$

$$= \int \frac{\left(\frac{2\,dt}{1+t^2}\right)}{\frac{5(1+t^2)+4(1-t^2)}{1+t^2}}$$

$$= 2\int \frac{dt}{t^2+9} = 2\int \frac{dt}{t^2+3^2}$$

$$= 2\left(\frac{1}{3}\tan^{-1}\frac{t}{3}\right) + c,$$

from 12 of Table 46.1, page 483. Hence

$$\int \frac{d\theta}{5+4\cos\theta} = \frac{2}{3}\tan^{-1}\left(\frac{1}{3}\tan\frac{\theta}{2}\right) + c$$

Now try the following Practice Exercise

Practice Exercise 231 The $t = \tan\frac{\theta}{2}$ substitution (Answers on page 727)

Integrate the following with respect to the variable:

1. $\int \frac{d\theta}{1+\sin\theta}$
2. $\int \frac{dx}{1-\cos x+\sin x}$
3. $\int \frac{d\alpha}{3+2\cos\alpha}$
4. $\int \frac{dx}{3\sin x-4\cos x}$

48.3 Further worked problems on the $t = \tan\frac{\theta}{2}$ substitution

Problem 5. Determine: $\int \frac{dx}{\sin x+\cos x}$

If $t = \tan\frac{x}{2}$ then $\sin x = \frac{2t}{1+t^2}$, $\cos x = \frac{1-t^2}{1+t^2}$ and $dx = \frac{2\,dt}{1+t^2}$ from equations (1), (2) and (3).

Thus

$$\int \frac{dx}{\sin x+\cos x} = \int \frac{\frac{2\,dt}{1+t^2}}{\left(\frac{2t}{1+t^2}\right)+\left(\frac{1-t^2}{1+t^2}\right)}$$

$$= \int \frac{\frac{2\,dt}{1+t^2}}{\frac{2t+1-t^2}{1+t^2}} = \int \frac{2\,dt}{1+2t-t^2}$$

$$= \int \frac{-2\,dt}{t^2-2t-1} = \int \frac{-2\,dt}{(t-1)^2-2}$$

$$= \int \frac{2\,dt}{(\sqrt{2})^2-(t-1)^2}$$

$$= 2\left[\frac{1}{2\sqrt{2}}\ln\left\{\frac{\sqrt{2}+(t-1)}{\sqrt{2}-(t-1)}\right\}\right] + c$$

(see problem 11, Chapter 47, page 494),

i.e. $\int \frac{dx}{\sin x+\cos x}$

$$= \frac{1}{\sqrt{2}}\ln\left\{\frac{\sqrt{2}-1+\tan\frac{x}{2}}{\sqrt{2}+1-\tan\frac{x}{2}}\right\} + c$$

Problem 6. Determine: $\int \frac{dx}{7 - 3\sin x + 6\cos x}$

From equations (1) and (3),

$$\int \frac{dx}{7 - 3\sin x + 6\cos x} = \int \frac{\frac{2\,dt}{1+t^2}}{7 - 3\left(\frac{2t}{1+t^2}\right) + 6\left(\frac{1-t^2}{1+t^2}\right)}$$

$$= \int \frac{\frac{2\,dt}{1+t^2}}{\frac{7(1+t^2) - 3(2t) + 6(1-t^2)}{1+t^2}}$$

$$= \int \frac{2\,dt}{7 + 7t^2 - 6t + 6 - 6t^2}$$

$$= \int \frac{2\,dt}{t^2 - 6t + 13} = \int \frac{2\,dt}{(t-3)^2 + 2^2}$$

$$= 2\left[\frac{1}{2}\tan^{-1}\left(\frac{t-3}{2}\right)\right] + c$$

from 12, Table 46.1, page 483. Hence

$$\int \frac{dx}{7 - 3\sin x + 6\cos x} = \mathbf{\tan^{-1}\left(\frac{\tan\frac{x}{2} - 3}{2}\right) + c}$$

Problem 7. Determine: $\int \frac{d\theta}{4\cos\theta + 3\sin\theta}$

From equations (1) to (3),

$$\int \frac{d\theta}{4\cos\theta + 3\sin\theta} = \int \frac{\frac{2\,dt}{1+t^2}}{4\left(\frac{1-t^2}{1+t^2}\right) + 3\left(\frac{2t}{1+t^2}\right)}$$

$$= \int \frac{2\,dt}{4 - 4t^2 + 6t} = \int \frac{dt}{2 + 3t - 2t^2}$$

$$= -\frac{1}{2}\int \frac{dt}{t^2 - \frac{3}{2}t - 1}$$

$$= -\frac{1}{2}\int \frac{dt}{\left(t - \frac{3}{4}\right)^2 - \frac{25}{16}}$$

$$= \frac{1}{2}\int \frac{dt}{\left(\frac{5}{4}\right)^2 - \left(t - \frac{3}{4}\right)^2}$$

$$= \frac{1}{2}\left[\frac{1}{2\left(\frac{5}{4}\right)}\ln\left\{\frac{\frac{5}{4} + \left(t - \frac{3}{4}\right)}{\frac{5}{4} - \left(t - \frac{3}{4}\right)}\right\}\right] + c$$

from problem 11, Chapter 47, page 494,

$$= \frac{1}{5}\ln\left\{\frac{\frac{1}{2} + t}{2 - t}\right\} + c$$

Hence $\int \frac{d\theta}{4\cos\theta + 3\sin\theta}$

$$= \mathbf{\frac{1}{5}\ln\left\{\frac{\frac{1}{2} + \tan\frac{\theta}{2}}{2 - \tan\frac{\theta}{2}}\right\} + c}$$

or $$\mathbf{\frac{1}{5}\ln\left\{\frac{1 + 2\tan\frac{\theta}{2}}{4 - 2\tan\frac{\theta}{2}}\right\} + c}$$

Now try the following Practice Exercise

Practice Exercise 232 The $t = \tan\frac{\theta}{2}$ substitution (Answers on page 727)

In Problems 1 to 4, integrate with respect to the variable.

1. $\int \frac{d\theta}{5 + 4\sin\theta}$
2. $\int \frac{dx}{1 + 2\sin x}$
3. $\int \frac{dp}{3 - 4\sin p + 2\cos p}$

4. $\displaystyle\int \frac{d\theta}{3-4\sin\theta}$

5. Show that

$$\int \frac{dt}{1+3\cos t} = \frac{1}{2\sqrt{2}}\ln\left\{\frac{\sqrt{2}+\tan\frac{t}{2}}{\sqrt{2}-\tan\frac{t}{2}}\right\}+c$$

6. Show that $\displaystyle\int_0^{\pi/3} \frac{3\,d\theta}{\cos\theta} = 3.95$, correct to 3 significant figures.

7. Show that $\displaystyle\int_0^{\pi/2} \frac{d\theta}{2+\cos\theta} = \frac{\pi}{3\sqrt{3}}$

For fully worked solutions to each of the problems in Practice Exercises 231 and 232 in this chapter, go to the website:

www.routledge.com/cw/bird

Chapter 49

Integration by parts

Why it is important to understand: Integration by parts

Integration by parts is a very important technique that is used often in engineering and science. It is frequently used to change the integral of a product of functions into an ideally simpler integral. It is the foundation for the theory of differential equations and is used with Fourier series. We have looked at standard integrals followed by various techniques to change integrals into standard ones; integration by parts is a particularly important technique for integrating a product of two functions.

At the end of this chapter, you should be able to:

- appreciate when integration by parts is required
- integrate functions using integration by parts
- evaluate definite integrals using integration by parts

49.1 Introduction

From the product rule of differentiation:

$$\frac{d}{dx}(uv) = v\frac{du}{dx} + u\frac{dv}{dx}$$

where u and v are both functions of x.

Rearranging gives: $u\dfrac{dv}{dx} = \dfrac{d}{dx}(uv) - v\dfrac{du}{dx}$

Integrating both sides with respect to x gives:

$$\int u\frac{dv}{dx}dx = \int \frac{d}{dx}(uv)dx - \int v\frac{du}{dx}dx$$

ie $$\int u\frac{dv}{dx}dx = uv - \int v\frac{du}{dx}dx$$

or $$\int u\,dv = uv - \int v\,du$$

This is known as the **integration by parts formula** and provides a method of integrating such products of simple functions as $\int xe^x dx$, $\int t\sin t\,dt$, $\int e^{\theta}\cos\theta\,d\theta$ and $\int x\ln x\,dx$.

Given a product of two terms to integrate the initial choice is: 'which part to make equal to u' and 'which part to make equal to dv'. The choice must be such that the 'u part' becomes a constant after successive differentiation and the 'dv part' can be integrated from standard integrals. Invariable, the following rule holds: 'If a product to be integrated contains an algebraic term (such as x, t^2 or 3θ) then this term is chosen as the u part. The one exception to this rule is when a '$\ln x$' term is involved; in this case $\ln x$ is chosen as the 'u part'.

49.2 Worked problems on integration by parts

Problem 1. Determine: $\int x\cos x\,dx$

From the integration by parts formula,

$$\int u\,dv = uv - \int v\,du$$

Let $u = x$, from which $\frac{du}{dx} = 1$, i.e. $du = dx$ and let $dv = \cos x\,dx$, from which $v = \int \cos x\,dx = \sin x$.

Expressions for u, du and v are now substituted into the 'by parts' formula as shown below.

$$\int \boxed{u}\ \boxed{dv} = \boxed{u}\ \boxed{v} - \int \boxed{v}\ \boxed{du}$$

$$\int \boxed{x}\ \boxed{\cos x\,dx} = \boxed{(x)}\ \boxed{(\sin x)} - \int \boxed{(\sin x)}\ \boxed{(dx)}$$

$$\text{i.e.} \int x\cos x\,dx = x\sin x - (-\cos x) + c$$

$$= \mathbf{x\sin x + \cos x + c}$$

[This result may be checked by differentiating the right hand side,

i.e. $\frac{d}{dx}(x\sin x + \cos x + c)$

$= [(x)(\cos x) + (\sin x)(1)] - \sin x + 0$ using the product rule

$= x\cos x$, which is the function being integrated.]

Problem 2. Find: $\int 3te^{2t}dt$

Let $u = 3t$, from which, $\frac{du}{dt} = 3$, i.e. $du = 3\,dt$ and let $dv = e^{2t}\,dt$, from which, $v = \int e^{2t}dt = \frac{1}{2}e^{2t}$

Substituting into $\int u\,dv = uv - \int v\,du$ gives:

$$\int 3te^{2t}dt = (3t)\left(\frac{1}{2}e^{2t}\right) - \int\left(\frac{1}{2}e^{2t}\right)(3\,dt)$$

$$= \frac{3}{2}te^{2t} - \frac{3}{2}\int e^{2t}dt$$

$$= \frac{3}{2}te^{2t} - \frac{3}{2}\left(\frac{e^{2t}}{2}\right) + c$$

Hence $$\int \mathbf{3te^{2t}\,dt = \frac{3}{2}e^{2t}\left(t - \frac{1}{2}\right) + c},$$

which may be checked by differentiating.

Problem 3. Evaluate: $\int_0^{\frac{\pi}{2}} 2\theta\sin\theta\,d\theta$

Let $u = 2\theta$, from which, $\frac{du}{d\theta} = 2$, i.e. $du = 2\,d\theta$ and let $dv = \sin\theta\,d\theta$, from which,

$$v = \int \sin\theta\,d\theta = -\cos\theta$$

Substituting into $\int u\,dv = uv - \int v\,du$ gives:

$$\int 2\theta\sin\theta\,d\theta = (2\theta)(-\cos\theta) - \int(-\cos\theta)(2\,d\theta)$$

$$= -2\theta\cos\theta + 2\int\cos\theta\,d\theta$$

$$= -2\theta\cos\theta + 2\sin\theta + c$$

Hence $\int_0^{\frac{\pi}{2}} 2\theta\sin\theta\,d\theta$

$$= \Big[2\theta\cos\theta + 2\sin\theta\Big]_0^{\frac{\pi}{2}}$$

$$= \left[-2\left(\frac{\pi}{2}\right)\cos\frac{\pi}{2} + 2\sin\frac{\pi}{2}\right] - [0 + 2\sin 0]$$

$$= (-0 + 2) - (0 + 0) = \mathbf{2}$$

$$\text{since} \cos\frac{\pi}{2} = 0 \quad\text{and}\quad \sin\frac{\pi}{2} = 1$$

Problem 4. Evaluate: $\int_0^1 5xe^{4x}dx$, correct to 3 significant figures

Let $u = 5x$, from which $\frac{du}{dx} = 5$, i.e. $du = 5\,dx$ and let $dv = e^{4x}\,dx$, from which, $v = \int e^{4x}dx = \frac{1}{4}e^{4x}$

Substituting into $\int u\,dv = uv - \int v\,du$ gives:

$$\int 5xe^{4x}dx = (5x)\left(\frac{e^{4x}}{4}\right) - \int\left(\frac{e^{4x}}{4}\right)(5\,dx)$$

$$= \frac{5}{4}xe^{4x} - \frac{5}{4}\int e^{4x}\,dx$$

$$= \frac{5}{4}xe^{4x} - \frac{5}{4}\left(\frac{e^{4x}}{4}\right) + c$$

$$= \frac{5}{4}e^{4x}\left(x - \frac{1}{4}\right) + c$$

Hence $\int_0^1 5xe^{4x}\,dx$

$$= \left[\frac{5}{4}e^{4x}\left(x - \frac{1}{4}\right)\right]_0^1$$

$$= \left[\frac{5}{4}e^4\left(1-\frac{1}{4}\right)\right] - \left[\frac{5}{4}e^0\left(0-\frac{1}{4}\right)\right]$$

$$= \left(\frac{15}{16}e^4\right) - \left(-\frac{5}{16}\right)$$

$$= 51.186 + 0.313 = 51.499 = \mathbf{51.5},$$

correct to 3 significant figures.

Problem 5. Determine: $\int x^2 \sin x\, dx$

Let $u = x^2$, from which, $\frac{du}{dx} = 2x$, i.e. $du = 2x\,dx$, and

let $dv = \sin x\, dx$, from which, $v = \int \sin x\,dx = -\cos x$

Substituting into $\int u\,dv = uv - \int v\,du$ gives:

$$\int x^2 \sin x\,dx = (x^2)(-\cos x) - \int(-\cos x)(2x\,dx)$$

$$= -x^2 \cos x + 2\left[\int x\cos x\,dx\right]$$

The integral, $\int x\cos x\,dx$, is not a 'standard integral' and it can only be determined by using the integration by parts formula again.

From Problem 1, $\int x\cos x\,dx = x\sin x + \cos x$

Hence $\int x^2 \sin x\,dx$

$$= -x^2\cos x + 2\{x\sin x + \cos x\} + c$$

$$= -x^2\cos x + 2x\sin x + 2\cos x + c$$

$$= \mathbf{(2 - x^2)\cos x + 2x \sin x + c}$$

In general, if the algebraic term of a product is of power n, then the integration by parts formula is applied n times.

Now try the following Practice Exercise

Practice Exercise 233 Integration by parts (Answers on page 727)

Determine the integrals in Problems 1 to 5 using integration by parts.

1. $\int xe^{2x}dx$
2. $\int \frac{4x}{e^{3x}}dx$
3. $\int x\sin x\,dx$
4. $\int 5\theta\cos 2\theta\,d\theta$
5. $\int 3t^2e^{2t}dt$

Evaluate the integrals in Problems 6 to 9, correct to 4 significant figures.

6. $\int_0^2 2xe^x\,dx$
7. $\int_0^{\frac{\pi}{4}} x\sin 2x\,dx$
8. $\int_0^{\frac{\pi}{2}} t^2\cos t\,dt$
9. $\int_1^2 3x^2e^{\frac{x}{2}}\,dx$

49.3 Further worked problems on integration by parts

Problem 6. Find: $\int x\ln x\,dx$

The logarithmic function is chosen as the 'u part' Thus when $u = \ln x$, then $\frac{du}{dx} = \frac{1}{x}$ i.e. $du = \frac{dx}{x}$

Letting $dv = x\,dx$ gives $v = \int x\,dx = \frac{x^2}{2}$

Substituting into $\int u\,dv = uv - \int v\,du$ gives:

$$\int x\ln x\,dx = (\ln x)\left(\frac{x^2}{2}\right) - \int\left(\frac{x^2}{2}\right)\frac{dx}{x}$$

$$= \frac{x^2}{2}\ln x - \frac{1}{2}\int x\,dx$$

$$= \frac{x^2}{2}\ln x - \frac{1}{2}\left(\frac{x^2}{2}\right) + c$$

Hence $\int x\ln x\,dx = \mathbf{\frac{x^2}{2}\left(\ln x - \frac{1}{2}\right) + c}$

or $\mathbf{\frac{x^2}{4}(2\ln x - 1) + c}$

Problem 7. Determine: $\int \ln x \, dx$

$\int \ln x dx$ is the same as $\int (1) \ln x dx$

Let $u = \ln x$, from which, $\frac{du}{dx} = \frac{1}{x}$ i.e. $du = \frac{dx}{x}$ and let $dv = 1\,dx$, from which, $v = \int 1\,dx = x$

Substituting into $\int u\,dv = uv - \int v\,du$ gives:

$$\int \ln x\,dx = (\ln x)(x) - \int x \frac{dx}{x}$$
$$= x \ln x - \int dx = x \ln x - x + c$$

Hence $\boldsymbol{\int \ln x\,dx = x(\ln x - 1) + c}$

Problem 8. Evaluate: $\int_1^9 \sqrt{x} \ln x\,dx$, correct to 3 significant figures

Let $u = \ln x$, from which $du = \frac{dx}{x}$ and let $dv = \sqrt{x}\,dx = x^{\frac{1}{2}}\,dx$, from which,

$$v = \int x^{\frac{1}{2}}\,dx = \frac{2}{3}x^{\frac{3}{2}}$$

Substituting into $\int u\,dv = uv - \int v\,du$ gives:

$$\int \sqrt{x} \ln x\,dx = (\ln x)\left(\frac{2}{3}x^{\frac{3}{2}}\right) - \int \left(\frac{2}{3}x^{\frac{3}{2}}\right)\left(\frac{dx}{x}\right)$$
$$= \frac{2}{3}\sqrt{x^3} \ln x - \frac{2}{3}\int x^{\frac{1}{2}}\,dx$$
$$= \frac{2}{3}\sqrt{x^3} \ln x - \frac{2}{3}\left(\frac{2}{3}x^{\frac{3}{2}}\right) + c$$
$$= \frac{2}{3}\sqrt{x^3}\left[\ln x - \frac{2}{3}\right] + c$$

Hence $\int_1^9 \sqrt{x} \ln x\,dx = \left[\frac{2}{3}\sqrt{x^3}\left(\ln x - \frac{2}{3}\right)\right]_1^9$

$$= \left[\frac{2}{3}\sqrt{9^3}\left(\ln 9 - \frac{2}{3}\right)\right] - \left[\frac{2}{3}\sqrt{1^3}\left(\ln 1 - \frac{2}{3}\right)\right]$$
$$= \left[18\left(\ln 9 - \frac{2}{3}\right)\right] - \left[\frac{2}{3}\left(0 - \frac{2}{3}\right)\right]$$
$$= 27.550 + 0.444 = 27.994 = \mathbf{28.0},$$

correct to 3 significant figures.

Problem 9. Find: $\int e^{ax} \cos bx\,dx$

When integrating a product of an exponential and a sine or cosine function it is immaterial which part is made equal to 'u'.

Let $u = e^{ax}$, from which $\frac{du}{dx} = ae^{ax}$, i.e. $du = ae^{ax}\,dx$ and let $dv = \cos bx\,dx$, from which,

$$v = \int \cos bx\,dx = \frac{1}{b}\sin bx$$

Substituting into $\int u\,dv = uv - \int v\,du$ gives:

$$\int e^{ax} \cos bx\,dx$$
$$= (e^{ax})\left(\frac{1}{b}\sin bx\right) - \int \left(\frac{1}{b}\sin bx\right)(ae^{ax}\,dx)$$
$$= \frac{1}{b}e^{ax}\sin bx - \frac{a}{b}\left[\int e^{ax}\sin bx\,dx\right] \qquad (1)$$

$\int e^{ax} \sin bx\,dx$ is now determined separately using integration by parts again:

Let $u = e^{ax}$ then $du = ae^{ax}\,dx$, and let $dv = \sin bx\,dx$, from which

$$v = \int \sin bx\,dx = -\frac{1}{b}\cos bx$$

Substituting into the integration by parts formula gives:

$$\int e^{ax} \sin bx\,dx = (e^{ax})\left(-\frac{1}{b}\cos bx\right) - \int \left(-\frac{1}{b}\cos bx\right)(ae^{ax}\,dx)$$
$$= -\frac{1}{b}e^{ax}\cos bx + \frac{a}{b}\int e^{ax}\cos bx\,dx$$

Substituting this result into equation (1) gives:

$$\int e^{ax} \cos bx\,dx = \frac{1}{b}e^{ax}\sin bx - \frac{a}{b}\left[-\frac{1}{b}e^{ax}\cos bx + \frac{a}{b}\int e^{ax}\cos bx\,dx\right]$$
$$= \frac{1}{b}e^{ax}\sin bx + \frac{a}{b^2}e^{ax}\cos bx - \frac{a^2}{b^2}\int e^{ax}\cos bx\,dx$$

The integral on the far right of this equation is the same as the integral on the left hand side and thus they may be combined.

$$\int e^{ax}\cos bx\,dx+\frac{a^2}{b^2}\int e^{ax}\cos bx\,dx$$
$$=\frac{1}{b}e^{ax}\sin bx+\frac{a}{b^2}e^{ax}\cos bx$$

i.e. $$\left(1+\frac{a^2}{b^2}\right)\int e^{ax}\cos bx\,dx$$
$$=\frac{1}{b}e^{ax}\sin bx+\frac{a}{b^2}e^{ax}\cos bx$$

i.e. $$\left(\frac{b^2+a^2}{b^2}\right)\int e^{ax}\cos bx\,dx$$
$$=\frac{e^{ax}}{b^2}(b\sin bx+a\cos bx)$$

Hence $$\int e^{ax}\cos bx\,dx$$
$$=\left(\frac{b^2}{b^2+a^2}\right)\left(\frac{e^{ax}}{b^2}\right)(b\sin bx+a\cos bx)$$
$$=\frac{e^{ax}}{a^2+b^2}(b\sin bx+a\cos bx)+c$$

Using a similar method to above, that is, integrating by parts twice, the following result may be proved:

$$\int e^{ax}\sin bx\,dx$$
$$=\frac{e^{ax}}{a^2+b^2}(a\sin bx-b\cos bx)+c \qquad (2)$$

Problem 10. Evaluate: $\int_0^{\frac{\pi}{4}} e^t\sin 2t\,dt$, correct to 4 decimal places

Comparing $\int e^t\sin 2t\,dt$ with $\int e^{ax}\sin bx\,dx$ shows that $x=t$, $a=1$ and $b=2$.
Hence, substituting into equation (2) gives:

$$\int_0^{\frac{\pi}{4}} e^t\sin 2t\,dt$$
$$=\left[\frac{e^t}{1^2+2^2}(1\sin 2t-2\cos 2t)\right]_0^{\frac{\pi}{4}}$$
$$=\left[\frac{e^{\frac{\pi}{4}}}{5}\left(\sin 2\left(\frac{\pi}{4}\right)-2\cos 2\left(\frac{\pi}{4}\right)\right)\right]$$
$$-\left[\frac{e^0}{5}(\sin 0-2\cos 0)\right]$$
$$=\left[\frac{e^{\frac{\pi}{4}}}{5}(1-0)\right]-\left[\frac{1}{5}(0-2)\right]=\frac{e^{\frac{\pi}{4}}}{5}+\frac{2}{5}$$
$= \mathbf{0.8387}$, correct to 4 decimal places

Now try the following Practice Exercise

Practice Exercise 234 Integration by parts (Answers on page 727)

Determine the integrals in Problems 1 to 5 using integration by parts.

1. $\int 2x^2\ln x\,dx$
2. $\int 2\ln 3x\,dx$
3. $\int x^2\sin 3x\,dx$
4. $\int 2e^{5x}\cos 2x\,dx$
5. $\int 2\theta\sec^2\theta\,d\theta$

Evaluate the integrals in Problems 6 to 9, correct to 4 significant figures.

6. $\int_1^2 x\ln x\,dx$
7. $\int_0^1 2e^{3x}\sin 2x\,dx$
8. $\int_0^{\frac{\pi}{2}} e^t\cos 3t\,dt$
9. $\int_1^4 \sqrt{x^3}\ln x\,dx$
10. In determining a Fourier series to represent $f(x)=x$ in the range $-\pi$ to π, Fourier coefficients are given by:
$$a_n=\frac{1}{\pi}\int_{-\pi}^{\pi} x\cos nx\,dx$$
and $$b_n=\frac{1}{\pi}\int_{-\pi}^{\pi} x\sin nx\,dx$$
where n is a positive integer. Show by using integration by parts that $a_n=0$ and $b_n=-\frac{2}{n}\cos n\pi$

11. The equations:

$$C = \int_0^1 e^{-0.4\theta} \cos 1.2\theta \, d\theta$$

and $$S = \int_0^1 e^{-0.4\theta} \sin 1.2\theta \, d\theta$$

are involved in the study of damped oscillations. Determine the values of C and S

Practice Exercise 235 Multiple-choice questions on integration by parts (Answers on page 728)

Each question has only one correct answer

1. $\int xe^{2x}\,dx$ is:

   (a) $\frac{x^2}{4}e^{2x}+c$ (b) $2e^{2x}+c$

   (c) $\frac{e^{2x}}{4}(2x-1)+c$ (d) $2e^{2x}(x-2)+c$

2. $\int \ln x\,dx$ is equal to:

   (a) $x(\ln x-1)+c$ (b) $\frac{1}{x}+c$

   (c) $x\ln x-1+c$ (d) $\frac{1}{x}+\frac{1}{x^2}+c$

3. The indefinite integral $\int x\ln x\,dx$ is equal to:

   (a) $x(\ln 2x-1)+c$

   (b) $x^2\left(\ln 2x-\frac{1}{4}\right)+c$

   (c) $\frac{x^2}{2}\left(\ln x-\frac{1}{2}\right)+c$

   (d) $\frac{x}{2}\ln\left(2x-\frac{1}{4}\right)+c$

4. $\int_0^{\frac{\pi}{2}} 3t\sin t\,dt$ is equal to:

   (a) -1.712 (b) 3 (c) 6 (d) -3

5. $\int_0^1 5xe^{3x}\,dx$ is equal to:

   (a) 1.67 (b) 22.32

   (c) -267.25 (d) 22.87

For fully worked solutions to each of the problems in Practice Exercises 233 and 234 in this chapter, go to the website:
www.routledge.com/cw/bird

Chapter 50

Numerical integration

Why it is important to understand: Numerical integration

There are two main reasons for why there is a need to do numerical integration – analytical integration may be impossible or unfeasible, or it may be necessary to integrate tabulated data rather than known functions. As has been mentioned before, there are many applications for integration. For example, Maxwell's equations can be written in integral form; numerical solutions of Maxwell's equations can be directly used for a huge number of engineering applications. Integration is involved in practically every physical theory in some way – vibration, distortion under weight or one of many types of fluid flow – be it heat flow, air flow (over a wing), or water flow (over a ship's hull, through a pipe, or perhaps even groundwater flow regarding a contaminant) and so on; all these things can be either directly solved by integration (for simple systems), or some type of numerical integration (for complex systems). Numerical integration is also essential for the evaluation of integrals of functions available only at discrete points; such functions often arise in the numerical solution of differential equations or from experimental data taken at discrete intervals. Engineers therefore often require numerical integration and this chapter explains the procedures available.

At the end of this chapter, you should be able to:

- appreciate the need for numerical integration
- evaluate integrals using the trapezoidal rule
- evaluate integrals using the mid-ordinate rule
- evaluate integrals using Simpson's rule
- apply numerical integration to practical situations

50.1 Introduction

Even with advanced methods of integration there are many mathematical functions which cannot be integrated by analytical methods and thus approximate methods have then to be used. In many cases, such as in modelling airflow around a car, an exact answer may not even be strictly necessary. Approximate methods of definite integrals may be determined by what is termed **numerical integration**.

It may be shown that determining the value of a definite integral is, in fact, finding the area between a curve, the horizontal axis and the specified ordinates. Three methods of finding approximate areas under curves are the trapezoidal rule, the mid-ordinate rule and Simpson's rule, and these rules are used as a basis for numerical integration.

50.2 The trapezoidal rule

Let a required definite integral be denoted by $\int_a^b y\,dx$ and be represented by the area under the graph of $y=f(x)$ between the limits $x=a$ and $x=b$ as shown in Fig. 50.1.

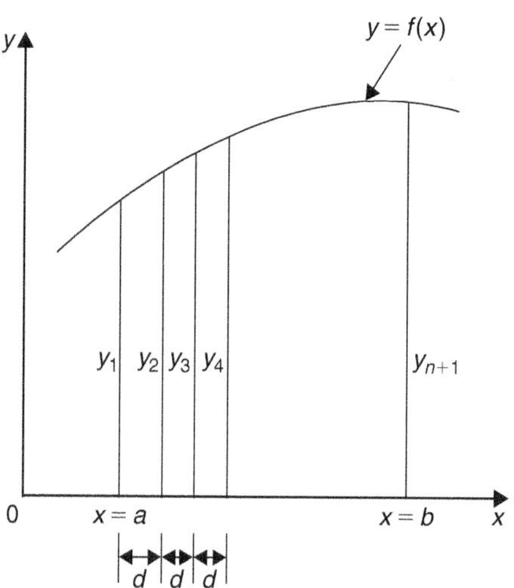

Figure 50.1

Let the range of integration be divided into n equal intervals each of width d, such that $nd=b-a$, i.e. $d=\dfrac{b-a}{n}$

The ordinates are labelled $y_1, y_2, y_3, \ldots, y_{n+1}$ as shown. An approximation to the area under the curve may be determined by joining the tops of the ordinates by straight lines. Each interval is thus a trapezium, and since the area of a trapezium is given by:

$$\text{area} = \frac{1}{2}(\text{sum of parallel sides}) (\text{perpendicular distance between them})$$

then

$$\int_a^b y\,dx \approx \frac{1}{2}(y_1+y_2)d + \frac{1}{2}(y_2+y_3)d + \frac{1}{2}(y_3+y_4)d + \cdots + \frac{1}{2}(y_n+y_{n+1})d$$

$$\approx d\left[\frac{1}{2}y_1 + y_2 + y_3 + y_4 + \cdots + y_n + \frac{1}{2}y_{n+1}\right]$$

i.e. **the trapezoidal rule states:**

$$\int_a^b y\,dx \approx \begin{pmatrix}\text{width of}\\ \text{interval}\end{pmatrix}\left\{\frac{1}{2}\begin{pmatrix}\text{first + last}\\ \text{ordinate}\end{pmatrix} + \begin{pmatrix}\text{sum of}\\ \text{remaining}\\ \text{ordinates}\end{pmatrix}\right\} \quad (1)$$

Problem 1. (a) Use integration to evaluate, correct to 3 decimal places, $\displaystyle\int_1^3 \frac{2}{\sqrt{x}}\,dx$
(b) Use the trapezoidal rule with 4 intervals to evaluate the integral in part (a), correct to 3 decimal places

(a) $$\int_1^3 \frac{2}{\sqrt{x}}\,dx = \int_1^3 2x^{-\frac{1}{2}}\,dx = \left[\frac{2x^{\left(-\frac{1}{2}\right)+1}}{-\frac{1}{2}+1}\right]_1^3 = \left[4x^{\frac{1}{2}}\right]_1^3$$

$$= 4\left[\sqrt{x}\right]_1^3 = 4\left[\sqrt{3}-\sqrt{1}\right]$$

$$= \mathbf{2.928}, \text{ correct to 3 decimal places.}$$

(b) The range of integration is the difference between the upper and lower limits, i.e. $3-1=2$. Using the trapezoidal rule with 4 intervals gives an interval width $d=\dfrac{3-1}{4}=0.5$ and ordinates situated at 1.0, 1.5, 2.0, 2.5 and 3.0. Corresponding values of $\dfrac{2}{\sqrt{x}}$ are shown in the table below, each correct to 4 decimal places (which is one more decimal place than required in the problem).

x	$\dfrac{2}{\sqrt{x}}$
1.0	2.0000
1.5	1.6330
2.0	1.4142
2.5	1.2649
3.0	1.1547

From equation (1):

$$\int_1^3 \frac{2}{\sqrt{x}}\,dx \approx (0.5)\left\{\frac{1}{2}(2.0000+1.1547) + 1.6330 + 1.4142 + 1.2649\right\}$$

$$= \mathbf{2.945}, \text{correct to 3 decimal places.}$$

This problem demonstrates that even with just 4 intervals a close approximation to the true value of 2.928 (correct to 3 decimal places) is obtained using the trapezoidal rule.

Problem 2. Use the trapezoidal rule with 8 intervals to evaluate $\displaystyle\int_1^3 \frac{2}{\sqrt{x}}\,dx$, correct to 3 decimal places

With 8 intervals, the width of each is $\frac{3-1}{8}$ i.e. 0.25 giving ordinates at 1.00, 1.25, 1.50, 1.75, 2.00, 2.25, 2.50, 2.75 and 3.00. Corresponding values of $\frac{2}{\sqrt{x}}$ are shown in the table below:

x	$\frac{2}{\sqrt{x}}$
1.00	2.000
1.25	1.7889
1.50	1.6330
1.75	1.5119
2.00	1.4142
2.25	1.3333
2.50	1.2649
2.75	1.2060
3.00	1.1547

From equation (1):

$$\int_1^3 \frac{2}{\sqrt{x}}dx \approx (0.25)\left\{\frac{1}{2}(2.000+1.1547)+1.7889 + 1.6330+1.5119+1.4142 +1.3333+1.2649+1.2060\right\}$$

$$= \mathbf{2.932}, \text{correct to 3 decimal places}$$

This problem demonstrates that the greater the number of intervals chosen (i.e. the smaller the interval width) the more accurate will be the value of the definite integral. The exact value is found when the number of intervals is infinite, which is what the process of integration is based upon.

Problem 3. Use the trapezoidal rule to evaluate $\int_0^{\pi/2} \frac{1}{1+\sin x}dx$ using 6 intervals. Give the answer correct to 4 significant figures

With 6 intervals, each will have a width of $\frac{\frac{\pi}{2}-0}{6}$ i.e. $\frac{\pi}{12}$ rad (or 15°) and the ordinates occur at 0, $\frac{\pi}{12}, \frac{\pi}{6}, \frac{\pi}{4}, \frac{\pi}{3}, \frac{5\pi}{12}$ and $\frac{\pi}{2}$. Corresponding values of $\frac{1}{1+\sin x}$ are shown in the table below:

x	$\frac{1}{1+\sin x}$
0	1.0000
$\frac{\pi}{12}$ (or 15°)	0.79440
$\frac{\pi}{6}$ (or 30°)	0.66667
$\frac{\pi}{4}$ (or 45°)	0.58579
$\frac{\pi}{3}$ (or 60°)	0.53590
$\frac{5\pi}{12}$ (or 75°)	0.50867
$\frac{\pi}{2}$ (or 90°)	0.50000

From equation (1):

$$\int_0^{\frac{\pi}{2}} \frac{1}{1+\sin x}dx \approx \left(\frac{\pi}{12}\right)\left\{\frac{1}{2}(1.00000+0.50000) + 0.79440+0.66667+0.58579 +0.53590+0.50867\right\}$$

$$= \mathbf{1.006}, \text{correct to 4 significant figures}$$

Now try the following Practice Exercise

Practice Exercise 236 Trapezoidal rule (Answers on page 728)

Evaluate the following definite integrals using the **trapezoidal rule**, giving the answers correct to 3 decimal places:

1. $\int_0^1 \frac{2}{1+x^2}dx$ (Use 8 intervals)
2. $\int_1^3 2\ln 3x\,dx$ (Use 8 intervals)
3. $\int_0^{\pi/3} \sqrt{\sin\theta}\,d\theta$ (Use 6 intervals)
4. $\int_0^{1.4} e^{-x^2}dx$ (Use 7 intervals)

50.3 The mid-ordinate rule

Let a required definite integral be denoted again by $\int_a^b y\,dx$ and represented by the area under the graph of $y=f(x)$ between the limits $x=a$ and $x=b$, as shown in Fig. 50.2.

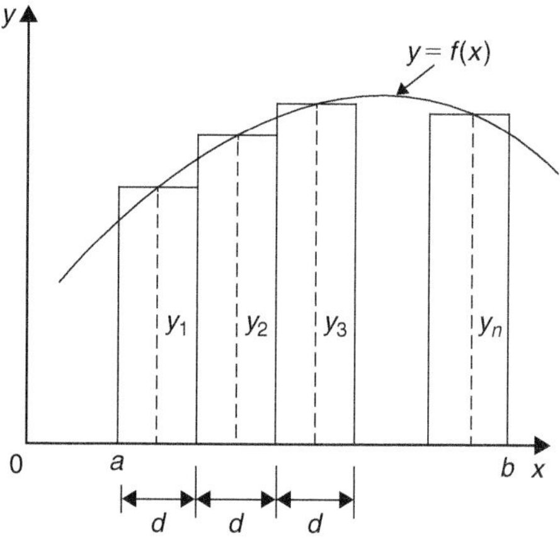

Figure 50.2

With the mid-ordinate rule each interval of width d is assumed to be replaced by a rectangle of height equal to the ordinate at the middle point of each interval, shown as $y_1, y_2, y_3, \ldots, y_n$ in Fig. 50.2.

Thus $$\int_a^b y\,dx \approx dy_1 + dy_2 + dy_3 + \cdots + dy_n$$
$$\approx d(y_1 + y_2 + y_3 + \cdots + y_n)$$

i.e. **the mid-ordinate rule states:**

$$\int_a^b y\,dx \approx \left(\begin{array}{c}\textbf{width of}\\ \textbf{interval}\end{array}\right)\left(\begin{array}{c}\textbf{sum of}\\ \textbf{mid-ordinates}\end{array}\right) \quad (2)$$

Problem 4. Use the mid-ordinate rule with (a) 4 intervals, (b) 8 intervals, to evaluate $\int_1^3 \frac{2}{\sqrt{x}}\,dx$, correct to 3 decimal places

(a) With 4 intervals, each will have a width of $\frac{3-1}{4}$, i.e. 0.5. and the ordinates will occur at 1.0, 1.5, 2.0, 2.5 and 3.0. Hence the mid-ordinates y_1, y_2, y_3 and y_4 occur at 1.25, 1.75, 2.25 and 2.75

Corresponding values of $\frac{2}{\sqrt{x}}$ are shown in the following table:

x	$\frac{2}{\sqrt{x}}$
1.25	1.7889
1.75	1.5119
2.25	1.3333
2.75	1.2060

From equation (2):
$$\int_1^3 \frac{2}{\sqrt{x}}\,dx \approx (0.5)[1.7889 + 1.5119 + 1.3333 + 1.2060]$$
$$= \mathbf{2.920}, \text{ correct to 3 decimal places}$$

(b) With 8 intervals, each will have a width of 0.25 and the ordinates will occur at 1.00, 1.25, 1.50, 1.75,... and thus mid-ordinates at 1.125, 1.375, 1.625, 1.875.... Corresponding values of $\frac{2}{\sqrt{x}}$ are shown in the following table:

x	$\frac{2}{\sqrt{x}}$
1.125	1.8856
1.375	1.7056
1.625	1.5689
1.875	1.4606
2.125	1.3720
2.375	1.2978
2.625	1.2344
2.875	1.1795

From equation (2):
$$\int_1^3 \frac{2}{\sqrt{x}}\,dx \approx (0.25)[1.8856 + 1.7056 + 1.5689 + 1.4606 + 1.3720 + 1.2978 + 1.2344 + 1.1795]$$
$$= \mathbf{2.926}, \text{correct to 3 decimal places}$$

As previously, the greater the number of intervals the nearer the result is to the true value of 2.928, correct to 3 decimal places.

Problem 5. Evaluate: $\int_0^{2.4} e^{-x^2/3}\,dx$, correct to 4 significant figures, using the mid-ordinate rule with 6 intervals

With 6 intervals each will have a width of $\frac{2.4-0}{6}$, i.e. 0.40 and the ordinates will occur at 0, 0.40, 0.80, 1.20, 1.60, 2.00 and 2.40 and thus mid-ordinates at 0.20, 0.60, 1.00, 1.40, 1.80 and 2.20

Corresponding values of $e^{-x^2/3}$ are shown in the following table:

x	$e^{-\frac{x^2}{3}}$
0.20	0.98676
0.60	0.88692
1.00	0.71653
1.40	0.52031
1.80	0.33960
2.20	0.19922

From equation (2):

$$\int_0^{2.4} e^{-\frac{x^2}{3}}\,dx \approx (0.40)[0.98676+0.88692+0.71653+0.52031+0.33960+0.19922]$$

$= \mathbf{1.460}$, correct to 4 significant figures.

Now try the following Practice Exercise

Practice Exercise 237 Mid-ordinate rule (Answers on page 728)

Evaluate the following definite integrals using the **mid-ordinate rule**, giving the answers correct to 3 decimal places.

1. $\int_0^2 \frac{3}{1+t^2}\,dt$ (Use 8 intervals)
2. $\int_0^{\pi/2} \frac{1}{1+\sin\theta}\,d\theta$ (Use 6 intervals)
3. $\int_1^3 \frac{\ln x}{x}\,dx$ (Use 10 intervals)
4. $\int_0^{\pi/3} \sqrt{\cos^3 x}\,dx$ (Use 6 intervals)

50.4 Simpson's rule

The approximation made with the trapezoidal rule is to join the top of two successive ordinates by a straight line, i.e. by using a linear approximation of the form $a+bx$. With **Simpson's* rule**, the approximation made is to join the tops of three successive ordinates by a parabola, i.e. by using a quadratic approximation of the form $a+bx+cx^2$.

Figure 50.3 shows a parabola $y=a+bx+cx^2$ with ordinates y_1, y_2 and y_3 at $x=-d$, $x=0$ and $x=d$ respectively.

Thus the width of each of the two intervals is d. The area enclosed by the parabola, the x-axis and ordinates $x=-d$ and $x=d$ is given by:

$$\int_{-d}^{d}(a+bx+cx^2)\,dx = \left[ax+\frac{bx^2}{2}+\frac{cx^3}{3}\right]_{-d}^{d}$$

$$= \left(ad+\frac{bd^2}{2}+\frac{cd^3}{3}\right) - \left(-ad+\frac{bd^2}{2}-\frac{cd^3}{3}\right)$$

$$= 2ad+\frac{2}{3}cd^3$$

$$\text{or } \frac{1}{3}d(6a+2cd^2) \qquad (3)$$

Since $y=a+bx+cx^2$

at $x=-d$, $y_1=a-bd+cd^2$

at $x=0$, $y_2=a$

and at $x=d$, $y_3=a+bd+cd^2$

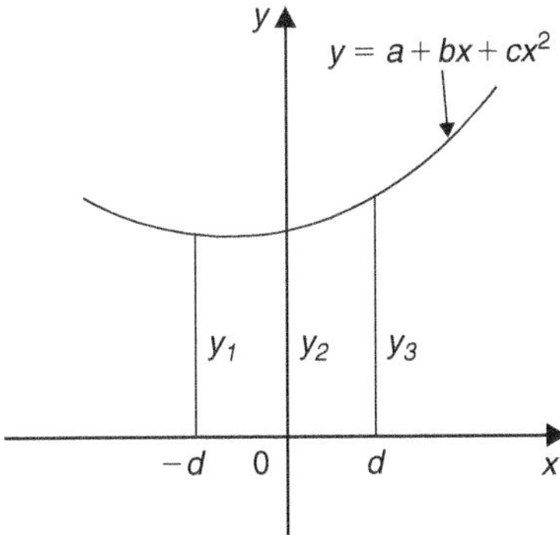

Figure 50.3

*Who was Simpson? – See page 277. To find out more go to www.routledge.com/cw/bird

Hence $\quad y_1 + y_3 = 2a + 2cd^2$

and $\quad y_1 + 4y_2 + y_3 = 6a + 2cd^2 \qquad (4)$

Thus the area under the parabola between $x = -d$ and $x = d$ in Fig. 50.3 may be expressed as $\frac{1}{3}d(y_1 + 4y_2 + y_3)$, from equation (3) and (4), and the result is seen to be independent of the position of the origin.

Let a definite integral be denoted by $\int_a^b y\,dx$ and represented by the area under the graph of $y = f(x)$ between the limits $x = a$ and $x = b$, as shown in Fig. 50.4. The range of integration, $b - a$, is divided into an **even** number of intervals, say $2n$, each of width d.

Since an even number of intervals is specified, an odd number of ordinates, $2n + 1$, exists. Let an approximation to the curve over the first two intervals be a parabola of the form $y = a + bx + cx^2$ which passes through the tops of the three ordinates y_1, y_2 and y_3. Similarly, let an approximation to the curve over the next two intervals be the parabola which passes through the tops of the ordinates y_3, y_4 and y_5 and so on. Then

$$\int_a^b y\,dx \approx \frac{1}{3}d(y_1 + 4y_2 + y_3) + \frac{1}{3}d(y_3 + 4y_4 + y_5) + \frac{1}{3}d(y_{2n-1} + 4y_{2n} + y_{2n+1})$$

$$\approx \frac{1}{3}d[(y_1 + y_{2n+1}) + 4(y_2 + y_4 + \cdots + y_{2n}) + 2(y_3 + y_5 + \cdots + y_{2n-1})]$$

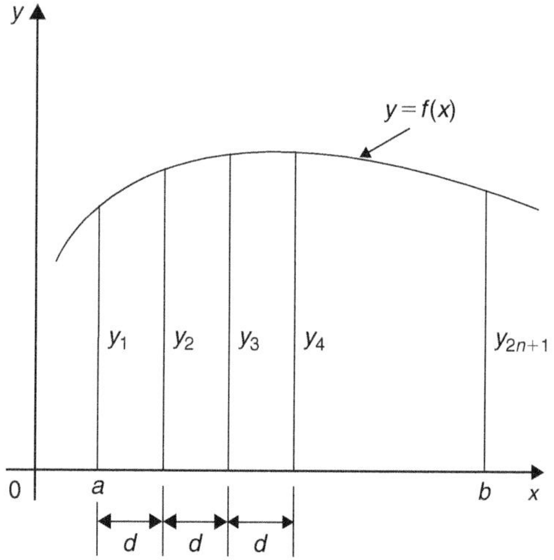

Figure 50.4

i.e. **Simpson's rule states:**

$$\int_a^b y\,dx \approx \frac{1}{3}\left(\begin{matrix}\text{width of}\\ \text{interval}\end{matrix}\right)\left\{\left(\begin{matrix}\text{first + last}\\ \text{ordinate}\end{matrix}\right) + 4\left(\begin{matrix}\text{sum of even}\\ \text{ordinates}\end{matrix}\right) + 2\left(\begin{matrix}\text{sum of remaining}\\ \text{odd ordinates}\end{matrix}\right)\right\} \qquad (5)$$

Note that Simpson's rule can only be applied when an **even** number of intervals is chosen, i.e. an odd number of ordinates.

Problem 6. Use Simpson's rule with (a) 4 intervals, (b) 8 intervals, to evaluate $\int_1^3 \frac{2}{\sqrt{x}}\,dx$, correct to 3 decimal places

(a) With 4 intervals, each will have a width of $\frac{3-1}{4}$ i.e. 0.5 and the ordinates will occur at 1.0, 1.5, 2.0, 2.5 and 3.0

The values of the ordinates are as shown in the table of Problem 1(b), page 508.

Thus, from equation (5):

$$\int_1^3 \frac{2}{\sqrt{x}}\,dx \approx \frac{1}{3}(0.5)[(2.0000 + 1.1547) + 4(1.6330 + 1.2649) + 2(1.4142)]$$

$$= \frac{1}{3}(0.5)[3.1547 + 11.5916 + 2.8284]$$

$= \mathbf{2.929}$, correct to 3 decimal places.

(b) With 8 intervals, each will have a width of $\frac{3-1}{8}$ i.e. 0.25 and the ordinates occur at 1.00, 1.25, 1.50, 1.75, ..., 3.0

The values of the ordinates are as shown in the table in Problem 2, page 509.

Thus, from equation (5):

$$\int_1^3 \frac{2}{\sqrt{x}}\,dx \approx \frac{1}{3}(0.25)[(2.0000 + 1.1547) + 4(1.7889 + 1.5119 + 1.3333 + 1.2060) + 2(1.6330 + 1.4142 + 1.2649)]$$

$$= \frac{1}{3}(0.25)[3.1547 + 23.3604 + 8.6242]$$

$$= \mathbf{2.928}, \text{ correct to 3 decimal places.}$$

It is noted that the latter answer is exactly the same as that obtained by integration. In general, Simpson's rule is regarded as the most accurate of the three approximate methods used in numerical integration.

Problem 7. Evaluate: $\int_0^{\pi/3} \sqrt{1 - \frac{1}{3}\sin^2\theta}\, d\theta$, correct to 3 decimal places, using Simpson's rule with 6 intervals

With 6 intervals, each will have a width of $\dfrac{\frac{\pi}{3} - 0}{6}$ i.e. $\frac{\pi}{18}$ rad (or 10°), and the ordinates will occur at 0, $\frac{\pi}{18}, \frac{\pi}{9}, \frac{\pi}{6}, \frac{2\pi}{9}, \frac{5\pi}{18}$ and $\frac{\pi}{3}$

Corresponding values of $\sqrt{1 - \frac{1}{3}\sin^2\theta}$ are shown in the table below:

θ	0	$\frac{\pi}{18}$ (or 10°)	$\frac{\pi}{9}$ (or 20°)	$\frac{\pi}{6}$ (or 30°)
$\sqrt{1 - \frac{1}{3}\sin^2\theta}$	1.0000	0.9950	0.9803	0.9574

θ	$\frac{2\pi}{9}$ (or 40°)	$\frac{5\pi}{18}$ (or 50°)	$\frac{\pi}{3}$ (or 60°)
$\sqrt{1 - \frac{1}{3}\sin^2\theta}$	0.9286	0.8969	0.8660

From equation (5):

$$\int_0^{\frac{\pi}{3}} \sqrt{1 - \frac{1}{3}\sin^2\theta}\, d\theta$$

$$\approx \frac{1}{3}\left(\frac{\pi}{18}\right)[(1.0000 + 0.8660) + 4(0.9950 + 0.9574 + 0.8969) + 2(0.9803 + 0.9286)]$$

$$= \frac{1}{3}\left(\frac{\pi}{18}\right)[1.8660 + 11.3972 + 3.8178]$$

$$= \mathbf{0.994}, \text{ correct to 3 decimal places.}$$

Problem 8. An alternating current i has the following values at equal intervals of 2.0 milliseconds:

Time (ms)	0	2.0	4.0	6.0	8.0	10.0	12.0
Current i (A)	0	3.5	8.2	10.0	7.3	2.0	0

Charge, q, in millicoulombs, is given by $q = \int_0^{12.0} i\, dt$. Use Simpson's rule to determine the approximate charge in the 12 ms period

From equation (5):

$$\text{Charge, } q = \int_0^{12.0} i\, dt$$

$$\approx \frac{1}{3}(2.0)[(0 + 0) + 4(3.5 + 10.0 + 2.0) + 2(8.2 + 7.3)]$$

$$= \mathbf{62\ mC}$$

Now try the following Practice Exercise

Practice Exercise 238 Simpson's rule (Answers on page 728)

In Problems 1 to 5, evaluate the definite integrals using **Simpson's rule**, giving the answers correct to 3 decimal places.

1. $\int_0^{\pi/2} \sqrt{\sin x}\, dx$ (Use 6 intervals)
2. $\int_0^{1.6} \frac{1}{1 + \theta^4}\, d\theta$ (Use 8 intervals)
3. $\int_{0.2}^{1.0} \frac{\sin\theta}{\theta}\, d\theta$ (Use 8 intervals)
4. $\int_0^{\pi/2} x\cos x\, dx$ (Use 6 intervals)
5. $\int_0^{\pi/3} e^{x^2}\sin 2x\, dx$ (Use 10 intervals)

In Problems 6 and 7 evaluate the definite integrals using (a) integration, (b) the trapezoidal rule, (c) the mid-ordinate rule, (d) Simpson's rule. Give answers correct to 3 decimal places.

6. $\int_1^4 \frac{4}{x^3}\, dx$ (Use 6 intervals)

7. $\int_2^6 \frac{1}{\sqrt{2x-1}}\,dx$ (Use 8 intervals)

In Problems 8 and 9 evaluate the definite integrals using (a) the trapezoidal rule, (b) the mid-ordinate rule, (c) Simpson's rule. Use 6 intervals in each case and give answers correct to 3 decimal places.

8. $\int_0^3 \sqrt{1+x^4}\,dx$

9. $\int_{0.1}^{0.7} \frac{1}{\sqrt{1-y^2}}\,dy$

10. A vehicle starts from rest and its velocity is measured every second for 8 seconds, with values as follows:

time t (s)	velocity v (ms^{-1})
0	0
1.0	0.4
2.0	1.0
3.0	1.7
4.0	2.9
5.0	4.1
6.0	6.2
7.0	8.0
8.0	9.4

The distance travelled in 8.0 seconds is given by $\int_0^{8.0} v\,dt$

Estimate this distance using Simpson's rule, giving the answer correct to 3 significant figures

11. A pin moves along a straight guide so that its velocity v (m/s) when it is a distance x (m) from the beginning of the guide at time t (s) is given in the table below:

t (s)	v (m/s)
0	0
0.5	0.052
1.0	0.082
1.5	0.125
2.0	0.162
2.5	0.175
3.0	0.186
3.5	0.160
4.0	0

Use Simpson's rule with 8 intervals to determine the approximate total distance travelled by the pin in the 4.0 second period

50.5 Accuracy of numerical integration

For a function with an increasing gradient, the trapezoidal rule will tend to over-estimate and the mid-ordinate rule will tend to under-estimate (but by half as much). The appropriate combination of the two in Simpson's rule eliminates this error term, giving a rule which will perfectly model anything up to a cubic, and have a proportionately lower error for any function of greater complexity.

In general, for a given number of strips, Simpson's rule is considered the most accurate of the three numerical methods.

Practice Exercise 239 Multiple-choice questions on numerical integration (Answers on page 728)

Each question has only one correct answer

1. An alternating current i has the following values at equal intervals of 2 ms:

Time t (ms)	0	2.0	4.0	6.0	8.0	10.0	12.0
Current $I(A)$	0	4.6	7.4	10.8	8.5	3.7	0

   Charge q (in millicoulombs) is given by $q = \int_0^{12.0} i\,dt$. Using the trapezoidal rule, the approximate charge in the 12 ms period is:
   (a) 70 mC (b) 72.1 mC
   (c) 35 mC (d) 216.4 mC

2. Using the mid-ordinate rule with 8 intervals, the value of the integral $\int_0^4 \left(\frac{5}{1+x^2}\right) dx$, correct to 3 decimal places is:
   (a) 13.258 (b) 6.630 (c) 9.000 (d) 4.276

3. Using Simpson's rule with 8 intervals, the value of $\int_0^8 \ln\left(x^2+3\right) dx$, correct to 4 significant figures is:
   (a) 25.88 (b) 67.03 (c) 22.34 (d) 75.04

For fully worked solutions to each of the problems in Practice Exercises 236 to 238 in this chapter, go to the website:
www.routledge.com/cw/bird

COMPANION @ WEBSITE

Revision Test 13 Further integration

This Revision Test covers the material contained in Chapters 47 to 50. *The marks for each question are shown in brackets at the end of each question.*

1. Determine: (a) $\int \frac{x-11}{x^2-x-2}dx$

   (b) $\int \frac{3-x}{(x^2+3)(x+3)}dx$ (21)

2. Evaluate: $\int_1^2 \frac{3}{x^2(x+2)}dx$ correct to 4 significant figures. (11)

3. Determine: $\int \frac{dx}{2\sin x+\cos x}$ (7)

4. Determine the following integrals:

   (a) $\int 5xe^{2x}dx$ (b) $\int t^2 \sin 2t\,dt$ (12)

5. Evaluate correct to 3 decimal places:

   $\int_1^4 \sqrt{x}\ln x\,dx$ (9)

6. Evaluate: $\int_1^3 \frac{5}{x^2}dx$ using

   (a) integration

   (b) the trapezoidal rule

   (c) the mid-ordinate rule

   (d) Simpson's rule.

   In each of the approximate methods use 8 intervals and give the answers correct to 3 decimal places. (16)

7. An alternating current i has the following values at equal intervals of 5 ms:

Time t (ms)	0	5	10	15	20	25	30
Current i (A)	0	4.8	9.1	12.7	8.8	3.5	0

   Charge q, in coulombs, is given by

$$q = \int_0^{30\times 10^{-3}} i\,dt$$

   Use Simpson's rule to determine the approximate charge in the 30 ms period. (4)

For lecturers/instructors/teachers, fully worked solutions to each of the problems in Revision Test 13, together with a full marking scheme, are available at the website:

www.routledge.com/cw/bird

Chapter 51

Areas under and between curves

Why it is important to understand: Areas under and between curves

One of the important applications of integration is to find the area bounded by a curve. Often such an area can have a physical significance like the work done by a motor or the distance travelled by a vehicle. Other examples where finding the area under a curve is important can involve position, velocity, force, charge density, resistivity and current density. Hence, finding the area under and between curves is of some importance in engineering and science.

At the end of this chapter, you should be able to:

- appreciate that the area under a curve of a known function is a definite integral
- state practical examples where areas under a curve need to be accurately determined
- calculate areas under curves using integration
- calculate areas between curves using integration

51.1 Area under a curve

The area shown shaded in Fig. 51.1 may be determined using approximate methods (such as the trapezoidal rule, the mid-ordinate rule or Simpson's rule) or, more precisely, by using integration.

(i) Let A be the area shown shaded in Fig. 51.1 and let this area be divided into a number of strips each of width δx. One such strip is shown and let the area of this strip be δA.

Then: $\delta A \approx y\delta x$ (1)

The accuracy of statement (1) increases when the width of each strip is reduced, i.e. area A is divided into a greater number of strips.

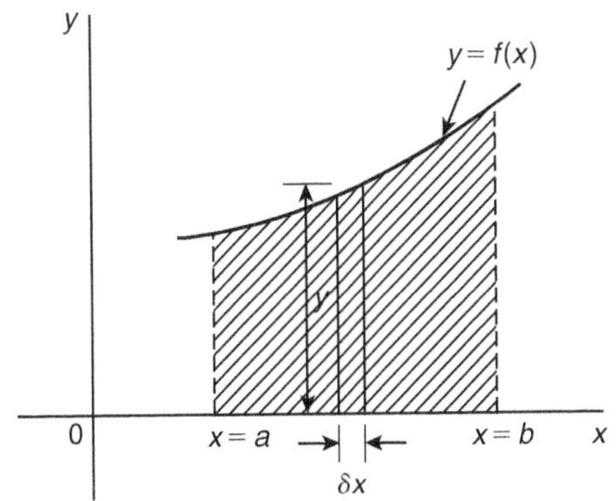

Figure 51.1

(ii) Area A is equal to the sum of all the strips from $x=a$ to $x=b$, i.e. $A = \lim_{\delta x \to 0} \sum_{x=a}^{x=b} y\,\delta x$ (2)

(iii) From statement (1), $\frac{\delta A}{\delta x} \approx y$ (3)

In the limit, as δx approaches zero, $\frac{\delta A}{\delta x}$ becomes the differential coefficient $\frac{dA}{dx}$

Hence $\lim_{\delta x \to 0} \left(\frac{\delta A}{\delta x}\right) = \frac{dA}{dx} = y$, from statement (3).

By integration,

$$\int \frac{dA}{dx} dx = \int y\,dx \quad \text{i.e.} \quad A = \int y\,dx$$

The ordinates $x=a$ and $x=b$ limit the area and such ordinate values are shown as limits. Hence

$$A = \int_a^b y\,dx \qquad (4)$$

(iv) Equating statements (2) and (4) gives:

$$\textbf{Area } A = \lim_{\delta x \to 0} \sum_{x=a}^{x=b} y\,\delta x = \int_a^b y\,dx$$

$$= \int_a^b f(x)\,dx$$

(v) If the area between a curve $x=f(y)$, the y-axis and ordinates $y=p$ and $y=q$ is required, then

$$\text{area} = \int_p^q x\,dy$$

Thus, determining the area under a curve by integration merely involves evaluating a definite integral.
There are several instances in engineering and science where the area beneath a curve needs to be accurately determined. For example, **the areas between limits of a:**

velocity/time graph gives distance travelled,
force/distance graph gives work done,
voltage/current graph gives power and so on.

Should a curve drop below the x-axis, then $y\ (=f(x))$ becomes negative and $f(x)\,dx$ is negative. When determining such areas by integration, a negative sign is placed before the integral. For the curve shown in Fig. 51.2, the total shaded area is given by (area $E+$ area $F+$area G).
By integration, **total shaded area**

$$= \int_a^b f(x)\,dx - \int_b^c f(x)\,dx + \int_c^d f(x)\,dx$$

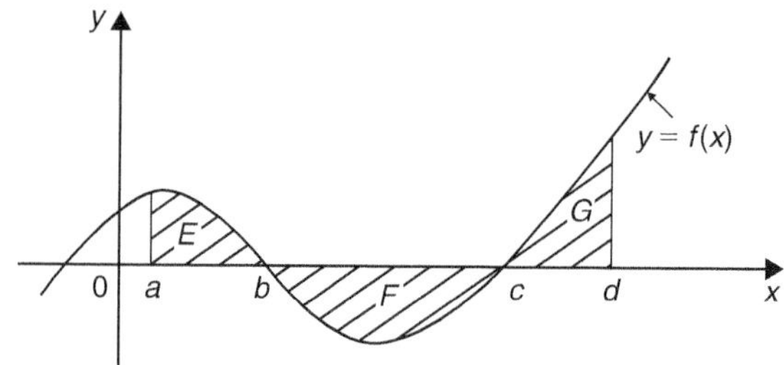

Figure 51.2

(Note that this is **not** the same as $\int_a^d f(x)\,dx$)
It is usually necessary to sketch a curve in order to check whether it crosses the x-axis.

51.2 Worked problems on the area under a curve

Problem 1. Determine the area enclosed by $y=2x+3$, the x-axis and ordinates $x=1$ and $x=4$

$y=2x+3$ is a straight line graph as shown in Fig. 51.3, where the required area is shown shaded.

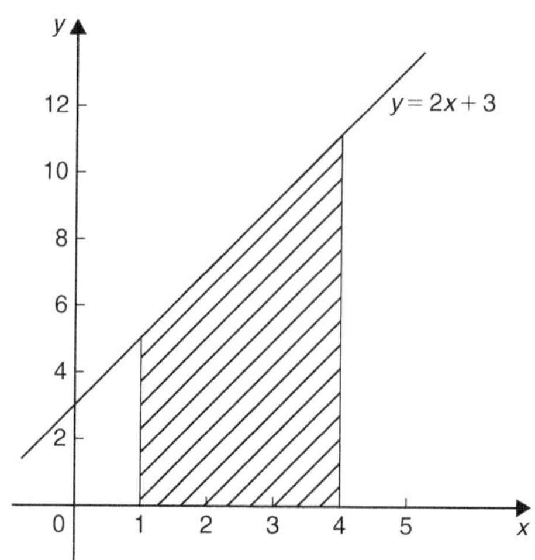

Figure 51.3

By integration,

$$\text{shaded area} = \int_1^4 y\,dx$$

$$= \int_1^4 (2x+3)\,dx$$

$$= \left[\frac{2x^2}{2} + 3x\right]_1^4$$

$$= [(16 + 12) - (1 + 3)]$$
$$= \textbf{24 square units}$$

[This answer may be checked since the shaded area is a trapezium.

Area of trapezium

$$= \frac{1}{2}\begin{pmatrix}\text{sum of parallel}\\ \text{sides}\end{pmatrix}\begin{pmatrix}\text{perpendicular distance}\\ \text{between parallel sides}\end{pmatrix}$$
$$= \frac{1}{2}(5 + 11)(3)$$
$$= \textbf{24 square units}]$$

Problem 2. The velocity v of a body t seconds after a certain instant is: $(2t^2 + 5)$ m/s. Find by integration how far it moves in the interval from $t = 0$ to $t = 4$ s

Since $2t^2 + 5$ is a quadratic expression, the curve $v = 2t^2 + 5$ is a parabola cutting the v-axis at $v = 5$, as shown in Fig. 51.4.
The distance travelled is given by the area under the v/t curve (shown shaded in Fig. 51.4).

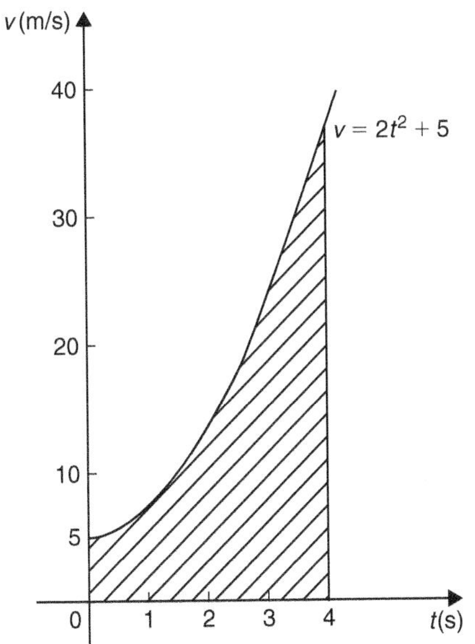

Figure 51.4

By integration,

$$\text{shaded area} = \int_0^4 v\,dt$$
$$= \int_0^4 (2t^2 + 5)\,dt$$
$$= \left[\frac{2t^3}{3} + 5t\right]_0^4$$
$$= \left(\frac{2(4^3)}{3} + 5(4)\right) - (0)$$

i.e. **distance travelled = 62.67 m**

Problem 3. Sketch the graph $y = x^3 + 2x^2 - 5x - 6$ between $x = -3$ and $x = 2$ and determine the area enclosed by the curve and the x-axis

A table of values is produced and the graph sketched as shown in Fig. 51.5 where the area enclosed by the curve and the x-axis is shown shaded.

x	-3	-2	-1	0	1	2
x^3	-27	-8	-1	0	1	8
$2x^2$	18	8	2	0	2	8
$-5x$	15	10	5	0	-5	-10
-6	-6	-6	-6	-6	-6	-6
y	0	4	0	-6	-8	0

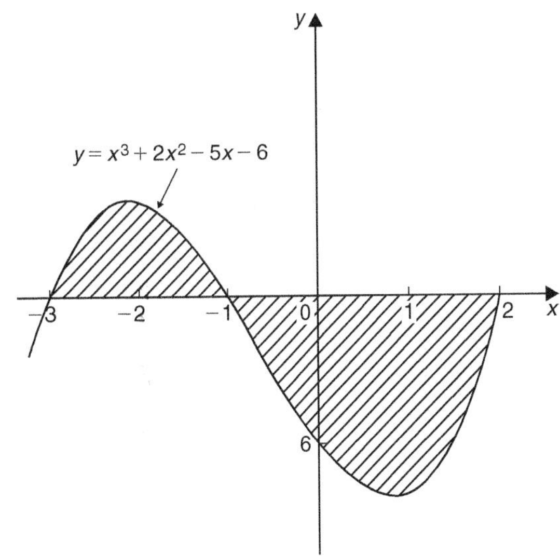

Figure 51.5

Shaded area $= \int_{-3}^{-1} y\,dx - \int_{-1}^{2} y\,dx$, the minus sign before the second integral being necessary since the enclosed area is below the x-axis.

Hence shaded area

$$= \int_{-3}^{-1} (x^3 + 2x^2 - 5x - 6)\, dx - \int_{-1}^{2} (x^3 + 2x^2 - 5x - 6)\, dx$$

$$= \left[\frac{x^4}{4} + \frac{2x^3}{3} - \frac{5x^2}{2} - 6x\right]_{-3}^{-1} - \left[\frac{x^4}{4} + \frac{2x^3}{3} - \frac{5x^2}{2} - 6x\right]_{-1}^{2}$$

$$= \left[\left\{\frac{1}{4} - \frac{2}{3} - \frac{5}{2} + 6\right\} - \left\{\frac{81}{4} - 18 - \frac{45}{2} + 18\right\}\right] - \left[\left\{4 + \frac{16}{3} - 10 - 12\right\} - \left\{\frac{1}{4} - \frac{2}{3} - \frac{5}{2} + 6\right\}\right]$$

$$= \left[\left\{3\frac{1}{12}\right\} - \left\{-2\frac{1}{4}\right\}\right] - \left[\left\{-12\frac{2}{3}\right\} - \left\{3\frac{1}{12}\right\}\right]$$

$$= \left[5\frac{1}{3}\right] - \left[-15\frac{3}{4}\right]$$

$$= \mathbf{21\frac{1}{12}} \quad \text{or} \quad \mathbf{21.08\ square\ units}$$

Problem 4. Determine the area enclosed by the curve $y = 3x^2 + 4$, the x-axis and ordinates $x = 1$ and $x = 4$ by (a) the trapezoidal rule, (b) the mid-ordinate rule, (c) Simpson's rule and (d) integration

The curve $y = 3x^2 + 4$ is shown plotted in Fig. 51.6.

(a) **By the trapezoidal rule,**

$$\text{Area} = \begin{pmatrix}\text{width of}\\ \text{interval}\end{pmatrix}\left[\frac{1}{2}\begin{pmatrix}\text{first + last}\\ \text{ordinate}\end{pmatrix} + \begin{pmatrix}\text{sum of}\\ \text{remaining}\\ \text{ordinates}\end{pmatrix}\right]$$

Selecting 6 intervals each of width 0.5 gives:

$$\text{Area} = (0.5)\left[\frac{1}{2}(7 + 52) + 10.75 + 16 + 22.75 + 31 + 40.75\right]$$

$$= \mathbf{75.375\ square\ units}$$

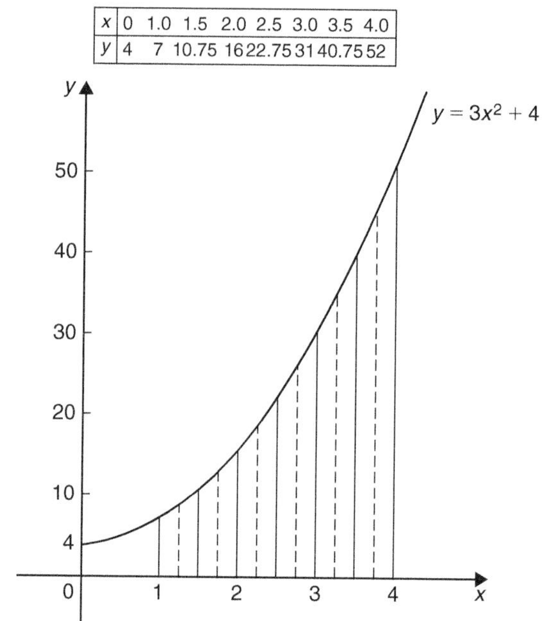

x	0	1.0	1.5	2.0	2.5	3.0	3.5	4.0
y	4	7	10.75	16	22.75	31	40.75	52

Figure 51.6

(b) **By the mid-ordinate rule,**
area = (width of interval) (sum of mid-ordinates).
Selecting 6 intervals, each of width 0.5 gives the mid-ordinates as shown by the broken lines in Fig. 51.6.

$$\text{Thus, area} = (0.5)(8.5 + 13 + 19 + 26.5 + 35.5 + 46)$$

$$= \mathbf{74.25\ square\ units}$$

(c) **By Simpson's rule,**

$$\text{area} = \frac{1}{3}\begin{pmatrix}\text{width of}\\ \text{interval}\end{pmatrix}\left[\begin{pmatrix}\text{first + last}\\ \text{ordinates}\end{pmatrix} + 4\begin{pmatrix}\text{sum of even}\\ \text{ordinates}\end{pmatrix} + 2\begin{pmatrix}\text{sum of remaining}\\ \text{odd ordinates}\end{pmatrix}\right]$$

Selecting 6 intervals, each of width 0.5, gives:

$$\text{area} = \frac{1}{3}(0.5)[(7 + 52) + 4(10.75 + 22.75 + 40.75) + 2(16 + 31)]$$

$$= \mathbf{75\ square\ units}$$

(d) **By integration**, shaded area

$$= \int_1^4 y\, dx$$

$$= \int_1^4 (3x^2 + 4)\, dx$$

$$= \left[x^3 + 4x\right]_1^4$$

$= \mathbf{75\ square\ units}$

Integration gives the precise value for the area under a curve. In this case Simpson's rule is seen to be the most accurate of the three approximate methods.

Problem 5. Find the area enclosed by the curve $y = \sin 2x$, the x-axis and the ordinates $x=0$ and $x=\pi/3$

A sketch of $y = \sin 2x$ is shown in Fig. 51.7.

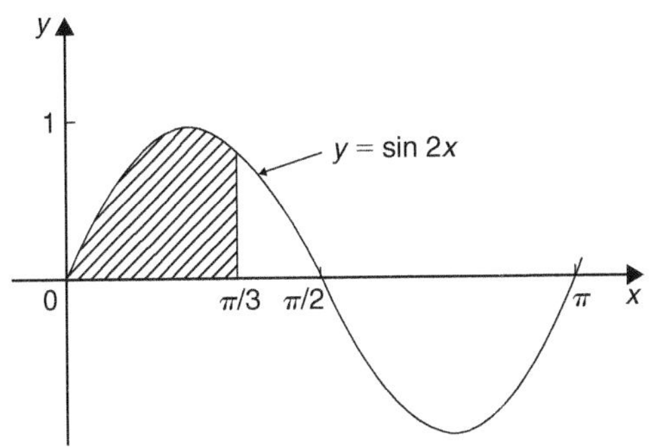

Figure 51.7

(Note that $y = \sin 2x$ has a period of $\frac{2\pi}{2}$, i.e. π radians.)

$$\text{Shaded area} = \int_0^{\pi/3} y\,dx$$

$$= \int_0^{\pi/3} \sin 2x\,dx$$

$$= \left[-\frac{1}{2}\cos 2x\right]_0^{\pi/3}$$

$$= \left\{-\frac{1}{2}\cos\frac{2\pi}{3}\right\} - \left\{-\frac{1}{2}\cos 0\right\}$$

$$= \left\{-\frac{1}{2}\left(-\frac{1}{2}\right)\right\} - \left\{-\frac{1}{2}(1)\right\}$$

$$= \frac{1}{4} + \frac{1}{2} = \frac{3}{4}\ \textbf{square units}$$

Now try the following Practice Exercise

Practice Exercise 240 Areas under curves (Answers on page 728)

Unless otherwise stated all answers are in square units.

1. Shown by integration that the area of the triangle formed by the line $y=2x$, the ordinates $x=0$ and $x=4$ and the x-axis is 16 square units.

2. Sketch the curve $y=3x^2+1$ between $x=-2$ and $x=4$. Determine by integration the area enclosed by the curve, the x-axis and ordinates $x=-1$ and $x=3$. Use an approximate method to find the area and compare your result with that obtained by integration.

In Problems 3 to 8, find the area enclosed between the given curves, the horizontal axis and the given ordinates.

3. $y=5x;\quad x=1, x=4$

4. $y=2x^2-x+1;\quad x=-1, x=2$

5. $y=2\sin 2\theta;\quad \theta=0, \theta=\frac{\pi}{4}$

6. $\theta=t+e^t;\quad t=0, t=2$

7. $y=5\cos 3t;\quad t=0, t=\frac{\pi}{6}$

8. $y=(x-1)(x-3);\quad x=0, x=3$

51.3 Further worked problems on the area under a curve

Problem 6. A gas expands according to the law $pv=$constant. When the volume is 3 m^3 the pressure is 150 kPa. Given that work done $= \int_{v_1}^{v_2} p\,dv$, determine the work done as the gas expands from 2 m^3 to a volume of 6 m^3

$pv=$constant. When $v=3$ m^3 and $p=150$ kPa the constant is given by $(3\times 150)=450$ kPa m^3 or 450 kJ.

Hence $pv=450$, or $p=\frac{450}{v}$

$$\text{Work done} = \int_2^6 \frac{450}{v}\,dv$$

$$= \left[450\ln v\right]_2^6 = 450[\ln 6 - \ln 2]$$

$$= 450\ln\frac{6}{2} = 450\ln 3 = \mathbf{494.4\ kJ}$$

Problem 7. Determine the area enclosed by the curve $y=4\cos\left(\frac{\theta}{2}\right)$, the θ-axis and ordinates $\theta=0$ and $\theta=\frac{\pi}{2}$

The curve $y=4\cos(\theta/2)$ is shown in Fig. 51.8.

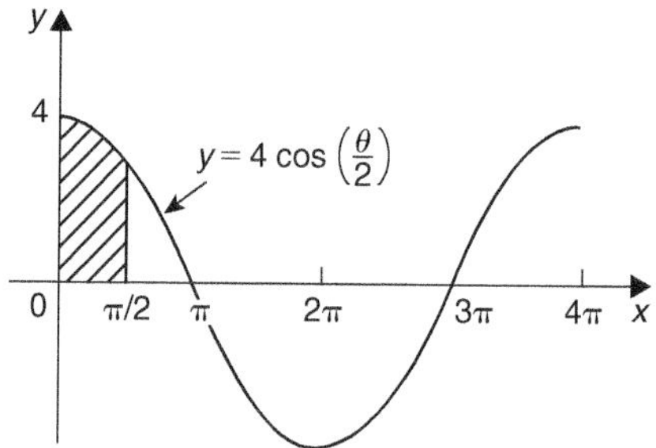

Figure 51.8

(Note that $y=4\cos\left(\frac{\theta}{2}\right)$ has a maximum value of 4 and period $2\pi/(1/2)$, i.e. 4π rads.)

$$\text{Shaded area} = \int_0^{\pi/2} y\,d\theta = \int_0^{\pi/2} 4\cos\frac{\theta}{2}\,d\theta$$

$$= \left[4\left(\frac{1}{\frac{1}{2}}\right)\sin\frac{\theta}{2}\right]_0^{\pi/2}$$

$$= \left(8\sin\frac{\pi}{4}\right) - (8\sin 0)$$

$$= \textbf{5.657 square units}$$

Problem 8. Determine the area bounded by the curve $y=3e^{t/4}$, the t-axis and ordinates $t=-1$ and $t=4$, correct to 4 significant figures

A table of values is produced as shown.

t	−1	0	1	2	3	4
$y=3e^{t/4}$	2.34	3.0	3.85	4.95	6.35	8.15

Since all the values of y are positive the area required is wholly above the t-axis.

$$\text{Hence area} = \int_1^4 y\,dt$$

$$= \int_1^4 3e^{t/4}dt = \left[\frac{3}{\left(\frac{1}{4}\right)}e^{t/4}\right]_{-1}^4$$

$$= 12\left[e^{t/4}\right]_{-1}^4 = 12(e^1 - e^{-1/4})$$

$$= 12(2.7183 - 0.7788)$$

$$= 12(1.9395) = \textbf{23.27 square units}$$

Problem 9. Sketch the curve $y=x^2+5$ between $x=-1$ and $x=4$. Find the area enclosed by the curve, the x-axis and the ordinates $x=0$ and $x=3$. Determine also, by integration, the area enclosed by the curve and the y-axis, between the same limits

A table of values is produced and the curve $y=x^2+5$ plotted as shown in Fig. 51.9.

x	−1	0	1	2	3
y	6	5	6	9	14

$$\text{Shaded area} = \int_0^3 y\,dx = \int_0^3 (x^2+5)\,dx$$

$$= \left[\frac{x^3}{5} + 5x\right]_0^3$$

$$= \textbf{24 square units}$$

When $x=3$, $y=3^2+5=14$, and when $x=0$, $y=5$.

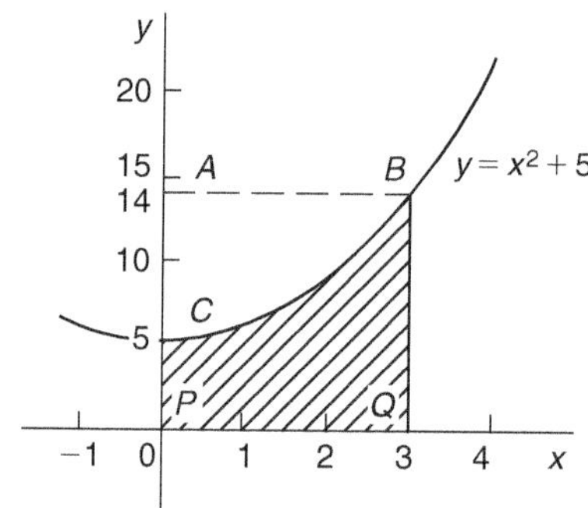

Figure 51.9

Since $y=x^2+5$ then $x^2=y-5$ and $x=\sqrt{y-5}$

The area enclosed by the curve $y=x^2+5$ (i.e. $x=\sqrt{y-5}$), the y-axis and the ordinates $y=5$ and $y=14$ (i.e. area ABC of Fig. 51.9) is given by:

$$\text{Area} = \int_{y=5}^{y=14} x\,dy = \int_5^{14} \sqrt{y-5}\,dy$$

$$= \int_5^{14} (y-5)^{1/2}\,dy$$

Let $u=y-5$, then $\frac{du}{dy}=1$ and $dy=du$

Hence $\int (y-5)^{1/2}dy=\int u^{1/2}du=\frac{2}{3}u^{3/2}$
(from algebraic substitutions, see Chapter 45)

Since $u=y-5$ then

$$\int_5^{14}\sqrt{y-5}\,dy=\frac{2}{3}\left[(y-5)^{3/2}\right]_5^{14}$$
$$=\frac{2}{3}[\sqrt{9^3}-0]$$
$$=\textbf{18 square units}$$

(Check: From Fig. 51.9, area $BCPQ$ + area $ABC=24+18=42$ square units, which is the area of rectangle $ABQP$)

Problem 10. Determine the area between the curve $y=x^3-2x^2-8x$ and the x-axis

$$y=x^3-2x^2-8x=x(x^2-2x-8)$$
$$=x(x+2)(x-4)$$

When $y=0$, then $x=0$ or $(x+2)=0$ or $(x-4)=0$, i.e. when $y=0$, $x=0$ or -2 or 4, which means that the curve crosses the x-axis at 0, -2 and 4. Since the curve is a continuous function, only one other co-ordinate value needs to be calculated before a sketch of the curve can be produced. When $x=1$, $y=-9$, showing that the part of the curve between $x=0$ and $x=4$ is negative. A sketch of $y=x^3-2x^2-8x$ is shown in Fig. 51.10. (Another method of sketching Fig. 51.10 would have been to draw up a table of values.)

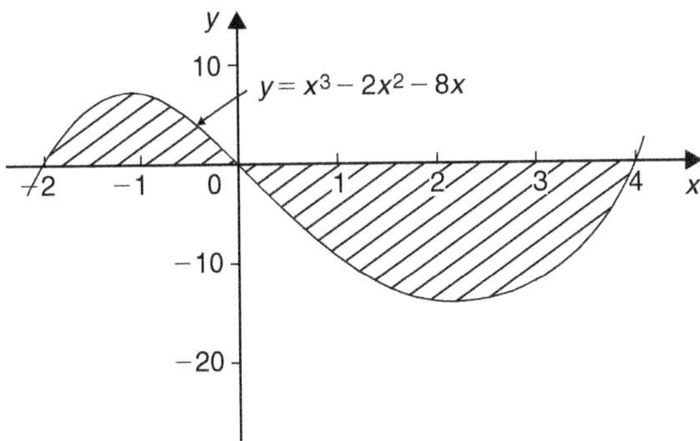

Figure 51.10

$$\text{Shaded area}=\int_{-2}^{0}(x^3-2x^2-8x)\,dx-\int_0^4(x^3-2x^2-8x)\,dx$$

$$=\left[\frac{x^4}{4}-\frac{2x^3}{3}-\frac{8x^2}{2}\right]_{-2}^{0}-\left[\frac{x^4}{4}-\frac{2x^3}{3}-\frac{8x^2}{2}\right]_0^4$$

$$=\left(6\frac{2}{3}\right)-\left(-42\frac{2}{3}\right)$$

$$=\mathbf{49\frac{1}{3}}\textbf{ square units}$$

Now try the following Practice Exercise

Practice Exercise 241 Areas under curves (Answers on page 728)

In Problems 1 and 2, find the area enclosed between the given curves, the horizontal axis and the given ordinates.

1. $y=2x^3$; $x=-2, x=2$
2. $xy=4$; $x=1, x=4$
3. The force F newtons acting on a body at a distance x metres from a fixed point is given by: $F=3x+2x^2$. If work done $=\int_{x_1}^{x_2}F\,dx$, determine the work done when the body moves from the position where $x=1$ m to that where $x=3$ m.
4. Find the area between the curve $y=4x-x^2$ and the x-axis.
5. Determine the area enclosed by the curve $y=5x^2+2$, the x-axis and the ordinates $x=0$ and $x=3$. Find also the area enclosed by the curve and the y-axis between the same limits.
6. Calculate the area enclosed between $y=x^3-4x^2-5x$ and the x-axis.
7. The velocity v of a vehicle t seconds after a certain instant is given by: $v=(3t^2+4)$ m/s. Determine how far it moves in the interval from $t=1$ s to $t=5$ s.
8. A gas expands according to the law $pv=$ constant. When the volume is 2 m^3 the pressure is 250 kPa. Find the work done as the gas expands from 1 m^3 to a volume of 4 m^3 given that work done $=\int_{v_1}^{v_2}p\,dv$

51.4 The area between curves

The area enclosed between curves $y=f_1(x)$ and $y=f_2(x)$ (shown shaded in Fig. 51.11) is given by:

$$\text{shaded area} = \int_a^b f_2(x)dx - \int_a^b f_1(x)dx$$

$$= \int_a^b [f_2(x) - f_2(x)]dx$$

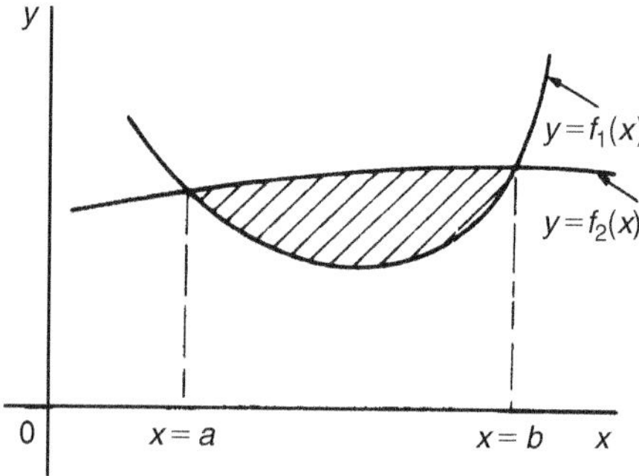

Figure 51.11

Problem 11. Determine the area enclosed between the curves $y=x^2+1$ and $y=7-x$

At the points of intersection, the curves are equal. Thus, equating the y-values of each curve gives: $x^2+1=7-x$, from which $x^2+x-6=0$. Factorising gives $(x-2)(x+3)=0$, from which, $x=2$ and $x=-3$. By firstly determining the points of intersection the range of x-values has been found. Tables of values are produced as shown below.

x	-3	-2	-1	0	1	2
$y=x^2+1$	10	5	2	1	2	5

x	-3	0	2
$y=7-x$	10	7	5

A sketch of the two curves is shown in Fig. 51.12.

$$\text{Shaded area} = \int_{-3}^{2} (7-x)dx - \int_{-3}^{2} (x^2+1)dx$$

$$= \int_{-3}^{2} [(7-x)-(x^2+1)]dx$$

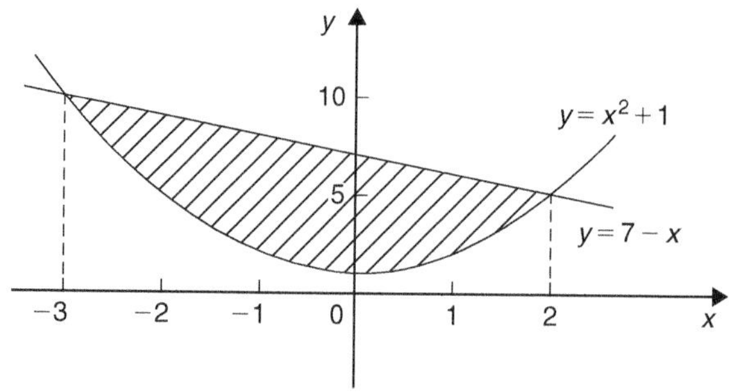

Figure 51.12

$$= \int_{-3}^{2} (6-x-x^2)dx$$

$$= \left[6x - \frac{x^2}{2} - \frac{x^3}{3}\right]_{-3}^{2}$$

$$= \left(12-2-\frac{8}{3}\right) - \left(-18-\frac{9}{2}+9\right)$$

$$= \left(7\frac{1}{3}\right) - \left(-13\frac{1}{2}\right)$$

$$= \mathbf{20\frac{5}{6}} \textbf{ square units}$$

Problem 12. (a) Determine the co-ordinates of the points of intersection of the curves $y=x^2$ and $y^2=8x$. (b) Sketch the curves $y=x^2$ and $y^2=8x$ on the same axes. (c) Calculate the area enclosed by the two curves

(a) At the points of intersection the co-ordinates of the curves are equal. When $y=x^2$ then $y^2=x^4$.

Hence at the points of intersection $x^4=8x$, by equating the y^2 values.

Thus $x^4-8x=0$, from which $x(x^3-8)=0$, i.e. $x=0$ or $(x^3-8)=0$

Hence at the points of intersection $x=0$ or $x=2$

When $x=0$, $y=0$ and when $x=2$, $y=2^2=4$

Hence the points of intersection of the curves $y=x^2$ and $y^2=8x$ are (0, 0) and (2, 4).

(b) A sketch of $y=x^2$ and $y^2=8x$ is shown in Fig. 51.13

(c) **Shaded area** $= \int_0^2 \{\sqrt{8x}-x^2\}dx$

$$= \int_0^2 \{(\sqrt{8})x^{1/2}-x^2\}dx$$

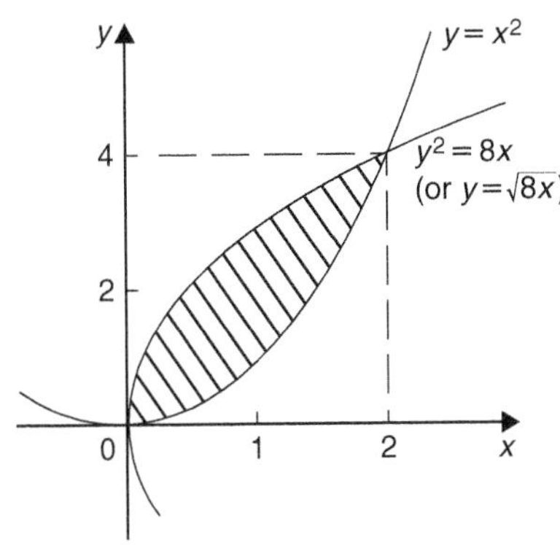

Figure 51.13

$$= \left[(\sqrt{8}) \frac{x^{3/2}}{\left(\frac{3}{2}\right)} - \frac{x^3}{3} \right]_0^2$$

$$= \left\{ \frac{\sqrt{8}\sqrt{8}}{\left(\frac{3}{2}\right)} - \frac{8}{3} \right\} - \{0\}$$

$$= \frac{16}{3} - \frac{8}{3} = \frac{8}{3}$$

$$= \mathbf{2\frac{2}{3}} \textbf{ square units}$$

Problem 13. Determine by integration the area bounded by the three straight lines $y = 4 - x$, $y = 3x$ and $3y = x$

Each of the straight lines is shown sketched in Fig. 51.14.

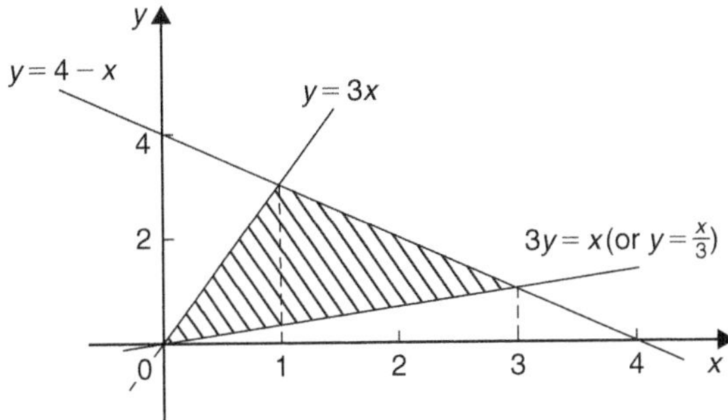

Figure 51.14

$$\text{Shaded area} = \int_0^1 \left(3x - \frac{x}{3}\right) dx + \int_1^3 \left[(4 - x) - \frac{x}{3}\right] dx$$

$$= \left[\frac{3x^2}{2} - \frac{x^2}{6} \right]_0^1 + \left[4x - \frac{x^2}{2} - \frac{x^2}{6} \right]_1^3$$

$$= \left[\left(\frac{3}{2} - \frac{1}{6} \right) - (0) \right]$$

$$+ \left[\left(12 - \frac{9}{2} - \frac{9}{6} \right) - \left(4 - \frac{1}{2} - \frac{1}{6} \right) \right]$$

$$= \left(1\frac{1}{3} \right) + \left(6 - 3\frac{1}{3} \right)$$

$$= \textbf{4 square units}$$

Now try the following Practice Exercise

Practice Exercise 242 Areas between curves (Answers on page 728)

1. Determine the co-ordinates of the points of intersection and the area enclosed between the parabolas $y^2 = 3x$ and $x^2 = 3y$
2. Sketch the curves $y = x^2 + 3$ and $y = 7 - 3x$ and determine the area enclosed by them.
3. Determine the area enclosed by the curves $y = \sin x$ and $y = \cos x$ and the y-axis.
4. Determine the area enclosed by the three straight lines $y = 3x$, $2y = x$ and $y + 2x = 5$

Practice Exercise 243 Multiple-choice questions on areas under and between curves (Answers on page 728)

Each question has only one correct answer

1. The area, in square units, enclosed by the curve $y = 2x + 3$, the x-axis and ordinates $x = 1$ and $x = 4$ is:
   (a) 28 (b) 2 (c) 24 (d) 39
2. The area enclosed by the curve $y = 3\cos 2\theta$, the ordinates $\theta = 0$ and $\theta = \frac{\pi}{4}$ and the θ axis is:
   (a) −3 (b) 6 (c) 1.5 (d) 3

3. The area enclosed between the curve $y = x^2$, the x-axis, and the lines $x = 1$ and $x = 4$ is:
(a) 63 square units (b) 21 square units
(c) 27 square units (d) 9 square units

4. The area under a force/distance graph gives the work done. The shaded area shown between p and q in Figure 51.15 is:
(a) $c\ln\frac{q}{p}$ (b) $-\frac{c}{2}\left(\frac{1}{q^2}-\frac{1}{p^2}\right)$
(c) $\frac{c}{2}(\ln q - \ln p)$ (d) $c(\ln p - \ln q)$

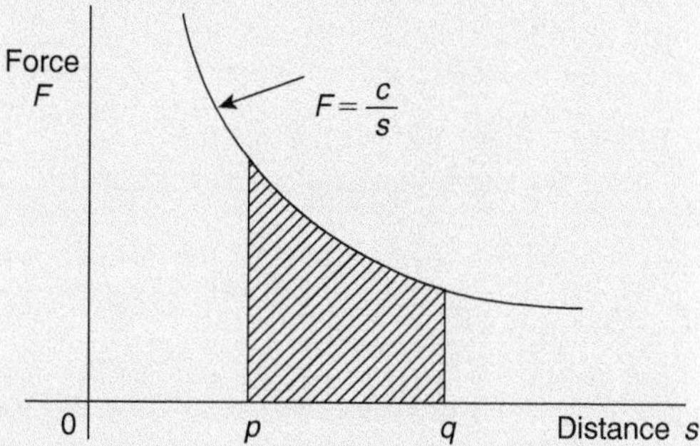

Figure 51.15

5. A vehicle has a velocity $v = (2 + 3t)$ m/s after t seconds. The distance travelled is equal to the area under the v/t graph. In the first 3 seconds the vehicle has travelled:
(a) 11 m (b) 33 m (c) 13.5 m (d) 19.5 m

For fully worked solutions to each of the problems in Practice Exercises 240 to 242 in this chapter, go to the website:
www.routledge.com/cw/bird

Chapter 52

Mean and root mean square values

Why it is important to understand: Mean and root mean square values

Electrical currents and voltages often vary with time and engineers may wish to know the average or mean value of such a current or voltage over some particular time interval. The mean value of a time-varying function is defined in terms of an integral. An associated quantity is the root mean square (r.m.s.) value of a current which is used, for example, in the calculation of the power dissipated by a resistor. Mean and r.m.s. values are required with alternating currents and voltages, pressure of sound waves and much more.

At the end of this chapter, you should be able to:

- determine the mean or average value of a function over a given range using integration
- define an r.m.s. value
- determine the r.m.s. value of a function over a given range using integration

52.1 Mean or average values

(i) The mean or average value of the curve shown in Fig. 52.1, between $x=a$ and $x=b$, is given by:

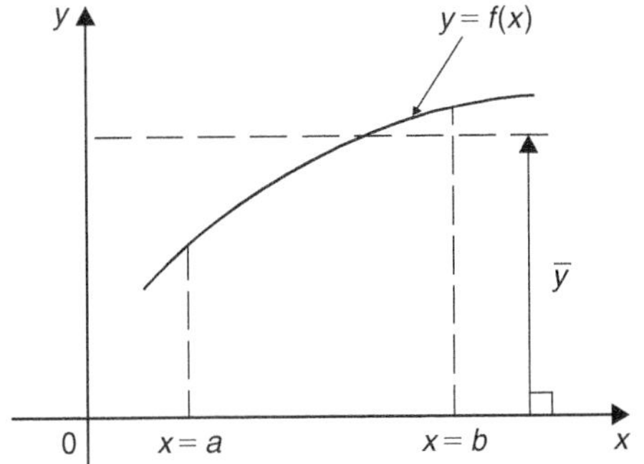

Figure 52.1

mean or average value,

$$\bar{y} = \frac{\textbf{area under curve}}{\textbf{length of base}}$$

(ii) When the area under a curve may be obtained by integration then:
mean or average value,

$$\bar{y} = \frac{\int_a^b y\,dx}{b-a}$$

i.e. $$\bar{y} = \frac{1}{b-a}\int_a^b f(x)\,dx$$

(iii) For a periodic function, such as a sine wave, the mean value is assumed to be 'the mean value over half a cycle', since the mean value over a complete cycle is zero.

Problem 1. Determine, using integration, the mean value of $y=5x^2$ between $x=1$ and $x=4$

Mean value,

$$\bar{y}=\frac{1}{4-1}\int_1^4 y\,dx=\frac{1}{3}\int_1^4 5x^2\,dx$$

$$=\frac{1}{3}\left[\frac{5x^3}{3}\right]_1^4=\frac{5}{9}[x^3]_1^4=\frac{5}{9}(64-1)=\mathbf{35}$$

Problem 2. A sinusoidal voltage is given by $v=100\sin\omega t$ volts. Determine the mean value of the voltage over half a cycle using integration

Half a cycle means the limits are 0 to π radians. Mean value,

$$\bar{v}=\frac{1}{\pi-0}\int_0^\pi v\,d(\omega t)$$

$$=\frac{1}{\pi}\int_0^\pi 100\sin\omega t\,d(\omega t)=\frac{100}{\pi}\left[-\cos\omega t\right]_0^\pi$$

$$=\frac{100}{\pi}[(-\cos\pi)-(-\cos 0)]$$

$$=\frac{100}{\pi}[(+1)-(-1)]=\frac{200}{\pi}$$

$$=\mathbf{63.66\ volts}$$

[Note that for a sine wave,

$$\textbf{mean value}=\frac{\mathbf{2}}{\pi}\times\textbf{maximum value}$$

In this case, mean value $=\frac{2}{\pi}\times 100=63.66$ V]

Problem 3. Calculate the mean value of $y=3x^2+2$ in the range $x=0$ to $x=3$ by (a) the mid-ordinate rule and (b) integration

(a) A graph of $y=3x^2$ over the required range is shown in Fig. 52.2 using the following table:

x	0	0.5	1.0	1.5	2.0	2.5	3.0
y	2.0	2.75	5.0	8.75	14.0	20.75	29.0

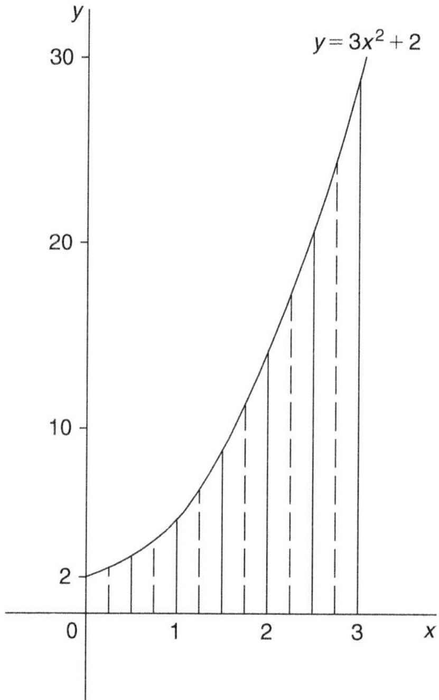

Figure 52.2

Using the mid-ordinate rule, mean value

$$=\frac{\text{area under curve}}{\text{length of base}}$$

$$=\frac{\text{sum of mid-ordinates}}{\text{number of mid-ordinates}}$$

Selecting 6 intervals, each of width 0.5, the mid-ordinates are erected as shown by the broken lines in Fig. 52.2.

$$\text{Mean value}=\frac{2.2+3.7+6.7+11.2+17.2+24.7}{6}$$

$$=\frac{65.7}{6}=\mathbf{10.95}$$

(b) By integration, mean value

$$=\frac{1}{3-0}\int_0^3 y\,dx=\frac{1}{3}\int_0^3(3x^2+2)dx$$

$$=\frac{1}{3}[x^3+2x]_0^3=\frac{1}{3}[(27+6)-(0)]$$

$$=\mathbf{11}$$

The answer obtained by integration is exact; greater accuracy may be obtained by the mid-ordinate rule if a larger number of intervals are selected.

Problem 4. The number of atoms, N, remaining in a mass of material during radioactive decay after time t seconds is given by: $N=N_0e^{-\lambda t}$, where N_0 and λ are constants. Determine the mean number of atoms in the mass of material for the time period $t=0$ and $t=\frac{1}{\lambda}$

Mean number of atoms

$$=\frac{1}{\frac{1}{\lambda}-0}\int_0^{1/\lambda} N\,dt=\frac{1}{\frac{1}{\lambda}}\int_0^{1/\lambda} N_0e^{-\lambda t}\,dt$$

$$=\lambda N_0\int_0^{1/\lambda} e^{-\lambda t}\,dt=\lambda N_0\left[\frac{e^{-\lambda t}}{-\lambda}\right]_0^{1/\lambda}$$

$$=-N_0[e^{-\lambda(1/\lambda)}-e^0]=-N_0[e^{-1}-e^0]$$

$$=+N_0[e^0-e^{-1}]=N_0[1-e^{-1}]=\mathbf{0.632\,N_0}$$

Problem 5. A current waveform, i, is given by: $i=(0.1\sin 20\pi t+0.02\sin 30\pi t)$ amperes, where t is the time in seconds. Calculate its mean value between the limits of $t=0$ to $t=200$ ms.

Since current, $i=(0.1\sin 20\pi t+0.02\sin 30\pi t)$ A then **the mean value**

$$=\frac{1}{200\times 10^{-3}-0}\times\int_0^{200\times 10^{-3}}(0.1\sin 20\pi t+0.02\sin 30\pi t)\,dt$$

$$=5\left[0.1\left(-\frac{1}{20\pi}\cos 20\pi t\right)+0.02\left(-\frac{1}{30\pi}\cos 30\pi t\right)\right]_0^{200\times 10^{-3}}$$

$$=5\left[\begin{array}{l}\left\{\left(-\frac{0.1}{20\pi}\cos 20\pi(200\times 10^{-3})\right)+\left(-\frac{0.02}{30\pi}\cos 30\pi(200\times 10^{-3})\right)\right\}\\ -\left\{\left(-\frac{0.1}{20\pi}\cos 20\pi(0)\right)+\left(-\frac{0.02}{30\pi}\cos 30\pi(0)\right)\right\}\end{array}\right]$$

$$=5\,[(-1.5915\times 10^{-3}-2.1221\times 10^{-4})-(-1.5915\times 10^{-3}-2.1221\times 10^{-4})]$$

$$=\mathbf{0}$$

i.e. the mean value of the current is zero

Now try the following Practice Exercise

Practice Exercise 244 Mean or average values (Answers on page 728)

1. Determine the mean value of (a) $y=3\sqrt{x}$ from $x=0$ to $x=4$ (b) $y=\sin 2\theta$ from $\theta=0$ to $\theta=\frac{\pi}{4}$ (c) $y=4e^t$ from $t=1$ to $t=4$
2. Calculate the mean val ue of $y=2x^2+5$ in the range $x=1$ to $x=4$ by (a) the mid-ordinate rule, and (b) integration
3. The speed v of a vehicle is given by: $v=(4t+3)$ m/s, where t is the time in seconds. Determine the average value of the speed from $t=0$ to $t=3$ s
4. Find the mean value of the curve $y=6+x-x^2$ which lies above the x-axis by using an approximate method. Check the result using integration
5. The vertical height h km of a missile varies with the horizontal distance d km, and is given by $h=4d-d^2$. Determine the mean height of the missile from $d=0$ to $d=4$ km
6. The velocity v of a piston moving with simple harmonic motion at any time t is given by: $v=c\sin\omega t$, where c is a constant. Determine the mean velocity between $t=0$ and $t=\frac{\pi}{\omega}$

52.2 Root mean square values

The **root mean square value** of a quantity is 'the square root of the mean value of the squared values of the quantity' taken over an interval. With reference to Fig. 52.1, the r.m.s. value of $y=f(x)$ over the range $x=a$ to $x=b$ is given by:

$$\textbf{r.m.s. value}=\sqrt{\frac{1}{b-a}\int_a^b y^2dx}$$

One of the principal applications of r.m.s. values is with alternating currents and voltages. The r.m.s. value of an alternating current is defined as that current which will give the same heating effect as the equivalent direct current.

Problem 6. Determine the r.m.s. value of $y=2x^2$ between $x=1$ and $x=4$

R.m.s. value

$$= \sqrt{\frac{1}{4-1}\int_1^4 y^2\,dx} = \sqrt{\frac{1}{3}\int_1^4 (2x^2)^2\,dx}$$

$$= \sqrt{\frac{1}{3}\int_1^4 4x^4\,dx} = \sqrt{\frac{4}{3}\left[\frac{x^5}{5}\right]_1^4}$$

$$= \sqrt{\frac{4}{15}(1024-1)} = \sqrt{272.8} = \mathbf{16.5}$$

Problem 7. A sinusoidal voltage has a maximum value of 100 V. Calculate its r.m.s. value

A sinusoidal voltage v having a maximum value of 100 V may be written as: $v=100\sin\theta$. Over the range $\theta=0$ to $\theta=\pi$,
r.m.s. value

$$= \sqrt{\frac{1}{\pi-0}\int_0^{\pi} v^2\,d\theta}$$

$$= \sqrt{\frac{1}{\pi}\int_0^{\pi} (100\sin\theta)^2\,d\theta}$$

$$= \sqrt{\frac{10000}{\pi}\int_0^{\pi} \sin^2\theta\,d\theta}$$

which is not a 'standard' integral. It is shown in Chapter 22 that $\cos 2A = 1-2\sin^2 A$ and this formula is used whenever $\sin^2 A$ needs to be integrated.
Rearranging $\cos 2A = 1-2\sin^2 A$ gives
$\sin^2 A = \frac{1}{2}(1-\cos 2A)$

Hence $\sqrt{\frac{10000}{\pi}\int_0^{\pi} \sin^2\theta\,d\theta}$

$$= \sqrt{\frac{10000}{\pi}\int_0^{\pi} \frac{1}{2}(1-\cos 2\theta)\,d\theta}$$

$$= \sqrt{\frac{10000}{\pi}\frac{1}{2}\left[\theta - \frac{\sin 2\theta}{2}\right]_0^{\pi}}$$

$$= \sqrt{\frac{10000}{\pi}\frac{1}{2}\left[\left(\pi - \frac{\sin 2\pi}{2}\right) - \left(0 - \frac{\sin 0}{2}\right)\right]}$$

$$= \sqrt{\frac{10000}{\pi}\frac{1}{2}[\pi]} = \sqrt{\frac{10000}{2}}$$

$$= \frac{100}{\sqrt{2}} = \mathbf{70.71\ volts}$$

[Note that for a sine wave,

$$\textbf{r.m.s. value} = \frac{\mathbf{1}}{\sqrt{\mathbf{2}}} \times \textbf{maximum value.}$$

In this case, r.m.s. value $= \frac{1}{\sqrt{2}} \times 100 = 70.71$ V]

Problem 8. In a frequency distribution the average distance from the mean, y, is related to the variable, x, by the equation $y=2x^2-1$. Determine, correct to 3 significant figures, the r.m.s. deviation from the mean for values of x from -1 to $+4$

R.m.s. deviation

$$= \sqrt{\frac{1}{4--1}\int_{-1}^4 y^2\,dx}$$

$$= \sqrt{\frac{1}{5}\int_{-1}^4 (2x^2-1)^2 dx}$$

$$= \sqrt{\frac{1}{5}\int_{-1}^4 (4x^4-4x^2+1) dx}$$

$$= \sqrt{\frac{1}{5}\left[\frac{4x^5}{5} - \frac{4x^3}{3} + x\right]_{-1}^4}$$

$$= \sqrt{\frac{1}{5}\left[\left(\frac{4}{5}(4)^5 - \frac{4}{3}(4)^3 + 4\right) - \left(\frac{4}{5}(-1)^5 - \frac{4}{3}(-1)^3 + (-1)\right)\right]}$$

$$= \sqrt{\frac{1}{5}[(737.87)-(-0.467)]}$$

$$= \sqrt{\frac{1}{5}[738.34]}$$

$$= \sqrt{147.67} = 12.152 = \mathbf{12.2},$$

correct to 3 significant figures.

Now try the following Practice Exercise

Practice Exercise 245 Root mean square values (Answers on page 728)

1. Determine the r.m.s. values of:
   (a) $y=3x$ from $x=0$ to $x=4$
   (b) $y=t^2$ from $t=1$ to $t=3$
   (c) $y=25\sin\theta$ from $\theta=0$ to $\theta=2\pi$

2. Calculate the r.m.s. values of:
   (a) $y = \sin 2\theta$ from $\theta = 0$ to $\theta = \frac{\pi}{4}$
   (b) $y = 1 + \sin t$ from $t = 0$ to $t = 2\pi$
   (c) $y = 3\cos 2x$ from $x = 0$ to $x = \pi$

   (Note that $\cos^2 t = \frac{1}{2}(1 + \cos 2t)$, from Chapter 22)

3. The distance, p, of points from the mean value of a frequency distribution are related to the variable, q, by the equation $p = \frac{1}{q} + q$. Determine the standard deviation (i.e. the r.m.s. value), correct to 3 significant figures, for values from $q = 1$ to $q = 3$

4. A current, $i = 30\sin 100\pi t$ amperes is applied across an electric circuit. Determine its mean and r.m.s. values, each correct to 4 significant figures, over the range $t = 0$ to $t = 10$ ms

5. A sinusoidal voltage has a peak value of 340 V. Calculate its mean and r.m.s. values, correct to 3 significant figures

6. Determine the form factor, correct to 3 significant figures, of a sinusoidal voltage of maximum value 100 volts, given that form factor $= \frac{\text{r.m.s. value}}{\text{average value}}$

7. A wave is defined by the equation:
   $$v = E_1 \sin \omega t + E_3 \sin 3\omega t$$
   where, E_1, E_3 and ω are constants.

   Determine the r.m.s. value of v over the interval $0 \le t \le \frac{\pi}{\omega}$

Practice Exercise 246 Multiple-choice questions on mean and root mean square values (Answers on page 728)

Each question has only one correct answer

1. The mean value of $y = 4x$ between $x = 1$ and $x = 4$ is:
   (a) 1.33 (b) 3.75 (c) 10 (d) 30
2. The mean value of $y = 2x^2$ between $x = 1$ and $x = 3$ is:
   (a) 2 (b) 4 (c) $4\frac{1}{3}$ (d) $8\frac{2}{3}$
3. The r.m.s. value of $y = x^2$ between $x = 1$ and $x = 3$, correct to 2 decimal places, is:
   (a) 2.08 (b) 4.92 (c) 6.96 (d) 24.2
4. The velocity v of a piston moving with SHM at any time t is given by $v = 3\sin \omega t$. The mean value between $t = 0$ and $t = \frac{\pi}{\omega}$ is:
   (a) $\frac{6}{\pi}$ (b) 0 (c) $\frac{\pi}{3}$ (d) $\frac{2}{\pi}$
5. The r.m.s. value of $y = 2\cos 3\theta$, between $x = 0$ and $x = \pi$, is:
   (a) 2 (b) 0.707 (c) 1.414 (d) 0.637

For fully worked solutions to each of the problems in Practice Exercises 244 and 245 in this chapter, go to the website:
www.routledge.com/cw/bird

COMPANION @ WEBSITE

Chapter 53

Volumes of solids of revolution

Why it is important to understand: Volumes of solids of revolution

Revolving a plane figure about an axis generates a volume. The solid generated by rotating a plane area about an axis in its plane is called a solid of revolution, and integration may be used to calculate such a volume. There are many applications of volumes of solids of revolution in engineering and particularly in manufacturing.

At the end of this chapter, you should be able to:

- understand the term volume of a solid of revolution
- determine the volume of a solid of revolution for functions between given limits using integration

53.1 Introduction

If the area under the curve $y=f(x)$, (shown in Fig. 53.1(a)), between $x=a$ and $x=b$ is rotated 360° about the x-axis, then a volume known as a **solid of revolution** is produced as shown in Fig. 53.1(b).
The volume of such a solid may be determined precisely using integration.

(i) Let the area shown in Fig. 53.1(a) be divided into a number of strips each of width δx. One such strip is shown shaded.

(ii) When the area is rotated 360° about the x-axis, each strip produces a solid of revolution approximating to a circular disc of radius y and thickness δx. Volume of disc = (circular cross-sectional area) (thickness) = $(\pi y^2)(\delta x)$

(iii) Total volume, V, between ordinates $x=a$ and $x=b$ is given by:

$$\textbf{Volume } V = \mathbf{\underset{\delta x \to 0}{limit}} \sum_{x=a}^{x=b} \pi \mathbf{y^2 \delta x} = \int_a^b \pi y^2\, dx$$

If a curve $x=f(y)$ is rotated about the y-axis 360° between the limits $y=c$ and $y=d$, as shown in Fig. 53.2, then the volume generated is given by:

$$\textbf{Volume } V = \mathbf{\underset{\delta y \to 0}{limit}} \sum_{y=c}^{y=d} \pi \mathbf{x^2 \delta y} = \int_c^d \pi x^2\, dy$$

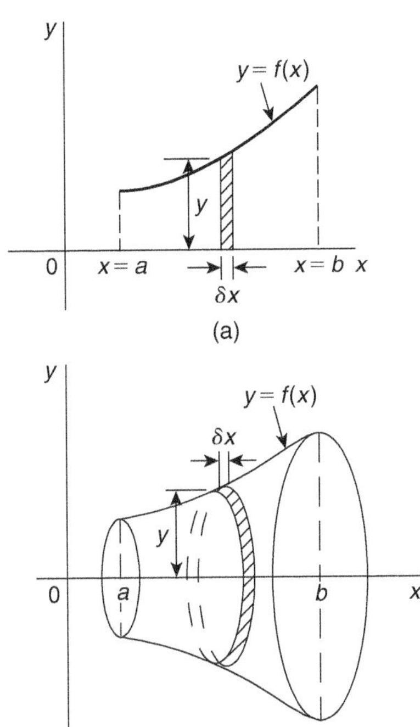

Figure 53.1

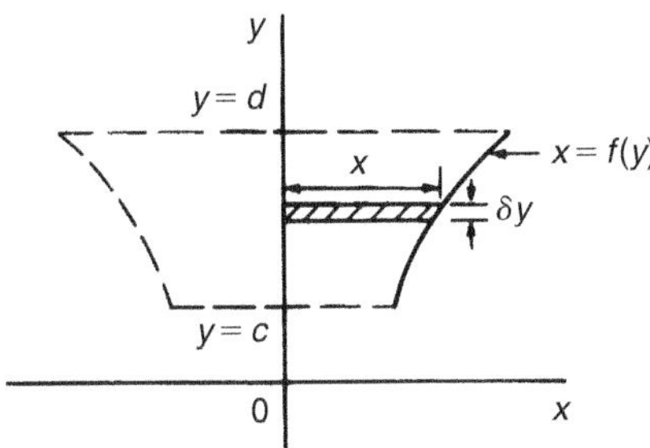

Figure 53.2

53.2 Worked problems on volumes of solids of revolution

Problem 1. Determine the volume of the solid of revolution formed when the curve $y = 2$ is rotated 360° about the x-axis between the limits $x = 0$ to $x = 3$

When $y = 2$ is rotated 360° about the x-axis between $x = 0$ and $x = 3$ (see Fig. 53.3):

Volume generated

$$= \int_0^3 \pi y^2\,dx = \int_0^3 \pi(2)^2\,dx$$

$$= \int_0^3 4\pi\,dx = 4\pi[x]_0^3 = \mathbf{12\pi\ cubic\ units}$$

[Check: The volume generated is a cylinder of radius 2 and height 3.
Volume of cylinder $= \pi r^2 h = \pi(2)^2(3) =$ **12π cubic units.**]

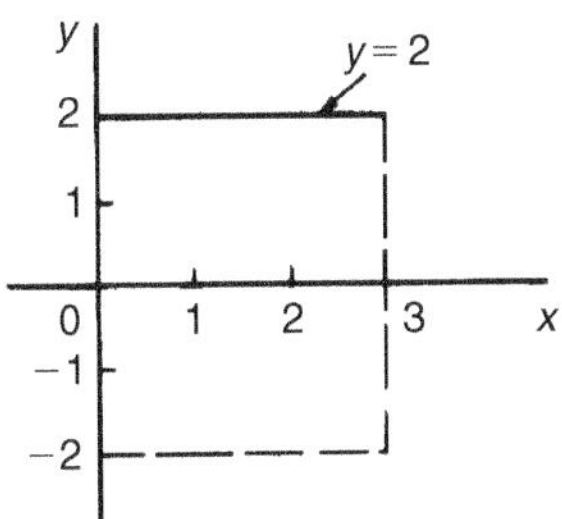

Figure 53.3

Problem 2. Find the volume of the solid of revolution when the curve $y = 2x$ is rotated one revolution about the x-axis between the limits $x = 0$ and $x = 5$

When $y = 2x$ is revolved one revolution about the x-axis between $x = 0$ and $x = 5$ (see Fig. 53.4) then:

Volume generated

$$= \int_0^5 \pi y^2 dx = \int_0^5 \pi(2x)^2 dx$$

$$= \int_0^5 4\pi x^2 dx = 4\pi\left[\frac{x^3}{3}\right]_0^5$$

$$= \frac{500\pi}{3} = \mathbf{166\frac{2}{3}\pi\ cubic\ units}$$

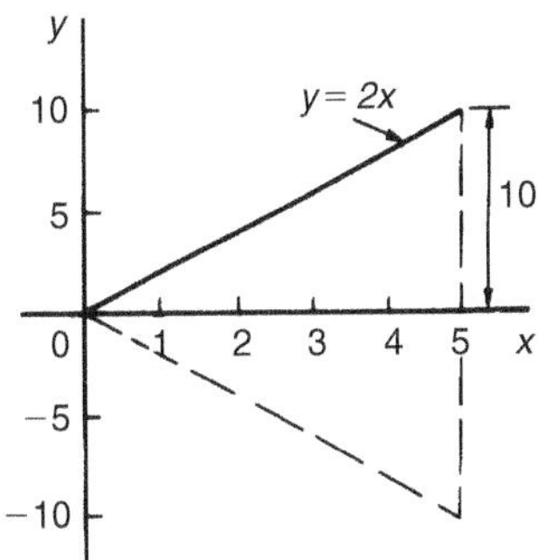

Figure 53.4

[Check: The volume generated is a cone of radius 10 and height 5. Volume of cone

$$= \frac{1}{3}\pi r^2 h = \frac{1}{3}\pi(10)^2 5 = \frac{500\pi}{3}$$

$$= \mathbf{166\frac{2}{3}\pi\ cubic\ units.}]$$

Problem 3. The curve $y = x^2 + 4$ is rotated one revolution about the x-axis between the limits $x = 1$ and $x = 4$. Determine the volume of the solid of revolution produced

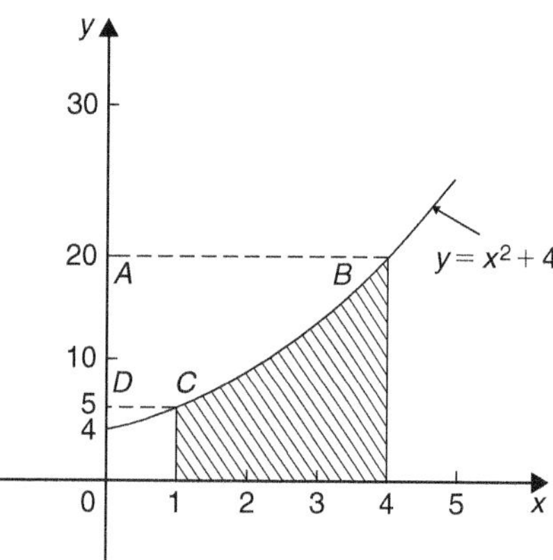

Figure 53.5

Revolving the shaded area shown in Fig. 53.5 about the x-axis 360° produces a solid of revolution given by:

$$\text{Volume} = \int_1^4 \pi y^2\, dx = \int_1^4 \pi (x^2+4)^2 dx$$

$$= \int_1^4 \pi(x^4 + 8x^2 + 16)\, dx$$

$$= \pi\left[\frac{x^5}{5} + \frac{8x^3}{3} + 16x\right]_1^4$$

$$= \pi[(204.8 + 170.67 + 64) - (0.2 + 2.67 + 16)]$$

$$= \mathbf{420.6\pi \text{ cubic units}}$$

Problem 4. If the curve in Problem 3 is revolved about the y-axis between the same limits, determine the volume of the solid of revolution produced

The volume produced when the curve $y = x^2 + 4$ is rotated about the y-axis between $y = 5$ (when $x = 1$) and $y = 20$ (when $x = 4$), i.e. rotating area ABCD of Fig. 53.5 about the y-axis is given by:

$$\text{Volume} = \int_5^{20} \pi x^2\, dy$$

Since $y = x^2 + 4$, then $x^2 = y - 4$

$$\text{Hence volume} = \int_5^{20} \pi(y-4)dy = \pi\left[\frac{y^2}{2} - 4y\right]_5^{20}$$

$$= \pi[(120) - (-7.5)]$$

$$= \mathbf{127.5\pi \text{ cubic units}}$$

Now try the following Practice Exercise

Practice Exercise 247 Volumes of solids of revolution (Answers on page 728)

In Problems 1 to 5, determine the volume of the solid of revolution formed by revolving the areas enclosed by the given functions, the x-axis and the given ordinates through one revolution about the x-axis.

1. $y = 5x;\quad x = 1, x = 4$
2. $y = x^2;\quad x = -2, x = 3$
3. $y = 2x^2 + 3;\quad x = 0, x = 2$
4. $\frac{y^2}{4} = x;\quad x = 1, x = 5$
5. $xy = 3;\quad x = 2, x = 3$

In Problems 6 to 8, determine the volume of the solid of revolution formed by revolving the areas enclosed by the given functions, the y-axis and the given ordinates through one revolution about the y-axis.

6. $y = x^2;\quad y = 1, y = 3$
7. $y = 3x^2 - 1;\quad y = 2, y = 4$
8. $y = \frac{2}{x};\quad y = 1, y = 3$
9. The curve $y = 2x^2 + 3$ is rotated about (a) the x-axis between the limits $x = 0$ and $x = 3$, and (b) the y-axis, between the same limits. Determine the volume generated in each case.

53.3 Further worked problems on volumes of solids of revolution

Problem 5. The area enclosed by the curve $y = 3e^{\frac{x}{3}}$, the x-axis and ordinates $x = -1$ and $x = 3$ is rotated 360° about the x-axis. Determine the volume generated

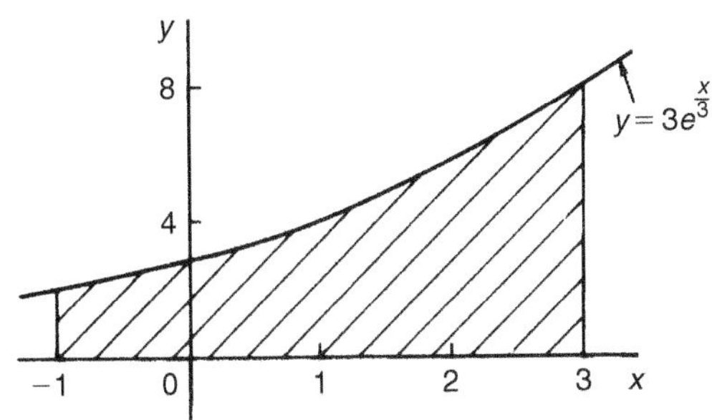

Figure 53.6

A sketch of $y=3e^{\frac{x}{3}}$ is shown in Fig. 53.6. When the shaded area is rotated 360° about the x-axis then:

$$\text{Volume generated} = \int_{-1}^{3} \pi y^2\, dx$$

$$= \int_{-1}^{3} \pi\left(3e^{\frac{x}{3}}\right)^2 dx$$

$$= 9\pi\int_{-1}^{3} e^{\frac{2x}{3}}\, dx$$

$$= 9\pi\left[\frac{e^{\frac{2x}{3}}}{\frac{2}{3}}\right]_{-1}^{3}$$

$$= \frac{27\pi}{2}\left(e^2 - e^{-\frac{2}{3}}\right)$$

$$= \mathbf{92.82\pi\ cubic\ units}$$

Problem 6. Determine the volume generated when the area above the x-axis bounded by the curve $x^2+y^2=9$ and the ordinates $x=3$ and $x=-3$ is rotated one revolution about the x-axis

Figure 53.7 shows the part of the curve $x^2+y^2=9$ lying above the x-axis. Since, in general, $x^2+y^2=r^2$ represents a circle, centre 0 and radius r, then $x^2+y^2=9$ represents a circle, centre 0 and radius 3. When the semi-circular area of Fig. 53.7 is rotated one revolution about the x-axis then:

$$\text{Volume generated} = \int_{-3}^{3} \pi y^2 dx$$

$$= \int_{-3}^{3} \pi(9-x^2)dx$$

$$= \pi\left[9x - \frac{x^3}{3}\right]_{-3}^{3}$$

$$= \pi[(18)-(-18)]$$

$$= \mathbf{36\pi\ cubic\ units}$$

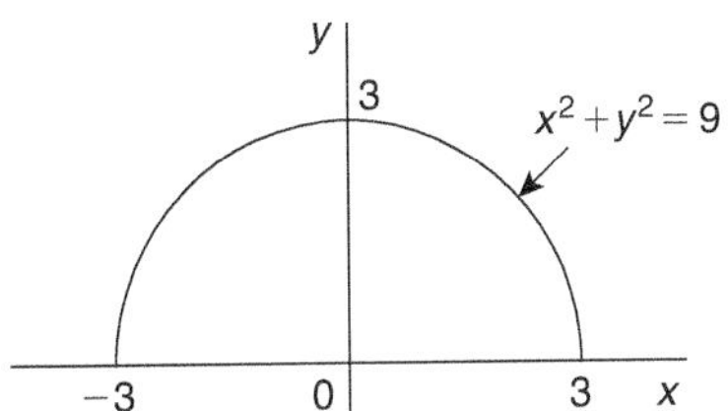

Figure 53.7

(Check: The volume generated is a sphere of radius 3. Volume of sphere $=\frac{4}{3}\pi r^3 = \frac{4}{3}\pi(3)^3 =$ **36π cubic units**.)

Problem 7. Calculate the volume of frustum of a sphere of radius 4 cm that lies between two parallel planes at 1 cm and 3 cm from the centre and on the same side of it

The volume of a frustum of a sphere may be determined by integration by rotating the curve $x^2+y^2=4^2$ (i.e. a circle, centre 0, radius 4) one revolution about the x-axis, between the limits $x=1$ and $x=3$ (i.e. rotating the shaded area of Fig. 53.8).

$$\text{Volume of frustum} = \int_{1}^{3} \pi y^2\, dx$$

$$= \int_{1}^{3} \pi(4^2 - x^2)dx$$

$$= \pi\left[16x - \frac{x^3}{3}\right]_{1}^{3}$$

$$= \pi\left[(39) - \left(15\frac{2}{3}\right)\right]$$

$$= \mathbf{23\frac{1}{3}\pi\ cubic\ units}$$

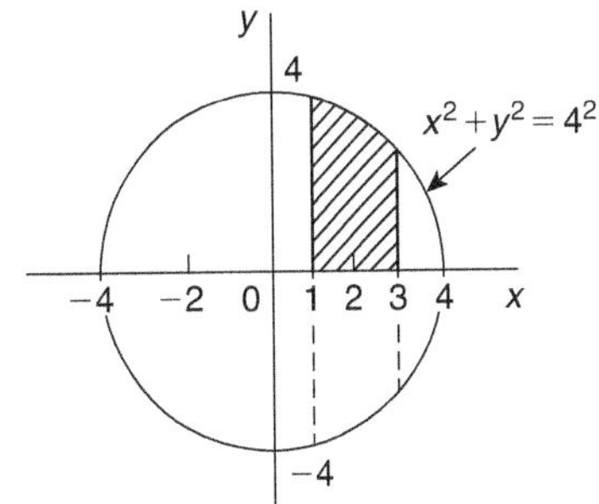

Figure 53.8

Problem 8. The area enclosed between the two parabolas $y = x^2$ and $y^2 = 8x$ of Problem 12, Chapter 51, page 524, is rotated 360° about the x-axis. Determine the volume of the solid produced

The area enclosed by the two curves is shown in Fig. 51.13, page 525. The volume produced by revolving the shaded area about the x-axis is given by: [(volume produced by revolving $y^2 = 8x$) − (volume produced by revolving $y = x^2$)]

i.e. $$\textbf{volume} = \int_0^2 \pi(8x)\,dx - \int_0^2 \pi(x^4)\,dx$$

$$= \pi\int_0^2 (8x - x^4)\,dx = \pi\left[\frac{8x^2}{2} - \frac{x^5}{5}\right]_0^2$$

$$= \pi\left[\left(16 - \frac{32}{5}\right) - (0)\right]$$

$$= \mathbf{9.6\pi\ cubic\ units}$$

Now try the following Practice Exercise

Practice Exercise 248 Volumes of solids of revolution (Answers on page 729)

In Problems 1 and 2, determine the volume of the solid of revolution formed by revolving the areas enclosed by the given functions, the x-axis and the given ordinates through one revolution about the x-axis.

1. $y = 4e^x$; $x = 0$; $x = 2$
2. $y = \sec x$; $x = 0$, $x = \frac{\pi}{4}$

In Problems 3 and 4, determine the volume of the solid of revolution formed by revolving the areas enclosed by the given functions, the y-axis and the given ordinates through one revolution about the y-axis.

3. $x^2 + y^2 = 16$; $y = 0$, $y = 4$
4. $x\sqrt{y} = 2$; $y = 2$, $y = 3$
5. Determine the volume of a plug formed by the frustum of a sphere of radius 6 cm which lies between two parallel planes at 2 cm and 4 cm from the centre and on the same side of it.

   (The equation of a circle, centre 0, radius r is $x^2 + y^2 = r^2$)
6. The area enclosed between the two curves $x^2 = 3y$ and $y^2 = 3x$ is rotated about the x-axis. Determine the volume of the solid formed
7. The portion of the curve $y = x^2 + \frac{1}{x}$ lying between $x = 1$ and $x = 3$ is revolved 360° about the x-axis. Determine the volume of the solid formed
8. Calculate the volume of the frustum of a sphere of radius 5 cm that lies between two parallel planes at 3 cm and 2 cm from the centre and on opposite sides of it
9. Sketch the curves $y = x^2 + 2$ and $y - 12 = 3x$ from $x = -3$ to $x = 6$. Determine (a) the co-ordinates of the points of intersection of the two curves, and (b) the area enclosed by the two curves. (c) If the enclosed area is rotated 360° about the x-axis, calculate the volume of the solid produced

Practice Exercise 249 Multiple-choice questions on volumes of solids of revolution (Answers on page 729)

Each question has only one correct answer

1. The volume of the solid of revolution when the curve $y = 2x$ is rotated one revolution about the x-axis between the limits $x = 0$ and $x = 4$ cm is:

   (a) $85\frac{1}{3}\pi$ cm^3 (b) 8 cm^3
   (c) $85\frac{1}{3}$ cm^3 (d) 64π cm^3
2. A volume of a solid of revolution is formed when the area under the curve $y = x^2$ is rotated once about the x-axis between $x = 0$ and $x = 2$. Leaving the answer as a multiple of π, the volume is:
   (a) 32π cubic units (b) 16π cubic units
   (c) 4π cubic units (d) 6.4π cubic units
3. The curve $y = x^2 + 1$ is rotated one revolution about the x-axis between the limits $x = 1$ and $x = 3$. The volume of the solid of revolution produced is:
   (a) 158.3 cu units (b) 67.7 cu units
   (c) 212.8 cu units (d) 307.9 cu units

4. The curve $y = x^2 + 3$ is rotated one revolution about the y-axis between the limits $y = 2$ and $y = 5$. The volume of the solid of revolution produced, in terms of π, is:
(a) 3π cu units (b) 1.5π cu units
(c) 304.5π cu units (d) 48π cu units

5. A volume of a solid of revolution is formed when the area under the curve $y = 3x + 4$ is rotated once about the x-axis between $x = 1$ and $x = 3$. Leaving the answer as a multiple of π, the volume is:
(a) 120π cubic units (b) 206π cubic units
(c) 20π cubic units (d) 120π cubic units

For fully worked solutions to each of the problems in Exercises 247 and 248 in this chapter, go to the website:
www.routledge.com/cw/bird

COMPANION @ WEBSITE

Chapter 54

Centroids of simple shapes

Why it is important to understand: Centroids of simple shapes

The centroid of an area is similar to the centre of mass of a body. Calculating the centroid involves only the geometrical shape of the area; the centre of gravity will equal the centroid if the body has constant density. Centroids of basic shapes can be intuitive – such as the centre of a circle; centroids of more complex shapes can be found using integral calculus – as long as the area, volume or line of an object can be described by a mathematical equation. Centroids are of considerable importance in manufacturing, and in mechanical, civil and structural design engineering.

At the end of this chapter, you should be able to:

- define a centroid
- determine the centroid of an area between a curve and the x-axis for functions between given limits using integration
- determine the centroid of an area between a curve and the y-axis for functions between given limits using integration
- state the theorem of Pappus
- determine the centroid of an area using the theorem of Pappus

54.1 Centroids

A **lamina** is a thin flat sheet having uniform thickness. The **centre of gravity** of a lamina is the point where it balances perfectly, i.e. the lamina's **centre of mass**. When dealing with an area (i.e. a lamina of negligible thickness and mass) the term **centre of area** or **centroid** is used for the point where the centre of gravity of a lamina of that shape would lie.

54.2 The first moment of area

The **first moment of area** is defined as the product of the area and the perpendicular distance of its centroid from a given axis in the plane of the area. In Fig. 54.1, the first moment of area A about axis XX is given by (Ay) cubic units.

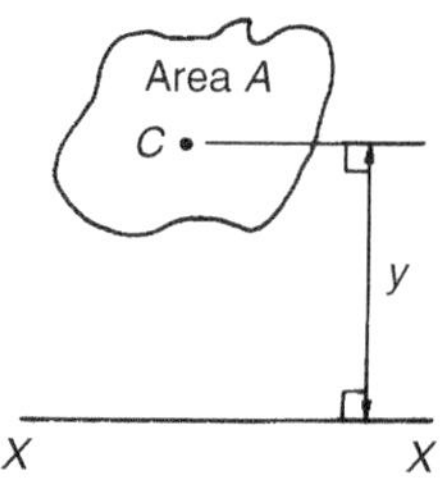

Figure 54.1

54.3 Centroid of area between a curve and the *x*-axis

(i) Figure 54.2 shows an area *PQRS* bounded by the curve $y=f(x)$, the *x*-axis and ordinates $x=a$ and $x=b$. Let this area be divided into a large number of strips, each of width δx. A typical strip is shown shaded drawn at point (x, y) on $f(x)$. The area of the strip is approximately rectangular and is given by $y\delta x$. The centroid, *C*, has co-ordinates $\left(x, \frac{y}{2}\right)$

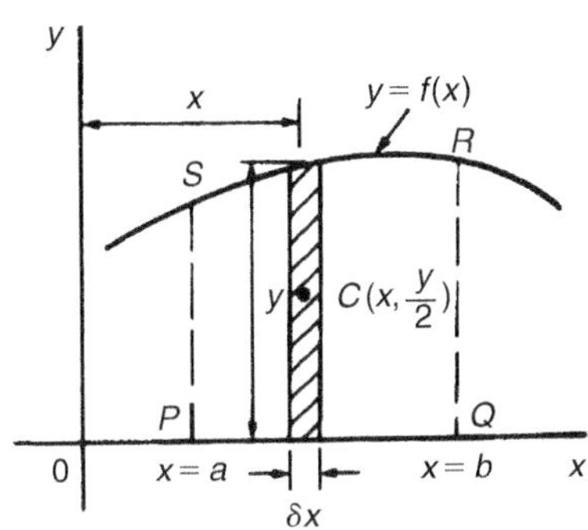

Figure 54.2

(ii) First moment of area of shaded strip about axis $Oy=(y\delta x)(x)=xy\delta x$
Total first moment of area *PQRS* about axis $Oy=\lim_{\delta x\to 0}\sum_{x=a}^{x=b} xy\delta x=\int_a^b xy\,dx$

(iii) First moment of area of shaded strip about axis $Ox=(y\delta x)\left(\frac{y}{2}\right)=\frac{1}{2}y^2\delta x$
Total first moment of area *PQRS* about axis $Ox=\lim_{\delta x\to 0}\sum_{x=a}^{x=b}\frac{1}{2}y^2\delta x=\frac{1}{2}\int_a^b y^2\,dx$

(iv) Area of *PQRS*, $A=\int_a^b y\,dx$ (from Chapter 51)

(v) Let $\bar{x}$ and $\bar{y}$ be the distances of the centroid of area *A* about *Oy* and *Ox* respectively then: $(\bar{x})(A)=$ total first moment of area *A* about axis $Oy=\int_a^b xy\,dx$

from which, $$\bar{x}=\frac{\int_a^b xy\,dx}{\int_a^b y\,dx}$$

and $(\bar{y})(A)=$ total moment of area *A* about axis $Ox=\frac{1}{2}\int_a^b y^2\,dx$

from which, $$\bar{y}=\frac{\frac{1}{2}\int_a^b y^2\,dx}{\int_a^b y\,dx}$$

54.4 Centroid of area between a curve and the *y*-axis

If $\bar{x}$ and $\bar{y}$ are the distances of the centroid of area *EFGH* in Fig. 54.3 from *Oy* and *Ox* respectively, then, by similar reasoning as above:

$$(\bar{x})(\text{total area})=\lim_{\delta y\to 0}\sum_{y=c}^{y=d} x\delta y\left(\frac{x}{2}\right)=\frac{1}{2}\int_c^d x^2dy$$

from which, $$\bar{x}=\frac{\frac{1}{2}\int_c^d x^2dy}{\int_c^d x\,dy}$$

and $$(\bar{y})(\text{total area})=\lim_{\delta y\to 0}\sum_{y=c}^{y=d}(x\delta y)y=\int_c^d xy\,dy$$

from which, $$\bar{y}=\frac{\int_c^d xy\,dy}{\int_c^d x\,dy}$$

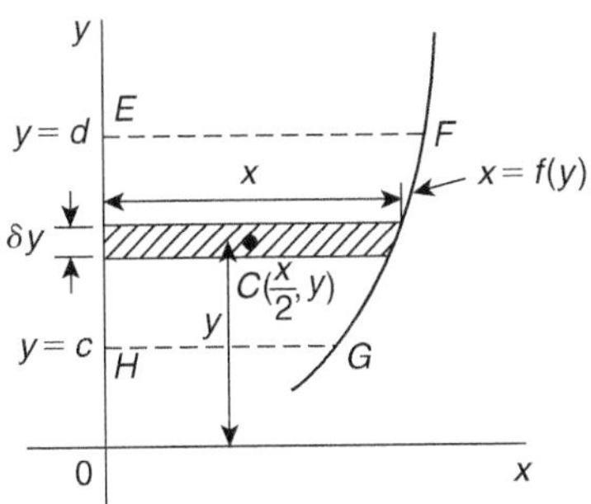

Figure 54.3

54.5 Worked problems on centroids of simple shapes

Problem 1. Show, by integration, that the centroid of a rectangle lies at the intersection of the diagonals

Let a rectangle be formed by the line $y=b$, the x-axis and ordinates $x=0$ and $x=l$ as shown in Fig. 54.4. Let the co-ordinates of the centroid C of this area be $(\bar{x},\bar{y})$.

By integration, $$\bar{x}=\frac{\int_0^l xy\,dx}{\int_0^l y\,dx}=\frac{\int_0^l (x)(b)\,dx}{\int_0^l b\,dx}$$

$$=\frac{\left[b\frac{x^2}{2}\right]_0^l}{[bx]_0^l}=\frac{\frac{bl^2}{2}}{bl}=\frac{l}{2}$$

and $$\bar{y}=\frac{\frac{1}{2}\int_0^l y^2\,dx}{\int_0^l y\,dx}=\frac{\frac{1}{2}\int_0^l b^2\,dx}{bl}$$

$$=\frac{\frac{1}{2}[b^2x]_0^l}{bl}=\frac{\frac{b^2l}{2}}{bl}=\frac{b}{2}$$

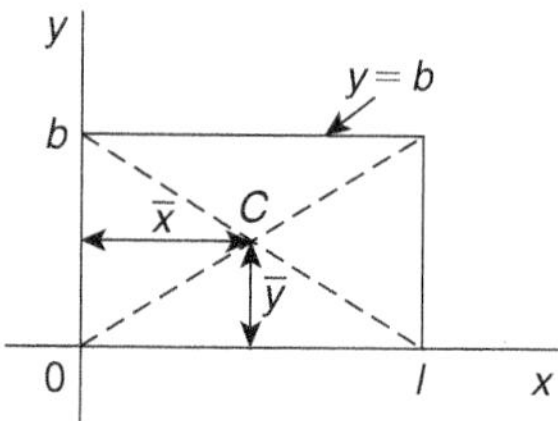

Figure 54.4

i.e. **the centroid lies at** $\left(\frac{l}{2},\frac{b}{2}\right)$ **which is at the intersection of the diagonals.**

Problem 2. Find the position of the centroid of the area bounded by the curve $y=3x^2$, the x-axis and the ordinates $x=0$ and $x=2$

If, $(\bar{x},\bar{y})$ are the co-ordinates of the centroid of the given area then:

$$\bar{x}=\frac{\int_0^2 xy\,dx}{\int_0^2 y\,dx}=\frac{\int_0^2 x(3x^2)dx}{\int_0^2 3x^2dx}$$

$$=\frac{\int_0^2 3x^3dx}{\int_0^2 3x^2dx}=\frac{\left[\frac{3x^4}{4}\right]_0^2}{[x^3]_0^2}=\frac{12}{8}=\mathbf{1.5}$$

$$\bar{y}=\frac{\frac{1}{2}\int_0^2 y^2dx}{\int_0^2 ydx}=\frac{\frac{1}{2}\int_0^2 (3x^2)^2dx}{8}$$

$$=\frac{\frac{1}{2}\int_0^2 9x^4dx}{8}=\frac{\frac{9}{2}\left[\frac{x^5}{5}\right]_0^2}{8}=\frac{\frac{9}{2}\left(\frac{32}{5}\right)}{8}$$

$$=\frac{18}{5}=\mathbf{3.6}$$

Hence the centroid lies at (1.5, 3.6)

Problem 3. Determine by integration the position of the centroid of the area enclosed by the line $y=4x$, the x-axis and ordinates $x=0$ and $x=3$

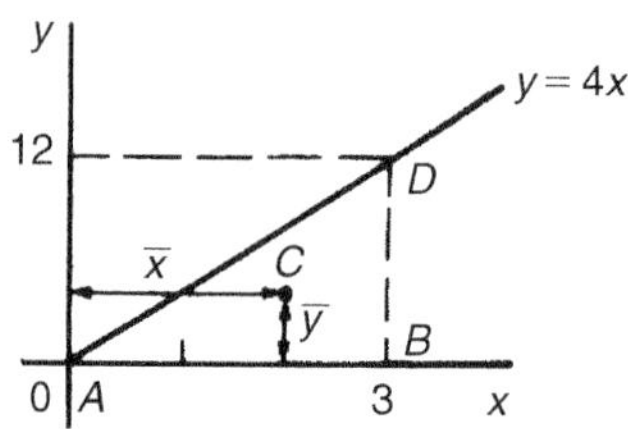

Figure 54.5

Let the co-ordinates of the area be $(\bar{x},\bar{y})$ as shown in Fig. 54.5.

Then $$\bar{x}=\frac{\int_0^3 xy\,dx}{\int_0^3 y\,dx}=\frac{\int_0^3 (x)(4x)dx}{\int_0^3 4x\,dx}$$

$$=\frac{\int_0^3 4x^2dx}{\int_0^3 4x\,dx}=\frac{\left[\frac{4x^3}{3}\right]_0^3}{[2x^2]_0^3}=\frac{36}{18}=2$$

$$\bar{y}=\frac{\frac{1}{2}\int_0^3 y^2dx}{\int_0^3 y\,dx}=\frac{\frac{1}{2}\int_0^3 (4x)^2dx}{18}$$

$$=\frac{\frac{1}{2}\int_0^3 16x^2dx}{18}=\frac{\frac{1}{2}\left[\frac{16x^3}{3}\right]_0^3}{18}=\frac{72}{18}=4$$

Hence the centroid lies at (2, 4).

In Fig. 54.5, ABD is a right-angled triangle. The centroid lies 4 units from AB and 1 unit from BD showing that the centroid of a triangle lies at one-third of the perpendicular height above any side as base.

Now try the following Practice Exercise

Practice Exercise 250 Centroids of simple shapes (Answers on page 729)

In Problems 1 to 5, find the position of the centroids of the areas bounded by the given curves, the x-axis and the given ordinates.

1. $y=2x;\ x=0, x=3$
2. $y=3x+2;\ x=0, x=4$
3. $y=5x^2;\ x=1, x=4$
4. $y=2x^3;\ x=0, x=2$
5. $y=x(3x+1);\ x=-1, x=0$

54.6 Further worked problems on centroids of simple shapes

Problem 4. Determine the co-ordinates of the centroid of the area lying between the curve $y=5x-x^2$ and the x-axis

$y=5x-x^2=x(5-x)$. When $y=0$, $x=0$ or $x=5$, Hence the curve cuts the x-axis at 0 and 5 as shown in Fig. 54.6. Let the co-ordinates of the centroid be $(\bar{x},\bar{y})$ then, by integration,

$$\bar{x}=\frac{\int_0^5 xy\,dx}{\int_0^5 y\,dx}=\frac{\int_0^5 x(5x-x^2)dx}{\int_0^5 (5x-x^2)dx}$$

$$=\frac{\int_0^5 (5x^2-x^3)dx}{\int_0^5 (5x-x^2)dx}=\frac{\left[\frac{5x^3}{3}-\frac{x^4}{4}\right]_0^5}{\left[\frac{5x^2}{2}-\frac{x^3}{3}\right]_0^5}$$

$$=\frac{\frac{625}{3}-\frac{625}{4}}{\frac{125}{2}-\frac{125}{3}}=\frac{\frac{625}{12}}{\frac{125}{6}}$$

$$=\left(\frac{625}{12}\right)\left(\frac{6}{125}\right)=\frac{5}{2}=\mathbf{2.5}$$

$$\bar{y}=\frac{\frac{1}{2}\int_0^5 y^2dx}{\int_0^5 y\,dx}=\frac{\frac{1}{2}\int_0^5 (5x-x^2)^2dx}{\int_0^5 (5x-x^2)\,dx}$$

$$=\frac{\frac{1}{2}\int_0^5 (25x^2-10x^3+x^4)\,dx}{\frac{125}{6}}$$

$$=\frac{\frac{1}{2}\left[\frac{25x^3}{3}-\frac{10x^4}{4}+\frac{x^5}{5}\right]_0^5}{\frac{125}{6}}$$

$$=\frac{\frac{1}{2}\left(\frac{25(125)}{3}-\frac{6250}{4}+625\right)}{\frac{125}{6}}=\mathbf{2.5}$$

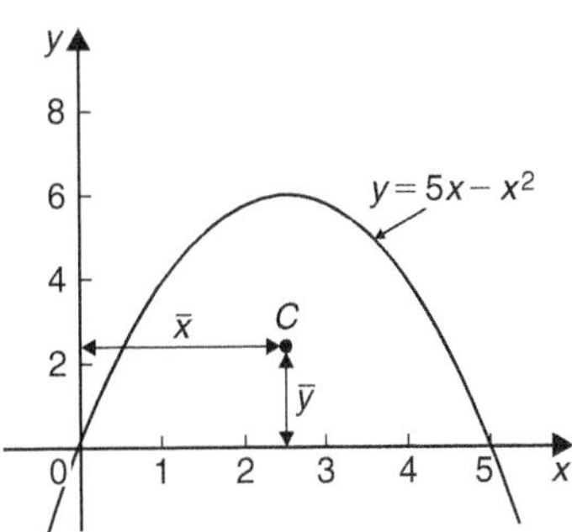

Figure 54.6

Hence the centroid of the area lies at (2.5, 2.5)
(Note from Fig. 54.6 that the curve is symmetrical about $x=2.5$ and thus $\bar{x}$ could have been determined 'on sight'.)

Problem 5. Locate the centroid of the area enclosed by the curve $y=2x^2$, the y-axis and ordinates $y=1$ and $y=4$, correct to 3 decimal places

From Section 54.4,

$$\bar{x} = \frac{\frac{1}{2}\int_1^4 x^2 dy}{\int_1^4 x dy} = \frac{\frac{1}{2}\int_1^4 \frac{y}{2} dy}{\int_1^4 \sqrt{\frac{y}{2}} dy}$$

$$= \frac{\frac{1}{2}\left[\frac{y^2}{4}\right]_1^4}{\left[\frac{2y^{3/2}}{3\sqrt{2}}\right]_1^4} = \frac{\frac{15}{8}}{\frac{14}{3\sqrt{2}}} = \mathbf{0.568}$$

and $$\bar{y} = \frac{\int_1^4 xy\,dy}{\int_1^4 x\,dy} = \frac{\int_1^4 \sqrt{\frac{y}{2}}(y)dy}{\frac{14}{3\sqrt{2}}}$$

$$= \frac{\int_1^4 \frac{y^{3/2}}{\sqrt{2}} dy}{\frac{14}{3\sqrt{2}}} = \frac{\frac{1}{\sqrt{2}}\left[\frac{y^{5/2}}{\frac{5}{2}}\right]_1^4}{\frac{14}{3\sqrt{2}}}$$

$$= \frac{\frac{2}{5\sqrt{2}}(31)}{\frac{14}{3\sqrt{2}}} = \mathbf{2.657}$$

Hence the position of the centroid is at (0.568, 2.657)

Problem 6. Locate the position of the centroid enclosed by the curves $y = x^2$ and $y^2 = 8x$

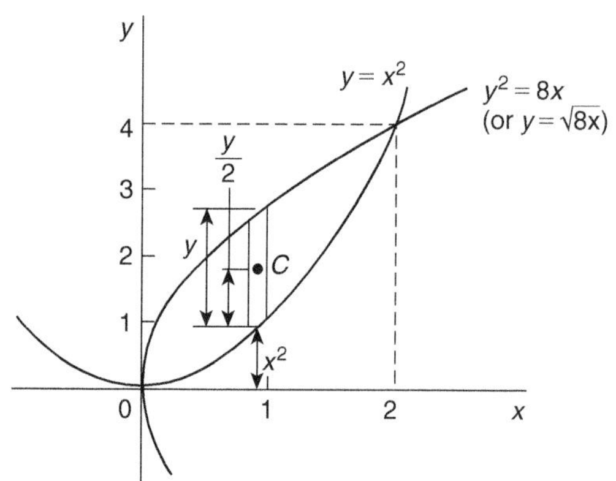

Figure 54.7

Figure 54.7 shows the two curves intersection at (0, 0) and (2, 4). These are the same curves as used in Problem 12, Chapter 51 where the shaded area was calculated as $2\frac{2}{3}$ square units. Let the co-ordinates of centroid C be $\bar{x}$ and $\bar{y}$.

By integration, $$\bar{x} = \frac{\int_0^2 xy\,dx}{\int_0^2 y\,dx}$$

The value of y is given by the height of the typical strip shown in Fig. 54.7, i.e. $y = \sqrt{8x} - x^2$. Hence,

$$\bar{x} = \frac{\int_0^2 x(\sqrt{8x} - x^2)dx}{2\frac{2}{3}} = \frac{\int_0^2 (\sqrt{8}x^{3/2} - x^3)dx}{2\frac{2}{3}}$$

$$= \frac{\left[\sqrt{8}\frac{x^{5/2}}{\frac{5}{2}} - \frac{x^4}{4}\right]_0^2}{2\frac{2}{3}} = \left(\frac{\sqrt{8}\frac{\sqrt{2^5}}{\frac{5}{2}} - 4}{2\frac{2}{3}}\right)$$

$$= \frac{2\frac{2}{5}}{2\frac{2}{3}} = \mathbf{0.9}$$

Care needs to be taken when finding $\bar{y}$ in such examples as this. From Fig. 54.7, $y = \sqrt{8x} - x^2$ and $\frac{y}{2} = \frac{1}{2}(\sqrt{8x} - x^2)$. The perpendicular distance from centroid C of the strip to Ox is $\frac{1}{2}(\sqrt{8x} - x^2) + x^2$. Taking moments about Ox gives:

(total area) ($\bar{y}$) = $\sum_{x=0}^{x=2}$(area of strip) (perpendicular distance of centroid of strip to Ox)

Hence (area) ($\bar{y}$)

$$= \int_0^2 \left[\sqrt{8x} - x^2\right]\left[\frac{1}{2}(\sqrt{8x} - x^2) + x^2\right]dx$$

i.e. $$\left(2\frac{2}{3}\right)(\bar{y}) = \int_0^2 \left[\sqrt{8x} - x^2\right]\left(\frac{\sqrt{8x}}{2} + \frac{x^2}{2}\right)dx$$

$$= \int_0^2 \left(\frac{8x}{2} - \frac{x^4}{2}\right)dx = \left[\frac{8x^2}{4} - \frac{x^5}{10}\right]_0^2$$

$$= \left(8 - 3\frac{1}{5}\right) - (0) = 4\frac{4}{5}$$

Hence $$\bar{y} = \frac{4\frac{4}{5}}{2\frac{2}{3}} = \mathbf{1.8}$$

Thus the position of the centroid of the enclosed area in Fig. 54.7 is at (0.9, 1.8)

Now try the following Practice Exercise

Practice Exercise 251 Centroids of simple shapes (Answers on page 729)

1. Determine the position of the centroid of a sheet of metal formed by the curve $y=4x-x^2$ which lies above the x-axis
2. Find the co-ordinates of the centroid of the area that lies between curve $\frac{y}{x}=x-2$ and the x-axis
3. Determine the co-ordinates of the centroid of the area formed between the curve $y=9-x^2$ and the x-axis
4. Determine the centroid of the area lying between $y=4x^2$, the y-axis and the ordinates $y=0$ and $y=4$
5. Find the position of the centroid of the area enclosed by the curve $y=\sqrt{5x}$, the x-axis and the ordinate $x=5$
6. Sketch the curve $y^2=9x$ between the limits $x=0$ and $x=4$. Determine the position of the centroid of this area
7. Calculate the points of intersection of the curves $x^2=4y$ and $\frac{y^2}{4}=x$, and determine the position of the centroid of the area enclosed by them
8. Sketch the curves $y=2x^2+5$ and $y-8=x(x+2)$ on the same axes and determine their points of intersection. Calculate the co-ordinates of the centroid of the area enclosed by the two curves

54.7 Theorem of Pappus

A theorem of Pappus* states:
'If a plane area is rotated about an axis in its own plane but not intersecting it, the volume of the solid formed is given by the product of the area and the distance moved by the centroid of the area'.

With reference to Fig. 54.8, when the curve $y=f(x)$ is rotated one revolution about the x-axis between the limits $x=a$ and $x=b$, the volume V generated is given by: volume $V=(A)(2\pi\bar{y})$, from which,

$$\bar{y} = \frac{V}{2\pi A}$$

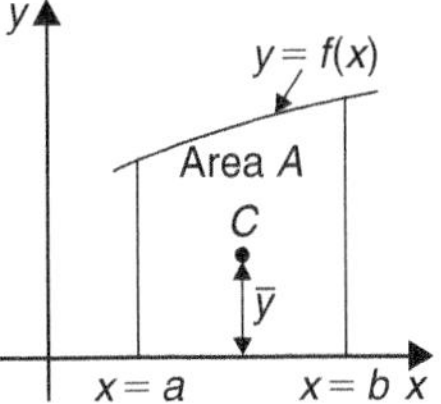

Figure 54.8

Problem 7. Determine the position of the centroid of a semicircle of radius r by using the theorem of Pappus. Check the answer by using integration (given that the equation of a circle, centre 0, radius r is $x^2+y^2=r^2$)

A semicircle is shown in Fig. 54.9 with its diameter lying on the x-axis and its centre at the origin. Area of semicircle $=\frac{\pi r^2}{2}$. When the area is rotated about the x-axis one revolution a sphere is generated of volume $\frac{4}{3}\pi r^3$.

*Who was Pappus? – **Pappus of Alexandria** (c. 290–c. 350) was one of the last great Greek mathematicians of Antiquity. *Collection*, his best-known work, is a compendium of mathematics in eight volumes. It covers a wide range of topics, including geometry, recreational mathematics, doubling the cube, polygons and polyhedra. To find out more go to www.routledge.com/cw/bird

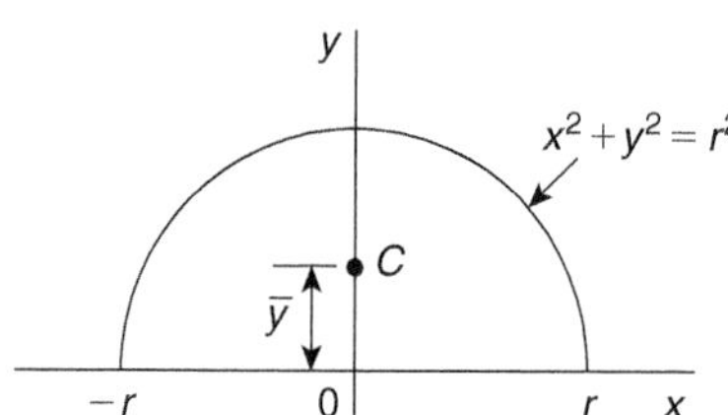

Figure 54.9

Let centroid C be at a distance $\bar{y}$ from the origin as shown in Fig. 54.9. From the theorem of Pappus, volume generated = area × distance moved through by centroid i.e.

$$\frac{4}{3}\pi r^3 = \left(\frac{\pi r^2}{2}\right)(2\pi\bar{y})$$

Hence $$\bar{y} = \frac{\frac{4}{3}\pi r^3}{\pi^2 r^2} = \frac{4r}{3\pi}$$

By integration,

$$\bar{y} = \frac{\frac{1}{2}\int_{-r}^{r} y^2 dx}{\text{area}}$$

$$= \frac{\frac{1}{2}\int_{-r}^{r}(r^2 - x^2)dx}{\frac{\pi r^2}{2}} = \frac{\frac{1}{2}\left[r^2 x - \frac{x^3}{3}\right]_{-r}^{r}}{\frac{\pi r^2}{2}}$$

$$= \frac{\frac{1}{2}\left[\left(r^3 - \frac{r^3}{3}\right) - \left(-r^3 + \frac{r^3}{3}\right)\right]}{\frac{\pi r^2}{2}} = \frac{4r}{3\pi}$$

Hence the centroid of a semicircle lies on the axis of symmetry, distance $\frac{4r}{3\pi}$ (or $0.424r$) from its diameter.

Problem 8. (a) Calculate the area bounded by the curve $y = 2x^2$, the x-axis and ordinates $x = 0$ and $x = 3$. (b) If this area is revolved (i) about the x-axis and (ii) about the y-axis, find the volumes of the solids produced. (c) Locate the position of the centroid using (i) integration, and (ii) the theorem of Pappus

(a) The required area is shown shaded in Fig. 54.10.

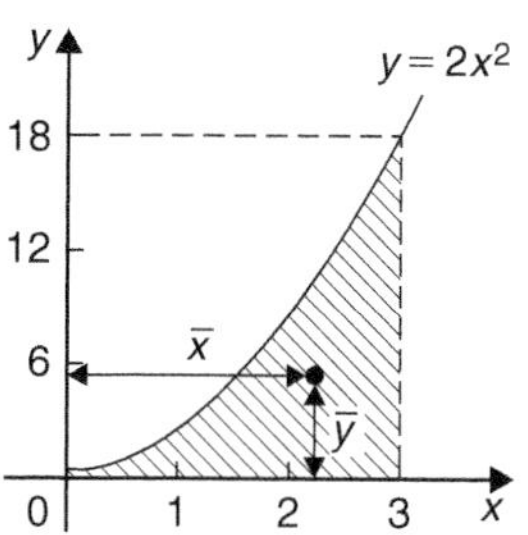

Figure 54.10

$$\text{Area} = \int_0^3 y\,dx = \int_0^3 2x^2\,dx = \left[\frac{2x^3}{3}\right]_0^3$$

$$= \mathbf{18\ square\ units}$$

(b) (i) When the shaded area of Fig. 54.10 is revolved 360° about the x-axis, the volume generated

$$= \int_0^3 \pi y^2 dx = \int_0^3 \pi(2x^2)^2 dx$$

$$= \int_0^3 4\pi x^4 dx = 4\pi\left[\frac{x^5}{5}\right]_0^3 = 4\pi\left(\frac{243}{5}\right)$$

$$= \mathbf{194.4\pi\ cubic\ units}$$

(ii) When the shaded area of Fig. 54.10 is revolved 360° about the y-axis, the volume generated = (volume generated by $x = 3$) − (volume generated by $y = 2x^2$)

$$= \int_0^{18} \pi(3)^2 dy - \int_0^{18} \pi\left(\frac{y}{2}\right)dy$$

$$= \pi\int_0^{18}\left(9 - \frac{y}{2}\right)dy = \pi\left[9y - \frac{y^2}{4}\right]_0^{18}$$

$$= \mathbf{81\pi\ cubic\ units}$$

(c) If the co-ordinates of the centroid of the shaded area in Fig. 54.10 are $(\bar{x}, \bar{y})$ then:

(i) by integration,

$$\bar{x} = \frac{\int_0^3 xy\,dx}{\int_0^3 y\,dx} = \frac{\int_0^3 x(2x^2)\,dx}{18}$$

$$= \frac{\int_0^3 2x^3\,dx}{18} = \frac{\left[\frac{2x^4}{4}\right]_0^3}{18} = \frac{81}{36} = \mathbf{2.25}$$

$$\bar{y} = \frac{\frac{1}{2}\int_0^3 y^2\,dx}{\int_0^3 y\,dx} = \frac{\frac{1}{2}\int_0^3 (2x^2)^2\,dx}{18}$$

$$= \frac{\frac{1}{2}\int_0^3 4x^4\,dx}{18} = \frac{\frac{1}{2}\left[\frac{4x^5}{5}\right]_0^3}{18} = \mathbf{5.4}$$

(ii) using the theorem of Pappus:

Volume generated when shaded area is revolved about $Oy = (\text{area})(2\pi\bar{x})$

i.e. $81\pi = (18)(2\pi\bar{x})$

from which, $\bar{x} = \dfrac{81\pi}{36\pi} = \mathbf{2.25}$

Volume generated when shaded area is revolved about $Ox = (\text{area})(2\pi\bar{y})$

i.e. $194.4\pi = (18)(2\pi\bar{y})$

from which, $\bar{y} = \dfrac{194.4\pi}{36\pi} = \mathbf{5.4}$

Hence the centroid of the shaded area in Fig. 54.10 is at (2.25, 5.4)

Problem 9. A cylindrical pillar of diameter 400 mm has a groove cut round its circumference. The section of the groove is a semicircle of diameter 50 mm. Determine the volume of material removed, in cubic centimetres, correct to 4 significant figures

A part of the pillar showing the groove is shown in Fig. 54.11.

The distance of the centroid of the semicircle from its base is $\dfrac{4r}{3\pi}$ (see Problem 7) $= \dfrac{4(25)}{3\pi} = \dfrac{100}{3\pi}$ mm.

The distance of the centroid from the centre of the pillar $= \left(200 - \dfrac{100}{3\pi}\right)$ mm.

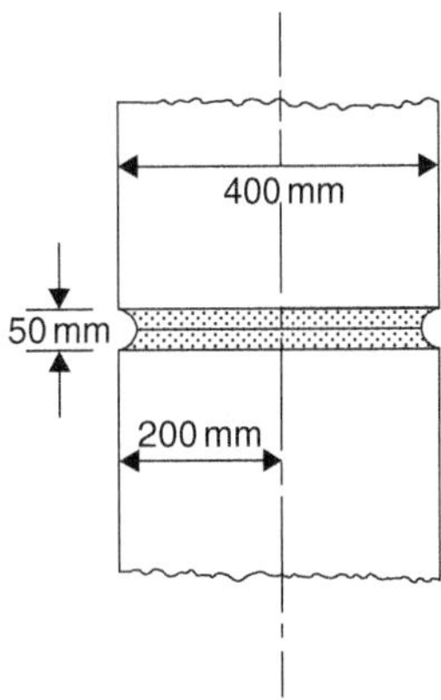

Figure 54.11

The distance moved by the centroid in one revolution

$$= 2\pi\left(200 - \frac{100}{3\pi}\right) = \left(400\pi - \frac{200}{3}\right)\text{mm.}$$

From the theorem of Pappus,
volume = area × distance moved by centroid

$$= \left(\frac{1}{2}\pi 25^2\right)\left(400\pi - \frac{200}{3}\right) = 1168250\,\text{mm}^3$$

Hence the volume of material removed is 1168 cm^3 correct to 4 significant figures.

Problem 10. A metal disc has a radius of 5.0 cm and is of thickness 2.0 cm. A semicircular groove of diameter 2.0 cm is machined centrally around the rim to form a pulley. Determine, using Pappus' theorem, the volume and mass of metal removed and the volume and mass of the pulley if the density of the metal is 8000 kg m^{-3}

A side view of the rim of the disc is shown in Fig. 54.12.

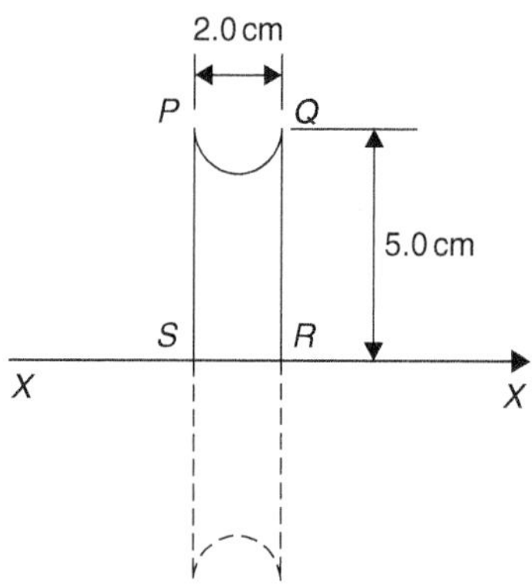

Figure 54.12

When area $PQRS$ is rotated about axis XX the volume generated is that of the pulley. The centroid of the semicircular area removed is at a distance of $\frac{4r}{3\pi}$ from its diameter (see Problem 7), i.e. $\frac{4(1.0)}{3\pi}$, i.e. 0.424 cm from PQ. Thus the distance of the centroid from XX is $(5.0-0.424)$, i.e. 4.576 cm. The distance moved through in one revolution by the centroid is $2\pi(4.576)$ cm.

Area of semicircle

$$= \frac{\pi r^2}{2} = \frac{\pi(1.0)^2}{2} = \frac{\pi}{2}\,\text{cm}^2$$

By the theorem of Pappus, volume generated

$$= \text{area} \times \text{distance moved by centroid}$$

$$= \left(\frac{\pi}{2}\right)(2\pi)(4.576)$$

i.e. **volume of metal removed = 45.16 cm³**

Mass of metal removed = density × volume

$$= 8000\,\text{kg m}^{-3} \times \frac{45.16}{10^6}\,\text{m}^3$$

$$= \mathbf{0.3613\,kg} \text{ or } \mathbf{361.3\,g}$$

Volume of pulley = volume of cylindrical disc − volume of metal removed

$$= \pi(5.0)^2(2.0) - 45.16 = \mathbf{111.9\,cm^3}$$

Mass of pulley = density × volume

$$= 8000\,\text{kg m}^{-3} \times \frac{111.9}{10^6}\,\text{m}^3$$

$$= \mathbf{0.8952\,kg} \text{ or } \mathbf{895.2\,g}$$

Now try the following Practice Exercise

Practice Exercise 252 The theorem of Pappus (Answers on page 729)

1. A right angled isosceles triangle having a hypotenuse of 8 cm is revolved one revolution about one of its equal sides as axis. Determine the volume of the solid generated using Pappus' theorem

2. A rectangle measuring 10.0 cm by 6.0 cm rotates one revolution about one of its longest sides as axis. Determine the volume of the resulting cylinder by using the theorem of Pappus

3. Using (a) the theorem of Pappus, and (b) integration, determine the position of the centroid of a metal template in the form of a quadrant of a circle of radius 4 cm. (The equation of a circle, centre 0, radius r is $x^2+y^2=r^2$)

4. (a) Determine the area bounded by the curve $y=5x^2$, the x-axis and the ordinates $x=0$ and $x=3$.

   (b) If this area is revolved 360° about (i) the x-axis, and (ii) the y-axis, find the volumes of the solids of revolution produced in each case.

   (c) Determine the co-ordinates of the centroid of the area using (i) integral calculus, and (ii) the theorem of Pappus

5. A metal disc has a radius of 7.0 cm and is of thickness 2.5 cm. A semicircular groove of diameter 2.0 cm is machined centrally around the rim to form a pulley. Determine the volume of metal removed using Pappus' theorem and express this as a percentage of the original volume of the disc. Find also the mass of metal removed if the density of the metal is 7800 kg m^{-3}

Practice Exercise 253 Multiple-choice questions on centroids of simple shapes (Answers on page 729)

Each question has only one correct answer

1. A metal template is bounded by the curve $y = x^2$, the x-axis and ordinates $x = 0$ and $x = 2$. The x-co-ordinate of the centroid of the area is:
(a) 2.5 (b) 1.5 (c) 2.0 (d) 1.0

2. Using the theorem of Pappus, the position of the centroid of a semicircle of radius r lies on the axis of symmetry at a distance from the diameter of:
(a) $\frac{3\pi}{4r}$ (b) $\frac{3r}{4\pi}$ (c) $\frac{4\pi}{3r}$ (d) $\frac{4r}{3\pi}$

3. The distance of the centroid from the y-axis of the surface area described by $y = 3x$ between $x = 0$ and $x = 3$ is:
(a) $\bar{x} = 1$ (b) $\bar{x} = 1.5$ (c) $\bar{x} = 2$ (d) $\bar{x} = 3$

4. A ceramic template is bounded by the curve $y = 2x^2$, the x-axis and ordinates $x = 0$ and $x = 2$. The y-co-ordinate of the centroid of the area is:
(a) 2.4 (b) 1.0 (c) 3.0 (d) 1.5

5. The x-co-ordinate of the centroid of the area lying between the curve $y = 4x - x^2$ and the x-axis is:
(a) 1.25 (b) 2.0 (c) 4.0 (d) 16.0

For fully worked solutions to each of the problems in Practice Exercises 250 to 252 in this chapter, go to the website:
www.routledge.com/cw/bird

COMPANION @ WEBSITE

Chapter 55

Second moments of area

Why it is important to understand: Second moments of area

The second moment of area is a property of a cross-section that can be used to predict the resistance of a beam to bending and deflection around an axis that lies in the cross-sectional plane. The stress in, and deflection of, a beam under load depends not only on the load but also on the geometry of the beam's cross-section; larger values of second moment cause smaller values of stress and deflection. This is why beams with larger second moments of area, such as I-beams, are used in building construction in preference to other beams with the same cross-sectional area. The second moment of area has applications in many scientific disciplines including fluid mechanics, engineering mechanics and biomechanics – for example to study the structural properties of bone during bending. The static roll stability of a ship depends on the second moment of area of the waterline section – short, fat ships are stable, long, thin ones are not. It is clear that calculations involving the second moment of area are very important in many areas of engineering.

At the end of this chapter, you should be able to:

- define second moment of area and radius of gyration
- derive second moments of area of regular sections – rectangle, triangle, circle and semicircle
- state the parallel and perpendicular axis theorems
- determine second moment of area and radius of gyration of regular sections using a table of derived results
- determine second moment of area and radius of gyration of composite areas using a table of derived results

55.1 Second moments of area and radius of gyration

The **first moment of area** about a fixed axis of a lamina of area A, perpendicular distance y from the centroid of the lamina is defined as Ay cubic units. The **second moment of area** of the same lamina as above is given by Ay^2, i.e. the perpendicular distance from the centroid of the area to the fixed axis is squared. Second moments of areas are usually denoted by I and have units of mm^4, cm^4 and so on.

Radius of gyration

Several areas, $a_1, a_2, a_3, \ldots$ at distances $y_1, y_2, y_3, \ldots$ from a fixed axis, may be replaced by a single area A, where $A = a_1 + a_2 + a_3 + \cdots$ at distance k from the axis, such that $Ak^2 = \sum ay^2$. k is called the **radius of gyration** of area A about the given axis. Since $Ak^2 = \sum ay^2 = I$ then the radius of gyration, $k = \sqrt{\dfrac{I}{A}}$

The second moment of area is a quantity much used in the theory of bending of beams, in the torsion of shafts, and in calculations involving water planes and centres of pressure.

55.2 Second moment of area of regular sections

The procedure to determine the second moment of area of regular sections about a given axis is (i) to find the second moment of area of a typical element and (ii) to sum all such second moments of area by integrating between appropriate limits.

For example, the second moment of area of the rectangle shown in Fig. 55.1 about axis PP is found by initially considering an elemental strip of width δx, parallel to and distance x from axis PP. Area of shaded strip $= b\delta x$. Second moment of area of the shaded strip about $PP = (x^2)(b\delta x)$

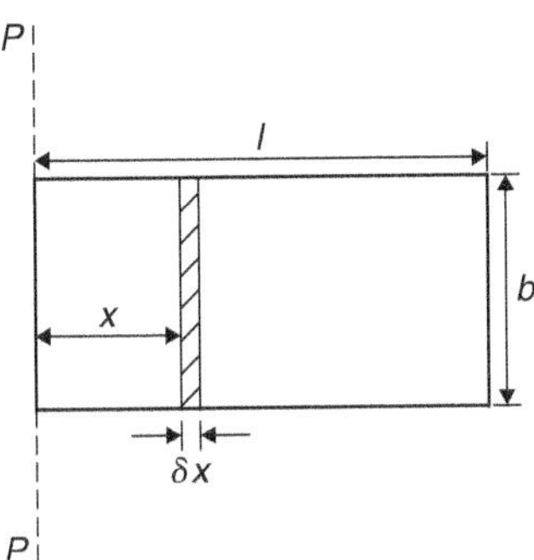

Figure 55.1

The second moment of area of the whole rectangle about PP is obtained by summing all such strips between $x=0$ and $x=l$, i.e. $\sum_{x=0}^{x=l} x^2 b\delta x$. It is a fundamental theorem of integration that

$$\lim_{\delta x \to 0} \sum_{x=0}^{x=l} x^2 b\delta x = \int_0^l x^2 b\, dx$$

Thus the second moment of area of the rectangle about $PP = b\int_0^l x^2 dx = b\left[\frac{x^3}{3}\right]_0^l = \mathbf{\frac{bl^3}{3}}$

Since the total area of the rectangle, $A = lb$, then

$$I_{pp} = (lb)\left(\frac{l^2}{3}\right) = \mathbf{\frac{Al^2}{3}}$$

$$I_{pp} = Ak_{pp}^2 \text{ thus } k_{pp}^2 = \frac{l^2}{3}$$

i.e. the radius of gyration about axes PP,

$$\mathbf{k_{pp}} = \sqrt{\frac{l^2}{3}} = \mathbf{\frac{l}{\sqrt{3}}}$$

55.3 Parallel axis theorem

In Fig. 55.2, axis GG passes through the centroid C of area A. Axes DD and GG are in the same plane, are parallel to each other and distance d apart. The parallel axis theorem states:

$$I_{DD} = I_{GG} + Ad^2$$

Using the parallel axis theorem the second moment of area of a rectangle about an axis through the centroid may be determined. In the rectangle shown in Fig. 55.3, $I_{pp} = \frac{bl^3}{3}$ (from above). From the parallel axis theorem

$$I_{pp} = I_{GG} + (bl)\left(\frac{l}{2}\right)^2$$

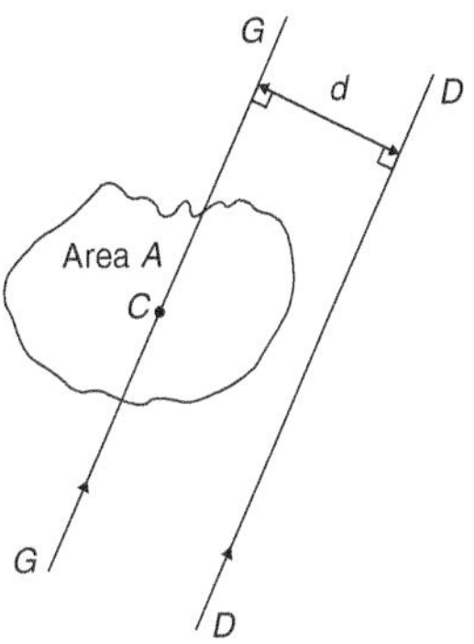

Figure 55.2

i.e. $\qquad \frac{bl^3}{3} = I_{GG} + \frac{bl^3}{4}$

from which, $\quad \mathbf{I_{GG}} = \frac{bl^3}{3} - \frac{bl^3}{4} = \mathbf{\frac{bl^3}{12}}$

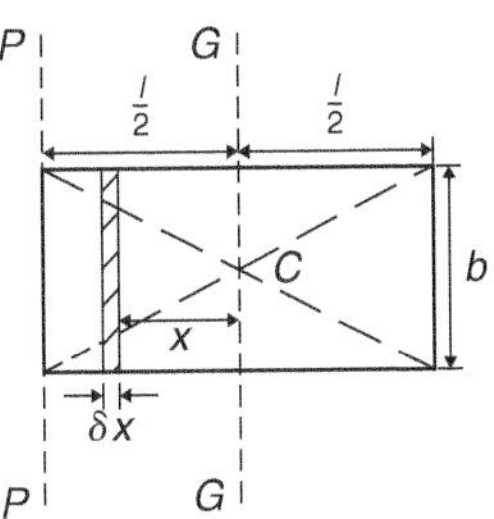

Figure 55.3

55.4 Perpendicular axis theorem

In Fig. 55.4, axes OX, OY and OZ are mutually perpendicular. If OX and OY lie in the plane

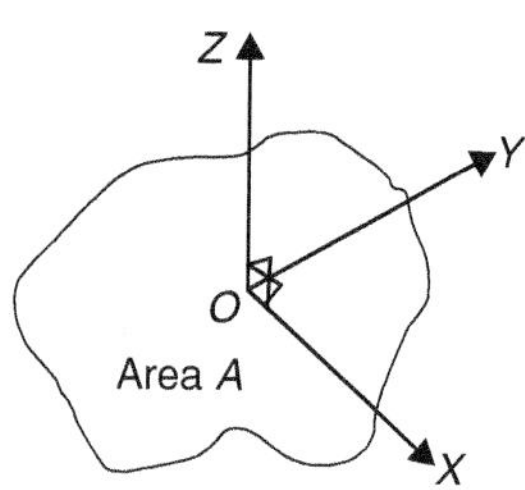

Figure 55.4

of area A then the perpendicular axis theorem states:

$$I_{OZ} = I_{OX} + I_{OY}$$

55.5 Summary of derived results

A summary of derive standard results for the second moment of area and radius of gyration of regular sections are listed in Table 55.1.

55.6 Worked problems on second moments of area of regular sections

Table 55.1 Summary of standard results of the second moments of areas of regular sections

Shape	Position of axis	Second moment of area, I	Radius of gyration, k
Rectangle length l breadth b	(1) Coinciding with b	$\frac{bl^3}{3}$	$\frac{l}{\sqrt{3}}$
	(2) Coinciding with l	$\frac{lb^3}{3}$	$\frac{b}{\sqrt{3}}$
	(3) Through centroid, parallel to b	$\frac{bl^3}{12}$	$\frac{l}{\sqrt{12}}$
	(4) Through centroid, parallel to l	$\frac{lb^3}{12}$	$\frac{b}{\sqrt{12}}$
Triangle Perpendicular height h base b	(1) Coinciding with b	$\frac{bh^3}{12}$	$\frac{h}{\sqrt{6}}$
	(2) Through centroid, parallel to base	$\frac{bh^3}{36}$	$\frac{h}{\sqrt{18}}$
	(3) Through vertex, parallel to base	$\frac{bh^3}{4}$	$\frac{h}{\sqrt{2}}$
Circle radius r	(1) Through centre, perpendicular to plane (i.e. polar axis)	$\frac{\pi r^4}{2}$	$\frac{r}{\sqrt{2}}$
	(2) Coinciding with diameter	$\frac{\pi r^4}{4}$	$\frac{r}{2}$
	(3) About a tangent	$\frac{5\pi r^4}{4}$	$\frac{\sqrt{5}}{2}r$
Semicircle radius r	Coinciding with diameter	$\frac{\pi r^4}{8}$	$\frac{r}{2}$

Problem 1. Determine the second moment of area and the radius of gyration about axes *AA*, *BB* and *CC* for the rectangle shown in Fig. 55.5

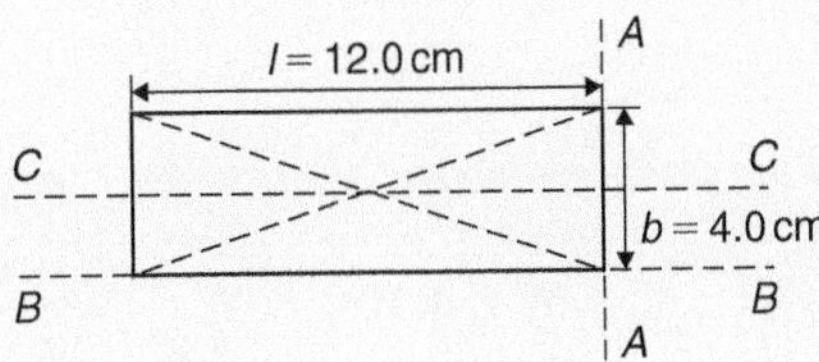

Figure 55.5

From Table 55.1, the second moment of area about axis *AA*, $I_{AA} = \frac{bl^3}{3} = \frac{(4.0)(12.0)^3}{3} = \mathbf{2304\,cm^4}$

Radius of gyration,

$$k_{AA} = \frac{l}{\sqrt{3}} = \frac{12.0}{\sqrt{3}} = \mathbf{6.93\,cm}$$

Similarly, $$I_{BB} = \frac{lb^3}{3} = \frac{(12.0)(4.0)^3}{3} = \mathbf{256\,cm^4}$$

and $$k_{BB} = \frac{b}{\sqrt{3}} = \frac{4.0}{\sqrt{3}} = \mathbf{2.31\,cm}$$

The second moment of area about the centroid of a rectangle is $\frac{bl^3}{12}$ when the axis through the centroid is parallel with the breadth, b. In this case, the axis *CC* is parallel with the length l.

Hence $$I_{CC} = \frac{lb^3}{12} = \frac{(12.0)(4.0)^3}{12} = \mathbf{64\,cm^4}$$

and $$k_{CC} = \frac{b}{\sqrt{12}} = \frac{4.0}{\sqrt{12}} = \mathbf{1.15\,cm}$$

Problem 2. Find the second moment of area and the radius of gyration about axis *PP* for the rectangle shown in Fig. 55.6

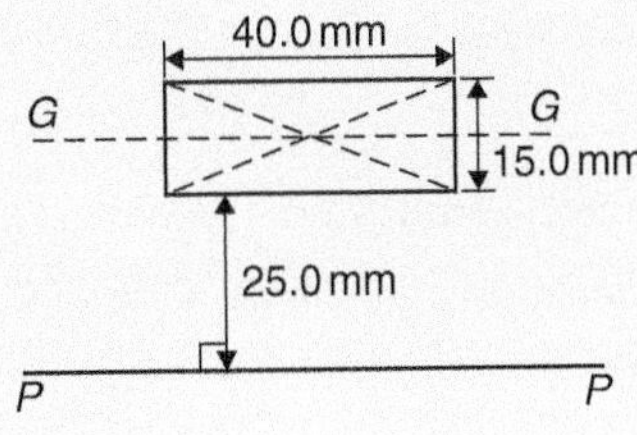

Figure 55.6

$I_{GG} = \frac{lb^3}{12}$ where $l = 40.0\,$mm and $b = 15.0\,$mm

Hence $I_{GG} = \frac{(40.0)(15.0)^3}{12} = 11\,250\,\text{mm}^4$

From the parallel axis theorem, $I_{PP} = I_{GG} + Ad^2$, where $A = 40.0 \times 15.0 = 600\,\text{mm}^2$ and $d = 25.0 + 7.5 = 32.5\,$mm, the perpendicular distance between *GG* and *PP*.

Hence, $$I_{PP} = 11\,250 + (600)(32.5)^2$$
$$= \mathbf{645\,000\,mm^4}$$

$$I_{PP} = Ak_{PP}^2$$

from which, $$k_{PP} = \sqrt{\frac{I_{PP}}{A}}$$
$$= \sqrt{\frac{645\,000}{600}} = \mathbf{32.79\,mm}$$

Problem 3. Determine the second moment of area and radius of gyration about axis *QQ* of the triangle *BCD* shown in Fig. 55.7

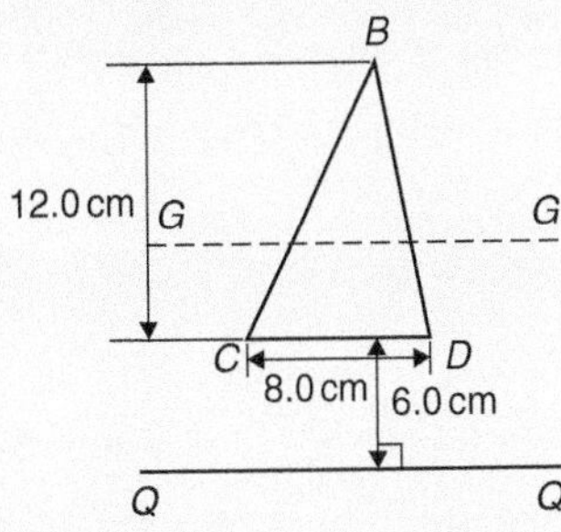

Figure 55.7

Using the parallel axis theorem: $I_{QQ} = I_{GG} + Ad^2$, where I_{GG} is the second moment of area about the centroid of the triangle,

i.e. $\frac{bh^3}{36} = \frac{(8.0)(12.0)^3}{36} = 384\,\text{cm}^4$, A is the area of the triangle $= \frac{1}{2}bh = \frac{1}{2}(8.0)(12.0) = 48\,\text{cm}^2$ and d is the distance between axes *GG* and $QQ = 6.0 + \frac{1}{3}(12.0) = 10\,$cm.

Hence the second moment of area about axis *QQ*,

$$I_{QQ} = 384 + (48)(10)^2 = \mathbf{5184\,cm^4}$$

Radius of gyration,

$$k_{QQ} = \sqrt{\frac{I_{QQ}}{A}} = \sqrt{\frac{5184}{48}} = \mathbf{10.4\,cm}$$

Problem 4. Determine the second moment of area and radius of gyration of the circle shown in Fig. 55.8 about axis YY

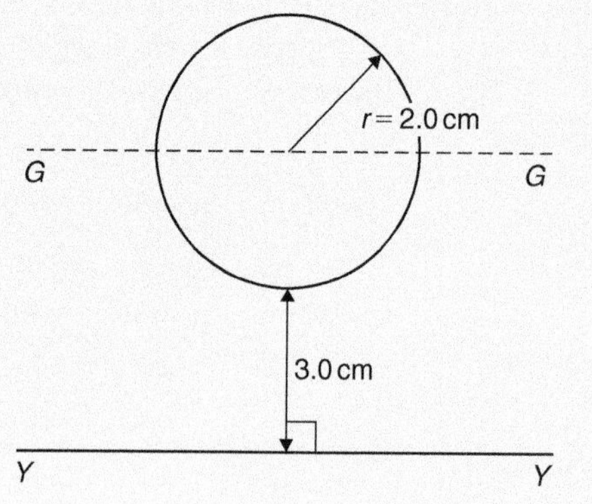

Figure 55.8

In Fig. 55.8, $I_{GG} = \frac{\pi r^4}{4} = \frac{\pi}{4}(2.0)^4 = 4\pi\,\text{cm}^4$. Using the parallel axis theorem, $I_{YY} = I_{GG} + Ad^2$, where $d = 3.0 + 2.0 = 5.0\,\text{cm}$.

Hence $\quad \boldsymbol{I_{YY}} = 4\pi + [\pi(2.0)^2](5.0)^2$

$= 4\pi + 100\pi = 104\pi = \mathbf{327\,cm^4}$

Radius of gyration,

$$k_{YY} = \sqrt{\frac{I_{YY}}{A}} = \sqrt{\frac{104\pi}{\pi(2.0)^2}} = \sqrt{26} = \mathbf{5.10\,cm}$$

Problem 5. Determine the second moment of area and radius of gyration for the semicircle shown in Fig. 55.9 about axis XX

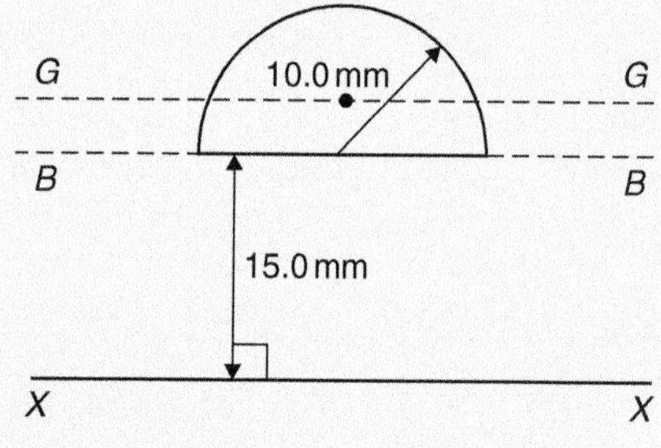

Figure 55.9

The centroid of a semicircle lies at $\frac{4r}{3\pi}$ from its diameter.

Using the parallel axis theorem: $I_{BB} = I_{GG} + Ad^2$,

where $\quad I_{BB} = \frac{\pi r^4}{8}$ (from Table 55.1)

$$= \frac{\pi(10.0)^4}{8} = 3927\,\text{mm}^4,$$

$$A = \frac{\pi r^2}{2} = \frac{\pi(10.0)^2}{2} = 157.1\,\text{mm}^2$$

and $\quad d = \frac{4r}{3\pi} = \frac{4(10.0)}{3\pi} = 4.244\,\text{mm}$

Hence $\quad 3927 = I_{GG} + (157.1)(4.244)^2$

i.e. $\quad 3927 = I_{GG} + 2830,$

from which, $\quad I_{GG} = 3927 - 2830 = 1097\,\text{mm}^4$

Using the parallel axis theorem again:

$I_{XX} = I_{GG} + A(15.0 + 4.244)^2$

i.e. $\boldsymbol{I_{XX}} = 1097 + (157.1)(19.244)^2$

$= 1097 + 58\,179 = 59\,276\,\text{mm}^4$ or $\mathbf{59\,280\,mm^4}$, correct to 4 significant figures.

Radius of gyration, $\boldsymbol{k_{XX}} = \sqrt{\frac{I_{XX}}{A}} = \sqrt{\frac{59\,276}{157.1}}$

$= \mathbf{19.42\,mm}$

Problem 6. Determine the polar second moment of area of the propeller shaft cross-section shown in Fig. 55.10

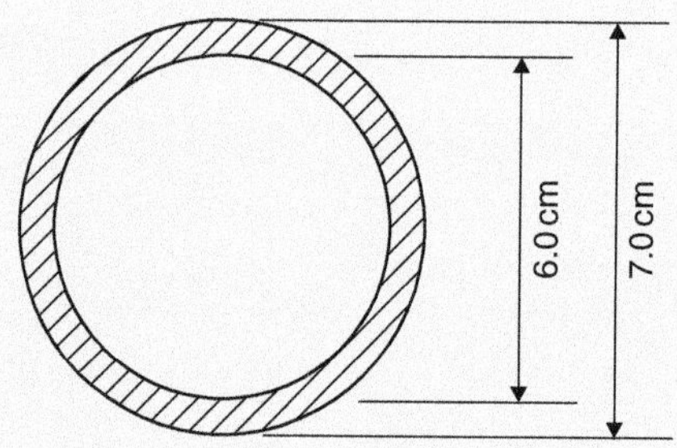

Figure 55.10

The polar second moment of area of a circle $= \frac{\pi r^4}{2}$. The polar second moment of area of the shaded area is given by the polar second moment of area of the 7.0 cm diameter circle minus the polar second moment of area of the 6.0 cm diameter circle. Hence the polar second moment of area of the

$$\text{cross-section shown} = \frac{\pi}{2}\left(\frac{7.0}{2}\right)^4 - \frac{\pi}{2}\left(\frac{6.0}{2}\right)^4$$

$$= 235.7 - 127.2 = \mathbf{108.5\ cm^4}$$

Problem 7. Determine the second moment of area and radius of gyration of a rectangular lamina of length 40 mm and width 15 mm about an axis through one corner, perpendicular to the plane of the lamina

The lamina is shown in Fig. 55.11.

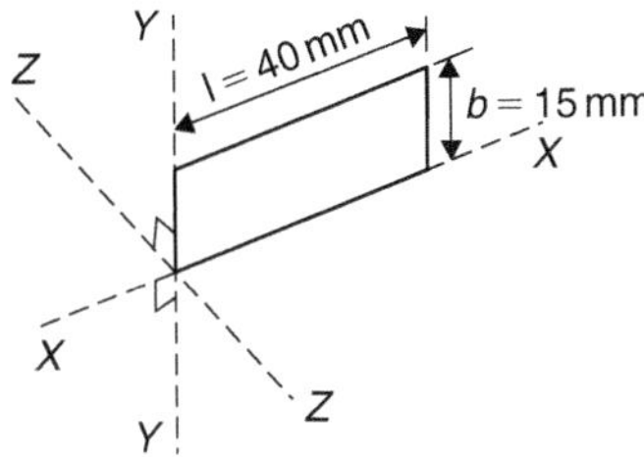

Figure 55.11

From the perpendicular axis theorem:

$$I_{ZZ} = I_{XX} + I_{YY}$$

$$I_{XX} = \frac{lb^3}{3} = \frac{(40)(15)^3}{3} = 45\,000\,\text{mm}^4$$

and $$I_{YY} = \frac{bl^3}{3} = \frac{(15)(40)^3}{3} = 320\,000\,\text{mm}^4$$

Hence $$I_{ZZ} = 45\,000 + 320\,000$$

$$= \mathbf{365\,000\ mm^4} \text{ or } \mathbf{36.5\ cm^4}$$

Radius of gyration,

$$k_{ZZ} = \sqrt{\frac{I_{ZZ}}{\text{area}}} = \sqrt{\frac{365\,000}{(40)(15)}}$$

$$= \mathbf{24.7\ mm} \text{ or } \mathbf{2.47\ cm}$$

Now try the following Practice exercise

Practice Exercise 254 Second moment of areas of regular sections (Answers on page 729)

1. Determine the second moment of area and radius of gyration for the rectangle shown in Fig. 55.12 about (a) axis AA (b) axis BB, and (c) axis CC

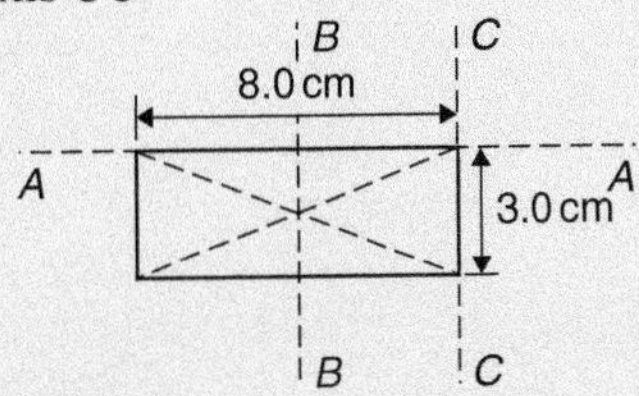

Figure 55.12

2. Determine the second moment of area and radius of gyration for the triangle shown in Fig. 55.13 about (a) axis DD (b) axis EE, and (c) an axis through the centroid of the triangle parallel to axis DD

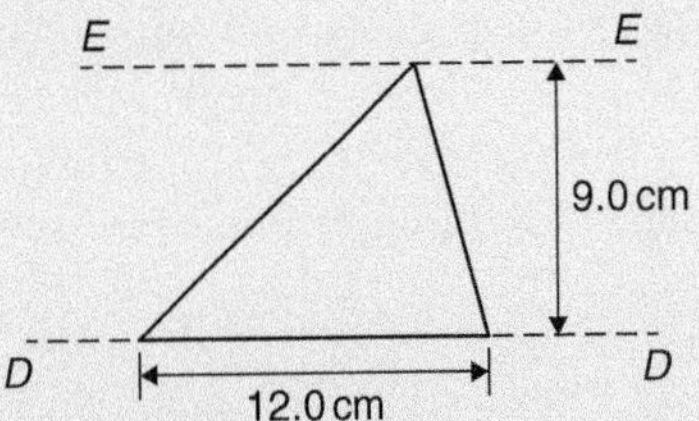

Figure 55.13

3. For the circle shown in Fig. 55.14, find the second moment of area and radius of gyration about (a) axis FF, and (b) axis HH

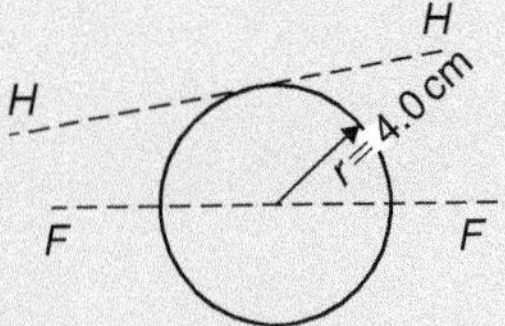

Figure 55.14

4. For the semicircle shown in Fig. 55.15, find the second moment of area and radius of gyration about axis JJ

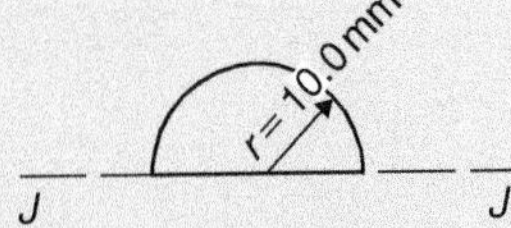

Figure 55.15

5. For each of the areas shown in Fig. 55.16 determine the second moment of area and

radius of gyration about axis LL, by using the parallel axis theorem

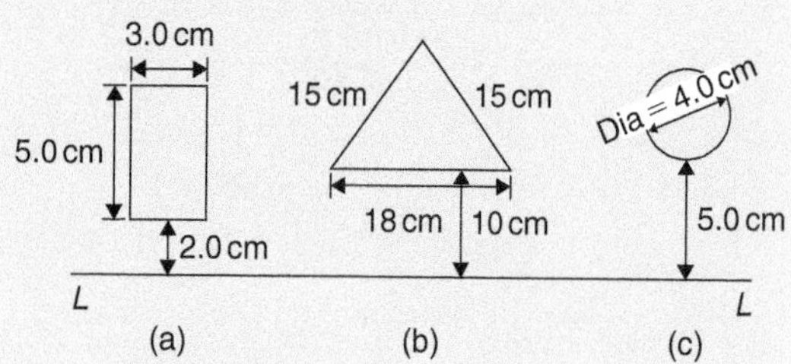

Figure 55.16

6. Calculate the radius of gyration of a rectangular door 2.0 m high by 1.5 m wide about a vertical axis through its hinge

7. A circular door of a boiler is hinged so that it turns about a tangent. If its diameter is 1.0 m, determine its second moment of area and radius of gyration about the hinge

8. A circular cover, centre 0, has a radius of 12.0 cm. A hole of radius 4.0 cm and centre X, where $OX = 6.0$ cm, is cut in the cover. Determine the second moment of area and the radius of gyration of the remainder about a diameter through 0 perpendicular to OX

55.7 Worked problems on second moments of area of composite areas

Problem 8. Determine correct to 3 significant figures, the second moment of area about XX for the composite area shown in Fig. 55.17

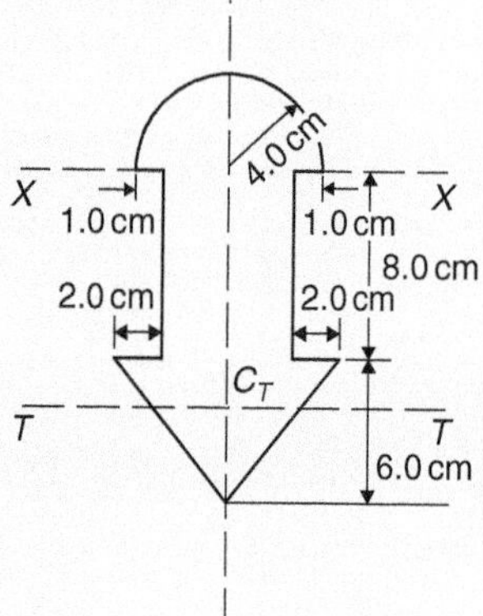

Figure 55.17

For the semicircle, $I_{XX} = \dfrac{\pi r^4}{8} = \dfrac{\pi(4.0)^4}{8}$

$= 100.5\,\text{cm}^4$

For the rectangle, $I_{XX} = \dfrac{bl^3}{3} = \dfrac{(6.0)(8.0)^3}{3}$

$= 1024\,\text{cm}^4$

For the triangle, about axis TT through centroid C_T,

$$I_{TT} = \frac{bh^3}{36} = \frac{(10)(6.0)^3}{36} = 60\,\text{cm}^4$$

By the parallel axis theorem, the second moment of area of the triangle about axis XX

$$= 60 + \left[\tfrac{1}{2}(10)(6.0)\right]\left[8.0 + \tfrac{1}{3}(6.0)\right]^2 = 3060\,\text{cm}^4$$

Total second moment of area about XX.

$$= 100.5 + 1024 + 3060 = 4184.5 = \mathbf{4180\,cm^4},$$

correct to 3 significant figures

Problem 9. Determine the second moment of area and the radius of gyration about axis XX for the I-section shown in Fig. 55.18

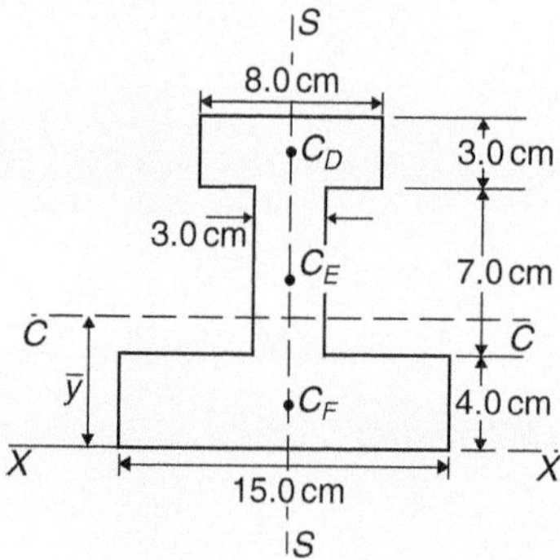

Figure 55.18

The I-section is divided into three rectangles, D, E and F and their centroids denoted by C_D, C_E and C_F respectively.

For rectangle D:
The second moment of area about C_D (an axis through C_D parallel to XX)

$$= \frac{bl^3}{12} = \frac{(8.0)(3.0)^3}{12} = 18\,\text{cm}^4$$

Using the parallel axis theorem: $I_{XX} = 18 + Ad^2$ where $A = (8.0)(3.0) = 24\,\text{cm}^2$ and $d = 12.5\,\text{cm}$

Hence $\boldsymbol{I_{XX}} = 18 + 24(12.5)^2 = \mathbf{3768\,cm^4}$

For rectangle E:

The second moment of area about C_E (an axis through C_E parallel to XX)

$$= \frac{bl^3}{12} = \frac{(3.0)(7.0)^3}{12} = 85.75\,\text{cm}^4$$

Using the parallel axis theorem:

$\boldsymbol{I_{XX}} = 85.75 + (7.0)(3.0)(7.5)^2 = \mathbf{1267\,cm^4}$

For rectangle F:

$$I_{XX} = \frac{bl^3}{3} = \frac{(15.0)(4.0)^3}{3} = \mathbf{320\,cm^4}$$

Total second moment of area for the *I*-section about axis *XX*,

$\boldsymbol{I_{XX}} = 3768 + 1267 + 320 = \mathbf{5355\,cm^4}$

Total area of I-section

$$= (8.0)(3.0) + (3.0)(7.0) + (15.0)(4.0) = 105\,\text{cm}^2.$$

Radius of gyration,

$$k_{XX} = \sqrt{\frac{I_{XX}}{A}} = \sqrt{\frac{5355}{105}} = \mathbf{7.14\,cm}$$

Now try the following Practice exercise

Practice Exercise 255 Second moment of areas of composite sections (Answers on page 729)

1. For the sections shown in Fig. 55.19, find the second moment of area and the radius of gyration about axis XX.

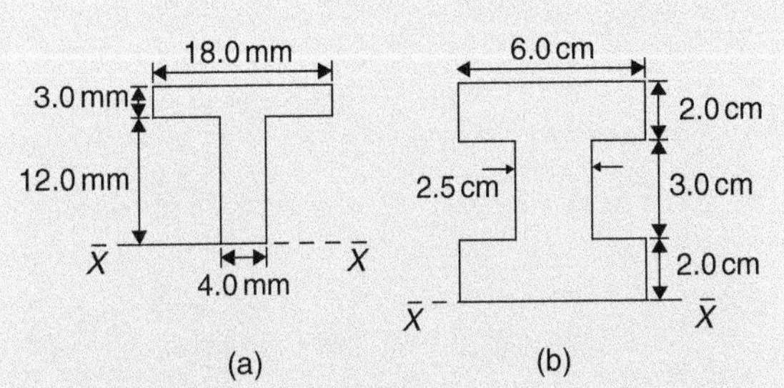

Figure 55.19

2. Determine the second moment of area about the given axes for the shapes shown in Fig. 55.20. (In Fig. 55.20(b), the circular area is removed.)

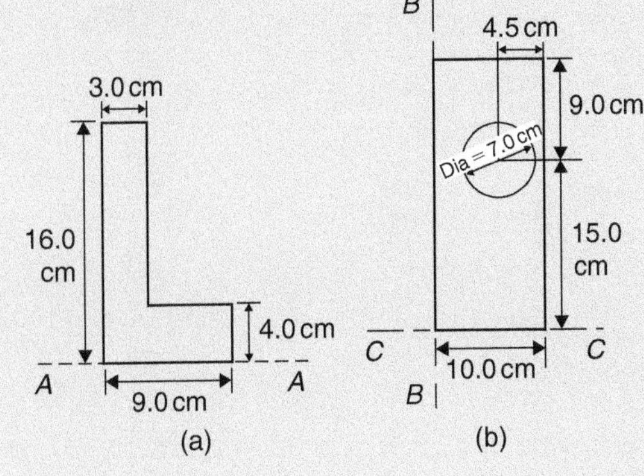

Figure 55.20

3. Find the second moment of area and radius of gyration about the axis XX for the beam section shown in Fig. 55.21.

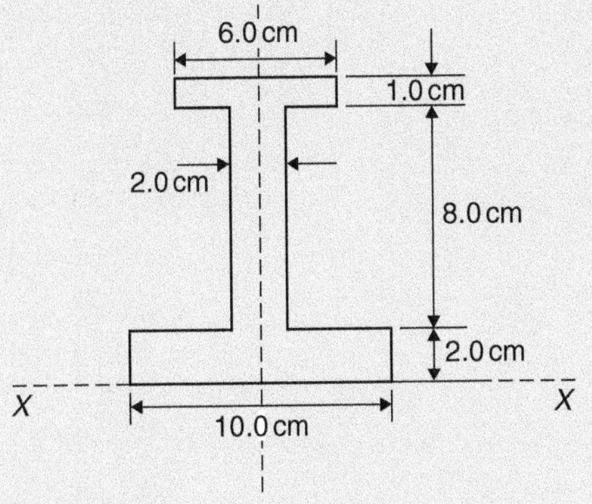

Figure 55.21

Practice Exercise 256 Multiple-choice questions on second moments of area (Answers on page 729)

(Answers on page 729)

Each question has only one correct answer

1. A rectangle has length x and breadth y. The second moment of area of the rectangle about its centroid parallel to y is:

   (a) $\frac{yx^3}{3}$ (b) $\frac{xy^3}{3}$ (c) $\frac{yx^3}{12}$ (d) $\frac{xy^3}{12}$

2. A circle has a diameter d. The second moment of area of the circle coinciding with the diameter is:

   (a) $\frac{\pi d^4}{32}$ (b) $\frac{\pi d^4}{64}$ (c) $\frac{5\pi d^4}{64}$ (d) $\frac{\pi d^2}{8}$

3. A triangle has perpendicular height h and base b. The second moment of area of the rectangle through the centroid parallel to the base is:

   (a) $\frac{bh^3}{12}$ (b) $\frac{bh^3}{24}$ (c) $\frac{bh^3}{36}$ (d) $\frac{bh^3}{4}$

4. A rectangle has length x and breadth y. The radius of gyration of the rectangle about its centroid parallel to y is:

   (a) $\frac{x}{\sqrt{3}}$ (b) $\frac{y}{\sqrt{12}}$ (c) $\frac{y}{\sqrt{3}}$ (d) $\frac{x}{\sqrt{12}}$

5. A circle has a diameter d. The radius of gyration of the circle coinciding about a tangent is:

   (a) $\frac{d}{4}$ (b) $\frac{\sqrt{5}d}{4}$ (c) $\frac{d}{2\sqrt{2}}$ (d) $\frac{\sqrt{5}d}{2}$

COMPANION WEBSITE

For fully worked solutions to each of the problems in Practice Exercises 254 and 255 in this chapter, go to the website:

www.routledge.com/cw/bird

Revision Test 14 Applications of integration

This Revision Test covers the material contained in Chapters 51 to 55. *The marks for each question are shown in brackets at the end of each question.*

1. The force F newtons acting on a body at a distance x metres from a fixed point is given by: $F = 2x + 3x^2$. If work done $= \int_{x_1}^{x_2} F\,dx$, determine the work done when the body moves from the position when $x = 1$ m to that when $x = 4$ m. (4)

2. Sketch and determine the area enclosed by the curve $y = 3\sin\frac{\theta}{2}$, the θ-axis and ordinates $\theta = 0$ and $\theta = \frac{2\pi}{3}$ (4)

3. Calculate the area between the curve $y = x^3 - x^2 - 6x$ and the x-axis. (10)

4. A voltage $v = 25\sin 50\pi t$ volts is applied across an electrical circuit. Determine, using integration, its mean and r.m.s. values over the range $t = 0$ to $t = 20$ ms, each correct to 4 significant figures. (12)

5. Sketch on the same axes the curves $x^2 = 2y$ and $y^2 = 16x$ and determine the co-ordinates of the points of intersection. Determine (a) the area enclosed by the curves, and (b) the volume of the solid produced if the area is rotated one revolution about the x-axis. (13)

6. Calculate the position of the centroid of the sheet of metal formed by the x-axis and the part of the curve $y = 5x - x^2$ which lies above the x-axis. (9)

7. A cylindrical pillar of diameter 500 mm has a groove cut around its circumference as shown in Fig. RT14.1. The section of the groove is a semicircle of diameter 40 mm. Given that the centroid of a semicircle from its base is $\frac{4r}{3\pi}$, use the theorem of Pappus to determine the volume of material removed, in cm^3, correct to 3 significant figures. (8)

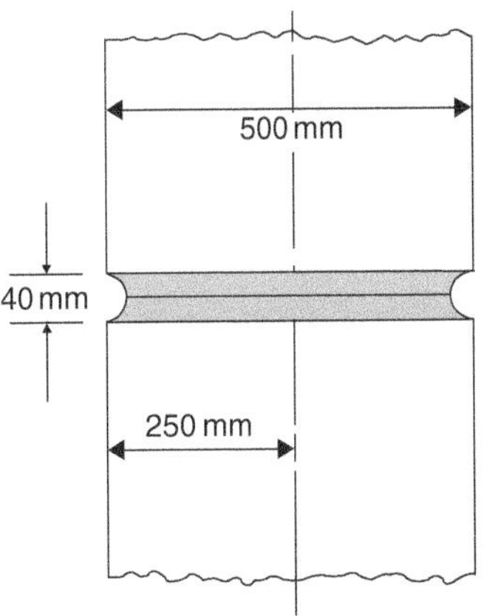

Figure RT14.1

8. For each of the areas shown in Fig. RT14.2 determine the second moment of area and radius of gyration about axis XX. (15)

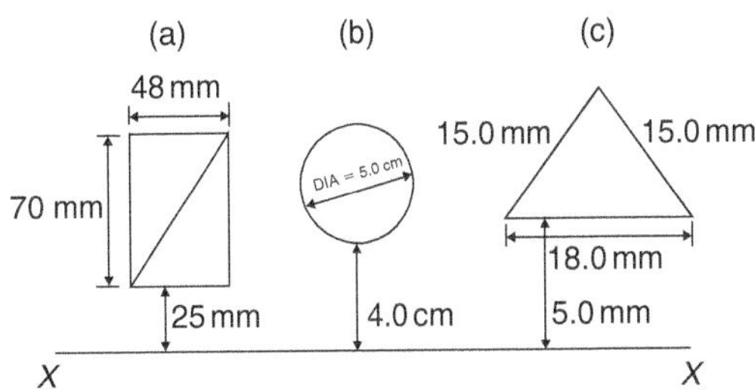

Figure RT14.2

9. A circular door is hinged so that it turns about a tangent. If its diameter is 1.0 m find its second moment of area and radius of gyration about the hinge. (5)

For lecturers/instructors/teachers, fully worked solutions to each of the problems in Revision Test 14, together with a full marking scheme, are available at the website:

www.routledge.com/cw/bird

Section 9

Differential equations

Chapter 56

Introduction to differential equations

Why it is important to understand: **Introduction to differential equations**

Differential equations play an important role in modelling virtually every physical, technical or biological process, from celestial motion, to bridge design, to interactions between neurons. Further applications are found in fluid dynamics with the design of containers and funnels, in heat conduction analysis with the design of heat spreaders in microelectronics, in rigid-body dynamic analysis, with falling objects and in exponential growth of current in an R-L circuit, to name but a few. This chapter introduces first order differential equations – the subject is clearly of great importance in many different areas of engineering.

At the end of this chapter, you should be able to:

- sketch a family of curves given a simple derivative
- define a differential equation – first order, second order, general solution, particular solution, boundary conditions
- solve a differential equation of the form $\frac{dy}{dx}=f(x)$
- solve a differential equation of the form $\frac{dy}{dx}=f(y)$
- solve a differential equation of the form $\frac{dy}{dx}=f(x)\cdot f(y)$

56.1 Family of curves

Integrating both sides of the derivative $\frac{dy}{dx}=3$ with respect to x gives $y=\int 3\,dx$, i.e. $y=3x+c$, where c is an arbitrary constant.

$y=3x+c$ represents a **family of curves**, each of the curves in the family depending on the value of c. Examples include $y=3x+8$, $y=3x+3$, $y=3x$ and $y=3x-10$ and these are shown in Fig. 56.1.

Each are straight lines of gradient 3. A particular curve of a family may be determined when a point on the curve is specified. Thus, if $y=3x+c$ passes through the point (1, 2) then $2=3(1)+c$, from which, $c=-1$. The equation of the curve passing through (1, 2) is therefore $y=3x-1$.

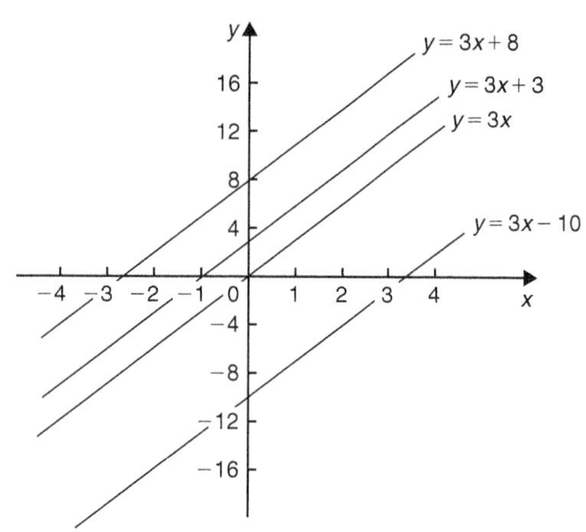

Figure 56.1

Problem 1. Sketch the family of curves given by the equations $\frac{dy}{dx}=4x$ and determine the equation of one of these curves which passes through the point (2, 3)

Integrating both sides of $\frac{dy}{dx}=4x$ with respect to x gives:

$$\int \frac{dy}{dx}\,dx = \int 4x\,dx, \quad \text{i.e. } y = 2x^2 + c$$

Some members of the family of curves having an equation $y=2x^2+c$ include $y=2x^2+15$, $y=2x^2+8$, $y=2x^2$ and $y=2x^2-6$, and these are shown in Fig. 56.2. To determine the equation of the curve passing through

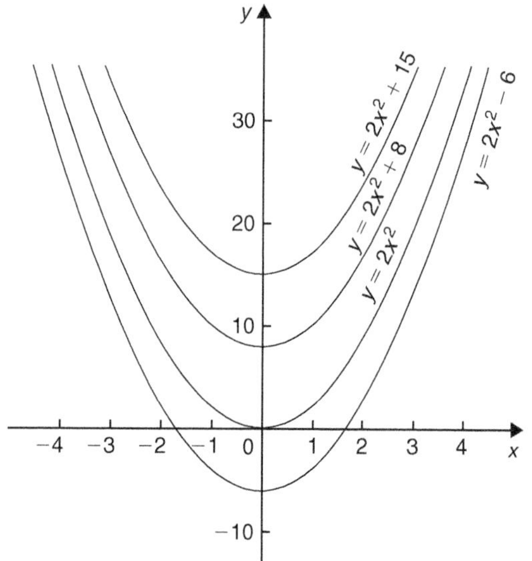

Figure 56.2

the point (2, 3), $x=2$ and $y=3$ are substituted into the equation $y=2x^2+c$

Thus $3=2(2)^2+c$, from which $c=3-8=-5$

Hence the equation of the curve passing through the point (2, 3) is $y=2x^2-5$

Now try the following Practice Exercise

Practice Exercise 257 Families of curves (Answers on page 729)

1. Sketch a family of curves represented by each of the following differential equations:

   (a) $\frac{dy}{dx}=6$ (b) $\frac{dy}{dx}=3x$ (c) $\frac{dy}{dx}=x+2$

2. Sketch the family of curves given by the equation $\frac{dy}{dx}=2x+3$ and determine the equation of one of these curves which passes through the point (1, 3)

56.2 Differential equations

A **differential equation** is one that contains differential coefficients.

Examples include

$$\text{(i)}\ \frac{dy}{dx}=7x \quad \text{and} \quad \text{(ii)}\ \frac{d^2y}{dx^2}+5\frac{dy}{dx}+2y=0$$

Differential equations are classified according to the highest derivative which occurs in them. Thus example (i) above is a **first order differential equation**, and example (ii) is a **second order differential equation**.

The **degree** of a differential equation is that of the highest power of the highest differential which the equation contains after simplification.

Thus $\left(\frac{d^2x}{dt^2}\right)^3+2\left(\frac{dx}{dt}\right)^5=7$ is a second order differential equation of degree three.

Starting with a differential equation it is possible, by integration and by being given sufficient data to determine unknown constants, to obtain the original function. This process is called **'solving the differential equation'**. A solution to a differential equation which contains one or more arbitrary constants of integration is called the **general solution** of the differential equation.

When additional information is given so that constants may be calculated the **particular solution** of the differential equation is obtained. The additional information is called **boundary conditions**. It was shown in Section 56.1 that $y=3x+c$ is the general solution of the differential equation $\frac{dy}{dx}=3$

Given the boundary conditions $x=1$ and $y=2$, produces the particular solution of $y=3x-1$. Equations which can be written in the form

$$\frac{dy}{dx}=f(x),\ \frac{dy}{dx}=f(y) \text{ and } \frac{dy}{dx}=f(x)\cdot f(y)$$

can all be solved by integration. In each case it is possible to separate the y's to one side of the equation and the x's to the other. Solving such equations is therefore known as solution by **separation of variables**.

56.3 The solution of equations of the form $\frac{dy}{dx}=f(x)$

A differential equation of the form $\frac{dy}{dx}=f(x)$ is solved by integrating both sides with respect to x

i.e. $$y=\int f(x)\,dx$$

Problem 2. Determine the general solution of:

$$x\frac{dy}{dx}=2-4x^3$$

Rearranging $x\frac{dy}{dx}=2-4x^3$ gives:

$$\frac{dy}{dx}=\frac{2-4x^3}{x}=\frac{2}{x}-\frac{4x^3}{x}=\frac{2}{x}-4x^2$$

Integrating both sides with respect to x gives:

$$y=\int\left(\frac{2}{x}-4x^2\right)dx$$

i.e. $$\mathbf{y=2\ln x-\frac{4}{3}x^3+c}$$

which is the general solution.

Problem 3. Find the particular solution of the differential equation $5\frac{dy}{dx}+2x=3$, given the boundary conditions $y=1\frac{2}{5}$ when $x=2$

Since $5\frac{dy}{dx}+2x=3$ then $\frac{dy}{dx}=\frac{3-2x}{5}=\frac{3}{5}-\frac{2x}{5}$

Hence $$y=\int\left(\frac{3}{5}-\frac{2x}{5}\right)dx$$

i.e. $$y=\frac{3x}{5}-\frac{x^2}{5}+c$$

which is the general solution.

Substituting the boundary conditions $y=1\frac{2}{5}$ and $x=2$ to evaluate c gives:

$1\frac{2}{5}=\frac{6}{5}-\frac{4}{5}+c$, from which, $c=1$

Hence the particular solution is $\mathbf{y=\frac{3x}{5}-\frac{x^2}{5}+1}$

Problem 4. Solve the equation:
$2t\left(t-\frac{d\theta}{dt}\right)=5$, given $\theta=2$ when $t=1$

Rearranging gives:

$$t-\frac{d\theta}{dt}=\frac{5}{2t} \quad\text{and}\quad \frac{d\theta}{dt}=t-\frac{5}{2t}$$

Integrating both sides with respect to t gives:

$$\theta=\int\left(t-\frac{5}{2t}\right)dt$$

i.e. $$\theta=\frac{t^2}{2}-\frac{5}{2}\ln t+c$$

which is the general solution.

When $\theta=2, t=1$, thus $2=\frac{1}{2}-\frac{5}{2}\ln 1+c$ from which, $c=\frac{3}{2}$

Hence the particular solution is:

$$\theta=\frac{t^2}{2}-\frac{5}{2}\ln t+\frac{3}{2}$$

i.e. $$\boldsymbol{\theta=\frac{1}{2}(t^2-5\ln t+3)}$$

Problem 5. The bending moment M of the beam is given by $\frac{dM}{dx}=-w(l-x)$, where w and x are constants. Determine M in terms of x given: $M=\frac{1}{2}wl^2$ when $x=0$

$$\frac{dM}{dx}=-w(l-x)=-wl+wx$$

Integrating both sides with respect to x gives:

$$M=-wlx+\frac{wx^2}{2}+c$$

which is the general solution.

When $M=\frac{1}{2}wl^2$, $x=0$.

Thus $\frac{1}{2}wl^2=-wl(0)+\frac{w(0)^2}{2}+c$

from which, $c=\frac{1}{2}wl^2$

Hence the particular solution is:

$$M=-wlx+\frac{wx^2}{2}+\frac{1}{2}wl^2$$

i.e. $\mathbf{M=\frac{1}{2}w(l^2-2lx+x^2)}$

or $\mathbf{M=\frac{1}{2}w(l-x)^2}$

Now try the following Practice Exercise

Practice Exercise 258 Solving equations of the form $\frac{dy}{dx}=f(x)$ (Answers on page 729)

In Problems 1 to 5, solve the differential equations.

1. $\frac{dy}{dx}=\cos 4x-2x$
2. $2x\frac{dy}{dx}=3-x^3$
3. $\frac{dy}{dx}+x=3$, given $y=2$ when $x=1$
4. $3\frac{dy}{d\theta}+\sin\theta=0$, given $y=\frac{2}{3}$ when $\theta=\frac{\pi}{3}$
5. $\frac{1}{e^x}+2=x-3\frac{dy}{dx}$, given $y=1$ when $x=0$
6. The gradient of a curve is given by:

   $$\frac{dy}{dx}+\frac{x^2}{2}=3x$$

   Find the equation of the curve if it passes through the point $\left(1,\frac{1}{3}\right)$
7. The acceleration, a, of a body is equal to its rate of change of velocity, $\frac{dv}{dt}$. Find an equation for v in term of t, given that when $t=0$, velocity $v=u$
8. An object is thrown vertically upwards with an initial velocity, u, of 20 m/s. The motion of the object follows the differential equation $\frac{ds}{dt}=u-gt$, where s is the height of the object in metres at time t seconds and $g=9.8\,\text{m/s}^2$. Determine the height of the object after 3 seconds if $s=0$ when $t=0$

56.4 The solution of equations of the form $\frac{dy}{dx}=f(y)$

A differential equation of the form $\frac{dy}{dx}=f(y)$ is initially rearranged to give $dx=\frac{dy}{f(y)}$ and then the solution is obtained by direct integration,

i.e. $$\int dx=\int\frac{dy}{f(y)}$$

Problem 6. Find the general solution of: $\frac{dy}{dx}=3+2y$

Rearranging $\frac{dy}{dx}=3+2y$ gives:

$$dx=\frac{dy}{3+2y}$$

Integrating both sides gives:

$$\int dx=\int\frac{dy}{3+2y}$$

Thus, by using the substitution $u=(3+2y)$ — see Chapter 45,

$$\mathbf{x=\tfrac{1}{2}\ln(3+2y)+c} \quad (1)$$

It is possible to give the general solution of a differential equation in a different form. For example, if $c=\ln k$, where k is a constant, then:

$$x=\tfrac{1}{2}\ln(3+2y)+\ln k,$$

i.e. $$x=\ln(3+2y)^{\frac{1}{2}}+\ln k$$

or $$\mathbf{x=\ln[k\sqrt{(3+2y)}]} \quad (2)$$

by the laws of logarithms, from which,

$$\mathbf{e^x=k\sqrt{(3+2y)}} \quad (3)$$

Equations (1), (2) and (3) are all acceptable general solutions of the differential equation

$$\frac{dy}{dx}=3+2y$$

Problem 7. Determine the particular solution of: $(y^2-1)\frac{dy}{dx}=3y$ given that $y=1$ when $x=2\frac{1}{6}$

Rearranging gives:

$$dx=\left(\frac{y^2-1}{3y}\right)dy=\left(\frac{y}{3}-\frac{1}{3y}\right)dy$$

Integrating gives:

$$\int dx=\int\left(\frac{y}{3}-\frac{1}{3y}\right)dy$$

i.e. $$x=\frac{y^2}{6}-\frac{1}{3}\ln y+c$$

which is the general solution.

When $y=1$, $x=2\frac{1}{6}$, thus $2\frac{1}{6}=\frac{1}{6}-\frac{1}{3}\ln 1+c$, from which, $c=2$

Hence the particular solution is:

$$\boldsymbol{x=\frac{y^2}{6}-\frac{1}{3}\ln y+2}$$

Problem 8. (a) The variation of resistance R ohms, of an aluminium conductor with temperature θ°C is given by $\frac{dR}{d\theta}=\alpha R$, where α is the temperature coefficient of resistance of aluminum. If $R=R_0$ when $\theta=0$°C, solve the equation for R. (b) If $\alpha=38\times10^{-4}$/°C, determine the resistance of an aluminum conductor at 50°C, correct to 3 significant figures, when its resistance at 0°C is 24.0 Ω

(a) $\frac{dR}{d\theta}=\alpha R$ is the form $\frac{dy}{dx}=f(y)$

Rearranging givens: $d\theta=\frac{dR}{\alpha R}$

Integrating both sides gives:

$$\int d\theta=\int\frac{dR}{\alpha R}$$

i.e. $$\theta=\frac{1}{\alpha}\ln R+c$$

which is the general solution.

Substituting the boundary conditions $R=R_0$ when $\theta=0$ gives:

$$0=\frac{1}{\alpha}\ln R_0+c$$

from which $c=-\frac{1}{\alpha}\ln R_0$

Hence the particular solution is

$$\theta=\frac{1}{\alpha}\ln R-\frac{1}{\alpha}\ln R_0=\frac{1}{\alpha}(\ln R-\ln R_0)$$

i.e. $\theta=\frac{1}{\alpha}\ln\left(\frac{R}{R_0}\right)$ or $\alpha\theta=\ln\left(\frac{R}{R_0}\right)$

Hence $e^{\alpha\theta}=\frac{R}{R_0}$ from which, $\boldsymbol{R=R_0e^{\alpha\theta}}$

(b) Substituting $\alpha=38\times10^{-4}$, $R_0=24.0$ and $\theta=50$ into $R=R_0e^{\alpha\theta}$ gives the resistance at 50°C, i.e. $\boldsymbol{R_{50}}=24.0e^{(38\times10^{-4}\times50)}=\textbf{29.0 ohms}$.

Now try the following Practice Exercise

Practice Exercise 259 Solving equations of the form $\frac{dy}{dx}=f(y)$ (Answers on page 730)

In Problems 1 to 3, solve the differential equations.

1. $\frac{dy}{dx}=2+3y$
2. $\frac{dy}{dx}=2\cos^2 y$
3. $(y^2+2)\frac{dy}{dx}=5y$, given $y=1$ when $x=\frac{1}{2}$
4. The current in an electric circuit is given by the equation
$$Ri+L\frac{di}{dt}=0$$
where L and R are constants. Shown that $i=Ie^{-\frac{Rt}{L}}$, given that $i=I$ when $t=0$
5. The velocity of a chemical reaction is given by $\frac{dx}{dt}=k(a-x)$. where x is the amount transferred in time t, k is a constant and a is the concentration at time $t=0$ when $x=0$. Solve the equation and determine x in terms of t

6. (a) Charge Q coulombs at time t seconds is given by the differential equation $R\frac{dQ}{dt}+\frac{Q}{C}=0$, where C is the capacitance in farads and R the resistance in ohms. Solve the equation for Q given that $Q=Q_0$ when $t=0$

   (b) A circuit possesses a resistance of $250\times10^3\,\Omega$ and a capacitance of 8.5×10^{-6} F, and after 0.32 seconds the charge falls to 8.0 C. Determine the initial charge and the charge after 1 second, each correct to 3 significant figures

7. A differential equation relating the difference in tension T, pulley contact angle θ and coefficient of friction μ is $\frac{dT}{d\theta}=\mu T$. When $\theta=0$, $T=150$ N, and $\mu=0.30$ as slipping starts. Determine the tension at the point of slipping when $\theta=2$ radians. Determine also the value of θ when T is 300 N

8. The rate of cooling of a body is given by $\frac{d\theta}{dt}=k\theta$, where k is a constant. If $\theta=60°$C when $t=2$ minutes and $\theta=50°$C when $t=5$ minutes, determine the time taken for θ to fall to 40°C, correct to the nearest second

56.5 The solution of equations of the form $\frac{dy}{dx}=f(x)\cdot f(y)$

A differential equation of the form $\frac{dy}{dx}=f(x)\cdot f(y)$, where $f(x)$ is a function of x only and $f(y)$ is a function of y only, may be rearranged as $\frac{dy}{f(y)}=f(x)dx$, and then the solution is obtained by direct integration, i.e.

$$\int\frac{dy}{f(y)}=\int f(x)dy$$

Problem 9. Solve the equation: $4xy\frac{dy}{dx}=y^2-1$

Separating the variables gives:

$$\left(\frac{4y}{y^2-1}\right)dy=\frac{1}{x}dx$$

Integrating both sides gives:

$$\int\left(\frac{4y}{y^2-1}\right)dy=\int\left(\frac{1}{x}\right)dx$$

Using the substituting $u=y^2-1$, the general solution is:

$$\mathbf{2\ln(y^2-1)=\ln x+c} \qquad (1)$$

or $\ln(y^2-1)^2-\ln x=c$

from which, $\ln\left\{\frac{(y^2-1)^2}{x}\right\}=c$

and $$\mathbf{\frac{(y^2-1)^2}{x}=e^c} \qquad (2)$$

If in equation (1), $c=\ln A$, where A is a different constant,

then $\ln(y^2-1)^2=\ln x+\ln A$

i.e. $\ln(y^2-1)^2=\ln Ax$

i.e. $\mathbf{(y^2-1)^2=Ax}$ (3)

Equations (1) to (3) are thus three valid solutions of the differential equation

$$4xy\frac{dy}{dx}=y^2-1$$

Problem 10. Determine the particular solution of $\frac{d\theta}{dt}=2e^{3t-2\theta}$, given that $t=0$ when $\theta=0$

$$\frac{d\theta}{dt}=2e^{3t-2\theta}=2(e^{3t})(e^{-2\theta})$$

by the laws of indices.

Separating the variables gives:

$$\frac{d\theta}{e^{-2\theta}}=2e^{2t}dt$$

i.e. $e^{2\theta}d\theta=2e^{3t}dt$

Integrating both sides gives:

$$\int e^{2\theta}d\theta=\int 2e^{3t}dt$$

Thus the general solution is:

$$\frac{1}{2}e^{2\theta}=\frac{2}{3}e^{3t}+c$$

When $t=0$, $\theta=0$, thus:

$$\frac{1}{2}e^0=\frac{2}{3}e^0+c$$

from which, $c=\frac{1}{2}-\frac{2}{3}=-\frac{1}{6}$

Hence the particular solution is:

$$\frac{1}{2}e^{2\theta}=\frac{2}{3}e^{3t}-\frac{1}{6}$$

or $\quad \mathbf{3e^{2\theta}=4e^{3t}-1}$

Problem 11. Find the curve which satisfies the equation $xy=(1+x^2)\frac{dy}{dx}$ and passes through the point (0, 1)

Separating the variables gives:

$$\frac{x}{(1+x^2)}dx=\frac{dy}{y}$$

Integrating both sides gives:

$$\frac{1}{2}\ln(1+x^2)=\ln y+c$$

When $x=0$, $y=1$ thus $\frac{1}{2}\ln 1=\ln 1+c$, from which, $c=0$

Hence the particular solution is $\frac{1}{2}\ln(1+x^2)=\ln y$

i.e. $\ln(1+x^2)^{\frac{1}{2}}=\ln y$, from which, $(1+x^2)^{\frac{1}{2}}=y$

Hence the equation of the curve is $y=\sqrt{(1+x^2)}$

Problem 12. The current i in an electric circuit containing resistance R and inductance L in series with a constant voltage source E is given by the differential equation $E-L\left(\frac{di}{dt}\right)=Ri$. Solve the equation and find i in terms of time t given that when $t=0$, $i=0$

In the $R-L$ series circuit shown in Fig. 56.3, the supply p.d., E, is given by

$$E=V_R+V_L$$

$$V_R=iR \quad \text{and} \quad V_L=L\frac{di}{dt}$$

Hence $\quad E=iR+L\frac{di}{dt}$

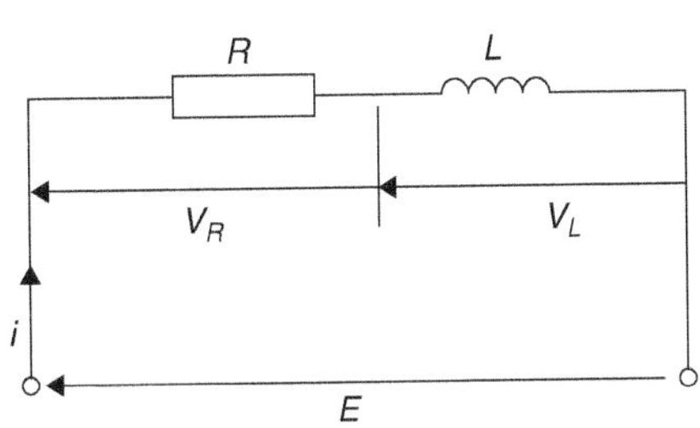

Figure 56.3

from which $\quad E-L\frac{di}{dt}=Ri$

Most electrical circuits can be reduced to a differential equation.

Rearranging $E-L\frac{di}{dt}=Ri$ gives $\frac{di}{dt}=\frac{E-Ri}{L}$

and separating the variables gives:

$$\frac{di}{E-Ri}=\frac{dt}{L}$$

Integrating both sides gives:

$$\int\frac{di}{E-Ri}=\int\frac{dt}{L}$$

Hence the general solution is:

$$-\frac{1}{R}\ln(E-Ri)=\frac{t}{L}+c$$

(by making a substitution $u=E-Ri$, see Chapter 45).

When $t=0$, $i=0$, thus $-\frac{1}{R}\ln E=c$

Thus the particular solution is:

$$-\frac{1}{R}\ln(E-Ri)=\frac{t}{L}-\frac{1}{R}\ln E$$

Transposing gives:

$$-\frac{1}{R}\ln(E-Ri)+\frac{1}{R}\ln E=\frac{t}{L}$$

$$\frac{1}{R}[\ln E-\ln(E-Ri)]=\frac{t}{L}$$

$$\ln\left(\frac{E}{E-Ri}\right)=\frac{Rt}{L}$$

from which $\dfrac{E}{E-Ri}=e^{\frac{Rt}{L}}$

Hence $\dfrac{E-Ri}{E}=e^{-\frac{Rt}{L}}$ and $E-Ri=Ee^{-\frac{Rt}{L}}$ and

$$Ri=E-Ee^{-\frac{Rt}{L}}$$

Hence current,

$$\boldsymbol{i=\frac{E}{R}\left(1-e^{-\frac{Rt}{L}}\right)}$$

which represents the law of growth of current in an inductive circuit as shown in Fig. 56.4.

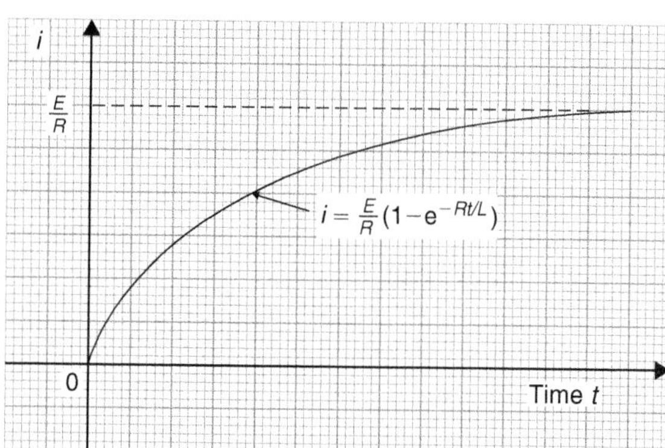

Figure 56.4

Problem 13. For an adiabatic expansion of a gas

$$C_v\frac{dp}{p}+C_p\frac{dV}{V}=0$$

where C_p and C_v are constants. Given $n=\dfrac{C_p}{C_v}$ show that $pV^n=$constant

Separating the variables gives:

$$C_v\frac{dp}{p}=-C_p\frac{dV}{V}$$

Integrating both sides gives:

$$C_v\int\frac{dp}{p}=-C_p\int\frac{dV}{V}$$

i.e. $$C_v\ln p=-C_p\ln V+k$$

Dividing throughout by constant C_v gives:

$$\ln p=-\frac{C_p}{C_v}\ln V+\frac{k}{C_v}$$

Since $\dfrac{C_p}{C_v}=n$, then $\ln p+n\ln V=K$,

where $K=\dfrac{k}{C_v}$

i.e. $\ln p+\ln V^n=K$ or $\ln pV^n=K$, by the laws of logarithms.

Hence $pV^n=e^K$ i.e. $\boldsymbol{pV^n=}$ **constant**.

Now try the following Practice Exercise

Practice Exercise 260 Solving equations of the form $\dfrac{dy}{dx}=f(x)\cdot f(y)$ (Answers on page 730)

In Problems 1 to 4, solve the differential equations.

1. $\dfrac{dy}{dx}=2y\cos x$
2. $(2y-1)\dfrac{dy}{dx}=(3x^2+1)$, given $x=1$ when $y=2$
3. $\dfrac{dy}{dx}=e^{2x-y}$, given $x=0$ when $y=0$
4. $2y(1-x)+x(1+y)\dfrac{dy}{dx}=0$, given $x=1$ when $y=1$
5. Show that the solution of the equation $\dfrac{y^2+1}{x^2+1}=\dfrac{y}{x}\dfrac{dy}{dx}$ is of the form $$\sqrt{\left(\frac{y^2+1}{x^2+1}\right)}=\text{constant}$$
6. Solve $xy=(1-x^2)\dfrac{dy}{dx}$ for y, given $x=0$ when $y=1$
7. Determine the equation of the curve which satisfies the equation $xy\dfrac{dy}{dx}=x^2-1$, and which passes through the point (1, 2)

8. The p.d., V, between the plates of a capacitor C charged by a steady voltage E through a resistor R is given by the equation

$$CR\frac{dV}{dt}+V=E$$

(a) Solve the equation for V given that at $t=0, V=0$
(b) Calculate V, correct to 3 significant figures, when $E=25\,\text{V}$, $C=20\times10^{-6}\,\text{F}$, $R=200\times10^{3}\,\Omega$ and $t=3.0\,\text{s}$

9. Determine the value of p, given that $x^3\frac{dy}{dx}=p-x$, and that $y=0$ when $x=2$ and when $x=6$

Practice Exercise 261 Multiple-choice questions on the introduction to differential equations (Answers on page 730)

Each question has only one correct answer

1. The order and degree of the differential equation $3x\frac{d^4y}{dx^4}+4x^3\left(\frac{dy}{dx}\right)^3-5xy=0$ is:
(a) third order, first degree
(b) fourth order, first degree
(c) first order, fourth degree
(d) first order, third degree.

2. Solving the differential equation $\frac{dy}{dx}=2x$ gives:
(a) $y=2\ln x+c$ (b) $y=2x^2+c$
(c) $y=2$ (d) $y=x^2+c$

3. Which of the following equations is a variable-separable differential equation?
(a) $(x+x^2y)dy=(3x+xy^2)dx$
(b) $(x+y)dx-5ydy=0$
(c) $3ydx=(x^2+1)dy$
(d) $y^2dx+(4x-3y)dy=0$

4. Given that $dy=x^2dx$, the equation of y in terms of x if the curve passes through $(1,1)$ is:
(a) $x^2-3y-3=0$ (b) $x^3-3y+2=0$
(c) $x^3+3y^2+2=0$ (d) $2y+x^3=2=0$

5. The solution of the differential equation $dy-xdx=0$, if the curve passes through $(1,0)$, is:
(a) $3x^2+2y-3=0$
(b) $2y^2+x^2-1=0$
(c) $x^2-2y-1=0$
(d) $2x^2+2y-2=0$

6. Solving the differential equation $\frac{dy}{dx}=2y$ gives:
(a) $\ln y=2x+c$ (b) $\ln x=y^2+c$
(c) $y=e^{2x+c}$ (d) $y=ce^{2x}$

7. The velocity of a particle moving in a straight line is described by the differential equation: $\frac{ds}{dt}=4t$. At time $t=2$, the position $s=12$. The particular solution is given by:
(a) $s=2t^2-12$ (b) $s=2t^2-280$
(c) $s=2t^2+4$ (d) $s=4$

8. Which of the following solutions to the differential equation $\frac{dy}{dx}=y(2x-3)$ is incorrect?
(a) $\ln y=x^2-3x+c$ (b) $y=e^{x^2}-3x+c$
(c) $y=e^{x^2-3x+c}$ (d) $y=ce^{x^2-3x}$

9. The particular solution of the differential equation $\frac{dy}{dx}+3x=4$, given that when $x=1, y=5$, is:
(a) $2y+3x^2=9$
(b) $2y=8x+3x^2-5$
(c) $2y-8x=3x^2-5$
(d) $2y+3x^2=8x+5$

10. Solving the differential equation $\frac{dy}{dx}=3e^{2x-3y}$, given that $x=0$ when $y=0$, gives:
(a) $2e^{3y}=9e^{2x}-7$ (b) $y=-e^{2x-3y}$
(c) $2e^{3y}-9e^{2x}=7$ (d) $2y=3(e^{2x-3y}-1)$

For fully worked solutions to each of the problems in Practice Exercises 257 to 260 in this chapter, go to the website:
www.routledge.com/cw/bird

COMPANION @ WEBSITE

Revision Test 15 Differential equations

This Revision Test covers the material contained in Chapter 56. *The marks for each question are shown in brackets at the end of each question.*

1. Solve the differential equation: $x\frac{dy}{dx} + x^2 = 5$ given that $y = 2.5$ when $x = 1$. (5)

2. Determine the equation of the curve which satisfies the differential equation $2xy\frac{dy}{dx} = x^2 + 1$ and which passes through the point (1, 2). (6)

3. A capacitor C is charged by applying a steady voltage E through a resistance R. The p.d. between the plates, V, is given by the differential equation:

$$CR\frac{dV}{dt} + V = E$$

   (a) Solve the equation for E given that when time $t = 0$, $V = 0$.

   (b) Evaluate voltage V when $E = 50$ V, $C = 10\,\mu$F, $R = 200\,$kΩ and $t = 1.2$ s.

(14)

For lecturers/instructors/teachers, fully worked solutions to each of the problems in Revision Test 15, together with a full marking scheme, are available at the website:

www.routledge.com/cw/bird

Section 10

Further number and algebra

Chapter 57

Boolean algebra and logic circuits

Why it is important to understand: Boolean algebra and logic circuits

Logic circuits are the basis for modern digital computer systems; to appreciate how computer systems operate an understanding of digital logic and Boolean algebra is needed. Boolean algebra (named after its developer, George Boole), is the algebra of digital logic circuits all computers use; Boolean algebra is the algebra of binary systems. A logic gate is a physical device implementing a Boolean function, performing a logical operation on one or more logic inputs, and produces a single logic output. Logic gates are implemented using diodes or transistors acting as electronic switches, but can also be constructed using electromagnetic relays, fluidic relays, pneumatic relays, optics, molecules or even mechanical elements. Learning Boolean algebra for logic analysis, learning about gates that process logic signals and learning how to design some smaller logic circuits is clearly of importance to computer engineers.

At the end of this chapter, you should be able to:

- draw a switching circuit and truth table for a 2-input and 3-input or-function and state its Boolean expression
- draw a switching circuit and truth table for a 2-input and 3-input and-function and state its Boolean expression
- produce the truth table for a 2-input not-function and state its Boolean expression
- simplify Boolean expressions using the laws and rules of Boolean algebra
- simplify Boolean expressions using de Morgan's laws
- simplify Boolean expressions using Karnaugh maps
- draw a circuit diagram symbol and truth table for a 3-input and-gate and state its Boolean expression
- draw a circuit diagram symbol and truth table for a 3-input or-gate and state its Boolean expression
- draw a circuit diagram symbol and truth table for a 3-input invert (or nor)-gate and state its Boolean expression
- draw a circuit diagram symbol and truth table for a 3-input nand-gate and state its Boolean expression
- draw a circuit diagram symbol and truth table for a 3-input nor-gate and state its Boolean expression
- devise logic systems for particular Boolean expressions
- use universal gates to devise logic circuits for particular Boolean expressions

57.1 Boolean algebra and switching circuits

A **two-state device** is one whose basic elements can only have one of two conditions. Thus, two-way switches, which can either be on or off, and the binary numbering system, having the digits 0 and 1 only, are two-state devices. In Boolean algebra (named after George Boole*), if A represents one state, then $\overline{A}$, called 'not-A', represents the second state.

The or-function

In Boolean algebra, the **or**-function for two elements A and B is written as $A+B$, and is defined as 'A, or B, or both A and B'. The equivalent electrical circuit for a two-input **or**-function is given by two switches connected in parallel. With reference to Fig. 57.1(a), the lamp will be on when A is on, when B is on, or when both A and B are on. In the table shown in Fig. 57.1(b), all the possible switch combinations are shown in columns 1 and 2, in which a 0 represents a switch being off and a 1 represents the switch being on, these columns being called the inputs. Column 3 is called the output and a 0 represents the lamp being off and a 1 represents the lamp being on. Such a table is called a **truth table**.

*Who was Boole? – **George Boole** (2 November 1815–8 December 1864) was an English mathematician, philosopher and logician that worked in the fields of differential equations and algebraic logic. Best known as the author of *The Laws of Thought*, Boole is also the inventor of the prototype of what is now called Boolean logic, which became the basis of the modern digital computer. To find out more go to www.routledge.com/cw/bird

The and-function

In Boolean algebra, the **and**-function for two elements A and B is written as $A \cdot B$ and is defined as 'both A and B'. The equivalent electrical circuit for a two-input **and**-function is given by two switches connected in series. With reference to Fig. 57.2(a) the lamp will be on only when both A and B are on. The truth table for a two-input **and**-function is shown in Fig. 57.2(b).

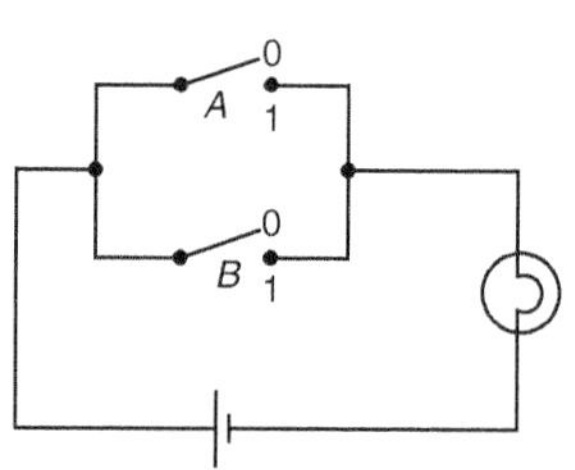

(a) Switching circuit for or-function

1	2	3
Input (switches)		Output (lamp)
A	B	$Z=A+B$
0	0	0
0	1	1
1	0	1
1	1	1

(b) Truth table for or-function

Figure 57.1

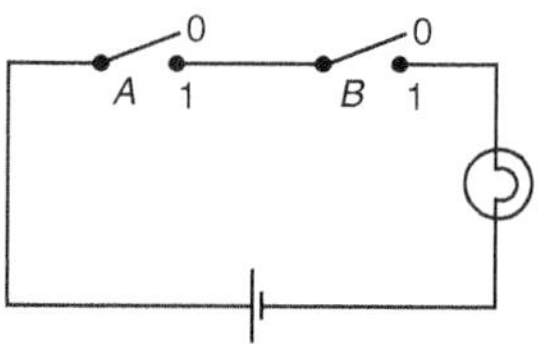

(a) Switching circuit for and-function

Input (switches)		Output (lamp)
A	B	$Z=A \cdot B$
0	0	0
0	1	0
1	0	0
1	1	1

(b) Truth table for and-function

Figure 57.2

The not-function

In Boolean algebra, the **not**-function for element A is written as $\overline{A}$, and is defined as 'the opposite to A'. Thus if A means switch A is on, $\overline{A}$ means that switch A is off. The truth table for the **not**-function is shown in Table 57.1.

In the above, the Boolean expressions, equivalent switching circuits and truth tables for the three functions used in Boolean algebra are given for a two-input

Table 57.1

Input A	Output $Z=\overline{A}$
0	1
1	0

system. A system may have more than two inputs and the Boolean expression for a three-input **or**-function having elements A, B and C is $A+B+C$. Similarly, a three-input **and**-function is written as $A \cdot B \cdot C$. The equivalent electrical circuits and truth tables for three-input **or** and **and**-functions are shown in Figs. 57.3(a) and (b) respectively.

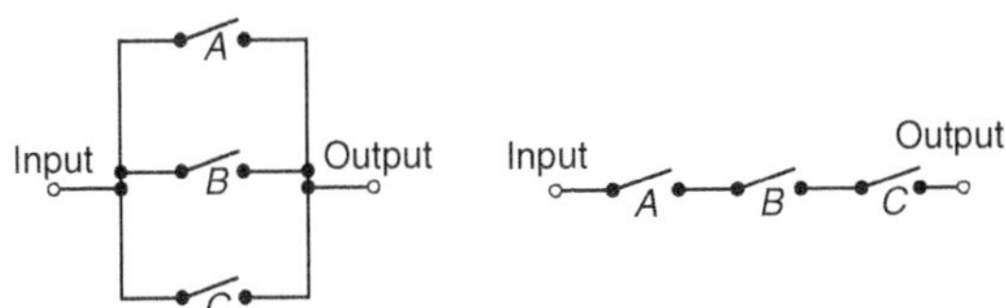

Input A B C	Output $Z=A+B+C$
0 0 0	0
0 0 1	1
0 1 0	1
0 1 1	1
1 0 0	1
1 0 1	1
1 1 0	1
1 1 1	1

(a) The or-function electrical circuit and truth table

Input A B C	Output $Z=A \cdot B \cdot C$
0 0 0	0
0 0 1	0
0 1 0	0
0 1 1	0
1 0 0	0
1 0 1	0
1 1 0	0
1 1 1	1

(b) The and-function electrical circuit and truth table

Figure 57.3

To achieve a given output, it is often necessary to use combinations of switches connected both in series and in parallel. If the output from a switching circuit is given by the Boolean expression $Z=A \cdot B+\overline{A} \cdot \overline{B}$, the truth table is as shown in Fig. 57.4(a). In this table, columns 1 and 2 give all the possible combinations of A and B. Column 3 corresponds to $A \cdot B$ and column 4 to $\overline{A} \cdot \overline{B}$, i.e. a 1 output is obtained when $A=0$ and when $B=0$. Column 5 is the **or**-function applied to columns 3 and 4 giving an output of $Z=A \cdot B+\overline{A} \cdot \overline{B}$. The corresponding switching circuit is shown in Fig. 57.4(b) in which A and B are connected in series to give $A \cdot B$, $\overline{A}$ and $\overline{B}$ are connected in series to give $\overline{A} \cdot \overline{B}$, and $A \cdot B$ and $\overline{A} \cdot \overline{B}$ are connected in parallel to give $A \cdot B+\overline{A} \cdot \overline{B}$. The circuit symbols

1	2	3	4	5
A	B	$A \cdot B$	$\overline{A} \cdot \overline{B}$	$Z=AB+\overline{A} \cdot \overline{B}$
0	0	0	1	1
0	1	0	0	0
1	0	0	0	0
1	1	1	0	1

(a) Truth table for $Z=A \cdot B+\overline{A} \cdot \overline{B}$

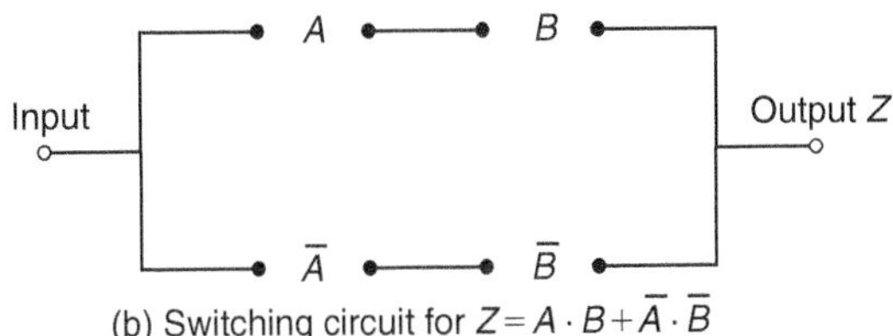

(b) Switching circuit for $Z=A \cdot B+\overline{A} \cdot \overline{B}$

Figure 57.4

used are such that A means the switch is on when A is 1, $\overline{A}$ means the switch is on when A is 0 and so on.

Problem 1. Derive the Boolean expression and construct a truth table for the switching circuit shown in Fig. 57.5.

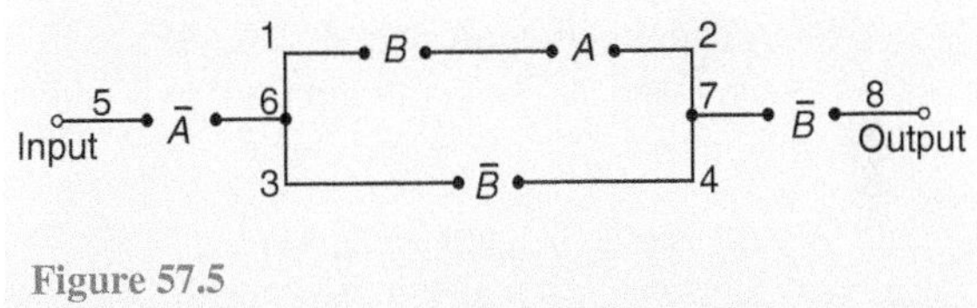

Figure 57.5

The switches between 1 and 2 in Fig. 57.5 are in series and have a Boolean expression of $B \cdot A$. The parallel circuit 1 to 2 and 3 to 4 have a Boolean expression of $(B \cdot A+\overline{B})$. The parallel circuit can be treated as a single switching unit, giving the equivalent of switches 5 to 6, 6 to 7 and 7 to 8 in series. Thus the output is given by:

$$\mathbf{Z=\overline{A} \cdot (B \cdot A+\overline{B}) \cdot \overline{B}}$$

The truth table is as shown in Table 57.2. Columns 1 and 2 give all the possible combinations of switches A and B. Column 3 is the **and**-function applied to columns 1 and 2, giving $B \cdot A$. Column 4 is $\overline{B}$, i.e., the opposite to column 2. Column 5 is the **or**-function applied to columns 3 and 4. Column 6 is $\overline{A}$, i.e. the opposite to column 1. The output is column 7 and is obtained by applying the **and**-function to columns 4, 5 and 6.

Table 57.2

1	2	3	4	5	6	7
A	B	$B\cdot A$	$\overline{B}$	$B\cdot A+\overline{B}$	$\overline{A}$	$Z=\overline{A}\cdot(B\cdot A+\overline{B})\cdot\overline{B}$
0	0	0	1	1	1	1
0	1	0	0	0	1	0
1	0	0	1	1	0	0
1	1	1	0	1	0	0

Problem 2. Derive the Boolean expression and construct a truth table for the switching circuit shown in Fig. 57.6

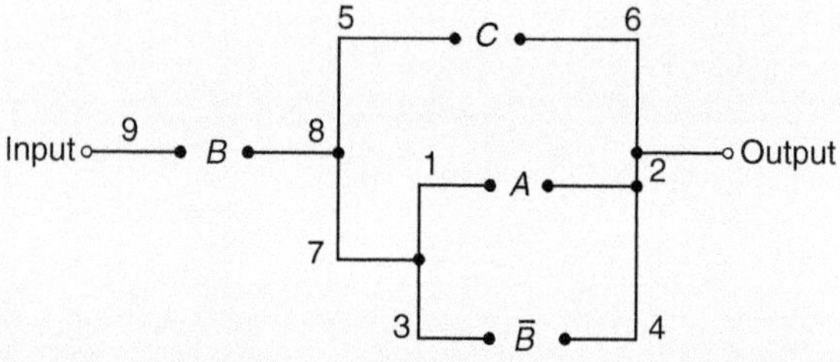

Figure 57.6

The parallel circuit 1 to 2 and 3 to 4 gives $(A+\overline{B})$ and this is equivalent to a single switching unit between 7 and 2. The parallel circuit 5 to 6 and 7 to 2 gives $C+(A+\overline{B})$ and this is equivalent to a single switching unit between 8 and 2. The series circuit 9 to 8 and 8 to 2 gives the output

$$\boldsymbol{Z=B\cdot[C+(A+\overline{B})]}$$

The truth table is shown in Table 57.3. Columns 1, 2 and 3 give all the possible combinations of A, B and C. Column 4 is $\overline{B}$ and is the opposite to column 2. Column 5 is the **or**-function applied to columns 1 and 4, giving $(A+\overline{B})$. Column 6 is the **or**-function applied to columns 3 and 5 giving $C+(A+\overline{B})$. The output is given in column 7 and is obtained by applying the **and**-function to columns 2 and 6, giving $Z=B\cdot[C+(A+\overline{B})]$

Problem 3. Construct a switching circuit to meet the requirements of the Boolean expression:

$$Z=A\cdot\overline{C}+\overline{A}\cdot B+\overline{A}\cdot B\cdot\overline{C}$$

Construct the truth table for this circuit

The three terms joined by **or**-functions, (+), indicate three parallel branches.

Table 57.3

1	2	3	4	5	6	7
A	B	C	$\overline{B}$	$A+\overline{B}$	$C+(A+\overline{B})$	$Z=B\cdot[C+(A+\overline{B})]$
0	0	0	1	1	1	0
0	0	1	1	1	1	0
0	1	0	0	0	0	0
0	1	1	0	0	1	1
1	0	0	1	1	1	0
1	0	1	1	1	1	0
1	1	0	0	1	1	1
1	1	1	0	1	1	1

having: branch 1 A **and** $\overline{C}$ in series

branch 2 $\overline{A}$ **and** B *in*series

and branch 3 $\overline{A}$ **and** B **and** $\overline{C}$ in series

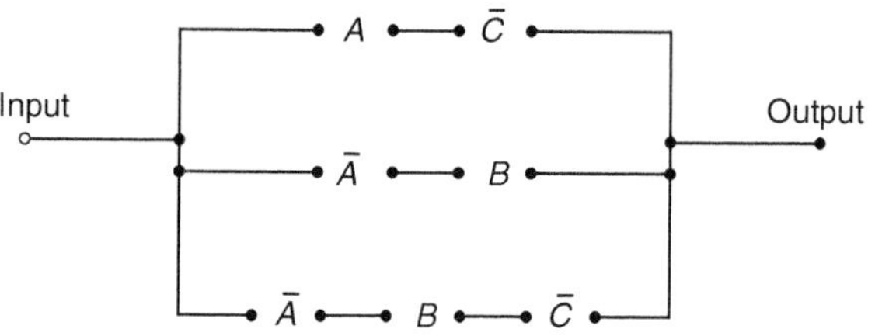

Figure 57.7

Hence the required switching circuit is as shown in Fig. 57.7. The corresponding truth table is shown in Table 57.4.

Table 57.4

1	2	3	4	5	6	7	8	9
A	B	C	$\overline{C}$	$A\cdot\overline{C}$	$\overline{A}$	$\overline{A}\cdot B$	$\overline{A}\cdot B\cdot\overline{C}$	$Z=A\cdot\overline{C}+\overline{A}\cdot B+\overline{A}\cdot B\cdot\overline{C}$
0	0	0	1	0	1	0	0	0
0	0	1	0	0	1	0	0	0
0	1	0	1	0	1	1	1	1
0	1	1	0	0	1	1	0	1
1	0	0	1	1	0	0	0	1
1	0	1	0	0	0	0	0	0
1	1	0	1	1	0	0	0	1
1	1	1	0	0	0	0	0	0

Column 4 is $\overline{C}$, i.e. the opposite to column 3

Column 5 is $A \cdot \overline{C}$, obtained by applying the **and**-function to columns 1 and 4

Column 6 is $\overline{A}$, the opposite to column 1

Column 7 is $\overline{A} \cdot B$, obtained by applying the **and**-function to columns 2 and 6

Column 8 is $\overline{A} \cdot B \cdot \overline{C}$, obtained by applying the **and**-function to columns 4 and 7

Column 9 is the output, obtained by applying the **or**-function to columns 5, 7 and 8

Problem 4. Derive the Boolean expression and construct the switching circuit for the truth table given in Table 57.5

Table 57.5

	A	B	C	Z
1	0	0	0	1
2	0	0	1	0
3	0	1	0	1
4	0	1	1	1
5	1	0	0	0
6	1	0	1	1
7	1	1	0	0
8	1	1	1	0

Examination of the truth table shown in Table 57.5 shows that there is a 1 output in the Z-column in rows 1, 3, 4 and 6. Thus, the Boolean expression and switching circuit should be such that a 1 output is obtained for row 1 **or** row 3 **or** row 4 **or** row 6. In row 1, A is 0 **and** B is 0 **and** C is 0 and this corresponds to the Boolean expression $\overline{A} \cdot \overline{B} \cdot \overline{C}$. In row 3, A is 0 **and** B is 1 **and** C is 0, i.e. the Boolean expression in $\overline{A} \cdot B \cdot \overline{C}$. Similarly in rows 4 and 6, the Boolean expressions are $\overline{A} \cdot B \cdot C$ and $A \cdot \overline{B} \cdot C$ respectively. Hence the Boolean expression is:

$$Z = \overline{A} \cdot \overline{B} \cdot \overline{C} + \overline{A} \cdot B \cdot \overline{C} + \overline{A} \cdot B \cdot C + A \cdot \overline{B} \cdot C$$

The corresponding switching circuit is shown in Fig. 57.8. The four terms are joined by **or**-functions, (+), and are represented by four parallel circuits. Each term has three elements joined by an **and**-function, and is represented by three elements connected in series.

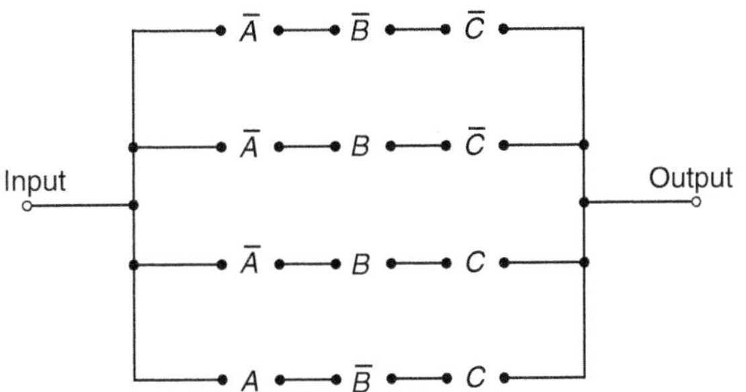

Figure 57.8

Now try the following Practice Exercise

Practice Exercise 262 Boolean algebra and switching circuits (Answers on page 730)

In Problems 1 to 4, determine the Boolean expressions and construct truth tables for the switching circuits given.

1. The circuit shown in Fig. 57.9

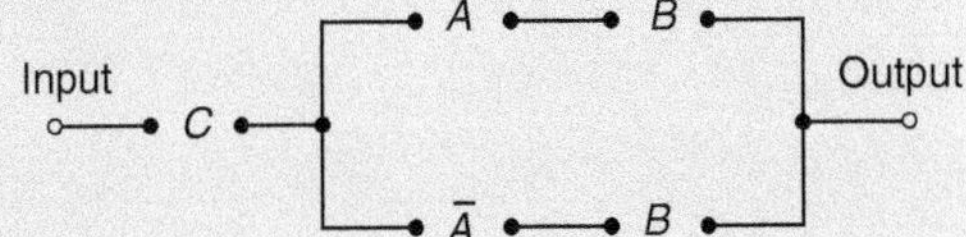

Figure 57.9

2. The circuit shown in Fig. 57.10

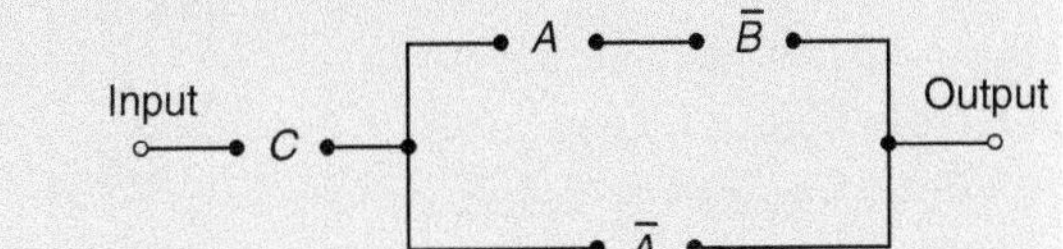

Figure 57.10

3. The circuit shown in Fig. 57.11

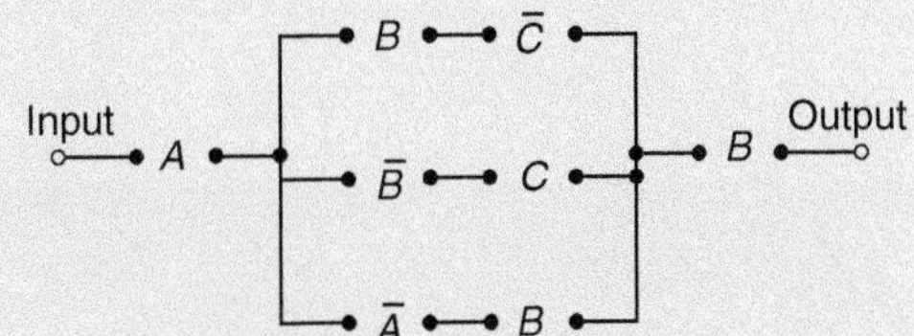

Figure 57.11

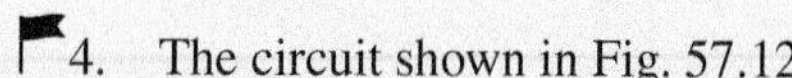

4. The circuit shown in Fig. 57.12

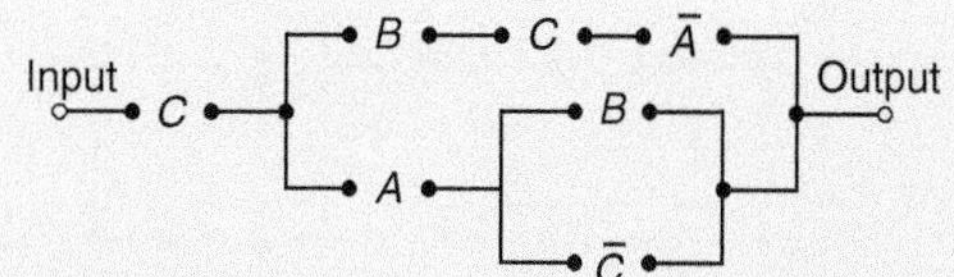

Figure 57.12

In Problems 5 to 7, construct switching circuits to meet the requirements of the Boolean expressions given.

5. $A\cdot C+A\cdot\overline{B}\cdot C+A\cdot B$

6. $A\cdot B\cdot C\cdot(A+B+C)$

7. $A\cdot(A\cdot\overline{B}\cdot C+B\cdot(A+\overline{C}))$

In Problems 8 to 10, derive the Boolean expressions and construct the switching circuits for the truth table stated.

8. Table 57.6, column 4

Table 57.6

1 A	2 B	3 C	4	5	6
0	0	0	0	1	1
0	0	1	1	0	0
0	1	0	0	0	1
0	1	1	0	1	0
1	0	0	0	1	1
1	0	1	0	0	1
1	1	0	1	0	0
1	1	1	0	0	0

9. Table 57.6, column 5

10. Table 57.6, column 6

57.2 Simplifying Boolean expressions

A Boolean expression may be used to describe a complex switching circuit or logic system. If the Boolean expression can be simplified, then the number of switches or logic elements can be reduced resulting in a saving in cost. Three principal ways of simplifying Boolean expressions are:

(a) by using the laws and rules of Boolean algebra (see Section 57.3),
(b) by applying de Morgan's laws (see Section 57.4), and
(c) by using Karnaugh maps (see Section 57.5).

57.3 Laws and rules of Boolean algebra

A summary of the principal laws and rules of Boolean algebra are given in Table 57.7. The way in which these laws and rules may be used to simplify Boolean expressions is shown in Problems 5 to 10.

Table 57.7

Ref.	Name	Rule or law
1	Commutative laws	$A+B=B+A$
2		$A\cdot B=B\cdot A$
3	Associative laws	$(A+B)+C=A+(B+C)$
4		$(A\cdot B)\cdot C=A\cdot(B\cdot C)$
5	Distributive laws	$A\cdot(B+C)=A\cdot B+A\cdot C$
6		$A+(B\cdot C)$ $=(A+B)\cdot(A+C)$
7	Sum rules	$A+0=A$
8		$A+1=1$
9		$A+A=A$
10		$A+\overline{A}=1$
11	Product rules	$A\cdot 0=0$
12		$A\cdot 1=A$
13		$A\cdot A=A$
14		$A\cdot\overline{A}=0$
15	Absorption rules	$A+A\cdot B=A$
16		$A\cdot(A+B)=A$
17		$A+\overline{A}\cdot B=A+B$

Problem 5. Simplify the Boolean expression:
$$\overline{P}\cdot\overline{Q}+\overline{P}\cdot Q+P\cdot\overline{Q}$$

With reference to Table 57.7:	*Reference*
$\overline{P}\cdot\overline{Q}+\overline{P}\cdot Q+P\cdot\overline{Q}$	
$=\overline{P}\cdot(\overline{Q}+Q)+P\cdot\overline{Q}$	5
$=\overline{P}\cdot 1+P\cdot\overline{Q}$	10
$=\mathbf{\overline{P}+P\cdot\overline{Q}}$	12

Problem 6. Simplify:

$$(P+\overline{P}\cdot Q)\cdot(Q+\overline{Q}\cdot P)$$

With reference to Table 57.7:	*Reference*
$(P+\overline{P}\cdot Q)\cdot(Q+\overline{Q}\cdot P)$	
$=P\cdot(Q+\overline{Q}\cdot P)$	
$+\overline{P}\cdot Q\cdot(Q+\overline{Q}\cdot P)$	5
$=P\cdot Q+P\cdot\overline{Q}\cdot P+\overline{P}\cdot Q\cdot Q$	
$+\overline{P}\cdot Q\cdot\overline{Q}\cdot P$	5
$=P\cdot Q+P\cdot\overline{Q}+\overline{P}\cdot Q$	
$+\overline{P}\cdot Q\cdot\overline{Q}\cdot P$	13
$=P\cdot Q+P\cdot\overline{Q}+\overline{P}\cdot Q+0$	14
$=P\cdot Q+P\cdot\overline{Q}+\overline{P}\cdot Q$	7
$=P\cdot(Q+\overline{Q})+\overline{P}\cdot Q$	5
$=P\cdot 1+\overline{P}\cdot Q$	10
$=\mathbf{P+\overline{P}\cdot Q}$	12

Problem 7. Simplify:

$$F\cdot G\cdot\overline{H}+F\cdot G\cdot H+\overline{F}\cdot G\cdot H$$

With reference to Table 57.7	*Reference*
$F\cdot G\cdot\overline{H}+F\cdot G\cdot H+\overline{F}\cdot G\cdot H$	
$=F\cdot G\cdot(\overline{H}+H)+\overline{F}\cdot G\cdot H$	5
$=F\cdot G\cdot 1+\overline{F}\cdot G\cdot H$	10
$=F\cdot G+\overline{F}\cdot G\cdot H$	12
$=\mathbf{G\cdot(F+\overline{F}\cdot H)}$	5

Problem 8. Simplify:
$\overline{F}\cdot\overline{G}\cdot H+\overline{F}\cdot G\cdot H+F\cdot\overline{G}\cdot H+F\cdot G\cdot H$

With reference to Table 57.7:	*Reference*
$\overline{F}\cdot\overline{G}\cdot H+\overline{F}\cdot G\cdot H+F\cdot\overline{G}\cdot H+F\cdot G\cdot H$	
$=\overline{G}\cdot H\cdot(\overline{F}+F)+G\cdot H\cdot(\overline{F}+F)$	5
$=\overline{G}\cdot H\cdot 1+G\cdot H\cdot 1$	10
$=\overline{G}\cdot H+G\cdot H$	12
$=H\cdot(\overline{G}+G)$	5
$=H\cdot 1=\mathbf{H}$	10 and 12

Problem 9. Simplify:

$$A\cdot\overline{C}+\overline{A}\cdot(B+C)+A\cdot B\cdot(C+\overline{B})$$

using the rules of Boolean algebra

With reference to Table 57.7	*Reference*
$A\cdot\overline{C}+\overline{A}\cdot(B+C)+A\cdot B\cdot(C+\overline{B})$	
$=A\cdot\overline{C}+\overline{A}\cdot B+\overline{A}\cdot C+A\cdot B\cdot C$	
$+A\cdot B\cdot\overline{B}$	5
$=A\cdot\overline{C}+\overline{A}\cdot B+\overline{A}\cdot C+A\cdot B\cdot C$	
$+A\cdot 0$	14
$=A\cdot\overline{C}+\overline{A}\cdot B+\overline{A}\cdot C+A\cdot B\cdot C$	7 and 11
$=A\cdot(\overline{C}+B)+\overline{A}\cdot B+\overline{A}\cdot C$	17
$=A\cdot\overline{C}+A\cdot B+\overline{A}\cdot B+\overline{A}\cdot C$	5
$=A\cdot\overline{C}+B\cdot(A+\overline{A})+\overline{A}\cdot C$	5
$=A\cdot\overline{C}+B\cdot 1+\overline{A}\cdot C$	10
$=\mathbf{A\cdot\overline{C}+B+\overline{A}\cdot C}$	12

Problem 10. Simplify the expression
$P\cdot\overline{Q}\cdot R+P\cdot Q\cdot(\overline{P}+R)+Q\cdot R\cdot(\overline{Q}+P)$
using the rules of Boolean algebra

With reference to Table 57.7:	*Reference*
$P\cdot\overline{Q}\cdot R+P\cdot Q\cdot(\overline{P}+R)+Q\cdot R\cdot(\overline{Q}+P)$	
$=P\cdot\overline{Q}\cdot R+P\cdot Q\cdot\overline{P}+P\cdot Q\cdot R$	
$+Q\cdot R\cdot\overline{Q}+Q\cdot R\cdot P$	5
$=P\cdot\overline{Q}\cdot R+0\cdot Q+P\cdot Q\cdot R+0\cdot R$	
$+P\cdot Q\cdot R$	14
$=P\cdot\overline{Q}\cdot R+P\cdot Q\cdot R+P\cdot Q\cdot R$	7 and 11
$=P\cdot\overline{Q}\cdot R+P\cdot Q\cdot R$	9
$=P\cdot R\cdot(Q+\overline{Q})$	5
$=P\cdot R\cdot 1$	10
$=\mathbf{P\cdot R}$	12

Now try the following Practice Exercise

Practice Exercise 263 Laws and rules of Boolean algebra (Answer on page 731)

Use the laws and rules of Boolean algebra given in Table 57.7 to simplify the following expressions:

1. $\overline{P}\cdot\overline{Q}+\overline{P}\cdot Q$
2. $\overline{P}\cdot Q+P\cdot Q+\overline{P}\cdot\overline{Q}$

3. $\overline{F}\cdot\overline{G}+F\cdot\overline{G}+\overline{G}\cdot(F+\overline{F})$
4. $F\cdot\overline{G}+F\cdot(G+\overline{G})+F\cdot G$
5. $(P+P\cdot Q)\cdot(Q+Q\cdot P)$
6. $\overline{F}\cdot\overline{G}\cdot H+\overline{F}\cdot G\cdot H+F\cdot\overline{G}\cdot H$
7. $F\cdot\overline{G}\cdot\overline{H}+F\cdot G\cdot H+\overline{F}\cdot G\cdot H$
8. $\overline{P}\cdot\overline{Q}\cdot\overline{R}+\overline{P}\cdot Q\cdot R+P\cdot\overline{Q}\cdot\overline{R}$
9. $\overline{F}\cdot\overline{G}\cdot\overline{H}+\overline{F}\cdot\overline{G}\cdot H+F\cdot\overline{G}\cdot\overline{H} + F\cdot\overline{G}\cdot H$
10. $F\cdot\overline{G}\cdot H+F\cdot G\cdot H+F\cdot G\cdot\overline{H} +\overline{F}\cdot G\cdot\overline{H}$
11. $R\cdot(P\cdot Q+P\cdot\overline{Q})+\overline{R}\cdot(\overline{P}\cdot\overline{Q}+\overline{P}\cdot Q)$
12. $\overline{R}\cdot(\overline{P}\cdot\overline{Q}+P\cdot Q+P\cdot\overline{Q}) +P\cdot(Q\cdot R+\overline{Q}\cdot R)$

57.4 De Morgan's laws

De Morgan's* **laws** may be used to simplify **not**-functions having two or more elements. The laws state that:

*Who was De Morgan? – **Augustus De Morgan** (27 June 1806–18 March 1871) was a British mathematician and logician. He formulated **De Morgan's laws** and introduced the term mathematical induction. To find out more go to www.routledge.com/cw/bird

$$\overline{A+B}=\overline{A}\cdot\overline{B} \quad \text{and} \quad \overline{A\cdot B}=\overline{A}+\overline{B}$$

and may be verified by using a truth table (see Problem 11). The application of De Morgan's laws in simplifying Boolean expressions is shown in Problems 12 and 13.

Problem 11. Verify that $\overline{A+B}=\overline{A}\cdot\overline{B}$

A Boolean expression may be verified by using a truth table. In Table 57.8, columns 1 and 2 give all the possible arrangements of the inputs A and B. Column 3 is the **or**-function applied to columns 1 and 2 and column 4 is the **not**-function applied to column 3. Columns 5 and 6 are the **not**-function applied to columns 1 and 2 respectively and column 7 is the **and**-function applied to columns 5 and 6.

Table 57.8

1 A	2 B	3 $A+B$	4 $\overline{A+B}$	5 $\overline{A}$	6 $\overline{B}$	7 $\overline{A}\cdot\overline{B}$
0	0	0	1	1	1	1
0	1	1	0	1	0	0
1	0	1	0	0	1	0
1	1	1	0	0	0	0

Since columns 4 and 7 have the same pattern of 0's and 1's this verifies that $\overline{A+B}=\overline{A}\cdot\overline{B}$

Problem 12. Simplify the Boolean expression $(\overline{\overline{A}\cdot B})+(\overline{\overline{A}+B})$ by using De Morgan's laws and the rules of Boolean algebra

Applying De Morgan's law to the first term gives:

$$\overline{\overline{A}\cdot B}=\overline{\overline{A}}+\overline{B}=A+\overline{B} \quad \text{since } \overline{\overline{A}}=A$$

Applying De Morgan's law to the second term gives:

$$\overline{\overline{A}+B}=\overline{\overline{A}}\cdot\overline{B}=A\cdot\overline{B}$$

Thus, $(\overline{\overline{A}\cdot B})+(\overline{\overline{A}+B})=(A+\overline{B})+A\cdot\overline{B}$

Removing the bracket and reordering gives:
$A+A\cdot\overline{B}+\overline{B}$
But, by rule 15, Table 57.7, $A+A\cdot B=A$. It follows that:
$A+A\cdot\overline{B}=A$

Thus: $\boldsymbol{(\overline{\overline{A}\cdot B})+(\overline{\overline{A}+B})=A+\overline{B}}$

Problem 13. Simplify the Boolean expression $(\overline{A\cdot\overline{B}+C})\cdot(\overline{A}+\overline{B\cdot\overline{C}})$ by using De Morgan's laws and the rules of Boolean algebra

Applying De Morgan's laws to the first term gives:

$$\overline{A\cdot\overline{B}+C}=\overline{A\cdot\overline{B}}\cdot\overline{C}=(\overline{A}+\overline{\overline{B}})\cdot\overline{C}$$
$$=(\overline{A}+B)\cdot\overline{C}=\overline{A}\cdot\overline{C}+B\cdot\overline{C}$$

Applying De Morgan's law to the second term gives:

$$\overline{A}+\overline{B\cdot\overline{C}}=\overline{A}+(\overline{B}+\overline{\overline{C}})=\overline{A}+(\overline{B}+C)$$

Thus $\quad(\overline{A\cdot\overline{B}+C})\cdot(\overline{A}+\overline{B\cdot\overline{C}})$

$$=(\overline{A}\cdot\overline{C}+B\cdot\overline{C})\cdot(\overline{A}+\overline{B}+C)$$
$$=\overline{A}\cdot\overline{A}\cdot\overline{C}+\overline{A}\cdot\overline{B}\cdot\overline{C}+\overline{A}\cdot\overline{C}\cdot C$$
$$+\overline{A}\cdot B\cdot\overline{C}+B\cdot\overline{B}\cdot\overline{C}+B\cdot\overline{C}\cdot C$$

But from Table 57.7, $\overline{A}\cdot\overline{A}=\overline{A}$ and $\overline{C}\cdot C=B\cdot\overline{B}=0$
Hence the Boolean expression becomes:

$$\overline{A}\cdot\overline{C}+\overline{A}\cdot\overline{B}\cdot\overline{C}+\overline{A}\cdot B\cdot\overline{C}$$
$$=\overline{A}\cdot\overline{C}(1+\overline{B}+B)$$
$$=\overline{A}\cdot\overline{C}(1+B)$$
$$=\overline{A}\cdot\overline{C}$$

Thus: $\quad\boldsymbol{(\overline{A\cdot\overline{B}+C})\cdot(\overline{A}+\overline{B\cdot\overline{C}})=\overline{A}\cdot\overline{C}}$

Now try the following Practice Exercise

Practice Exercise 264 Simplifying Boolean expressions using De Morgan's laws (Answers on page 731)

Use De Morgan's laws and the rules of Boolean algebra given in Table 57.7 on page 578 to simplify the following expressions.

1. $(\overline{A}\cdot\overline{B})\cdot(\overline{\overline{A}\cdot B})$
2. $(\overline{A+\overline{B\cdot C}})+(\overline{A\cdot B+C})$
3. $(\overline{\overline{A}\cdot B+B\cdot\overline{C}})\cdot\overline{A\cdot\overline{B}}$
4. $(\overline{A\cdot\overline{B}}+\overline{B\cdot\overline{C}})+(\overline{\overline{A}\cdot B})$
5. $(\overline{P\cdot\overline{Q}}+\overline{\overline{P}\cdot R})\cdot(\overline{P}\cdot\overline{Q\cdot R})$

57.5 Karnaugh maps

(i) Two-variable Karnaugh maps

A truth table for a two-variable expression is shown in Table 57.9(a), the '1' in the third row output showing that $Z=A\cdot\overline{B}$. Each of the four possible Boolean expressions associated with a two-variable function can be depicted as shown in Table 57.9(b) in which one cell is allocated to each row of the truth table. A matrix similar to that shown in Table 57.9(b) can be used to depict $Z=A\cdot\overline{B}$, by putting a 1 in the cell corresponding to $A\cdot\overline{B}$ and 0's in the remaining cells. This method of depicting a Boolean expression is called a two-variable **Karnaugh*** **map**, and is shown in Table 57.9(c).
To simplify a two-variable Boolean expression, the Boolean expression is depicted on a Karnaugh map, as outlined above. Any cells on the map having either a common vertical side or a common horizontal side are grouped together to form a **couple**. (This is a coupling

*Who is Karnaugh? – **Maurice Karnaugh** (4 October 1924 in New York City) is an American physicist, famous for the Karnaugh map used in Boolean algebra. To find out more go to www.routledge.com/cw/bird

Table 57.9

Inputs A	B	Output Z	Boolean expression
0	0	0	$\overline{A}\cdot\overline{B}$
0	1	0	$\overline{A}\cdot B$
1	0	1	$A\cdot\overline{B}$
1	1	0	$A\cdot B$

(a)

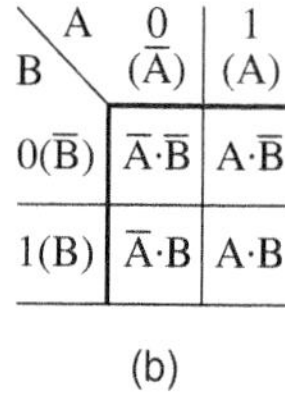

B \ A	0 ($\overline{A}$)	1 (A)
0($\overline{B}$)	$\overline{A}\cdot\overline{B}$	$A\cdot\overline{B}$
1(B)	$\overline{A}\cdot B$	$A\cdot B$

(b)

B \ A	0	1
0	0	1
1	0	0

(c)

together of cells, not just combining two together.) The simplified Boolean expression for a couple is given by those variables common to all cells in the couple. See Problem 14.

(ii) Three-variable Karnaugh maps

A truth table for a three-variable expression is shown in Table 57.10(a), the 1's in the output column showing that:

$$Z=\overline{A}\cdot\overline{B}\cdot C+\overline{A}\cdot B\cdot C+A\cdot B\cdot\overline{C}$$

Each of the eight possible Boolean expressions associated with a three-variable function can be depicted as shown in Table 57.10(b) in which one cell is allocated to each row of the truth table. A matrix similar to that shown in Table 57.10(b) can be used to depict: $Z=\overline{A}\cdot\overline{B}\cdot C+\overline{A}\cdot B\cdot C+A\cdot B\cdot\overline{C}$, by putting 1's in the cells corresponding to the Boolean terms on the right of the Boolean equation and 0's in the remaining cells. This method of depicting a three-variable Boolean expression is called a three-variable Karnaugh map, and is shown in Table 57.10(c).

To simplify a three-variable Boolean expression, the Boolean expression is depicted on a Karnaugh map as outlined above. Any cells on the map having common edges either vertically or horizontally are grouped together to form couples of four cells or two cells. During coupling the horizontal lines at the top and bottom of the cells are taken as a common edge, as are the vertical lines on the left and right of the cells. The simplified Boolean expression for a couple is given by those variables common to all cells in the couple. See Problems 15 and 17.

Table 57.10

Inputs A	B	C	Output Z	Boolean expression
0	0	0	0	$\overline{A}\cdot\overline{B}\cdot\overline{C}$
0	0	1	1	$\overline{A}\cdot\overline{B}\cdot C$
0	1	0	0	$\overline{A}\cdot B\cdot\overline{C}$
0	1	1	1	$\overline{A}\cdot B\cdot C$
1	0	0	0	$A\cdot\overline{B}\cdot\overline{C}$
1	0	1	0	$A\cdot\overline{B}\cdot C$
1	1	0	1	$A\cdot B\cdot\overline{C}$
1	1	1	0	$A\cdot B\cdot C$

(a)

C \ A·B	00 ($\overline{A}\cdot\overline{B}$)	01 ($\overline{A}\cdot B$)	11 (A·B)	10 ($A\cdot\overline{B}$)
0($\overline{C}$)	$\overline{A}\cdot\overline{B}\cdot\overline{C}$	$\overline{A}\cdot B\cdot\overline{C}$	$A\cdot B\cdot\overline{C}$	$A\cdot\overline{B}\cdot\overline{C}$
1(C)	$\overline{A}\cdot\overline{B}\cdot C$	$\overline{A}\cdot B\cdot C$	$A\cdot B\cdot C$	$A\cdot\overline{B}\cdot C$

(b)

C \ A.B	00	01	11	10
0	0	0	1	0
1	1	1	0	0

(c)

(iii) Four-variable Karnaugh maps

A truth table for a four-variable expression is shown in Table 57.11(a), the 1's in the output column showing that:

$$Z=\overline{A}\cdot\overline{B}\cdot C\cdot\overline{D}+\overline{A}\cdot B\cdot C\cdot\overline{D}$$
$$+A\cdot\overline{B}\cdot C\cdot\overline{D}+A\cdot B\cdot C\cdot\overline{D}$$

Each of the sixteen possible Boolean expressions associated with a four-variable function can be depicted as shown in Table 57.11(b), in which one cell is allocated to each row of the truth table. A matrix similar to that shown in Table 57.11(b) can be used to depict

$$Z=\overline{A}\cdot\overline{B}\cdot C\cdot\overline{D}+\overline{A}\cdot B\cdot C\cdot\overline{D}$$
$$+A\cdot\overline{B}\cdot C\cdot\overline{D}+A\cdot B\cdot C\cdot\overline{D}$$

by putting 1's in the cells corresponding to the Boolean terms on the right of the Boolean equation and 0's in the remaining cells. This method of depicting a

Table 57.11

Inputs *A*	*B*	*C*	*D*	Output *Z*	Boolean expression
0	0	0	0	0	$\overline{A}\cdot\overline{B}\cdot\overline{C}\cdot\overline{D}$
0	0	0	1	0	$\overline{A}\cdot\overline{B}\cdot\overline{C}\cdot D$
0	0	1	0	1	$\overline{A}\cdot\overline{B}\cdot C\cdot\overline{D}$
0	0	1	1	0	$\overline{A}\cdot\overline{B}\cdot C\cdot D$
0	1	0	0	0	$\overline{A}\cdot B\cdot\overline{C}\cdot\overline{D}$
0	1	0	1	0	$\overline{A}\cdot B\cdot\overline{C}\cdot D$
0	1	1	0	1	$\overline{A}\cdot B\cdot C\cdot\overline{D}$
0	1	1	1	0	$\overline{A}\cdot B\cdot C\cdot D$
1	0	0	0	0	$A\cdot\overline{B}\cdot\overline{C}\cdot\overline{D}$
1	0	0	1	0	$A\cdot\overline{B}\cdot\overline{C}\cdot D$
1	0	1	0	1	$A\cdot\overline{B}\cdot C\cdot\overline{D}$
1	0	1	1	0	$A\cdot\overline{B}\cdot C\cdot D$
1	1	0	0	0	$A\cdot B\cdot\overline{C}\cdot\overline{D}$
1	1	0	1	0	$A\cdot B\cdot\overline{C}\cdot D$
1	1	1	0	1	$A\cdot B\cdot C\cdot\overline{D}$
1	1	1	1	0	$A\cdot B\cdot C\cdot D$

(a)

C·D \ A·B	00 ($\overline{A}\cdot\overline{B}$)	01 ($\overline{A}\cdot B$)	11 ($A\cdot B$)	10 ($A\cdot\overline{B}$)
00 ($\overline{C}\cdot\overline{D}$)	$\overline{A}\cdot\overline{B}\cdot\overline{C}\cdot\overline{D}$	$\overline{A}\cdot B\cdot\overline{C}\cdot\overline{D}$	$A\cdot B\cdot\overline{C}\cdot\overline{D}$	$A\cdot\overline{B}\cdot\overline{C}\cdot\overline{D}$
01 ($\overline{C}\cdot D$)	$\overline{A}\cdot\overline{B}\cdot\overline{C}\cdot D$	$\overline{A}\cdot B\cdot\overline{C}\cdot D$	$A\cdot B\cdot\overline{C}\cdot D$	$A\cdot\overline{B}\cdot\overline{C}\cdot D$
11 ($C\cdot D$)	$\overline{A}\cdot\overline{B}\cdot C\cdot D$	$\overline{A}\cdot B\cdot C\cdot D$	$A\cdot B\cdot C\cdot D$	$A\cdot\overline{B}\cdot C\cdot D$
10 ($C\cdot\overline{D}$)	$\overline{A}\cdot\overline{B}\cdot C\cdot\overline{D}$	$\overline{A}\cdot B\cdot C\cdot\overline{D}$	$A\cdot B\cdot C\cdot\overline{D}$	$A\cdot\overline{B}\cdot C\cdot\overline{D}$

(b)

C·D \ A·B	0.0	0.1	1.1	1.0
0.0	0	0	0	0
0.1	0	0	0	0
1.1	0	0	0	0
1.0	1	1	1	1

(c)

four-variable expression is called a four-variable Karnaugh map, and is shown in Table 57.11(c).

To simplify a four-variable Boolean expression, the Boolean expression is depicted on a Karnaugh map as outlined above. Any cells on the map having common edges either vertically or horizontally are grouped together to form couples of eight cells, four cells or two cells. During coupling, the horizontal lines at the top and bottom of the cells may be considered to be common edges, as are the vertical lines on the left and the right of the cells. The simplified Boolean expression for a couple is given by those variables common to all cells in the couple. See Problems 18 and 19.

Summary of procedure when simplifying a Boolean expression using a Karnaugh map

(a) Draw a four, eight or sixteen-cell matrix, depending on whether there are two, three or four variables.

(b) Mark in the Boolean expression by putting 1's in the appropriate cells.

(c) Form couples of 8, 4 or 2 cells having common edges, forming the largest groups of cells possible. (Note that a cell containing a 1 may be used more than once when forming a couple. Also note that each cell containing a 1 must be used at least once.)

(d) The Boolean expression for the couple is given by the variables which are common to all cells in the couple.

Problem 14. Use Karnaugh map techniques to simplify the expression $\overline{P}\cdot\overline{Q}+\overline{P}\cdot Q$

Using the above procedure:

(a) The two-variable matrix is drawn and is shown in Table 57.12.

Table 57.12

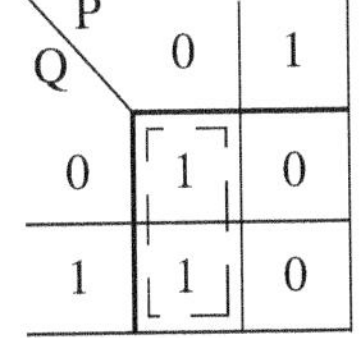

(b) The term $\overline{P}\cdot\overline{Q}$ is marked with a 1 in the top left-hand cell, corresponding to $P=0$ and $Q=0$; $\overline{P}\cdot Q$ is marked with a 1 in the bottom left-hand cell corresponding to $P=0$ and $Q=1$.

(c) The two cells containing 1's have a common horizontal edge and thus a vertical couple, can be formed.

(d) The variable common to both cells in the couple is $P=0$, i.e. $\overline{P}$ thus

$$\overline{P}\cdot\overline{Q}+\overline{P}\cdot Q=\overline{P}$$

Problem 15. Simplify the expression $\overline{X}\cdot Y\cdot\overline{Z}+\overline{X}\cdot\overline{Y}\cdot Z+X\cdot Y\cdot\overline{Z}+X\cdot\overline{Y}\cdot Z$ by using Karnaugh map techniques

Using the above procedure:

(a) A three-variable matrix is drawn and is shown in Table 57.13.

Table 57.13

Z \ X·Y	0·0	0·1	1·1	1·0
0	0	1	1	0
1	1	0	0	1

(b) The 1's on the matrix correspond to the expression given, i.e. for $\overline{X}\cdot Y\cdot\overline{Z}$, $X=0$, $Y=1$ and $Z=0$ and hence corresponds to the cell in the two row and second column and so on.

(c) Two couples can be formed as shown. The couple in the bottom row may be formed since the vertical lines on the left and right of the cells are taken as a common edge.

(d) The variables common to the couple in the top row are $Y=1$ and $Z=0$, that is, $\boldsymbol{Y\cdot\overline{Z}}$ and the variables common to the couple in the bottom row are $Y=0$, $Z=1$, that is, $\boldsymbol{\overline{Y}\cdot Z}$. Hence:

$$\boldsymbol{\overline{X}\cdot Y\cdot\overline{Z}+\overline{X}\cdot\overline{Y}\cdot Z+X\cdot Y\cdot\overline{Z}}$$
$$\boldsymbol{+X\cdot\overline{Y}\cdot Z=Y\cdot\overline{Z}+\overline{Y}\cdot Z}$$

Problem 16. Using a Karnaugh map technique to simplify the expression $(\overline{\overline{A}\cdot B})\cdot(\overline{\overline{A}+B})$

Using the procedure, a two-variable matrix is drawn and is shown in Table 57.14.

$\overline{A}\cdot B$ corresponds to the bottom left-hand cell and $(\overline{\overline{A}\cdot B})$ must therefore be all cells except this one, marked with a 1 in Table 57.14. $(\overline{A}+B)$ corresponds to all the cells except the top right-hand cell marked with a 2 in Table 57.14. Hence $(\overline{\overline{A}+B})$ must correspond to the cell marked with a 2. The expression $(\overline{\overline{A}\cdot B})\cdot(\overline{\overline{A}+B})$ corresponds to the cell having both 1 and 2 in it, i.e.

Table 57.14

B \ A	0	1
0	1	1 2
1		1

$$\boldsymbol{(\overline{\overline{A}\cdot B})\cdot(\overline{\overline{A}+B})=A\cdot\overline{B}}$$

Problem 17. Simplify $(\overline{P+\overline{Q}\cdot R})+(\overline{P\cdot Q+\overline{R}})$ using a Karnaugh map technique

The term $(P+\overline{Q}\cdot R)$ corresponds to the cells marked 1 on the matrix in Table 57.15(a), hence $(\overline{P+\overline{Q}\cdot R})$ corresponds to the cells marked 2. Similarly, $(P\cdot Q+\overline{R})$ corresponds to the cells marked 3 in Table 57.15(a), hence $(\overline{P\cdot Q+\overline{R}})$ corresponds to the cells marked 4. The expression $(\overline{P+\overline{Q}\cdot R})+(\overline{P\cdot Q+\overline{R}})$ corresponds to cells marked with either a 2 or with a 4 and is shown in Table 57.15(b) by X's. These cells may be coupled as shown. The variables common to the group of four cells is $P=0$, i.e. $\boldsymbol{\overline{P}}$, and those common to the group of two cells are $Q=0$, $R=1$, i.e. $\boldsymbol{\overline{Q}\cdot R}$

Thus: $\boldsymbol{(\overline{P+\overline{Q}\cdot R})+(\overline{P\cdot Q+\overline{R}})=\overline{P}+\overline{Q}\cdot R}$

Table 57.15

(a)

R \ P·Q	0·0	0·1	1·1	1·0
0	3 2	3 2	3 1	3 1
1	4 1	4 2	3 1	4 1

(b)

R \ P·Q	0·0	0·1	1·1	1·0
0	X	X		
1	X	X		X

Problem 18. Use Karnaugh map techniques to simplify the expression:

$$A\cdot B\cdot\overline{C}\cdot\overline{D}+A\cdot B\cdot C\cdot D+\overline{A}\cdot B\cdot C\cdot D$$
$$+A\cdot B\cdot C\cdot\overline{D}+\overline{A}\cdot B\cdot C\cdot\overline{D}$$

Using the procedure, a four-variable matrix is drawn and is shown in Table 57.16. The 1's marked on the matrix correspond to the expression given. Two couples can be formed as shown. The four-cell couple has $B=1$, $C=1$, i.e. $\boldsymbol{B \cdot C}$ as the common variables to all four cells and the two-cell couple has $\boldsymbol{A \cdot B \cdot \overline{D}}$ as the common variables to both cells. Hence, the expression simplifies to:

$$\boldsymbol{B \cdot C + A \cdot B \cdot \overline{D}} \quad \text{i.e.} \quad \boldsymbol{B \cdot (C + A \cdot \overline{D})}$$

Table 57.16

C·D \ A·B	0.0	0.1	1.1	1.0
0.0			1	
0.1				
1.1		1	1	
1.0		1	1	

Problem 19. Simplify the expression $\overline{A} \cdot \overline{B} \cdot \overline{C} \cdot \overline{D} + A \cdot \overline{B} \cdot \overline{C} \cdot \overline{D} + \overline{A} \cdot \overline{B} \cdot C \cdot \overline{D} +$ $A \cdot \overline{B} \cdot C \cdot \overline{D} + A \cdot B \cdot C \cdot D$ by using Karnaugh map techniques

The Karnaugh map for the expression is shown in Table 57.17. Since the top and bottom horizontal lines are common edges and the vertical lines on the left and right of the cells are common, then the four corner cells form a couple, $\overline{B} \cdot \overline{D}$, (the cells can be considered as if they are stretched to completely cover a sphere, as far as common edges are concerned). The cell $A \cdot B \cdot C \cdot D$ cannot be coupled with any other. Hence the expression simplifies to

$$\boldsymbol{\overline{B} \cdot \overline{D} + A \cdot B \cdot C \cdot D}$$

Table 57.17

C·D \ A·B	0.0	0.1	1.1	1.0
0.0	1			1
0.1				
1.1			1	
1.0	1			1

Now try the following Practice Exercise

Practice Exercise 265 Simplifying Boolean expressions using Karnaugh maps (Answers on page 731)

In Problems 1 to 11 use Karnaugh map techniques to simplify the expressions given.

1. $\overline{X} \cdot Y + X \cdot Y$
2. $\overline{X} \cdot \overline{Y} + \overline{X} \cdot Y + X \cdot Y$
3. $(\overline{P} \cdot \overline{Q}) \cdot (\overline{\overline{P} \cdot Q})$
4. $A \cdot \overline{C} + \overline{A} \cdot (B + C) + A \cdot B \cdot (C + \overline{B})$
5. $\overline{P} \cdot \overline{Q} \cdot \overline{R} + \overline{P} \cdot Q \cdot \overline{R} + P \cdot Q \cdot \overline{R}$
6. $\overline{P} \cdot \overline{Q} \cdot \overline{R} + P \cdot Q \cdot \overline{R} + P \cdot Q \cdot R + P \cdot \overline{Q} \cdot R$
7. $\overline{A} \cdot \overline{B} \cdot \overline{C} \cdot \overline{D} + \overline{A} \cdot B \cdot \overline{C} \cdot \overline{D} + \overline{A} \cdot B \cdot \overline{C} \cdot D$
8. $\overline{A} \cdot \overline{B} \cdot C \cdot D + \overline{A} \cdot \overline{B} \cdot C \cdot \overline{D} + A \cdot \overline{B} \cdot C \cdot \overline{D}$
9. $\overline{A} \cdot B \cdot \overline{C} \cdot D + A \cdot B \cdot \overline{C} \cdot D + A \cdot B \cdot C \cdot D$ $+ A \cdot \overline{B} \cdot \overline{C} \cdot D + A \cdot \overline{B} \cdot C \cdot D$
10. $\overline{A} \cdot \overline{B} \cdot \overline{C} \cdot D + A \cdot B \cdot \overline{C} \cdot \overline{D} + A \cdot \overline{B} \cdot \overline{C} \cdot \overline{D}$ $+ A \cdot B \cdot C \cdot \overline{D} + A \cdot \overline{B} \cdot C \cdot D$
11. $A \cdot B \cdot \overline{C} \cdot \overline{D} + \overline{A} \cdot \overline{B} \cdot \overline{C} \cdot \overline{D} + \overline{A} \cdot B \cdot C \cdot D$ $+ \overline{A} \cdot \overline{B} \cdot C \cdot D + A \cdot \overline{B} \cdot \overline{C} \cdot \overline{D} + \overline{A} \cdot \overline{B} \cdot C \cdot \overline{D}$ $+ \overline{A} \cdot B \cdot C \cdot \overline{D}$

57.6 Logic circuits

In practice, logic gates are used to perform the **and, or** and **not**-functions introduced in Section 57.1. Logic gates can be made from switches, magnetic devices or fluidic devices, but most logic gates in use are electronic devices. Various logic gates are available. For example, the Boolean expression $(A \cdot B \cdot C)$ can be produced using a three-input, **and**-gate and $(C + D)$ by using a two-input **or**-gate. The principal gates in common use are introduced below. The term 'gate' is used in the same sense as a normal gate, the open state being indicated by a binary '1' and the closed state by a binary '0'. A gate will only open when the requirements of the gate are met and, for example, there will only be a '1' output on a two-input **and**-gate when both the inputs to the gate are at a '1' state.

The and-gate

The different symbols used for a three-input, **and**-gate are shown in Fig. 57.13(a) and the truth table is shown in Fig. 57.13(b). This shows that there will only be a '1' output when A is 1, or B is 1, or C is 1, written as:

$$Z = A \cdot B \cdot C$$

The or-gate

The different symbols used for a three-input **or**-gate are shown in Fig. 57.14(a) and the truth table is shown in Fig. 57.14(b). This shows that there will be a '1' output

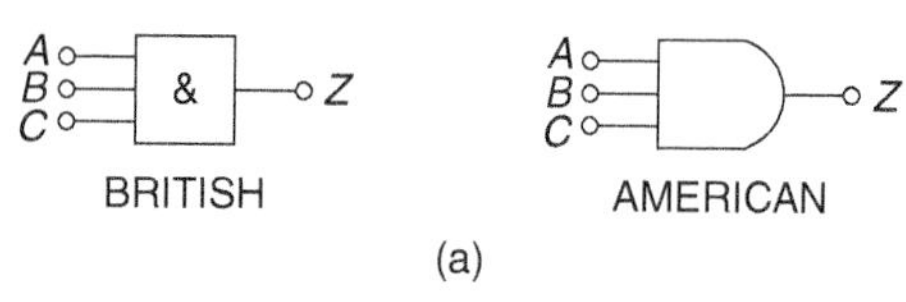

(a)

INPUTS A	B	C	OUTPUT $Z=A\cdot B\cdot C$
0	0	0	0
0	0	1	0
0	1	0	0
0	1	1	0
1	0	0	0
1	0	1	0
1	1	0	0
1	1	1	1

(b)

Figure 57.13

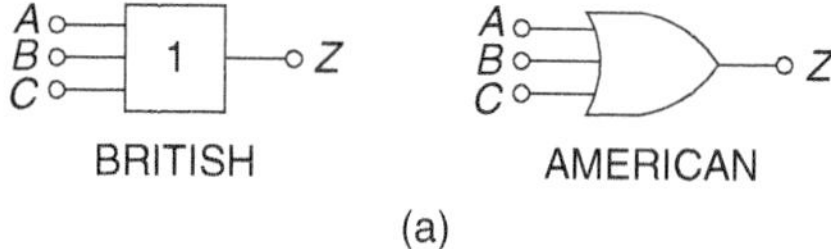

(a)

INPUTS A	B	C	OUTPUT $Z=A+B+C$
0	0	0	0
0	0	1	1
0	1	0	1
0	1	1	1
1	0	0	1
1	0	1	1
1	1	0	1
1	1	1	1

(b)

Figure 57.14

when A is 1, or B is 1, or C is 1, or any combination of A, B or C is 1, written as:

$$Z=A+B+C$$

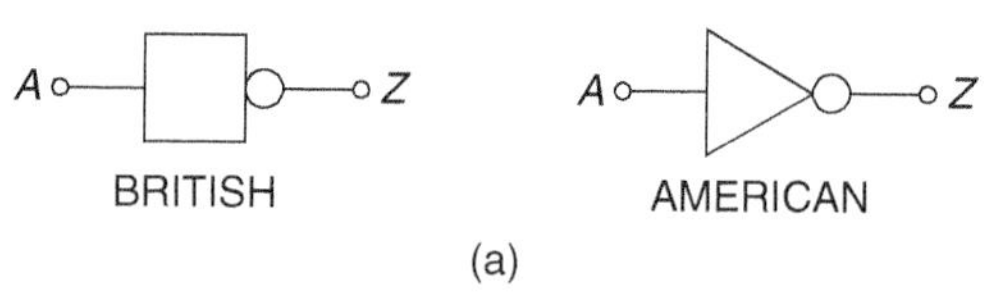

(a)

INPUT A	OUTPUT $Z=\overline{A}$
0	1
1	0

(b)

Figure 57.15

The invert-gate or not-gate

The different symbols used for an **invert**-gate are shown in Fig. 57.15(a) and the truth table is shown in Fig. 57.15(b). This shows that a '0' input gives a '1' output and vice versa, i.e. it is an 'opposite to' function. The invert of A is written $\overline{A}$ and is called 'not-A'.

The nand-gate

The different symbols used for a **nand**-gate are shown in Fig. 57.16(a) and the truth table is shown in Fig. 57.16(b). This gate is equivalent to an **and**-gate and an **invert**-gate in series (not-and = nand) and the output is written as:

$$Z=\overline{A\cdot B\cdot C}$$

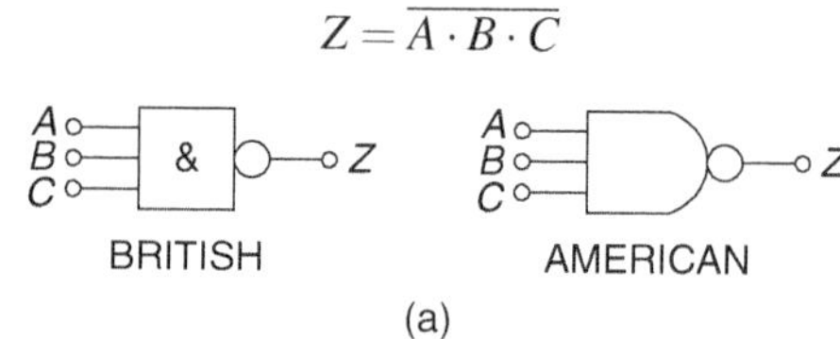

(a)

INPUTS A	B	C	A.B.C.	OUTPUT $Z=\overline{A.B.C.}$
0	0	0	0	1
0	0	1	0	1
0	1	0	0	1
0	1	1	0	1
1	0	0	0	1
1	0	1	0	1
1	1	0	0	1
1	1	1	1	0

(b)

Figure 57.16

The nor-gate

The different symbols used for a **nor**-gate are shown in Fig. 57.17(a) and the truth table is shown in Fig. 57.17(b). This gate is equivalent to an **or**-gate and an **invert**-gate in series, (not-or = nor), and the output is written as:

$$Z=\overline{A+B+C}$$

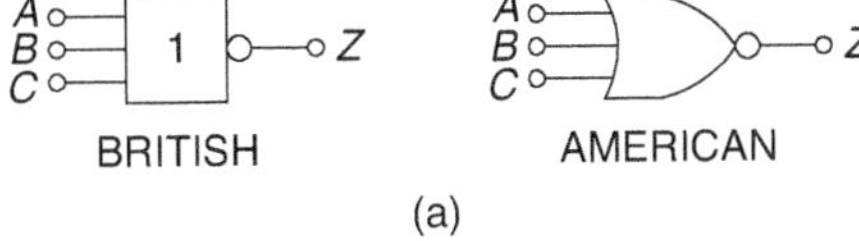

(a)

INPUTS A	B	C	$A+B+C$	OUTPUT $Z=\overline{A+B+C}$
0	0	0	0	1
0	0	1	1	0
0	1	0	1	0
0	1	1	1	0
1	0	0	1	0
1	0	1	1	0
1	1	0	1	0
1	1	1	1	0

(b)

Figure 57.17

Combinational logic networks

In most logic circuits, more than one gate is needed to give the required output. Except for the **invert**-gate, logic gates generally have two, three or four inputs and are confined to one function only. Thus, for example, a two-input, **or**-gate or a four-input **and**-gate can be used when designing a logic circuit. The way in which logic gates are used to generate a given output is shown in Problems 20 to 23.

Problem 20. Devise a logic system to meet the requirements of: $Z=A\cdot\overline{B}+C$

With reference to Fig. 57.18 an **invert**-gate, shown as (1), gives $\overline{B}$. The **and**-gate, shown as (2), has inputs of A and $\overline{B}$, giving $A\cdot\overline{B}$. The **or**-gate, shown as (3), has inputs of $A\cdot\overline{B}$ and C, giving:

$$Z=A\cdot\overline{B}+C$$

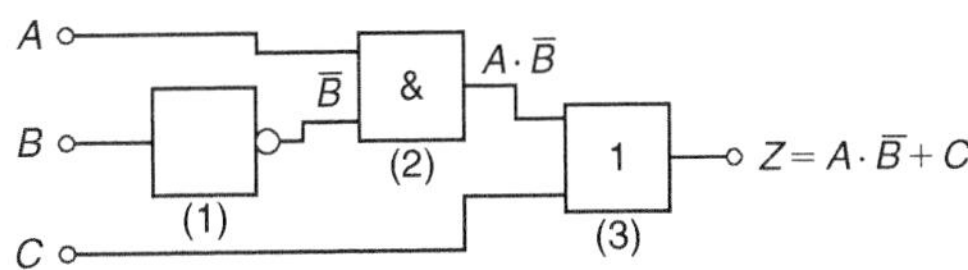

Figure 57.18

Problem 21. Devise a logic system to meet the requirements of $(P+\overline{Q})\cdot(\overline{R}+S)$

The logic system is shown in Fig. 57.19. The given expression shows that two **invert**-functions are needed to give $\overline{Q}$ and $\overline{R}$ and these are shown as gates (1) and (2). Two **or**-gates, shown as (3) and (4), give $(P+\overline{Q})$ and $(\overline{R}+S)$ respectively. Finally, an **and**-gate, shown as (5), gives the required output.

$$Z=(P+\overline{Q})\cdot(\overline{R}+S)$$

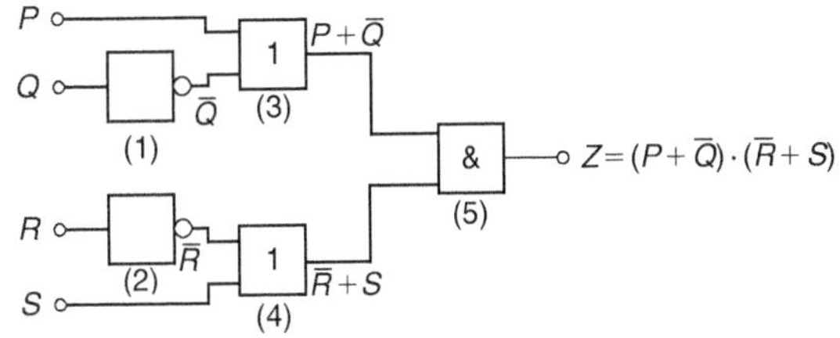

Figure 57.19

Problem 22. Devise a logic circuit to meet the requirements of the output given in Table 57.18, using as few gates as possible

Table 57.18

Inputs A	B	C	Output Z
0	0	0	0
0	0	1	0
0	1	0	0
0	1	1	0
1	0	0	0
1	0	1	1
1	1	0	1
1	1	1	1

The '1' outputs in rows 6, 7 and 8 of Table 57.18 show that the Boolean expression is:

$$Z = A\cdot\overline{B}\cdot C + A\cdot B\cdot\overline{C} + A\cdot B\cdot C$$

The logic circuit for this expression can be built using three, 3-input **and**-gates and one, 3-input **or**-gate, together with two **invert**-gates. However, the number of gates required can be reduced by using the techniques introduced in Sections 57.3 to 57.5, resulting in the cost of the circuit being reduced. Any of the techniques can be used, and in this case, the rules of Boolean algebra (see Table 57.7) are used.

$$\begin{aligned} Z &= A\cdot\overline{B}\cdot C + A\cdot B\cdot\overline{C} + A\cdot B\cdot C \\ &= A\cdot[\overline{B}\cdot C + B\cdot\overline{C} + B\cdot C] \\ &= A\cdot[\overline{B}\cdot C + B(\overline{C}+C)] = A\cdot[\overline{B}\cdot C + B] \\ &= A\cdot[B+\overline{B}\cdot C] = \mathbf{A\cdot[B+C]} \end{aligned}$$

The logic circuit to give this simplified expression is shown in Fig. 57.20.

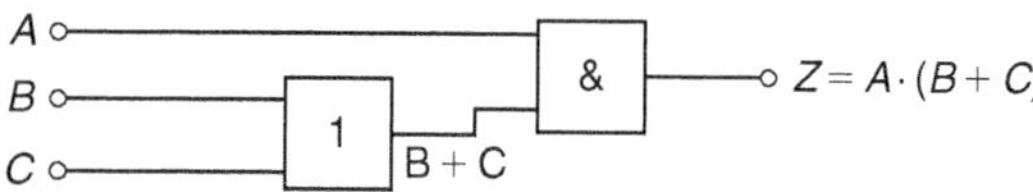

Figure 57.20

Problem 23. Simplify the expression:

$$Z = \overline{P}\cdot\overline{Q}\cdot\overline{R}\cdot\overline{S} + \overline{P}\cdot\overline{Q}\cdot\overline{R}\cdot S + \overline{P}\cdot Q\cdot\overline{R}\cdot\overline{S} + \overline{P}\cdot Q\cdot\overline{R}\cdot S + P\cdot\overline{Q}\cdot\overline{R}\cdot\overline{S}$$

and devise a logic circuit to give this output

The given expression is simplified using the Karnaugh map techniques introduced in Section 57.5. Two couples are formed as shown in Fig. 57.21(a) and the simplified expression becomes:

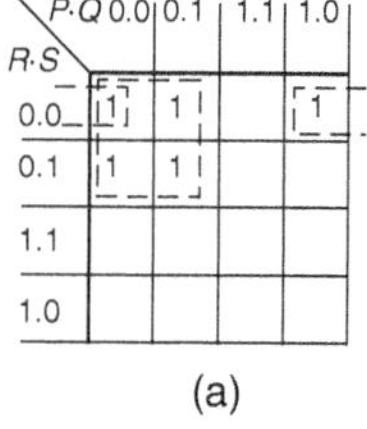

(a)

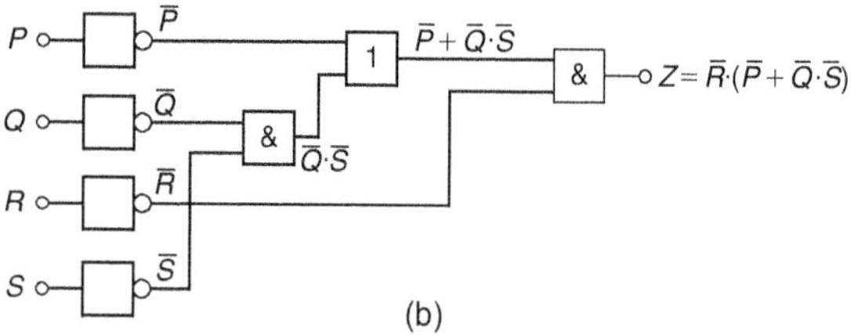

(b)

Figure 57.21

$$Z = \overline{Q}\cdot\overline{R}\cdot\overline{S} + \overline{P}\cdot\overline{R}$$

i.e. $\mathbf{Z = \overline{R}\cdot(\overline{P} + \overline{Q}\cdot\overline{S})}$

The logic circuit to produce this expression is shown in Fig. 57.21(b).

Now try the following Practice Exercise

Practice Exercise 266 Logic circuits (Answers on page 731)

In Problems 1 to 4, devise logic systems to meet the requirements of the Boolean expressions given.

1. $Z = \overline{A} + B\cdot C$
2. $Z = A\cdot\overline{B} + B\cdot\overline{C}$
3. $Z = A\cdot B\cdot\overline{C} + \overline{A}\cdot\overline{B}\cdot C$
4. $Z = (\overline{A} + B)\cdot(\overline{C} + D)$

In Problems 5 to 7, simplify the expression given in the truth table and devise a logic circuit to meet the requirements stated.

5. Column 4 of Table 57.19
6. Column 5 of Table 57.19

Table 57.19

1 A	2 B	3 C	4 Z_1	5 Z_2	6 Z_3
0	0	0	0	0	0
0	0	1	1	0	0
0	1	0	0	0	1
0	1	1	1	1	1
1	0	0	0	1	0
1	0	1	1	1	1
1	1	0	1	0	1
1	1	1	1	1	1

7. Column 6 of Table 57.19

In Problems 8 to 12, simplify the Boolean expressions given and devise logic circuits to give the requirements of the simplified expressions.

8. $\overline{P}\cdot\overline{Q} + \overline{P}\cdot Q + P\cdot Q$
9. $\overline{P}\cdot\overline{Q}\cdot\overline{R} + P\cdot Q\cdot\overline{R} + P\cdot\overline{Q}\cdot\overline{R}$

10. $P\cdot\overline{Q}\cdot R+P\cdot\overline{Q}\cdot\overline{R}+\overline{P}\cdot\overline{Q}\cdot\overline{R}$

11. $\overline{A}\cdot\overline{B}\cdot\overline{C}\cdot\overline{D}+A\cdot\overline{B}\cdot\overline{C}\cdot\overline{D}+\overline{A}\cdot\overline{B}\cdot C\cdot\overline{D}$
$+\overline{A}\cdot B\cdot C\cdot\overline{D}+A\cdot\overline{B}\cdot C\cdot\overline{D}$

12. $\overline{(\overline{P}\cdot Q\cdot R)}\cdot\overline{(P+Q\cdot R)}$

57.7 Universal logic gates

The function of any of the five logic gates in common use can be obtained by using either **nand**-gates or **nor**-gates and when used in this manner, the gate selected in called a **universal gate**. The way in which a universal **nand**-gate is used to produce the **invert, and, or** and **nor**-functions is shown in Problem 24. The way in which a universal **nor**-gate is used to produce the **invert, or, and** and **nand**-function is shown in Problem 25.

Problem 24. Show how **invert, and, or** and **nor**-functions can be produced using nand-gates only

A single input to a **nand**-gate gives the **invert**-function, as shown in Fig. 57.22(a). When two **nand**-gates are connected, as shown in Fig. 57.22(b), the output from the first gate is $\overline{A\cdot B\cdot C}$ and this is inverted by the second gate, giving $Z=\overline{\overline{A\cdot B\cdot C}}=A\cdot B\cdot C$ i.e. the **and**-function is produced. When $\overline{A}$, $\overline{B}$ and $\overline{C}$ are the inputs to a **nand**-gate, the output is $\overline{\overline{A}\cdot\overline{B}\cdot\overline{C}}$.

By de Morgan's law, $\overline{\overline{A}\cdot\overline{B}\cdot\overline{C}}=\overline{\overline{A}}+\overline{\overline{B}}+\overline{\overline{C}}=A+B+C$, i.e. a **nand**-gate is used to produce **or**-function. The logic circuit shown in Fig. 57.22(c). If the output from the logic circuit in Fig. 57.22(c) is inverted by adding an additional **nand**-gate, the output becomes the invert of an **or**-function, i.e. the **nor**-function, as shown in Fig. 57.22(d).

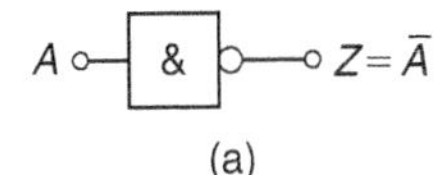

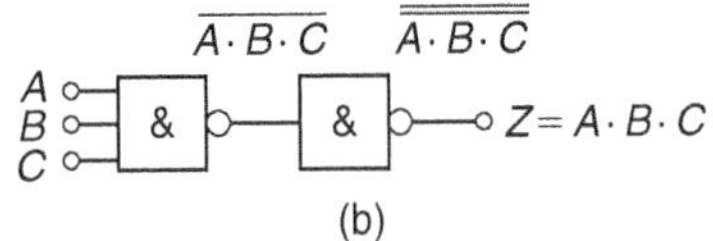

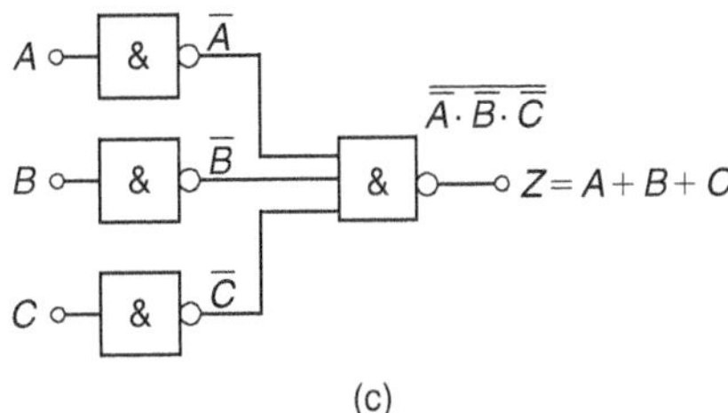

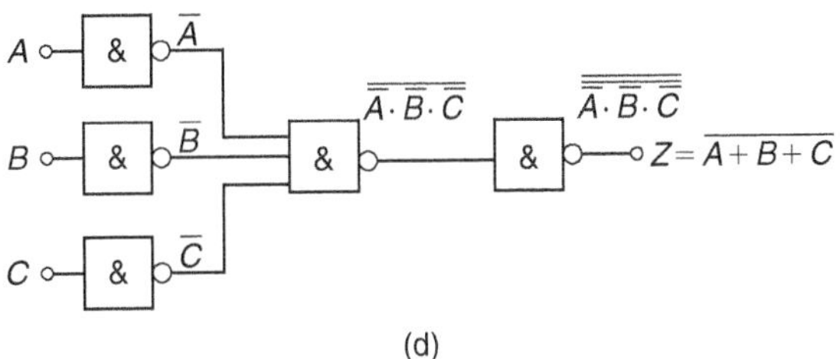

Figure 57.22

Problem 25. Show how **invert, or, and** and **nand**-functions can be produced by using **nor**-gates only

A single input to a **nor**-gate gives the **invert**-function, as shown in Fig. 57.23(a). When two **nor**-gates are connected, as shown in Fig. 57.23(b), the output from the first gate is $\overline{A+B+C}$ and this is inverted by the second gate, giving $Z=\overline{\overline{A+B+C}}=A+B+C$, i.e. the **or**-function is produced. Inputs of $\overline{A}$, $\overline{B}$, and $\overline{C}$ to a **nor**-gate give an output of $\overline{\overline{A}+\overline{B}+\overline{C}}$

By de Morgan's law, $\overline{\overline{A}+\overline{B}+\overline{C}}=\overline{\overline{A}}\cdot\overline{\overline{B}}\cdot\overline{\overline{C}}=A\cdot B\cdot C$, i.e. the **nor**-gate can be used to produce the **and**-function. The logic circuit is shown in Fig. 57.23(c). When the output of the logic circuit, shown in Fig. 57.23(c), is inverted by adding an additional **nor**-gate, the output then becomes the invert of an **or**-function, i.e. the **nor**-function as shown in Fig. 57.23(d).

Problem 26. Design a logic circuit, using **nand**-gates having not more than three inputs, to meet the requirements of the Boolean expression:

$$Z=\overline{A}+\overline{B}+C+\overline{D}$$

When designing logic circuits, it is often easier to start at the output of the circuit. The given expression shows there are four variables joined by **or**-functions. From the principles introduced in Problem 24, if a four-input **nand**-gate is used to give the expression given, the input are $\overline{\overline{A}}$, $\overline{\overline{B}}$, $\overline{C}$ and $\overline{\overline{D}}$ that is A, B, $\overline{C}$ and D. However, the problem states that three-inputs are not to be exceeded so two of the variables are joined, i.e. the inputs to the

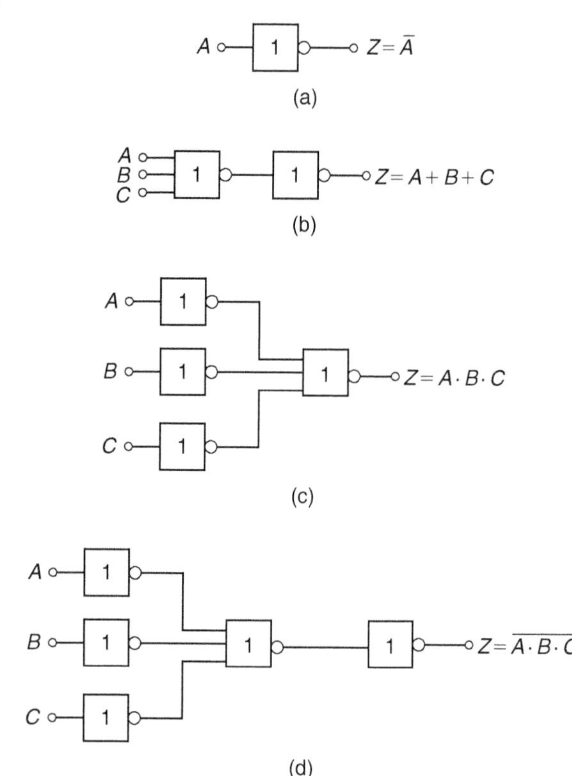

Figure 57.23

three-input **nand**-gate, shown as gate (1) Fig. 57.24, is A, B, $\overline{C}$, and D. From Problem 24, the **and**-function is generated by using two **nand**-gates connected in series, as shown by gates (2) and (3) in Fig. 57.24. The logic circuit required to produce the given expression is as shown in Fig. 57.24.

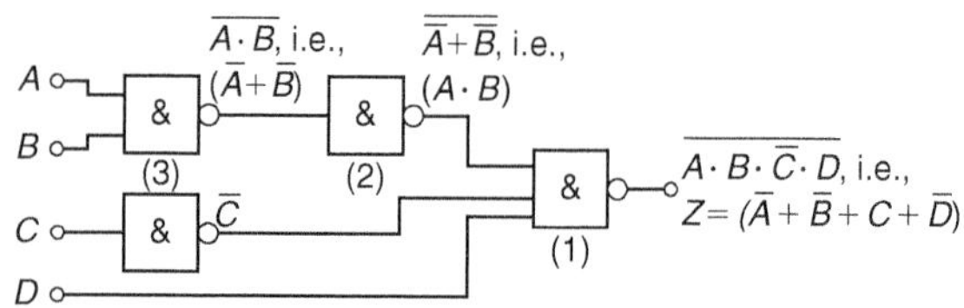

Figure 57.24

Problem 27. Use **nor**-gates only to design a logic circuit to meet the requirements of the expressions:

$$Z = \overline{D} \cdot (\overline{A} + B + \overline{C})$$

It is usual in logic circuit design to start the design at the output. From Problem 25, the **and**-function between $\overline{D}$ and the terms in the bracket can be produced by using inputs of $\overline{\overline{D}}$ and $\overline{\overline{A} + B + \overline{C}}$ to a **nor**-gate, i.e. by de Morgan's law, inputs of D and $A \cdot \overline{B} \cdot C$. Again, with reference to Problem 25, inputs of $\overline{A} \cdot B$ and $\overline{C}$ to a **nor**-gate give an output of $\overline{\overline{A} + B + \overline{C}}$, which by de Morgan's law is $A \cdot \overline{B} \cdot C$. The logic circuit to produce the required expression is as shown in Fig. 57.25

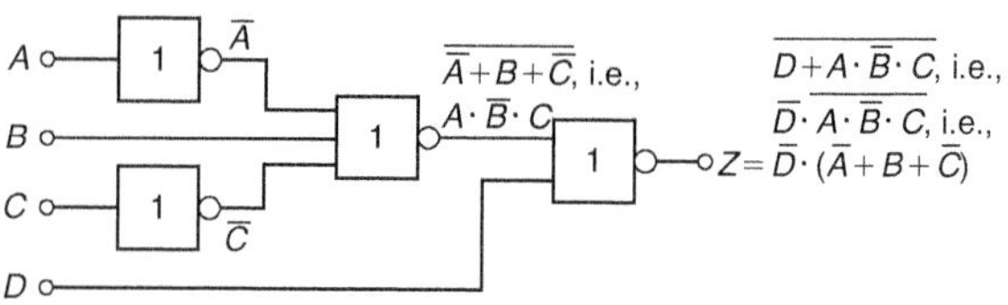

Figure 57.25

Problem 28. An alarm indicator in a grinding mill complex should be activated if (a) the power supply to all mills is off and (b) the hopper feeding the mills is less than 10% full, and (c) if less than two of the three grinding mills are in action. Devise a logic system to meet these requirements

Let variable A represent the power supply on to all the mills, then $\overline{A}$ represents the power supply off. Let B represent the hopper feeding the mills being more than 10% full, then $\overline{B}$ represents the hopper being less than 10% full. Let C, D and E represent the three mills respectively being in action, then $\overline{C}$, $\overline{D}$ and $\overline{E}$ represent the three mills respectively not being in action. The required expression to activate the alarm is:

$$Z = \overline{A} \cdot \overline{B} \cdot (\overline{C} + \overline{D} + \overline{E})$$

There are three variables joined by **and**-functions in the output, indicating that a three-input **and**-gate is required, having inputs of $\overline{A}$, $\overline{B}$ and $(\overline{C} + \overline{D} + \overline{E})$.

The term $(\overline{C} + \overline{D} + \overline{E})$ is produced by a three-input **nand**-gate. When variables C, D and E are the inputs to a **nand**-gate, the output is $\overline{C \cdot D \cdot E}$ which, by de Morgan's law is $\overline{C} + \overline{D} + \overline{E}$. Hence the required logic circuit is as shown in Fig. 57.26.

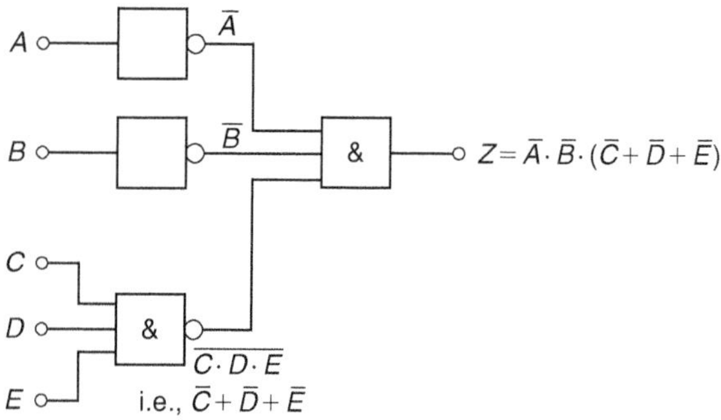

Figure 57.26

Now try the following Practice Exercise

Practice Exercise 267 Universal logic circuits (Answers on page 732)

In Problems 1 to 3, use **nand**-gates only to devise the logic systems stated.

1. $Z=A+B\cdot C$
2. $Z=A\cdot\overline{B}+B\cdot\overline{C}$
3. $Z=A\cdot B\cdot\overline{C}+\overline{A}\cdot\overline{B}\cdot C$

In Problems 4 to 6, use **nor**-gates only to devise the logic systems stated.

4. $Z=(\overline{A}+B)\cdot(\overline{C}+D)$
5. $Z=A\cdot\overline{B}+B\cdot\overline{C}+C\cdot\overline{D}$
6. $Z=\overline{P}\cdot Q+P\cdot(Q+R)$
7. In a chemical process, three of the transducers used are P, Q and R, giving output signals of either 0 or 1. Devise a logic system to give a 1 output when:
   (a) P and Q and R all have 0 outputs, or when:
   (b) P is 0 and (Q is 1 or R is 0)
8. Lift doors should close, (Z), if:
   (a) the master switch, (A), is on and either
   (b) a call, (B), is received from any other floor, or
   (c) the doors, (C), have been open for more than 10 seconds, or
   (d) the selector push within the lift (D), is pressed for another floor

   Devise a logic circuit to meet these requirements
9. A water tank feeds three separate processes. When any two of the processes are in operation at the same time, a signal is required to start a pump to maintain the head of water in the tank. Devise a logic circuit using **nor**-gates only to give the required signal
10. A logic signal is required to give an indication when:
    (a) the supply to an oven is on, and
    (b) the temperature of the oven exceeds 210°C, or
    (c) the temperature of the oven is less than 190°C

    Devise a logic circuit using **nand**-gates only to meet these requirements

Practice Exercise 268 Multiple-choice questions on Boolean algebra and logic circuits (Answers on page 732)

Each question has only one correct answer

1. Boolean algebra can be used:
   (a) in building logic symbols
   (b) for designing digital computers
   (c) in circuit theory
   (d) in building algebraic functions
2. The universal logic gates are:
   (a) NOT (b) OR and NOR
   (c) AND (d) NAND and NOR
3. According to the Boolean law, $A+1$ is equal to:
   (a) A (b) 0 (c) 1 (d) $\overline{A}$
4. In Boolean algebra $A.(A+B)$ is equal to:
   (a) $A.B$ (b) A (c) 1 (d) $(1+A.B)$
5. The Boolean expression $A+\overline{A}.B$ is equivalent to:
   (a) A (b) B (c) $A+B$ (d) $A+\overline{A}$
6. If the input to an invertor is A, the output will be:
   (a) $\frac{1}{A}$ (b) 1 (c) A (d) $\overline{A}$
7. The Boolean expression $\overline{P}.\overline{Q}+\overline{P}.Q$ is equivalent to:
   (a) $\overline{P}$ (b) $\overline{Q}$ (c) P (d) Q
8. A 4-variable Karnaugh map has:
   (a) 4 cells (b) 8 cells
   (c) 16 cells (d) 32 cells
9. The Boolean expression $\overline{F}.\overline{G}.\overline{H}+\overline{F}.\overline{G}.H$ is equivalent to:
   (a) $F.G$ (b) $F.\overline{G}$ (c) $\overline{F}.H$ (d) $\overline{F}.\overline{G}$
10. In Boolean algebra, $A+B.C$ is equivalent to:
    (a) $(A+B).(A+C)$ (b) $A.B.C$
    (c) $A+B$ (d) $A.B+A.C$

For fully worked solutions to each of the problems in Practice Exercises 262 to 267 in this chapter, go to the website:
www.routledge.com/cw/bird

COMPANION @ WEBSITE

Chapter 58

The theory of matrices and determinants

Why it is important to understand: The theory of matrices and determinants

Matrices are used to solve problems in electronics, optics, quantum mechanics, statics, robotics, linear programming, optimisation, genetics and much more. Matrix calculus is a mathematical tool used in connection with linear equations, linear transformations, systems of differential equations and so on, and is vital for calculating forces, vectors, tensions, masses, loads and a lot of other factors that must be accounted for in engineering to ensure safe and resource-efficient structure. Electrical and mechanical engineers, chemists, biologists and scientists all need knowledge of matrices to solve problems. In computer graphics, matrices are used to project a 3-dimensional image on to a 2-dimentional screen, and to create realistic motion. Matrices are therefore very important in solving engineering problems.

At the end of this chapter, you should be able to:

- understand matrix notation
- add, subtract and multiply 2 by 2 and 3 by 3 matrices
- recognise the unit matrix
- calculate the determinant of a 2 by 2 matrix
- determine the inverse (or reciprocal) of a 2 by 2 matrix
- calculate the determinant of a 3 by 3 matrix
- determine the inverse (or reciprocal) of a 3 by 3 matrix

58.1 Matrix notation

Matrices and determinants are mainly used for the solution of linear simultaneous equations. The theory of matrices and determinants is dealt with in this chapter and this theory is then used in Chapter 59 to solve simultaneous equations.

The coefficients of the variables for linear simultaneous equations may be shown in matrix form. The coefficients of x and y in the simultaneous equations

$$x + 2y = 3$$

$$4x - 5y = 6$$

become $\begin{pmatrix} 1 & 2 \\ 4 & -5 \end{pmatrix}$ in matrix notation.

Similarly, the coefficients of p, q and r in the equations

$$1.3p - 2.0q + r = 7$$
$$3.7p + 4.8q - 7r = 3$$
$$4.1p + 3.8q + 12r = -6$$

become $\begin{pmatrix} 1.3 & -2.0 & 1 \\ 3.7 & 4.8 & -7 \\ 4.1 & 3.8 & 12 \end{pmatrix}$ in matrix form.

The numbers within a matrix are called an **array** and the coefficients forming the array are called the **elements** of the matrix. The number of rows in a matrix is usually specified by m and the number of columns by n and a matrix referred to as an 'm by n' matrix. Thus, $\begin{pmatrix} 2 & 3 & 6 \\ 4 & 5 & 7 \end{pmatrix}$ is a '2 by 3' matrix. Matrices cannot be expressed as a single numerical value, but they can often be simplified or combined, and unknown element values can be determined by comparison methods. Just as there are rules for addition, subtraction, multiplication and division of numbers in arithmetic, rules for these operations can be applied to matrices and the rules of matrices are such that they obey most of those governing the algebra of numbers.

58.2 Addition, subtraction and multiplication of matrices

(i) Addition of matrices

Corresponding elements in two matrices may be added to form a single matrix.

Problem 1. Add the matrices

(a) $\begin{pmatrix} 2 & -1 \\ -7 & 4 \end{pmatrix}$ and $\begin{pmatrix} -3 & 0 \\ 7 & -4 \end{pmatrix}$

(b) $\begin{pmatrix} 3 & 1 & -4 \\ 4 & 3 & 1 \\ 1 & 4 & -3 \end{pmatrix}$ and $\begin{pmatrix} 2 & 7 & -5 \\ -2 & 1 & 0 \\ 6 & 3 & 4 \end{pmatrix}$

(a) Adding the corresponding elements gives:

$$\begin{pmatrix} 2 & -1 \\ -7 & 4 \end{pmatrix} + \begin{pmatrix} -3 & 0 \\ 7 & -4 \end{pmatrix}$$
$$= \begin{pmatrix} 2+(-3) & -1+0 \\ -7+7 & 4+(-4) \end{pmatrix}$$
$$= \begin{pmatrix} \mathbf{-1} & \mathbf{-1} \\ \mathbf{0} & \mathbf{0} \end{pmatrix}$$

(b) Adding the corresponding elements gives:

$$\begin{pmatrix} 3 & 1 & -4 \\ 4 & 3 & 1 \\ 1 & 4 & -3 \end{pmatrix} + \begin{pmatrix} 2 & 7 & -5 \\ -2 & 1 & 0 \\ 6 & 3 & 4 \end{pmatrix}$$
$$= \begin{pmatrix} 3+2 & 1+7 & -4+(-5) \\ 4+(-2) & 3+1 & 1+0 \\ 1+6 & 4+3 & -3+4 \end{pmatrix}$$
$$= \begin{pmatrix} \mathbf{5} & \mathbf{8} & \mathbf{-9} \\ \mathbf{2} & \mathbf{4} & \mathbf{1} \\ \mathbf{7} & \mathbf{7} & \mathbf{1} \end{pmatrix}$$

(ii) Subtraction of matrices

If A is a matrix and B is another matrix, then $(A - B)$ is a single matrix formed by subtracting the elements of B from the corresponding elements of A.

Problem 2. Subtract

(a) $\begin{pmatrix} -3 & 0 \\ 7 & -4 \end{pmatrix}$ from $\begin{pmatrix} 2 & -1 \\ -7 & 4 \end{pmatrix}$

(b) $\begin{pmatrix} 2 & 7 & -5 \\ -2 & 1 & 0 \\ 6 & 3 & 4 \end{pmatrix}$ from $\begin{pmatrix} 3 & 1 & -4 \\ 4 & 3 & 1 \\ 1 & 4 & -3 \end{pmatrix}$

To find matrix A minus matrix B, the elements of B are taken from the corresponding elements of A. Thus:

(a) $$\begin{pmatrix} 2 & -1 \\ -7 & 4 \end{pmatrix} - \begin{pmatrix} -3 & 0 \\ 7 & -4 \end{pmatrix}$$
$$= \begin{pmatrix} 2-(-3) & -1-0 \\ -7-7 & 4-(-4) \end{pmatrix}$$
$$= \begin{pmatrix} \mathbf{5} & \mathbf{-1} \\ \mathbf{-14} & \mathbf{8} \end{pmatrix}$$

(b) $$\begin{pmatrix} 3 & 1 & -4 \\ 4 & 3 & 1 \\ 1 & 4 & -3 \end{pmatrix} - \begin{pmatrix} 2 & 7 & -5 \\ -2 & 1 & 0 \\ 6 & 3 & 4 \end{pmatrix}$$
$$= \begin{pmatrix} 3-2 & 1-7 & -4-(-5) \\ 4-(-2) & 3-1 & 1-0 \\ 1-6 & 4-3 & -3-4 \end{pmatrix}$$
$$= \begin{pmatrix} \mathbf{1} & \mathbf{-6} & \mathbf{1} \\ \mathbf{6} & \mathbf{2} & \mathbf{1} \\ \mathbf{-5} & \mathbf{1} & \mathbf{-7} \end{pmatrix}$$

Problem 3. If

$A=\begin{pmatrix}-3 & 0\\ 7 & -4\end{pmatrix}$, $B=\begin{pmatrix}2 & -1\\ -7 & 4\end{pmatrix}$ and

$C=\begin{pmatrix}1 & 0\\ -2 & -4\end{pmatrix}$ find $A+B-C$

$$A+B=\begin{pmatrix}-1 & -1\\ 0 & 0\end{pmatrix}$$

(from Problem 1)

Hence, $A+B-C=\begin{pmatrix}-1 & -1\\ 0 & 0\end{pmatrix}-\begin{pmatrix}1 & 0\\ -2 & -4\end{pmatrix}$

$$=\begin{pmatrix}-1-1 & -1-0\\ 0-(-2) & 0-(-4)\end{pmatrix}$$

$$=\begin{pmatrix}\mathbf{-2} & \mathbf{-1}\\ \mathbf{2} & \mathbf{4}\end{pmatrix}$$

Alternatively $A+B-C$

$$=\begin{pmatrix}-3 & 0\\ 7 & -4\end{pmatrix}+\begin{pmatrix}2 & -1\\ -7 & 4\end{pmatrix}-\begin{pmatrix}1 & 0\\ -2 & -4\end{pmatrix}$$

$$=\begin{pmatrix}-3+2-1 & 0+(-1)-0\\ 7+(-7)-(-2) & -4+4-(-4)\end{pmatrix}$$

$$=\begin{pmatrix}\mathbf{-2} & \mathbf{-1}\\ \mathbf{2} & \mathbf{4}\end{pmatrix}$$ as obtained previously

(iii) Multiplication

When a matrix is multiplied by a number, called **scalar multiplication**, a single matrix results in which each element of the original matrix has been multiplied by the number.

Problem 4. If $A=\begin{pmatrix}-3 & 0\\ 7 & -4\end{pmatrix}$, $B=\begin{pmatrix}2 & -1\\ -7 & 4\end{pmatrix}$ and $C=\begin{pmatrix}1 & 0\\ -2 & -4\end{pmatrix}$ find $2A-3B+4C$

For scalar multiplication, each element is multiplied by the scalar quantity, hence

$$2A=2\begin{pmatrix}-3 & 0\\ 7 & -4\end{pmatrix}=\begin{pmatrix}-6 & 0\\ 14 & -8\end{pmatrix}$$

$$3B=3\begin{pmatrix}2 & -1\\ -7 & 4\end{pmatrix}=\begin{pmatrix}6 & -3\\ -21 & 12\end{pmatrix}$$

and $$4C=4\begin{pmatrix}1 & 0\\ -2 & -4\end{pmatrix}=\begin{pmatrix}4 & 0\\ -8 & -16\end{pmatrix}$$

Hence $2A-3B+4C$

$$=\begin{pmatrix}-6 & 0\\ 14 & -8\end{pmatrix}-\begin{pmatrix}6 & -3\\ -21 & 12\end{pmatrix}+\begin{pmatrix}4 & 0\\ -8 & -16\end{pmatrix}$$

$$=\begin{pmatrix}-6-6+4 & 0-(-3)+0\\ 14-(-21)+(-8) & -8-12+(-16)\end{pmatrix}$$

$$=\begin{pmatrix}\mathbf{-8} & \mathbf{3}\\ \mathbf{27} & \mathbf{-36}\end{pmatrix}$$

When a matrix A is multiplied by another matrix B, a single matrix results in which elements are obtained from the sum of the products of the corresponding rows of A and the corresponding columns of B.

Two matrices A and B may be multiplied together, provided the number of elements in the rows of matrix A are equal to the number of elements in the columns of matrix B. In general terms, when multiplying a matrix of dimensions (m by n), by a matrix of dimensions (n by r), the resulting matrix has dimensions (m by r). Thus a 2 by 3 matrix multiplied by a 3 by 1 matrix gives a matrix of dimensions 2 by 1.

Problem 5. If $A=\begin{pmatrix}2 & 3\\ 1 & -4\end{pmatrix}$ and $B=\begin{pmatrix}-5 & 7\\ -3 & 4\end{pmatrix}$ find $A\times B$

Let $A\times B=C$ where $C=\begin{pmatrix}C_{11} & C_{12}\\ C_{21} & C_{22}\end{pmatrix}$

C_{11} is the sum of the products of the first row elements of A and the first column elements of B taken one at a time,

i.e. $C_{11}=(2\times(-5))+(3\times(-3))=-19$

C_{12} is the sum of the products of the first row elements of A and the second column elements of B, taken one at a time,

i.e. $C_{12}=(2\times 7)+(3\times 4)=26$

C_{21} is the sum of the products of the second row elements of A and the first column elements of B, taken one at a time.

i.e. $C_{21}=(1\times(-5))+(-4)\times(-3))=7$

Finally, C_{22} is the sum of the products of the second row elements of A and the second column elements of B, taken one at a time

i.e. $C_{22}=(1\times 7)+((-4)\times 4)=-9$

Thus, $\boldsymbol{A\times B}=\begin{pmatrix}\mathbf{-19} & \mathbf{26}\\ \mathbf{7} & \mathbf{-9}\end{pmatrix}$

Problem 6. Simplify

$$\begin{pmatrix} 3 & 4 & 0 \\ -2 & 6 & -3 \\ 7 & -4 & 1 \end{pmatrix} \times \begin{pmatrix} 2 \\ 5 \\ -1 \end{pmatrix}$$

The sum of the products of the elements of each row of the first matrix and the elements of the second matrix, (called a **column matrix**), are taken one at a time. Thus:

$$\begin{pmatrix} 3 & 4 & 0 \\ -2 & 6 & -3 \\ 7 & -4 & 1 \end{pmatrix} \times \begin{pmatrix} 2 \\ 5 \\ -1 \end{pmatrix}$$

$$= \begin{pmatrix} (3\times 2)+(4\times 5)+(0\times(-1)) \\ (-2\times 2)+(6\times 5)+(-3\times(-1)) \\ (7\times 2)+(-4\times 5)+(1\times(-1)) \end{pmatrix}$$

$$= \begin{pmatrix} \mathbf{26} \\ \mathbf{29} \\ \mathbf{-7} \end{pmatrix}$$

Problem 7. If $A = \begin{pmatrix} 3 & 4 & 0 \\ -2 & 6 & -3 \\ 7 & -4 & 1 \end{pmatrix}$ and $B = \begin{pmatrix} 2 & -5 \\ 5 & -6 \\ -1 & -7 \end{pmatrix}$ find $A\times B$

The sum of the products of the elements of each row of the first matrix and the elements of each column of the second matrix are taken one at a time. Thus:

$$\begin{pmatrix} 3 & 4 & 0 \\ -2 & 6 & -3 \\ 7 & -4 & 1 \end{pmatrix} \times \begin{pmatrix} 2 & -5 \\ 5 & -6 \\ -1 & -7 \end{pmatrix}$$

$$= \begin{pmatrix} [(3\times 2) + (4\times 5) + (0\times(-1))] & [(3\times(-5)) + (4\times(-6)) + (0\times(-7))] \\ [(-2\times 2) + (6\times 5) + (-3\times(-1))] & [(-2\times(-5)) + (6\times(-6)) + (-3\times(-7))] \\ [(7\times 2) + (-4\times 5) + (1\times(-1))] & [(7\times(-5)) + (-4\times(-6)) + (1\times(-7))] \end{pmatrix}$$

$$= \begin{pmatrix} \mathbf{26} & \mathbf{-39} \\ \mathbf{29} & \mathbf{-5} \\ \mathbf{-7} & \mathbf{-18} \end{pmatrix}$$

Problem 8. Determine

$$\begin{pmatrix} 1 & 0 & 3 \\ 2 & 1 & 2 \\ 1 & 3 & 1 \end{pmatrix} \times \begin{pmatrix} 2 & 2 & 0 \\ 1 & 3 & 2 \\ 3 & 2 & 0 \end{pmatrix}$$

The sum of the products of the elements of each row of the first matrix and the elements of each column of the second matrix are taken one at a time. Thus:

$$\begin{pmatrix} 1 & 0 & 3 \\ 2 & 1 & 2 \\ 1 & 3 & 1 \end{pmatrix} \times \begin{pmatrix} 2 & 2 & 0 \\ 1 & 3 & 2 \\ 3 & 2 & 0 \end{pmatrix}$$

$$= \begin{pmatrix} [(1\times 2) + (0\times 1) + (3\times 3)] & [(1\times 2) + (0\times 3) + (3\times 2)] & [(1\times 0) + (0\times 2) + (3\times 0)] \\ [(2\times 2) + (1\times 1) + (2\times 3)] & [(2\times 2) + (1\times 3) + (2\times 2)] & [(2\times 0) + (1\times 2) + (2\times 0)] \\ [(1\times 2) + (3\times 1) + (1\times 3)] & [(1\times 2) + (3\times 3) + (1\times 2)] & [(1\times 0) + (3\times 2) + (1\times 0)] \end{pmatrix}$$

$$= \begin{pmatrix} \mathbf{11} & \mathbf{8} & \mathbf{0} \\ \mathbf{11} & \mathbf{11} & \mathbf{2} \\ \mathbf{8} & \mathbf{13} & \mathbf{6} \end{pmatrix}$$

In algebra, the commutative law of multiplication states that $a\times b = b\times a$. For matrices, this law is only true in a few special cases, and in general $A\times B$ is **not** equal to $B\times A$.

Problem 9. If $A = \begin{pmatrix} 2 & 3 \\ 1 & 0 \end{pmatrix}$ and $B = \begin{pmatrix} 2 & 3 \\ 0 & 1 \end{pmatrix}$ show that $A\times B \neq B\times A$

$$A\times B = \begin{pmatrix} 2 & 3 \\ 1 & 0 \end{pmatrix} \times \begin{pmatrix} 2 & 3 \\ 0 & 1 \end{pmatrix}$$

$$= \begin{pmatrix} [(2\times 2)+(3\times 0)] & [(2\times 3)+(3\times 1)] \\ [(1\times 2)+(0\times 0)] & [(1\times 3)+(0\times 1)] \end{pmatrix}$$

$$= \begin{pmatrix} 4 & 9 \\ 2 & 3 \end{pmatrix}$$

$$B \times A = \begin{pmatrix} 2 & 3 \\ 0 & 1 \end{pmatrix} \times \begin{pmatrix} 2 & 3 \\ 1 & 0 \end{pmatrix}$$

$$= \begin{pmatrix} [(2\times 2)+(3\times 1)] & [(2\times 3)+(3\times 0)] \\ [(0\times 2)+(1\times 1)] & [(0\times 3)+(1\times 0)] \end{pmatrix}$$

$$= \begin{pmatrix} 7 & 6 \\ 1 & 0 \end{pmatrix}$$

Since $\begin{pmatrix} 4 & 9 \\ 2 & 3 \end{pmatrix} \neq \begin{pmatrix} 7 & 6 \\ 1 & 0 \end{pmatrix}$ then $\boldsymbol{A \times B \neq B \times A}$

Problem 10. A T-network is a cascade connection of a series impedance, a shunt admittance and then another series impedance. The transmission matrix for such a network is given by:

$$\begin{pmatrix} A & B \\ C & D \end{pmatrix} = \begin{pmatrix} 1 & Z_1 \\ 0 & 1 \end{pmatrix}\begin{pmatrix} 1 & 0 \\ Y & 1 \end{pmatrix}\begin{pmatrix} 1 & Z_2 \\ 0 & 1 \end{pmatrix}$$

Determine expressions for A, B, C and D in terms of Z_1, Z_2 and Y.

$$\begin{pmatrix} \boldsymbol{A} & \boldsymbol{B} \\ \boldsymbol{C} & \boldsymbol{D} \end{pmatrix} = \begin{pmatrix} 1 & Z_1 \\ 0 & 1 \end{pmatrix}\begin{pmatrix} 1 & 0 \\ Y & 1 \end{pmatrix}\begin{pmatrix} 1 & Z_2 \\ 0 & 1 \end{pmatrix}$$

$$= \begin{pmatrix} (1+YZ_1) & Z_1 \\ Y & 1 \end{pmatrix}\begin{pmatrix} 1 & Z_2 \\ 0 & 1 \end{pmatrix}$$

$$= \begin{pmatrix} (1+YZ_1) & [(1+YZ_1)(Z_2)+Z_1] \\ Y & (YZ_2+1) \end{pmatrix}$$

$$= \begin{pmatrix} \boldsymbol{(1+YZ_1)} & \boldsymbol{Z_1+Z_2+YZ_1Z_2} \\ \boldsymbol{Y} & \boldsymbol{(1+YZ_2)} \end{pmatrix}$$

Now try the following Practice Exercise

Practice Exercise 269 Addition, subtraction and multiplication of matrices (Answers on page 733)

In Problems 1 to 13, the matrices A to K are:

$$A = \begin{pmatrix} 3 & -1 \\ -4 & 7 \end{pmatrix} \quad B = \begin{pmatrix} 5 & 2 \\ -1 & 6 \end{pmatrix}$$

$$C = \begin{pmatrix} -1.3 & 7.4 \\ 2.5 & -3.9 \end{pmatrix}$$

$$D = \begin{pmatrix} 4 & -7 & 6 \\ -2 & 4 & 0 \\ 5 & 7 & -4 \end{pmatrix}$$

$$E = \begin{pmatrix} 3 & 6 & 2 \\ 5 & -3 & 7 \\ -1 & 0 & 2 \end{pmatrix}$$

$$F = \begin{pmatrix} 3.1 & 2.4 & 6.4 \\ -1.6 & 3.8 & -1.9 \\ 5.3 & 3.4 & -4.8 \end{pmatrix} \quad G = \begin{pmatrix} 6 \\ -2 \end{pmatrix}$$

$$H = \begin{pmatrix} -2 \\ 5 \end{pmatrix} \quad J = \begin{pmatrix} 4 \\ -11 \\ 7 \end{pmatrix} \quad K = \begin{pmatrix} 1 & 0 \\ 0 & 1 \\ 1 & 0 \end{pmatrix}$$

In Problems 1 to 12, perform the matrix operation stated.

1. $A+B$
2. $D+E$
3. $A-B$
4. $A+B-C$
5. $5A+6B$
6. $2D+3E-4F$
7. $A\times H$
8. $A\times B$
9. $A\times C$
10. $D\times J$
11. $E\times K$
12. $D\times F$
13. Show that $A\times C \neq C\times A$

58.3 The unit matrix

A **unit matrix, *I***, is one in which all elements of the leading diagonal (\) have a value of 1 and all other elements have a value of 0. Multiplication of a matrix by I is the equivalent of multiplying by 1 in arithmetic.

58.4 The determinant of a 2 by 2 matrix

The determinant of a 2 by 2 matrix, $\begin{pmatrix} a & b \\ c & d \end{pmatrix}$ **is defined as** $(ad-bc)$

The elements of the determinant of a matrix are written between vertical lines. Thus, the determinant of $\begin{pmatrix} 3 & -4 \\ 1 & 6 \end{pmatrix}$ is written as $\begin{vmatrix} 3 & -4 \\ 1 & 6 \end{vmatrix}$ and is equal to $(3\times 6)-(-4\times 1)$, i.e. $18-(-4)$ or 22. Hence the determinant of a matrix can be expressed as a single numerical value, i.e. $\begin{vmatrix} 3 & -4 \\ 1 & 6 \end{vmatrix}=22$

Problem 11. Determine the value of $\begin{vmatrix} 3 & -2 \\ 7 & 4 \end{vmatrix}$

$$\begin{vmatrix} 3 & -2 \\ 7 & 4 \end{vmatrix} = (3\times 4)-(-2\times 7)$$
$$= 12-(-14) = \mathbf{26}$$

Problem 12. Evaluate: $\begin{vmatrix} (1+j) & j2 \\ -j3 & (1-j4) \end{vmatrix}$

$$\begin{vmatrix} (1+j) & j2 \\ -j3 & (1-j4) \end{vmatrix} = (1+j)(1-j4)-(j2)(-j3)$$
$$= 1-j4+j-j^2 4+j^2 6$$
$$= 1-j4+j-(-4)+(-6)$$

since from Chapter 32, $j^2=-1$

$$= 1-j4+j+4-6$$
$$= \mathbf{-1-j3}$$

Problem 13. Evaluate: $\begin{vmatrix} 5\angle 30^\circ & 2\angle -60^\circ \\ 3\angle 60^\circ & 4\angle -90^\circ \end{vmatrix}$

$$\begin{vmatrix} 5\angle 30^\circ & 2\angle -60^\circ \\ 3\angle 60^\circ & 4\angle -90^\circ \end{vmatrix} = (5\angle 30^\circ)(4\angle -90^\circ) - (2\angle -60^\circ)(3\angle 60^\circ)$$
$$= (20\angle -60^\circ)-(6\angle 0^\circ)$$
$$= (10-j17.32)-(6+j0)$$
$$= \mathbf{(4-j17.32)} \text{ or } \mathbf{17.78\angle -77^\circ}$$

Now try the following Practice Exercise

Practice Exercise 270 2 by 2 determinants (Answers on page 733)

1. Calculate the determinant of $\begin{pmatrix} 3 & -1 \\ -4 & 7 \end{pmatrix}$
2. Calculate the determinant of $\begin{pmatrix} -2 & 5 \\ 3 & -6 \end{pmatrix}$
3. Calculate the determinant of $\begin{pmatrix} -1.3 & 7.4 \\ 2.5 & -3.9 \end{pmatrix}$
4. Evaluate $\begin{vmatrix} j2 & -j3 \\ (1+j) & j \end{vmatrix}$
5. Evaluate $\begin{vmatrix} 2\angle 40^\circ & 5\angle -20^\circ \\ 7\angle -32^\circ & 4\angle -117^\circ \end{vmatrix}$
6. Given matrix $A = \begin{pmatrix} (x-2) & 6 \\ 2 & (x-3) \end{pmatrix}$, determine values of x for which $|A| = 0$

58.5 The inverse or reciprocal of a 2 by 2 matrix

The inverse of matrix A is A^{-1} such that $A\times A^{-1}=I$, the unit matrix.

Let matrix A be $\begin{pmatrix} 1 & 2 \\ 3 & 4 \end{pmatrix}$ and let the inverse matrix, A^{-1} be $\begin{pmatrix} a & b \\ c & d \end{pmatrix}$

Then, since $A\times A^{-1}=I$,

$$\begin{pmatrix} 1 & 2 \\ 3 & 4 \end{pmatrix}\times\begin{pmatrix} a & b \\ c & d \end{pmatrix}=\begin{pmatrix} 1 & 0 \\ 0 & 1 \end{pmatrix}$$

Multiplying the matrices on the left hand side, gives

$$\begin{pmatrix} a+2c & b+2d \\ 3a+4c & 3b+4d \end{pmatrix}=\begin{pmatrix} 1 & 0 \\ 0 & 1 \end{pmatrix}$$

Equating corresponding elements gives:

$$b+2d=0, \quad \text{i.e.} \quad b=-2d$$

and $\quad 3a+4c=0, \quad$ i.e. $\quad a=-\frac{4}{3}c$

Substituting for a and b gives:

$$\begin{pmatrix} -\frac{4}{3}c+2c & -2d+2d \\ 3\left(-\frac{4}{3}c\right)+4c & 3(-2d)+4d \end{pmatrix}=\begin{pmatrix} 1 & 0 \\ 0 & 1 \end{pmatrix}$$

i.e.

$$\begin{pmatrix} \frac{2}{3}c & 0 \\ 0 & -2d \end{pmatrix}=\begin{pmatrix} 1 & 0 \\ 0 & 1 \end{pmatrix}$$

showing that $\frac{2}{3}c = 1$, i.e. $c = \frac{3}{2}$ and $-2d = 1$, i.e. $d = -\frac{1}{2}$

Since $b = -2d$, $b = 1$ and since $a = -\frac{4}{3}c, a = -2$

Thus the inverse of matrix $\begin{pmatrix} 1 & 2 \\ 3 & 4 \end{pmatrix}$ is $\begin{pmatrix} a & b \\ c & d \end{pmatrix}$ that is, $\begin{pmatrix} -2 & 1 \\ \frac{3}{2} & -\frac{1}{2} \end{pmatrix}$

There is, however, **a quicker method of obtaining the inverse** of a 2 by 2 matrix.

For any matrix $\begin{pmatrix} p & q \\ r & s \end{pmatrix}$ the inverse may be obtained by:

(i) interchanging the positions of p and s,
(ii) changing the signs of q and r, and
(iii) multiplying this new matrix by the reciprocal of the determinant of $\begin{pmatrix} p & q \\ r & s \end{pmatrix}$.

Thus the inverse of matrix $\begin{pmatrix} 1 & 2 \\ 3 & 4 \end{pmatrix}$ is

$$\frac{1}{4-6}\begin{pmatrix} 4 & -2 \\ -3 & 1 \end{pmatrix} = \begin{pmatrix} -2 & 1 \\ \frac{3}{2} & -\frac{1}{2} \end{pmatrix}$$

as obtained previously.

Problem 14. Determine the inverse of $\begin{pmatrix} 3 & -2 \\ 7 & 4 \end{pmatrix}$

The inverse of matrix $\begin{pmatrix} p & q \\ r & s \end{pmatrix}$ is obtained by interchanging the positions of p and s, changing the signs of q and r and multiplying by the reciprocal of the determinant $\begin{vmatrix} p & q \\ r & s \end{vmatrix}$. Thus, the inverse of $\begin{pmatrix} 3 & -2 \\ 7 & 4 \end{pmatrix}$

$$= \frac{1}{(3\times 4)-(-2\times 7)}\begin{pmatrix} 4 & 2 \\ -7 & 3 \end{pmatrix}$$

$$= \frac{1}{26}\begin{pmatrix} 4 & 2 \\ -7 & 3 \end{pmatrix} = \begin{pmatrix} \mathbf{\frac{2}{13}} & \mathbf{\frac{1}{13}} \\ \mathbf{\frac{-7}{26}} & \mathbf{\frac{3}{26}} \end{pmatrix}$$

Now try the following Practice Exercise

Practice Exercise 271 The inverse of 2 by 2 matrices (Answers on page 733)

1. Determine the inverse of $\begin{pmatrix} 3 & -1 \\ -4 & 7 \end{pmatrix}$
2. Determine the inverse of $\begin{pmatrix} \frac{1}{2} & \frac{2}{3} \\ -\frac{1}{3} & -\frac{3}{5} \end{pmatrix}$
3. Determine the inverse of $\begin{pmatrix} -1.3 & 7.4 \\ 2.5 & -3.9 \end{pmatrix}$

58.6 The determinant of a 3 by 3 matrix

(i) The **minor** of an element of a 3 by 3 matrix is the value of the 2 by 2 determinant obtained by covering up the row and column containing that element.

Thus for the matrix $\begin{pmatrix} 1 & 2 & 3 \\ 4 & 5 & 6 \\ 7 & 8 & 9 \end{pmatrix}$ the minor of element 4 is obtained by covering the row (4 5 6) and the column $\begin{pmatrix} 1 \\ 4 \\ 7 \end{pmatrix}$, leaving the 2 by determinant $\begin{vmatrix} 2 & 3 \\ 8 & 9 \end{vmatrix}$, i.e. the minor of element 4 is $(2\times 9)-(3\times 8) = -6$

(ii) The sign of a minor depends on its position within the matrix, the sign pattern being $\begin{pmatrix} + & - & + \\ - & + & - \\ + & - & + \end{pmatrix}$

Thus the signed-minor of element 4 in the matrix $\begin{pmatrix} 1 & 2 & 3 \\ 4 & 5 & 6 \\ 7 & 8 & 9 \end{pmatrix}$ is $-\begin{vmatrix} 2 & 3 \\ 8 & 9 \end{vmatrix} = -(-6) = 6$

The signed-minor of an element is called the **cofactor** of the element.

(iii) **The value of a 3 by 3 determinant is the sum of the products of the elements and their cofactors of any row or any column of the corresponding 3 by 3 matrix.**

There are thus six different ways of evaluating a 3×3 determinant — and all should give the same value.

Problem 15. Find the value of

$$\begin{vmatrix} 3 & 4 & -1 \\ 2 & 0 & 7 \\ 1 & -3 & -2 \end{vmatrix}$$

The value of this determinant is the sum of the products of the elements and their cofactors, of any row or of any column. If the second row or second column is selected, the element 0 will make the product of the element and its cofactor zero and reduce the amount of arithmetic to be done to a minimum.

Supposing a second row expansion is selected.

The minor of 2 is the value of the determinant remaining when the row and column containing the 2 (i.e. the second row and the first column), is covered up. Thus the cofactor of element 2 is $\begin{vmatrix} 4 & -1 \\ -3 & -2 \end{vmatrix}$ i.e. -11. The sign of element 2 is minus, (see (ii) above), hence the cofactor of element 2, (the signed-minor) is $+11$. Similarly the minor of element 7 is $\begin{vmatrix} 3 & 4 \\ 1 & -3 \end{vmatrix}$ i.e. -13, and its cofactor is $+13$. Hence the value of the sum of the products of the elements and their cofactors is $2 \times 11 + 7 \times 13$, i.e.,

$$\begin{vmatrix} 3 & 4 & -1 \\ 2 & 0 & 7 \\ 1 & -3 & -2 \end{vmatrix} = 2(11) + 0 + 7(13) = \mathbf{113}$$

The same result will be obtained whichever row or column is selected. For example, the third column expansion is

$$(-1)\begin{vmatrix} 2 & 0 \\ 1 & -3 \end{vmatrix} - 7\begin{vmatrix} 3 & 4 \\ 1 & -3 \end{vmatrix} + (-2)\begin{vmatrix} 3 & 4 \\ 2 & 0 \end{vmatrix}$$

$= 6 + 91 + 16 = \mathbf{113}$, as obtained previously.

Problem 16. Evaluate $\begin{vmatrix} 1 & 4 & -3 \\ -5 & 2 & 6 \\ -1 & -4 & 2 \end{vmatrix}$

Using the first row: $\begin{vmatrix} 1 & 4 & -3 \\ -5 & 2 & 6 \\ -1 & -4 & 2 \end{vmatrix}$

$$= 1\begin{vmatrix} 2 & 6 \\ -4 & 2 \end{vmatrix} - 4\begin{vmatrix} -5 & 6 \\ -1 & 2 \end{vmatrix} + (-3)\begin{vmatrix} -5 & 2 \\ -1 & -4 \end{vmatrix}$$

$$= (4 + 24) - 4(-10 + 6) - 3(20 + 2)$$

$$= 28 + 16 - 66 = \mathbf{-22}$$

Using the second column: $\begin{vmatrix} 1 & 4 & -3 \\ -5 & 2 & 6 \\ -1 & -4 & 2 \end{vmatrix}$

$$= -4\begin{vmatrix} -5 & 6 \\ -1 & 2 \end{vmatrix} + 2\begin{vmatrix} 1 & -3 \\ -1 & 2 \end{vmatrix} - (-4)\begin{vmatrix} 1 & -3 \\ -5 & 6 \end{vmatrix}$$

$$= -4(-10 + 6) + 2(2 - 3) + 4(6 - 15)$$

$$= 16 - 2 - 36 = \mathbf{-22}$$

Problem 17. Determine the value of

$$\begin{vmatrix} j2 & (1+j) & 3 \\ (1-j) & 1 & j \\ 0 & j4 & 5 \end{vmatrix}$$

Using the first column, the value of the determinant is:

$$(j2)\begin{vmatrix} 1 & j \\ j4 & 5 \end{vmatrix} - (1-j)\begin{vmatrix} (1+j) & 3 \\ j4 & 5 \end{vmatrix} + (0)\begin{vmatrix} (1+j) & 3 \\ 1 & j \end{vmatrix}$$

$$= j2(5 - j^2 4) - (1-j)(5 + j5 - j12) + 0$$

$$= j2(9) - (1-j)(5 - j7)$$

$$= j18 - [5 - j7 - j5 + j^2 7]$$

$$= j18 - [-2 - j12]$$

$$= j18 + 2 + j12 = \mathbf{2 + j30} \text{ or } \mathbf{30.07\angle 86.19^\circ}$$

Now try the following Practice Exercise

Practice Exercise 272 3 by 3 determinants (Answers on page 733)

1. Find the matrix of minors of
$$\begin{pmatrix} 4 & -7 & 6 \\ -2 & 4 & 0 \\ 5 & 7 & -4 \end{pmatrix}$$
2. Find the matrix of cofactors of
$$\begin{pmatrix} 4 & -7 & 6 \\ -2 & 4 & 0 \\ 5 & 7 & -4 \end{pmatrix}$$
3. Calculate the determinant of
$$\begin{pmatrix} 4 & -7 & 6 \\ -2 & 4 & 0 \\ 5 & 7 & -4 \end{pmatrix}$$

4. Evaluate $\begin{vmatrix} 8 & -2 & -10 \\ 2 & -3 & -2 \\ 6 & 3 & 8 \end{vmatrix}$

5. Calculate the determinant of
$\begin{pmatrix} 3.1 & 2.4 & 6.4 \\ -1.6 & 3.8 & -1.9 \\ 5.3 & 3.4 & -4.8 \end{pmatrix}$

6. Evaluate $\begin{vmatrix} j2 & 2 & j \\ (1+j) & 1 & -3 \\ 5 & -j4 & 0 \end{vmatrix}$

7. Evaluate $\begin{vmatrix} 3\angle 60^\circ & j2 & 1 \\ 0 & (1+j) & 2\angle 30^\circ \\ 0 & 2 & j5 \end{vmatrix}$

58.7 The inverse or reciprocal of a 3 by 3 matrix

The **adjoint** of a matrix A is obtained by:

(i) forming a matrix B of the cofactors of A, and
(ii) **transposing** matrix B to give B^T, where B^T is the matrix obtained by writing the rows of B as the columns of B^T. Then $\textbf{adj } \boldsymbol{A}=\boldsymbol{B}^T$

The **inverse of matrix** $\boldsymbol{A}$, A^{-1} is given by

$$A^{-1}=\frac{\text{adj}A}{|A|}$$

where adj A is the adjoint of matrix A and $|A|$ is the determinant of matrix A.

Problem 18. Determine the inverse of the matrix $\begin{pmatrix} 3 & 4 & -1 \\ 2 & 0 & 7 \\ 1 & -3 & -2 \end{pmatrix}$

The inverse of matrix A, $A^{-1}=\dfrac{\text{adj}A}{|A|}$

The adjoint of A is found by:

(i) obtaining the matrix of the cofactors of the elements, and
(ii) transposing this matrix.

The cofactor of element 3 is $+\begin{vmatrix} 0 & 7 \\ -3 & -2 \end{vmatrix}=21$

The cofactor of element 4 is $-\begin{vmatrix} 2 & 7 \\ 1 & -2 \end{vmatrix}=11$ and so on.

The matrix of cofactors is $\begin{pmatrix} 21 & 11 & -6 \\ 11 & -5 & 13 \\ 28 & -23 & -8 \end{pmatrix}$

The transpose of the matrix of cofactors, i.e. the adjoint of the matrix, is obtained by writing the rows as columns, and is $\begin{pmatrix} 21 & 11 & 28 \\ 11 & -5 & -23 \\ -6 & 13 & -8 \end{pmatrix}$

From Problem 14, the determinant of $\begin{vmatrix} 3 & 4 & -1 \\ 2 & 0 & 7 \\ 1 & -3 & -2 \end{vmatrix}$ is 113

Hence the inverse of $\begin{pmatrix} 3 & 4 & -1 \\ 2 & 0 & 7 \\ 1 & -3 & -2 \end{pmatrix}$ is

$$\frac{\begin{pmatrix} 21 & 11 & 28 \\ 11 & -5 & -23 \\ -6 & 13 & -8 \end{pmatrix}}{113} \text{ or } \boldsymbol{\frac{1}{113}\begin{pmatrix} 21 & 11 & 28 \\ 11 & -5 & -23 \\ -6 & 13 & -8 \end{pmatrix}}$$

Problem 19. Find the inverse of
$\begin{pmatrix} 1 & 5 & -2 \\ 3 & -1 & 4 \\ -3 & 6 & -7 \end{pmatrix}$

$$\text{Inverse}=\frac{\text{adjoint}}{\text{determinant}}$$

The matrix of cofactors is $\begin{pmatrix} -17 & 9 & 15 \\ 23 & -13 & -21 \\ 18 & -10 & -16 \end{pmatrix}$

The transpose of the matrix of cofactors (i.e. the adjoint) is $\begin{pmatrix} -17 & 23 & 18 \\ 9 & -13 & -10 \\ 15 & -21 & -16 \end{pmatrix}$

The determinant of $\begin{pmatrix} 1 & 5 & -2 \\ 3 & -1 & 4 \\ -3 & 6 & -7 \end{pmatrix}$

$=1(7-24)-5(-21+12)-2(18-3)$

$=-17+45-30=-2$

Hence the inverse of $\begin{pmatrix} 1 & 5 & -2 \\ 3 & -1 & 4 \\ -3 & 6 & -7 \end{pmatrix}$

$$= \frac{\begin{pmatrix} -17 & 23 & 18 \\ 9 & -13 & -10 \\ 15 & -21 & -16 \end{pmatrix}}{-2}$$

$$= \begin{pmatrix} \mathbf{8.5} & \mathbf{-11.5} & \mathbf{-9} \\ \mathbf{-4.5} & \mathbf{6.5} & \mathbf{5} \\ \mathbf{-7.5} & \mathbf{10.5} & \mathbf{8} \end{pmatrix}$$

Now try the following Practice Exercise

Practice Exercise 273 The inverse of a 3 by 3 matrix (Answers on page 733)

1. Write down the transpose of $\begin{pmatrix} 4 & -7 & 6 \\ -2 & 4 & 0 \\ 5 & 7 & -4 \end{pmatrix}$
2. Write down the transpose of $\begin{pmatrix} 3 & 6 & \frac{1}{2} \\ 5 & -\frac{2}{3} & 7 \\ -1 & 0 & \frac{3}{5} \end{pmatrix}$
3. Determine the adjoint of $\begin{pmatrix} 4 & -7 & 6 \\ -2 & 4 & 0 \\ 5 & 7 & -4 \end{pmatrix}$
4. Determine the adjoint of $\begin{pmatrix} 3 & 6 & \frac{1}{2} \\ 5 & -\frac{2}{3} & 7 \\ -1 & 0 & \frac{3}{5} \end{pmatrix}$
5. Find the inverse of $\begin{pmatrix} 4 & -7 & 6 \\ -2 & 4 & 0 \\ 5 & 7 & -4 \end{pmatrix}$
6. Find the inverse of $\begin{pmatrix} 3 & 6 & \frac{1}{2} \\ 5 & -\frac{2}{3} & 7 \\ -1 & 0 & \frac{3}{5} \end{pmatrix}$

Practice Exercise 274 Multiple-choice questions on the theory of matrices and determinants (Answers on page 734)

Each question has only one correct answer

1. The vertical lines of elements in a matrix are called:
   (a) rows (b) row matrix
   (c) columns (d) a column matrix
2. If a matrix P has the same number of rows and columns, then matrix P is called a:
   (a) row matrix (b) square matrix
   (c) column matrix (d) rectangular matrix
3. A matrix of order $m \times 1$ is called a:
   (a) row matrix (b) column matrix
   (c) square matrix (d) rectangular matrix
4. $\begin{vmatrix} j2 & j \\ j & -j \end{vmatrix}$ is equal to:
   (a) 5 (b) 4 (c) 3 (d) 2
5. The matrix product $\begin{pmatrix} 2 & 3 \\ -1 & 4 \end{pmatrix}\begin{pmatrix} 1 & -5 \\ -2 & 6 \end{pmatrix}$ is equal to:
   (a) $\begin{pmatrix} -13 \\ 26 \end{pmatrix}$ (b) $\begin{pmatrix} 3 & -2 \\ -3 & 10 \end{pmatrix}$
   (c) $\begin{pmatrix} 1 & -2 \\ -3 & -2 \end{pmatrix}$ (d) $\begin{pmatrix} -4 & 8 \\ -9 & 29 \end{pmatrix}$
6. If $\begin{vmatrix} 2k & -1 \\ 4 & 2 \end{vmatrix} = \begin{vmatrix} 3 & 0 \\ 2 & 1 \end{vmatrix}$ then the value of k is:
   (a) $\frac{2}{3}$ (b) 3 (c) $-\frac{1}{4}$ (d) $\frac{3}{2}$
7. The value of $\begin{vmatrix} 6 & 0 & -1 \\ 2 & 1 & 4 \\ 1 & 1 & 3 \end{vmatrix}$ is:
   (a) −7 (b) 8 (c) 7 (d) 10
8. Solving $\begin{vmatrix} (2k+5) & 3 \\ (5k+2) & 9 \end{vmatrix} = 0$ for k gives:
   (a) $-\frac{17}{11}$ (b) −17 (c) 17 (d) −13
9. If a matrix has only one row, then it is called a:
   (a) row matrix (b) column matrix
   (c) square matrix (d) rectangular matrix

10. If $P = \begin{pmatrix} 5 & 3 & 2 \\ 0 & 4 & 1 \\ 0 & 0 & 3 \end{pmatrix}$ then $|P|$ is equal to:
(a) 30 (b) 40 (c) 50 (d) 60

11. If the order of a matrix P is $a \times b$ and the order of a matrix Q is $b \times c$ then the order of $P \times Q$ is:
(a) $a \times c$ (b) $c \times a$ (c) $c \times b$ (d) $a \times b$

12. The value of $\begin{vmatrix} (1+j2) & j \\ j & -j \end{vmatrix}$ is:
(a) $-3-j$ (b) $-1-j$ (c) $3-j$ (d) $1-j$

13. $\begin{pmatrix} 1 & 0 & 0 \\ 0 & 1 & 0 \\ 0 & 0 & 1 \end{pmatrix}$ is an example of a:
(a) null matrix (b) triangular matrix
(c) rectangular matrix (d) unit matrix

14. The inverse of the matrix $\begin{pmatrix} 5 & -3 \\ -2 & 1 \end{pmatrix}$ is:
(a) $\begin{pmatrix} -5 & -3 \\ 2 & -1 \end{pmatrix}$ (b) $\begin{pmatrix} -1 & -3 \\ -2 & -5 \end{pmatrix}$
(c) $\begin{pmatrix} -1 & 3 \\ 2 & -5 \end{pmatrix}$ (d) $\begin{pmatrix} 1 & 3 \\ 2 & 5 \end{pmatrix}$

15. The value of the determinant $\begin{vmatrix} 2 & -1 & 4 \\ 0 & 1 & 5 \\ 6 & 0 & -1 \end{vmatrix}$ is:
(a) -56 (b) 52 (c) 4 (d) 8

16. The value of $\begin{vmatrix} j2 & -(1+j) \\ (1-j) & 1 \end{vmatrix}$ is:
(a) $-j2$ (b) $2(1+j)$ (c) 2 (d) $-2+j2$

17. The inverse of the matrix $\begin{pmatrix} 2 & -3 \\ 1 & -4 \end{pmatrix}$ is:
(a) $\begin{pmatrix} 0.8 & 0.6 \\ -0.2 & -0.4 \end{pmatrix}$
(b) $\begin{pmatrix} -4 & 3 \\ -1 & 2 \end{pmatrix}$
(c) $\begin{pmatrix} -0.4 & -0.6 \\ 0.2 & 0.8 \end{pmatrix}$
(d) $\begin{pmatrix} 0.8 & -0.6 \\ 0.2 & -0.4 \end{pmatrix}$

18. Matrix A is given by: $A = \begin{pmatrix} 3 & -2 \\ 4 & -1 \end{pmatrix}$ The determinant of matrix A is:
(a) -11 (b) 5 (c) 4 (d) $\begin{pmatrix} -1 & 2 \\ -4 & 3 \end{pmatrix}$

19. The inverse of the matrix $\begin{pmatrix} -1 & 4 \\ -3 & 2 \end{pmatrix}$ is:
(a) $\begin{pmatrix} \frac{1}{10} & -\frac{3}{10} \\ \frac{2}{5} & -\frac{1}{5} \end{pmatrix}$ (b) $\begin{pmatrix} -\frac{1}{7} & \frac{2}{7} \\ -\frac{3}{14} & \frac{1}{14} \end{pmatrix}$
(c) $\begin{pmatrix} \frac{1}{5} & -\frac{2}{5} \\ \frac{3}{10} & -\frac{1}{10} \end{pmatrix}$ (d) $\begin{pmatrix} \frac{1}{14} & -\frac{3}{14} \\ \frac{2}{7} & -\frac{1}{7} \end{pmatrix}$

20. The value(s) of λ given $\begin{vmatrix} (2-\lambda) & -1 \\ -4 & (-1-\lambda) \end{vmatrix} = 0$ is:
(a) -2 and $+3$ (b) $+2$ and -3
(c) -2 (d) 3

For fully worked solutions to each of the problems in Practice Exercises 269 to 273 in this chapter, go to the website:
www.routledge.com/cw/bird

Chapter 59

The solution of simultaneous equations by matrices and determinants

Why it is important to understand: **Applications of matrices and determinants**

As mentioned previously, matrices are used to solve problems, for example, in electrical circuits, optics, quantum mechanics, statics, robotics, genetics and much more, and for calculating forces, vectors, tensions, masses, loads and a lot of other factors that must be accounted for in engineering. In the main, matrices and determinants are used to solve a system of simultaneous linear equations. The simultaneous solution of multiple equations finds its way into many common engineering problems. In fact, modern structural engineering analysis techniques are all about solving systems of equations simultaneously.

At the end of this chapter, you should be able to:

- solve simultaneous equations in two and three unknowns using matrices
- solve simultaneous equations in two and three unknowns using determinants
- solve simultaneous equations using Cramer's rule
- solve simultaneous equations using Gaussian elimination

59.1 Solution of simultaneous equations by matrices

(a) The procedure for solving linear simultaneous equations in **two unknowns using matrices** is:

(i) write the equations in the form

$$a_1x + b_1y = c_1$$

$$a_2x + b_2y = c_2$$

(ii) write the matrix equation corresponding to these equations,

i.e. $\begin{pmatrix} a_1 & b_1 \\ a_2 & b_2 \end{pmatrix} \times \begin{pmatrix} x \\ y \end{pmatrix} = \begin{pmatrix} c_1 \\ c_2 \end{pmatrix}$

(iii) determine the inverse matrix of $\begin{pmatrix} a_1 & b_1 \\ a_2 & b_2 \end{pmatrix}$

i.e. $\dfrac{1}{a_1b_2 - b_1a_2}\begin{pmatrix} b_2 & -b_1 \\ -a_2 & a_1 \end{pmatrix}$

(from Chapter 58)

(iv) multiply each side of (ii) by the inverse matrix, and

(v) solve for x and y by equating corresponding elements.

Problem 1. Use matrices to solve the simultaneous equations:

$$3x+5y-7=0 \qquad (1)$$

$$4x-3y-19=0 \qquad (2)$$

(i) Writing the equations in the $a_1x+b_1y=c$ form gives:

$$3x+5y=7$$

$$4x-3y=19$$

(ii) The matrix equation is

$$\begin{pmatrix}3 & 5\\ 4 & -3\end{pmatrix}\times\begin{pmatrix}x\\ y\end{pmatrix}=\begin{pmatrix}7\\ 19\end{pmatrix}$$

(iii) The inverse of matrix $\begin{pmatrix}3 & 5\\ 4 & -3\end{pmatrix}$ is

$$\frac{1}{3\times(-3)-5\times 4}\begin{pmatrix}-3 & -5\\ -4 & 3\end{pmatrix}$$

i.e. $\begin{pmatrix}\frac{3}{29} & \frac{5}{29}\\ \frac{4}{29} & \frac{-3}{29}\end{pmatrix}$

(iv) Multiplying each side of (ii) by (iii) and remembering that $A\times A^{-1}=I$, the unit matrix, gives:

$$\begin{pmatrix}1 & 0\\ 0 & 1\end{pmatrix}\begin{pmatrix}x\\ y\end{pmatrix}=\begin{pmatrix}\frac{3}{29} & \frac{5}{29}\\ \frac{4}{29} & \frac{-3}{29}\end{pmatrix}\times\begin{pmatrix}7\\ 19\end{pmatrix}$$

Thus $\begin{pmatrix}x\\ y\end{pmatrix}=\begin{pmatrix}\frac{21}{29}+\frac{95}{29}\\ \frac{28}{29}-\frac{57}{29}\end{pmatrix}$

i.e. $\begin{pmatrix}x\\ y\end{pmatrix}=\begin{pmatrix}4\\ -1\end{pmatrix}$

(v) By comparing corresponding elements:

$$\boldsymbol{x=4} \quad \text{and} \quad \boldsymbol{y=-1}$$

Checking:
equation (1),

$$3\times 4+5\times(-1)-7=0=\text{RHS}$$

equation (2),

$$4\times 4-3\times(-1)-19=0=\text{RHS}$$

(b) The procedure for solving linear simultaneous equations in **three unknowns using matrices** is:

(i) write the equations in the form

$$a_1x+b_1y+c_1z=d_1$$

$$a_2x+b_2y+c_2z=d_2$$

$$a_3x+b_3y+c_3z=d_3$$

(ii) write the matrix equation corresponding to these equations, i.e.

$$\begin{pmatrix}a_1 & b_1 & c_1\\ a_2 & b_2 & c_2\\ a_3 & b_3 & c_3\end{pmatrix}\times\begin{pmatrix}x\\ y\\ z\end{pmatrix}=\begin{pmatrix}d_1\\ d_2\\ d_3\end{pmatrix}$$

(iii) determine the inverse matrix of

$$\begin{pmatrix}a_1 & b_1 & c_1\\ a_2 & b_2 & c_2\\ a_3 & b_3 & c_3\end{pmatrix}\text{ (see Chapter 58)}$$

(iv) multiply each side of (ii) by the inverse matrix, and

(v) solve for x, y and z by equating the corresponding elements.

Problem 2. Use matrices to solve the simultaneous equations:

$$x+y+z-4=0 \qquad (1)$$

$$2x-3y+4z-33=0 \qquad (2)$$

$$3x-2y-2z-2=0 \qquad (3)$$

(i) Writing the equations in the $a_1x+b_1y+c_1z=d_1$ form gives:

$$x+y+z=4$$

$$2x-3y+4z=33$$

$$3x-2y-2z=2$$

(ii) The matrix equation is

$$\begin{pmatrix}1 & 1 & 1\\ 2 & -3 & 4\\ 3 & -2 & -2\end{pmatrix}\times\begin{pmatrix}x\\ y\\ z\end{pmatrix}=\begin{pmatrix}4\\ 33\\ 2\end{pmatrix}$$

(iii) The inverse matrix of

$$A=\begin{pmatrix}1 & 1 & 1\\ 2 & -3 & 4\\ 3 & -2 & -2\end{pmatrix}$$

is given by

$$A^{-1}=\frac{\text{adj}\,A}{|A|}$$

The adjoint of A is the transpose of the matrix of the cofactors of the elements (see Chapter 58). The matrix of cofactors is

$$\begin{pmatrix} 14 & 16 & 5 \\ 0 & -5 & 5 \\ 7 & -2 & -5 \end{pmatrix}$$

and the transpose of this matrix gives

$$\text{adj}\,A = \begin{pmatrix} 14 & 0 & 7 \\ 16 & -5 & -2 \\ 5 & 5 & -5 \end{pmatrix}$$

The determinant of A, i.e. the sum of the products of elements and their cofactors, using a first row expansion is

$$1\begin{vmatrix} -3 & 4 \\ -2 & -2 \end{vmatrix} - 1\begin{vmatrix} 2 & 4 \\ 3 & -2 \end{vmatrix} + 1\begin{vmatrix} 2 & -3 \\ 3 & -2 \end{vmatrix}$$

$$= (1\times 14) - (1\times(-16)) + (1\times 5) = 35$$

Hence the inverse of A,

$$A^{-1} = \frac{1}{35}\begin{pmatrix} 14 & 0 & 7 \\ 16 & -5 & -2 \\ 5 & 5 & -5 \end{pmatrix}$$

(iv) Multiplying each side of (ii) by (iii), and remembering that $A\times A^{-1} = I$, the unit matrix, gives

$$\begin{pmatrix} 1 & 0 & 0 \\ 0 & 1 & 0 \\ 0 & 0 & 1 \end{pmatrix} \times \begin{pmatrix} x \\ y \\ z \end{pmatrix}$$

$$= \frac{1}{35}\begin{pmatrix} 14 & 0 & 7 \\ 16 & -5 & -2 \\ 5 & 5 & -5 \end{pmatrix} \times \begin{pmatrix} 4 \\ 33 \\ 2 \end{pmatrix}$$

$$\begin{pmatrix} x \\ y \\ z \end{pmatrix} = \frac{1}{35}\begin{pmatrix} (14\times 4)+(0\times 33)+(7\times 2) \\ (16\times 4)+((-5)\times 33)+((-2)\times 2) \\ (5\times 4)+(5\times 33)+((-5)\times 2) \end{pmatrix}$$

$$= \frac{1}{35}\begin{pmatrix} 70 \\ -105 \\ 175 \end{pmatrix}$$

$$= \begin{pmatrix} 2 \\ -3 \\ 5 \end{pmatrix}$$

(v) By comparing corresponding elements, $\boldsymbol{x = 2}$, $\boldsymbol{y = -3}$, $\boldsymbol{z = 5}$, which can be checked in the original equations.

Now try the following Practice Exercise

Practice Exercise 275 Solving simultaneous equations using matrices (Answers on page 734)

In Problems 1 to 5 use **matrices** to solve the simultaneous equations given.

1. $3x + 4y = 0$
   $2x + 5y + 7 = 0$

2. $2p + 5q + 14.6 = 0$
   $3.1p + 1.7q + 2.06 = 0$

3. $x + 2y + 3z = 5$
   $2x - 3y - z = 3$
   $-3x + 4y + 5z = 3$

4. $3a + 4b - 3c = 2$
   $-2a + 2b + 2c = 15$
   $7a - 5b + 4c = 26$

5. $p + 2q + 3r + 7.8 = 0$
   $2p + 5q - r - 1.4 = 0$
   $5p - q + 7r - 3.5 = 0$

6. In two closed loops of an electrical circuit, the currents flowing are given by the simultaneous equations:

   $I_1 + 2I_2 + 4 = 0$
   $5I_1 + 3I_2 - 1 = 0$

   Use matrices to solve for I_1 and I_2

7. The relationship between the displacement, s, velocity, v, and acceleration, a, of a piston is given by the equations:

   $s + 2v + 2a = 4$
   $3s - v + 4a = 25$
   $3s + 2v - a = -4$

   Use matrices to determine the values of s, v and a

8. In a mechanical system, acceleration $\ddot{x}$, velocity $\dot{x}$ and distance x are related by the simultaneous equations:

   $3.4\ddot{x} + 7.0\dot{x} - 13.2x = -11.39$
   $-6.0\ddot{x} + 4.0\dot{x} + 3.5x = 4.98$
   $2.7\ddot{x} + 6.0\dot{x} + 7.1x = 15.91$

   Use matrices to find the values of $\ddot{x}$, $\dot{x}$ and x

59.2 Solution of simultaneous equations by determinants

(a) When solving linear simultaneous equations in **two unknowns using determinants**:

(i) write the equations in the form

$$a_1x + b_1y + c_1 = 0$$
$$a_2x + b_2y + c_2 = 0$$

and then

(ii) the solution is given by

$$\frac{x}{D_x} = \frac{-y}{D_y} = \frac{1}{D}$$

where $D_x = \begin{vmatrix} b_1 & c_1 \\ b_2 & c_2 \end{vmatrix}$

i.e. the determinant of the coefficients left when the x-column is covered up,

$$D_y = \begin{vmatrix} a_1 & c_1 \\ a_2 & c_2 \end{vmatrix}$$

i.e. the determinant of the coefficients left when the y-column is covered up,

and $D = \begin{vmatrix} a_1 & b_1 \\ a_2 & b_2 \end{vmatrix}$

i.e. the determinant of the coefficients left when the constants-column is covered up.

Problem 3. Solve the following simultaneous equations using determinants:

$$3x - 4y = 12$$
$$7x + 5y = 6.5$$

Following the above procedure:

(i) $3x - 4y - 12 = 0$
$7x + 5y - 6.5 = 0$

(ii) $$\frac{x}{\begin{vmatrix} -4 & -12 \\ 5 & -6.5 \end{vmatrix}} = \frac{-y}{\begin{vmatrix} 3 & -12 \\ 7 & -6.5 \end{vmatrix}} = \frac{1}{\begin{vmatrix} 3 & -4 \\ 7 & 5 \end{vmatrix}}$$

i.e. $$\frac{x}{(-4)(-6.5) - (-12)(5)} = \frac{-y}{(3)(-6.5) - (-12)(7)} = \frac{1}{(3)(5) - (-4)(7)}$$

i.e. $$\frac{x}{26 + 60} = \frac{-y}{-19.5 + 84} = \frac{1}{15 + 28}$$

i.e. $$\frac{x}{86} = \frac{-y}{64.5} = \frac{1}{43}$$

Since $\frac{x}{86} = \frac{1}{43}$ then $\boldsymbol{x} = \frac{86}{43} = \mathbf{2}$

and since $\frac{-y}{64.5} = \frac{1}{43}$ then

$$\boldsymbol{y} = -\frac{64.5}{43} = \mathbf{-1.5}$$

Problem 4. The velocity of a car, accelerating at uniform acceleration a between two points, is given by $v = u + at$, where u is its velocity when passing the first point and t is the time taken to pass between the two points. If $v = 21$ m/s when $t = 3.5$ s and $v = 33$ m/s when $t = 6.1$ s, use determinants to find the values of u and a, each correct to 4 significant figures

Substituting the given values in $v = u + at$ gives:

$$21 = u + 3.5a \quad (1)$$
$$33 = u + 6.1a \quad (2)$$

(i) The equations are written in the form

$$a_1x + b_1y + c_1 = 0$$

i.e. $u + 3.5a - 21 = 0$

and $u + 6.1a - 33 = 0$

(ii) The solution is given by

$$\frac{u}{D_u} = \frac{-a}{D_a} = \frac{1}{D}$$

where D_u is the determinant of coefficients left when the u column is covered up, i.e. $D_u = \begin{vmatrix} 3.5 & -21 \\ 6.1 & -33 \end{vmatrix}$

$= (3.5)(-33) - (-21)(6.1)$

$= 12.6$

Similarly, $D_a = \begin{vmatrix} 1 & -21 \\ 1 & -33 \end{vmatrix}$

$= (1)(-33) - (-21)(1)$

$= -12$

and $D = \begin{vmatrix} 1 & 3.5 \\ 1 & 6.1 \end{vmatrix}$

$= (1)(6.1) - (3.5)(1) = 2.6$

Thus $\dfrac{u}{12.6} = \dfrac{-a}{-12} = \dfrac{1}{2.6}$

i.e. $u = \dfrac{12.6}{2.6} = \mathbf{4.846\ m/s}$

and $a = \dfrac{12}{2.6} = \mathbf{4.615\ m/s^2}$,

each correct to 4 significant figures

Problem 5. Applying Kirchhoff's laws to an electric circuit results in the following equations:

$(9 + j12)I_1 - (6 + j8)I_2 = 5$

$-(6 + j8)I_1 + (8 + j3)I_2 = (2 + j4)$

Solve the equations for I_1 and I_2

Following the procedure:

(i) $(9 + j12)I_1 - (6 + j8)I_2 - 5 = 0$

$-(6 + j8)I_1 + (8 + j3)I_2 - (2 + j4) = 0$

(ii) $\dfrac{I_1}{\begin{vmatrix} -(6+j8) & -5 \\ (8+j3) & -(2+j4) \end{vmatrix}}$

$= \dfrac{-I_2}{\begin{vmatrix} (9+j12) & -5 \\ -(6+j8) & -(2+j4) \end{vmatrix}}$

$= \dfrac{I}{\begin{vmatrix} (9+j12) & -(6+j8) \\ -(6+j8) & (8+j3) \end{vmatrix}}$

$\dfrac{I_1}{(-20 + j40) + (40 + j15)}$

$= \dfrac{-I_2}{(30 - j60) - (30 + j40)}$

$= \dfrac{1}{(36 + j123) - (-28 + j96)}$

$\dfrac{I_1}{20 + j55} = \dfrac{-I_2}{-j100} = \dfrac{1}{64 + j27}$

Hence $I_1 = \dfrac{20 + j55}{64 + j27}$

$= \dfrac{58.52\angle 70.02^\circ}{69.46\angle 22.87^\circ} = \mathbf{0.84\angle 47.15^\circ\ A}$

and $I_2 = \dfrac{100\angle 90^\circ}{69.46\angle 22.87^\circ} = \mathbf{1.44\angle 67.13^\circ\ A}$

(b) When solving simultaneous equations in **three unknowns using determinants:**

(i) Write the equations in the form

$a_1x + b_1y + c_1z + d_1 = 0$

$a_2x + b_2y + c_2z + d_2 = 0$

$a_3x + b_3y + c_3z + d_3 = 0$

and then

(ii) the solution is given by

$\dfrac{x}{D_x} = \dfrac{-y}{D_y} = \dfrac{z}{D_z} = \dfrac{-1}{D}$

where $D = x \begin{vmatrix} b_1 & c_1 & d_1 \\ b_2 & c_2 & d_2 \\ b_3 & c_3 & d_3 \end{vmatrix}$

i.e. the determinant of the coefficients obtained by covering up the x column.

$D_y = \begin{vmatrix} a_1 & c_1 & d_1 \\ a_2 & c_2 & d_2 \\ a_3 & c_3 & d_3 \end{vmatrix}$

i.e., the determinant of the coefficients obtained by covering up the y column.

$D_z = \begin{vmatrix} a_1 & b_1 & d_1 \\ a_2 & b_2 & d_2 \\ a_3 & b_3 & d_3 \end{vmatrix}$

i.e. the determinant of the coefficients obtained by covering up the z column.

and $D = \begin{vmatrix} a_1 & b_1 & c_1 \\ a_2 & b_2 & c_2 \\ a_3 & b_3 & c_3 \end{vmatrix}$

i.e. the determinant of the coefficients obtained by covering up the constants column.

Problem 6. A d.c. circuit comprises three closed loops. Applying Kirchhoff's laws to the closed loops gives the following equations for current flow in milliamperes:

$$2I_1 + 3I_2 - 4I_3 = 26$$
$$I_1 - 5I_2 - 3I_3 = -87$$
$$-7I_1 + 2I_2 + 6I_3 = 12$$

Use determinants to solve for I_1, I_2 and I_3

(i) Writing the equations in the $a_1x + b_1y + c_1z + d_1 = 0$ form gives:

$$2I_1 + 3I_2 - 4I_3 - 26 = 0$$
$$I_1 - 5I_2 - 3I_3 + 87 = 0$$
$$-7I_1 + 2I_2 + 6I_3 - 12 = 0$$

(ii) The solution is given by

$$\frac{I_1}{D_{I_1}} = \frac{-I_2}{D_{I_2}} = \frac{I_3}{D_{I_3}} = \frac{-1}{D}$$

where D_{I_1} is the determinant of coefficients obtained by covering up the I_1 column, i.e.

$$D_{I_1} = \begin{vmatrix} 3 & -4 & -26 \\ -5 & -3 & 87 \\ 2 & 6 & -12 \end{vmatrix}$$

$$= (3)\begin{vmatrix} -3 & 87 \\ 6 & -12 \end{vmatrix} - (-4)\begin{vmatrix} -5 & 87 \\ 2 & -12 \end{vmatrix} + (-26)\begin{vmatrix} -5 & -3 \\ 2 & 6 \end{vmatrix}$$

$$= 3(-486) + 4(-114) - 26(-24)$$

$$= \mathbf{-1290}$$

$$D_{I_2} = \begin{vmatrix} 2 & -4 & -26 \\ 1 & -3 & 87 \\ -7 & 6 & -12 \end{vmatrix}$$

$$= (2)(36 - 522) - (-4)(-12 + 609) + (-26)(6 - 21)$$

$$= -972 + 2388 + 390$$

$$= \mathbf{1806}$$

$$D_{I_3} = \begin{vmatrix} 2 & 3 & -26 \\ 1 & -5 & 87 \\ -7 & 2 & -12 \end{vmatrix}$$

$$= (2)(60 - 174) - (3)(-12 + 609) + (-26)(2 - 35)$$

$$= -228 - 1791 + 858 = \mathbf{-1161}$$

and $D = \begin{vmatrix} 2 & 3 & -4 \\ 1 & -5 & -3 \\ -7 & 2 & 6 \end{vmatrix}$

$$= (2)(-30 + 6) - (3)(6 - 21) + (-4)(2 - 35)$$

$$= -48 + 45 + 132 = \mathbf{129}$$

Thus

$$\frac{I_1}{-1290} = \frac{-I_2}{1806} = \frac{I_3}{-1161} = \frac{-1}{129}$$

giving

$$I_1 = \frac{-1290}{-129} = \mathbf{10\,mA},$$

$$I_2 = \frac{1806}{129} = \mathbf{14\,mA}$$

and $I_3 = \frac{1161}{129} = \mathbf{9\,mA}$

Now try the following Practice Exercise

Practice Exercise 276 Solving simultaneous equations using determinants (Answers on page 734)

In Problems 1 to 5 use **determinants** to solve the simultaneous equations given.

1. $3x - 5y = -17.6$
   $7y - 2x - 22 = 0$
2. $2.3m - 4.4n = 6.84$
   $8.5n - 6.7m = 1.23$
3. $3x + 4y + z = 10$
   $2x - 3y + 5z + 9 = 0$
   $x + 2y - z = 6$
4. $1.2p - 2.3q - 3.1r + 10.1 = 0$
   $4.7p + 3.8q - 5.3r - 21.5 = 0$
   $3.7p - 8.3q + 7.4r + 28.1 = 0$

5. $\frac{x}{2} - \frac{y}{3} + \frac{2z}{5} = -\frac{1}{20}$

$\frac{x}{4} + \frac{2y}{3} - \frac{z}{2} = \frac{19}{40}$

$x + y - z = \frac{59}{60}$

6. In a system of forces, the relationship between two forces F_1 and F_2 is given by:

$$5F_1 + 3F_2 + 6 = 0$$

$$3F_1 + 5F_2 + 18 = 0$$

Use determinants to solve for F_1 and F_2

7. Applying mesh-current analysis to an a.c. circuit results in the following equations:

$$(5 - j4)I_1 - (-j4)I_2 = 100\angle 0^\circ$$

$$(4 + j3 - j4)I_2 - (-j4)I_1 = 0$$

Solve the equations for I_1 and I_2

8. Kirchhoff's laws are used to determine the current equations in an electrical network and show that

$$i_1 + 8i_2 + 3i_3 = -31$$

$$3i_1 - 2i_2 + i_3 = -5$$

$$2i_1 - 3i_2 + 2i_3 = 6$$

Use determinants to solve for i_1, i_2 and i_3

9. The forces in three members of a framework are F_1, F_2 and F_3. They are related by the simultaneous equations shown below

$$1.4F_1 + 2.8F_2 + 2.8F_3 = 5.6$$

$$4.2F_1 - 1.4F_2 + 5.6F_3 = 35.0$$

$$4.2F_1 + 2.8F_2 - 1.4F_3 = -5.6$$

Find the values of F_1, F_2 and F_3 using determinants

10. Mesh-current analysis produces the following three equations:

$$20\angle 0^\circ = (5 + 3 - j4)I_1 - (3 - j_4)I_2$$

$$10\angle 90^\circ = (3 - j4 + 2)I_2 - (3 - j_4)I_1 - 2I_3$$

$$-15\angle 0^\circ - 10\angle 90^\circ = (12 + 2)I_3 - 2I_2$$

Solve the equations for the loop currents I_1, I_2 and I_3

59.3 Solution of simultaneous equations using Cramer's rule

Cramer's* rule states that if

$$a_{11}x + a_{12}y + a_{13}z = b_1$$

$$a_{21}x + a_{22}y + a_{23}z = b_2$$

$$a_{31}x + a_{32}y + a_{33}z = b_3$$

then $\quad x = \frac{D_x}{D}, y = \frac{D_y}{D}$ **and** $z = \frac{D_z}{D}$

where $\quad D = \begin{vmatrix} a_{11} & a_{12} & a_{13} \\ a_{21} & a_{22} & a_{23} \\ a_{31} & a_{32} & a_{33} \end{vmatrix}$

$$D_x = \begin{vmatrix} b_1 & a_{12} & a_{13} \\ b_2 & a_{22} & a_{23} \\ b_3 & a_{32} & a_{33} \end{vmatrix}$$

i.e. the x-column has been replaced by the R.H.S. b column

$$D_y = \begin{vmatrix} a_{11} & b_1 & a_{13} \\ a_{21} & b_2 & a_{23} \\ a_{31} & b_3 & a_{33} \end{vmatrix}$$

*Who was Cramer? – **Gabriel Cramer** (31 July 1704–4 January 1752) was a Swiss mathematician, born in Geneva. His articles cover a wide range of subjects including the study of geometric problems, the history of mathematics, philosophy and the date of Easter. Cramer's most famous book is a work which Cramer modelled on Newton's memoir on cubic curves and he highly praises a commentary on Newton's memoir written by Stirling. To find out more go to **www.routledge.com/cw/bird**

i.e. the y-column has been replaced by the R.H.S. b column

$$D_z = \begin{vmatrix} a_{11} & a_{12} & b_1 \\ a_{21} & a_{22} & b_2 \\ a_{31} & a_{32} & b_3 \end{vmatrix}$$

i.e. the z-column has been replaced by the R.H.S. b column.

Problem 7. Solve the following simultaneous equations using Cramer's rule

$$x+y+z=4$$
$$2x-3y+4z=33$$
$$3x-2y-2z=2$$

(This is the same as Problem 2 and a comparison of methods may be made.) Following the above method:

$$\boldsymbol{D} = \begin{vmatrix} 1 & 1 & 1 \\ 2 & -3 & 4 \\ 3 & -2 & -2 \end{vmatrix}$$

$$= 1(6-(-8)) - 1((-4)-12) + 1((-4)-(-9)) = 14+16+5 = \mathbf{35}$$

$$\boldsymbol{D} = \begin{vmatrix} 4 & 1 & 1 \\ 33 & -3 & 4 \\ 2 & -2 & -2 \end{vmatrix}$$

$$= 4(6-(-8)) - 1((-66)-8) + 1((-66)-(-6)) = 56+74-60 = \mathbf{70}$$

$$\boldsymbol{D_y} = \begin{vmatrix} 1 & 4 & 1 \\ 2 & 33 & 4 \\ 3 & 2 & -2 \end{vmatrix}$$

$$= 1((-66)-8) - 4((-4)-12) + 1(4-99) = -74+64-95 = \mathbf{-105}$$

$$\boldsymbol{D_z} = \begin{vmatrix} 1 & 1 & 4 \\ 2 & -3 & 33 \\ 3 & -2 & 2 \end{vmatrix}$$

$$= 1((-6)-(-66)) - 1(4-99) + 4((-4)-(-9)) = 60+95+20 = \mathbf{175}$$

Hence

$$\boldsymbol{x} = \frac{D_x}{D} = \frac{70}{35} = \mathbf{2},\ y = \frac{D_y}{D} = \frac{-105}{35} = \mathbf{-3}$$

and $z = \dfrac{D_z}{D} = \dfrac{175}{35} = \mathbf{5}$

Now try the following Practice Exercise

Practice Exercise 277 Solving simultaneous equations using Cramer's rule (Answers on page 734)

1. Repeat problems 3, 4, 5, 7 and 8 of Exercise 275 on page 605, using Cramer's rule
2. Repeat problems 3, 4, 8 and 9 of Exercise 276 on page 608, using Cramer's rule

59.4 Solution of simultaneous equations using the Gaussian elimination method

Consider the following simultaneous equations:

$$x+y+z=4 \quad (1)$$
$$2x-3y+4z=33 \quad (2)$$
$$3x-2y-2z=2 \quad (3)$$

Leaving equation (1) as it is gives:

$$x+y+z=4 \quad (1)$$

Equation (2) $-\ 2\times$ equation (1) gives:

$$0-5y+2z=25 \quad (2')$$

and equation (3) $-\ 3\times$ equation (1) gives:

$$0-5y-5z=-10 \quad (3')$$

Leaving equations (1) and (2′) as they are gives:

$$x+y+z=4 \quad (1)$$
$$0-5y+2z=25 \quad (2')$$

Equation (3′) − equation (2′) gives:

$$0+0-7z=-35 \quad (3'')$$

By appropriately manipulating the three original equations we have deliberately obtained zeros in the positions shown in equations (2′) and (3″).
Working backwards, from equation (3″),

$$z = \frac{-35}{-7} = \mathbf{5},$$

from equation (2′),

$$-5y+2(5)=25,$$

from which,

$$y=\frac{25-10}{-5}=\mathbf{-3}$$

and from equation (1),

$$x+(-3)+5=4,$$

from which,

$$\boldsymbol{x}=4+3-5=\mathbf{2}$$

(This is the same example as Problems 2 and 7, and a comparison of methods can be made). The above method is known as the **Gaussian elimination method**.

We conclude from the above example that if

$$a_{11}x+a_{12}y+a_{13}z=b_1$$
$$a_{21}x+a_{22}y+a_{23}z=b_2$$
$$a_{31}x+a_{32}y+a_{33}z=b_3$$

the three-step **procedure** to solve simultaneous equations in three unknowns using the **Gaussian*** **elimination method** is:

*Who was Gauss? – **Johann Carl Friedrich Gauss** (30 April 1777–23 February 1855) was a German mathematician and physical scientist who contributed significantly to many fields, including number theory, statistics, electrostatics, astronomy and optics. To find out more go to www.routledge.com/cw/bird

1. Equation (2) $-\dfrac{a_{21}}{a_{11}}\times$ equation (1) to form equation (2′) and equation (3) $-\dfrac{a_{31}}{a_{11}}\times$ equation (1) to form equation (3′).
2. Equation (3′) $-\dfrac{a_{32}}{a_{22}}\times$ equation (2′) to form equation (3″).
3. Determine z from equation (3″), then y from equation (2′) and finally, x from equation (1).

Problem 8. A d.c. circuit comprises three closed loops. Applying Kirchhoff's laws to the closed loops gives the following equations for current flow in milliamperes:

$$2I_1+3I_2-4I_3=26 \quad (1)$$
$$I_1-5I_2-3I_3=-87 \quad (2)$$
$$-7I_1+2I_2+6I_3=12 \quad (3)$$

Use the Gaussian elimination method to solve for I_1, I_2 and I_3

(This is the same example as Problem 6 on pages 607/608, and a comparison of methods may be made)

Following the above procedure:

1. $2I_1+3I_2-4I_3=26 \quad (1)$

   Equation (2) $-\dfrac{1}{2}\times$ equation (1) gives:

   $0-6.5I_2-I_3=-100 \quad (2')$

   Equation (3) $-\dfrac{-7}{2}\times$ equation (1) gives:

   $0+12.5I_2-8I_3=103 \quad (3')$

2. $2I_1+3I_2-4I_3=26 \quad (1)$

   $0-6.5I_2-I_3=-100 \quad (2')$

   Equation (3′) $-\dfrac{12.5}{-6.5}\times$ equation (2′) gives:

   $0+0-9.923I_3=-89.308 \quad (3'')$

3. From equation (3″),

$$\boldsymbol{I_3}=\frac{-89.308}{-9.923}=\mathbf{9\,mA},$$

from equation (2′), $-6.5I_2-9=-100$,

from which, $\boldsymbol{I_2}=\dfrac{-100+9}{-6.5}=\mathbf{14\,mA}$

and from equation (1), $2I_1+3(14)-4(9)=26$,

from which, $\boldsymbol{I_1} = \dfrac{26-42+36}{2} = \dfrac{20}{2}$

$= \mathbf{10\,mA}$

Now try the following Practice Exercise

Practice Exercise 278 Solving simultaneous equations using Gaussian elimination (Answers on page 734)

1. In a mass-spring-damper system, the acceleration $\ddot{x}$ m/s^2, velocity $\dot{x}$ m/s and displacement x m are related by the following simultaneous equations:

$$6.2\ddot{x} + 7.9\dot{x} + 12.6x = 18.0$$
$$7.5\ddot{x} + 4.8\dot{x} + 4.8x = 6.39$$
$$13.0\ddot{x} + 3.5\dot{x} - 13.0x = -17.4$$

By using Gaussian elimination, determine the acceleration, velocity and displacement for the system, correct to 2 decimal places.

2. The tensions, T_1, T_2 and T_3 in a simple framework are given by the equations:

$$5T_1 + 5T_2 + 5T_3 = 7.0$$
$$T_1 + 2T_2 + 4T_3 = 2.4$$
$$4T_1 + 2T_2 = 4.0$$

Determine T_1, T_2 and T_3 using Gaussian elimination

3. Repeat problems 3, 4, 5, 7 and 8 of Exercise 275 on page 605, using the Gaussian elimination method

4. Repeat problems 3, 4, 8 and 9 of Exercise 276 on page 608, using the Gaussian elimination method

Practice Exercise 279 Multiple-choice questions on the solution of simultaneous equations by matrices and determinants (Answers on page 734)

Each question has only one correct answer

1. For the following simultaneous equations:

$$3x - 4y + 10 = 0$$
$$5y - 2x = 9$$

using matrices, the value of x is:
(a) -1 (b) 1 (c) 2 (d) -2

2. For the following simultaneous equations:

$$3x - 4y + 2z = 11$$
$$5x + 3y - 2z = -17$$
$$2x - 3y + 5z = 19$$

using Cramer's rule, the value of y is:
(a) -2 (b) 3 (c) 1 (d) -1

3. If $x + 2y = 1$ and $3x - y = -11$ then, using determinants, the value of x is:
(a) -3 (b) 4.2 (c) 2 (d) -1.6

4. If $4x - 3y = 7$ and $2x + 5y = -1$ then, using matrices, the value of y is:
(a) -1 (b) $-\dfrac{9}{13}$ (c) 2 (d) $\dfrac{59}{26}$

5. For the following simultaneous equations:

$$2a - 3b + 6c = -17$$
$$7a + 2b - 4c = 28$$
$$a - 5b + 2c = 7$$

using Cramer's rule, the value of c is:
(a) -3 (b) 2 (c) -5 (d) 3

COMPANION @ WEBSITE

For fully worked solutions to each of the problems in Practice Exercises 275 to 278 in this chapter, go to the website:

www.routledge.com/cw/bird

Revision Test 16 Boolean algebra, logic circuits, matrices and determinants

This Revision Test covers the material contained in Chapters 57 to 59. *The marks for each question are shown in brackets at the end of each question.*

1. Use the laws and rules of Boolean algebra to simplify the following expressions:

   (a) $B \cdot (A+\overline{B})+A \cdot \overline{B}$

   (b) $\overline{A} \cdot \overline{B} \cdot \overline{C}+\overline{A} \cdot B \cdot \overline{C}+\overline{A} \cdot B \cdot C+\overline{A} \cdot \overline{B} \cdot C$

   (9)

2. Simplify the Boolean expression: $\overline{A \cdot \overline{B}+A \cdot B \cdot \overline{C}}$ using de Morgan's laws. (5)

3. Use a Karnaugh map to simplify the Boolean expression:

   $$\overline{A} \cdot \overline{B} \cdot \overline{C}+\overline{A} \cdot B \cdot \overline{C}+\overline{A} \cdot B \cdot C+A \cdot \overline{B} \cdot C$$

   (6)

4. A clean room has two entrances, each having two doors, as shown in Fig. RT16.1. A warning bell must sound if both doors A and B or doors C and D are open at the same time. Write down the Boolean expression depicting this occurrence, and devise a logic network to operate the bell using NAND-gates only. (8)

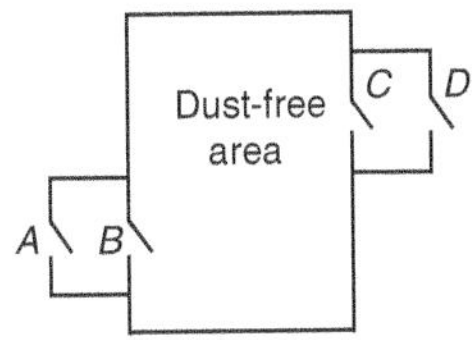

Figure RT16.1

In questions 5 to 9, the matrices stated are:

$$A=\begin{pmatrix}-5 & 2\\ 7 & -8\end{pmatrix} \quad B=\begin{pmatrix}1 & 6\\ -3 & -4\end{pmatrix}$$

$$C=\begin{pmatrix}j3 & (1+j2)\\ (-1-j4) & -j2\end{pmatrix}$$

$$D=\begin{pmatrix}2 & -1 & 3\\ -5 & 1 & 0\\ 4 & -6 & 2\end{pmatrix} \quad E=\begin{pmatrix}-1 & 3 & 0\\ 4 & -9 & 2\\ -5 & 7 & 1\end{pmatrix}$$

5. Determine A × B. (4)

6. Calculate the determinant of matrix C. (4)

7. Determine the inverse of matrix A. (4)

8. Determine E × D. (9)

9. Calculate the determinant of matrix D. (5)

10. Use matrices to solve the following simultaneous equations:

    $$4x-3y=17$$
    $$x+y+1=0$$

    (6)

11. Use determinants to solve the following simultaneous equations:

    $$4x+9y+2z=21$$
    $$-8x+6y-3z=41$$
    $$3x+y-5z=-73$$

    (10)

12. The simultaneous equations representing the currents flowing in an unbalanced, three-phase, star-connected, electrical network are as follows:

    $$2.4I_1+3.6I_2+4.8I_3=1.2$$
    $$-3.9I_1+1.3I_2-6.5I_3=2.6$$
    $$1.7I_1+11.9I_2+8.5I_3=0$$

    Using matrices, solve equations for I_1, I_2 and I_3 (10)

For lecturers/instructors/teachers, fully worked solutions to each of the problems in Revision Test 16, together with a full marking scheme, are available at the website:

www.routledge.com/cw/bird

Section 11

Statistics

Chapter 60

Presentation of statistical data

Why it is important to understand: Presentation of statistical data

Statistics is the study of the collection, organisation, analysis and interpretation of data. It deals with all aspects of this, including the planning of data collection in terms of the design of surveys and experiments. Statistics is applicable to a wide variety of academic disciplines, including natural and social sciences, engineering, government and business. Statistical methods can be used for summarising or describing a collection of data. Engineering statistics combines engineering and statistics. Design of experiments is a methodology for formulating scientific and engineering problems using statistical models. Quality control and process control use statistics as a tool to manage conformance to specifications of manufacturing processes and their products. Time and methods engineering use statistics to study repetitive operations in manufacturing in order to set standards and find optimum manufacturing procedures. Reliability engineering measures the ability of a system to perform for its intended function (and time) and has tools for improving performance. Probabilistic design involves the use of probability in product and system design. System identification uses statistical methods to build mathematical models of dynamical systems from measured data. System identification also includes the optimal design of experiments for efficiently generating informative data for fitting such models. This chapter introduces the presentation of statistical data.

At the end of this chapter, you should be able to:

- distinguish between discrete and continuous data
- present data diagrammatically – pictograms, horizontal and vertical bar charts, percentage component bar charts, pie diagrams
- produce a tally diagram for a set of data
- form a frequency distribution from a tally diagram
- construct a histogram from a frequency distribution
- construct a frequency polygon from a frequency distribution
- produce a cumulative frequency distribution from a set of grouped data
- construct an ogive from a cumulative frequency distribution

60.1 Some statistical terminology

Data are obtained largely by two methods:

(a) by counting — for example, the number of stamps sold by a post office in equal periods of time, and

(b) by measurement — for example, the heights of a group of people.

When data are obtained by counting and only whole numbers are possible, the data are called **discrete**. Measured data can have any value within certain limits and are called **continuous** (see Problem 1).

A **set** is a group of data and an individual value within the set is called a **member** of the set. Thus, if the masses of five people are measured correct to the nearest 0.1 kilogram and are found to be 53.1 kg, 59.4 kg, 62.1 kg, 77.8 kg and 64.4 kg, then the set of masses in kilograms for these five people is:

$$\{53.1, 59.4, 62.1, 77.8, 64.4\}$$

and one of the members of the set is 59.4

A set containing all the members is called a **population**. A **sample** is a set of data collected and/or selected from a statistical population by a defined procedure. Typically, the population is very large, making a census or a complete enumeration of all the values in the population impractical or impossible. The sample usually represents a subset of manageable size. Samples are collected and statistics are calculated from the samples so that inferences or extrapolations can be made from the sample to the population. Thus all car registration numbers form a population, but the registration numbers of, say, 20 cars taken at random throughout the country are a sample drawn from that population.

The number of times that the value of a member occurs in a set is called the **frequency** of that member. Thus in the set: {2, 3, 4, 5, 4, 2, 4, 7, 9}, member 4 has a frequency of three, member 2 has a frequency of 2 and the other members have a frequency of one.

The **relative frequency** with which any member of a set occurs is given by the ratio:

$$\frac{\text{frequency of member}}{\text{total frequency of all members}}$$

For the set: {2, 3, 5, 4, 7, 5, 6, 2, 8}, the relative frequency of member 5 is $\frac{2}{9}$

Often, relative frequency is expressed as a percentage and the **percentage relative frequency** is: (relative frequency × 100)%

Problem 1. Data are obtained on the topics given below. State whether they are discrete or continuous data.

(a) The number of days on which rain falls in a month for each month of the year.
(b) The mileage travelled by each of a number of salesmen.
(c) The time that each of a batch of similar batteries lasts.
(d) The amount of money spent by each of several families on food.

(a) The number of days on which rain falls in a given month must be an integer value and is obtained by **counting** the number of days. Hence, these data are **discrete**.

(b) A salesman can travel any number of miles (and parts of a mile) between certain limits and these data are **measured**. Hence the data are **continuous**.

(c) The time that a battery lasts is **measured** and can have any value between certain limits. Hence these data are **continuous**.

(d) The amount of money spent on food can only be expressed correct to the nearest pence, the amount being **counted**. Hence, these data are **discrete**.

Now try the following Practice Exercise

Practice Exercise 280 Discrete and continuous data (Answers on page 734)

In Problems 1 and 2, state whether data relating to the topics given are discrete or continuous.

1. (a) The amount of petrol produced daily, for each of 31 days, by a refinery
   (b) The amount of coal produced daily by each of 15 miners
   (c) The number of bottles of milk delivered daily by each of 20 milkmen
   (d) The size of 10 samples of rivets produced by a machine
2. (a) The number of people visiting an exhibition on each of 5 days
   (b) The time taken by each of 12 athletes to run 100 metres

(c) The value of stamps sold in a day by each of 20 post offices
(d) The number of defective items produced in each of 10 one-hour periods by a machine

60.2 Presentation of ungrouped data

Ungrouped data can be presented diagrammatically in several ways and these include:

(a) **pictograms**, in which pictorial symbols are used to represent quantities (see Problem 2),
(b) **horizontal bar charts**, having data represented by equally spaced horizontal rectangles (see Problem 3), and
(c) **vertical bar charts**, in which data are represented by equally spaced vertical rectangles (see Problem 4).

Trends in ungrouped data over equal periods of time can be presented diagrammatically by a **percentage component bar chart**. In such a chart, equally spaced rectangles of any width, but whose height corresponds to 100%, are constructed. The rectangles are then subdivided into values corresponding to the percentage relative frequencies of the members (see Problem 5).
A **pie diagram** is used to show diagrammatically the parts making up the whole. In a pie diagram, the area of a circle represents the whole, and the areas of the sectors of the circle are made proportional to the parts which make up the whole (see Problem 6).

Problem 2. The number of television sets repaired in a workshop by a technician in six, one-month periods is as shown below. Present these data as a pictogram.

Month	January	February	March
Number repaired	11	6	15

Month	April	May	June
Number repaired	9	13	8

Each symbol shown in Fig. 60.1 represents two television sets repaired. Thus, in January, $5\frac{1}{2}$ symbols are used to represents the 11 sets repaired, in February, 3 symbols are used to represent the 6 sets repaired and so on.

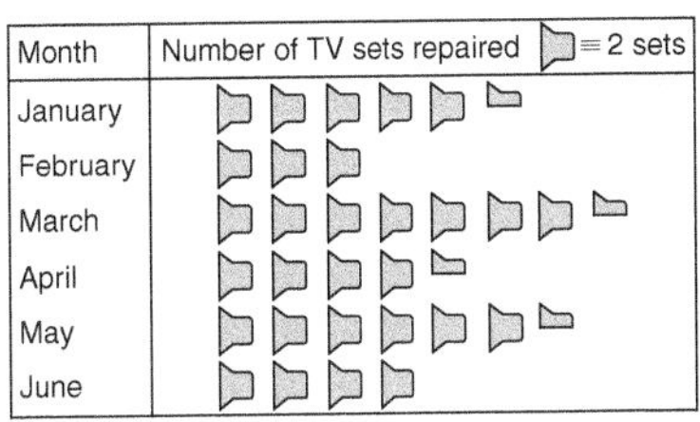

Figure 60.1

Problem 3. The distance in miles travelled by four salesmen in a week are as shown below.

Salesmen	*P*	*Q*	*R*	*S*
Distance travelled (miles)	413	264	597	143

Use a horizontal bar chart to represent these data diagrammatically

Equally spaced horizontal rectangles of any width, but whose length is proportional to the distance travelled, are used. Thus, the length of the rectangle for salesman P is proportional to 413 miles and so on. The horizontal bar chart depicting these data is shown in Fig. 60.2.

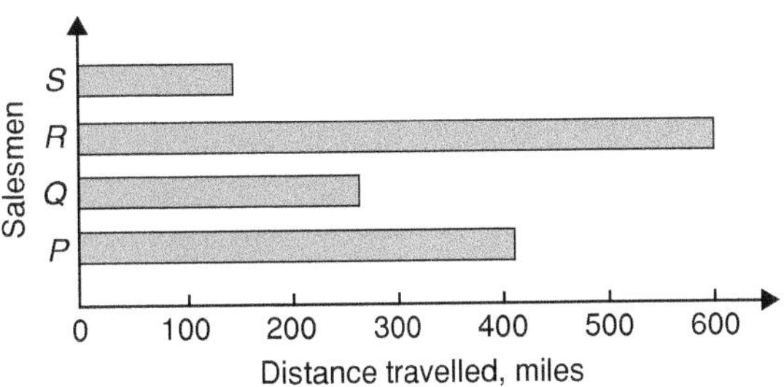

Figure 60.2

Problem 4. The number of issues of tools or materials from a store in a factory is observed for seven, one-hour periods in a day, and the results of the survey are as follows:

Period	1	2	3	4	5	6	7
Number of issues	34	17	9	5	27	13	6

Present these data on a vertical bar chart.

In a vertical bar chart, equally spaced vertical rectangles of any width, but whose height is proportional to the quantity being represented, are used. Thus the height of the rectangle for period 1 is proportional to 34 units and so on. The vertical bar chart depicting these data is shown in Fig. 60.3.

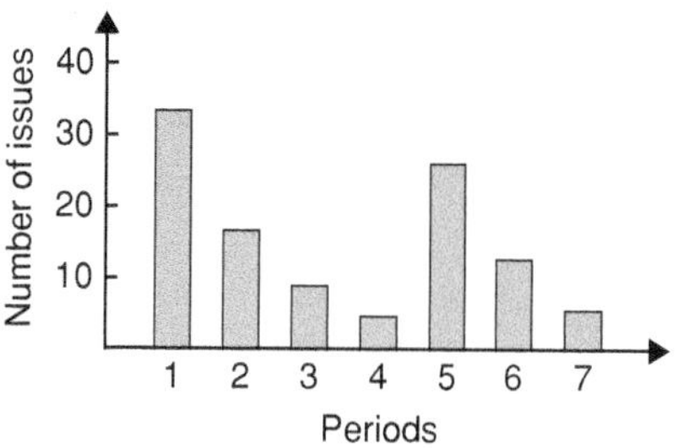

Figure 60.3

Problem 5. The number of various types of dwellings sold by a company annually over a three-year period are as shown below. Draw percentage component bar charts to present these data.

	Year 1	Year 2	Year 3
4-roomed bungalows	24	17	7
5-roomed bungalows	38	71	118
4-roomed houses	44	50	53
5-roomed houses	64	82	147
6-roomed houses	30	30	25

A table of percentage relative frequency values, correct to the nearest 1%, is the first requirement. Since,

$$\text{percentage relative frequency} = \frac{\text{frequency of member} \times 100}{\text{total frequency}}$$

then for 4-roomed bungalows in year 1:

$$\text{percentage relative frequency} = \frac{24 \times 100}{24 + 38 + 44 + 64 + 30} = 12\%$$

The percentage relative frequencies of the other types of dwellings for each of the three years are similarly calculated and the results are as shown in the table below.

	Year 1	Year 2	Year 3
4-roomed bungalows	12%	7%	2%
5-roomed bungalows	19%	28%	34%
4-roomed houses	22%	20%	15%
5-roomed houses	32%	33%	42%
6-roomed houses	15%	12%	7%

The percentage component bar chart is produced by constructing three equally spaced rectangles of any width, corresponding to the three years. The heights of the rectangles correspond to 100% relative frequency, and are subdivided into the values in the table of percentages shown above. A key is used (different types of shading or different colour schemes) to indicate corresponding percentage values in the rows of the table of percentages. The percentage component bar chart is shown in Fig. 60.4.

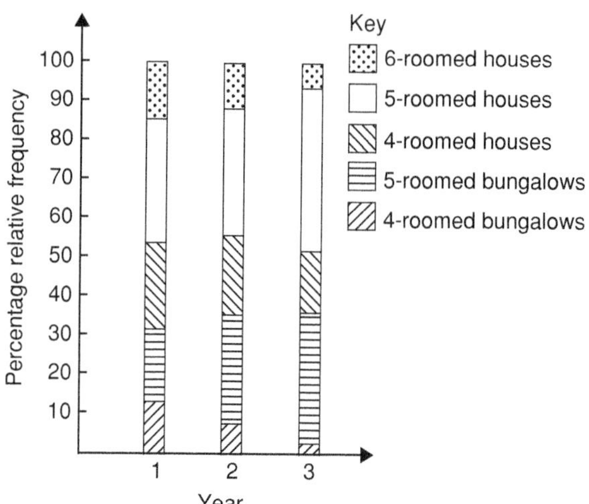

Figure 60.4

Problem 6. The retail price of a product costing £2 is made up as follows: materials 10 p, labour 20 p, research and development 40 p, overheads 70 p, profit 60 p. Present these data on a pie diagram

A circle of any radius is drawn, and the area of the circle represents the whole, which in this case is £2. The circle is subdivided into sectors so that the areas of the sectors are proportional to the parts, i.e. the parts which make up the total retail price. For the area of a sector to be proportional to a part, the angle at the centre of the circle must be proportional to that part. The whole, £2 or 200 p, corresponds to 360°. Therefore,

$$10\,\text{p corresponds to } 360 \times \frac{10}{200} \text{ degrees, i.e. } 18^\circ$$

$$20\,\text{p corresponds to } 360 \times \frac{20}{200} \text{ degrees, i.e. } 36^\circ$$

and so on, giving the angles at the centre of the circle for the parts of the retail price as: 18°, 36°, 72°, 126° and 108°, respectively.
The pie diagram is shown in Fig. 60.5.

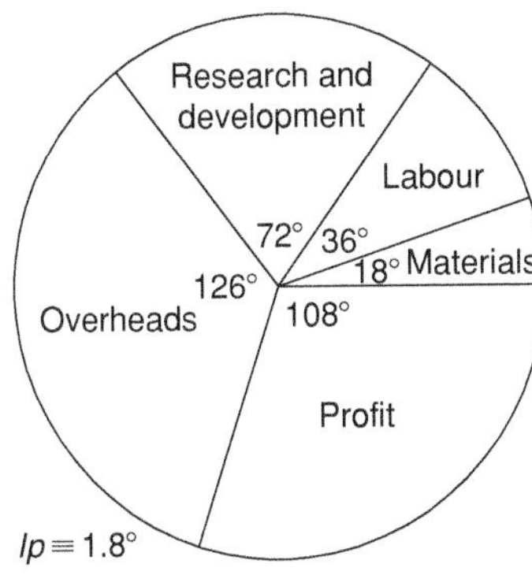

Figure 60.5

Problem 7.
(a) Using the data given in Fig. 60.2 only, calculate the amount of money paid to each salesman for travelling expenses, if they are paid an allowance of 37 p per mile
(b) Using the data presented in Fig. 60.4, comment on the housing trends over the three-year period.
(c) Determine the profit made by selling 700 units of the product shown in Fig. 60.5

(a) By measuring the length of rectangle P the mileage covered by salesman P is equivalent to 413 miles. Hence salesman P receives a travelling allowance of

$$\frac{£413 \times 37}{100} \text{ i.e. } \mathbf{£152.81}$$

Similarly, for salesman Q, the miles travelled are 264 and this allowance is

$$\frac{£264 \times 37}{100} \text{ i.e. } \mathbf{£97.68}$$

Salesman R travels 597 miles and he receives

$$\frac{£597 \times 37}{100} \text{ i.e. } \mathbf{£220.89}$$

Finally, salesman S receives

$$\frac{£143 \times 37}{100} \text{ i.e. } \mathbf{£52.91}$$

(b) An analysis of Fig. 60.4 shows that 5-roomed bungalows and 5-roomed houses are becoming more popular, the greatest change in the three years being a 15% increase in the sales of 5-roomed bungalows.

(c) Since 1.8° corresponds to 1 p and the profit occupies 108° of the pie diagram, then the profit per unit is $\frac{108 \times 1}{1.8}$, that is, 60 p.
The profit when selling 700 units of the product is $£\frac{700 \times 60}{100}$, that is, **£420.**

Now try the following Practice Exercise

Practice Exercise 281 Presentation of ungrouped data (Answers on page 734)

1. The number of vehicles passing a stationary observer on a road in six ten-minute intervals is as shown. Draw a pictogram to represent these data

Period of Time	1	2	3	4	5	6
Number of Vehicles	35	44	62	68	49	41

2. The number of components produced by a factory in a week is as shown below

Day	Mon	Tues	Wed
Number of Components	1580	2190	1840

Day	Thurs	Fri
Number of Components	2385	1280

Show these data on a pictogram

3. For the data given in Problem 1 above, draw a horizontal bar chart

4. Present the data given in Problem 2 above on a horizontal bar chart

5. For the data given in Problem 1 above, construct a vertical bar chart

6. Depict the data given in Problem 2 above on a vertical bar chart

7. A factory produces three different types of components. The percentages of each of these components produced for three, one-month periods are as shown below. Show this information on percentage component bar charts and comment on the changing trend in the percentages of the types of component produced

Month	1	2	3
Component *P*	20	35	40
Component *Q*	45	40	35
Component *R*	35	25	25

8. A company has five distribution centres and the mass of goods in tonnes sent to each centre during four, one-week periods, is as shown

Week	1	2	3	4
Centre *A*	147	160	174	158
Centre *B*	54	63	77	69
Centre *C*	283	251	237	211
Centre *D*	97	104	117	144
Centre *E*	224	218	203	194

Use a percentage component bar chart to present these data and comment on any trends

9. The employees in a company can be split into the following categories: managerial 3, supervisory 9, craftsmen 21, semi-skilled 67, others 44. Show these data on a pie diagram

10. The way in which an apprentice spent his time over a one-month period is a follows: drawing office 44 hours, production 64 hours, training 12 hours, at college 28 hours.
Use a pie diagram to depict this information

11. (a) With reference to Fig. 60.5, determine the amount spent on labour and materials to produce 1650 units of the product
(b) If in year 2 of Fig. 60.4, 1% corresponds to 2.5 dwellings, how many bungalows are sold in that year

12. (a) If the company sell 23 500 units per annum of the product depicted in Fig. 60.5, determine the cost of their overheads per annum
(b) If 1% of the dwellings represented in year 1 of Fig. 60.4 corresponds to 2 dwellings, find the total number of houses sold in that year

60.3 Presentation of grouped data

When the number of members in a set is small, say ten or less, the data can be represented diagrammatically without further analysis, by means of pictograms, bar charts, percentage components bar charts or pie diagrams (as shown in Section 60.2).

For sets having more than ten members, those members having similar values are grouped together in **classes** to form a **frequency distribution**. To assist in accurately counting members in the various classes, a **tally diagram** is used (see Problems 8 and 12).

A frequency distribution is merely a table showing classes and their corresponding frequencies (see Problems 8 and 12).

The new set of values obtained by forming a frequency distribution is called **grouped data**.

The terms used in connection with grouped data are shown in Fig. 60.6(a). The size or range of a class is given by the **upper class boundary value** minus the **lower class boundary value**, and in Fig. 60.6 is 7.65 − 7.35, i.e. 0.30. The **class interval** for the class shown in Fig. 60.6(b) is 7.4 to 7.6 and the class mid-point value is given by:

$$\frac{\begin{pmatrix}\text{upper class}\\ \text{boundary value}\end{pmatrix}+\begin{pmatrix}\text{lower class}\\ \text{boundary value}\end{pmatrix}}{2}$$

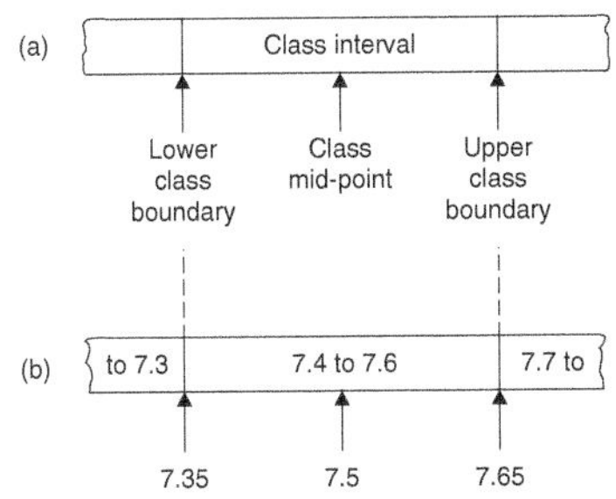

Figure 60.6

and in Fig. 60.6 is $\frac{7.65+7.35}{2}$, i.e. 7.5

One of the principal ways of presenting grouped data diagrammatically is by using a **histogram**, in which the **areas** of vertical, adjacent rectangles are made proportional to frequencies of the classes (see Problem 9). When class intervals are equal, the heights of the rectangles of a histogram are equal to the frequencies of the classes. For histograms having unequal class intervals, the area must be proportional to the frequency. Hence, if the class interval of class A is twice the class interval of class B, then for equal frequencies, the height of the rectangle representing A is half that of B (see Problem 11).

Another method of presenting grouped data diagrammatically is by using a **frequency polygon**, which is the graph produced by plotting frequency against class mid-point values and joining the co-ordinates with straight lines (see Problem 12).

A **cumulative frequency distribution** is a table showing the cumulative frequency for each value of upper class boundary. The cumulative frequency for a particular value of upper class boundary is obtained by adding the frequency of the class to the sum of the previous frequencies. A cumulative frequency distribution is formed in Problem 13.

The curve obtained by joining the co-ordinates of cumulative frequency (vertically) against upper class boundary (horizontally) is called an **ogive** or a **cumulative frequency distribution curve** (see Problem 13).

Problem 8. The data given below refer to the gain of each of a batch of 40 transistors, expressed correct to the nearest whole number. Form a frequency distribution for these data having seven classes

81	83	87	74	76	89	82	84
86	76	77	71	86	85	87	88
84	81	80	81	73	89	82	79
81	79	78	80	85	77	84	78
83	79	80	83	82	79	80	77

The **range** of the data is the value obtained by taking the value of the smallest member from that of the largest member. Inspection of the set of data shows that, range $=89-71=18$. The size of each class is given approximately by range divided by the number of classes. Since 7 classes are required, the size of each class is 18/7, that is, approximately 3. To achieve seven equal classes spanning a range of values from 71 to 89, the class intervals are selected as: 70–72, 73–75 and so on.

To assist with accurately determining the number in each class, a **tally diagram** is produced, as shown in Table 60.1(a). This is obtained by listing the classes in the left-hand column, and then inspecting each of the 40 members of the set in turn and allocating them to the appropriate classes by putting '1*s*' in the appropriate rows. Every fifth '1' allocated to a particular row is

Table 60.1(a)

Class	Tally
70–72	1
73–75	11
76–78	~~1111~~ 11
79–81	~~1111~~ ~~1111~~ 11
82–84	~~1111~~ 1111
85–87	~~1111~~ 1
88–90	111

Table 60.1(b)

Class	Class mid-point	Frequency
70–72	71	1
73–75	74	2
76–78	77	7
79–81	80	12
82–84	83	9
85–87	86	6
88–90	89	3

shown as an oblique line crossing the four previous '1s', to help with final counting.

A **frequency distribution** for the data is shown in Table 60.1(b) and lists classes and their corresponding frequencies, obtained from the tally diagram. (Class mid-point values are also shown in the table, since they are used for constructing the histogram for these data (see Problem 9).)

Problem 9. Construct a histogram for the data given in Table 60.1(b)

The histogram is shown in Fig. 60.7. The width of the rectangles correspond to the upper class boundary values minus the lower class boundary values and the heights of the rectangles correspond to the class frequencies. The easiest way to draw a histogram is to mark the class mid-point values on the horizontal scale and draw the rectangles symmetrically about the appropriate class mid-point values and touching one another.

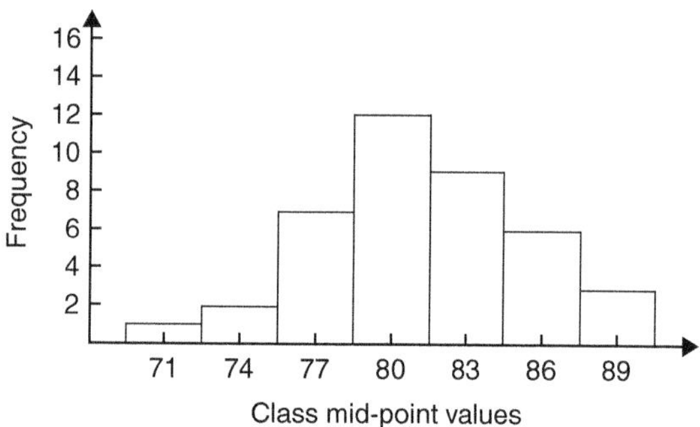

Figure 60.7

Problem 10. The amount of money earned weekly by 40 people working part-time in a factory, correct to the nearest £10, is shown below. Form a frequency distribution having 6 classes for these data.

80	90	70	110	90	160	110	80
140	30	90	50	100	110	60	100
80	90	110	80	100	90	120	70
130	170	80	120	100	110	40	110
50	100	110	90	100	70	110	80

Inspection of the set given shows that the majority of the members of the set lie between £80 and £110 and that there are a much smaller number of extreme values ranging from £30 to £170. If equal class intervals are selected, the frequency distribution obtained does not give as much information as one with unequal class intervals. Since the majority of members are between £80 and £100, the class intervals in this range are selected to be smaller than those outside of this range. There is no unique solution and one possible solution is shown in Table 60.2.

Table 60.2

Class	Frequency
20–40	2
50–70	6
80–90	12
100–110	14
120–140	4
150–170	2

Problem 11. Draw a histogram for the data given in Table 60.2

When dealing with unequal class intervals, the histogram must be drawn so that the areas, (and not the heights), of the rectangles are proportional to the frequencies of the classes. The data given are shown in columns 1 and 2 of Table 60.3. Columns 3 and 4 give the upper and lower class boundaries, respectively. In column 5, the class ranges (i.e. upper class boundary minus lower class boundary values) are listed. The heights of the rectangles are proportional to the ratio $\frac{\text{frequency}}{\text{class range}}$, as shown in column 6. The histogram is shown in Fig. 60.8.

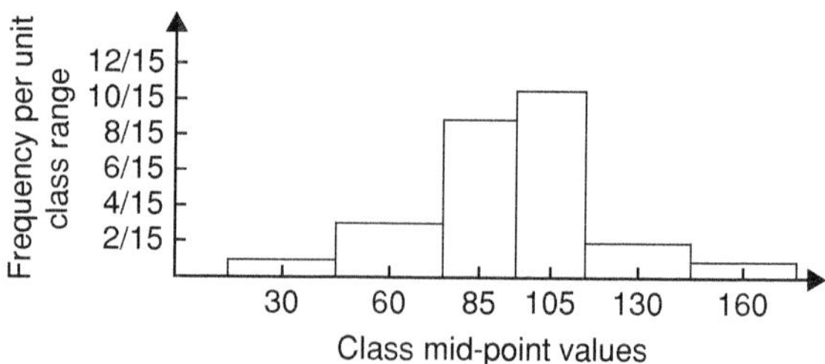

Figure 60.8

Table 60.3

1 Class	2 Frequency	3 Upper class boundary	4 Lower class boundary	5 Class range	6 Height of rectangle
20–40	2	45	15	30	$\frac{2}{30}=\frac{1}{15}$
50–70	6	75	45	30	$\frac{6}{30}=\frac{3}{15}$
80–90	12	95	75	20	$\frac{12}{20}=\frac{9}{15}$
100–110	14	115	95	20	$\frac{14}{20}=\frac{10\frac{1}{2}}{15}$
120–140	4	145	115	30	$\frac{4}{30}=\frac{2}{15}$
150–170	2	175	145	30	$\frac{2}{30}=\frac{1}{15}$

Problem 12. The masses of 50 ingots in kilograms are measured correct to the nearest 0.1 kg and the results are as shown below. Produce a frequency distribution having about 7 classes for these data and then present the grouped data as (a) a frequency polygon and (b) histogram.

8.0	8.6	8.2	7.5	8.0	9.1	8.5	7.6	8.2	7.8
8.3	7.1	8.1	8.3	8.7	7.8	8.7	8.5	8.1	8.5
7.7	8.4	7.9	8.8	7.2	8.1	7.8	8.2	7.7	7.5
8.1	7.4	8.8	8.0	8.4	8.5	8.1	7.3	9.0	8.6
7.4	8.2	8.4	7.7	8.3	8.2	7.9	8.5	7.9	8.0

The **range** of the data is the member having the largest value minus the member having the smallest value. Inspection of the set of data shows that:

$$\text{range} = 9.1 - 7.1 = 2.0$$

The size of each class is given approximately by

$$\frac{\text{range}}{\text{number of classes}}$$

Since about seven classes are required, the size of each class is 2.0/7, that is approximately 0.3, and thus the **class limits** are selected as 7.1 to 7.3, 7.4 to 7.6, 7.7 to 7.9 and so on.

The **class mid-point** for the 7.1 to 7.3 class is $\frac{7.35+7.05}{2}$, i.e. 7.2, for the 7.4 to 7.6 class is $\frac{7.65+7.35}{2}$, i.e. 7.5 and so on.

To assist with accurately determining the number in each class, a **tally diagram** is produced as shown in Table 60.4. This is obtained by listing the classes in the left-hand column and then inspecting each of the 50 members of the set of data in turn and allocating it to the appropriate class by putting a '1' in the appropriate row. Each fifth '1' allocated to a particular row is marked as an oblique line to help with final counting.

A **frequency distribution** for the data is shown in Table 60.5 and lists classes and their corresponding frequencies. Class mid-points are also shown in this table, since they are used when constructing the frequency polygon and histogram.

A **frequency polygon** is shown in Fig. 60.9, the co-ordinates corresponding to the class mid-point/frequency values, given in Table 60.5. The co-ordinates are joined by straight lines and the polygon is 'anchored-down' at each end by joining to the next class mid-point value and zero frequency.

Table 60.4

Class	Tally
7.1 to 7.3	111
7.4 to 7.6	𝍸
7.7 to 7.9	𝍸 1111
8.0 to 8.2	𝍸 𝍸 1111
8.3 to 8.5	𝍸 𝍸 1
8.6 to 8.8	𝍸 1
8.9 to 9.1	11

Table 60.5

Class	Class mid-point	Frequency
7.1 to 7.3	7.2	3
7.4 to 7.6	7.5	5
7.5 to 7.9	7.8	9
8.0 to 8.2	8.1	14
8.3 to 8.5	8.4	11
8.6 to 8.8	8.7	6
8.9 to 9.1	9.0	2

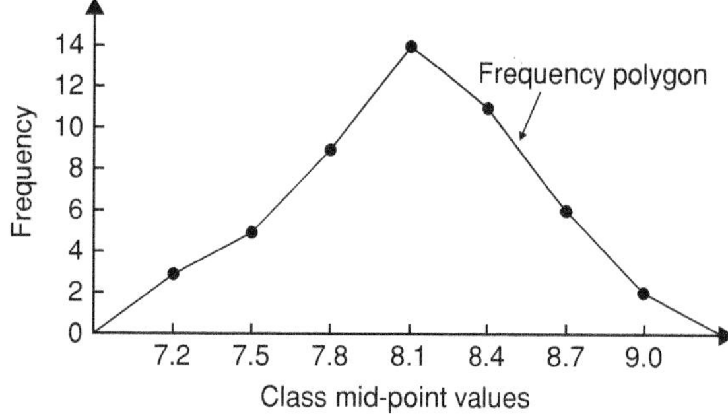

Figure 60.9

A **histogram** is shown in Fig. 60.10, the width of a rectangle corresponding to (upper class boundary value — lower class boundary value) and height corresponding

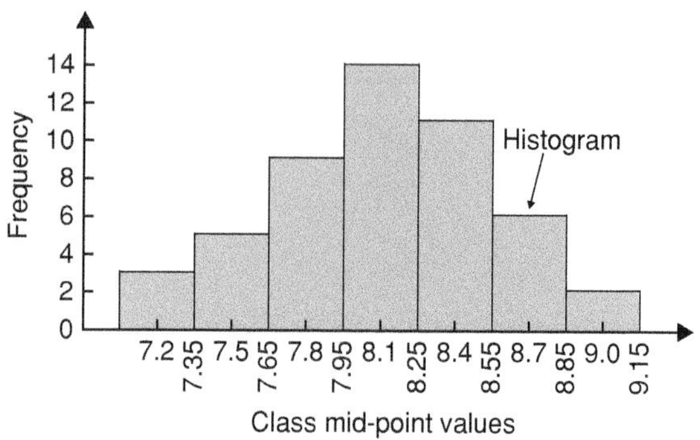

Figure 60.10

to the class frequency. The easiest way to draw a histogram is to mark class mid-point values on the horizontal scale and to draw the rectangles symmetrically about the appropriate class mid-point values and touching one another. A histogram for the data given in Table 60.5 is shown in Fig. 60.10.

Problem 13. The frequency distribution for the masses in kilograms of 50 ingots is:

7.1 to 7.3 3, 7.4 to 7.6 5, 7.7 to 7.9 9,
8.0 to 8.2 14, 8.3 to 8.5 11, 8.6 to 8.8, 6,
8.9 to 9.1 2,

Form a cumulative frequency distribution for these data and draw the corresponding ogive

A **cumulative frequency distribution** is a table giving values of cumulative frequency for the values of upper class boundaries, and is shown in Table 60.6. Columns 1 and 2 show the classes and their frequencies. Column 3 lists the upper class boundary values for the classes given in column 1. Column 4 gives the cumulative frequency values for all frequencies less than the upper class boundary values given in column 3. Thus, for example, for the 7.7 to 7.9 class shown in row 3, the cumulative frequency value is the sum of all frequencies having values of less than 7.95, i.e. $3+5+9=17$ and so on. The **ogive** for the cumulative frequency distribution given in Table 60.6 is shown in Fig. 60.11. The co-ordinates corresponding to each upper class boundary/cumulative frequency value are plotted and the co-ordinates are joined by straight lines (— not the best curve drawn through the co-ordinates as in experimental work). The ogive is 'anchored' at its start by adding the co-ordinate (7.05, 0).

Table 60.6

1 Class	2 Frequency	3 Upper Class boundary	4 Cumulative frequency
		Less than	
7.1–7.3	3	7.35	3
7.4–7.6	5	7.65	8
7.7–7.9	9	7.95	17
8.0–8.2	14	8.25	31
8.3–8.5	11	8.55	42
8.6–8.8	6	8.85	48
8.9–9.1	2	9.15	50

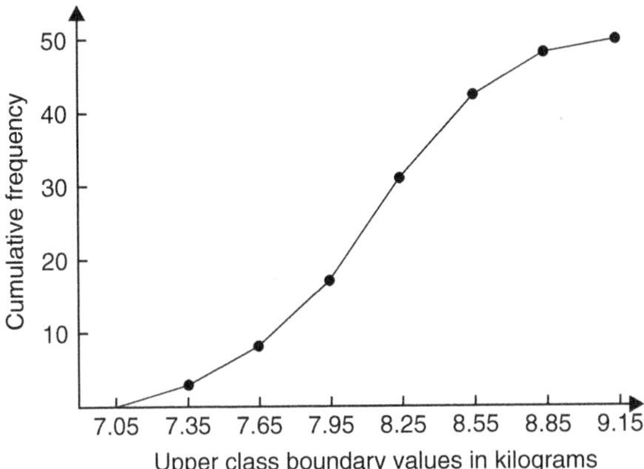

Figure 60.11

Now try the following Practice Exercise

Practice Exercise 282 Presentation of grouped data (Answers on page 735)

1. The mass in kilograms, correct to the nearest one-tenth of a kilogram, of 60 bars of metal are as shown. Form a frequency distribution of about 8 classes for these data

39.8	40.3	40.6	40.0	39.6
39.6	40.2	40.3	40.4	39.8
40.2	40.3	39.9	39.9	40.0
40.1	40.0	40.1	40.1	40.2
39.7	40.4	39.9	40.1	39.9
39.5	40.0	39.8	39.5	39.9
40.1	40.0	39.7	40.4	39.3
40.7	39.9	40.2	39.9	40.0
40.1	39.7	40.5	40.5	39.9
40.8	40.0	40.2	40.0	39.9
39.8	39.7	39.5	40.1	40.2
40.6	40.1	39.7	40.2	40.3

2. Draw a histogram for the frequency distribution given in the solution of Problem 1

3. The information given below refers to the value of resistance in ohms of a batch of 48 resistors of similar value. Form a frequency distribution for the data, having about 6 classes and draw a frequency polygon and histogram to represent these data diagrammatically

21.0	22.4	22.8	21.5	22.6	21.1	21.6	22.3
22.9	20.5	21.8	22.2	21.0	21.7	22.5	20.7
23.2	22.9	21.7	21.4	22.1	22.2	22.3	21.3
22.1	21.8	22.0	22.7	21.7	21.9	21.1	22.6
21.4	22.4	22.3	20.9	22.8	21.2	22.7	21.6
22.2	21.6	21.3	22.1	21.5	22.0	23.4	21.2

4. The time taken in hours to the failure of 50 specimens of a metal subjected to fatigue failure tests are as shown. Form a frequency distribution, having about 8 classes and unequal class intervals, for these data

28	22	23	20	12	24	37	28	21	25
21	14	30	23	27	13	23	7	26	19
24	22	26	3	21	24	28	40	27	24
20	25	23	26	47	21	29	26	22	33
27	9	13	35	20	16	20	25	18	22

5. Form a cumulative frequency distribution and hence draw the ogive for the frequency distribution given in the solution to Problem 3

6. Draw a histogram for the frequency distribution given in the solution to Problem 4

7. The frequency distribution for a batch of 50 capacitors of similar value, measured in microfarads, is:

10.5–10.9	2,	11.0–11.4	7,
11.5–11.9	10,	12.0–12.4	12,
12.5–12.9	11,	13.0–13.4	8

Form a cumulative frequency distribution for these data

8. Draw an ogive for the data given in the solution of Problem 7

9. The diameter in millimetres of a reel of wire is measured in 48 places and the results are as shown

2.10	2.29	2.32	2.21	2.14	2.22
2.28	2.18	2.17	2.20	2.23	2.13
2.26	2.10	2.21	2.17	2.28	2.15
2.16	2.25	2.23	2.11	2.27	2.34
2.24	2.05	2.29	2.18	2.24	2.16
2.15	2.22	2.14	2.27	2.09	2.21
2.11	2.17	2.22	2.19	2.12	2.20
2.23	2.07	2.13	2.26	2.16	2.12

(a) Form a frequency distribution of diameters having about 6 classes

(b) Draw a histogram depicting the data

(c) Form a cumulative frequency distribution

(d) Draw an ogive for the data

Practice Exercise 283 Multiple-choice questions on presentation of statistical data (Answers on page 735)

Each question has only one correct answer

1. A pie diagram is shown in Figure 60.12 where P, Q, R and S represent the salaries of four employees of a construction company. P earns £24000 p.a. Employee S earns:
(a) £20000 (b) £36000
(c) £40000 (d) £24000

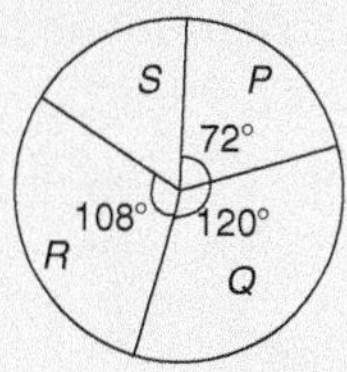

Figure 60.12

2. The curve obtained by joining the co-ordinates of cumulative frequency against upper class boundary values is called;
(a) a histogram (b) a frequency polygon
(c) a tally diagram (d) an ogive

3. Which of the following is a discrete variable?
(a) The diameter of a workpiece
(b) The temperature of a furnace
(c) The number of people working in a factory
(d) The volume of air in an office

4. A histogram is drawn for a frequency distribution with unequal class intervals. The frequency of the class is represented by the:
(a) height of the rectangles
(b) perimeter of the rectangles
(c) area of the rectangles
(d) width of the rectangles

5. An ogive is a:
(a) measure of dispersion
(b) cumulative frequency diagram
(c) measure of location
(d) type of histogram

COMPANION @ WEBSITE **For fully worked solutions to each of the problems in Practice Exercises 280 to 282 in this chapter, go to the website:**
www.routledge.com/cw/bird

Chapter 61

Mean, median, mode and standard deviation

Why it is important to understand: Mean, median, mode and standard deviation

Statistics is a field of mathematics that pertains to data analysis. In many real-life situations, it is helpful to describe data by a single number that is most representative of the entire collection of numbers. Such a number is called a measure of central tendency; the most commonly used measures are mean, median, mode and standard deviation, the latter being the average distance between the actual data and the mean. Statistics is important in the field of engineering since it provides tools to analyse collected data. For example, a chemical engineer may wish to analyse temperature measurements from a mixing tank. Statistical methods can be used to determine how reliable and reproducible the temperature measurements are, how much the temperature varies within the data set, what future temperatures of the tank may be and how confident the engineer can be in the temperature measurements made. When performing statistical analysis on a set of data, the mean, median, mode and standard deviation are all helpful values to calculate; this chapter explains how to determine these measures of central tendency.

At the end of this chapter, you should be able to:

- determine the mean, median and mode for a set of ungrouped data
- determine the mean, median and mode for a set of grouped data
- draw a histogram from a set of grouped data
- determine the mean, median and mode from a histogram
- calculate the standard deviation from a set of ungrouped data
- calculate the standard deviation from a set of grouped data
- determine the quartile values from an ogive
- determine quartile, decile and percentile values from a set of data

61.1 Measures of central tendency

A single value, which is representative of a set of values, may be used to give an indication of the general size of the members in a set, the word **'average'** often being used to indicate the single value.

The statistical term used for 'average' is the arithmetic mean or just the **mean**. Other measures of central tendency may be used and these include the **median** and the **modal** values.

61.2 Mean, median and mode for discrete data

Mean

The **arithmetic mean value** is found by adding together the values of the members of a set and dividing by the number of members in the set. Thus, the mean of the set of numbers: {4, 5, 6, 9} is:

$$\frac{4+5+6+9}{4} \quad \text{i.e.} \quad 6$$

In general, the mean of the set: $\{x_1, x_2, x_3, \ldots, x_n\}$ is

$$\bar{x} = \frac{x_1 + x_2 + x_3 + \cdots + x_n}{n}, \text{written as } \frac{\sum x}{n}$$

where Σ is the Greek letter 'sigma' and means 'the sum of', and $\bar{x}$ (called x-bar) is used to signify a mean value.

Median

The **median value** often gives a better indication of the general size of a set containing extreme values. The set: {7, 5, 74, 10} has a mean value of 24, which is not really representative of any of the values of the members of the set. The median value is obtained by:

(a) **ranking** the set in ascending order of magnitude, and

(b) selecting the value of the **middle member** for sets containing an odd number of members, or finding the value of the mean of the two middle members for sets containing an even number of members.

For example, the set: {7, 5, 74, 10} is ranked as {5, 7, 10, 74}, and since it contains an even number of members (four in this case), the mean of 7 and 10 is taken, giving a median value of 8.5. Similarly, the set: {3, 81, 15, 7, 14} is ranked as {3, 7, 14, 15, 81} and the median value is the value of the middle member, i.e. 14.

Mode

The **modal value**, or **mode**, is the most commonly occurring value in a set. If two values occur with the same frequency, the set is 'bi-modal'. The set: {5, 6, 8, 2, 5, 4, 6, 5, 3} has a modal value of 5, since the member having a value of 5 occurs three times.

Problem 1. Determine the mean, median and mode for the set:

$$\{2,3,7,5,5,13,1,7,4,8,3,4,3\}$$

The mean value is obtained by adding together the values of the members of the set and dividing by the number of members in the set.

Thus, **mean value**,

$$\bar{x} = \frac{2+3+7+5+5+13+1+7+4+8+3+4+3}{13} = \frac{65}{13} = \mathbf{5}$$

To obtain the median value the set is ranked, that is, placed in ascending order of magnitude, and since the set contains an odd number of members the value of the middle member is the median value. Ranking the set gives:

$$\{1,2,3,3,3,4,4,5,5,7,7,8,13\}$$

The middle term is the seventh member, i.e. 4, thus the **median value is 4**.

The **modal value** is the value of the most commonly occurring member and is **3**, which occurs three times, all other members only occurring once or twice.

Problem 2. The following set of data refers to the amount of money in £s taken by a news vendor for 6 days. Determine the mean, median and modal values of the set:

$$\{27.90, 34.70, 54.40, 18.92, 47.60, 39.68\}$$

$$\textbf{Mean value} = \frac{27.90+34.70+54.40+18.92+47.60+39.68}{6} = \textbf{£37.20}$$

The ranked set is:

$$\{18.92, 27.90, 34.70, 39.68, 47.60, 54.40\}$$

Since the set has an even number of members, the mean of the middle two members is taken to give the median value, i.e.

$$\textbf{median value} = \frac{34.70+39.68}{2} = \textbf{£37.19}$$

Since no two members have the same value, this set has **no mode**.

Now try the following Practice Exercise

Practice Exercise 284 Mean, median and mode for discrete data (Answers on page 735)

In Problems 1 to 4, determine the mean, median and modal values for the sets given.

1. {3, 8, 10, 7, 5, 14, 2, 9, 8}

2. $\{26, 31, 21, 29, 32, 26, 25, 28\}$
3. $\{4.72, 4.71, 4.74, 4.73, 4.72, 4.71, 4.73, 4.72\}$
4. $\{73.8, 126.4, 40.7, 141.7, 28.5, 237.4, 157.9\}$

61.3 Mean, median and mode for grouped data

The mean value for a set of grouped data is found by determining the sum of the (frequency × class mid-point values) and dividing by the sum of the frequencies,

i.e. mean value $\bar{x} = \dfrac{f_1x_1 + f_2x_2 + \cdots + f_nx_n}{f_1 + f_2 + \cdots + f_n}$

$$= \frac{\sum(fx)}{\sum f}$$

where f is the frequency of the class having a mid-point value of x and so on.

Problem 3. The frequency distribution for the value of resistance in ohms of 48 resistors is as shown. Determine the mean value of resistance.

20.5–20.9 3, 21.0–21.4 10, 21.5–21.9 11,
22.0–22.4 13, 22.5–22.9 9, 23.0–23.4 2

The class mid-point/frequency values are:

20.7 3, 21.2 10, 21.7 11, 22.2 13,
22.7 9 and 23.2 2

For grouped data, the mean value is given by:

$$\bar{x} = \frac{\sum(fx)}{\sum f}$$

where f is the class frequency and x is the class mid-point value. Hence mean value,

$$\bar{x} = \frac{(3 \times 20.7) + (10 \times 21.2) + (11 \times 21.7) + (13 \times 22.2) + (9 \times 22.7) + (2 \times 23.2)}{48}$$

$$= \frac{1052.1}{48} = 21.919\ldots$$

i.e. **the mean value is 21.9 ohms**, correct to 3 significant figures.

Histogram

The mean, median and modal values for grouped data may be determined from a **histogram**. In a histogram, frequency values are represented vertically and variable values horizontally. The mean value is given by the value of the variable corresponding to a vertical line drawn through the centroid of the histogram. The median value is obtained by selecting a variable value such that the area of the histogram to the left of a vertical line drawn through the selected variable value is equal to the area of the histogram on the right of the line. The modal value is the variable value obtained by dividing the width of the highest rectangle in the histogram in proportion to the heights of the adjacent rectangles. The method of determining the mean, median and modal values from a histogram is shown in Problem 4.

Problem 4. The time taken in minutes to assemble a device is measured 50 times and the results are as shown. Draw a histogram depicting this data and hence determine the mean, median and modal values of the distribution.

14.5–15.5 5, 16.5–17.5 8, 18.5–19.5 16,
20.5–21.5 12, 22.5–23.5 6, 24.5–25.5 3

The histogram is shown in Fig. 61.1. The mean value lies at the centroid of the histogram. With reference to any arbitrary axis, say YY shown at a time of 14 minutes, the position of the horizontal value of the centroid can be obtained from the relationship $AM = \Sigma(am)$, where A is the area of the histogram, M is the horizontal distance of the centroid from the axis YY, a is the area of a rectangle of the histogram and m is the distance of the centroid of the rectangle from YY. The areas of

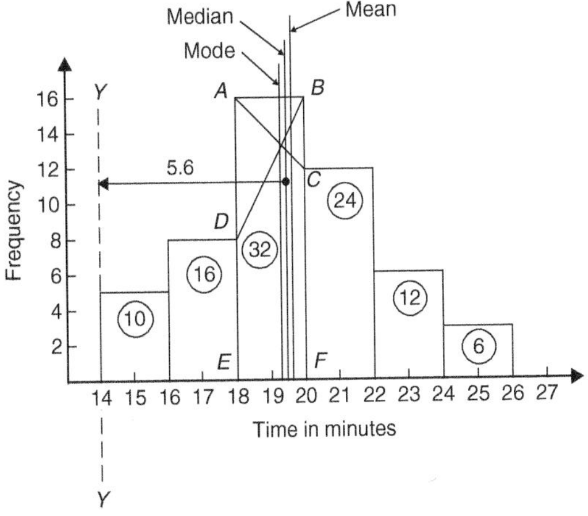

Figure 61.1

the individual rectangles are shown circled on the histogram giving a total area of 100 square units. The positions, m, of the centroids of the individual rectangles are 1, 3, 5, ... units from YY. Thus

$$100M = (10\times 1)+(16\times 3)+(32\times 5) + (24\times 7)+(12\times 9)+(6\times 11)$$

i.e. $$M=\frac{560}{100}=5.6 \text{ units from } YY$$

Thus the position of the **mean** with reference to the time scale is 14+5.6, i.e. **19.6 minutes**.

The median is the value of time corresponding to a vertical line dividing the total area of the histogram into two equal parts. The total area is 100 square units, hence the vertical line must be drawn to give 50 square units of area on each side. To achieve this with reference to Fig. 61.1, rectangle $ABFE$ must be split so that $50-(10+16)$ square units of area lie on one side and $50-(24+12+6)$ square units of area lie on the other. This shows that the area of $ABFE$ is split so that 24 square units of area lie to the left of the line and 8 square units of area lie to the right, i.e. the vertical line must pass through 19.5 minutes. Thus the **median value** of the distribution is **19.5 minutes**.

The mode is obtained by dividing the line AB, which is the height of the highest rectangle, proportionally to the heights of the adjacent rectangles. With reference to Fig. 61.1, this is done by joining AC and BD and drawing a vertical line through the point of intersection of these two lines. This gives the **mode** of the distribution and is **19.3 minutes**.

Now try the following Practice Exercise

Practice Exercise 285 Mean, median and mode for grouped data (Answers on page 735)

1. 21 bricks have a mean mass of 24.2 kg, and 29 similar bricks have a mass of 23.6 kg. Determine the mean mass of the 50 bricks

2. The frequency distribution given below refers to the heights in centimetres of 100 people. Determine the mean value of the distribution, correct to the nearest millimetre

   150–156 5, 157–163 18, 164–170 20
   171–177 27, 178–184 22, 185–191 8

3. The gain of 90 similar transistors is measured and the results are as shown

   83.5–85.5 6, 86.5–88.5 39, 89.5–91.5 27, 92.5–94.5 15, 95.5–97.5 3

   By drawing a histogram of this frequency distribution, determine the mean, median and modal values of the distribution

4. The diameters, in centimetres, of 60 holes bored in engine castings are measured and the results are as shown. Draw a histogram depicting these results and hence determine the mean, median and modal values of the distribution

   2.011–2.014 7, 2.016–2.019 16,
   2.021–2.024 23, 2.026–2.029 9,
   2.031–2.034 5

61.4 Standard deviation

(a) Discrete data

The standard deviation of a set of data gives an indication of the amount of dispersion, or the scatter, of members of the set from the measure of central tendency. Its value is the root-mean-square value of the members of the set and for discrete data is obtained as follows:

(a) determine the measure of central tendency, usually the mean value, (occasionally the median or modal values are specified),

(b) calculate the deviation of each member of the set from the mean, giving

$$(x_1-\overline{x}),(x_2-\overline{x}),(x_3-\overline{x}),\ldots,$$

(c) determine the squares of these deviations, i.e.

$$(x_1-\overline{x})^2,(x_2-\overline{x})^2,(x_3-\overline{x})^2,\ldots,$$

(d) find the sum of the squares of the deviations, that is

$$(x_1-\overline{x})^2+(x_2-\overline{x})^2+(x_3-\overline{x})^2,\ldots,$$

(e) divide by the number of members in the set, n, giving

$$\frac{(x_1-\overline{x})^2+(x_2-\overline{x})^2+(x_3-\overline{x})^2+\cdots}{n}$$

(f) determine the square root of (e).

The standard deviation is indicated by σ (the Greek letter small 'sigma') and is written mathematically as:

$$\textbf{standard deviation}, \sigma = \sqrt{\frac{\sum(x-\overline{x})^2}{n}}$$

where x is a member of the set, $\overline{x}$ is the mean value of the set and n is the number of members in the set. The value of standard deviation gives an indication of the distance of the members of a set from the mean value. The set: {1, 4, 7, 10, 13} has a mean value of 7 and a standard deviation of about 4.2. The set {5, 6, 7, 8, 9} also has a mean value of 7, but the standard deviation is about 1.4. This shows that the members of the second set are mainly much closer to the mean value than the members of the first set. The method of determining the standard deviation for a set of discrete data is shown in Problem 5.

Note that in the above formula for standard deviation, $\frac{\sum(x-\overline{x})^2}{n}$ is referred to as the **variance**, i.e. $\sigma = \sqrt{\text{variance}}$ or variance $= \sigma^2$

Problem 5. Determine the standard deviation from the mean of the set of numbers: {5, 6, 8, 4, 10, 3}, correct to 4 significant figures.

The arithmetic mean, $\overline{x} = \frac{\sum x}{n} = \frac{5+6+8+4+10+3}{6} = 6$

Standard deviation, $\sigma = \sqrt{\frac{\sum(x-\overline{x})^2}{n}}$

The $(x-\overline{x})^2$ values are: $(5-6)^2$, $(6-6)^2$, $(8-6)^2$, $(4-6)^2$, $(10-6)^2$ and $(3-6)^2$.

The sum of the $(x-\overline{x})^2$ values,

i.e. $\sum(x-\overline{x})^2 = 1+0+4+4+16+9 = 34$

and $\frac{\sum(x-\overline{x})^2}{n} = \frac{34}{6} = 5.\dot{6}$

since there are 6 members in the set.

Hence, **standard deviation,**

$$\sigma = \sqrt{\frac{\sum(x-\overline{x}^2)}{n}} = \sqrt{5.\dot{6}} = \mathbf{2.380}$$

correct to 4 significant figures.

(b) Grouped data

For **grouped data, standard deviation**

$$\sigma = \sqrt{\frac{\sum\{f(x-\overline{x})^2\}}{\sum f}}$$

where f is the class frequency value, x is the class mid-point value and $\overline{x}$ is the mean value of the grouped data. The method of determining the standard deviation for a set of grouped data is shown in Problem 6.

Problem 6. The frequency distribution for the values of resistance in ohms of 48 resistors is as shown. Calculate the standard deviation from the mean of the resistors, correct to 3 significant figures.

20.5–20.9 3, 21.0–21.4 10, 21.5–21.9 11,
22.0–22.4 13, 22.5–22.9 9, 23.0–23.4 2

The standard deviation for grouped data is given by:

$$\sigma = \sqrt{\frac{\sum\{f(x-\overline{x})^2\}}{\sum f}}$$

From Problem 3, the distribution mean value, $\overline{x} = 21.92$, correct to 4 significant figures.

The 'x-values' are the class mid-point values, i.e. 20.7, 21.2, 21.7, ...,

Thus the $(x-\overline{x})^2$ values are $(20.7-21.92)^2$, $(21.2-21.92)^2$, $(21.7-21.92)^2$, ...

and the $f(x-\overline{x})^2$ values are $3(20.7-21.92)^2$, $10(21.2-21.92)^2$, $11(21.7-21.92)^2$, ...

The $\sum f(x-\overline{x})^2$ values are

$$4.4652+5.1840+0.5324+1.0192+5.4756+3.2768 = 19.9532$$

$$\frac{\sum\{f(x-\overline{x})^2\}}{\sum f} = \frac{19.9532}{48} = 0.41569$$

and **standard deviation,**

$$\sigma = \sqrt{\frac{\sum\{f(x-\overline{x})^2\}}{\sum f}} = \sqrt{0.41569} = \mathbf{0.645},$$

correct to 3 significant figures.

Now try the following Practice Exercise

Practice Exercise 286 Standard deviation (Answers on page 735)

1. Determine the standard deviation from the mean of the set of numbers:

$$\{35, 22, 25, 23, 28, 33, 30\}$$

correct to 3 significant figures.

2. The values of capacitances, in microfarads, of ten capacitors selected at random from a large batch of similar capacitors are:

$$34.3, 25.0, 30.4, 34.6, 29.6, 28.7, 33.4, 32.7, 29.0 \text{ and } 32.3$$

Determine the standard deviation from the mean for these capacitors, correct to 3 significant figures.

3. The tensile strength in megapascals for 15 samples of tin were determined and found to be:

$$34.61, 34.57, 34.40, 34.63, 34.63, 34.51, 34.49, 34.61, 34.52, 34.55, 34.58, 34.53, 34.44, 34.48 \text{ and } 34.40$$

Calculate the mean and standard deviation from the mean for these 15 values, correct to 4 significant figures.

4. Calculate the standard deviation from the mean for the mass of the 50 bricks given in Problem 1 of Exercise 285, page 632, correct to 3 significant figures.

5. Determine the standard deviation from the mean, correct to 4 significant figures, for the heights of the 100 people given in Problem 2 of Exercise 285, page 632.

6. Calculate the standard deviation from the mean for the data given in Problem 4 of Exercise 285, page 632, correct to 3 significant figures.

61.5 Quartiles, deciles and percentiles

Other measures of dispersion, which are sometimes used, are the quartile, decile and percentile values. The **quartile values** of a set of discrete data are obtained by selecting the values of members that divide the set into four equal parts. Thus for the set: {2, 3, 4, 5, 5, 7, 9, 11, 13, 14, 17} there are 11 members and the values of the members dividing the set into four equal parts are 4, 7, and 13. These values are signified by Q_1, Q_2 and Q_3 and called the first, second and third quartile values, respectively. It can be seen that the second quartile value, Q_2, is the value of the middle member and hence is the median value of the set.

For grouped data the ogive may be used to determine the quartile values. In this case, points are selected on the vertical cumulative frequency values of the ogive, such that they divide the total value of cumulative frequency into four equal parts. Horizontal lines are drawn from these values to cut the ogive. The values of the variable corresponding to these cutting points on the ogive give the quartile values (see Problem 7).

When a set contains a large number of members, the set can be split into ten parts, each containing an equal number of members. These ten parts are then called **deciles**. For sets containing a very large number of members, the set may be split into one hundred parts, each containing an equal number of members. One of these parts is called a **percentile**.

Problem 7. The frequency distribution given below refers to the overtime worked by a group of craftsmen during each of 48 working weeks in a year.

25–29 5, 30–34 4, 35–39 7, 40–44 11, 45–49 12, 50–54 8, 55–59 1

Draw an ogive for this data and hence determine the quartile values

The cumulative frequency distribution (i.e. upper class boundary/cumulative frequency values) is:

29.5 5, 34.5 9, 39.5 16,
44.5 27, 49.5 39, 54.5 47,
59.5 48

The ogive is formed by plotting these values on a graph, as shown in Fig. 61.2. The total frequency is divided into four equal parts, each having a range of 48/4, i.e. 12. This gives cumulative frequency values of 0 to

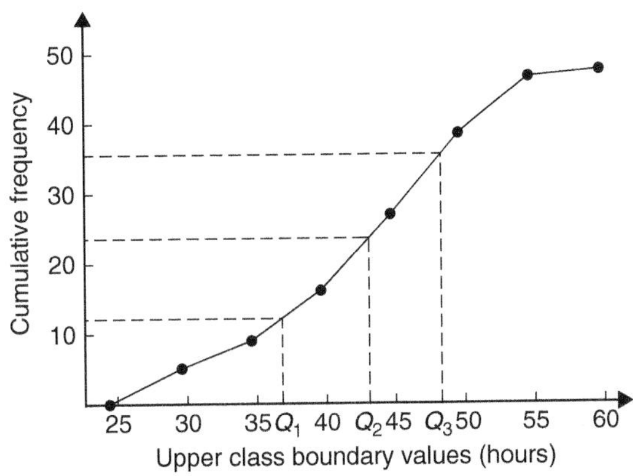

Figure 61.2

12 corresponding to the first quartile, 12 to 24 corresponding to the second quartile, 24 to 36 corresponding to the third quartile and 36 to 48 corresponding to the fourth quartile of the distribution, i.e. the distribution is divided into four equal parts. The quartile values are those of the variable corresponding to cumulative frequency values of 12, 24 and 36, marked Q_1, Q_2 and Q_3 in Fig. 61.2. These values, correct to the nearest hour, are **37 hours, 43 hours and 48 hours**, respectively. The Q_2 value is also equal to the median value of the distribution. One measure of the dispersion of a distribution is called the **semi-interquartile range** and is given by: $(Q_3 - Q_1)/2$, and is $(48 - 37)/2$ in this case, i.e. $\mathbf{5\frac{1}{2}}$ **hours**.

Problem 8. Determine the numbers contained in the (a) 41st to 50th percentile group, and (b) 8th decile group of the set of numbers shown below:

14	22	17	21	30	28	37	7	23	32
24	17	20	22	27	19	26	21	15	29

The set is ranked, giving:

7	14	15	17	17	19	20	21	21	22
22	23	24	26	27	28	29	30	32	37

(a) There are 20 numbers in the set, hence the first 10% will be the two numbers 7 and 14, the second 10% will be 15 and 17 and so on. Thus the 41st to 50th percentile group will be the numbers **21 and 22**

(b) The first decile group is obtained by splitting the ranked set into 10 equal groups and selecting the first group, i.e. the numbers 7 and 14. The second decile group are the numbers 15 and 17 and so on. Thus the 8th decile group contains the numbers **27 and 28**

Now try the following Practice Exercise

Practice Exercise 287 Quartiles, deciles and percentiles (Answers on page 735)

1. The number of working days lost due to accidents for each of 12 one-monthly periods are as shown. Determine the median and first and third quartile values for this data

27	37	40	28	23	30
35	24	30	32	31	28

2. The number of faults occurring on a production line in a nine-week period are as shown below. Determine the median and quartile values for the data

30	27	25	24	27
37	31	27	35	

3. Determine the quartile values and semi-interquartile range for the frequency distribution given in Problem 2 of Exercise 285, page 632.

4. Determine the numbers contained in the 5th decile group and in the 61st to 70th percentile groups for the set of numbers:

40 46 28 32 37 42 50 31 48 45
32 38 27 33 40 35 25 42 38 41

5. Determine the numbers in the 6th decile group and in the 81st to 90th percentile group for the set of numbers:

43 47 30 25 15 51 17 21 37 33 44 56 40 49 22
36 44 33 17 35 58 51 35 44 40 31 41 55 50 16

Practice Exercise 288 Multiple-choice questions on mean, median, mode and standard deviation (Answers on page 735)

Each question has only one correct answer

Questions 1 and 2 relate to the following information.

A set of measurements (in mm) is as follows:
$\{4, 5, 2, 11, 7, 6, 5, 1, 5, 8, 12, 6\}$

1. The median is:

   (a) 6 mm (b) 5 mm

   (c) 72 mm (d) 5.5 mm

2. The mean is:

   (a) 6 mm (b) 5 mm

   (c) 72 mm (d) 5.5 mm

3. The mean of the numbers 5, 13, 9, x and 10 is 11. The value of x is:

   (a) 20 (b) 18

   (c) 17 (d) 16

Questions 4 to 7 relate to the following information:
The capacitance (in pF) of six capacitors is as follows: $\{5, 6, 8, 5, 10, 2\}$

4. The median value is:

   (a) 36 pF (b) 6 pF

   (c) 5.5 pF (d) 5 pF

5. The modal value is:

   (a) 36 pF (b) 6 pF

   (c) 5.5 pF (d) 5 pF

6. The mean value is:

   (a) 36 pF (b) 6 pF

   (c) 5.5 pF (d) 5 pF

7. The standard deviation is:

   (a) 2.66 pF (b) 2.52 pF

   (c) 2.45 pF (d) 6.33 pF

8. The number of faults occurring on a production line in a 11-week period are as shown:

   32 29 27 26 29 39 33 29 37 28 35

   The third quartile value is:

   (a) 29 (b) 31

   (c) 35 (d) 33

Questions 9 and 10 relate to the following information:
The frequency distribution for the values of resistance in ohms of 40 transistors is as follows:

15.5 – 15.9 3 16.0 – 16.4 10

16.5 – 16.9 13 17.0 – 17.4 8

17.5 – 17.9 6

9. The mean value of the resistance is:

   (a) 16.75 Ω (b) 1.0 Ω

   (c) 15.85 Ω (d) 16.95 Ω

10. The standard deviation is:

    (a) 0.335 Ω (b) 0.251 Ω

    (c) 0.682 Ω (d) 0.579 Ω

COMPANION @ WEBSITE

For fully worked solutions to each of the problems in Practice Exercises 284 to 287 in this chapter, go to the website:

www.routledge.com/cw/bird

Chapter 62

Probability

Why it is important to understand: Probability

Engineers deal with uncertainty in their work, often with precision and analysis, and probability theory is widely used to model systems in engineering and scientific applications. There are a number of examples of where probability is used in engineering. For example, with electronic circuits, scaling down the power and energy of such circuits reduces the reliability and predictability of many individual elements, but the circuits must nevertheless be engineered so that the overall circuit is reliable. Centres for disease control need to decide whether to institute massive vaccination or other preventative measures in the face of globally threatening, possibly mutating diseases in humans and animals. System designers must weigh the costs and benefits of measures for reliability and security, such as levels of backups and firewalls, in the face of uncertainty about threats from equipment failures or malicious attackers. Models incorporating probability theory have been developed and are continuously being improved for understanding the brain, gene pools within populations, weather and climate forecasts, microelectronic devices and imaging systems such as computer aided tomography (CAT) scan and radar. The electric power grid, including power generating stations, transmission lines and consumers, is a complex system; however, breakdowns occur, and guidance for investment comes from modelling the most likely sequences of events that could cause outage. Similar planning and analysis is done for communication networks, transportation networks, water and other infrastructure. Probabilities, permutations and combinations are used daily in many different fields that range from gambling and games, to mechanical or structural failure rates, to rates of detection in medical screening. Uncertainty is clearly all around us, in our daily lives and in many professions. Use of standard deviation is widely used when results of opinion polls are described. The language of probability theory lets people break down complex problems, and to argue about pieces of them with each other, and then aggregate information about subsystems to analyse a whole system. This chapter briefly introduces the important subject of probability.

At the end of this chapter, you should be able to:

- define probability
- define expectation, dependent event, independent event and conditional probability
- state the addition and multiplication laws of probability
- use the laws of probability in simple calculations
- use the laws of probability in practical situations
- determine permutations and combinations

62.1 Introduction to probability

Probability and statistics are quite extensively linked. For instance, when a scientist performs a measurement, the outcome of that measurement has a certain amount of 'chance' associated with it: factors such as electronic noise in the equipment, minor fluctuations in environmental conditions and even human error have a random effect on the measurement. Often, the variations caused by these factors are minor, but they do have a significant effect in many cases. As a result, the scientist cannot expect to get the exact same measurement result in every case, and this variation requires that he describe his measurements *statistically* (for instance, using a mean and standard deviation). Likewise, when an anthropologist considers a small group of people from a larger population, the results of the study (assuming they involve numerical data) will involve some random variations that must be taken into account using statistics.

The **probability** of something happening is the likelihood or chance of it happening. Values of probability lie between 0 and 1, where 0 represents an absolute impossibility and 1 represents an absolute certainty. The probability of an event happening usually lies somewhere between these two extreme values and is expressed either as a proper or decimal fraction. Examples of probability are:

that a length of copper wire has zero resistance at 100°C	0
that a fair, six-sided dice will stop with a 3 upwards	$\frac{1}{6}$ or 0.1667
that a fair coin will land with a head upwards	$\frac{1}{2}$ or 0.5
that a length of copper wire has some resistance at 100°C	1

If p is the probability of an event happening and q is the probability of the same event not happening, then the total probability is $p+q$ and is equal to unity, since it is an absolute certainty that the event either does or does not occur, i.e. $p+q=1$

Expectation

The **expectation**, E, of an event happening is defined in general terms as the product of the probability p of an event happening and the number of attempts made, n, i.e. $\boldsymbol{E=pn}$.

Thus, since the probability of obtaining a 3 upwards when rolling a fair dice is $\frac{1}{6}$, the expectation of getting a 3 upwards on four throws of the dice is $\frac{1}{6}\times 4$, i.e. $\frac{2}{3}$

Thus expectation is the average occurrence of an event.

Dependent event

A **dependent event** is one in which the probability of an event happening affects the probability of another event happening. Let 5 transistors be taken at random from a batch of 100 transistors for test purposes, and the probability of there being a defective transistor, p_1, be determined. At some later time, let another 5 transistors be taken at random from the 95 remaining transistors in the batch and the probability of there being a defective transistor, p_2, be determined. The value of p_2 is different from p_1 since batch size has effectively altered from 100 to 95, i.e. probability p_2 is dependent on probability p_1. Since transistors are drawn, and then another 5 transistors drawn without replacing the first 5, the second random selection is said to be **without replacement**.

Independent event

An independent event is one in which the probability of an event happening does not affect the probability of another event happening. If 5 transistors are taken at random from a batch of transistors and the probability of a defective transistor p_1 is determined and the process is repeated after the original 5 have been replaced in the batch to give p_2, then p_1 is equal to p_2. Since the 5 transistors are replaced between draws, the second selection is said to be **with replacement**.

Conditional probability

Conditional probability is concerned with the probability of say event B occurring, given that event A has already taken place. If A and B are independent events, then the fact that event A has already occurred will not affect the probability of event B. If A and B are dependent events, then event A having occurred will effect the probability of event B.

62.2 Laws of probability

The addition law of probability

The addition law of probability is recognised by the word **'or'** joining the probabilities. If p_A is the probability of event A happening and p_B is the probability of event B happening, the probability of **event A or event B** happening is given by $p_A + p_B$ (provided events A and B are **mutually exclusive**, i.e. A and B are events which cannot occur together). Similarly, the probability of events **A or B or C or** ... N happening is given by

$$p_A + p_B + p_C + \cdots + p_N$$

Suppose that we draw a card from a well-shuffled standard deck of cards. We want to determine the probability that the card drawn is a two or a face card. The event 'a face card is drawn' is mutually exclusive with the event 'a two is drawn,' so we will simply need to add the probabilities of these two events together.

There is a total of 12 face cards (king, queen and jack for four suits), and so the probability of drawing a face card is 12/52. There are four twos in the deck, and so the probability of drawing a two is 4/52. This means that the probability of drawing a two or a face card is $12/52 + 4/52 = 16/52$ or $8/26$ or $4/13$.

The multiplication law of probability

The multiplication law of probability is recognized by the word **'and'** joining the probabilities. If p_A is the probability of event A happening and p_B is the probability of event B happening, the probability of **event A and event B** happening is given by $p_A \times p_B$. Similarly, the probability of events **A and B and C and** ... N happening is given by:

$$p_A \times p_B \times p_C \times \cdots \times p_N$$

62.3 Worked problems on probability

Problem 1. Determine the probabilities of selecting at random (a) a man, and (b) a woman from a crowd containing 20 men and 33 women

(a) The probability of selecting at random a man, p, is given by the ratio

$$\frac{\text{number of men}}{\text{number in crowd}}$$

i.e. $p = \dfrac{20}{20+33} = \mathbf{\dfrac{20}{53}}$ or **0.3774**

(b) The probability of selecting at random a woman, q, is given by the ratio

$$\frac{\text{number of women}}{\text{number in crowd}}$$

i.e. $q = \dfrac{33}{20+33} = \mathbf{\dfrac{33}{53}}$ or **0.6226**

(Check: the total probability should be equal to 1;

$$p = \frac{20}{53} \text{ and } q = \frac{33}{53},$$

thus the total probability,

$$p + q = \frac{20}{53} + \frac{33}{53} = 1$$

hence no obvious error has been made.)

Problem 2. A fair coin is tossed three times. What is the probability of obtaining two heads and one tail?

There are 8 possible ways the coin can land:
(H, H, H), (H, H, T), (H, T, H), (H, T, T), (T, H, H), (T, H, T), (T, T, H) and (T, T, T) where H represents a head and T represents a tail

Of these, 3 have two heads and one tail: (H, H, T), (H, T, H) and (T, H, H)

Thus, the number of ways it can happen is 3 and the total number of outcomes is 8

The probability of an event happening

$$= \frac{\text{number of ways it can happen}}{\text{total number of outcomes}}$$

Therefore, the probability of obtaining one head and two tails $= \mathbf{\dfrac{3}{8}}$

Problem 3. A committee of three is chosen from five candidates - A, B, C, D and E. What is the probability that B will be on the committee?

There are 10 possible committees: (A, B, C), (A, B, D), (A, B, E), (A, C, D), (A, C, E), (A, D, E), (B, C, D), (B, C, E), (B, D, E) and (C, D, E)

Of these, B is included in 6: (A, B, C), (A, B, D), (A, B, E), (B, C, D), (B, C, E) and (B, D, E)

Thus, the number of ways it can happen is 6 and the total number of outcomes is 10

The probability of an event happening

$$= \frac{\text{number of ways it can happen}}{\text{total number of outcomes}}$$

Therefore, the probability of B being on the committee

$$= \frac{6}{10} = \mathbf{\frac{3}{5}}$$

Problem 4. Find the expectation of obtaining a 4 upwards with 3 throws of a fair dice

Expectation is the average occurrence of an event and is defined as the probability times the number of attempts. The probability, p, of obtaining a 4 upwards for one throw of the dice, is $\frac{1}{6}$

Also, 3 attempts are made, hence $n=3$ and the expectation, E, is pn, i.e.

$$E=\frac{1}{6}\times 3=\frac{1}{2} \text{ or } \mathbf{0.50}$$

Problem 5. Calculate the probabilities of selecting at random:

(a) the winning horse in a race in which 10 horses are running

(b) the winning horses in both the first and second races if there are 10 horses in each race

(a) Since only one of the ten horses can win, the probability of selecting at random the winning horse is

$$\frac{\text{number of winners}}{\text{number of horses}} \text{ i.e. } \frac{\mathbf{1}}{\mathbf{10}} \text{ or } \mathbf{0.10}$$

(b) The probability of selecting the winning horse in the first race is $\frac{1}{10}$. The probability of selecting the winning horse in the second race is $\frac{1}{10}$. The probability of selecting the winning horses in the first **and** second race is given by the multiplication law of probability,

$$\text{i.e. } \textbf{probability}=\frac{1}{10}\times\frac{1}{10}=\frac{\mathbf{1}}{\mathbf{100}} \text{ or } \mathbf{0.01}$$

Problem 6. There are 30 buttons in total in a bag; 13 are blue and 17 are red. What is the probability of taking two red buttons in a row out of the bag without looking?

The probability of taking a red button from the bag the first time is17/30. Note that the numerator is 17, the total number of red buttons and the denominator is 30, the total number of buttons. If this button is not put back into the bag, then the probability of taking a red button from the bag now is 16/29. The numerator is 16, the total number of red buttons remaining in the bag and the denominator is 29, the total number of buttons remaining. The probability of taking two red buttons in a row without looking is given by the multiplication rule of probability, i.e. **probability**

$$=\frac{17}{30}\times\frac{16}{29}=\frac{272}{870}=\frac{136}{435}=\mathbf{0.313=31.3\%}$$

Problem 7. The probability of a component failing in one year due to excessive temperature is $\frac{1}{20}$, due to excessive vibration is $\frac{1}{25}$ and due to excessive humidity is $\frac{1}{50}$. Determine the probabilities that during a one-year period a component: (a) fails due to excessive (b) fails due to excessive vibration or excessive humidity, and (c) will not fail because of both excessive temperature and excessive humidity

Let p_A be the probability of failure due to excessive temperature, then

$$p_A=\frac{1}{20} \text{ and } \overline{p_A}=\frac{19}{20}$$

(where $\overline{p_A}$ is the probability of not failing).

Let p_B be the probability of failure due to excessive vibration, then

$$p_B=\frac{1}{25} \text{ and } \overline{p_B}=\frac{24}{25}$$

Let p_C be the probability of failure due to excessive humidity, then

$$p_C=\frac{1}{50} \text{ and } \overline{p_C}=\frac{49}{50}$$

(a) The probability of a component failing due to excessive temperature **and** excessive vibration is given by:

$$p_A\times p_B=\frac{1}{20}\times\frac{1}{25}=\frac{\mathbf{1}}{\mathbf{500}} \text{ or } \mathbf{0.002}$$

(b) The probability of a component failing due to excessive vibration **or** excessive humidity is:

$$p_B+p_C=\frac{1}{25}+\frac{1}{50}=\frac{\mathbf{3}}{\mathbf{50}} \text{ or } \mathbf{0.06}$$

(c) The probability that a component will not fail due to excessive temperature **and** will not fail due to excessive humidity is:

$$\overline{p_A}\times\overline{p_C}=\frac{19}{20}\times\frac{49}{50}=\frac{\mathbf{931}}{\mathbf{1000}} \text{ or } \mathbf{0.931}$$

Problem 8. A batch of 100 capacitors contains 73 that are within the required tolerance values, 17 which are below the required tolerance values, and the remainder are above the required tolerance values. Determine the probabilities that when randomly selecting a capacitor and then a second capacitor: (a) both are within the required tolerance values when selecting with replacement, and (b) the first one drawn is below and the second one drawn is above the required tolerance value, when selection is without replacement

(a) The probability of selecting a capacitor within the required tolerance values is $\frac{73}{100}$. The first capacitor drawn is now replaced and a second one is drawn from the batch of 100. The probability of this capacitor being within the required tolerance values is also $\frac{73}{100}$.

This an **independent event**. Thus, the probability of selecting a capacitor within the required tolerance values for both the first **and** the second draw is:

$$\frac{73}{100}\times\frac{73}{100}=\mathbf{\frac{5329}{10000}} \quad \text{or} \quad \mathbf{0.5329}$$

(b) The probability of obtaining a capacitor below the required tolerance values on the first draw is $\frac{17}{100}$. There are now only 99 capacitors left in the batch, since the first capacitor is not replaced. The probability of drawing a capacitor above the required tolerance values on the second draw is $\frac{10}{99}$, since there are $(100-73-17)$, i.e. 10 capacitors above the required tolerance value. Thus, the probability of randomly selecting a capacitor below the required tolerance values and followed by randomly selecting a capacitor above the tolerance values is

$$\frac{17}{100}\times\frac{10}{99}=\frac{170}{9900}=\mathbf{\frac{17}{990}} \quad \text{or} \quad \mathbf{0.0172}$$

Now try the following Practice Exercise

Practice Exercise 289 Probability (Answers on page 735)

1. In a batch of 45 lamps there are 10 faulty lamps. If one lamp is drawn at random, find the probability of it being (a) faulty and (b) satisfactory
2. A box of fuses are all of the same shape and size and comprises 23 2 A fuses, 47 5 A fuses and 69 13 A fuses. Determine the probability of selecting at random (a) a 2 A fuse, (b) a 5 A fuse and (c) a 13 A fuse
3. (a) Find the probability of having a 2 upwards when throwing a fair 6-sided dice. (b) Find the probability of having a 5 upwards when throwing a fair 6-sided dice. (c) Determine the probability of having a 2 and then a 5 on two successive throws of a fair 6-sided dice
4. There are 10 counters in a bag, 3 are red, 2 are blue and 5 are green. The contents of the bag are shaken before one is randomly chosen from the bag. What is the probability that a red counter is not chosen?
5. Determine the probability that the total score is 8 when two like dice are thrown
6. The probability of event A happening is $\frac{3}{5}$ and the probability of event B happening is $\frac{2}{3}$. Calculate the probabilities of (a) both A and B happening, (b) only event A happening, i.e. event A happening and event B not happening, (c) only event B happening, and (d) either A, or B, or A and B happening
7. When testing 1000 soldered joints, 4 failed during a vibration test and 5 failed due to having a high resistance. Determine the probability of a joint failing due to (a) vibration, (b) high resistance, (c) vibration or high resistance and (d) vibration and high resistance

62.4 Further worked problems on probability

Problem 9. A batch of 40 components contains 5 which are defective. A component is drawn at random from the batch and tested and then a second component is drawn. Determine the probability that neither of the components is defective when drawn (a) with replacement, and (b) without replacement

(a) With replacement

The probability that the component selected on the first draw is satisfactory is $\frac{35}{40}$, i.e. $\frac{7}{8}$. The component is now replaced and a second draw is made. The probability that this component is also satisfactory is $\frac{7}{8}$. Hence, the probability that both the first component drawn **and** the second component drawn are satisfactory is:

$$\frac{7}{8}\times\frac{7}{8}=\mathbf{\frac{49}{64}} \quad \text{or} \quad \mathbf{0.7656}$$

(b) Without replacement

The probability that the first component drawn is satisfactory is $\frac{7}{8}$. There are now only 34 satisfactory components left in the batch and the batch number is 39. Hence, the probability of drawing a satisfactory component on the second draw is $\frac{34}{39}$. Thus the probability that the first component drawn **and** the second component drawn are satisfactory, i.e. neither is defective, is:

$$\frac{7}{8}\times\frac{34}{39}=\mathbf{\frac{238}{312}} \quad \text{or} \quad \mathbf{0.7628}$$

Problem 10. A batch of 40 components contains 5 that are defective. If a component is drawn at random from the batch and tested and then a second component is drawn at random, calculate the probability of having one defective component, both with and without replacement

The probability of having one defective component can be achieved in two ways. If p is the probability of drawing a defective component and q is the probability of drawing a satisfactory component, then the probability of having one defective component is given by drawing a satisfactory component and then a defective component **or** by drawing a defective component and then a satisfactory one, i.e. by $q\times p+p\times q$

With replacement:

$$p=\frac{5}{40}=\frac{1}{8} \quad \text{and} \quad q=\frac{35}{40}=\frac{7}{8}$$

Hence, probability of having one defective component is:

$$\frac{1}{8}\times\frac{7}{8}+\frac{7}{8}\times\frac{1}{8}$$

i.e. $$\frac{7}{64}+\frac{7}{64}=\mathbf{\frac{7}{32}} \quad \text{or} \quad \mathbf{0.2188}$$

Without replacement:

$p_1=\frac{1}{8}$ and $q_1=\frac{7}{8}$ on the first of the two draws. The batch number is now 39 for the second draw, thus,

$$p_2=\frac{5}{39} \quad \text{and} \quad q_2=\frac{35}{39}$$

$$p_1q_2+q_1p_2=\frac{1}{8}\times\frac{35}{39}+\frac{7}{8}\times\frac{5}{39}$$

$$=\frac{35+35}{312}$$

$$=\mathbf{\frac{70}{312}} \quad \text{or} \quad \mathbf{0.2244}$$

Problem 11. A box contains 74 brass washers, 86 steel washers and 40 aluminium washers. Three washers are drawn at random from the box without replacement. Determine the probability that all three are steel washers

Assume, for clarity of explanation, that a washer is drawn at random, then a second, then a third (although this assumption does not affect the results obtained). The total number of washers is 74+86+40, i.e. 200.

The probability of randomly selecting a steel washer on the first draw is $\frac{86}{200}$. There are now 85 steel washers in a batch of 199. The probability of randomly selecting a steel washer on the second draw is $\frac{85}{199}$. There are now 84 steel washers in a batch of 198. The probability of randomly selecting a steel washer on the third draw is $\frac{84}{198}$. Hence the probability of selecting a steel washer on the first draw **and** the second draw **and** the third draw is:

$$\frac{86}{200}\times\frac{85}{199}\times\frac{84}{198}=\frac{614040}{7880400}$$

$$=\mathbf{0.0779}$$

Problem 12. For the box of washers given in Problem 11 above, determine the probability that there are no aluminium washers drawn, when three washers are drawn at random from the box without replacement

The probability of not drawing an aluminium washer on the first draw is $1-\left(\frac{40}{200}\right)$, i.e. $\frac{160}{200}$. There are now 199 washers in the batch of which 159 are not aluminium washers. Hence, the probability of not drawing an aluminium washer on the second draw is $\frac{159}{199}$. Similarly, the probability of not drawing an aluminium washer on the third draw is $\frac{158}{198}$. Hence the probability of not drawing an aluminium washer on the first **and** second **and** third draw is

$$\frac{160}{200}\times\frac{159}{199}\times\frac{158}{198}=\frac{4019520}{7880400}$$

$$=\mathbf{0.5101}$$

Problem 13. For the box of washers in Problem 11 above, find the probability that there are two brass washers and either a steel or an aluminium washer when three are drawn at random, without replacement

Two brass washers (A) and one steel washer (B) can be obtained in any of the following ways:

1st draw	2nd draw	3rd draw
A	A	B
A	B	A
B	A	A

Two brass washers and one aluminium washer (C) can also be obtained in any of the following ways:

1st draw	2nd draw	3rd draw
A	A	C
A	C	A
C	A	A

Thus there are six possible ways of achieving the combinations specified. If A represents a brass washer, B a steel washer and C an aluminium washer, then the combinations and their probabilities are as shown:

First	Draw Second	Third	Probability
A	A	B	$\frac{74}{200}\times\frac{73}{199}\times\frac{86}{198}=0.0590$
A	B	A	$\frac{74}{200}\times\frac{86}{199}\times\frac{73}{198}=0.0590$
B	A	A	$\frac{86}{200}\times\frac{74}{199}\times\frac{73}{198}=0.0590$
A	A	C	$\frac{74}{200}\times\frac{73}{199}\times\frac{40}{198}=0.0274$
A	C	A	$\frac{74}{200}\times\frac{40}{199}\times\frac{73}{198}=0.0274$
C	A	A	$\frac{40}{200}\times\frac{74}{199}\times\frac{73}{198}=0.0274$

The probability of having the first combination **or** the second, **or** the third and so on, is given by the sum of the probabilities,

i.e. by $3\times 0.0590+3\times 0.0274$, that is **0.2592**

Now try the following Practice Exercise

Practice Exercise 290 Probability (Answers on page 736)

1. The probability that component A will operate satisfactorily for 5 years is 0.8 and that B will operate satisfactorily over that same period of time is 0.75. Find the probabilities that in a 5 year period: (a) both components operate satisfactorily, (b) only component A will operate satisfactorily, and (c) only component B will operate satisfactorily
2. In a particular street, 80% of the houses have telephones. If two houses selected at random are visited, calculate the probabilities that (a) they both have a telephone and (b) one has a telephone but the other does not have a telephone
3. Veroboard pins are packed in packets of 20 by a machine. In a thousand packets, 40 have less than 20 pins. Find the probability that if 2 packets are chosen at random, one will contain less than 20 pins and the other will contain 20 pins or more

4. A batch of 1 kW fire elements contains 16 which are within a power tolerance and 4 which are not. If 3 elements are selected at random from the batch, calculate the probabilities that (a) all three are within the power tolerance and (b) two are within but one is not within the power tolerance

5. An amplifier is made up of three transistors, A, B and C. The probabilities of A, B or C being defective are $\frac{1}{20}$, $\frac{1}{25}$ and $\frac{1}{50}$, respectively. Calculate the percentage of amplifiers produced (a) which work satisfactorily and (b) which have just one defective transistor

6. A box contains 14 40 W lamps, 28 60 W lamps and 58 25 W lamps, all the lamps being of the same shape and size. Three lamps are drawn at random from the box, first one, then a second, then a third. Determine the probabilities of: (a) getting one 25 W, one 40 W and one 60 W lamp, with replacement, (b) getting one 25 W, one 40 W and one 60 W lamp without replacement, and (c) getting either one 25 W and two 40 W or one 60 W and two 40 W lamps with replacement

62.5 Permutations and combinations

Permutations

If n different objects are available and r objects are selected from n, they can be arranged in different orders of selection. Each different ordered arrangement is called a **permutation**. For example, permutations of the three letters X, Y and Z taken together are:

$$XYZ, XZY, YXZ, YZX, ZXY \text{ and } ZYX$$

This can be expressed as ${}^3P_3 = 6$, the upper 3 denoting the number of items from which the arrangements are made, and the lower 3 indicating the number of items used in each arrangement.

If we take the same three letters XYZ two at a time the permutations

$$XY, YX, XZ, ZX, YZ, ZY$$

can be found, and denoted by ${}^3P_2 = 6$

(Note that the order of the letters matter in permutations, i.e. YX is a different permutation from XY). In general, ${}^nP_r = n(n-1)(n-2)\ldots(n-r+1)$ or

$${}^nP_r = \frac{n!}{(n-r)!} \text{ as stated in Chapter 15}$$

For example, ${}^5P_4 = (5)(4)(3)(2) = 120$ or

${}^5P_4 = \frac{5!}{(5-4)!} = \frac{5!}{1!} = (5)(4)(3)(2) = 120$

Also, ${}^3P_3 = 6$ from above; using ${}^nP_r = \frac{n!}{(n-r)!}$ gives ${}^3P_3 = \frac{3!}{(3-3)!} = \frac{6}{0!}$. Since this must equal 6, then $0! = 1$ (check this with your calculator).

Combinations

If selections of the three letters X, Y, Z are made without regard to the order of the letters in each group, i.e. XY is now the same as YX for example, then each group is called a **combination**. The number of possible combinations is denoted by ${}^nC_r$, where n is the number of different items and r is the number in each selection.

In general,

$${}^nC_r = \frac{n!}{r!(n-r)!}$$

For example,

$${}^5C_4 = \frac{5!}{4!(5-4)!} = \frac{5!}{4!} = \frac{5 \times 4 \times 3 \times 2 \times 1}{4 \times 3 \times 2 \times 1} = 5$$

Problem 14. Calculate the number of permutations there are of: (a) 5 distinct objects taken 2 at a time, (b) 4 distinct objects taken 2 at a time

(a) ${}^5P_2 = \frac{5!}{(5-2)!} = \frac{5!}{3!} = \frac{5 \times 4 \times 3 \times 2}{3 \times 2} = \mathbf{20}$

(b) ${}^4P_2 = \frac{4!}{(4-2)!} = \frac{4!}{2!} = \mathbf{12}$

Problem 15. Calculate the number of combinations there are of: (a) 5 distinct objects taken 2 at a time, (b) 4 distinct objects taken 2 at a time

(a) ${}^5C_2 = \frac{5!}{2!(5-2)!} = \frac{5!}{2!3!}$

$$= \frac{5 \times 4 \times 3 \times 2 \times 1}{(2 \times 1)(3 \times 2 \times 1)} = \mathbf{10}$$

(b) $^4C_2 = \frac{4!}{2!(4-2)!} = \frac{4!}{2!2!} = \mathbf{6}$

Problem 16. A class has 24 students. 4 can represent the class at an exam board. How many combinations are possible when choosing this group?

Number of combinations possible,

$$^nC_r = \frac{n!}{r!(n-r!)}$$

i.e. $^{24}C_4 = \frac{24!}{4!(24-4)!} = \frac{24!}{4!20!} = \mathbf{10\,626}$

Problem 17. In how many ways can a team of eleven be picked from sixteen possible players?

Number of ways $= {}^nC_r = {}^{16}C_{11}$

$$= \frac{16!}{11!(16-11)!} = \frac{16!}{11!5!} = \mathbf{4368}$$

Now try the following Practice Exercise

Practice Exercise 291 Permutations and combinations (Answers on page 736)

1. Calculate the number of permutations there are of: (a) 15 distinct objects taken 2 at a time, (b) 9 distinct objects taken 4 at a time
2. Calculate the number of combinations there are of: (a) 12 distinct objects taken 5 at a time, (b) 6 distinct objects taken 4 at a time
3. In how many ways can a team of six be picked from ten possible players?
4. 15 boxes can each hold one object. In how many ways can 10 identical objects be placed in the boxes?
5. Six numbers between 1 and 49 are chosen when playing the National lottery. Determine the probability of winning the top prize (i.e. 6 correct numbers!) if 10 tickets were purchased and six different numbers were chosen on each ticket.

62.6 Bayes' theorem

Bayes' theorem is one of probability theory (originally stated by the Reverend Thomas Bayes), and may be seen as a way of understanding how the probability that a theory is true is affected by a new piece of evidence. The theorem has been used in a wide variety of contexts, ranging from marine biology to the development of 'Bayesian' spam blockers for email systems; in science, it has been used to try to clarify the relationship between theory and evidence. Insights in the philosophy of science involving confirmation, falsification and other topics can be made more precise, and sometimes extended or corrected, by using Bayes' theorem.
Bayes' theorem may be stated mathematically as:

$$P(A_1|B) = \frac{P(B|A_1)P(A_1)}{P(B|A_1)P(A_1) + P(B|A_2)P(A_2) + \cdots}$$

$$\text{or } P(A_i|B) = \frac{P(B|A_i)P(A_i)}{\sum_{j=1}^{n} P(B|A_j)P(A_j)} \qquad (i = 1, 2, \ldots, n)$$

where $P(A|B)$ is the probability of A *given* B, i.e. *after* B is observed

$P(A)$ and $P(B)$ are the probabilities of A and B without regard to each other

and $P(B|A)$ is the probability of observing event B given that A is true

In the Bayes' therorem formula, 'A' represents a theory or hypothesis that is to be tested, and 'B' represents a new piece of evidence that seems to confirm or disprove the theory.
Bayes' theorem is demonstrated in the following worked problem.

Problem 18. An outdoor degree ceremony is taking place tomorrow, 5 July, in the hot climate of Dubai. In recent years it has rained only 2 days in the four-month period June to September. However, the weather forecaster has predicted rain for tomorrow. When it actually rains, the weatherman correctly forecasts rain 85% of the time. When it doesn't rain, he incorrectly forecasts rain 15% of the time. Determine the probability that it will rain tomorrow

There are two possible mutually-exclusive events occurring here – it either rains or it does not rain.
Also, a third event occurs when the weatherman predicts rain.

Let the notation for these events be:

Event A_1 It rains at the ceremony

Event A_2 It does not rain at the ceremony

Event B The weatherman predicts rain

The probability values are:

$P(A_1) = \frac{2}{30+31+31+30} = \frac{1}{61}$ (i.e. it rains 2 days in the months June to September)

$P(A_2) = \frac{120}{30+31+31+30} = \frac{60}{61}$ (i.e. it does not rain for 120 of the 122 days in the months June to September)

$P(B|A_1) = 0.85$ (i.e. when it rains, the weatherman predicts rain 85% of the time)

$P(B|A_2) = 0.15$ (i.e. when it does not rain, the weatherman predicts rain 15% of the time)

Using Bayes' theorem to determine the probability that it will rain tomorrow, given the forecast of rain by the weatherman:

$$\boldsymbol{P(A_1|B) = \frac{P(B|A_1)P(A_1)}{P(B|A_1)P(A_1) + P(B|A_2)P(A_2)}}$$

$$= \frac{(0.85)\left(\frac{1}{61}\right)}{0.85 \times \frac{1}{61} + 0.15 \times \frac{60}{61}} = \frac{0.0139344}{0.1614754}$$

$$= \mathbf{0.0863} \text{ or } \mathbf{8.63\%}$$

Even when the weatherman predicts rain, it rains only between 8% and 9% of the time. **Hence, there is a good chance it will not rain tomorrow in Dubai for the degree ceremony**.

Now try the following Practice Exercise

Practice Exercise 292 Bayes' theorem (Answers on page 736)

1. Machines X, Y and Z produce similar vehicle engine parts. Of the total output, machine X produces 35%, machine Y 20% and machine Z 45%. The proportions of the output from each machine that do not conform to the specification are 7% for X, 4% for Y and 3% for Z. Determine the proportion of those parts that do not conform to the specification that are produced by machine X.
2. A doctor is called to see a sick child. The doctor has prior information that 85% of sick children in that area have the 'flu, while the other 15% are sick with measles. For simplicity, assume that there are no other maladies in that area. A symptom of measles is a rash. The probability of children developing a rash and having measles is 94% and the probability of children with flu occasionally also developing a rash is 7%. Upon examining the child, the doctor finds a rash. Determine the probability that the child has measles.
3. In a study, 100 oncologists were asked what the odds of breast cancer would be in a woman who was initially thought to have a 1% risk of cancer but who ended up with a positive mammogram result (a mammogram accurately classifies about 80% of cancerous tumours; 10% of mammograms detect breast cancer when it's not there, and therefore 90% correctly return a negative result). 95 of oncologists estimated the probability of cancer to be about 75%. Use Bayes' theorem to determine the probability of cancer.

Practice Exercise 293 Multiple-choice questions on probability (Answers on page 736)

Each question has only one correct answer

1. In a box of 50 nails, 4 are faulty. One nail is taken from the box at random. The probability that the nail is faulty is:
(a) $\frac{1}{25}$ (b) $\frac{2}{27}$ (c) $\frac{2}{25}$ (d) $\frac{23}{25}$
2. Using a standard deck of 52 cards, the percentage probability of drawing a three or a king or a queen is:
(a) 5.77% (b) 1.18%
(c) 17.31% (d) 23.08%
3. The percentage probability of selecting at random the winning horses in all of three consecutive races if there are 8 horses in each race is:
(a) 12.50% (b) 1.563%
(c) 0.195% (d) 37.5%

4. In a bag are 12 red balls and 8 blue balls. If one ball is drawn from the bag, then a second ball drawn without returning the first ball to the bag, the percentage probability of drawing two blue balls is:
   (a) 76.84% (b) 14.74%
   (c) 80% (d) 16%

Questions 5 to 7 relate to the following information.
The probability of a component failing in one year due to excessive temperature is $\frac{1}{16}$, due to excessive vibration is $\frac{1}{20}$ and due to excessive humidity is $\frac{1}{40}$

5. The probability that a component fails due to excessive temperature and excessive vibration is:
   (a) $\frac{285}{320}$ (b) $\frac{1}{320}$ (c) $\frac{9}{80}$ (d) $\frac{1}{800}$
6. The probability that a component fails due to excessive vibration or excessive humidity is:
   (a) 0.00125 (b) 0.00257
   (c) 0.1125 (d) 0.0750
7. The probability that a component will not fail because of both excessive temperature and excessive humidity is:
   (a) 0.914 (b) 1.913
   (c) 0.00156 (d) 0.0875

Questions 8 to 10 relate to the following information:
A box contains 35 brass washers, 40 steel washers and 25 aluminium washers. 3 washers are drawn at random from the box without replacement.

8. The probability that all three are steel washers is:
   (a) 0.0611 (b) 1.200
   (c) 0.0640 (d) 1.182
9. The probability that there are no aluminium washers is:
   (a) 2.250 (b) 0.418
   (c) 0.014 (d) 0.422
10. The probability that there are two brass washers and either a steel or an aluminium washer is:
   (a) 0.071 (b) 0.687
   (c) 0.239 (d) 0.343

For fully worked solutions to each of the problems in Practice Exercises 289 to 292 in this chapter, go to the website:
www.routledge.com/cw/bird

COMPANION @ WEBSITE

Revision Test 17 Presentation of statistical data, mean, median, mode, standard deviation and probability

This Revision Test covers the material in Chapters 60 to 62. *The marks for each question are shown in brackets at the end of each question.*

1. A company produces five products in the following proportions:
   Product A 24 Product B 16 Product C 15 Product D 11 Product E 6
   Present these data visually by drawing (a) a vertical bar chart, (b) a percentage bar chart, (c) a pie diagram. (13)

2. The following lists the diameters of 40 components produced by a machine, each measured correct to the nearest hundredth of a centimetre:

1.39	1.36	1.38	1.31	1.33	1.40	1.28
1.40	1.24	1.28	1.42	1.34	1.43	1.35
1.36	1.36	1.35	1.45	1.29	1.39	1.38
1.38	1.35	1.42	1.30	1.26	1.37	1.33
1.37	1.34	1.34	1.32	1.33	1.30	1.38
1.41	1.35	1.38	1.27	1.37		

   (a) Using 8 classes form a frequency distribution and a cumulative frequency distribution
   (b) For the above data draw a histogram, a frequency polygon and an ogive (21)

3. Determine for the 10 measurements of lengths shown below:

   (a) the arithmetic mean, (b) the median, (c) the mode and (d) the standard deviation.

   28 m, 20 m, 32 m, 44 m, 28 m, 30 m, 30 m, 26 m, 28 m and 34 m (9)

4. The heights of 100 people are measured correct to the nearest centimetre with the following results:

150–157 cm	5	158–165 cm	18
166–173 cm	42	174–181 cm	27
182–189 cm	8		

   Determine for the data (a) the mean height and (b) the standard deviation. (10)

5. Determine the probabilities of:
   (a) drawing a white ball from a bag containing 6 black and 14 white balls
   (b) winning a prize in a raffle by buying 6 tickets when a total of 480 tickets are sold
   (c) selecting at random a female from a group of 12 boys and 28 girls
   (d) winning a prize in a raffle by buying 8 tickets when there are 5 prizes and a total of 800 tickets are sold (8)

6. In a box containing 120 similar transistors 70 are satisfactory, 37 give too high a gain under normal operating conditions and the remainder give too low a gain.

   Calculate the probability that when drawing two transistors in turn, at random, **with replacement**, of having (a) two satisfactory, (b) none with low gain, (c) one with high gain and one satisfactory, (d) one with low gain and none satisfactory.

   Determine the probabilities in (a), (b) and (c) above if the transistors are drawn **without replacement**. (14)

For lecturers/instructors/teachers, fully worked solutions to each of the problems in Revision Test 17, together with a full marking scheme, are available at the website:
www.routledge.com/cw/bird

Chapter 63

The binomial and Poisson distribution

Why it is important to understand: **The binomial and Poisson distributions**

The binomial distribution is used only when both of two conditions are met – the test has only two possible outcomes, and the sample must be random. If both of these conditions are met, then this distribution may be used to predict the probability of a desired result. For example, a binomial distribution may be used in determining whether a new drug being tested has or has not contributed to alleviating symptoms of a disease. Common applications of this distribution range from scientific and engineering applications to military and medical ones, in quality assurance, genetics and in experimental design. A Poisson distribution has several applications, and is essentially a derived limiting case of the binomial distribution. It is most applicably relevant to a situation in which the total number of successes is known, but the number of trials is not. An example of such a situation would be if the mean expected number of cancer cells present per sample is known and it was required to determine the probability of finding 1.5 times that amount of cells in any given sample; this is an example when the Poisson distribution would be used. The Poisson distribution has widespread applications in analysing traffic flow, in fault prediction on electric cables, in the prediction of randomly occurring accidents and in reliability engineering.

At the end of this chapter, you should be able to:

- define the binomial distribution
- use the binomial distribution
- apply the binomial distribution to industrial inspection
- draw a histogram of probabilities
- define the Poisson distribution
- apply the Poisson distribution to practical situations

63.1 The binomial distribution

The binomial distribution deals with two numbers only, these being the probability that an event will happen, p, and the probability that an event will not happen, q. Thus, when a coin is tossed, if p is the probability of the coin landing with a head upwards, q is the probability of the coin landing with a tail upwards. $p+q$ must always be equal to unity. A binomial distribution can be used for finding, say, the probability of getting three heads in seven tosses of the coin, or in industry for determining defect rates as a result of sampling. One way of defining a binomial distribution is as follows:

> *'if p is the probability that an event will happen and q is the probability that the event will not happen, then the probabilities that the event will happen 0, 1, 2, 3, ..., n times in n trials are given by the successive terms of the expansion of $(q+p)^n$ taken from left to right'.*

The binomial expansion of $(q+p)^n$ is:

$$q^n + nq^{n-1}p + \frac{n(n-1)}{2!}q^{n-2}p^2 + \frac{n(n-1)(n-2)}{3!}q^{n-3}p^3 + \cdots + p^n$$

from Chapter 16.
This concept of a binomial distribution is used in Problems 1 and 2.

Problem 1. Determine the probabilities of having (a) at least 1 girl and (b) at least 1 girl and 1 boy in a family of 4 children, assuming equal probability of male and female birth

The probability of a girl being born, p, is 0.5 and the probability of a girl not being born (male birth), q, is also 0.5. The number in the family, n, is 4. From above, the probabilities of 0, 1, 2, 3, 4 girls in a family of 4 are given by the successive terms of the expansion of $(q+p)^4$ taken from left to right.

From the binomial expansion:

$$(q+p)^4 = q^4 + 4q^3p + 6q^2p^2 + 4qp^3 + p^4$$

Hence the probability of no girls is q^4,

i.e. $\quad 0.5^4 = 0.0625$

the probability of 1 girl is $4q^3p$,

i.e. $\quad 4 \times 0.5^3 \times 0.5 = 0.2500$

the probability of 2 girls is $6q^2p^2$,

i.e. $\quad 6 \times 0.5^2 \times 0.5^2 = 0.3750$

the probability of 3 girls is $4qp^3$,

i.e. $\quad 4 \times 0.5 \times 0.5^3 = 0.2500$

the probability of 4 girls is p^4,

i.e. $\quad 0.5^4 = 0.0625$

Total probability, $(q+p)^4 = 1.0000$

(a) The probability of having at least one girl is the sum of the probabilities of having 1, 2, 3 and 4 girls, i.e.

$$0.2500 + 0.3750 + 0.2500 + 0.0625 = \mathbf{0.9375}$$

(Alternatively, the probability of having at least 1 girl is: 1 – (the probability of having no girls), i.e. $1 - 0.0625$, giving **0.9375**, as obtained previously.)

(b) The probability of having at least 1 girl and 1 boy is given by the sum of the probabilities of having: 1 girl and 3 boys, 2 girls and 2 boys and 3 girls and 1 boy, i.e.

$$0.2500 + 0.3750 + 0.2500 = \mathbf{0.8750}$$

(Alternatively, this is also the probability of having $1 - $(probability of having no girls + probability of having no boys), i.e. $1 - 2 \times 0.0625 = \mathbf{0.8750}$, as obtained previously.)

Problem 2. A dice is rolled 9 times. Find the probabilities of having a 4 upwards (a) 3 times and (b) less than 4 times

Let p be the probability of having a 4 upwards. Then $p = 1/6$, since dice have six sides.
Let q be the probability of not having a 4 upwards. Then $q = 5/6$. The probabilities of having a 4 upwards 0, 1, $2 \ldots n$ times are given by the successive terms of the expansion of $(q+p)^n$, taken from left to right. From the binomial expansion:

$$(q+q)^9 = q^9 + 9q^8p + 36q^7p^2 + 84q^6p^3 + \cdots$$

The probability of having a 4 upwards no times is

$$q^9 = (5/6)^9 = 0.1938$$

The probability of having a 4 upwards once is

$$9q^8p = 9(5/6)^8(1/6) = 0.3489$$

The probability of having a 4 upwards twice is

$$36q^7p^2 = 36(5/6)^7(1/6)^2 = 0.2791$$

The probability of having a 4 upwards 3 times is

$$84q^6p^3 = 84(5/6)^6(1/6)^3 = 0.1302$$

(a) The probability of having a 4 upwards 3 times is **0.1302**

(b) The probability of having a 4 upwards less than 4 times is the sum of the probabilities of having a 4 upwards 0, 1, 2, and 3 times, i.e.

$$0.1938 + 0.3489 + 0.2791 + 0.1302 = \mathbf{0.9520}$$

Industrial inspection

In industrial inspection, p is often taken as the probability that a component is defective and q is the probability that the component is satisfactory. In this case, a binomial distribution may be defined as:

> *'the probabilities that 0, 1, 2, 3, …, n components are defective in a sample of n components, drawn at random from a large batch of components, are given by the successive terms of the expansion of $(q+p)^n$, taken from left to right'.*

This definition is used in Problems 3 and 4.

Problem 3. A machine is producing a large number of bolts automatically. In a box of these bolts, 95% are within the allowable tolerance values with respect to diameter, the remainder being outside of the diameter tolerance values. Seven bolts are drawn at random from the box. Determine the probabilities that (a) two and (b) more than two of the seven bolts are outside of the diameter tolerance values

Let p be the probability that a bolt is outside of the allowable tolerance values, i.e. is defective, and let q be the probability that a bolt is within the tolerance values, i.e. is satisfactory. Then $p = 5\%$, i.e. 0.05 per unit and $q = 95\%$, i.e. 0.95 per unit. The sample number is 7.

The probabilities of drawing $0, 1, 2, \ldots, n$ defective bolts are given by the successive terms of the expansion of $(q+p)^n$, taken from left to right. In this problem

$$(q+p)^n = (0.95 + 0.05)^7$$
$$= 0.95^7 + 7 \times 0.95^6 \times 0.05$$
$$+ 21 \times 0.95^5 \times 0.05^2 + \cdots$$

Thus the probability of no defective bolts is:

$$0.95^7 = 0.6983$$

The probability of 1 defective bolt is:

$$7 \times 0.95^6 \times 0.05 = 0.2573$$

The probability of 2 defective bolts is:

$$21 \times 0.95^5 \times 0.05^2 = 0.0406 \text{ and so on.}$$

(a) The probability that two bolts are outside of the diameter tolerance values is **0.0406**

(b) To determine the probability that more than two bolts are defective, the sum of the probabilities of 3 bolts, 4 bolts, 5 bolts, 6 bolts and 7 bolts being defective can be determined. An easier way to find this sum is to find 1 − (sum of 0 bolts, 1 bolt and 2 bolts being defective), since the sum of all the terms is unity. Thus, the probability of there being more than two bolts outside of the tolerance values is:

$$1 - (0.6983 + 0.2573 + 0.0406), \text{ i.e. } \mathbf{0.0038}$$

Problem 4. A package contains 50 similar components and inspection shows that four have been damaged during transit. If six components are drawn at random from the contents of the package determine the probabilities that in this sample (a) one and (b) less than three are damaged

The probability of a component being damaged, p, is 4 in 50, i.e. 0.08 per unit. Thus, the probability of a component not being damaged, q, is $1 - 0.08$, i.e. 0.92 The probabilities of there being $0, 1, 2, \ldots, 6$ damaged components are given by the successive terms of $(q+p)^6$, taken from left to right.

$$(q+p)^6 = q^6 + 6q^5p + 15q^4p^2 + 20q^3p^3 + \cdots$$

(a) The probability of one damaged component is

$$6q^5p = 6 \times 0.92^5 \times 0.08 = \mathbf{0.3164}$$

(b) The probability of less than three damaged components is given by the sum of the probabilities of 0, 1 and 2 damaged components.

$$q^6 + 6q^5p + 15q^4p^2$$
$$= 0.92^6 + 6 \times 0.92^5 \times 0.08$$
$$+ 15 \times 0.92^4 \times 0.08^2$$
$$= 0.6064 + 0.3164 + 0.0688 = \mathbf{0.9916}$$

Histogram of probabilities

The terms of a binomial distribution may be represented pictorially by drawing a histogram, as shown in Problem 5.

Problem 5. The probability of a student successfully completing a course of study in three years is 0.45. Draw a histogram showing the probabilities of 0, 1, 2, ..., 10 students successfully completing the course in three years

Let p be the probability of a student successfully completing a course of study in three years and q be the probability of not doing so. Then $p=0.45$ and $q=0.55$. The number of students, n, is 10.

The probabilities of 0, 1, 2, ..., 10 students successfully completing the course are given by the successive terms of the expansion of $(q+p)^{10}$, taken from left to right.

$$(q+p)^{10} = q^{10} + 10q^9p + 45q^8p^2 + 120q^7p^3 + 210q^6p^4 + 252q^5p^5 + 210q^4p^6 + 120q^3p^7 + 45q^2p^8 + 10qp^9 + p^{10}$$

Substituting $q=0.55$ and $p=0.45$ in this expansion gives the values of the success terms as:

0.0025, 0.0207, 0.0763, 0.1665, 0.2384, 0.2340, 0.1596, 0.0746, 0.0229, 0.0042 and 0.0003. The histogram depicting these probabilities is shown in Fig. 63.1.

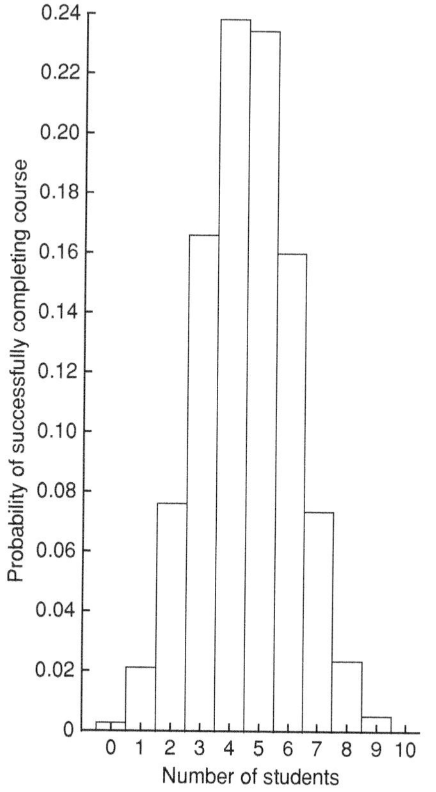

Figure 63.1

Now try the following Practice Exercise

Practice Exercise 294 The binomial distribution (Answers on page 736)

1. Concrete blocks are tested and it is found that, on average, 7% fail to meet the required specification. For a batch of 9 blocks, determine the probabilities that (a) three blocks and (b) less than four blocks will fail to meet the specification
2. If the failure rate of the blocks in Problem 1 rises to 15%, find the probabilities that (a) no blocks and (b) more than two blocks will fail to meet the specification in a batch of 9 blocks
3. The average number of employees absent from a firm each day is 4%. An office within the firm has seven employees. Determine the probabilities that (a) no employee and (b) three employees will be absent on a particular day
4. A manufacturer estimates that 3% of his output of a small item is defective. Find the probabilities that in a sample of 10 items (a) less than two and (b) more than two items will be defective
5. Five coins are tossed simultaneously. Determine the probabilities of having 0, 1, 2, 3, 4 and 5 heads upwards, and draw a histogram depicting the results
6. If the probability of rain falling during a particular period is 2/5, find the probabilities of having 0, 1, 2, 3, 4, 5, 6 and 7 wet days in a week. Show these results on a histogram
7. An automatic machine produces, on average, 10% of its components outside of the tolerance required. In a sample of 10 components from this machine, determine the probability of having three components outside of the tolerance required by assuming a binomial distribution

63.2 The Poisson distribution

When the number of trials, n, in a binomial distribution becomes large (usually taken as larger than 10), the calculations associated with determining the values of the terms become laborious. If n is large and p is small, and the product np is less than 5, a very good approximation to a binomial distribution is given by the corresponding **Poisson*** **distribution**, in which calculations are usually simpler.

The Poisson approximation to a binomial distribution may be defined as follows:

'the probabilities that an event will happen 0, 1, 2, 3, …, n times in n trials are given by the successive terms of the expression

$$e^{-\lambda}\left(1+\lambda+\frac{\lambda^2}{2!}+\frac{\lambda^3}{3!}+\cdots\right)$$

taken from left to right'

The symbol λ is the expectation of an event happening and is equal to np.

Problem 6. If 3% of the gearwheels produced by a company are defective, determine the probabilities that in a sample of 80 gearwheels (a) two and (b) more than two will be defective

*Who was Poisson? – **Siméon Denis Poisson** (21 June 1781-25 April 1840), was a French mathematician, geometer and physicist. His work on the theory of electricity and magnetism virtually created a new branch of mathematical physics, and his study of celestial mechanics discussed the stability of the planetary orbits. To find out more go to www.routledge.com/cw/bird

The sample number, n, is large, the probability of a defective gearwheel, p, is small and the product np is 80×0.03, i.e. 2.4, which is less than 5. Hence a Poisson approximation to a binomial distribution may be used. The expectation of a defective gearwheel, $\lambda = np = 2.4$

The probabilities of 0, 1, 2, … defective gearwheels are given by the successive terms of the expression

$$e^{-\lambda}\left(1+\lambda+\frac{\lambda^2}{2!}+\frac{\lambda^3}{3!}+\cdots\right)$$

taken from left to right, i.e. by

$$e^{-\lambda}, \lambda e^{-\lambda}, \frac{\lambda^2 e^{-\lambda}}{2!}, \ldots \quad \text{Thus:}$$

probability of no defective gearwheels is

$$e^{-\lambda} = e^{-2.4} = 0.0907$$

probability of 1 defective gearwheel is

$$\lambda e^{-\lambda} = 2.4e^{-2.4} = 0.2177$$

probability of 2 defective gearwheels is

$$\frac{\lambda^2 e^{-\lambda}}{2!} = \frac{2.4^2 e^{-2.4}}{2 \times 1} = 0.2613$$

(a) The probability of having 2 defective gearwheels is **0.2613**

(b) The probability of having more than 2 defective gearwheels is 1 – (the sum of the probabilities of having 0, 1, and 2 defective gearwheels), i.e.

$$1 - (0.0907 + 0.2177 + 0.2613),$$

that is, **0.4303**

The principal use of a Poisson distribution is to determine the theoretical probabilities when p, the probability of an event happening, is known, but q, the probability of the event not happening is unknown. For example, the average number of goals scored per match by a football team can be calculated, but it is not possible to quantify the number of goals that were not scored. In this type of problem, a Poisson distribution may be defined as follows:

'the probabilities of an event occurring 0, 1, 2, 3…times are given by the successive terms of the expression

$$e^{-\lambda}\left(1+\lambda+\frac{\lambda^2}{2!}+\frac{\lambda^3}{3!}+\cdots\right),$$

taken from left to right'

The symbol λ is the value of the average occurrence of the event.

Problem 7. A production department has 35 similar milling machines. The number of breakdowns on each machine averages 0.06 per week. Determine the probabilities of having (a) one, and (b) less than three machines breaking down in any week

Since the average occurrence of a breakdown is known but the number of times when a machine did not break down is unknown, a Poisson distribution must be used.

The expectation of a breakdown for 35 machines is 35×0.06, i.e. 2.1 breakdowns per week. The probabilities of a breakdown occurring 0, 1, 2, ... times are given by the successive terms of the expression

$$e^{-\lambda}\left(1+\lambda+\frac{\lambda^2}{2!}+\frac{\lambda^3}{3!}+\cdots\right),$$

taken from left to right. Hence:

probability of no breakdowns

$$e^{-\lambda}=e^{-2.1}=0.1225$$

probability of 1 breakdown is

$$\lambda e^{-\lambda}=2.1e^{-2.1}=0.2572$$

probability of 2 breakdowns is

$$\frac{\lambda^2 e^{-\lambda}}{2!}=\frac{2.1^2 e^{-2.1}}{2\times 1}=0.2700$$

(a) The probability of 1 breakdown per week is **0.2572**

(b) The probability of less than 3 breakdowns per week is the sum of the probabilities of 0, 1 and 2 breakdowns per week,

i.e. $0.1225+0.2572+0.2700=\mathbf{0.6497}$

Histogram of probabilities

The terms of a Poisson distribution may be represented pictorially by drawing a histogram, as shown in Problem 8.

Problem 8. The probability of a person having an accident in a certain period of time is 0.0003. For a population of 7500 people, draw a histogram showing the probabilities of 0, 1, 2, 3, 4, 5 and 6 people having an accident in this period

The probabilities of 0, 1, 2, ... people having an accident are given by the successive terms of the expression

$$e^{-\lambda}\left(1+\lambda+\frac{\lambda^2}{2!}+\frac{\lambda^3}{3!}+\cdots\right),$$

taken from left to right.

The average occurrence of the event, λ, is 7500×0.0003, i.e. 2.25

The probability of no people having an accident is

$$e^{-\lambda}=e^{-2.25}=0.1054$$

The probability of 1 person having an accident is

$$\lambda e^{-\lambda}=2.25e^{-2.25}=0.2371$$

The probability of 2 people having an accident is

$$\frac{\lambda^2 e^{-\lambda}}{2!}=\frac{2.25^2 e^{-2.25}}{2!}=0.2668$$

and so on, giving probabilities of 0.2001, 0.1126, 0.0506 and 0.0190 for 3, 4, 5 and 6 respectively having an accident. The histogram for these probabilities is shown in Fig. 63.2.

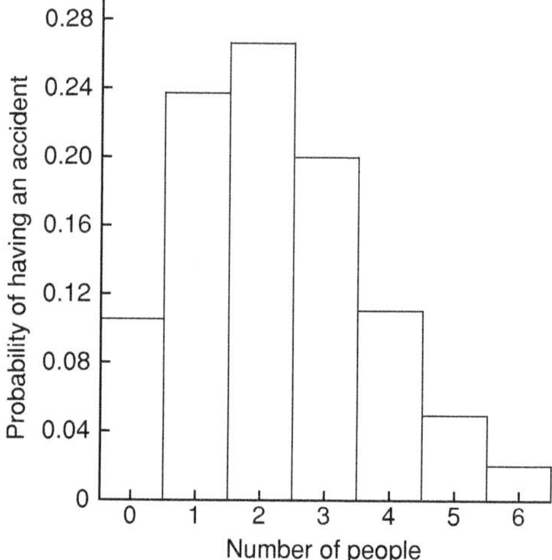

Figure 63.2

Now try the following Practice Exercise

Practice Exercise 295 The Poisson distribution (Answers on page 736)

1. In problem 7 of Exercise 294, page 652, determine the probability of having three components outside of the required tolerance using the Poisson distribution
2. The probability that an employee will go to hospital in a certain period of time is 0.0015. Use a Poisson distribution to determine the probability of more than two employees going to hospital during this period of time if there are 2000 employees on the payroll
3. When packaging a product, a manufacturer finds that one packet in twenty is underweight. Determine the probabilities that in a box of 72 packets (a) two and (b) less than four will be underweight
4. A manufacturer estimates that 0.25% of his output of a component are defective. The components are marketed in packets of 200. Determine the probability of a packet containing less than three defective components
5. The demand for a particular tool from a store is, on average, five times a day and the demand follows a Poisson distribution. How many of these tools should be kept in the stores so that the probability of there being one available when required is greater than 10%?
6. Failure of a group of particular machine tools follows a Poisson distribution with a mean value of 0.7. Determine the probabilities of 0, 1, 2, 3, 4 and 5 failures in a week and present these results on a histogram

Practice Exercise 296 Multiple-choice questions on the binomial and Poisson distribution (Answers on page 736)

Each question has only one correct answer

1. 2% of the components produced by a manufacturer are defective. Using the Poisson distribution, the percentage probability that more than two will be defective in a sample of 100 components is:

   (a) 13.5% (b) 32.3% (c) 27.1% (d) 59.4%
2. A manufacturer estimates that 4% of components produced are defective. Using the binomial distribution, the percentage probability that more than two components will be defective in a sample of 10 components is:

   (a) 0.40% (b) 5.19% (c) 0.63% (d) 99.4%
3. A box contains 60 similar resistors and inspection shows that 3 have been damaged in transit. Five resistors are drawn at random from the box. Using the binomial distribution, the probability that in this sample one is damaged is:

   (a) 0.2036 (b) 0.7738 (c) 0.0214 (d) 0.9988
4. A machine shop has 40 similar lathes. The number of breakdowns on each machine averages 0.05 per week. Using the Poisson distribution, the probability of one machine breaking down in any week is:

   (a) 0.1353 (b) 0.5413 (c) 0.2707 (d) 0.1804
5. In a large car plant the probability of a worker falling ill in a certain period of time is 0.00012. For a workforce of 6500, the percentage probability of two workers falling ill in that time, using the Poisson distribution, is:

   (a) 27.89% (b) 1.39% (c) 35.76% (d) 13.94%

For fully worked solutions to each of the problems in Practice Exercises 294 and 295 in this chapter, go to the website:
www.routledge.com/cw/bird

Chapter 64

The normal distribution

Why it is important to understand: The normal distribution

A normal distribution is a very important statistical data distribution pattern occurring in many natural phenomena, such as height, blood pressure, lengths of objects produced by machines, marks in a test, errors in measurements and so on. In general, when data is gathered, we expect to see a particular pattern to the data, called a *normal distribution.* This is a distribution where the data is evenly distributed around the mean in a very regular way, which when plotted as a histogram will result in a *bell curve.* The normal distribution is the most important of all probability distributions; it is applied directly to many practical problems in every engineering discipline. There are two principal applications of the normal distribution to engineering and reliability. One application deals with the analysis of items which exhibit failure to wear, such as mechanical devices – frequently the wear-out failure distribution is sufficiently close to normal that the use of this distribution for predicting or assessing reliability is valid. Another application is in the analysis of manufactured items and their ability to meet specifications. No two parts made to the same specification are exactly alike; the variability of parts leads to a variability in systems composed of those parts. The design must take this variability into account, otherwise the system may not meet the specification requirement due to the combined effect of part variability.

At the end of this chapter, you should be able to:

- recognise a normal curve
- use the normal distribution in calculations
- test for a normal distribution using probability paper

64.1 Introduction to the normal distribution

When data is obtained, it can frequently be considered to be a sample (i.e. a few members) drawn at random from a large population (i.e. a set having many members). If the sample number is large, it is theoretically possible to choose class intervals which are very small, but which still have a number of members falling within each class. A frequency polygon of this data then has a large number of small line segments and approximates to a continuous curve. Such a curve is called a **frequency** or a **distribution curve**.

An extremely important symmetrical distribution curve is called the **normal curve** and is as shown in Fig. 64.1. This curve can be described by a mathematical equation and is the basis of much of the work done in more advanced statistics. Many natural occurrences such as the heights or weights of a group of people, the sizes of components produced by a particular machine and the life length of certain components approximate to a normal distribution.

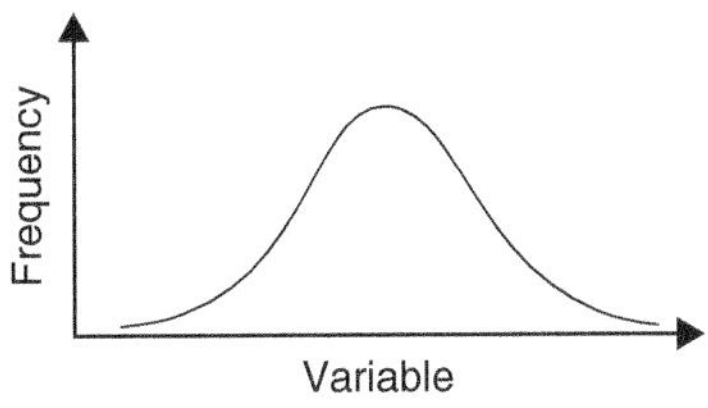

Figure 64.1

Normal distribution curves can differ from one another in the following four ways:

(a) by having different mean values
(b) by having different values of standard deviations
(c) the variables having different values and different units
(d) by having different areas between the curve and the horizontal axis.

A normal distribution curve is **standardised** as follows:

(a) The mean value of the unstandardised curve is made the origin, thus making the mean value, $\bar{x}$, zero.
(b) The horizontal axis is scaled in standard deviations. This is done by letting $z = \frac{x - \bar{x}}{\sigma}$, where z is called the **normal standard variate**, x is the value of the variable, $\bar{x}$ is the mean value of the distribution and σ is the standard deviation of the distribution.
(c) The area between the normal curve and the horizontal axis is made equal to unity.

When a normal distribution curve has been standardised, the normal curve is called a **standardised normal curve** or a **normal probability curve**, and any normally distributed data may be represented by the **same** normal probability curve.

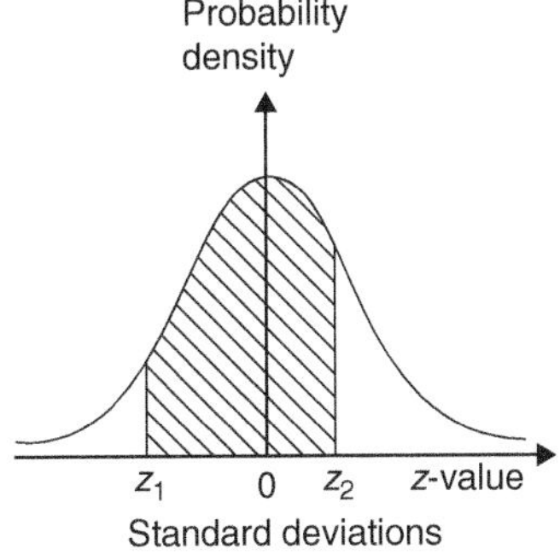

Figure 64.2

The area under part of a normal probability curve is directly proportional to probability and the value of the shaded area shown in Fig. 64.2 can be determined by evaluating:

$$\int \frac{1}{\sqrt{2\pi}} e^{(z^2/2)} dz, \quad \text{where } z = \frac{x - \bar{x}}{\sigma}$$

To save repeatedly determining the values of this function, tables of partial areas under the standardised normal curve are available in many mathematical formulae books, and such a table is shown in Table 64.1.

Problem 1. The mean height of 500 people is 170 cm and the standard deviation is 9 cm. Assuming the heights are normally distributed, determine the number of people likely to have heights between 150 cm and 195 cm

The mean value, $\bar{x}$, is 170 cm and corresponds to a normal standard variate value, z, of zero on the standardised normal curve. A height of 150 cm has a z-value given by $z = \frac{x - \bar{x}}{\sigma}$ standard deviations, i.e. $\frac{150 - 170}{9}$ or -2.22 standard deviations. Using a table of partial areas beneath the standardised normal curve (see Table 64.1), a z-value of -2.22 corresponds to an area of 0.4868 between the mean value and the ordinate $z = -2.22$. The negative z-value shows that it lies to the left of the $z = 0$ ordinate.
This area is shown shaded in Fig. 64.3(a) on page 660.
Similarly, 195 cm has a z-value of $\frac{195 - 170}{9}$ that is 2.78 standard deviations. From Table 64.1, this value of z corresponds to an area of 0.4973, the positive value of z showing that it lies to the right of the $z = 0$ ordinate. This area is shown shaded in Fig. 64.3(b). The total area shaded in Fig. 64.3(a) and (b) is shown in Fig. 64.3(c) and is $0.4868 + 0.4973$, i.e. 0.9841 of the total area beneath the curve.
However, the area is directly proportional to probability. Thus, the probability that a person will have a height of between 150 and 195 cm is 0.9841. For a group of 500 people, 500×0.9841, i.e. **492 people are likely to have heights in this range**. The value of 500×0.9841 is 492.05, but since answers based on a normal probability distribution can only be approximate, results are usually given correct to the nearest whole number.

Table 64.1 Partial areas under the standardised normal curve

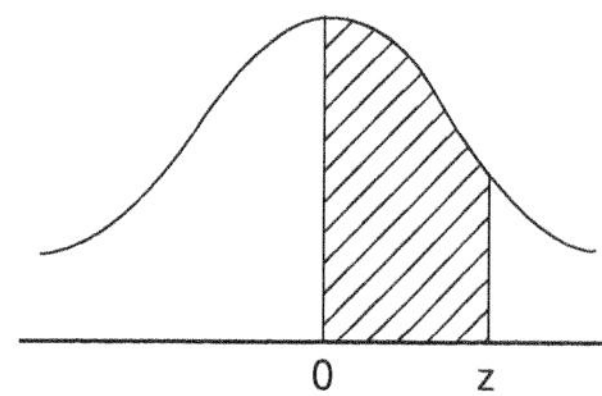

$z = \frac{x - \bar{x}}{\sigma}$	0	1	2	3	4	5	6	7	8	9
0.0	0.0000	0.0040	0.0080	0.0120	0.0159	0.0199	0.0239	0.0279	0.0319	0.0359
0.1	0.0398	0.0438	0.0478	0.0517	0.0557	0.0596	0.0636	0.0678	0.0714	0.0753
0.2	0.0793	0.0832	0.0871	0.0910	0.0948	0.0987	0.1026	0.1064	0.1103	0.1141
0.3	0.1179	0.1217	0.1255	0.1293	0.1331	0.1388	0.1406	0.1443	0.1480	0.1517
0.4	0.1554	0.1591	0.1628	0.1664	0.1700	0.1736	0.1772	0.1808	0.1844	0.1879
0.5	0.1915	0.1950	0.1985	0.2019	0.2054	0.2086	0.2123	0.2157	0.2190	0.2224
0.6	0.2257	0.2291	0.2324	0.2357	0.2389	0.2422	0.2454	0.2486	0.2517	0.2549
0.7	0.2580	0.2611	0.2642	0.2673	0.2704	0.2734	0.2760	0.2794	0.2823	0.2852
0.8	0.2881	0.2910	0.2939	0.2967	0.2995	0.3023	0.3051	0.3078	0.3106	0.3133
0.9	0.3159	0.3186	0.3212	0.3238	0.3264	0.3289	0.3315	0.3340	0.3365	0.3389
1.0	0.3413	0.3438	0.3451	0.3485	0.3508	0.3531	0.3554	0.3577	0.3599	0.3621
1.1	0.3643	0.3665	0.3686	0.3708	0.3729	0.3749	0.3770	0.3790	0.3810	0.3830
1.2	0.3849	0.3869	0.3888	0.3907	0.3925	0.3944	0.3962	0.3980	0.3997	0.4015
1.3	0.4032	0.4049	0.4066	0.4082	0.4099	0.4115	0.4131	0.4147	0.4162	0.4177
1.4	0.4192	0.4207	0.4222	0.4236	0.4251	0.4265	0.4279	0.4292	0.4306	0.4319
1.5	0.4332	0.4345	0.4357	0.4370	0.4382	0.4394	0.4406	0.4418	0.4430	0.4441
1.6	0.4452	0.4463	0.4474	0.4484	0.4495	0.4505	0.4515	0.4525	0.4535	0.4545
1.7	0.4554	0.4564	0.4573	0.4582	0.4591	0.4599	0.4608	0.4616	0.4625	0.4633
1.8	0.4641	0.4649	0.4656	0.4664	0.4671	0.4678	0.4686	0.4693	0.4699	0.4706
1.9	0.4713	0.4719	0.4726	0.4732	0.4738	0.4744	0.4750	0.4756	0.4762	0.4767
2.0	0.4772	0.4778	0.4783	0.4785	0.4793	0.4798	0.4803	0.4808	0.4812	0.4817
2.1	0.4821	0.4826	0.4830	0.4834	0.4838	0.4842	0.4846	0.4850	0.4854	0.4857

(*Continued*)

Table 64.1 (*Continued*)

$z=\frac{x-\bar{x}}{\sigma}$	0	1	2	3	4	5	6	7	8	9
2.2	0.4861	0.4864	0.4868	0.4871	0.4875	0.4878	0.4881	0.4884	0.4887	0.4890
2.3	0.4893	0.4896	0.4898	0.4901	0.4904	0.4906	0.4909	0.4911	0.4913	0.4916
2.4	0.4918	0.4920	0.4922	0.4925	0.4927	0.4929	0.4931	0.4932	0.4934	0.4936
2.5	0.4938	0.4940	0.4941	0.4943	0.4945	0.4946	0.4948	0.4949	0.4951	0.4952
2.6	0.4953	0.4955	0.4956	0.4957	0.4959	0.4960	0.4961	0.4962	0.4963	0.4964
2.7	0.4965	0.4966	0.4967	0.4968	0.4969	0.4970	0.4971	0.4972	0.4973	0.4974
2.8	0.4974	0.4975	0.4976	0.4977	0.4977	0.4978	0.4979	0.4980	0.4980	0.4981
2.9	0.4981	0.4982	0.4982	0.4983	0.4984	0.4984	0.4985	0.4985	0.4986	0.4986
3.0	0.4987	0.4987	0.4987	0.4988	0.4988	0.4989	0.4989	0.4989	0.4990	0.4990
3.1	0.4990	0.4991	0.4991	0.4991	0.4992	0.4992	0.4992	0.4992	0.4993	0.4993
3.2	0.4993	0.4993	0.4994	0.4994	0.4994	0.4994	0.4994	0.4995	0.4995	0.4995
3.3	0.4995	0.4995	0.4995	0.4996	0.4996	0.4996	0.4996	0.4996	0.4996	0.4997
3.4	0.4997	0.4997	0.4997	0.4997	0.4997	0.4997	0.4997	0.4997	0.4997	0.4998
3.5	0.4998	0.4998	0.4998	0.4998	0.4998	0.4998	0.4998	0.4998	0.4998	0.4998
3.6	0.4998	0.4998	0.4999	0.4999	0.4999	0.4999	0.4999	0.4999	0.4999	0.4999
3.7	0.4999	0.4999	0.4999	0.4999	0.4999	0.4999	0.4999	0.4999	0.4999	0.4999
3.8	0.4999	0.4999	0.4999	0.4999	0.4999	0.4999	0.4999	0.4999	0.4999	0.4999
3.9	0.5000	0.5000	0.5000	0.5000	0.5000	0.5000	05000	0.5000	0.5000	0.5000

Problem 2. For the group of people given in Problem 1, find the number of people likely to have heights of less than 165 cm

A height of 165 cm corresponds to $\frac{165-170}{9}$, i.e. -0.56 standard deviations. The area between $z=0$ and $z=-0.56$ (from Table 64.1) is 0.2123, shown shaded in Fig. 64.4(a). The total area under the standardised normal curve is unity and since the curve is symmetrical, it follows that the total area to the left of the $z=0$ ordinate is 0.5000. Thus the area to the left of the $z=-0.56$ ordinate ('left' means 'less than', 'right' means 'more than') is $0.5000-0.2123$, i.e. 0.2877 of the total area, which is shown shaded in Fig. 64.4(b). The area is directly proportional to probability and since the total area beneath the standardised normal curve is unity, the probability of a person's height being less than 165 cm is 0.2877. For a group of 500 people, 500×0.2877, i.e. **144 people are likely to have heights of less than 165 cm**.

Problem 3. For the group of people given in Problem 1 find how many people are likely to have heights of more than 194 cm

194 cm correspond to a z-value of $\frac{194-170}{9}$ that is, 2.67 standard deviations. From Table 64.1, the area between $z=0$, $z=2.67$ and the standardised normal

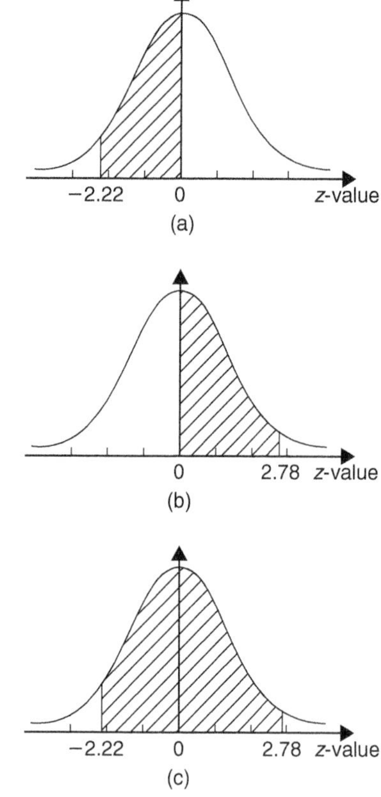

Figure 64.3

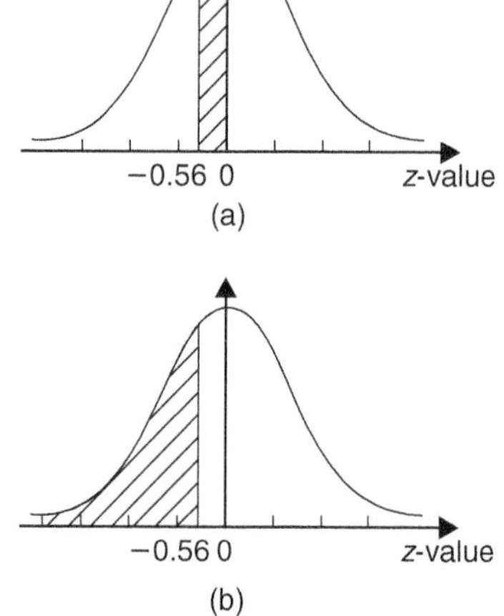

Figure 64.4

curve is 0.4962, shown shaded in Fig. 64.5(a). Since the standardised normal curve is symmetrical, the total area to the right of the $z = 0$ ordinate is 0.5000, hence the shaded area shown in Fig. 64.5(b) is $0.5000 - 0.4962$, i.e. 0.0038. This area represents the probability of a person having a height of more than 194 cm,

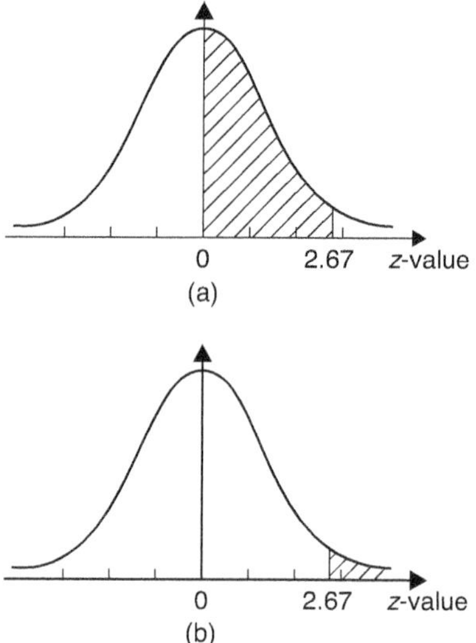

Figure 64.5

and for 500 people, the number of people likely to have a height of more than 194 cm is 0.0038×500, i.e. **2 people**.

Problem 4. A batch of 1500 lemonade bottles have an average contents of 753 ml and the standard deviation of the contents is 1.8 ml. If the volumes of the content are normally distributed, find the number of bottles likely to contain: (a) less than 750 ml, (b) between 751 and 754 ml, (c) more than 757 ml, and (d) between 750 and 751 ml

(a) The z-value corresponding to 750 ml is given by $\frac{x - \bar{x}}{\sigma}$ i.e. $\frac{750 - 753}{1.8} = -1.67$ standard deviations. From Table 64.1, the area between $z = 0$ and $z = -1.67$ is 0.4525. Thus the area to the left of the $z = -1.67$ ordinate is $0.5000 - 0.4525$ (see Problem 2), i.e. 0.0475. This is the probability of a bottle containing less than 750 ml. Thus, for a batch of 1500 bottles, it is likely that 1500×0.0475, i.e. **71 bottles will contain less than 750 ml**.

(b) The z-value corresponding to 751 and 754 ml are $\frac{751 - 753}{1.8}$ and $\frac{754 - 753}{1.8}$ i.e. -1.11 and 0.56 respectively. From Table 64.1, the areas corresponding to these values are 0.3665 and 0.2123 respectively. Thus the probability of a bottle containing between 751 and 754 ml is $0.3665 + 0.2123$ (see Problem 1), i.e. 0.5788. For 1500 bottles, it is likely that 1500×0.5788, i.e. **868 bottles will contain between 751 and 754 ml**.

(c) The z-value corresponding to 757 ml is $\frac{757-753}{1.8}$, i.e. 2.22 standard deviations. From Table 64.1, the area corresponding to a z-value of 2.22 is 0.4868. The area to the right of the $z=2.22$ ordinate is $0.5000-0.4868$ (see Problem 3), i.e. 0.0132. Thus, for 1500 bottles, it is likely that 1500×0.0132, i.e. **20 bottles will have contents of more than 750 ml**.

(d) The z-value corresponding to 750 ml is -1.67 (see part (a)), and the z-value corresponding to 751 ml is -1.11 (see part (b)). The areas corresponding to these z-values area 0.4525 and 0.3665 respectively, and both these areas lie on the left of the $z=0$ ordinate. The area between $z=-1.67$ and $z=-1.11$ is $0.4525-0.3665$, i.e. 0.0860 and this is the probability of a bottle having contents between 750 and 751 ml. For 1500 bottles, it is likely that 1500×0.0860, i.e. **129 bottles will be in this range**.

Now try the following Practice Exercise

Practice Exercise 297 Introduction to the normal distribution (Answers on page 736)

1. A component is classed as defective if it has a diameter of less than 69 mm. In a batch of 350 components, the mean diameter is 75 mm and the standard deviation is 2.8 mm. Assuming the diameters are normally distributed, determine how many are likely to be classed as defective
2. The masses of 800 people are normally distributed, having a mean value of 64.7 kg, and a standard deviation of 5.4 kg. Find how many people are likely to have masses of less than 54.4 kg
3. 500 tins of paint have a mean content of 1010 ml and the standard deviation of the contents is 8.7 ml. Assuming the volumes of the contents are normally distributed, calculate the number of tins likely to have contents whose volumes are less than (a) 1025 ml (b) 1000 ml and (c) 995 ml
4. For the 350 components in Problem 1, if those having a diameter of more than 81.5 mm are rejected, find, correct to the nearest component, the number likely to be rejected due to being oversized
5. For the 800 people in Problem 2, determine how many are likely to have masses of more than (a) 70 kg, and (b) 62 kg
6. The mean diameter of holes produced by a drilling machine bit is 4.05 mm and the standard deviation of the diameters is 0.0028 mm. For twenty holes drilled using this machine, determine, correct to the nearest whole number, how many are likely to have diameters of between (a) 4.048 and 4.0553 mm, and (b) 4.052 and 4.056 mm, assuming the diameters are normally distributed
7. The I.Q.s of 400 children have a mean value of 100 and a standard deviation of 14. Assuming that I.Q.s are normally distributed, determine the number of children likely to have I.Q.s of between (a) 80 and 90, (b) 90 and 110, and (c) 110 and 130
8. The mean mass of active material in tablets produced by a manufacturer is 5.00 g and the standard deviation of the masses is 0.036 g. In a bottle containing 100 tablets, find how many tablets are likely to have masses of (a) between 4.88 and 4.92 g, (b) between 4.92 and 5.04 g, and (c) more than 5.04 g

64.2 Testing for a normal distribution

It should never be assumed that because data is continuous it automatically follows that it is normally distributed. One way of checking that data is normally distributed is by using **normal probability paper**, often just called **probability paper**. This is special graph paper which has linear markings on one axis and percentage probability values from 0.01 to 99.99 on the other axis (see Figs. 64.6 and 64.7). The divisions on the probability axis are such that a straight line graph results for normally distributed data when percentage cumulative frequency values are plotted against upper class boundary values. If the points do not lie in a reasonably straight line, then the data is not normally distributed. The method used to test the normality of a

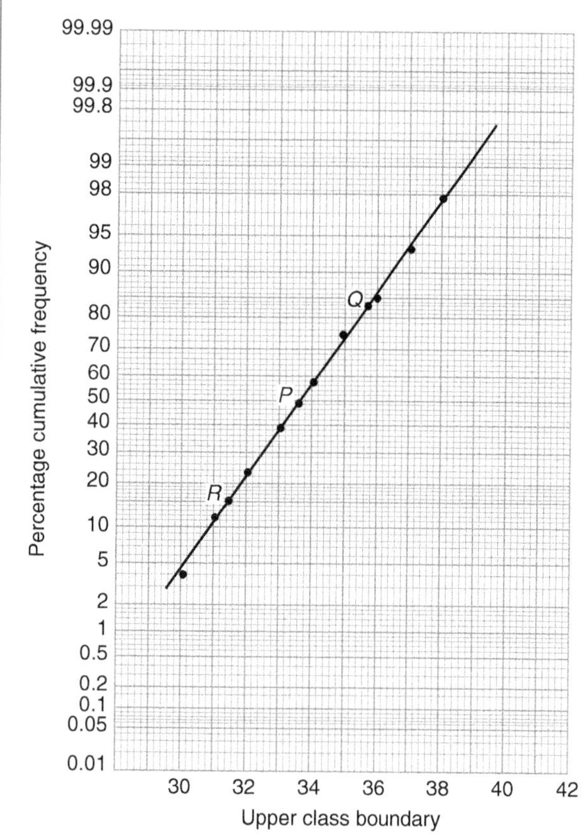

Figure 64.6

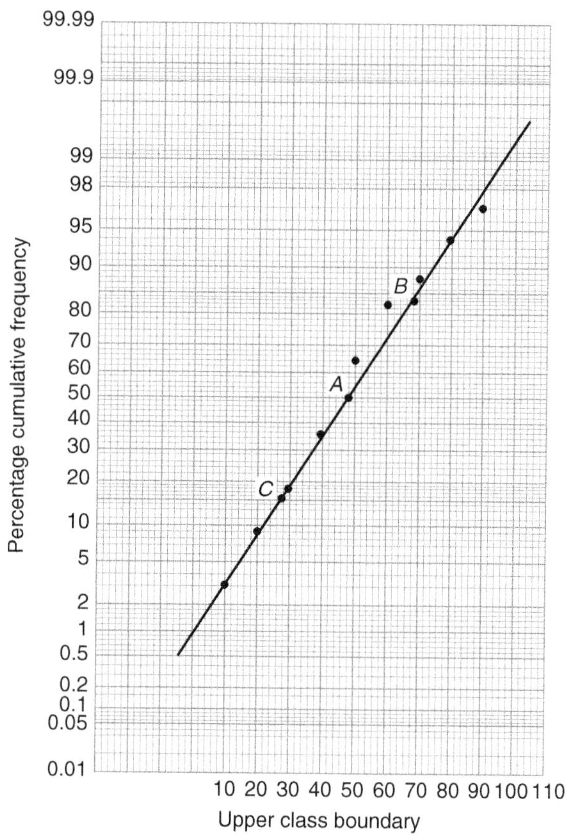

Figure 64.7

distribution is shown in Problems 5 and 6. The mean value and standard deviation of normally distributed data may be determined using normal probability paper. For normally distributed data, the area beneath the standardised normal curve and a z-value of unity (i.e. one standard deviation) may be obtained from Table 64.1. For one standard deviation, this area is 0.3413, i.e. 34.13%. An area of ± 1 standard deviation is symmetrically placed on either side of the $z=0$ value, i.e. is symmetrically placed on either side of the 50 per cent cumulative frequency value. Thus an area corresponding to ± 1 standard deviation extends from percentage cumulative frequency values of $(50+34.13)\%$ to $(50-34.13)\%$, i.e. from 84.13% to 15.87%. For most purposes, these values are taken as 84% and 16%. Thus, when using normal probability paper, the standard deviation of the distribution is given by:

$$\frac{\text{(variable value for 84\% cumulative frequency)} - \text{(variable value for 16\% cumulative frequency)}}{2}$$

Problem 5. Use normal probability paper to determine whether the data given below, which refers to the masses of 50 copper ingots, is approximately normally distributed. If the data is normally distributed, determine the mean and standard deviation of the data from the graph drawn

Class mid-point value (kg)	29.5	30.5	31.5	32.5	33.5
Frequency	2	4	6	8	9
Class mid-point value (kg)	34.5	35.5	36.5	37.5	38.5
Frequency	8	6	4	2	1

To test the normality of a distribution, the upper class boundary/percentage cumulative frequency values are plotted on normal probability paper. The upper class boundary values are: 30, 31, 32, …, 38, 39. The corresponding cumulative frequency values (for 'less than' the upper class boundary values) are: 2, $(4+2)=6$,

$(6+4+2)=12$, 20, 29, 37, 43, 47, 49 and 50. The corresponding percentage cumulative frequency values are $\frac{2}{50}\times 100=4$, $\frac{6}{50}\times 100=12$, 24, 40, 58, 74, 86, 94, 98 and 100%.

The co-ordinates of upper class boundary/percentage cumulative frequency values are plotted as shown in Fig. 64.6. When plotting these values, it will always be found that the co-ordinate for the 100% cumulative frequency value cannot be plotted, since the maximum value on the probability scale is 99.99. **Since the points plotted in Fig. 64.6 lie very nearly in a straight line, the data is approximately normally distributed**.

The mean value and standard deviation can be determined from Fig. 64.6. Since a normal curve is symmetrical, the mean value is the value of the variable corresponding to a 50% cumulative frequency value, shown as point P on the graph. This shows that **the mean value is 33.6 kg**. The standard deviation is determined using the 84% and 16% cumulative frequency values, shown as Q and R in Fig. 64.6. The variable values for Q and R are 35.7 and 31.4 respectively; thus two standard deviations correspond to $35.7-31.4$, i.e. 4.3, showing that the standard deviation of the distribution is approximately $\frac{4.3}{2}$ i.e. **2.15 kg**.

The mean value and standard deviation of the distribution can be calculated using mean, $\bar{x}=\frac{\sum fx}{\sum f}$ and standard deviation, $\sigma=\sqrt{\frac{\sum[f(x-\bar{x})^2]}{\sum f}}$ where f is the frequency of a class and x is the class mid-point value. Using these formulae gives a mean value of the distribution of 33.6 kg (as obtained graphically) and a standard deviation of 2.12 kg, showing that the graphical method of determining the mean and standard deviation give quite realistic results.

Problem 6. Use normal probability paper to determine whether the engineering data given below is normally distributed. Use the graph and assume a normal distribution whether this is so or not, to find approximate values of the mean and standard deviation of the distribution.

Class mid-point Values	5	15	25	35	45	55	65	75	85	95
Frequency	1	2	3	6	9	6	2	2	1	1

To test the normality of a distribution, the upper class boundary/percentage cumulative frequency values are plotted on normal probability paper. The upper class boundary values are: 10, 20, 30, …, 90 and 100. The corresponding cumulative frequency values are 1, $1+2=3$, $1+2+3=6$, 12, 21, 27, 29, 31, 32 and 33. The percentage cumulative frequency values are $\frac{1}{33}\times 100=3$, $\frac{3}{33}\times 100=9$, 18, 36, 64, 82, 88, 94, 97 and 100.

The co-ordinates of upper class boundary values/percentage cumulative frequency values are plotted as shown in Fig. 64.7. Although six of the points lie approximately in a straight line, three points corresponding to upper class boundary values of 50, 60 and 70 are not close to the line and indicate that **the distribution is not normally** distributed. However, if a normal distribution is assumed, the **mean value** corresponds to the variable value at a cumulative frequency of 50% and, from Fig. 64.7, point A is **48**. The value of the standard deviation of the distribution can be obtained from the variable values corresponding to the 84% and 16% cumulative frequency values, shown as B and C in Fig. 64.7 and give: $2\sigma=69-28$, i.e. **the standard deviation $\sigma=20.5$**. The calculated values of the mean and standard deviation of the distribution are 45.9 and 19.4 respectively, showing that errors are introduced if the graphical method of determining these values is used for data that is not normally distributed.

Now try the following Practice Exercise

Practice Exercise 298 Testing for a normal distribution (Answers on page 736)

1. A frequency distribution of 150 measurements is as shown:

Class mid-point value	26.4	26.6	26.8	27.0
Frequency	5	12	24	36

Class mid-point value	27.2	27.4	27.6
Frequency	36	25	12

Use normal probability paper to show that this data approximates to a normal distribution and hence determine the approximate values of the mean and standard deviation of the distribution. Use the formula for mean and standard deviation to verify the results obtained

2. A frequency distribution of the class mid-point values of the breaking loads for 275 similar fibres is as shown below

Load (kN)	17	19	21	23
Frequency	9	23	55	78

Load (kN)	25	27	29	31
Frequency	64	28	14	4

Use normal probability paper to show that this distribution is approximately normally distributed and determine the mean and standard deviation of the distribution (a) from the graph and (b) by calculation

Practice Exercise 299 Multiple-choice questions on the normal distribution (Answers on page 736)

Each question has only one correct answer

Questions 1 to 3 relate to the following information.
The mean height of 400 people is 170 cm and the standard deviation is 8 cm. Assume a normal distribution.

1. The number of people likely to have heights of between 154 cm and 186 cm is:
(a) 390 (b) 382 (c) 191 (d) 185
2. The number of people likely to have heights less than 162 cm is:
(a) 126 (b) 273 (c) 63 (d) 137
3. The number of people likely to have a height of more than 186 cm is:
(a) 9 (b) 95 (c) 191 (d) 18

Questions 4 and 5 relate to the following information.
A batch of 2000 bottles of drinks have an average content of 600 ml and the standard deviation is 2 ml. Assume a normal distribution.

4. The number of bottles likely to contain less than 595 ml is:
(a) 92 (b) 24 (c) 46 (d) 12
5. The number of bottles likely to contain between 596 ml and 604 ml is:
(a) 1909 (b) 1974 (c) 974 (d) 937

COMPANION WEBSITE

For fully worked solutions to each of the problems in Practice Exercises 297 and 298 in this chapter, go to the website:
www.routledge.com/cw/bird

Revision Test 18 The binomial, Poisson and normal distributions

This Revision Test covers the material contained in Chapters 63 and 64. *The marks for each question are shown in brackets at the end of each question.*

1. A machine produces 15% defective components. In a sample of 5, drawn at random, calculate, using the binomial distribution, the probability that:
   (a) there will be 4 defective items
   (b) there will be not more than 3 defective items
   (c) all the items will be non-defective (14)

2. 2% of the light bulbs produced by a company are defective. Determine, using the Poisson distribution, the probability that in a sample of 80 bulbs:

   (a) 3 bulbs will be defective, (b) not more than 3 bulbs will be defective, (c) at least 2 bulbs will be defective. (13)

3. Some engineering components have a mean length of 20 mm and a standard deviation of 0.25 mm. Assume that the data on the lengths of the components is normally distributed.

   In a batch of 500 components, determine the number of components likely to:
   (a) have a length of less than 19.95 mm
   (b) be between 19.95 mm and 20.15 mm
   (c) be longer than 20.54 mm (12)

4. In a factory, cans are packed with an average of 1.0 kg of a compound and the masses are normally distributed about the average value. The standard deviation of a sample of the contents of the cans is 12 g. Determine the percentage of cans containing (a) less than 985 g, (b) more than 1030 g, (c) between 985 g and 1030 g. (11)

For lecturers/instructors/teachers, fully worked solutions to each of the problems in Revision Test 18, together with a full marking scheme, are available at the website:
www.routledge.com/cw/bird

Chapter 65

Linear correlation

Why it is important to understand: Linear correlation

Correlation coefficients **measure the strength of association between two variables. The most common correlation coefficient, called the** ***product-moment correlation coefficient*****, measures the strength of the** ***linear association*** **between variables. A positive value indicates a positive correlation and the higher the value, the stronger the correlation. Similarly, a negative value indicates a negative correlation and the lower the value the stronger the correlation. This chapter explores linear correlation and the meaning of values obtained calculating the coefficient of correlation.**

At the end of this chapter, you should be able to:

- recognise linear correlation
- state the product-moment formula
- appreciate the significance of a coefficient of correlation
- determine the linear coefficient of correlation between two given variables

65.1 Introduction to linear correlation

Correlation is a measure of the amount of association existing between two variables. For linear correlation, if points are plotted on a graph and all the points lie on a straight line, then **perfect linear correlation** is said to exist. When a straight line having a positive gradient can reasonably be drawn through points on a graph, **positive or direct linear correlation** exists, as shown in Fig. 65.1(a). Similarly, when a straight line having a negative gradient can reasonably be drawn through points on a graph, **negative or inverse linear correlation** exists, as shown in Fig. 65.1(b). When there is no apparent relationship between co-ordinate values plotted on a graph then no **correlation** exists between the points, as shown in Fig. 65.1(c). In statistics, when two variables are being investigated, the location of the co-ordinates on a rectangular co-ordinate system is called a **scatter diagram** — as shown in Fig. 65.1.

65.2 The Pearson product-moment formula for determining the linear correlation coefficient

The Pearson product-moment correlation coefficient (or **Pearson correlation coefficient**, for short) is a measure of the strength of a linear association between two variables and is denoted by the symbol r. A Pearson product-moment correlation attempts to draw a line of best fit through the data of two variables, and the Pearson correlation coefficient, r, indicates how far away all these data points are to this line of best fit (how well the data points fit this new model/line of best fit). The **product-moment formula**, states:

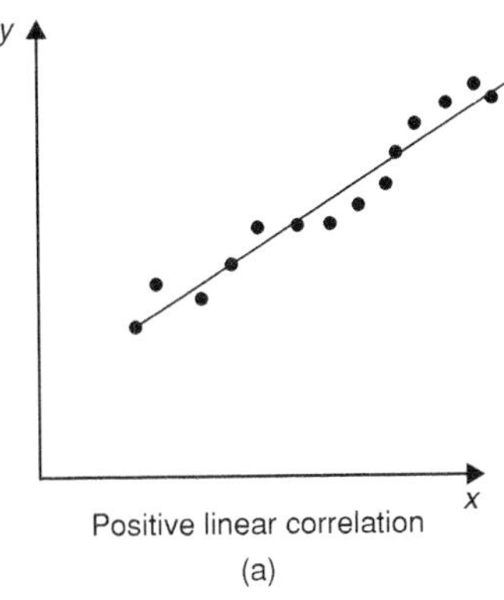

Positive linear correlation
(a)

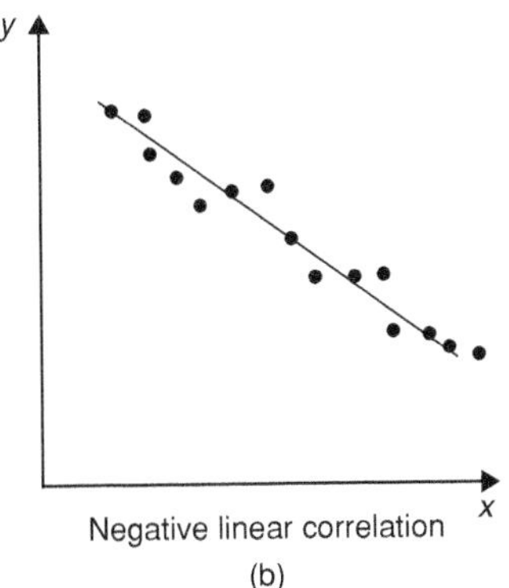

Negative linear correlation
(b)

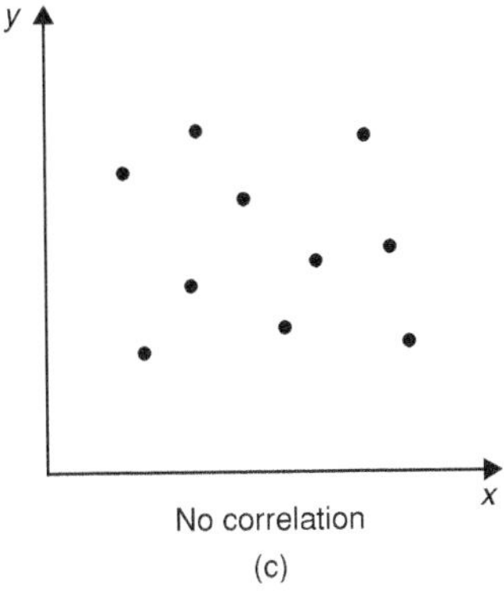

No correlation
(c)

Figure 65.1

coefficient of correlation,

$$r = \frac{\sum xy}{\sqrt{\left(\sum x^2\right)\left(\sum y^2\right)}} \qquad (1)$$

where the x-values are the values of the deviations of co-ordinates X from $\overline{X}$, their mean value and the y-values are the values of the deviations of co-ordinates Y from $\overline{Y}$, their mean value. That is, $x=(X-\overline{X})$ and $y=(Y-\overline{Y})$. The results of this determination give values of r lying between $+1$ and -1, where $+1$ indicates perfect direct correlation, -1 indicates perfect inverse correlation and 0 indicates that no correlation exists. Between these values, the smaller the value of r, the less is the amount of correlation which exists. Generally, values of r in the ranges 0.7 and 1 and -0.7 to -1 show that a fair amount of correlation exists.

65.3 The significance of a coefficient of correlation

When the value of the coefficient of correlation has been obtained from the product moment formula, some care is needed before coming to conclusions based on this result. Checks should be made to ascertain the following two points:

(a) that a 'cause and effect' relationship exists between the variables; it is relatively easy, mathematically, to show that some correlation exists between, say, the number of ice creams sold in a given period of time and the number of chimneys swept in the same period of time, although there is no relationship between these variables;

(b) that a linear relationship exists between the variables; the product-moment formula given in Section 65.2 is based on linear correlation. Perfect non-linear correlation may exist (for example, the co-ordinates exactly following the curve $y=x^3$), but this gives a low value of coefficient of correlation since the value of r is determined using the product-moment formula, based on a linear relationship.

65.4 Worked problems on linear correlation

Problem 1. In an experiment to determine the relationship between force on a wire and the resulting extension, the following data are obtained:

Force (N)	10	20	30	40	50	60	70
Extension (mm)	0.22	0.40	0.61	0.85	1.20	1.45	1.70

Determine the linear coefficient of correlation for this data

Let X be the variable force values and Y be the dependent variable extension values. The coefficient of correlation is given by:

$$r = \frac{\sum xy}{\sqrt{\left(\sum x^2\right)\left(\sum y^2\right)}}$$

where $x=(X-\overline{X})$ and $y=(Y-\overline{Y})$, $\overline{X}$ and $\overline{Y}$ being the mean values of the X and Y values respectively. Using a tabular method to determine the quantities of this formula gives:

X	Y	$x=(X-\overline{X})$	$y=(Y-\overline{Y})$
10	0.22	−30	−0.699
20	0.40	−20	−0.519
30	0.61	−10	−0.309
40	0.85	0	−0.069
50	1.20	10	0.281
60	1.45	20	0.531
70	1.70	30	0.781

$\sum X=280,\quad \overline{X}=\frac{280}{7}=40$

$\sum Y=6.43,\quad \overline{Y}=\frac{6.43}{7}=0.919$

xy	x^2	y^2
20.97	900	0.489
10.38	400	0.269
3.09	100	0.095
0	0	0.005
2.81	100	0.079
10.62	400	0.282
23.43	900	0.610

$\sum xy=71.30 \quad \sum x^2=2800 \quad \sum y^2=1.829$

Thus $r=\frac{71.3}{\sqrt{2800\times 1.829}}=\mathbf{0.996}$

This shows that a **very good direct correlation exists** between the values of force and extension.

Problem 2. The relationship between expenditure on welfare services and absenteeism for similar periods of time is shown below for a small company.

Expenditure (£'000)	3.5	5.0	7.0	10	12	15	18
Days lost	241	318	174	110	147	122	86

Determine the coefficient of linear correlation for this data

Let X be the expenditure in thousands of pounds and Y be the days lost.

The coefficient of correlation,

$$r = \frac{\sum xy}{\sqrt{\left(\sum x^2\right)\left(\sum y^2\right)}}$$

where $x=(X-\overline{X})$ and $y=(Y-\overline{Y})$, $\overline{X}$ and $\overline{Y}$ being the mean values of X and Y respectively. Using a tabular approach:

X	Y	$x=(X-\overline{X})$	$y=(Y-\overline{Y})$
3.5	241	−6.57	69.9
5.0	318	−5.07	146.9
7.0	174	−3.07	2.9
10	110	−0.07	−61.1
12	147	1.93	−24.1
15	122	4.93	−49.1
18	86	7.93	−85.1

$\sum X=70.5,\quad \overline{X}=\frac{70.5}{7}=10.07$

$\sum Y=1198,\quad \overline{Y}=\frac{1198}{7}=171.1$

xy	x^2	y^2
−459.2	43.2	4886
−744.8	25.7	21580
−8.9	9.4	8
4.3	0	3733
−46.5	3.7	581

xy	x^2	y^2
−242.1	24.3	2411
−674.8	62.9	7242
$\sum xy = -2172$	$\sum x^2 = 169.2$	$\sum y^2 = 40441$

Thus $r = \dfrac{-2172}{\sqrt{169.2 \times 40441}} = \mathbf{-0.830}$

This shows that there is **fairly good inverse correlation** between the expenditure on welfare and days lost due to absenteeism.

Problem 3. The relationship between monthly car sales and income from the sale of petrol for a garage is as shown:

Cars sold	2	5	3	12	14	7
Income from petrol sales (£'000)	12	9	13	21	17	22
Cars sold	3	28	14	7	3	13
Income from petrol sales (£'000)	31	47	17	10	9	11

Determine the linear coefficient of correlation between these quantities

Let X represent the number of cars sold and Y the income, in thousands of pounds, from petrol sales. Using the tabular approach:

X	Y	$x=(X-\overline{X})$	$y=(Y-\overline{Y})$
2	12	−7.25	−6.25
5	9	−4.25	−9.25
3	13	−6.25	−5.25
12	21	2.75	2.75
14	17	4.75	−1.25
7	22	−2.25	3.75
3	31	−6.25	12.75
28	47	18.75	28.75
14	17	4.75	−1.25
7	10	−2.25	−8.25
3	9	−6.25	−9.25
13	11	3.75	−7.25

$\sum X = 111$, $\overline{X} = \dfrac{111}{12} = 9.25$

$\sum Y = 219$, $\overline{Y} = \dfrac{219}{12} = 18.25$

xy	x^2	y^2
45.3	52.6	39.1
39.3	18.1	85.6
32.8	39.1	27.6
7.6	7.6	7.6
−5.9	22.6	1.6
−8.4	5.1	14.1
−79.7	39.1	162.6
539.1	351.6	826.6
−5.9	22.6	1.6
18.6	5.1	68.1
57.8	39.1	85.6
−27.2	14.1	52.6
$\sum xy = 613.4$	$\sum x^2 = 616.7$	$\sum y^2 = 1372.7$

The coefficient of correlation,

$$r = \frac{\sum xy}{\sqrt{\left(\sum x^2\right)\left(\sum y^2\right)}}$$

$$= \frac{613.4}{\sqrt{(616.7)(1372.7)}} = \mathbf{0.667}$$

Thus, there is **no appreciable correlation** between petrol and car sales.

Now try the following Practice Exercise

Practice Exercise 300 Linear correlation (Answers on page 737)

In Problems 1 to 3, determine the coefficient of correlation for the data given, correct to 3 decimal places.

1.

X	14	18	23	30	50
Y	900	1200	1600	2100	3800

2.

X	2.7	4.3	1.2	1.4	4.9
Y	11.9	7.10	33.8	25.0	7.50

3.

X	24	41	9	18	73
Y	39	46	90	30	98

4. In an experiment to determine the relationship between the current flowing in an electrical circuit and the applied voltage, the results obtained are:

Current (mA)	5	11	15	19	24	28	33
Applied voltage (V)	2	4	6	8	10	12	14

Determine, using the product-moment formula, the coefficient of correlation for these results.

5. A gas is being compressed in a closed cylinder and the values of pressures and corresponding volumes at constant temperature are as shown:

Pressure (kPa)	160	180	200	220
Volume (m^3)	0.034	0.036	0.030	0.027

Pressure (kPa)	240	260	280	300
Volume (m^3)	0.024	0.025	0.020	0.019

Find the coefficient of correlation for these values.

6. The relationship between the number of miles travelled by a group of engineering salesmen in ten equal time periods and the corresponding value of orders taken is given below. Calculate the coefficient of correlation using the product-moment formula for these values.

Miles travelled	1370	1050	980	1770	1340
Orders taken (£'000)	23	17	19	22	27

Miles travelled	1560	2110	1540	1480	1670
Orders taken (£'000)	23	30	23	25	19

7. The data shown below refer to the number of times machine tools had to be taken out of service, in equal time periods, due to faults occurring and the number of hours worked by maintenance teams. Calculate the coefficient of correlation for this data.

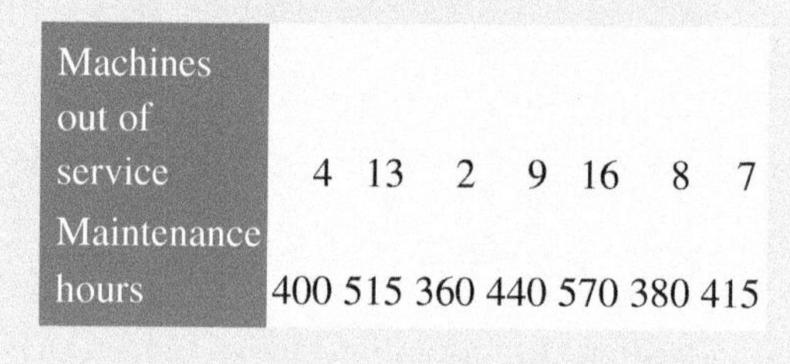

Machines out of service	4	13	2	9	16	8	7
Maintenance hours	400	515	360	440	570	380	415

For fully worked solutions to each of the problems in Practice Exercise 300 in this chapter, go to the website:
www.routledge.com/cw/bird

Chapter 66

Linear regression

Why it is important to understand: Linear regression

The general process of fitting data to a linear combination of basic functions is termed linear regression. Linear least squares regression is by far the most widely used modelling method; it is what most people mean when they say they have used 'regression', 'linear regression' or 'least squares' to fit a model to their data. Not only is linear least squares regression the most widely used modelling method, but it has been adapted to a broad range of situations that are outside its direct scope. It plays a strong underlying role in many other modelling methods. This chapter explains how regression lines are determined.

At the end of this chapter, you should be able to:

- explain linear regression
- understand least-squares regression lines
- determine, for two variables X and Y, the equations of the regression lines of X on Y and Y on X

66.1 Introduction to linear regression

Regression analysis, usually termed **regression**, is used to draw the line of 'best fit' through co-ordinates on a graph. The techniques used enable a mathematical equation of the straight line form $y=mx+c$ to be deduced for a given set of co-ordinate values, the line being such that the sum of the deviations of the co-ordinate values from the line is a minimum, i.e. it is the line of 'best fit'. When a regression analysis is made, it is possible to obtain two lines of best fit, depending on which variable is selected as the dependent variable and which variable is the independent variable. For example, in a resistive electrical circuit, the current flowing is directly proportional to the voltage applied to the circuit. There are two ways of obtaining experimental values relating the current and voltage. Either, certain voltages are applied to the circuit and the current values are measured, in which case the voltage is the independent variable and the current is the dependent variable; or, the voltage can be adjusted until a desired value of current is flowing and the value of voltage is measured, in which case the current is the independent value and the voltage is the dependent value.

66.2 The least-squares regression lines

For a given set of co-ordinate values, (X_1,Y_1), $(X_2,Y_2),\ldots,(X_n,Y_n)$ let the X values be the independent variables and the Y-values be the dependent values. Also let $D_1,\ldots,D_n$ be the vertical distances between the line shown as PQ in Fig. 66.1 and the points representing the co-ordinate values. The least-squares regression line, i.e. the line of best fit, is the line which makes the value of $D_1^2+D_2^2+\cdots+D_n^2$ a minimum value.

The equation of the least-squares regression line is usually written as $Y=a_0+a_1X$, where a_0 is the Y-axis intercept value and a_1 is the gradient of the line (analogous to c and m in the equation $y=mx+c$). The

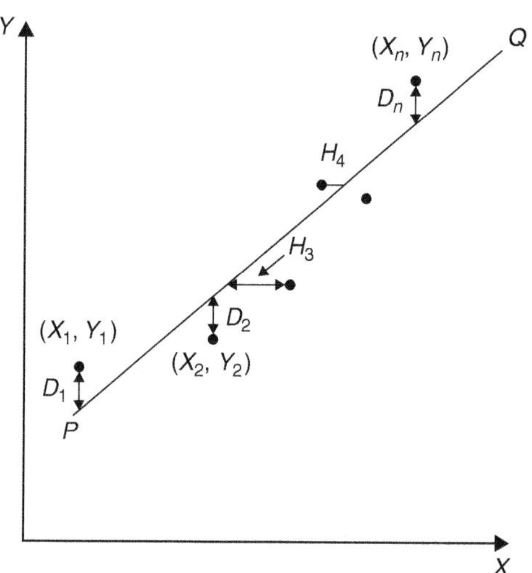

Figure 66.1

values of a_0 and a_1 to make the sum of the 'deviations squared' a minimum can be obtained from the two equations:

$$\sum Y = a_0 N + a_1 \sum X \quad (1)$$

$$\sum (XY) = a_0 \sum X + a_1 \sum X^2 \quad (2)$$

where X and Y are the co-ordinate values, N is the number of co-ordinates and a_0 and a_1 are called the **regression coefficients** of Y on X. Equations (1) and (2) are called the **normal equations** of the regression lines of Y on X. The regression line of Y on X is used to estimate values of Y for given values of X.

If the Y-values (vertical axis) are selected as the independent variables, the horizontal distances between the line shown as PQ in Fig. 66.1 and the co-ordinate values (H_3, H_4, etc.) are taken as the deviations. The equation of the regression line is of the form: $X = b_0 + b_1 Y$ and the normal equations become:

$$\sum X = b_0 N + b_1 \sum Y \quad (3)$$

$$\sum (XY) = b_0 \sum Y + b_1 \sum Y^2 \quad (4)$$

where X and Y are the co-ordinate values, b_0 and b_1 are the regression coefficients of X on Y and N is the number of co-ordinates. These normal equations are of the regression line of X on Y, which is slightly different to the regression line of Y on X. The regression line of X on Y is used to estimate values of X for given values of Y. The regression line of Y on X is used to determine any value of Y corresponding to a given value of X. If the value of Y lies within the range of Y-values of the extreme co-ordinates, the process of finding the corresponding value of X is called **linear interpolation**. If it lies outside of the range of Y-values of the extreme co-ordinates than the process is called **linear extrapolation** and the assumption must be made that the line of best fit extends outside of the range of the co-ordinate values given.

By using the regression line of X on Y, values of X corresponding to given values of Y may be found by either interpolation or extrapolation.

66.3 Worked problems on linear regression

Problem 1. In an experiment to determine the relationship between frequency and the inductive reactance of an electrical circuit, the following results were obtained:

Frequency (Hz)	50	100	150
Inductive reactance (ohms)	30	65	90

Frequency (Hz)	200	250	300	350
Inductive reactance (ohms)	130	150	190	200

Determine the equation of the regression line of inductive reactance on frequency, assuming a linear relationship.

Since the regression line of inductive reactance on frequency is required, the frequency is the independent variable, X, and the inductive reactance is the dependent variable, Y. The equation of the regression line of Y on X is:

$$Y = a_0 + a_1 X$$

and the regression coefficients a_0 and a_1 are obtained by using the normal equations

$$\sum Y = a_0 N + a_1 \sum X$$

$$\text{and} \quad \sum XY = a_0 \sum X + a_1 \sum X^2$$

(from equations (1) and (2))

A tabular approach is used to determine the summed quantities.

Frequency, X	Inductive reactance, Y	X^2
50	30	2500
100	65	10 000
150	90	22 500
200	130	40 000
250	150	62 500
300	190	90 000
350	200	122 500
$\sum X=1400$	$\sum Y=855$	$\sum X^2=350\,000$

XY	Y^2
1500	900
6500	4225
13 500	8100
26 000	16 900
37 500	22 500
57 000	36 100
70 000	40 000
$\sum XY=212\,000$	$\sum Y^2=128\,725$

The number of co-ordinate values given, N is 7. Substituting in the normal equations gives:

$$855 = 7a_0 + 1400a_1 \quad (1)$$

$$212\,000 = 1400a_0 + 350\,000a_1 \quad (2)$$

$1400 \times (1)$ gives:

$$1197\,000 = 9800a_0 + 1\,960\,000a_1 \quad (3)$$

$7 \times (2)$ gives:

$$1\,484\,000 = 9800a_0 + 2\,450\,000a_1 \quad (4)$$

$(4) - (3)$ gives:

$$287\,000 = 0 + 490\,000a_1 \quad (5)$$

from which, $a_1 = \frac{287\,000}{490\,000} = 0.586$

Substituting $a_1 = 0.586$ in equation (1) gives:

$$855 = 7a_0 + 1400(0.586)$$

i.e. $$a_0 = \frac{855 - 820.4}{7} = 4.94$$

Thus the equation of the regression line of inductive reactance on frequency is:

$$\mathbf{Y = 4.94 + 0.586X}$$

Problem 2. For the data given in Problem 1, determine the equation of the regression line of frequency on inductive reactance, assuming a linear relationship

In this case, the inductive reactance is the independent variable Y and the frequency is the dependent variable X. From equations 3 and 4, the equation of the regression line of X on Y is:

$$X = b_0 + b_1 Y$$

and the normal equations are

$$\sum X = b_0 N + b_1 \sum Y$$

and $$\sum XY = b_0 \sum Y + b_1 \sum Y^2$$

From the table shown in Problem 1, the simultaneous equations are:

$$1400 = 7b_0 + 855b_1$$

$$212\,000 = 855b_0 + 128\,725b_1$$

Solving these equations in a similar way to that in Problem 1 gives:

$b_0 = -6.15$

and $b_1 = 1.69$, correct to 3 significant figures.

Thus the equation of the regression line of frequency on inductive reactance is:

$$\mathbf{X = -6.15 + 1.69Y}$$

Problem 3. Use the regression equations calculated in Problems 1 and 2 to find (a) the value of inductive reactance when the frequency is 175 Hz, and (b) the value of frequency when the inductive reactance is 250 ohms, assuming the line of best fit extends outside of the given co-ordinate values. Draw a graph showing the two regression lines

(a) From Problem 1, the regression equation of inductive reactance on frequency is: $Y = 4.94 + 0.586X$. When the frequency, X, is 175 Hz, $Y = 4.94 + 0.586(175) = 107.5$, correct to 4 significant figures, i.e. the inductive reactance is **107.5 ohms** when the frequency is 175 Hz.

(b) From Problem 2, the regression equation of frequency on inductive reactance is: $X = -6.15 + 1.69Y$. When the inductive reactance, Y, is 250 ohms, $X = -6.15 + 1.69(250) = 416.4$ Hz, correct to 4 significant figures, i.e. the frequency is **416.4 Hz** when the inductive reactance is 250 ohms.

The graph depicting the two regression lines is shown in Fig. 66.2. To obtain the regression line of inductive reactance on frequency the regression line equation $Y = 4.94 + 0.586X$ is used, and X (frequency) values of 100 and 300 have been selected in order to find the corresponding Y values. These values gave the co-ordinates as (100, 63.5) and (300, 180.7), shown as points A and B in Fig. 66.2. Two co-ordinates for the regression line of frequency on inductive reactance are calculated using the equation $X = -6.15 + 1.69Y$, the values of inductive reactance of 50 and 150 being used to obtain the co-ordinate values. These values gave co-ordinates (78.4, 50) and (247.4, 150), shown as points C and D in Fig. 66.2.

It can be seen from Fig. 66.2 that to the scale drawn, the two regression lines coincide. Although it is not necessary to do so, the co-ordinate values are also shown to indicate that the regression lines do appear to be the lines of best fit. A graph showing co-ordinate values is called a **scatter diagram** in statistics.

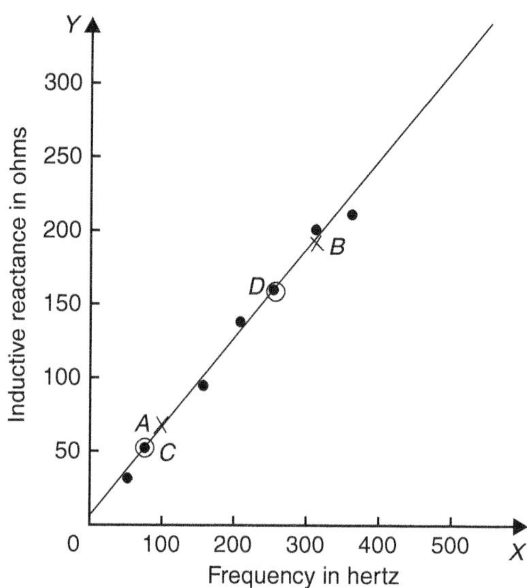

Figure 66.2

Problem 4. The experimental values relating centripetal force and radius, for a mass travelling at constant velocity in a circle, are as shown:

Force (N)	5	10	15	20	25	30	35	40
Radius (cm)	55	30	16	12	11	9	7	5

Determine the equations of (a) the regression line of force on radius and (b) the regression line of force on radius. Hence, calculate the force at a radius of 40 cm and the radius corresponding to a force of 32 N

Let the radius be the independent variable X, and the force be the dependent variable Y. (This decision is usually based on a 'cause' corresponding to X and an 'effect' corresponding to Y)

(a) The equation of the regression line of force on radius is of the form $Y = a_0 + a_1X$ and the constants a_0 and a_1 are determined from the normal equations:

$$\sum Y = a_0N + a_1\sum X$$

$$\text{and} \quad \sum XY = a_0\sum X + a_1\sum X^2$$

(from equations (1) and (2))

Using a tabular approach to determine the values of the summations gives:

Radius, X	Force, Y	X^2
55	5	3025
30	10	900
16	15	256
12	20	144
11	25	121
9	30	81
7	35	49
5	40	25
$\sum X = 145$	$\sum Y = 180$	$\sum X^2 = 4601$

XY	Y^2
275	25
300	100
240	225
240	400
275	625
270	900
245	1225
200	1600
$\sum XY = 2045$	$\sum Y^2 = 5100$

Thus $180 = 8a_0 + 145a_1$

and $2045 = 145a_0 + 4601a_1$

Solving these simultaneous equations gives $a_0 = 33.7$ and $a_1 = -0.617$, correct to 3 significant figures. Thus the equation of the regression line of force on radius is:

$$Y = 33.7 - 0.617X$$

(b) The equation of the regression line of radius on force is of the form $X = b_0 + b_1 Y$ and the constants b_0 and b_1 are determined from the normal equations:

$$\sum X = b_0 N + b_1 \sum Y$$

and $\sum XY = b_0 \sum Y + b_1 \sum Y^2$

(from equations (3) and (4))

The values of the summations have been obtained in part (a) giving:

$$145 = 8b_0 + 180b_1$$

and $2045 = 180b_0 + 5100b_1$

Solving these simultaneous equations gives $b_0 = 44.2$ and $b_1 = -1.16$, correct to 3 significant figures. Thus the equation of the regression line of radius on force is:

$$X = 44.2 - 1.16Y$$

The force, Y, at a radius of 40 cm, is obtained from the regression line of force on radius, i.e. $y = 33.7 - 0.617(40) = 9.02$,

i.e. **the force at a radius of 40 cm is 9.02 N**

The radius, X, when the force is 32 Newton's is obtained from the regression line of radius on force, i.e. $X = 44.2 - 1.16(32) = 7.08$,

i.e. **the radius when the force is 32 N is 7.08 cm**

Now try the following Practice Exercise

Practice Exercise 301 Linear regression (Answers on page 737)

In Problems 1 and 2, determine the equation of the regression line of Y on X, correct to 3 significant figures.

1.

X	14	18	23	30	50
Y	900	1200	1600	2100	3800

2.

X	6	3	9	15	2	14	21	13
Y	1.3	0.7	2.0	3.7	0.5	2.9	4.5	2.7

In Problems 3 and 4, determine the equations of the regression lines of X on Y for the data stated, correct to 3 significant figures.

3. The data given in Problem 1.
4. The data given in Problem 2.
5. The relationship between the voltage applied to an electrical circuit and the current flowing is as shown:

Current (mA)	2	4	6	8	10	12	14
Applied voltage (V)	5	11	15	19	24	28	33

Assuming a linear relationship, determine the equation of the regression line of applied voltage, Y, on current, X, correct to 4 significant figures.

6. For the data given in Problem 5, determine the equation of the regression line of current

on applied voltage, correct to 3 significant figures.

7. Draw the scatter diagram for the data given in Problem 5 and show the regression lines of applied voltage on current and current on applied voltage. Hence determine the values of (a) the applied voltage needed to give a current of 3 mA and (b) the current flowing when the applied voltage is 40 volts, assuming the regression lines are still true outside of the range of values given.

8. In an experiment to determine the relationship between force and momentum, a force X, is applied to a mass, by placing the mass on an inclined plane, and the time, Y, for the velocity to change from u m/s to v m/s is measured. The results obtained are as follows:

Force (N)	11.4	18.7	11.7
Time (s)	0.56	0.35	0.55

Force (N)	12.3	14.7	18.8	19.6
Time (s)	0.52	0.43	0.34	0.31

Determine the equation of the regression line of time on force, assuming a linear relationship between the quantities, correct to 3 significant figures.

9. Find the equation for the regression line of force on time for the data given in Problem 8, correct to 3 decimal places.

10. Draw a scatter diagram for the data given in Problem 8 and show the regression lines of time on force and force on time. Hence find (a) the time corresponding to a force of 16 N, and (b) the force at a time of 0.25 s, assuming the relationship is linear outside of the range of values given.

For fully worked solutions to each of the problems in Practice Exercise 301 in this chapter, go to the website:

www.routledge.com/cw/bird

Chapter 67

Sampling and estimation theories

Why it is important to understand: Sampling and estimation theories

Estimation theory is a branch of statistics and signal processing that deals with estimating the values of parameters based on measured/empirical data that has a random component. Estimation theory can be found at the heart of many electronic signal processing systems designed to extract information; these systems include radar, sonar, speech, image, communications, control and seismology. This chapter introduces some of the principles involved with sampling and estimation theories.

At the end of this chapter, you should be able to:

- understand sampling distributions
- determine the standard error of the means
- understand point and interval estimates and confidence intervals
- calculate confidence limits
- estimate the mean and standard deviation of a population from sample data
- estimate the mean of a population based on a small sample data using a Student's t distribution

67.1 Introduction

The concepts of elementary sampling theory and estimation theories introduced in this chapter will provide the basis for a more detailed study of inspection, control and quality control techniques used in industry. Such theories can be quite complicated; in this chapter a full treatment of the theories and the derivation of formulae have been omitted for clarity–basic concepts only have been developed.

67.2 Sampling distributions

In statistics, it is not always possible to take into account all the members of a set and in these circumstances, a **sample**, or many samples, are drawn from a population. Usually when the word sample is used, it means that a **random sample** is taken. If each member of a population has the same chance of being selected, then a sample taken from that population is called random. A sample which is not random is said to be **biased** and this usually occurs when some influence affects the selection.

When it is necessary to make predictions about a population based on random sampling, often many samples of, say, N members are taken, before the predictions are made. If the mean value and standard deviation of each of the samples is calculated, it is found that the results vary from sample to sample, even though the samples are all taken from the same population. In the theories introduced in the following sections, it is important to know whether the differences in the values obtained are due to chance or whether the differences obtained are related in some way. If M samples of N members are drawn at random from a population, the mean values for the M samples together form a set of data. Similarly, the standard deviations of the M samples collectively form a set of data. Sets of data based on many samples drawn from a population are called **sampling distributions**. They are often used to describe the chance fluctuations of mean values and standard deviations based on random sampling.

67.3 The sampling distribution of the means

Suppose that it is required to obtain a sample of two items from a set containing five items. If the set is the five letters A, B, C, D and E, then the different samples which are possible are:

AB, AC, AD, AE, BC, BD, BE,

CD, CE and DE,

that is, ten different samples. The number of possible different samples in this case is given by ${}^5C_2 = \frac{5!}{2!3!} = 10$, from combination on pages 154 and 644. Similarly, the number of different ways in which a sample of three items can be drawn from a set having ten members, ${}^{10}C_3 = \frac{10!}{3!7!} = 120$. It follows that when a small sample is drawn from a large population, there are very many different combinations of members possible. With so many different samples possible, quite a large variation can occur in the mean values of various samples taken from the same population.

Usually, the greater the number of members in a sample, the closer will be the mean value of the sample to that of the population. Consider the set of numbers 3, 4, 5, 6, and 7. For a sample of 2 members, the lowest value of the mean is $\frac{3+4}{2}$, i.e. 3.5; the highest is $\frac{6+7}{2}$, i.e. 6.5, giving a range of mean values of $6.5 - 3.5 = 3$. For a sample of 3 members, the range is, $\frac{3+4+5}{3}$ to $\frac{5+6+7}{3}$ that is, 2. As the number in the sample increases, the range decreases until, in the limit, if the sample contains all the members of the set, the range of mean values is zero. When many samples are drawn from a population and a sample distribution of the mean values is small provided the number in the sample is large. Because the range is small it follows that the standard deviation of all the mean values will also be small, since it depends on the distance of the mean values from the distribution mean. The relationship between the standard deviation of the mean values of a sampling distribution and the number in each sample can be expressed as follows:

Theorem 1

'If all possible samples of size N are drawn from a finite population, N_p, without replacement, and the standard deviation of the mean values of the sampling distribution of means is determined then:

$$\sigma_{\bar{x}} = \frac{\sigma}{\sqrt{N}}\sqrt{\frac{N_p - N}{N_p - 1}}$$

where $\sigma_{\bar{x}}$ is the standard deviation of the sampling distribution of means and σ is the standard deviation of the population.'

The standard deviation of a sampling distribution of mean values is called the **standard error of the means**, thus

standard error of the means,

$$\sigma_{\bar{x}} = \frac{\sigma}{\sqrt{N}}\sqrt{\frac{N_p - N}{N_p - 1}} \qquad (1)$$

Equation (1) is used for a finite population of size N_p and/or for sampling without replacement. The word 'error' in the 'standard error of the means' does not mean that a mistake has been made but rather that there is a degree of uncertainty in predicting the mean value of a population based on the mean values of the samples. The formula for the standard error of the means is true for all values of the number in the sample, N. When N_p is very large compared with N or when the population is infinite (this can be considered to be the case when sampling is done with replacement), the correction factor $\sqrt{\frac{N_p - N}{N_p - 1}}$ approaches unit and equation (1) becomes

$$\sigma_{\bar{x}} = \frac{\sigma}{\sqrt{N}} \qquad (2)$$

Equation (2) is used for an infinite population and/or for sampling with replacement.

Theorem 2

'If all possible samples of size N are drawn from a population of size N_p and the mean value of the sampling distribution of means $\mu_{\bar{x}}$ is determined then

$$\mu_{\bar{x}} = \mu \qquad (3)$$

where μ is the mean value of the population'

In practice, all possible samples of size N are not drawn from the population. However, if the sample size is large (usually taken as 30 or more), then the relationship between the mean of the sampling distribution of means and the mean of the population is very near to that shown in equation (3). Similarly, the relationship between the standard error of the means and the standard deviation of the population is very near to that shown in equation (2).

Another important property of a sampling distribution is that when the sample size, N, is large, **the sampling distribution of means approximates to a normal distribution** (see Chapter 64), of mean value $\mu_{\overline{x}}$ and standard deviation $\sigma_{\overline{x}}$. This is true for all normally distributed populations and also for populations that are not normally distributed provided the population size is at least twice as large as the sample size. This property of normality of a sampling distribution is based on a special case of the 'central limit theorem', an important theorem relating to sampling theory. Because the sampling distribution of means and standard deviations is normally distributed, the table of the partial areas under the standardised normal curve (shown in Table 64.1 on page 658) can be used to determine the probabilities of a particular sample lying between, say, ± 1 standard deviation and so on. This point is expanded in Problem 3.

Problem 1. The heights of 3000 people are normally distributed with a mean of 175 cm, and a standard deviation of 8 cm. If random samples are taken of 40 people. predict the standard deviation and the mean of the sampling distribution of means if sampling is done (a) with replacement, and (b) without replacement

For the population: number of members, $N_p = 3000$; standard deviation, $\sigma = 8$ cm; mean, $\mu = 175$ cm

For the samples: number in each sample, $N = 40$

(a) When sampling is done **with replacement**, the total number of possible samples (two or more can be the same) is infinite. Hence, from equation (2) the **standard error of the mean (i.e. the standard deviation of the sampling distribution of means)**

$$\sigma_{\overline{x}} = \frac{\sigma}{\sqrt{N}} = \frac{8}{\sqrt{40}} = \mathbf{1.265\,cm}$$

From equation (3), **the mean of the sampling distribution,** $\mu_{\overline{x}} = \mu = \mathbf{175\,cm}$.

(b) When sampling is done **without replacement**, the total number of possible samples is finite and hence equation (1) applies. Thus **the standard error of the means**

$$\sigma_{\overline{x}} = \frac{\sigma}{\sqrt{N}}\sqrt{\frac{N_p - N}{N_p - 1}}$$

$$= \frac{8}{\sqrt{40}}\sqrt{\frac{3000 - 40}{3000 - 1}}$$

$$= (1.265)(0.9935) = \mathbf{1.257\,cm}$$

As stated, following equation (3), provided the sample size is large, the mean of the sampling distribution of means is the same for both finite and infinite populations. Hence, from equation (3),

$$\mu_{\overline{x}} = \mathbf{175\,cm}$$

Problem 2. 1500 ingots of a metal have a mean mass of 6.5 kg and a standard deviation of 0.5 kg. Find the probability that a sample of 60 ingots chosen at random from the group, without replacement, will have a combined mass of (a) between 378 and 396 kg, and (b) more than 399 kg

For the population: numbers of members, $N_p = 1500$; standard deviation, $\sigma = 0.5$ kg; mean $\mu = 6.5$ kg.

For the sample: number in sample, $N = 60$

If many samples of 60 ingots had been drawn from the group, then the mean of the sampling distribution of means, $\mu_{\overline{x}}$ would be equal to the mean of the population. Also, the standard error of means is given by

$$\sigma_{\overline{x}} = \frac{\sigma}{\sqrt{N}}\sqrt{\frac{N_p - N}{N_p - 1}}$$

In addition, the sample distribution would have been approximately normal. Assume that the sample given in the problem is one of many samples. For many (theoretical) samples:

the mean of the sampling distribution of means, $\mu_{\overline{x}} = \mu = 6.5$ kg

Also, the standard error of the means,

$$\sigma_{\overline{x}} = \frac{\sigma}{\sqrt{N}}\sqrt{\frac{N_p - N}{N_p - 1}}$$

$$= \frac{0.5}{\sqrt{60}}\sqrt{\frac{1500 - 60}{1500 - 1}}$$

$$= 0.0633\,\text{kg}$$

Thus, the sample under consideration is part of a normal distribution of mean value 6.5 kg and a standard error of the means of 0.0633 kg.

(a) If the combined mass of 60 ingots is between 378 and 396 kg, then the mean mass of each of the 60 ingots lies between $\frac{378}{60}$ and $\frac{396}{60}$ kg, i.e. between 6.3 kg and 6.6 kg.

Since the masses are normally distributed, it is possible to use the techniques of the normal distribution to determine the probability of the mean mass lying between 6.3 and 6.6 kg. The normal standard variate value, z, is given by

$$z = \frac{x - \overline{x}}{\sigma}$$

hence for the sampling distribution of means, this becomes,

$$z = \frac{x - \mu_{\overline{x}}}{\sigma_{\overline{x}}}$$

Thus, 6.3 kg corresponds to a z-value of $\frac{6.3 - 6.5}{0.0633} = -3.16$ standard deviations.

Similarly, 6.6 kg corresponds to a z-value of $\frac{6.6 - 6.5}{0.0633} = 1.58$ standard deviations.

Using Table 64.1 (page 658), the areas corresponding to these values of standard deviations are 0.4992 and 0.4430 respectively. Hence **the probability of the mean mass lying between 6.3 kg and 6.6 kg is** $0.4992 + 0.4430 = \mathbf{0.9422}$. (This means that if 10 000 samples are drawn, 9422 of these samples will have a combined mass of between 378 and 396 kg.)

(b) If the combined mass of 60 ingots is 399 kg, the mean mass of each ingot is $\frac{399}{60}$, that is, 6.65 kg.

The z-value for 6.65 kg is $\frac{6.65 - 6.5}{0.0633}$, i.e. 2.37 standard deviations. From Table 64.1 (page 658), the area corresponding to this z-value is 0.4911. But this is the area between the ordinate $z = 0$ and ordinate $z = 2.37$. The 'more than' value required is the total area to the right of the $z = 0$ ordinate, less the value between $z = 0$ and $z = 2.37$, i.e. $0.5000 - 0.4911$

Thus, since areas are proportional to probabilities for the standardised normal curve, **the probability of the mean mass being more than 6.65 kg** is $0.5000 - 0.4911$, i.e. **0.0089**. (This means that only 89 samples in 10 000, for example, will have a combined mass exceeding 399 kg.)

Now try the following Practice Exercise

Practice Exercise 302 The sampling distribution of means (Answers on page 737)

1. The lengths of 1500 bolts are normally distributed with a mean of 22.4 cm and a standard deviation of 0.0438 cm. If 30 samples are drawn at random from this population, each sample being 36 bolts, determine the mean of the sampling distribution and standard error of the means when sampling is done with replacement.

2. Determine the standard error of the means in Problem 1, if sampling is done without replacement, correct to four decimal places.

3. A power punch produces 1800 washers per hour. The mean inside diameter of the washers is 1.70 cm and the standard deviation is 0.013 cm. Random samples of 20 washers are drawn every 5 minutes. Determine the mean of the sampling distribution of means and the standard error of the means for the one hour's output from the punch, (a) with replacement and (b) without replacement, correct to three significant figures.

A large batch of electric light bulbs have a mean time to failure of 800 hours and the standard deviation of the batch is 60 hours. Use this data and also Table 64.1 on page 658 to solve Problems 4 to 6.

4. If a random sample of 64 light bulbs is drawn from the batch, determine the probability that the mean time to failure will be less than 785 hours, correct to three decimal places.

5. Determine the probability that the mean time to failure of a random sample of 16 light bulbs will be between 790 hours and 810 hours, correct to three decimal places.

6. For a random sample of 64 light bulbs, determine the probability that the mean time to failure will exceed 820 hours, correct to two significant figures.

7. The contents of a consignment of 1200 tins of a product have a mean mass of 0.504 kg and a standard deviation of 2.3 g. Determine the probability that a random sample of 40 tins drawn from the consignment will have a combined mass of (a) less than 20.13 kg, (b) between 20.13 kg and 20.17 kg, and (c) more than 20.17 kg, correct to three significant figures.

67.4 The estimation of population parameters based on a large sample size

When a population is large, it is not practical to determine its mean and standard deviation by using the basic formulae for these parameters. In fact, when a population is infinite, it is impossible to determine these values. For large and infinite populations the values of the mean and standard deviation may be estimated by using the data obtained from samples drawn from the population.

Point and interval estimates

An estimate of a population parameter, such as mean or standard deviation, based on a single number is called a **point estimate**. An estimate of a population parameter given by two numbers between which the parameter may be considered to lie is called an **interval estimate**. Thus if an estimate is made of the length of an object and the result is quoted as 150 cm, this is a point estimate. If the result is quoted as 150 ± 10 cm, this is an interval estimate and indicates that the length lies between 140 and 160 cm. Generally, a point estimate does not indicate how close the value is to the true value of the quantity and should be accompanied by additional information on which its merits may be judged. A statement of the error or the precision of an estimate is often called its **reliability**. In statistics, when estimates are made of population parameters based on samples, usually interval estimates are used. The word estimate does not suggest that we adopt the approach 'let's guess that the mean value is about ..', but rather that a value is carefully selected and the degree of confidence which can be placed in the estimate is given in addition.

Confidence intervals

It is stated in Section 67.3 that when samples are taken from a population, the mean values of these samples are approximately normally distributed, that is, the mean values forming the sampling distribution of means is approximately normally distributed. It is also true that if the standard deviations of each of the samples is found, then the standard deviations of all the samples are approximately normally distributed, that is, the standard deviations of the sampling distribution of standard deviations are approximately normally distributed. Parameters such as the mean or the standard deviation of a sampling distribution are called **sampling statistics**, S. Let μ_S be the mean value of a sampling statistic of the sampling distribution, that is, the mean value of the means of the samples or the mean value of the standard deviations of the samples. Also let σ_S be the standard deviation of a sampling statistic of the sampling distribution, that is, the standard deviation of the means of the samples or the standard deviation of the standard deviations of the samples. Because the sampling distribution of the means and of the standard deviations are normally distributed, it is possible to predict the probability of the sampling statistic lying in the intervals:

mean ± 1 standard deviation,

mean ± 2 standard deviations,

or mean ± 3 standard deviations,

by using tables of the partial areas under the standardised normal curve given in Table 64.1 on page 658. From this table, the area corresponding to a z-value of $+1$ standard deviation is 0.3413, thus the area corresponding to ± 1 standard deviation is 2×0.3413, that is, 0.6826. Thus the percentage probability of a sampling statistic lying between the mean ± 1 standard deviation is 68.26%. Similarly, the probability of a sampling statistic lying between the mean ± 2 standard deviations is 95.44% and of lying between the mean ± 3 standard deviations is 99.74%

The values 68.26%, 95.44% and 99.74% are called the **confidence levels** for estimating a sampling statistic. A confidence level of 68.26% is associated with two distinct values, these being, $S-$ (1 standard deviation), i.e. $S-\sigma_S$ and $S+$(1 standard deviation), i.e. $S+\sigma_S$. These two values are called the **confidence limits** of the estimate and the distance between the confidence limits is called the **confidence interval**. A confidence interval indicates the expectation or confidence of finding an estimate of the population statistic in that interval, based on a sampling statistic. The list in Table 67.1 is based on values given in Table 64.1, and gives some of the confidence levels used in practice and their associated z-values (some of the values given are based on interpolation). When the table is used in this context, z-values are usually indicated by 'z_C' and are called the **confidence coefficients**.

Table 67.1

Confidence level, %	Confidence coefficient, z_C
99	2.58
98	2.33
96	2.05
95	1.96
90	1.645
80	1.28
50	0.6745

Any other values of confidence levels and their associated confidence coefficients can be obtained using Table 64.1.

Problem 3. Determine the confidence coefficient corresponding to a confidence level of 98.5%

98.5% is equivalent to a per unit value of 0.9850. This indicates that the area under the standardised normal curve between $-z_C$ and $+z_C$, i.e. corresponding to $2z_C$, is 0.9850 of the total area. Hence the area between the mean value and z_C is $\frac{0.9850}{2}$ i.e. 0.4925 of the total area. The z-value corresponding to a partial area of 0.4925 is 2.43 standard deviations from Table 64.1. Thus, **the confidence coefficient corresponding to a confidence level of 98.5% is 2.43**

(a) Estimating the mean of a population when the standard deviation of the population is known

When a sample is drawn from a large population whose standard deviation is known, the mean value of the sample, $\bar{x}$, can be determined. This mean value can be used to make an estimate of the mean value of the population, μ. When this is done, the estimated mean value of the population is given as lying between two values, that is, lying in the confidence interval between the confidence limits. If a high level of confidence is required in the estimated value of μ, then the range of the confidence interval will be large. For example, if the required confidence level is 96%, then from Table 67.1 the confidence interval is from $-z_C$ to $+z_C$, that is, $2\times 2.05=4.10$ standard deviations wide. Conversely, a low level of confidence has a narrow confidence interval and a confidence level of, say, 50%, has a confidence interval of 2×0.6745, that is 1.3490 standard deviations. The 68.26% confidence level for an estimate of the population mean is given by estimating that the population mean, μ, is equal to the same mean, $\bar{x}$, and then stating the confidence interval of the estimate. Since the 68.26% confidence level is associated with '± 1 standard deviation of the means of the sampling distribution', then the 68.26% confidence level for the estimate of the population mean is given by:

$$\bar{x}\pm 1\sigma_{\bar{x}}$$

In general, any particular confidence level can be obtained in the estimate, by using $\bar{x}\pm z_C\sigma_{\bar{x}}$, where z_C is the confidence coefficient corresponding to the particular confidence level required. Thus for a 96% confidence level, the confidence limits of the population mean are given by $\bar{x}\pm 2.05\sigma_{\bar{x}}$. Since only one sample has been drawn, the standard error of the means, $\sigma_{\bar{x}}$, is not known. However, it is shown in Section 67.3 that

$$\sigma_{\bar{x}}=\frac{\sigma}{\sqrt{N}}\sqrt{\frac{N_p-N}{N_p-1}}$$

Thus, **the confidence limits of the mean of the population are**:

$$\bar{x}\pm\frac{z_C\sigma}{\sqrt{N}}\sqrt{\frac{N_p-N}{N_p-1}} \qquad (4)$$

for a finite population of size N_p

The **confidence limits for the mean of the population are**:

$$\bar{x}\pm\frac{z_C\sigma}{\sqrt{N}} \qquad (5)$$

for an infinite population.

Thus for a sample of size N and mean $\bar{x}$, drawn from an infinite population having a standard deviation of σ, the mean value of the population is estimated to be, for example,

$$\bar{x}\pm\frac{2.33\sigma}{\sqrt{N}}$$

for a confidence level of 98%. This indicates that the mean value of the population lies between

$$\bar{x}-\frac{2.33\sigma}{\sqrt{N}} \text{ and } \bar{x}+\frac{2.33\sigma}{\sqrt{N}}$$

with 98% confidence in this prediction.

Problem 4. It is found that the standard deviation of the diameters of rivets produced by a certain machine over a long period of time is 0.018 cm. The diameters of a random sample of 100 rivets produced by this machine in a day have a mean value of 0.476 cm. If the machine produces 2500 rivets per day, determine (a) the 90% confidence limits, and (b) the 97% confidence limits for an estimate of the mean diameter of all the rivets produced by the machine in a day

For the population:

standard deviation, $\sigma = 0.018\,\text{cm}$

number in the population, $N_p = 2500$

For the sample:

number in the sample, $N = 100$

mean, $\bar{x} = 0.476\,\text{cm}$

There is a finite population and the standard deviation of the population is known, hence expression (4) is used for determining an estimate of the confidence limits of the population mean, i.e.

$$\bar{x} \pm \frac{z_C\sigma}{\sqrt{N}}\sqrt{\frac{N_p - N}{N_p - 1}}$$

(a) For a 90% confidence level, the value of z_C, the confidence coefficient, is 1.645 from Table 67.1. Hence, the estimate of the confidence limits of the population mean,

$$\mu = 0.476 \pm \left(\frac{(1.645)(0.018)}{\sqrt{100}}\right)\sqrt{\frac{2500 - 100}{2500 - 1}}$$

$$= 0.476 \pm (0.00296)(0.9800)$$

$$= 0.476 \pm 0.0029\,\text{cm}$$

Thus, **the 90% confidence limits are 0.473 cm and 0.479 cm**.

This indicates that if the mean diameter of a sample of 100 rivets is 0.476cm, then it is predicted that the mean diameter of all the rivets will be between 0.473cm and 0.479cm and this prediction is made with confidence that it will be correct nine times out of ten.

(b) For a 97% confidence level, the value of z_C has to be determined from a table of partial areas under the standardised normal curve given in Table 64.1, as it is not one of the values given in Table 67.1. The total area between ordinates drawn at $-z_C$ and $+z_C$ has to be 0.9700. Because the is $\frac{0.9700}{2}$, i.e. 0.4850. From Table 64.1 an area of 0.4850 corresponds to a z_C value of 2.17. Hence, the estimated value of the confidence limits of the population mean is between

$$\bar{x} \pm \frac{z_C\sigma}{\sqrt{N}}\sqrt{\frac{N_p - N}{N_p - 1}}$$

$$= 0.476 \pm \left(\frac{(2.17)(0.018)}{\sqrt{100}}\right)\sqrt{\frac{2500 - 100}{2500 - 1}}$$

$$= 0.476 \pm (0.0039)(0.9800)$$

$$= 0.476 \pm 0.0038$$

Thus, **the 97% confidence limits are 0.472 cm and 0.480 cm**.

It can be seen that the higher value of confidence level required in part (b) results in a larger confidence interval.

Problem 5. The mean diameter of a long length of wire is to be determined. The diameter of the wire is measured in 25 places selected at random throughout its length and the mean of these values is 0.425 mm. If the standard deviation of the diameter of the wire is given by the manufacturers as 0.030 mm, determine (a) the 80% confidence interval of the estimated mean diameter of the wire, and (b) with what degree of confidence it can be said that 'the mean diameter is 0.425 ± 0.012 mm'

For the population: $\sigma = 0.030\,\text{mm}$
For the sample: $N = 25$, $\bar{x} = 0.425\,\text{mm}$
Since an infinite number of measurements can be obtained for the diameter of the wire, the population is infinite and the estimated value of the confidence interval of the population mean is given by expression (5).

(a) For an 80% confidence level, the value of z_C is obtained from Table 67.1 and is 1.28.

The 80% confidence level estimate of the confidence interval is

$$\mu = \bar{x} \pm \frac{z_C\sigma}{\sqrt{N}} = 0.425 \pm \frac{(1.28)(0.030)}{\sqrt{25}}$$

$$= 0.425 \pm 0.0077\,\text{mm}$$

i.e. **the 80% confidence interval is from 0.417 mm to 0.433 mm**.

This indicates that the estimated mean diameter of the wire is between 0.417mm and 0.433mm and that this prediction is likely to be correct 80 times out of 100

(b) To determine the confidence level, the given data is equated to expression (5), giving

$$0.425 \pm 0.012 = \overline{x} \pm z_C \frac{\sigma}{\sqrt{N}}$$

But $\overline{x} = 0.425$, therefore

$$\pm z_C \frac{\sigma}{\sqrt{N}} = \pm 0.012$$

i.e. $z_C = \dfrac{0.012\sqrt{N}}{\sigma} = \pm\dfrac{(0.012)(5)}{0.030} = \pm 2$

Using Table 64.1 of partial areas under the standardised normal curve, a z_C value of 2 standard deviations corresponds to an area of 0.4772 between the mean value ($z_C = 0$) and +2 standard deviations. Because the standardised normal curve is symmetrical, the area between the mean and ±2 standard deviations is 0.4772 × 2, i.e. 0.9544

Thus the confidence level corresponding to 0.425 ± 0.012 mm is 95.44%.

(b) Estimating the mean and standard deviation of a population from sample data

The standard deviation of a large population is not known and, in this case, several samples are drawn from the population. The mean of the sampling distribution of means, $\mu_{\overline{x}}$ and the standard deviation of the sampling distribution of means (i.e. the standard error of the means), $\sigma_{\overline{x}}$, may be determined. The confidence limits of the mean value of the population, μ, are given by:

$$\mu_{\overline{x}} \pm z_C \sigma_{\overline{x}} \qquad (6)$$

where z_C is the confidence coefficient corresponding to the confidence level required.

To make an estimate of the standard deviation, σ, of a normally distributed population:

(i) a sampling distribution of the standard deviations of the samples is formed, and

(ii) the standard deviation of the sampling distribution is determined by using the basic standard deviation formula.

This standard deviation is called the standard error of the standard deviations and is usually signified by σ_s. If s is the standard deviation of a sample, then the confidence limits of the standard deviation of the population are given by:

$$s \pm z_C \sigma_S \qquad (7)$$

where z_C is the confidence coefficient corresponding to the required confidence level.

Problem 6. Several samples of 50 fuses selected at random from a large batch are tested when operating at a 10% overload current and the mean time of the sampling distribution before the fuses failed is 16.50 minutes. The standard error of the means is 1.4 minutes. Determine the estimated mean time to failure of the batch of fuses for a confidence level of 90%

For the sampling distribution: the mean, $\mu_{\overline{x}} = 16.50$, the standard error of the means, $\sigma_{\overline{x}} = 1.4$

The estimated mean of the population is based on sampling distribution data only and so expression (6) is used, i.e. the confidence limits of the estimated mean of the population are $\mu_{\overline{x}} \pm z_C \sigma_{\overline{x}}$

For a 90% confidence level, $z_C = 1.645$ (from Table 67.1), thus

$$\mu_{\overline{x}} \pm z_C \sigma_{\overline{x}} = 16.50 \pm (1.645)(1.4)$$
$$= 16.50 \pm 2.30 \text{ minutes}$$

Thus, **the 90% confidence level of the mean time to failure is from 14.20 minutes to 18.80 minutes**.

Problem 7. The sampling distribution of random samples of capacitors drawn from a large batch is found to have a standard error of the standard deviations of 0.12 µF. Determine the 92% confidence interval for the estimate of the standard deviation of the whole batch, if in a particular sample, the standard deviation is 0.60 µF. It can be assumed that the values of capacitance of the batch are normally distributed

For the sample: the standard deviation, $s = 0.60\,\mu\text{F}$.
For the sampling distribution: the standard error of the standard deviations,

$$\sigma_S = 0.12\,\mu\text{F}$$

When the confidence level is 92%, then by using Table 64.1 of partial areas under the standardised normal curve,

$$\text{area} = \frac{0.9200}{2} = 0.4600,$$

giving z_C as ±1.751 standard deviations (by interpolation)

Since the population is normally distributed, the confidence limits of the standard deviation of the population may be estimated by using expression (7), i.e. $s \pm z_C \sigma_S = 0.60 \pm (1.751)(0.12)$

$$= 0.60 \pm 0.21\,\mu\text{F}$$

Thus, **the 92% confidence interval for the estimate of the standard deviation for the batch is from 0.39 μF to 0.81 μF**

Now try the following Practice Exercise

Practice Exercise 303 Estimation of population parameters based on a large sample size (Answers on page 737)

1. Measurements are made on a random sample of 100 components drawn from a population of size 1546 and having a standard deviation of 2.93 mm. The mean measurement of the components in the sample is 67.45 mm. Determine the 95% and 99% confidence limits for an estimate of the mean of the population.

2. The standard deviation of the masses of 500 blocks is 150 kg. A random sample of 40 blocks has a mean mass of 2.40 Mg.

   (a) Determine the 95% and 99% confidence intervals for estimating the mean mass of the remaining 460 blocks.

   (b) With what degree of confidence can it be said that the mean mass of the remaining 460 blocks is 2.40 ± 0.035 Mg?

3. In order to estimate the thermal expansion of a metal, measurements of the change of length for a known change of temperature are taken by a group of students. The sampling distribution of the results has a mean of $12.81 \times 10^{-4}\,\text{m}\,^\circ\text{C}^{-1}$ and a standard error of the means of $0.04 \times 10^{-4}\,\text{m}\,^\circ\text{C}^{-1}$. Determine the 95% confidence interval for an estimate of the true value of the thermal expansion of the metal, correct to two decimal places.

4. The standard deviation of the time to failure of an electronic component is estimated as 100 hours. Determine how large a sample of these components must be, in order to be 90% confident that the error in the estimated time to failure will not exceed (a) 20 hours, and (b) 10 hours.

5. A sample of 60 slings of a certain diameter, used for lifting purposes, are tested to destruction (that is, loaded until they snapped). The mean and standard deviation of the breaking loads are 11.09 tonnes and 0.73 tonnes respectively. Find the 95% confidence interval for the mean of the snapping loads of all the slings of this diameter produced by this company.

6. The time taken to assemble a servo-mechanism is measured for 40 operatives and the mean time is 14.63 minutes with a standard deviation of 2.45 minutes. Determine the maximum error in estimating the true mean time to assemble the servo-mechanism for all operatives, based on a 95% confidence level.

67.5 Estimating the mean of a population based on a small sample size

The methods used in Section 67.4 to estimate the population mean and standard deviation rely on a relatively large sample size, usually taken as 30 or more. This is because when the sample size is large the sampling distribution of a parameter is approximately normally distributed. When the sample size is small, usually taken as less than 30, the techniques used for estimating the population parameters in Section 67.4 become more and more inaccurate as the sample size becomes smaller, since the sampling distribution no longer approximates to a normal distribution. Investigations were carried out into the effect of small sample sizes on the estimation theory by W. S. Gosset in the early twentieth century and, as a result of his work, tables are available which enable a realistic estimate to be made, when sample sizes are small. In these tables, the t-value is determined from the relationship

$$t = \frac{(\bar{x} - \mu)}{s}\sqrt{N-1}$$

where $\bar{x}$ is the mean value of a sample, μ is the mean value of the population from which the sample is drawn, s is the standard deviation of the sample and N is the number of independent observations in the sample. He published his findings under the pen name of 'Student', and these tables are often referred to as the **'Student's t distribution'**.

The confidence limits of the mean value of a population based on a small sample drawn at random from

Table 67.2 Percentile values (t_p) for Student's t distribution with ν degrees of freedom (shaded area $= p$)

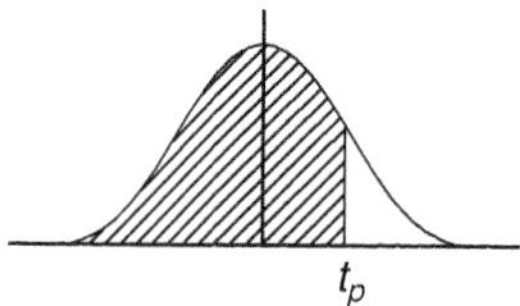

ν	$t_{0.995}$	$t_{0.99}$	$t_{0.975}$	$t_{0.95}$	$t_{0.90}$	$t_{0.80}$	$t_{0.75}$	$t_{0.70}$	$t_{0.60}$	$t_{0.55}$
1	63.66	31.82	12.71	6.31	3.08	1.376	1.000	0.727	0.325	0.158
2	9.92	6.96	4.30	2.92	1.89	1.061	0.816	0.617	0.289	0.142
3	5.84	4.54	3.18	2.35	1.64	0.978	0.765	0.584	0.277	0.137
4	4.60	3.75	2.78	2.13	1.53	0.941	0.741	0.569	0.271	0.134
5	4.03	3.36	2.57	2.02	1.48	0.920	0.727	0.559	0.267	0.132
6	3.71	3.14	2.45	1.94	1.44	0.906	0.718	0.553	0.265	0.131
7	3.50	3.00	2.36	1.90	1.42	0.896	0.711	0.549	0.263	0.130
8	3.36	2.90	2.31	1.86	1.40	0.889	0.706	0.546	0.262	0.130
9	3.25	2.82	2.26	1.83	1.38	0.883	0.703	0.543	0.261	0.129
10	3.17	2.76	2.23	1.81	1.37	0.879	0.700	0.542	0.260	0.129
11	3.11	2.72	2.20	1.80	1.36	0.876	0.697	0.540	0.260	0.129
12	3.06	2.68	2.18	1.78	1.36	0.873	0.695	0.539	0.259	0.128
13	3.01	2.65	2.16	1.77	1.35	0.870	0.694	0.538	0.259	0.128
14	2.98	2.62	2.14	1.76	1.34	0.868	0.692	0.537	0.258	0.128
15	2.95	2.60	2.13	1.75	1.34	0.866	0.691	0.536	0.258	0.128
16	2.92	2.58	2.12	1.75	1.34	0.865	0.690	0.535	0.258	0.128
17	2.90	2.57	2.11	1.74	1.33	0.863	0.689	0.534	0.257	0.128
18	2.88	2.55	2.10	1.73	1.33	0.862	0.688	0.534	0.257	0.127
19	2.86	2.54	2.09	1.73	1.33	0.861	0.688	0.533	0.257	0.127
20	2.84	2.53	2.09	1.72	1.32	0.860	0.687	0.533	0.257	0.127
21	2.83	2.52	2.08	1.72	1.32	0.859	0.686	0.532	0.257	0.127
22	2.82	2.51	2.07	1.72	1.32	0.858	0.686	0.532	0.256	0.127
23	2.81	2.50	2.07	1.71	1.32	0.858	0.685	0.532	0.256	0.127
24	2.80	2.49	2.06	1.71	1.32	0.857	0.685	0.531	0.256	0.127
25	2.79	2.48	2.06	1.71	1.32	0.856	0.684	0.531	0.256	0.127
26	2.78	2.48	2.06	1.71	1.32	0.856	0.684	0.531	0.256	0.127
27	2.77	2.47	2.05	1.70	1.31	0.855	0.684	0.531	0.256	0.127
28	2.76	2.47	2.05	1.70	1.31	0.855	0.683	0.530	0.256	0.127
29	2.76	2.46	2.04	1.70	1.31	0.854	0.683	0.530	0.256	0.127
30	2.75	2.46	2.04	1.70	1.31	0.854	0.683	0.530	0.256	0.127
40	2.70	2.42	2.02	1.68	1.30	0.851	0.681	0.529	0.255	0.126
60	2.66	2.39	2.00	1.67	1.30	0.848	0.679	0.527	0.254	0.126
120	2.62	2.36	1.98	1.66	1.29	0.845	0.677	0.526	0.254	0.126
∞	2.58	2.33	1.96	1.645	1.28	0.842	0.674	0.524	0.253	0.126

the population are given by:

$$\bar{x} \pm \frac{t_C s}{\sqrt{N-1}} \quad (8)$$

In this estimate, t_C is called the confidence coefficient for small samples, analogous to z_C for large samples, s is the standard deviation of the sample, $\bar{x}$ is the mean value of the sample and N is the number of members in the sample. Table 67.2 is called 'percentile values for Student's t distribution'. The columns are headed t_p where p is equal to $0.995, 0.99, 0.975, \ldots, 0.55$. For a confidence level of, say, 95%, the column headed $t_{0.95}$ is selected and so on. The rows are headed with the Greek letter 'nu', ν, and are numbered from 1 to 30 in steps of 1, together with the numbers 40, 60, 120 and ∞. These numbers represent a quantity called the **degrees of freedom**, which is defined as follows:

'the sample number, N, minus the number of population parameters which must be estimated for the sample.'

When determining the t-value, given by

$$t = \frac{(\bar{x}-\mu)}{s}\sqrt{N-1}$$

it is necessary to know the sample parameters $\bar{x}$ and s and the population parameter μ. $\bar{x}$ and s can be calculated for the sample, but usually an estimate has to be made of the population mean μ, based on the sample mean value. The number of degrees of freedom, ν, is given by the number of independent observations in the sample, N, minus the number of population parameters which have to be estimated, k, i.e. $\nu = N - k$. For the equation

$$t = \frac{(\bar{x}-\mu)}{s}\sqrt{N-1}$$

only μ has to be estimated, hence $k = 1$, and $\nu = N - 1$.

When determining the mean of a population based on a small sample size, only one population parameter is to be estimated, and hence ν can always be taken as $(N-1)$. The method used to estimate the mean of a population based on a small sample is shown in Problems 8 to 10.

Problem 8. A sample of 12 measurements of the diameter of a bar are made and the mean of the sample is 1.850 cm. The standard deviation of the sample is 0.16 mm. Determine (a) the 90% confidence limits, and (b) the 70% confidence limits for an estimate of the actual diameter of the bar

For the sample: the sample size, $N = 12$;
mean, $\bar{x} = 1.850$ cm;
standard deviation $s = 0.16$ mm $= 0.016$ cm
Since the sample number is less than 30, the small sample estimate as given in expression (8) must be used. The number of degrees of freedom, i.e. sample size minus the number of estimations of population parameters to be made, is $12 - 1$, i.e. 11

(a) The confidence coefficient corresponding to a percentile value of $t_{0.90}$ and a degree of freedom value of $\nu = 11$ can be found by using Table 67.2, and is 1.36, that is, $t_C = 1.36$. The estimated value of the mean of the population is given by

$$\bar{x} \pm \frac{t_C s}{\sqrt{N-1}} = 1.850 \pm \frac{(1.36)(0.016)}{\sqrt{11}} = 1.850 \pm 0.0066\,\text{cm}.$$

Thus, **the 90% confidence limits are 1.843 cm and 1.857 cm**.
This indicates that the actual diameter is likely to lie between 1.843 cm and 1.857 cm and that this prediction stands a 90% chance of being correct.

(b) The confidence coefficient corresponding to $t_{0.70}$ and to $\nu = 11$ is obtained from Table 67.2, and is 0.540, that is, $t_C = 0.540$

The estimated value of the 70% confidence limits is given by:

$$\bar{x} \pm \frac{t_C s}{\sqrt{N-1}} = 1.850 \pm \frac{(0.540)(0.016)}{\sqrt{11}} = 1.850 \pm 0.0026\,\text{cm}$$

Thus, **the 70% confidence limits are 1.847 cm and 1.853 cm**, i.e. the actual diameter of the bar is between 1.847 cm and 1.853 cm and this result has a 70% probability of being correct.

Problem 9. A sample of 9 electric lamps are selected randomly from a large batch and are tested until they fail. The mean and standard deviations of the time to failure are 1210 hours and 26 hours respectively. Determine the confidence level based on an estimated failure time of 1210 ± 6.5 hours

For the sample: sample size, $N = 9$; standard deviation, $s = 26$ hours; mean, $\bar{x} = 1210$ hours. The confidence limits are given by:

$$\bar{x} \pm \frac{t_C s}{\sqrt{N-1}}$$

and these are equal to 1210 ± 6.5

Since $\bar{x}$ = 1210 hours, then

$$\pm\frac{t_C s}{\sqrt{N-1}} = \pm 6.5$$

i.e. $$t_C = \pm\frac{65\sqrt{N-1}}{s} = \pm\frac{(65)\sqrt{8}}{26} = \pm 0.707$$

From Table 67.2, a t_C value of 0.707, having a value of $N-1$, i.e. 8, gives a t_p value of $t_{0.75}$

Hence, **the confidence level of an estimated failure time of 1210 ± 6.5 hours is 75%**, i.e. it is likely that 75% of all of the lamps will fail between 1203.5 and 1216.5 hours.

Problem 10. The specific resistance of some copper wire of nominal diameter 1 mm is estimated by determining the resistance of 6 samples of the wire. The resistance values found (in ohms per metre) were:

2.16, 2.14, 2.17, 2.15, 2.16 and 2.18

Determine the 95% confidence interval for the true specific resistance of the wire

For the sample: sample size, $N=6$ mean,

$$\bar{x} = \frac{2.16+2.14+2.17+2.15+2.16+2.18}{6} = 2.16\,\Omega\,\text{m}^{-1}$$

standard deviation,

$$s = \sqrt{\frac{(2.16-2.16)^2+(2.14-2.16)^2+(2.17-2.16)^2+(2.15-2.16)^2+(2.16-2.16)^2+(2.18-2.16)^2}{6}}$$

$$= \sqrt{\frac{0.001}{6}} = 0.0129\,\Omega\,\text{m}^{-1}$$

The confidence coefficient corresponding to a percentile value of $t_{0.95}$ and a degree of freedom value of $N-1$, i.e. $6-1=5$ is 2.02 from Table 67.2. The estimated value of the 95% confidence limits is given by:

$$\bar{x} \pm \frac{t_C s}{\sqrt{N-1}} = 2.16 \pm \frac{(2.02)(0.0129)}{\sqrt{5}} = 2.16 \pm 0.01165\,\Omega\,\text{m}^{-1}$$

Thus, **the 95% confidence limits are 2.148 Ω m^{-1} and 2.172 Ω m^{-1}** which indicatesthat there is a 95% chance that the true specific resistance of the wire lies between 2.148 Ω m^{-1} and 2.172 Ω m^{-1}.

Now try the following Practice Exercise

Practice Exercise 304 Estimating the mean of population based on a small sample size (Answers on page 737)

1. The value of the ultimate tensile strength of a material is determined by measurements on 10 samples of the materials. The mean and standard deviation of the results are found to be 5.17 MPa and 0.06 MPa respectively. Determine the 95% confidence interval for the mean of the ultimate tensile strength of the material.
2. Use the data given in Problem 1 above to determine the 97.5% confidence interval for the mean of the ultimate tensile strength of the material.
3. The specific resistance of a reel of German silver wire of nominal diameter 0.5 mm is estimated by determining the resistance of 7 samples of the wire. These were found to have resistance values (in ohms per metre) of:

   1.12 1.15 1.10 1.14 1.15 1.10 and 1.11

   Determine the 99% confidence interval for the true specific resistance of the reel of wire.
4. In determining the melting point of a metal, five determinations of the melting point are made. The mean and standard deviation of the five results are 132.27°C and 0.742°C. Calculate the confidence with which the prediction 'the melting point of the metal is between 131.48°C and 133.06°C' can be made.

For fully worked solutions to each of the problems in Practice Exercises 302 to 304 in this chapter, go to the website:

www.routledge.com/cw/bird

Revision Test 19 Linear correlation and regression, sampling and estimation theories

This Revision Test covers the material contained in Chapters 65 to 67. *The marks for each question are shown in brackets at the end of each question.*

1. The data given below gives the experimental values obtained for the torque output, X, from an electric motor and the current, Y, taken from the supply.

Torque X	Current Y
0	3
1	5
2	6
3	6
4	9
5	11
6	12
7	12
8	14
9	13

Determine the linear coefficient of correlation for this data. (16)

2. Some results obtained from a tensile test on a steel specimen are shown below:

Tensile force (kN)	4.8	9.3	12.8	17.7	21.6	26.0
Extension (mm)	3.5	8.2	10.1	15.6	18.4	20.8

Assuming a linear relationship:

(a) determine the equation of the regression line of extension on force,
(b) determine the equation of the regression line of force on extension,
(c) estimate (i) the value of extension when the force is 16 kN, and (ii) the value of force when the extension is 17 mm. (23)

3. 1200 metal bolts have a mean mass of 7.2 g and a standard deviation of 0.3 g. Determine the standard error of the means. Calculate also the probability that a sample of 60 bolts chosen at random, without replacement, will have a mass of (a) between 7.1 g and 7.25 g, and (b) more than 7.3 g. (11)

4. A sample of 10 measurements of the length of a component are made and the mean of the sample is 3.650 cm. The standard deviation of the samples is 0.030 cm. Determine (a) the 99% confidence limits, and (b) the 90% confidence limits for an estimate of the actual length of the component. (10)

For lecturers/instructors/teachers, fully worked solutions to each of the problems in Revision Test 19, together with a full marking scheme, are available at the website:
www.routledge.com/cw/bird

List of essential formulae

Number and algebra

Length in metric units:

$$1\text{ m} = 100\text{ cm} = 1000\text{ mm}$$

Areas in metric units:

$1\text{ m}^2 = 10^4\text{ cm}^2$ $\quad$ $1\text{ cm}^2 = 10^{-4}\text{m}^2$

$1\text{ m}^2 = 10^6\text{ mm}^2$ $\quad$ $1\text{ mm}^2 = 10^{-6}\text{ m}^2$

$1\text{ cm}^2 = 10^2\text{ mm}^2$ $\quad$ $1\text{ mm}^2 = 10^{-2}\text{ cm}^2$

Volumes in metric units:

$1\text{ m}^3 = 10^6\text{ cm}^2$ $\quad$ $1\text{ cm}^3 = 10^{-6}\text{m}^3$

$1\text{ litre} = 1000\text{ cm}^3$

$1\text{ m}^3 = 10^9\text{ mm}^3$ $\quad$ $1\text{ mm}^3 = 10^{-9}\text{ m}^3$

$1\text{ cm}^3 = 10^3\text{ mm}^3$ $\quad$ $1\text{ mm}^3 = 10^{-3}\text{ cm}^3$

Laws of indices:

$a^m \times a^n = a^{m+n}$ $\quad$ $\dfrac{a^m}{a^n} = a^{m-n}$ $\quad$ $(a^m)^n = a^{mn}$

$a^{\frac{m}{n}} = \sqrt[n]{a^m}$ $\quad$ $a^{-n} = \dfrac{1}{a^n}$ $\quad$ $a^0 = 1$

Factor theorem

If $x = a$ is a root of the equation $f(x) = 0$, then $(x - a)$ is a factor of $f(x)$

Remainder theorem

If $(ax^2 + bx + c)$ is divided by $(x - p)$, the remainder will be: $ap^2 + bp + c$

or if $(ax^3 + bx^2 + cx + d)$ is divided by $(x - p)$, the remainder will be: $ap^3 + bp^2 + cp + d$

Partial fractions

Provided that the numerator $f(x)$ is of less degree than the relevant denominator, the following identities are typical examples of the form of partial fractions used:

$$\frac{f(x)}{(x+a)(x+b)(x+c)} \equiv \frac{A}{(x+a)} + \frac{B}{(x+b)} + \frac{C}{(x+c)}$$

$$\frac{f(x)}{(x+a)^3(x+b)} \equiv \frac{A}{(x+a)} + \frac{B}{(x+a)^2} + \frac{C}{(x+a)^3} + \frac{D}{(x+b)}$$

$$\frac{f(x)}{(ax^2+bx+c)(x+d)} \equiv \frac{Ax+B}{(ax^2+bx+c)} + \frac{C}{(x+d)}$$

Quadratic formula:

If $ax^2 + bx + c = 0$ then $x = \dfrac{-b \pm \sqrt{b^2 - 4ac}}{2a}$

Definition of a logarithm:

If $y = a^x$ then $x = \log_a y$

Laws of logarithms:

$$\log(A \times B) = \log A + \log B$$

$$\log\left(\frac{A}{B}\right) = \log A - \log B$$

$$\log A^n = n \times \log A$$

Exponential series:

$$e^x = 1 + x + \frac{x^2}{2!} + \frac{x^3}{3!} + \ldots$$

(valid for all values of x)

Arithmetic progression:

If a = first term and d = common difference, then the arithmetic progression is: $a, a+d, a+2d, \ldots$

The n'th term is : $a + (n-1)d$

Sum of n terms, $S_n = \frac{n}{2}[2a + (n-1)d]$

Geometric progression:

If a = first term and r = common ratio, then the geometric progression is: $a, ar, ar^2, \ldots$

The n'th term is: ar^{n-1}

Sum of n terms, $S_n = \frac{a(1-r^n)}{(1-r)}$ or $\frac{a(r^n-1)}{(r-1)}$

If $-1 < r < 1$, $S_\infty = \frac{a}{(1-r)}$

Binomial series:

$$(a+b)^n = a^n + na^{n-1}b + \frac{n(n-1)}{2!}a^{n-2}b^2 + \frac{n(n-1)(n-2)}{3!}a^{n-3}b^3 + \ldots$$

$$(1+x)^n = 1 + nx + \frac{n(n-1)}{2!}x^2 + \frac{n(n-1)(n-2)}{3!}x^3 + \ldots$$

Newton Raphson iterative method

If r_1 is the approximate value for a real root of the equation $f(x) = 0$, then a closer approximation to the root, r_2, is given by:

$$r_2 = r_1 - \frac{f(r_1)}{f'(r_1)}$$

Trigonometry

Theorem of Pythagoras:

$$b^2 = a^2 + c^2$$

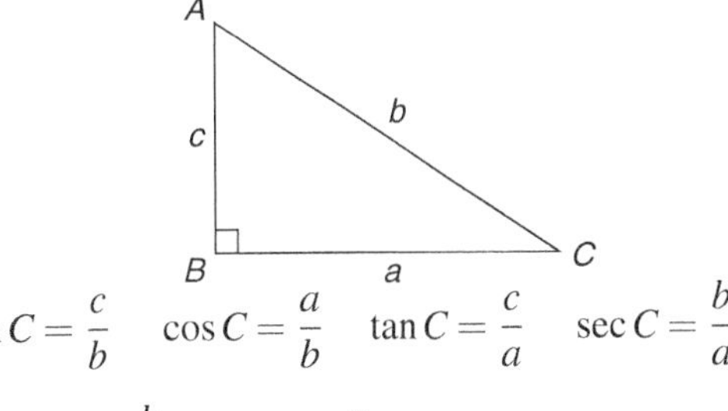

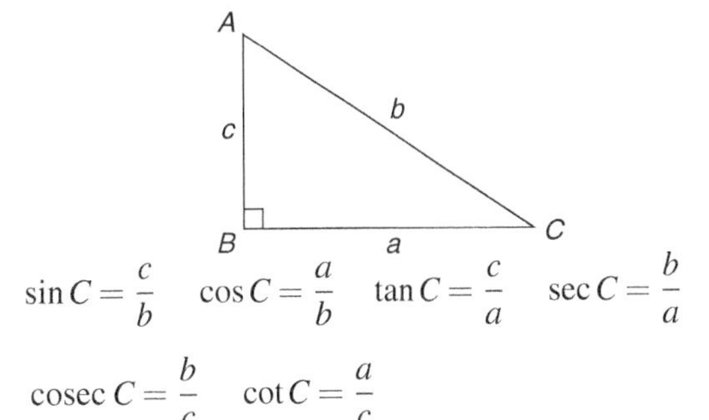

$\sin C = \frac{c}{b}$ $\cos C = \frac{a}{b}$ $\tan C = \frac{c}{a}$ $\sec C = \frac{b}{a}$

$\operatorname{cosec} C = \frac{b}{c}$ $\cot C = \frac{a}{c}$

Identities:

$\sec\theta = \frac{1}{\cos\theta}$ $\operatorname{cosec}\theta = \frac{1}{\sin\theta}$ $\cot\theta = \frac{1}{\tan\theta}$

$\tan\theta = \frac{\sin\theta}{\cos\theta}$ $\cos^2\theta + \sin^2\theta = 1$ $1 + \tan^2\theta = \sec^2\theta$

$\cot^2\theta + 1 = \operatorname{cosec}^2\theta$

Triangle formulae:

Sine rule $\frac{a}{\sin A} = \frac{b}{\sin B} = \frac{c}{\sin C}$

Cosine rule $a^2 = b^2 + c^2 - 2bc\cos A$

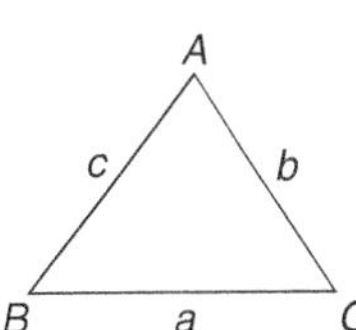

Area of any triangle

(i) $\frac{1}{2} \times$ base $\times$ perpendicular height

(ii) $\frac{1}{2}ab\sin C$ or $\frac{1}{2}ac\sin B$ or $\frac{1}{2}bc\sin A$

(iii) $\sqrt{[s(s-a)(s-b)(s-c)]}$ where $s = \frac{a+b+c}{2}$

Compound angle formulae

$$\sin(A \pm B) = \sin A \cos B \pm \cos A \sin B$$

$$\cos(A \pm B) = \cos A \cos B \mp \sin A \sin B$$

$$\tan(A \pm B) = \frac{\tan A \pm \tan B}{1 \mp \tan A \tan B}$$

If $\mathbf{R \sin(\omega t + \alpha) = a \sin\omega t + b \cos\omega t}$, then $a = R\cos\alpha$, $b = R\sin\alpha$, $R = \sqrt{(a^2 + b^2)}$ and $\alpha = \tan^{-1}\frac{b}{a}$

Double angles

$$\sin 2A = 2\sin A \cos A$$

$$\cos 2A = \cos^2 A - \sin^2 A = 2\cos^2 A - 1 = 1 - 2\sin^2 A$$

$$\tan 2A = \frac{2\tan A}{1 - \tan^2 A}$$

Products of sines and cosines into sums or differences:

$$\sin A \cos B = \frac{1}{2}[\sin(A + B) + \sin(A - B)]$$

$$\cos A \sin B = \frac{1}{2}[\sin(A + B) - \sin(A - B)]$$

$$\cos A \cos B = \frac{1}{2}[\cos(A + B) + \cos(A - B)]$$

$$\sin A \sin B = -\frac{1}{2}[\cos(A + B) - \cos(A - B)]$$

Sums or differences of sines and cosines into products:

$$\sin x + \sin y = 2\sin\left(\frac{x+y}{2}\right)\cos\left(\frac{x-y}{2}\right)$$

$$\sin x - \sin y = 2\cos\left(\frac{x+y}{2}\right)\sin\left(\frac{x-y}{2}\right)$$

$$\cos x + \cos y = 2\cos\left(\frac{x+y}{2}\right)\cos\left(\frac{x-y}{2}\right)$$

$$\cos x - \cos y = -2\sin\left(\frac{x+y}{2}\right)\sin\left(\frac{x-y}{2}\right)$$

For a **general sinusoidal function** $\mathbf{y = A \sin(\omega t \pm \alpha)}$, then
A = amplitude, ω = angular velocity $= 2\pi f$ rad/s, periodic time, $T = \frac{2\pi}{\omega}$ seconds, frequency, $f = \frac{\omega}{2\pi}$ hertz, α = angle of lead or lag (compared with $y = A\sin\omega t$)

Cartesian and polar co-ordinates

If co-ordinate $(x, y) = (r, \theta)$ then $r = \sqrt{x^2 + y^2}$ and $\theta = \tan^{-1}\frac{y}{x}$

If co-ordinate $(r, \theta) = (x, y)$ then

$x = r\cos\theta$ and $y = r\sin\theta$

Areas and volumes

Areas of plane figures:

Rectangle Area $= l \times b$

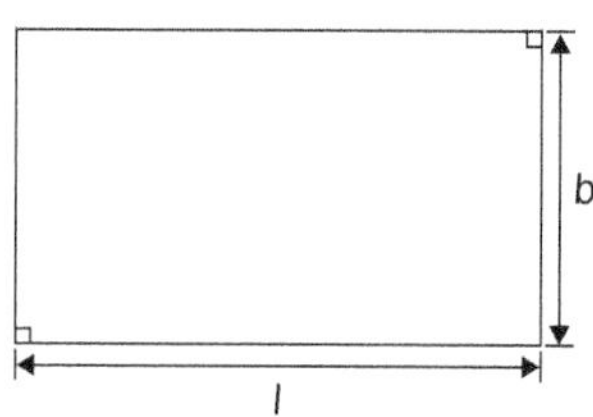

Parallelogram Area $= b \times h$

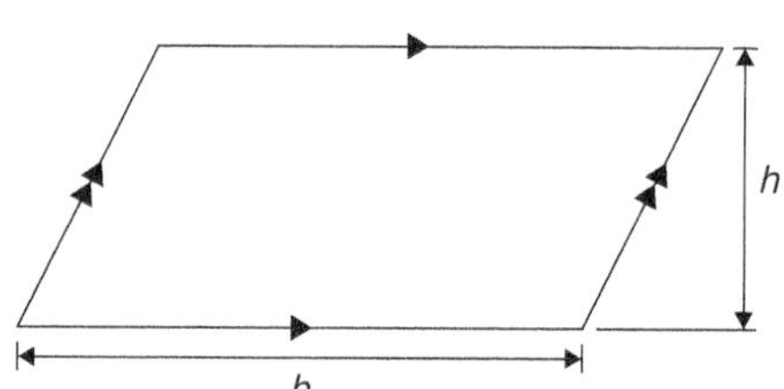

Trapezium Area $= \frac{1}{2}(a + b)h$

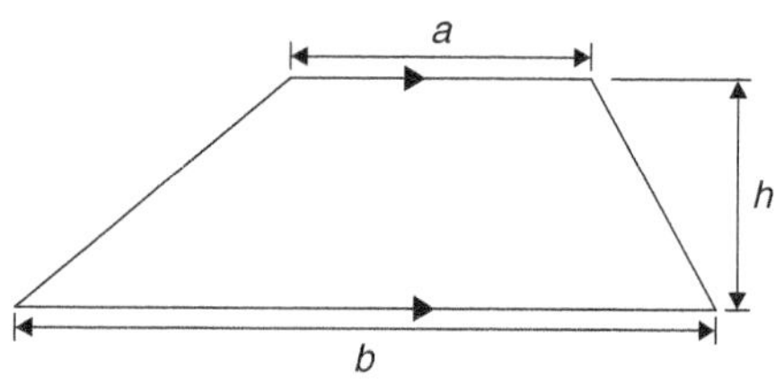

Triangle Area $= \frac{1}{2} \times b \times h$

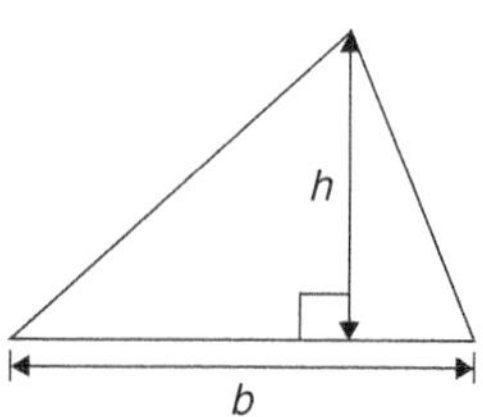

Circle Area $= \pi r^2$ Circumference $= 2\pi r$

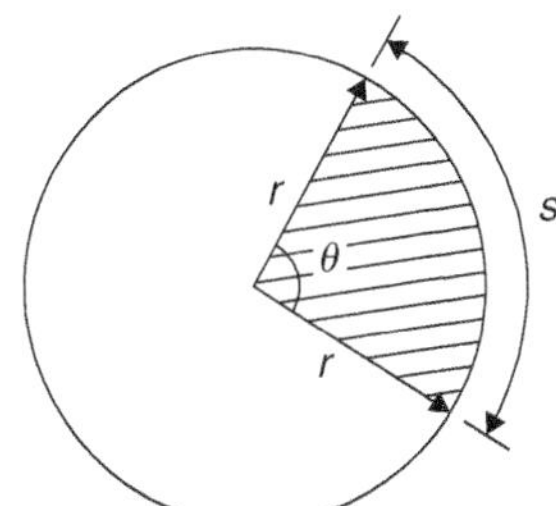

Radian measure: 2π radians = 360 degrees
For a sector of circle:

arc length, $s = \frac{\theta^\circ}{360}(2\pi r) = r\theta$ (θ in rad)

shaded area $= \frac{\theta^\circ}{360}(\pi r^2) = \frac{1}{2}r^2\theta$ (θ in rad)

Equation of a circle, centre at origin, radius r:
$x^2 + y^2 = r^2$

Equation of a circle, centre at (a, b), radius r:
$(x - a)^2 + (y - b)^2 = r^2$

Ellipse Area $= \pi ab$ Perimeter $\approx \pi(a + b)$

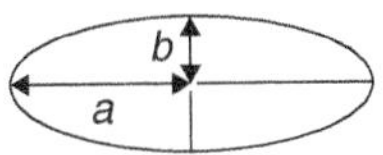

Volumes and surface areas of regular solids:

Rectangular prism (or cuboid)

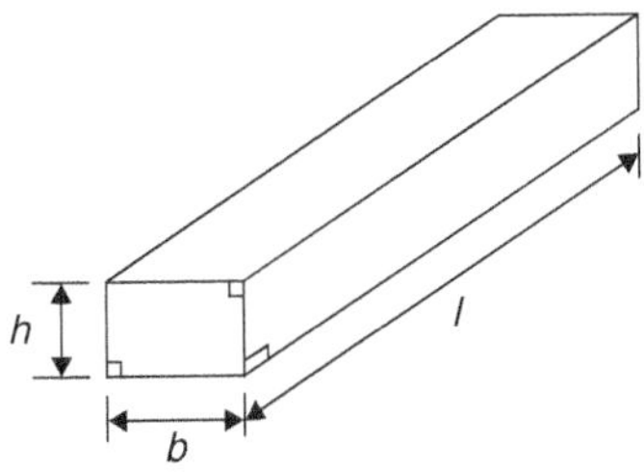

Volume $= l \times b \times h$
Surface area $= 2(bh + hl + lb)$

Cylinder

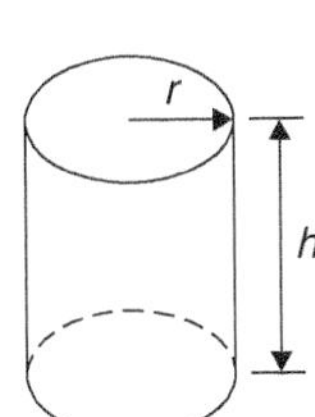

Volume $= \pi r^2 h$
Total surface area $= 2\pi rh + 2\pi r^2$

Pyramid

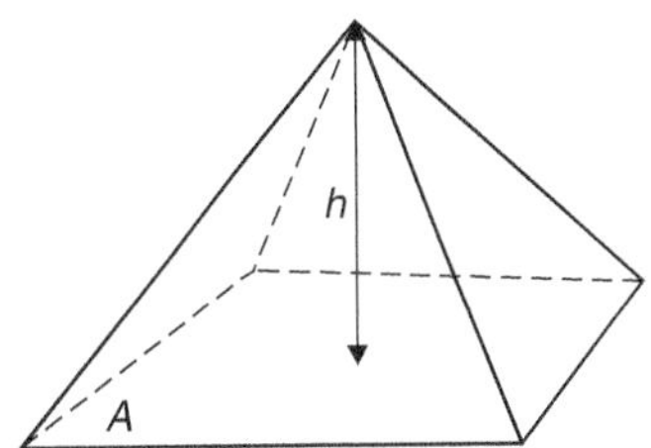

If area of base = A and perpendicular height $= h$ then:

Volume $= \frac{1}{3} \times A \times h$
Total surface area = sum of areas of triangles forming sides + area of base

Cone

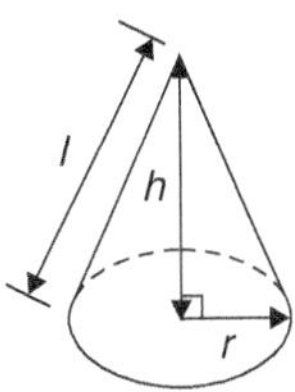

Volume $= \frac{1}{3}\pi r^2 h$

Curved surface area $= \pi rl$
Total surface area $= \pi rl + \pi r^2$

Sphere

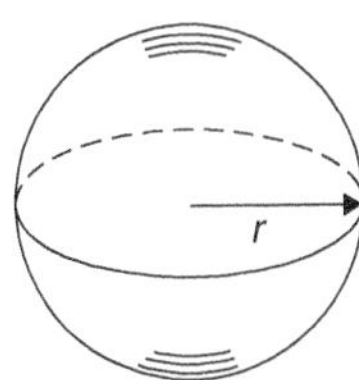

Volume $= \frac{4}{3}\pi r^3$ Surface area $= 4\pi r^2$

Frustum of sphere

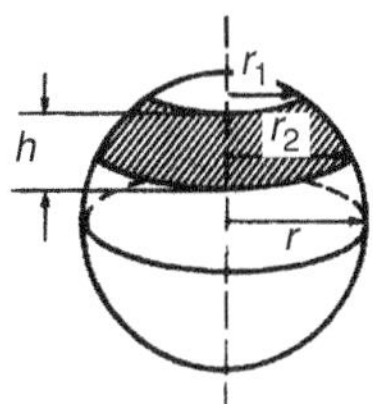

Surface area of zone of a sphere $= 2\pi rh$

Volume of frustum of sphere $= \frac{\pi h}{6}\left(h^2 + 3r_1^2 + 3r_2^2\right)$

Prismoidal rule

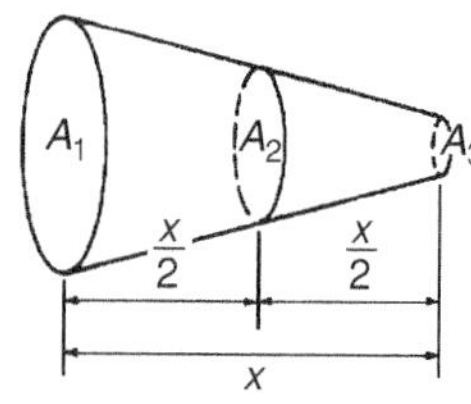

$$\text{Volume} = \frac{x}{6}\left(A_1 + 4A_2 + A_3\right)$$

Irregular areas

Trapezoidal rule

$$\text{Area} \approx \begin{pmatrix}\text{width of}\\ \text{interval}\end{pmatrix}\left[\frac{1}{2}\begin{pmatrix}\text{first + last}\\ \text{ordinates}\end{pmatrix} + \text{sum of remaining ordinates}\right]$$

Mid-ordinate rule

$$\text{Area} \approx \begin{pmatrix}\text{width of}\\ \text{interval}\end{pmatrix}\begin{pmatrix}\text{sum of}\\ \text{mid-ordinates}\end{pmatrix}$$

Simpson's rule

$$\text{Area} \approx \frac{1}{3}\begin{pmatrix}\text{width of}\\ \text{interval}\end{pmatrix}\left[\begin{pmatrix}\text{first + last}\\ \text{ordinate}\end{pmatrix} + 4\begin{pmatrix}\text{sum of even}\\ \text{ordinates}\end{pmatrix} + 2\begin{pmatrix}\text{sum of remaining}\\ \text{odd ordinates}\end{pmatrix}\right]$$

Volume of irregular solids

A1 A2 A3 A4 A5 A6 A7

d d d d d d

$$\text{Volume} = \frac{d}{3}\left\{\left(A_1 + A_7\right) + 4\left(A_2 + A_4 + A_6\right) + 2\left(A_3 + A_5\right)\right\}$$

For a sine wave:

over half a cycle, **mean value** $= \frac{2}{\pi} \times$ **maximum value**

of a full-wave rectified waveform, **mean value** $= \frac{2}{\pi} \times$ **maximum value**

of a half-wave rectified waveform, **mean value** $= \frac{1}{\pi} \times$ **maximum value**

Graphs

Equations of functions

Equation of a straight line: $y = mx + c$ where m is the gradient and c is the y-axis intercept

If $y = ax^n$ then $\lg y = n \lg x + \lg a$
If $y = ab^x$ then $\lg y = (\lg b)x + \lg a$
If $y = ae^{kx}$ then $\ln y = kx + \ln a$

Equation of a parabola: $y = ax^2 + bx + c$

Circle, centre (a, b), radius r: $(x-a)^2 + (y-b)^2 = r^2$

Equation of an ellipse, centre at origin, semi-axes a and b: $\frac{x^2}{a^2} + \frac{y^2}{b^2} = 1$

Equation of a hyperbola: $\frac{x^2}{a^2} - \frac{y^2}{b^2} = 1$

Equation of a rectangular hyperbola: $xy = c$

Odd and even functions

A function $y = f(x)$ is **odd** if $f(-x) = -f(x)$ for all values of x
Graphs of odd functions are always symmetrical about the origin.
A function $y = f(x)$ is **even** if $f(-x) = f(x)$ for all values of x
Graphs of even functions are always symmetrical about the y-axis.

Complex numbers

$z = a + jb = r(\cos\theta + j\sin\theta) = r\angle\theta = re^{j\theta}$ where $j^2 = -1$
Modulus $r = |z| = \sqrt{(a^2 + b^2)}$
Argument $\theta = \arg z = \tan^{-1}\frac{b}{a}$

Addition: $(a+jb)+(c+jd)=(a+c)+j(b+d)$

Subtraction: $(a+jb)-(c+jd)=(a-c)+j(b-d)$

Complex equations: If $m+jn=p+jq$ then $m=p$ and $n=q$

Multiplication: $z_1z_2=r_1\,r_2\angle(\theta_1+\theta_2)$

Division: $\frac{z_1}{z_2}=\frac{r_1}{r_2}\angle(\theta_1-\theta_2)$

De Moivre's theorem:
$[r\angle\theta]^n=r^n\angle n\theta=r^n(\cos n\theta+j\sin n\theta)=re^{j\theta}$

Differential calculus

Standard derivatives

y or $f(x)$	$\frac{dy}{dx}$ or $f'(x)$
ax^n	anx^{n-1}
$\sin ax$	$a\cos ax$
$\cos ax$	$-a\sin ax$
$\tan ax$	$a\sec^2 ax$
$\sec ax$	$a\sec ax\tan ax$
$\operatorname{cosec} ax$	$-a\operatorname{cosec} ax\cot ax$
$\cot ax$	$-a\operatorname{cosec}^2 ax$
e^{ax}	ae^{ax}
$\ln ax$	$\frac{1}{x}$
$\sinh ax$	$a\cosh ax$
$\cosh ax$	$a\sinh ax$
$\tanh ax$	$a\operatorname{sech}^2 ax$
$\operatorname{sech} ax$	$-a\operatorname{sech} ax\tanh ax$
$\operatorname{cosech} ax$	$-a\operatorname{cosech} ax\coth ax$
$\coth ax$	$-a\operatorname{cosech}^2 ax$

Product rule: When $y=uv$ and u and v are functions of x then:

$$\frac{dy}{dx}=u\frac{dv}{dx}+v\frac{du}{dx}$$

Quotient rule: When $y=\frac{u}{v}$ and u and v are functions of x then:

$$\frac{dy}{dx}=\frac{v\frac{du}{dx}-u\frac{dv}{dx}}{v^2}$$

Function of a function:

If u is a function of x then: $\frac{dy}{dx}=\frac{dy}{du}\times\frac{du}{dx}$

Parametric differentiation

If x and y are both functions of θ, then:

$$\frac{dy}{dx}=\frac{\frac{dy}{d\theta}}{\frac{dx}{d\theta}} \quad\text{and}\quad \frac{d^2y}{dx^2}=\frac{\frac{d}{d\theta}\left(\frac{dy}{dx}\right)}{\frac{dx}{d\theta}}$$

Implicit function:

$$\frac{d}{dx}[f(y)]=\frac{d}{dy}[f(y)]\times\frac{dy}{dx}$$

Maximum and minimum values:

If $y=f(x)$ then $\frac{dy}{dx}=0$ for stationary points.

Let a solution of $\frac{dy}{dx}=0$ be $x=a$; if the value of $\frac{d^2y}{dx^2}$ when $x=a$ is: *positive*, the point is a *minimum*, *negative*, the point is a *maximum*

Points of inflexion

(i) Given $y=f(x)$, determine $\frac{dy}{dx}$ and $\frac{d^2y}{dx^2}$

(ii) Solve the equation $\frac{d^2y}{dx^2}=0$

(iii) Test whether there is a change of sign occurring in $\frac{d^2y}{dx^2}$. This is achieved by substituting into the expression for $\frac{d^2y}{dx^2}$ firstly a value of x just less than the solution and then a value just greater than the solution.

(iv) A point of inflexion has been found if $\frac{d^2y}{dx^2}=0$ **and** there is a change of sign.

Velocity and acceleration

If distance $x=f(t)$, then velocity $v=f'(t)$ or $\frac{dx}{dt}$

and acceleration $a=f''(t)$ or $\frac{d^2x}{dt^2}$

Tangents and normals

Equation of tangent to curve $y=f(x)$ at the point (x_1, y_1) is:

$$y - y_1 = m(x - x_1)$$

where m = gradient of curve at (x_1, y_1)

Equation of normal to curve $y=f(x)$ at the point (x_1, y_1) is:

$$y - y_1 = -\frac{1}{m}(x - x_1)$$

Maclaurin's series

$$f(x) = f(0) + xf'(0) + \frac{x^2}{2!}f''(0) + \frac{x^3}{3!}f'''(0) + \ldots$$

Integral calculus

Standard integrals

y	$\int y\,dx$
ax^n	$a\frac{x^{n+1}}{n+1} + c$ (except where $n=-1$)
$\cos ax$	$\frac{1}{a}\sin ax + c$
$\sin ax$	$-\frac{1}{a}\cos ax + c$
$\sec^2 ax$	$\frac{1}{a}\tan ax + c$
$\operatorname{cosec}^2 ax$	$-\frac{1}{a}\cot ax + c$
$\operatorname{cosec} ax \cot ax$	$-\frac{1}{a}\operatorname{cosec} ax + c$
$\sec ax \tan ax$	$\frac{1}{a}\sec ax + c$
e^{ax}	$\frac{1}{a}e^{ax} + c$
$\frac{1}{x}$	$\ln x + c$
$\tan ax$	$\frac{1}{a}\ln(\sec ax) + c$
$\cos^2 x$	$\frac{1}{2}\left(x + \frac{\sin 2x}{2}\right) + c$
$\sin^2 x$	$\frac{1}{2}\left(x - \frac{\sin 2x}{2}\right) + c$
$\tan^2 x$	$\tan x - x + c$
$\cot^2 x$	$-\cot x - x + c$

$t = \tan\frac{\theta}{2}$ substitution

To determine $\int \frac{1}{a\cos\theta + b\sin\theta + c}\,d\theta$ let

$$\sin\theta = \frac{2t}{(1+t^2)} \qquad \cos\theta = \frac{1-t^2}{1+t^2}$$

$$\text{and} \qquad d\theta = \frac{2\,dt}{(1+t^2)}$$

Integration by parts If u and v are both functions of x then:

$$\int u\frac{dv}{dx}\,dx = uv - \int v\frac{du}{dx}\,dx$$

Numerical integration

Trapezoidal rule

$$\int y\,dx \approx \left(\begin{array}{c}\text{width of}\\ \text{interval}\end{array}\right)\left[\frac{1}{2}\left(\begin{array}{c}\text{first + last}\\ \text{ordinates}\end{array}\right) + \text{sum of remaining ordinates}\right]$$

Mid-ordinate rule

$$\int y\,dx \approx \left(\begin{array}{c}\text{width of}\\ \text{interval}\end{array}\right)\left(\begin{array}{c}\text{sum of}\\ \text{mid-ordinates}\end{array}\right)$$

Simpson's rule

$$\int y\,dx \approx \frac{1}{3}\left(\begin{array}{c}\text{width of}\\ \text{interval}\end{array}\right)\left[\left(\begin{array}{c}\text{first + last}\\ \text{ordinate}\end{array}\right) + 4\left(\begin{array}{c}\text{sum of even}\\ \text{ordinates}\end{array}\right) + 2\left(\begin{array}{c}\text{sum of remaining}\\ \text{odd ordinates}\end{array}\right)\right]$$

Second moment of area and radius of gyration

Shape	Position of axis	Second moment of area, I	Radius of gyration, k
Rectangle length l breadth b	(1) Coinciding with b	$\frac{bl^3}{3}$	$\frac{l}{\sqrt{3}}$
	(2) Coinciding with l	$\frac{lb^3}{3}$	$\frac{b}{\sqrt{3}}$
	(3) Through centroid, parallel to b	$\frac{bl^3}{12}$	$\frac{l}{\sqrt{12}}$
	(4) Through centroid, parallel to l	$\frac{lb^3}{12}$	$\frac{b}{\sqrt{12}}$
Triangle Perpendicular height h base b	(1) Coinciding with b	$\frac{bh^3}{12}$	$\frac{h}{\sqrt{6}}$
	(2) Through centroid, parallel to base	$\frac{bh^3}{36}$	$\frac{h}{\sqrt{18}}$
	(3) Through vertex, parallel to base	$\frac{bh^3}{4}$	$\frac{h}{\sqrt{2}}$
Circle radius r	(1) Through centre, perpendicular to plane (i.e. polar axis)	$\frac{\pi r^4}{2}$	$\frac{r}{\sqrt{2}}$
	(2) Coinciding with diameter	$\frac{\pi r^4}{4}$	$\frac{r}{2}$
	(3) About a tangent	$\frac{5\pi r^4}{4}$	$\frac{\sqrt{5}}{2}r$
Semicircle radius r	Coinciding with diameter	$\frac{\pi r^4}{8}$	$\frac{r}{2}$

Parallel axis theorem:
If C is the centroid of area A in the diagram shown then

$$A\,k_{BB}^2 = A\,k_{GG}^2 + A\,d^2 \quad \text{or} \quad k_{BB}^2 = k_{GG}^2 + d^2$$

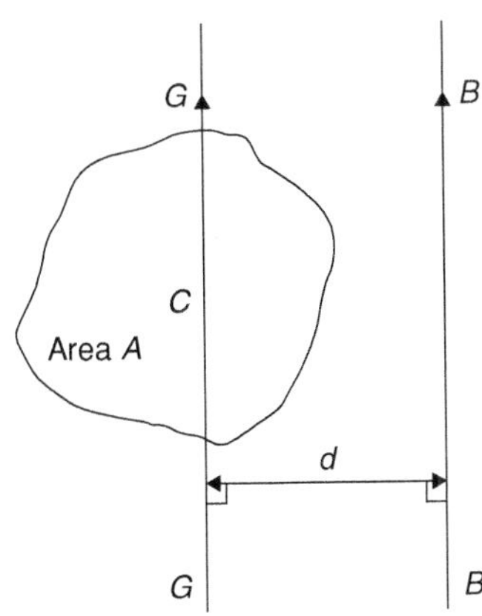

Perpendicular axis theorem:
If OX and OY lie in the plane of area A in the diagram shown then

$$A\,k_{OZ}^2 = A\,k_{OX}^2 + A\,k_{OY}^2 \quad \text{or} \quad k_{OZ}^2 = k_{OX}^2 + k_{OY}^2$$

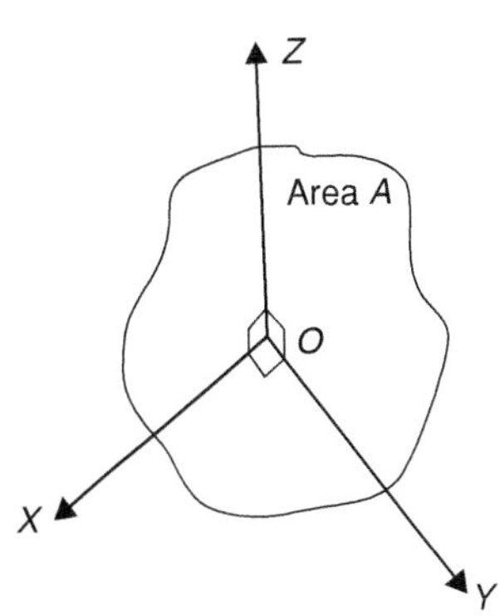

Area under a curve:

$$\text{area } A = \int_a^b y\,dx$$

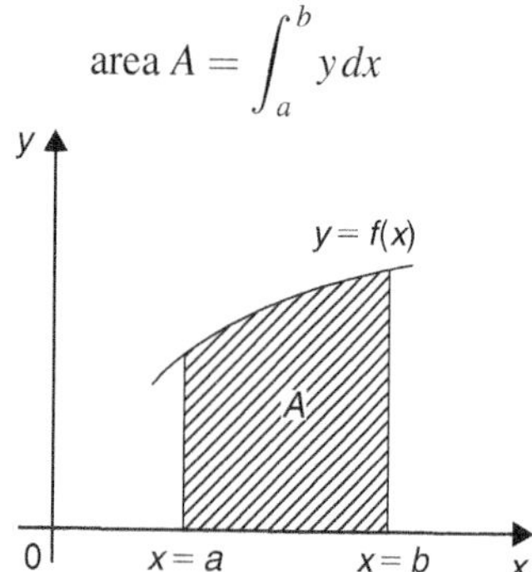

Mean value:

$$\text{mean value} = \frac{1}{b-a}\int_a^b y\,dx$$

R.m.s. value:

$$\text{r.m.s. value} = \sqrt{\left\{\frac{1}{b-a}\int_a^b y^2\,dx\right\}}$$

Volume of solid of revolution:

$$\text{volume} = \int_a^b \pi y^2\,dx \text{ about the } x\text{-axis}$$

Centroids

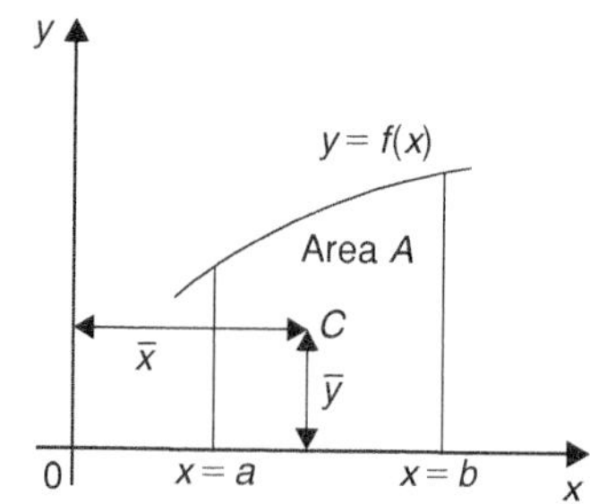

$$\bar{x} = \frac{\int_a^b xy\,dx}{\int_a^b y\,dx} \quad \text{and} \quad \bar{y} = \frac{\frac{1}{2}\int_a^b y^2\,dx}{\int_a^b y\,dx}$$

Theorem of Pappus With reference to the above diagram, when the curve is rotated one revolution about the x-axis between the limits $x = a$ and $x = b$, the volume V generated is given by:

$$V = 2\pi A\bar{y}$$

Differential equations

First order differential equations

Separation of variables

If $\frac{dy}{dx} = f(x)$ then $y = \int f(x)\,dx$

If $\frac{dy}{dx} = f(y)$ then $\int dx = \int \frac{dy}{f(y)}$

If $\frac{dy}{dx} = f(x)\cdot f(y)$ then $\int \frac{dy}{f(y)} = \int f(x)\,dx$

Further number and algebra

Boolean algebra

Laws and rules of Boolean algebra

Commutative Laws: $A + B = B + A$, $A \cdot B = B \cdot A$

Associative Laws: $A + B + C = (A + B) + C$, $A \cdot B \cdot C = (A \cdot B) \cdot C$

Distributive Laws: $A \cdot (B + C) = A \cdot B + A \cdot C$, $A + (B \cdot C) = (A + B) \cdot (A + C)$

Sum rules: $A + \overline{A} = 1$, $A + 1 = 1$, $A + 0 = A$, $A + A = A$

Product rules: $A \cdot \overline{A} = 0$, $A \cdot 0 = 0$, $A \cdot 1 = A$, $A \cdot A = A$

Absorption rules: $A + A \cdot B = A$, $A \cdot (A + B) = A$, $A + \overline{A} \cdot B = A + B$

De Morgan's Laws: $\overline{A + B} = \overline{A} \cdot \overline{B}$, $\overline{A \cdot B} = \overline{A} + \overline{B}$

Matrices:

If $A = \begin{pmatrix} a & b \\ c & d \end{pmatrix}$ and $B = \begin{pmatrix} e & f \\ g & h \end{pmatrix}$ then

$$A + B = \begin{pmatrix} a+e & b+f \\ c+g & d+h \end{pmatrix}$$

$$A - B = \begin{pmatrix} a-e & b-f \\ c-g & d-h \end{pmatrix}$$

$$A \times B = \begin{pmatrix} ae+bg & af+bh \\ ce+dg & cf+dh \end{pmatrix}$$

$$A^{-1} = \frac{1}{ad-bc}\begin{pmatrix} d & -b \\ -c & a \end{pmatrix}$$

If $A = \begin{pmatrix} a_1 & b_1 & c_1 \\ a_2 & b_2 & c_2 \\ a_3 & b_3 & c_3 \end{pmatrix}$ then $A^{-1} = \frac{B^T}{|A|}$ where

B^T = transpose of cofactors of matrix A

Determinants: $\begin{vmatrix} a & b \\ c & d \end{vmatrix} = ad - bc$

$$\begin{vmatrix} a_1 & b_1 & c_1 \\ a_2 & b_2 & c_2 \\ a_3 & b_3 & c_3 \end{vmatrix} = a_1\begin{vmatrix} b_2 & c_2 \\ b_3 & c_3 \end{vmatrix} - b_1\begin{vmatrix} a_2 & c_2 \\ a_3 & c_3 \end{vmatrix} + c_1\begin{vmatrix} a_2 & b_2 \\ a_3 & b_3 \end{vmatrix}$$

Statistics and probability

Mean, median, mode and standard deviation

If x = variate and f = frequency then :

mean $\bar{x} = \frac{\sum fx}{\sum f}$

The **median** is the middle term of a ranked set of data.
The **mode** is the most commonly occurring value in a set of data

Standard deviation $\sigma = \sqrt{\left[\frac{\sum\{f(x-\bar{x})^2\}}{\sum f}\right]}$ for a population

Bayes' theorem

$$P(A_1|B) = \frac{P(B|A_1)P(A_1)}{P(B|A_1)P(A_1) + P(B|A_2)P(A_2) + \cdots}$$

$$\text{or } P(A_i|B) = \frac{P(B|A_i)P(A_i)}{\sum_{j=1}^{n} P(B|A_j)P(A_j)} \quad (i = 1, 2, \ldots, n)$$

Binomial probability distribution

If n = number in sample, p = probability of the occurrence of an event and $q = 1 - p$, then the probability of 0, 1, 2, 3, .. occurrences is given by:

$$q^n,\ nq^{n-1}p,\ \frac{n(n-1)}{2!}q^{n-2}p^2,\ \frac{n(n-1)(n-2)}{3!}q^{n-3}p^3, ..$$

(i.e. successive terms of the $(q+p)^n$ expansion)

Normal approximation to a binomial distribution:
Mean = np Standard deviation $\sigma = \sqrt{(npq)}$

Poisson distribution

If λ is the expectation of the occurrence of an event then the probability of 0, 1, 2, 3, .. occurrences is given by:

$$e^{-\lambda}, \lambda e^{-\lambda}, \lambda^2\frac{e^{-\lambda}}{2!}, \lambda^3\frac{e^{-\lambda}}{3!}, ..$$

Product-moment formula for the linear correlation coefficient:

Coefficient of correlation $r = \frac{\sum xy}{\sqrt{[(\sum x^2)(\sum y^2)]}}$

where $x = X - \overline{X}$ and $y = Y - \overline{Y}$ and (X_1, Y_1), (X_2, Y_2), .. denote a random sample from a bivariate normal distribution and $\overline{X}$ and $\overline{Y}$ are the means of the X and Y values respectively

Normal probability distribution – Partial areas under the standardised normal curve – see Table 64.1 on page 658.

Student's *t* distribution – Percentile values (t_p) for Student's t distribution with ν degrees of freedom – see Table 67.2 on page 686.

Symbols:

Population
number of members N_p, mean μ, standard deviation σ.

Sample
number of members N, mean $\bar{x}$, standard deviation s.

Sampling distributions
mean of sampling distribution of means $\mu_{\bar{x}}$
standard error of means $\sigma_{\bar{x}}$
standard error of the standard deviations σ_s

Standard error of the means

Standard error of the means of a sample distribution, i.e. the standard deviation of the means of samples, is:

$\sigma_{\bar{x}} = \dfrac{\sigma}{\sqrt{N}}\sqrt{\left(\dfrac{N_p - N}{N_p - 1}\right)}$ for a finite population and/or for sampling without replacement, and

$\sigma_{\bar{x}} = \dfrac{\sigma}{\sqrt{N}}$

for an infinite population and/or for sampling with replacement

The relationship between sample mean and population mean

$\mu_{\bar{x}} = \mu$ for all possible samples of size N are drawn from a population of size N_p

Estimating the mean of a population (σ known)

The confidence coefficient for a large sample size, ($N \geq 30$) is z_c where:

Confidence level %	Confidence coefficient z_c
99	2.58
98	2.33
96	2.05
95	1.96
90	1.645
80	1.28
50	0.6745

The confidence limits of a population mean based on sample data are given by:

$\bar{x} \pm \dfrac{z_c\sigma}{\sqrt{N}}\sqrt{\left(\dfrac{N_p - N}{N_p - 1}\right)}$ for a finite population of size N_p, and by

$\bar{x} \pm \dfrac{z_c\sigma}{\sqrt{N}}$ for an infinite population

Estimating the mean of a population (σ unknown)

The confidence limits of a population mean based on sample data are given by:

$$\mu_{\bar{x}} \pm z_c\sigma_{\bar{x}}$$

Estimating the standard deviation of a population

The confidence limits of the standard deviation of a population based on sample data are given by:

$$s \pm z_c\sigma_s$$

Estimating the mean of a population based on a small sample size

The confidence coefficient for a small sample size ($N < 30$) is t_c which can be determined using Table 67.1. The confidence limits of a population mean based on sample data given by:

$$\bar{x} \pm \frac{t_c\, s}{\sqrt{(N-1)}}$$

These formulae are available for download at the website:
www.routledge.com/cw/bird

Answers to Practice Exercises

Chapter 1

Exercise 1 (page 6)

1. (a) $\frac{9}{10}$ (b) $\frac{3}{16}$
2. (a) $\frac{43}{77}$ (b) $\frac{47}{63}$
3. (a) $1\frac{16}{21}$ (b) $\frac{17}{60}$
4. (a) $\frac{5}{12}$ (b) $\frac{3}{49}$
5. (a) $\frac{3}{5}$ (b) 11
6. (a) $\frac{8}{15}$ (b) $\frac{12}{23}$
7. $1\frac{7}{24}$
8. $5\frac{4}{5}$
9. $-\frac{13}{126}$
10. $2\frac{28}{55}$
11. 14
12. (a) £60 (b) £36, £16

Exercise 2 (page 7)

1. 81 cm, 189 cm, 351 cm
2. 17 g
3. 72 kg : 27 kg
4. 5 men
5. (a) 2 h 10 min (b) 4 h 20 min
6. 12 kg, 20 kg and 24 kg

Exercise 3 (page 9)

1. 11.989
2. −31.265
3. 10.906
4. 2.2446
5. (a) 24.81 (b) 24.812
6. (a) 0.00639 (b) 0.0064
7. (a) $\frac{13}{20}$ (b) $\frac{21}{25}$ (c) $\frac{1}{80}$ (d) $\frac{141}{500}$ (e) $\frac{3}{125}$
8. (a) $1\frac{41}{50}$ (b) $4\frac{11}{40}$ (c) $14\frac{1}{8}$ (d) $15\frac{7}{20}$ (e) $16\frac{17}{80}$
9. 0.444
10. 0.62963
11. 1.563
12. 13.84
13. 12.52 mm
14. 2400 tins

Exercise 4 (page 11)

1. (a) 5.7% (b) 37.4% (c) 128.5%
2. (a) 21.2% (b) 79.2% (c) 169%
3. (a) 496.4 t (b) 8.657 g (c) 20.73 s
4. 2.25%
5. (a) 14% (b) 15.67% (c) 5.36%
6. 37.8 g
7. 7.2%
8. A 0.6 kg B 0.9 kg C 0.5 kg
9. 54%, 31%, 15%, 0.3 t
10. 20,000 kg
11. 13.5 mm, 11.5 mm
12. 600 kW
13. 126

Exercise 5 (page 11)

1. (a) **2.** (a) **3.** (c) **4.** (b) **5.** (c) **6.** (c) **7.** (b) **8.** (b) **9.** (a) **10.** (a) **11.** (a) **12.** (d) **13.** (d) **14.** (c) **15.** (c) **16.** (d) **17.** (d) **18.** (d) **19.** (c) **20.** (c) **21.** (b) **22.** (a) **23.** (b) **24.** (d) **25.** (b)

Chapter 2

Exercise 6 (page 15)

1. (a) 3^7 (b) 4^9 **2.** (a) 2^6 (b) 7^{10}
3. (a) 2 (b) 3^5 **4.** (a) 5^3 (b) 7^3
5. (a) 7^6 (b) 3^6 **6.** (a) 2 (b) 3^6
7. (a) 5^2 (b) 13^2 **8.** (a) 3^4 (b) 1
9. (a) 5^2 (b) $\frac{1}{3^5}$ **10.** (a) 7^2 (b) $\frac{1}{2}$

Exercise 7 (page 17)

1. (a) $\frac{1}{3 \times 5^2}$ (b) $\frac{1}{7^3 \times 3^7}$

2. (a) $\frac{3^2}{2^5}$ (b) $\frac{1}{2^{10} \times 5^2}$

3. (a) 9 (b) 3 (c) $\frac{1}{2}$ (d) $\frac{2}{3}$

4. $\frac{147}{148}$ **5.** $\frac{1}{9}$

6. $-5\frac{65}{72}$ **7.** 64

8. $4\frac{1}{2}$

Exercise 8 (page 18)

1. (a) 7.39×10 (b) 2.84×10 (c) 1.9772×10^2
2. (a) 2.748×10^3 (b) 3.317×10^4
(c) 2.74218×10^5
3. (a) 2.401×10^{-1} (b) 1.74×10^{-2}
(c) 9.23×10^{-3}
4. (a) 5×10^{-1} (b) 1.1875×10 (c) 1.306×10^2
(d) 3.125×10^{-2}
5. (a) 1010 (b) 932.7 (c) 54100 (d) 7
6. (a) 0.0389 (b) 0.6741 (c) 0.008
7. (a) $2.71 \times 10^3\,kg\,m^{-3}$ (b) 4.4×10^{-1}
(c) $3.7673 \times 10^2\,\Omega$ (d) 5.11×10^{-1} MeV
(e) $9.57897 \times 10^7\,C\,kg^{-1}$
(f) $2.241 \times 10^{-2}\,m^3\,mol^{-1}$

Exercise 9 (page 19)

1. (a) 100 kW (b) 0.54 mA or 540 μA
(c) 1.5 MΩ (d) 22.5 mV
(e) 35 GHz (f) 15 pF
(g) 17 μA (h) 46.2 kΩ
2. (a) 25 μA (b) 1 nF (c) 620 kV (d) 1.25 MΩ
3. (a) 13.5×10^{-3} (b) 4×10^3
4. 720 g

Exercise 10 (page 20)

1. 2450 mm **2.** 167.5 cm **3.** 658 mm
4. 25.4 m **5.** 56.32 m **6.** 4356 mm
7. 87.5 cm
8. (a) 4650 mm (b) 4.65 m
9. (a) 504 cm (b) 5.04 m
10. 148.5 mm to 151.5 mm

Exercise 11 (page 21)

1. 8×10^4 cm^2 **2.** 240×10^{-4} m^2 or 0.024 m^2
3. 3.6×10^6 mm^2 **4.** 350×10^{-6} m^2 or 0.35×10^{-3}m^2
5. 5000 mm^2 **6.** 2.5 cm^2
7. (a) 288000 mm^2 (b) 2880 cm^2 (c) 0.288 m^2

Exercise 12 (page 23)

1. 2.5×10^6 cm^3
2. 400×10^{-6} m^3
3. 0.87×10^9 mm^3 or 870×10^6 mm^3
4. 2.4 m^3
5. 1500×10^{-9} m^3 or 1.5×10^{-6} m^3
6. 400×10^{-3} cm^3 or 0.4 cm^3
7. 6.4×10^3 mm^3
8. 7500×10^{-3} cm^3 or 7.5 cm^3
9. (a) 0.18 m^3 (b) 180000 cm^3 (c) 180 litres

Exercise 13 (page 27)

1. 18.74 in **2.** 82.28 in **3.** 37.95 yd
4. 18.36 mile **5.** 41.66 cm **6.** 1.98 m
7. 14.36 m **8.** 5.91 km
9. (a) 22,745 yd (b) 20.81 km

10. 9.688 in^2 **11.** 18.18 yd^2 **12.** 30.89 acres
13. 21.89 $mile^2$ **14.** 41 cm^2 **15.** 28.67 m^2
16. 10117 m^2 **17.** 55.17 km^2 **18.** 12.24 in^3
19. 7.572 ft^3 **20.** 17.59 yd^3
21. 31.7 fluid pints **22.** 35.23 cm^3 **23.** 4.956 m^3
24. 3.667 litre **25.** 47.32 litre **26.** 34.59 oz
27. 121.3 lb **28.** 4.409 short tons
29. 220.6 g **30.** 1.630 kg
31. (a) 280 lb (b) 127.2 kg
32. 131°F **33.** 75°C

Exercise 14 (page 28)

1. (d) **2.** (b) **3.** (d) **4.** (b) **5.** (b) **6.** (c) **7.** (c) **8.** (a) **9.** (a) **10.** (c) **11.** (d) **12.** (a) **13.** (b) **14.** (d) **15.** (b) **16.** (c) **17.** (a) **18.** (a) **19.** (a) **20.** (c) **21.** (d) **22.** (c) **23.** (d) **24.** (d) **25.** (b)

Chapter 3

Exercise 15 (page 32)

1. (a) 6_{10} (b) 11_{10} (c) 14_{10} (d) 9_{10}
2. (a) 21_{10} (b) 25_{10} (c) 45_{10} (d) 51_{10}
3. (a) 42_{10} (b) 56_{10} (c) 65_{10} (d) 184_{10}
4. (a) 0.8125_{10} (b) 0.78125_{10} (c) 0.21875_{10} (d) 0.34375_{10}
5. (a) 26.75_{10} (b) 23.375_{10} (c) 53.4375_{10} (d) 213.71875_{10}

Exercise 16 (page 33)

1. (a) 101_2 (b) 1111_2 (c) 10011_2 (d) 11101_2
2. (a) 11111_2 (b) 101010_2 (c) 111001_2 (d) 111111_2
3. (a) 101111_2 (b) 111100_2 (c) 1001001_2 (d) 1010100_2
4. (a) 0.01_2 (b) 0.00111_2 (c) 0.01001_2 (d) 0.10011_2
5. (a) 101111.01101_2 (b) 11110.1101_2 (c) 110101.11101_2 (d) 111101.10101_2

Exercise 17 (page 34)

1. 101 **2.** 1011 **3.** 10100
4. 101100 **5.** 1001000 **6.** 100001010
7. 1010110111 **8.** 1001111101 **9.** 111100
10. 110111 **11.** 110011 **12.** 1101110

Exercise 18 (page 36)

1. (a) 101010111_2 (b) 1000111100_2 (c) 10011110001_2
2. (a) 0.01111_2 (b) 0.1011_2 (c) 0.10111_2
3. (a) 11110111.00011_2 (b) 1000000010.0111_2 (c) 11010110100.11001_2
4. (a) 7.4375_{10} (b) 41.25_{10} (c) 7386.1875_{10}

Exercise 19 (page 38)

1. 231_{10} **2.** 44_{10} **3.** 152_{10} **4.** 753_{10}
5. 36_{16} **6.** $C8_{16}$ **7.** $5B_{16}$ **8.** EE_{16}

Exercise 20 (page 39)

1. $D7_{16}$ **2.** EA_{16} **3.** $8B_{16}$
4. $A5_{16}$ **5.** 110111_2 **6.** 11101101_2
7. 10011111_2 **8.** 101000100001_2

Exercise 21 (page 39)

1. (c) **2.** (b) **3.** (d) **4.** (c) **5.** (b) **6.** (c) **7.** (d) **8.** (c) **9.** (d) **10.** (a) **11.** (c) **12.** (a) **13.** (d) **14.** (b) **15.** (c) **16.** (c) **17.** (a) **18.** (a) **19.** (a) **20.** (b) **21.** (a) **22.** (b) **23.** (c) **24.** (b) **25.** (d)

Chapter 4

Exercise 22 (page 42)

1. order of magnitude error – should be 2.1
2. Rounding-off error – should add 'correct to 4 significant figures' or 'correct to 1 decimal place'
3. Blunder
4. Measured values, hence $c = 55800\,\text{Pa}\,\text{m}^3$
5. Order of magnitude error and rounding-off error – should be 0.0225, correct to 3 significant figures or 0.0225, correct to 4 decimal places
6. ≈ 30 (29.61 by calculator)
7. ≈ 2 (1.988, correct to 4 s.f., by calculator)
8. ≈ 10 (9.481, correct to 4 s.f., by calculator)

Exercise 23 (page 44)

1. (a) 10.56 (b) 5443 (c) 96970 (d) 0.004083
2. (a) 2.176 (b) 5.955 (c) 270.7 (d) 0.1600
3. (a) 0.1287 (b) 0.02064 (c) 12.25 (d) 0.8945
4. (a) 109.1 (b) 3.641
5. (a) 0.2489 (b) 500.5
6. (a) 2515 (b) 146.0 (c) 0.00002932
7. (a) 0.005559 (b) 1.900
8. (a) 6.248 (b) 0.9630
9. (a) 1.605 (b) 11.74
10. (a) 6.874×10^{-3} (b) 87.31×10^{-3}

Exercise 24 (page 46)

1. (a) 38.03 euros (b) \$119.68 (c) £104 (d) £155.20 (e) 93.50 dollars
2. (a) 381 mm (b) 56.35 km/h (c) 378.35 km (d) 52 lb 13 oz (e) 6.82 kg (f) 54.55 litre (g) 5.5 gallon
3. (a) 7.09 a.m. (b) 52 minutes, 31.15 m.p.h. (c) 7.04 a.m.

Exercise 25 (page 50)

1. $R = 37.5$
2. 159 m/s
3. 0.00502 m or 5.02 mm
4. 0.144 J
5. 224.5
6. 14230 kg/m^3
7. 281.1 m/s
8. 2.526 Ω
9. 508.1 W
10. $V = 2.61\,V$
11. $F = 854.5$
12. $I = 3.81\,A$
13. $t = 14.79$ s
14. $E = 3.96\,J$
15. $I = 12.77\,A$
16. $s = 17.25$ m
17. $A = 7.184$ cm^2
18. $v = 7.327$
19. 0.50 kg m^2
20. 7.063 MN

Exercise 26 (page 51)

1. (b) **2.** (d) **3.** (a) **4.** (c) **5.** (b) **6.** (c) **7.** (b) **8.** (a) **9.** (d) **10.** (d)

Chapter 5

Exercise 27 (page 56)

1. -16
2. -8
3. 4a
4. $a - 4b - c$
5. 9d – 2e
6. $3x - 5y + 5z$
7. $-5\frac{1}{2}a + \frac{5}{6}b - 4c$
8. $3x^2 - xy - 2y^2$
9. $6a^2 - 13ab + 3ac - 5b^2 + bc$
10. (i) $\frac{1}{3b}$ (ii) 2ab

Exercise 28 (page 58)

1. $x^5y^4z^3$, $13\frac{1}{2}$
2. $a^2b^{\frac{1}{2}}c^{-2}$, $\pm 4\frac{1}{2}$
3. $a^3b^{-2}c$, 9
4. $x^{\frac{7}{10}}y^{\frac{1}{6}}z^{\frac{1}{2}}$
5. $\frac{1+a}{b}$
6. $\frac{p^2q}{q-p}$
7. $ab^6c^{\frac{3}{2}}$
8. $a^{-4}b^5c^{11}$
9. $xy^3\sqrt[6]{z^{13}}$
10. $a^{\frac{11}{6}}b^{\frac{1}{3}}c^{-\frac{3}{2}}$ or $\frac{\sqrt[6]{a^{11}}\sqrt[3]{b}}{\sqrt{c^3}}$

Exercise 29 (page 60)

1. $3x + y$
2. $5(x - y)$
3. $-5p + 10q - 6r$
4. $a^2 + 3ab + 2b^2$
5. $3p^2 + pq - 2q^2$
6. (i) $x^2 - 4xy + 4y^2$ (ii) $9a^2 - 6ab + b^2$
7. $4 - a$
8. $2 + 5b^2$
9. $11q - 2p$
10. (i) $p(b + 2c)$ (ii) $2q(q + 4n)$
11. (i) $7ab(3ab - 4)$ (ii) $2xy(y + 3x + 4x^2)$
12. (i) $(a + b)(y + 1)$ (ii) $(p + q)(x + y)$ (iii) $(a - 2b)(2x + 3y)$

Exercise 30 (page 62)

1. $\frac{1}{2} + 6x$
2. $\frac{1}{5}$
3. $4a(1 - 2a)$
4. $a(3 - 10a)$
5. $\frac{2}{3y} - 3y + 12$
6. $\frac{2}{3y} + 12 - 13y$
7. $\frac{5}{y} + 1$
8. pq
9. $\frac{1}{2}(x - 4)$
10. $y\left(\frac{1}{2} + 3y\right)$

Exercise 31 (page 64)

1. (a) 15 (b) 78
2. (a) 0.0075 (b) 3.15 litres (c) 350 K
3. (a) 0.00008 (b) 4.16×10^{-3} A (c) 45 V
4. (a) 9.18 (b) 6.12 (c) 0.3375
5. (a) 300×10^3 (b) 0.375 m^3 (c) 240×10^3 Pa

Exercise 32 (page 65)

1. (d) **2.** (a) **3.** (b) **4.** (b) **5.** (a) **6.** (d) **7.** (c) **8.** (c) **9.** (a) **10.** (b) **11.** (b) **12.** (d) **13.** (d) **14.** (c) **15.** (c)

Chapter 6

Exercise 33 (page 68)

1. $2x - y$
2. $3x - 1$
3. $5x - 2$
4. $7x + 1$
5. $x^2 + 2xy + y^2$
6. $5x + 4 + \frac{8}{x-1}$
7. $3x^2 - 4x + 3 - \frac{2}{x+2}$
8. $5x^3 + 18x^2 + 54x + 160 + \frac{481}{x-3}$

Exercise 34 (page 70)

1. $(x-1)(x+3)$
2. $(x+1)(x+2)(x-2)$
3. $(x+1)(2x^2+3x-7)$
4. $(x-1)(x+3)(2x-5)$
5. $x^3+4x^2+x-6=(x-1)(x+2)(x+3)$
   $x=1$, $x=-2$ and $x=-3$
6. $x=1$, $x=2$ and $x=-1$

Exercise 35 (page 71)

1. (a) 6 (b) 9
2. (a) −39 (b) −29
3. $(x-1)(x-2)(x-3)$
4. $x=-1$, $x=-2$ and $x=-4$
5. $a=-3$
6. $x=1$, $x=-2$ and $x=1.5$

Exercise 36 (page 72)

1. (d) **2.** (b) **3.** (c) **4.** (a) **5.** (d)

Chapter 7

Exercise 37 (page 76)

1. $\frac{2}{(x-3)}-\frac{2}{(x+3)}$
2. $\frac{5}{(x+1)}-\frac{1}{(x-3)}$
3. $\frac{3}{x}+\frac{2}{(x-2)}-\frac{4}{(x-1)}$
4. $\frac{7}{(x+4)}-\frac{3}{(x+1)}-\frac{2}{(2x-1)}$
5. $1+\frac{2}{(x+3)}+\frac{6}{(x-2)}$
6. $1+\frac{3}{(x+1)}-\frac{2}{(x-3)}$
7. $3x-2+\frac{1}{(x-2)}-\frac{5}{(x+2)}$

Exercise 38 (page 77)

1. $\frac{4}{(x+1)}-\frac{7}{(x+1)^2}$
2. $\frac{2}{x}+\frac{1}{x^2}-\frac{1}{(x+3)}$
3. $\frac{5}{(x-2)}-\frac{10}{(x-2)^2}+\frac{4}{(x-2)^3}$
4. $\frac{2}{(x-5)}-\frac{3}{(x+2)}+\frac{4}{(x+2)^2}$

Exercise 39 (page 78)

1. $\frac{2x+3}{(x^2+7)}-\frac{1}{(x-2)}$
2. $\frac{1}{(x-4)}+\frac{2-x}{(x^2+3)}$
3. $\frac{1}{x}+\frac{3}{x^2}+\frac{2-5x}{(x^2+5)}$
4. $\frac{3}{(x-1)}+\frac{2}{(x-1)^2}+\frac{1-2x}{(x^2+8)}$
5. Proof

Chapter 8

Exercise 40 (page 82)

1. 1 **2.** 2 **3.** −4 **4.** $1\frac{2}{3}$
5. 2 **6.** $\frac{1}{2}$ **7.** 0 **8.** −10
9. 6 **10.** −2 **11.** 2.5 **12.** 2
13. $6\frac{1}{4}$ **14.** −3

Exercise 41 (page 84)

1. −2 **2.** $-4\frac{1}{2}$ **3.** 2 **4.** 12
5. 15 **6.** −4 **7.** 2 **8.** 13
9. 2 **10.** 3 **11.** −11 **12.** −6
13. 9 **14.** 4 **15.** 10 **16.** ± 12
17. $-3\frac{1}{3}$ **18.** ±4

Exercise 42 (page 85)

1. 10^{-7} **2.** 8 m/s^2 **3.** 3.472
4. (a) 1.8 Ω (b) 30 Ω **5.** 800 Ω **6.** 176 MPa

Exercise 43 (page 86)

1. 0.004 **2.** 30 **3.** 45°C
4. 50 **5.** 3.5 N **6.** 12 m, 8 m
7. 7.2 m

Exercise 44 (page 87)

1. (c) **2.** (b) **3.** (a) **4.** (c) **5.** (b) **6.** (a) **7.** (d) **8.** (c) **9.** (d) **10.** (a)

Chapter 9

Exercise 45 (page 90)

1. $d = c - e - a - b$
2. $y = \frac{1}{3}(t - x)$
3. $r = \frac{c}{2\pi}$
4. $x = \frac{y - c}{m}$
5. $T = \frac{I}{PR}$
6. $R = \frac{E}{I}$
7. $c = \frac{Q}{m\Delta T}$
8. $r = \frac{S - a}{S}$ or $r = 1 - \frac{a}{S}$
9. $C = \frac{5}{9}(F - 32)$
10. $R = \frac{pV}{mT}$
11. $a = N^2y - x$
12. $T_2 = \frac{P_2V_2T_1}{P_1V_1}$
13. $L = \frac{\sqrt{Z^2 - R^2}}{2\pi f}$
14. $v = \sqrt{\frac{2L}{\rho ac}}$
15. $V = \frac{k^2H^2L^2}{\theta^2}$
16. $x = \sqrt{\left(a^2 - \left(\frac{V}{W}\right)^2\right)}$
17. 750 GHz

Exercise 46 (page 92)

1. $x = \frac{d}{\lambda}(y + \lambda)$ or $x = d + \frac{yd}{\lambda}$
2. $f = \frac{3F - AL}{3}$ or $f = F - \frac{AL}{3}$
3. $E = \frac{M\ell^2}{8yI}$
4. $t = \frac{R - R_0}{R_0\alpha}$
5. $R_2 = \frac{RR_1}{R_1 - R}$
6. $R = \frac{E - e - Ir}{I}$ or $R = \frac{E - e}{I} - r$
7. $b = \sqrt{\left(\frac{y}{4ac^2}\right)}$
8. $V_2 = \frac{P_1V_1T_2}{P_2T_1}$
9. $x = \frac{ay}{\sqrt{(y^2 - b^2)}}$
10. $\ell = \frac{gt^2}{4\pi^2}$
11. $u = \sqrt{v^2 - 2as}$
12. $R = \sqrt{\left(\frac{360A}{\pi\theta}\right)}$

Exercise 47 (page 94)

1. $a = \sqrt{\left(\frac{xy}{m - n}\right)}$
2. $R = \sqrt[4]{\left(\frac{M}{\pi} + r^4\right)}$
3. $r = \frac{3(x + y)}{(1 - x - y)}$
4. $L = \frac{mrCR}{\mu - m}$
5. $b = \frac{c}{\sqrt{1 - a^2}}$
6. $r = \sqrt{\left(\frac{x - y}{x + y}\right)}$
7. $b = \frac{a(p^2 - q^2)}{2(p^2 + q^2)}$
8. $v = \frac{uf}{u - f}$, 30
9. $t_2 = t_1 + \frac{Q}{mc}$, 55
10. $v = \sqrt{\left(\frac{2dgh}{0.03L}\right)}$, 0.965
11. $l = \frac{8S^2}{3d} + d$, 2.725
12. $C = \frac{1}{\omega\left\{\omega L - \sqrt{Z^2 - R^2}\right\}}$, 63.1×10^{-6}
13. 64 mm
14. $\lambda = \sqrt[5]{\left(\frac{a\mu}{\rho CZ^4n}\right)^2}$
15. $w = \frac{2R - F}{L}$, 3 kN/m
16. $t_2 = t_1 - \frac{Qd}{kA}$
17. $r = \frac{v}{\omega}\left(1 - \frac{s}{100}\right)$
18. $F = EI\left(\frac{n\pi}{L}\right)^2$, 13.61 MN

19. $v_2 = \sqrt{2g\left(\frac{p_1}{\rho g} - \frac{p_2}{\rho g} + \frac{v_1^2}{2g} + h_1 + h_2\right)}$
20. 500 MN/m^2
21. P = 37.0

Exercise 48 (page 96)

1. (c) **2.** (b) **3.** (d) **4.** (a) **5.** (a) **6.** (b) **7.** (c) **8.** (d) **9.** (a) **10.** (c) **11.** (b) **12.** (a) **13.** (d) **14.** (d) **15.** (c) **16.** (b) **17.** (b) **18.** (a) **19.** (d) **20.** (c)

Chapter 10

Exercise 49 (page 100)

1. $a = 5, b = 2$ **2.** $x = 1, y = 1$
3. $s = 2, t = 3$ **4.** $x = 3, y = -2$
5. $x = 2, y = 5$ **6.** $c = 2, d = -3$

Exercise 50 (page 102)

1. $p = -1, q = -2$ **2.** $x = 4, y = 6$
3. $a = 2, b = 3$ **4.** $x = 3, y = 4$
5. $x = 10, y = 15$ **6.** $a = 0.30, b = 0.40$

Exercise 51 (page 103)

1. $x = \frac{1}{2}, y = \frac{1}{4}$ **2.** $a = \frac{1}{3}, b = -\frac{1}{2}$
3. $p = \frac{1}{4}, q = \frac{1}{5}$ **4.** $c = 3, d = 4$
5. $r = 3, s = \frac{1}{2}$ **6.** 1

Exercise 52 (page 107)

1. $a = 0.2, b = 4$
2. $I_1 = 6.47, I_2 = 4.62$
3. $u = 12, a = 4, v = 26$
4. $m = -0.5, c = 3$
5. $\alpha = 0.00426, R_0 = 22.56\ \Omega$
6. $a = 15, b = 0.50$
7. $R_1 = 5.7\,kN, R_2 = 6.3\,kN$
8. $r = 0.258, \omega = 32.3$

Exercise 53 (page 107)

1. (c) **2.** (d) **3.** (c) **4.** (d) **5.** (b) **6.** (a) **7.** (b) **8.** (a) **9.** (b) **10.** (a)

Chapter 11

Exercise 54 (page 111)

1. 4 or −8 **2.** 4 or −4
3. 2 or −6 **4.** 1.5 or −1
5. $\frac{1}{2}$ or $\frac{1}{3}$ **6.** $\frac{1}{2}$ or $-\frac{4}{5}$
7. 2 **8.** $1\frac{1}{3}$ or $-\frac{1}{7}$
9. $1\frac{1}{3}$ or $-\frac{1}{2}$ **10.** $\frac{5}{4}$ or $-\frac{3}{2}$
11. $x^2 - 4x + 3 = 0$ **12.** $x^2 + 3x - 10 = 0$
13. $x^2 + 5x + 4 = 0$ **14.** $4x^2 - 8x - 5 = 0$
15. $x^2 - 36 = 0$ **16.** $x^2 - 1.7x - 1.68 = 0$

Exercise 55 (page 113)

1. −0.268 or −3.732 **2.** 0.637 or −3.137
3. 1.468 or −1.135 **4.** 1.290 or 0.310
5. 2.443 or 0.307

Exercise 56 (page 114)

1. 0.637 or −3.137 **2.** 0.296 or −0.792
3. 2.781 or 0.719 **4.** 0.443 or −1.693
5. 3.608 or −1.108

Exercise 57 (page 116)

1. 1.191 s **2.** 0.905 A or 0.345 A
3. 19.38 m or 0.619 m **4.** 0.0133
5. 1.066 m **6.** 86.78 cm
7. 18.165 m or 1.835 m **8.** 7 m
9. 12 ohms, 28 ohms **10.** 5.73 s or 0.52 s
11. 400 rad/s **12.** 2.602 m
13. 1229 m or 238.9 m

Exercise 58 (page 117)

1. $x = 1, y = 3$ and $x = -3, y = 7$
2. $x = \frac{2}{5}, y = -\frac{1}{5}$ and $-1\frac{2}{3}, y = -4\frac{1}{3}$
3. $x = 0, y = 4$ and $x = 3, y = 1$

Exercise 59 (page 118)

1. (d) **2.** (d) **3.** (a) **4.** (c) **5.** (a) **6.** (c) **7.** (b) **8.** (d) **9.** (c) **10.** (b) **11.** (c) **12.** (b) **13.** (a) **14.** (d) **15.** (a) **16.** (a) **17.** (b) **18.** (b) **19.** (c) **20.** (d)

Chapter 12

Exercise 60 (page 121)

1. (a) $t>2$ (b) $x<5$

2. (a) $x>3$ (b) $x\geq 3$

3. (a) $t\leq 1$ (b) $x\leq 6$

4. (a) $k\geq \frac{3}{2}$ (b) $z>\frac{1}{2}$

5. (a) $y\geq -\frac{4}{3}$ (b) $x\geq -\frac{1}{2}$

Exercise 61 (page 122)

1. $-5<t<3$ **2.** $-5\leq y\leq -1$

3. $-\frac{3}{2}<x<\frac{5}{2}$ **4.** $t>3$ and $t<\frac{1}{3}$

5. $k\leq -2$ and $k\geq 4$

Exercise 62 (page 123)

1. $-4\leq x\leq 3$ **2.** $t>5$ or $t<-9$

3. $-5<z\leq 14$ **4.** $-3<x\leq -2$

Exercise 63 (page 124)

1. $z>4$ or $z<-4$ **2.** $-4<z<4$

3. $x\geq\sqrt{3}$ or $x\leq -\sqrt{3}$ **4.** $-2\leq k\leq 2$

5. $-5\leq t\leq 7$ **6.** $t\geq 7$ or $t\leq -5$

7. $y\geq 2$ or $y\leq -2$ **8.** $k>-\frac{1}{2}$ or $k<-2$

Exercise 64 (page 125)

1. $x>3$ or $x<-2$ **2.** $-4\leq t\leq 2$

3. $-2<x<\frac{1}{2}$ **4.** $y\geq 5$ or $y\leq -4$

5. $-4\leq z\leq 0$

6. $(-\sqrt{3}-3)\leq x\leq(\sqrt{3}-3)$

7. $t\geq\left(2+\sqrt{11}\right)$ or $t\leq\left(2-\sqrt{11}\right)$

8. $k\geq\left(\sqrt{\frac{13}{4}}+\frac{1}{2}\right)$ or $k\leq\left(-\sqrt{\frac{13}{4}}+\frac{1}{2}\right)$

Alternatively, $k\geq\frac{1}{2}(1+\sqrt{13})$ or $k\leq\frac{1}{2}(1-\sqrt{13})$

Exercise 65 (Page 125)

1. (b) **2.** (c) **3.** (a) **4.** (d) **5.** (b)

Chapter 13

Exercise 66 (page 128)

1. 4 **2.** 4 **3.** 3

4. -3 **5.** $\frac{1}{3}$ **6.** 3

7. 2 **8.** -2 **9.** $1\frac{1}{2}$

10. $\frac{1}{3}$ **11.** 2 **12.** 10,000

13. 100,000 **14.** 9 **15.** $\frac{1}{32}$

16. 0.01 **17.** $\frac{1}{16}$ **18.** e^3

Exercise 67 (page 130)

1. log 6 **2.** log 15 **3.** log 2

4. log 3 **5.** log 12 **6.** log 500

7. log 100 **8.** log 6 **9.** log 10

10. log 1 = 0 **11.** log 2

12. log 243 or log 3^5 or 5 log 3

13. log 16 or log 2^4 or 4 log 2

14. log 64 or log 2^6 or 6 log 2

15. 0.5 **16.** 1.5 **17.** $x=2.5$

18. $t=8$ **19.** $b=2$ **20.** $x=2$

21. $a=6$ **22.** $x=5$

Exercise 68 (page 132)

1. 1.690 **2.** 3.170 **3.** 0.2696 **4.** 6.058

5. 2.251 **6.** 3.959 **7.** 2.542 **8.** -0.3272

9. 316.2 **10.** 0.057 m^3

Exercise 69 (page 133)

1. (c) **2.** (c) **3.** (d) **4.** (b) **5.** (a) **6.** (b) **7.** (c) **8.** (b) **9.** (d) **10.** (c) **11.** (d) **12.** (d) **13.** (a) **14.** (a) **15.** (b)

Chapter 14

Exercise 70 (page 136)

1. (a) 0.1653 (b) 0.4584 (c) 22030

2. (a) 5.0988 (b) 0.064037 (c) 40.446

3. (a) 4.55848 (b) 2.40444 (c) 8.05124
4. (a) 48.04106 (b) 4.07482 (c) −0.08286
5. 2.739 **6.** 120.7 m

Exercise 71 (page 138)

1. 2.0601 **2.** (a) 7.389 (b) 0.7408
3. $1 - 2x^2 - \frac{8}{3}x^3 - 2x^4$
4. $2x^{1/2} + 2x^{5/2} + x^{9/2} + \frac{1}{3}x^{13/2} + \frac{1}{12}x^{17/2} + \frac{1}{60}x^{21/2}$

Exercise 72 (page 139)

1. 3.95, 2.05 **2.** 1.65, −1.30
3. (a) 28 cm^3 (b) 116 min
4. (a) 70°C (b) 5 minutes

Exercise 73 (page 142)

1. (a) 0.55547 (b) 0.91374 (c) 8.8941
2. (a) 2.2293 (b) −0.33154 (c) 0.13087
3. 8.166 **4.** 1.522
5. 1.485 **6.** −0.4904
7. −0.5822 **8.** 2.197
9. 816.2 **10.** 0.8274
11. 1.962 **12.** 3
13. 4 **14.** 147.9
15. 4.901 **16.** 3.095
17. $t = e^{b+a\ln D} = e^b e^{a\ln D} = e^b e^{\ln D^a}$ i.e. $t = e^b D^a$
18. 500 **19.** $W = PV\ln\left(\frac{U_2}{U_1}\right)$
20. $p_2 = 348.5$ Pa **21.** 992 m/s
22. 25 min

Exercise 74 (page 145)

1. (a) 150°C (b) 100.5°C **2.** 99.21 kPa
3. (a) 29.32 volts (b) 71.31×10^{-6} s
4. (a) 2.038×10^{-4} (b) 2.293 m
5. (a) 50°C (b) 55.45 s **6.** 30.4 N, 0.807 rad
7. (a) 3.04 A (b) 1.46 s **8.** 2.45 mol/cm^3
9. (a) 7.07 A (b) 0.966 s **10.** £1760
11. (a) 100% (b) 67.03% (c) 1.83%
12. 2.45 mA **13.** 142 ms
14. 99.752% **15.** 20 min 38 s

Exercise 75 (page 146)

1. (b) **2.** (b) **3.** (a) **4.** (c) **5.** (c) **6.** (a) **7.** (b) **8.** (d) **9.** (d) **10.** (c)

Chapter 15

Exercise 76 (page 149)

1. 68 **2.** 6.2 **3.** 85.25 **4.** 23.5
5. 11th **6.** 209 **7.** 346.5

Exercise 77 (page 150)

1. −0.5 **2.** 1.5, 3, 4.5
3. 7808 **4.** 25
5. 8.5, 12, 15.5, 19
6. (a) 120 (b) 26070 (c) 250.5
7. £19840, £223,680 **8.** £8720

Exercise 78 (page 152)

1. 2560 **2.** 273.25 **3.** 512, 4096
4. 812.5 **5.** 8 **6.** $1\frac{2}{3}$

Exercise 79 (page 154)

1. (a) 3 (b) 2 (c) 59022 **2.** 10th
3. £1566, 11 years **4.** 57.57 M
5. 71.53 g
6. (a) £346.50 (b) 22 years
7. 100, 139, 193, 268, 373, 518, 720, 1000 rev/min

Exercise 80 (page 155)

1. (a) 84 (b) 3 **2.** (a) 15 (b) 56
3. (a) 12 (b) 840 **4.** (a) 720 (b) 6720

Exercise 81 (page 155)

1. (d) **2.** (a) **3.** (b) **4.** (b) **5.** (d) **6.** (c) **7.** (d) **8.** (b) **9.** (c) **10.** (a)

Chapter 16

Exercise 82 (page 158)

1. $x^7 - 7x^6y + 21x^5y^2 - 35x^4y^3 + 35x^3y^4 - 21x^2y^5 + 7xy^6 - y^7$
2. $32a^5 + 240a^4b + 720a^3b^2 + 1080a^2b^3 + 810ab^4 + 243b^5$

Exercise 83 (page 159)

1. $a^4+8a^3x+24a^2x^2+32ax^3+16x^4$
2. $64-192x+240x^2-160x^3+60x^4-12x^5+x^6$
3. $16x^4-96x^3y+216x^2y^2-216xy^3+81y^4$
4. $32x^5+160x^3+320x+\frac{320}{x}+\frac{160}{x^3}+\frac{32}{x^5}$
5. $p^{11}+22p^{10}q+210p^9q^2+1320p^8q^3+5280p^7q^4$
6. $34749p^8q^5$
7. $700000a^4b^4$

Exercise 84 (page 161)

1. $1+x+x^2+x^3+\ldots,\ |x|<1$
2. $1-2x+3x^2-4x^3+\ldots,\ |x|<1$
3. $\frac{1}{8}\left[1-\frac{3}{2}x+\frac{3}{2}x^2-\frac{5}{4}x^3+\ldots\right],\ |x|<2$
4. $\sqrt{2}\left(1+\frac{x}{4}-\frac{x^2}{32}+\frac{x^3}{128}-..\right),\ |x|<2$ or $-2<x<2$
5. $1-\frac{3}{2}x+\frac{27}{8}x^2-\frac{135}{16}x^3,\ |x|<\frac{1}{3}$
6. $\frac{1}{64}\left[1-9x+\frac{189}{4}x^2+\ldots\right],\ |x|<\frac{2}{3}$
7. Proofs 8. $4-\frac{31}{15}x$
9. (a) $1-x+\frac{x^2}{2},\ |x|<1$ (b) $1-x-\frac{7}{2}x^2,\ |x|<\frac{1}{3}$

Exercise 85 (page 163)

1. 0.6% decrease 2. 3.5% decrease
3. (a) 4.5% increase (b) 3.0% increase
4. 2.2% increase 5. 4.5% increase
6. Proof 7. 7.5% decrease
8. 2.5% increase 9. 0.9% too small
10. +7% 11. Proof
12. 5.5% 13. +5%
14. +1.5%

Exercise 86 (page 164)

1. (d) **2.** (b) **3.** (b) **4.** (c) **5.** (a)

Chapter 17

Exercise 87 (page 170)

1. 24.11 mm 2. (a) 27.20 cm each (b) 45°
3. 20.81 km 4. 3.35 m, 10 cm
5. 132.7 naut. mile 6. 2.94 mm
7. 24 mm

Exercise 88 (page 172)

1. $\sin A=\frac{3}{5}$, $\cos A=\frac{4}{5}$, $\tan A=\frac{3}{4}$, $\sin B=\frac{4}{5}$, $\cos B=\frac{3}{5}$, $\tan B=\frac{4}{3}$
2. (a) $\frac{15}{17}$ (b) $\frac{15}{17}$ (c) $\frac{8}{15}$
3. $\sin A=\frac{5}{13}$, $\tan A=\frac{5}{12}$
4. (a) 9.434 (b) −0.625 (c) 32°

Exercise 89 (page 174)

1. $\frac{1}{2}$ 2. $\frac{7}{2}\sqrt{3}$
3. 1 4. $2-\sqrt{3}$
5. $\frac{1}{\sqrt{3}}$

Exercise 90 (page 177)

1. (a) 0.4540 (b) 0.1321 (c) −0.8399
2. (a) −0.5592 (b) 0.9307 (c) 0.2447
3. (a) −0.7002 (b) −1.1671 (c) 1.1612
4. (a) 3.4203 (b) 3.5313 (c) −1.0974
5. (a) −1.8361 (b) 3.7139 (c) −1.3421
6. (a) 0.3443 (b) −1.8510 (c) −1.2519
7. (a) 0.8660 (b) −0.1100 (c) 0.5865
8. (a) 1.0824 (b) 5.5675 (c) −1.7083
9. 13.54°, 13°32′, 0.236 rad
10. 34.20°, 34°12′, 0.597 rad
11. 39.03°, 39°2′, 0.681 rad
12. 51.92°, 51°55′, 0.906 rad
13. 23.69°, 23°41′, 0.413 rad
14. 27.01°, 27°1′, 0.471 rad
15. 29.05° 16. 20°21′
17. 1.097 18. 0.7199
19. 21°42′ 20. 1.8258, 1.1952, 0.6546
21. 0.07448 22. 12.85
23. (a) −0.8192 (b) −1.8040 (c) 0.6528
24. (a) −1.6616 (b) −0.32492 (c) 2.5985
25. −68.37°

Exercise 91 (page 179)

1. $BC = 3.50\,\text{cm}$, $AB = 6.10\,\text{cm}$, $\angle B = 55^\circ$
2. $FE = 5\,\text{cm}$, $\angle E = 53.13^\circ$, $\angle F = 36.87^\circ$
3. $GH = 9.841\,\text{mm}$, $GI = 11.32\,\text{mm}$, $\angle H = 49^\circ$
4. $KL = 5.43\,\text{cm}$, $JL = 8.62\,\text{cm}$, $\angle J = 39^\circ$, area $= 18.19\,\text{cm}^2$
5. $MN = 28.86\,\text{mm}$, $NO = 13.82\,\text{mm}$, $\angle O = 64^\circ 25'$, area $= 199.4\,\text{mm}^2$
6. $PR = 7.934\,\text{m}$, $\angle Q = 65.06^\circ$, $\angle R = 24.94^\circ$, area $= 14.64\,\text{m}^2$
7. 6.54 m
8. 9.40 mm

Exercise 92 (page 181)

1. 48 m
2. 110.1 m
3. 53.0 m
4. 9.50 m
5. 107.8 m
6. 9.43 m, 10.56 m
7. 60 m

Exercise 93 (page 182)

1. (d) **2.** (a) **3.** (b) **4.** (a) **5.** (b) **6.** (c) **7.** (c) **8.** (a) **9.** (c) **10.** (d) **11.** (d) **12.** (a) **13.** (b) **14.** (a) **15.** (b) **16.** (c) **17.** (d) **18.** (b) **19.** (c) **20.** (d)

Chapter 18

Exercise 94 (page 187)

1. (a) 42.78° and 137.22° (b) 188.53° and 351.47°
2. (a) 29.08° and 330.92° (b) 123.86° and 236.14°
3. (a) 44.21° and 224.21° (b) 113.12° and 293.12°
4. $t = 122^\circ 7'$ and $237^\circ 53'$
5. $\alpha = 218^\circ 41'$ and $321^\circ 19'$
6. $\theta = 39^\circ 44'$ and $219^\circ 44'$

Exercise 95 (page 191)

1. 1, 120°
2. 2, 144°
3. 3, 90°
4. 5, 720°
5. 3.5, 960°
6. 6, 360°
7. 4, 180°

Exercise 96 (page 194)

1. (a) 40 mA (b) 25 Hz (c) 0.04 s or 40 ms (d) 0.29 rad (or 16.62°) leading 40 sin 50πt
2. (a) 75 cm (b) 6.37 Hz (c) 0.157 s (d) 0.54 rad (or 30.94°) lagging 75 sin 40t
3. (a) 300 V (b) 100 Hz (c) 0.01 s or 10 ms (d) 0.412 rad (or 23.61°) lagging 300 sin 200πt
4. (a) $v = 120\sin 100\pi t$ volts (b) $v = 120\sin(100\pi t + 0.43)$ volts
5. $i = 20\sin(80\pi t - \frac{\pi}{6})$ A or $i = 20\sin(80\pi t - 0.524)$ A
6. $3.2\sin(100\pi t + 0.488)$ m
7. (a) 5 A, 50 Hz, 20 ms, 24.75° lagging (b) −2.093 A (c) 4.363 A (d) 6.375 ms (e) 3.423 ms

Exercise 97 (page 195)

1. (d) **2.** (a) **3.** (a) **4.** (c) **5.** (b)

Chapter 19

Exercise 98 (page 198)

1. (5.83, 59.04°) or (5.83, 1.03 rad)
2. (6.61, 20.82°) or (6.61, 0.36 rad)
3. (4.47, 116.57°) or (4.47, 2.03 rad)
4. (6.55, 145.58°) or (6.55, 2.54 rad)
5. (7.62, 203.20°) or (7.62, 3.55 rad)
6. (4.33, 236.31°) or (4.33, 4.12 rad)
7. (5.83, 329.04°) or (5.83, 5.74 rad)
8. (15.68, 307.75°) or (15.68, 5.37 rad)

Exercise 99 (page 199)

1. (1.294, 4.830)
2. (1.917, 3.960)
3. (−5.362, 4.500)
4. (−2.884, 2.154)
5. (−9.353, −5.400)
6. (−2.615, −3.027)
7. (0.750, −1.299)
8. (4.252, −4.233)
9. (a) $40\angle 18^\circ$, $40\angle 90^\circ$, $40\angle 162^\circ$, $40\angle 234^\circ$, $40\angle 306^\circ$ (b) (38.04, 12.36), (0, 40), (−38.04, 12.36), (−23.51, −32.36), (23.51, −32.36) (c) 47.02 mm

Exercise 100 (page 201)

1. (a) **2.** (d) **3.** (b) **4.** (a) **5.** (c)

Chapter 20

Exercise 101 (page 205)

1. $C = 83°$, $a = 14.1$ mm, $c = 28.9$ mm, area $= 189$ mm^2
2. $A = 52°2'$, $c = 7.568$ cm, $a = 7.152$ cm, area $= 25.65$ cm^2
3. $D = 19°48'$, $E = 134°12'$, $e = 36.0$ cm, area $= 134$ cm^2
4. $E = 49°0'$, $F = 26°38'$, $f = 15.09$ mm, area $= 185.6$ mm^2
5. $J = 44°29'$, $L = 99°31'$, $l = 5.420$ cm, area $= 6.133$ cm^2
   OR $J = 135°31'$, $L = 8°29'$, $l = 0.811$ cm, area $= 0.917$ cm^2
6. $K = 47°8'$, $J = 97°52'$, $j = 62.2$ mm, area $= 820.2$ mm^2
   OR $K = 132°52'$, $J = 12°8'$, $j = 13.19$ mm, area $= 174.0$ mm^2

Exercise 102 (page 207)

1. $p = 13.2$ cm, $Q = 47.34°$, $R = 78.66°$, area $= 77.7$ cm^2
2. $p = 6.127$ m, $Q = 30.83°$, $R = 44.17°$, area $= 6.938$ m^2
3. $X = 83.33°$, $Y = 52.62°$, $Z = 44.05°$, area $= 27.8$ cm^2
4. $X = 29.77°$, $Y = 53.50°$, $Z = 96.73°$, area $= 355$ mm^2

Exercise 103 (page 209)

1. 193 km
2. (a) 122.6 m (b) 44.54°, 40.66°, 94.80°
3. (a) 11.4 m (b) 17.55°
4. 163.4 m
5. $BF = 3.9$ m, $BE = 4.0$ m
6. 6.35 m, 5.37 m
7. 32.48 A, 14.31°

Exercise 104 (page 211)

1. 40.20°, 80.42°, 59.38°
2. (a) 15.23 m (b) 50.07°
3. 40.25 cm, 126.05°
4. 19.8 cm
5. 36.2 m
6. $x = 69.3$ mm, $y = 142$ mm
7. 130°
8. 13.66 mm

Exercise 105 (page 212)

1. (d) **2.** (d) **3.** (a) **4.** (a) **5.** (c) **6.** (c) **7.** (b) **8.** (c) **9.** (a) **10.** (b)

Chapter 21

Exercise 106 (page 216)

1–6. Proofs

Exercise 107 (page 218)

1. $\theta = 34.85°$ or 145.15°
2. $A = 213.06°$ or 326.94°
3. $t = 66.75°$ or 246.75°
4. 60°, 300°
5. 59°, 239°
6. 41.81°, 138.19°
7. $\pm 131.81°$
8. 39.81°, −140.19°
9. −30°, −150°
10. 33.69°, 213.69°
11. 101.31°, 281.31°

Exercise 108 (page 218)

1. $y = 50.77°$, 129.23°, 230.77° or 309.23°
2. $\theta = 60°$, 120°, 240° or 300°
3. $\theta = 60°$, 120°, 240° or 300°
4. $D = 90°$ or 270°
5. $\theta = 32.31°$, 147.69°, 212.31° or 327.69°

Exercise 109 (page 219)

1. $A = 19.47°$, 160.53°, 203.58° or 336.42°
2. $\theta = 51.34°$, 123.69°, 231.34° or 303.69°
3. $t = 14.48°$, 165.52°, 221.81° or 318.19°
4. $\theta = 60°$ or 300°

Exercise 110 (page 220)

1. $\theta = 90°$, 210°, 330°
2. $t = 190.10°$, 349.90°
3. $\theta = 38.67°$, 321.33°
4. $\theta = 0°$, 60°, 300°, 360°
5. $\theta = 48.19°$, 138.59°, 221.41° or 311.81°
6. $x = 52.94°$ or 307.06°
7. $A = 90°$
8. $t = 107.83°$ or 252.17°
9. $a = 27.83°$ or 152.17°
10. $\beta = 60.17°$, 161.02°, 240.17° or 341.02°
11. $\theta = 51.83°$, 308.17°
12. $\theta = 30°$, 150°

Exercise 111 (page 220)

1. (c) **2.** (a) **3.** (c) **4.** (a) **5.** (b) **6.** (d) **7.** (b) **8.** (d)

Chapter 22

Exercise 112 (page 224)

1. (a) sin 58° (b) sin 4t
2. (a) cos 104° (b) $\cos\frac{\pi}{12}$
3. Proofs **4.** Proofs
5. (a) 0.3136 (b) 0.9495 (c) −2.4678
6. 64.72° or 244.72°
7. 67.52° or 247.52°

Exercise 113 (page 227)

1. $9.434\sin(\omega t + 1.012)$ **2.** $5\sin(\omega t - 0.644)$
3. $8.062\sin(\omega t + 2.622)$ **4.** $6.708\sin(\omega t - 2.034)$
5. (a) 74.44° or 338.70° (b) 64.69° or 189.05°
6. (a) 72.74° or 354.64° (b) 11.15° or 311.98°
7. (a) 90° or 343.74° (b) 0° or 53.14°
8. (a) 82.92° or 296° (b) 32.36°, 97°, 152.36°, 217°, 272.36° or 337°
9. $8.13\sin(3\theta + 2.584)$
10. $x = 4.0\sin(\omega t + 0.927)$ m
11. $9.434\sin(\omega t + 2.583)$ V
12. $x = 7.07\sin\left(2t + \frac{\pi}{4}\right)$ cm

Exercise 114 (page 229)

1. $\frac{V^2}{2R}(1 + \cos 2t)$
2. Proofs
3. $\cos 3\theta = 4\cos^3\theta - 3\cos\theta$
4. −90°, 30°, 150°
5. −160.47°, −90°, −19.47°, 90°
6. −150°, −90°, −30°, 90°
7. −90°
8. 45°, −135°

Exercise 115 (page 230)

1. $\frac{1}{2}[\sin 9t + \sin 5t]$ **2.** $\frac{1}{2}[\sin 10x - \sin 6x]$
3. $\cos 4t - \cos 10t$ **4.** $2[\cos 4\theta + \cos 2\theta]$
5. $\frac{3}{2}\left[\sin\frac{\pi}{2} + \sin\frac{\pi}{6}\right]$ **6.** $-\frac{\cos 4t}{4} - \frac{\cos 2t}{2} + c$
7. $-\frac{20}{21}$ **8.** 30°, 90° and 150°

Exercise 116 (page 231)

1. $2\sin 2x\cos x$ **2.** $\cos 8\theta\sin\theta$
3. $2\cos 4t\cos t$ **4.** $-\frac{1}{4}\sin 3t\sin 2t$
5. $\cos\frac{7\pi}{24}\cos\frac{\pi}{24}$
6. Proofs
7. 22.5°, 45°, 67.5°, 112.5°, 135°, 157.5°
8. 0°, 45°, 135°, 180° **9.** 21.47° or 158.53°
10. 0°, 60°, 90°, 120°, 180°, 240°, 270°, 300°, 360°
11. $20\sin 400t\cos 10t$

Exercise 117 (page 232)

1. (c) **2.** (d) **3.** (b) **4.** (c) **5.** (d) **6.** (a)

Chapter 23

Exercise 118 (page 241)

1. 35.7 cm^2
2. (a) 80 m (b) 170 m
3. (a) 29 cm^2 (b) 650 mm^2
4. 1136 cm^2 **5.** 482 m^2
6. 3.4 cm
7. $p = 105°$, $q = 35°$, $r = 142°$, $s = 95°$, $t = 146°$
8. (i) rhombus (a) 14 cm^2 (b) 16 cm
(ii) parallelogram (a) 180 mm^2 (b) 80 mm
(iii) rectangle (a) 3600 mm^2 (b) 300 mm
(iv) trapezium (a) 190 cm^2 (b) 62.91 cm
9. 6750 mm^2

Exercise 119 (page 242)

1. (a) 50.27 cm^2 (b) 706.9 mm^2 (c) 3183 mm^2
2. 2513 mm^2 **3.** (a) 20.19 mm (b) 63.41 mm
4. (a) 53.01 cm^2 (b) 129.9 mm^2 (c) 6.84 cm^2
5. 5773 mm^2 **6.** 1.89 m^2
7. 15710 mm^2 (b) 471 mm
8. £4712 **9.** 6597 m^2
10. 0.196 m^2, 1.785 m **11.** 586.9 cm^2

Exercise 120 (page 244)

1. 1932 mm^2 **2.** 1624 mm^2
3. (a) 0.918 ha (b) 456 m **4.** 32
5. (a) 306 **6.** 0.416 m^2

Exercise 121 (page 245)

1. 80 ha **2.** 80 m^2 **3.** 3.14 ha

Exercise 122 (page 246)

1. (b) **2.** (a) **3.** (b) **4.** (d) **5.** (b) **6.** (c) **7.** (a) **8.** (c) **9.** (c) **10.** (d)

Chapter 24

Exercise 123 (page 249)

1. 45.24 cm **2.** 259.5 mm **3.** 2.629 cm
4. 47.68 cm **5.** 38.73 cm **6.** 12730 km
7. 97.13 mm

Exercise 124 (page 250)

1. (a) $\frac{\pi}{6}$ (b) $\frac{5\pi}{12}$ (c) $\frac{5\pi}{4}$
2. (a) 0.838 (b) 1.481 (c) 4.054
3. (a) 210° (b) 80° (c) 105°
4. (a) 0°43′ (b) 154°8′ (c) 414°53′
5. 104.72 rad/s

Exercise 125 (page 252)

1. 113 cm^2 **2.** 2376 mm^2
3. 1790 mm^2 **4.** 802 mm^2
5. 1709 mm^2 **6.** 1269 m^2
7. 1548 m^2 **8.** 17.80 cm, 74.07 cm^2
9. (a) 59.86 mm (b) 197.8 mm
10. 26.2 cm **11.** 8.67 cm, 54.48 cm
12. 82°30′ **13.** 19.63 m^2
14. 2107 mm^2 **15.** 2880 mm^2
16. 4.49 m^2 **17.** 8.48 m
18. 9.55 m **19.** 60 mm
20. 748
21. (a) 0.698 rad (b) 804.2 m^2
22. 10.47 m^2 **23.** (a) 396 mm^2 (b) 42.24%
24. 701.8 mm **25.** 7.74 mm

Exercise 126 (page 255)

1. (a) 2 (b) (3, −4)
2. Centre at (3, −2), radius 4
3. Circle, centre (0, 1), radius 5
4. Circle, centre (0, 0), radius 6

Exercise 127 (page 255)

1. (d) **2.** (b) **3.** (c) **4.** (c) **5.** (c) **6.** (d) **7.** (a) **8.** (b) **9.** (a) **10.** (b)

Chapter 25

Exercise 128 (page 260)

1. 15 cm^3, 135 g **2.** 500 litre
3. 1.44 m^3 **4.** 8796 cm^3
5. 4.709 cm, 153.9 cm^2 **6.** 201.1 cm^3, 159.0 cm^2
7. 2.99 cm **8.** 28060 cm^3, 1.099 m^2
9. 7.68 cm^3, 25.81 cm^2 **10.** 113.1 cm^3, 113.1 cm^2
11. 5890 mm^3 or 58.90 cm^2 **12.** 62.5 minutes
13. (a) 17.36 m^2 (b) 4.778 m^3, 4778 litre
14. 8.48 m^3 **15.** 272

Exercise 129 (page 264)

1. 13.57 kg **2.** 5.131 cm
3. 29.32 cm^3 **4.** 393.4 m^2
5. (i) (a) 670 cm^3 (b) 523 cm^2
(ii) (a) 180 cm^3 (b) 154 cm^2
(iii) (a) 56.5 cm^3 (b) 84.8 cm^2
(iv) (a) 10.4 cm^3 (b) 32.0 cm^2
(v) (a) 96.0 cm^3 (b) 146 cm^2
(vi) (a) 86.5 cm^3 (b) 142 cm^2
(vii) (a) 805 cm^3 (b) 539 cm^2
6. 8.53 cm **7.** (a) 17.9 cm (b) 38.0 cm
8. 125 cm^3 **9.** 10.3 m^3, 25.5 m^2
10. 6560 litre **11.** 657.1 cm^3, 1027 cm^2
12. 220.7 cm^3
13. (a) 1458 litre (b) 9.77 m^2 (c) £140.45
14. 92 m^3, 92,000 litre
15. 7.5 m^3 **16.** 31.13 kg
17. (a) 201.1 mm^3 (b) 1250 (c) 1.965 kg
(d) 49.73% (e) 1.588 g (f) 58.73%

Exercise 130 (page 269)

1. 147 cm^3, 164 cm^2
2. 403 cm^3, 337 cm^2
3. 10480 m^3, 1852 m^2
4. 1707 cm^2 **5.** 10.69 cm
6. 55910 cm^3, 6051 cm^2
7. 5.14 m **8.** 72790 cm^3, 0.07279 m^3

Exercise 131 (page 272)

1. $11210\,cm^3$, $1503\,cm^2$
2. $259.2\,cm^3$, $118.3\,cm^2$
3. $1150\,cm^3$, $531\,cm^2$, 2.60 cm, $326.7\,cm^3$
4. $14.84\,cm^3$ 5. 35.34 litres

Exercise 132 (page 274)

1. $1500\,cm^3$ 2. $418.9\,cm^3$
3. 31.20 litres 4. 1.267×10^6 litres

Exercise 133 (page 275)

1. 8 : 125 2. 137.2 g

Exercise 134 (page 275)

1. (d) **2.** (c) **3.** (a) **4.** (b) **5.** (c) **6.** (a) **7.** (b) **8.** (c) **9.** (d) **10.** (a)

Chapter 26

Exercise 135 (page 278)

1. (a) 4.375 sq units (b) 4.563 sq units (c) 4.5 sq units
2. 54.7 square units 3. 63.33 m
4. 4.70 ha 5. $143\,m^2$

Exercise 136 (page 280)

1. $42.59\,m^3$ 2. $147\,m^3$ 3. $20.42\,m^3$

Exercise 137 (page 283)

1. (a) 2 A (b) 50 V (c) 2.5 A
2. (a) 2.5 mV (b) 3 A
3. 0.093 As, 3.1 A
4. (a) 31.83 V (b) 0
5. $49.13\,cm^2$, 368.5 kPa

Exercise 138 (page 284)

1. (c) **2.** (d) **3.** (b) **4.** (c) **5.** (a)

Chapter 27

Exercise 139 (page 294)

1. 14.5
2. Missing values: −0.75, 0.25, 0.75, 1.75, 2.25, 2.75 Gradient = $\frac{1}{2}$
3. (a) 4, −2 (b) −1, 0 (c) −3, −4 (d) 0, 4
4. (a) $2, \frac{1}{2}$ (b) $3, -2\frac{1}{2}$ (c) $\frac{1}{24}, \frac{1}{2}$
5. (a) 6, −3 (b) 3, 0 (c) 0, 7 (d) $-\frac{2}{3}, -1\frac{2}{3}$
6. (a) $\frac{3}{5}$ (b) −4 (c) $-1\frac{5}{6}$
7. (a) and (c), (b) and (e)
8. (a) −1.1 (b) −1.4
9. (2, 1)
10. (1.5, 6)

Exercise 140 (page 300)

1. (a) 40°C (b) 128 Ω
2. (a) 850 rev/min (b) 77.5 V
3. (a) 0.25 (b) 12 N (c) $F = 0.25L + 12$ (d) 89.5 N (e) 592 N (f) 212 N
4. −0.003, $8.73\,N/cm^2$
5. (a) 22.5 m/s (b) 6.5 s (c) $v = 0.7t + 15.5$
6. $m = 26.8L$
7. (a) $1.25t$ (b) 21.6% (c) $F = -0.095w + 2.2$
8. (a) 96×10^9 Pa (b) 0.00022 (c) 29×10^6 Pa
9. (a) 0.2 (b) 6 (c) $E = 0.2L + 6$ (d) 12 N (e) 65 N
10. $a = 0.85$, $b = 12$, 254.3 kPa, 275.5 kPa, 280 K

Exercise 141 (Page 301)

1. (d) **2.** (a) **3.** (c) **4.** (d) **5.** (b) **6.** (d) **7.** (b) **8.** (a) **9.** (c) **10.** (c) **11.** (d) **12.** (b) **13.** (a) **14.** (b) **15.** (a)

Chapter 28

Exercise 142 (page 307)

1. (a) y (b) x^2 (c) c (d) d
2. (a) y (b) $\sqrt{x}$ (c) b (d) a

3. (a) y (b) $\frac{1}{x}$ (c) f (d) e
4. (a) $\frac{y}{x}$ (b) x (c) b (d) c
5. (a) $\frac{y}{x}$ (b) $\frac{1}{x^2}$ (c) a (d) b
6. $a = 1.5, b = 0.4$, 11.78 mm^2
7. $y = 2x^2 + 7$, 5.15
8. (a) 950 (b) 317 kN
9. $a = 0.4, b = 8.6$ (i) 94.4 (ii) 11.2

Exercise 143 (page 311)

1. (a) $\lg y$ (b) x (c) $\lg a$ (d) $\lg b$
2. (a) $\lg y$ (b) $\lg x$ (c) l (d) $\lg k$
3. (a) $\ln y$ (b) x (c) n (d) $\ln m$
4. $I = 0.0012\ V^2$, 6.75 candelas
5. $a = 3.0, b = 0.5$
6. $a = 5.6, b = 2.6$, 37.86, 3.0
7. $R_0 = 25.1, c = 1.42$
8. $y = 0.08e^{0.24x}$
9. $T_0 = 35.3$ N, $\mu = 0.27$, 64.8 N, 1.29 radians

Exercise 144 (page 312)

1. (a) **2.** (c) **3.** (d) **4.** (b) **5.** (c) **6.** (b) **7.** (d) **8.** (a)

Chapter 29

Exercise 145 (page 316)

1. $a = 12, n = 1.8$, 451, 28.5
2. $k = 1.5, n = -1$
3. $m = 3, n = 2.5$

Exercise 146 (page 318)

1. (1) $a = -8, b = 5.3, p = -8(5.3)^q$ (ii) -224.7 (iii) 3.31

Exercise 147 (page 319)

1. $a = 76, k = -7 \times 10^{-5}, p = 76e^{-7\times10^{-5}h}$, 37.74 kPa
2. $\theta_0 = 152, k = -0.05$

Exercise 148 (Page 320)

1. (c) **2.** (a) **3.** (d) **4.** (a) **5.** (c) **6.** (d) **7.** (b) **8.** (b) **9.** (d) **10.** (b)

Chapter 30

Exercise 149 (page 322)

1. $x = 1, y = 1$
2. $x = 3.5, y = 1.5$
3. $x = -1, y = 2$
4. $x = 2.3, y = -1.2$
5. $x = -2, y = -3$
6. $a = 0.4, b = 1.6$

Exercise 150 (page 326)

1. (a) Minimum (0, 0) (b) Minimum (0, −1) (c) Maximum (0, 3) (d) Maximum (0, −1)
2. −0.4 or 0.6
3. −3.9 or 6.9
4. −1.1 or 4.1
5. −1.8 or 2.2
6. $x = -1.5$ or -2, Minimum at $(-1.75, -0.1)$
7. $x = -0.7$ or 1.6
8. (a) ± 1.63 (b) 1 or −0.3
9. (−2.6, 13.2), (0.6, 0.8); $x = -2.6$ or 0.6
10. $x = -1.2$ or 2.5 (a) −30 (b) 2.75 and −1.50 (c) 2.3 or −0.8

Exercise 151 (page 327)

1. $x = 4, y = 8$ and $x = -0.5, y = -5.5$
2. (a) $x = -1.5$ or 3.5 (b) $x = -1.24$ or 3.24 (c) $x = -1.5$ or 3.0

Exercise 152 (page 328)

1. $x = -2.0$, −0.5 or 1.5
2. $x = -2$, 1 or 3, Minimum at (2.1, −4.1), Maximum at (−0.8, 8.2)
3. $x = 1$
4. $x = -2.0$, 0.4 or 2.6
5. $x = -1.2$, 0.7 or 2.5
6. $x = -2.3$, 1.0 or 1.8
7. $x = -1.5$

Exercise 153 (page 329)

1. (b) **2.** (b) **3.** (c) **4.** (d) **5.** (a)

Chapter 31

Exercise 154 (page 337)

1.

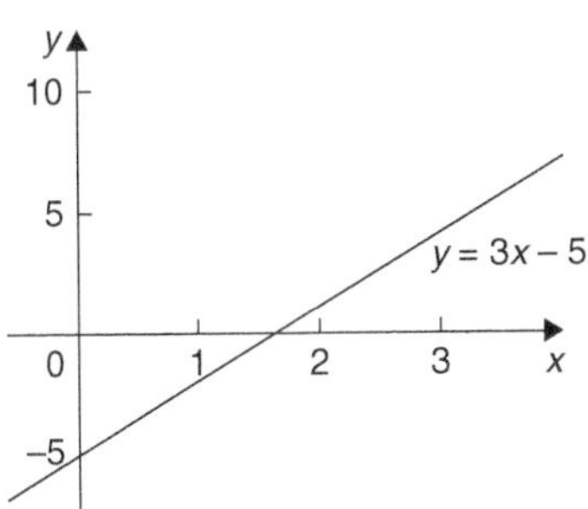

2.

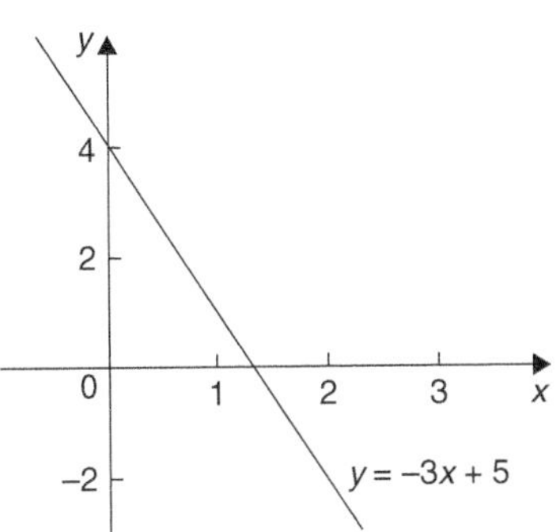

3.

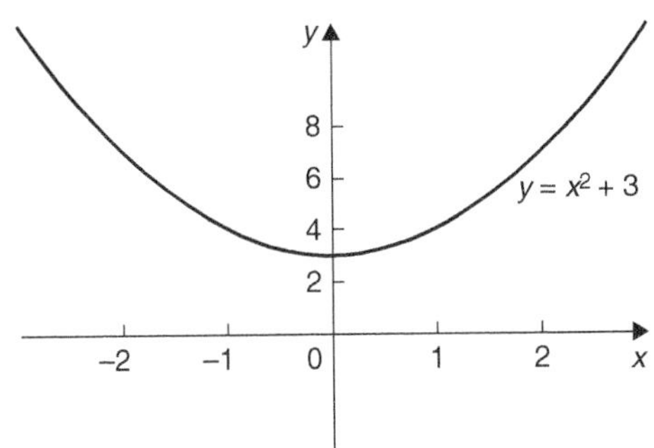

4.

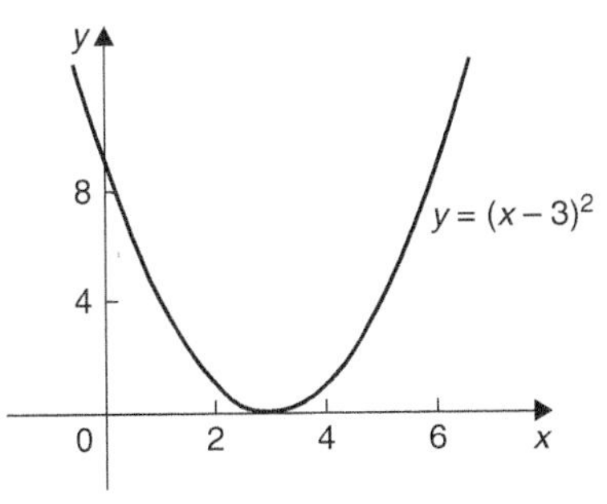

5.

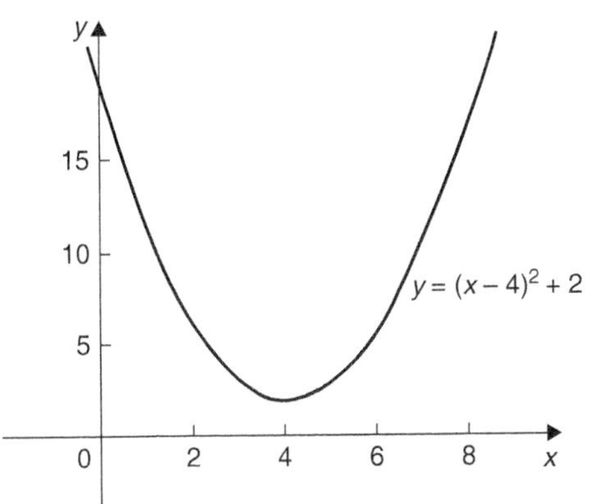

6.

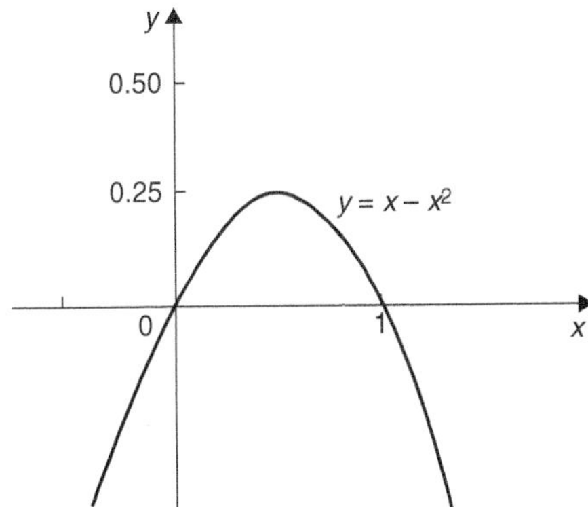

7.

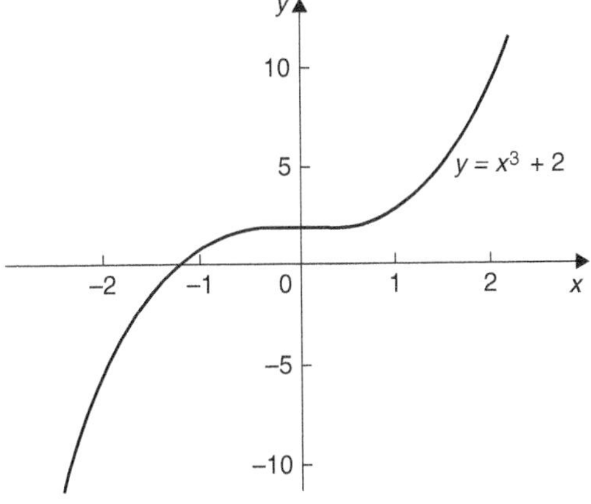

8.

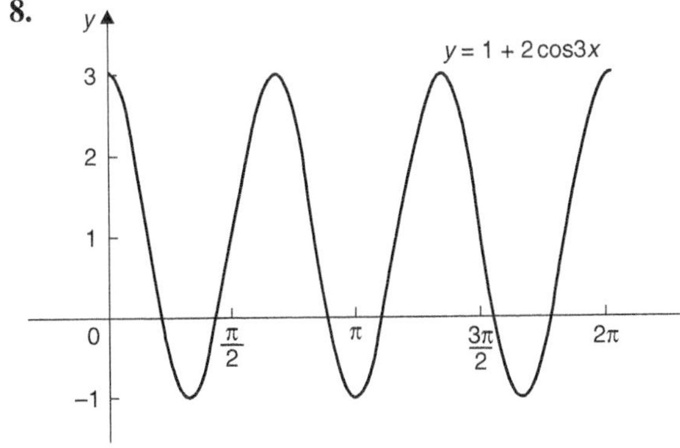

9. $y = 3 - 2\sin(x + \frac{\pi}{4})$

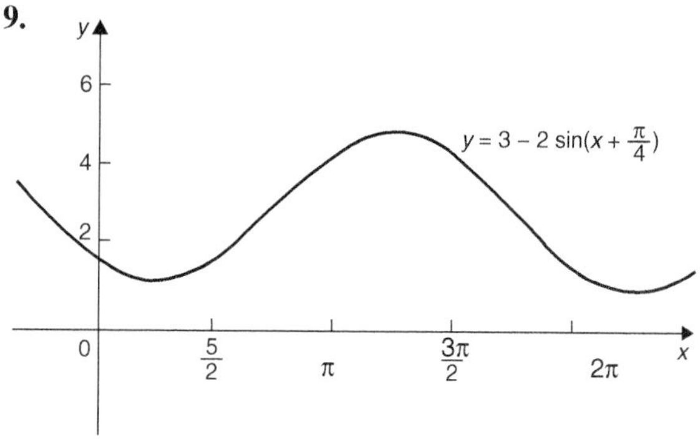

10. $y = 2 \ln x$

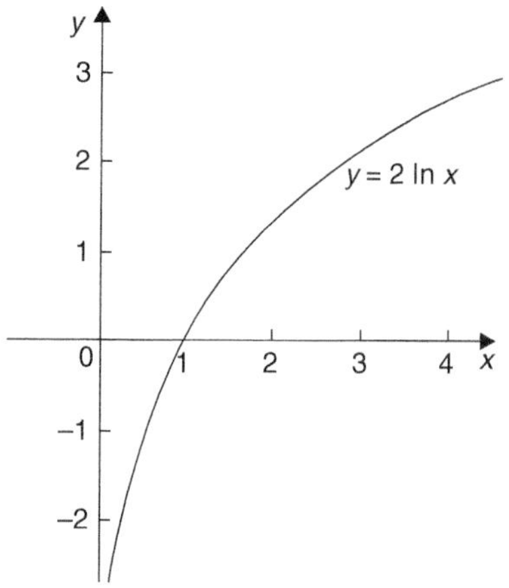

Exercise 155 (page 339)

1. (a) even (b) odd (c) neither (d) even
2. (a) odd (b) even (c) odd (d) neither
3. (a) even (b) odd

Exercise 156 (page 341)

1. $f^{-1}(x) = x - 1$ **2.** $f^{-1}(x) = \frac{1}{5}(x + 1)$
3. $f^{-1}(x) = \sqrt[3]{x - 1}$ **4.** $f^{-1}(x) = \frac{1}{x - 2}$
5. $-\frac{\pi}{2}$ or -1.5708 rad **6.** $\frac{\pi}{3}$ or 1.0472 rad
7. $\frac{\pi}{4}$ or 0.7854 rad **8.** 0.4636 rad
9. 0.4115 rad **10.** 0.8411 rad
11. $\frac{\pi}{4}$ or 0.7854 rad **12.** 0.257 rad
13. 1.533 rad

Exercise 157 (page 341)

1. (c) **2.** (b) **3.** (a) **4.** (b) **5.** (d)

Chapter 32

Exercise 158 (page 348)

1. $x = \pm j5$ **2.** $x = 1 \pm j$
3. $x = 2 \pm j$ **4.** $x = 3 \pm j$
5. $x = 0.5 \pm j0.5$ **6.** $x = 2 \pm j2$
7. $x = 0.2 \pm j0.2$
8. $x = -\frac{3}{4} \pm j\frac{\sqrt{23}}{4}$ or $x = -0.750 \pm j1.199$
9. $x = \frac{5}{8} \pm j\frac{\sqrt{87}}{8}$ or $x = 0.625 \pm j1.166$
10. (a) 1 (b) $-j$ (c) $-j2$

Exercise 159 (page 351)

1. (a) $8 + j$ (b) $-5 + j8$
2. (a) $3 - j4$ (b) $2 + j$
3. (a) 5 (b) $1 - j2$ (c) $j5$ (d) $16 + j3$ (e) 5 (f) $3 + j4$
4. (a) $7 - j4$ (b) $-2 - j6$
5. (a) $10 + j5$ (b) $13 - j13$
6. (a) $-13 - j2$ (b) $-35 + j20$
7. (a) $-\frac{2}{25} + j\frac{11}{25}$ (b) $-\frac{19}{85} + j\frac{43}{85}$
8. (a) $\frac{3}{26} + j\frac{41}{26}$ (b) $\frac{45}{26} - j\frac{9}{26}$
9. (a) $-j$ (b) $\frac{1}{2} - j\frac{1}{2}$
10. Proof

Exercise 160 (page 352)

1. $a = 8, b = -1$ **2.** $x = \frac{3}{2}, y = -\frac{1}{2}$
3. $a = -5, b = -12$ **4.** $x = 3, y = 1$
5. $10 + j13.75$

Exercise 161 (page 355)

1. (a) 4.472, 63.43° (b) 5.385, −158.20° (c) 2.236, 63.43°
2. (a) $\sqrt{13}\angle 56.31°$ (b) $4\angle 180°$ (c) $\sqrt{37}\angle 170.54°$

3. (a) $3\angle-90°$ (b) $\sqrt{125}\angle 100.30°$
(c) $\sqrt{2}\angle-135°$
4. (a) $4.330+j2.500$ (b) $1.500+j2.598$
(c) $4.950+j4.950$
5. (a) $-3.441+j4.915$ (b) $-4.000+j0$
(c) $-1.750-j3.031$
6. (a) $45\angle 65°$ (b) $10.56\angle 44°$
7. (a) $3.2\angle 42°$ (b) $2\angle 150°$
8. (a) $6.986\angle 26.79°$ (b) $7.190\angle 85.77°$

Exercise 162 (page 358)

1. (a) $R=3\,\Omega$, $L=25.5\,\text{mH}$
(b) $R=2\,\Omega$, $C=1061\,\mu\text{F}$
(c) $R=0$, $L=44.56\,\text{mH}$
(d) $R=4\,\Omega$, $C=459.4\,\mu\text{F}$
2. 15.76 A, 23.20° lagging
3. 27.25 A, 3.37° lagging
4. 14.42 A, 43.83° lagging, 0.721
5. 14.6 A, 2.51° leading
6. 8.394 N, 208.68° from force A
7. $(10+j20)\Omega$, $22.36\angle 63.43°\,\Omega$
8. $\pm\dfrac{mh}{2\pi}$
9. (a) 922 km/h at 77.47° (b) 922 km/h at −102.53°
10. (a) $3.770\angle 8.17°$ (b) $1.488\angle 100.37°$
11. Proof
12. $353.6\angle-45°$
13. $275\angle-36.87°$ mA

Exercise 163 (page 359)

1. (b) **2.** (d) **3.** (c) **4.** (d) **5.** (a) **6.** (d) **7.** (a) **8.** (c) **9.** (b) **10.** (a) **11.** (b) **12.** (c) **13.** (b) **14.** (d) **15.** (a) **16.** (a) **17.** (b) **18.** (c) **19.** (c) **20.** (d)

Chapter 33

Exercise 164 (page 361)

1. (a) $7.594\angle 75°$ (b) $125\angle 20.61°$
2. (a) $81\angle 164°$, $-77.86+j22.33$
(b) $55.90\angle-47.18°$, $38-j41$
3. $\sqrt{10}\angle-18.43°$, $3162\angle-129°$
4. $476.4\angle 119.42°$, $-234+j415$
5. $45530\angle 12.78°$, $44400+j10070$
6. $2809\angle 63.78°$, $1241+j2520$
7. $(38.27\times 10^6)\angle 176.15°$, $10^6(-38.18+j2.570)$

Exercise 165 (page 363)

1. (a) $\pm(1.099+j0.455)$ (b) $\pm(0.707+j0.707)$
2. (a) $\pm(2-j)$ (b) $\pm(0.786-j1.272)$
3. (a) $\pm(2.291+j1.323)$ (b) $\pm(-2.449+j2.449)$
4. Modulus 1.710, arguments 17.71°, 137.71° and 257.71°
5. Modulus 1.223, arguments 38.36°, 128.36°, 218.36° and 308.36°
6. Modulus 2.795, arguments 109.90° and 289.90°
7. Modulus 0.3420, arguments 24.58°, 144.58° and 264.58°
8. $Z_0=390.2\angle-10.43°\,\Omega$, $\gamma=0.1029\angle 61.92°$

Exercise 166 (page 363)

1. (a) **2.** (b) **3.** (c) **4.** (d) **5.** (b)

Chapter 34

Exercise 167 (page 368)

1. A scalar quantity has magnitude only; a vector quantity has both magnitude and direction.
2. Scalar 3. Scalar 4. Vector
5. Scalar 6. Scalar 7. Vector
8. Scalar 9. Vector

Exercise 168 (page 375)

1. 17.35 N at 18.00° to the 12 N vector
2. 13 m/s at 22.62° to the 12 m/s velocity
3. 16.40 N at 37.57° to the 13 N force
4. 28.43 N at 129.29° to the 18 N force
5. 32.31 m at 21.80° to the 30 m displacement
6. 14.72 N at −14.72° to the 5 N force
7. 29.15 m/s at 29.04° to the horizontal
8. 9.28 N at 16.70° to the horizontal
9. 6.89 m/s at 159.56° to the horizontal
10. 15.62 N at 26.33° to the 10 N force
11. 21.07 knots, E 9.22° S

Exercise 169 (page 379)

1. (a) 54.0 N at 78.16° (b) 45.64 N at 4.66°
2. (a) 31.71 m/s at 121.81° (b) 19.55 m/s at 8.63°

Exercise 170 (page 380)

1. 83.5 km/h at 71.6° to the vertical
2. 4 minutes 55 seconds, 60°
3. 22.79 km/h at E 9.78° N

Exercise 171 (page 380)

1. $i - j - 4k$ 2. $4i + j - 6k$
3. $-i + 7j - k$ 4. $5i - 10k$
5. $-3i + 27j - 8k$ 6. $-5i + 10k$
7. $i + 7.5j - 4k$ 8. $20.5j - 10k$
9. $3.6i + 4.4j - 6.9k$ 10. $2i + 40j - 43k$

Exercise 172 (page 381)

1. (c) **2.** (c) **3.** (a) **4.** (a) **5.** (d) **6.** (b) **7.** (d) **8.** (a) **9.** (b) **10.** (c)

Chapter 35

Exercise 173 (page 385)

1. $4.5\sin(A + 63.5°)$
2. (a) $20.9\sin(\omega t + 0.63)$ volts
   (b) $12.5\sin(\omega t - 1.36)$ volts
3. $13\sin(\omega t + 0.393)$

Exercise 174 (page 386)

1. $4.5\sin(\theta + 63.5°)$
2. (a) $20.9\sin(\omega t + 0.62)$ volts
   (b) $12.5\sin(\omega t - 1.33)$ volts
3. $13\sin(\omega t + 0.40)$

Exercise 175 (page 388)

1. $4.472\sin(A + 63.44°)$
2. (a) $20.88\sin(\omega t + 0.62)$ volts
   (b) $12.50\sin(\omega t - 1.33)$ volts
3. $13\sin(\omega t + 0.395)$
4. $11.11\sin(\omega t + 0.324)$
5. $8.73\sin(\omega t - 0.173)$

Exercise 176 (page 389)

1. $11.11\sin(\omega t + 0.324)$A
2. $8.73\sin(\omega t - 0.173)$V
3. $i = 21.79\sin(\omega t - 0.639)$A
4. $x = 14.38\sin(\omega t + 1.444)$ m
5. (a) $305.3\sin(314.2t - 0.233)$ V (b) 50 Hz
6. (a) $10.21\sin(628.3t + 0.818)$ V
   (b) 100 Hz (c) 10 ms
7. (a) $79.83\sin(300\pi t + 0.352)$ V
   (b) 150 Hz (c) 6.667 ms
8. $150.6\sin(\omega t - 0.247)$ V

Exercise 177 (page 392)

1. $12.07\sin(\omega t + 0.297)$ 2. $14.51\sin(\omega t - 0.315)$
3. $9.173\sin(\omega t + 0.396)$ 4. $16.168\sin(\omega t + 1.451)$
5. (a) $371.95\sin(314.2t - 0.239)$ V (b) 50 Hz
6. (a) $11.44\sin(200\pi t + 0.715)$ V
   (b) 100 Hz (c) 10 ms
7. (a) $79.73\sin(300\pi t - 0.536)$ V
   (b) 150 Hz (c) 6.667 ms (d) 56.37 V
8. $I_N = 354.6\angle 32.41°$A
9. $s = 85\sin(\omega t + 0.49)$ mm
10. 15 V, 0.927 rad or 53.13°

Exercise 178 (page 392)

1. (b) **2.** (d) **3.** (c) **4.** (b) **5.** (a)

Chapter 36

Exercise 179 (page 398)

1. 1, 5, 21, 9, 61 **2.** 0, 11, −10, 21
3. proof **4.** $8, -a^2 - a + 8, -a^2 - a, -a - 1$

Exercise 180 (page 399)

1. 16, 8

Exercise 181 (page 401)

1. 1	**2.** 7	**3.** $8x$
4. $15x^2$	**5.** $-4x + 3$	**6.** 0
7. 9	**8.** $\frac{2}{3}$	**9.** $18x$
10. $-21x^2$	**11.** $2x + 15$	**12.** 0
13. $12x^2$	**14.** $6x$	

Exercise 182 (page 403)

1. $28x^3$

2. $\frac{1}{2\sqrt{x}}$

3. $\frac{3}{2}\sqrt{t}$

4. $-\frac{3}{x^4}$

5. $3+\frac{1}{2\sqrt{x^3}}-\frac{1}{x^2}$

6. $-\frac{10}{x^3}+\frac{7}{2\sqrt{x^9}}$

7. $6t-12$

8. $3x^2+6x+3$

9. See answers above

10. $12x-3$ (a) -15 (b) 21

11. $6x^2+6x-4$, 32

12. $-6x^2+4$, -9.5

Exercise 183 (page 405)

1. (a) $12\cos 3x$ (b) $-12\sin 6x$

2. $6\cos 3\theta+10\sin 2\theta$

3. 270.2 A/s

4. 1393.4 V/s

5. $12\cos(4t+0.12)+6\sin(3t-0.72)$

Exercise 184 (page 406)

1. (a) $15e^{3x}$ (b) $-\frac{4}{7e^{2x}}$

2. $\frac{5}{\theta}-\frac{4}{\theta}=\frac{1}{\theta}$

3. 16

4. 664

Exercise 185 (page 406)

1. (b) **2.** (d) **3.** (b) **4.** (c) **5.** (a) **6.** (b) **7.** (c) **8.** (d) **9.** (a) **10.** (c)

Chapter 37

Exercise 186 (page 410)

1. (a) $25x^4$ (b) $8.4x^{2.5}$ (c) $-\frac{1}{x^2}$

2. (a) $\frac{8}{x^3}$ (b) 0 (c) 2

3. (a) $\frac{1}{\sqrt{x}}$ (b) $5\sqrt[3]{x^2}$ (c) $-\frac{2}{\sqrt{x^3}}$

4. (a) $\frac{1}{\sqrt[3]{x^4}}$ (b) $2(x-1)$ (c) $6\cos 3x$

5. (a) $8\sin 2x$ (b) $12e^{6x}$ (c) $-\frac{15}{e^{5x}}$

6. (a) $\frac{4}{x}$ (b) $\frac{e^x+e^{-x}}{2}$ (c) $-\frac{1}{x^2}+\frac{1}{2\sqrt{x^3}}$

7. -1, 16

8. $\left(\frac{1}{2},\frac{3}{4}\right)$

9. (a) $-\frac{4}{\theta^3}+\frac{2}{\theta}+10\sin 5\theta-12\cos 2\theta+\frac{6}{e^{3\theta}}$ (b) 22.30

10. 3.29

11. $x=\frac{mg}{k}$

12. 27.0 volts

Exercise 187 (page 411)

1. $x\cos x+\sin x$

2. $2xe^{2x}(x+1)$

3. $x(1+2\ln x)$

4. $6x^2\left(\cos 3x-x\sin 3x\right)$

5. $\sqrt{x}\left(1+\frac{3}{2}\ln 3x\right)$

6. $e^{3t}\left(4\cos 4t+3\sin 4t\right)$

7. $e^{4\theta}\left(\frac{1}{\theta}+4\ln 3\theta\right)$

8. $e^t\left\{\left(\frac{1}{t}+\ln t\right)\cos t-\ln t\sin t\right\}$

9. 8.732

10. 32.31

Exercise 188 (page 413)

1. $\frac{x\cos x-\sin x}{x^2}$

2. $-\frac{6}{x^4}\left(x\sin 3x+\cos 3x\right)$

3. $\frac{2(1-x^2)}{\left(x^2+1\right)^2}$

4. $\frac{\frac{\cos x}{2\sqrt{x}}+\sqrt{x}\sin x}{\cos^2 x}$

5. $\frac{3\sqrt{\theta}\left\{3\sin 2\theta-4\theta\cos 2\theta\right\}}{4\sin^2 2\theta}$

6. $\frac{1}{\sqrt{t^3}}\left(1-\frac{1}{2}\ln 2t\right)$

7. $\frac{2e^{4x}}{\sin^2 x}\left\{\left(1+4x\right)\sin x-x\cos x\right\}$

8. -18

9. 3.82

Exercise 189 (page 414)

1. $12(2x-1)^5$ 2. $5(2x^3-5x)^4(6x^2-5)$
3. $6\cos(3\theta-2)$
4. $-10\cos^4\alpha\sin\alpha$
5. $\dfrac{5(2-3x^2)}{(x^3-2x+1)^6}$
6. $10e^{2t+1}$
7. $-20t\,\mathrm{cosec}^2(5t^2+3)$
8. $18\sec^2(3y+1)$
9. $2\sec^2\theta\, e^{\tan\theta}$ 10. 1.86
11. (a) 24.21 mm/s (b) −70.46 mm/s

Exercise 190 (page 415)

1. (a) $36x^2+12x$ (b) $72x+12$
2. (a) $\dfrac{4}{5}-\dfrac{12}{t^5}+\dfrac{6}{t^3}+\dfrac{1}{4\sqrt{t^3}}$ (b) −4.95
3. (a) $-\dfrac{V}{R}e^{-\frac{t}{CR}}$ (b) $\dfrac{V}{CR^2}e^{-\frac{t}{CR}}$
4. (a) $-(12\sin 2t+\cos t)$ (b) $-\dfrac{2}{\theta^2}$
5. (a) $4(\sin^2 x-\cos^2 x)$ (b) $48(2x-3)^2$
6. 18 7. Proof
8. Proof 9. Proof
10. $M=-\left(PL+\dfrac{WL^2}{2}\right)+x(P-WL)-\dfrac{Wx^2}{2}$

Exercise 191 (page 416)

1. (a) 2. (d) 3. (a) 4. (b) 5. (c) 6. (b) 7. (d) 8. (a) 9. (b) 10. (c)

Chapter 38

Exercise 192 (page 418)

1. 3000π A/s
2. (a) 0.24 cd/V (b) 250 V
3. (a) −625 V/s (b) −220.5 V/s 4. −1.635 Pa/m
5. $-390\,\text{m}^3/\text{min}$

Exercise 193 (page 421)

1. (a) 100 m/s (b) 4 s (c) 200 m (d) −100 m/s
2. (a) 90 km/h (b) 62.5 m
3. (a) 4 s (b) 3 rads
4. (a) 3 m/s, $-1\,\text{m/s}^2$ (b) 6 m/s, $-4\,\text{m/s}^2$ (c) 0.75 s
5. (a) $\omega=1.40$ rad/s (b) $\alpha=-0.37\,\text{rad/s}^2$ (c) $t=6.28$ s
6. (a) 6 m/s, $-23\,\text{m/s}^2$ (b) 117 m/s, $97\,\text{m/s}^2$ (c) 0.75 s or 0.4 s (d) 1.5 s (e) $75\frac{1}{6}$ m
7. 3 s
8. (a) 12 rad/s (b) 48 rad/s^2 (c) 0 or $\frac{1}{3}$ s
9. 162 kJ
10. 2.378 m/s
11. (a) 959.2 m (b) 132.3 m/s (c) 3.97 m/s^2

Exercise 194 (page 425)

1. (3, −9) Minimum
2. (1, 9) Maximum
3. (2, −1) Minimum
4. (0, 3) Minimum, (2, 7) Maximum
5. Minimum at $\left(\frac{2}{3},\frac{2}{3}\right)$
6. (3, 9) Maximum
7. (2, −88) Minimum, (−2.5, 94.25) Maximum
8. (0.4000, 3.8326) Minimum
9. (0.6931, − 0.6137) Maximum
10. (1, 2.5) Minimum, $\left(-\frac{2}{3},4\frac{22}{27}\right)$ Maximum
11. (0.5, 6) Minimum
12. Maximum of 13 at 337.38°, minimum of −13 at 157.38°
13. Proof

Exercise 195 (page 428)

1. 54 km/h
2. 90,000 m^2
3. 48 m
4. 11.42 m^2
5. Radius = 4.607 cm, height = 9.212 cm
6. 6.67 cm
7. Proof
8. Height = 5.42 cm, radius = 2.71 cm
9. 44.72
10. 42.72 volts
11. 50.0 miles/gallon, 52.6 miles/hour

12. 45°
13. 0.607
14. 1.028

Exercise 196 (page 430)

1. $\left(\frac{1}{2},-1\right)$ **2.** $\left(-\frac{1}{4},4\right)$
3. (0,0) **4.** (3, −100)

5. (2, 0.541)

6. Max at (0, 10), Min at (2,−2), point of inflexion at (1, 4)

Exercise 197 (page 432)

1.(a) $y = 4x - 2$ (b) $4y + x = 9$
2.(a) $y = 10x - 12$ (b) $10y + x = 82$
3.(a) $y = \frac{3}{2}x + 1$ (b) $6y + 4x + 7 = 0$
4.(a) $y = 5x + 5$ (b) $5y + x + 27 = 0$
5.(a) $9\theta + t = 6$ (b) $\theta = 9t - 26\frac{2}{3}$ or $3\theta = 27t - 80$

Exercise 198 (page 433)

1. (a) −0.03 (b) −0.008
2. −0.032, −1.6%
3. (a) 60 cm³ (b) 12 cm²
4. (a) −6.03 cm² (b) −18.10 cm³
5. 12.5%

Exercise 199 (page 433)

1. (c) **2.** (d) **3.** (c) **4.** (b) **5.** (c) **6.** (d) **7.** (b) **8.** (a) **9.** (a) **10.** (b)

Chapter 39

Exercise 200 (page 438)

1. 4.742, −2.742 **2.** 2.313
3. 2.648, −1.721 **4.** −1.386, 1.491
5. 1.147 **6.** 0.744, −0.846, −1.693
7. 2.05 **8.** 0.0399
9. 4.19 **10.** 2.9143

Exercise 201 (page 438)

1. (d) **2.** (c) **3.** (b) **4.** (c) **5.** (a)

Chapter 40

Exercise 202 (page 444)

1. $2x - \frac{4}{3}x^3 + \frac{4}{15}x^5 - \frac{8}{315}x^7$

2. $1 + 2x + 2x^2 + \frac{4}{3}x^3 + \frac{2}{3}x^4$

3. $\ln 2 + x^2 + \frac{x^2}{8}$

4. $1 - 8t^2 + \frac{32}{3}t^4 - \frac{256}{45}t^6$

5. $1 + \frac{3}{2}x^2 + \frac{9}{8}x^4 + \frac{9}{16}x^6$

6. $1 + 2x^2 + \frac{10}{3}x^4$

7. $1 + 2\theta - \frac{5}{2}\theta^2$

8. $x^2 - \frac{1}{3}x^4 + \frac{2}{45}x^6$

9. $81 + 216t + 216t^2 + 96t^3 + 16t^4$

Exercise 203 (page 444)

1. (b) **2.** (d) **3.** (c) **4.** (a) **5.** (b)

Chapter 41

Exercise 204 (page 449)

1. $\frac{1}{3}(2t-1)$ **2.** 2

3. (a) $-\frac{1}{4}\cot\theta$ (b) $-\frac{1}{16}\operatorname{cosec}^3\theta$

4. 4 **5.** −6.25

6. $y = -1.155x + 4$ **7.** $y = -\frac{1}{4}x + 5$

Exercise 205 (page 451)

1. (a) 3.122 (b) −14.43 **2.** $y = -2x + 3$
3. $y = -x + \pi$ **4.** 0.02975
5. (a) 13.14 (b) 5.196

Exercise 206 (page 452)

1. (a) **2.** (d) **3.** (c) **4.** (b) **5.** (a)

Chapter 42

Exercise 207 (page 454)

1. (a) $15y^4\frac{dy}{dx}$ (b) $-8\sin 4\theta\frac{d\theta}{dx}$ (c) $\frac{1}{2\sqrt{k}}\frac{dk}{dx}$

2. (a) $\frac{5}{2t}\frac{dt}{dx}$ (b) $\frac{3}{2}e^{2y+1}\frac{dy}{dx}$ (c) $6\sec^2 3y\frac{dy}{dx}$

3. (a) $6\cos 2\theta\frac{d\theta}{dy}$ (b) $6\sqrt{x}\frac{dx}{dy}$ (c) $-\frac{2}{e^t}\frac{dt}{dy}$

4. (a) $-\frac{6}{(3x+1)^2}\frac{dx}{du}$ (b) $6\sec 2\theta\tan 2\theta\frac{d\theta}{du}$

(c) $-\frac{1}{\sqrt{y^3}}\frac{dy}{du}$

Exercise 208 (page 455)

1. $3xy^2\left(3x\frac{dy}{dx}+2y\right)$
2. $\frac{2}{5x^2}\left(x\frac{dy}{dx}-y\right)$
3. $\frac{3}{4v^2}\left(v-u\frac{dv}{du}\right)$
4. $3\left(\frac{\cos 3x}{2\sqrt{y}}\right)\frac{dy}{dx}-9\sqrt{y}\sin 3x$
5. $2x^2\left(\frac{x}{y}+3\ln y\frac{dx}{dy}\right)$

Exercise 209 (page 457)

1. $\frac{2x+4}{3-2y}$ 2. $\frac{3}{1-6y^2}$
3. $-\frac{\sqrt{5}}{2}$ 4. $\frac{-(x+\sin 4y)}{4x\cos 4y}$
5. $\frac{4x-y}{3y+x}$ 6. $\frac{x(4y+9x)}{\cos y-2x^2}$
7. $\frac{1-2\ln y}{3+\frac{2x}{y}-4y^3}$ 8. 5
9. ± 0.5774 10. ± 1.5
11. -6

Exercise 210 (page 458)

1. (d) **2.** (a) **3.** (c) **4.** (b) **5.** (c)

Chapter 43

Exercise 211 (page 460)

1. $\frac{2}{2x-5}$ 2. $-3\tan 3x$
3. $\frac{9x^2+1}{3x^3+x}$ 4. $\frac{10(x+1)}{5x^2+10x-7}$
5. $\frac{1}{x}$ 6. $\frac{2x}{x^2-1}$
7. $\frac{3}{x}$ 8. $2\cot x$
9. $\frac{12x^2-12x+3}{4x^3-6x^2+3x}$

Exercise 212 (page 462)

1. $\frac{(x-2)(x+1)}{(x-1)(x+3)}\left\{\frac{1}{(x-2)}+\frac{1}{(x+1)}-\frac{1}{(x-1)}-\frac{1}{(x+3)}\right\}$

2. $\frac{(x+1)(2x+1)^3}{(x-3)^2(x+2)^4}\left\{\frac{1}{(x+1)}+\frac{6}{(2x+1)}-\frac{2}{(x-3)}-\frac{4}{(x+2)}\right\}$

3. $\frac{(2x-1)\sqrt{(x+2)}}{(x-3)\sqrt{(x+1)^3}}\left\{\frac{2}{(2x+1)}+\frac{1}{2(x+2)}-\frac{1}{(x-3)}-\frac{3}{2(x+1)}\right\}$

4. $\frac{e^{2x}\cos 3x}{\sqrt{(x-4)}}\left\{2-3\tan 3x-\frac{1}{2(x-4)}\right\}$

5. $3\theta\sin\theta\cos\theta\left\{\frac{1}{\theta}+\cos\theta-\tan\theta\right\}$

6. $\frac{2x^4\tan x}{e^{2x}\ln 2x}\left\{\frac{4}{x}+\frac{1}{\sin x\cos x}-2-\frac{1}{x\ln 2x}\right\}$

7. $\frac{13}{16}$

8. -6.71

Exercise 213 (page 463)

1. $2x^{2x}(1+\ln x)$
2. $(2x-1)^x\left\{\dfrac{2x}{2x-1}+\ln(2x-1)\right\}$
3. $\sqrt[x]{(x+3)}\left\{\dfrac{1}{x(x+3)}-\dfrac{\ln(x+3)}{x^2}\right\}$
4. $3x^{4x+1}\left(4+\dfrac{1}{x}+4\ln x\right)$
5. Proof 6. $\dfrac{1}{3}$
7. Proof

Exercise 214 (page 464)

1. (b) **2.** (a) **3.** (a) **4.** (d) **5.** (c)

Chapter 44

Exercise 215 (page 472)

1. (a) $4x+c$ (b) $\dfrac{7x^2}{2}+c$
2. (a) $\dfrac{2}{15}x^3+c$ (b) $\dfrac{5}{24}x^4+c$
3. (a) $\dfrac{3x^2}{2}-5x+c$ (b) $4\theta+2\theta^2+\dfrac{\theta^3}{3}+c$
4. (a) $-\dfrac{4}{3x}+c$ (b) $-\dfrac{1}{4x^3}+c$
5. (a) $\dfrac{4}{5}\sqrt{x^5}+c$ (b) $\dfrac{1}{9}\sqrt[4]{x^9}+c$
6. (a) $\dfrac{10}{\sqrt{t}}+c$ (b) $\dfrac{15}{7}\sqrt[5]{x}+c$
7. (a) $\dfrac{3}{2}\sin 2x+c$ (b) $-\dfrac{7}{3}\cos 3\theta+c$
8. (a) $\dfrac{1}{4}\tan 3x+c$ (b) $-\dfrac{1}{2}\cot 4\theta+c$
9. (a) $-\dfrac{5}{2}\operatorname{cosec} 2t+c$ (b) $\dfrac{1}{3}\sec 4t+c$
10. (a) $\dfrac{3}{8}e^{2x}+c$ (b) $\dfrac{-2}{15e^{5x}}+c$
11. (a) $\dfrac{2}{3}\ln x+c$ (b) $\dfrac{u^2}{2}-\ln u+c$
12. (a) $8\sqrt{x}+8\sqrt{x^3}+\dfrac{18}{5}\sqrt{x^5}+c$
    (b) $-\dfrac{1}{t}+4t+\dfrac{4t^3}{3}+c$

Exercise 216 (page 474)

1. (a) 105 (b) −0.5
2. (a) 6 (b) $-1\dfrac{1}{3}$
3. (a) 0 (b) 4
4. (a) 1 (b) 4.248
5. (a) 0.2352 (b) 2.598
6. (a) 0.2527 (b) 2.638
7. (a) 19.09 (b) 2.457
8. (a) 0.2703 (b) 9.099
9. 55.65 10. Proof
11. 7.26 12. 77.7 m^3

Exercise 217 (page 475)

1. (d) **2.** (c) **3.** (a) **4.** (a) **5.** (c) **6.** (d) **7.** (b) **8.** (b) **9.** (b) **10.** (d)

Chapter 45

Exercise 218 (page 478)

1. $-\dfrac{1}{2}\cos(4x+9)+c$ 2. $\dfrac{3}{2}\sin(2\theta-5)+c$
3. $\dfrac{4}{3}\tan(3t+1)+c$ 4. $\dfrac{1}{70}(5x-3)^7+c$
5. $-\dfrac{3}{2}\ln(2x-1)+c$ 6. $e^{3\theta+5}+c$
7. 227.5 8. 4.333
9. 0.9428 10. 0.7369
11. 1.6 years

Exercise 219 (page 480)

1. $\dfrac{1}{12}(2x^2-3)^6+c$
2. $-\dfrac{5}{6}\cos^6 t+c$
3. $\dfrac{1}{2}\tan^2 3x+c$ or $\dfrac{1}{2}\sec^2 3x+c$
4. $\dfrac{2}{9}\sqrt{(3t^2-1)^3}+c$
5. $\dfrac{1}{2}(\ln\theta)^2+c$
6. $\dfrac{3}{2}\ln(\sec 2t)+c$

7. $4\sqrt{(e^4+4)}+c$

8. 1.763 **9.** 0.6000

10. 0.09259

11. $2\pi\sigma\left\{\sqrt{(9^2+r^2)}-r\right\}$

12. $\dfrac{8\pi^2 IkT}{h^2}$

13. Proof **14.** 11 min 50 s

Exercise 220 (page 481)

1. (d) **2.** (d) **3.** (a) **4.** (b) **5.** (c)

Chapter 46

Exercise 221 (page 484)

1. $\dfrac{1}{2}\left(x-\dfrac{\sin 4x}{4}\right)+c$

2. $\dfrac{3}{2}\left(t+\dfrac{\sin 2t}{2}\right)+c$

3. $5\left(\dfrac{1}{3}\tan 3\theta-\theta\right)+c$

4. $-(\cot 2t+2t)+c$

5. $\dfrac{\pi}{2}$ or 1.571 **6.** $\dfrac{\pi}{8}$ or 0.3927

7. -4.185 **8.** 0.6311

Exercise 222 (page 486)

1. $-\cos\theta+\dfrac{\cos^3\theta}{3}+c$

2. $\sin 2x-\dfrac{\sin^3 2x}{3}+c$

3. $-\dfrac{2}{3}\cos^3 t+\dfrac{2}{5}\cos^5 t+c$

4. $-\dfrac{\cos^5 x}{5}+\dfrac{\cos^7 x}{7}+c$

5. $\dfrac{3\theta}{4}-\dfrac{1}{4}\sin 4\theta+\dfrac{1}{32}\sin 8\theta+c$

6. $\dfrac{t}{8}-\dfrac{1}{32}\sin 4t+c$

Exercise 223 (page 487)

1. $-\dfrac{1}{2}\left(\dfrac{\cos 7t}{7}+\dfrac{\cos 3t}{3}\right)+c$

2. $\dfrac{\sin 2x}{2}-\dfrac{\sin 4x}{4}+c$

3. $\dfrac{3}{2}\left[\dfrac{\sin 7x}{7}+\dfrac{\sin 5x}{5}\right]+c$

4. $\dfrac{1}{4}\left(\dfrac{\cos 2\theta}{2}-\dfrac{\cos 6\theta}{6}\right)+c$

5. $\dfrac{3}{7}$ or 0.4286 **6.** 0.5973

7. 0.2474 **8.** -0.1999

Exercise 224 (page 488)

1. $5\sin^{-1}\dfrac{t}{2}+c$

2. $3\sin^{-1}\dfrac{x}{3}+c$

3. $2\sin^{-1}\dfrac{x}{2}+\dfrac{x}{2}\sqrt{(4-x^2)}+c$

4. $\dfrac{8}{3}\sin^{-1}\dfrac{3t}{4}+\dfrac{t}{2}\sqrt{(16-9t^2)}+c$

5. $\dfrac{\pi}{2}$ or 1.571

6. 2.760

Exercise 225 (page 489)

1. $\dfrac{3}{2}\tan^{-1}\dfrac{t}{2}+c$ **2.** $\dfrac{5}{12}\tan^{-1}\dfrac{3\theta}{4}+c$

3. 2.356 **4.** 2.457

Exercise 226 (page 489)

1. (a) **2.** (a) **3.** (b) **4.** (c) **5.** (d)

Chapter 47

Exercise 227 (page 492)

1. $2\ln(x-3)-2\ln(x+3)+c$ or $\ln\left(\dfrac{x-3}{x+3}\right)^2+c$

2. $5\ln(x+1)-\ln(x-3)+c$ or $\ln\left\{\dfrac{(x+1)^5}{(x-3)}\right\}+c$

3. $7\ln(x+4)-3\ln(x+1)-\ln(2x-1)+c$

or $\ln\left(\dfrac{(x+4)^7}{(x+1)^3(2x-1)}\right)+c$

4. $x+2\ln(x+3)+6\ln(x-2)+c$

or $x+\ln\left\{(x+3)^2(x-2)^6\right\}+c$

5. $\dfrac{3x^2}{2}-2x+\ln(x-2)-5\ln(x+2)+c$

6. 0.6275 7. 0.8122

8. 19.05 ms

Exercise 228 (page 493)

1. $4\ln(x+1)+\dfrac{7}{(x+1)}+c$

2. $5\ln(x-2)+\dfrac{10}{(x-2)}-\dfrac{2}{(x-2)^2}+c$

3. 1.663 4. 1.089

Exercise 229 (page 495)

1. $\ln\left(x^2+7\right)+\dfrac{3}{\sqrt{7}}\tan^{-1}\dfrac{x}{\sqrt{7}}-\ln(x-2)+c$

2. 0.5880 3. 0.2939

4. 0.1865

Exercise 230 (page 495)

1. (a) **2.** (c) **3.** (d) **4.** (b) **5.** (c)

Chapter 48

Exercise 231 (page 498)

1. $-\dfrac{2}{1+\tan\dfrac{\theta}{2}}+c$

2. $\ln\left\{\dfrac{\tan\dfrac{x}{2}}{1+\tan\dfrac{x}{2}}\right\}+c$

3. $\dfrac{2}{\sqrt{5}}\tan^{-1}\left(\dfrac{1}{\sqrt{5}}\tan\dfrac{\alpha}{2}\right)+c$

4. $\dfrac{1}{5}\ln\left\{\dfrac{2\tan\dfrac{x}{2}-1}{\tan\dfrac{x}{2}+2}\right\}+c$

Exercise 232 (page 499)

1. $\dfrac{2}{3}\tan^{-1}\left(\dfrac{5\tan\dfrac{\theta}{2}+4}{3}\right)+c$

2. $\dfrac{1}{\sqrt{3}}\ln\left(\dfrac{\tan\dfrac{x}{2}+2-\sqrt{3}}{\tan\dfrac{x}{2}+2+\sqrt{3}}\right)+c$

3. $\dfrac{1}{\sqrt{11}}\ln\left\{\dfrac{\tan\dfrac{p}{2}-4-\sqrt{11}}{\tan\dfrac{p}{2}-4+\sqrt{11}}\right\}+c$

4. $\dfrac{1}{\sqrt{7}}\ln\left(\dfrac{3\tan\dfrac{\theta}{2}-4-\sqrt{7}}{3\tan\dfrac{\theta}{2}-4+\sqrt{7}}\right)+c$

5. $\dfrac{1}{2\sqrt{2}}\ln\left\{\dfrac{\sqrt{2}+\tan\dfrac{t}{2}}{\sqrt{2}-\tan\dfrac{t}{2}}\right\}+c$

6. Proof 7. Proof

Chapter 49

Exercise 233 (page 503)

1. $\dfrac{e^{2x}}{2}\left(x-\dfrac{1}{2}\right)+c$

2. $-\dfrac{4}{3}e^{-3x}\left(x+\dfrac{1}{3}\right)+c$

3. $-x\cos x+\sin x+c$

4. $\dfrac{5}{2}\left(\theta\sin 2\theta+\dfrac{1}{2}\cos 2\theta\right)+c$

5. $\dfrac{3}{2}e^{2t}\left(t^2-t+\dfrac{1}{2}\right)+c$

6. 16.78 7. 0.2500

8. 0.4674 9. 15.78

Exercise 234 (page 505)

1. $\dfrac{2}{3}x^3\left(\ln x-\dfrac{1}{3}\right)+c$

2. $2x(\ln 3x-1)+c$

3. $\dfrac{\cos 3x}{27}\left(2-9x^2\right)+\dfrac{2}{9}x\sin 3x+c$

4. $\frac{2}{29}e^{5x}(2\sin 2x + 5\cos 2x) + c$

5. $2[\theta \tan\theta - \ln(\sec\theta)] + c$

6. 0.6363 **7.** 11.31

8. −1.543 **9.** 12.78

10. Proof **11.** $C = 0.66, S = 0.41$

Exercise 235 (page 506)

1. (c) **2.** (a) **3.** (c) **4.** (b) **5.** (d)

Chapter 50

Exercise 236 (page 509)

1. 1.569 **2.** 6.979 **3.** 0.672 **4.** 0.843

Exercise 237 (page 511)

1. 3.323 **2.** 0.997 **3.** 0.605 **4.** 0.799

Exercise 238 (page 513)

1. 1.187 **2.** 1.034 **3.** 0.747

4. 0.571 **5.** 1.260

6. (a) 1.875 (b) 2.107 (c) 1.765 (d) 1.916

7. (a) 1.585 (b) 1.588 (c) 1.583 (d) 1.585

8. (a) 10.194 (b) 10.007 (c) 10.070

9. (a) 0.677 (b) 0.674 (c) 0.675

10. 28.8 m **11.** 0.485 m

Exercise 239 (page 515)

1. (a) **2.** (b) **3.** (c)

Chapter 51

Exercise 240 (page 521)

1. Proof **2.** 32 sq units

3. 37.5 sq units **4.** 7.5 sq units

5. 1 sq units **6.** 8.389 sq units

7. 1.67 sq units **8.** 2.67 sq units

Exercise 241 (page 523)

1. 16 sq units **2.** 5.545 sq units

3. 29.33 N m **4.** 10.67 sq units

5. 51 sq units, 90 sq units **6.** 73.83 sq units

7. 140 m **8.** 693.1 kJ

Exercise 242 (page 525)

1. (0, 0) and (3, 3), 3 sq units **2.** 20.83 sq units

3. 0.4142 sq units **4.** 2.5 sq units

Exercise 243 (page 525)

1. (c) **2.** (c) **3.** (b) **4.** (a) **5.** (d)

Chapter 52

Exercise 244 (page 529)

1. (a) 4 (b) $\frac{2}{\pi}$ or 0.637 (c) 69.17

2. 19 **3.** 9 m/s

4. 4.17 **5.** 2.67 km

6. $\frac{2c}{\pi}$

Exercise 245 (page 530)

1. (a) 6.928 (b) 4.919 (c) $\frac{25}{\sqrt{2}}$ or 17.68

2. (a) $\frac{1}{\sqrt{2}}$ or 0.707 (b) 1.225 (c) 2.121

3. 2.58 **4.** 19.10 A, 21.21 A

5. 216 V, 240 V **6.** 1.11

7. $\sqrt{\frac{E_1^2 + E_3^2}{2}}$

Exercise 246 (page 531)

1. (c) **2.** (d) **3.** (b) **4.** (a) **5.** (c)

Chapter 53

Exercise 247 (page 534)

1. 525π cubic units **2.** 55π cubic units

3. 75.6π cubic units **4.** 48π cubic units

5. 1.5π cubic units **6.** 4π cubic units

7. 2.67π cubic units **8.** 2.67π cubic units

9. (a) 329.4π cubic units (b) 81π cubic units

Exercise 248 (page 536)

1. 428.8π cubic units
2. π cubic units
3. 42.67π cubic units
4. 1.622π cubic units
5. 53.33π cubic units
6. 8.1π cubic units
7. 57.07π cubic units
8. 113.33π cubic units
9. (a) $(-2, 6)$ and $(5, 27)$ (b) 57.17 sq units (c) 1326π cubic units

Exercise 249 (page 536)

1. (a) **2.** (d) **3.** (c) **4.** (b) **5.** (b)

Chapter 54

Exercise 250 (page 541)

1. (2, 2)
2. (2.50, 4.75)
3. (3.036, 24.36)
4. (1.60, 4.57)
5. (−0.833, 0.633)

Exercise 251 (page 543)

1. (2, 1.6)
2. (1, −0.4)
3. (0, 3.6)
4. (0.375, 2.40)
5. (3.0, 1.875)
6. (2.4, 0)
7. (0, 0) and (4, 4), (1.80, 1.80)
8. (−1, 7) and (3, 23), (1, 10.20)

Exercise 252 (page 546)

1. $189.6\,\text{cm}^3$
2. $1131\,\text{cm}^3$
3. On the centre line, distance 2.40 cm from the centre, i.e. at co-ordinates (1.70, 1.70)
4. (a) 45 sq units (b) (i) 1215π cubic units (ii) 202.5π cubic units (c) (2.25, 13.5)
5. $64.90\,\text{cm}^3$, 16.86%, 506.2 g

Exercise 253 (page 547)

1. (b) **2.** (d) **3.** (c) **4.** (a) **5.** (b)

Chapter 55

Exercise 254 (page 553)

1. (a) $72\,\text{cm}^4$, 1.73 cm (b) $128\,\text{cm}^4$, 2.31 cm (c) $512\,\text{cm}^4$, 4.62 cm
2. (a) $729\,\text{cm}^4$, 3.67 cm (b) $2187\,\text{cm}^4$, 6.36 cm (c) $243\,\text{cm}^4$, 2.12 cm
3. (a) $201\,\text{cm}^4$, 2.0 cm (b) $1005\,\text{cm}^4$, 4.47 cm
4. $3927\,\text{mm}^4$, 5.0 mm
5. (a) $335\,\text{cm}^4$, 4.73 cm (b) $22030\,\text{cm}^4$, 14.3 cm (c) $628\,\text{cm}^4$, 7.07 cm
6. 0.866 m
7. $0.245\,\text{m}^4$, 0.559 m
8. $14280\,\text{cm}^4$, 5.96 cm

Exercise 255 (page 555)

1. (a) $12190\,\text{mm}^4$, 10.9 mm (b) $549.5\,\text{cm}^4$, 4.18 cm
2. $I_{AA} = 4224\,\text{cm}^4$, $I_{BB} = 6718\,\text{cm}^4$, $I_{CC} = 37300\,\text{cm}^4$
3. $1350\,\text{cm}^4$, 5.67 cm

Exercise 256 (page 556)

1. (a) **2.** (b) **3.** (c) **4.** (d) **5.** (b)

Chapter 56

Exercise 257 (page 562)

1. Sketches
2. $y = x^2 + 3x - 1$

Exercise 258 (page 564)

1. $y = \frac{1}{4}\sin 4x - x^2 + c$
2. $y = \frac{3}{2}\ln x - \frac{x^3}{6} + c$
3. $y = 3x - \frac{x^2}{2} - \frac{1}{2}$
4. $y = \frac{1}{3}\cos\theta + \frac{1}{2}$
5. $y = \frac{1}{6}\left(x^2 - 4x + \frac{2}{e^x} + 4\right)$
6. $y = \frac{3}{2}x^2 - \frac{x^3}{6} - 1$
7. $v = u + at$
8. 15.9 m

Exercise 259 (page 565)

1. $x = \frac{1}{3}\ln(2+3y)+c$ **2.** $\tan y = 2x + c$

3. $\frac{y^2}{2} + 2\ln y = 5x - 2$ **4.** Proof

5. $x = a\left(1 - e^{-kt}\right)$

6. (a) $Q = Q_0 e^{-\frac{t}{CR}}$ (b) 9.30 C, 5.81 C

7. 273.3 N, 2.31 rad **8.** 8 min 40 s

Exercise 260 (page 568)

1. $\ln y = 2\sin x + c$ **2.** $y^2 - y = x^3 + x$

3. $e^y = \frac{1}{2}e^{2x} + \frac{1}{2}$ **4.** $\ln\left(x^2 y\right) = 2x - y - 1$

5. Proof **6.** $y = \frac{1}{\sqrt{\left(1 - x^2\right)}}$

7. $y^2 = x^2 - 2\ln x + 3$

8. (a) $V = E\left(1 - e^{-\frac{t}{CR}}\right)$ (b) 13.2 V

9. 3

Exercise 261 (page 569)

1. (b) **2.** (d) **3.** (c) **4.** (b) **5.** (c) **6.** (a) **7.** (c) **8.** (b) **9.** (d) **10.** (a)

Chapter 57

Exercise 262 (page 577)

1. $Z = C.\left(A.B + \overline{A}.B\right)$

A	B	C	$A.B$	$\overline{A}$	$\overline{A}.B$	$A.B+\overline{A}.B$	$Z=C.(A.B+\overline{A}.B)$
0	0	0	0	1	0	0	0
0	0	1	0	1	0	0	0
0	1	0	0	1	1	1	0
0	1	1	0	1	1	1	1
1	0	0	0	0	0	0	0
1	0	1	0	0	0	0	0
1	1	0	1	0	0	1	0
1	1	1	1	0	0	1	1

2. $Z = C.\left(A.\overline{B} + \overline{A}\right)$

A	B	C	$\overline{A}$	$\overline{B}$	$A.\overline{B}$	$A.\overline{B}+\overline{A}$	$Z=C.(A.\overline{B}+\overline{A})$
0	0	0	1	1	0	1	0
0	0	1	1	1	0	1	1
0	1	0	1	0	0	1	0
0	1	1	1	0	0	1	1
1	0	0	0	1	1	1	0
1	0	1	0	1	1	1	1
1	1	0	0	0	0	0	0
1	1	1	0	0	0	0	0

3. $Z = A.B.\left(B.\overline{C} + \overline{B}.C + \bar{A}.B\right)$

A	B	C	$\overline{A}$	$\overline{B}$	$\overline{C}$	$B.\overline{C}$	$\overline{B}.C$	$\bar{A}.B$	$(B.\overline{C}+\overline{B}.C+\bar{A}.B)$	$Z=A.B.(B.\overline{C}+\overline{B}.C+\bar{A}.B)$
0	0	0	1	1	1	0	0	0	0	0
0	0	1	1	1	0	0	1	0	1	0
0	1	0	1	0	1	1	0	1	1	0
0	1	1	1	0	0	0	0	1	1	0
1	0	0	0	1	1	0	0	0	0	0
1	0	1	0	1	0	0	1	0	1	0
1	1	0	0	0	1	1	0	0	1	1
1	1	1	0	0	0	0	0	0	0	0

4. $Z = C.\left(B.C.\overline{A} + A.(B + \overline{C})\right)$

A	B	C	$\overline{A}$	$\overline{C}$	$B.C.\overline{A}$	$B+\overline{C}$	$A.(B+\overline{C})$	$B.C.\overline{A}+A.(B+\overline{C})$	Z
0	0	0	1	1	0	1	0	0	0
0	0	1	1	0	0	0	0	0	0
0	1	0	1	1	0	1	0	0	0
0	1	1	1	0	1	1	0	1	1
1	0	0	0	1	0	1	1	1	0
1	0	1	0	0	0	0	0	0	0
1	1	0	0	1	0	1	1	1	0
1	1	1	0	0	0	1	1	1	1

5.

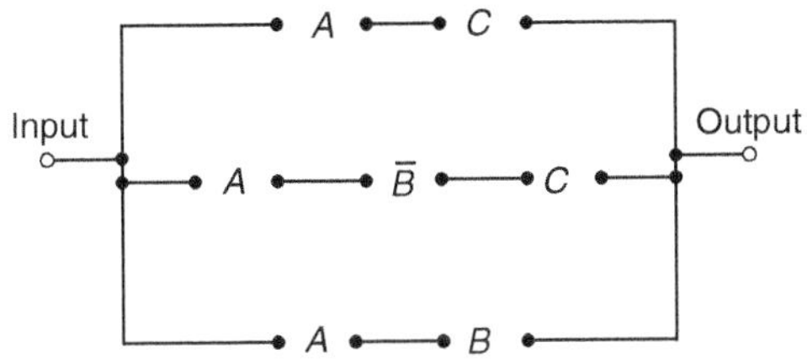

6.

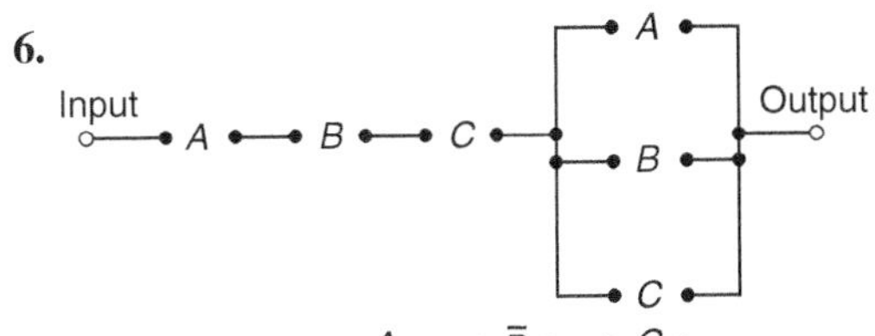

7.

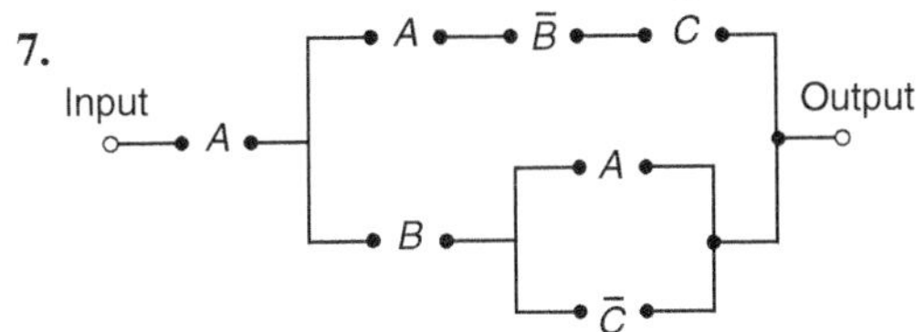

8. $\overline{A}.\overline{B}.C + A.B.\overline{C}$

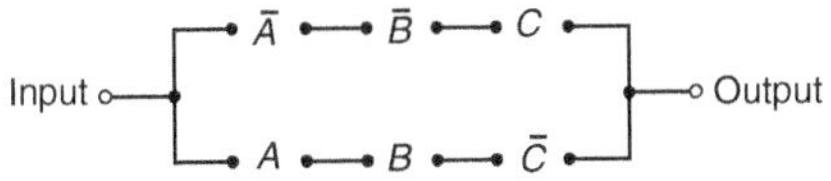

9. $\overline{A}.\overline{B}.\overline{C} + \overline{A}.B.C + A.\overline{B}.\overline{C}$

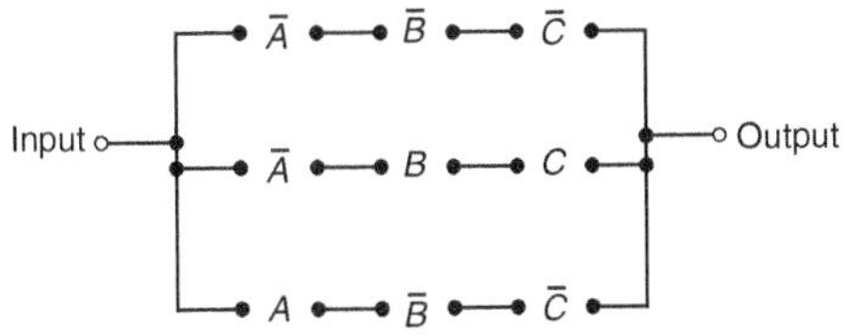

10. $\overline{A}.\overline{B}.\overline{C} + \overline{A}.B.\overline{C} + A.\overline{B}.\overline{C} + A.\overline{B}.C$

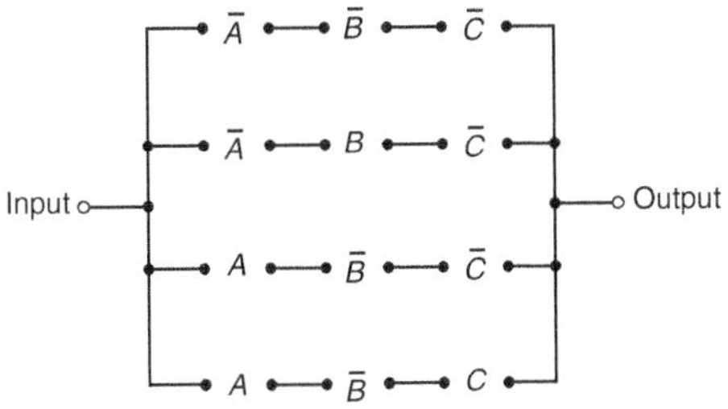

Exercise 263 (page 579)

1. $\overline{P}$
2. $\overline{P} + P.Q$
3. $\overline{G}$
4. F
5. $P.Q$
6. $H.\left(\overline{F} + F.\overline{G}\right)$
7. $F.\overline{G}.\overline{H} + G.H$
8. $\overline{Q}.\overline{R} + \overline{P}.Q.R$
9. $\overline{G}$
10. $F.H + G.\overline{H}$
11. $P.R + \overline{P}.\overline{R}$
12. $P + \overline{R}.\overline{Q}$

Exercise 264 (page 581)

1. $\overline{A}.\overline{B}$
2. $\overline{A} + \overline{B} + C$
3. $\overline{A}.\overline{B} + A.B.C$
4. 1
5. $\overline{P}.\left(\overline{Q} + \overline{R}\right)$

Exercise 265 (page 585)

1. Y
2. $\overline{X} + Y$
3. $\overline{P}.\overline{Q}$
4. $B + A.\overline{C} + \overline{A}.C$
5. $\overline{R}.\left(\overline{P} + Q\right)$
6. $P.(Q + R) + \overline{P}.\overline{Q}.\overline{R}$
7. $\overline{A}.\overline{C}.\left(B + \overline{D}\right)$
8. $\overline{B}.C.\left(\overline{A} + \overline{D}\right)$
9. $D.\left(A + B.\overline{C}\right)$
10. $A.\overline{D} + \overline{A}.\overline{B}.\overline{C}.D$
11. $\overline{A}.C + A.\overline{C}.\overline{D} + \overline{B}.\overline{D}.\left(\overline{A} + \overline{C}\right)$

Exercise 266 (page 588)

1.

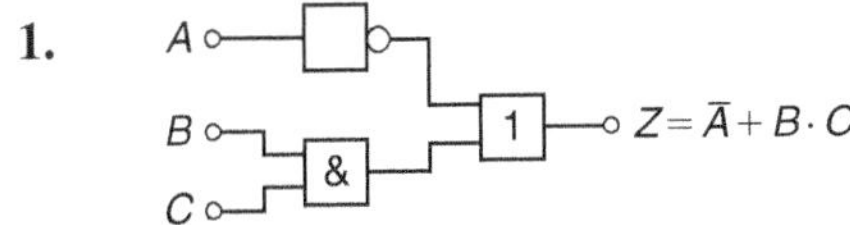

2.

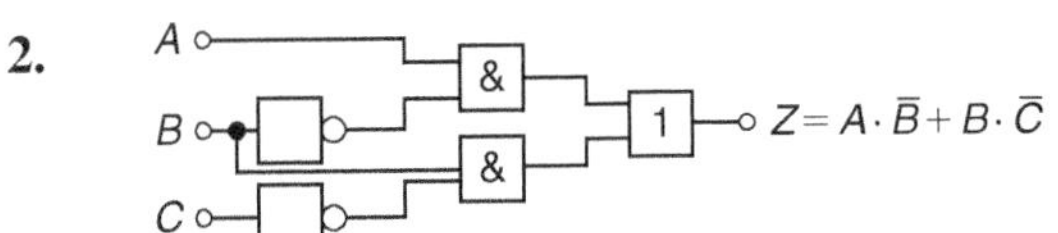

3.

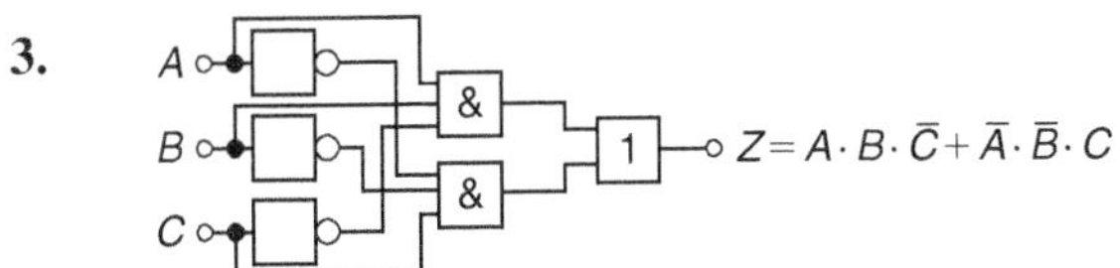

4.

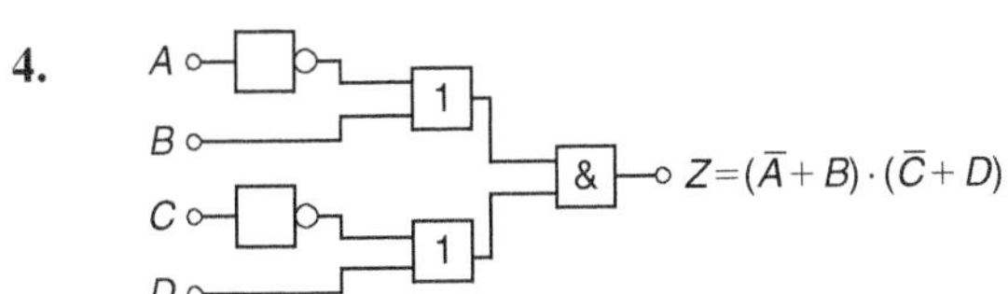

5. $Z_1 = A.B + C$

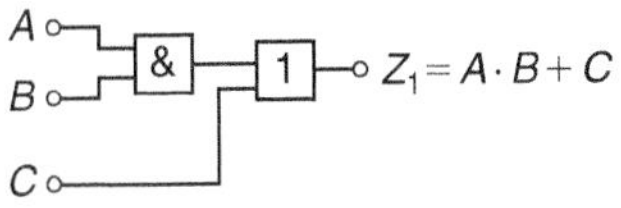

6. $Z_2 = A.\bar{B} + B.C$

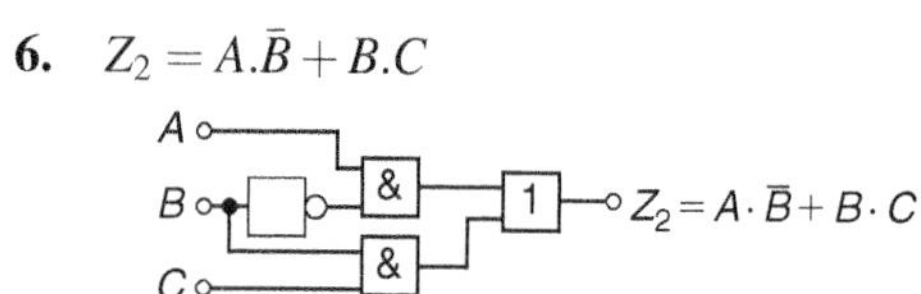

7. $Z_3 = A.C + B$

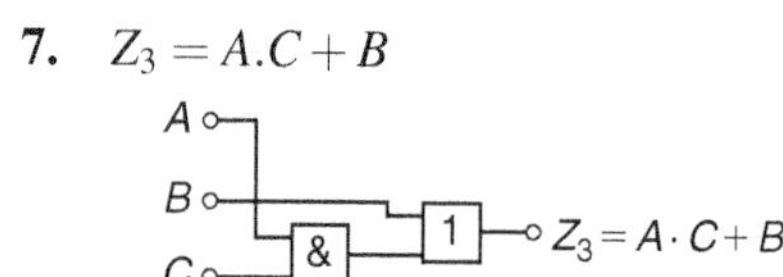

8. $Z = \bar{P} + Q$

P
Q
$Z = \bar{P} + Q$

9. $\bar{R}.\left(P + \bar{Q}\right)$

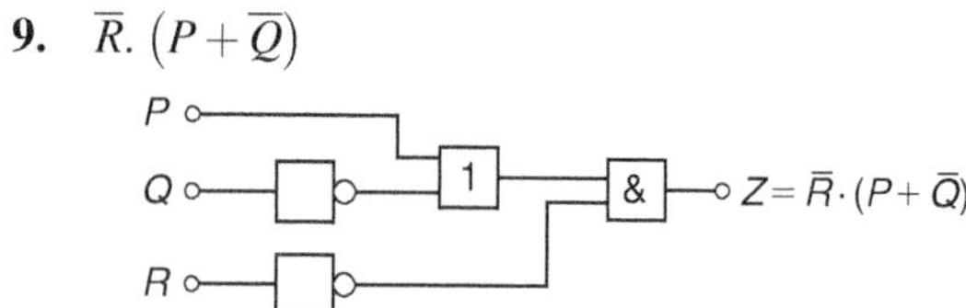

10. $\bar{Q}.\left(P + \bar{R}\right)$

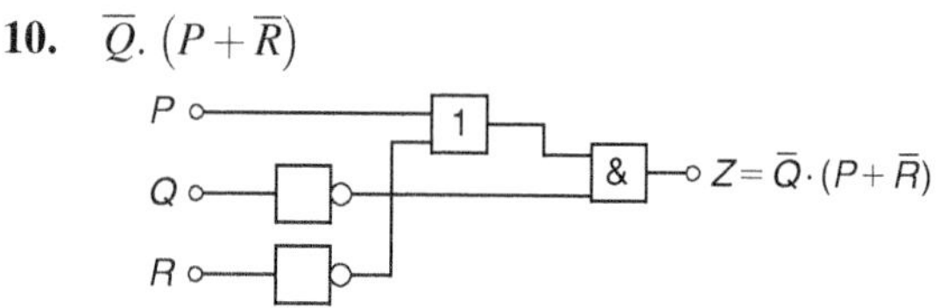

11. $\bar{D}.\left(\bar{A}.C + \bar{B}\right)$

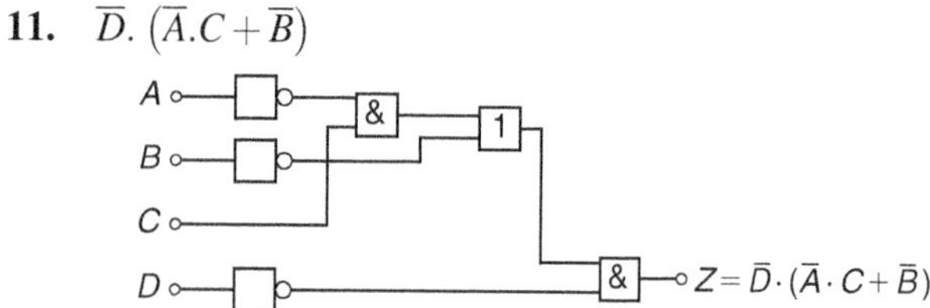

12. $\bar{P}.\left(\bar{Q} + \bar{R}\right)$

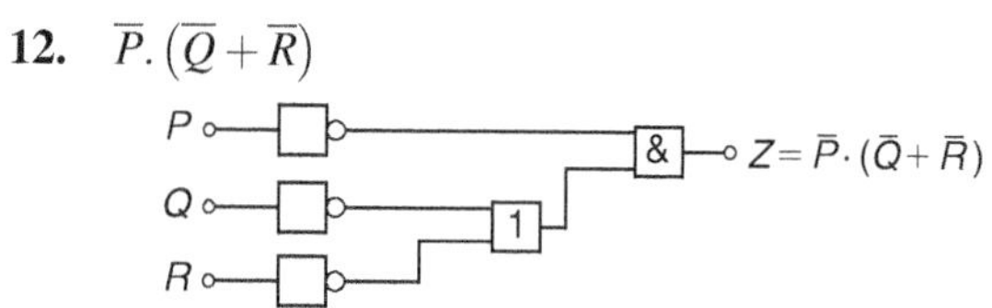

Exercise 267 (page 591)

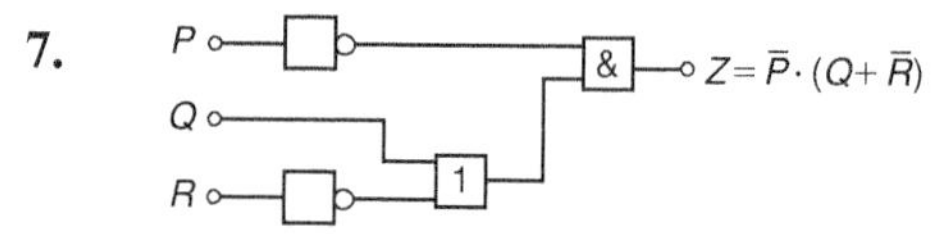

1.

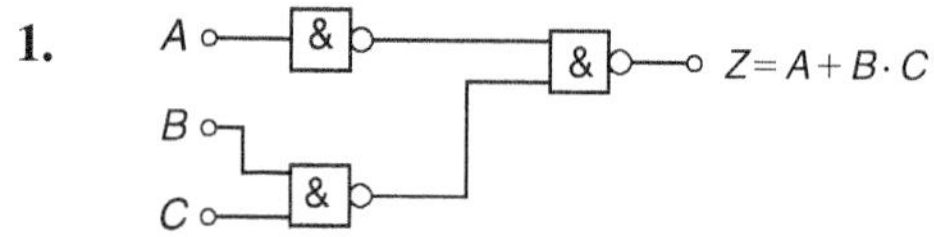

2.

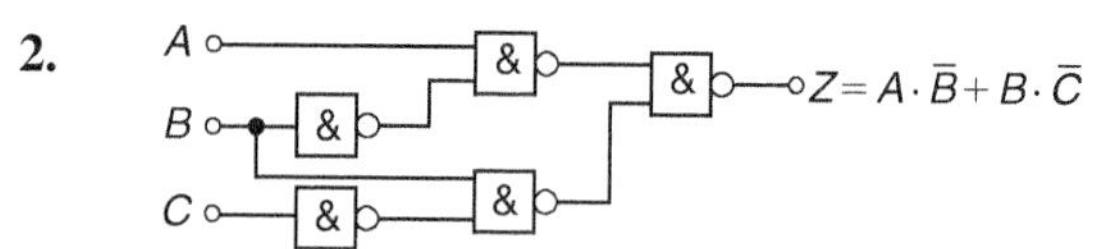

3.

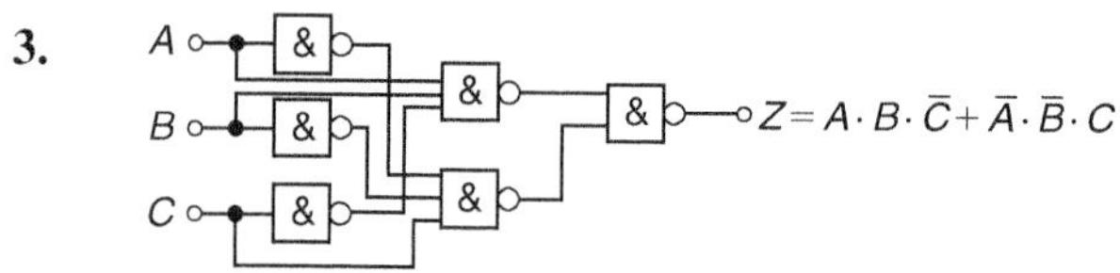

4.

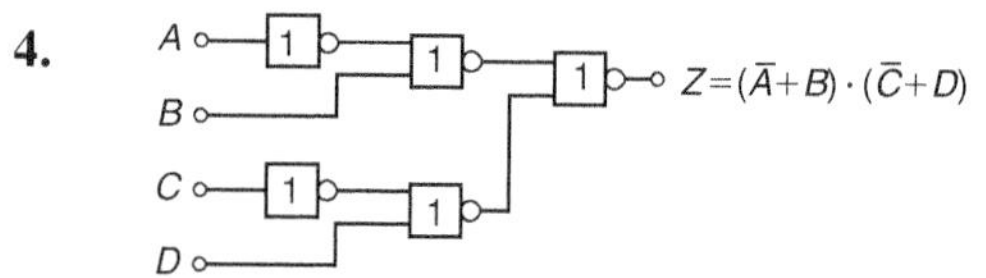

5.

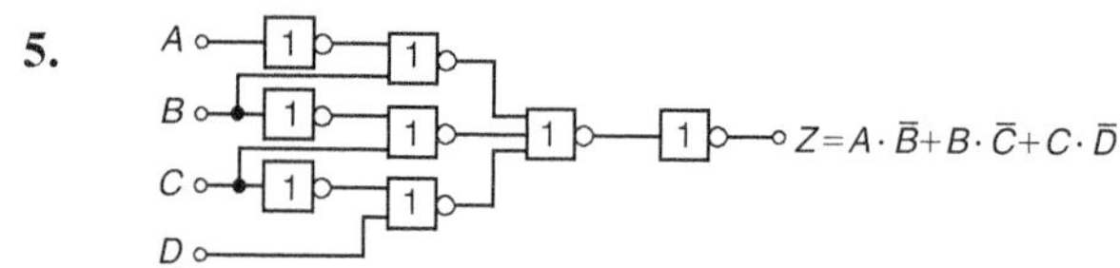

6.

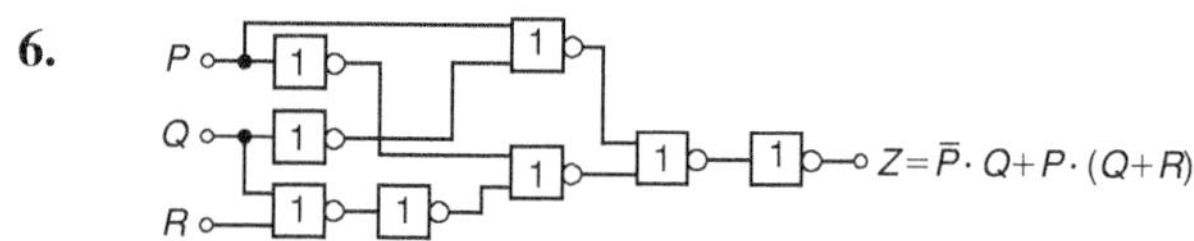

7.

P
Q
R
$Z = \bar{P} \cdot (Q + \bar{R})$

8.

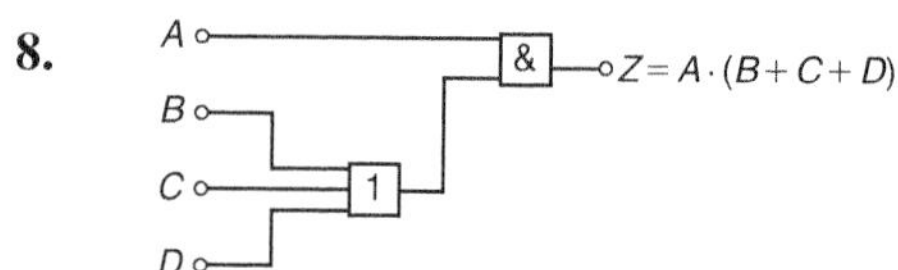

9.

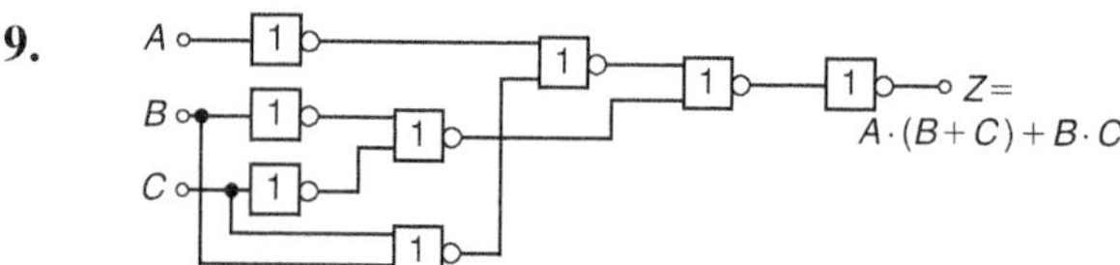

10.

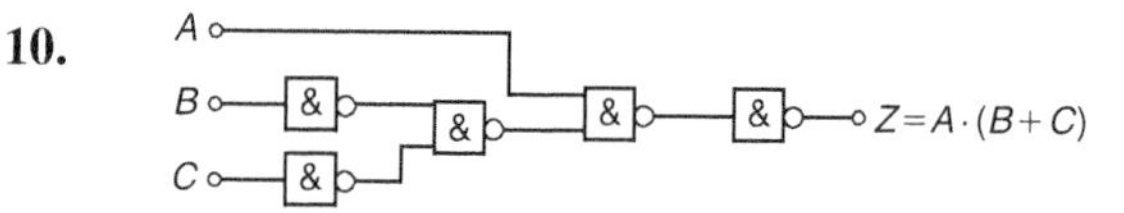

Exercise 268 (page 591)

1. (b) **2.** (d) **3.** (c) **4.** (b) **5.** (c) **6.** (d) **7.** (a) **8.** (c) **9.** (d) **10.** (a)

Chapter 58

Exercise 269 (page 596)

1. $\begin{pmatrix} 8 & 1 \\ -5 & 13 \end{pmatrix}$

2. $\begin{pmatrix} 7 & -1 & 8 \\ 3 & 1 & 7 \\ 4 & 7 & -2 \end{pmatrix}$

3. $\begin{pmatrix} -2 & -3 \\ -3 & 1 \end{pmatrix}$

4. $\begin{pmatrix} 9.3 & -6.4 \\ -7.5 & 16.9 \end{pmatrix}$

5. $\begin{pmatrix} 45 & 7 \\ -26 & 71 \end{pmatrix}$

6. $\begin{pmatrix} 4.6 & -5.6 & -7.6 \\ 17.4 & -16.2 & 28.6 \\ -14.2 & 0.4 & 17.2 \end{pmatrix}$

7. $\begin{pmatrix} -11 \\ 43 \end{pmatrix}$

8. $\begin{pmatrix} 16 & 0 \\ -27 & 34 \end{pmatrix}$

9. $\begin{pmatrix} -6.4 & 26.1 \\ 22.7 & -56.9 \end{pmatrix}$

10. $\begin{pmatrix} 135 \\ -52 \\ -85 \end{pmatrix}$

11. $\begin{pmatrix} 5 & 6 \\ 12 & -3 \\ 1 & 0 \end{pmatrix}$

12. $\begin{pmatrix} 55.4 & 3.4 & 10.1 \\ -12.6 & 10.4 & -20.4 \\ -16.9 & 25.0 & 37.9 \end{pmatrix}$

13. $A \times C = \begin{pmatrix} -6.4 & 26.1 \\ 22.7 & -56.9 \end{pmatrix}$

$C \times A = \begin{pmatrix} -33.5 & 53.1 \\ 23.1 & -29.8 \end{pmatrix}$

Hence, $A \times C \neq C \times A$

Exercise 270 (page 597)

1. 17
2. -3
3. -13.43
4. $-5 + j3$
5. $(-19.75 + j19.79)$ or $27.96\angle 134.94°$
6. $x = 6$ or $x = -1$

Exercise 271 (page 598)

1. $\begin{pmatrix} \frac{7}{17} & \frac{1}{17} \\ \frac{4}{17} & \frac{3}{17} \end{pmatrix}$

2. $\begin{pmatrix} 7\frac{5}{7} & 8\frac{4}{7} \\ -4\frac{2}{7} & -6\frac{3}{7} \end{pmatrix}$

3. $\begin{pmatrix} 0.290 & 0.551 \\ 0.186 & 0.097 \end{pmatrix}$

Exercise 272 (page 599)

1. $\begin{pmatrix} -16 & 8 & -34 \\ -14 & -46 & 63 \\ -24 & 12 & 2 \end{pmatrix}$

2. $\begin{pmatrix} -16 & -8 & -34 \\ 14 & -46 & -63 \\ -24 & -12 & 2 \end{pmatrix}$

3. -212
4. -328
5. -242.83
6. $-2 - j$
7. $26.94\angle -139.52°$ or $(-20.49 - j17.49)$

Exercise 273 (page 601)

1. $\begin{pmatrix} 4 & -2 & 5 \\ -7 & 4 & 7 \\ 6 & 0 & -4 \end{pmatrix}$

2. $\begin{pmatrix} 3 & 5 & -1 \\ 6 & -\frac{2}{3} & 0 \\ \frac{1}{2} & 7 & \frac{3}{5} \end{pmatrix}$

3. $\begin{pmatrix} -16 & 14 & -24 \\ -8 & -46 & -12 \\ -34 & -63 & 2 \end{pmatrix}$

4. $\begin{pmatrix} -\frac{2}{5} & -3\frac{3}{5} & 42\frac{1}{3} \\ -10 & 2\frac{3}{10} & -18\frac{1}{2} \\ -\frac{2}{3} & -6 & -32 \end{pmatrix}$

5. $-\frac{1}{212}\begin{pmatrix} -16 & 14 & -24 \\ -8 & -46 & -12 \\ -34 & -63 & 2 \end{pmatrix}$

6. $-\frac{15}{923}\begin{pmatrix} -\frac{2}{5} & -3\frac{3}{5} & 42\frac{1}{3} \\ -10 & 2\frac{3}{10} & -18\frac{1}{2} \\ -\frac{2}{3} & -6 & -32 \end{pmatrix}$

Exercise 274 (page 601)

1. (c) **2.** (b) **3.** (b) **4.** (c) **5.** (d) **6.** (c) **7.** (a) **8.** (d) **9.** (a) **10.** (d) **11.** (a) **12.** (c) **13.** (d) **14.** (b) **15.** (a) **16.** (b) **17.** (d) **18.** (b) **19.** (c) **20.** (a)

Chapter 59

Exercise 275 (page 605)

1. $x = 4, y = -3$
2. $p = 1.2, q = -3.4$
3. $x = 1, y = -1, z = 2$
4. $a = 2.5, b = 3.5, c = 6.5$
5. $p = 4.1, q = -1.9, r = -2.7$
6. $I_1 = 2, I_2 = -3$
7. $s = 2, v = -3, a = 4$
8. $\ddot{x} = 0.5, \dot{x} = 0.77, x = 1.4$

Exercise 276 (page 608)

1. $x = -1.2, y = 2.8$
2. $m = -6.4, n = -4.9$
3. $x = 1, y = 2, z = -1$
4. $p = 1.5, q = 4.5, r = 0.5$
5. $x = \frac{7}{20}, y = \frac{17}{40}, z = -\frac{5}{24}$
6. $F_1 = 1.5, F_2 = -4.5$
7. $I_1 = 10.77\angle 19.23°$ A, $I_2 = 10.45\angle -56.73°$A
8. $i_1 = -5, i_2 = -4, i_3 = 2$
9. $F_1 = 2, F_2 = -3, F_3 = 4$
10. $I_1 = 3.317\angle 22.57°$A, $I_2 = 1.963\angle 40.97°$A, $I_3 = 1.010\angle -148.32°$A

Exercise 277 (page 610)

Answers to Exercises 275 and 276 are as above

Exercise 278 (page 612)

1. $\ddot{x} = -0.30, \dot{x} = 0.60, x = 1.20$
2. $T_1 = 0.8, T_2 = 0.4, T_3 = 0.2$
3. Answers to Exercise 275 are as above
4. Answers to Exercise 276 are as above

Exercise 279 (page 612)

1. (d) **2.** (a) **3.** (a) **4.** (b) **5.** (c)

Chapter 60

Exercise 280 (page 618)

1. (a) continuous (b) continuous (c) discrete (d) continuous
2. (a) discrete (b) continuous (c) discrete (d) discrete

Exercise 281 (page 621)

1. If one symbol is used to represent 10 vehicles, working correct to the nearest 5 vehicles, gives 3.5, 4.5, 6, 7, 5 and 4 symbols respectively.
2. If one symbol represents 200 components, working correct to the nearest 100 components gives: Mon 8, Tues 11, Wed 9, Thurs 12 and Fri 6.5
3. 6 equally spaced horizontal rectangles, whose lengths are proportional to 35, 44, 62, 68, 49 and 41 respectively
4. 5 equally spaced horizontal rectangles, whose lengths are proportional to 1580, 2190, 1840, 2385 and 1280 units, respectively
5. 6 equally spaced vertical rectangles, whose heights are proportional to 35, 44, 62, 68, 49 and 41 units, respectively
6. 6 equally spaced vertical rectangles, whose heights are proportional to 1580, 2190, 1840, 2385 and 1280 units, respectively
7. 3 rectangles of equal height, subdivided in the percentages shown; P increases by 20% at the expense of Q and R
8. 4 rectangles of equal heights, subdivided as follows: **Week 1**: 18%, 7%, 35%, 12%, 28% **Week 2**: 20%, 8%, 32%, 13%, 27% **Week 3**: 22%, 10%, 29%, 14%, 25% **Week 4**: 20%, 9%, 27%, 19%, 25%. Little change in centres A and B, there is a reduction of around 8% in centre C, an increase of around 7% in centre D and a reduction of about 3% in centre E.

9. A circle of any radius, subdivided into sectors, having angles of 7.5°, 22.5°, 52.5°, 167.5° and 110°, respectively.
10. A circle of any radius, subdivided into sectors, having angles of 107°, 156°, 29° and 68°, respectively.
11. (a) £495 (b) 88
12. (a) £16450 (b) 138

Exercise 282 (page 627)

1. There is no unique solution, but one solution is: 39.3 – 39.4 1; 39.5 – 39.6 5; 39.7 – 39.8 9; 39.9 – 40.0 17; 40.1 – 40.2 15; 40.3 – 40.4 7; 40.5 – 40.6 4; 40.7 – 40.8 2;
2. Rectangles, touching one another, having mid-points of 39.35, 39.55, 39.75, 39.95..... and heights of 1, 5, 9, 17, ...
3. There is no unique solution, but one solution is: 20.5 – 20.9 3; 21.0 – 21.4 10; 21.5 – 21.9 11; 22.0 – 22.4 13; 22.5 – 22.9 9; 23.0 – 23.4 2
4. There is no unique solution, but one solution is: 1 – 10 3; 11 – 19 7; 20 – 22 12; 23 – 25 11; 26 – 28 10; 29 – 38 5; 39 – 48 2
5. 20.95 3; 21.45 13; 21.95 24; 22.45 37; 22.95 46; 23.45 48
6. Rectangles, touching one another, having mid-points of 5.5, 15, 21, 24, 27, 33.5 and 43.5. The heights the rectangles (frequency per unit class range) are 0.3, 0.78, 4, 3.67, 3.33, 0.5 and 0.2
7. (10.95 2), (11.45 9), (11.95 19), (12.45 31), (12.95 42), (13.45 50)
8. Ogive
9. (a) There is no unique solution, but one solution is: 2.05 – 2.09 3; 2.10 – 2.14 10; 2.15 – 2.19 11; 2.20 – 2.24 13; 2.25 – 2.29 9; 2.30 – 2.34 2. (b) Rectangles, touching one another, having mid-points of 2.07, 2.12 .... and heights of 3, 10 ... (c) Using the frequency distribution given in the solution to part (a) gives: 2.095 3; 2.145 13; 2.195 24; 2.245 37; 2.295 46; 2.345 48. (d) A graph of cumulative frequency against upper class boundary having the co-ordinates given in part (c).

Exercise 283 (page 628)

1. (a) **2.** (d) **3.** (c) **4.** (c) **5.** (b)

Chapter 61

Exercise 284 (page 630)

1. mean 7.33, median 8, mode 8
2. mean 27.25, median 27, mode 26
3. mean 4.7225, median 4.72, mode 4.72
4. mean 115.2, median 126.4, no mode

Exercise 285 (page 632)

1. 23.85 kg
2. 171.7 cm
3. mean 89.5, median 89, mode 88.2
4. mean 2.02158 cm, median 2.02152 cm, mode 2.02167 cm

Exercise 286 (page 634)

1. 4.60
2. 2.83 μF
3. mean 34.53 MPa, standard deviation 0.07474 MPa
4. 0.296 kg
5. 9.394 cm
6. 0.00544 cm

Exercise 287 (page 635)

1. 30, 27.5, 33.5 days
2. 27, 26, 33 faults
3. $Q_1 = 164.5$ cm, $Q_2 = 172.5$ cm, $Q_3 = 179$ cm, 7.25 cm
4. 37 and 38; 40 and 41
5. 40, 40, 41; 50, 51, 51

Exercise 288 (page 635)

1. (d) **2.** (a) **3.** (b) **4.** (c) **5.** (d) **6.** (b) **7.** (b) **8.** (c) **9.** (a) **10.** (d)

Chapter 62

Exercise 289 (page 641)

1. (a) $\frac{2}{9}$ or 0.2222 (b) $\frac{7}{9}$ or 0.7778

2. (a) $\frac{23}{139}$ or 0.1655 (b) $\frac{47}{139}$ or 0.3381

(c) $\frac{69}{139}$ or 0.4964

3. (a) $\frac{1}{6}$ (b) $\frac{1}{6}$ (c) $\frac{1}{36}$

4. 0.7 or 70% **5.** $\frac{5}{36}$

6. (a) $\frac{2}{5}$ (b) $\frac{1}{5}$ (c) $\frac{4}{15}$ (d) $\frac{13}{15}$

7. (a) $\frac{1}{250}$ (b) $\frac{1}{200}$ (c) $\frac{9}{1000}$ (d) $\frac{1}{50000}$

Exercise 290 (page 643)

1. (a) 0.6 (b) 0.2 (c) 0.15
2. (a) 0.64 (b) 0.32
3. 0.0768
4. (a) 0.4912 (b) 0.4211
5. (a) 89.38% (b) 10.25%
6. (a) 0.0227 (b) 0.0234 (c) 0.0169

Exercise 291 (page 645)

1. (a) 210 (b) 3024 **2.** (a) 792 (b) 15
3. 210 **4.** 3003
5. $\frac{10}{^{49}C_6} = \frac{10}{13983816} = \frac{1}{1398382}$ or 715×10^{-9}

Exercise 292 (page 646)

1. 53.26% **2.** 70.32 % **3.** 7.48 %

Exercise 293 (page 646)

1. (c) **2.** (d) **3.** (c) **4.** (b) **5.** (b) **6.** (d) **7.** (a) **8.** (a) **9.** (b) **10.** (c)

Chapter 63

Exercise 294 (page 652)

1. (a) 0.0186 (b) 0.9976
2. (a) 0.2316 (b) 0.1408
3. (a) 0.7514 (b) 0.0019
4. (a) 0.9655 (b) 0.0028
5. Vertical adjacent rectangles, whose heights are proportional to 0.03125, 0.15625, 0.3125, 0.3125, 0.15625 and 0.03125
6. Vertical adjacent rectangles, whose heights are proportional to 0.0280, 0.1306, 0.2613, 0.2903, 0.1935, 0.0774, 0.0172 and 0.0016
7. 0.0574

Exercise 295 (page 655)

1. 0.0613
2. 0.5768
3. (a) 0.1771 (b) 0.5153
4. 0.9856
5. The probabilities of the demand for 0, 1, 2,..... tools are 0.0067, 0.0337, 0.0842, 0.1404, 0.1755, 0.1755, 0.1462, 0.1044, 0.0653, ... This shows that the probability of wanting a tool 8 times a day is 0.0653, i.e. less than 10%. Hence 7 should be kept in the store
6. Vertical adjacent rectangles having heights proportional to 0.4966, 0.3476, 0.1217, 0.0284, 0.0050 and 0.0007

Exercise 296 (page 655)

1. (b) **2.** (c) **3.** (a) **4.** (c) **5.** (d)

Chapter 64

Exercise 297 (page 661)

1. 6 **2.** 22
3. (a) 479 (b) 63 (c) 21 **4.** 4
5. (a) 131 (b) 553 **6.** (a) 15 (b) 4
7. (a) 65 (b) 209 (c) 89
8. (a) 1 (b) 85 (c) 13

Exercise 298 (page 663)

1. Graphically, $\bar{x} = 27.1$, $\sigma = 0.3$ By calculation, $\bar{x} = 27.079$, $\sigma = 0.3001$
2. (a) $\bar{x} = 23.5$ kN, $\sigma = 2.9$ kN (b) $\bar{x} = 23.364$ kN, $\sigma = 2.917$ kN

Exercise 299 (page 664)

1. (b) **2.** (c) **3.** (a) **4.** (d) **5.** (a)

Chapter 65

Exercise 300 (page 670)

1. 0.999 **2.** −0.916 **3.** 0.422
4. 0.999 **5.** −0.962 **6.** 0.632
7. 0.937

Chapter 66

Exercise 301 (page 675)

1. $Y = -256 + 80.6X$
2. $Y = 0.0477 + 0.216X$
3. $X = 3.20 + 0.0124Y$
4. $X = -0.056 + 4.56Y$
5. $Y = 1.142 + 2.268X$
6. $X = -0.483 + 0.440Y$
7. (a) 7.95 V (b) 17.1 mA
8. $Y = 0.881 - 0.0290X$
9. $X = 30.194 - 34.039Y$
10. (a) 0.417 s (b) 21.7 N

Chapter 67

Exercise 302 (page 680)

1. $\mu_{\bar{x}} = \mu = 22.4\,\text{cm}$ $\sigma_{\bar{x}} = 0.0080\,\text{cm}$
2. $\sigma_{\bar{x}} = 0.0079\,\text{cm}$
3. (a) $\mu_{\bar{x}} = 1.70\,\text{cm}$, $\sigma_{\bar{x}} = 2.91 \times 10^{-3}\,\text{cm}$
(b) $\mu_{\bar{x}} = 1.70\,\text{cm}$, $\sigma_{\bar{x}} = 2.89 \times 10^{-3}\,\text{cm}$
4. 0.023 **5.** 0.497
6. 0.0038
7. (a) 0.0179 (b) 0.740 (c) 0.242

Exercise 303 (page 685)

1. 66.89 and 68.01 mm, 66.72 and 68.18 mm
2. (a) 2.355 Mg to 2.445 Mg; 2.341 Mg to 2.459 Mg
(b) 86%
3. $12.73 \times 10^{-4}\,\text{m}^\circ\text{C}^{-1}$ to $12.89 \times 10^{-4}\,\text{m}^\circ\text{C}^{-1}$
4. (a) at least 68 (b) at least 271
5. 10.91 t to 11.27 t **6.** 45.6 seconds

Exercise 304 (page 688)

1. 5.133 MPa to 5.207 MPa
2. 5.125 MPa to 5.215 MPa
3. $1.10\,\Omega\text{m}^{-1}$ to $1.15\,\Omega\text{m}^{-1}$ **4.** 95%

Index